International Conference on Engineering Vibration (ICoEV 2017)

MATEC Web of Conferences Volume 148 (2018)

Sofia, Bulgaria
4 - 7 September 2017

Editors:

Emil Manoach
Stanislav Stoykov
Marian Wiercigroch

ISBN: 978-1-5108-5726-1

Printed from e-media with permission by:

Curran Associates, Inc.
57 Morehouse Lane
Red Hook, NY 12571

Some format issues inherent in the e-media version may also appear in this print version.

Printed by Curran Associates, Inc. (2018)

For additional information, please contact EDP Sciences – Web of Conferences
at the address below.

EDP Sciences – Web of Conferences
17, Avenue du Hoggar
Parc d'Activité de Courtabœuf
BP 112
F-91944 Les Ulis Cedex A
France

Phone: +33 (0) 1 69 18 75 75
Fax: +33 (0) 1 69 28 84 91

contact-edps@webofconferences.org

Additional copies of this publication are available from:

Curran Associates, Inc.
57 Morehouse Lane
Red Hook, NY 12571 USA
Phone: 845-758-0400
Fax: 845-758-2633
Email: curran@proceedings.com
Web: www.proceedings.com

TABLE OF CONTENTS

International Conference of Engineering Vibration (ICoEV-2017)

Preface

The International Conference of Engineering Vibration (ICoEV) originates from a long-running series called the International Conference of Vibrational Problems (ICoVP), which was founded in India over 20 years ago. The last conference organised in Lubljana in 2015, was renamed as International Conference of Engineering Vibration (ICoEV). The ICoEV-2017 was organised by the Institute of Information and Communication Technology and by the Institute of Mechanics at the Bulgarian Academy of Sciences and held between 4 and 7 September 2017 in Sofia, the capital of Bulgaria. The conference has attracted more than 150 participants from 34 countries including senior scientists, PhD students and practicing vibration engineers.

The conference was based on two full days and two half days of technical presentations, structured into four parallel sessions. Six invited lectures gave plenary talks on important and state of the art problems on dynamics and vibrations.

The topics of the conference included Structural Dynamics, Wave Propagation, Fluid-Structure Interactions, Dynamics of Rotating Systems, Structural Health Monitoring, Vibration Control and Isolation, Experimental Techniques, Energy harvesting systems, Vibration and control in technological process, Biomechanics and other vibration themes.

The scientific program was organized into 16 mini-symposia and one other with general topics:

- Mitigation of Vibrations of Mechanical and Structural Systems;
- Nonlinear Oscillations and Controls of Mechanical, Civil, Aerospace and Naval Structures
- Modelling, Methodologies and Engineering Applications of Nonlinear Dynamical Systems;
- Vibration of Solids and Structures under Moving Loads: Modelling and Analysis;
- Dynamics of Composite and Smart Structures
- Nonlinear Dynamics of MEMS & NEMS
- Vibration of Beams, Plates and Shells, from Nano to Macro
- Nonlinearity and Stochasticity in Vibrating Systems
- Vibration in Mechanical and Biomechanical Systems
- Modelling of Friction and Dynamics of Frictional Oscillators
- Active Vibration Control
- Nonlinear Effects in Broadband Energy Harvesting from Mechanical Vibrations
- New Trends in Analytical Approaches to Nonlinear Vibration
- Vibration-Based Structural Health Monitoring Data Analysis and Time Series Methods
- Wave Mechanics: Generation and Propagation of Waves in Fluids, Solids and Structures
- Vibration and Control in Downhole Drilling Processes

During the conference the advanced scientific achievements in the field of dynamics and stability of mechanical systems were presented. We believe that the scientific results presented at ICoEV-2017 will contribute to the clarification of many scientific issues related to vibrations and the further development of analytical, numerical and experimental modelling of dynamic processes in mechanical systems.
In the current Proceedings of ICoEV-2017 a part of the contribution presented at the conference are published.

E. Manoach S. Stoykov M. Wiercigroch

edp sciences

Web of Conferences

Statement of Peer review

In submitting conference proceedings to *Web of Conferences*, the editors of the proceedings certify to the Publisher that

1. They adhere to its **Policy on Publishing Integrity** in order to safeguard good scientific practice in publishing.
2. All articles have been subjected to peer review administered by the proceedings editors.
3. Reviews have been conducted by expert referees, who have been requested to provide unbiased and constructive comments aimed, whenever possible, at improving the work.
4. Proceedings editors have taken all reasonable steps to ensure the quality of the materials they publish and their decision to accept or reject a paper for publication has been based only on the merits of the work and the relevance to the journal.

Title, date and place of the conference

International Conference on Engineering Vibration - 2017
4-7 September 2017, Sofia, Bulgaria

Proceedings editor(s):

Emil Manoach

Stanislav Stoykov

Marian Wiercigroch

Date and editor's signature

15.01.2018

Motorised momentum exchange space tethers: the dynamics of asymmetrical tethers, and some recent new applications

Matthew Cartmell[1,2,*], *Olga* Ganilova[1,2], *Eoin* Lennon[1], and *Gavin* Shuttleworth[1]

[1]Department of Mechanical & Aerospace Engineering, University of Strathclyde, Glasgow, G1 1XJ, Scotland, UK
[2]Strathclyde Space Institute, University of Strathclyde, Glasgow, G1 1XJ, Scotland, UK

Abstract. This paper reports on a first attempt to model the dynamics of an asymmetrical motorised momentum exchange tether for spacecraft payload propulsion, and it also provides some interesting summary results for two novel applications for motorised momentum exchange tethers. The asymmetrical tether analysis is very important because it represents the problematic scenario when payload mass unbalance intrudes, due to unexpected payload loss or failure to retrieve. Mass symmetry is highly desirable both dynamically and logistically, but it is shown in this paper that there is still realistic potential for mission rescue should an asymmetry condition arise. Conceptual designs for tethered payload release from LEO and lunar tether delivery and retrieval are also presented as options for future development.

1 Introduction

Momentum exchange tether dynamics have been extensively studied in recent years, and are of current considerable interest internationally as a technology for cost effective and environmentally clean propulsion of payload mass from Low Earth Orbit (LEO). A conceptual design for a motorised momentum exchange tether is given in Figure 1 where the central facility motor drive shaft attaches to the propulsion sub-spans by mean of a gantry, mounted so that it can be rotated by the shaft. Typical performance figures for an operational motorised MET on a circular LEO predict an orbit velocity in the region of 7.6 km/s and a tether generated increment of around 3.1 km/s, which has the potential to accelerate the outer payload, at an optimal release position, to around 10.7 km/s. This is fast enough for Earth escape, using a tether made of *Spectra2000*™. It should be noted that a fundamental requirement for motorised spin is shown in Figure 1 comprising a counter-inertia in the form of an additional pair of tethers and counter-masses attached to the motor stator [1]. This sub-system is required to satisfy Newton's third law of motion and is not analysed here but is assumed to be present, coupling through the drive motor electrodynamics in order to provide the requisite operational configuration.

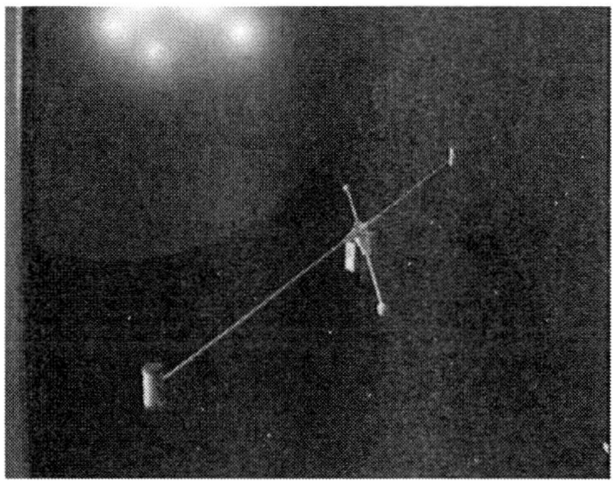

Fig. 1. A motorised momentum exchange tether in LEO.

A fairly comprehensive review of space tether research up to 2008 is given in [2] and a detailed study of fundamental rigid-body motorised tether system dynamics is given in [3, 4]. Tether flexure, accommodating various forms of vibration, is discussed in full in [5]. Important mission scenarios for motorised tethers are discussed in [6, 7]. The fundamental issue with asymmetrical payload mass distribution is that the centre of mass location of the tether will then change on orbit, introducing dynamic perturbations that will de-orbit the tether, with potentially disastrous consequences for the mission. In order to maintain the orbit, and also to capitalise on opportunities offered by the exploitation of orbital harmonics, payload mass symmetry is required. In terms of mission architecture this allows a simultaneous in and out flow of payload from the location of the host planet to the destination planet [6]. On the basis that symmetry (defined as the tether being laden with two payloads of identical mass properties

*Corresponding author : matthew.cartmell@strath.ac.uk

located at both ends, or, conversely, no payloads on either end of the tether) cannot be absolutely guaranteed, a study of the asymmetrical mass distribution problem was seen to be necessary. In this paper an initial model is proposed, with some calculations that suggest that a pragmatic level of reserve chemical propulsion on board each payload could be used to rescue a mission in which asymmetry has arisen for some reason. This is investigated in the form of the complete loss of one of the payloads. This is followed by discussions of two further investigations in which the tethered release of small payloads from an Earth orbiter and the use of a specialised *lunavator* tether with an end reeling facility for lunar touchdown are investigated. These two application scenarios in fact reflect different levels of mass symmetry requirement and on that basis they provide interesting platforms for further research into robust solutions to the tether asymmetry problem.

2. Reference model for a planar motorised tether on an elliptical Low Earth Orbit

We start with the kinetic and potential energies for a mass-discretised dumb-bell tether (i.e. in symmetrical rigid-body form), and these can be stated as follows [3] for the system shown diagrammatically in Figure 2, with an Earth-fixed frame EXY and a tether-fixed frame Cx_0y_0. The tether is assumed to be symmetrical in length and therefore point C, the location of the central facility, is exactly half-way between each payload, M_p. Note that the time argument has been dropped from $R(t)$, $\theta(t)$, and $\psi(t)$ in both equations for clarity and compactness. Quantities are defined in full after the equations.

$$T = \left[M_p + \rho A l + \frac{M_m}{2}\right]\left(\dot{R}^2 + R^2\dot{\theta}^2\right) + \left[\begin{array}{c} M_p\left(l^2 + \frac{r_p^2}{2}\right) + M_m\frac{r_m^2}{4} + \\ \frac{\rho A l}{12}\left(4l^2 + 3r_T^2\right)\end{array}\right]\left(\dot{\psi} + \dot{\theta}\right)^2 \quad (1)$$

$$U = -\sum_{j=1}^{n}\frac{A\mu l\rho}{n\sqrt{\frac{(2j-1)^2 l^2}{4n^2} - \frac{(2j-1)lR\cos\psi}{n} + R^2}} - \sum_{j=1}^{n}\frac{A\mu l\rho}{n\sqrt{\frac{(2j-1)^2 l^2}{4n^2} + \frac{(2j-1)lR\cos\psi}{n} + R^2}} - \frac{\mu M_p}{\sqrt{l^2 - 2lR\cos\psi + R^2}} - \frac{\mu M_p}{\sqrt{l^2 + 2lR\cos\psi + R^2}} - \frac{\mu M_m}{R} \quad (2)$$

*Corresponding author : matthew.cartmell@strath.ac.uk

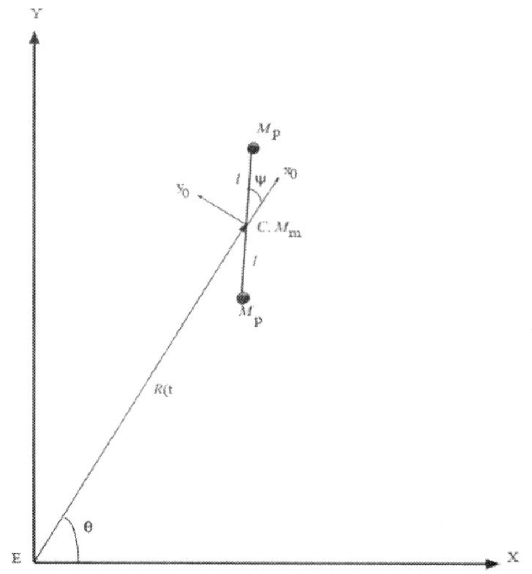

Fig. 2. A planar dumb-bell tether in LEO where E represents the centre of the Earth and EXY is an Earth fixed frame.

Lagrange's equations are used to obtain three differential equations of motion based on generalised coordinates $R(t)$, $\theta(t)$, and $\psi(t)$. These are intricately coupled, nonlinear, and lengthy, and are not reproduced here but are obtainable in full in [3]. They represent the *planar* orbital dynamics of the tether, accommodating orbital eccentricity. The data used for the study is as follows. Tether sub-span length $l = 50$ km, payload masses $M_p = 1000$ kg, central facility mass $M_m = 5000$ kg, tether density $\rho = 970$ kg/m^3, cross-sectional area $A = 62.83*10^{-6}$ m^2, equal radii of central facility and payloads $r_m = r_p = 0.5$m, standard gravitational parameter $\mu = 3.9877848*10^{14}$ m^3/s^2, drive torque $\tau = 5$MNm, and integration time = 86400 s. Initial conditions are $R(0) = 6870000$ m, $\theta(0) = 97.1877$ rad, and $\psi(0) = 0$ rad, and $R'(0) = 0$ m/s, $\theta'(0) = 0.00121485$ rad/s, and $\psi'(0) = 0$ rad/s. These initial conditions accommodate an elliptical LEO ($e = 0.20004$) on which the tether's long-term motion is predicted to be in the form of a monotonic spin. The tangential velocity of the payloads relative to the COM at the centre of the system builds up usefully. We note that the centripetal stress is proportional to the angular velocity squared. The system responses in the time domain are given in the following Figures. Figure 3 shows a continuous build-up in the angular position of the tether over time, with Figures 4 and 5 indicating the increasing angular velocity and the tangential velocity of the tip of each sub-span relative to the spin axis located at the CoM of the tether. The increase in centripetal stress in the tether over time is shown in Figure 6 where it can be seen that this is still within reasonable limits for *Spectra2000*™ at the end of the integration time of 24 hours. The orbital elements are given in Figures 7-10 with clear indication of perigee and apogee locations for R in Figure 7, and a commensurate ripple in θ shown in Figure 9.

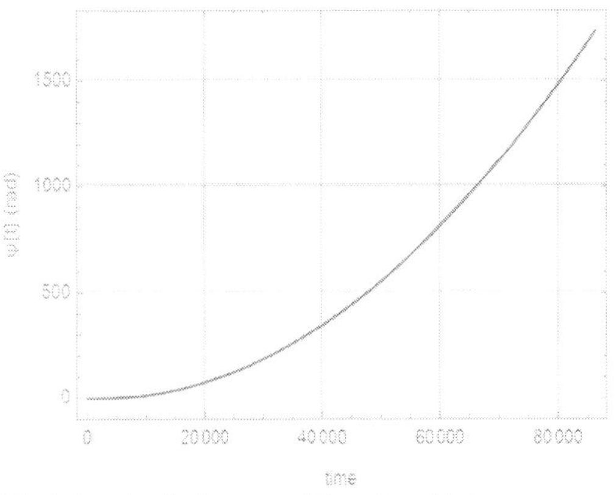

Fig. 3. Angular displacement of the tether with time.

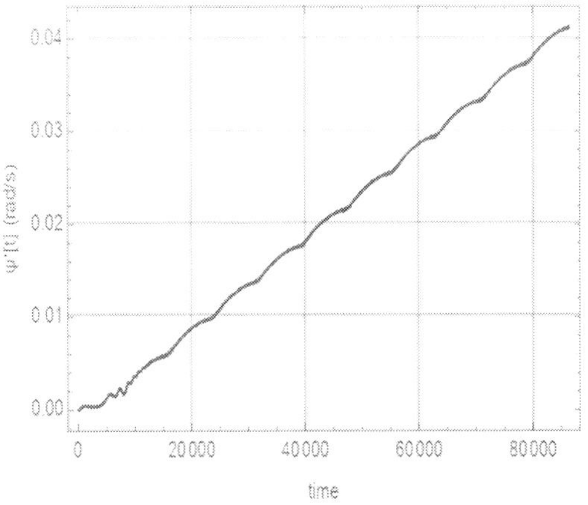

Fig. 4. Angular rate of the tether with time.

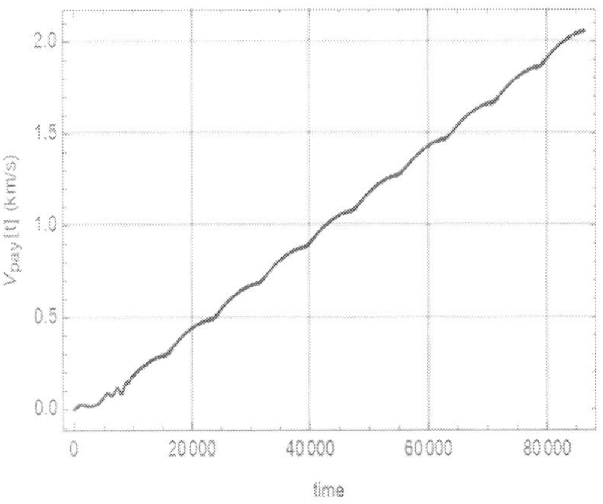

Fig. 5. Tangential velocity of the sub-spans relative to the spin axis with time.

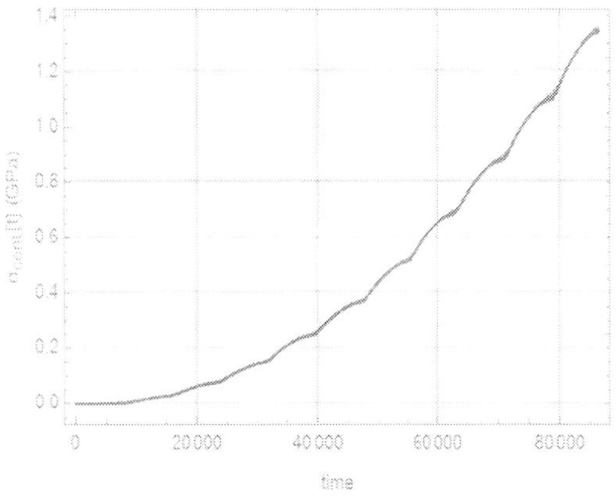

Fig. 6. Sub-span centripetal stress with time.

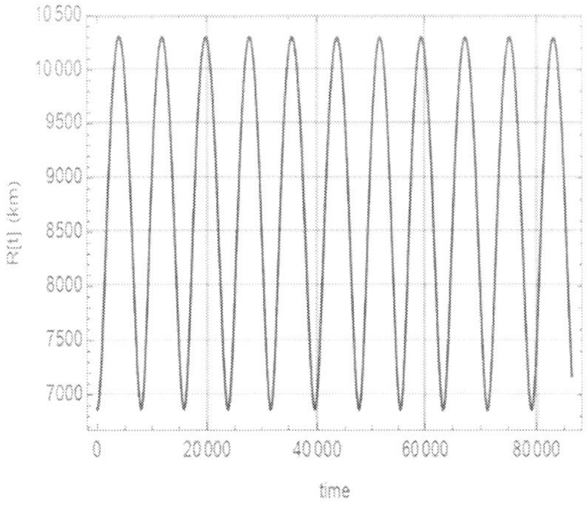

Fig. 7. Magnitude of position vector **R** with time.

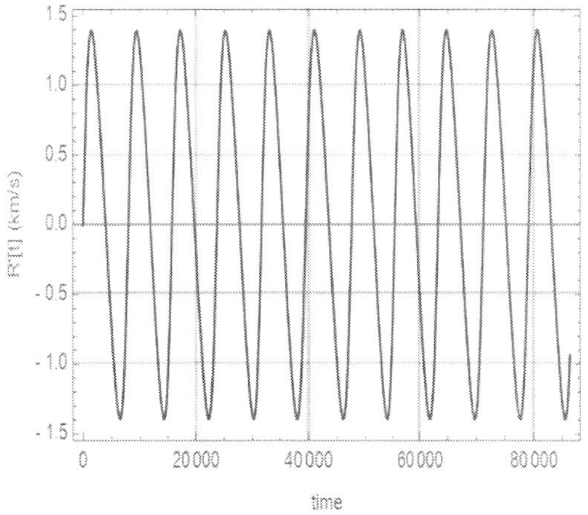

Fig. 8. Time rate of change of magnitude of **R**.

*Corresponding author : matthew.cartmell@strath.ac.uk

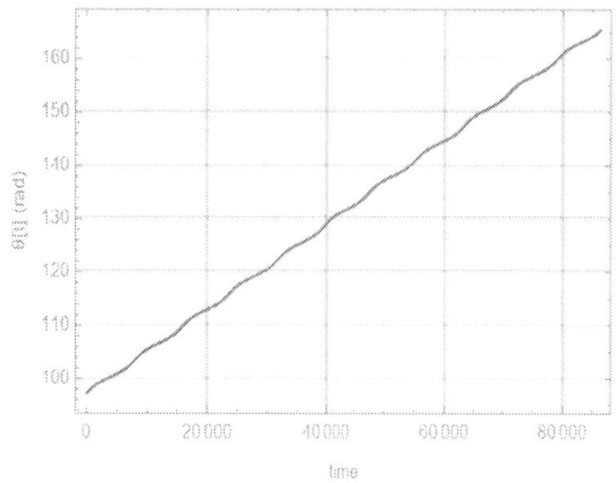

Fig. 9. True anomaly with time.

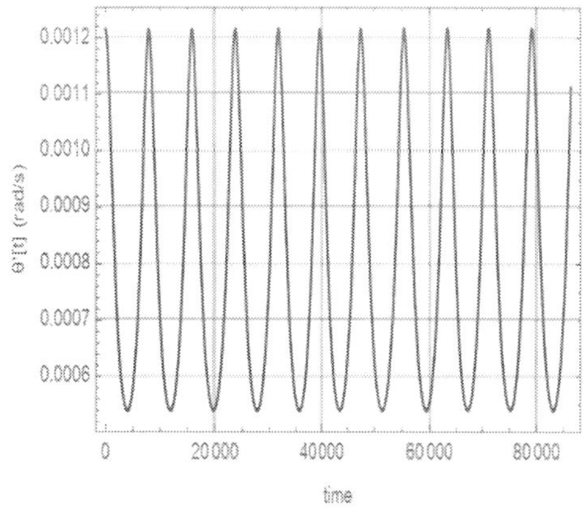

Fig. 10. Rate of change of true anomaly with time.

This model shows some interesting and typical tether phenomena on orbit and was useful for an initial data scoping study for the more complex scenario of the asymmetry case in the next section.

3. Conditions for Mass Asymmetry

It was shown by [8, Fu] that even small mass asymmetry will quickly lead to catastrophic failure of the tether as the propulsion side proceeds to collide with the outrigger counter-inertia system. A typical motorised MET system is shown in side-elevation in Figure 11.

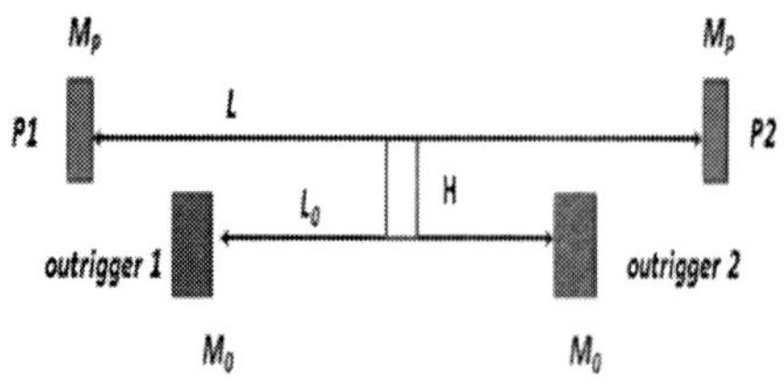

Fig. 11. Side elevation of the tether showing the propulsion and outrigger tethers and the separation distance H between them.

The separation distance H is assumed to comprise the height of the motor drive system. The centripetal load in a sub-span due to the mass of a payload and the tether sub-span can readily be calculated from,

$$F_{cent} = \left(M_p l + \frac{\rho A l^2}{2}\right)\dot{\psi}^2 \tag{3}$$

and shown against time, as in Figure 12.

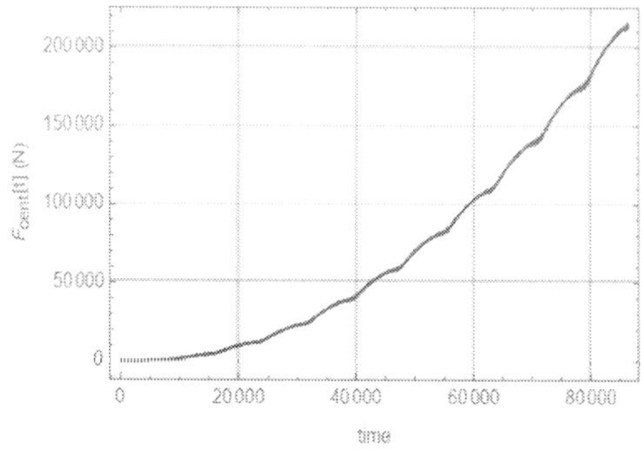

Fig. 12. Sub-span centripetal load as a function of time.

4. Asymmetry model

Under normal operating conditions the two sub-spans each develop equal centripetal loads, and although the moment arms, about the half-way point, of $H/2$ mean that there are moments associated with the centripetal loads in each sub-span these are equal and opposite and therefore mutually balancing. These loads can reach 200kN for typical design data, and would normally be balanced by each identical sub-span. But in the asymmetrical configuration the tether will obviously be de-stabilised so that it is pushed out of the spin-and-orbit plane. The precise effect depends on the physical design adopted but it cannot be ignored. For the moment arm of $H/2$, where H could be up to 3 m, the corresponding moment could reach up to 300 kNm. A corrective counter-moment, but exerted instead over a moment arm comprising the full length of the sub-span, (taken to be 50 km in this example) would typically require a normal thrust of up to 6N and this could easily be obtained from simple cold-gas thrusters, and provided as emergency propulsion in each payload. So, although the out of balance moment generated by asymmetry could probably be inexpensively corrected using low thrust propulsion on board the payloads there will also be the problem of whole-body translation of the tether to consider, due to subsequent re-positioning of the CoM. It is this that provides the focus of the remaining analysis, with an asymmetry scenario shown in Figure 13.

*Corresponding author : matthew.cartmell@strath.ac.uk

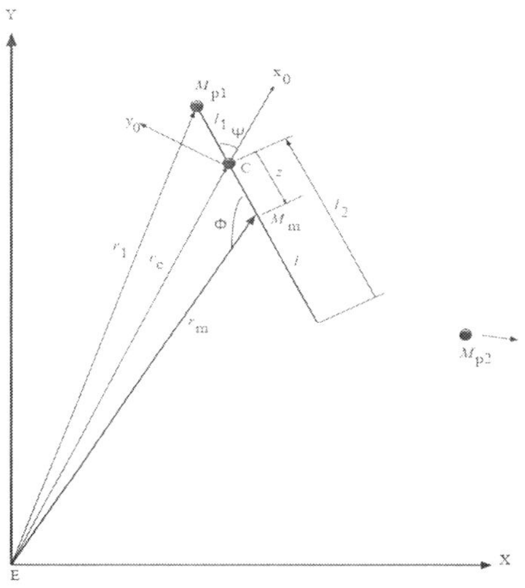

Fig. 13. Asymmetrical tether in the Earth-fixed frame with asymmetry due to the unplanned loss of payload M_{p2}.

The CoM is now no longer co-located with the central facility and moves closer to the remaining payload. This means that the tether will translate through $z(t)$ to stay on orbit. We can relate the position vector magnitudes through simple geometry to get,

$$r_m(-2lr_c z\cos\psi + lr_c^2 + 2r_c z^2\cos\psi - r_c^2 z + lz^2 - z^3) - (l-z)r_m^3 = 0 \quad (4)$$

noting that we have dropped the arguments from $r_m(t)$ and $z(t)$ in equations (4) and (5) for compactness. We also note that $R(t)$ in Figure 2 for the general elliptical orbit case is represented instead by r_c in Figure 13, where r_c is a constant radius for the circular orbit test case considered here. The procedure is to calculate $r_m(t)$ in terms of r_c and $z(t)$ and then re-derive the equations of motion for a circular orbit condition, and in terms of generalised coordinates $\psi(t)$ and $z(t)$ to show the tether spin and also the translational movement of the tether from its initially symmetrical location on orbit. The revised kinetic energy is now given by equation (5),

$$T = \frac{1}{2}\dot{z}^2(2Al\rho + M_m + M_p) + \frac{1}{2}A\rho(l-z)(\dot{x}_{Tl1}^2 + \dot{y}_{Tl1}^2) + \frac{1}{2}A\rho(l+z)(\dot{x}_{Tl2}^2 + \dot{y}_{Tl2}^2) + \frac{1}{2}M_m(\dot{x}_m^2 + \dot{y}_m^2) + \left(\frac{I_m}{2} + \frac{I_p}{2} + I_T\right)(\dot{\theta} + \dot{\psi})^2 + \frac{1}{2}M_p(\dot{x}_{p1}^2 + \dot{y}_{p1}^2) \quad (5)$$

*Corresponding author : matthew.cartmell@strath.ac.uk

The potential energy remains as expressed in equation (2) except that the contribution due to M_{p2} is no longer present and $r_m(t)$ is re-expressed in terms of r_c and $z(t)$. The tether will re-balance itself by sliding along to a new CoM position on the orbit. An axial propulsive force can be introduced into the differential equation as a generalised force for $z(t)$ in order to counteract the displacement of the CoM. The effect of an additional 5.5kN axial propulsive force introduced to the asymmetrical tether, for the same numerical data as previously used, can be seen in Figure 14.

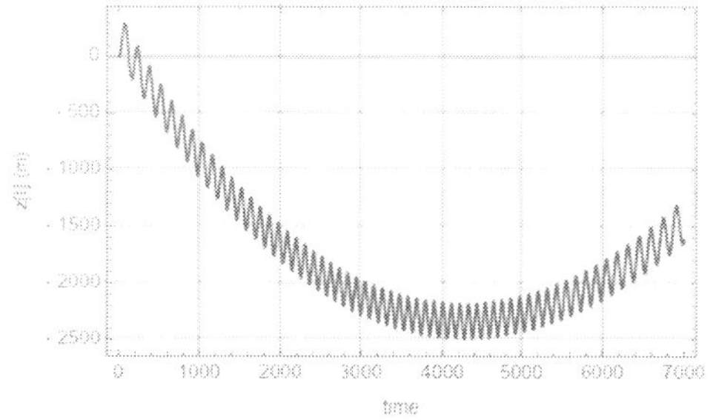

Fig. 14. Effect of axial propulsive force of 5.5kN on $z(t)$ with time.

It can be clearly seen that the whole tether starts to retract axially back from its new position from around 4000 s, thus moving to restore the CoM to its original position over time. We also note the appearance of an axial oscillation, to which we return later. The monotonicity of the spin-up of the tether is not adversely affected but numerically the spin-up is clearly somewhat different for the specific initial conditions of the circular orbit case considered here, as shown in Figure 15.

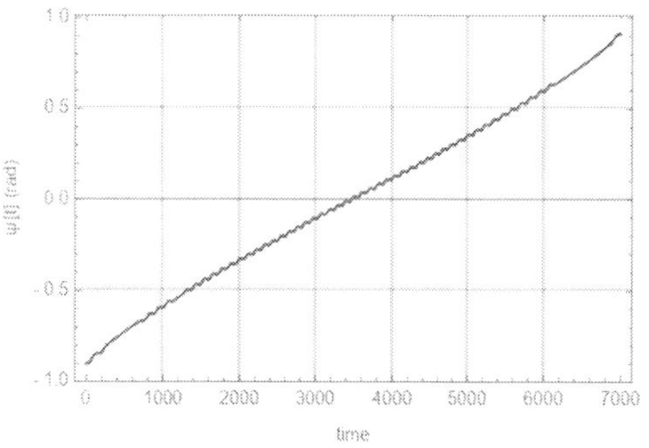

Fig. 15. Spin-up of tether on a circular orbit, with activated axial position restoration force as well as motor torque.

In summary, the tether has lost one payload, M_{p2} and from that point onwards the problem has had to be treated differently – as one that is asymmetrically configured. The motor torque is applied as usual and tether spins monotonically from the initial displacement condition. It is assumed that in practice the centripetal moment unbalance is simply compensated for by a cold-gas thruster, but that the tether then translates so that the new CoM is no longer coincident with the location of the central facility M_m. For the data chosen here this displacement is up to around 2.3 km. An axial thrust, taken on observation to be 5.5 kN in this example, is then applied (potentially from another cold gas thruster) to pull the tether back, noting that a simple *uncontrolled* system is shown here in which the force is simply switched on when the tether asymmetry condition arises. Obviously it will subsequently have to be reduced and then a phase of controlled thrust and counter-thrusts provided to decelerate and stabilise the tether, and also remove the axial oscillation, when it is close to its desired on-orbit configuration. This restoration force system would be highly useful for maintaining the tether spin radius so that the remaining payload can receive its expected ΔV, as well as for maintaining the expected configurational geometry of the tether on orbit so that the mission can continue to some extent as planned. It can be seen in Figure 16 what happens if the axial restoration force is not applied from the onset of asymmetry.

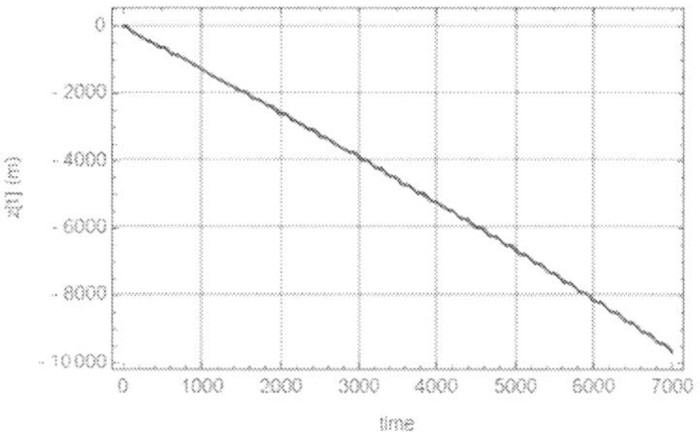

Fig. 16. The axial translation of the torqued tether with time for no axial restoration force.

It can be seen that z(t) grows with time and eventually it can be shown that the tether no longer spins, and de-orbits. It is conjectured that all asymmetrical tether system dynamics will degenerate into some form of irregular behaviour unless a suitable controlled axial thrust is provided.

5. Design of a Payload Transfer System for a Lunavator Tether

A lunavator is a MMET in orbit about the moon which can be used to transfer payloads to and from the lunar surface. Payloads are captured from LEO or EEO as they enter the LSI. No previous research has considered the design of a lunavator which can both deliver and collect payloads at the lunar surface.

The system proposed herein comprises a MMET equipped with grappled payloads at either end. As the tether approaches instantaneous tidal stabilisation, the payload is extended towards the lunar surface, with the cable providing greater in-plane, and out-of-plane, positional control, and extended ground time. Crawler masses along the tether reposition to maintain lunavator dynamic stability.

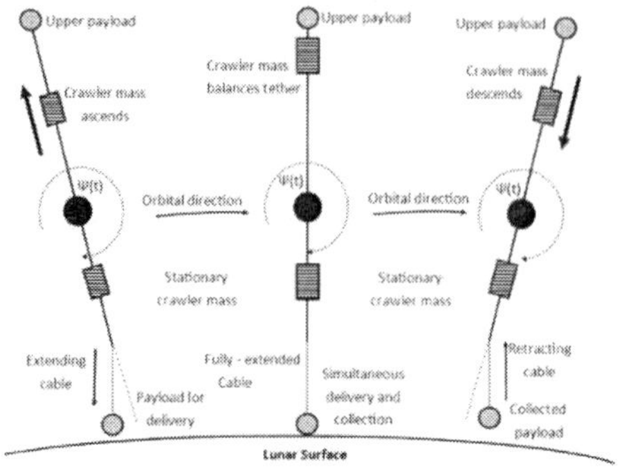

Fig. 17. Proposed lunavator design and payload transfer method

Component positions are defined relative to the orbit plane. Individual mass moments of inertia (MOI) are determined locally, and with the parallel axis theorem to the centre of mass. Due to its significant cross-sectional area, the tether is discretised when deriving its mass moment of inertia. Total kinetic, T_i and potential, U_i energies of each component, and propulsion braid, are also defined.

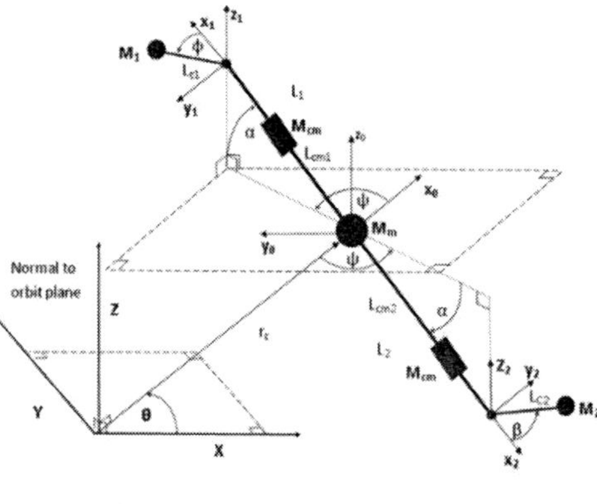

Fig. 18. 3D schematic representation of proposed *lunavator*

Seven representative equations of motion can be derived by substituting the potential and kinetic energies into Lagrange's Equation.

$$\frac{d}{dt}\left[\frac{\delta T}{\delta q_i^{\prime}}\right]-\frac{\delta T}{\delta q_i}+\frac{\delta U}{\delta q_i}=Q_i \qquad (6)$$

where $q_i = \{L_{c1}, L_{c2}, \emptyset, \beta, L_{cm1}, L_{cm2}, \varphi\}$

By defining cable extension as a simple quadratic function, all seven dependent variables can be related to φ, and the model reduced to a single equation of motion for subsequent dynamic study.

*Corresponding author : matthew.cartmell@strath.ac.uk

It was found that a proposed 580km tether (with 2 tonne payloads) required a motor torque of 1000 MNm, with a motor mass of 1000 tonnes, to result in continuous spin (Figure 19). Smaller torque values cause in-plane tether libration wherein the lunavator cannot overcome its inertial elements (Figure 20). Alternatively, a 100km tether is found to perform adequately with 25 MNm of torque and a motor mass of 25 tonnes. Differential positioning of crawler masses may be used to ascend the tether during normal orbit, and descend during the payload transfer phases. Further work will determine the viability of such an approach.

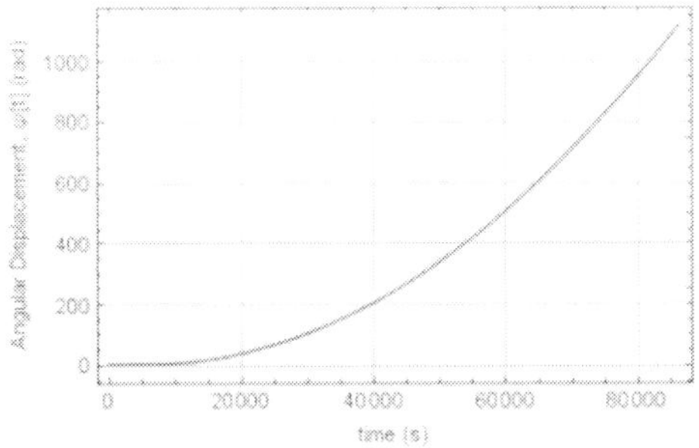

Fig. 19. Continuous spin of 580km tether with 2,000kg payloads, 1,000 tonne motor mass, 1000 MNm torque

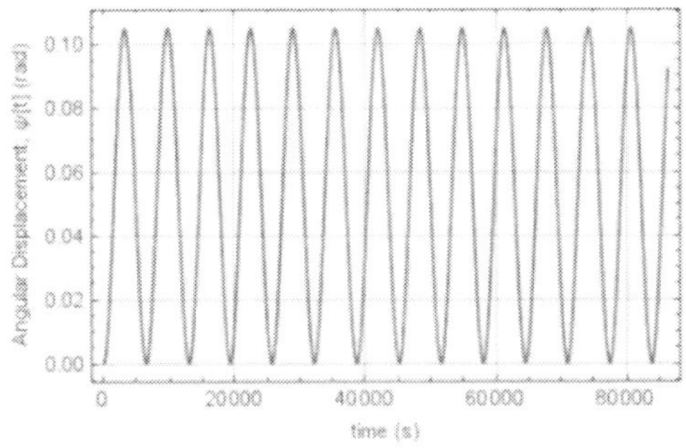

Fig. 20. In-plane libration of 580km tether with 2,000kg payloads, 1,000 tonne motor mass, 145 MNm torque

6. Deployment of a Short Tether from an Earth Orbiter for Sample Return

Momentum exchange space tethers can be used to deorbit payloads from a vehicle travelling in a low Earth orbit. As demonstrated by the YES2 mission of 2007 [9], a payload can be swung in libration from an Earth orbiter and if released at the right moment a retrograde velocity can be transferred to it causing it to have insufficient velocity to remain in orbit, and therefore fall to the Earth.

In order to analyse accurately the requirements for deorbiting payloads of different mass and shape, a computer program was created which was capable of analysing the libration of an Earth orbiter in LEO connected to a payload via a tether. The program aimed to analyse the movements of a planar representation of a tether system comprising two masses and a solid cylinder representing a tether. This planar system is shown in Figure 21. The equation of motion for the system can be derived using the Lagrange equation as shown in Equation (6), and just as in previous sections.

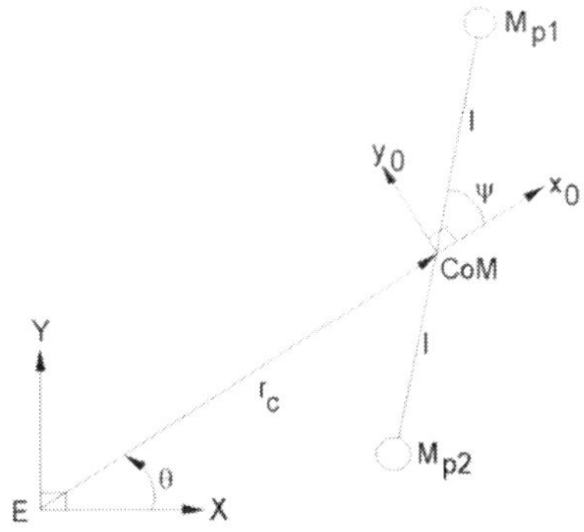

Fig. 21. Planar tether system

The equation of motion is inherently nonlinear and can be numerically integrated for different initial conditions. The angular displacement, angular velocity and tangential velocity of the payload are plotted so that the behaviour of a librating tether system in LEO can be analysed. Examples of the plots from this exercise are shown in Figures 22, 23 and 24.

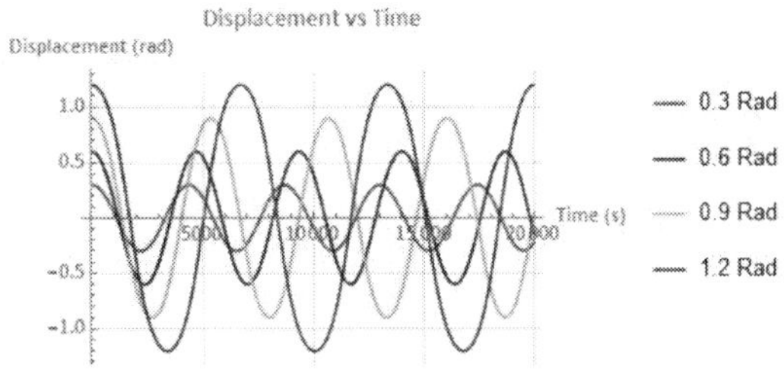

Fig. 22. An example of the angular displacement vs time chart for a tether system.

Further work is required to realise fully the deorbit of payloads using tethers. However, tethers offer a far more financially profitable and environmentally friendly solution to current payload deorbit techniques and could revolutionise payload transfer in space if suitably developed. This numerical study offers a means for the analysis of how the initial conditions of a system can effect its libration, and thus the deorbit of the payload.

*Corresponding author : matthew.cartmell@strath.ac.uk

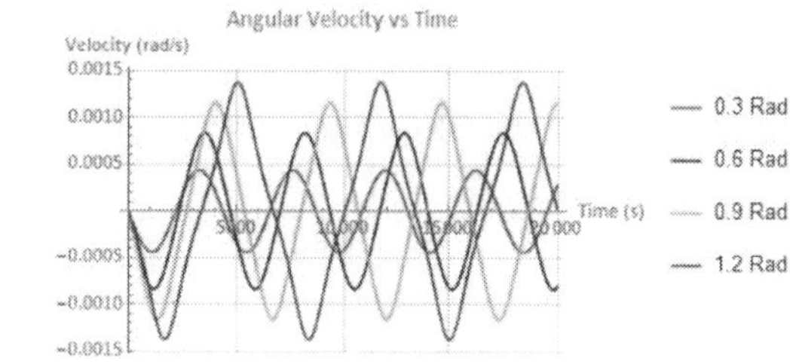

Fig. 23. An example of the angular velocity vs time chart for a tether system

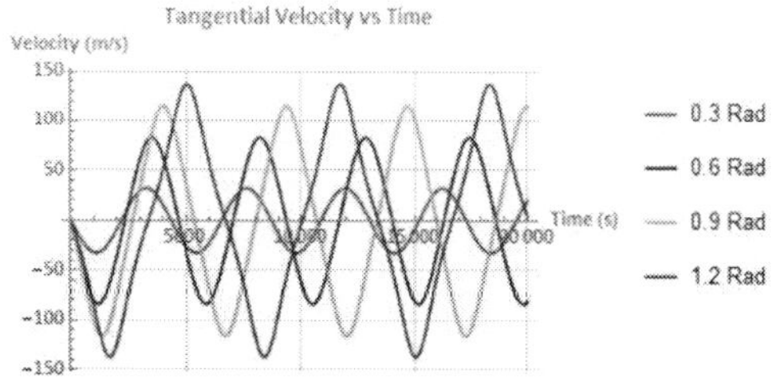

Fig. 24. An example of the tangential velocity vs time chart for a tether system.

Conclusions

Motorised tethers potentially offer clean, green, reusable propulsion in space. Many practical (equatorial and even polar) orbits could in principle be accommodated. The dynamics of space tethers are complicated, and so robust GNC and compensation propulsion will be required in all practical scenarios, particularly in asymmetrical conditions, to safeguard against failure. It may be still possible to operate a motorised momentum exchange tether on a pre-specified orbit after catastrophic payload loss - if controlled axial thrust is provided by suitable emergency on-board payload propellant. METs can almost certainly be used for small sample return from an Earth orbiter. A Lunavator with *reeled end-cables* may offer a practical solution for smooth lunar payload delivery and retrieval. Both these applications are currently receiving further attention, with the goal of integrating them within practical mission architectures in the near future.

References

1. M.P. Cartmell, *Generating Velocity Increments by means of a Spinning Motorised Tether*, AIAA-98-3739, Cleveland, Ohio, USA (1998)
2. M.P. Cartmell, D.J. McKenzie, Progress in Aerospace Sciences, **44**, 1-21 (2008)
3. S.W.Ziegler, *The Rigid Body Dynamics of Tethers in Space*, PhD Thesis, University of Glasgow (2003)
4. D.J. McKenzie, *The Dynamics of Tethers and Space-Webs* PhD Thesis, University of Glasgow (2010)
5. N.A. Ismail, *The Dynamics of a Flexible Motorised Momentum Exchange Tether (MMET)*, PhD Thesis, University of Glasgow (2011)
6. M.P. Cartmell, C.R. McInnes, *Modern Astrodynamics*, Chapter 7, Elsevier Astrodynamics Series, Edited by P.Gurfil (2006)
7. C. Murray, *Continuous Earth-Moon Payload Exchange using Motorised Tethers with Associated Dynamics*, PhD Thesis, University of Glasgow (2011)
8. G. Fu, *Research on Asymmetrical Motorised Momentum Exchange Tether*, Final Research Report, Department of Mechanical Engineering, University of Sheffield, (2016)
9. Y. Chen, R. Huang, X. Ren, L. He, Y. He, *History of the Tether Concept and Tether Missions: A Review*, ISRN Astronomy and Astrophysics, vol. **2013**, Article ID 502973, 7 pages, 2013. doi:10.1155/2013/502973

*Corresponding author : matthew.cartmell@strath.ac.uk

Dynamics of nonlinear waves in the tubes filled with aerosol

Damir Gubaidullin[1,*], *Rinat* Zaripov[1], and *Liudmila* Tkachenko[1]

[1]Institute of Mechanics and Engineering, Kazan Science Center, Russian Academy of Sciences, 420111 Kazan, Russia

Abstract. The results of experimental investigations of nonlinear oscillations of finely dispersed aerosol in the tube with various geometry on the end in the shock-wave, the shock-free wave modes and in the mode of transition to shock waves near the resonance frequency are presented. The time dependences of the numerical concentration of the oscillating aerosol droplets are presented. The effect of the frequency and amplitude of the piston displacement and the influence of the diaphragm internal diameter on the time coagulation and sedimentation of aerosol were studied. An increase in the amplitude of the piston displacement in all modes results in acceleration of the process of coagulation and sedimentation of aerosol. The dependence of time of coagulation and sedimentation of aerosol on the excitation frequency was found to be of a nonmonotonic character with the minimum value upon the resonance frequency.

1 INTRODUCTION

Various inhomogeneous media with droplets or solid particles are the main working bodies in many technological processes. The study of the wave processes in such media is one of important problems in the dynamics and thermophysics of multiphase systems [1]. Technologies created on the basis of the wave principle can be used for cleaning of gases, including high-temperature (flue, smoke) and radioactive from solid fine-dispersed inclusions; for sedimentation of aerosols, mists, including toxic, chemically reactive in limited volumes; for trapping of water vapor and other liquids in thermal plants, in particular, in cooling towers. The study of coagulation and sedimentation of aerosols under influence of wave acoustic fields is of great importance for the development of such technologies [2-4]. Some results on the coagulation and sedimentation of aerosols at resonance oscillations in tubes with different conditions on the ends are considered in experimental works [5-15].

In this paper, detailed results of experimental studies of the dynamics of nonlinear waves in closed and open tubes filled with aerosol are presented in the shock-wave and the shock-free wave modes and in the mode of transition to shock waves near the resonance frequencies.

2 NONLINEAR OSCILLATIONS OF AEROSOL IN TUBES IN THE SHOCK-WAVE MODE

2.1 Experimental setup

The experimental setup consisted of tube-resonator, generator of aerosol oscillations, oscillation frequency measuring unit containing frequency transducer and frequency meter, photo-transducer registering light-penetrability of medium, aerosol generator, laser spectrometer, combined multimeter and digital chamber. Longitudinal oscillations of aerosol column inside the tube with different length of L_0 = 2.7, 3.7, 4.7, 6.7 m and inner diameter $2R_0$ = 0.048 m were created by the flat piston. The tube is made of stainless sections and was joined to the cylinder head with inner diameter $2R$ = 0.077 m through a cone reducer of height h = 0.22 m. The cone reducer serves to obtain greater forced aerosol oscillations in comparison with the tube of uniform cross-section. One section in length of 1.2 m has been made of a quartz glass and established in the middle of the tube in experiment with the closed tube and on distance of 1.87 m from cone reducer in experiment with the open tube. The photo-transducer attached to the quartz section of the tube permitted us to register the aerosol transparency at different particle concentration of droplets in a medium. The excitation frequency ν varied from 0 to 18 Hz and was controlled by the frequency meter with accuracy of 0.1 Hz. Aerosol is turned out from the liquid Di-ethyl-hexyl-sebacate $C_{26}H_{50}O_4$ (DEHS) by means of ATM 225 (TOPAS) aerosol generator. Average geometrical diameter droplets of aerosol in undisturbed state were measured by LAP 320 laser spectrometer (TOPAS) and equalled 0.863 μm (Figure 1).

2.2 Experimental results

As a result of experimental researches dependences of numerical concentration of droplets of aerosol on time have been received. They are approximated by dependence

$$N = A_0 + A_1 \exp(-K_1 t) + A_2 \exp(-K_2 t),$$

* Corresponding author: damirgubaidullin@gmail.com

 https://doi.org/10.1051/matecconf/201814801002

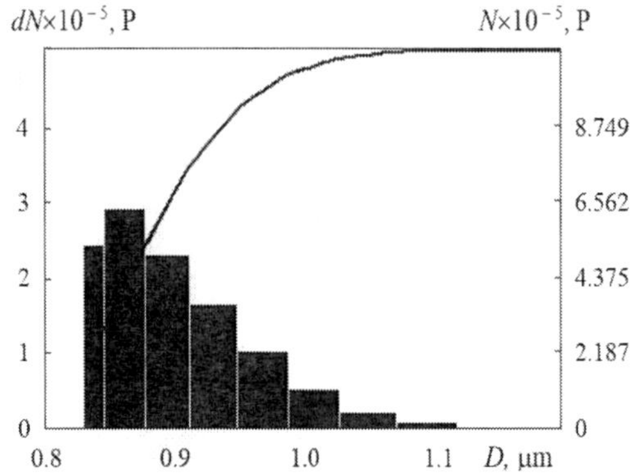

Fig. 1. Distribution of the unperturbed weighed phase of a dispersion medium on the geometrical diameter.

where coefficients are from experiment. First two terms well describe experimental data at oscillations of aerosol far from a subharmonic resonance. The third term takes into account nonlinear effects in a resonance (shock waves, secondary flows, pulsing jet of aerosol from the open end). Concentration of droplets of aerosol for all experiments monotonously decreases in due course. The above frequency of excitation, the occurs this process faster. In the closed tube, the process of decrease of droplets concentration of aerosol in due course includes coagulation of droplets and their sedimentation on walls of the tube. In the open tube, the process of decrease of concentration of the droplets of aerosol in due course includes basically coagulation of droplets, their sedimentation on walls of the tube, and also non-stationary emission of aerosol from the open end of the tube as a pulsing jet. The time of coagulation and sedimentation of aerosol in the closed tube in the shock-wave mode on subharmonic resonances is up to 5 times less, than at the natural sedimentation. In the open tube, the natural sedimentation lasts up to 30 times more slowly than time of coagulation and sedimentation of aerosol in the shock-wave mode.

Dependences of time of coagulation and sedimentation of aerosol on the excitation frequency for different values of length of the closed tube are given in Figure 2. Every point in the curves corresponds to the certain experiment. It is seen from the Figure 2 that for tubes of $L_0 = 2.7$, 3.7, 4.7 m length with increasing frequency in pre-resonant regime, time of coagulation and sedimentation of aerosol decreases. It takes place due to the fact that as the frequency approaches the resonance aerosol oscillation amplitude increases. For the tube of 6.7 m length, the minimum of time of coagulation and sedimentation of aerosol occurs at the 12 Hz frequency, afterwards it starts to increase. It is a matter of fact that at this frequency, subharmonic resonance occurs in the tube with a frequency two times lower as the fundamental one of a medium $\nu_{1/2} = c_0/4L$ (c_0 is the equilibrium velocity of sound in aerosol [16], and $L = L_0 + l + (m^2 + m + 1)h/3$ is the reduced tube length, where $l = m^2 l_0$, $m = R/R_0$), the oscillation amplitude reaching maximum at this frequency. For tubes of $L_0 = 2.7$, 3.7, 4.7 m length, the resonant frequencies lie higher than 18 Hz, therefore they were not reached in the present experiment.

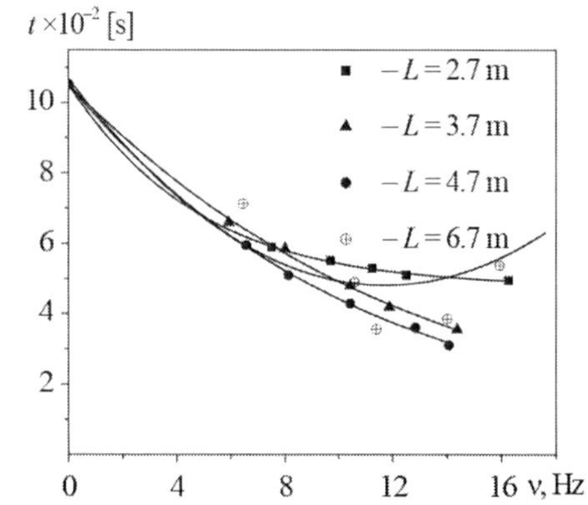

Fig. 2. Dependences of time of coagulation and sedimentation of the aerosol on frequency in the closed tube. The points indicate experimental data, and the solid lines indicate the polynomial approximation.

On Figure 3 dependencies of time of coagulation and sedimentation of aerosol on frequency of excitation for various lengths of the open tube are given. Note that the curve for the tube of $L_0 = 4.7$ m length has a minimum and a maximum. The minimum corresponds to frequency 9 Hz of a subharmonic resonance $\nu_{1/2} = c_0/8L$ ($L = L_0 + l + (m^2 + m + 1)h/3 + \sigma R_0$, where σR_0 Rayleigh correction with $\sigma = 0.6133$). Behind the resonance time of coagulation and sedimentation of aerosol increases up to the maximum and then quickly decreases. Thus, time of coagulation and sedimentation of aerosol in the open tube decreases in two and more times in comparison in due course coagulations and sedimentation in the closed tube.

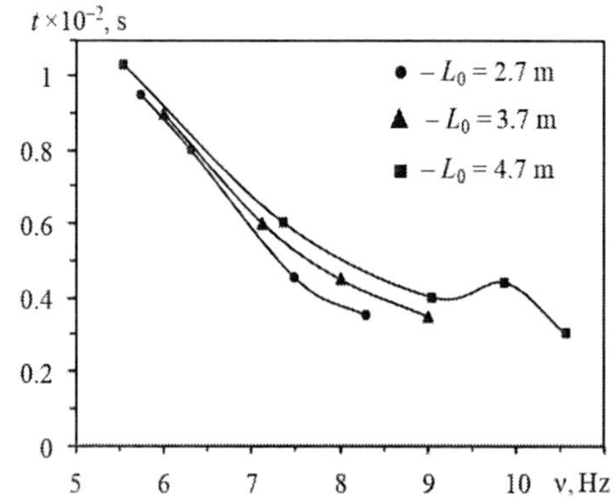

Fig. 3. Dependences of time of coagulation and sedimentation of aerosol on frequency in the open tube. The points indicate the experimental data, and the solid lines indicate the polynomial approximation.

Also, influence of the diaphragms at the passive end of the tube on the process of coagulation and sedimentation of aerosol is studied. In Figure 4 the dependence of time of coagulation and sedimentation of aerosol on the excitation frequency for the closed and the open tubes without diaphragm, as well as the open tube with diaphragms of different internal diameters is given. Figure 4 shows that decrease in the internal diameter of the diaphragm $2R_{inf}$ results in the increase of time of coagulation and sedimentation of aerosol, which reaches the maximum value in the case of the closed tube. This is due, firstly, to significant influence of the unsteady discharge of aerosol from the open end of the tube, and secondly, with the approach of the investigated excitation frequencies of aerosol in the open tube to its resonance frequency [17, 18]. Note that time of coagulation and sedimentation of aerosol in the open tube with the diaphragm is longer than in the open tube without a diaphragm. This is due to the fact that the open tube with the diaphragm which is in connection with the tube forms a right angle, the consumption on the formation of vortexes near the outlet section is greater than in the tube with a rounded open end [19].

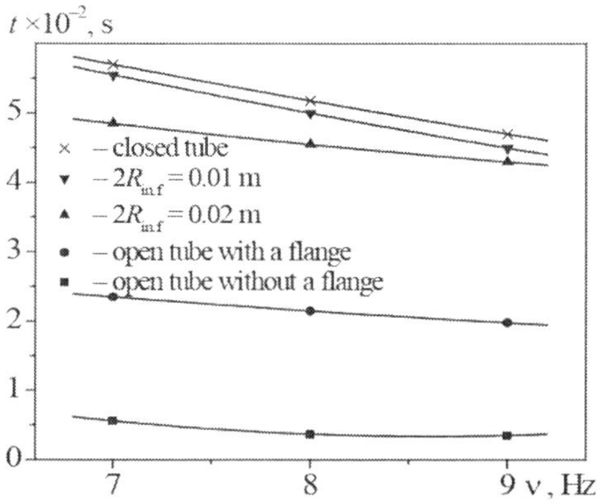

Fig.4. Dependence of time of coagulation and sedimentation of aerosol on frequency in to-resonance regime for the tube of L_0= 2.7 m length with diaphragms of different internal diameters. The points indicate the experimental data, and the solid lines indicate the polynomial approximation.

3 DYNAMICS OF AEROSOL AT OSCILLATIONS IN THE SHOCK-FREE WAVE MODE AND IN THE MODE OF TRANSITION TO SHOCK WAVES

3.1 Experimental setup

The investigation was carried out on an experimental setup containing the tube, the system of oscillations excitation, and the system of detection of process parameters. The oscillations of aerosol in closed and open tubes were excited with the flat piston. Studies in the shock-free wave mode were carried out in the tube L = 1.06 m long and R_0 = 0.0365 m in radius using TIRAvib 51075 vibrostend, and in the mode of transition to shock waves in the tube L = 0.938 m long and R_0 = 0.1 m in radius using TIRAvib S 5220/LS vibrostend. The measurements and control of sinusoidal oscillations were performed using software module installed on computer, by means of the integrated electronic piezoelectric accelerometer and aerosol, we used 8530C-15 (Bruel and Kjaer) pressure gauge with ENDEVCO-136 (Bruel and Kjaer) three-channel bridge voltage amplifier and DSO 3062A (Agilent Technologies) digital oscillograph. For determining of aerosol transparency, we used the complex consisting of the light source and ATT-1505 (Aktakom) digital luxmeter. As before, DEHS was used as the operating liquid.

3.2 Experimental results

The time dependences of pressure oscillations and numerical concentration of aerosol droplets were obtained for various values of amplitude of the piston displacement near the first eigenfrequency. In the shock-free wave mode, this frequency is equal to 160.7 Hz in the closed tube and 78.6 Hz in the open tube and in the mode of transition to shock waves, 182 Hz and 90.3 Hz, respectively. In all these cases, numerical concentration of aerosol decreases monotonically with time, as in the case of the shock-wave mode. This decrease is faster with increase of the amplitude. For the natural sedimentation, a smoother decrease in time for aerosol concentration is observed. The process time exceeds 2-4 and 6-12 times that in the presence of oscillations in the shock-free wave mode and exceeds 2.5-5 and 3-18 times that in the presence of oscillations in the mode of transition to shock waves, in the closed and the open tubes, respectively. As in before, we accepted that time of change in the numerical concentration in the tubes is the time of coagulation and sedimentation of aerosol.

In Figure 5, we plot the dependences of the specified times on the relative frequency of excitation (a) and also present the diagrams of oscillations of aerosol pressure for the closed and the open tubes at the first eigenfrequency (b) in the shock-free wave mode for amplitude of the piston displacement l = 0.35 mm. Figure 6 shows the similar dependence for the mode of transition to shock waves for amplitude of the piston displacement l = 0.3 mm.

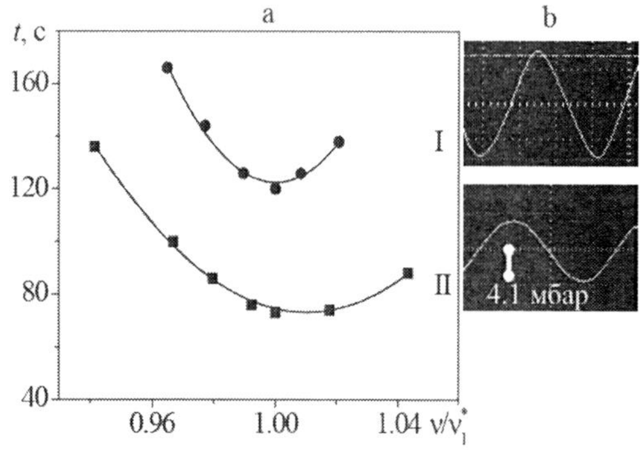

Fig.5. Dependences of time of coagulation and sedimentation of aerosol on the frequency (a) and aerosol pressure oscillations in time near the piston (b) for (I) closed and (II) open tubes in the shock-free wave mode. The lines show the polynomial approximation.

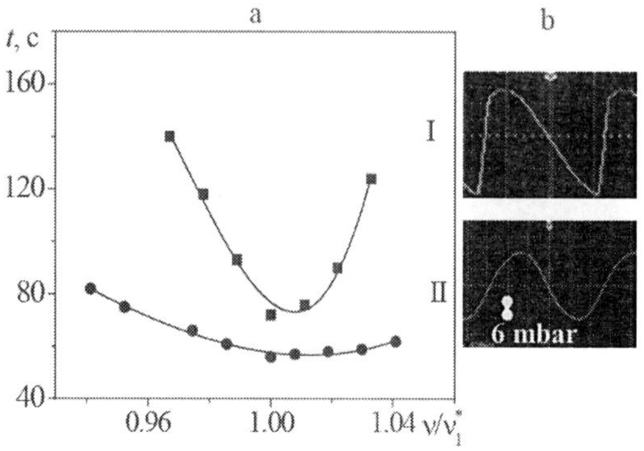

Fig.6. Dependences of time of coagulation and sedimentation of aerosol on the frequency (a) and aerosol pressure oscillations in time near the piston (b) for (I) closed and (II) open tubes in the mode of transition to shock. The lines show the polynomial approximation.

The dependences of time of coagulation and sedimentation of aerosol on the frequency have a nonmonotonic character with the minimum at the first eigenfrequency, when the amplitude of aerosol pressure oscillations is the largest. In the closed tube at the shock-free wave mode deformation of the pressure-wave shape (compression time is less the rarefaction time) is observed. In the open tube, both in the shock-free wave and in the mode of transition to shock waves the pressure-wave shape is close to a sine wave even in the resonance. In these cases, the change in time of coagulation and sedimentation of aerosol upon approaching the first eigenfrequency is insignificant. At the same time, a sharp decrease in time of coagulation and sedimentation of aerosol in the closed tube in the mode of transition to shock waves, when the pressure-wave shape is close to discontinuous, occurs. In both cases, this time in the open tube is approximately 1.5 times less than in the closed tube. Such an insignificant difference in times of the processes is associated with the fact that, although a part of aerosol is thrown out into the medium in the open tube, as was already noted above, distortion of the aerosol-pressure-wave shape takes place in the closed tube. It is possible to note that, in the mode of transition to shock waves, the time of coagulation and sedimentation of aerosol in the closed and the open tubes is 1.5 times less than the corresponding time in the shock-free wave mode of oscillations for a practically identical amplitude of the piston displacement.

5 CONCLUSIONS

Time of coagulation and sedimentation of aerosol decreases with increase in the intensity of oscillations caused by increase in the amplitude of the piston displacement or decrease in the length of the tube and with growth in the internal diameter of the diaphragm at the passive end of the tube. The dependence of time of coagulation and sedimentation of aerosol on the excitation frequency was found to be of a nonmonotonic character with the minimum value upon the resonance frequency. At nonlinear resonance oscillations of aerosol in the mode of transition to shock waves, more efficient coagulation and sedimentation of aerosol are observed than in the shock-free wave mode at the same small amplitudes of the piston displacement. In the shock-wave mode (at the large amplitudes of the piston displacement), the process is accelerated even more.

The study was performed by a grant from the Russian Science Foundation (project No. 15-11-20022).

References

1. R. I. Nigmatulin, *Dynamics of Multiphase Media* (Hemisphere, New York, 1990)
2. E. P. Mednikov, *Acoustic Coagulation and Precipitation of Aerosols* (Consultant Bureau, New York, 1965)
3. N. A. Fuks, *Aerosol Mechanics* (USSR Acad. Sci., Moscow, 1955)
4. S. Temkin, *Suspension Acoustics: an Introduction to the Physics of Suspensions* (Cambridge University Press, New York, 2005)
5. A. I. Gulyaev, V. M. Kuznetsov, Sov. Phys. Acoust. **8**, 4, 236–238 (1963)
6. S. Temkin, Phys. Fluids **13**, 1639–1641 (1970)
7. K. Shuster, M. Fichman, A. Goldshtein, C. Gutfinger, Phys. Fluids **14**, 5, 1802–1805 (2002)
8. D. A. Gubaidullin, R. G. Zaripov, R. G. Galiullin, E. R. Galiullina, L. A. Tkachenko, High Temperature **42**, 5, 794–802 (2004)
9. D. A. Gubaidullin, R. G. Zaripov, L. A. Tkachenko, High Temperature **50**, 4, 564–566 (2012)
10. D. A. Gubaidullin, R. G. Zaripov, L. A. Tkachenko, High Temperature **51**, 6, 955–957 (2013)
11. D. A. Gubaidullin, R. G. Zaripov, L. A. Tkachenko, Doklady Physics **58**, 9, 392–395 (2013)
12. D. A. Gubaidullin, R. G. Zaripov, L. A. Tkachenko, J. of Engin. Physics and Thermophysics **85**, 2, 259–264 (2012)
13. D. A. Gubaidullin, R. G. Zaripov, L. A. Tkachenko, High Temperature **52**, 6, 895–899 (2014)
14. D. A. Gubaidullin, R. G. Zaripov, L. A. Tkachenko, L. R. Shaidullin, High Temperature **55**, 3, 463–465 (2017)
15. D. A. Gubaidullin, R. G. Zaripov, L. A. Tkachenko, Doklady Physics **62**, 7, 363–365 (2017)
16. M. A. Ilgamov, R. G. Zaripov, R. G. Galiullin, V. B. Repin, Appl. Mech. Rev. **49**, 3, 137–154 (1996)
17. J. Jimenez, Fluid Mech. **59**. 23–46 (1973)
18. B. B. Sturtevant, J. Fluid Mech. **63**, 97–120 (1974)
19. B. B. Sturtevant and J. J. Keller, Z. Angew. Math. Physics **29**, 3, 473–485 (1978)

Experimental Determination of Rigid Body Properties: An Evaluation on the Use of Piezoelectric or MEMS Tri-Axial Accelerometers

António Urgueira, *Nuno* Venâncio, *Pedro* Riscado, *Raquel* Almeida, *Tiago* Silva*

NOVA UNIDEMI, Department of Mechanical and Industrial Engineering, FCT-NOVA, 2829-516 Caparica, Portugal

Abstract. The development of new sensors that are available at more accessible prices may lead to the spread of their use on common studies in structural dynamics. One of areas of interest is the experimental determination of rigid body properties that are mandatory when the vibration response is to be calculated at low frequency ranges. In this work, a comparison of the experimental determination of rigid body properties is carried out to evaluate the performance of the commonly used tri-axial piezoelectric accelerometers and their equivalent MEMS sensors. Although their prices are quite different, both sensors can measure the inertia restraint line that is related to the inertia properties of the tested object. An identification algorithm is applied to the frequency response functions obtained by using both sensors, leading to the estimation of the body mass value, as well as the three coordinates of the centre of mass and the six elements of the inertia tensor. An experimental example supports the use of the referred low-cost sensors.

1 Introduction

Whenever the dynamic behaviour of a structure at very low frequencies is important, the ten rigid body properties of a mechanical system is of vital importance. These rigid body properties are mass, position of the centre of mass and inertial moments and they may be estimated using numerical or analytical approaches or using experimental driven procedures. The latter have been addressed by several authors [1-10], considering different methods that can be divided into two main categories: Frequency domain methods [1-5] and Time domain methods [6,7].

The Time domain methods [6,7] present advantages mainly due to the fact that there is no need to transform the acquired data into the frequency domain, although care should be taken whenever the structure does not behave like a rigid body in the frequency range of interest. This difficulty can be circumvented by using Frequency domain methods because it is possible to separate the rigid and the elastic system behaviour. The methods based on the Frequency domain are mainly divided into three groups: Inertia Restraint methods (IRM) [1,3], Methods of Direct Physical Parameter Identification [2] and the Modal methods (MM) [4, 11-14].

In this work, the authors focused their attention on the use of IRM or the so-called Mass Line methods (MLM) that are based on the principle that the dynamic response of structures in free-free conditions are characterized, in the low frequency region, by a constant term designated as inertial restraint or mass line. This approach is suitable for structures where rigid and flexible modes are well separated. Note that the mass line is the modal constant at zero Hz of a typical frequency response function (FRF) and can be determined according to the methods described in [15,16]. The inertia properties can then be estimated using direct or iterative approaches.

The authors recall here an IRM that was used by several authors [2-4] that is based on several measured accelerations at different locations due to applied excitations at different points. As discussed by Lee [9] the accuracy of this method heavily depends on the selection of the measuring locations, as well as, the conditions of the excitation.

The IRM is here applied to a case study where one has considered the use of two types of sensors, tri-axial piezoelectric accelerometers and their equivalent MEMS sensors, to acquire the measurements at different test configurations. Doing so, one aims at assessing the accuracy of an acquisition system significantly affordable, in economic terms, in comparison to the traditional use of piezoelectric accelerometers. The rigid body properties are experimentally estimated, and the results are discussed.

Note that the MEMS sensor type used is capacitive based displacement accelerometers, which measures the difference of capacitance by the displacement of a proof-mass between reference plates [17]. These sensors are lighter and have a very low power consumption when compared to the equivalent piezoelectric accelerometers. The latter have a broader bandwidth, which may be very important for modal studies, but it is not paramount

* Corresponding author: tan.silva@fct.unl.pt

regarding the application, where one is only interest in the rigid body motion.

2 Theoretical background

As referred, the IRM is based on a set of measured FRFs, specifically accelerance (acceleration/force). In the present work, one uses tri-axial piezoelectric accelerometers in a first stage and after their equivalent MEMS sensors to measure up to three accelerations at a point p_i of the structure, in local coordinates. Hence, the vector of measured translational responses at a given location is,

$$\left\{\ddot{x}_{m_{p_i}}\right\}=\left[\ddot{u}_{m_{p_i}} \quad \ddot{v}_{m_{p_i}} \quad \ddot{w}_{m_{p_i}}\right]^T \tag{1}$$

where u, v and w are the displacements in the three axis, and the double dot represents the second time derivative. Considering k sensors, the vector containing all the responses at different locations can be cast as,

$$\left\{\ddot{x}_m\right\}=\left[\left\{\ddot{x}_{m_{p1}}\right\}^T \quad \left\{\ddot{x}_{m_{p2}}\right\}^T \quad \cdots \quad \left\{\ddot{x}_{m_{pk}}\right\}^T\right]^T \tag{2}$$

However, one needs to express all the measured responses regarding a global referential for which the rigid body properties are estimated. Thus, using the transformation matrix $\left[G_{m0}\right]$, one can transform the measured responses to the origin of a given global referential 0 as

$$\left\{\ddot{x}_0\right\}=\left[G_{m0}\right]^+\left\{\ddot{x}_m\right\} \tag{3}$$

where the superscript + is the pseudo-inverse operator. Recalling the Newton's second law, one can relate the exciting forces to the measured responses, being

$$\left\{f_0\right\}=\left[M_0\right]\left\{\ddot{x}_0\right\} \tag{4}$$

where $\left[M_0\right]$ is a inertia matrix given by

$$\left[M_0\right]=\left[\begin{array}{ccc|ccc} m & 0 & 0 & 0 & mz_g & -my_g \\ 0 & m & 0 & -mz_g & 0 & mx_g \\ 0 & 0 & m & my_g & -mx_g & 0 \\ \hline 0 & -mz_g & my_g & J_{0_{xx}} & -J_{0_{xy}} & -J_{0_{xz}} \\ mz_g & 0 & -mx_g & -J_{0_{yx}} & J_{0_{yy}} & -J_{0_{yz}} \\ -my_g & mx_g & 0 & -J_{0_{zx}} & -J_{0_{zy}} & J_{0_{zz}} \end{array}\right] \tag{5}$$

where m is the mass, x_g, y_g and z_g are the coordinates of the centre of mass, and $J_{0_{ij}}$ are the elements of the inertia tensor.

From eq. (4), it is possible to compute all the referred rigid body properties, although one must assure that the number of measurements and exciting forces is sufficient to avoid an ill-posed problem and to fulfil the recommendations of [9]. Note that the measured accelerations of eq. (2) are the mass line constant or the modal constant at zero Hz, that may be obtained using a modal identification or a curve-fitting method.

To demonstrate the applicability of the presented method and to evaluate the use of the MEMS tri-axial accelerometers, in comparison to the traditional piezoelectric ones, one presents the following experimental case study.

Fig. 1. Rigid body under study.

3 Case study

In this work, one has selected a structure to be studied that has important property (Fig. 1). This structure does not present any plane of symmetry and therefore it exhibits moments and products of inertia when one considers the coordinate system of reference aligned to

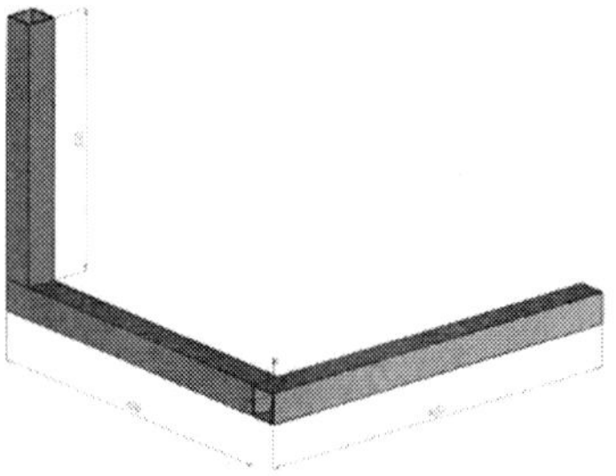

Fig. 2. Model of the rigid body under study (see Fig. 1).

Table 1. Theoretical rigid body properties, regarding the origin of the referential in Figs. 1 and 2.

$mass = 1.23\ kg$	$\rho = 7200 kgm^{-3}$	
$x_g = 0.169m$	$y_g = -0.089m$	$z_g = 0.020m$
$J_{g_{xx}} = 0.023029kgm^2$	$J_{g_{yy}} = 0.023029kgm^2$	$J_{g_{zz}} = 0.023029kg$
$J_{g_{xy}} = 0.023029kgm^2$	$J_{g_{xz}} = 0.023029kgm^2$	$J_{g_{yz}} = 0.023029kg$

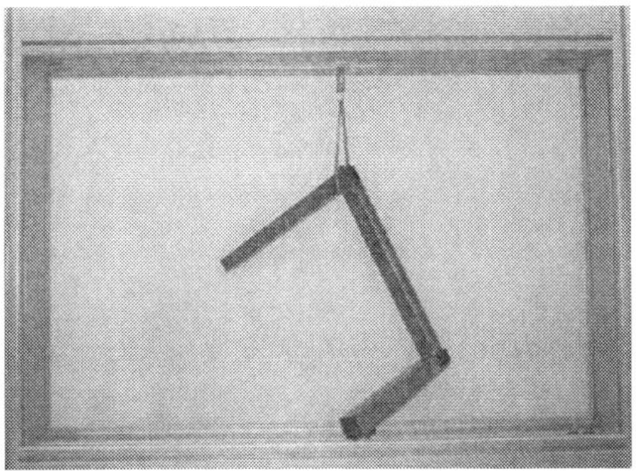

Fig. 3. Soft suspension of the tested structure.

the three components that form the overall structure. These six elements of the inertia tensor together with the three coordinates of the centre of mass and the value of the mass, form the ten unknowns of this experimental method (section 2).

The theoretical model of the structure is built using a CAD software (Fig. 2). The rigid body properties of the model are given in Table 1. These properties will be taken as reference for comparison with the values obtained by the experimental application of the IRM. Note that the reference coordinate system is marked both in Figs. 1 and 2.

The physical structure, shown in Fig. 1 was built up based on the needs of the present experimental work. This structure is made of steel and it was obtained by welding to each other, three rectilinear components.

The experimental work was carried out by assuming the structure to be freely suspended. The soft suspension of the tested structure is shown in Fig. 3, allowing the free movement of the rigid structure, as it is required by the IRM.

The experimental FRFs at different points and directions on the structure were obtained using both type of tri-axial translational sensors, the piezoelectrical and MEMS ones, and an instrumented impact hammer. The experimental data was acquired using an acquisition system built at our lab, mainly due to the work of Riscado [18], in the frequency range of 0-400 Hz.

The acquisition system for both types of sensors was built by using a NI DAQ 9172 chassis, as it is shown in Fig. 4. It collects data from the accelerometers triggered by an instrumented impact hammer (PBC 086C03 model). In this work, one has used the tri-axial translational piezoelectric accelerometers B&K 4506 B and MEMS sensors ADXL326 (lighter and cheaper).

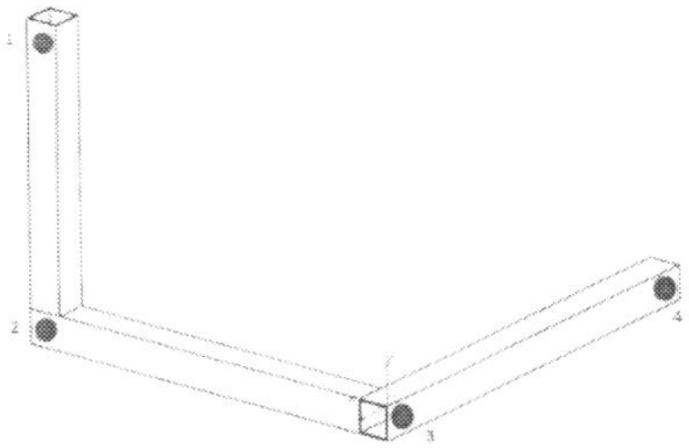

Fig. 5. Reference points on the structure.

Table 2. List of applied forces and sensors positions for each test case (according to Fig. 5).

	Measuring configuration	Exciting configuration
Test 1	**Conf. A** (locations: 1,2,3)	**Conf. I** (1Y, 2Z, 4X)
Test 2	**Conf. B** (locations: 1,2,4)	**Conf. I**
Test 3	**Conf. A**	**Conf. II** (1X, 2Z, 2Y)
Test 4	**Conf. B**	**Conf. II**

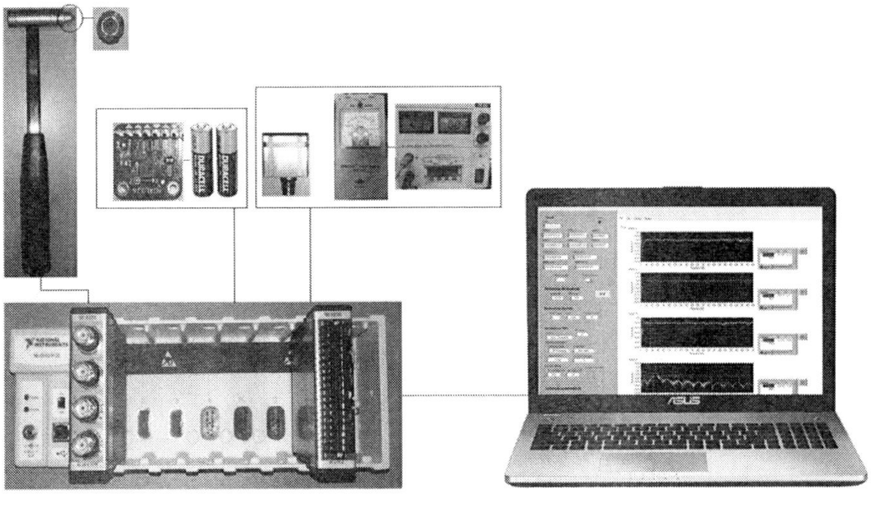

Fig. 4. Schematic representation of the acquisition system used with both types of sensors.

This acquisition system is implemented using LabView that allows the experimental data to be exported and treated in MatLab, for further developments.

In terms of the experimental procedure, different test cases were considered to evaluate the influence, on the IRM accuracy, of the position and orientation of the sensors and set of applied forces, as described in Table 2 in conjunction with the representation given in Fig. 5.

For each test case (1-4 in Table 2), one measures 9 FRFs per applied force. Note that the exciting configurations given in Table 2 consider a single excitation at a given location and direction, accordingly to a certain axis, and that one uses three different excitations per test.

The entire set of FRFs for each test case is then used of compute the rigid body properties, that cannot be measured directly.

Note that all the pairs sensor/impact hammer were calibrated using a suspended block of known mass.

4 Results and discussion

In this section, one shows the most significant results, aiming at evaluating the accuracy of the estimated rigid body properties. Note that the values of the referred properties are shown as a relative error regarding the values of the CAD model (Table 1). Moreover, one also aims at evaluating the performance of the MEMS accelerometers in comparison to the piezoelectric ones.

Hence, from the set of measured FRFs for each test case, here accelerance FRFs (acceleration/force) one obtains the rigid body properties. Figs. 6 and 7 show the values of the relative error for each estimated property, considering the data from both sensors (MEMS in red and piezoelectric in blue). Note that the error at 0 is the one obtained for the comparison of the model value against itself.

From Figs. 6 and 7, it can be inferred that the location of the sensors and of the location and orientation of the exciting forces have a moderated impact on the estimated quantities. Moreover, the use of the MEMS accelerometers is as accurate as the traditional, well-known, piezoelectrical accelerometers. This is a very important observation, regarding the cost difference between them. From the results, there are no evidences to support a decision towards the choice for a particular type of sensor. It appears that the results accuracy is more related to the configuration of each test case, besides the measurement noise and errors from the identification of the mass line, as there are always some differences in the structure in each test case.

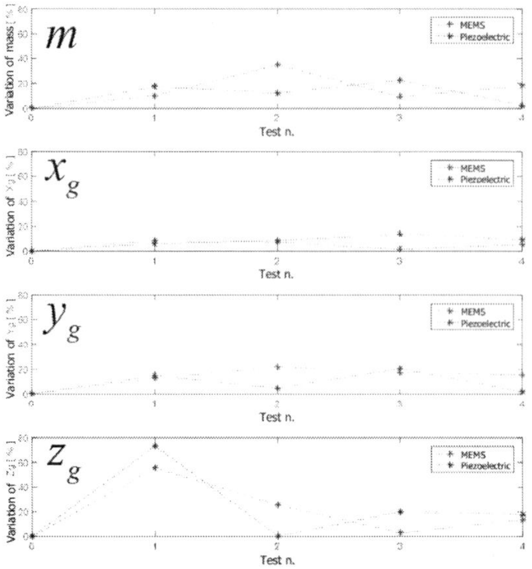

Fig. 6. Relative errors for the estimated rigid body properties for each test case: mass and coordinates of the centre of mass.

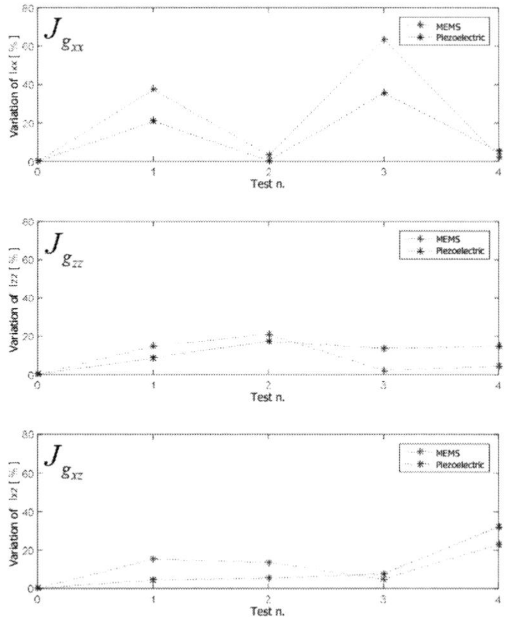

Fig. 7.a. Relative errors for the estimated rigid body properties for each test case: elements of the inertia tensor.

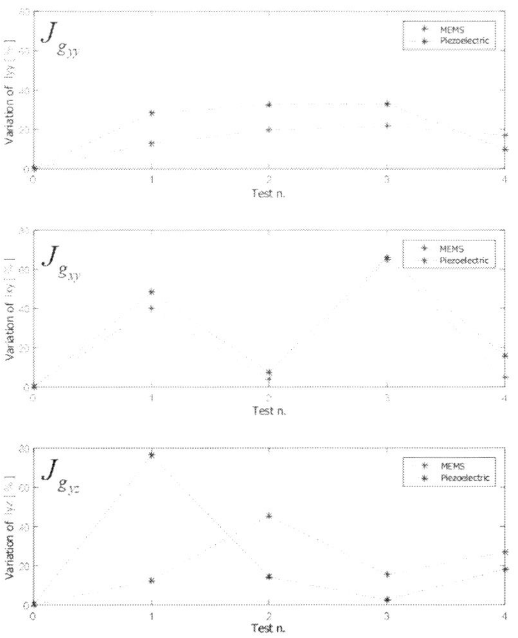

Fig. 7.b. Relative errors for the estimated rigid body properties for each test case: elements of the inertia tensor.

5 Conclusion

The IRM is here used to determine the rigid body properties of a non-symmetric structure. The mass line values are extracted from experimental FRFs acquired by two types of tri-axial translational accelerometers, piezoelectric and MEMS. There is no evidence to conclude that the well disseminated piezoelectric sensors have advantage over the MEMS ones. It is important to note that the cost difference between the two types of sensors is very significant. However, as referred in the literature, a special attention to the relative location of the sensors during the testing process is paramount for the accuracy of the estimated rigid body properties, for both types of sensors.

Acknowledgements

The authors acknowledge the support of the Portuguese Foundation for Science and Technology, FCT, under the strategic project PEst-OE/EME/UI0667/2014 (UNIDEMI, FCT-NOVA)

References

[1] J. Bretl, P. Conti, *in Proc. Int. Modal Analysis Conference IMAC V* (1987)

[2] Y. S. Wei, J. Reis, *in Proc. Int. Modal Analysis Conference IMAC VII* (1989)

[3] A. P. V Urgueira, *in Proc. Int. Modal Analysis Conference IMAC XIII* (1995)

[4] P. Conti, J. Bretl, J. Vib. Acoust. Stress. Reliab. Des. **111**, 2 (1989)

[5] J. Toivola, O. Nuutila, *in Proc. Int. Modal Analysis Conference IMAC XI* (1993)

[6] S. M. Pandit, Y.-X. Yao, Z.-Q. Hu, J. Vib. Acoust. **116**, 3 (1994)

[7] S. M. Pandit, Z.-Q. Hu, J. Sound Vib. **177**, 1 (1994)

[8] A. Fregolent, A. Sestieri, Mech. Syst. Signal Pr. **10**, 6 (1996)

[9] H. Lee, Y.-B. Lee, Y.-S. Park, Mech. Syst. Signal Pr. **13**, 4 (1999)

[10] C. Schedlinski, M. Link, Mech Syst Signal Pr. **15**, 1 (2001)

[11] R. A. B. Almeida, A. P. V. Urgueira, N. M. M. Maia, Mech. Syst. Signal Pr. **24**, 5(2010)

[12] R. A. B. Almeida, A. P. V. Urgueira, N. M. M. Maia, J. Sound Vib. **299**, 4-5 (2007)

[13] R. A. B. Almeida, A. P. V. Urgueira, N. M. M. Maia, Shock Vib. **15**, 3-4 (2008)

[14] A. Malekjafarian, M. R. Ashory, M. M. Khatibi, *in Proc. Int. Modal Analysis Conference IMAC XXVII* (2010)

[15] Silva, Maia, He, Lieven, Lin, Skingle, To, Urgueira, *Theoretical and Experimental Modal Analysis*, RSP (1997)

[16] D. J. Ewins, *Modal Testing: Theory and Practice*, RSP (1999)

[17] S. E. Lyshevski, *MEMS and NEMS: Systems, Devices, and Structures*, CRC Press (2002)

[18] P. Riscado, MSc Thesis, UNL-FCT, Portugal, (2016)

Theoretical Study on the Dynamic Behavior of Pipes Conveying Gas-Liquid Flow

L. Enrique Ortiz-Vidal[1,*], *David G.* Castillo[2], and *Quino* Valverde[2]

[1]Mechanical Engineering Department, Sao Carlos School of Engineering, University of Sao Paulo, 13563-120, São Carlos/SP, Brazil
[2]Engineering Department, Mechanical Engineering Section, Pontifical Catholic University of Peru, Av. Universitaria 1801, Lima, Peru

Abstract. The dynamic behavior of clamped-clamped straight pipes conveying gas-liquid two-phase flow is theoretically investigated, specifically the effect of the flow parameters on the frequency of the system. First, the equation of motion is derived based on the classic Païdoussis formulation. Assuming Euler-Bernoulli beam theory, small-deflection approximation and no-slip homogeneous model, a coupled fluid-structure fourth-order partial differential equation (PDE) is obtained. Then, the equation of motion is rendered dimensionless and discretized through Galerkin's method. That method transforms the PDE into a set of Ordinary Differential Equations (ODEs). The system frequency is obtained by solving the system of ODEs by allowing the determinant to be equal to zero. System frequencies for different geometries, structural properties and flow conditions have been calculated. The results show that the system frequency decreases with increasing two-phase flow velocity. By contrast, the former increases with increasing homogeneous void fraction. These theoretical results are in agreement with experimental findings reported in the literature. Furthermore, even for typical two phase flow conditions, the system can become unstable for inadequate chooses of geometry or material of the pipe.

1 Introduction

Pipes conveying gas-liquid two-phase flow are common in nuclear, process and oil&gas industries, where the flow generates dynamics forces. Depending on the system conditions these forces may induce moderate or excessive structural vibrations, leading to structural instability. Then, a proper knowledge of the dynamic behavior of this kind of systems is relevant and has motivated relevant researches over the last five decades, see *e.g.* [1–4]. Fundamental experimental studies have shown a strong influence of two-phase flow parameters, such as mixture velocity, void fraction, flow pattern and slip, on the structural vibration response [5–9]. In the specific case of the frequency of the system, recently experimental results [8] indicate this parameter increases with increasing void fraction due to the effect of hydrodynamic mass (added mass). On the other hand, system frequency decreases with increasing two-phase mixture flow velocity. Theoretical studies on system frequency in order to imitate the effect of those two-phase flow parameters are needed. In general, the reported studies focused on determining the critical mixture velocity for reaching instability. In this paper, the dynamic stability of clamped-clamped straight pipes conveying gas-liquid two-phase flow is theoretically investigated, specifically the system frequency for typical two-phase flow conditions.

2 Theory

2.1 Equation of the motion

The mathematical formulation for pipes conveying fluid presented here is based on the classic Païdoussis formulation [10]. The system is composed by (i) a uniform pipe of length L_{span}, inner cross-sectional area A, internal perimeter S, linear mass density m, and flexural rigidity EI and (ii) a two-phase fluid flow of linear mass density M with mixture flow velocity J. The equation of motion is derived applying Newton's second law over free-body diagrams of the both fluid and pipe differential elements of Fig. 1. Euler-Bernoulli beam theory and small-deflection approximation are adopted. For the fluid element, in the X- and Z- directions, the following equations are obtained, respectively,

$$-A\frac{\partial p}{\partial x} + qS - F\frac{\partial w}{\partial x} = M\frac{\partial J}{\partial t} \tag{1}$$

$$\begin{aligned} &A\frac{\partial}{\partial x}\left(p\frac{\partial w}{\partial x}\right) - qS\frac{\partial w}{\partial x} - F + Mg \\ &= M\left[\frac{\partial^2 w}{\partial t^2} + J^2\frac{\partial^2 w}{\partial x^2} + 2J\frac{\partial^2 w}{\partial x \partial t} + \frac{\partial J}{\partial t}\frac{\partial w}{\partial x}\right] \end{aligned} \tag{2}$$

where w is the deflection of the pipe in the Z-direction. p and g are pressure and gravity, respectively. qS and F are related to reaction forces of the pipe on the two-phase fluid, in tangential and normal directions, respectively.

* Corresponding author: leortiz@sc.usp.br | enrique.ortizvidal@outlook.com

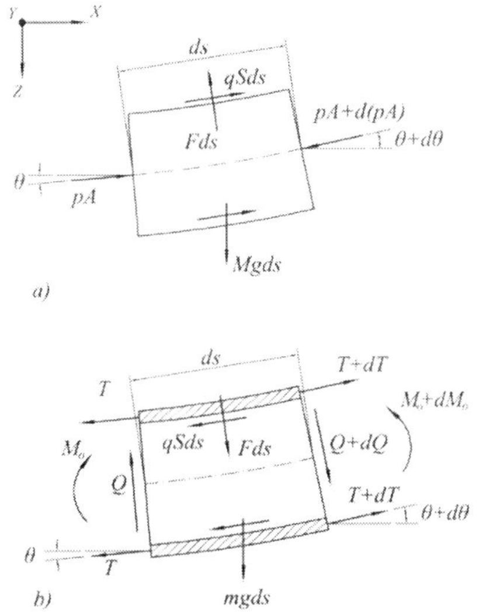

Fig. 1. Free-body diagrams of a differential element ds of a horizontal pipe conveying fluid, (a) fluid and (b) pipe.

In the same form, for the pipe element, it is obtained,

$$\frac{\partial T}{\partial x} - qS + F\frac{\partial w}{\partial x} = 0 \tag{3}$$

$$-\frac{\partial}{\partial x}\left(T\frac{\partial w}{\partial x}\right) + qS\frac{\partial w}{\partial x} + F + \frac{\partial Q}{\partial x} + mg = m\frac{\partial^2 w}{\partial t^2} \tag{4}$$

where T and Q are longitudinal tension and transverse shear force, respectively. Combining Eqs. (1)-(4), and using Païdoussis' procedure for calculation of T and Q, gives the equation of the motion,

$$\begin{aligned} &EI\frac{\partial^4 w}{\partial x^4} + (M+m)g\left((x-L)\frac{\partial^2 w}{\partial x^2} + \frac{\partial w}{\partial x}\right) \\ &+MJ^2\frac{\partial^2 w}{\partial x^2} + 2MJ\frac{\partial^2 w}{\partial t\,\partial x} + (M+m)\frac{\partial^2 w}{\partial t^2} \\ &+EI\frac{\mu}{\Omega}\frac{\partial}{\partial t}\left(\frac{\partial^4 w}{\partial x^4}\right) = 0 \end{aligned} \tag{5}$$

where the terms represent respectively (from left to rigth): flexural force, pressure and tension, centrifugal force, Coriolis force, inertia force and structural damping force. Ω is the system frequency.

2.2 Two-phase flow

For the case of two-phase flow, the mixture flow velocity J, and the two-phase fluid mass M, can be represented by

$$J = J_G + J_L \tag{6}$$

$$M = \rho_M A \tag{7}$$

where J_G and J_L represent the superficial velocities of the gas and liquid, respectively. Superficial velocity of the phase is defined as the volumetric flow rate divided by inner cross-sectional area of the pipe. ρ_M is the mixture density. This parameter depends on the mass density and volume fraction of each phase in the flow. Assuming no-slip homogeneous model [11], the mixture density can be calculated from homogeneous void fraction (λ) by,

$$\rho_M = \rho_L(1-\lambda) + \rho_G\lambda \tag{8}$$

$$\lambda = \frac{J_G}{J} \tag{9}$$

For the purposes of this study, a two-phase flow condition is totally defined when both the mixture flow velocity J and the homogeneous void fraction λ are established.

2.3 Nondimensional equation of motion

The equation of the motion may be rendered dimensionless introducing the following parameters,

$$\xi = \frac{x}{L}, \quad \eta = \frac{w}{L}, \quad \tau = \left(\frac{EI}{M+m}\right)^{1/2}\frac{t}{L^2} \tag{10}$$

Thus, Eq. (5) becomes

$$\eta'''' + \eta'' j^2 + 2\beta^{1/2} j\,\dot{\eta}' + \ddot{\eta} = 0 \tag{11}$$

where $(\dot{})$ and $(\,)'$ indicates time (τ) and space (ξ) derivate of parameter η. Eq. (11) represents the dimensionless equation of motion with neglected gravity and structural damping. The following expressions relate both dimensional and nondimensional parameters,

$$\begin{aligned} j &= \left(\frac{M}{EI}\right)^{1/2} JL_{span}, \quad \beta = \frac{M}{M+m}, \\ \omega &= \left(\frac{M+m}{EI}\right)^{1/2}\Omega L_{span}^2 \end{aligned} \tag{12}$$

where j and β represent dimensionless velocity and mass ratio, respectively. ω is the dimensionless frequency of the system.

2.4 Solution method

Galerkin's method [10] is used to solve Eq. (11). This method transforms the distributed system into discrete one. Thus, the partial differential equation of the system becomes a set of ordinary differential equations. According to the method, the solution may be written as,

$$\eta(\xi,\tau) = \sum_{r=1}^{N}\phi_r(\xi)q_r(\tau), \qquad r = 1,2,\ldots,N \tag{13}$$

where the summation considers the first $r = N$ modes. $\phi_r(\xi)$ are the dimensionless eigenfunctions of a beam with clamped-clamped boundary conditions. $q_r(\tau)$ represent the generalized coordinates of the discretized system. Substituting Eq. (13) in Eq. (11), multiplying by $\phi_s(\xi)$ and integrating over the domain [0,1], is obtained,

$$\sum_{r=1}^{N}\left\{\begin{matrix}\delta_{sr}\ddot{q}_r + \left[2\beta^{1/2} j\, b_{sr}\right]\dot{q}_r \\ +\left[\delta_{sr}\lambda_r^4 + j^2 c_{sr}\right]q_r\end{matrix}\right\} = 0 \qquad (14)$$

where

$$\delta_{sr} = \int_0^1 \phi_s \phi_r d\xi \qquad (15)$$

$$b_{sr} = \int_0^1 \phi_s \phi_r' d\xi, \qquad c_{sr} = \int_0^1 \phi_s \phi_r'' d\xi \qquad (16)$$

The parameter δ_{sr} represents the Kronecker delta. This is equal to 0 for $r \neq s$ and equal to 1 for $r = s$. For a clamped-clamped pipe, the constants b_{sr} y c_{sr} can be expressed by [12],

$$b_{sr} = \begin{cases} \dfrac{4\lambda_r^2\lambda_s^2}{\lambda_r^4 - \lambda_s^4}\left\{(-1)^{r+s} - 1\right\}, & s \neq r \\ 0, & s = r \end{cases} \qquad (17)$$

$$c_{sr} = \begin{cases} \dfrac{4\lambda_r^2\lambda_s^2}{\lambda_r^4 - \lambda_s^4}(\lambda_r\sigma_r - \lambda_s\sigma_s)\left\{(-1)^{r+s} + 1\right\}, & s \neq r \\ \lambda_r\sigma_r(2 - \lambda_r\sigma_r), & s = r \end{cases} \qquad (18)$$

where λ_r, λ_s, σ_r and σ_s are parameters of the eigenproblem of a beam [13]. Eq. (14) can be expressed in standard form by,

$$[\mathbf{M}]\ddot{q} + [\mathbf{C}]\dot{q} + [\mathbf{K}]q = 0 \qquad (19)$$

where **M**, **C** and **K** are, respectively, mass, damping and stiffness matrices. In this paper, the system of Eq. (19) is solved adopting oscillatory solutions, *i.e.* $q = q_r e^{i\omega\tau}$. Dimensionless frequencies of the system ω, or eigenvalues, are found by allowing the determinant to be equal to zero for each specific flow condition.

3 Results and Discussion

Theoretical results for the dimensionless frequency of pipes conveying gas-liquid flow are presented. The two pipe systems of Table 1 are tested. CASE 1 and CASE 2 correspond to air-water two-phase flow in ¾" and 1-1/2" commercial PVC pipes, respectively. In the case of the span length L_{span}, two lengths L1 = 1.5 m and L2 = 2.5 m are used.

3.1 Verification of the formulation

The CASE 1 pipe system with $\lambda = 0$ is used to verify the presented formulation. This is equivalent to a pipe conveying liquid single-phase flow with $\beta = 0.5$ (substituting the values of Table 1 on Eq. (12)). The results of dimensionless frequency of the system as a function of dimensionless velocity for the first three modes are shown in Fig 2. It can be observed that the Argand diagram of the complex frequencies is identical to the diagram presented in Païdoussis's book [10]. Some important matters should be pointed. First, the

Table 1. Tested pipe-systems conveying fluid.

	CASE 1	CASE 2
Outer diameter (m)	0.02667	0.04826
Inner diameter (m)	0.02071	0.04078
Pipe mass (kg/m)	0.3362	0.7989
Liquid density (kg/m³)	998.2	998.2
Gas density (kg/m³)	1.49	1.49
EI (N·m2)	37.93	313.3

instability of the system occurs for the first mode, when, for the lowest dimensionless velocity, the imaginary part of frequency is different to zero. Then, the dimensionless frequency for $j = 0$ (no-flow) is 22.3733. Finally, the dimensionless critical velocity (where instability occurs) is 6.28; approximately 2π as pointed by [10].

3.2 Frequency for two-phase flow

Dimensionless frequency for two-phase flow has been calculated using the flow conditions of Fig. 3. We use the parameter Stability Level (Eq. (20)) to illustrate the results,

$$\text{Stability Level} = \left(1 - \frac{\omega_{ref} - \omega_i}{\omega_{ref}}\right) \cdot 100\% \qquad (20)$$

where ω_i is the calculated dimensionless frequency of a specific point of Fig. 3. ω_{ref} represents the dimensionless frequency for no-flow liquid-filled pipe system and is equal to 22.3733. A diminution in the Stability Level value is associated to the reduction of dimensionless frequency of the system, and vice-versa. Moreover, Stability Level equals to 0 indicates that the system became unstable.

Fig. 4 shows Stability Level results as a function of two-phase flow parameters, for the CASE 1-L1 pipe system. Each one of the 42-points corresponds to the flow conditions of Fig. 3. For constant values of homogeneous void fraction, it is observed that Stability Level decreases with increasing the superficial velocity.

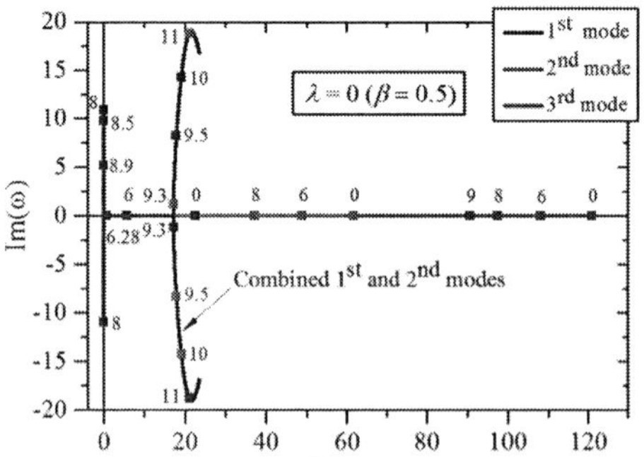

Fig. 2. Argand diagram of the complex frequencies of CASE 1 pipe system, considering $\lambda = 0$ $(\beta = 0.5)$. Numbers indicate the values of dimensionless velocity j.

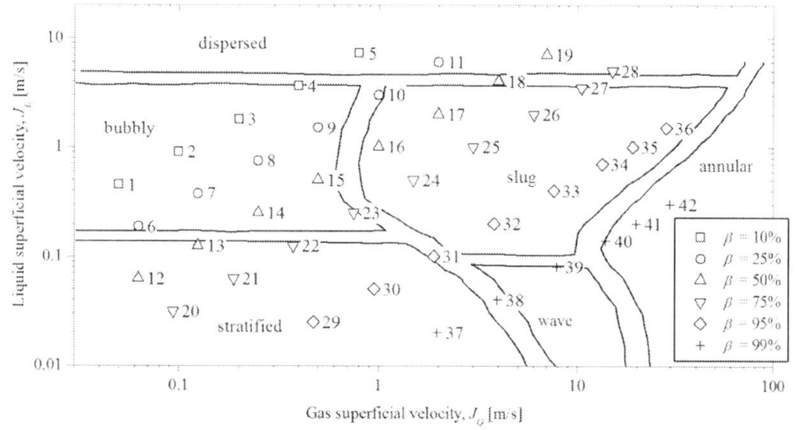

Fig. 3. Typical gas-liquid two-phase flow conditions on Mandhane *et al.* flow map [14]. The symbols are classified by homogeneous void fraction β.

For constant values of superficial velocity, an opposite behaviour of Stability Level is observed, *i.e.* Stability Level increases with increasing of homogeneous void fraction. This behaviour is due to the effect of hydrodynamic mass (added mass). Some values of Stability Level are presented in parenthesis. In general, the effect of velocity on the Stability Level is stronger than the effect of the homogeneous void fraction. These results are in agreement with experimental results reported by [8].

Fig 5 shows the Stability Level due to two-phase flow for three pipe systems. The effect of the pipe length L_{span} is observed when compared results for CASE1-L1 and CASE1-L2. On the other hand, the effect of both geometry and mechanical properties of the pipe are taken into account when compared results for CASE 1-L2 and CASE 2-L2. It can be observed that an increment on the pipe length reduces the Stability Level of the system, making it unstable for two conditions. This fact is due to the pipe system becoming more slender. The opposite happens with the rigidity *EI*; for the same pipe length, *i.e.* CASE 1-L2 versus CASE 2-L2, the higher rigidity value, the larger Stability Level. The strong effect of the velocity on Stability Level is also evident in Fig. 5.

4 Conclusions and Future Works

The classical equation of pipes conveying fluid, incorporating the no-slip homogeneous two-phase model, captures properly the trend between two-phase flow parameters and the Stability Level of the system. The lower Stability Level, the smaller frequency of the system. Furthermore, frequency equals to 0 represents an unstable system. In agree with reported experimental results in the literature, theoretical simulations show that Stability level decreases with increasing mixture velocity; and, by contrast, it increases with increasing homogeneous void fraction. Also, inadequate combinations of two-phase flow conditions, geometry and mechanical properties of the pipe can lead to unstable systems.

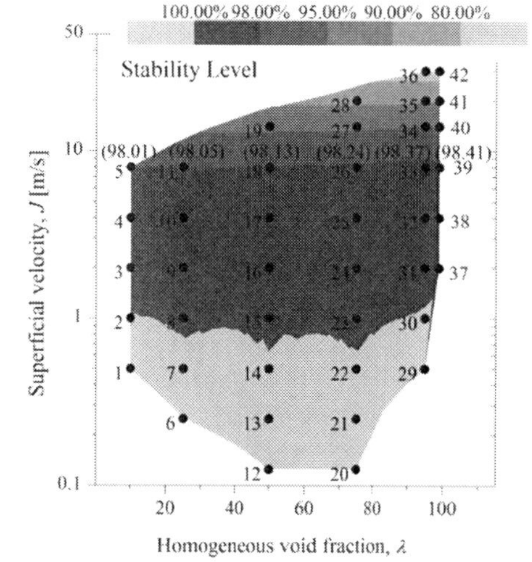

Fig. 4. Stability Level for the CASE 1-L1 pipe system. The numbers represent flow conditions of Fig. 3. The values in parenthesis indicate the Stability Level for two-phase flow conditions with $J = 8$ m/s.

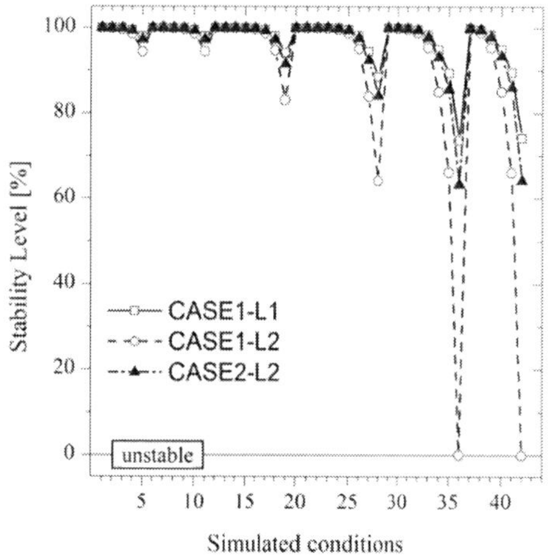

Fig. 5. Stability Level versus two-phase conditions of Fig. 3 for three pipe systems.

Future works should perform comparisons with experimental results and analyse the influence of slip between phases, gravity and structural damping on the Stability level.

This research has been supported by CIENCIACTIVA del CONCYTEC. L. Enrique Ortiz Vidal is grateful to FIPAI.

References

1. K. Fujita, J. Wind Eng. Ind. Aerodyn. **33**, 405 (1990).

2. M. J. Pettigrew, C. E. Taylor, N. J. Fisher, M. Yetisir, and B. A. W. Smith, Nucl. Eng. Des. **185**, 249 (1998).

3. C. Monette and M. J. Pettigrew, J. Fluids Struct. **19**, 943 (2004).

4. H. Chung, in *Offshore Mech. Arct. Eng. Symp.* (Dallas, TX, USA, 1985), pp. 398–402.

5. M. J. Pettigrew and C. E. Taylor, J. Press. Vessel Technol. Trans. ASME **116**, 233 (1994).

6. C. Zhang, M. J. Pettigrew, and N. W. Mureithi, J. Press. Vessel Technol. **130**, 11301 (2008).

7. T. Hibiki and M. Ishii, Nucl. Eng. Des. **185**, 113 (1998).

8. L. E. Ortiz-Vidal, N. W. Mureithi, and O. M. H. Rodriguez, Nucl. Eng. Des. **313**, 214 (2017).

9. Y. Geng, F. Ren, and C. Hua, Flow Meas. Instrum. **27**, 113 (2012).

10. M. P. Païdoussis, *Fluid-Structure Interactions: Slender Structures and Axial Flow, Vol. 1* (Elsevier Academic Press, 1998).

11. G. Wallis, *One Dimensional Two-Phase Flow* (McGraw-Hill, New York, 1969).

12. R. E. D. Bishop and D. C. Johnson, *The Mechanics of Vibration* (Cambridge University Press, 1979).

13. M. P. Paidoussis, *Fluid-Structure Interacions: Slender Structures and Axial Flow* (1998).

14. J. M. Mandhane, G. A. Gregory, and K. Aziz, Int. J. Multiph. Flow **1**, 537 (1974).

An interferometric radar sensor for monitoring the vibrations of structures at short ranges

Guido Luzi[1]*, *Michele* Crosetto[1], *Eduard* Angelats[1], *Enric* Fernández[1]

[1] Centre Tecnològic de Telecomunicacions de Catalunya, CTTC/CERCA, 08860 Castelldefels (BCN), Spain

Abstract. The Real-Aperture-Radar (RAR) interferometry technique consolidated in the last decade as an operational tool for the monitoring of large civil engineering structures as bridges, towers, and buildings. In literature, experimental campaigns collected through a well-known commercial equipment have been widely documented, while the cases where different types of sensors have been tested are a few. On the bases of some experimental tests, a new sensor working at high frequency, providing some improved performances, is here discussed. The core of the proposed system is an off-the-shelf, linear frequency modulated continuous wave device. The development of this apparatus is aimed at achieving a proof-of-concept, tackling operative aspects related to the development of a low cost and reliable system. The capability to detect the natural frequencies of a lightpole has been verified; comparing the results of the proposed sensor with those ones obtained through a commercial system based on the same technique, a more detailed description of the vibrating structure has been achieved. The results of this investigation confirmed that the development of sensors working at higher frequencies, although deserving deeper studies, is very promising and could open new applications demanding higher spatial resolutions at close ranges.

1 Introduction

The use of interferometric radar sensors, namely Real-Aperture-Radar (RAR) interferometry, consolidated in the last decade as an operational tool for the monitoring of large civil engineering structures as bridges, towers, and buildings: for an exhaustive bibliography on the topic see [1]. Although the accuracy of a radar estimate of displacement, usually of the order of tens of microns, cannot be comparable with that of optical systems, as Laser Doppler vibrometers, a microwave sensor is able to provide data with a negligible effect of the atmospheric propagation, weather conditions, and from distance up to one kilometre. Furthermore an interferometric radar, with respect to velocimeters, provides directly estimates of the displacement of the vibrating object, allowing to obtain direct inputs for modal analysis. The main limitation of this technique is that only the component along the radar-target line of sight (LOS) can be measured. The spatial resolution available for the target monitoring are mainly dictated by radar parameters as the swept bandwidth, and the achievable signal to noise ratio (SNR), which affect the range resolution, i.e. the capability to distinguish two different targets or parts of a surface, and the available accuracy. In literature, data collected through a well-known commercial apparatus working at Ku band are well documented, while the cases where different sensors have been tested are a few. The most popular system, namely Ibis-S™, is manufactured by the IDS (Ingegneria dei Sistemi SpA) company, and is available in the market. The system was born more than ten years ago, and it represents a trade-off after some experimental pioneer tests carried out during this period.

In the last years the costs of microwave technology at higher frequencies drastically lowered, and the performances of the on-the shelf devices have strongly improved. More recently, in the last few years, new prototypes have been presented in literature (see for example [2-6]); systems with different characteristics, operating at different bands have been developed to improve the performances of Ibis-S and aiming at reducing the costs of the apparatus. On the bases of some basic experiments, a novel sensor working at a higher frequency, providing some improved performances, has been developed at CTTC; in this paper some preliminary results are discussed. The core of the proposed system is an off-the-shelf, linear frequency modulated continuous wave (FMCW) K-band (centre frequency: 24 GHz) sensor. This device can provide a lower maximum operating range in comparison with respect to the commercial available systems, but a better range resolution. The development of this system is devoted to a proof-of-concept aimed at tackling operative aspects, including a reduction of the costs. In the present version, the data acquired through the radar are processed in two steps and not in real time. The main test carried out at CTTC is focused to evaluate its capability to detect the first natural frequency of a lightpole, comparing the results of the novel sensor, from here on referred as "NeoRAR", with those ones obtained through Ibis-S. An outstanding outcome of this test is the achievement of a

* Corresponding author: gluzi@cttc.es

better range resolution, and hence a more detailed description of the vibrating structure.
The paper contains a brief introduction of the working principle followed by a recall of the main features of a commercial microwave interferometer. The novel apparatus is then described and some experimental results are discussed.

1.1 The working principle

Radar is a time of flight device, which uses the time, elapsed between the transmitting and the receiving of an electromagnetic waveform, to locate targets included in the illuminated area. The basic output from a radar survey is a 1D signal, representing the amplitude of the echoes coming from the target included in the field of view (FOV) of the radar antenna; this signal is usually called "range profile" because its different peaks correspond to contributions coming from targets located at different distances. Standard radar is based on the transmission of signals composed of temporal periodic pulses of the amplitude. In a different way radar interferometers, which need coherent signal detection, are usually based on the frequency modulation of the transmitted signal, and specific waveforms sweeping a finite band composed of different frequencies, B, (Frequency Modulated Continuous Wave, FMCW, or Step Frequency Continuous Wave, SFCW) are used to assure coherent signals. The coherence of the radar sensor allows interferometric processing providing the capability of a sub-meter range resolution [7]. When different targets are present, the radar is able to provide their displacement history. The characteristics of the used antennas determine the volume of a spatial cell of the radar measurement, usually called radar bin.

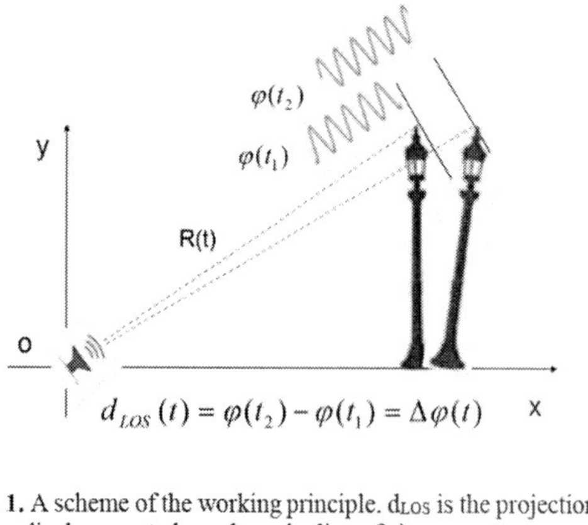

Fig. 1. A scheme of the working principle. d_{LOS} is the projection of the displacement along the radar line of site.

The use of interferometric radar to detect the vibration of an object is based on the capability of a coherent radar working at microwave frequency, to measure variations in time of the differential phase of the received echo with respect to the transmitted signal. As simply shown in fig.1 the variation of the LOS distance can be measured calculating the temporal variation of the interferometric phase. A higher operating frequency means in general a higher sensitivity to displacement variation. Considering that presently available apparatus uses a wavelength of the order of two centimetres, we can appreciate variations down to tens of microns. The vibration is seen by the radar as a range variation between the radar and the vibrating objects. The achievable accuracy is mainly determined by the signal to noise ratio of the acquisition which, in turn, depends on the intensity of the reflected signal. The radar reflecting properties of a target, the transmitted power, the distance, geometric factors (shape and orientation) and finally the dielectric characteristics of the target are the main factors affecting the strength of the received signal and hence the accuracy of the phase measurement.

1.2 A commercial system

The CTTC owns a commercial radar with interferometric capability: the IBIS-S™, manufactured and marketed by IDS [8]. The system consists of a sensor module, a control PC and a power supply unit and data processing software. A summary of its main characteristics is reported in Table 1. It operates at Ku band (17 GHz), i.e. with a corresponding wavelength lower than 2 centimetres, and with a maximum band of 300 MHz (model available in Europe). The system can be used with various antennas to better adapt the measuring conditions to different geometries and monitored surfaces. The radar has been used since many years by many researches to carry out monitoring different structures This apparatus has been used to measure the dynamic response of a variety of structures, such as bridges [9], wind farms [10], buildings [11] and towers [12], [13].

Table 1. Ibis-S Characteristics.

Parameters	Value
Operating frequency	17.2 GHz (Ku band)
Bandwidth (CWSF)	300 MHz
Max. operational distance	1000 m
Range resolution	0.5 m
Nominal sensitivity	0.00001 m
Max. acquisition rate	200 Hz
Polarization	Linear (HH or VV)

2 The novel apparatus

2.1. The radar sensor

The transceiver used to implement the radar sensor here discussed, is based on the use of a on the shelf device marketed by Sivers IMA AB (Sweden). Its main characteristics are resumed in Table 2. With respect to Ibis-S™, the operating frequency is higher, and its maximum bandwidth is larger, 1500 MHz in comparison to 300 MHz; this allows achieving a nominal range resolution of 10 cm. It is a monostatic radar with a 20dB standard gain horn antenna. The system is powered by a single 12 V battery.

In fig. 2, a picture of the system, including the RF sensor mounted on a small tripod, a laptop and a battery, is shown. In the present configuration, the sampling frequency is 3.5 Hz but we expect to increase it up to 12 Hz

Fig. 2. Picture of the NeoRAR system.

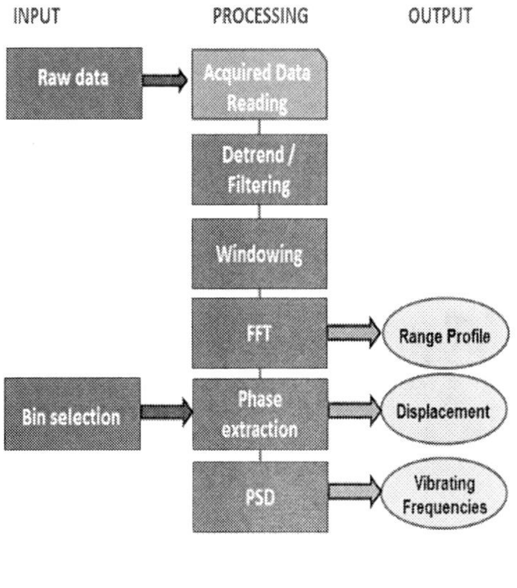

Fig. 3. A scheme of the data processing chain used to elaborate the NeoRAR data

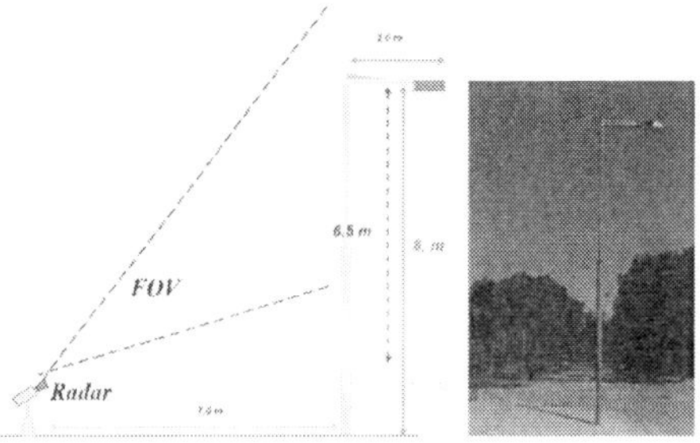

Fig. 4 . A scheme and a picture of the target used for testing the sensor.

Table 2. Sivers IMA transceiver sensor characteristics.

Parameters	Value
Operating frequency	24.75 GHz (K band)
Bandwidth (CWFM)	1500 MHz
Max. operational distance	75 m
Range resolution	0.1 m
Max. acquisition rate	3.5 Hz
Polarization	VV
Battery Autonomy	12 h
Antenna Field of View (°) H	18.6
Antenna Field of View (°) E	16.1

2.2 The data processing

The data acquired through radar are usually processed in two steps: the one necessary to transform the radar data to a range profile where we can identity the different parts of the monitored surface or target. This step is based on an inverse Fourier transform after a previous filtering and windowing of the raw signal. In particular a Kaiser function is used for windowing the signal before the DFT, and a high pass Butterworth filter of third order, with 0.1 Hz cut-off frequency, is applied. The second step consists in the extraction of the phase temporal series and their spectral analysis. DFT or power spectral density of the displacement stories are usually calculated to detect the main frequencies. In fig. 3 a scheme of the processing steps is shown.

3 Experimental Results

3.1. The target

The target selected for the test of the novel sensor performances is a metal lightpole located in the Parc Mediterrani de la Tecnologia of Castelldefels (Spain), where CTTC is located. In fig. 5a, a picture and a scheme of the size of the target are shown. The measurements were carried out on 29th September 2016; the day was affected by irregular wind shears. Micro tremor and wind are the main causes of target vibration; vehicular traffic in this area is very sporadic and it was absent during the data

https://doi.org/10.1051/matecconf/201814801005

acquisition. A first test consisted in verifying the actual range resolution: moving the NeoRAR of known displacement steps, a 10 cm value was obtained as expected using the whole 1500 MHz bandwidth. The results of this test are shown in fig. 5 where the displacement obtained from the bin identification and the target displacement are compared.

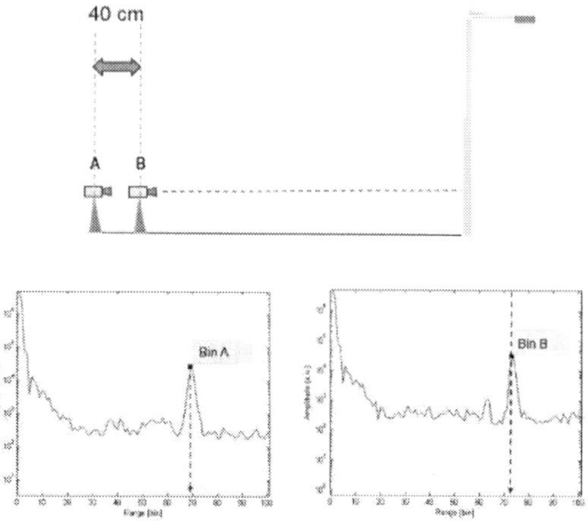

Fig. 5. Test to verify the range resolution of the NeoRAR: a displacement of 40 cm corresponds to a 4 bins (BinA=69, BinB=73) variation.

3.2. The data acquisition

To test the capability in vibrations monitoring of the proposed system, data acquisitions have been acquired in front of the lightpole with the geometry shown in fig. 4 to sample different parts of the pole. In fig. 6 we show the section of the range profile where four bins corresponding to different heights of the lightpole are marked. To retrieve the distance from the radar bin number note that each bin to a 0.1 m step, and due to an instrumental constant, we have to take off an offset (3 bins).

Table 3. Bin identification and detected frequencies obtained from the NeoRAR monitoring.

Bin	Range(m)	Corresponding Height (m)	Detected Frequency (Hz)
76	7.3	4.2	1.512
88	8.5	6.8	1.519
93	9.0	7.6	1.519
95	9.2	7.8	0.467/ 1.512

3.3 Data analysis

In fig. 7 we show the spectral response obtained processing the LOS displacements corresponding to the four bins identified in fig. 6. The amplitude of the displacement increases with the height, as expected for the first vibrating mode of the pole. In table 3 the height of the lightpole part sampled by the radar measurement is also reported. These results can be compared to those obtained from the monitoring of the same lightpole carried out using Ibis-S only two bins were analysed, as can be seen in fig. 8.

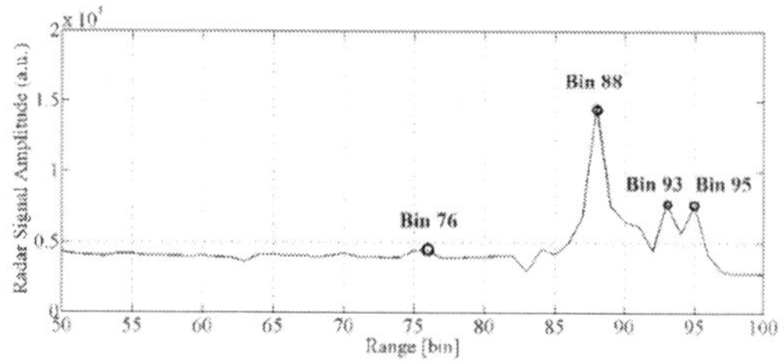

Fig. 6 Range profile and bin identification.

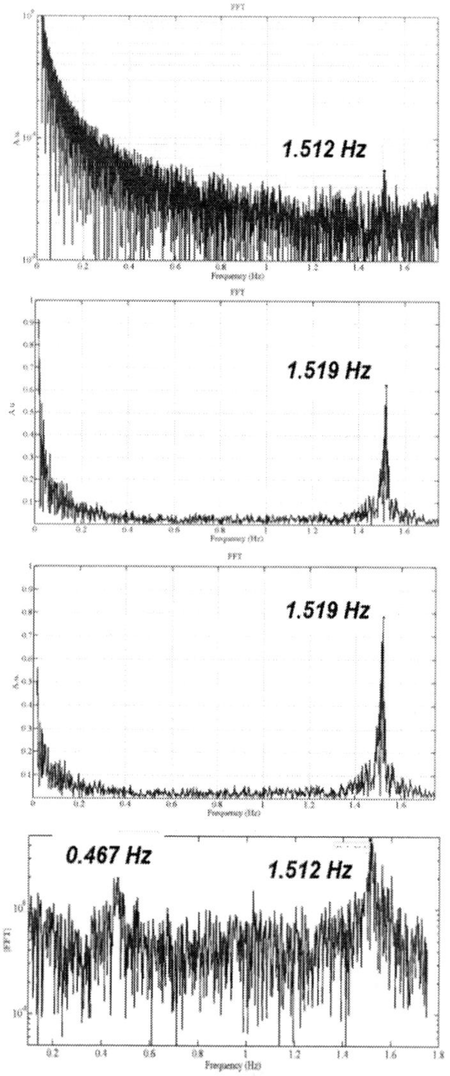

Fig. 7 FFT calculated from the displacement samples obtained from the four selected bins; from top to bottom: Bin76, Bin88, Bin94 and Bin95. In the first and the last plots logarithmic y axis scale is used to improve the peaks' detection.

4 Conclusions

In this paper after introducing the basic of radar interferometry applied to detect civil structures' vibrating frequencies, preliminary results of a test aimed at testing a novel microwave interferometer have been discussed. On the bases of a simple experimental test, the main performances of a new FMCW sensor, working at K band (24 GHz), as a monitoring tool of mechanical vibration, have been analysed. Results obtained through the new sensor are in good agreement with that detected by the Ibis-S measurement: the power spectral density obtained from Ibis-S monitoring is shown in fig. 8. In addition, the higher spatial resolution available from NeoRAR allowed obtaining of a second frequency not detected by Ibis-S. The analysed sensor is very promising and deserves deeper studies.

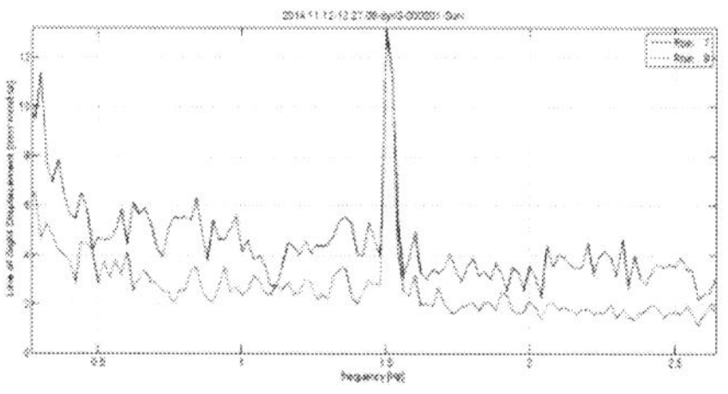

Fig. 8 LOS displacement PSD calculated from data acquired through Ibis-S.

References

1. M. Pieraccini, Monitoring of Civil Infrastructures by Interferometric Radar: A Review. The Scientific World Journal (2013), Article ID 786961, http://dx.doi.org/10.1155/2013/786961. (2013)
2. G. Grazzini., M. Pieraccini, D. Dei, C. Atzeni, Simple Microwave sensor for remote detection of structural vibration. Electronics Letters 2009, **45** 11, 567-569 (2009).
3. J.Abril, E. Nova., A. Broquetas, A. Aguasca, J. Romeu, L. Jofre, Micrometric deformation imaging at W-Band". In *Proceedings of the 2011 IEEE International Microwave Workshop. IMWS 2011.* pp. 65-68 (2011).
4. F. Papi, N. Donati, M. Pieraccini, Handy Microwave Sensor for Remote Detection of Structural Vibration. EWSHM - *7th European Workshop on Structural Health Monitoring*, Jul 2014, Nantes, France (2014)
5. Li Cunlong, Chen Weimin, Liu Gang, Yan Rong, Xu Hengyi and Qi Yi, A Noncontact FMCW Radar Sensor for Displacement Measurement in Structural Health Monitoring. Sensors 2015, **15** 7412-7433; doi:10.3390/s150407412. (2015)
6. G. Giunta, A.Monti-Guarnieri, D. D.'Aria, F.Speziali, P.Falcone, L.Maggi, G. Amoroso, A novel technique for very accurate three-dimensional monitoring of structural vibrations and displacement by remote radar sensors, *Radar Conference IET.* (2015)
7. C. Gentile, G. Bernardini. An interferometric radar for non-contact measurement of deflections on civil engineering structures: laboratory and full-scale tests. Structure and Infrastructure Engineering **6** , 521-534. DOI: 10.1080/15732470903068557 (2010)
8. F. Coppi, C. Gentile, P. Ricci, A software tool for processing the displacement time series extracted from raw radar data. *Proceedings of the 9th Int. Conference on Vibration Measurements by Laser and non-contact Techniques,* Ancona, Italy; 22-25 June 2010. AIP Conference Proceedings 1253, ed. E.P. Tomasini (2010)
9. M. Pieraccini, M. Fratini, F. Parrini, C. Atzeni, Dynamic monitoring of bridges using a high-speed coherent radar. 2006, IEEE TGRS, **40** 11, 3284–3288 (2006)
10. M. Pieraccini, F. Parrini, M. Fratini, C. Atzeni, P. Spinelli. In-service testing of wind turbine towers using a microwave sensor. Renewable Energy 2008, **33**, 1 pp. 13-21, ISSN:0960-1481 URL:http://dx.medra.org/10.1016/j.renene.2007.02.001 (2008)
11. C. Negulescu, G. Luzi, M. Crosetto, D. Raucoules, A. Roullé, D. Monfort, L. Pujades, B. Colas, T. Dewez, Comparison of seismometer and radar measurements for the modal identification of civil engineering structures. Engineering Structures, **51**, 10–22 (2013)
12. G. Luzi, M. Crosetto, M. Cuevas-González, A radar-based monitoring of the Collserola Tower (Barcelona). Mechanical Systems and Signal Processing 2014, **49**, 234-248 (2014)
13. C. Atzeni, A. Bicci, D. Dei, M. Fratini, M. Pieraccini. Remote survey of the leaning tower of Pisa by interferometric sensing. IEEE GRSL 2010, **7** 1, 185–189.

Influence of dash-pot with controllable damping coefficient on damping efficiency of TMDI

Mateusz Lazarek[1,*], *Piotr* Brzeski[1,**], and *Przemyslaw* Perlikowski[1,***]

[1] *Division of Dynamics, Lodz University of Technology, Stefanowskiego 1/15, 90-924 Lodz, Poland*

Abstract. In this paper we study the dynamics of two degree freedom system, which is consist of main body and tuned mass damper with inerter (TMDI). We add the dash-pot with variable damping coefficient to TMDI to study the overall efficiency of the device. The shape of non-linear characteristic of the dash-pot is dependent on one control parameter which governs the steepness of the function and the value of damping coefficient changes according to the relative displacement or velocity between main mass and tuned mass damper. We show the two parameters diagrams showing the maximum amplitude of the main body versus frequency of excitation of main body and controlling parameter.

1 Introduction

All mechanical systems are susceptible to oscillate according to its natural frequency [1]. Therefore, matching of the external excitation forcing frequency with the natural frequency can lead to resonance phenomenon what is one of the reasons of structure's destruction [2]. Assuming inability to remove external excitation one can move the work point far from the natural frequency or use am external devices which absorb the energy. Limitation of frequencies' spectra corresponding to the first alternative is useful mainly for systems working under known and restricted conditions. In a more general case with low predictability of external forcing frequency, additional devices become the best solution [3]. In 1909 Frahm patented the classical tuned mass damper (TMD) [4]. The TMD is consists of a body of mass m and spring with stiffness k where the mass m is smaller than the mass of damped system but their natural frequencies are the same. Frahm device is extremely effective in reducing the response of the damped structure only in resonance, but outside the resonance it causes the amplification of oscillations' magnitude. To overcome this problem Den Hartog [3] proposed to add a viscous damper to Frahm's system design.

One of possible solution to extend the efficiency of TMD is addition of an inerter [6]. Inerter is a mechanical device generating force proportional to acceleration between its ends [5]. Initially inerters were successfully applied in sports cars' suspensions [7]. Now we observe growing number of studies on other possible applications [8–12].

This work is devoted to the examination of the efficiency of a tuned mass damper with ineter (TMDI) with a non-linear dash-pot. We use the non-linear dash-pot which damping function is dependent on one control parameter and relative displacement or velocity between the TMDI

*e-mail: mateusz.lazarek@p.lodz.pl
**e-mail: piotr.brzeski@p.lodz.pl
***e-mail: przemyslaw.perlikowski@p.lodz.pl

and main mass. We show how such dash-pot is influencing the mitigation properties of TMDI using different shapes of damping functions.

2 The model of the tuned mass damper with inerter and non-linear dash-pot

Let us describe the physical model of the system. It has two degrees of freedom and consists of two coupled oscillators that can move in the vertical direction. The first one serves as a model of the main oscillator which motion we are going to mitigate. It is connected with the support and forced via kinematic excitation realized with the crank mechanism. The second oscillator is connected to the first one and represents the TMDI. It is connected with the main oscillator via linear spring, non-linear viscous damper and inerter. The motion of the system is described by two generalized coordinates: the position of the main oscillator by coordinate x, while the displacement of the TMDI by coordinate y. The main oscillator is characterized with the following parameters: M is its mass, K is the stiffness of the single spring that connects the main mass to the ground and C is the viscous damping coefficient of dash-pot that links mass M and the support. Formula $A\cos\left(\frac{s\pi}{30}t\right)$ expresses the kinematic excitation of the main oscillator with the frequency $\frac{s\pi}{30}$ that corresponds to the speed $s\,[rpm]$ of the forcing servomotor and the crank on the exciting plate set to A. During the investigation we analyse the response of the structure for fixed crank $A = 0.02955\,[\mathrm{m}]$ and different frequencies of excitation that corresponds to the following range of rotational speeds $s\epsilon\langle 150, 250\rangle$ [rpm]. To characterize TMDI we use the following parameters values: the moving mass is given by m, the stiffness of the spring that connects it to the main oscillator is described with parameter k. The viscous damping function of the non-linear dash-pot that connect mass m with M is given by c. Values of parameters are following: $M = 102.66\,[\mathrm{kg}]$, $K = 8181.0\,[\frac{\mathrm{N}}{\mathrm{m}}]$, $C = 20.0\,[\frac{\mathrm{Ns}}{\mathrm{m}}]$, $A = 0.02955\,[\mathrm{m}]$, $s\epsilon\langle 150, 300\rangle$ [rpm], $m = 12.81\,[\mathrm{kg}]$, $k = 10985.1\,[\frac{\mathrm{N}}{\mathrm{m}}]$ and $I = 6.877\,[\mathrm{kg}]$. Damping function c is variable and we introduce it in each subsection separately with corresponding equations of system's motion .

3 Influence of a linear viscous dash-pot on the response of the system

In this section we study the response of a linear system with varying of linear viscous damping coefficient. The system is given by the following equations of motion:

$$M\ddot{x} + 7Kx + C\dot{x} + I(\ddot{x} - \ddot{y}) + k(x - y) + \\ + c(\dot{x} - \dot{y}) = KA\cos\left(\frac{s\pi}{30}t\right), \quad (1)$$

$$m\ddot{y} - I(\ddot{x} - \ddot{y}) - k(x - y) - c(\dot{x} - \dot{y}) = 0, \quad (2)$$

where the c is varied in the following range $c\epsilon\langle 30, 330\rangle$ [Nm/s]. As aforementioned, all our investigations are possible to realize experimentally and the chosen values of all parameters are based on the characteristics of real devices. We present obtained results in Fig. 2(a) as a colour diagram. The plot is calculated numerically with Auto07p, we divide the range of parameter c in 100 steps and for each we compute the FRC. The blue colour corresponds to low amplitude and it dominates in the plot. The increase of amplitude is observed in neighbourhood of resonances. For low value of damping $c = 30\,[\mathrm{Nm/s}]$ two resonances occur for $s = 194\,[\mathrm{rpm}]$ and $s = 257\,[\mathrm{rpm}]$. For maximum value of damping coefficient we observe one resonance for $s = 208\,[\mathrm{rpm}]$. The transition is caused by increase of value of damping coefficient which cause the stiffening of connection between main body and damper. Hence, the independent motion of the TMD is no more possible and the system is reduced to one DoF.

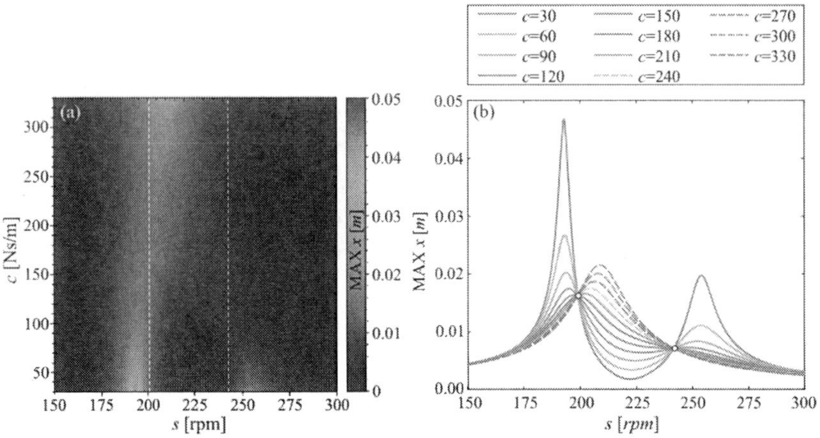

Figure 2. In panel (a) we show two dimensional plot of maximum amplitude of damped system (max x[m]) in two parameters space: c and s. Plot shows influence of linear viscus damper with constant damping coefficient (Eqs (1,2)) on system response. Panel (b) shows eleven exemplary FRCs of damped system. Points P and Q are marked as circles in panel (b).

The corresponding FRCs are shown in Fig. 2(b) for eleven values of damping coefficients starting with $c = 30\,[\mathrm{Nm/s}]$ and step $\Delta c = 30\,[\mathrm{Nm/s}]$. The amplitude of second resonance is deceasing rapidly and finally for $c = 150\,[\mathrm{Nm/s}]$ the increase of amplitude is not more present. As it is well known all FRCs are crossing two points - P and Q where the response of system is nearly independent on value of damping coefficient.

4 Influence of a non-linear viscous dash-pot on the response of the system

In this subsection we show the influence of non-linear damping on the response of the main system. The equations of system are as follow:

$$M\ddot{x} + 7Kx + C\dot{x} + I(\ddot{x} - \ddot{y}) + k(x - y) + \\ + c(x, y, \dot{x}, \dot{y})(\dot{x} - \dot{y}) = KA\cos\left(\frac{s\pi}{30}t\right), \quad (3)$$

$$m\ddot{y} - I(\ddot{x} - \ddot{y}) - k(x - y) + \\ - c(x, y, \dot{x}, \dot{y})(\dot{x} - \dot{y}) = 0, \quad (4)$$

where $c(x, y, \dot{x}, \dot{y})$ is a continuous damping function which depends on the retaliative displacement or velocity between the main system and the TMD. For sake of generality, we include all state variables in the general form of damping function. However, we consider two types of damping function, i.e., one is dependent on the retaliative displacement and second one on the relative velocity. The system is smooth, hence we perform all calculations in Auto07p.

The first case corresponds to the damping dependent on the retaliative displacement between the main mass and the TMD. For $(x - y) = 0.0$ the damping function has minimum value: $c_{MIN} = 30\,[\mathrm{Nm/s}]$ and it is increasing according to following formula:

$$c(x, y, \dot{x}, \dot{y}) = c(x, y) = \\ = c_{MIN}\left(1 + 10\tanh\left(a(x - y)^2\right)\right), \quad (5)$$

to its maximum value: $c_{MAX} = 330$ [Nm/s]. The controlling parameter of the damping function is a and it varies in range $a \in \langle 10, 30000 \rangle$ [m^{-2}] which let us change the value of relative displacement for which function achieves its maximum. Ten exemplary plots of the damping functions are presented in Fig. 3(a), we restricted the range of relative displacement to $|x - y| < 0.15$ [m].

Results are shown in Fig. 3(b, c). In panel (b) we present a two dimensional colour plot of maximum amplitude of main system as a function of rotational speed s (frequency of excitation) and parameter of damping function a. White dash lines indicate the location of points $P = 205$ [rpm] and $Q = 240$ [rpm] for system with linear viscous damping. In Fig. 3(c) we show eleven exemplary FRCs for chosen values of parameter a within the considered range. As it is easy to see points P and Q do not change their locations with change of shape of damping function, while FRCs are significantly different. For low value of a we observe two resonances and for maximum value of a the main mass and TMD behave like one DoF system.

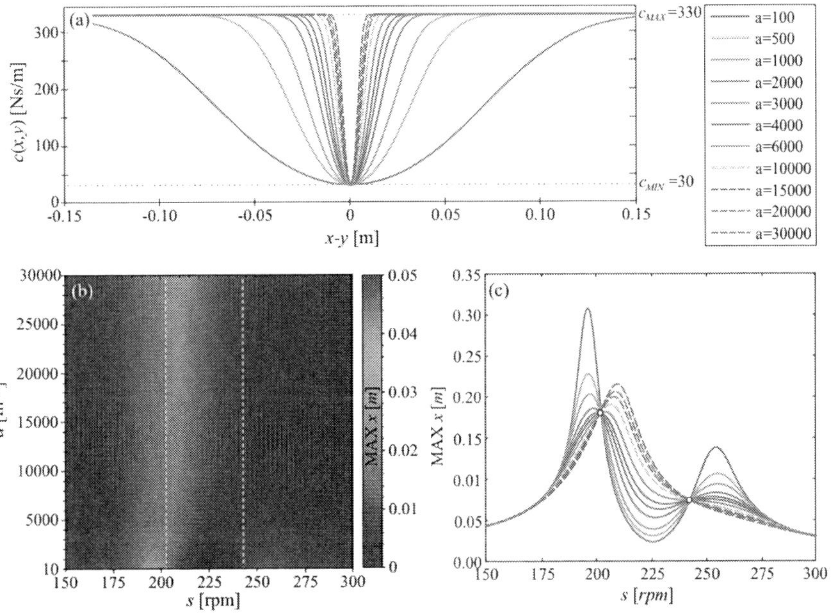

Figure 3. In panel (a) we show the damping functions $c(x, y)$ (Eq. 5) for selected values of a. In panel (b) we show two dimensional plot of maximum amplitude of damped system (max x[m]) in two parameters space: a and s. The last panel (c) shows eleven exemplary FRCs of damped system. Points P and Q are marked as circles in panel (b).

In second case we consider the same shape of damping function as in previous case but it is dependent on retaliative velocity:

$$c(x, y, \dot{x}, \dot{y}) = c(\dot{x}, \dot{y}) =$$
$$= c_{MIN}\left(1 + 10\tanh\left(a(\dot{x} - \dot{y})^2\right)\right), \quad (6)$$

where parameter a is in the range $a\epsilon\langle 0.1, 7\rangle$ [s^2/m^2]. Eleven exemplary plots are shown in Fig. 4(a) and as in previous case the value of a is governing the slope of function.

The corresponding two dimensional colour plots of amplitude versus rotational speed of servomotor and parameter a are shown in Fig. 4(b). The dashed white lines indicate the position of pointspoints $P = 205$ [rpm] and $Q = 240$ [rpm]. With increase of parameter a the system is behaving as one DoF system. All periodic solutions are stable and location of points P and Q along the FRC is not affected by non-linear damping (see Fig. 4(c)).

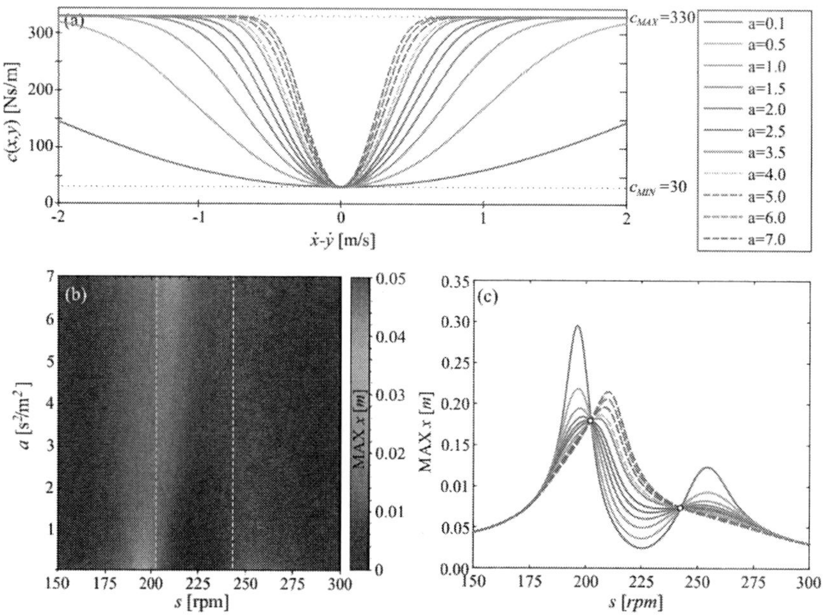

Figure 4. In panel (a) we show the damping functions $c(\dot{x}, \dot{y})$ (Eq. 6) for selected values of a. In panel (b) we show two dimensional plot of maximum amplitude of damped system (max x[m]) in two parameters space: a and s. The last panel (c) shows eleven exemplary FRCs of damped system. Points P and Q are marked as circles in panel (b).

5 Conclusions

In this paper we present the description of the influence of a non-linear dash-pot with variable coefficient of damping placed between the main mass and the TMD. In case of linear dashpot there are two points along the FRC that do not dependent on the value damping coefficient known as points P and Q. We study how using the non-linear dash-pot and the inerter we can decrease the maximum amplitude of damped body. Analysis proves that suspicion of stationary character of this points is true even with non-linear damper. However, we find ranges where one can observe the decrease of amplitude of main system.

Acknowledgement

This work has been supported by National Science Centre, Poland - Project No. 2015/17/B/ST8/03325.

References

[1] S.S. Rao, F.F. Yap, *Mechanical vibrations*, Vol. 4 (Prentice Hall Upper Saddle River, 2011)

[2] K.Y. Billah, R.H. Scanlan, American Journal of Physics **59**, 118 (1991)

[3] J.P. Den Hartog, *Mechanical Vibrations* (McGraw-Hill, New York, 1934)

[4] H. Frahm, *Device for damping vibrations of bodies* (1909)

[5] F.C.W. M. C. Smith, Proceedings of the IEEE Conference on Decision and Control **3**, 2258 (2004)

[6] M. Smith, Automatic Control, IEEE Transactions on **47**, 1648 (2002)

[7] M. Chen, C. Papageorgiou, F. Scheibe, F. cheng Wang, M. Smith, Circuits and Systems Magazine, IEEE **9**, 10 (2009)

[8] F.C. Wang, M.K. Liao, Vehicle System Dynamics **48**, 619 (2010)

[9] R. Faraj, J. Holnicki-Szulc, L. Knap, J. Senko, Smart Materials and Structures **25**, 035031 (2016)

[10] I. Takewaki, S. Murakami, S. Yoshitomi, M. Tsuji, Structural Control and Health Monitoring **19**, 590 (2012)

[11] Y.C. Chen, J.Y. Tu, F.C. Wang, *Earthquake vibration control for buildings with inerter networks*, in *Control Conference (ECC), 2015 European* (2015), pp. 3137–3142

[12] M Lazarek, P Brzeski, P Perlikowski, Mechanism and Machine Theory **119**, 161-173 (2017)

The use of Classical Rolling Pendulum Bearings (CRPB) for vibration control of stay-cables

Georgia Papastergiou[1,*], and *Ioannis* Raftoyiannis[2]

[1,2]National Technical University of Athens, Dept. of Civil Engineering, Greece

Abstract. Cables are efficient structural elements that are used in cable-stayed bridges, suspension bridges and other cable structures. A significant problem which arose from the praxis is the cables' rain-wind induced vibrations as these cables are subjected to environmental excitations. Rain-wind induced stay-cable vibrations may occur at different cable eigenfrequencies. Large amplitude Rain-Wind-Induced-Vibrations (RWIV) of stay cables are a challenging problem in the design of cable-stayed bridges. Several methods, including aerodynamic or structural means, have been investigated in order to control the vibrations of bridge's stay-cables. The present research focuses on the effectiveness of a movable anchorage system with a Classical Rolling Pendulum Bearing (CRPB) device. An analytical model of cable-damper system is developed based on the taut string representation of the cable. The gathered integral-differential equations are solved through the use of the Lagrange transformation. . Finally, a case study with realistic geometrical parameters is also presented to establish the validity of the proposed system.

1 Introduction

Cable-stayed (C-S) bridges have been known since the beginning of the 18th century, but there were difficulties in their static and dynamic analysis. A significant problem, which arose from the practice, is the cables rain-wind induced vibrations. Large amplitude Rain-Wind-Induced-Vibrations (RWIV) of stay-cables constitutes a challenging problem in the design of C-S bridges. Several methods, have been investigated in order to control the vibrations of bridge's stay cables.

This paper investigates the effectiveness of a movable anchorage system with a Classical Rolling Pendulum Bearing (CRPB) device. An analytical model of the cable-damper system is developed herein based on the taut string representation of the cable. The gathered integral-differential equations are solved through the use of the Lagrange transformation. Finally, a case study with realistic geometrical parameters is also presented to establish the validity of the proposed system, while the required device for the studied case is designed (see Fig. 1).

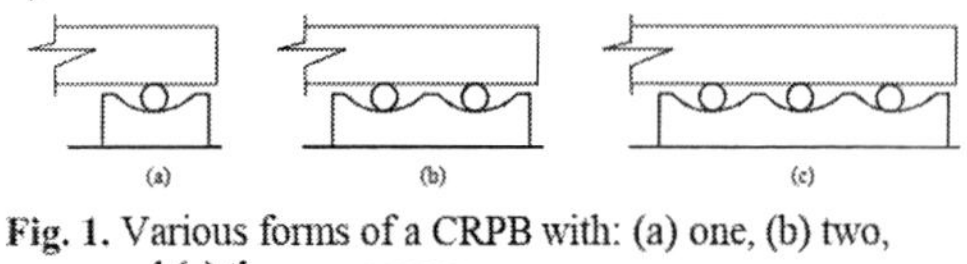

Fig. 1. Various forms of a CRPB with: (a) one, (b) two, and (c) three concaves

2 Basic assumptions

The deformed shape of the cables under dead and live loads is a catenary curve, with displacements wo and tensile forces To (see Fig.2), which because of its very shallow form can be approximated by a second-order parabola.

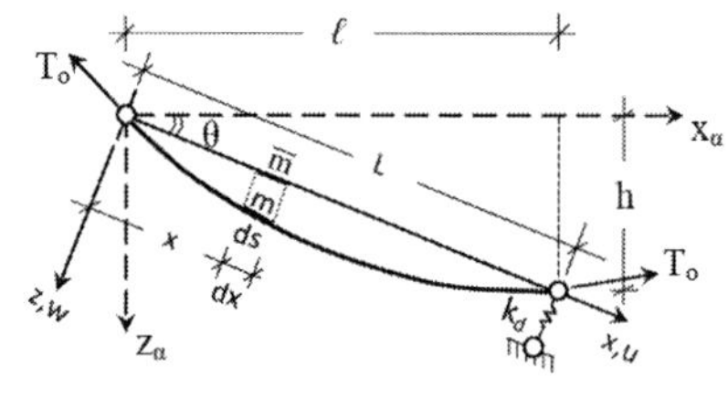

Fig. 2. Cable and reference axes

Under the action of the dynamic loads py(x,t) and pz(x,t), the cable takes the form of Fig. 3, with additional displacements u_d, υ_d, w_d and tensile forces T_d.

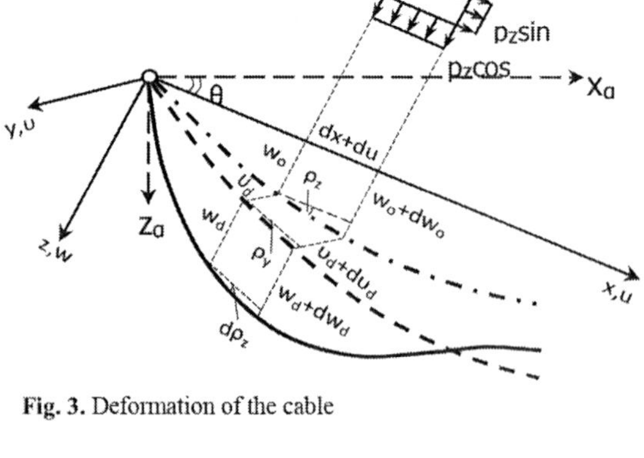

Fig. 3. Deformation of the cable

* Corresponding author: papastergiou.georgina@gmail.com

3 Equilibrium equations of a taut cable

3.1 Projection on xoz-plane

3.1.1 Equilibrium of horizontal forces

Projecting on xoz-plane and taking the equilibrium of horizontal forces, we obtain:

$$\frac{\partial H}{\partial s} - c\dot{u} - m\ddot{u} = -p_x(x,t) \tag{1}$$

3.1.2 Equilibrium of vertical forces

Projecting on xoz-plane and taking the equilibrium of vertical forces, we obtain:

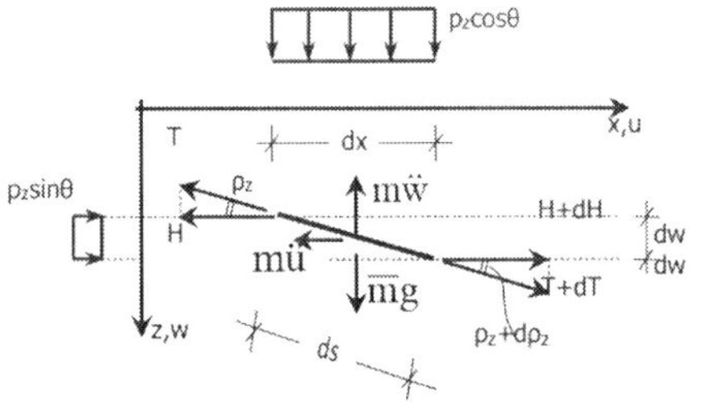

Fig. 4. Projection on xoz-plane

$$T_o \frac{\partial^2 w_d}{\partial x^2} + T_d\left(\frac{\partial^2 w_o}{\partial x^2} + \frac{\partial^2 w_d}{\partial x^2}\right) - c\dot{w}_d - m\ddot{w}_d = -p_z(x,t) \tag{2}$$

3.2 Projection on xoy-plane

Projecting on xoy-plane, we obtain:

$$T_o \frac{\partial^2 \upsilon_d}{\partial x^2} + T_d \frac{\partial^2 \upsilon_d}{\partial x^2} - c\dot{\upsilon}_d - m\ddot{\upsilon}_d = -p_y(x,t) \tag{3}$$

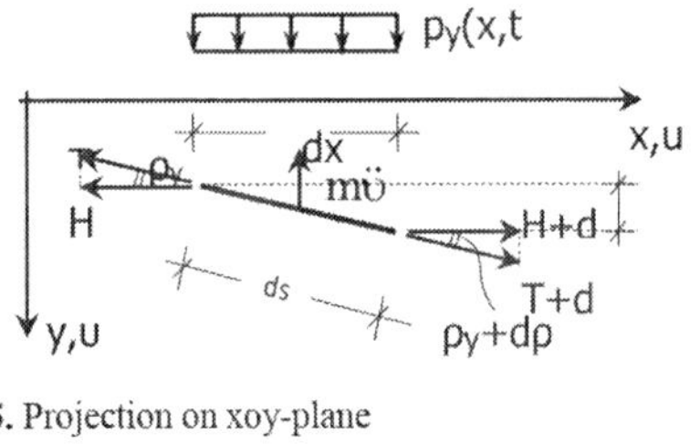

Fig. 5. Projection on xoy-plane

3.3 The cables' deformation

With boundary condition $w_d(0) = 0$ we obtain:

$$T_d = \frac{1}{L_o}\left(w_o'(L)w_d(L) - w_o''\int_0^L w_d dx\right) \text{with: } L_o = \int_0^L \frac{dx}{EA\cos^3 \rho_z} \tag{4}$$

3.4 Catenary and the parabola approach

It is usual to use the parabola as a curve that is very close to the catenary one, especially for shallow forms of cables. For a cable's shallow form the equation of a parabola passing from the points (0,0), (L,0) and having $w_o'' = -\frac{\overline{m}\,g}{H_o}$ is given by the following formula:

$$w_o(x) = \frac{\overline{m}\,g}{2H_o} \cdot x \cdot (L - x) \tag{5}$$

4 Analysis

4.1. The cable

The stay-cable model with the considered anchorage-bearing system is shown in Fig. 6.

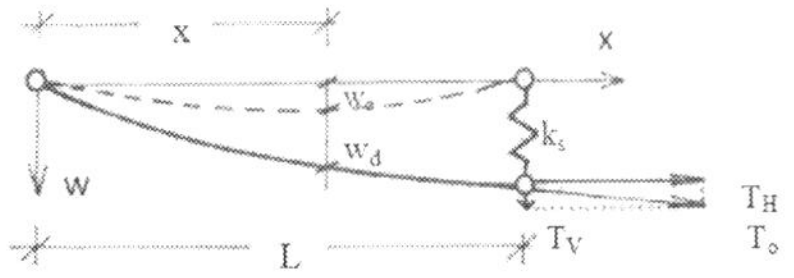

Fig. 6. The stay-cable model

4.2 The rolling pendulum bearing system

Let us consider a C.R.P.B. device with one concave rolling, like the one of Fig.7. The C.R.P.B. system is made from material like the one of the classical ball-bearings having surfaces elaborated wery diligently, with coefficient of rolling friction ranging from 0.002 to 0.005. Therefore, the developed friction forces can be neglected in this preliminary study. On the other hand, the angle of the friction cone amounts up to 0.34o, which corresponds to a very small static friction.

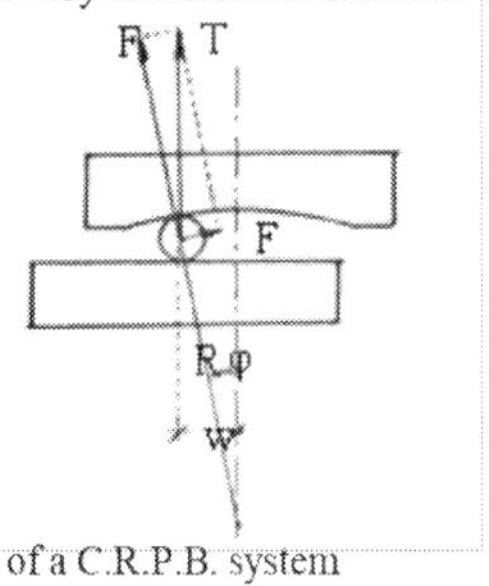

Fig. 7. Geometry of a C.R.P.B. system

4.3 The equation of motion

We obtain the following equation of motion:

$$\left.\begin{aligned} &T_o w_d'' + \frac{w_o'(L)w_d(L)}{L_o} w_o'' - \frac{w_o''^2}{L_o}\int_0^L w_d dx - c\dot{w}_d - m\ddot{w}_d = p(x,t) + \frac{T_o}{R} w_d \delta(x-L) \\ &\textit{boundary conditions}: \quad w_d(0) = 0, T_o w_d'(L) + \frac{T_o}{R} w_d(L) = 0 \end{aligned}\right\} \tag{6}$$

4.4 The free vibrating cable

The equation of motion of a free vibrating cable with movable anchorage is:

$$T_o w_d'' - c\dot{w}_d - m\ddot{w}_d = -\frac{w_o'(L)w_d(L)}{L_o} w_o'' + \frac{w_o''^2}{L_o}\int_0^L w_d dx \tag{7}$$

We are searching for a solution of separated variables of the form:

$$w_d(x,t) = W(x)\Phi(t) \qquad (8)$$

One can determine the following form of the shape functions:

$$W_n(x) = c_1\left[(\sin\lambda_n x + D_1) + \frac{\lambda_n T_o \cos\lambda_n L + \frac{T_o}{R}(\sin\lambda_n L + D_2)}{\lambda_n T_n \sin\lambda_n L - \frac{T_o}{R}(\cos\lambda_n L + D_2)}\cdot(\cos\lambda_n x + D_2)\right] \qquad (9)$$

4.5 The forced vibrating cable

We are searching for a solution of the form:

$$w_d(x,t) = \sum_n W_n(x) Z_n(t) \qquad (10)$$

and we conclude finally to: $Z_k(t) = L^{-1}\zeta_k(s)$ (11)

5 Numerical results and discussion

5.1 The cables

Let us consider a C-S bridge with dense distribution of cables from which we study a cable having tension To=300000 dN/cable, cross-sectional area F=7.5·10-3m2, diameter D=0.13 m, weight G=70 dN/m, mass per unit length m=7kg/m, and variable length L=150, 250, and 350m.

5.2 The rain-wind combination

It has been observed that the rain-wind-induced vibration in bridge cables usually occurs in a frequency range from 0.5 to 4 sec-1. For the study of the vibration of a cable under the action of a rain-wind combination we choose the following loading: $p = p(x)\cdot f(t) = 20\cdot\sin\omega t$, where ω=1, 2, 3, 4 sec-1 (for the study of the above cables without a damping system) and ω=3 sec-1 (for the study of the above cables with the proposed C.R.P.B. devise).

5.3 Behavior of the cables without a damping system

Table 1. Eigenfrequencies of the cables

	m=7kg/m ,	To=300000 dN	
	L=150m	L=250m	L=350m
ω 1	4.4697	2.8186	2.1507
ω 2	8.6716	5.2029	3.7164
ω 3	13.0125	7.8130	5.5868
ω 4	17.3432	10.4059	7.4328
ω 5	21.6801	13.0093	9.2936

In Figs 8 to 10, one can see the oscillations of the mid-length of the studied cables with length L=150, L=250 and L=350m, subjected to loadings acting with frequencies ω=1, 2, 3, 4 sec-1.

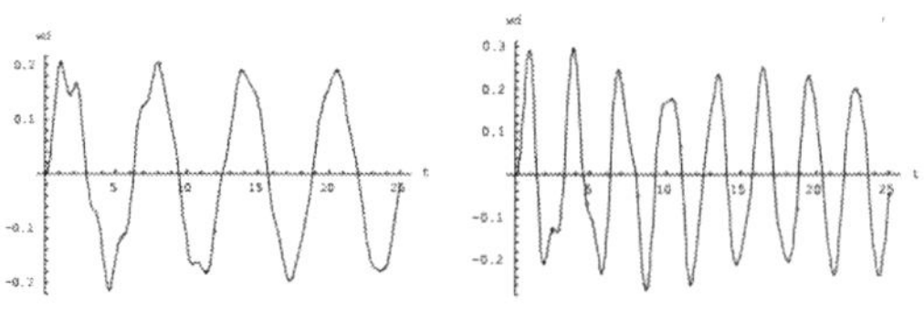

p=20·Sin(1·t) p=20·Sin(2·t)

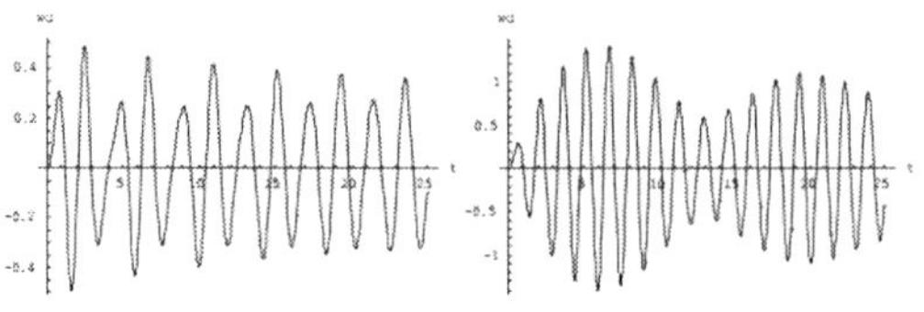

p=20·Sin(3·t) p=20·Sin(4·t)

In this case, big deformations appear, because the eigenfrequency of the external load is near to the first one of the cable (ω1 =4.47 sec-1)

Fig. 8. Oscillations of the mid-length of a cable of length L=150m, without any damping system

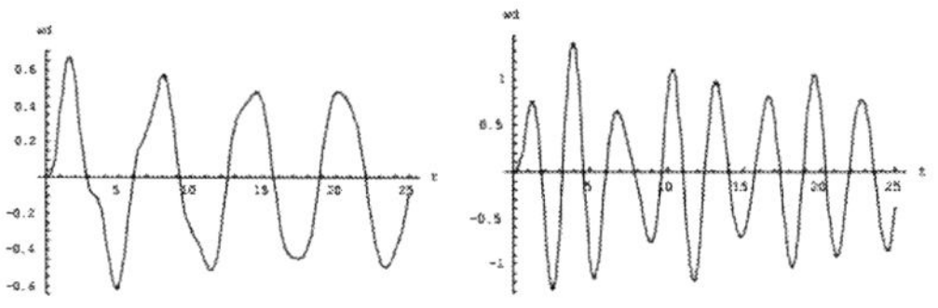

p=20·Sin(1·t) p=20·Sin(2·t)

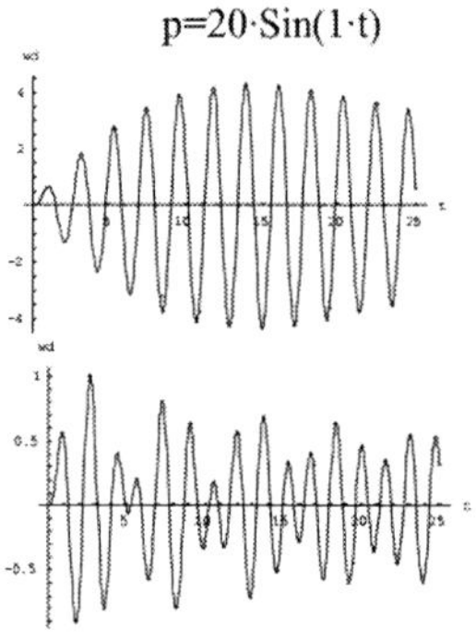

p=20·Sin(3·t) p=20·Sin(4·t)

In this case, big deformations appear, because the eigenfrequency of the external load is near to the first one of the cable (ω1 =2.82 sec-1).

Fig. 9. Oscillations of the mid-length of a cable of length L=250m, without any damping system

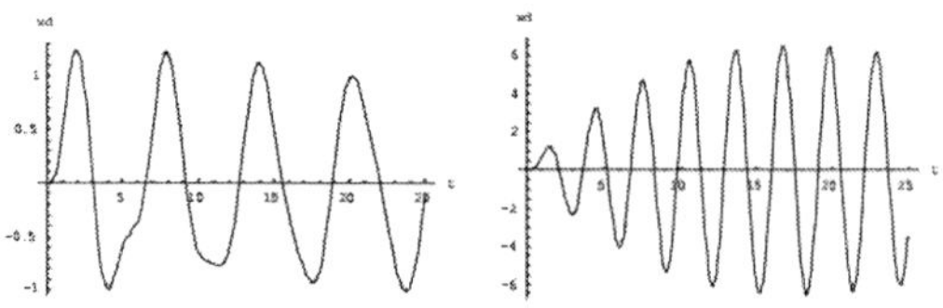

p=20·Sin(1·t) p=20·Sin(2·t)

In this case, big deformations appear, because the eigenfrequency of the external load is near to the first one of the cable (ω1 =2.15 sec-1).

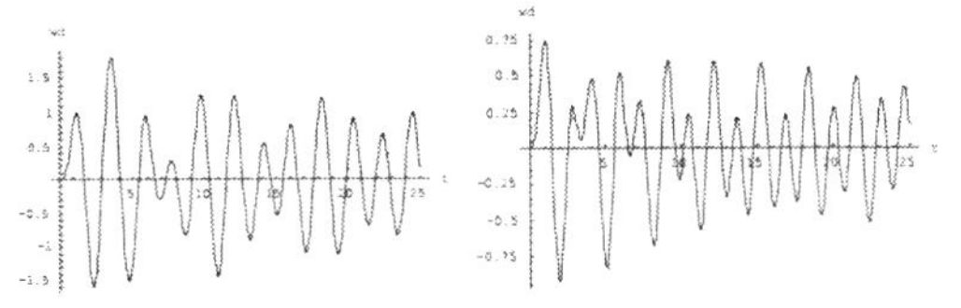

p=20·Sin(3·t) p=20·Sin(4·t)

Fig. 10. Oscillations of the mid-length of a cable of length L=350m, without any damping system

5.4 The damping system

In the followings, we will use a C.R.P.B. device, based on the operation principle of the simple system of Fig. 7. We will consider devices with concave radii R=1, 2, and 3 meters.

In the plots of Figs 12, 13, 14, we see the oscillations of the middle and of the anchor head of a cable of length 150m, tension 300000dN, and for different values of R.

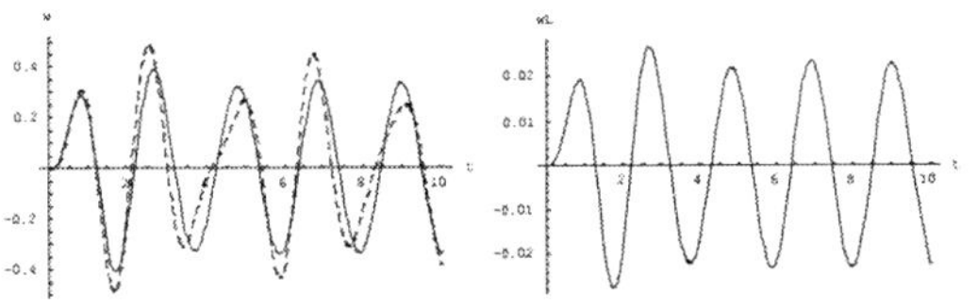

Fig. 11. Oscillations of the mid-length and of its anchor head of a cable of L=150m, R=3m ___ with , - - - without C.R.P.B

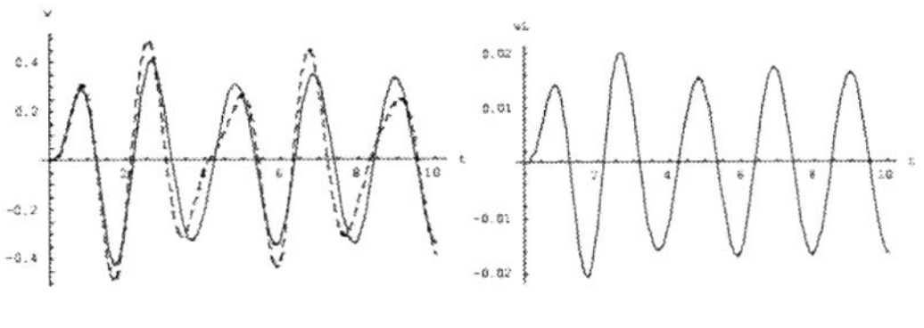

Fig. 12. Oscillations of the mid-length and of its anchor head of a cable of L=150m, R=2m ___ with , - - - without C.R.P.B

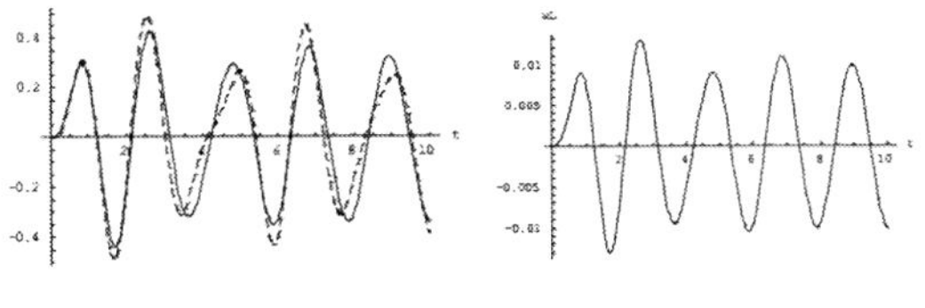

Fig. 13. Oscillations of the mid-length and of its anchor head of a cable of L=150m, R=1m ___ with,- - - without C.R.P.B

From the above plots of Figs 11 to 13, we ascertain that smaller radii are more effective than the greater ones. The above results are valid for both the cables' deformations and the anchorages' motion.

In the plots of Figs 14, 15, and 16, we see the oscillations of the mid-length and of the anchor head of a cable of length 250m, tension 300000dN, and for different values of R.

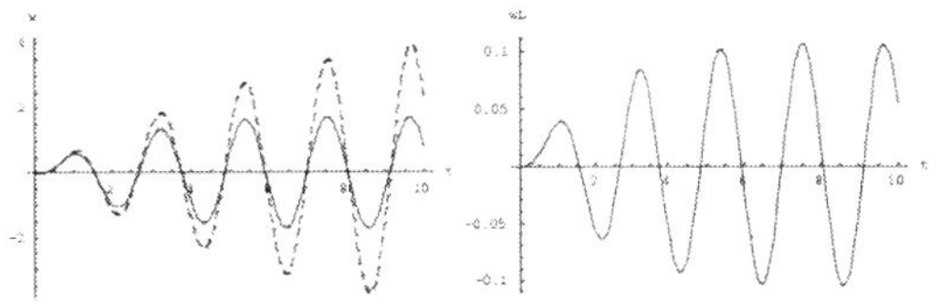

Fig. 14. Oscillations of the mid-length and of its anchor head of a cable of L=250m, R=3m ___ with, - - - without C.R.P.B

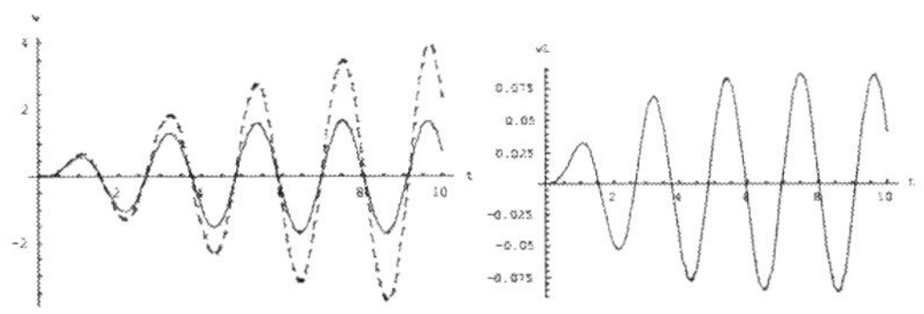

Fig. 15. Oscillations of the mid-length and of its anchor head of a cable of L=250m, R=2m ___ with, - - - without C.R.P.B

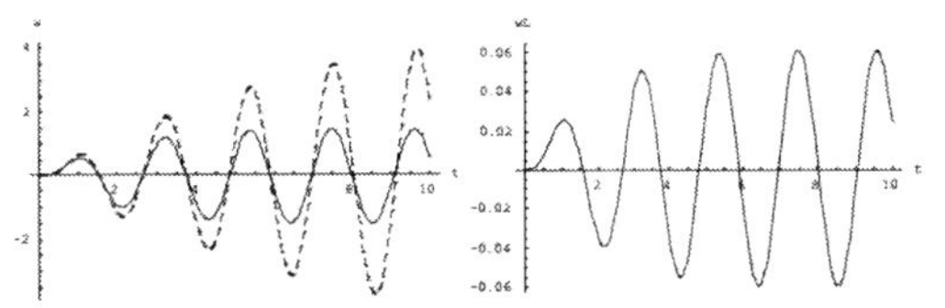

Fig. 16. Oscillations of the mid-length and of its anchor head of a cable of L=250m, R=1m ___ with, - - - without C.R.P.B

From the above plots of Figs 14 to 16, we observe that although the oscillations' amplitude is remarkable large the effectiveness of the system is obvious. We ascertain, again, that smaller radii are more effective than the greater ones. The above results are valid for both the cables' deformations and the anchorages' motion.

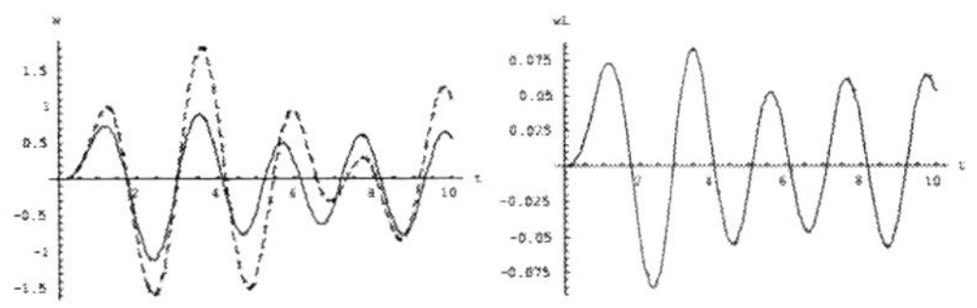

Fig. 17. Oscillations of the mid-length and of its anchor head of a cable of L=350m, R=3m ___ with, - - - without C.R.P.B

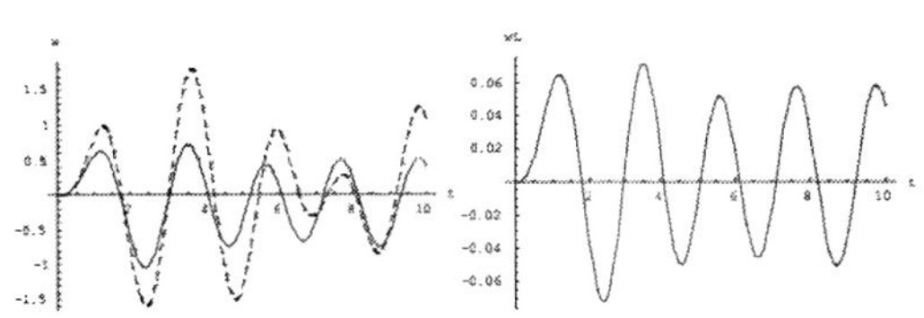

Fig. 18. Oscillations of the mid-length and of its anchor head of a cable of L=350m, R=2m ___ with, - - - without C.R.P.B

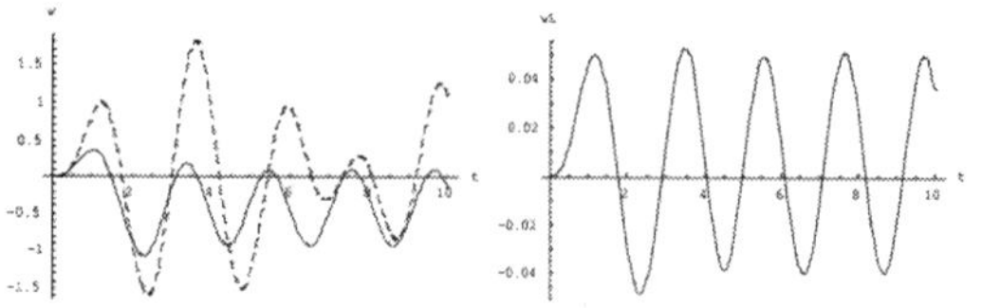

Fig. 19. Oscillations of the mid-length and of its anchor head of a cable of L=350m, R=1m ___ with, - - - without C.R.P.B

In the plots of Figs 17, 18, and 19, we see the oscillations of the mid-length and of the anchor head of a cable with length 350m, tension 300000dN, and for different values of R.

From the above plots of Figs 17 to 19, we ascertain, once again, that smaller radii are more effective than the greater ones. The above results are valid for both the cables' deformations and the anchorages' motion.

6 Design of a C.R.P.B. device

6.1 Bearing capability

Manufactures of ball bearings typically publish "LOAD RATINGS" for each bearing that they produce. Both ABMA and ISO have published standards related to load ratings. ABMA std. 9 - Load Ratings and Fatigue life for Ball Bearings. ABMA std. 12.1 and 12.2 - Instrument Ball Bearings. ISO 76 - Static Load Ratings. ISO 281- Dynamic Load Ratings and Rating Life. Static load ratings and dynamic load ratings are calculated on completely different ways and there is not direct relationship to one another. The Basic Static Load Rating applies to bearings where motion does not occur or occurs only infrequently. The basic load ratings and calculation methods are based on methods described by the above-mentioned ISO recommendations.

As a standard of permissible static load, the basic load rating is specified as follows: Maximum contact pressure at the contact point 4200 MPa (1 MPa=100 N/cm2)

Total permanent deformation of the compressed zone can be, approximately, 1/10000th of the rolling elementary diameter. The basic load rating for stainless steel is 80% of that for standard bearing steel.

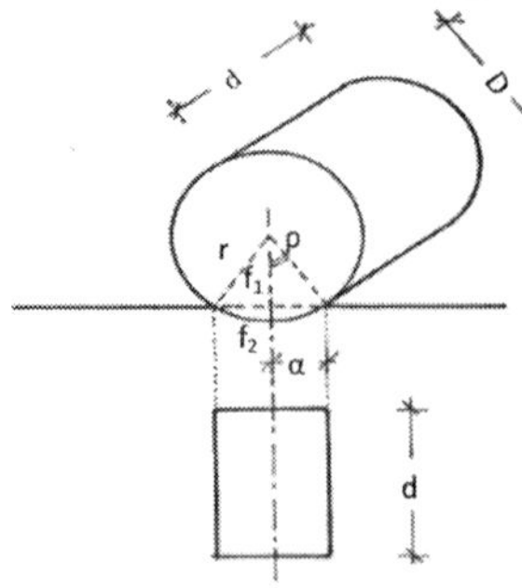

Fig. 20. Geometry of a cylinder bearing

According to the above recommendations, one can proceed as follows (see Fig. 20):

$f_2 = \frac{2r}{10000} = r \cdot 2 \cdot 10^{-4}$ and $f_1 = r - f_2 = 0.9998 \cdot r$

From the above, we find: $\cos\rho = \frac{f_1}{r} = 0.9998 = \text{constant}$ or finally:

$\rho = 0.020\,\text{rad}\ (\sim 1.146^\circ) = \text{constant}$

From the sketch of Fig. 20, we get: $\alpha \cong 2\pi r \frac{\rho}{2\pi} = \rho r$

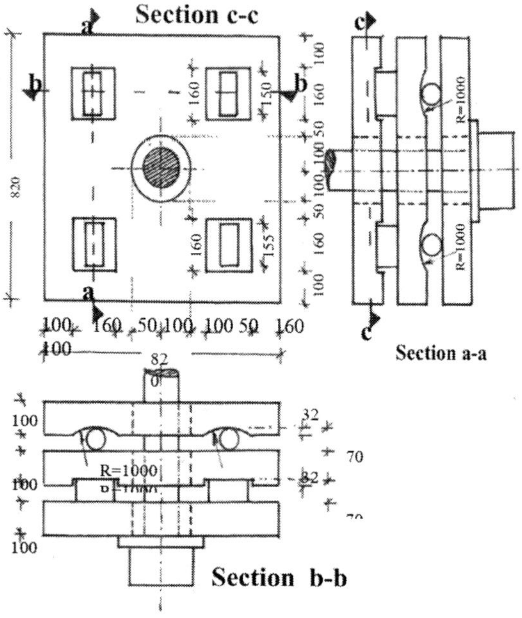

Fig. 21. The required C.R.P.B.

Therefore, for the safe undertaking of a load To, the required length d of the cylinder of Fig. 21, can be determined by the following relations:

$$d = \frac{T_o}{2\alpha\sigma_{per}}, \quad \text{or for stainless steel: } d = \frac{1.25 \cdot T_o}{2\alpha\sigma_{per}}$$

6.2 Selection of the appropriate C.R.P.B.

We select a device with R=1m, and D=2r=7cm.

From §6.1, we have $\alpha = \rho r = 0.020 \cdot 3.5 = 0.07\,\text{cm}$ and $d = \frac{T_o}{2\alpha\sigma_{per}} = \frac{300000}{2 \cdot 0.07 \cdot 42000} = 51\text{cm}$ or, equivalently, 4 cylinders with length 15cm.

The designed C.R.P.B. device shown in Fig. 22 can undertake loads acting to any direction because the rolling cylinders operate along two perpendicular axes.

7 Conclusions

In this paper, a movable anchorage system with Classical Rolling Pendulum Bearing (C.R.P.B.) for vibration control of stay cables has been proposed and investigated. A model for the control system has been formulated, based on the taut string representation in which the proposed device has been incorporated.

From the studied cases, one can conclude to the followings:

a) The constant of the equivalent spring of the CRPB system has been assessed.

b) The results of cable response show that the proposed CRPB device can effectively reduce the oscillation magnitude of the cable, proving the efficiency of the system.

c) The observed decrease of the cable's oscillations amounts from 15% to 50%, while the motion of the anchor-head of the selected and designed device amounts from 1.5 to 6 cm. One must note that the CRPB device is very effective even for external loads acting with frequencies equal or near to one of the eigenfrequencies of the strained cable.

d) The design parameters of the CRPB system for the selected cables are identified and the proper device has been designed.

e) The proposed anchorage system is shown to perform more efficiently than the conventional passive external dampers, presenting a better solution from aesthetics point of view.

References

[1] J.F. Fleming, *"Non-linear static analysis of cable-stayed bridge structures"*, Comp. Struct., (1979), **10**, pp. 986-1000

[2] C.F. Kollbruner, N. Hajdin, and B. Stipanic, *"Contribution to the analysis of cable-stayed Bridges"*, Inst. For Engineering Research Editions, (1980), N. **48**, Schulthess Verlag, Zürich

[3] M.S. Khalil, *"Non-linear analysis of cable-stayed bridges at ultimate load level"*, Canadian J. of Civil Engineering, (1999), **23 (5)**, pp. 1111-1117

[4] G.T Michaltsos, J.C. Ermopoulos, and T.G. Konstantakopoulos, *"Preliminary design of cable-stayed bridges for vertical static loads"*, J. of Structural Engineering and Mechanics, (2003), **16(1)**, pp. 1-15

[5] J.F. Fleming, and E.A. Egeseli, *"Dynamic behavior of a cable-stayed bridge"*, Int. Journal of Earthquake Eng. & Structural Dynamics, (1980), **8(1)**, pp. 1-16

[6] D. Bruno, and V. Colotti, *"Vibration analysis of cable-stayed bridges"*, Journal of Structures Ing. International, (1994), **1**, pp. 23-28

[7] G.T. Michaltsos, *"A simplified model for the dynamic analysis of cable-stayed bridges"*, Facta Universitatis, (2001), **3(11)**, pp. 185-204

[8] T.G. Konstantakopoulos, G.T. Michaltsos, and D.S. Sophianopoulos, *"A simplified model for the study of the lateral-torsional vibration of cable-stayed bridges"*, Proceedings of the Eurodyn 2002, Munich, (2002)

[9] D.Q. Cao, M.T. Song, W.D. Zhu, R.W. Tucker, and C.H.-T. Wang, *"Modeling and analysis of the in-plane vibration of a complex cable-stayed bridge"*, J. of Sound and Vibration, (2012), **331(26)**, pp. 5685-5714

[10] I.G. Raftoyiannis, and G.T. Michaltsos, *"Movable anchorage systems for vibration control of stay-cables in bridges"*,Engineering Structures,(2016),**112**,pp. 162-171

[11] I.G. Raftoyiannis, and G.T. Michaltsos, *"Curved-in-plane cable-stayed bridges. A mathematical model"*, Int. Journal of Structural Stability and Dynamics, (2012), **12(3)**, DOI: 10.1142/S0219455412500113

[12] J. Ermopoulos, A. Vlahinos, and Y.-C. Wang, *"Stability analysis of cable-stayed bridges"*, Comp. Structures, (1992), **44(5)**, pp. 1083-1089

[13] A. Bosdogianni, and D. Olivari, *"Wind-reduced and Rain-induced oscillations of cable-stayed bridges"*, Journal of Wind Eng. and Industrial Aerodynamics, (1997), **64(2-3)**, pp. 171-185

[14] G.T. Michaltsos, *"Stability of a cable-stayed bridge's pylon, under time-depended loading"*, Collection of papers in memory of Academician P.S. Theocharis. Inst. of Mechanics Problems and Academy of Sciences, Russia, Inst. of Mechanics of Nat. Acad. of Sciences, Armenia, (2005)

[15] G.T. Michaltsos, I.G. Raftoyiannis, and T.G. Konstantakopoulos, *"Dynamic Stability of cable-stayed bridge pylons"*, Int. Journal of Structural Stability and Dynamics, (2008), **8(4)**, pp. 627-643

[16] Y. Hikami, and N. Shiraishi, *"Rain-wind-induced vibrations of cables in cable stayed bridges"*, Journal of Wind Engineering and Industrial Aerodynamics, (1988), **29(1-3)**, pp. 409-418

[17] K. Wilde, and W. Witkowski, *"Simple model of rain-wind-induced vibrations of stayed cables"*, Journal of Wind Engineering and Industrial Aerodynamics, (2003), **91(7)**, pp. 873-891

[18] D. Sophianopoulos, *"Rain-wind induced vibrations in stay cables – Response characteristics, excitation mechanism and modeling."* Int. Journal of Bridge Engineering, (2013), **1(1)**, pp. 29-36

Optimal design of power frames for special purpose vehicles' cockpits with regard to their eigenfrequencies and shock resistance

Aleksandr Leontev[1,*] *Mikhail* Aleshin[1,*] *Oleg* Klyavin[1,*] *Alexey* Borovkov[1,*]

[1]ST Petersburg State Polytechnical University, 195220 St.-Petersburg Gzhatskaya str. 21/ 2, Russia

Abstract. The paper is focused on development of a methodology for designing special purpose vehicles' cockpits minimizing vibration and impacts in the workspace. Finite elements method is utilized as the main tool in assessment of the structural design. The method also forms a foundation for further topology optimization, which allows for obtaining lightweight adaptable to production design fulfilling both vibration and impact requests, standards and regulations. The method has been used in designing experimental cockpit for a prospective tractor. Thus it is shown that the methodology being presented in the article can be successfully implemented in design of special purpose vehicles.

1 Introduction

Nowadays there is a trend of increasing interest to the problems of acoustics and vibration during the vehicle design. These are the factors, which a person directly interacts with. The rising concern on the topic is dictated by many factors:

• Desire to increase consumer satisfaction and, as a result, the vehicle competitiveness;

• Increase the computing power of modern computers, which allows solving large-scale problems, which were too consuming in terms of computation before.

2 Problem description

Tractor production is not an exception and the area has a rising interest to the NVH problems. The machine operator spends 80% of the working time in the cabin and, therefore, they are subjected to a constant affliction of vibrations and noise. As a result, when buying new equipment, the customers of the tractor pay special attention to the vibro-acoustic comfort of the cabin. Therefore, manufacturers of special equipment have to take into account the wishes of consumers and strive to reduce the noise and vibration.

Another important requirement for the cabin of special machinery is the presence of a protective structure. Satisfaction of this requirement ensures the safety of the tractor operator in the case of dangerous and emergency situations: tractor roll-over and heavy objects drop on the roof. These requirements are regulated by international standards, technical regulations of the Eurasian Customs Union, GOST standards. Its fulfillment is mandatory for all manufacturers of special vehicles.

Therefore, the developers of new technology face the need to meet the mutually exclusive requirements imposed on the design in attempts to increase vibro-acoustic comfort, as well as passive safety. For example, the tractor cabin must be rigid so the lowest natural frequency would be high enough, and, at the same time, it must be sufficiently compliant to maximize impact energy absorption during the roll-over.

In accordance with this, it becomes necessary to develop an instrument that will provide an opportunity of obtaining the optimal solution that will satisfy mutually exclusive requirements for vibro-acoustics and passive safety in an automatic mode.

3 Goals and objectives

Within the framework of the project funded by the Ministry of Education of the Russian Federation, we set the goal to develop such a tool while creating a next-gen tractor with improved design, visibility, ergonomics, vibro-acoustic and climatic comfort.

At the first stages of the project, a digital model of the original tractor design was created in order to determine the layout constraints on the design. Next, an updated style surface and the power frame of the cabin was developed were designed. The current focus is on fine-tuning of the cabin design and layout (Fig. 1).

* Corresponding author: Leontev@compmechlab.ru; aleshin@compmechlab.com; klyavin@compmechlab.com; Borovkov@compmechlab.com

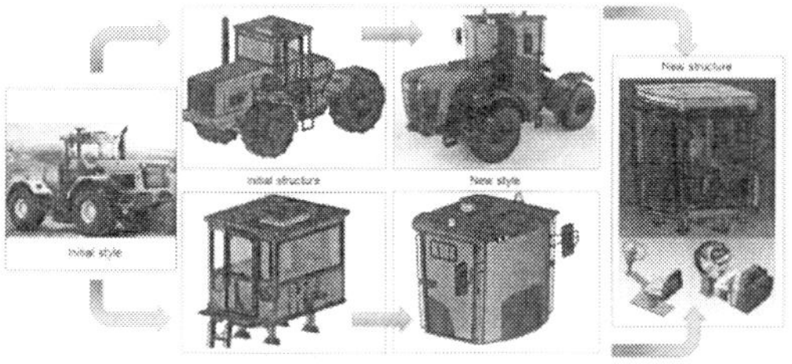

Fig. 1. The process of the development of the new cabin design.

To achieve this goal, it was necessary to solve the following tasks:
• Considering the classical approaches to the design of special equipment cabins;
• Set up the goals and limitations for optimization by cascading targets and constraints from the subjective wishes of the customer to the technical requirements for the nodes and components;
• Development of an algorithm for multidisciplinary cross-platform structural optimization;
• Carry out a multi-disciplinary cross-platform optimization of the power frame of the tractor cab.

4 The classical approach to the design process

The classical approach to design is based on the results of topology optimization [1]. First of all, the analysis of layout and style surface is carried out and the scope of the design is determined, that is, the volume in which the structural elements of the structure can potentially be located (Fig. 2).

Fig. 2. Design volume.

After this, topology optimization is carried out. Its results determine the main force paths of the power frame (Fig. 3).

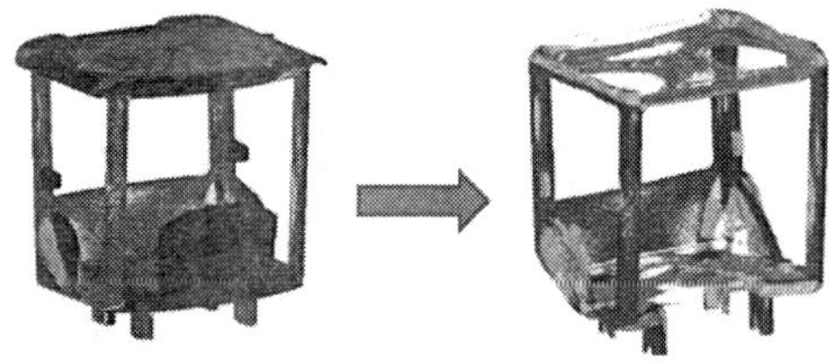

Fig. 3. Results of topology optimization.

Then a beam frame model is constructed and parametric optimization of the beam cross-sections is performed (Fig. 4).

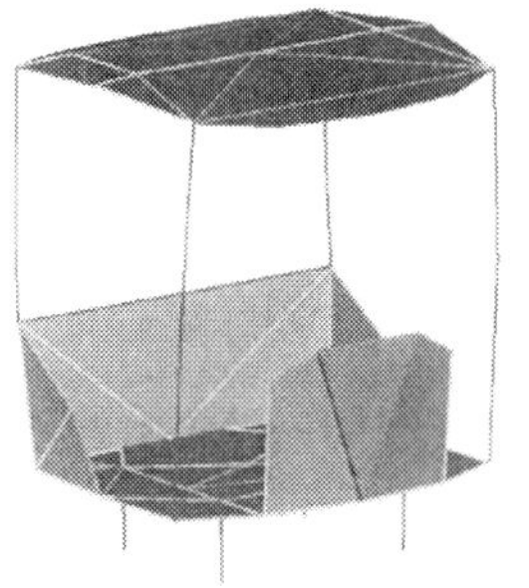

Fig. 4. Beam model of the tractor cab.

As a result, a section map is constructed. Taking is as a basis a mathematical model of the prototype of the power frame is developed. Further, a multi-variant modification of the structure is performed separately for vibro-acoustic comfort and for passive safety. Often, those design solutions that have a positive effect on vibro-acoustic comfort have a negative effect on passive safety and vice versa. Thus, the process of creating a design that satisfies the set goals is multi-iterative and takes a long time.

All in all, the classical approach requires a large number of highly qualified stuff, it takes a lot of time and does not give the most effective result.

5 Algorithm of multidisciplinary cross-platform optimization

The process of obtaining the optimal solution using multidisciplinary cross-platform optimization for the most part resembles the process of the classical approach, but it has drastic differences at the stage of the design finalization. Instead of manual refinement of the design, optimization is used at each iteration to analyze both vibro-acoustic parameters, such as the global natural frequencies of the design and the parameters of passive safety, such as the penetration of structural elements into the cabin inner space. In the end of each iteration the engineer determines whether the targets are met. If the requirements are not fulfilled, the optimization variables are altered and the next iteration is performed (Figure 5).

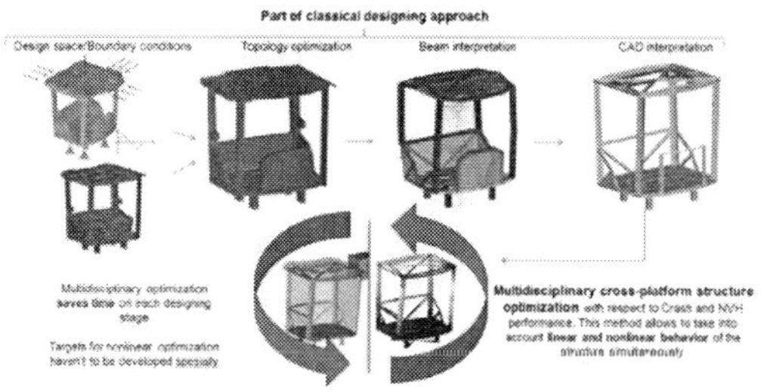

Fig. 5. The process of multidisciplinary cross-platform optimization.

Optimization is considered converged if all targets are met.

Despite its advantages, multidisciplinary cross-platform optimization has a number of limitations:

• The topology of the design does not change during optimization;

• The shape change in the optimization process is limited

6 Variables, boundary conditions and constraints for optimization

Analysis of the NVH main requirements and their further cascading showed that one of the main parameters are global natural frequencies of the cabin on its own suspension. Requirements for these indicators are directly related to the excitation frequencies from the power unit. Own frequencies of the cabin should square root of two larger than the frequency of the installed engine idle.

Target values for natural frequencies were calculated based on the frequency of the dominant motor harmonics for the internal combustion engine according to the formula (1). The project took into account the characteristics of the engines YaMZ 238 and TMZ 84;

For V-type engine:

$$f = \frac{0.5 \cdot n \cdot i}{60 \cdot \tau}, \quad (1)$$

i is the number of cylinders, $\tau = 1$ for a two-stroke engine, $\tau = 2$ for a four-stroke engine.

The selected target values are guaranteed to ensure compliance with the requirements for vibro-acoustic comfort:

• target values of the frequencies of the torsional vibrations of the assembled tractor cab roof relative to the floor and supports around the Y axis were set in the range of more than 25 Hz and more than 30 Hz, respectively;

• frequency of natural oscillations of the assembled tractor cab in the longitudinal direction - in the range from 11 to 18 Hz and more than 25 Hz;

• natural frequencies of the assembled tractor cab in the transverse direction - in the range from 16 to 18 Hz and more than 25 Hz.

Analysis of the basic requirements for passive safety and their cascading showed that the main indicators are the maximum displacement of the structural elements of the cabin under the impact of shock loads.

Thus, the optimization goal is to reduce the mass, while limiting the lowest global natural frequency of the cabin's power frame, according to the maximum penetration of the pendulum with lateral and rear impacts.

The design optimization variables are the thicknesses of all the beams (Figure 6) that make up the power frame of the cabin. In addition to this, the shape of the cabin is a design variable as well. Variables are width, length, height of the cabin, the angle of inclination of the windshield, as well as the angle of inclination of the side walls of the cabin.

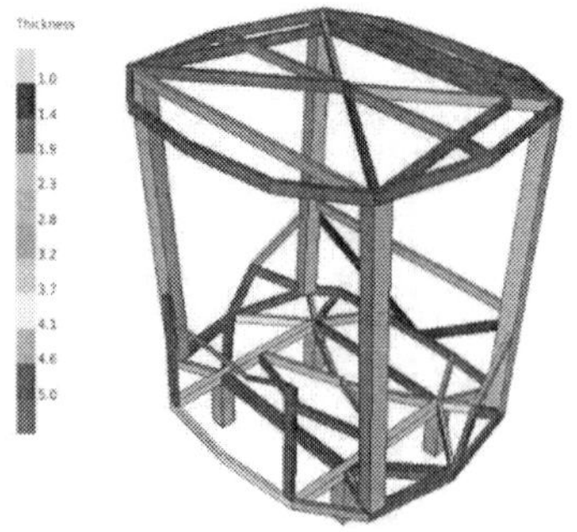

Fig. 6. Parts whose thicknesses are alerting during optimization.

7 Results of multidisciplinary cross-platform optimization

During the optimization process 57 iterations were carried out. It took 136 hours. As a result, the design of the tractor cab which satisfies all the requirements was obtained (Fig. 7,8):

• The first global natural frequency increased from 19.8 Hz to 25.2 Hz;

• The second global natural frequency increased from 22.6 Hz to 28.5 Hz;

• The third global natural frequency increased from 26.3 Hz to 32 Hz;

• Penetration of the pendulum with a lateral impact decreased from 261 mm to 215 mm;

• Penetration of the pendulum with a rear impact dropped from 204 mm to 194 mm;

• The weight of the power frame decreased from 709.6 kg to 656.2 kg.

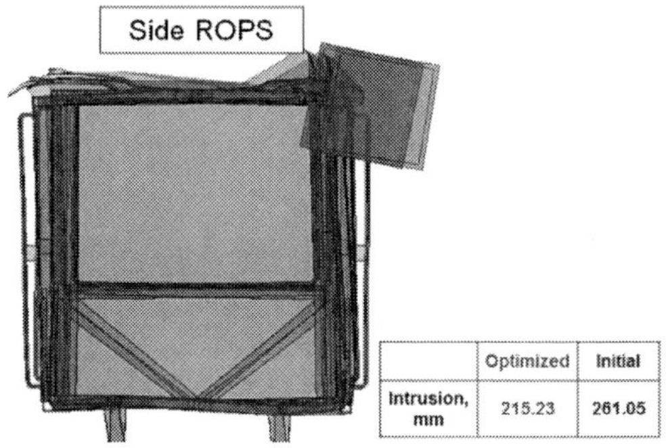

Fig. 7. Comparison of optimized and non-optimized side impact structures.

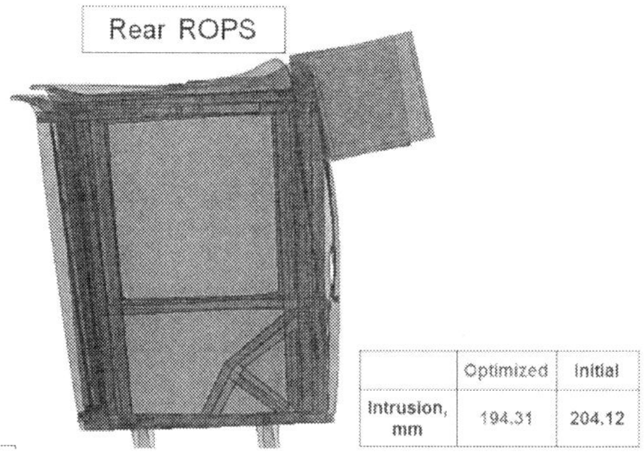

Fig. 8. Comparison of optimized and non-optimized designs in the case of the rear impact.

8 Specifications of the computing server

Optimization was carried out on the HPC of SPBPU, which is on the third place in Russia in terms of performance and is the most powerful computer in Russia, which is used to solve applied problems of various industries of the Russian Federation. The performance of the HPC is 1.3 Pflops (Figure 10).

Fig. 9. The used HPC.

9 Main conclusions

The outcome of the performed actions is the tractor cab structure, which:

- Has a reduced mass
- Satisfies the vibro-acoustics rquirements
- Satisfies Passive Safety Requirements

Thus, it can be concluded that the approach of multidisciplinary cross-platform optimization for the designing procedure of power frames of special vehicles leads to reduction of production costs (be the means of material expenses reduction), design time reduction, and it also allows to obtain an optimal result while taking both requirements for vibro-acoustics and passive safety into account.

The work is carried out with financial support of the Ministry of Education and Science of the Russian Federation within the framework of the Federal Program "Research and Development in Priority Areas for the Development of the Russian Science and Technology Complex for 2014-2020", Activity 1.3., Agreement on the Grant No. 14.578.21.0206 of 03.10.2016, the unique identifier of the program of research and experimental works: RFMEFI57816X0206..

References

1. Bendsoe M.P., Sigmund O Theory, Methods and Applications. 2nd ed. Berlin: Springer-Verlag, 2004. **370** p.

https://doi.org/10.1051/matecconf/201814803001

Non-linear finite element analysis of the dynamics of a slender cable stayed tower

Eliot Pezo, *Paulo* Gonçalves*, and *Deane* Roehl

Civil and Environmental Engineering Department, PUC-Rio, 22451-900, Rio de Janeiro, RJ, Brazil

Abstract. The aim of the present work is to investigate the static and dynamic nonlinear behaviour of a cable-stayed tower. A continuous structural model of a slender tower is discretized by the finite element method. First the buckling and post-buckling behaviour of the tower under axial load is explored, in order to understand the influence of the cable stiffness and lateral restrain on the load carrying capacity of the tower. Then, the linear vibration modes and frequencies are obtained. Due to the inherent symmetries of the tower, coincident buckling loads and vibration frequencies are obtained. This may lead to interactive buckling and internal resonance, increasing the effect of the geometric nonlinearities on the response. The results show that the tower exhibits a highly nonlinear response, which must be considered with care in the design stage.

1 Introduction

Slender guyed towers are used to support broadcasting antennas, for meteorological measurements, power transmission, etc. Nowadays, modern tall guyed towers constructed of high-strength and lightweight materials are highly flexible and lightly damped. These structures present non-linear behaviour even under working conditions [1]. The main source of nonlinearity is the change of stiffness of the guys due to the variation of stresses. Also they may exhibit undesired vibrations under environmental loads such as wind and earthquakes. Considering the strong structure sensitivity to imperfections, the structure's work conditions must follow severe requirements in terms of maximum displacements and rotations. In addition, static and dynamic instabilities and non-linear vibration of these structures are important design concerns.

2 Phenomenological discrete model

A 2dof simplified model of a guyed tower, supported laterally by three linear springs, as illustrated in Fig. 1, was proposed by Thompson and Gaspar [1] as an example of interactive buckling in the context of the catastrophe theory. The nonlinear static and dynamic behaviour of this model was recently explored by Orlando et al. [2] and Gavassoni et al. [3]. Their results show that the inherent symmetries and post-buckling solutions have a marked influence on the underlying potential function and, consequently, on the nonlinear dynamics of the system. The model is composed by an inverted spatial pendulum with a tip-mass m and a slender, rigid (but massless) bar of a length l pinned at the base, connected to three linear springs. The springs are symmetrically positioned around the tower.

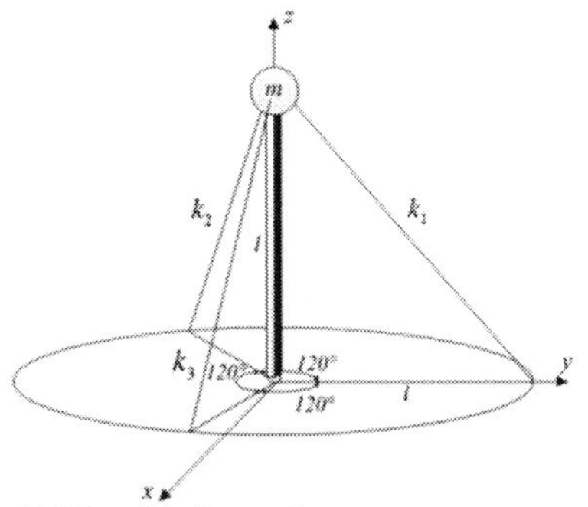

Fig. 1. 2DOF model of a guyed tower

3 Finite element model

In order to understand the behaviour of the continuous system and infer if it exhibits the same type and degree of nonlinearity of the discrete model, a 100m tall tower clamped at the base and supported laterally by three guys inclined at 60^0 and equally distributed around the tower is considered. The numerical analysis of the elastic guyed tower is carried out with a finite element model composed by a tower discretized with 10 two-node cubic spatial beam elements and three cables. The cables are discretized with linear three-dimensional truss elements. Figure 2 presents the geometry and finite element model of the structure.

The tower presents the following properties: cross section 0.0191 m^2, inertial moment 5.69E-4 m^4, and Young's modulus 1.18E+8 kN/m^2. The cable properties

* Corresponding author: paulo@puc-rio.br

are: diameter 0.018 m, cross section 2.5446E-4 m^2 and Young's modulus 1.0E+8 kN/m^2.
An initial pretension of 10 kN is applied to the cables defining an initial stress and deformation state. Figure 3 and Figure 4 show the guyed tower with cables loaded at 10 kN, before and after equilibrium, respectively. Unloading of stresses in the cables and the compression in the tower can be observed.

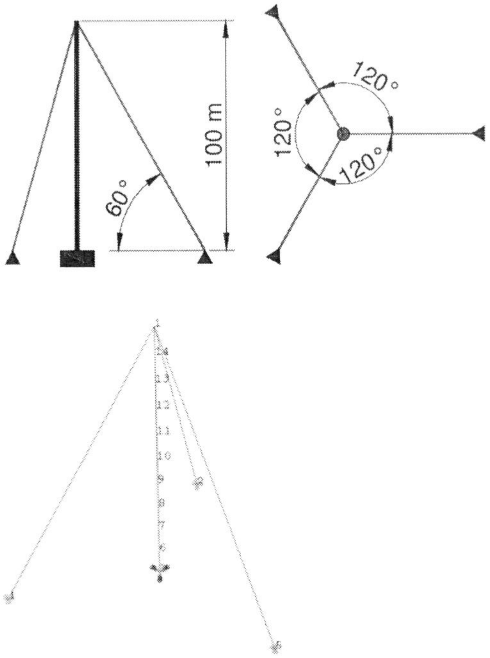

Fig. 2 Guyed tower geometry and finite element model.

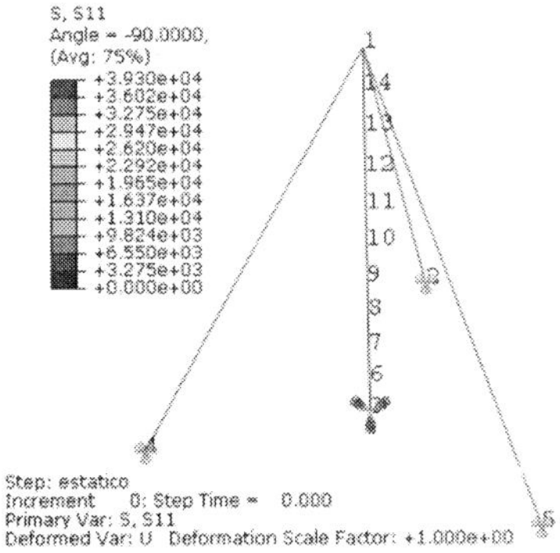

Fig. 3 Initial applied stresses- undeformed state. Stresses in kPa.

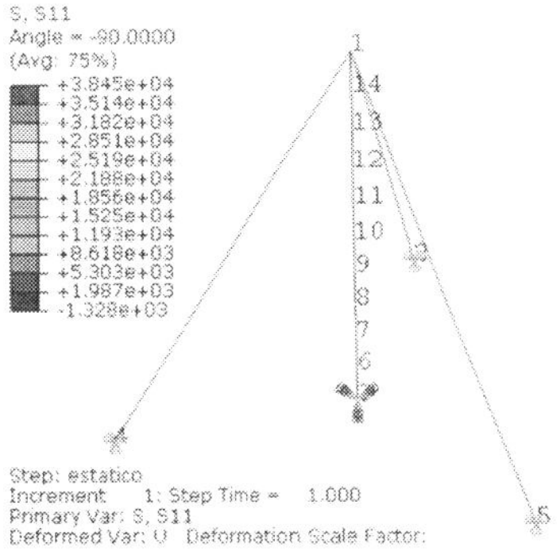

Fig. 4 Equilibrium in the initial – deformed state. Stresses in kPa.

4 Stability analysis – modal coupling

4.1 Buckling FE analysis

Fist, the buckling loads and modes of the finite element model of the elastic guyed tower are obtained. As expected, due to the system symmetry, the first two bifurcation loads are the same, indicating the possibility of modal interaction. Table 1 provides the first four bifurcation loads for the finite element model. The buckling load of a clamped-free column with the same physical and geometrical parameters is equal to 16.6 kN. Thus, as shown in Table 1, a considerable increase in the static buckling load is observed, due to the cable stiffness, approaching the critical load of the associated clamped-pinned column. Figure 5 presents the critical loads and configurations for the first three buckling modes, where the similarity of the first two modes with the buckling mode of a clamped-pinned column is observed.

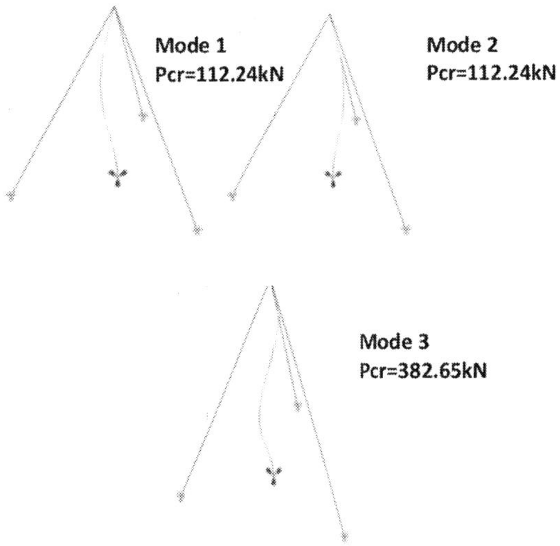

Fig. 5 Buckling modes of the finite element model

Table 1 Bifurcation loads (kN)

Mode	Pcr
1	112.24
2	112.24
3	382.65
4	382.65

4.2 Post buckling FE analysis

Next, a post buckling analysis is carried out for the tower model. The post-buckling response is shown in Figure 6, where the lateral normalized displacement components at the top of the tower are plotted as a function of the nondimensional load parameter PL^2/EI. The initial unstable response suggests high imperfection sensitivity, with the post-buckling minimum approaching the critical load of a clamped-free column.

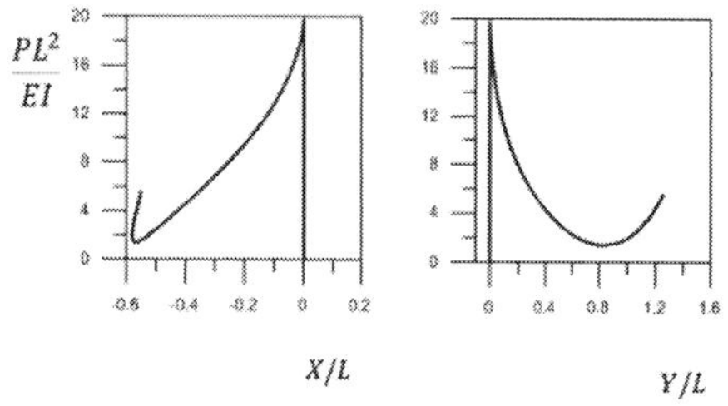

Fig. 6 Post critical analysis showing an unstable response

5 Dynamic analysis

The dynamic response is investigated for a system with a perturbation introduced through a horizontal load applied at an angle β with respect to the x axis, as illustrated in Figure 7.

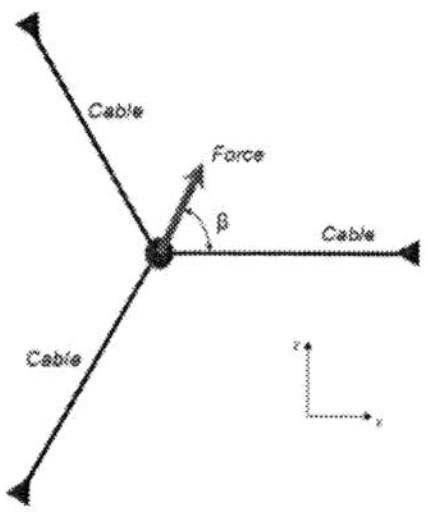

Fig. 7 Top view of the guyed tower model with load perturbation at an angle β with the vertical plane.

The first four vibration modes and frequencies are shown in Fig. 8. Again, due to the structure' symmetries, the natural frequencies occur in pairs, which may lead to 1:1 internal resonance and energy transfer between these modes.

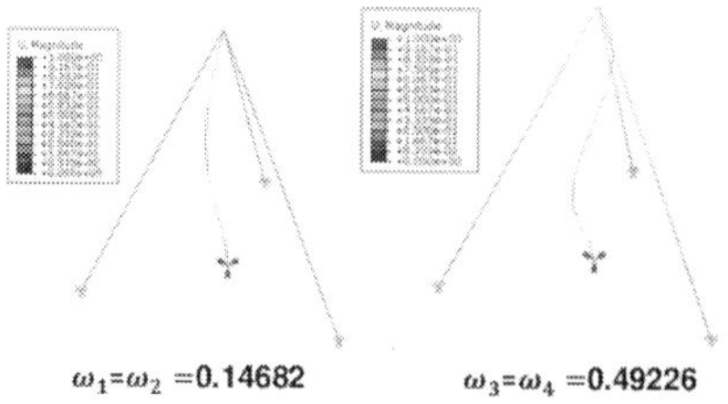

Fig. 3 Vibration modes and frequencies (Hz).

5.1 Free vibration with initial tip perturbation

The discrete model shows the coexisting of several similar and nonsimilar nonlinear normal modes [2, 3]. In order to investigate the existence of such modes on the continuous model, a nonlinear free vibration analysis is conducted. Figure 9 shows the nonlinear damped free vibration response considering an initial perturbation at β=0°, while Fig. 10 shows the response for β=60°. The initial conditions at t=0s are also shown in Fig. 9. A response similar to that in Fig. 10 is obtained for β=-60°. Thus, the same three similar nonlinear normal modes obtained for the discrete model are here obtained [2, 3].

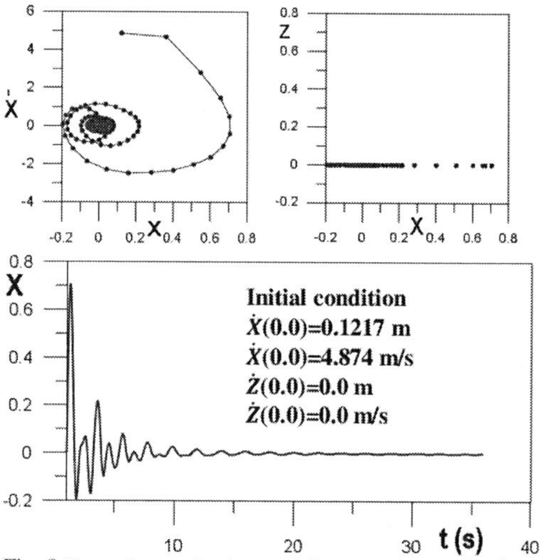

Fig. 9 Phase plane projections and time response for β =0°.

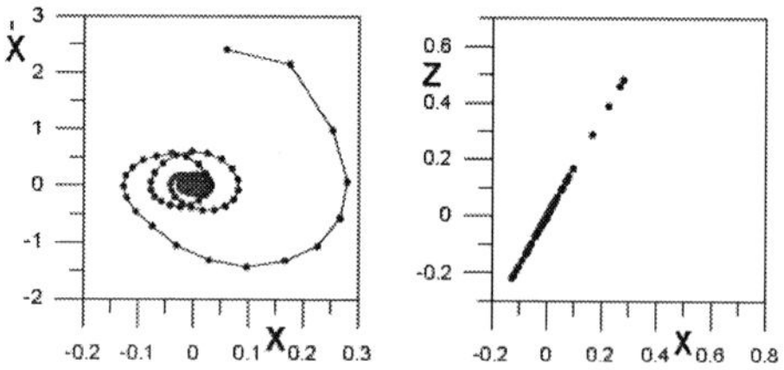

Fig. 10 Phase plane projections, displacement of top of the tower for β =60°.

For any other value of β the nonlinear interaction among these NNMs is observed.

Figure 11 shows the undamped free vibration response considering again an initial perturbation at β=0°, where energy is transferred from the x to the z direction after a high number of cycles.

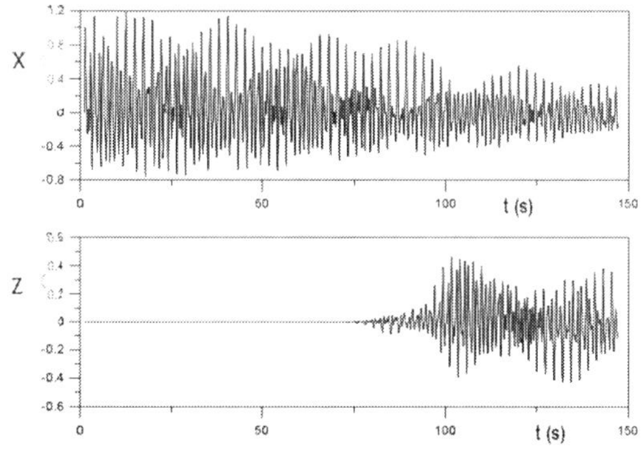

Fig. 11 Tower tip horizontal displacements in time

5.2 Damped forced vibration

The tower is then subjected to a distributed periodic horizontal load at an angle β, as illustrated in Figure 12 where the structure and loading are depicted.

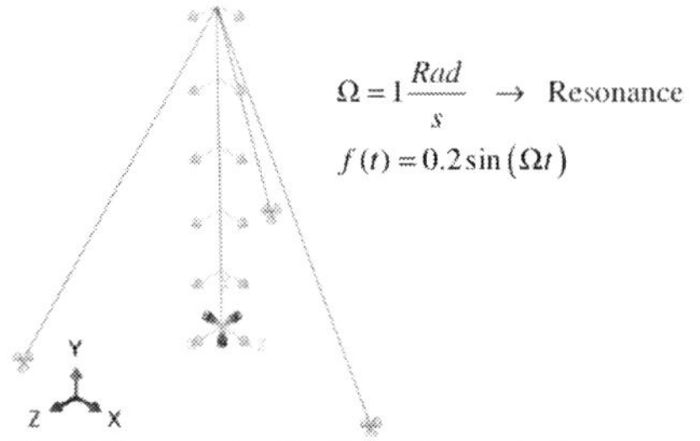

Fig. 4 Tower under distributed harmonic force

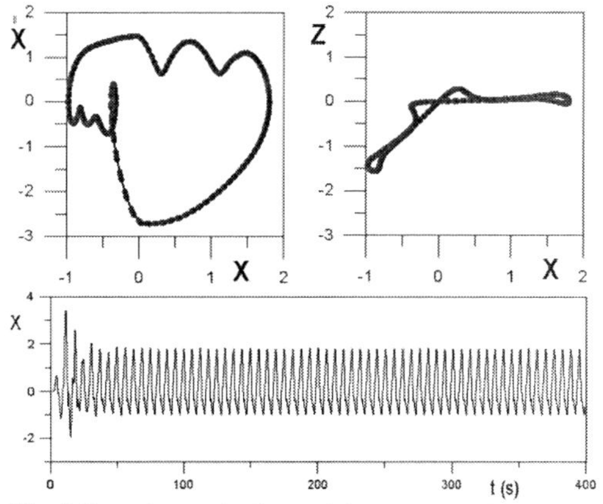

Fig. 5 Phase plane projections and time response of the tower. Inclination of the excitation force β= 30°.

Figure 13 shows two phase plane projections of the forced steady-state response and the time response of the x coordinate, which displays a small transient. The response in configuration space of the tower top (x-z plane) show clearly the coupling between the two nonlinear normal modes depicted in Figs. 9 and 10.

Final remarks

In this work, a non-linear finite element model of a guyed tower was used to investigate the stability characteristics and nonlinear vibrations of this widely employed system. Due to their inherent symmetries, these structures display coincident or nearly coincident buckling loads leading to coupled post-buckling paths. Further, the existence of coincident and nearly coincident natural frequencies may induce even more complex behaviours with the presence of internal resonance. Again, the FE model is able to provide the expected non-linear oscillations with several nonlinear normal modes which have a fundamental importance for the understanding of the tower forced response.

References

1. M. Grey, Finite element seismic analysis of guyed mast, Master thesis, University of Oxford, Oxford, UK (2006)
2. J.M.T. Thompson, Z. Gaspar, A buckling model for the set of umbilic catastrophes, Math. Proc. Camb. Phil. Soc.. 82 (1977) 497
3. Orlando, D., Gonçalves, P.B., Rega, G., Lenci, S. Influence of symmetries and imperfections on the non-linear vibration modes of archetypal structural systems, International Journal of Non-Linear Mechanics Vol. 49, 175–195, 2013
4. Gavassoni, E., Gonçalves, P.B., Roehl, D. Nonlinear Vibration Modes of an Offshore Articulated Tower, Ocean Engineering, Vol. 109, pp. 226-242, 2015

Energy balance analysis in non linear dynamic equivalent systems.

Iturregui Arranz Carlos[1,*], *Soria Herrera* Jose Manuel[1], *Muñoz Díaz* Ivan[1], *García Palacios* Jaime Higinio[1].

[1] Department of Continuum Mechanics and Theory of Structures, Escuela Técnica Superior de Ingenieros de Caminos, Canales y Puertos- Technical University of Madrid, 28040 Profesor Aranguren, Spain

Abstract. The aim of the paper is to present a critical analysis for nonlinear dynamic vibrations. It is applicable to single degree of freedom -SDOF- of reinforced concrete -RC- structures, revealing its multi degree of freedom -MDOF- performance, showing contrast using the balance of energy, using six accelerograms based on type I and II spectrum, according with Eurocode-8. The degradation curve was obtained applying a new formulation, based on the system work and complementary work, into the pushover analysis. A new method incorporating the Bouc-Wen-Baben-Noori theory and global damage was used for the analysis, adding relevance to the: energy balance in its dissipative part, analysis of the structure's fundamental parameters, relation's effective period, equivalent damping and global ductility.
The powers, energies and works developed are analyzed, creating a precise balance since energy enters selectively. Hence, an equivalent damping containing a viscous and hysteretic part is predictable, accordingly to the variation of the building's nonlinear properties. Evaluation of the adequateness and safety level are also obtainable. The controlled parameters contrasted with the balance predicts the structure's MDOF situation, at any moment related with seismic events. This methodology can be used to stablish a systematic control of nonlinearities for other structural schemes.

1. Introduction.

In an SDOF system, the general model, normalized to mass, is:

$$\ddot{x}(t)+g(x,\dot{x})=f(t) \qquad (1)$$

In this equation the first term corresponds to the inertial forces due to the acceleration of the system, the second $g(x,\dot{x})$ to the forces of restoration and loss of energy-velocity by damping, and the right hand side is the external action $f(t)$

In case of a seismic action the linear equation is:

$$m\ddot{x}(t)+c\dot{x}(t)+kx(t)=-ma(t) \qquad (2)$$

In the case of a non-linear system, the energy balance can be established in a similar way:

$$m\ddot{x}(t)+\int_0^t G(t-\tau)\,\dot{x}(\tau)d\tau+f(x,t)=-ma(t) \quad (3)$$

With this analysis, it is possible to do a general balance of the system, which will be extended to the MDOF systems, since as demonstrated by Akiyama the energy input of an MDOF system is the sum of the energy input of the SDOF that composes it [1].

$$W_{ek}+W_d+W_e+W_p=E \qquad (4)$$

Where W_{ek}, W_d, W_e, W_p are the kinetic, viscous damping, elastic and plastic work develop by the system's forces, and E is the energy input.

As an example, 6 synthetic accelerograms, created with the software acelsin [2], will be used. In the figures 1.1. and 1.2 are shown the numbers 1 and 4.

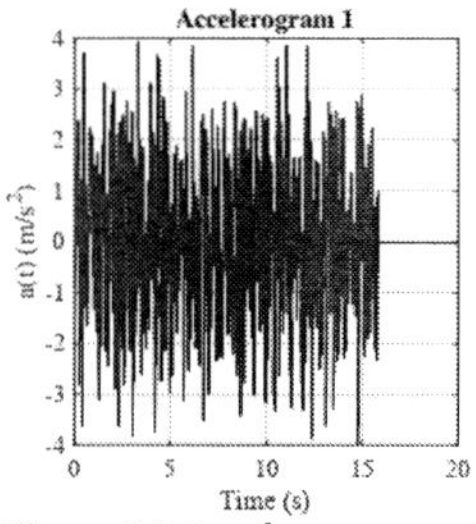

Fig 1.1. Accelerogram 1. Example of application. Spectrum type II.

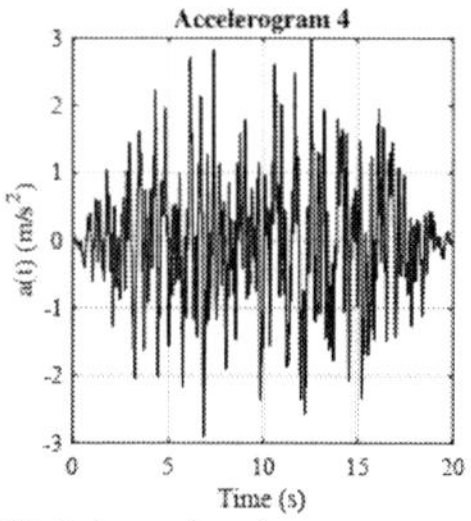

Fig1.2. Accelerogram 4. Example of application. Spectrum type I.

In the figure 2 it is shown the structure under analysis:

Fig 2. Building GF+3F.

m*=2518 t. Γ=1.2956. According with EC-8 [3].

* Corresponding author: carlos.iturregui@upm.es

K_O=74074 $\frac{kN}{m}$ T_{eqSDOF}*=1.135 sec.

The building has been designed conventionally with the ETABS program. [4] and EC 2 and 8. The symbols m*, Γ, K_O and T_{eqSDOF}* are refer to those codes.

2 Relevant parameters to consider in the energy evaluation of non linear systems in concrete structures.

2.1. Damage.

The mechanical properties of damaged differs from those initially conceived in the elastic design. This paper will concentrate the inelastic parameters, besides the lateral deformation in the case of concrete buildings considering they are indicative of the structure´s damage.

2.2. Effective period- T_e.

- Akiyama proposed a general expression for an SDOF system as follows:

$$T_e = \psi(\mu) T_o \left(\frac{\mu}{q}\right)^{0.5} \qquad (5)$$

μ is the coefficient of ductility, q is the instantaneous force normalized to the yield shear force. $\psi(\mu)$ is a function of ductility and the period of the undamaged structure.

- Kowlasky, Calvi and Priestley [5] proposed a formulation for a bilinear inelastic solid, whose value is

$$T_e = T_o \left(\frac{\mu}{1 + r(\mu - 1)}\right)^{0.5} \qquad (6)$$

r is the ratio of inelastic to elastic stiffness in the hardening section, and μ is the coefficient of deformational ductility of the system, that is the ratio between maximum displacement and elastic yield displacement.

In this paper is proposed a formula based on the work and complementary work developed by the system in the general case of inelasticity, not necessarily linear, developed by any SDOF or equivalent SDOF system.

The figure 3 shows building push over curve, being the shear force F and the displacement Δ.

Then, with this notation:

$$W = \int_0^{\Delta} F(\Delta) d\Delta \text{ and } W^* = \int_0^{F} \Delta(F) dF \text{ ;}$$

with $\mu(\Delta) = \frac{\Delta}{\Delta_y}$, $T_e = T_o\left(1 + \frac{W - W^*}{W + W^*}\mu\right)^{0.5}$ *(7)*

The factor that regulates the function is f=$\frac{W - W^*}{W + W^*}$*, which is the ratio between the difference and the sum of work and complementary work of a structural system.*

The figure 4 shows the frequency evolution with the displacement, obtained with the secant period proposed in the formula (7).

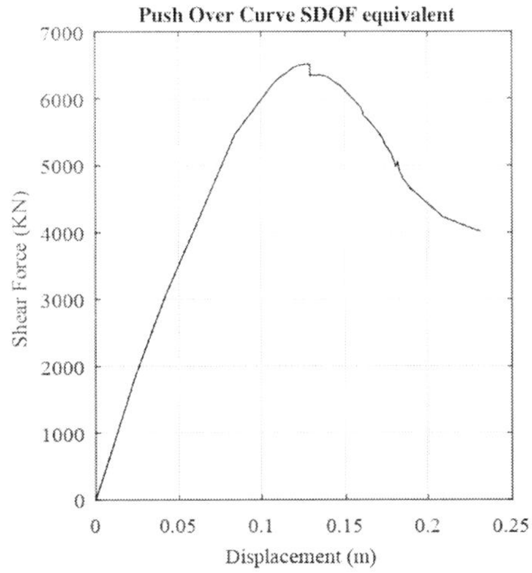

Fig 3. Push-Over curve for a GF+3F building

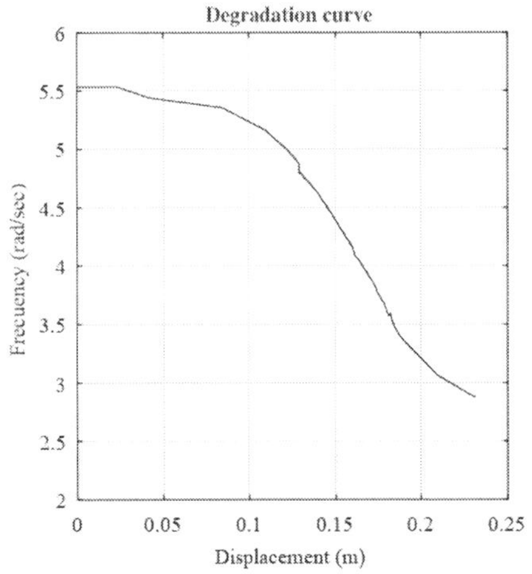

Fig 4. ***Degradation curve*** obtained with equation 7. Application to a GF+3F building with first natural frequency 5.543 rad / s for the SDOF equivalent.

2.3. Inelastic damping.

The energy dissipation is one of the most relevant aspects of the behaviour of a structure versus external action in civil construction, especially when the load is a base acceleration.

Akiyama [6] made numerous studies in elastic-perfectly plastic systems, arriving at a conclusion on the hysteretic work developed by the system.

Hysteretic damping coefficient is defined by Jacobsen's [7] as:

$$\zeta_{eq} = \frac{1}{4\pi}\frac{A_{hyst}}{F_0 A_0} \quad (8)$$

It is a ratio between hysteretic energy in one complete loop in comparation with the maximum elastic energy.

Gulkan and Sozen [8] proposed an evolution of Jacobsen's methodology

$$\zeta_{eq} = \frac{T_{eq}\int_0^t a(t)\,x(t)dt}{4\pi\int_0^t \dot{x}^2(t)dt} \quad (9)$$

The formulation is critically analysed:

Integrating in (2) the powers:

$\int_0^t a(t)\,x(t)dt = \frac{E}{m}$ and

$\int_0^t \dot{x}^2(t)dt = \frac{W_d}{c} = \frac{mW_d}{2\xi_0\omega}$, being ξ_0 the elastic viscous damping coefficient.

Substituting in the equation of Gulkan and Sozen:

$$\zeta_{eq} = \frac{T_{eq}\frac{E}{m}}{4\pi\frac{W_h}{c}} = \frac{T_{eq}Ec}{4\pi m W_h} = \frac{E\zeta_o\omega}{2\pi W_h}T_{eq} = \frac{E}{W_h}\zeta_o \quad (10)$$

In elastic system $W_d \approx E$, so $\zeta_{eq} \approx \zeta_o$

In the case of an inelastic system, the same formulation can also be considered, simply replacing c by c(ω), although some precision may be lost.

Other formulations have been developed by Kowalsky, Calvi and Priestley [5] depending on the ductility required by the system, with or without period dependence.

$$\zeta_{eq} = \zeta_0 + a\left(1 - \frac{1}{\mu^b}\right)\left(1 + \frac{1}{\left(T_{eq} + c\right)^d}\right)\frac{1}{N} \quad (11)$$

This formulation greatly improved the one raised by Gulkan and Sozen, since it takes into account the degree of damage of the structure as well as damage model, with the parameters a, b, c and d.

All aspects of this point are schematically summarized in the figure 5, by superposition of the last 4 seconds in the Inelastic Time History Analysis (ITHA) with linear and Bouc-Wen-Baber-Noori (BWBN) theory (look 3.2):

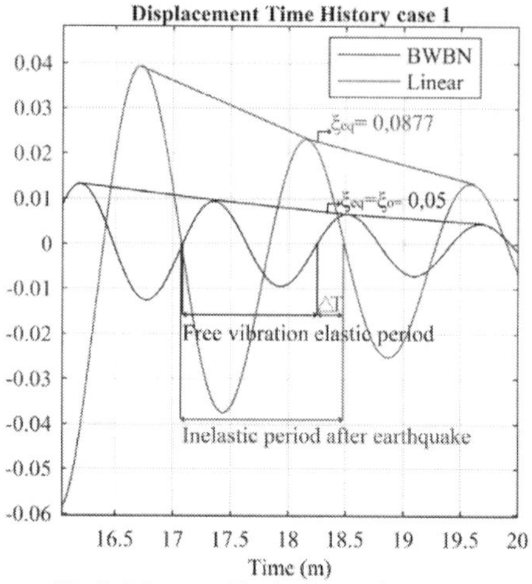

Fig 5. Schemes of inelastic quasi-period, equivalent damping caused by damage to concrete structures. Example figure 1.1. Case I.

The energy balance based design is based on, once the inelastic energy input is defined, the formulation of the viscous and plastic working developed by the structural system.

In the case of an unknown external action clearly known as a project element, the energy balance based design becomes a much more stable than forces method.

In addition to this, the energy input can be established in the balance if the equivalent damping is proportional to the viscous work.

3. Analysis of nonlinear equivalent models.

3.1. Nonlinear model with accumulated degradation associated to an incremental push curve. Analysis of damping problem and effective period of vibration.

Over the last few years the incremental push-over procedure has been implemented in different regulations, especially in seismic design standards. It consists of obtaining a SDOF substitute structure from the incremental monotonous push until breaking, it is then analyzed in a static way.

Many studies of this methodology have been done by Chopra [9], although they have been restricted to the adaptive push over methodology.

In concrete structures it could be assumed that there is a monotonous decreasing of the natural frequency of vibration, since, except the crack close by loop change, a cracked piece can not gain vibration frequency. Even in the case of close of cracks the only "distortion" is the lack of symmetry of the load loop.

From the non-linear static analysis curve it is possible to obtain the Inelastic Time History Analysis (ITHA), in comparation with the elastic Time History Analysis (THA) with the equation:

$$\ddot{x}(t) + 2\xi_o \omega(x,t)\dot{x}(t) + \omega^2(x,t)x(t) = -a(t) \quad (12)$$

In case of THA $\omega(x,t) = \omega_0$, which is supposed constant value.

For to obtain the ITHA an algorithm has been created in Matlab using the push-over curve (fig 3), so that the frequency is derived from the degradation curve (fig 4) in function of the secant inelastic period, obtained with (7).

The results shown for case 1 and 4 are representative of all 6.

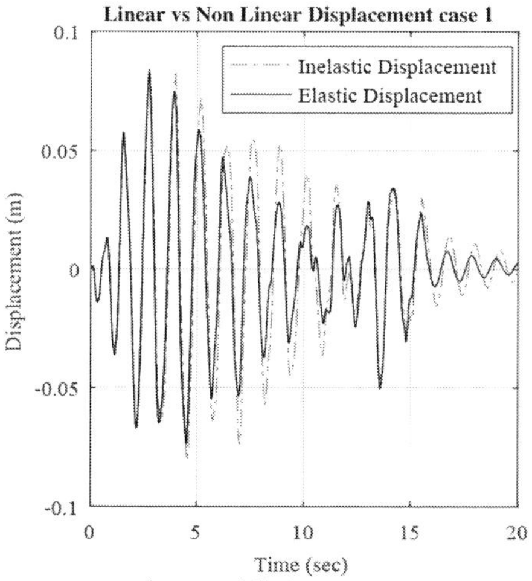

Fig 6. THA and ITHA of displacements case 1.

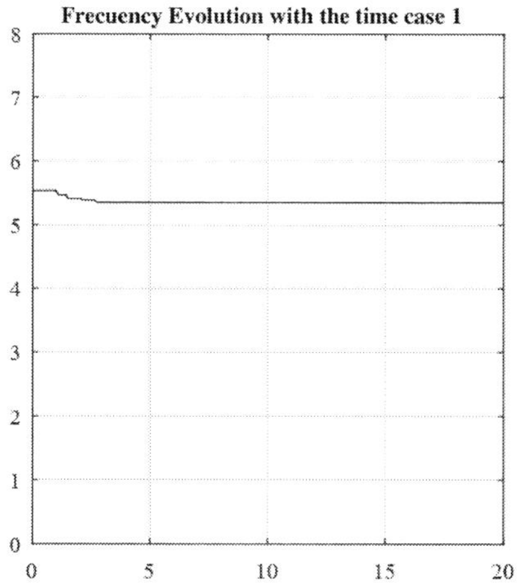

Fig 7. Evolution of circular frequency during the earthquake, case 1. (Y-Frecuency rad/s); (X,time)

In the case of type II spectrum accelerograms the degradation is quite low (fig 7). In spite of this, the parameters are observed quite well. Conversely, in type I spectrum, the degradation is much more pronounced (fig 9).

It is expected to have more damage and inelasticity in accelerograms obtained with type I spectrum, even if the acceleration is smaller, precisely because the energy introduced in the system is greater for periods over 0.7 s, like it is shown in the present paper example.

Table 1. Maximum values of the spectrums type I and II. EC-8. 0.25g, soil IV. S_v is cuasi-spectrum velocity.

Spectrum	$S_a(m/s^2)$	S_v (m/s)	S_d(m)
I	8.268	1.05	0.335
II	11.02	0.562	0.1

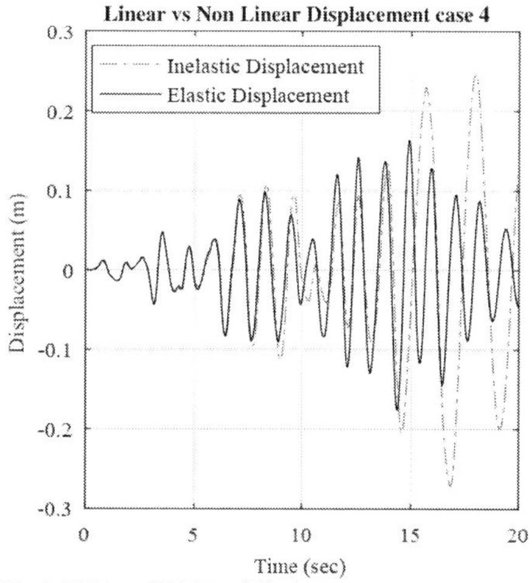

Fig 8. THA and ITHA of displacements case 4.

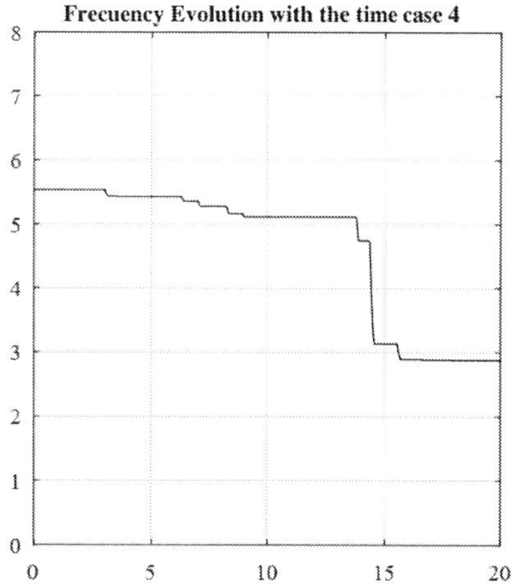

Fig 9. Evolution of circular frequency during the earthquake, case 4. (Y-Frecuency rad/s); (X,time)

https://doi.org/10.1051/matecconf/201814803002

3.2. Bouc Wen Baber Noori (BWBN).

In the BWBN the general equation of the SDOF model is established as follow:

$$\ddot{x}(t) + 2\xi_o\omega_o \dot{x}(t) + \alpha\omega_o^2 x(t) + ... \quad (13)$$
$$..... + (1-\alpha)\omega_o^2 z(t) = -a(t)$$

The parameters has been adapted according with [10] and [11] ; the last term of the power integrated equation is the Hysteretic Energy.

Tabel 2. Summary of maximum abs displacements (m).

TH	Spectrum	Elastic	Push	BWBN	DDBD
1	II	0.08420	0.08420	0.06260	0.04905
2	I	0.14560	0.19080	0.15360	0.16453
3	II	0.08510	0.07910	0.06710	0.04151
4	I	0.17610	0.27130	0.01497	0.13063
5	II	0.09170	0.09290	0.07890	0.04126
6	I	0.16100	0.37110	0.20880	0.13507

Note: The displacements according with DDBD methodology are explained in 3.3.

3.3. Methods of synthesis. Direct Displacement Based Design (DDBD) example.

Generally the interest to know the evolution during the time of the parameters that regulate the behaviour of a structural system is irrelevant This lack of interest promotes certain methods such as the dynamic modal spectral analysis and the direct displacement design method (DDBD), which overall, have had a great implementation in professional practice.
The DDBD method is summarised by calculating an equivalent SDOF structure with an elastic plus an inelastic phase. In a reinforced concrete frame building:

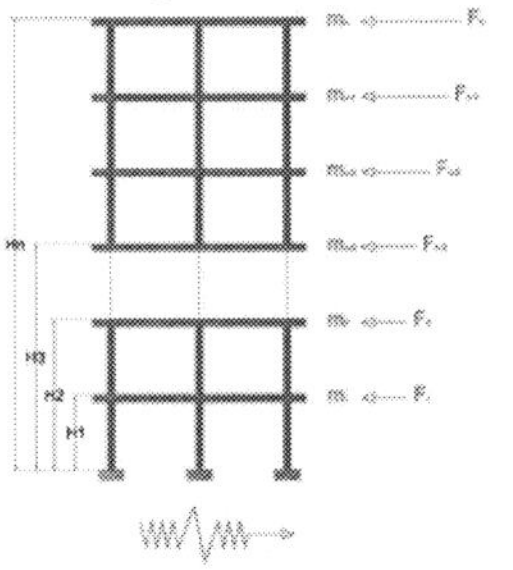

Fig 10.1 MDOF real Structure

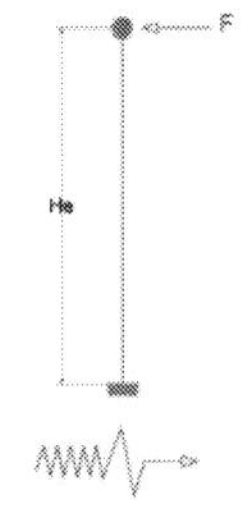

Fig 10.2 Equivalent SDOF scheme

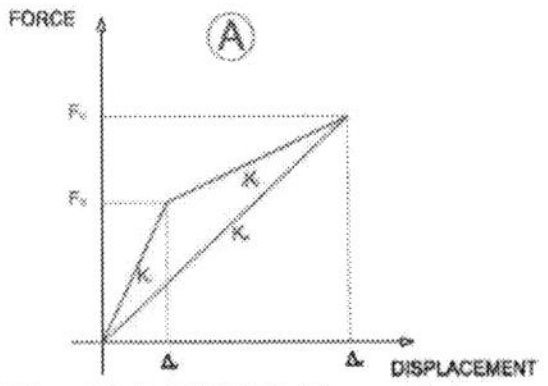

Fig 10.3 DDBD Force-Balance Scheme.

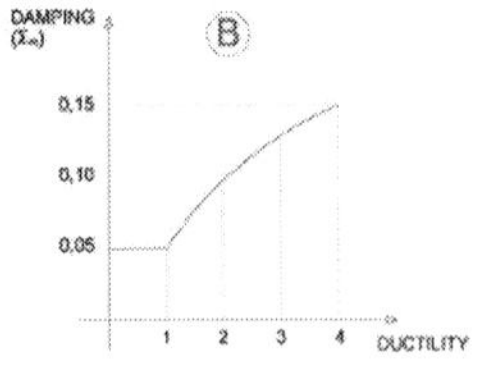

Fig 10.4. R Damping-Ductility curve.

$$\Delta_i = \delta_i\left(\frac{\Delta_c}{\delta_c}\right) = \left(\frac{\Delta_c}{\delta_c}\right)\left[\frac{4}{3}\left(\frac{H_i}{H_n}\right)\left(1 - \frac{H_i}{4H_n}\right)\right] \quad (14)$$

The equivalent displacement is: $\Delta_d = \dfrac{\sum_1^n m_i \Delta_i^2}{\sum_1^n m_i \Delta_i}$ (15)

The equivalent mass is: $m_e = \dfrac{\sum_1^n m_i \Delta_i}{\Delta_d}$ (16)

The equivalent height: $H_e = \dfrac{\sum_1^n m_i \Delta_i H_i}{\sum_1^n m_i \Delta_i}$ (17)

Whenever the equivalent period T_e is known the parameters of the system could be fixed:

Equivalent Rigidity: $K_e = \dfrac{F}{\Delta_d} = \dfrac{4\pi^2 m_e}{T^2_e}$ (18)

The equivalent period of the system is obtained by introducing the deformation demand of the equivalent system Δ_d (15) in the spectrum, based on equivalent damping. The method introduces some conceptual improvements such as the association of a displacement ductility proportionally to the damage (fig 10.4) and coherence for the damping balance (11).
In this case, the building GF+3F is introduced in its equivalent SDOF response. Using the formulation of the energy balance the elastic work part is:

$$W_e = K_o \frac{\Delta^2_{dy}}{2} \quad \textit{(19)}$$

And the plastic work:

$$W_p = \frac{K_o\Delta_{dy} + K_e\Delta_d}{2}(\Delta_d - \Delta_{dy}) \quad \textit{(20)}$$

In case of considering an inelastic section without hardening $W_p = F(\Delta_d - \Delta_{dy}) = F\Delta d$ ***(21)***
whereby it is possible to obtain the increment of plastic displacement associated with the energy contributing to damage, deducing the elastic component.

Tabel 4. Summary of energy balance DDBD.

TH	Spec	E-W_d	W_e	W_p	Δ_d	Total
1	II	0.130	0.597	0.000	0.000	0.049
2	I	0.881	0.597	0.284	0.060	0.165
3	II	0.093	0.597	0.000	0.000	0.042
4	I	0.719	0.597	0.122	0.026	0.131
5	II	0.092	0.597	0.000	0.000	0.041
6	I	0.741	0.597	0.144	0.030	0.135

4. Conclusions.

The substitution of a MDOF system for a SDOF one is very common to simplify the seismic design projects. In this sense, due to the natural dispersion derived in the formulation and the action, the energy balance represents a parameter of fundamental contrast for the methodology of nonlinear vibrations, whatever its formulation may be.
In particular, in the case of an equivalent SDOF system in concrete structures, the energy balance allows it to be to critically discussed and contrasted with the suitability between actions, inelastic response, damage, damping and effective vibration period.
This paper has shown the following aspects:
- A formulation has been developed to obtain the secant period of a structural system from the curve derived from the nonlinear static analysis, general form and including geometric and material nonlinearity. This formulation, based on the energy balance of the shear-displacement curve, may be more adequate than that which binds the modal mass, as it evolves during the development of inelasticity. This does not happen in the DDBD method, which always implies the same inelastic profile, with only the magnitude varying.
- It has been shown how, with some care in the parameter delimitation which regulate the concrete damage model, it is possible to evaluate with a high degree of accuracy its values, even in extreme cases where is not considered the energy dissipated in hysteresis cycles.
The energy balance, despite the difficulties of its approach, with a correct definition and calibration, can suppose a fundamental tool for professional practice in nonlinear vibrations and specifically in seismic engineering, as they are currently introduced in some design regulations.

ACKNOWLEDGEMENTS

The authors acknowledge the financial support provided by Ministry of Economy and Competitiveness (Government of Spanish Kingdom) by funding the Research Project REVES-P (DPI2013-47441). The authors also acknowledge the partial financial support provided by the Research Net VIBRASTRUNET (BIA2015-71942-REDT).

References

[1] H. Akiyama, "Collapse modes of structures under strong motions of earthquake," *Ann. Geophys.*, vol. 45, no. 6, pp. 791–798, 2002.
[2] Don Jose Antonio Agudelo Zapata, *User manual of Acelsin. Synthetic accelerograms*. Granada (Spain)., 2017.
[3] Comité Técnico AEN/CTN 140 Eurocódigos estructurales, "Eurocode 8: Design of structures for earthquake resistance. Part 1: General r ules, seismic actions and rules for buildings," p. 208, 2006.
[4] Computers and Structures Inc., "CSi Analysis Reference Manual," p. 556, 2016.
[5] T. J. Sullivan, G. M. Calvi, and M. J. N. Priestley, "Initial Stiffness Versus Secant Stiffness in Displacement Based Design," *13th World Conf. Earthq. Eng.*, no. 2888, 2004.
[6] H. Akiyama, *Metodología de proyecto sismorresistente de edificios basada en el balance energético*. Reverté, 2003.
[7] S. Superiori and S. Pavia, "Reduction of Seismic Risk Equiva L Ent Viscous Damping Equations for Direct," 2004.
[8] R. Zaharia and F. Taucer, "Equivalent period and damping for EC8 spectral response of SDOF ring-spring hysteretic models," *Eur. Comm. Jt. Res. Centre, Sci. Tech. Reports*, vol. JRC 45403, no. m, 2008.
[9] A. K. Chopra and R. K. Goel, "A modal pushover analysis procedure for estimating seismic demands for buildings," vol. 582, no. January 2001, pp. 561–582, 2002.
[10] P. Sengupta and B. Li, "Modified Bouc-Wen model for hysteresis behavior of RC beam-column joints with limited transverse reinforcement," *Eng. Struct.*, vol. 46, pp. 392–406, 2013.
[11] A. E. Charalampakis and V. K. Ã. Koumousis, "ARTICLE IN PRESS On the response and dissipated energy of Bouc – Wen hysteretic model," *J. Sound Vib.*, vol. 309, pp. 887–895, 2008.

Jump Attenuation in a Non-Ideal System Using Shape Memory Element

Adriano Kossoski[1], *Angelo* M. Tusset[1,*], *Frederic* C. Janzen[1], *Rodrigo* T. Rocha[1], *Jose* M. Balthazar[2], *Reyolando* M.L.R.F. Brasil[3], and *Airton* Nabarrete[2]

[1]Dept. of Electronics, Federal University of Technology – Parana, Brazil
[2]Aeronautics Division, Aeronautics Technological Institute, Brazil
[3]CECS – Aerospace Engineering, Federal University of ABC, Brazil

Abstract. The studies on the so-called smart materials have grown in the last years due to the diverse possibilities that these materials can provide. These types of materials have the ability to respond to an external excitation, altering its physical form, and being able to be considered as actuators. Among these materials are the Shape Memory Alloys (SMA), several metal alloys that are able to memorize a shape and recover it after a deformation through an increase of its thermal energy. In this paper, an actuator consisting of an SMA wire is used to attenuate the vibration and Sommerfeld effect of a non-ideal type oscillator. The temperature control of the actuator was carried out through the application of an electric current in the wire. Results are presented for different currents, with the objective of investigating the temperature variation for vibration control applications. The results showed that it is possible to apply SMA actuators to the attenuation of the Sommerfeld Effect as well as in the reduction of the total vibration of the system.

1 Introduction

Among the most common configurations in engineering designs, there are systems composed of a mechanical structure and some rotating excitation, e.g., a DC (Direct Current) motor. All these systems that have some rotating device are subject to mechanical vibrations that may be induced by an imbalance or poor alignment [1].

Sometimes, the methods of balance a system can be difficult or even unfeasible due to presented critical speeds. These velocities emerge when the angular frequency of the excitation shaft equals the natural frequency of the mechanical structure, being a characteristic effect of a non-ideal type system, where the motor (responsible for the vibration) is influenced by the behavior of the system, thus altering its normal functioning [1, 2].

In 1907, Sommerfeld performed an experiment composed of a cantilever beam with an unbalanced DC motor at its tip. Sommerfeld realized that when the angular frequency of the motor came close to the natural frequency of the system, the applied electrical voltage would stop corresponding to an increase of the rotation of the motor, where only the amplitudes of vibration increased. However, after the motor have passed through the resonance region, its rotation returned to function normally. This effect was called by Sommerfeld Effect and is one of the most characteristic effects of a non-ideal type system [2, 3].

The behavior of a non-ideal system implies an increase in the vibration amplitudes and loss of energy, which can cause a reduction in useful life or a critical system failure, thus, such systems should be studied in advance to prevent this kind of situation in a real project [1, 3].

However, an opportunity that have emerged is the use of new materials that were previously unavailable, materials that little alter the project and can act in a way to attenuate, control or remove certain behavior of the structure, allowing the project barriers to be expanded. These are the so-called smart materials, which are a class composed of several materials that have the capacity to respond in a specific way to some type of excitation, and can be used as actuators and sensors. Among these materials, there are the Shape Memory Alloys (SMA), a group of metal alloys that can change and return to a memorized form by increasing their thermal energy [4-6].

The SMA materials have two different crystallographic phases. A phase called Austenite, referring to high temperatures and a phase called martensite, referring to the low temperatures. In the SMA material, the so-called martensitic transformation effect occurs, which is a transition between the two phases. This transition occurs at high speed and in solid state and the effects of shape memory and pseudoelasticity come from this transformation, which occurs due to temperature change and/or mechanical loading [4-6].

As stated, the effect of shape memory consists of the structural return of the material to a previous form through the increase of its thermal energy. The effect of

* Corresponding author: a.m.tusset@gmail.com

pseudoelasticity occurs when the system is at a temperature above its austenite temperature and, in this condition, the system tends to return after deformed, performing mechanical work in the return [5-6].

In this way, this work presents the pseudoelastic effect of a SMA actuator used in a non-ideal oscillator in order to reduce its vibration and the jump effect that occurs in these types of systems. The built acquisition system was explained and experimental results were obtained for different electric currents applied in the actuator.

2 Experimental procedures

3.1 System built and instrumentation

The system built for the tests consists of a cantilever beam made of aluminum with an unbalanced DC (Direct Current) motor attached to its free end. For the motor unbalance, on its shaft was fixed an extension with a certain mass at its free end. When the motor works, this mass destabilizes and makes the system vibrate vertically (flexural vibration). Figure 1 presents the non-ideal system built for the tests.

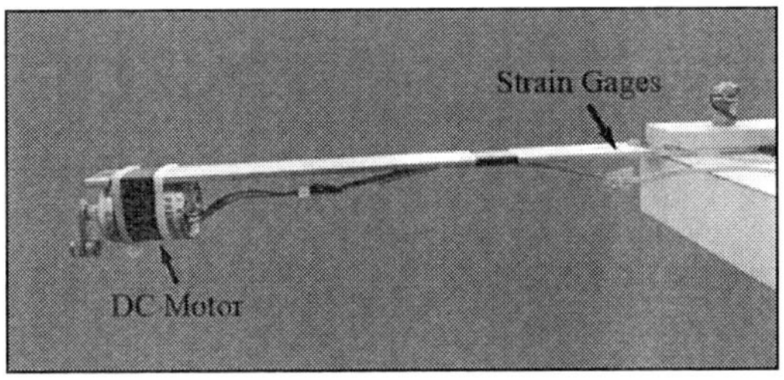

Fig. 1. Non-ideal oscillator.

In order to acquire the vibration of the system, it was used a strain gauge type sensor. This kind of sensor has an extremely low mass, not physically influencing the system and can be used for both static and dynamic system measurements. The principle of operation of a strain gauge occurs by the variation of its electrical resistance. When variations of its dimensions occur, the changes of its dimensions are transformed into electrical signals that can be measured and related to its real time physical deformation through a Wheatstone Bridge. The Wheatstone bridge is a common circuit used to discover an unknown electric resistance applying a determined voltage at one of its ends and reading the output voltage at another end. In this paper, two of the resistors of the bridge are strain gauges and the other two are normal resistors, all of then having the same electric resistance (120Ω). If the four resistances are equal (the strain gauges are not deformed), the voltage read is equal to the voltage applied, and it is said that the bridge is in equilibrium. If the electric voltage is different, then this variation can be used to relate to the deformation suffered by the structure.

As the electric signal of the Wheatstone bridge has a low magnitude, a signal amplifier circuit was built using an INA128P low power instrumentation amplifier of Texas Instruments®. All instrumentation data were obtained with the use of a National Instruments Data Acquisition (DAq) model 6212. electric current and rotation of the DC motor.

As power source, a nominal 24V DC motor was used. Figure 2 shows the torque/speed curve of the motor used.

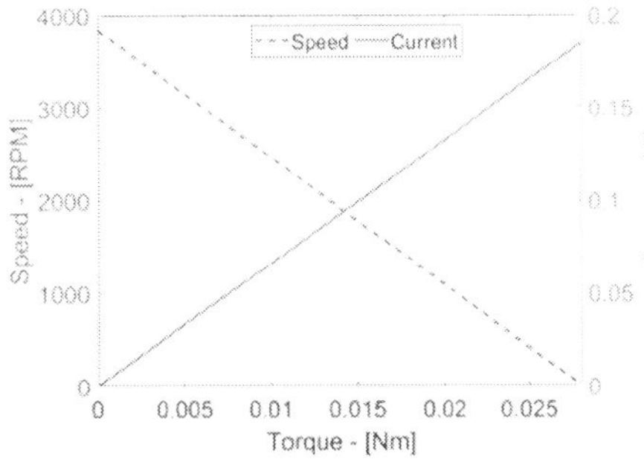

Fig. 2. Torque versus speed curve of the DC motor.

The Labview® software was used to process and save all the data acquired from the vibration, electric current in the SMA element and the rotation of the motor. Figure 3 shows the scheme of the data acquisition.

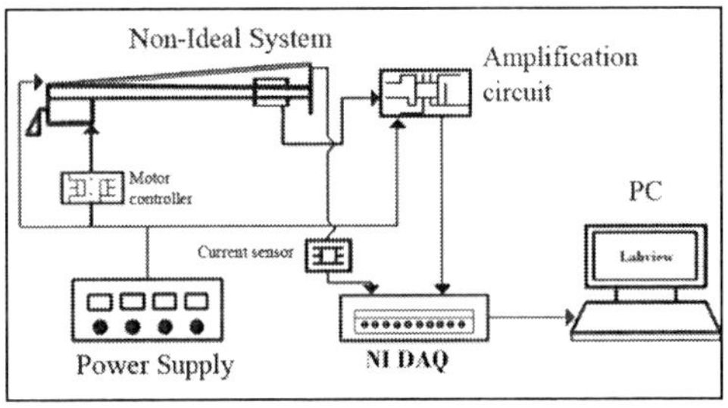

Fig. 3. Scheme used to obtain the results.

3.2 Tests performed

As previously stated, the objective of this paper is to use an actuator made of SMA to attenuate the Sommerfeld Effect (jump phenomenon) in a non-ideal type system. First, the natural frequency was found through a free vibration type excitation. In sequence was obtained the two types of graphs: the graphics of the jump phenomenon based on the angular frequency of the motor and the graphics of the jump due to the applied voltage in the DC motor. With these types of graphics, it is possible to analyze the amplitudes of the Sommerfeld Effect and the behavior of the motor during the vibration.

A SMA actuator was used as a tendon, applying a force in order to hold the system in vibration. The actuator was a Flexinol® wire with the properties shown in Table 1.

For activation of the SMA actuator, an electric current passed through the wire, thus, due to the Joule's effect, the actuator tends to heat proportionally to the current that crosses it. However, due to vibration, the actuator changes its shape dynamically, hence, there are variations in its temperature during the dynamic operation of the system. This fact makes it difficult to obtain the exact temperature of the actuator, and then the electric current was used as the control parameter of the shape memory alloy actuator for the tests.

Table 1. Actuator parameters.

Properties	Value
SMA wire diameter (μm)	375
Austenite start temp. (°C)	68
Austenite finish temp. (°C)	78
Martensite start temp. (°C)	52
Martensite finish temp. (°C)	42

Seven different tests were carried out. The first one was the non-ideal system without the actuator. The other six consist of different electric currents applied to the actuator, whose tested electric current are of 500mA, 600mA, 700mA, 800mA, 900mA and 1A. These values were chosen because, for this actuator, 500mA was the first value that activated the actuator (temperature above the Austenite Start) and 1A was the last value that was able to activate the actuator at the same time that it did not physically deform the system before the tests. To avoid this behavior, electric currents that deformed the system before the tests were discarded.

4 Results and discussion

In order to obtain the natural frequency, the system was excited and then its vibration frequencies were analyzed with the use of a Fast Fourier Transform (FFT). Figure 4 shows the frequency that is most active in the system; this is one of the natural frequencies of the system.

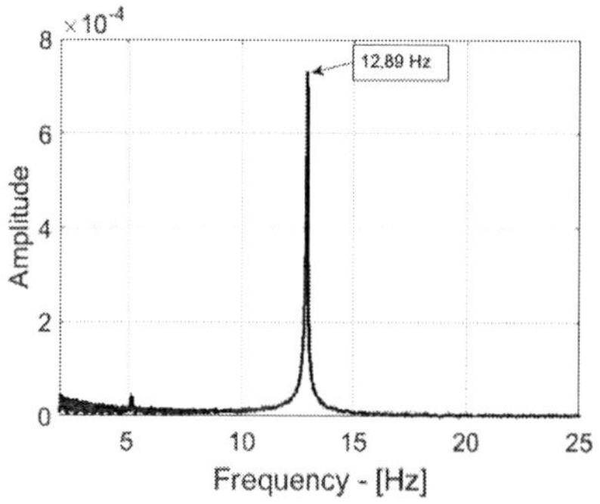

Fig. 4. Natural frequency of the electromechanical system built.

This frequency will be the expected point where the Sommerfeld effect occurs.

4.1 System without the SMA actuator

In this subsection, results for the jump effect are presented. Two types of graphs are displayed as results. The first type shows the response diagram frequency for the Sommerfeld effect. With these results, it is possible to observe the amplitude and angular frequency of the motor where the jump is occurring. The second type shows the jump effect in function of the voltage applied to the motor.

Figure 5 shows the jump according to the angular frequency of the motor for the system without actuator.

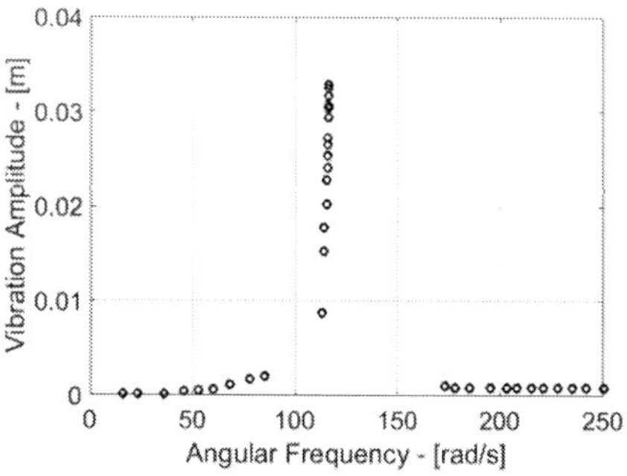

Fig. 5. Frequency response diagram for the Sommerfeld Effect - Without the SMA actuator.

Figure 6 shows the results for the jump phenomenon based on the voltage applied to the motor.

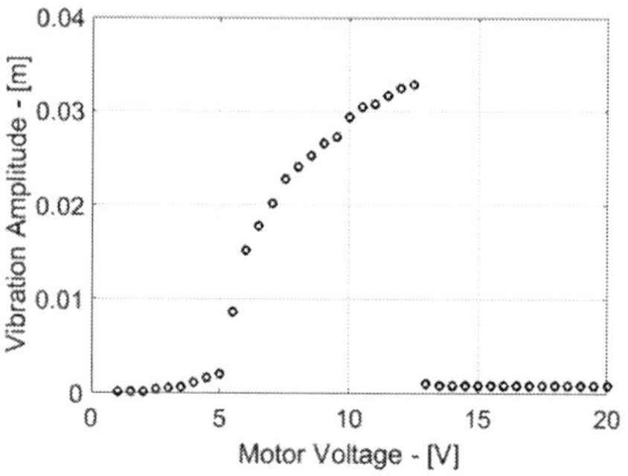

Fig. 6. Jump phenomenon due to the applied voltage in the DC motor - Without the SMA actuator.

The jump phenomenon occurs when the motor reaches 115 rad/s and 12.5 volts, after this point the system returns to normal operation. Due to the effect of the coupling between the motor and the structure, the system has its operation weakened and it is realized that between 6 and 12.5 volts all the increase of voltage in the motor serves almost solely to increase the amplitudes of vibration and not the rotation of the motor. This is the biggest problem in a non-ideal system.

4.2 System with the SMA actuator

Figure 7 shows the results for different parameters of the SMA actuator based in the angular frequency of the DC motor.

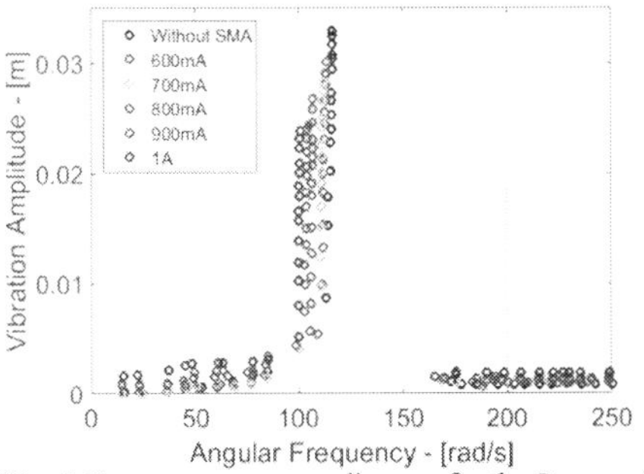

Fig. 7. Frequency response diagram for the Sommerfeld Effect.

Figure 8 shows the results for different parameters of the SMA actuator based in the motor voltage.

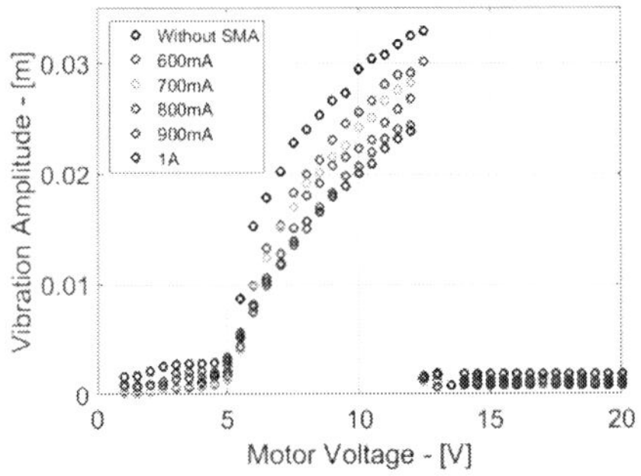

Fig. 8. Jump phenomenon due to the applied voltage in the DC motor.

Due to the force exerted by the actuator, the system has reduced its amplitudes of vibration. It is also observed that due to the mechanical alteration of the oscillator, the jump occurs each time before, with a lower motor voltage and consequently with a lower rotation. In these tests performed, even with the actuator with its maximum current, was not enough for the Sommerfeld effect to disappear.

Table 2 presents a summary of the results obtained for each parameter of the actuator used.

Table 2. Results summary.

Parameter in the actuator	Result in the system
600mA	Reduction of 16.20% in the positive peak of vibration
700mA	Reduction of 18.16% in the positive peak of vibration
800mA	Reduction of 23.18% in the positive peak of vibration
900mA	Reduction of 30.72% in the positive peak of vibration
1A	Reduction of 32.12% in the positive peak of vibration

4 Conclusions

This work presented an experimental method of attenuating the Sommerfeld effect of a non-ideal oscillator type system by coupling a Shape Memory Alloy (SMA) actuator. Non-ideal systems are systems where the source of excitation is influenced by its own performance, losing energy in the process that serves only to amplify the amplitudes of vibration.

As a control parameter, an electric current was passed through the actuator, which, through the Joule's effect, heated the wire and modified its physical properties. Five electric currents were tested: 600mA, 700mA, 800mA, 900mA and 1A. The results showed that the actuator does not need to reach austenite temperature to act on the system, although this is the temperature with the best results due to it is the temperature where the system has the highest energy available to perform work.

According to the force applied to the system through the actuator, the jump effect tends to occur at a lower motor voltage. This is an effect similar to change the stiffness and damping of the system. With the used actuator, it was possible to reduce the amplitude peak of the Sommerfeld effect by up to 32.12%. This is equivalent to reduce the vibration amplitudes of the system by the same value.

In all the performed tests, it was not possible to completely remove the Sommerfeld effect from the system. Future research may focus on the use of greater quantity and optimum positioning of the actuators, noting that the design of the system has a great influence on the Sommerfeld effect.

References

1. P. J. P. Gonçalves, M. Silveira, B. P. Junior, J. M. Balthazar, The dynamic behavior of a cantilever beam coupled to a non-ideal unbalanced motor through numerical and experimental analysis. J. Sound and Vibration, **333(20)**, 5115-5129 (2014)
2. J. M. Balthazar, D. T. Mook, H. I. Weber, R. M. Brasil, A. Fenili, D. Belato, J. L. P. Felix, An overview on non-ideal vibrations. Meccanica, **38(6)**, 613-621 (2003)
3. V. Piccirillo, A. M. Tusset, J. M. Balthazar, Dynamical jump attenuation in a non-ideal system through a magnetorheological damper. J. of Theoretical and Applied Mechanics, **52(3)**, 595-604 (2014)
4. L. Sun, W. M. Huang, Z. Ding, Y. Zhao,C. C. Wang, H. Purnawali, C. Tang, Stimulus-responsive shape memory materials: a review. Materials & Design, **33**, 577-640 (2012)
5. D. C. Lagoudas, *Shape memory alloys: modeling and engineering applications*. Springer Science & Business Media (2008).
6. M. V. Gandhi, B. D. Thompson, *Smart materials and structures*. Springer Science & Business Media (1992)

Motorcycle control-oriented dynamics modeling for accelerated maneuvers on slippy terrains

Elżbieta Jarzębowska[1*], *Michał* Cieśluk[1]

[1]Warsaw University of Technology, Power and Aeronautical Engineering Dept., 00-665 Warsaw, Nowowiejska 24 st., Poland

Abstract. The paper presents a motorcycle dynamics model developed for testing and future model based controller designs for accelerated maneuvers on variable slip terrains. The dynamics is simplified, yet it captures real motorcycle behaviors when it accelerates and decelerates on a curvy trajectory and contact forces due to variable ground properties change and allow to generate slip. The tire - terrain model is based upon a simplified Pacejka model. The paper objective is not to obtain complex dynamical systems of equations like those required for high-fidelity simulations. Instead, the aim is to derive simple but reliable and manageable models that enable designing and implementing, and verifying control laws, as well as maintaining their capability to capture the main behaviors of real systems. The paper applies the adopted assumptions to develop the motorcycle model, and presents simulation tests for the motorcycle acceleration and deceleration during turn maneuvers, and during changes of the ground the vehicle moves on.

1 Introduction

Motorcycling is a popular means of transportation in urban areas, used for sports and entertainment, so a number of drivers increase. Many road accidents with motorcyclists involved are reported. The reports show that major accidents are due to human driving errors. Many motorcyclists have little knowledge about the dynamic behavior of their vehicles, about the driver-motorcycle interaction. Driving assistance systems could be the key to better control vehicles. In this context, several active safety systems for motorcycles were developed [1]. Also, the motorcycle is itself an interesting object from dynamics and control point of view.

There are many dynamic models of motorcycles in the literature starting from very simple to very sophisticated and specialized to some modeling goals. A high-fidelity model may accurately reflect the response of a vehicle, but added details may complicate motion planning and control designs. It is then a trade-off between the accuracy of the dynamics model and difficulties of controller designs and their performance. An extensive overview of general modeling concepts and a survey of models used for motion planning and control can be found in [1, 2]. Also, there are many popular models of mobility for car-like vehicles [3-6]. They are widely used in control and motion planning to approximate a vehicle's behavior in response to control actions. Many of these models are kinematical ones [1]. Many dynamic models are complex, they focus upon the vehicle itself and use either very simple tire – terrain interaction models or assume no slip motion conditions, i.e. introduce nonholonomic constraints into the model. Formally, the nonholonomic constraint limits maneuverability of the vehicle model. A real vehicle due to slip has more degrees of freedom. Since forces that develop between tires and the ground play the significant role in fast motorcycle maneuvers, the main goal of this research is to develop and test a form of simplified control oriented dynamics model with reliable tire-ground interaction model is welcome.

The paper presents a motorcycle dynamics model developed for testing and future model based controller designs for accelerated maneuvers on variable slip terrains. The dynamics is simplified, yet it captures real motorcycle behaviors when accelerates and decelerates on a curvy trajectory and contact forces due to variable ground properties are allowed to change, and allow to generate slip. The tire - terrain model is based upon a Pacejka model [2] but it is simplified. The adopted simplifications proved to be satisfactory for reproducing the motorcycle behaviors. The motivations for this research are multiple. When designing control strategies and algorithms for wheeled vehicles like mobile robots, cars or motorcycles, the development of mathematical models that describe the dynamics of real systems is of a significant importance. It allows using model-based approaches to controller designs, which achieve better implementation of control laws and their better performance. In these cases, the objective of modeling is not to obtain complex dynamical systems of equations like those required for high-fidelity simulations. Instead, the aim is to derive simple but reliable and manageable models that enable designing and implementing, and verifying control laws, as well as maintaining, at the

* Corresponding author: elajarz@meil.pw.edu.pl

same time, their capability to capture the main behaviors of real systems.

The research contribution is twofold. It is a motorcycle dynamics model developed for testing and future model based controller designs for accelerated maneuvers on variable slip terrains and a simplified tire - terrain Pacejka model, which satisfactory reproduce the motorcycle behaviors.

The paper is organized as follows. Section 1 provides some introductory comments about the motorcycle reported research. Section 2 presents the adopted assumptions to the motorcycle model and the dynamics model development. Simulation tests for the motorcycle acceleration and deceleration during turn maneuvers and during the change of the ground the vehicle moves on are detailed in Section 3. The paper closes with conclusions and a list of references.

2 Motorcycle dynamics model

The motorcycle dynamics model is developed for a vehicle model presented in Fig. 1. It is simplified, yet captures the main bodies the model consists of as well as the contact force due to the change of the mass center of the motorcyclist- motorcycle system.
The following simulation model inputs were taken into account:
- m_k - handle bar mass,
- m_r - motorcycle frame mass (includes the drives mass),
- m_p, m_t - front and rear wheel masses,
- l_{rosi} - distance between the vehicle axes,
- l_{siedz} - distance between the driver seat and the rear wheel axis,
- l_{rama} - distance between the frame mass centre and the rear wheel axis,
- R_p - radius of the front wheel,
- R_t - radius of the rear wheel,
- h_{siedz} - height of a driver seat (measured from the ground),
- h_{rama} - distance between the frame mass centre and the ground,
- ε - yaw angle of the frame,
- g_r - prism width, which models the frame, the motorcyclist and the rear wheel,
- a- distance between the front wheel contact point and the vehicle mass center,
- b - distance between the rear wheel contact point and the vehicle mass center,
- h_m- height of the mass center location,
- I_{px0}, I_{mx0}, I_{pz0}, I_{mz0}, I_{mxz0}, I_{wp}, I_{wt}- moments and products of inertia of the motorcycle and its wheels.

Fig. 1. Motorcycle model.

2.1. Motorcycle tire-ground interaction forces

The significant part of the model is the tire – ground interaction model. It is based upon the Pacejka model [2] and its components are presented at a scheme in Fig. 2.

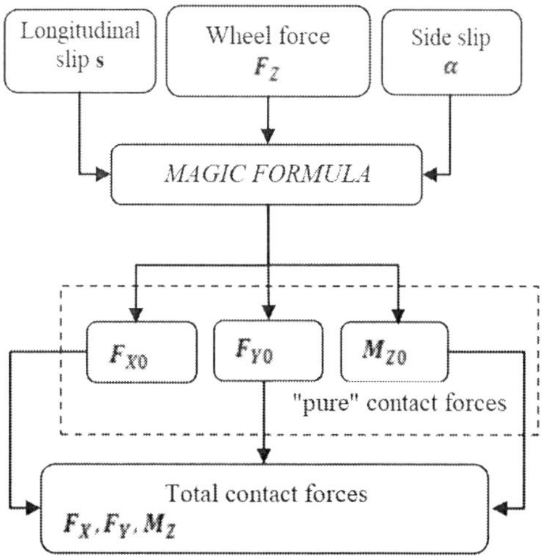

Fig. 2. Tire-ground interaction forces modeling.

The "pure contact forces" mean the forces determined for one slip type. The reference inertial frame *X, Y, Z* is the right hand side Euclidian.

2.2 Motorcycle dynamic equations

The simplified motorcycle dynamics model is developed for the seven degrees of freedom represented by the quasi velocity vector $\dot{q} = [u\ v\ r\ \dot{\varphi}\ \dot{\delta}\ \dot{\theta}_P\ \dot{\theta}_T]^T$.
The velocities *u*, *v*, *r*, are the velocities along the *x*, *y*, *z*, body-fixed frame, and the angles φ, δ, θ_P, θ_T are the motorcycle bank angle, handle bar heading angle, yaw angles of the front and rear wheels, respectively. The Lagrange equations of motion developed for the motorcycle model have the form

$$M(q)\cdot\ddot{q} = V_{COR}(q,\dot{q}) + F_{ACT} + F_{PAC}(q,\dot{q}) \quad (1)$$

where the matrices in (1) are as follows:

$$F_{PAC} = \begin{Bmatrix} F_{xp} + F_{xt} - F_{yp}\cdot\delta\cdot\cos\varepsilon \\ F_{yp} + F_{yt} + F_{xp}\cdot\delta\cdot\cos\varepsilon \\ -bF_{yt} + aF_{yp} + M_{zt} + M_{zp} + (t_c + a\cos\varepsilon)\delta F_{xp} \\ -bF_{yt} + aF_{yp} + M_{xt} + M_{xp} + t_c\delta F_{zp} \\ -F_{PX}\cdot r_T \\ -F_{LX}\cdot r_T \\ -F_{FX}\cdot r_F \end{Bmatrix}; \quad (2)$$

is the force vector of interaction between the tire and the ground and is determined according to Pacejka model,

$$F_{ACT} = [0\ 0\ 0\ 0\ M_T\ M_{Tnap}\ M_{Pnap}]^T \quad (3)$$

is the external force vector due to a driver,

$$M(q) = \begin{bmatrix} m_m + m_p & 0 & -m_m h_m \varphi - m_p r_p \varphi & 0 & 0 & 0 & 0 \\ 0 & m_m + m_p & m_p a & m_m h_m \varphi + m_p r_p & 0 & 0 & 0 \\ -m_m h_m \varphi - m_p r_p \varphi & m_p r_p \varphi + m_p a & m_m h_m^2 \varphi^2 + m_p r_p^2 \varphi^2 + m_p a^2 + I_{px0} sin^2\varepsilon + I_{pz0} cos^2\varepsilon & m_p r_p a + (I_{pz0} - I_{px0}) cos\varepsilon \cdot sin\varepsilon - I_{Mx0} & I_{pz0} cos\varepsilon & 0 & 0 \\ 0 & m_m h_m + m_p r_p & m_p r_p a - I_{px0} sin\varepsilon + I_{pz0} cos\varepsilon & I_{pz0} sin\varepsilon + I_{px0} cos\varepsilon + m_p r_p^2 + m_m h_m^2 + I_{mx0} & I_{pz0} & 0 & 0 \\ 0 & 0 & I_{pz0} cos\varepsilon & I_{pz0} & I_{pz0} & 0 & 0 \\ 0 & 0 & 0 & 0 & 0 & I_{py} & 0 \\ 0 & 0 & 0 & 0 & 0 & 0 & I_{ty} \end{bmatrix} \quad (4)$$

$$V_{COR}(q, \dot{q}) = \begin{Bmatrix} \delta \cdot cos\varepsilon + (m_m + m_p) rv + m_p a r^2 + 2(m_m h_m + m_p r_p) \cdot r\dot{\varphi} \\ -(m_m + m_p) ur + (m_m h_m + m_p r_p) \varphi r^2 \\ -2r\varphi\dot{\varphi}(r_p^2 + h_m^2) - (m_m h_m + m_p r_p) \varphi vr - a m_p ur + \varphi g (m_m h_m + m_p r_p) \\ 0 \\ 0 \\ 0 \\ 0 \end{Bmatrix} \quad (5)$$

In Eq. (1) adhesion forces are projected onto the vehicle principle axes. Forces that act upon the wheels are adhesive force and moments of forces regulated by a motorcyclist. The moments are due to the handle bar rotation, propulsion from the rear wheel and the braking torque of the front wheel. The dynamics model (1) encounters for the motorcycle stabilizing effect due to the so called dragging effect for the front wheel. It is measured by the distance between the contact point of the front wheel and the point of intersection of the ground plane with the handlebar rotation axis.

3 Simulation studies – accelerated motorcycle maneuvers on slippy roads

Three driving scenarios are tested in simulation studies. They are selected as "everyday driving maneuvers". They demonstrate both the effectiveness of the developed motorcycle dynamics model and effects of accelerated motion when variable terrain – tire interactions are accounted for.

3.1. Scenario 1 – constant velocity motion along a circle

Motorcycle moves along a circle with a constant velocity and a constant handle bar angle.

In all scenarios the following data is used (all are SI units): m_r+m_k=278, m_p=10, m_t=11, I_{px0}=0.44, I_{pz0}=0.44, I_{mx0}=170.02, I_{mz0}=5.55, I_{mxz0}=2.59, I_{wp}=0.88, I_{wt}=1.16.

The trajectories of the front and rear wheels, and of the mass center are shown in Fig. 3.

Figure 4 presents aligning torques (top) and the camber torques for the front and rear wheels.

Figure 5 presents longitudinal slips to the front and rear wheels (top) and the side slip angle to these wheels for the scenario 1 conditions.

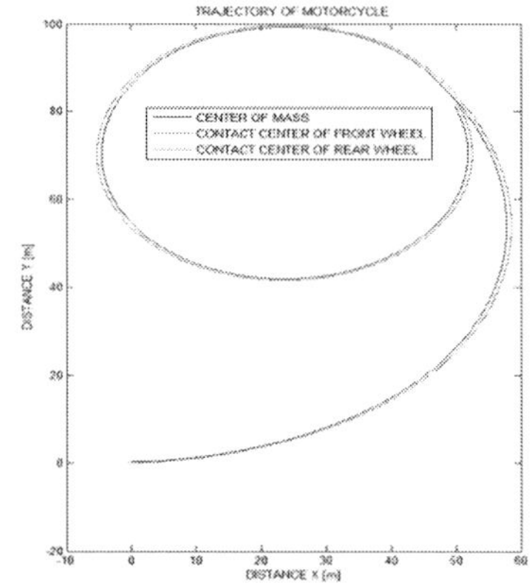

Fig. 3. Trajectories of the front, rear wheels and the mass center in scenario 1.

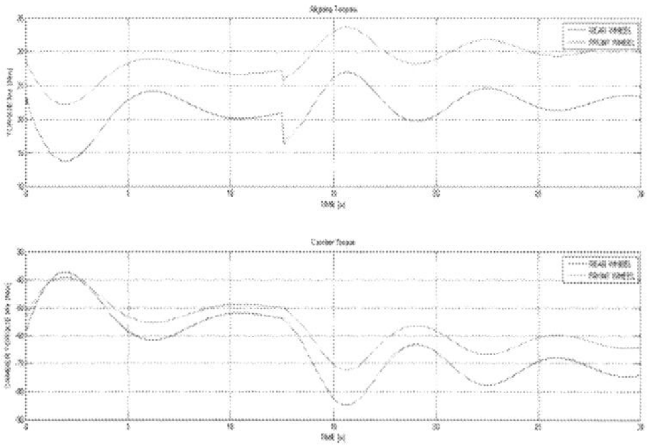

Fig. 4. Aligning and camber torques for a motorcycle.

* Corresponding author: elajarz@meil.pw.edu.pl

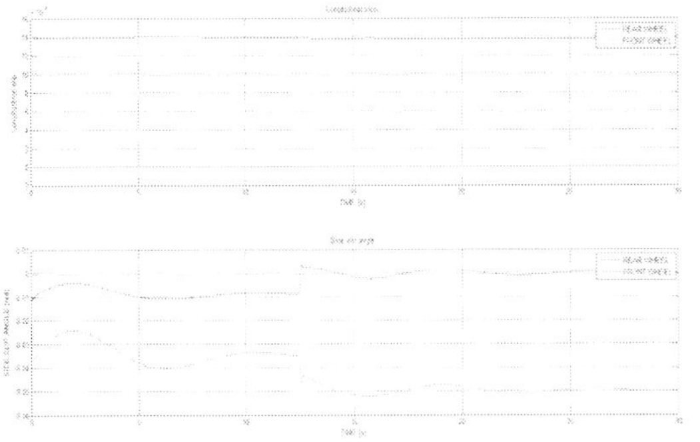

Fig. 5. Longitudinal slips to the front and rear wheels (top) and the side slip angle to the wheels in scenario 1.

3.2. Scenario 2 – accelerated turning and the change of a surface friction

In this case, the motorcycle turns and accelerates when turning. Suddenly, a driver notices an obstacle and starts decelerating after 1s (the reaction time for a motorcyclist is assumed to be 1 s). But the surface changes in this 1s (friction coefficient decreases about 0.5). This scenario is for example for a case of oil spot, which was not seen by a motorcyclist.

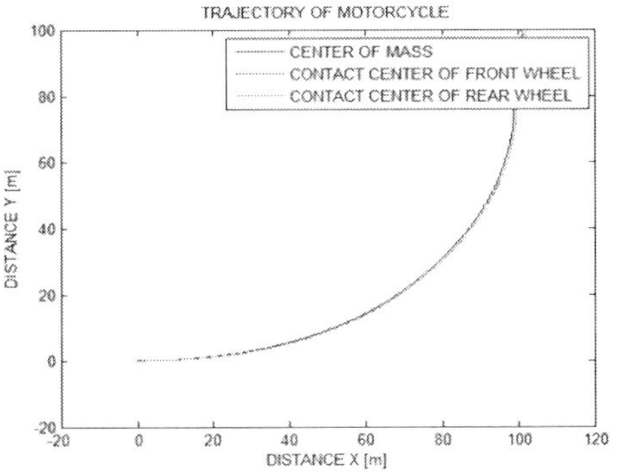

Fig. 6. Trajectories of the front, rear wheels and the mass center in scenario 2.

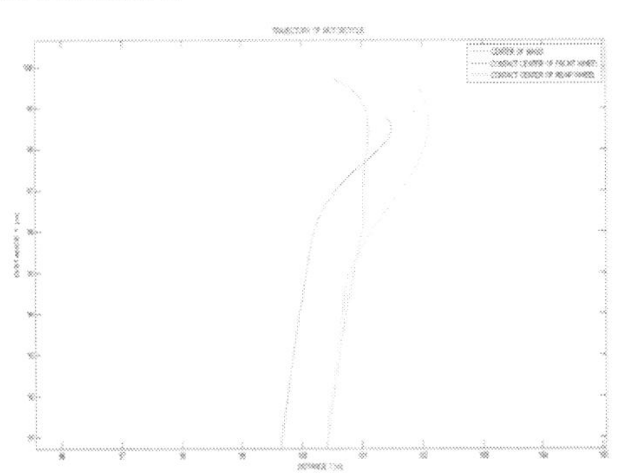

Fig. 7. Trajectories of the front, rear wheels and the mass center in scenario 2 - magnified.

The motorcycle falls; see fig. 6 and 7. Notice that the adhesive side forces almost vanishes during braking, and it contributes to falling.

Figures 8 to 11 present side and longitudinal forces as well as side slip angle and longitudinal slips, vertical loads on the front and rear wheels, and side and longitudinal velocities associated to the maneuver.

Each motorcycle due to its design and performance properties has a specified maneuver related longitudinal velocity, handle bar heading angle and its bank angle. Simulation studies presented in these three scenarios start from setting these variables at some initial magnitudes. Between 2 to 7 seconds of motion, the motorcycle accelerates what manifests in the growing longitudinal force on the rear wheel. At the same time, side forces on both wheels grow; see Fig. 8. Between 7 to 9 seconds, the motorcycle does not accelerate. Braking starts at the second 9. Next, according to our scenario, after 1 second the friction coefficient drops from 1.2 to 0.7. It causes significant delay of the braking wheel. It results in increase of longitudinal slip magnitude; see Fig.9. The adhesion force is directed along the wheel axis and the lack of the side component of this force results in the motorcycle falling; see Fig. 11.

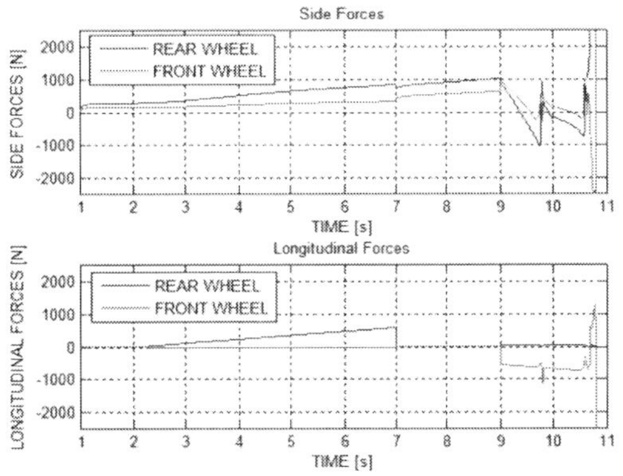

Fig. 8. Side and longitudinal forces on the wheels in scenario 2.

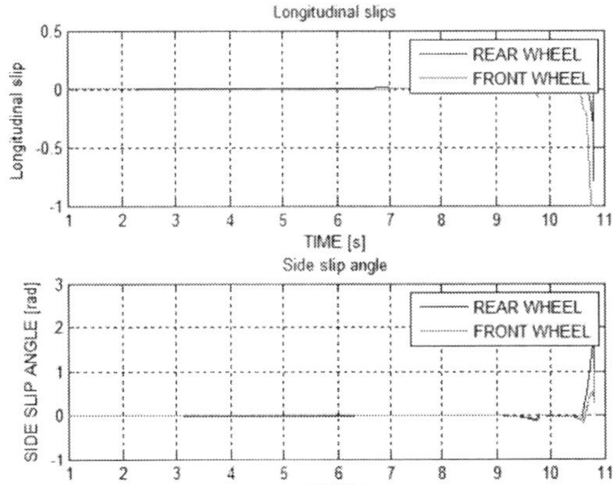

Fig. 9. Longitudinal slip to the front and rear wheels (top) and the side slip angle to the wheels in scenario 2.

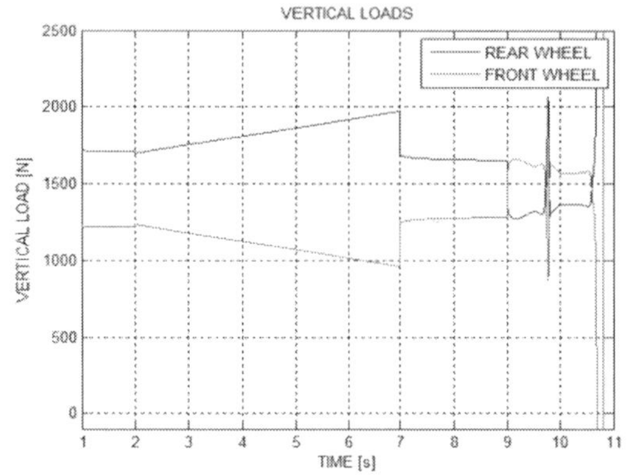

Fig. 10. Vertical loads on the motorcycle wheels in scenario 2.

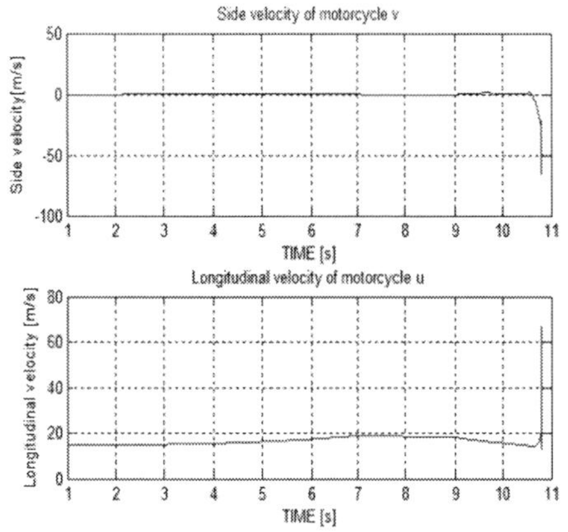

Fig. 11. Side and longitudinal velocities of the motorcycle.

3.3. Scenario 3 – accelerated turning and no change of a surface friction

The scenario 3 looks like scenario 2 but the surface the motorcycle drives on does not change. The motorcyclist can survive this time and does not fall. Figures 12 and 13 present the trajectories of the wheels and the vehicle mass center.

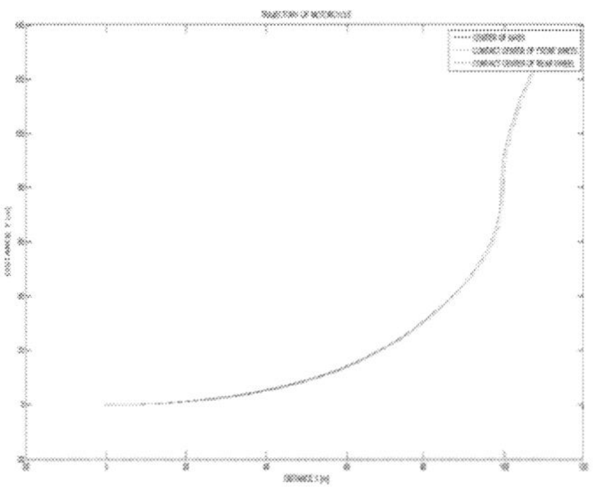

Fig. 12. Trajectories of the front, rear wheels and the mass center in scenario 3.

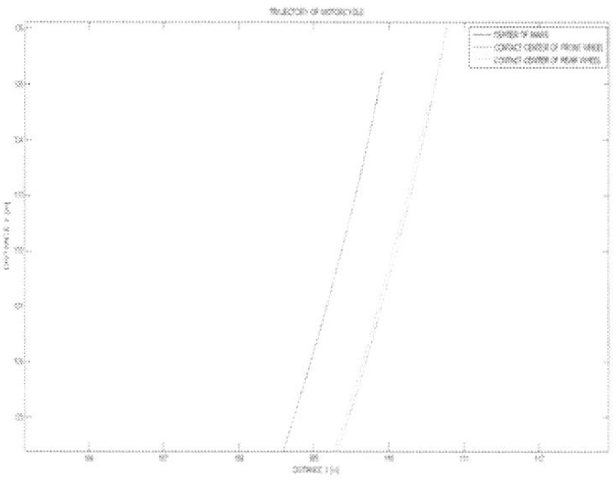

Fig. 13. Trajectories of the front, rear wheels and the mass center in scenario 3 - maginified.

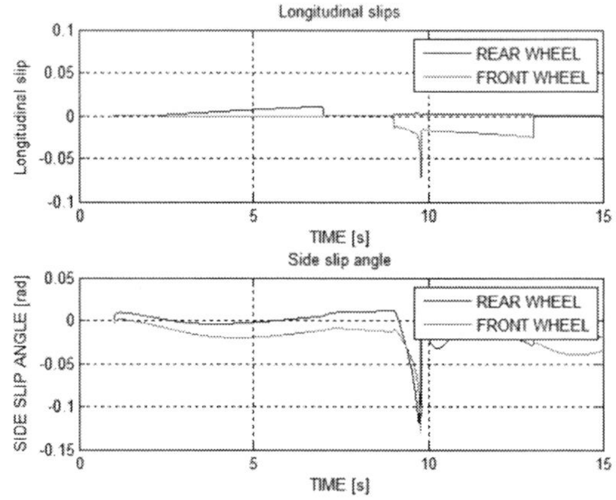

Fig. 14. Longitudinal slips to the front and rear wheels (top) and the side slip angle to the wheels in scenario 3.

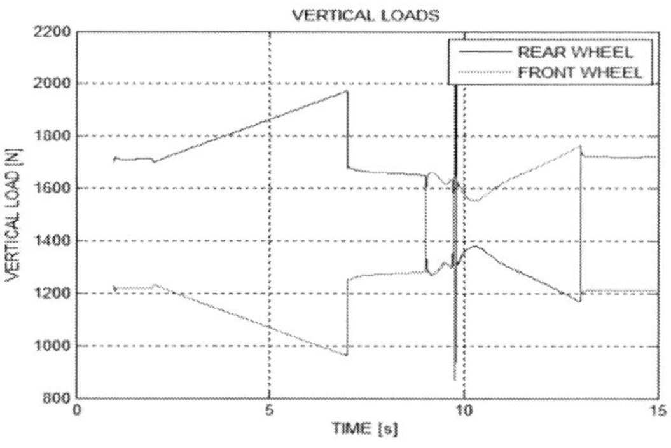

Fig. 15. Vertical loands on the wheels in the driving scenario 3.

Comparing scenarios 2 and 3, the motorcycle accelerates and then brakes in the region, at which the longitudinal slip does not increase. Also, adhesion forces are uniformly distributed. Longitudinal acceleration magnitude, force exerted by the wheels onto the ground and adhesion forces are all related. Braking causes the force increase on the front wheel and accelerating on the the rear wheel. The vertical load on the motorcycle wheels is always symmetric with respect to half of its total weight; see Fig. 15.

Conclusions

The paper presents a motorcycle simplified dynamics model oriented to test accelerated maneuvers on variable slip terrains and for future model based controller designs for such maneuvers. The dynamics is simplified, yet it captures real motorcycle behaviors when it accelerates and decelerates on slippy terrains. Three simulation scenarios are presented. They reproduce real situations motorcyclists may experience. The simulation tests for the motorcycle model acceleration and deceleration during turn maneuvers, and during the change of the ground the vehicle moves on confirmed good agreement of the model outcomes with real motorcycle behaviors. Future research is planned to design a controller for motorcycle driving and training, and test it experimentally.

References

1. B. Paden, M. Cáp, S. Z. Yong, D. Yershov, and E. Frazzoli, A Survey of Motion Planning and Control Techniques for Self-driving Urban Vehicles, IEEE Trans. Intel. Vehicles, **1(1)**, pp. 33-55 (2016).
2. H. Pacejka, 2012, *Tire and Vehicle Dynamics* (SAE Intern. and Butterworth Heinemann, 2012)
3. G. Franke, W. Suhr, and F Rieß, An Advanced Model of Bicycle Dynamics, European J. Physics, **11 (2)**, pp. 116–121 (2000).
4. U. Kumar and N. Sukavanam, Dynamic Modeling and Tracking Control of a Four Wheeled Nonholonomic Mobile Robot, *Proc. 13th Int. Conf. Mechanics and Machines*, (NaCoMM07), pp. 127-134 (2007).
5. D. Limebeer and R.S. Sharp, Bicycles, Motorcycles and Models, IEEE Control Syst., **26 (5)**, pp. 34-61 (2006).
6. JP. Meijaard, J.M. Papadopoulos, A. Ruina, and A.L. Schwab, Linearized Dynamics Equations for the Balance and Steer of a Bicycle: A Benchmark and Review, Proc. Royal Soc. A: Math., Phys. Eng. Sci., 463 (2084), pp. 1955-1982 (2007).

Dynamics modeling and performance analysis of underwater vehicles based on the Boltzmann-Hamel equations approach

Elżbieta Jarzębowska[1,*], *Michał* Cichowski[1]

[1]Warsaw University of Technology, Power and Aeronautical Engineering Dept., 00-665 Warsaw, Nowowiejska 24st, Poland

Abstract. The paper addresses dynamics modeling of underwater vehicles, which are typical examples of variable mass and configuration mechanical systems. Specifically, we address inertia–based propelled vehicles, whose mass changes due to change of water amount in their water tanks and their configuration changes due to movable mass enabling maneuvers. They can be additionally subjected to constraints, which are imposed due to their desired performance, tracking pre-specified motions or are control constraints implying underactuation. The paper presents a new approach to variable mass and configuration systems modeling, which is based upon Boltzmann-Hamel equations derived in quasi-coordinates. The dynamics framework based upon quasi-coordinates results in dynamic models suitable for performance analysis and model-based control applications. It serves free or constrained variable mass and configuration systems modeling. The example researched in the paper is an underwater vehicle, which is purely inertia–based propelled. The theoretical development is illustrated with simulation studies of dynamical behavior and performance of the underwater vehicle model moving a specified trajectory.

1 Introduction

The paper addresses dynamics modeling of underwater vehicles (UV), which are variable mass and configuration mechanical systems. They can be additionally subjected to constraints, which are imposed due to their desired performance, tracking pre-specified motions or are control constraints implying underactuation. An UV is a typical example of variable mass and configuration system.

There are many systems whose mass or configuration change when they operate or a change of mass is their way of propulsion. The examples of such systems of engineering significance are ground mobile robots and manipulators, whose mass or configuration change due to carrying loads, excavating machines or others transporting various materials. Also, space vehicles are of interest in this context, since their service missions like refueling or debris removal are strongly related to control of their variable mass or configuration dynamics. If sloshing is to be taken into account for spacecraft, then a variable configuration dynamics is also welcome. All these variable mass systems mentioned above operate and deliver services so they are controlled systems for which reliable and computationally efficient dynamic models are needed.

Almost all modeling approaches to obtain either dynamics or control models are methods based upon Newton-Euler or Lagrange equations; see for example [1-3] and references there. Variable mass systems are of interest in mechanics for a long time and the beginning of its theory is attributed to Mieszczerskii [4] and the beginning of analytical mechanics of these systems is dated when Lagrange equations were derived for them [5]. Nowadays, research on variable mass systems is intensive and latest results with references to other works can be found in [6]. There, as in other works, holonomic systems with mass being mostly a time function are considered. Also, in e.g. like [1-3] and many other works, rigid body mechanics approach to variable mass and configuration systems is applied. Some preliminary results in modeling variable mass systems are presented in [7]. The results are promising so the main motivation of the paper is to develop a modeling framework for variable mass and configuration systems.

The paper presents a new approach to variable mass and configuration systems modeling, which is based upon Boltzmann-Hamel equations derived in quasi-coordinates. The dynamics framework based upon quasi-coordinates results in dynamics models suitable for performance analysis and model-based control applications. The paper contribution is then a development of dynamics modeling method based upon Boltzmann-Hamel equations derived in quasi-coordinates. It serves free or constrained variable mass and configuration systems modeling. The example researched herein is an UV, which can be powered by internal batteries, is a purely inertia–based propelled or hybrid propelled, when additional batteries are added to facilitate or speed up its maneuvers. The theoretical development is illustrated with simulation studies of dynamical behavior and performance of an UV moving a specified trajectory.

The paper is organized as follows. Section 1 presents briefly an introduction about variable mass systems modeling methods. The theoretical framework, i.e. the variable mass and configuration system Boltzmann-Hamel motion equations are derived in Section 2. Section 3 presents simulation studies of the new equations applications to modeling an UV. The paper closes with conclusions and a list of references.

* Corresponding author: elajarz@meil.pw.edu.pl

2 The Boltzmann-Hamel equations for variable mass and configuration systems

2.1 Underwater vehicle and external forces modeling

The Boltzmann-Hamel based modeling framework is derived in this Section for a variable mass and configuration system. The system is represented herein by a UV model, which is inertia based propelled. It is quite a general example of a system of variable mass – water amount in its two water tanks changes. Also, it is equipped with a movable mass that may move back and forth and asides enabling the vehicle maneuvers.

The UV has 9 degrees of freedom, i.e. 3 translations and 3 rotations for the vehicle, and 3 translations for movable ballast mass. It is equipped with two water tanks, the front one and the rear one, whose water amount can be controlled due to the vehicle motion mode. The scheme of a vehicle is presented in Fig. 1.

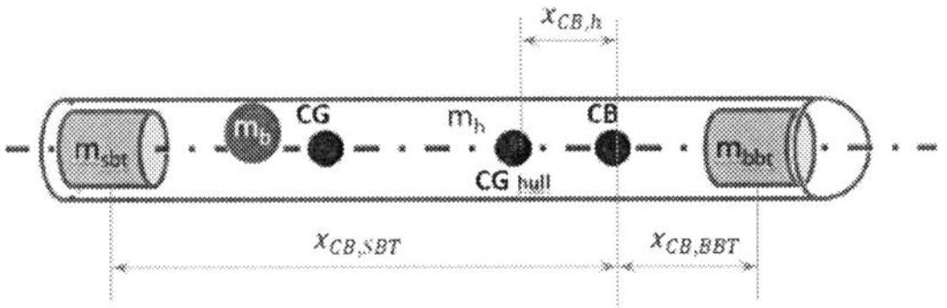

Fig. 1. Inertia and geometry parameters of the UV model.

The original Boltzmann-Hamel motion equations for constant mass systems are derived in quasi-coordinates and quasi-velocities [7]. They are non-inertial coordinates and their selection is presented in Fig. 2. They are velocities U, V, W along the body fixed frame. Rotations about these axes are described by angular velocities P, Q, R.

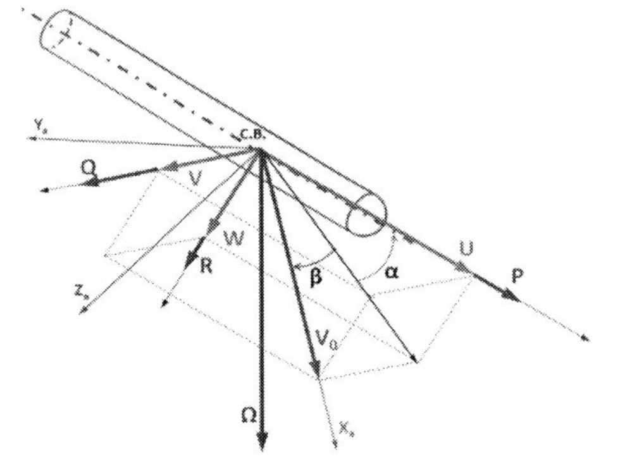

Fig. 2. The body fixed frame and quasi-velocities describing motion of the UV.

The UV model geometry and mass data is gathered in Table 1.

Table 1. UV geometry and mass data.

$x_{CB,SBT} = 0.5m$	distance from CB to stern ballast tank
$x_{CB,BBT} = 0.5m$	distance from CB tom bow ballast tank
$x_{CB,h} = 0$	distance from CB to center of gravity of hull
$m_h = 34.6kg$	mass of hull
$m_p = 0.4kg$	mass of movable ballast
$m_{disp} = 35kg$	displacement

Moments of inertia of hull (principal)

$I_{xx} = 0.172\ kg \cdot m^2$
$I_{yy} = 5.919\ kg \cdot m^2$
$I_{zz} = 5.919\ kg \cdot m^2$

The buoyancy and gravity forces that act upon the vehicle are presented by equation (1). It is assumed that the UV can move with moderate velocities so hydrodynamic forces are determined according to classical fluid dynamics as in (2) and Table 2.

$$F_{G,CB} + F_{B,CB} = \begin{bmatrix} 1_{3x3} & 0_{3x3} & 0_{3x3} \\ 0_{3x3} & \tilde{r}_{CB,CM} & 0_{3x3} \\ 0_{3x3} & 0_{3x3} & 0_{3x3} \end{bmatrix} \left(\begin{bmatrix} \Lambda_v \\ \Lambda_v \\ 0_{3x3} \end{bmatrix} \begin{bmatrix} 0 \\ 0 \\ Mg \end{bmatrix} + \begin{bmatrix} \Lambda_v \\ 0_{3x3} \\ 0_{3x3} \end{bmatrix} \begin{bmatrix} 0 \\ 0 \\ -\rho vg \end{bmatrix} \right) =$$
$$= \begin{bmatrix} 1_{3x3} & 0_{3x3} & 0_{3x3} \\ 0_{3x3} & \tilde{r}_{CB,CM} & 0_{3x3} \\ 0_{3x3} & 0_{3x3} & 0_{3x3} \end{bmatrix} \begin{bmatrix} -\sin\theta(Mg-\rho vg) \\ \cos\theta\sin\phi(Mg-\rho vg) \\ \sin\theta\sin\phi(Mg-\rho vg) \\ -\sin\theta Mg \\ \cos\theta\sin\phi Mg \\ \sin\theta\sin\phi Mg \\ 0 \\ 0 \\ 0 \end{bmatrix} = \begin{bmatrix} \sin\theta(Mg-\rho vg) \\ \cos\theta\sin\phi(Mg-\rho vg) \\ \cos\theta\cos\phi(Mg-\rho vg) \\ \cos\theta(y_{CB,CM}\cos\phi - z_{CB,CM}\sin\phi)Mg \\ -(z_{CB,CM}\sin\theta + x_{CB,CM}\cos\theta\cos\phi)Mg \\ (y_{CB,CM}\sin\theta + x_{CB,CM}\cos\theta\sin\phi)Mg \\ 0 \\ 0 \\ 0 \end{bmatrix} \quad (1)$$

$$F^V_{H,CB} = \begin{bmatrix} X^V_{H,CB} \\ Y^V_{H,CB} \\ Z^V_{H,CB} \\ L^V_{H,CB} \\ M^V_{H,CB} \\ N^V_{H,CB} \\ 0 \\ 0 \\ 0 \end{bmatrix} = \begin{bmatrix} -K_{D0} - K_D(\alpha^2+\beta^2) \\ K_\beta\beta \\ -K_\alpha\alpha \\ -K_P P \\ --K_M\alpha - K_Q Q \\ K_{MY}\beta - K_R R \\ 0 \\ 0 \\ 0 \end{bmatrix} V_0^2 \quad (2)$$

Table 2. Hydrodynamic force coefficients for the UV model in the velocity related frame.

$K_{d0} = 7.19 \frac{kg}{m}$
$K_d = 386.29 \frac{kg}{m \cdot rad^2}$
$K_\beta = 115.65 \frac{kg}{m \cdot rad}$
$K_\alpha = 440.99 \frac{kg}{m \cdot rad}$
$K_p = 190.83 \frac{kg \cdot s}{rad}$
$K_m = 65.84 \frac{kg}{rad}$
$K_q = 205.64 \frac{kg \cdot s}{rad^2}$
$K_{my} = 34.1 \frac{kg}{ra} d$
$K_r = 389.3 \frac{kg \cdot s}{rad^2}$

2.2 Boltzmann-Hamel equations in the matrix form

The original Boltzmann-Hamel equations for constant mass systems have the form, see e.g. [7] and references there

$$\frac{d}{dt}\left(\frac{\partial T}{\partial \omega_j}\right) - \frac{\partial T}{\partial \pi_j} + \sum_{l=1}^{n}\sum_{\alpha=1}^{n} \gamma_j{}^l{}_\alpha \frac{\partial T}{\partial \omega_l}\omega_\alpha = F_j \tag{3}$$

with the Boltzmann-Hamel symbols of the form

$$\gamma_j{}^l{}_\alpha = \sum_{\sigma=1}^{n}\sum_{\lambda=1}^{n}\left(\frac{\partial a_{l,\sigma}}{\partial q_\lambda} - \frac{\partial a_{l,\lambda}}{\partial q_\sigma}\right) b_{\sigma,j} b_{\lambda,\alpha}$$

Advantages of the Boltzmann-Hamel equations are multiple. They are first order differential equations in quasi-velocities, which do not have to possess physical interpretations and can be selected by a designer or a control engineer. Also, the kinematic constraint equations can be easily eliminated via the quasi-velocities selection and the forces acting and interacting in the system can be easily determined [7].

Let us write the Boltzmann-Hamel equations (3) in a matrix form applying the following notation

$$\frac{d}{dt}\begin{bmatrix}\frac{\partial T}{\partial \omega_1}\\ \frac{\partial T}{\partial \omega_2}\\ \vdots \\ \frac{\partial T}{\partial \omega_n}\end{bmatrix} - \begin{bmatrix}\frac{\partial T}{\partial \pi_1}\\ \frac{\partial T}{\partial \pi_2}\\ \vdots \\ \frac{\partial T}{\partial \pi_n}\end{bmatrix} + \sum_{\alpha=1}^{n}\omega_\alpha \begin{bmatrix}\sum_{l=1}^{n}\gamma_1{}^l{}_\alpha \frac{\partial T}{\partial \omega_l}\\ \sum_{l=1}^{n}\gamma_2{}^l{}_\alpha \frac{\partial T}{\partial \omega_l}\\ \vdots \\ \sum_{l=1}^{n}\gamma_n{}^l{}_\alpha \frac{\partial T}{\partial \omega_l}\end{bmatrix} = \begin{bmatrix}F_1\\ F_2\\ \vdots \\ F_n\end{bmatrix} \tag{4}$$

$$\sum_{\alpha=1}^{n}\omega_\alpha \begin{bmatrix}\sum_{l=1}^{n}\gamma_1{}^l{}_\alpha \frac{\partial T}{\partial \omega_l}\\ \sum_{l=1}^{n}\gamma_2{}^l{}_\alpha \frac{\partial T}{\partial \omega_l}\\ \vdots \\ \sum_{l=1}^{n}\gamma_n{}^l{}_\alpha \frac{\partial T}{\partial \omega_l}\end{bmatrix} = \sum_{\alpha=1}^{n}\left(\omega_\alpha \begin{bmatrix}\gamma_{1\alpha}^1 & \gamma_{1\alpha}^2 & \cdots & \gamma_{1\alpha}^n\\ \gamma_{2\alpha}^1 & \gamma_{2\alpha}^2 & \cdots & \gamma_{2\alpha}^n\\ \vdots & \vdots & \ddots & \vdots\\ \gamma_{n\alpha}^1 & \gamma_{n\alpha}^2 & \cdots & \gamma_{n\alpha}^n\end{bmatrix}\begin{bmatrix}\frac{\partial T}{\partial \omega_1}\\ \frac{\partial T}{\partial \omega_2}\\ \vdots \\ \frac{\partial T}{\partial \omega_n}\end{bmatrix}\right) \tag{5}$$

Also, introducing the notation

$$\Gamma_\alpha = \begin{bmatrix}\gamma_{1\alpha}^1 & \gamma_{1\alpha}^2 & \cdots & \gamma_{1\alpha}^n\\ \gamma_{2\alpha}^1 & \gamma_{2\alpha}^2 & \cdots & \gamma_{2\alpha}^n\\ \vdots & \vdots & \ddots & \vdots\\ \gamma_{n\alpha}^1 & \gamma_{n\alpha}^2 & \cdots & \gamma_{n\alpha}^n\end{bmatrix}$$

and

$$G(\omega) = \sum_{\alpha=1}^{n}\Gamma_\alpha \omega_\alpha$$

the matrix form of equations (3) can be presented as

$$\frac{d}{dt}\left(\frac{\partial T}{\partial \bar{\omega}}\right) - \frac{\partial T}{\partial \bar{\pi}} + G(\bar{\omega})\frac{\partial T}{\partial \bar{\omega}} = \bar{F}_{ext} \tag{6}$$

The matrix form (6) of the Boltzmann-Hamel equations is needed for derivation of the general modeling framework based upon these equations, for which mass and configuration can change or be constant.

2.3 Boltzmann-Hamel equations for variable mass and configuration systems

The derivation of the Boltzmann-Hamel equations for variable mass and configuration systems starts from the presentation of the system kinetic energy using matrix and vector notation. It enables easy handling of mass and inertia terms in the system model equations. The general form of the kinetic energy for the variable mass and configuration system is given by (7), where the mass matrix is composed of

$$M = m_h + m_p + m_{bbc} + m_{sbs}$$

$$M\bar{r}_{CB,CM} = m_h\bar{r}_{CB,h} + m_p\bar{r}_{CB,p} + m_{bbc}\bar{r}_{CB,bbc} + m_{sbc}\bar{r}_{CB,sbc} \tag{8}$$

$$\mathbb{I} = \mathbb{I}_{h,CM} + m_h\tilde{r}_{CB,h}\tilde{r}_{CB,h}^T + m_p\tilde{r}_{CB,p}\tilde{r}_{CB,p}^T + m_{bbc}\tilde{r}_{CB,bbc}\tilde{r}_{CB,bbc}^T + m_{sbc}\tilde{r}_{CB,sbc}\tilde{r}_{CB,sbc}^T \tag{9}$$

Vectors $\bar{\omega}_v$ and $\bar{\omega}_\Omega$ stand for the linear velocity vector and the angular velocity vector, both in the body fixed frame. The other vectors are:

- $\bar{r}_{CB,p} = \begin{bmatrix} r_{px} & r_{py} & r_{pz}\end{bmatrix}^T$ is a distance of a movable mass from the buoyancy center,
- $\bar{r}_{CB,bbc} = \begin{bmatrix} x_{CB,bbc} & 0 & 0\end{bmatrix}^T$ is a position vector of the front water tank,
- $\bar{r}_{CB,sbc} = \begin{bmatrix} x_{CB,sbc} & 0 & 0\end{bmatrix}^T$ is a position vector of the rear water tank,
- vectors with the top wave symbol denote the skew-symmetry matrices, e.g.:

$$\tilde{\omega}_\Omega = \begin{bmatrix} 0 & -\omega_{\Omega z} & \omega_{\Omega y}\\ \omega_{\Omega z} & 0 & -\omega_{\Omega x}\\ -\omega_{\Omega y} & \omega_{\Omega x} & 0\end{bmatrix}$$

The matrix of the Boltzmann-Hamel coefficients can be written as

$$G = \begin{bmatrix} 0 & -R & Q & 0 & 0 & 0 & 0 & 0 & 0\\ R & 0 & -P & 0 & 0 & 0 & 0 & 0 & 0\\ -Q & P & 0 & 0 & 0 & 0 & 0 & 0 & 0\\ 0 & -W & V & 0 & -R & Q & 0 & 0 & 0\\ W & 0 & -U & R & 0 & -P & 0 & 0 & 0\\ -V & U & 0 & -Q & P & 0 & 0 & 0 & 0\\ 0 & 0 & 0 & 0 & 0 & 0 & 0 & 0 & 0\\ 0 & 0 & 0 & 0 & 0 & 0 & 0 & 0 & 0\\ 0 & 0 & 0 & 0 & 0 & 0 & 0 & 0 & 0\end{bmatrix} = \begin{bmatrix}\tilde{\omega}_\Omega & 0_{3x3} & 0_{3x3}\\ \tilde{\omega}_v & \tilde{\omega}_\Omega & 0_{3x3}\\ 0_{3x3} & 0_{3x3} & 0_{3x3}\end{bmatrix} \tag{10}$$

Finally, determining

$$\frac{d}{dt}\left(\frac{\partial T}{\partial \bar{\omega}}\right) = \frac{d}{dt}\frac{\partial}{\partial \bar{\omega}}\left(\frac{1}{2}\bar{\omega}^T \mathbb{M}\bar{\omega}\right) = \frac{d}{dt}\left(\mathbb{M}\bar{\omega}\right) = \mathbb{M}\dot{\bar{\omega}} + \dot{\mathbb{M}}\bar{\omega} \tag{11}$$

with the inertia matrix derivative given by (12), the Boltzmann-Hamel equations have the form (13). They can be rewritten in the form (14), which clearly shows the inertia components in the system. Equations (14) are the Boltzmann-Hamel equations for variable mass and configuration systems. They can be also easily applied to constants mass systems and variable configuration or constant mass and configuration systems.

$$T = \frac{1}{2}\begin{bmatrix}\bar{\omega}_v^T & \bar{\omega}_\Omega^T & \dot{\bar{r}}_{CB,p}^T\end{bmatrix}\begin{bmatrix} M\mathbb{I}_{3x3} & M\tilde{r}^T_{CB,CM} & m_p\mathbb{I}_{3x3} \\ M\tilde{r}_{CB,CM} & \mathbb{I} & m_p\tilde{r}_{CB,p} \\ m_p\mathbb{I}_{3x3} & m_p\tilde{r}^T_{CB,p} & m_p\mathbb{I}_{3x3}\end{bmatrix}\begin{bmatrix}\bar{\omega}_v \\ \bar{\omega}_\Omega \\ \dot{\bar{r}}_{CB,p}\end{bmatrix} = \frac{1}{2}\bar{\omega}^T\mathbb{M}\bar{\omega} \tag{7}$$

$$\dot{\mathbb{M}} = \begin{bmatrix} (\dot{m}_{bbc} + \dot{m}_{sbc})\,\mathbb{I}_{3x3} & \dot{m}_{bbc}\tilde{r}^T_{CB,bbc} + \dot{m}_{sbc}\tilde{r}^T_{CB,sbc} & 0_{3x3} \\ \dot{m}_{bbc}\tilde{r}_{CB,bbc} + \dot{m}_{sbc}\tilde{r}_{CB,sbc} & \dot{m}_{bbc}\tilde{r}_{CB,bbc}\tilde{r}^T_{CB,bbc} + \dot{m}_{sbc}\tilde{r}_{CB,sbc}\tilde{r}^T_{CB,sbc} & 0_{3x3} \\ 0_{3x3} & 0_{3x3} & 0_{3x3}\end{bmatrix} + \\ + \begin{bmatrix} 0_{3x3} & m_p\dot{\tilde{r}}^T_{CB,p} & 0_{3x3} \\ m_p\dot{\tilde{r}}_{CB,p} & m_p\left(\dot{\tilde{r}}_{CB,p}\tilde{r}^T_{CB,p} + \tilde{r}_{CB,p}\dot{\tilde{r}}^T_{CB,p}\right) & m_p\dot{\tilde{r}}_{CB,p} \\ 0_{3x3} & m_p\dot{\tilde{r}}^T_{CB,p} & 0_{3x3}\end{bmatrix} \tag{12}$$

$$\mathbb{M}\dot{\bar{\omega}} + \left(\begin{bmatrix}\tilde{\omega}_\Omega & 0_{3x3} & 0_{3x3} \\ \tilde{\omega}_v & \tilde{\omega}_\Omega & \dot{\tilde{r}}_{CB,p} \\ 0_{3x3} & 0_{3x3} & \tilde{\omega}_\Omega\end{bmatrix}\mathbb{M} + \mathbb{M}\begin{bmatrix}0_{3x3} & 0_{3x3} & 0_{3x3} \\ 0_{3x3} & 0_{3x3} & 0_{3x3} \\ 0_{3x3} & \dot{\tilde{r}}^T_{CB,p} & 0_{3x3}\end{bmatrix} + \dot{\mathbb{M}}_{bbc} + \dot{\mathbb{M}}_{sbc}\right)\bar{\omega} = \bar{F} \tag{13}$$

$$\mathbb{M}\dot{\bar{\omega}} + \begin{bmatrix}\tilde{\omega}_\Omega M & \tilde{\omega}_\Omega M\tilde{r}^T_{CB,CM} & 2\tilde{\omega}_\Omega m_p \\ M\tilde{r}_{CB,CM}\tilde{\omega}_\Omega & \tilde{\omega}_\Omega\mathbb{I} & 2m_p\tilde{r}_{CB,p}\tilde{\omega}_\Omega \\ \tilde{\omega}_\Omega m_p & \tilde{\omega}_\Omega m_p\tilde{r}^T_{CB,p} & 2\tilde{\omega}_\Omega m_p\end{bmatrix}\bar{\omega} + \left(\dot{\mathbb{M}}_{bbc} + \dot{\mathbb{M}}_{sbc}\right)\bar{\omega} = \bar{F}_{ext} \tag{14}$$

3 Simulation studies – inertia propelled UV dynamics and performance

In this Section, using the general modeling framework for variable mass and configuration systems based upon the Boltzmann-Hamel equations, dynamics and performance of inertia propelled UV is presented. Two UV motion scenarios are selected. One presents the typical saw-tooth motion of the UV and the second, a spiral motion. Both motions are controlled in the open-loop through the control of the movable mass motion and the regulation of water amount in the water tanks.

3.1 Scenario 1- saw-tooth motion of the UV

The saw-tooth UV motion is its typical motion pattern. The movable mass does not move but water in the water tanks is regulated in a sinusoidal pattern with a period T=30s as shown in Fig. 3. The UV initial motion conditions are U=0.05m/s.

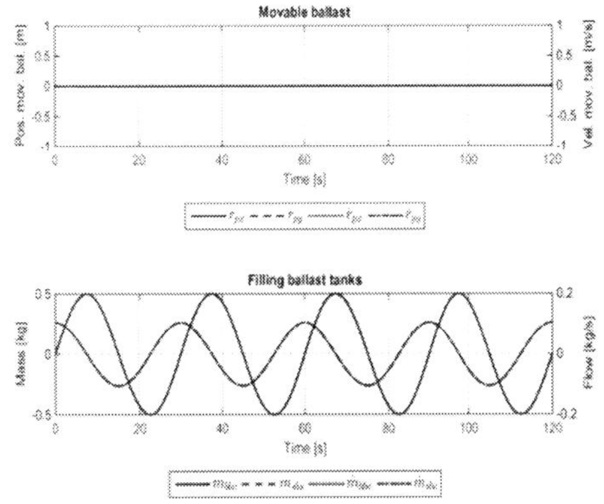

Fig. 3. The movable mass and changes of water amount in the tanks in the saw-tooth UV motion.

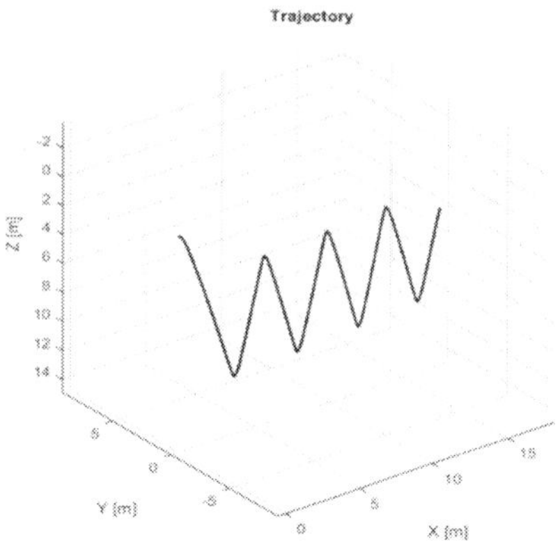

Fig. 4. The UV saw-tooth motion.

The UV velocities *U, V, W*, the total velocity V_0 and the angular velocities P, Q, R are depicted in Fig. 5.

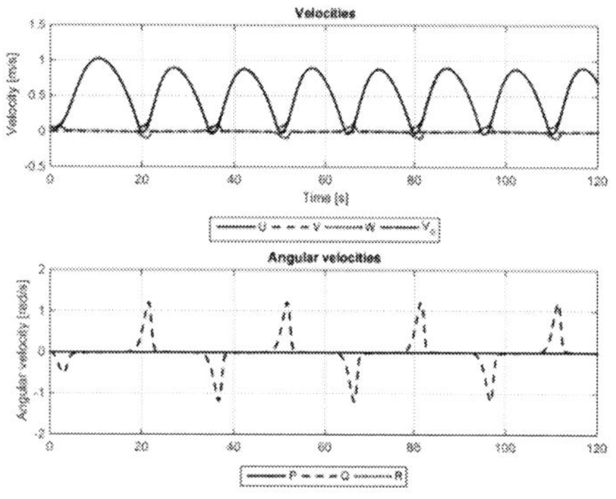

Fig. 5. The body fixed frame and quasi-velocities describing motion of the UV.

3.2 Scenario 2- spiral motion of the UV

In this scenario the UV is to move a spiral motion down into the water. The UV configuration is as follows: motion starts from rest, the water tanks are filled with water of 0.1kg each. The movable mass is located at 0.05m to stern and 0.1m to starboard. Initial motion conditions are selected to be: U=0.05m/s, simulation time is t = 1200 sec.

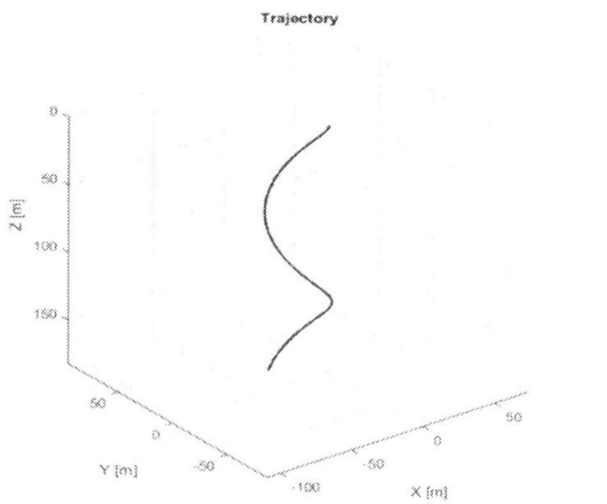

Fig. 6. The body fixed frame and quasi-velocities describing motion of the UV.

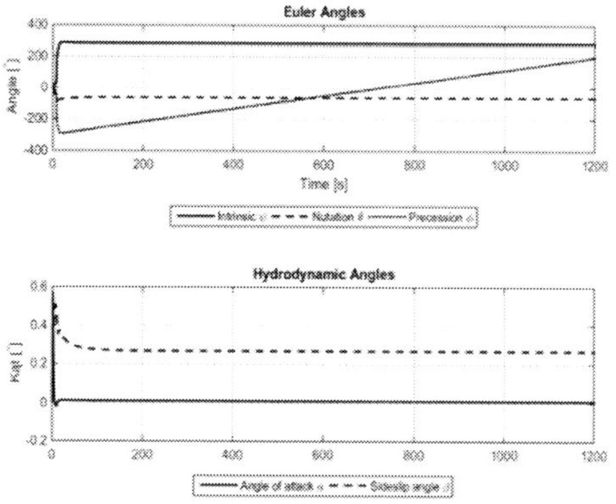

Fig. 7. The body fixed frame and quasi-velocities describing motion of the UV.

Other motions along other trajectories can be simulated for the UV in the same way.

Conclusions

The paper presents a new method of variable mass and configuration systems modeling, which is based upon the Boltzmann-Hamel equations derived in quasi-coordinates.

The presented dynamics framework based upon quasi-coordinates results in dynamic models suitable for performance analysis and model-based control applications. The dynamics framework serves free or constrained constant and/or variable mass and configuration systems modeling. The example of an underwater, inertia–based propelled vehicle model illustrates the theoretical development of variable mass and configuration systems dynamics. Simulation studies of the dynamical behavior and performance of the underwater vehicle moving a specified trajectory illustrate the framework applications.

Future research plans to incorporate kinematic constraints into the framework and apply it to design model-based control strategies for the UV.

References

1. A. Bender, *Analysis of an Autonomous Underwater Glider, Australian Centre for Field Robotics*, Univ. Sydney, Sydney, NSW (2006).
2. J. G. Graver, *Underwater Gliders: Dynamics, Control and Design*, Ph.D. Thesis, Princeton University (2005).
3. S. Guo, X. Lin, Development of a vectored water-jet-based spherical underwater vehicle, in: Nuno Cruz (Ed.), *Autonomous Underwater Vehicles*, Intech (2011).
4. I. V. Mieszczerskii, Motion Equations of a Variable Mass Particle in a General Case - Works on Mechanics of Variable Mass Bodies. Gostiechisdat, (in Russian), Moscow (1949).
5. A. A. Kosmodemianskii, *Lectures in Mechanics of Bodies with Variable Mass,* Uczionnye zapiski M.G.U. 154, Mechanics (1951).
6. L. Cveticanin, *Dynamics of Bodies with Time-Variable Mass*. Springer, Heidelberg (2016).
7. E. Jarzębowska, *Model Based Tracking Control of Nonlinear Systems*, Taylor and Francis Group, Boca Raton, 2012.

Numerical analysis of non-linear vibrations of a fractionally damped cylindrical shell under the conditions of combinational internal resonance

Yury A. Rossikhin[1], *Marina V.* Shitikova[1], and *Basem* Ajarmah[1,2*]

[1]Voronezh State Technical University, Research Center on Dynamics of Solids and Structures, 394006 Voronezh, Russia
[2]Al-istiqlal University Jericho, Palestine

Abstract. Non-linear damped vibrations of a cylindrical shell embedded into a fractional derivative medium are investigated for the case of the combinational internal resonance, resulting in modal interaction, using two different numerical methods with further comparison of the results obtained. The damping properties of the surrounding medium are described by the fractional derivative Kelvin-Voigt model utilizing the Riemann-Liouville fractional derivatives. Within the first method, the generalized displacements of a coupled set of nonlinear ordinary differential equations of the second order are estimated using numerical solution of nonlinear multi-term fractional differential equations by the procedure based on the reduction of the problem to a system of fractional differential equations. According to the second method, the amplitudes and phases of nonlinear vibrations are estimated from the governing nonlinear differential equations describing amplitude-and-phase modulations for the case of the combinational internal resonance. A good agreement in results is declared.

1. Introduction

Shells are in the art of many structural and building elements due to their several applications. Shell elements are in pipes and ducts, bodies of cars, space shuttles, aircraft fuselages, ship hulls, submarines and building structures. Recent investigations on dynamic response of cylindrical shells are based on nonlinear theories [1-5], and a comprehensive review could be found in [6, 7].

Recently Rossikhin and Shitikova [8] proposed a new approach for studying nonlinear dynamic response of thin fractionally damped cylindrical shells embedded into a fractional derivative viscoelastic medium. The dynamic behavior of the shell is described by a set of three coupled nonlinear differential equations The displacement functions are determined in terms of eigenfunctions of linear vibrations. The procedure resulting in decoupling linear parts of equations is proposed with the further use of the method of multiple scales for solving nonlinear governing equations of motion.

This procedure has been utilized for the analytical analysis of free vibrations of cylindrical shell subjected to the conditions of the different internal resonances, resulting in the interaction of two (in the case of two-to-one, one-to-one or three-to-one internal resonance [6, 9]) or three modes corresponding to the mutually orthogonal displacements (in the case of combinational internal resonances of additive or difference type [7]).

It has been shown that that the phenomenon of the internal resonance could be very critical, since in a circular cylindrical shell the internal resonance is always present. The influence of viscosity on the energy exchange mechanism has been analyzed. It has been shown that each mode is characterized by its damping coefficient dependent on the natural frequency by the exponential relationship with a negative fractional exponent.

In this paper, we are going to verify parameter values corresponding to the nonlinear vibrations of a fractionally damped cylindrical shell under the conditions of a combinational internal resonance, when a certain natural frequency of vibrations is equal to the sum of two other natural frequencies, using two different numerical methods. In the first method, the generalized displacements of a coupled set of nonlinear ordinary differential equations of the second order are estimated using the numerical solution of nonlinear multi-term fractional differential equations by the procedure based on the reducing of the problem to a system of fractional differential equations [10, 11]. According to the second method, the amplitudes and phases of nonlinear vibrations are estimated from the governing nonlinear differential equations describing amplitude-and-phase modulations for the case of the combinational internal resonance [7] using the Runga-Kutta fourth order method.

* Corresponding author: basem@pass.ps

2. Problem formulation and governing equations

Let us consider the dynamic response of a free supported nonlinear elastic circular cylindrical shell of radius R and length l, vibrations of which in the cylindrical system of coordinates described by the Donnell–Mushtari–Vlasov equations in terms of three mutually orthogonal displacements [12]:

$$\frac{\partial^2 u}{\partial x^2}+\frac{1-\nu}{2}\frac{1}{R^2}\frac{\partial^2 u}{\partial \varphi^2}+\frac{1+\nu}{2}\frac{1}{R}\frac{\partial^2 v}{\partial x\partial \varphi}-\nu\frac{1}{R}\frac{\partial w}{\partial x}+\frac{\partial w}{\partial x}\frac{\partial^2 w}{\partial x^2}$$
$$+\frac{1+\nu}{2}\frac{1}{R^2}\frac{\partial w}{\partial \varphi}\frac{\partial^2 w}{\partial x\partial \varphi}+\frac{1-\nu}{2}\frac{1}{R^2}\frac{\partial w}{\partial x}\frac{\partial^2 w}{\partial \varphi^2}=\frac{\rho\left(1-\nu^2\right)}{E}\frac{\partial^2 u}{\partial t^2} \quad (1)$$

$$\frac{1}{R^2}\frac{\partial^2 v}{\partial \varphi^2}+\frac{1-\nu}{2}\frac{\partial^2 v}{\partial x^2}+\frac{1+\nu}{2}\frac{1}{R}\frac{\partial^2 u}{\partial x\partial \varphi}-\frac{1}{R^2}\frac{\partial w}{\partial \varphi}+\frac{1}{R^3}\frac{\partial w}{\partial \varphi}\frac{\partial^2 w}{\partial \varphi^2}$$
$$+\frac{1+\nu}{2}\frac{1}{R}\frac{\partial w}{\partial x}\frac{\partial^2 w}{\partial x\partial \varphi}+\frac{1-\nu}{2}\frac{1}{R}\frac{\partial w}{\partial \varphi}\frac{\partial^2 w}{\partial x^2}=\frac{\rho\left(1-\nu^2\right)}{E}\frac{\partial^2 v}{\partial t^2} \quad (2)$$

$$\frac{h^2}{12}\nabla^4 w+\frac{1}{R^2}w-\nu\frac{1}{R}\frac{\partial u}{\partial x}-\frac{1}{R^2}\frac{\partial v}{\partial \varphi}-\frac{1}{2}\frac{\nu}{R}\left(\frac{\partial w}{\partial x}\right)^2-\frac{1}{2}\frac{1}{R^3}\left(\frac{\partial w}{\partial \varphi}\right)^2$$
$$-\frac{\partial}{\partial x}\left[\frac{\partial w}{\partial x}\left(\frac{\partial u}{\partial x}+\frac{\nu}{R}\frac{\partial v}{\partial \varphi}-\frac{\nu}{R}w\right)+\frac{1-\nu}{2}\frac{1}{R}\frac{\partial w}{\partial \varphi}\left(\frac{1}{R}\frac{\partial u}{\partial \varphi}+\frac{\partial v}{\partial x}\right)\right]$$
$$-\frac{1}{R}\frac{\partial}{\partial \varphi}\left[\frac{1}{R}\frac{\partial w}{\partial \varphi}\left(\nu\frac{\partial u}{\partial x}+\frac{1}{R}\frac{\partial v}{\partial \varphi}-\frac{1}{R}w\right)+\frac{1-\nu}{2}\frac{\partial w}{\partial x}\left(\frac{1}{R}\frac{\partial u}{\partial \varphi}+\frac{\partial v}{\partial x}\right)\right] \quad (3)$$
$$=\frac{\rho\left(1-\nu^2\right)}{E}\frac{\partial^2 w}{\partial t^2}$$

where x-axis is directed along the axis of the cylinder, φ is the polar angle in the plane perpendicular to the x-axis, $u=u(x,\varphi,t)$, $v=v(x,\varphi,t)$, and $w=w(x,\varphi,t)$ are the displacements of points located in the shell's middle surface in three mutually orthogonal directions x,φ,r, r is the polar radius, h is the thickness, ρ is the density, E and ν are the elastic modulus and Poisson's ratio, respectively, t is the time, and

$$\nabla^4=\nabla^2\nabla^2=\frac{\partial^4}{\partial x^4}+2\frac{1}{R^2}\frac{\partial^4}{\partial x^2\partial \varphi^2}+\frac{1}{R^4}\frac{\partial^4}{\partial \varphi^4}.$$

The initial conditions

$$u|_{t=0}=v|_{t=0}=w|_{t=0}, \quad (4)$$

$$\dot{u}|_{t=0}=\varepsilon V_1^0(x,\varphi), \quad \dot{v}|_{t=0}=\varepsilon V_2^0(x,\varphi), \quad \dot{w}|_{t=0}=\varepsilon V_3^0(x,\varphi) \quad (5)$$

where $V_\alpha^0(x,\varphi)$ are the corresponding initial velocities, and ε is a small value, should be added to Eqs. (1)-(3). Hereafter over dots denote time-derivatives.

The boundary conditions for a simply supported shell (Navier conditions for the edges free supported in the x-direction) have the form [12]:

$$w|_{x=0}=w|_{x=l}=0, \quad v|_{x=0}=v|_{x=l}=0,$$
$$\frac{\partial^2 w}{\partial x^2}|_{x=0}=\frac{\partial^2 w}{\partial x^2}|_{x=l}=0, \quad \frac{\partial u}{\partial x}|_{x=0}=\frac{\partial u}{\partial x}|_{x=l}=0. \quad (6)$$

Internal resonances, which could occur during free vibrations of elastic circular cylindrical shells described by partial differential equations (1)-(3) with initial and boundary conditions (4)-(6), were revealed and studied in [13] utilizing the solution of the Navier type

$$u(x,\varphi,t)=\sum_{m=1}^{\infty}\sum_{n=1}^{\infty}x_{1mn}(t)\eta_{1mn}(x,\varphi),$$
$$v(x,\varphi,t)=\sum_{m=1}^{\infty}\sum_{n=1}^{\infty}x_{2mn}(t)\eta_{2mn}(x,\varphi),$$
(7)
$$w(x,\varphi,t)=\sum_{m=1}^{\infty}\sum_{n=1}^{\infty}x_{3mn}(t)\eta_{3mn}(x,\varphi),$$

where $x_{imn}(t)$ and $\eta_{imn}(x,\varphi)$ $(i=1,2,3)$ are, respectively, the generalized time-dependent displacements and natural functions corresponding to the boundary conditions (6).

During the last decade, fractional calculus entered the mainstream of engineering analysis, and it has been widely applied to structural dynamics problems both in discrete and continuous equations. The results obtained in the field critically estimated in the light of the present view of the place and role of the fractional calculus in engineering problems and practice in the state-of-the-art article by Rossikhin and Shitikova [8].

Rossikhin and Shitikova [8] suggested a new idea to model nonlinear damped vibrations of plates and shells in a viscoelastic medium representing viscous resistance forces via fractional order time derivatives, as distinct to the traditional way [14] when damping forces are assumed to be proportional to first order time-derivatives of displacements. It has been shown in [15] that viscoelastic fractional derivative models used for dynamic response of suspension bridges produce better fit of theoretical investigation with experimental results.

Thus, utilizing the fractional derivative Kelvin-Voigt model and the procedure proposed in [8] for decoupling the linear parts of differential equations of motion of the cylindrical shell (1)-(3), the system of equations of three predominating modes of vibrations, which could be coupled by some conditions of internal resonance, have been derived in [6, 7, 9] and written in the following dimensionless form:

$$\ddot{X}_{1mn}+æ_1 D^{\gamma}X_{1mn}+\Omega_{1mn}^2X_{1mn}=-\sum_{i=1}^{3}F_{imn}L_{imn}^{I} \quad (7)$$

$$\ddot{X}_{2mn}+æ_2 D^{\gamma}X_{2mn}+\Omega_{2mn}^2X_{2mn}=-\sum_{i=1}^{3}F_{imn}L_{imn}^{II} \quad (8)$$

$$\ddot{X}_{3mn}+æ_3 D^{\gamma}X_{3mn}+\Omega_{3mn}^2X_{3mn}=-\sum_{i=1}^{3}F_{imn}L_{imn}^{III} \quad (9)$$

where m and n are integers defining the number of the natural mode of vibration, $D^{\gamma}=(d/dt)^{\gamma}$ is the fractional order ($0<\gamma<1$) of the operator of differentiation [8, 15], $æ_i=\varepsilon\mu_i\tau_i^{\gamma}$ $(i=1,2,3)$ are coefficients of small viscosity, μ_i are modal damping coefficients, τ_i are modal retardation times, F_{imn} are nonlinear terms [6].

In equations (8)-(10), X_{imn} are new generalized displacements which are connected with $x_{imn}(t)$ via eigenvectors L_{imn}^{I}, L_{imn}^{II}, L_{imn}^{III}

$$x_{imn}(t) = X_{1mn}L^{I}_{imn} + X_{2mn}L^{II}_{imn} + X_{3mn}L^{III}_{imn}$$

of the matrix S^{mn}_{ij} with the corresponding eigenvalues Ω_{1mn}, Ω_{2mn}, and Ω_{3mn}, the elements of which are the following:

$$S^{mn}_{11} = \left(\pi^2 m^2 + \frac{1-\nu}{2}\beta_1^2 n^2\right),\ S^{mn}_{12} = S^{mn}_{21} = \frac{1+\nu}{2}\beta_1 \pi mn\,,$$
$$S^{mn}_{13} = S^{mn}_{31} = \nu\beta_1\pi m,\ \ S^{mn}_{23} = S^{mn}_{32} = \beta_1^2 n\,,$$
$$S^{mn}_{22} = \left(\frac{1-\nu}{2}\pi^2 m^2 + \beta_1^2 n^2\right), \tag{10}$$
$$S^{mn}_{33} = \frac{\beta_2^2}{12}(\pi^2 m^2 + \beta_1^2 n^2)^2 + \beta_1\,,$$

where $\beta_1 = l/R$ and $\beta_2 = h/l$ are the parameters defining the dimensions of the shell.

Method of multiple time scales [16] has been directly applied to the governing differential equations (8)-(10), resulting in six first-order nonlinear ordinary-differential equations governing the modulation of the amplitudes and phases of the three interacting modes in case of the combinational additive-difference internal resonance $\Omega_1 = \Omega_2 + \Omega_3$:

$$\left(a_1^2\right)^{\cdot} + s_1 a_1^2 = \Omega_1^{-1} a^{I}_{23} a_1 a_2 a_3 \sin\delta\,, \tag{11}$$

$$\left(a_2^2\right)^{\cdot} + s_2 a_2^2 = -\Omega_2^{-1} a^{II}_{13} a_1 a_2 a_3 \sin\delta\,, \tag{12}$$

$$\left(a_3^2\right)^{\cdot} + s_3 a_3^2 = -\Omega_3^{-1} a^{III}_{12} a_1 a_2 a_3 \sin\delta\,, \tag{13}$$

$$\dot{\varphi}_1 - \frac{1}{2}\sigma_1 - \frac{1}{2}\frac{a^{I}_{23}}{\Omega_1}\frac{a_2 a_3}{a_1}\cos\delta = 0\,, \tag{14}$$

$$\dot{\varphi}_2 - \frac{1}{2}\sigma_2 - \frac{1}{2}\frac{a^{II}_{13}}{\Omega_2}\frac{a_1 a_3}{a_2}\cos\delta = 0\,, \tag{15}$$

$$\dot{\varphi}_3 - \frac{1}{2}\sigma_3 - \frac{1}{2}\frac{a^{III}_{12}}{\Omega_3}\frac{a_1 a_2}{a_3}\cos\delta = 0\,, \tag{16}$$

where a_i and φ_i $(i = 1,2,3)$ are, respectively, amplitudes and phases, $\delta = \varphi_1 - (\varphi_2 + \varphi_3)$ is the phase difference, $\psi = \frac{1}{2}\pi\gamma$, $s_i = \mu_i \tau_i^{\gamma}\Omega_i^{\gamma-1}\sin\psi$, $\sigma_i = \mu_i \tau_i^{\gamma}\Omega_i^{\gamma-1}\cos\psi$, and $a^{I}_{23}, a^{II}_{13}, a^{III}_{12}$ are the coefficients depending on the shell's dimensions, the properties of material it is made of, as well as on the numbers of the interacting modes, which could be found in [6, 7].

Thus, the solution of equations (12-17) could provide an approximation solution for small but finite amplitudes weakly varying with time based on a third-order uniform expansion in terms of different time scales. This set of the differential equations in amplitudes and phases approximates the previous system of equations (8-10) in the case of the additive combinational resonance $\Omega_1 = \Omega_2 + \Omega_3$.

3. Method of solution

3.1. Defining shell's parameters that could satisfy the combinational resonance conditions

To find the parameters which could satisfy the conditions of the combinational internal resonance $\Omega_1 = \Omega_2 + \Omega_3$, we should use the property of the symmetric matrix S^{mn}_{ij} possessing three real eigenvalues Ω_{imn} $(i$=1,2,3) corresponding to the three mutually orthogonal eigenvectors:

$$S^{mn}_{ij} = \Omega^2_{1mn}L^{I}_{imn}L^{I}_{jmn} + \Omega^2_{2mn}L^{II}_{imn}L^{II}_{jmn} + \Omega^2_{3mn}L^{III}_{imn}L^{III}_{jmn}\,.$$

The values for m, n, R, l, h, ν which satisfy the equality $\Omega_1 = \Omega_2 + \Omega_3$ for every m, n are presented below in Table 1.

Reference to Table (1) it shows that the condition of the combinational internal resonance could occur very often, resulting in coupling of three interacting modes of vibrations.

Table 1. Shell parameters that satisfy resonance conditions

Ω_1	Ω_2	Ω_3	m	n	ν	l	h	R
6.55	3.83	2.72	2	3	0.32	0.24	0.05	0.39
6.43	3.81	2.62	2	3	0.3	0.23	0.05	0.5
6.30	3.71	2.60	2	3	0.3	0.23	0.05	1.17
6.28	3.64	2.64	2	3	0.32	0.22	0.05	2.59
6.28	3.72	2.57	2	3	0.3	0.24	0.05	2.89
6.28	3.72	2.57	2	3	0.3	0.2	0.05	2.41

3.2. Numerical solution of general multi-term linear equations

Using the method proposed in [10, 11, 17, 18], the procedure based on the reduction of the problem to a system of fractional differential equations to estimate numerically the solution of equations (8-10) is as follows:

let

$$Y_1 = X_1\,,$$
$$Y_2 = D^{\gamma}X_1 = D^{\gamma}Y_1\,,$$
$$Y_3 = DX_1 = DY_1\,,$$
$$\ddot{X}_1 = DDX_1 = DY_3$$

and then substitute these equalities in equation (8), resulting in

$$DY_3 = -\sum_{i=1}^{3} F_{1mn}L^{I}_{imn} - æ_1 Y_2 - \Omega_1^2 Y_1\,, \tag{17}$$
$$Y_4 = X_2\,,$$
$$Y_5 = D^{\gamma}X_2 = D^{\gamma}Y_4\,,$$
$$Y_6 = DX_2 = DY_4\,,$$
$$\ddot{X}_2 = DDX_2 = DY_6\,,$$

then in equation (9), resulting in

$$DY_6 = -\sum_{i=1}^{3} F_{2\,mn} L^{II}_{i\,mn} - \text{æ}_2 Y_5 - \Omega_2^2 Y_4 ,$$

$$Y_7 = X_3 ,$$

$$Y_8 = D^{\gamma} X_3 = D^{\gamma} Y_7 ,$$

$$Y_9 = DX_3 = DY_7 ,$$

$$\ddot{X}_3 = DDX_3 = DY_9 ,$$

and finally in equation (10), resulting in

$$DY_9 = -\sum_{i=1}^{3} F_{3\,mn} L^{II}_{i\,mn} - \text{æ}_3 Y_8 - \Omega_3^2 Y_7 .$$

Thus, the governing set of nine equations in nine unknown values Y_i in the matrix form could be written as

$$\begin{bmatrix} D^{\gamma} & 0 & 0 & 0 & 0 & 0 & 0 & 0 & 0 \\ D & 0 & 0 & 0 & 0 & 0 & 0 & 0 & 0 \\ 0 & 0 & D & 0 & 0 & 0 & 0 & 0 & 0 \\ 0 & 0 & 0 & D^{\gamma} & 0 & 0 & 0 & 0 & 0 \\ 0 & 0 & 0 & D & 0 & 0 & 0 & 0 & 0 \\ 0 & 0 & 0 & 0 & 0 & D & 0 & 0 & 0 \\ 0 & 0 & 0 & 0 & 0 & 0 & D^{\gamma} & 0 & 0 \\ 0 & 0 & 0 & 0 & 0 & 0 & D & 0 & 0 \\ 0 & 0 & 0 & 0 & 0 & 0 & 0 & 0 & D \end{bmatrix} * \begin{bmatrix} Y_1 \\ Y_2 \\ Y_3 \\ Y_4 \\ Y_5 \\ Y_6 \\ Y_7 \\ Y_8 \\ Y_9 \end{bmatrix}$$

$$= \begin{bmatrix} Y_2 \\ Y_3 \\ -\sum_{i=1}^{3} F_{1\,mn} L^{I}_{i\,mn} - \text{æ}_1 Y_2 - \Omega_1^2 Y_1 \\ Y_5 \\ Y_6 \\ -\sum_{i=1}^{3} F_{2\,mn} L^{II}_{i\,mn} - \text{æ}_2 Y_5 - \Omega_2^2 Y_4 \\ Y_8 \\ Y_9 \\ -\sum_{i=1}^{3} F_{3\,mn} L^{III}_{i\,mn} - \text{æ}_3 Y_8 - \Omega_3^2 Y_7 \end{bmatrix} \qquad (18)$$

Two different types of discretization of derivatives in (19) could be utilized [10, 17]. For the first order differential equation, usually the Trapezium rule is used:

$$DY = f \xrightarrow{yields} Y_i = Y_{i-1} + \frac{1}{2} h \left(f_i + f_{i-1} \right) . \qquad (19)$$

So in our problem, the discrete derivatives (D) will take the form

$$Y_i - \frac{1}{2} h f_i = Y_{i-1} + \frac{1}{2} h \left(f_{i-1} \right) . \qquad (20)$$

To discretize the fractional derivative, the Diethelm's method could be used [10]:

$$D^{\gamma} Y = \frac{1}{{}^{\alpha}\gamma_i} \left(\sum_{k=0}^{i} {}^{\alpha}\omega_{k,i} Y_{i-k} + \frac{Y_0}{\alpha} \right) , \qquad (21)$$

where ${}^{\alpha}\gamma_i = (ih)^{\alpha} \Gamma(-\alpha)$ and ${}^{\alpha}\omega_{k,0}, \ldots, {}^{\alpha}\omega_{k,i}$ [10, 11] are convolution weights derived from the fact that the fractional operator defined in terms of a convolution integral. We will use the weights of the quadrature formula [10]

$$f(\alpha) = \alpha (1-\alpha) j^{-\alpha}\, {}^{\alpha}\omega_{k,0} =$$

$$= \begin{cases} -1 & \text{for } k = 0 \\ 2k^{1-\alpha} - (k-1)^{1-\alpha} - (k+1)^{1-\alpha} & \text{for } k = 1, 2, .., j-1 \\ (\alpha - 1) k^{-\alpha} - (k-1)^{1-\alpha} + k^{1-\alpha} & \text{for } k = j \end{cases}$$

(22)

Discretization of the equation

$$D^{\gamma} Y_1 = Y_2 .$$

(23)

results in the following relationships (note $\alpha = \gamma$):

$$\frac{1}{{}^{\alpha}\gamma_i} \left(\sum_{k=0}^{i} {}^{\alpha}\omega_{k,i} Y_{1i-k} + \frac{Y_{10}}{\alpha} \right) = Y_2 , \qquad (24)$$

$$\left({}^{\alpha}\omega_{0,i} Y_{1i} + \sum_{k=1}^{i} {}^{\alpha}\omega_{k,i} Y_{1i-k} + \frac{Y_{10}}{\alpha} \right) = {}^{\alpha}\gamma_i Y_{2i} . \qquad (25)$$

Let $\sum_{k=1}^{i} {}^{\alpha}\omega_{k,i} Y_{1i-k} + \frac{Y_{10}}{\alpha} = s_{1i-1}$, so we have

$$\left({}^{\alpha}\omega_{0,i} Y_{1i} + s_{1i-1} \right) = {}^{\alpha}\gamma_i Y_{2i} , \qquad (26)$$

$$s_{1i} = -{}^{\alpha}\omega_{0,i} Y_{1i} + {}^{\alpha}\gamma_i Y_{2i} . \qquad (27)$$

By the trapezium rule we can discretize $Y_3 = DY_1$ as

$$Y_{1i} = Y_{1i-1} + \frac{h}{2} (Y_{3\,i} + Y_{3\,i-1}) . \qquad (28)$$

So rearranging terms

$$Y_{1i} - \frac{h}{2} Y_{3\,i} = Y_{1i-1} + \frac{h}{2} Y_{3\,i-1} = s_{2i-1} \qquad (29)$$

and utilizing the trapezium rule, we can discretize

$$DY_3 = -\sum_{j=1}^{3} F_{1\,mn} L^{I}_{j\,mn} - \text{æ}_1 Y_2 - \Omega_1^2 Y_1 , \qquad (30)$$

$$Y_{3\,i} = Y_{3\,i-1} + \frac{h}{2} \Big[-\sum_{j=1}^{3} (F_{1\,mn\,i} + F_{1\,mn\,i-1}) L^{I}_{j\,mn} - \text{æ}_1 (Y_{2i} + Y_{2i-1}) - \Omega_1^2 (Y_{1i} + Y_{1i-1}) \Big] \qquad (31)$$

Rearranging terms, we have

$$\left(Y_{3\,i} + \frac{h}{2} (\text{æ}_1 Y_{2i} + \Omega_1^2 Y_{1i}) \right) = Y_{3\,i-1} + \frac{h}{2} \left(-\sum_{j=1}^{3} (F_{1\,mn\,i} + F_{1\,mn\,i-1}) L^{I}_{j\,mn} - \text{æ}_1 Y_{2i-1} - \Omega_1^2 Y_{1i-1} \right) = s_{3i-1} \qquad (32)$$

Repeating these steps (as we have done in Eqs. (24)-(33)) for all other values (Y_4-Y_9), and arranging them in matrix form, we obtain

$$\begin{bmatrix} -^{\alpha}\omega_{0,i} & ^{\alpha}\gamma_i & 0 & 0 & 0 & 0 & 0 & 0 & 0 \\ 1 & 0 & -\frac{h}{2} & 0 & 0 & 0 & 0 & 0 & 0 \\ \frac{\text{h}}{2}\Omega_1^2 & \frac{\text{h}}{2}æ_1 & 1 & 0 & 0 & 0 & 0 & 0 & 0 \\ 0 & 0 & 0 & -^{\alpha}\omega_{0,i} & ^{\alpha}\gamma_i & 0 & 0 & 0 & 0 \\ 0 & 0 & 0 & 1 & 0 & -\frac{h}{2} & 0 & 0 & 0 \\ 0 & 0 & 0 & \frac{\text{h}}{2}\Omega_2^2 & \frac{\text{h}}{2}æ_2 & 1 & 0 & 0 & 0 \\ 0 & 0 & 0 & 0 & 0 & 0 & -^{\alpha}\omega_{0,i} & ^{\alpha}\gamma_i & 0 \\ 0 & 0 & 0 & 0 & 0 & 0 & 1 & 0 & -\frac{h}{2} \\ 0 & 0 & 0 & 0 & 0 & 0 & \frac{\text{h}}{2}\Omega_3^2 & \frac{\text{h}}{2}æ_3 & 1 \end{bmatrix} *$$

$$\begin{bmatrix} Y_{1i} \\ Y_{2i} \\ Y_{3i} \\ Y_{4i} \\ Y_{5i} \\ Y_{6i} \\ Y_{7i} \\ Y_{8i} \\ Y_{9i} \end{bmatrix} = \begin{bmatrix} s_{1i-1} \\ s_{2i-1} \\ s_{3i-1} \\ s_{4i-1} \\ s_{5i-1} \\ s_{6i-1} \\ s_{7i-1} \\ s_{8i-1} \\ s_{9i-1} \end{bmatrix} \qquad (33)$$

It is quite straightforward to solve (34).

3.3. Numerical solution of the combinational additive-difference internal resonance using the Runga-Kutta fourth order method

To utilize the Runga-Kutta fourth order method to estimate numerically the solution of equations (12)-(17), we first rewrite these equations in the first order form as follows:

$$\dot{a}_1 = \frac{1}{2}(\Omega_1^{-1} a_{23}^{I}\, a_2 a_3 \sin\delta - s_1 a_1), \qquad (34)$$

$$\dot{a}_2 = \frac{1}{2}(\Omega_2^{-1} a_{13}^{II}\, a_1 a_3 \sin\delta - s_2 a_2), \qquad (35)$$

$$\dot{a}_3 = \frac{1}{2}(\Omega_3^{-1} a_{12}^{III}\, a_1 a_2 \sin\delta - s_3 a_3), \qquad (36)$$

$$\dot{\varphi}_1 = \frac{1}{2}\sigma_1 + \frac{1}{2}\frac{a_{23}^{I}}{\Omega_1}\frac{a_2 a_3}{a_1}\cos\delta, \qquad (37)$$

$$\dot{\varphi}_2 = \frac{1}{2}\sigma_2 + \frac{1}{2}\frac{a_{13}^{II}}{\Omega_2}\frac{a_1 a_3}{a_2}\cos\delta, \qquad (38)$$

$$\dot{\varphi}_3 = \frac{1}{2}\sigma_3 + \frac{1}{2}\frac{a_{12}^{III}}{\Omega_3}\frac{a_1 a_2}{a_3}\cos\delta. \qquad (39)$$

4. Numerical results

The numerical solution of equation (34) has been carried out at the dimensionless parameters presented in Table 2.

Table 2. Shell's parameters.

ν	*R*	*l*	*h*	γ	*m*	*N*
0.3	2	0.28	0.042	$0 \longrightarrow 1$	3	1

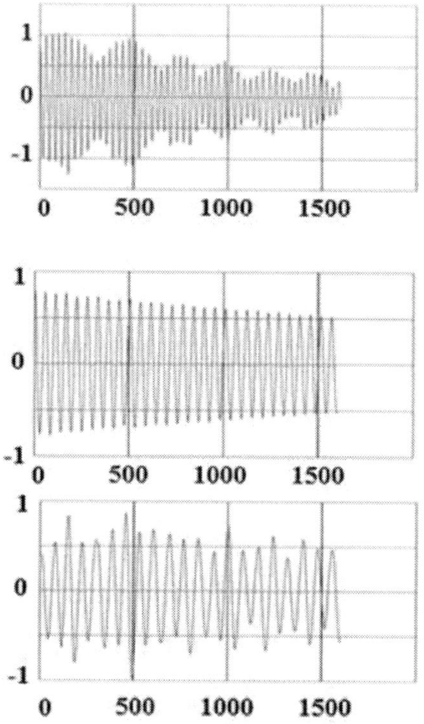

Fig. 1 Plot of numerical solution of equations (34) X_1, X_2, X_3

Variation of the fractional parameter from $\gamma \to 0$ to $\gamma \to 1$ allows one to investigate vibrations of cylindrical shells in surrounding media with different viscous properties, including the pure elastic case at $\gamma = 0$ and conventional Kelvin-Voigt model at $\gamma \to 1$.

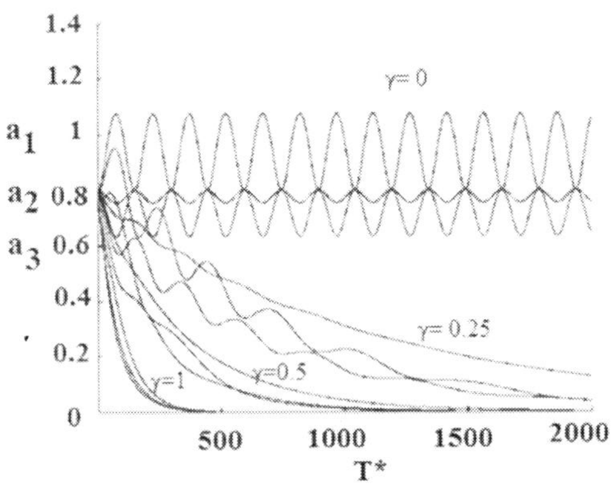

Fig. 2 Dimensionless amplitude vs. dimensionless time as the solution of equations (35-40).

The dynamic behavior of a cylindrical shell in a viscous medium at $\gamma = 0.25$, 0.5 and 1, which is found by using the first and second methods, is shown in Fig. 1 and 2, respectively, for the parameters taken from Table 2, which correspond to the combinational internal resonance at $\Omega_1 = \Omega_2 + \Omega_3 = 9.4259 = 3.8494 + 5.5764$.

The behavior of amplitudes of vibrations reveals the exchange of energy between the generalized displacements of the system under the considered case of the combinational internal resonance.

5. Conclusion

Free damped vibrations of a shallow nonlinear thin cylindrical shell in a fractional derivative viscoelastic medium are investigated numerically by two different methods based on the new approach proposed in [6].

The numerical solutions of the damped vibrations of the nonlinear cylindrical shell subjected to the conditions of the internal resonance have been estimated, and good agreement between the two methods has been achieved. Within the first method, the generalized displacements of a coupled set of nonlinear ordinary differential equations of the second order are calculated using the numerical solution of nonlinear multi-term fractional differential equations by the procedure based on the reducing the problem to a system of fractional differential equations.

According to the second method, the amplitudes and phases of nonlinear vibrations are estimated from the governing nonlinear differential equations describing amplitude-and-phase modulations for the case of the combinational internal resonance.

It has been shown that, as in [6,7], the phenomenon of the internal resonance could be very critical, since in a circular cylindrical shell the internal additive and difference combinational resonances are always present. The effect of viscosity on the energy exchange mechanism is analyzed.

This research was made possible by Grant No. 9.5138.2017/8.9 as a Government task from the Ministry of Education and Science of the Russian Federation.

References

1. V.D. Kubenko, P.S. Koval'chuk, Int. Appl. Mech. **34**, 703–728 (1998)
2. M. Amabili, M.P. Paidoussis, Appl. Mech. Rev. **56**, 349–381 (2003)
3. Y.-S. Lee, Trans. KSME **33**, 1–26 (2009)
4. I.D. Breslavsky, K.V. Avramov, Meccanica **46**, 817–832 (2011)
5. K.V. Avramov, Yu.V. Mikhin, E. Kurilov, Nonlinear Dyn. **47**, 331–352 (2012)
6. Yu.A. Rossikhin, M.V. Shitikova, Chapter 17 in: *Shell and Membrane Theories in Mechanics and Biology: From Macro- to Nanoscale Structures* (H. Altenbach, G.I. Mikhasev, eds., Springer, pp. 259–280, 2015) DOI 10.1007/978-3-319-02535-3_17
7. Yu.A. Rossikhin, M.V. Shitikova, Chapter 3 in: *Computational Problems in Science and Engineering* (N. Mastorakis et al. eds.), *Lecture Notes in Electrical Engineering* **343**, 59-107 (Springer Int. Publ., Switzerland, 2015) DOI 10.1007/978-3-319-15765-8_3
8. Yu.A. Rossikhin, M.V. Shitikova, Appl. Mech. Rev. **63**, 010801 (2010)
9. Yu.A. Rossikhin, M.V. Shitikova, Appl. Math. Comp. **257**, 498–525 (2015)
10. K. Diethelm, Elect. Trans. Num. Anal. **5**, 1–6 (1997)
11. J.T. Edwards, N.J. Ford, A.C. Simpson, J. Comp. Appl. Math. **148**, 401–418 (2002)
12. A. Volmir, *Nonlinear dynamics of plates and shells* (Nauka, Moscow, 1972)
13. Yu.A. Rossikhin, M.V. Shitikova, In: *Proceedings of the 18th International Congress on Sound and Vibration* **4**, 2775-2782, July 11-14, 2011, Rio de Janeiro, Brazil (2011)
14. R.W. Clough, J. Penzien, *Dynamics of structures* (McGraw-Hill, New York, 1975)
15. Yu.A. Rossikhin, M.V. Shitikova, ASCE J. Eng. Mech, **124**, 1029–1036 (1998)
16. A. Nayfeh, *Perturbation methods* (Wiley, New York 1973)
17. D. Baleanu, K. Diethelm, E. Scalas, J.J. Trujillo, *Fractional calculus: Models and numerical methods* (World Scientific, Singapore, pp. 88-114, 2012)
18. K. Diethelm, *Hildesheimer Informatikberichte* (1995)

 https://doi.org/10.1051/matecconf/201814803007

Exploiting nonlinearity in a flapping wing mechanism of a bio inspired micro air vehicle to enhance energy efficiency

Ali Abolfathi[1,*]

[1]Dept. of Mechanical Engineering, University College London (UCL), Roberts Engineering Building, London, UK WC1E 7JE

Abstract. In recent years, there was a great interest in developing flying drones with similar capabilities as flying insects. It is suggested that the flapping frequency of insects coincides with the resonance frequency of their flight mechanism to enhance the power consumptions. In this paper, the effect of nonlinearity in the flight mechanism on the power consumption is investigated. A simple nonlinear model of the insect flight mechanism is developed and normalised to study the effect of different parameters on its performance. Both bistable and hardening nonlinearity are considered. It is shown that for a harmonic loading, the bistable systems reach their peak power at lower frequencies when compared to the corresponding linear system. The maximum power factor of nonlinear oscillator would be lower than the liner one. It is also shown that the peak active power of the bistable system has a higher value than the linear system if the loading function is a pulse square signal.

1 Introduction

Flying insects have evolved over thousands of years and have become efficient flying machines. They served as a source of inspiration for mini and micro unmanned air vehicles [1,2]. According to Greenewalt [3], the frequency of flapping is constant at different flight regimes. The flying birds or insects are modelled as a damped linear oscillator which operates at its resonance frequency to minimise the energy consumption. Fischer et. al. [4] proposed mechanisms for flapping wing robots that operate at their resonance frequency to minimise the power consumption. Madangopal et. al. [5] used four bar mechanisms for a flapping drone and used linear springs to mimic the elastic energy storage of the insects. They optimised the geometrical parameters and spring constant to achieve the minimum energy consumption. Baek et. al. [6] designed a flapping wing drone and optimised its performance considering the effect of battery and electrical motor. Bronson et. al. [7] and Bao et. al. [8] developed flapping wings that can be excited at the resonance frequency of the wing.

Goosen et. al. [9] proposed a new design for a flapping wing drone. They listed nonlinearity as one of the difficulties in implementing a flying machine at its resonance frequency in practice. Ramananarivo et. al. [10] developed a nonlinear model of a beam like wing having cubic stiffness nonlinearity and quadratic damping term. They concluded that the resonance is not providing the maximum efficiency, but the temporal evolution of the wing is optimised to increase the aerodynamic efficiency.

In insects with indirect flight muscles, the flight muscles are not directly connected to the wings but they cause a deformation of the thorax [11]. The movement of the thorax transfers to the wings through a complex mechanism. Different models are suggested for the wing-muscle interface [12] but its exact function yet still to be explored. The common theme between different models is the nonlinearity in restoring force of the "flight mechanism". The benefit of such a nonlinear mechanism is attributed to an increase in the kinetic energy of the wings [13,14], modulation of the flapping amplitude [15] and an increase in the flapping speed during half of the flapping cycle [16].

In this paper, the energy efficiency of a flight mechanism is investigated by examining the active and reactive powers of a nonlinear oscillatory system. The power that conduct useful work in a system is called active power and the power that is used to accelerate masses and deform elastic elements of the system is the reactive power [17]. In theory, the energy attributed to the reactive power returns to the source in a cycle. However, it increases the energy loss as most mechanical actuators do not have an energy recovery capability. A larger actuator is also required to compensate the reactive power requirement. In a linear oscillator at the resonance, the stiffness and inertia term cancel out and the reactive power become zero which is not the case in a nonlinear oscillator.

A lumped parameter model in form of a Duffing oscillator is used in this study as a simplified representation of a flapping wing aerial vehicle. Both bistable and monostable oscillator are considered. The generated lift is modelled by a linear damping term for the simplicity. The performance of the system for a range of parameters is evaluated and it is compared with a linear counterpart.

* Corresponding author: a.abolfathi@ucl.ac.uk

2 A nonlinear model

A schematic view of the "click mechanism" which was proposed in Ref. [18] is shown in Fig. 1(a). A mechanical model of this mechanism is proposed by Brenan et. al. [13] and is shown in Fig. 1(b). The length $2b$ is the distances between the supports when springs are unloaded. The linear horizontal springs model the stiffness of notum and mesopleural process. The mechanism will snap through the unstable static equilibrium position if a force applies downward at point C. There are two static stable equilibrium positions which are shown by solid and dashed line in Fig. 1(b). There is also an unstable equilibrium position where the levers are horizontal. Such a system is called bistable as there are two equilibrium points.

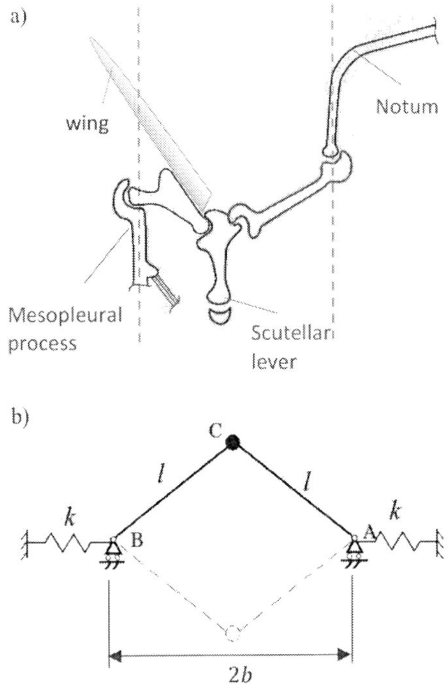

Fig. 1. a) A schematic view of insect click mechanism reproduced from Ref. [18]. b) Mechanical model of the click mechanism.

To account for the stiffness of scutellar lever (Fig. 1(a)) a vertical spring can be added to the mechanical model at point C in parallel to a damper as shown in Fig. 2. The damper is to model the aerodynamic load. The stiffness of the vertical spring can compensate the negative stiffness resulted from the lever mechanism and depending on the ratio between the vertical and horizontal springs $\alpha = k/k_v$ such a system can be bistable or only possess a single stable equilibrium position, hereafter will be referred to as *monostable* system. Different force-deflection curves for different values of α are shown in Fig. 3 for a ratio of $b/l = 0.7$.

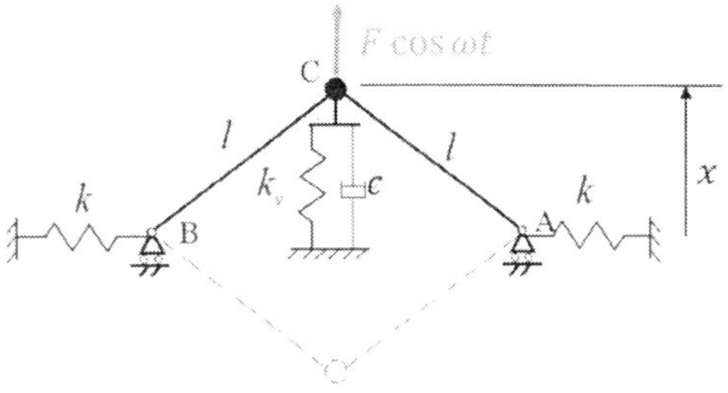

Fig. 2. Complete model of the insect flight mechanism.

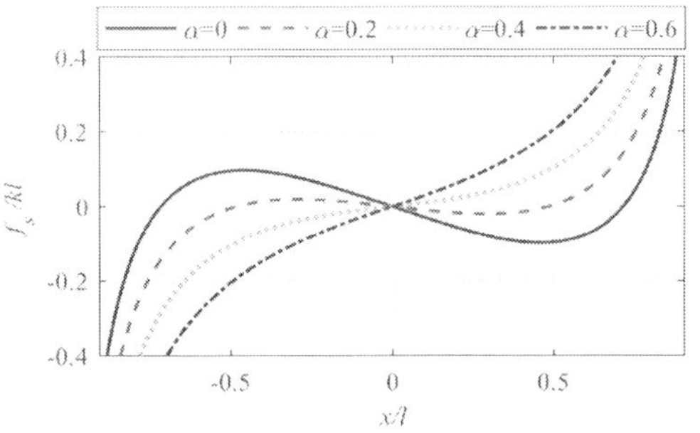

Fig. 3. Nondimensional force as a function of displacement of the flight mechanism model for different ratio of α and $b/l = 0.7$.

The equation of motion of the single degree of freedom model of the flapping mechanism can be obtained,

$$m\ddot{x} + c\dot{x} + f_s(x) = F\cos\omega t \qquad (1)$$

where $f_s(x)$ is the nonlinear spring force. The spring reaction force can be approximated by a third order polynomial and the equation of motion would have the form of,

$$m\ddot{x} + c\dot{x} + k_1x + k_3x^3 = F\cos\omega t \qquad (2)$$

where k_1 and k_3 are stiffness coefficients and can be obtained as function of system parameters [13]. Depending on the chosen values for α and β, linear stiffness term k_1 can have a positive or a negative value. In this paper, it is tried to understand how flapping can be benefitted from nonlinearity. The equation of motion is normalised to make it possible to analyse the effect of different parameters on flapping performance,

$$\hat{x}'' + 2\zeta\hat{x}' + \kappa\hat{x} + \gamma\hat{x}^3 = \cos\Omega t \qquad (3)$$

where $k = |k_1|$, $\omega_n = \sqrt{k/m}$, $\tau = \omega_n t$, $\hat{x} = x/x_0$, $x_0 = F/k$, $\zeta = c/(2m\omega_n)$, $\Omega = \omega/\omega_n$, $\kappa = k_1/k$, $\gamma = x_0^2 k_3/k_1$ and dash $(\blacksquare)'$ represents differentiation with respect to non-dimensional time τ. The normalised linear stiffness term κ can have a value of 1 or -1 depending on the type of the system, i.e. bistable or monostable. The normalised nonlinear stiffness term γ can be considered as the ratio between nonlinear spring force and linear spring force at the displacement x_0. It can also be linked to the applied force and thus is a function of the amplitude of response.

3 Power

The power that is used to produced aerodynamic lift is used as a measure of the performance here. This is called active or true power [17] and its average can be obtained over a period by calculating the work of the external force over a period,

$$P = \frac{1}{T}\int_0^T F(t)\,\dot{x}\,dt \tag{4}$$

where $F(t)$ is the applied force. The power can be nondimensionalised as $\hat{P} = P/(Fx_0\,\omega_n)$. The active power is the result of work done on the damping element and the same result can be obtained if external force is substituted by the damping force in Eq. (4).

There is no net transfer of energy in a cycle through the work conducted on mass and stiffness elements of the system, and the part of the power that is required to move them is called reactive power. Its average can be obtained over a cycle similarly,

$$Q = \frac{1}{T}\int_0^T \left| f_s(x(t)) - m\ddot{x} \right| \dot{x}\,dt \tag{5}$$

where Q is the average reactive power and can be nondimensionalised as $\hat{Q} = Q/(Fx_0\,\omega_n)$. It is important to note that although the net energy transfer due to reactive power is zero over a cycle, but it need to be considered in selection of the actuator as it should be sized based on the apparent power $S = \sqrt{P^2 + Q^2}$. The ratio between active power and apparent power is called power ratio and is considered as a measure of the performance of the system in this study. A power ratio of one would result when all the power is used to conduct useful work.

4 Harmonic loading

The active and reactive powers are obtained in this section for a range of parameters to compare the performance of nonlinear and linear system in term of energy efficiency. A purely harmonic loading is assumed to be applied to the mechanism. The aerodynamic force is modelled through the damping force and as the purpose of the oscillator is to produce lift, it is expected that the damping coefficient be high. By comparing the ratio of the powers reported in Ref. [19], a damping ratio of 0.9 can be obtained for a linear oscillator.

4.1 Bistable oscillator

The nondimensional active and reactive power for the bistable mechanism is shown in Fig. 4 as a function of excitation frequency for three different value of γ. The frequency of the peak active power increases by increasing γ. The drop in active power immediately after the peak corresponds to the bifurcation where the inter-well oscillation become chaotic before converting to intra-well oscillation. The peak in active power of the linear oscillator occurs at a normalised frequency of 1 where the reactive power is equal to zero. For a linear oscillator at the natural frequency, the inertia and stiffness forces cancel out and the input energy dissipates completely through damping term. However, for a nonlinear oscillator, the reactive power never become zero because multiple harmonics exist in the response.

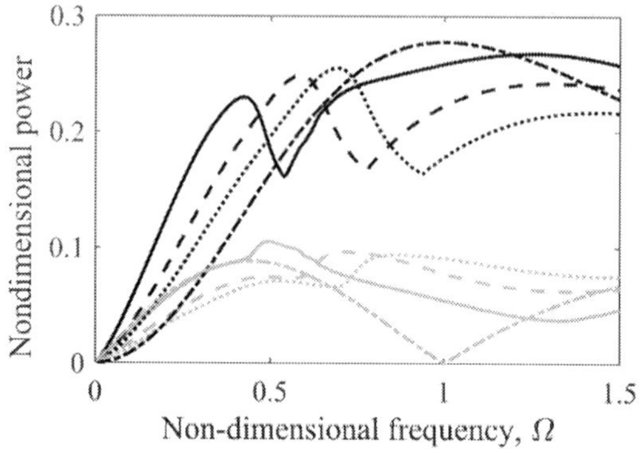

Fig. 4. Nondimensional power as a function of frequency for harmonic loading, $\kappa = -1$ and $\zeta = 0.9$. black lines: active power $\hat{P}$, cyan lines: reactive power $\hat{Q}$. Solid line: $\gamma = 1$; dashed line: $\gamma = 2$; dotted line: $\gamma = 3$; dashed-dotted line: linear oscillator.

Power factor is plotted in the Fig. 5 as a function of frequency for the bistable system and the linear oscillator. The power factor is equal to unity for the linear oscillator at the natural frequency of the system. The peak in power factor is lower for nonlinear oscillator and is about %95 but it is higher at low frequencies compared to the linear oscillator.

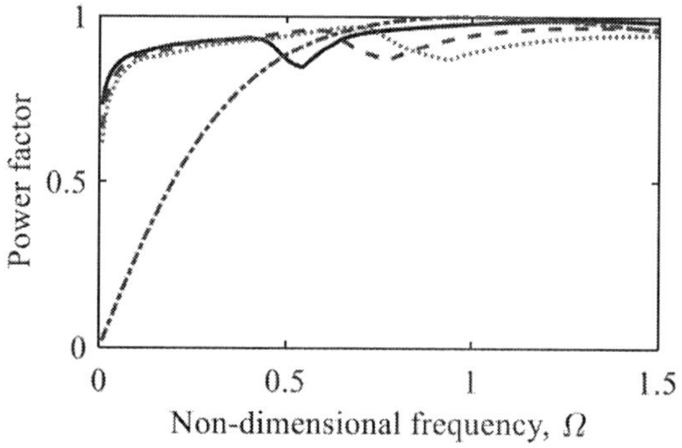

Fig. 5: Power factor as a function of frequency for harmonic loading, $\kappa = -1$ and $\zeta = 0.9$. Solid line: $\gamma = 1$; dashed line: $\gamma = 2$; dotted line: $\gamma = 3$; dashed-dotted line: linear oscillator.

Although the damping factor is expected to be high due the nature of the oscillatory system, its exact value is not known. To evaluate the effect of damping, the active power for three different level of damping is shown in Fig. 6 for the bistable oscillator with $\gamma = 0.2$. The lower damping level cause an increase in the peak of the nondimensional active power as well as a shift in its frequency to a higher frequency. Reducing the damping also cause a slight reduction in the power factor which is not shown here for brevity.

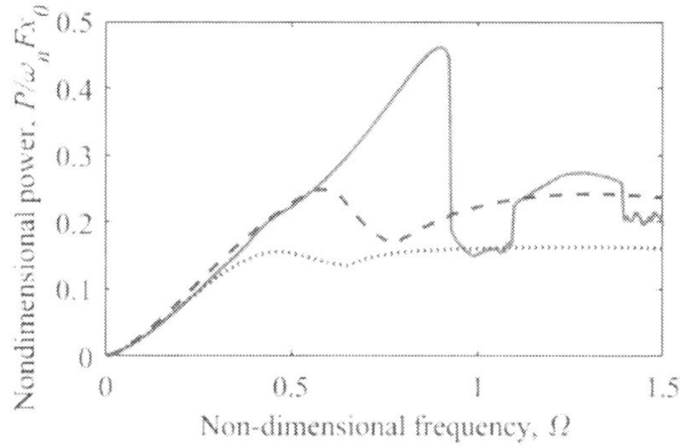

Fig. 6. Nondimensional power as a function of frequency for harmonic loading, $\kappa = -1$ and $\gamma = 2$. Solid line: $\zeta = 0.5$; dashed line: $\zeta = 0.9$; dotted line: $\zeta = 1.5$.

4.2 Monostable oscillator

The active and reactive power for a monostable oscillator ($\kappa = 1$) is shown in Fig. 7 for three different value of γ. Similar to Fig. 4, damping ratio is assumed to be equal to 0.9. The hardening nonlinearity of the oscillator cause a shift in the peak frequency to higher frequencies compared to the linear oscillator. Furthermore, the reactive power has a minimum almost equal to zero similar to the linear system. The resulted power factor is shown in Fig. 8. The nonlinear oscillator has a maximum power factor close to unity. The high damping ratio makes the effect of nonlinearity on the dynamics of the monostable oscillator minimal.

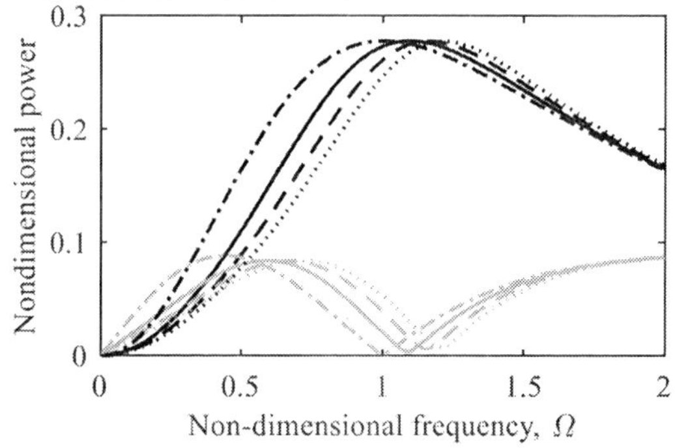

Fig. 7. Nondimensional power as a function of frequency for harmonic loading, $\kappa = 1$ and $\zeta = 0.9$. black lines: active power $\hat{P}$, cyan lines: reactive power $\hat{Q}$. Solid line: $\gamma = 1$; dashed line: $\gamma = 2$; dotted line: $\gamma = 3$; dashed-dotted line: linear oscillator.

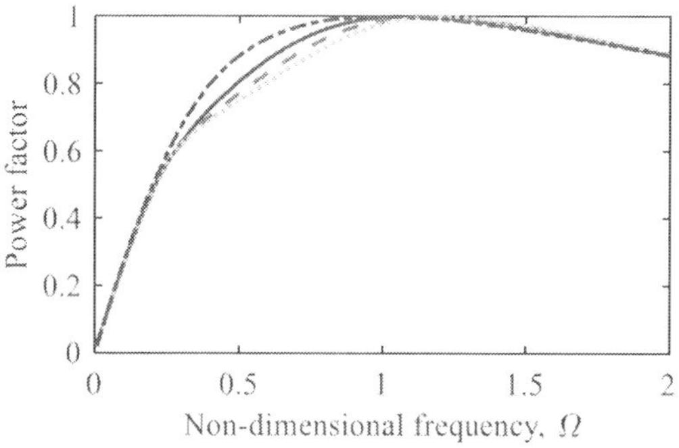

Fig. 8. Power factor as a function of frequency for $\kappa = 1$ and $\zeta = 0.9$. Solid line: $\gamma = 1$; dashed line: $\gamma = 2$; dotted line: $\gamma = 3$; dashed-dotted line: linear oscillator.

5 Non-harmonic loading

The forcing function of insect muscles are not most likely a simple harmonic. The existence of higher harmonics in a non-sinusoidal loading function can be beneficial to the performance of a nonlinear oscillator by increasing its power factor.

5.1 Bistable oscillator

The active and reactive power for a bistable system when subjected to a square loading function is shown in Fig. 9. The amplitude of the peak power is almost constant for three different values of nonlinearity and is higher than the linear oscillator. The amplitude of peak active power is almost two times of that for the same oscillator when harmonically loaded (Fig. 4). The power factor is shown in Fig. 10. The power factor is very close to one in all frequencies due to existence of multiple harmonics in the forcing function.

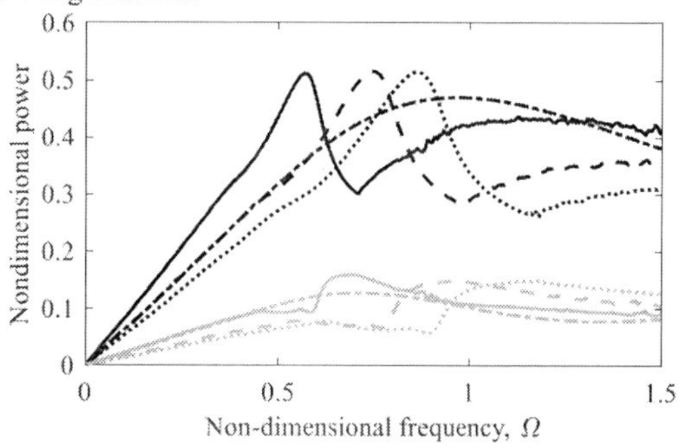

Fig. 9. Nondimensional power as a function of frequency for square loading, $\kappa = -1$ and $\zeta = 0.9$. black lines: active power $\hat{P}$, cyan lines: reactive power $\hat{Q}$. Solid line: $\gamma = 1$; dashed line: $\gamma = 2$; dotted line: $\gamma = 3$; dashed-dotted line: linear oscillator.

 https://doi.org/10.1051/matecconf/201814803007

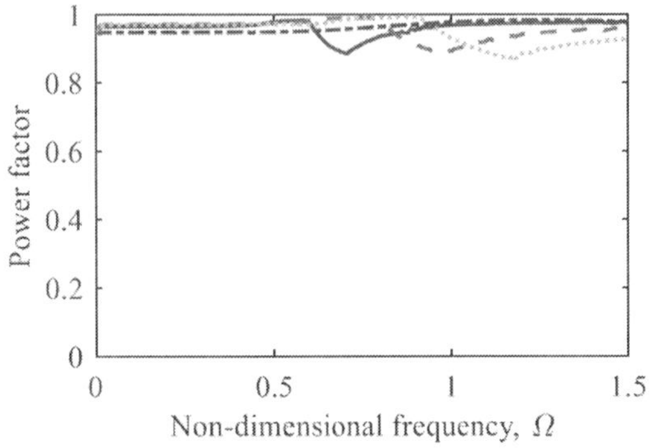

Fig. 10. Power factor as a function of frequency for square loading, $\kappa = -1$ and $\zeta = 0.9$. Solid line: $\gamma = 1$; dashed line: $\gamma = 2$; dotted line: $\gamma = 3$; dashed-dotted line: linear oscillator.

5.1 Monostable oscillator

The power curves of the monostable oscillator alongside a linear one are shown in Fig. 11 when subjected to a square pulse forcing function. The amplitude of active power is larger than the harmonic loading shown in Fig. 7. The hardening nonlinearity cause the peaks of active power to shift to higher frequencies. Similar to the bistable system, the power factor would be close to unity which is not shown here for brevity.

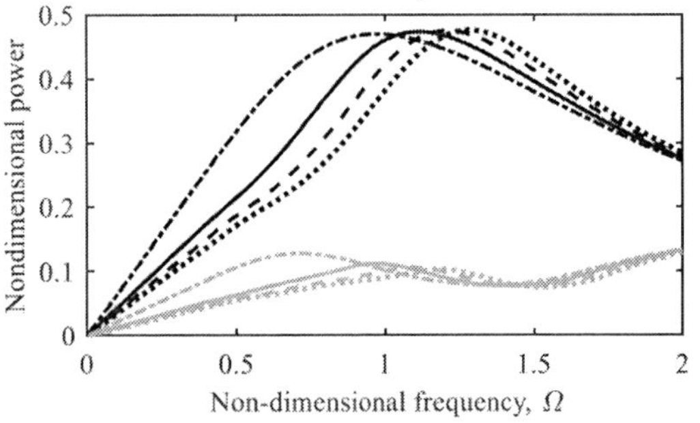

Fig. 11. Nondimensional power as a function of frequency for square loading, $\kappa = 1$ and $\zeta = 0.9$. Black lines: active power $\hat{P}$, cyan lines: reactive power $\hat{Q}$. Solid line: $\gamma = 1$; dashed line: $\gamma = 2$; dotted line: $\gamma = 3$; dashed-dotted line: linear oscillator.

6 Conclusions

The nonlinearity in insect flight mechanism is modelled by a cubic nonlinearity. The mechanism can be bistable or monostable depending on the values of system parameters. The aerodynamic lift generation is modelled by a linear damper and the corresponding active power is obtained. The magnitude and the frequency at which the active power has its maximum value and the power factor are obtained in order to compare the effect of different parameters.

For a bistable mechanism, the active power has a peak value at a lower frequency than the natural frequency of a comparable linear oscillator. The power factor of a linear oscillator is unity at its natural frequency but for a nonlinear oscillator it has a value smaller than one. For a monostable system with hardening stiffness the peak power shifts to the higher frequencies. The effect of nonlinearity is minimum in this case as the damping ratio is high.

A square pulse forcing function is examined to assess a nonharmonic forcing function. The peak active power for bistable nonlinear system is higher than the linear one which implies a higher lift generation. Furthermore, the power factor has a value close to unity for the frequency range that is examined here.

References

1. D. Floreano and R.J. Wood, Nature **521** 7553, 460–466 (2015).
2. G. Cai, J. Dias, and L. Seneviratne, Unmanned Syst. **2** 2, 175–199 (2014).
3. C.H. Greenewalt, Proc. Am. Philos. Soc. **104** 6, 605–611 (1960).
4. G. Fischer, A.G. Cox, M.A. Gogola, M.K. Gordon, N.O. Lobontiu, D.J. Monopoli, E. Garcia, and M. Goldfarb, Elastodynamic locomotion in mesoscale robotic insects, in: Ed. by Y. Bar-Cohen, Smart Struct. Mater. 1999 Electroact. Polym. Actuators Devices, (International Society for Optics and Photonics, Newport Beach, CA, US 1999), pp. 362–368.
5. R. Madangopal, Z.A. Khan, and S.K. Agrawal, J. Mech. Des. **127** 4, 809 (2005).
6. S.S. Baek, K.Y. Ma, and R.S. Fearing, Efficient resonant drive of flapping-wing robots, in: 2009 IEEE/RSJ Int. Conf. Intell. Robot. Syst., (IEEE, St. Louis, MO, USA 2009), pp. 2854–2860.
7. J.R. Bronson, J.S. Pulskamp, R.G. Polcawich, C.M. Kroninger, and E.D. Wetzel, PZT MEMS Actuated Flapping Wings for Insect-Inspired Robotics, in: 2009 IEEE 22nd Int. Conf. Micro Electro Mech. Syst., (IEEE, Sorrento, Italy 2009), pp. 1047–1050.
8. X.Q. Bao, T. Dargent, S. Grondel, J.B. Paquet, and E. Cattan, Microelectron. Eng. **88** 8, 2218–2224 (2011).
9. J.F.L. Goosen, H.J. Peters, Q. Wang, P. Tiso, and F. Van Keulen, Resonance Based Flapping Wing Micro Air Vehicle, in: Int. Micro Air Veh. Conf. Flight Compet., Toulouse, France 2013).
10. S. Ramananarivo, R. Godoy-Diana, and B. Thiria, Proc. Natl. Acad. Sci. U. S. A. **108** 15, 5964–9 (2011).
11. R.F. (Reginald F. Chapman, S.J. Simpson, and A.E. (Angela E. Douglas, *The insects : structure and function* n.d.).
12. A.R. ENNOS, J. Exp. Biol. **127** 1, (1987).
13. M.J. Brennan, S.J. Elliott, P. Bonello, and J.F. V Vincent, J. Theor. Biol. **224** 2, 205–213 (2003).
14. B. Tang and M.J. Brennan, J. Theor. Biol. **289**173–180 (2011).
15. R.L. Harne and K.W. Wang, J. R. Soc. Interface

12 104, 20141367 (2015).

16. A. Abolfathi, M.A. Changizi, and M.J. Brennan, J. Phys. Conf. Ser. **744** 1, (2016).
17. S.M. Kaplan, *Wiley electrical and electronics engineering dictionary* (IEEE Press, 2004).
18. A. Thomson and W. Thompson, Acta Biotheor. **26** 1, 19–29 (1977).
19. Y.-W. Chin, J.T.-W. Goh, and G.-K. Lau, Insect-inspired thoracic mechanism with non-linear stiffness for flapping-wing micro air vehicles, in: 2014 IEEE Int. Conf. Robot. Autom., (IEEE, 2014), pp. 3544–3549.

A computational study on the influence of the delayed yielding phenomenon in magnetorheological oils on the steady state vibration and forces transmitted between the rotor and its frame

Jaroslav Zapoměl[1,2,*], *Petr* Ferfecki[2,3], and *Jan* Kozánek[1]

[1]Institute of Thermomechanics, Department of Dynamics and Vibrations, Dolejškova 5, 182 00 Prague, Czech Republic
[2]VŠB - TU of Ostrava, Department of Applied Mechanics, 17. listopadu 15, 708 33 Ostrava, Czech Republic
[3]VŠB - TU of Ostrava, IT4Innovations National Supercomputing Center, 17. listopadu 15, 708 33 Ostrava, Czech Republic

Abstract. The theoretical analyses and practical experience show that only the damping effect adaptable to the current operating conditions makes it possible to achieve optimum performance of damping devices inserted in the supports of rotating machines. This is offered by magnetorheological squeeze film dampers. The magnetorheological oils are liquids sensitive to magnetic induction. Their response to the change of a magnetic field is not instantaneous, but it is a process called the delayed yielding. The research was focused on enhancement of the mathematical model of the magnetorheological squeeze film damper by considering the delayed yielding phenomenon and on its application for the study of the influence of the delayed yielding on the force transmission between the rotor and its stationary part. The results of the computational simulations show that rising value of the delayed yielding time constant that characterizes the delayed yielding process reduces the damping effect and efficiency of the magnetorheological damping devices.

1 Introduction

Unbalance is the main sources of excitation on vibration of rotors of rotating machines and of increase of forces transmitted between the rotor and its casing. A technological solution frequently used for suppressing these undesirable effects consists in inserting damping devices in the rotor supports. The analysis carried out in [1] shows that to achieve their high efficiency their damping effect must be adaptable to the operating speed. In the field of rotordynamics this is offered by squeeze film dampers lubricated by magnetorheological oils.

Wang et al. [2] used the experimental approach to study the vibration attenuation of a flexible rotor by means of a magnetorheological squeeze film damper. The design and work principles of magnetorheological dampers for rotordynamic applications are reported in [3-5]. The corresponding mathematical models applicable for computational simulations were developed and presented by Kim et al. [6], Zapoměl et al. [7-9], and other authors.

A great deal of effort of researchers was focused on the physical substance of the magnetorheological phenomenon, which consists in forming the inner structure in the flow of magnetorheological fluids [10].

The response of magnetorheological liquid materials on the change of a magnetic field is not instantaneous but it takes some time [11], which is known as the delayed yielding phenomenon. Zapoměl and Ferfecki extended mathematical models of the magnetorheological squeeze film dampers by taking this phenomenon into the consideration. [12-13] and performed the simulations.

The presented article deals with the study of the influence of the delayed yielding phenomenon on supressing amplitude of the time variable force transmitted between the rotor and its stationary part.

The principal contributions of the performed research consist in the description of the delayed yielding effect, in the development of a novel mathematical model of the magnetorheological squeeze film damper, and in learning more on behaviour of rotor systems equipped with controllable damping devices.

2 The damping device model

The damping effect of the magnetorheological squeeze film damper (Fig. 1) is produced by squeezing a thin layer of magnetorheological oil between two concentric rings. The magnetic flux generated in the coils embedded in the damper housing passes through the lubricant and as its resistance against the flow depends on magnetic induction, the change of the applied current can be used to change the damping force.

Magnetorheological oils are liquids composed of carrying fluid, tiny ferromagnetic particles, and of some additive materials. By application of a magnetic field the particles start to form a chain structure in the direction of the magnetic flux lines that are oriented in the damper gap in the radial direction. This effects the flow of the

* Corresponding author: jaroslav.zapomel@vsb.cz

magnetorheological oil and is the physical substance of occurrence of the yielding shear stress.

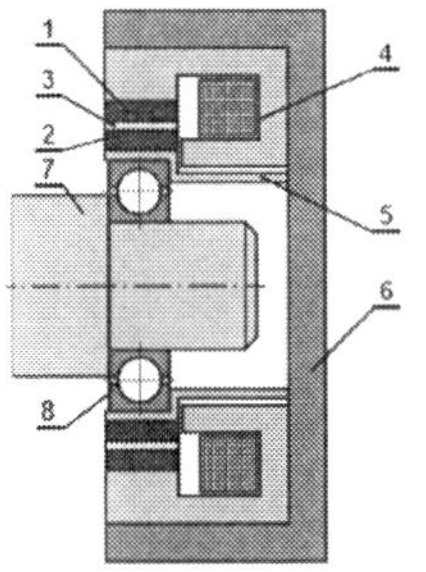

Fig. 1. The magnetorheological squeeze film damper.

The developed mathematical model of the magnetorheological damper is based on assumptions of the classical theory of lubrication [14, 15], only the magnetorheological oil is represented by bilinear material. In addition, it is assumed that the geometric and design parameters allow to consider the damper as short [14, 15].

Then the pressure distribution in the full oil film is governed by the modified Reynolds equation [9]

$$\frac{\partial}{\partial Z}\left(\frac{1}{\eta_C}h^3 p'\right) = 12\dot{h} \tag{1}$$

$$\frac{\partial}{\partial Z}\left[\frac{1}{\eta}\left(h^3 p' + 3h^2\tau_y + 8\frac{\tau_C^3}{p'^2} - 12\frac{\tau_y \tau_C^2}{p'^2}\right) - \frac{8}{\eta_C}\frac{\tau_C^3}{p'^2}\right] = 12\dot{h} \tag{2}$$

$$Z_C = -\frac{\tau_C h^2}{6\eta_C \dot{h}} \tag{3}$$

$$p'_C = -\frac{2\tau_C}{h} \tag{4}$$

p is the pressure, p' stands for the pressure gradient in the axial direction, h is the oil film thickness, Z is the axial coordinate defining position in the oil film, τ_y is the yielding shear stress, τ_C is the shear stress at the core border of bilinear materials, and η_C, η are the dynamic viscosities of the oil in and outside of the core region, respectively, p'_C denotes the pressure gradient in the axial direction at location given by the axial coordinate Z_C, and ($\dot{}$) reads for the first derivative with respect to time.

A cavitation is assumed in those parts of the damper gap where the thickness oil the film rises with time. In cavitated areas the pressure of the medium remains constant and equal to the pressure in the ambient space.

The damper housing is considered to be composed of a set of meridian segments and each segment as a divided core of an electromagnet with the gap filled with the magnetorheological oil. This makes it possible to express magnetic induction B in the oil film

$$B = k_B \mu_0 \mu_r \frac{I}{h} \tag{5}$$

k_B is the damper design parameter [16], μ_0 is the vacuum permeability, μ_r is the relative permeability of the magnetorheological oil, and I is the applied current.

The dependence of the stationary value of the yielding shear stress on magnetic induction can be approximated by a power function

$$\tau_y = k_y B^{n_y} \tag{6}$$

k_y and n_y are the material constants of the magnetorheological oil.

The components of the magnetorheological damping force acting on the rotor journal are given by integration of the oil pressure distribution around the damper circumference and along its length.

The physical substance of the yielding shear stress of magnetorheological fluids is a chain structure of ferromagnetic particles dispersed in the carrying liquid if a magnetic field is applied. Its forming or changing is a process that takes some time. It implies the yield shear stress depends not only on instantaneous magnitude of the magnetic induction but also on its history in the past. This is well described by a convolutory integral. Utilizing the exponential kernel and relation (6) the time history of the yielding shear stress accounting for the delayed yielding phenomenon reads

$$\tau_y = \frac{k_y}{T_y}\int_0^t e^{-\frac{1}{T_y}(t-\vartheta)} B^{n_y} \mathrm{d}\vartheta \tag{7}$$

T_y is the delayed yielding time constant characterizing rapidity of the process of forming the particle structure inside the magnetorheological oil layer and ϑ denotes the time in the past before the moment of time t.

Consequently, relation (7) is transformed to the differential form

$$T_y \dot{\tau}_y + \tau_y = k_y B^{n_y} \tag{8}$$

In areas, in which the cavitation takes place, it is assumed that the particle structure is completely disrupted and as a result, the yielding shear stress is considered to become zero there.

3 The investigated rotor system

The developed novel mathematical model of the magnetorheological squeeze film damper was applied for investigation of influence of the delayed yielding phenomenon on vibration attenuation and on the force transmission between the rotor and its stationary part during the steady state operating regime.

The studied rotor (Fig. 2) is rigid. At both its ends it is coupled with a rigid casing by two magnetorheological squeeze film dampers. The rotor rotates at constant angular speed and is loaded by its weight and by the disc unbalance. The deflection of the cage springs caused by the rotor weight is eliminated by their prestressing. The whole system can be considered as symmetric with respect to the middle plane of the disc.

The dampers are implemented in the computational model by springs and force couplings.

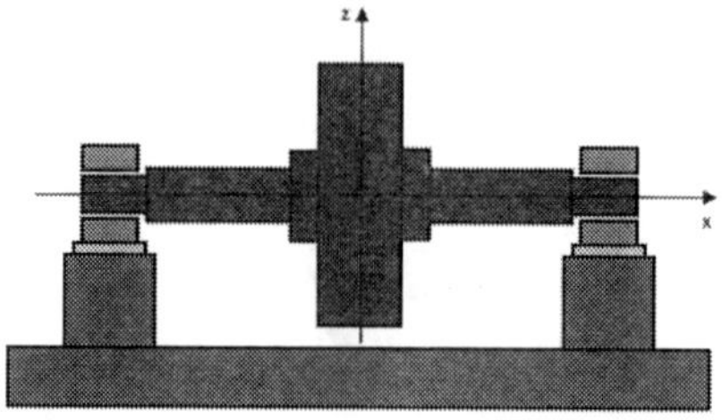

Fig. 2. The investigated rotor system.

Because of the system symmetry the lateral vibration of the rotor is governed by a set of two nonlinear motion equations

$$m_R\ddot{y} + b_P\dot{y} + 2k_D y = 2F_{mry} + m_R e_T \omega^2 \cos(\omega t + \psi_o) \quad (9)$$

$$m_R\ddot{z} + b_P\dot{z} + 2k_D z = 2F_{mrz} + m_R e_T \omega^2 \sin(\omega t + \psi_o) - m_R g + 2F_{PS} \quad (10)$$

m_R is the rotor mass, b_P is the coefficient of the rotor external damping, k_D is the bending stiffness of one cage spring, e_T is the eccentricity of the rotor centre of gravity, F_{mry}, F_{mrz} are the y and z components of the magnetorheological damping force acting on the rotor journal, F_{PS} is the prestressing force, ω is the angular speed of the rotor rotation, ψ_o is the phase shift, g is the gravity acceleration, y, z are the displacements of the rotor centre in the horizontal and vertical directions, and $(\ddot{\ })$ notates the second derivative with respect to time.

4 Results of the simulations

The technological and operating parameters of the system are: the rotor mass 450 kg, the bending stiffness of each damper cage spring 5.0 MN/m, the viscous damping coefficient (external damping) 10 Nm/s, the disc unbalance 45 kgmm, the oil viscosity (if not affected by a magnetic field) 0.3 Pas, the damper gap mean diameter 150 mm, the damper length 50 mm, the damper clearance 1.0 mm, the damper design parameter 60, the magnetorheological oil relative permeability 5, and the proportional and exponential constants 10000 PaT$^{-1.1}$, 1.1. The applied current is 0.2 A.

A simple dynamical analysis shows that the critical speed is 149.1 rad/s

Fig. 3 and 4 are related to the rotor angular speed of 100 rad/s, which is lower than the critical one. The figures show the steady state orbit of the rotor centre and the time history of the force transmitted between the rotor and the stationary part in the horizontal direction. The cage springs of both magnetorheological dampers are prestressed, which results in a circular shape of the orbit. It is evident that the application of a fluid with higher delayed yielding time constant arrives at increase of the size of the orbit and amplitude of the transmitted force (Fig. 4).

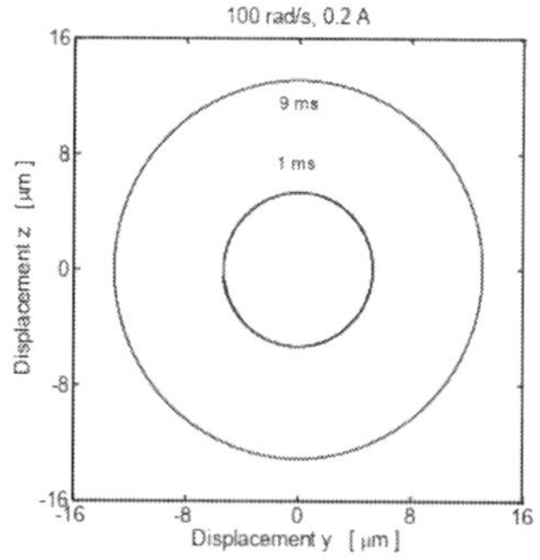

Fig. 3. Orbit of the rotor centre (100 rad/s).

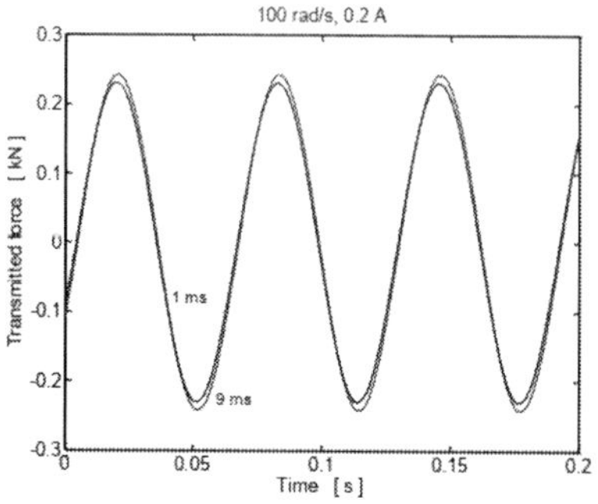

Fig. 4. Transmitted force (100 rad/s).

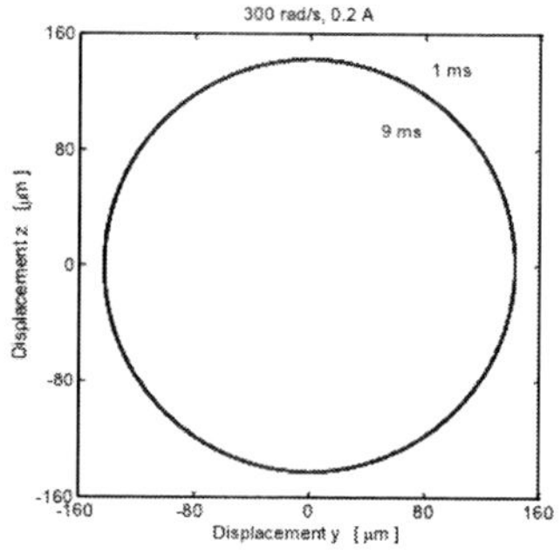

Fig. 5. Orbit of the rotor centre (300 rad/s).

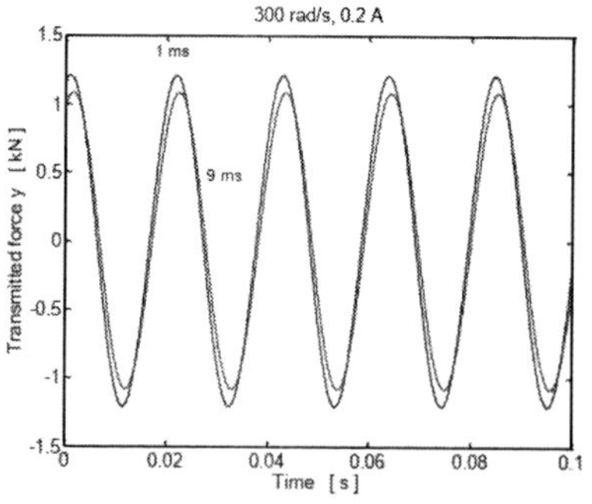

Fig. 6. Transmitted force (300 rad/s).

The orbit and the time history of the force transmitted to the rotor casing drawn in Fig. 5 and 6 are related to the operating regime when the rotor turns at angular speed of 300 rad/s, which is higher than the critical velocity. The results show that magnitude of the delayed yielding time constant has almost no influence on the orbit size but its lower value leads to increase of the transmitted force.

Fig. 7 and 8 show the time history of the yielding shear stress at a specified location in the damper gap for two magnitudes of the delayed yielding time constant. The rotor turns at angular speed of 300 rad/s. As evident, lower value of the time constant arrives at steeper increase of the yielding shear stress and at reaching its higher magnitude. During the periods when a cavitation takes place the yielding shear stress falls to zero.

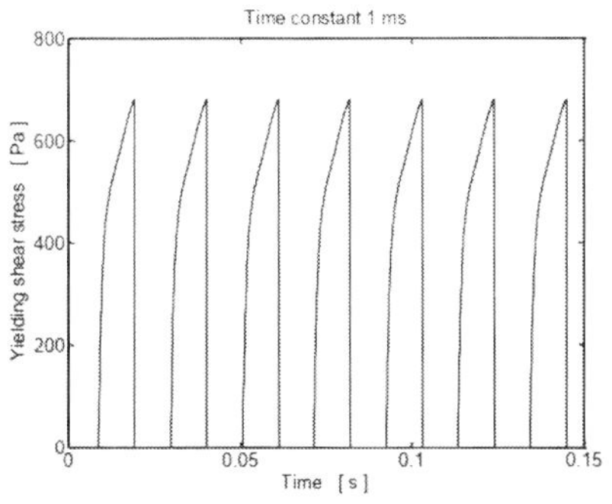

Fig. 7. Time history of the yielding shear stress (1 ms).

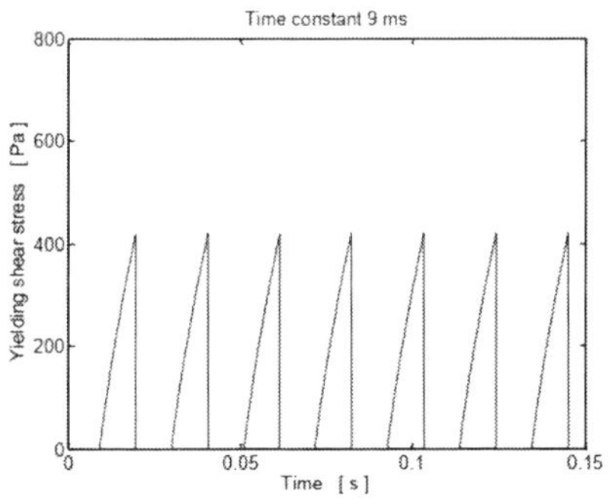

Fig. 8. Time history of the yielding shear stress (9 ms).

5 Conclusions

The research was aimed at investigation of the influence of the delayed yielding phenomenon on reducing the force transmitted from the rotor to its stationary part.

The results of the computational simulations show that lower value of the delayed yielding time constant leads to a smaller orbit and lower amplitude of the transmitted force for the velocity lower than the critical speed. For the velocity higher than the critical one lower value of the time constant increases the maximum transmitted force and has almost no influence of the orbit size. Such behaviour of the rotor shows that rising value of the delayed yielding time constant reduces the damping effect and efficiency of the damping device.

A more accurate description of behaviour of magnetorheological oils, the enhancement of the mathematical model of a short squeeze film damper by taking into consideration the delayed yielding phenomenon, and learning more on effect of magnetorheological dampers on oscillation of rotating machines are the main contributions of the carried out research.

The research work reported in this paper was made possible by the Czech Science Foundation (project no. 15-06621S) and the national Programme of Sustainability (project LQ1602 - IT4Innovations excellence in science). The support is highly acknowledged.

References

1. J. Zapoměl, P. Ferfecki, J. Kozánek, Appl. Comput. Mech., **7** (2013)
2. J. Wang, G. Meng, E. J. Hahn, *Proceedings of the 2003 ASME Design Engineering Technical Conference & Computers and Information in Engineering Conference*, 955-959 (2003)
3. J. Wang, G. Meng, Proc. Inst. Mech. Eng. Part L J. Mater. Des. Appl., **215** (2001)
4. X. Gong, X. Ruan, S. Xuan, Q. Yan, H. Deng, Adv. Mech. Eng., **6** (2014)
5. P. Forte, M. Paterno, E. Rustighi, Int. J. Rotat. Mach., **10** (2004)
6. K. J. Kim, C. W. Lee, J. H. Koo, Smart Mater. Struct., **17** (2008)
7. J. Zapoměl, P. Ferfecki, P. Forte, Smart Mater. Struct., **21** (2012)
8. J. Zapoměl, P. Ferfecki, Model. Opt. Phys. Syst., **9** (2010)
9. J. Zapoměl, P. Ferfecki, Appl. Math. Model., **52** (2017)
10. R. Tao, J. Phys. Condens. Matter, **13** (2001)
11. H. Li, X. Peng, W. Chen, J. Intell. Material Syst. Struct., **16** (2005)
12. J. Zapoměl, P. Ferfecki, *Proceedings of the 9th IFToMM International Conference on Rotor Dynamics*, 1021-1032 (2014)
13. J. Zapoměl, P. Ferfecki, *Proceedings of the 5th ECCOMAS Thematic Conference on Computational Methods in Structural Dynamics and Earthquake Engineering*, 1941-1948 (2015)
14. J. Zapoměl, *Computer Modelling of Lateral Vibration of Rotors Supported by Hydrodynamical Bearings and Squeeze Film Dampers*, VSB-Technical University of Ostrava, (2007, in Czech)
15. A. Z. Szeri, *Tribology: Friction, Lubrication, and Wear*, (1980)
16. P. Ferfecki, J. Zapoměl, J. Kozánek, Adv. Eng. Soft., **104** (2017)

Fluid nonlinearities effect on wake oscillator model performance

Victoria Kurushina[1,2,*] and *Ekaterina* Pavlovskaia[1,**]

[1] *Centre for Applied Dynamics Research, School of Engineering, University of Aberdeen, United Kingdom*
[2] *Department of Transport of Hydrocarbon Resources, Industrial University of Tyumen, Russia*

Abstract. Vortex-induced vibrations (VIV) need to be accounted for in the design of marine structures such as risers and umbilicals. If a resonance state of the slender structure develops due to its interaction with the surrounding fluid flow, the consequences can be severe resulting in the accelerated fatigue and structural damage. Wake oscillator models allow to estimate the fluid force acting on the structure without complex and time consuming CFD analysis of the fluid domain. However, contemporary models contain a number of empirical coefficients which are required to be tuned using experimental data. This is often left for the future work with the opened question on how to calibrate a model for a wide range of cases and find out what is working and is not.

The current research is focused on the problem of the best choice of the fluid nonlinearities for the base wake oscillator model [1] in order to improve the accuracy of prediction for the cases with mass ratios around 6.0. The paper investigates six nonlinear damping types for two fluid equations of the base model. The calibration is conducted using the data by Stappenbelt and Lalji [2] for 2 degrees-of-freedom rigid structure for mass ratio 6.54. The conducted analysis shows that predicted in-line and cross-flow displacements are more accurate if modelled separately using different damping types than using only one version of the model. The borders of application for each found option in terms of mass ratio are discussed in this work, and appropriate recommendations are provided.

1 Introduction

Vortex-induced vibrations (VIVs) is a form of fluid - structure interaction when fluctuations of the fluid forces due to growing vortices lead to increased fatigue and motion of the slender structure. Flexible structures in shear flows are in the main interest for industries, for example, subsea production systems, but the complex VIV phenomenon is still not explored fully for simpler cases of rigid structures with one and two degrees-of-freedom in the uniform flow.

Increased displacement amplitudes of a structure are observed during the so-called lock-in state (or synchronization) - condition of resonance between the fluid and the structure. Lock-in condition is achieved by matching the natural frequency of the structure and the frequency of vortex shedding.

The current work is performed using wake oscillator approach to VIV phenomenon. This method implies a very simplified representation of the fluid around the structure as fluctuations of lift and drag forces only. This allows to reduce computational time, but requires prior calibration and validation work in order to take empirical parameters correctly.

It was proved by Jauvtis and Williamson [3] in 2004 that development of the resonance state in terms of displacement amplitudes varies for rigid structures with 2 degrees-of-freedom depending on the mass ratio. The "border" between two types of lock-in peak was identified around mass ratio 6.0. This is why, the region of mass ratios 5.0-7.0 seems particularly important and challenging for modelling.

Previous work towards improving fluid oscillators for VIV problem included the study of different combinations of Van der Pol and Rayleigh equations [4]. The wider range of oscillators and the developed optimisation procedure are given in details in [5]. The focus of these previous works was on "low" mass ratio around 2.0-4.0, which left other cases for future exploration. In this study, the authors consider the most suitable oscillators and corresponding sets of coefficients for 2DOFs rigid structure with mass ratio around 6.0 with the focus on maximum possible accuracy of prediction for in-line and cross-flow displacement amplitudes. The basis for modification is the model presented in [1].

The current paper is structured as follows. Section 1 introduces the basic ideas that allow to narrow the aim of this study. Section 2 contains details of the improved model and the list of applied fluid nonlinearities of wake equations. Section 3 gives the summary of calibration procedure and presents three groups of calibration results. Section 4 observes advantages and drawbacks of the achieved precision of displacement prediction. And Section 5 gives concluding remarks and perspectives of the future studies.

*e-mail: v.kurushina@abdn.ac.uk
**e-mail: e.pavlovskaia@abdn.ac.uk

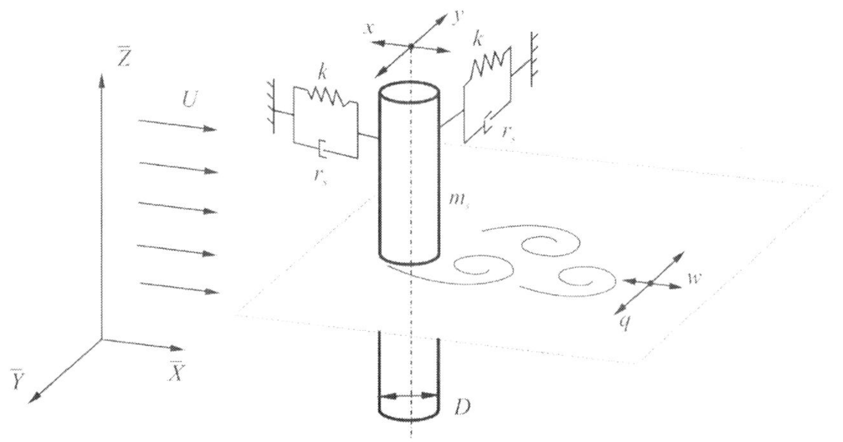

Figure 1. The model of two degrees-of-freedom elastically-supported rigid structure interacting with the uniform flow of velocity U [5]. Here, m_s is structural mass, r_s is structural damping, and K is stiffness of support.

2 Fluid nonlinearities

The considered case is 2 degrees-of-freedom rigid structure shown in Fig. 1, and the flow is assumed to be uniform. The structure can move in the direction of the flow (or in-line direction), and the direction perpendicular to the flow (or cross-flow direction). The base model for modifications [1] in dimensionless form is as follows:

$$\ddot{X} + 2\zeta\dot{X} + \omega_{st}^2 X = \frac{a}{2\pi St}\Omega_R^2 + \frac{b}{4\pi St}\Omega_R^2 w - 2a\Omega_R\dot{X} + \quad (1)$$
$$+ \frac{c}{2}\Omega_R q\dot{Y} + a\pi St\dot{Y}^2 + 2a\pi St\dot{X}^2 - b\Omega_R w\dot{X};$$

$$\ddot{Y} + 2\zeta\dot{Y} + \omega_{st}^2 Y = \frac{c\Omega_R^2}{4\pi St}q - a\Omega_R\dot{Y} + 2a\pi St\dot{X}\dot{Y} - \quad (2)$$
$$- \frac{b}{2}w\dot{Y}\Omega_R - cq\dot{X}\Omega_R;$$

$$\ddot{w} + f_{dampX} + 4\Omega_R^2 w = A_x\ddot{X}; \quad (3)$$

$$\ddot{q} + f_{dampY} + \Omega_R^2 q = A_y\ddot{Y}, \quad (4)$$

where

$$a = \frac{C_{D0}\rho_f D^2}{4\pi m_* St}; \quad b = \frac{C_{D0}^{fl}\rho_f D^2}{4\pi m_* St}; \quad c = \frac{C_{L0}\rho_f D^2}{4\pi m_* St}; \quad (5)$$

$$f_{dampX} = 2\varepsilon_x\Omega_R(w^2 - 1)\dot{w}; \quad (6)$$
$$f_{dampY} = \varepsilon_y\Omega_R(q^2 - 1)\dot{q}.$$

Here, X, Y are dimensionless structural displacements in in-line and cross-flow directions respectively, defined as $X = \frac{x}{D}$ and $Y = \frac{y}{D}$, where x and y are dimensional structural displacements, and D is diameter of the structure; w, q are fluctuating parts of drag and lift forces as wake coefficients; ω_{st} is natural frequency of the structure; Ω_R is vortex shedding frequency; St is Strouhal number; ζ is damping ratio; a, b, c are dimensionless complexes; C_{D0} is initial drag coefficient; C_{D0}^{fl} is initial fluctuating drag coefficient; C_{L0} is initial lift coefficient; m_* is mass per unit length which contains mass of the structure per unit length and the fluid added mass per unit length; ρ_f is fluid density; A_x, A_y are empirical coupling coefficients connecting wake oscillators with equations of structural motion; f_{dampX} and f_{dampY} are nonlinear damping terms of fluid equations; $\varepsilon_x, \varepsilon_y$ are damping coefficients (Van der Pol parameters).

Mass ratio μ is defined in this research as suggested in [2] as relationship of the structural mass per unit length m_s and the mass of displaced fluid per unit length m_f:

$$\mu = \frac{m_s}{m_f} = \frac{4m_s}{\pi\rho_f D^2}. \quad (7)$$

In this study, the accuracy of prediction with the Van der Pol damping terms (see Eq.(6)), applied in the model [1] for in-line and cross-flow wake equations, is compared with the accuracy provided by possible alternative fluid nonlinearities, as proposed in [5]:

- Modified Van der Pol terms $f_{dampX} = 2\varepsilon_{x1}\Omega_R\dot{w}w^2 - 2\varepsilon_{x2}\Omega_R\dot{w}$ and $f_{dampY} = \varepsilon_{y1}\Omega_R\dot{q}q^2 - \varepsilon_{y2}\Omega_R\dot{q}$;
- Modified Rayleigh terms $f_{dampX} = -2\varepsilon_{x1}\Omega_R\dot{w} + 2\frac{\varepsilon_{x2}}{\Omega_R}\dot{w}^3$ and $f_{dampY} = -\varepsilon_{y1}\Omega_R\dot{q} + \frac{\varepsilon_{y2}}{\Omega_R}\dot{q}^3$;
- Krenk-Nielsen terms $f_{dampX} = -2\varepsilon_{x1}\Omega_R\dot{w} + 2\varepsilon_{x2}\Omega_R\dot{w}w^2 + 2\frac{\varepsilon_{x3}}{\Omega_R}\dot{w}^3$ and $f_{dampY} = -\varepsilon_{y1}\Omega_R\dot{q} + \varepsilon_{y2}\Omega_R\dot{q}q^2 + \frac{\varepsilon_{y3}}{\Omega_R}\dot{q}^3$ suggested by Krenk and Nielsen [7];
- and Landl terms $f_{dampX} = 2\Omega_R\dot{w}(\varepsilon_{x1} - \varepsilon_{x2}w^2 + \varepsilon_{x3}w^4)$ and $f_{dampY} = \Omega_R\dot{q}(\varepsilon_{y1} - \varepsilon_{y2}q^2 + \varepsilon_{y3}q^4)$ based on the work by Landl [6].

The alternative models developed using these terms are calibrated with the experimental data [2] in the next section.

https://doi.org/10.1051/matecconf/201814804002

Table 1. Versions of the model calibrated for mass ratio 6.54 [2]

Modification	Set of coefficients
Prediction for both in-line and cross-flow displacement	
Rayleigh - Van der Pol	$C_{L0} = 1.83, C_{D0} = 1.73, C_{D0}^{fl} = 0.30, \varepsilon_x = 1.8598, \varepsilon_y = 0.3614, A_x = 12.33, A_y = 2.50, C_A = 0.67, K = 1.09$
Prediction for in-line displacement only	
Modified Rayleigh - Modified Van der Pol	$C_{L0} = 1.17, C_{D0} = 1.18, C_{D0}^{fl} = 0.64, \varepsilon_{x1} = 1.8976, \varepsilon_{x2} = 0.5211, \varepsilon_{y1} = 1.6508, \varepsilon_{y2} = 1.0386, A_x = 12.56, A_y = 3.12, C_A = 0.34, K = 3.90$
Modified Rayleigh - Modified Rayleigh	$C_{L0} = 0.21, C_{D0} = 1.46, C_{D0}^{fl} = 0.87, \varepsilon_{x1} = 1.9601, \varepsilon_{x2} = 0.5841, \varepsilon_{y1} = 0.6688, \varepsilon_{y2} = 1.3768, A_x = 12.10, A_y = 2.09, C_A = 1.04, K = 3.61$
Modified Van der Pol - Krenk-Nielsen	$C_{L0} = 1.14, C_{D0} = 1.62, C_{D0}^{fl} = 0.19, \varepsilon_{x1} = 0.0242, \varepsilon_{x2} = 0.5452, \varepsilon_{y1} = 1.9493, \varepsilon_{y2} = 0.0214, \varepsilon_{y3} = 0.2491, A_x = 1.02, A_y = 3.2, C_A = 1.96, K = 0.17$
Landl - Modified Rayleigh	$C_{L0} = 0.26, C_{D0} = 0.58, C_{D0}^{fl} = 0.64, \varepsilon_{x1} = 0.8290, \varepsilon_{x2} = 0.8493, \varepsilon_{x3} = 0.8411, \varepsilon_{y1} = 1.4904, \varepsilon_{y2} = 0.0192, A_x = 5.05, A_y = 5.02, C_A = 1.00, K = 1.09$
Prediction for cross-flow displacement only	
Landl - Van der Pol	$C_{L0} = 0.38, C_{D0} = 0.37, C_{D0}^{fl} = 0.51, \varepsilon_{x1} = 0.8519, \varepsilon_{x2} = 0.7962, \varepsilon_{x3} = 0.9258, \varepsilon_y = 0.3479, A_x = 6.36, A_y = 4.49, C_A = 1.03, K = 1.80$
Landl - Landl	$C_{L0} = 1.03, C_{D0} = 0.59, C_{D0}^{fl} = 0.02, \varepsilon_{x1} = 0.3058, \varepsilon_{x2} = 0.5479, \varepsilon_{x3} = 0.0558, \varepsilon_{y1} = 0.0009, \varepsilon_{y2} = 0.1228, \varepsilon_{y3} = 0.3494, A_x = 4.72, A_y = 1.62, C_A = 1.97, K = 1.7$
Landl - Modified Van der Pol	$C_{L0} = 0.33, C_{D0} = 0.31, C_{D0}^{fl} = 0.47, \varepsilon_{x1} = 0.7922, \varepsilon_{x2} = 0.8399, \varepsilon_{x3} = 0.8535, \varepsilon_{y1} = 0.5322, \varepsilon_{y2} = 0.6653, A_x = 9.81, A_y = 5.77, C_A = 1.19, K = 1.51$
Landl - Krenk-Nielsen	$C_{L0} = 0.63, C_{D0} = 0.51, C_{D0}^{fl} = 0.34, \varepsilon_{x1} = 1.2909, \varepsilon_{x2} = 1.4159, \varepsilon_{x3} = 1.0042, \varepsilon_{y1} = 0.2581, \varepsilon_{y2} = 0.2273, \varepsilon_{y3} = 0.2260, A_x = 5.39, A_y = 4.38, C_A = 1.00, K = 1.59$

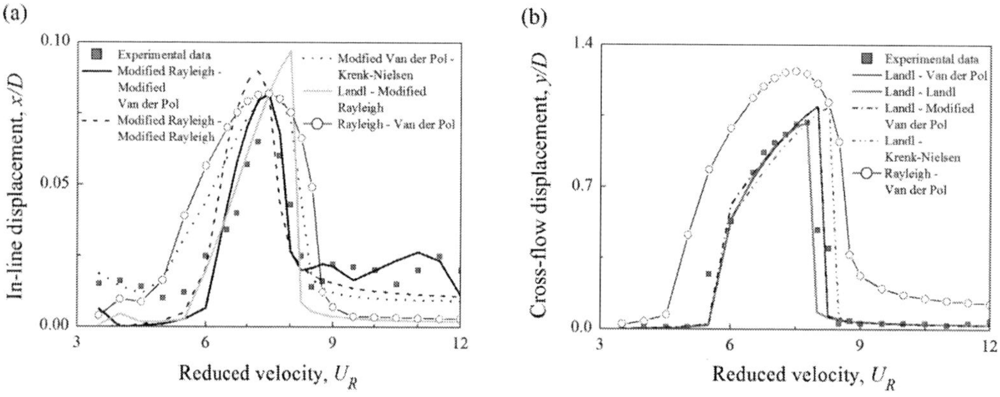

Figure 2. Displacement amplitudes generated by the model versions in comparison with experimental data [2]: a) calibrated in-line responses of considered models for mass ratio 6.54; b) generated cross-flow response of the Rayleigh - Van der Pol system in comparison with calibrated cross-flow responses of other models for mass ratio 6.54.

3 Calibration results

Calibration of the dimensionless coefficients for the aformentioned wake oscillator models is performed following the algorithm in [5]. The main priority during calibration is given to the highest displacement amplitudes developed during the resonance. Experimental data [2] are for mass ratio 6.54. The highest displacement amplitudes here are approximately 0.082 for in-line direction and 1.02 for cross-flow direction. The highest amplitudes correspond to the reduced velocity 7.5 and 7.75 respectively.

The models are calibrated in respect to either in-line or cross-flow data, and the standard deviation is applied as the statistics of the signal. Eleven control points of experimental data [2] in either in-line or cross-flow direction are utilized to estimate the error of the model prediction. Calibration is conducted using the constrained nonlinear minimization tool in Matlab as detailed in [5].

The calibrated sets for medium mass ratio are presented in Table 1 in three groups. The first part of Table 1 is the combination of Rayleigh - Van der Pol dampings which allows relatively suitable prediction for both in-line and cross-flow displacement amplitudes at the same time as illustrated in Fig. 2. In this case, Fig. 2a demonstrates the calibration result for in-line direction, and Fig. 2b, Figs 3 and 4 show the results of validation on the same exper-

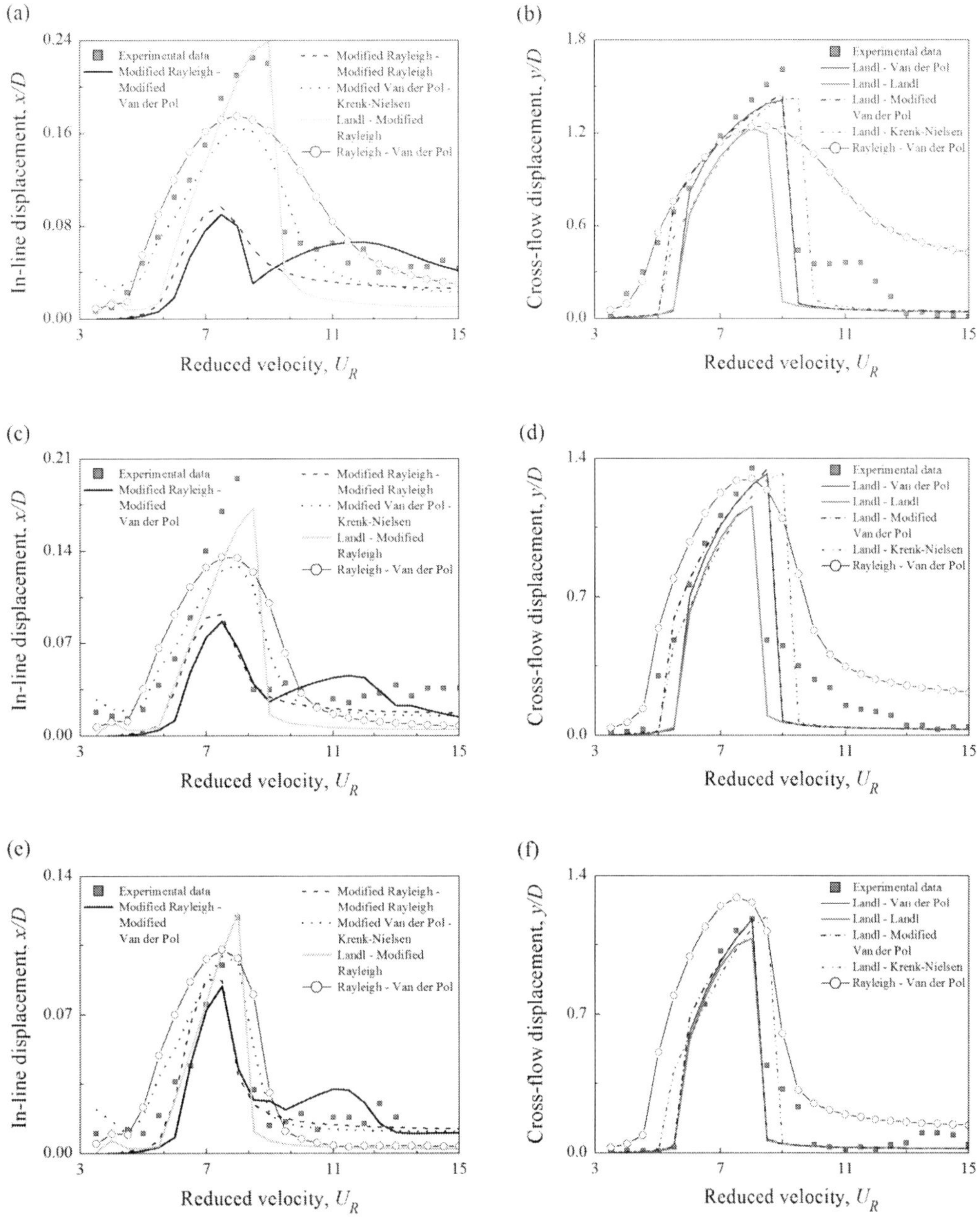

Figure 3. Displacement amplitudes generated by the model versions in comparison with experimental data [2]: a) in-line response for mass ratio 2.36; b) cross-flow response for mass ratio 2.36; c) in-line response for mass ratio 3.68; d) cross-flow response for mass ratio 3.68; e) in-line response for mass ratio 5.19; f) cross-flow response for mass ratio 5.19.

imental set-up [2]. Hence, Figs 2a, 3a, 3c, 3e, 4a, and 4c show the comparison of fit by the models calibrated with in-line displacement record for mass ratio 6.54; and Figs 2b, 3b, 3d, 3f, 4b, and 4d provide the comparison of the combination of Rayleigh - Van der Pol dampings (calibrated with the in-line displacement record) with the options calibrated with cross-flow displacement record for mass ratio 6.54.

The second part of Table 1 present options of the fluid nonlinearities which are able to predict in-line displacement only when calibrated with mass ratio 6.54 [2], and they are shown in Figs 2a, 3a, 3c, 3e, 4a, and 4c for the range of mass ratios. Those are options of Modified

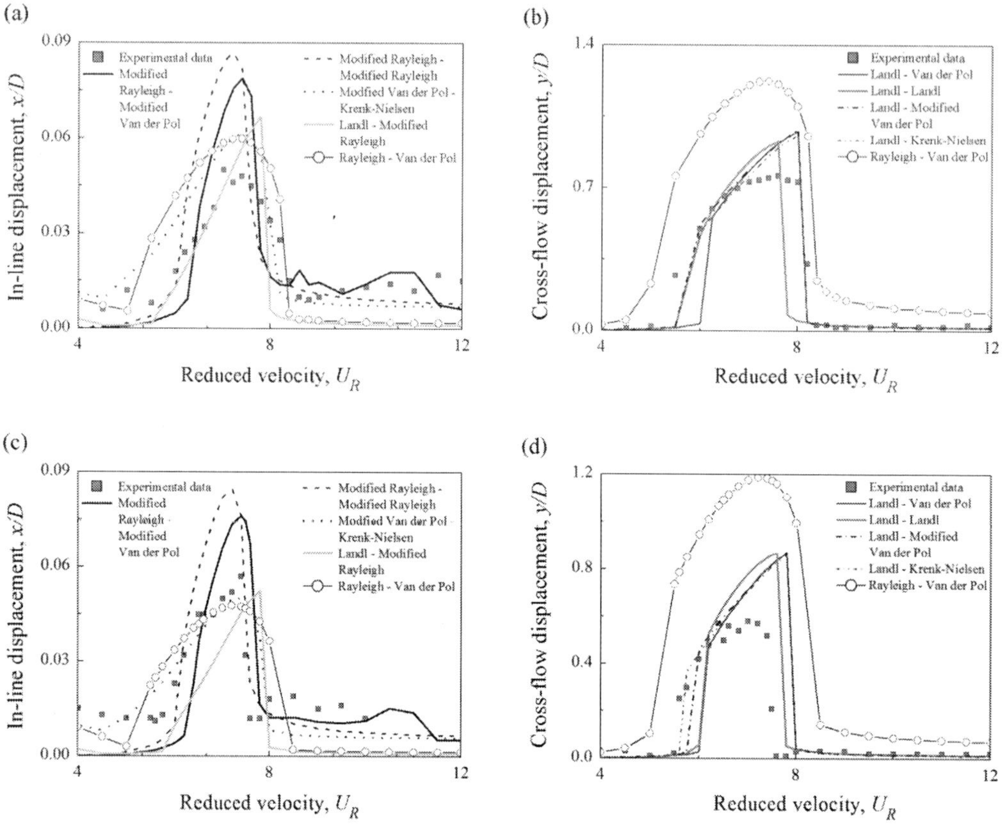

Figure 4. Displacement amplitudes generated by the model versions in comparison with experimental data [2]: a) in-line response for mass ratio 8.76; b) cross-flow response for mass ratio 8.76; c) in-line response for mass ratio 10.63; d) cross-flow response for mass ratio 10.63.

Rayleigh - Modified Van der Pol, Modified Rayleigh - Modified Rayleigh, Modified Van der Pol - Krenk-Nielsen, and Landl - Modified Rayleigh fluid nonlinearities. Fig. 2a demonstrates the result of calibration with the in-line displacement record for mass ratio 6.54, and Figs 3a, 3c, 3e, 4a, 4c are the results of validation on the same experimental set-up [2].

The third block of Table 1 contains the fluid nonlinearities to forecast cross-flow displacement only when calibrated with mass ratio 6.54 [2]: Landl - Van der Pol, Landl - Landl, Landl - Modified Van der Pol, and Landl - Krenk-Nielsen fluid nonlinearities. Accuracy of their fit is illustrated in Figs 2b, 3b, 3d, 3f, 4b, and 4d for the range of mass ratios. Fig. 2b is the result of calibration with mass ratio 6.54; and Figs 3b, 3d, 3f, 4b, and 4d are the results of validation on the same experimental set-up [2].

4 Discussion

Fig. 2a demonstrates the advantageous fit provided by fluid nonlinearities Modified Rayleigh - Modified Van der Pol calibrated with the in-line displacement amplitudes at mass ratio 6.54 [2]. It manifests in the correct position of the resonance peak, the correct highest displacement amplitude, and the best representation of the lower branch among considered options. Figs 3a, 3c, 3e, 4a, and 4c, however, illustrate that the best quality of the fit (in-line peak height and width) throughout the range of mass ratio from 2.0 to 9.0 belongs to the combination of Landl - Modified Rayleigh dampings.

Figs 2a, 3e and 4a prove that the range of the correct predictions by options with Modified Rayleigh - Modified Van der Pol and Modified Rayleigh - Modified Rayleigh dampings is very narrow - around mass ratio 6.0-7.0. The combination of nonlinearities Modified Rayleigh - Modified Van der Pol can be seen as a more interesting due to providing relatively safe versions of lower branch at different mass ratios.

At the same time, options with Rayleigh - Van der Pol and Modified Van der Pol - Krenk-Nielsen dampings, as appears in Figs 2a, 3e, 4a, and 4c, provide a suitable, safe fit for in-line displacement amplitudes over the mass ratio range approximately from 6.0 to 10.0.

Fig. 2b shows very accurate predictions by Landl - Van der Pol, Landl - Landl, Landl - Modified Van der Pol dampings of the cross-flow peak at mass ratio 6.54. Variation of mass ratio in Figs 2b, 3b, 3d, 3f, 4b, and 4d reveals the advantageous prediction by Landl - Van der Pol, Landl - Modified Van der Pol, and also Landl - Krenk-Nielsen dampings from approximately mass ratio 3.7 to mass ratio 9.0. Rayleigh - Van der Pol damping (from the first part of Table 1) significantly overestimates the cross-flow displacement amplitudes when mass ratio exceeds 6.0. Landl - Landl fluid nonlinearities with identified set of coefficients tend to predict cross-flow peak to be narrower than it appears in experiments [2].

5 Conclusions

The considered modifications of the base model [1] show that predicted in-line and cross-flow displacements are more accurate if modelled separately using different damping types than using only one version of the model, as follows from the analysis of Figs 2, 3 and 4.

Option of Landl - Modified Rayleigh damping calibrated with the in-line displacement amplitudes allows to predict in-line displacement in the largest range of mass ratio: approximately from 2.0 to 9.0, as illustrated in Figs 2a, 3a, 3c, 3e, 4a, and 4c.

Options of Landl - Van der Pol, Landl - Modified Van der Pol, Landl - Krenk-Nielsen dampings calibrated with cross-flow displacement amplitudes are proved reliable to predict cross-flow displacement only in the range of mass ratio from approximately 3.7 until 9.0. It is substantiated by Figs 2b, 3b, 3d, 3f, 4b, and 4d.

Option of Rayleigh - Van der Pol damping calibrated with the in-line displacement amplitudes predicts in-line displacement safe and relatively correctly for the range of mass ratio from approximately 6.0 to 10.0, and cross-flow displacement - from 3.7 to 6.0, as revealed by Figs 2, 3 and 4.

Analysis of Table 1 allows to suggest that Landl damping in in-line fluid equation enables relatively accurate prediction of cross-flow displacement amplitudes. This makes the combination of Landl - Modified Rayleigh fluid nonlinearities the most perspective option for the future development. Hence, the next stages of this research should focus on the synthetic oscillators specifically designed for the models of 2 degrees-of-freedom structures, and definitely involve more experimental data for calibration and validation.

Acknowledgements

V.K. would like to acknowledge the support of the Industrial University of Tyumen, Tyumen, Russia, and the State Program "Global Education", Russia.

This work was performed using the Maxwell High Performance Computing Cluster funded by the University of Aberdeen.

Authors also would like to express their gratitude to Dr. Andrew Starkey for advices regarding optimization procedures and to Naveed Khan for technical advices on accelerating computations.

References

[1] Postnikov A., Pavlovskaia E., Wiercigroch M. 2DOF CFD calibrated wake oscillator model to investigate vortex-induced vibrations, International Journal of Mechanical Sciences **127**, 176–190 (2017).

[2] Stappenbelt B., Lalji F. Vortex-induced vibration super-upper response branch boundaries, International Journal of Offshore and Polar Engineering **2(18)**, 99–105 (2008).

[3] Jauvtis N., Williamson C.H.K. The effect of two degrees-of-freedom on vortex-induced vibration at low mass and damping, Journal of Fluid Mechanics **509**, 23–62 (2008).

[4] Kurushina V., Pavlovskaia E. Wake oscillator equations in modelling vortex-induced vibrations at low mass ratio, OCEANS 2017 - Aberdeen, 1–6 (2017). IEEE. DOI: 10.1109/OCEANSE.2017.8084644

[5] Kurushina V., Pavlovskaia E., Postnikov A., Wiercigroch M. Calibration and comparison of VIV wake oscillator models for low mass ratio structures. Submitted for publication.

[6] Landl R. A mathematical model for vortex-excited vibrations of bluff bodies, Journal of Sound and Vibration **2(42)**, 219–234 (1975).

[7] Krenk S., Nielsen S.R.K. Energy balanced double oscillator model for vortex-induced vibrations, Journal of Engineering Mechanics **3(125)**, 263–271 (1999).

Dynamic behaviour of rotors supported by fluid-film bearings operated close to fluid-induced instability

Luboš Smolík[1,*], *Miroslav* Byrtus[1,**], *Štěpán* Dyk[1,***], and *Michal* Hajžman[1,****]

[1] *Department of Mechanics, University of West Bohemia, Czech Republic*

Abstract. This paper is focused on an analysis of rotating systems with fluid-film bearings, especially on their nonlinear behaviour in the course of developing fluid-induced instability. The studied system consists of Jeffcott rotor supported by a fluid-film bearing characterised by the Reynolds equation. The steady state response of the system is investigated by means of an approximate analytical solution of the Reynolds equation while the transient response of the system is investigated using a complex numerical solution. Results suggest that the rotor exercise a bounded chaotic motion if it becomes unstable. If the fluid-induced instability further develops, the motion actually becomes less chaotic and can be characterised as quasi-periodic.

1 Introduction

Fluid film bearings are widely used for supporting of various rotating machines because of their favourable properties, specifically low friction and wear and vibration-reducing capabilities. However, it has been known for more than 90 years that a fluid-induced instability can develop in a poorly designed bearing [1]. The quality of the bearing design can be expressed by Sommerfeld number $S_0 = {}^{\bar{p}\psi^2}/_{\eta\omega_0}$, where $\bar{p}$, ψ are a rated load relative clearance of the bearing, η is a dynamic viscosity of a fluid film and ω_0 is a rotor speed [2]. In general, bearings with low S_0 – i.e. not enough loaded bearings, bearings with high-speed rotors or bearings with highly viscous films – are prone to the the fluid-induced instability.

The instability, also known as oil whirl, was mathematically described at the turn of the 50s and 60s [3, 4]. Both authors also expressed a stability criterion for a linearised rotor-bearing system which is based on the Routh-Hurwitz criterion. The occurrence of the fluid-induced instability and the processes accompanying it were studied experimentally in the early 80s [5–7].

The oil whirl starts when hydrodynamical (HD) forces in the fluid film become higher than loading forces, which are predominantly caused by static loads. When this occurs, the HD forces start to push a journal – a part of a shaft, which rests in a bearing – in a forward circular motion around the bearing. The journal runs about the bearing with a period which in general lasts slightly longer than two revolutions. Therefore the oil whirl can be easily recognized by frequency which is 0.42–0.49 to journal speed [6].

*e-mail: carlist@kme.zcu.cz
**e-mail: mbyrtus@kme.zcu.cz
***e-mail: sdyk@kme.zcu.cz
****e-mail: mhajzman@kme.zcu.cz

HD forces are determined by a pressure distribution of the fluid film, which is governed by a partial differential equation known as the Reynolds equation (RE) [8]. The exact analytical solution of RE has not been known until 2012 [9], however, approximate analytical solutions for infinitely short [10, 11] and infinitely long bearings [11] have been known since the 50's and later many researchers used perturbation methods in order to obtain approximate analytical solutions of full RE [12, 13]. As the computing power of personal computers had been rising, robust numerical methods including finite differential [14], finite element [14, 15], finite volume [16] and element-based finite volume methods [17] were also employed.

Although all mentioned methods capture the phenomenon of oil whirl, there are relatively small number of studies which focus on nonlinear behaviours of rotors operated close to the instability threshold speed. A transient state which occurs in rotor-bearing system when a journal pass the threshold speed is studied in [13], where small amplitude perturbations are assumed in the bearing. The transient state of Jeffcott rotor supported by a gas bearing and accompanying bifurcations are studied in [18].

The main focus of this work is to perform an in-depth analysis of the transient state for Jeffcott rotor supported by journal bearing with consideration of all terms which are presented in RE. This may lead to better understanding of the oil whirl instability, and hence to understand how to design rotor systems which are operated in the unstable regime.

The rest of the article is outlined as follows. Section 2 presents a simple model on which computations of steady-state response were performed. Section 3 contains a brief description of an extended model for transient analysis. Model and simulation parameters, and results are discussed in sections 4 and 5, respectively. Finally, section 6 draws some conclusions.

2 Model for steady-state analysis

A general mathematical model of a simple rigid rotor which rotates at a constant angular velocity and is supported by a generic journal bearing is described in this section.

2.1 Equation of motion for a rigid rotor

We assume the *rigid rotor* of mass m which rotates at a constant angular velocity ω_0 and is subject to constant gravitational and out-of-balance forces $\mathbf{f}_g$ and $\mathbf{f}_u$ and to a vector of hydrodynamic forces $\mathbf{f}_{hd}$ (see fig. 1). The equilibrium of acting and inertia forces can be written as [2]

$$\mathbf{M}\,\ddot{\mathbf{q}} = \mathbf{f}_g + \mathbf{f}_u + \mathbf{f}_{hd}, \tag{1}$$

where $\mathbf{M}$ is a mass matrix and $\ddot{\mathbf{q}}(t)$ is an acceleration vector. If we further assume that the rotor does not perform yaw and pitch movements then its motion is given by horizontal displacement and vertical displacement $x = x(t)$ and $y = y(t)$ and Eq. (1) can be written in the form

$$\begin{bmatrix} m & 0 \\ 0 & m \end{bmatrix}\begin{bmatrix} \ddot{x} \\ \ddot{y} \end{bmatrix} = \begin{bmatrix} 0 \\ -m\,g \end{bmatrix} + U_{st}\,\omega_0^2 \begin{bmatrix} \cos\omega_0 t \\ \sin\omega_0 t \end{bmatrix} + \begin{bmatrix} F_x^{hd} \\ F_y^{hd} \end{bmatrix}, \tag{2}$$

where g is the gravitational constant, U_{st} is the static unbalance and F_x^{hd} and F_y^{hd} are horizontal and vertical components of vector $\mathbf{f}_{hd}$ [2].

Alternatively, the motion of the rotor can be characterised by eccentricity $e = e(t)$ and angle to the eccentricity $\phi = \phi(t)$, which are related to x and y as follows

$$e = \sqrt{x^2 + y^2}, \tag{3}$$

$$\phi = \begin{cases} \arctan\left(\frac{y}{x}\right) & x > 0 \wedge y > 0, \\ \arctan\left(\frac{y}{x}\right) + \pi & x < 0, \\ \arctan\left(\frac{y}{x}\right) + 2\pi & x > 0 \wedge y < 0. \end{cases} \tag{4}$$

Time derivatives $\dot{e}$ and $\dot{\phi}$ of e and ϕ are obviously

$$\dot{e} = \frac{x\,\dot{x} + y\,\dot{y}}{\sqrt{x^2 + y^2}}, \tag{5}$$

$$\dot{\phi} = \frac{x\,\dot{y} - y\,\dot{x}}{x^2 + y^2}. \tag{6}$$

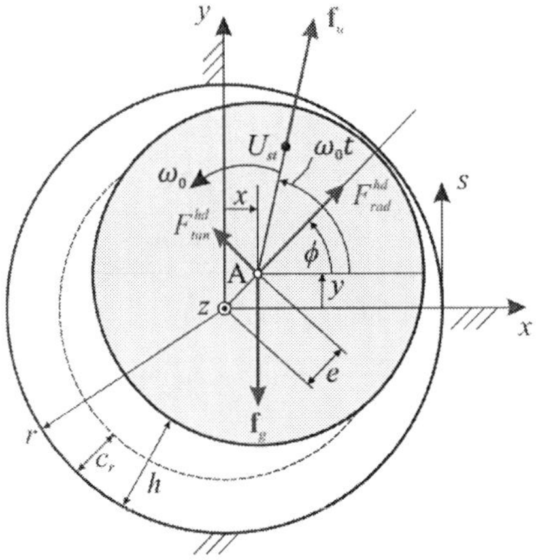

Figure 1. A rigid rotor supported by a journal bearing.

2.2 Hydrodynamic forces in a journal bearing

Hydrodynamic forces are obtained by integrating hydrodynamic pressure in an oil film over a surface of the bearing. Hydrodynamic pressure $p = p(s, z, t)$ is given by the Reynolds equation [8]

$$\frac{\partial}{\partial s}\left(\frac{h^3}{\mu}\frac{\partial p}{\partial s}\right) + \frac{\partial}{\partial z}\left(\frac{h^3}{\mu}\frac{\partial p}{\partial z}\right) = 6\,\omega_0\,r\,\frac{\partial h}{\partial s} + 12\,\frac{\partial h}{\partial t}, \tag{7}$$

where s and z are circumferential and axial coordinates, respectively, $h = h(s, z, t)$ is a gap between the journal and the bearing and μ is dynamic viscosity of the oil film.

The Reynolds equation (7) can be solved under various assumptions. Let us assume the half Sommerfeld conditions [10] and consider the following relations

$$h = c_r - e\,\cos\left(\frac{s}{r} - \phi\right) \quad \text{and} \quad l \to 0, \tag{8}$$

where l is the bearing length. Then radial and tangential components of hydrodynamic forces $F_{rad}^{hd,sb}$ and $F_{tan}^{hd,sb}$ shown in Fig. 1 can be expressed as [11]

$$F_{rad}^{hd,sb} = -\frac{\mu\,r\,l^3}{c_r^2}\left[\frac{|\omega_0 - 2\,\dot{\phi}|\,\varepsilon^2}{(1-\varepsilon^2)^2} + \frac{\pi\left(1 + 2\,\varepsilon^2\right)\dot{\varepsilon}}{2\,(1-\varepsilon^2)^{2.5}}\right], \tag{9}$$

$$F_{tan}^{hd,sb} = \frac{\mu\,r\,l^3}{c_r^2}\left[\frac{\left(\omega_0 - 2\,\dot{\phi}\right)\pi\,\varepsilon}{4\,(1-\varepsilon^2)^{1.5}} + \frac{2\,\varepsilon\,\dot{\varepsilon}}{(1-\varepsilon^2)^2}\right], \tag{10}$$

where $\varepsilon = {}^{e}/_{c_r}$ is the relative eccentricity. Equations (9) and (10) approximate the analytical solution of Reynolds equation (7) reasonably accurately if $l < r$ holds [11]. If l is from interval $\langle r, 4\,r\rangle$, the components of hydrodynamic forces have to be corrected. Bastani proposed [11] corrective functions in the form

$$F_{rad}^{hd} = \left(f_1\,\varepsilon^3 + f_2\,\varepsilon^2 + f_3\,\varepsilon + f_4\right)F_{rad}^{hd,sb}, \tag{11}$$

$$F_{tan}^{hd} = \left(g_1\,\varepsilon^2 + g_2\,\varepsilon + g_3\right)F_{tan}^{hd,sb}, \tag{12}$$

where f_i, g_i are cubic or quartic polynomial functions which are dependant on ${}^{l}/_{r}$ and are expressed in [11].

Hydrodynamic forces F_{rad}^{hd} and F_{tan}^{hd} are derived in a rotated coordinate system shown in fig. 1 and can be transformed to a fixed coordinate system from equation of motion (2) using relations

$$F_x^{hd} = F_{rad}^{hd}\,\cos\phi - F_{tan}^{hd}\,\sin\phi, \tag{13}$$

$$F_y^{hd} = F_{rad}^{hd}\,\sin\phi + F_{tan}^{hd}\,\cos\phi. \tag{14}$$

2.3 Solution strategy

Equation of motion (2) can be formally rewritten as

$$\mathbf{M}\,\ddot{\mathbf{q}} = \mathbf{f}, \tag{15}$$

where $\mathbf{f} = \mathbf{f}(\dot{\mathbf{q}}, \mathbf{q}, t)$ is a vector of all acting forces. Eq. (15) can be further transformed to the state-space; the resulting equation is then

$$\dot{\mathbf{u}} = \begin{bmatrix} \dot{\mathbf{q}} \\ \mathbf{M}^{-1}\,\mathbf{f} \end{bmatrix}, \quad \text{where} \quad \dot{\mathbf{u}} = \begin{bmatrix} \dot{\mathbf{q}} \\ \ddot{\mathbf{q}} \end{bmatrix}. \tag{16}$$

The equation of motion in this form can be integrated numerically by means of the Runge-Kutta method.

3 Model for transient analysis

An extension to the previous model allowing a transient analysis for $\omega_0 \neq$ const. is described in this section.

3.1 Equation of motion

Let us assume that the angular speed of the rigid rotor from section 2.1 is time-dependant variable $\dot{\varphi} = \dot{\varphi}(t)$. Since we want to be able to control $\dot{\varphi}$, we introduce a controller and couple the controlle and the rotor with a rotational spring as shown in Fig. 2. The rotation of the rotor in such configuration is given by the equation

$$I\,\ddot{\varphi} + d_r\,(\dot{\varphi} - \Omega) + k_r\,(\varphi - \Phi) = M_{fric}, \qquad (17)$$

where I is a moment of inertia of the rotor to the axis of rotation, d_r, k_r are rotational damping and stiffness of the rotational spring, respectively, $\Phi = \Phi(t)$ and $\Omega = \Omega(t)$ are arbitrary functions of controller's angular displacement and angular velocity and M_{fric} is a frictional moment [2].

Since frictional moment M_{fric} is of no concern because we do not analyse losses due friction nor thermodynamics of the oil film, M_{fric} is further neglected. With respect to the fact that angular velocity of the rotor $\dot{\varphi}$ is not constant, equations of motion are

$$m\,\ddot{y} = U_{st}\,\dot{\varphi}^2\,\cos\varphi + {}^{FEM}F_y^{hd}, \qquad (18)$$

$$m\,\ddot{z} = -m\,g + U\,\dot{\varphi}^2\,\sin\varphi + {}^{FEM}F_z^{hd}, \qquad (19)$$

$$I\,\ddot{\varphi} + d_r\,\dot{\varphi} + k_r\,\varphi = d_r\,\Omega + k_r\,\Phi, \qquad (20)$$

where the components of hydrodynamic forces ${}^{FEM}F_x^{hd}$ and ${}^{FEM}F_y^{hd}$ are obtained from finite-element-based numerical solution of the Reynolds equation [15].

In the first step of the solution process, hydrodynamic pressure p is obtained from Reynolds equation (7) under the assumptions tha i) the bearing is fully-flooded, ii) there is constant ambient pressure p_a at edges of the bearing, and c) if hydrodynamic pressure p in any node of FE mesh drops below the predefined value p_s, then $p = p_s$ is set in that particular node.

In the second step, p is integrated over the bearing surface. This integral yields ${}^{FEM}F_x^{hd}$ and ${}^{FEM}F_y^{hd}$ [9].

Note that the approximate analytical solution of Eq. (7) which is described in section 2.2 is derived under the assumption that the angular speed of the journal is constant and therefore cannot be used for transient analysis.

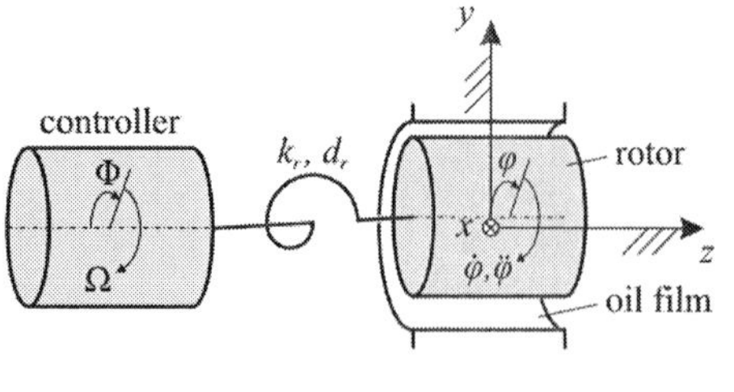

Figure 2. A rigid rotor coupled to a controller.

3.2 Solution strategy

The system of eqs. (18) has to be transformed to the state-space in order to be integrable by standard numerical solvers. The transformation was briefly described in section 2.3.

4 Model and simulation parameters

Parameters of the system and their respective symbols for both steady-state and transient analysis can be found in tab. 1. The parameters are chosen so that the fluid-induced instability arises at ca. 2300 rpm.

The rotor was considered perfectly balanced in the case of the steady-state analysis. The simulations were performed in the speed interval $n \in \langle 2000, 2900\rangle$ with the step 5 rpm. Time integration was performed in the interval $t \in \langle 0, 1.5\rangle$ s with a variable time step. The results were then transformed to time series with the constant time step 10^{-4} s. Only the time interval $t \in \langle 0.5, 1.5\rangle$ s was subjected to subsequent post-processing.

The transient analysis was performed in the speed interval $n \in \langle 800, 4000\rangle$ rpm with several values of constant angular acceleration and static unbalance U_{st}. The selected values of U_{st} correspond with balance quality grades G0, G6.3 and G16 per ISO 1940. Time integration was performed with a variable time step $\Delta t \in \langle 5\times10^{-7}, 5\times10^{-5}\rangle$ s and the results were then transformed to time series with the constant time step 10^{-4} s.

Table 1. Parameters of analysed system. Items denoted with asterisk symbol (*) are used only during transient analysis.

Parameter	*Symbol (unit)*	*Value*
rotor mass	m (kg)	3
moment of inertia*	I (kg·m^2)	0.03
rotor speed	n (rpm)	variable
gravitational constant	g (m·s^{-2})	9.81
static unbalance G0	U_{st} (g·mm)	0
static unbalance G6.3*	U_{st} (g·mm)	45
static unbalance G16*	U_{st} (g·mm)	120
bearing radius	r (mm)	23.69
bearing length	l (mm)	47.37
radial clearance	c_r (mm)	0.9
oil viscosity	μ (Pa·s)	0.07
ambient pressure*	p_a (bar)	0
saturation pressure*	p_c (bar)	0
rotational stiffness*	k_r (kN·m·rad)	1
rotational damping*	d_r (N·m·rad^{-1}·s^{-1})	1

Table 2. Summary of performed simulations.

Analysis:	steady-state	transient
Static unbalance:	0 g·mm	0, 45, 120 g·mm
Speed:	const.	const.
		const. angular acc.
Min. speed:	2000 rpm	800 rpm
Max. speed:	2900 rpm	4000 rpm
Simulation length:	1.5 s	15, 30, 60, 120 s

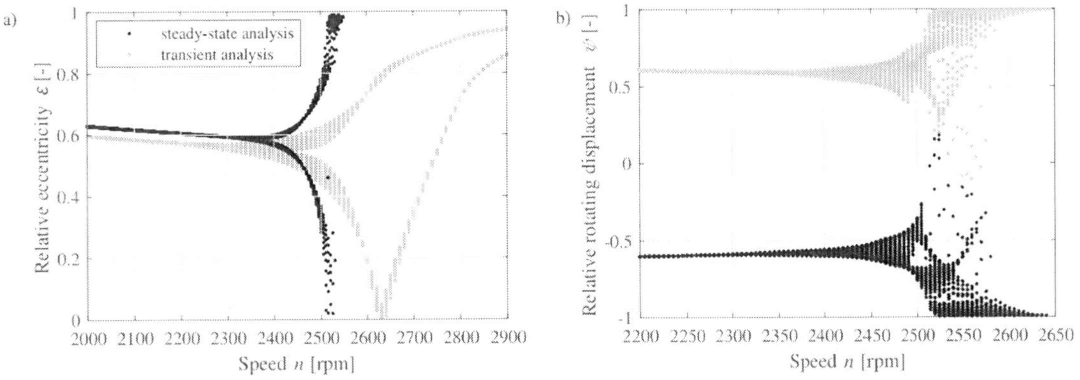

Figure 3. Results of brute force bifurcation. Depicted are a) local extremes of steady-state responses and b) values of relative rotating displacement ψ if relative rotating displacement χ is 0. Relative rotating displacements are defined in Eqs. (21) and (22). Fig. 3b can be also perceived as the intersection of rotating trajectories with $n\psi$ plane. The intersection is highlighted in fig. 4b.

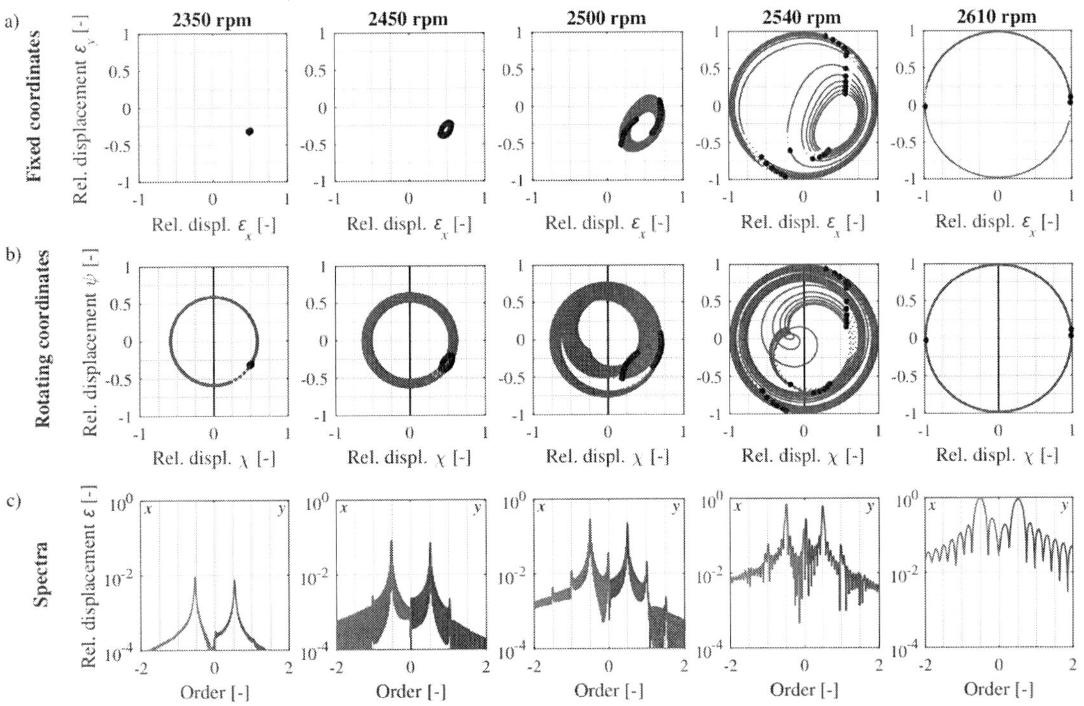

Figure 4. Trajectories (obtained from steady-state analysis) of the rotor at selected speeds are depicted in a) fixed coordinates and b) rotating coordinates. Black dots show the position of the rotor at the beginning of each turn. There are also shown responses in the frequency domain in fig. 4c.

5 Results and discussion

The results of the aforementioned steady-state and transient analysis are presented and discussed in this section.

5.1 Steady-state analysis

Fig. 3a shows the results of the steady-state response analysis, more precisely there are plotted local maxima and minima of relative eccentricity ε. We assume that depicted trajectories are reasonably close to trajectories in respective limit cycles.

It is apparent that until ca. 2350 rpm the system is working at its equilibrium. If the rotor speed further grows, the trajectory will become elliptic as shown in fig. 4a. This elliptic motion is happening at a frequency which is slightly lower than half of the rotor speed and might actually be chaotic because significant broadband noise is present [19]

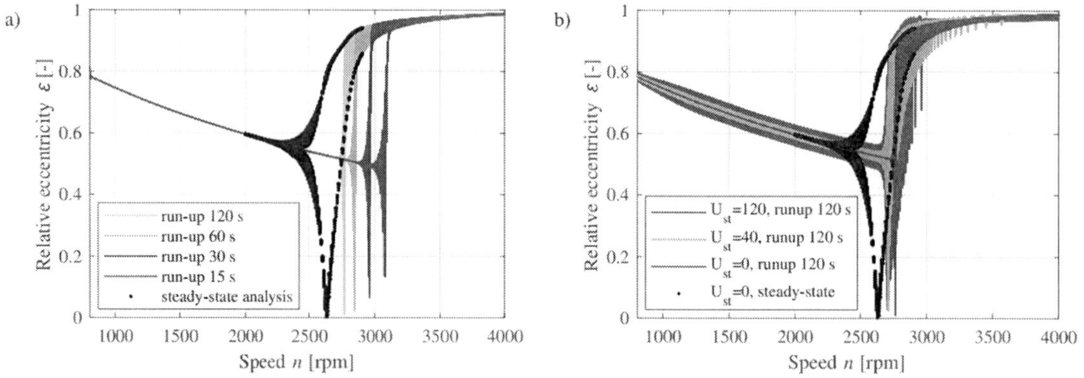

Figure 5. The results of transient analysis show that a) the threshold speed for the fluid-induced instability is dependent on the angular acceleration of the system and b) the static unbalance introduces new local extremes but almost does not influence the threshold speed.

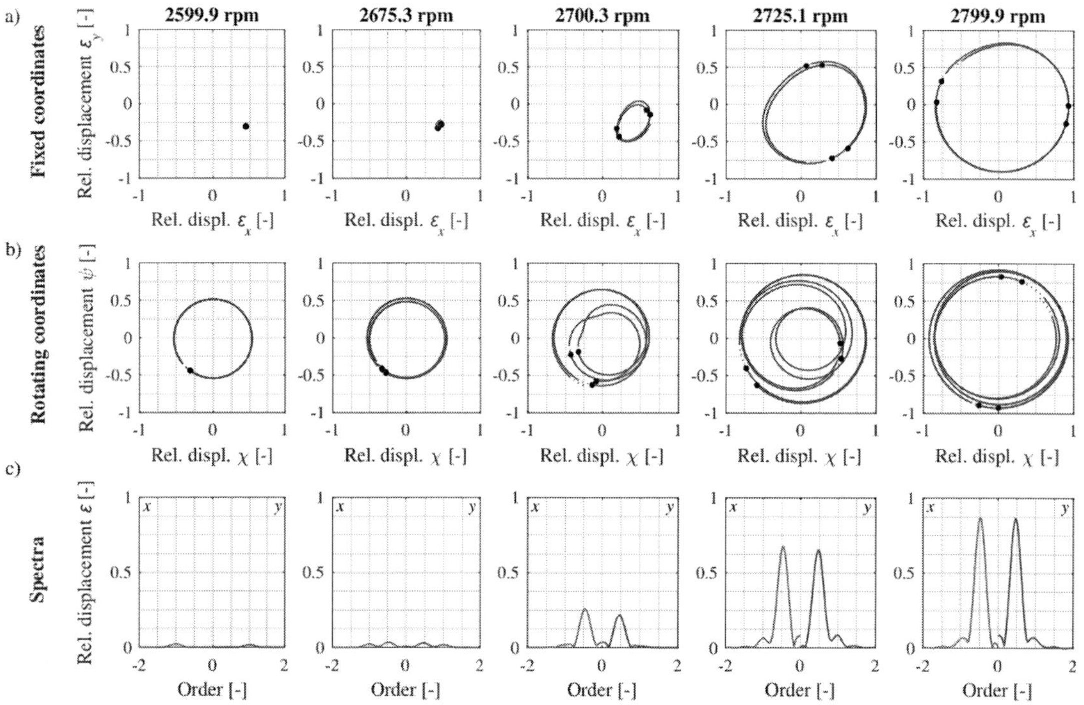

Figure 6. Trajectories (obtained from transient analysis) of the balanced rotor at selected speeds through 4 turns are depicted in a) fixed coordinates and b) rotating coordinates. Black dots show the position of the rotor at the beginning of each turn. There are also shown responses in the frequency domain through 6 turns in fig. 6c.

as can be seen in fig. 4c. It is safe to assume that around 2350 rpm a threshold speed of the fluid-induced instability can be found. After 2500 rpm mark is passed, the system comes into quasi-periodic motion with apparent response at the rotor speed and at even higher speed the motion is clearly limited by bearing clearance and the fluid-induced instability becomes fully developed.

We can study the trajectories more rigorously, if we transform them to a coordinate system (χ, ψ) which rotates around z axis at the same speed as the rotor [19]. $\chi = \chi(t)$ and $\psi = \psi(t)$ are defined as follows

$$\chi = x \cos\varphi - y \sin\varphi, \tag{21}$$

$$\psi = x \sin\varphi + y \cos\varphi. \tag{22}$$

We further refer to the transformed trajectories as *rotating trajectories*. Fig. 4b shows that there are three distinct types of the rotating trajectories in the zone delimited by the threshold speed of the instability and by the speed at which the instability is fully-developed (we call this zone the *transient zone* in the following text). Firstly, there is almost circular trajectory which shows properties of a bounded chaotic motion [19]. If the speed increases, the system comes into stable period doubling and tripling motion with less significant noise, possibly indicating quasi-periodic solution [13]. Finally, when the instability fully develops, the rotating trajectory is circular and the rotor covers roughly 540° in one cycle.

All described phenomena are exhibited in the results of both steady-state and transient analysis with zero angular acceleration. However, all qualitative changes happen at higher speeds in the case of the transient analysis.

5.2 Transient analysis

Fig. 5a shows how the threshold speed of the instability is influenced by angular acceleration of the rotor. Apparently, if the angular acceleration is higher, the system will became unstable at higher speed. Fig. 5b shows that the influence of static unbalance on the threshold speed is rather small.

Fig. 6 demonstrates there are three distinct types of rotating trajectories not only if the rotor is spinning at constant speed but also if its speed is increasing. Note that trajectories and spectra depicted in fig. 6 are taken from the simulation of the perfectly balanced rotor with the lowest angular acceleration because this simulation offers the best frequency resolution of frequency analysis.

Similar trajectories are to be found in all simulated cases. However, if the rotor is unbalanced, there is an additional synchronous (1X) component present in the response.

6 Conclusions

This paper provided a brief yet coherent theory for dynamic behaviour of rigid rotors supported by fluid-film bearings occurring in the so called *transient zone* delimited by the threshold speed of the fluid-induced instability and by the speed at which the instability is fully-developed.

There are three distinct types of the rotor motion in the transient zone. The first type appears as a bounded chaotic motion. Interestingly, later two types are less chaotic and tend to be quasi-periodic. This behaviour is presumably caused by the fact that these motions are more limited by the geometry of the bearing. The later two types of motion are also characteristic by the response with synchronous (1X) components even if the rotor is perfectly balanced.

These three types of motion can be find not only if the rotor is analysed at constant speed but also if the angular speed is being constantly increased and if the rotor is unbalanced.

Preliminary studies suggest that the presented conclusions are valid also for journals of flexible rotors which are supported by multiple journal bearings. However, to prove that such phenomena happen in general is left to future work.

Acknowledgement

This publication was supported by project 17-15915S of Czech Science Foundation entitled *Nonlinear dynamics of rotating systems considering fluid film instabilities with the emphasis on local effects* and by project SGS-2016-038 of University of West Bohemia.

The transient analysis was performed in the AVL Excite software. The AVL Excite software is available in the framework of the University Partnership Program of AVL List GmbH and its usage is greatly acknowledged.

References

[1] B.L. Newkirk, H.D. Taylor, General Electric Review, **28**, 559-568 (1925)

[2] E. Krämer, *Dynamics of Rotors and Foundations* (Springer, Berlin, 1993)

[3] Y. Hori, Trans ASME Ser E, **26**, 189-198 (1959)

[4] A. Tondl, *Some Problems of Rotor Dynamics* (Chapman and Hall, London, 1966)

[5] M. Akkok, G. M. M. Ettles, ASLE Trans, **22**, 175-184 (1980)

[6] J.S. Mitchell, *An Introduction to Machinery Analysis and Monitoring* (Pennwell Publishing, Tulsa, 1981)

[7] D.E. Bently, A. Muszynska, *NASA Conference Publication No. 2250* (Texas A&M Univ., College Station, 1982), 307-322

[8] O. Reynolds, Proc R Soc Lond, **40**, 191-203 (1886)

[9] D. Sfyris, A. Chasalevris, Tribol Int, **55**, 46-58 (2012)

[10] G.B. DuBois, F.W. and Ocvirk, *NACA Report No. 1157* (NACA, 1953), 1199–1206

[11] Y. Bastani, M. de Queiroz, J Tribol, **132**, 014502-1-9 (2010)

[12] F.K. Choy, M.J. Braun, Y. Hu, J Tribol, **113**, 555-561 (1991)

[13] L.-H. Yang, W.-M. Wang, S.-Q. Zhao, Y.-H. Sun, L. Yu, Appl Math Model, **38**, 5239-5255 (2014)

[14] L.R. Gero, C.M.M. Ettles, Tribol T, **29**, 166-172 (1986)

[15] H.N. Chandrawat, R. Sinhasan, Wear, **119**, 77-87 (1987)

[16] A.T. Prata, R.T.S.S. Ferriera, D.E.B. Mile, M.G.D. Bortoli, *International Compressor Engineering Conference* (Purdue University, Purdue, 1988), 34-41

[17] F.J. Profito, M. Giacopini, D.C. Zachariadis, D. Dini, Tribol Lett, **60**, 1-21 (2015)

[18] J.Z. Zhang, W. Kang, Y. Liu, ASME J Comput Nonlinear Dyn, **4**, 011007-1-9 (2009)

[19] A.D. Shaw, A.R. Champneys, M.I. Friswell, P Roy Soc Lond Ser-A, **472**, (2016)

Normal form analysis of stator rub in rotating machinery

Alexander Shaw[1,*], *Alan* Champneys[2], and *Michael* Friswell[1]

[1] *College of Engineernig, Swansea Uninversity, United Kingdom*
[2] *Department of Engineering Mathematics, The University of Bristol, United Kingdom*

Abstract. This work considers analysis of sustained impacting cycles of rotating shafts with potentially many disks. The insight that this is an internal resonance phenomena makes this an ideal system to be studied with the method of normal forms. However, the presence of arbitrary non smooth nonlinearities due to impact and rub mean that the method must be extended by incorporating an Alternating Frequency/Time (AFT) step to capture nonlinear forces. The process results in an elegant formulation that can model a very wide variety of rotor systems and is demonstrated by comparing against simulation of a contacting overhung rotor.

1 Introduction

Contact between rotating machinery and surrounding stators is an issue that can effect a wide variety of engineered systems, from pumps to turbines and drilling rigs. It has also been shown to lead to some highly complex dynamical behaviour [1–3] including internal resonance [4]. Recent work by the authors identified that sustained bouncing cycles can be seen as an internal resonance that is identifiable in the rotating coordinate system [5], unlike the phenomena in [4], which are explained in the stationary frame.

The work in [5] offers a useful insight for analysis of such asynchronous bouncing motions, because it identifies that bouncing cycles can primarily be seen as an interaction between just two modes of the system; this suggests that analysis can be performed by reducing the system to just its resonant modes. The method of normal forms [6] embodies this approach, because it works by reducing systems to just the resonant components; all nonresonant components are held within a near-identity transformation. The systems under consideration have the following form in a coordinate system that rotates with the shaft:

$$\mathbf{M}\ddot{\mathbf{q}} + \Omega\mathbf{G}\dot{\mathbf{q}} + \mathbf{K}\mathbf{q} + \mathbf{C}\dot{\mathbf{q}} + \mathbf{K}_c\mathbf{q} + \mathbf{n}_q(\mathbf{q}, \dot{\mathbf{q}}) = \mathbf{b} \quad (1)$$

However after transformation into complex modal rotating system coordinates, and then the normal forms transformation, the system typically becomes represented by just two complex modal amplitudes, which are solved with two transformed harmonic balance equations. Hence this is a very powerful and elegant method to reduce nonlinear problems in rotating shafts that can potentially handle large problems with complex nonlinearities.

*e-mail: a.d.shaw@swansea.ac.uk

2 Solution overview

Firstly, system (1) is transformed into first order form:

$$\dot{\mathbf{y}} = \mathbf{A}\mathbf{y} + \mathbf{n}_y(\mathbf{y}) \quad (2)$$

where $\mathbf{y} = \{\mathbf{q}, \dot{\mathbf{q}}\}^T$, $\mathbf{A}$ contains the underlying linear conservative terms, and therefore all nonconservative and nonlinear terms are contained within $\mathbf{n}_y(\mathbf{y})$. The system is then transformed to complex modal form using the eigensolutions of $\mathbf{A}$:

$$\dot{\mathbf{p}} = \mathbf{\Lambda}\mathbf{p} + \mathbf{n}_p(\mathbf{p}) \quad (3)$$

where $\mathbf{\Lambda}$ is a diagonal matrix of eigenvalues and $\mathbf{y} = \Re(\mathbf{\Phi}\mathbf{p})$ and $\mathbf{\Phi}$ contains eigenvalues. Note that this system has the same number of degrees of freedom as system (1) because only one of each conjugate pair of eigensolutions is used. Furthermore, the retained eigenvalues have form $\pm j\omega_i$ where it is important that the sign reflects the direction of the physical whirling i.e. positive if it is the same as the shaft rotation.

The method of normal forms seeks a near-identity transformation $\mathbf{p} = \mathbf{u} + \mathbf{h}(\mathbf{u})$ such that the variable $\mathbf{u}$ results in a system

$$\dot{\mathbf{u}} = \mathbf{\Lambda}\mathbf{u} + \mathbf{n}_u(\mathbf{u}) \quad (4)$$

that is much simpler equation than (3) to solve, and can be solved exactly. Note that form of the actual form of the transformation is decided in the frequency domain as a trial solution is implemented. If we subtract (4) from (3) we obtain:

$$\dot{\mathbf{p}} - \dot{\mathbf{u}} = \dot{\mathbf{h}} = \mathbf{\Lambda}\mathbf{h} + \mathbf{n}_p(\mathbf{p}) - \mathbf{n}_u(\mathbf{u}) \quad (5)$$

A problem occurs in trying to eliminate $\mathbf{p}$ from this equation; this has to be done by approximation. Many texts apply the approximation $\mathbf{n}_p(\mathbf{p}) \approx \mathbf{n}_p(\mathbf{u})$ to accuracy $O(\mathbf{h})$. In this method the more accurate approximation

$$\mathbf{n}_p(\mathbf{p}) \approx \mathbf{n}_p(\mathbf{u} + \mathbf{h}_{-1}) = \mathbf{n}_p(\mathbf{p}_{-1}) \quad (6)$$

https://doi.org/10.1051/matecconf/201814804004

may be used with accuracy of $O(\mathbf{h_0} - \mathbf{h_{-1}})$ where $\mathbf{h_{-1}}$ is an estimate from a previous iteration or initial guess and $\mathbf{h_0}$ is the true value for $\mathbf{h}$. After incorporating this approximation equation can be rearranged to

$$\dot{\mathbf{h}} - \mathbf{\Lambda h} - \mathbf{n_p}(\mathbf{u} + \mathbf{h_{-1}}) + \mathbf{n_u}(\mathbf{u}) = \mathbf{0} \quad (7)$$

The general solution of (4) is assumed to be of form:

$$\mathbf{u} = \sum_{i=-n_f/2}^{n_f/2-1} \mathbf{U}_i e^{ij\omega_r t} = \mathbf{Ut} \quad (8)$$

i.e. a summation of vectors multiplied by exponential time functions, so that each transformed modal variable is complex Fourier series with unknown fundamental frequency ω_r and signed harmonics. In the matrix form $\mathbf{Ut}$, $\mathbf{U}$ will have dimensions $N \times n_f$ where n_f indicates the length of the Fourier transform, and $\mathbf{t}$ will be a vector of all the terms $e^{ij\omega_r t}$ with length n_f. The aim is a system where $\mathbf{u}$ is simple, and therefore it is desirable that $\mathbf{U}$ is as sparse as possible. Similar representations are made for all other variables:

$$\begin{gathered}\mathbf{h} = \mathbf{Ht}\,,\ \mathbf{n_u}(\mathbf{u}) = \mathbf{N_u t}\,,\ \mathbf{n_p}(\mathbf{u} + \mathbf{h_{-1}}) = \mathbf{N_p t}\,, \\ \mathbf{h_{-1}} = \mathbf{H_{-1} t}\,,\ \mathbf{p_{-1}} = \mathbf{P_{-1} t}\end{gathered} \quad (9)$$

Differentiation can also be achieved by noting that

$$\dot{\mathbf{h}} = \mathbf{H}\dot{\mathbf{t}} = \mathbf{H\Psi t} \quad (10)$$

where $\mathbf{\Psi}$ is an $n_f \times n_f$ diagonal matrix where the diagonal entries are all of form $ij\omega_r$. Thus equation (7) may be written:

$$\mathbf{H\Psi t} - \mathbf{\Lambda Ht} - \mathbf{N_p t} + \mathbf{N_u t} = \mathbf{0} \quad (11)$$

In order to evaluate the trial nonlinear frequency components $\mathbf{N_p}$ an Alternating Frequency/Time (AFT) is used. This consists of firstly approximating the time series of $\mathbf{p}$ at all time steps using $\mathbf{p} = (\mathbf{U} + \mathbf{H_{-1}})\,\mathbf{t}$ (obtained from using the forms of (9) in (6)). Then, all nonlinear and nonconservative forces $\mathbf{n_p}(\mathbf{p})$ are evaluated based on this time series. Finally a Fast Fourier Transform is used to return the frequency components of the forces $\mathbf{N_p} = \mathbf{F}(\mathbf{n_p}(\mathbf{p}))$. Note that because our state variables give velocity as well as displacement, there is no need to know ω_r to evaluate this stage. Note that $\mathbf{U}$ is to be very sparse, whereas $\mathbf{H_{-1}}$ is a matrix of constants from a previous solution or initial guess, and is in general full except where $\mathbf{U}$ is nonzero.

The corresponding elements k, ℓ of the matrices in (11) can be compared separately because they each relate to a different harmonic of a different modal variable. Hence equation (11) (with $\mathbf{t}$ eliminated) can be considered term by term:

$$H_{k,\ell}\Psi_{\ell,\ell} - \Lambda_{k,k}H_{k,\ell} - N_{p(k,\ell)} + N_{u(k,\ell)} = 0 \quad (12)$$

In general, because we want to simplify equation (4), equation (12) is solved by choosing:

$$N_{u(k,\ell)} = 0 \quad , \quad H_{k,\ell} = \frac{N_{p(k,\ell)}}{\Psi_{\ell,\ell} - \Lambda_{k,k}} \quad (13)$$

However, if $\Psi_{\ell,\ell} \approx \Lambda_{k,k}$, this will cause $H_{k,\ell}$ to be large, violating the assumption of a near identity transformation. These terms are known as resonant terms and must be solved by choosing:

$$N_{u(k,\ell)} = N_{p(k,\ell)} \quad , \quad H_{k,\ell} = 0 \quad (14)$$

Typically interesting solutions (i.e. not simply synchronous whirling in or out of contact) occur when exactly two modes become internally resonant [5]. Therefore, expressing the harmonic components of the resonant equation of motion (4) will give a harmonic balance problem of the form:

$$\begin{gathered}U_{i,j}\Psi_{j,j} - \Lambda_{i,i}U_{i,j} - N_{u(i,j)} = 0 \\ U_{k,\ell}\Psi_{\ell,\ell} - \Lambda_{k,k}U_{k,\ell} - N_{u(k,\ell)} = 0\end{gathered} \quad (15)$$

Hence our system of N degrees of freedom results in this greatly reduced form. Equation (15) has 5 unknowns; the real and imaginary parts of $U_{i,j}$ and $U_{k,\ell}$ and also the unknown fundamental frequency ω_r which is within $\Psi_{j,j}$ and $\Psi_{\ell,\ell}$ (see the definition of $\mathbf{\Psi}$ following equation (10)). However, we may impose that one of the transformed modal variables is purely real; this constrains the phase of this modal variable and therefore locks the solution in time. Therefore, with this imposition equation (15) is solvable.

The method outlined is very general with regard to the form of nonlinearity in $\mathbf{n_q}(\mathbf{q}, \dot{\mathbf{q}})$ due to the use of the AFT step. Furthermore, it is also very general with regard to the size of the system matrices, so should scale to rotor systems with many degrees of freedom with appropriate modal truncations.

3 A Snubbed Overhung Rotor

3.1 System description

As shown in Fig. 1 (a), a disc of mass m, polar moment of inertia I_p, and diametral moment inertia I_d, is mounted on an inertialess rigid shaft of length ℓ. The shaft is pinned at point O, and rotations around this point are resisted by a linear isotropic rotational viscously damped spring, with rate k and damping coefficient c. The disc rotates at a constant angular speed Ω about its centre point C; however imperfections in its geometry cause its centre of mass M to be at a distance ε from C, resulting in out of balance forces. This rotor has two degrees of freedom which are the displacements of the centre point u and v as shown in Fig. 1 (b), which are directly coupled to rotations ψ and θ assuming small angles.

A snubber ring with clearance δ exists at a distance a along the shaft. This generates additional restoring forces when the displacement at this point is greater than δ, represented by the nonlinear contact force $\mathbf{N_x}$ given by

$$\mathbf{N_x} = \begin{cases} -k_s\left(\frac{a}{\ell}\right)^2 (r - \delta)\,\{u, v\}^T / r, & \text{if } r \geq \delta \\ \{0, 0\}^T, & \text{if } r < \delta, \end{cases} \quad (16)$$

where k_s is the stator contact stiffness, δ is the clearance of the stator and $r = \sqrt{x_c^2 + y_c^2}$.

The system is nondimensionalised using the undamped natural frequency of the underlying nonrotating

 https://doi.org/10.1051/matecconf/201814804004

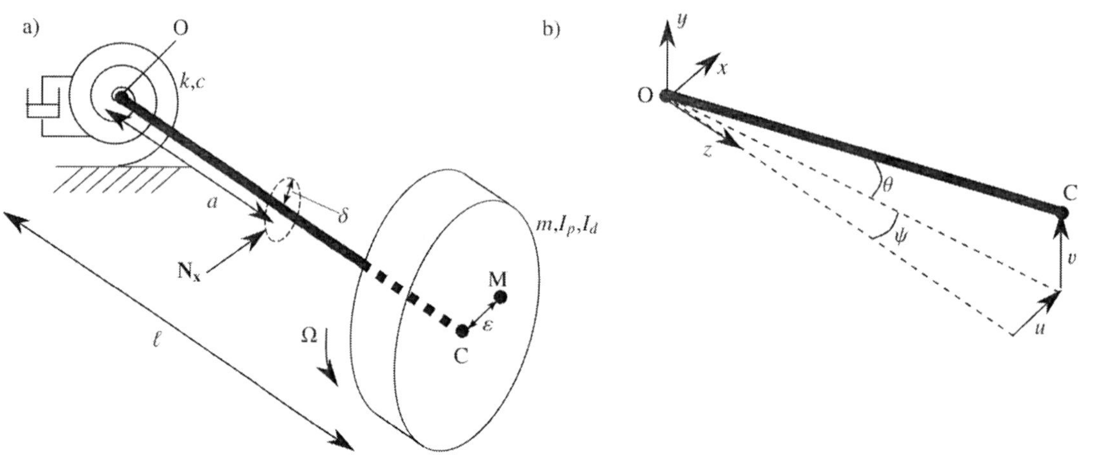

Figure 1. a) A snubbed overhung rotor. b) Assumed kinematics of point C.

system $\omega_n = \sqrt{\frac{k}{m_{eff}\ell}}$ to scale time where $m_{eff} = m + \frac{I_d}{\ell^2}$ is the effective inertia, and the displacement required to make contact with the stator $\Delta = \frac{\delta\ell}{a}$ is used to scale displacement:

$$\begin{aligned} \tau = \omega_n t \quad , \quad \hat{u} = \tfrac{u}{\Delta} \quad , \quad \hat{v} = \tfrac{v}{\Delta} \quad , \quad \hat{I}_p = \tfrac{I_p}{m_{eff}\ell^2} \\ \hat{\Omega} = \tfrac{\Omega}{\omega_n} \quad , \quad \hat{f} = \tfrac{\varepsilon}{\Delta} \quad , \quad \zeta = \tfrac{c}{2\ell^2\omega_n m_{eff}} \end{aligned} \tag{17}$$

This leads to equations of motion of the form of equation (1) with matrices as follows:

$$\begin{aligned} \mathbf{M} = \mathbf{I}_{2\times2} \ , \ \mathbf{G} = -(\hat{I}_p - 2)\mathbf{J} \ , \ \mathbf{K} = (1 - \hat{\Omega}^2(1 - \hat{I}_p))\mathbf{I}_{2\times2} \\ \mathbf{C} = 2\zeta\mathbf{I}_{2\times2} \ , \ \mathbf{K_c} = -2\zeta\hat{\Omega}\mathbf{J} \ , \ \mathbf{b} = \left\{\hat{f}\hat{\Omega}^2, 0\right\}^T \end{aligned} \tag{18}$$

and

$$\mathbf{N_q}(\mathbf{q}) = \begin{cases} -\beta(|\mathbf{q}| - 1)\{q_1, q_2\}^T/|\mathbf{q}|, & \text{if } |\mathbf{q}| \geq 1 \\ \{0, 0\}^T, & \text{if } |\mathbf{q}| < 1, \end{cases} \tag{19}$$

where $\beta = \frac{k_s a}{k}$ expresses the ratio of stator stiffness to the underlying linear stiffness.

3.2 Results

The above analysis was performed for a system with $\beta = 13.2$, $\hat{f} = 0.065$, $\hat{I}_p = 0.14$. The initial value for $\mathbf{h}_{-1}$ is obtained from the linear synchronous whirl solution. The solution was then iterated twice, each time using the current values of $\mathbf{U}$ to refine the estimate for $\mathbf{h}_{-1}$ by using equations (13) and (14). At each speed, four initial guesses where chosen - with the 2nd resonant component at initial phase differences of 0, $\pi/2$, π and $3\pi/2$, and the amplitude sufficient to give some contact. A 1024 point FFT was used for the AFT stage.

Figure 2 presents the underlying linear whirl velocities - although note that these are represented in the rotating coordinate system and hence show a downward trend, due to the subtraction of the drive speed. It can be seen that the linear whirl speeds form a 2:1 ratio at $\Omega \approx 3.3$; hence this is a good region in which to search for 2:1 internally resonant solutions. Figure 3 shows the results of the analysis, in terms of the amplitudes of the resonant components. At slightly above the drive speed for the onset of internal resonance as found from Fig. 2, the nontrivial solutions begin to be found.

These solution come in two branches, an upper and a lower for each resonant component. Figure 4 allows us to compare the two solutions at $\hat{\Omega} = 3.8$. It may be seen from Fig. 4 (a) that the upper branch solution is stable - the numerical simulation, initiated with the analytical solution, follows it perfectly (30 orbit cycles are simulated). Figure 4 shows that the lower branch solution is similar in magnitude and character to the upper branch, although the 'loop' in the orbit is oriented differently. However, while the numerical simulation follows the analytical solution initially, but then diverges away from it. This suggests that this solution branch consists of unstable solutions.

4 Conclusions and future work

This work has shown how the insight of internal resonance leads to the application of the method of normal forms to rotor stator contact problems. The analysis has been demonstrated for a simple rotor system, and it has been shown to match time simulation data and find both stable and unstable solutions. In principle the method extends to more complex systems and this is the focus of future work.

This work has received funding from EPSRC grant number **EP/K003836** *Engineering Nonlinearity*.

References

[1] F.F. Ehrich, Journal of System Design and Dynamics **2**, 641 (2008)

[2] S. Edwards, A.W. Lees, M.I. Friswell, Journal of Sound and Vibration **225**, 767 (1999)

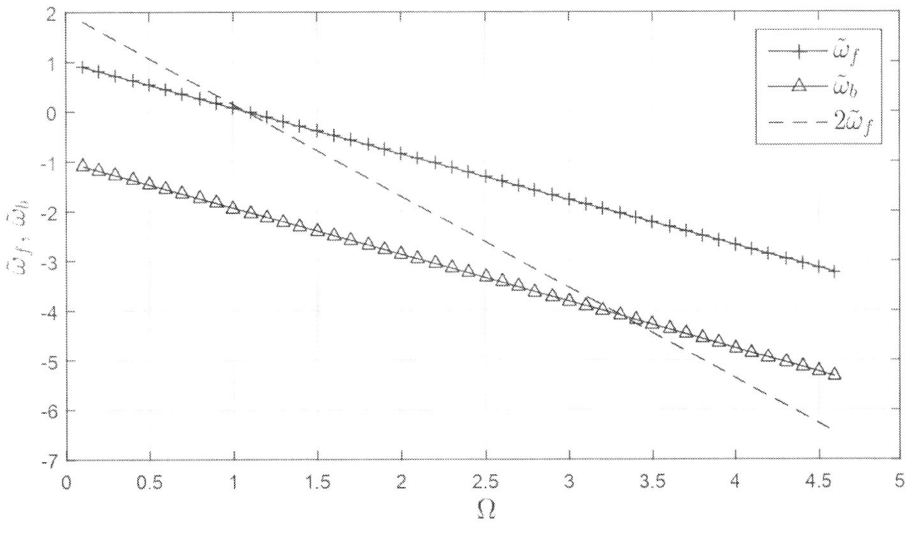

Figure 2. Linear whirl speeds of overhung rotor in rotating coordinate frame.

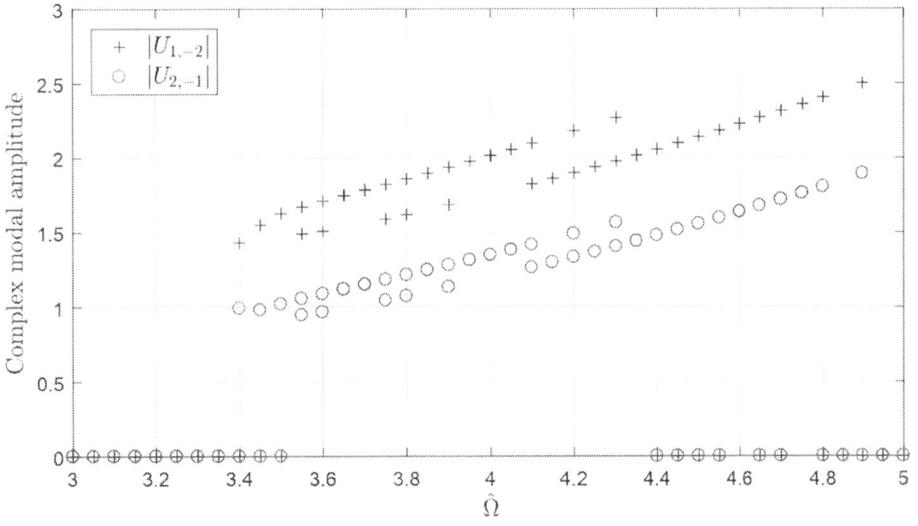

Figure 3. Solution plot from normal form analysis.

[3] A. Zilli, R.J. Williams, D.J. Ewins, Journal of Engineering for Gas Turbines and Power **137**, 065001 (2015)

[4] T. Inoue, Y. Ishida, Journal of Vibration and Acoustics **128**, 156 (2006)

[5] A.D. Shaw, A.R. Champneys, M.I. Friswell, Proc. R. Soc. A **472**, 20160303 (2016)

[6] S.A. Neild, A.R. Champneys, D.J. Wagg, T.L. Hill, A. Cammarano, Phil. Trans. R. Soc. A **373**, 20140404 (2015)

 https://doi.org/10.1051/matecconf/201814804004

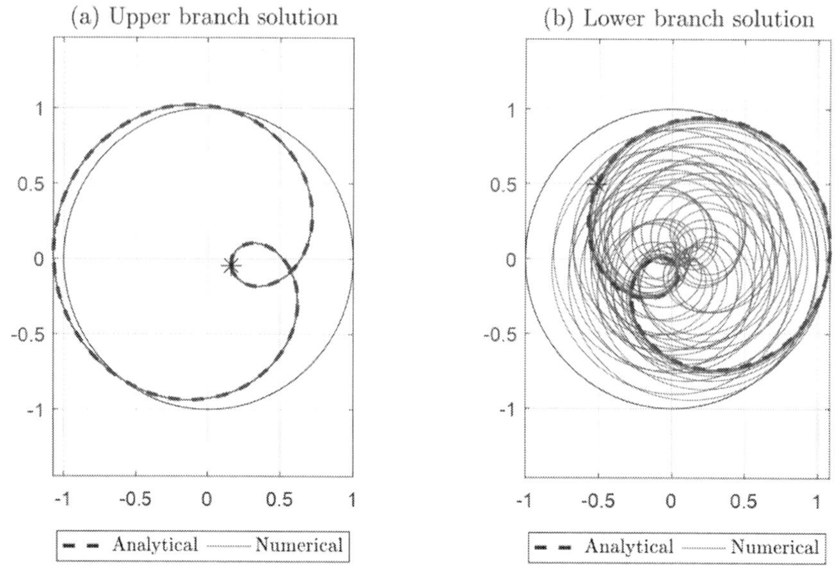

Figure 4. Orbits from solutions for $\hat{\Omega} = 3.8$. Star indicates start point of time solution.

A new autogenous mobile system driven by vibration without impacts, excited by an impulse periodic force

The-Hung Duong[1], Van-Du Nguyen[2,*] and Ngoc-Tuan La[3]

[1]Department of Civil Engineering, Thai Nguyen University of Technology, Viet Nam
[2]Department of Mechanical Engineering, Thai Nguyen University of Technology, Viet Nam
[3]Department of Manufacturing Engineering, Vinh University of Technology Education, Viet Nam

Abstract. This report describes a new proposed design for autogenous mobile systems which can move without any external mechanisms such as legs or wheels. A Duffing oscillator with a cubic spring, which is excited by an impulse periodic force, is utilized to drive the whole system. The rectilinear motion of the system is performed employing the periodically oscillation of the internal mass interacting without collisions with the main body. Utilizing the nonlinear restoring force of the cubic spring, the system can move in desired directions. When the ratio between the excitation force and the friction force is smaller than 2.5, backward or forward motion can be easily achieved by applying an excitation force in the same desired direction. Different from other vibro-impact drifting devices, no impact needed to drive the new proposed system. This novel structure allows to miniaturize the device as well as to simplify the control algorithm thus can significantly expand applicability of the proposed system.

1. Introduction

Mobile systems have been widely employed both in large- and micro-scale sizes. Such systems are preferable for special engineering applications in remote and hostile environments such as engineering diagnosis to disaster rescues or pipeline inspection [1-3] and impact moling [4-6]. In micro-scale biomechanics, locomotion systems can enhance capabilities of the capsule endoscopy to deal with problems of traditional systems such as inability to control and stop, and risk of capsule retention [7]. Of such movable dynamical systems, vibration-driven platforms are a new kind which is propelled by the periodic motion of internal masses or the periodic deformation of their bodies. Using vibration-driven mechanism, several complications induced by external driving mechanism (wheels, chain tracks, or legs) can be avoided [8,9]. The vibration-driven locomotion systems have been widely investigated from many aspects, including modeling and theoretical analysis and control [10-13], design and experimental implementation [14-17]. Several experimental platforms have been developed, employing centripetal forces generated by platform-mounted vibration micro-motors [14], electromagnetic force induced in voice coil motor [15,16], centrifugal forces of a couple-pendulum [17]. The common issue of such vibration-driven systems is that, a complex control for the inertial mass motion is necessary to obtain the desired progression of the system [10,15]. Moreover, the problem for the large-scale mobile platforms also make difficult to miniaturize the platform. The capsule model using vibro-impact device proposed by Liu et al. [9] addressed these issues. A similar mobile mechanism in a large-scale platform of size which can produce strong impact force for moling machines was also reported [6]. Those models are inspired by a vibro-impact drifting oscillator studied by Pavlovskaia et al. [18]. Experimental verification [19-21] and dynamic response analysis of such systems [22-23] have been investigated in order to carry out the best performance conditions, including the system stability and maximum progression rate. For capsule endoscopy applications, the system abilities for moving forwards and backwards are required. In order to address this issue, a position feedback control method suitable to control the vibro-impact capsule system which uses a soft impact and harmonic excitation was proposed [24]. However, requirement of constraint setup in vibo-impact oscillator still make it difficult for minimizing the device for capsule applications. This paper propose a new design of a drifting actuator which can drive the autogenous mobile system without any impacts. The actuator is a kind of Duffing oscillator, as similar to the one proposed [25], containing a cubic spring but excited by impulse force. Our study revealed that by choosing proper values of design and operational parameters, the system can move in a desired direction with a simple control strategy. Nonlinear spring is a prerequisite condition for the movable ability of the system.

2. Mathematical model

A physical model of the proposed system is depicted in Figure 1, where a movable mass m_1 is connected with the system frame m_2 by a cubic spring F_{spr} and a viscous damper with damping coefficient c. The system is excited by a periodic impulse force F with amplitude A and frequency Ω exerts on both masses in opposite directions. The excitation amplitude, A is assigned as a positive value if it acts on the mass m_1 in the forward direction and vice versa.

In this study, the excitation force is a 50% duty cycle force and thus can be expressed in a form of the Heaviside function as:

$$\begin{cases} F = \begin{cases} A, \sin(\Omega t) > 0 \\ 0, \sin(\Omega t) \le 0 \end{cases} \\ \text{or } F = Heavisde(\sin(\Omega t)) \end{cases} \tag{1}$$

The friction force F_f has direction opposite to the velocity of the frame body. X_1 and X_2 are the displacement of the inertial mass m_1 and the system body m_2, respectively.

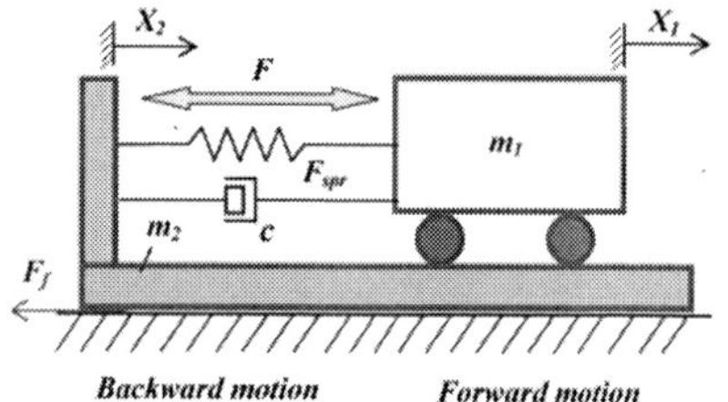

Fig. 1. *Physical model of the system*

The equations describing the motion of the system can be written as

$$\begin{cases} m_1 \dfrac{d^2 X_1}{dt^2} = F - F_{spr} - c\left(\dfrac{dX_1}{dt} - \dfrac{dX_2}{dt}\right) \\ m_2 \dfrac{d^2 X_2}{dt^2} = -F + F_{spr} + c\left(\dfrac{dX_1}{dt} - \dfrac{dX_2}{dt}\right) - F_f \\ F = \begin{cases} A, \sin(\Omega t) > 0 \\ 0, \sin(\Omega t) \le 0 \end{cases} \\ F_f = f.(\mathrm{m}_1 + \mathrm{m}_2) \cdot g.\mathrm{sgn}\left(\dfrac{dX_2}{dt}\right) \end{cases} \tag{2}$$

where $d()/dt$ denotes differentiation with respect to time, t; sgn() denotes the sign of velocity (dX_2/dt); f is the friction coefficient between the frame body and the environment; g is the gravitational acceleration.

An illustration of the 50% duty cycle forcing function with amplitude of 2 N and frequency of 5 Hz (Ω=31,416 rad/s) is depicted in Fig.2.

The spring force, F_{spr} is expressed as;

$$F_{spr} = k_1(X_1 - X_2) + k_2(X_1 - X_2)^3 \tag{3}$$

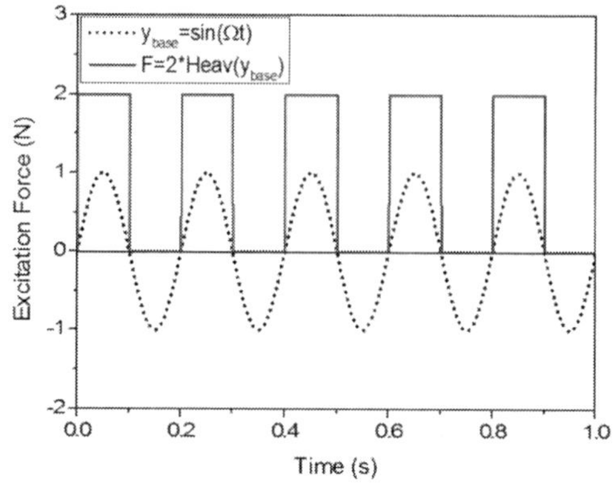

Fig. 2. *Time history of the excitation force (red solid line) with frequency of 5 Hz, and the corresponding sine function (black dots)*

The Equations (1) can be rewritten in a dimensionless form as

$$\begin{cases} x_1'' = \alpha H - (x_1 - x_2) - \beta(x_1 - x_2)^3 - 2\zeta(x_1' - x_2') \\ x_2'' = \begin{bmatrix} -\alpha H + (x_1 - x_2) + \beta(x_1 - x_2)^3 \\ + 2\zeta(x_1' - x_2') - \mathrm{sgn}(x_2') \end{bmatrix} / \mu \end{cases} \tag{4}$$

where

$$\begin{cases} \tau = \Omega_0 t;\ x_1 = \dfrac{k_1}{F_f} X_1; x_2 = \dfrac{k_1}{F_f} X_2; \Omega_0 = \sqrt{\dfrac{k_1}{m_1}}; \omega = \dfrac{\Omega}{\Omega_0}; \\ \zeta = \dfrac{c}{2m_1\Omega_0}; \alpha = \dfrac{A}{F_f}; \beta = \dfrac{k_2}{k_1}\left(\dfrac{F_f}{k_1}\right)^2; \mu = \dfrac{m_2}{m_1}; \\ H = heav(\sin(\omega t)) \end{cases} \tag{5}$$

There are two response issues are usually considered for a mobile system: the progression rate and directional motion of the system. As can be seen, the Equations (5) contains five parameters which can be classified into two groups: β, ζ and μ can be considered as design parameters; α and ω are the parameters of operation task. Several effects of such parameters on the system response will be elicited in the following sections.

3. Design parameters

In this study, the progression rate, P, is considered as the displacement of the system frame, x_2 for the dimensionless time of 100. The mass ratio is kept as constant μ=1, e.g. the two masses have the same value. Firstly, the coupled effect of the stiffness, β and the damping, ζ of the nonlinear spring on the progression rate, x_2 is considered. Figure 2 presents a contour plot of x_2 for $\beta \in [0,100]$ and $\zeta \in [0,1.0]$. All other parameters are fixed as μ=1, α=1 and ω=2.

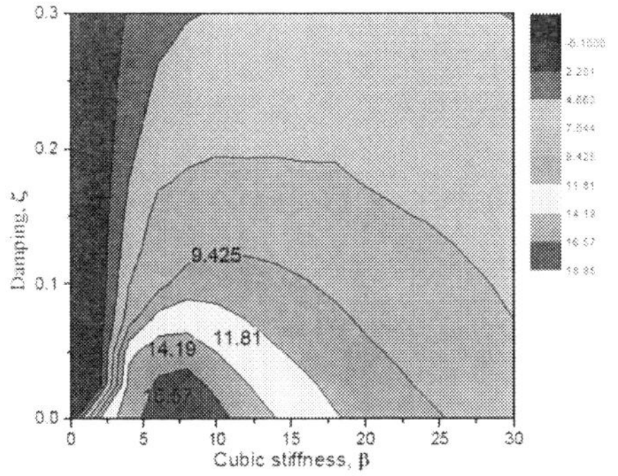

Fig. 3. *(Color online) A contour plot of P with respect to β and ζ: The values obtained for α=1, ω=2 and μ=1*

As can be seen on Figure 3, the progression rate reached values higher than 16.57 for $\beta\in[5, 10]$ and $\zeta\in[0, 0.3]$. Our numerical data revealed that, P get peak value $P=17.571$ at $\beta=8$ and $\zeta=0.025$. Figure 3 depicts x_2 as a function of the cubic stiffness, β, for four different levels of the damping factor, ζ to support this remark.

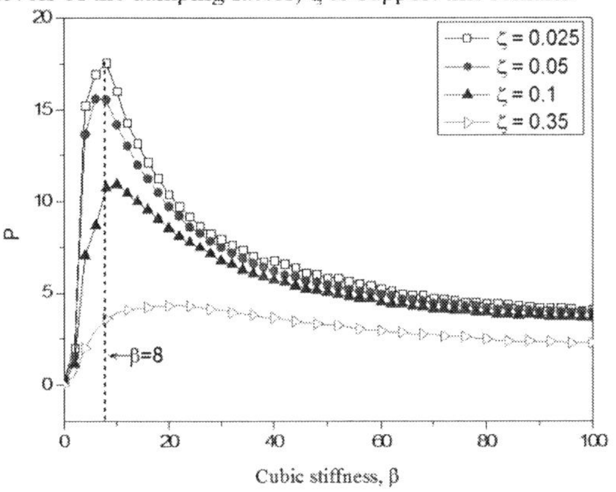

Fig. 4. *A plot of P as a function of β for for four levels of ζ: values obtained for α=1, ω=2 and μ=1*

As can be seen in Figure 4, the progression rate is significant when β increases from 0 to 8. It is worth to highlight that, the system is not able to move (P=0) for β=0. However, the progression rate trends down when β>8. In other words, the spring inside the system must be a nonlinear spring in order to provide a movable ability. Besides, the cubic term, β, should be chosen carefully to obtain the fastest progression rate. It is also noted that there have three apexes of x_2 appeared remarkably around $\beta\approx8$ for small damping values (ζ=0.025; ζ=0.05 and ζ=0.1). For higher values of damping, such as ζ=0.35 as shown in the Figure, changing β would not have significant effect on the progression rate.

It is also observed from Figure 3 and Figure 4 that, the system with smaller damping values would provide higher progression rate. The next section presents some primary results obtained from system analysis under the effects of operational parameters.

4. Operational parameters

In order to understand the system behavior under different values of operational parameters, two design parameters are kept as constants as: μ=1, ζ=0.025. Figure 5 presents the bifurcation diagram of the progression rate, P, and the relative velocity, v_1-v_2, with respects to the excitation force, α.

From Figure 5(a), it can be observed that the progression rate for positive α is exactly as the same as that for negative α but in opposite sign. The system can move with the highest progression rate for $|\alpha|\approx1$. For $\alpha\in[-2.5, 2.5]$, the system move forward when α>0 and backward when α<0. When increasing the force ratio, $|\alpha|$>2.5, the system moves back if α>0 and vice versa, i.e. the system trends to move in opposite direction of the excitation force.

The bifurcation diagram of the relative velocity, v_1-v_2 shown in Figure 5(b) is a projection of the Poincaré map on the vertical axis. The calculations were run for 400 cycles of the excitation force. The first 100 cycles were taken to be transient and not included so as to plot the steady state of next 300 cycles.

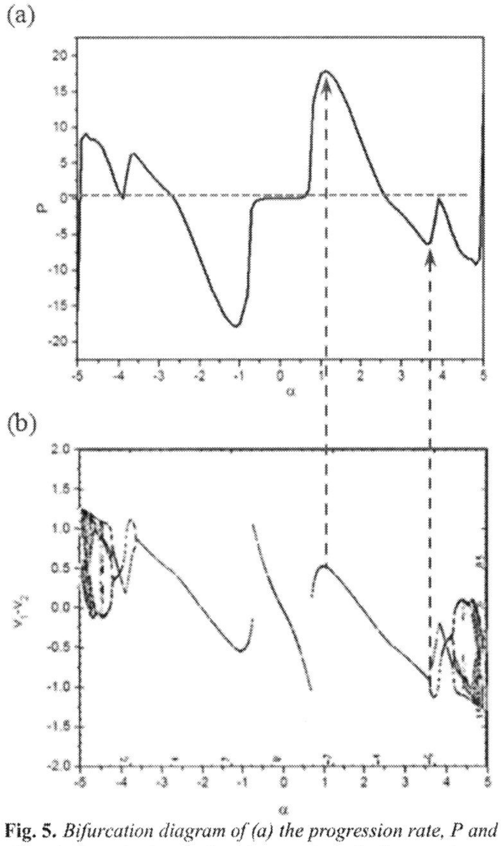

Fig. 5. *Bifurcation diagram of (a) the progression rate, P and (b) relative velocity v1-v2 with respects to the force ratio, α. The results obtained for μ=1, ζ=0.025 and ω=2.0*

As can be seen, the highest progression rate appeared when the mass motion has a period-one response. Besides, a peak of reverse movement (the progress movement is opposite to the excitation force) appeared where the relative motion is a period-two and higher. Figure 6 depicts these phenomena.

(a)

(b)

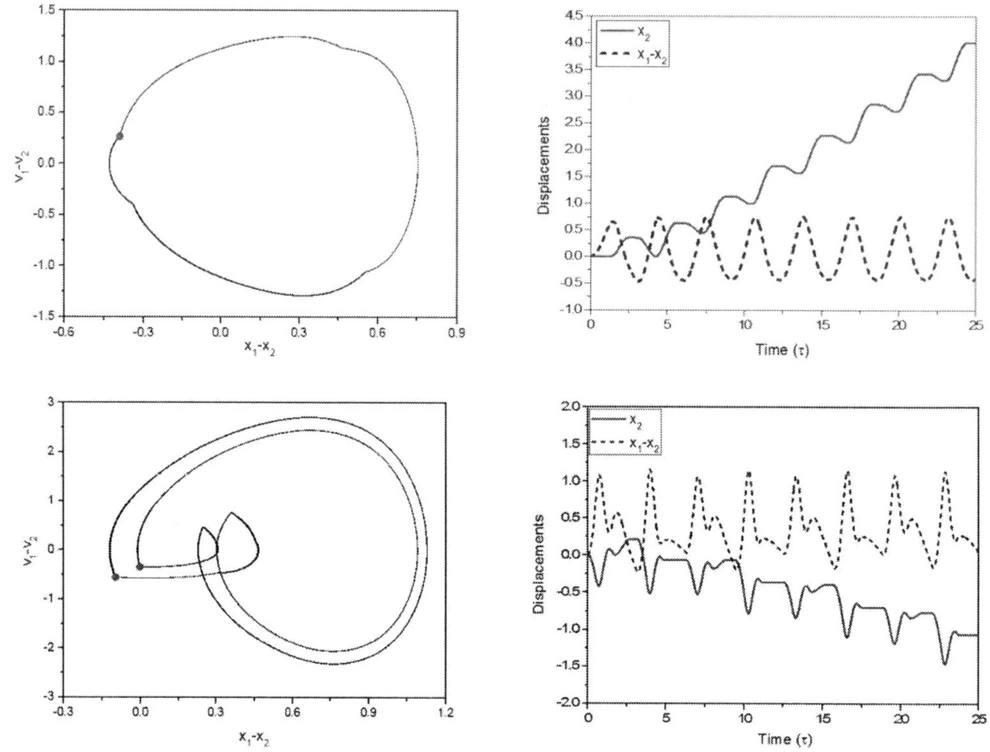

Fig. 6. *(Color online) Left hand-side sub-plots: Trajectories presented by gray lines on the phase plane (x_1-x_2, v_1-v_2), where Poincaré sections are marked by red dots; Right hand-side sub-plots: Time histories of the relative motion, x_1-x_2 (black dashed lines) and of the system displacement, x_2 (red solid lines). The calculations obtained for β=8, μ=1, ω=2, ζ=0.025 and (a) α=1, (b) α=3.75*

In Figure 6(a), the relative motion of the inertial mass is period-one, the system frame has a steady state movement in forward direction. When the mass has period-two response as shown in Figure 6(b), the system progressed in backward direction.

The influence of excitation frequency on the system progression is scrutinized by checking with the bifurcation diagram with respect to $\omega\in$[0.1, 5.1], as shown in Figure 7.

In overall, for all the values of the investigated excitation frequency, ω, the system always progresses in forward direction, as the same as the direction of the excited force. There appears one significant local maximum of forward progression for ω=2. For $\omega\in$[1, 2], increasing the excitation frequency would result in raising up the progression rate. However, for ω>2, increasing the excitation frequency would reduce the progression rate. When the motion of the inertial mass change from period-one to period-two motion at ω=3.1, the progression rate increase slightly. There appears a small peak of progression rate in the range of $\omega\in$[3.1, 3.75] but much smaller than the peak at ω=2.

(a)

(b)

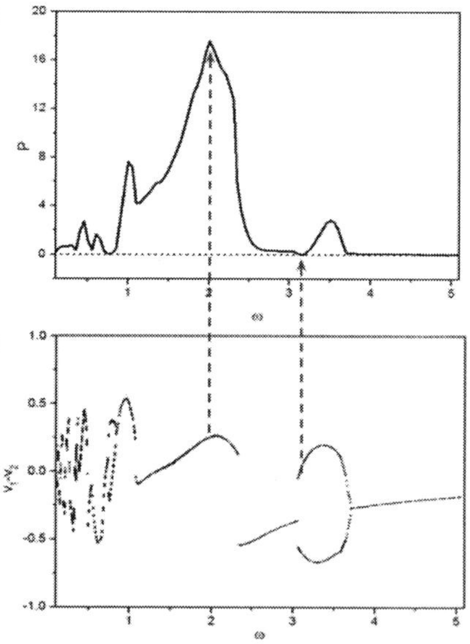

Fig. 7. *Bifurcation diagram of (a) the progression rate, P, and (b) relative velocity v_1-v_2 with respects to the excitation frequency, ω. The results obtained for μ=1, ζ=0.025 and α=1*

5. Conclusion

A new vibration-driven device which is able to drive an autogenous mobile system in desired directions has been developed. The most advantage of the new design is that the system can move in desired direction without impacts, thus can simplify the device structure. Moreover, the desired direction of the system movement can be obtained simply by applied an excitation force with the same direction as the desired motion.

The heart of the new system is a Duffing oscillator including a nonlinear spring with a cubic term of stiffness. For the investigated range of the parameters, our numerical study revealed that, the system can obtain the highest progression rate for the set of following parameters: cubic spring stiffness β=8, amplitude of the excitation force α=1 and the excitation frequency ω=2. For this set of parameters, the system not only reaches the highest progression rate, but also moves in the same direction as the direction of applied force.

The mathematical model with dimensionless parameters allows extending the results to both large- and micro-scale of different mobile systems. Further study should be paid to the system response for wider ranges of both design and operation parameters so as to optimize system structure and enhance its working effectiveness.

Acknowledgements: This research is funded by Vietnam National Foundation for Science and Technology Development (NAFOSTED) under grant number 107.01-2012.33.

References

[1]. Y. Zhang, S. Jiang, X. Zhang, X. Ruan, D. Guo, A variable-diameter capsule robot based on multiple wedge effects, *IEEE/ASME Trans. Mechatron.* **16** (2011) 241–254, DOI: 10.1109/TMECH.2009.2039942;

[2]. A. Shukla and H. Karki, Application of robotics in onshore oil and gas industry – a review Part I, Robot. Auton. Syst. **75**, 490-507, (2016) DOI:10.1016/j.robot.2015.09.012.

[3]. A. Shukla and H. Karki, Application of robotics in onshore oil and gas industry – a review Part II, Robot. Auton. Syst. **75**, pp508-524, (2016) DOI:10.1016/j.robot.2015.09.013.

[4]. J. Warminski, S. Lenci, M.P. Cartmell, G. Rega, M. Wiercigroch, Nonlinear Dynamic Phenomena in Mechanics,Vol **181**; *Chapter 5:Engineering Applications of Non-smooth Dyn., Springer, pp. 211-273, (2012) DOI:10.1007/978-94-007-2473-0_5;*

[5]. E. Pavlovskaia, M. Wiercigroch, K.C. Woo and A.A. Rodger, Modelling of Ground Moling Dynamics by an Impact Oscillator with a Frictional Slider, *Meccanica* ***38*** *(1), pp85–97, (2003), DOI:10.1023/A:1022023502199;*

[6]. V.D. Nguyen, K.C. Woo and E. Pavlovskaia, Experimental study and Mathematical Modelling of a New of Vibro-impact moling device, *Int. J. Nonlinear. Mech.* ***43*** *(6), pp542-550, (2008) DOI:10.1016/j.ijnonlinmec.2007.10.003;*

[7]. L. Liu, S. Towfighian and A. Hila, A review of locomotion systems for capsule endoscopy, *IEEE Reviews in Biomedical Engineering (2015), DOI: 10.1109/RBME.2015.2451031;*

[8]. N. A. Sobolev and K. S. Sorokin, Experimental Investigation of a Model of a Vibration-Driven Robot with Rotating Masses, *J. Comput. Sys. Sc. Int.* ***46****(5), pp826–835, (2007); DOI: 10.1134/S1064230707050140;*

[9]. Y. Liu, M. Wiercigroch, E. Pavlovskaia and H. Yu, Modeling of a vibro-impact capsule system, *Int. J. Mech. Sci.,* ***66****, 2-11 (2013), DOI:10.1016/j.ijmecsci.2012.09.012;*

[10]. H. Li, K. Furuta and F. L. Chernous'ko, Motion Generation of the Capsubot Using Internal Force and Static Friction; *Proceedings of the 45th IEEE Conference on Decision & Control, pp6575-6580, (2006) DOI: 10.1109/CDC.2006.377472;*

[11]. F.L. Chernous'ko, The optimal periodic motions of a two-mass system in a resistant medium, *J. Appl. Math. Mech.* ***72****, pp 116–125, (2008) DOI:10.1016/j.jappmathmech.2008.04.014.*

[12]. N.N. Bolotnik, T. Figurina, Optimal control of the rectilinear motion of a rigid body on a rough plane by means of the motion of two internal masses, *J. Appl. Math. Mech.* ***72,*** *pp126–135, (2008) DOI:10.1016/j.jappmathmech.2008.04.013;*

[13]. N.N. Bolotnik, F.L. Chernous'ko, T. Figurina, Optimal Control of a Two-body Vibration-driven Locomotion System in a Resistive Environment, *IFAC-PapersOnLine 48-25 (2015) 091–096, DOI:10.1016/j.ifacol.2015.11.065 ;*

[14]. P. Vartholomeos and E. Papadopoulos, Dynamics, Design and Simulation of a Novel Microrobotic Platform Employing Vibration Microactuators, *J. Dyn. Sys., Meas., Control* ***128****(1), 122-133 (2005) DOI:10.1115/1.2168472;*

[15]. G. Su, C. Zhang C, R. Tan and H. Li, A Design of the Electromagnetic Driver for the "Internal Force-Static Friction" Capsubot, *International Conference on Intelligent Robots and Systems, , (IEEE/RSJ 2009), DOI: 10.1109/IROS.2009.5354587;*

[16]. M.N. Huda, H.N. Yu and S.O. Wane, Self-contained Capsubot Propulsion Mechanism, *International Journal of Automation and Computing,* ***8****(3), 348-356, (2011), DOI: 10.1007/s11633-011-0591-3;*

[17]. Y. Liu, H. Yu and S. Cang, Modelling and motion control of a double-pendulum driven cart, *Proc. IMechE Vol.* ***226*** *Part I: J. Systems and Control Engineering, 226 No. 2 175-187, (2012), DOI: 10.1177/0959651811414507;*

[18]. E. Pavlovkskaia, M. Wiercigroch and C. Grebogi, Modelling of an impact system with a drift, *Phys. Rev. E,* ***64****, 056224 (2001), DOI: 10.1103/PhysRevE.64.056224;*

[19]. V.D. Nguyen and K.C. Woo, New electro-vibroimpact system, *Proc. IMechE, Part C: J. Mech. Eng. Sc., Vol.* ***222****, (4), 629-642 (2008), DOI: 10.1243/09544062JMES833;*

[20]. V.D. Nguyen and K.C. Woo, Nonlinear dynamic responses of new electro-vibroimpact system, *J. Sound. Vib.;* ***310****(4-5), pp. 769-775 (2008), DOI: 10.1016/j.jsv.2007.10.032;*

[21]. Y. Liu, E. Pavlovskaia and M. Wiercigroch, Experimental verification of the vibro-impact capsule model, *Nonlinear Dyn.,* ***83****(1), 1029-1041 (2016), DOI: 10.1007/s11071-015-2385-6;*

[22]. J.H. Ho, V.D. Nguyen and K.C. Woo, Nonlinear dynamics of a new electro-vibroimpact system; *Nonlinear Dyn.*, ***63****(1-2), 35-49 (2011), DOI: 10.1007/s11071-010-9783-6;*

[23]. Y. Liu, S. Islam, E. Pavlovskaia and M. Wiercigroch, Optimization of the Vibro-Impact Capsule System, *Strojniški vestnik - Journal of Mechanical Engineering* ***62****, 7-8, 430-439 (2016), DOI: 10.5545/sv-jme.2016.3754;*

[24]. Y. Liu, E. Pavlovskaia, M. Wiercigroch, Z. Peng, Forward and backward motion control of a vibro-impact capsule system, *Int. J. Nonlinear. Mech.* ***70,*** *30–46, (2015) DOI: 10.1016/j.ijnonlinmec.2014.10.009;*

[25]. V-D. Nguyen, T-H. Duong, N-H. Chu and Q-H. Ngo, The effect of inertial mass and excitation frequency on a Duffing vibro-impact drifting system, *Int. J. Mech. Sci., 124-125, pp. 9-21 (2017), DOI: 10.1016/j.ijmecsci.2017.02.023.*

Moving mass problem: Complete solution with the effect of initial conditions

Zuzana Dimitrovová[1,2,*]

[1]Departamento de Engenharia Civil, Faculdade de Ciências e Tecnologia, Universidade Nova de Lisboa, Portugal
[2]IDMEC, Instituto Superior Técnico, Universidade de Lisboa, Lisboa, Portugal

Abstract. In this contribution, recently published new semi-analytical solution for the moving mass problem [1] is extended to account for the transient terms that adapt the initial part of the complete solution in a way to match the initial conditions. It is assumed that a mass and a vertical force with harmonic component move by constant velocity along a horizontal infinite beam posted on a two-parameter visco-elastic foundation. The new semi-analytical solution is presented as a sum of truly steady-state terms, harmonic terms induced by the moving mass and transient terms adapting the initial conditions. Closed-form formula is given for the first two types of vibrations. It is concluded that transient terms have in most cases almost negligible effect on the full solution and that the initial conditions can significantly affect the amplitudes of the induced harmonic vibrations, but the induced frequencies are kept without any changes.

1 Introduction

Vibration analyses of beam structures under moving loads undoubtedly contributed to the design of modern railway lines. Deep understanding of dynamic phenomena related to train-track-soil interactions, and therefore, questions regarding the moving load and moving mass problems still attract the scientific community. New modelling approaches and solution methods are always welcomed to underline the necessary understanding. In this context, analytical and semi-analytical solutions have the unquestionable advantages of closed form solutions and quickly obtainable high-precision results solely in places of interest without the necessity to test numerical parameters ensuring the result convergence.

When the train passage is modelled by moving forces, then usually the critical velocity is the important feature to analyse. When inertia of the moving load is included, then additional feature as instability comes into account. Several authors have dedicated significant part of their research to instability of moving objects. Nevertheless, usually the main concern was on the identification of the instability interval, [2].

Considering the fact that someone may want to calibrate a numerical model, or control the unstable region, then it is pertinent not only to identify the instability velocity interval, but also determine the exact vibration pattern that will lead to instability. With this in mind the new semi-analytical solution was derived in closed form. The exact evolution of vibrations allows determining not only the onset of instability, but also its severity.

In Section 2 the problem is defined and solved. In Section 3 some examples and their validation are shown. The paper is concluded in Section 4.

2 Problem statement and solution

2.1. Assumptions

A uniform motion of a mass traversing a horizontal infinite beam posted on a two-parameter visco-elastic foundation is assumed. The beam has uniform cross section, its material is homogeneous and isotropic and it obeys linear elastic Euler-Bernoulli theory. It is further assumed that the moving mass is always in contact with the beam and no friction is considered at the contact point. The aim of the analysis is to determine full deflection shapes of the beam as a function of space and time, describing entirely the vertical vibrations induced by the moving mass and force. The problem at hand is depicted in Figure 1.

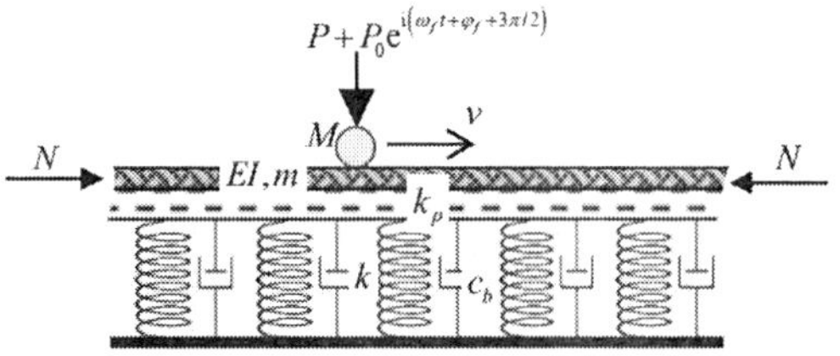

Fig. 1. Infinite beam on a visco-elastic two-parameter foundation subjected to a moving load and a normal force.

* Corresponding author: zdim@fct.unl.pt

In Figure 1 the following symbols are used: EI, m, and N stand for the bending stiffness and mass per unit length of the beam, and a normal force acting on the beam normal axis. k, k_p and c_b are Winkler's and Pasternak's moduli of the foundation and the coefficient of viscous damping of the foundation. Further in Figure 1, M designates the moving mass, P is the constant part of the moving force and the effect of the surface irregularity is approximated by a harmonic component with sine evolution in time t, which has an amplitude P_0, a frequency ω_f and a phase angle φ_f. The harmonic part of the moving force is given more conveniently in the complex domain, thus the phase angle $\varphi_f + 3\pi/2$ is used to ensure the correspondence with $\sin(\omega_f t + \varphi_f)$. Finally, v is the constant velocity of the mass/force system.

2.2 Solution

At first, the deflection of the loading point is derived. Next, the full deflection shapes can be obtained by joining two semi-infinite beams. The solution procedure for the determination of the loading point displacement follows these steps: moving coordinate r is introduced; dimensionless parameters are substituted; Laplace transform in time is applied; Fourier transform in space is applied; and then the analytical form of the displacement image can be obtained and inverse Fourier transform can be applied in an analytical way to determine the Laplace image of the loading point displacement.

To simplify the expressions, a leading polynomial form can be designated as

$$D(p,q) = p^4 - 4p^2(\eta_N - \eta_S + \alpha^2) \\ -4q^2 + 8\alpha pq + 8\mathrm{i}\eta_b q - 8\mathrm{i}\eta_b \alpha p + 4 \quad (1)$$

where p and q are transformed variables: iq was used in Laplace transform in dimensionless time τ, thus corresponds to frequency; p was used in Fourier transform in dimensionless spatial coordinate ξ, thus corresponds to the transformed moving space coordinate. Other parameters are specified as

$$\eta_b = \frac{c_b}{2\sqrt{mk}},\ \eta_N = \frac{N}{2\sqrt{kEI}},\ \eta_S = \frac{k_p}{2\sqrt{kEI}},\ \xi = \chi r \quad (2)$$

$$\tau = \chi v_{cr} t,\ \chi = \sqrt[4]{\frac{k}{4EI}},\ \alpha = \frac{v}{v_{cr}},\ v_{cr} = \sqrt[4]{\frac{4kEI}{m^2}} \quad (3)$$

where v_{cr} is the critical velocity of a constant force moving uniformly on the beam on Winkler's foundation k. Then the Laplace image of the loading point displacement is given by

$$\tilde{W}(0,\mathrm{i}q) = -\frac{4\mathrm{i}\left((q-\hat{\omega}_f)\eta_P + \eta_{P_0} q \mathrm{e}^{\mathrm{i}(\varphi_f + 3\pi/2)}\right)K(q)}{q(q-\hat{\omega}_f)(\pi - 2\eta_M q^2 K(q))} \quad (4)$$

where

$$\eta_M = \frac{M\chi}{m},\ \eta_P = \frac{P}{P} = 1,\ \eta_{P_0} = \frac{P_0}{P},\ \hat{\omega}_f = \frac{\omega_f}{\chi v_{cr}} \quad (5)$$

and

$$K(q) = \int_{-\infty}^{\infty} \frac{dp}{D(p,q)},\ C(q) = \frac{4}{\pi}K(q) \quad (6)$$

where $C(q)$ is the frequency dependent flexibility of the foundation.

As the last step, the inverse Laplace transform is accomplished. This starts with the definition

$$f(t) = \frac{1}{2\pi \mathrm{i}} \lim_{T\to\infty} \int_{a-\mathrm{i}T}^{a+\mathrm{i}T} e^{st} F(s)\, ds \quad (7)$$

where a is positive, real and greater than the real part of all singularities. In the expression above, the Laplace variable s=iq. Further, the variable s is switched to q and the final result is obtained by contour integration.

For correct application of the methods of the contour integration, it is necessary to analyse the behaviour of function $K(q)$. It is possible to conclude that the polynomial expression from Eq. (1) has always four complex roots and thus $K(q)$ can be obtained by the contour integration as well. The only exception is when $D(p,q)$ has real and multiple roots, which can only happen when $D(p,q)$ has real coefficients, which only occurs when Im(q)=iη_b. By further analysis it can be concluded that if certain two values of q have the same real parts and the imaginary parts are at the same distance from iη_b, then the corresponding $K(q)$ values are complex conjugates of each other. This also means that there is a discontinuity, or in other words, multiple value of $K(q)$ in complex q-plane along the horizontal line which crosses the imaginary axis at iη_b. In fact, by analysing the nature of the roots of $D(p,q)$ with real coefficients, it can be concluded, that these is a certain frequency, which can be called the "cutting frequency", q_C, for which $D(p,q)$ has multiple roots. The region with four complex roots is then limited by a certain velocity ratio α_C. For velocities lower than α_C there is an interval ($-q_C$+iη_b, $+q_C$+iη_b) where the imaginary part of $K(q)$ is zero, which ensures continuity along the part of the line designated above as discontinuity line. Such region is, however, not identified for velocity ratio higher than α_C, and discontinuity in the imaginary part of $K(q)$ is satisfied along the full discontinuity line.

Due to the occurrence of multiple values as described above, the contour integration must be adapted in the way to avoid these regions. Thus

$$\tilde{w}_0(\tau) = \sum \mathrm{res}\left(\mathrm{i}\tilde{W}(0,q)\mathrm{e}^{\mathrm{i}q\tau}, q\right) + I_{bc} \quad (8)$$

where I_{bc} is the contribution obtained by the integration along the discontinuity regions, which can be named as branch cuts, however, their meaning is slightly different, because their position is not arbitrary. Other parts of the final result expressed in Eq. (8) are harmonic functions that are obtained by the sum of all residues. The type of residues additionally separates the solution to purely steady state solution, that could be obtained by double Fourier transform, [3]. In this case it is composed by the residue at $q=0$, which represents the constant force contribution in form of

$$\bar{w}_{0,1}(\tau)=\frac{4K(0)}{\pi}\eta_P \tag{9}$$

and by the residue at $q=\bar{\omega}_f$, which represent the harmonic force contribution as

$$\bar{w}_{0,2}(\tau)=\frac{4K(\bar{\omega}_f)\eta_{P_0}e^{i(\varphi_f+3\pi/2)}}{\left(\pi-2K(\bar{\omega}_f)\eta_M\bar{\omega}_f^2\right)}e^{i\bar{\omega}_f\tau} \tag{10}$$

Other harmonic parts must be added by the remaining residues. This is, however, not an easy task, because the poles must be determined as complex roots of complex equation, which reads

$$\pi=2\eta_M q^2 K(q) \tag{11}$$

It is possible to conclude that in such a case the roots will always come in pairs, connected by

$$q_{M_2}=(-q_{M_1})^* \tag{12}$$

where * stand for complex conjugate value. Most commonly, there is only one pair of these roots, but there can be also two pairs or none. Having a root, q_M, the corresponding harmonic part is given by

$$\bar{w}_{0,3}(\tau)=\frac{-2K(q_M)\eta_P}{q_M^3 K_{,q}(q_M)\eta_M+\pi}e^{iq_M\tau} \tag{13}$$

for the constant force and by

$$\bar{w}_{0,4}(\tau)=\frac{-2q_M K(q_M)\eta_{P_0}e^{i(\varphi_f+3\pi/2)}}{\left(q_M^3\eta_M K_{,q}(q_M)+\pi\right)\left(q_M-\bar{\omega}_f\right)}e^{iq_M\tau} \tag{14}$$

for the harmonic one.

2.3 Influence of the initial conditions

The solution presented in the previous section was directly derived for the homogeneous initial conditions. When these conditions are not homogeneous, then the application of the Laplace transform must be adapted accordingly. Changes must be introduced on both sides of the transformed governing equation. On the right-hand side, it must be simply added

$$4\eta_M\left(\bar{w}_{,\tau}(0,0)+\bar{q}\bar{w}(0,0)\right) \tag{15}$$

The other term, located on the left-hand side is not so easy to implement:

$$H(p,0)= -4\bar{W}_{,\tau}(p,0)-4\bar{W}(p,0)(iq-2i\alpha p+2\eta_b) \tag{16}$$

Nevertheless, the leading effect can be attributed to the terms in Eq. (15) that directly express the initial conditions of the loading point and can significantly affect the amplitudes of the induced harmonics. For instance, the function for which residues are calculated in Eq. (8) for non-zero initial displacement and constant force changes to

$$\frac{2\left(2-\eta_M q^2\bar{w}(0,0)\right)K(q)\eta_P}{q\left(\pi-2\eta_M q^2 K(q)\right)}e^{iq\tau} \tag{17}$$

which alteres the contribution in Eq. (13) to

$$\bar{w}_{0,3}(\tau)=\frac{-\left(2-\eta_M q^2\bar{w}(0,0)\right)K(q_M)\eta_P}{q_M^3 K_{,q}(q_M)\eta_M+\pi}e^{iq_M\tau} \tag{18}$$

3 Examples

Examples presented in this section are related to railway applications. Only quite soft foundation is assumed, and the effect of Pasternak foundation, normal force and harmonic force is neglected. More details about these additional influences can be found in [1]. Moving force will be considered as a typical axle load, and for simplicity, the moving mass is assumed to have a value for which the moving force would represent the associated weight, according to approximate value of the gravity acceleration 10m/s^2. Nevertheless, dimensionless parameters are used and thus same results would be possible to obtain by different combinations of the input data. Values are listed in Table 1.

Table 1. Values used in numerical examples.

Property	Value
EI (10^6N m^2)	6.4
m (kg m^{-1})	60
k (10^6N m^{-2})	1
P (10^3N)	100
M (kg)	10000

The associated characteristics and dimensionless values for this case are

$$v_{cr}=290.4\text{m/s},\ \chi=0.445\text{m}^{-1},\ \eta_M=74.1 \tag{15}$$

Further, w_{st}=0.022m, which is the static displacement exerted by the constant part of the force on the beam on a Winkler foundation, that is used to obtain dimensionless displacements.

Firstly, the induced frequencies for the case of damping η_b = 0.05 are shown. These can be obtained by the iterative procedure described in [1]. Their evolution with velocity ratio is plotted in Figure 2. To each value shown there, another one exists in conformity with Eq. (12). It is seen that the values smoothly progress and the negative imaginary part, as the onset of instability is reached at α=1.07. It is also seen that when discontinuity in $K(p,q)$ is reached, the frequency lines are cut. This happens when the imaginary part of the frequency reaches η_b, *i.e.* 0.05. This example was selected as an illustrative case, because there are certain velocity ratios for which none, one or two pairs of induced frequencies exist.

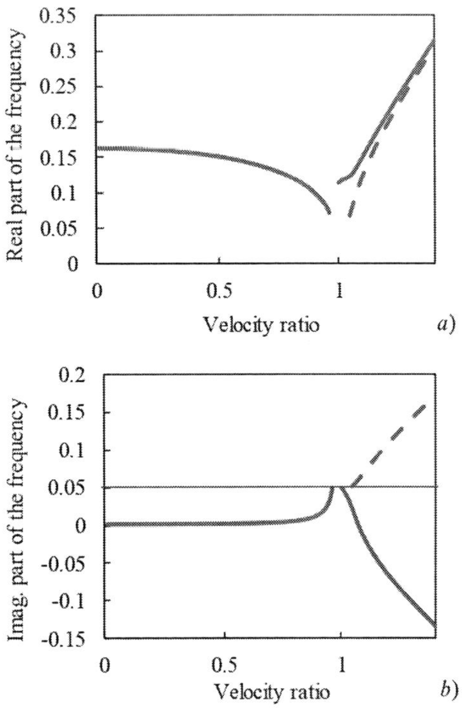

Fig. 2. Mass induced frequencies: a) the real part, b) the imaginary part.

Results for displacement evolution under the load are shown next. In all further figures validation on finite beams is presented as black dotted line. It is seen that the agreement between the results is excellent. In Figure 3 the case where only one pair of induced frequencies exists is shown. α = 0.344 is selected (100m/s) and comparison is done for η_b = 0 and 0.1. It is seen that when there is no damping, there are large oscillations around the steady-state force displacement, with amplitude practically exactly equal to this steady state value. Such oscillations theoretically last for ever. When damping is included, these oscillations are gradually damped and after some time only the truly steady-state stage exists. In Figure 3 only the harmonic parts are plotted, namely the ones from Eq. (9) and (13), because the transient part is completely negligible. It was shown in [4] that the transient part is important in this case only for low mass ratios, approximately equal to 10 or lower.

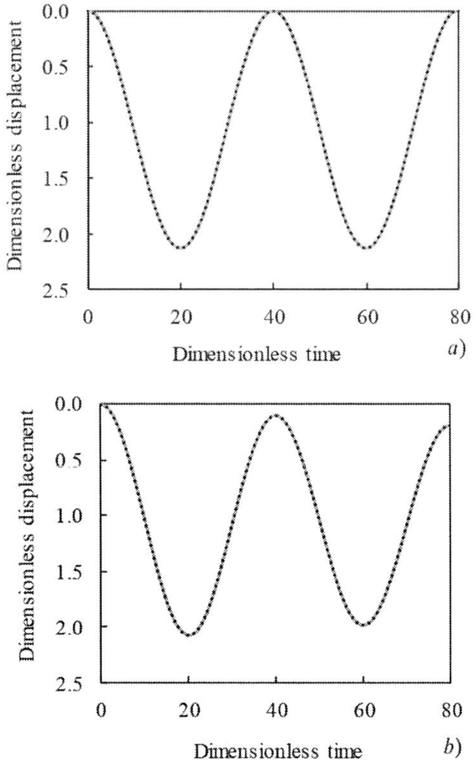

Fig. 3. Displacement under the load: a) η_b = 0, b) η_b = 0.1.

Next case is for η_b = 0.05. It is seen in Figure 2 that last one pair of frequencies occurs for α = 0.96. Then for α = 0.97 - 0.99 there are no induced frequencies, and one pair is gained again for α = 1. This means that these cases must have significant contribution of the transient part, that needs to adjust the initial conditions. Results are shown in Figures 4, 5 and 6. In Figure 4 the case of α = 0.96 is plotted.

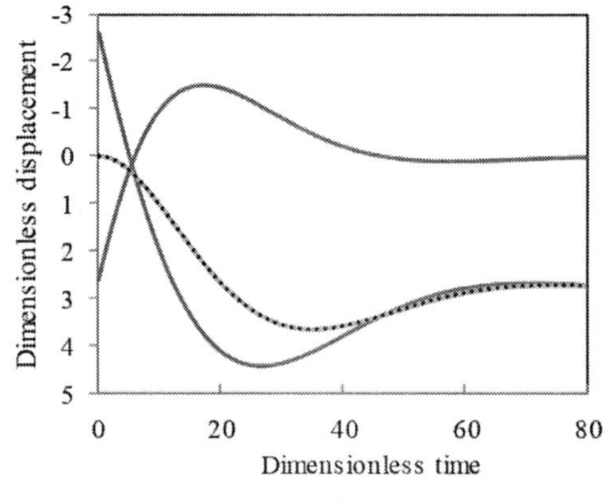

Fig. 4. Displacement under the load for α = 0.96: harmonic part (red), transient part (violet), full solution (green).

It is seen in Figure 4 that the harmonic part starts at values very different from the initial conditions, therefore quite large values of the transient part must correct this fact. But the influence of the transient solution rapidly disappears.

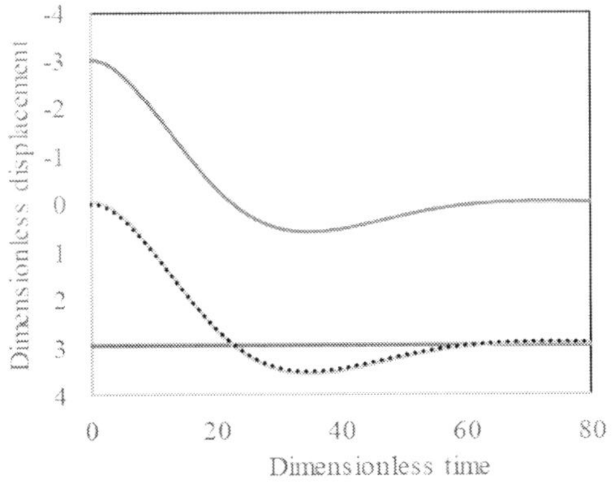

Fig. 5. Displacement under the load for $\alpha = 0.97$: harmonic part (red), transient part (violet), full solution (green).

In Figure 5 the case for $\alpha = 0.97$ is plotted. As there is no induced frequency, the harmonic part is constant, as it represents the steady-state solution only. Large influence of the transient solution corrects the initial conditions in the same way as before, and rapidly disappears.

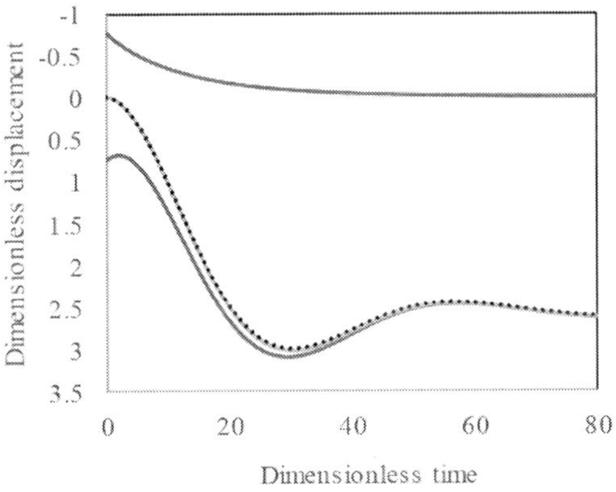

Fig. 6. Displacement under the load for $\alpha = 1$: harmonic part (red), transient part (violet), full solution (green).

When α reaches 1, there is again one pair of induced frequencies. This case is plotted in Figure 6. It is seen that as in all other cases, the transient part adjusts the initial conditions and then its effect over the solution is negligible.

Last example shown in this paper is the case when two pairs of induced frequencies exist. In such cases the second pair has always small influence on the solution, because it has large imaginary part, that is positive and thus quickly damp the vibration. Nevertheless, it is necessary to account for such solution, otherwise the initial conditions would be violated. The case selected has $\eta_b = 0.2$ and $\alpha = 1.26$, which is the first velocity ratio where two pairs of induced frequencies occur for such level of damping. Results are shown in Figure 7.

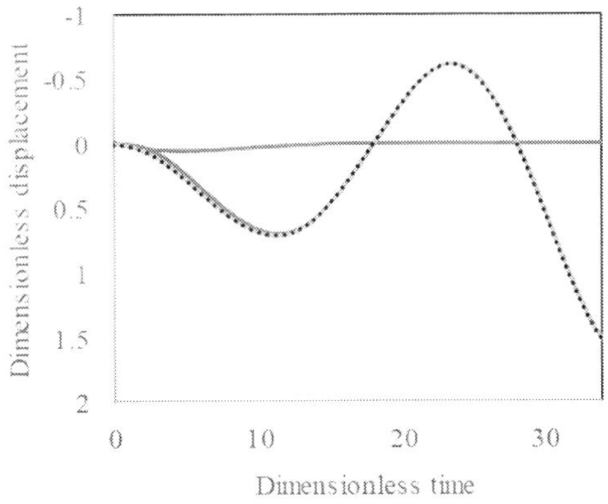

Fig. 7. Displacement under the load for $\alpha = 1.26$: harmonic part (red), transient part (violet), full solution (green).

In Figure 7 it is again seen that the transient part has negligible effect. The case selected belongs already to unstable vibrations, as can be verified by increasing amplitude of the vibrations. It can thus be concluded that the validity of the derived solution is not restricted to some specific velocity range.

4 Conclusion

In this paper closed form solution for moving mass problem was derived and validated. Several cases were discussed with respect to the number of induced pairs of frequencies and dominances of the distinct parts that compose the full solution.

This work was supported by FCT, through IDMEC, under LAETA, project UID/EMS/50022/2013.

References

1. Z. Dimitrovová, Int. J. Mech. Sci. **127**, 142–162 (2017).
2. A.V. Metrikine, H.A. Dieterman, J. Sound Vib., **201**, 567–576 (1997).
3. S. Mackertich, J. Acoust. Soc. Am. **101**, 337-340 (1997).
4. Z. Dimitrovová, *Procedia Engineering* **199**, 2537–2542 (2017).

Analytical approximation of rail bending stress

Piotr Koziol[1,*]

[1]Cracow University of Technology, Faculty of Civil Engineering, Kraków, Poland

Abstract. Phenomena associated with railway dynamics are usually analysed by using numerical approaches due to high computational complexity of such systems. However, classical methods based on analytical modelling are still highly valued and desirable by researchers and railway industry. This paper presents analytical solution representing dynamic response of railway track due to moving train in the case of nonlinear foundation. In published papers, one can find analyses of various characteristics such as velocity and acceleration of vibrations of track layers or bending moments of rails. The approach applied in this paper uses the Fourier transform combined with wavelet based approximation applied to the systems of infinitely long beams. The system of Euler-Bernoulli beams resting on viscoelastic foundation represents two-layer model (or one-layer model) of railway track, commonly used in engineering studies. It is shown that although both methods give good results for displacements, analysis of other characteristics, involving derivatives of higher orders, might lead to wrong results, even in the linear case. Possible reasons of this problem are pointed out. Some modifications of the known dynamic railway track models are proposed for further work.

1 Introduction

Railway track structures can be modelled with a use of many different approaches. FEM based techniques which are widely used in the literature give possibility of discussion of many dynamic effects appearing in rail tracks [1, 2]. Many published results show, however, that they might be burdened with quite significant error and, additionally, need high computational power. This makes them difficult to apply in many cases, especially in dynamical systems parametrical analysis. Alternative approaches are based on analytical computations, mainly using mathematical formulations, i.e. differential equations with initial and boundary conditions [3, 4]. Such methods allow to analyse desirable features of railway track components, if their description by analytical formulas is possible. Usefulness of such a model depends in turn on ability of introduction of additional assumptions to the system, e.g. nonlinear or stochastic properties. The approximation theory gives reliable tools helping to solve these complex systems [3, 5, 6]. Procedures arising from these techniques are usually effective when one deals only directly with exact solution without further computations involving differentiation or integration [7].

In the literature, one can observe a lack of analytical methods for a proper rail stress evaluation. This paper shows two analytical models of railway tracks: one-layer model and two-layer model. Both these models, although suitable for evaluation of vertical vibrations (or their velocity and acceleration), lead to wrong results with regard to characteristics using derivatives in relation to space variable (e.g. rail bending stress). Question of applicability of the analysed models and methods is partially discussed. Some possible ways of models improvements are mentioned as future work.

2 Railway track modelled by differential equations

Two analytical models of railway track are considered in this paper. The one-layer model representing rail vertical vibrations is described by the Euler-Bernoulli beam resting on viscoelastic foundation with nonlinear stiffness. This model is represented by the following equation:

$$EI\frac{\partial^4 y_r}{\partial x^4}+N\frac{\partial^2 y_r}{\partial x^2}+m_r\frac{\partial^2 y_r}{\partial t^2}+c_r\frac{\partial y_r}{\partial t} + k_r y_r + k_{rN} y_r^{\,3} = q(x,t). \quad (1)$$

The two-layer model can be described by the following system of two differential equations:

$$\begin{aligned} &EI\frac{\partial^4 y_r}{\partial x^4}+N\frac{\partial^2 y_r}{\partial x^2}+m_r\frac{\partial^2 y_r}{\partial t^2}+c_r\left(\frac{\partial y_r}{\partial t}-\frac{\partial y_s}{\partial t}\right) \\ &\quad +k_r(y_r-y_s)-k_{sN}y_s^{\,3}=q(x,t) \\ &m_s\frac{\partial^2 y_s}{\partial t^2}+c_s\frac{\partial y_s}{\partial t}+k_s y_s+k_{sN}y_s^{\,3} \\ &\quad =c_r\left(\frac{\partial y_r}{\partial t}-\frac{\partial y_s}{\partial t}\right)+k_r(y_r-y_s) \end{aligned} \quad (2)$$

The first equation represents rail and the second one describes sleepers layer. The used parameters are: EI [Nm^2] – bending stiffness of rail steel, N [N] – axial force, m_r [kg/m] – rail unit mass, k_r [N/m^2] – stiffness of

* Corresponding author: pkoziol@pk.edu.pl

fasteners, c_r [Ns/m^2] – viscous damping of fasteners, m_s [kg/m] – unit mass of sleepers uniformly distributed along the track. Effect of sleepers bending is neglected. The two parameters: k_s [N/m^2] and c_s [Ns/m^2] denote stiffness and viscous damping of sleepers foundation, respectively. Geometrical imperfections of rolling surface of the rail head are represented by regular cosine wave with frequency Ω [rad/m] and amplitude s_0 [m]. The load q [N/m] is considered as a set of distributed forces produced by wheels of train and associated with train axles configuration. In the case of constant speed v [km/h], geometrical imperfections produce the varying forces with circular frequency $\omega=\Omega v$ and the amplitude $F_s=s_0k_c$. These forces are distributed on the wheel-rail contact area and the contact is modelled by a single spring k_c. The terms $k_{rN}y_r^3$ (eq. (1)) and $k_{sN}y_s^3$ (eq. (2)) represent nonlinear factor associated with nonlinear part of rail foundation stiffness k_{rN} [N/m^4] and rail track foundation stiffness k_{sN} [N/m^4], respectively [6, 8, 9].

One can apply the moving coordinate system ($\eta=x$, $\xi=x-vt$) to equations (1-2) and then these formulas can be transformed to the following form:

$$EI\frac{\partial^4 y_r}{\partial\xi^4}+N\frac{\partial^2 y_r}{\partial\xi^2}+m_r(\frac{\partial^2 y_r}{\partial t^2}-2v\frac{\partial^2 y_r}{\partial t\partial\xi}+v^2\frac{\partial^2 y_r}{\partial\xi^2})+c_r(\frac{\partial y_r}{\partial t}-v\frac{\partial y_r}{\partial\xi})+k_r y_r+k_{rN}y_r^3=q(\xi,t) \quad (3)$$

and

$$EI\frac{\partial^4 y_r}{\partial\xi^4}+N\frac{\partial^2 y_r}{\partial\xi^2}+m_r(\frac{\partial^2 y_r}{\partial t^2}-2v\frac{\partial^2 y_r}{\partial t\partial\xi}+v^2\frac{\partial^2 y_r}{\partial\xi^2})+c_r(\frac{\partial y_r}{\partial t}-v\frac{\partial y_r}{\partial\xi}-\frac{\partial y_p}{\partial t}+v\frac{\partial y_s}{\partial\xi})+k_r(y_r-y_s)-k_{sN}y_s^3=q(\xi,t) \quad (4a)$$

$$m_s(\frac{\partial^2 y_s}{\partial t^2}-2v\frac{\partial^2 y_s}{\partial t\partial\xi}+v^2\frac{\partial^2 y_s}{\partial\xi^2})+c_s(\frac{\partial y_r}{\partial t}-\frac{\partial y_s}{\partial\xi})+k_s y_s-c_r(\frac{\partial y_r}{\partial t}-v\frac{\partial y_r}{\partial\xi}-\frac{\partial y_s}{\partial t}+v\frac{\partial y_s}{\partial\xi})-k_r(y_r-y_s)+k_{sN}y_s^3=0 \quad (4b)$$

These models can be solved by applying either the Fourier series or the Fourier transform, depending on assumptions with regard to the E-B beams and the decision about the nonlinearity investigation.

The load representing moving train can be described as follows:

$$q(\xi,t)=P_S(\xi,t)+P_D(\xi,t)=\sum_{l=0}^{l_T}(\frac{P_0}{a}+\frac{P_1}{a}e^{i\omega t})\cos^2(\frac{\pi(\xi-s_l)}{2a})H(a^2-(\xi-s_l)^2) \quad (5)$$

where

$$P_S(\xi,t)=\sum_{l=0}^{l_T}\frac{P_0}{a}\cos^2(\frac{\pi(\xi-s_l)}{2a})H(a^2-(\xi-s_l)^2) \quad (6a)$$

is the stationary part of load, i.e. constant in time, and

$$P_D(\tilde{x},t)=\sum_{l=0}^{l_T}\frac{P_1}{a}e^{i\omega t}\cos^2(\frac{\pi(\xi-s_l)}{2a})H(a^2-(\xi-s_l)^2) \quad (6b)$$

represents the part of load varying in time, produced by rail head surface imperfections of cosine form and constant amplitude. $H(.)$, $2a$, s_l, l_T, ω and V are the Heaviside function, the span of each load related to single wheel, the distance of consecutive forces produced by wheels, the number of axles, the frequency and the velocity of the moving load, respectively. The axles configuration can be also introduced by inclusion of the phase shift associated with the wheel position on the cosine shape of irregularity [3, 10].

3 Solution

The following method of solution can be applied to both systems, (3) and (4a-4b), in the case of infinitely long beams. It deals also with the assumed nonlinearity of foundation stiffness.

The cubic term representing nonlinearity can be expanded by Adomian polynomials [5, 6, 10]:

$$y^3(x)=\sum_{j=0}^{\infty}Q_j(x)=\sum_{j=0}^{\infty}\frac{1}{j!}\left[\frac{d^j}{ds^j}\left(\sum_{k=0}^{\infty}\lambda^k y_k(x)\right)^3\right]_{s=0} \quad (7)$$

Both models, the one-layer and the two-layer can be solved by applying the Fourier transform

$$\tilde{y}(\omega)=F[y(\xi)]=\int_{-\infty}^{\infty}y(\xi)e^{-i\omega\xi}d\xi \quad (8)$$

Because the form of solution in the transform domain becomes quite complicated, classical methods of retransformation are inefficient when applied to parametrical analysis. The response in the physical domain can be effectively obtained from the transform domain by applying the coiflet based approximation of the inverse Fourier transform [5, 6, 10]:

$$\begin{aligned} y(\xi) &= \lim_{n\to\infty} y_n(\xi) \\ &= \lim_{n\to\infty} \frac{1}{2^{n+1}\pi} \prod_{k=1}^{k_p} \Big(\sum_{j=0}^{3N-1} p_j e^{ij\xi/2^{n+k}} \Big) \\ &\quad \cdot \sum_{k=-\infty}^{+\infty} \tilde{y}\big((k + \sum_{k=0}^{3N-1} kp_k) 2^{-n}\big) e^{i\xi k 2^{-n}} \end{aligned} \tag{9}$$

where N is the degree of the used coiflet filter (p_k) and k_p is a number evaluated on the basis of stabilization of the obtained solution [5, 6, 11].

4 Examples and discussion

In the case of rail and sleepers displacement, one can find sufficiently exact solutions by applying the technique described in section 3. These solutions can be successfully used in parametrical analysis when one takes into account engineering applications. They have been also verified by experimental measurements [10, 12]. In the case of stresses produced by dynamic excitations, the problem of analytical modelling becomes more difficult.

The following system of parameters is taken in preparation of examples:
$EI = 6.4*10^6$ Nm^2, $N = 0$, $m_r = 60$ kg/m, $k_r = 8.8*10^7$ N/m^2, $c_r = 3950$ Ns/m^2, $m_s = 267$ kg/m, $k_s = 8.5*10^7$ N/m^2, $c_s = 8.2*10^3$ Ns/m^2 and $v = 150$ km/h. The nonlinear stiffness coefficients k_{rN} and k_{sN} can be taken similarly to values considered in past publications [6, 10-12], although it is neglected in this paper for calculation of normal bending stress. It is assumed that the part of load P_D (see eq. (6b)) is related to the rail deflection between sleepers and there are no other imperfections on rolling surface of rail head. The axles configuration of fast train Pendolino EMU250 is also considered.

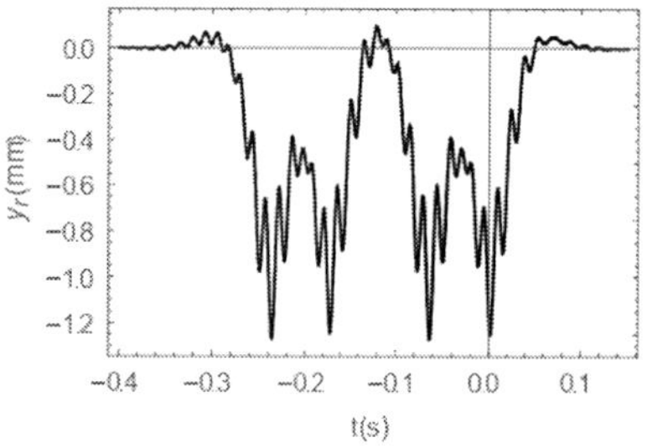

Fig. 1. Vertical vibrations of rail.

Figure 1 presents the dynamic response of railway track to moving train in terms of vertical displacements of rails obtained analytically by using the two-layer model and the coiflet based technique in the case of linear foundation stiffness.

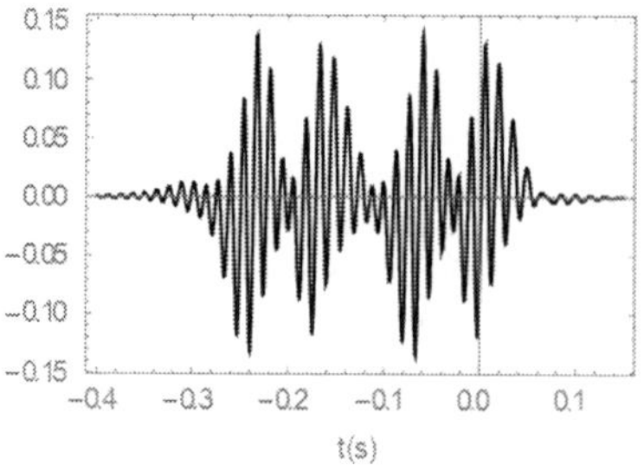

Fig. 2. Velocity of rail vertical vibrations.

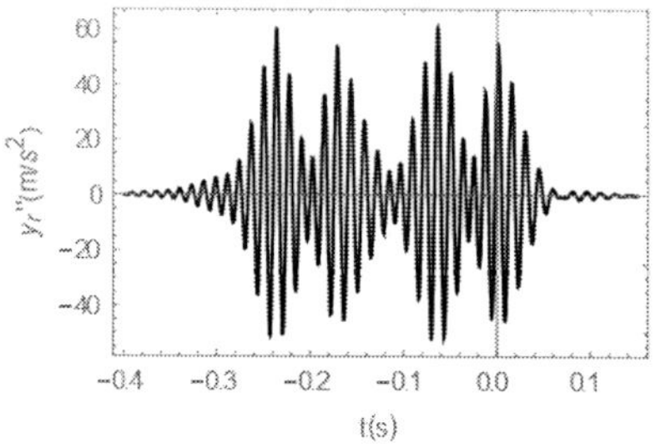

Fig. 3. Vertical vibrations acceleration.

Figures 2-3 show vibration velocity and vibration acceleration for the same system of parameters. As mentioned before, the results for vertical vibrations have been verified experimentally [10, 12]. When one deals with vibrations velocity or acceleration, the obtained results need to be verified by experimental measurements or other methods of solution for extended range of parameters, although maximal values are close to real ones, at least in the case of the track in relatively good condition. Some doubts arise from the analysis of the normal bending stresses, involving derivatives of higher order with respect to the space variable. It can be observed that analytically obtained results significantly exceed limit values of the strength of rail steel R260 (around 800 MPa).

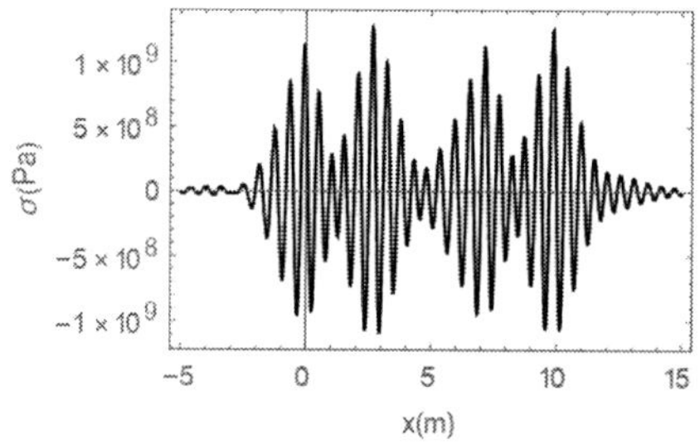

Fig. 4. Distribution of normal stress in rail obtained by using two-layer model and wavelet approximation.

Figure 4 presents the normal stress distribution in rail with irregularity related to the sleeper spacing, showing the method ability to analytically calculate stresses.

https://doi.org/10.1051/matecconf/201814805002

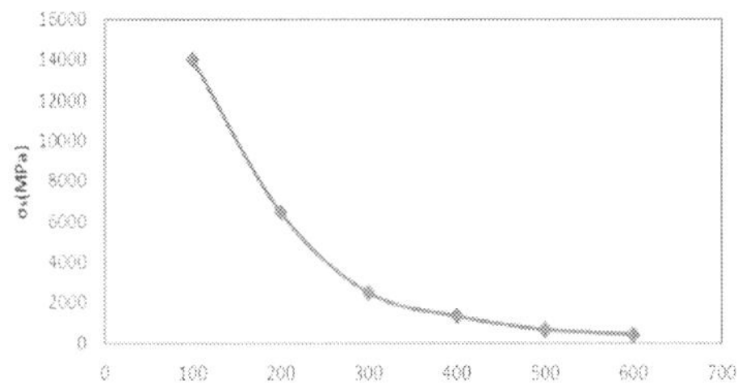

Fig. 5. Maximal normal stress in rail axis depending on the length [mm] of cosine imperfection [7].

For some lengths of waves, associated with the considered type of irregularity appearing on rail head rolling surface, the maximal value of the normal stress can significantly exceed the limit value (a few hundred percent and more). For example, for the length of irregularity equal to 30 cm (cosine type of imperfection), which is twice less than the distance between sleepers, and the train speed V=200 km/h, the obtained maximal value of the normal stress in rail axis can be estimated around 1500 MPa. This value is obtained under assumption that the force generated by the considered irregularity is the same as for a length of 60 cm, which corresponds to the sleepers spacing. In reality, this additional dynamic force can be stronger and therefore the resulting normal stress should reach higher values. Figure 5 shows maximal normal stress in rail axis depending on the length [mm] of cosine imperfection, under described above assumption about forces.
The project [7] was aimed at showing examples of analytically derived stresses exceeding limit values when using models validated and known from the literature. The analysis was carried out for a wide range of physical parameters and train speeds showing that a lack of appropriate methods (models) for the normal stress calculation should be underlined.

5 Conclusions

One can observe that the obtained results for normal bending stress in rail axis exceed, for some systems of parameters, limit values independently of model and method of analytical solution. One can say that applied approaches (models and/or methods) are insufficient for stresses analysis. In order to improve this analysis, one might suggest consideration of additional mechanical features of the investigated systems. These features remain unrecognised so far. In future, they should lead to reformulation of railway track models. Among them, one should mention a more detailed approach to rail modelling. This modification should avoid treating the rail as a homogeneous beam. Some preliminary results are obtained for "head on web" problem or rail internal damping included in the dynamic beam equation. These results are valid for some specific range of parameters only. They are especially strongly dependent on the load frequency. Therefore the undertaken analysis is not satisfactory and should be extended. One should underline that all analytical models and the wavelet based method considered in this paper are valid for rails and sleepers vertical displacement analysis.

References

1. D. Tran, Adv. Struct. Eng., **12**, 3, 399-409 (2009)
2. P. Amado-Mendes, P.A. Costa, L.M.C. Godinho, P. Lopes, Eng. Struct., **104**, 141-154 (2015)
3. W. Czyczula, P. Koziol, D. Kudla, S. Lisowski, J. Vib. Control., **23**, 18, 2989-3006 (2017)
4. P. Koziol, *Proceedings of ICoEV2015, International Conference on Engineering Vibration, Ljubljana, Slovenia, 7-10 September 2015*, ISBN 978-961-6536-97-4, paper 273, (2015)
5. P. Koziol, VDM Verlag Dr. Müller, Saarbrucken, (2010)
6. P. Koziol, J. Theor. Appl. Mech., **52**, 3, 687-697 (2014)
7. P. Bulinski, MSc thesis (in Polish), Cracow University of Technology (2016)
8. T. Dahlberg, *Proceedings of Fifth International Conference on Structural Dynamics*, Munich (2002)
9. M.H. Kargarnovin, D. Younesian, D.J. Thompson, C.J.C. Jones, Comput. Struct., **83**, 1865-1877 (2005)
10. P. Koziol, Mech. Syst. Signal Process., **79**, 174-181, (2016)
11. P. Koziol, Z. Hryniewicz, Shock Vib., **19**, 995-1007, (2012)
12. P. Koziol, *Proceedings of the Fifteenth International Conference on Civil, Structural and Environmental Engineering Computing*, Civil-Comp Press, Stirlingshire, UK, Paper 199 (2015)

The shaker parameters estimation, a first step to virtual testing

Jonathan Martino[1,*], *Kristof* Harri[1]

[1]Royal Military Academy, 30 Av. de la Renaissance, 1000 Bruxelles, Belgique

Abstract. The increase in computing resources makes simulation methods more affordable. The processing speed allows real time analysis or even more rapid tests analysis offering a real tool for test prediction and design process optimization. Vibration tests are no exception to this trend.
The so called 'Virtual Vibration Testing' offers solution among others to study the influence of specific loads such as optimized signals, to better anticipate the boundary conditions between the exciter and the structure under test or to study the influence of small changes in the structure under test.
This article will first present a virtual vibration test modeling with a main focus on the shaker model and will afterwards present the experimental parameters determination.
The classical way of modeling a shaker is to consider the shaker as a simple mechanical structure augmented by an electrical circuit that makes the shaker move. The developed model is a two degrees of freedom mechanical lumped parameters model and the electrical circuit model takes the coil impedance and the dynamic back-electromagnetic force into account. The establishment of the equations of this model, describing the dynamics of the shaker, is presented in this article and is strongly related to the internal physical quantities of the shaker. Those quantities will be reduced into global parameters which will be estimated through experiments in order to get a fully functional shaker model.
An experimental modal analysis will also be carried out to extract the modal parameters of the shaker and to combine them with the electrical measurements.
Finally this article will conclude with an experimental validation of the model.

1 Introduction

Vibration tests are an important part of the development process of a product and the optimization of these products in order to satisfy the imposed vibration requirements is essential. A constantly renewed aim is undoubtedly to push the limits of the possible. Is it possible to improve a product before it even exists? Is it possible to perform a physical test in less time or with less needed infrastructure? How to do more with less?

Virtual vibration testing aims to propose an approach to handle those aspects.

Increased computing capabilities make simulation tools more affordable. It is now possible to obtain a response in real time or even more rapidly from the response of a structure to a vibration test, to optimize the physical test or the structure to be tested, or to study the influence of specific loads, to implement the optimized signals in order to get more from the test infrastructure, to better anticipate the boundary conditions between the exciter and the structure under test, etc.

The simulation of a physical vibration test is carried out by modeling each of the constituent elements of this test [1, 2]. The shaker plays a primordial role and the validity of the simulation depends almost exclusively on the validity of the model of the shaker. The latter is based on the physical characteristics of the constituent elements. A good determination of the parameters of the shaker will therefore be essential.

Figure 1 shows a cross-sectional view of an electrodynamic shaker. It essentially operates as a loudspeaker – vibration is generated by a current flowing in a coil immersed in a magnetic field – but where guides and suspensions have been added, making this system an electro-mechanical system.

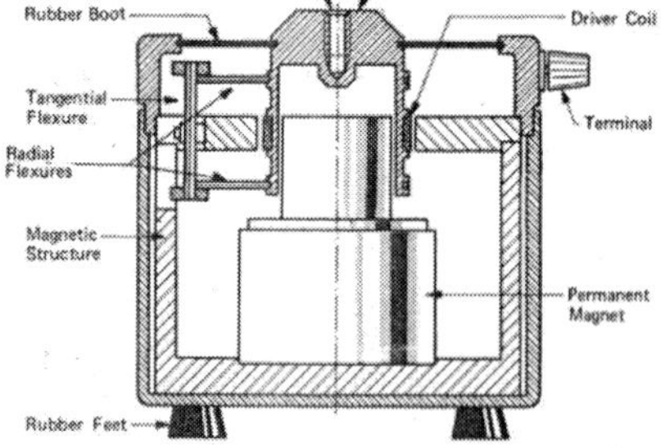

Fig. 1. Cross-sectional view of an electrodynamic shaker.

* Corresponding author: jonathan.martino@rma.ac.be

2 Shaker modeling

The shaker can be modeled by mechanical lumped parameters but an electrodynamic relationship has to be added to represent the behavior of the coil when a signal is applied to its terminals.

Olsem [3] and McConnell [4] propose a model with two degrees of freedom (DOFs) while Lang and Snyder [5] propose a four DOFs model. Unlike Olsem and McConnel, this four DOFs model makes it possible to observe all the characteristic modes of a shaker – isolation, suspension and coil mode.

Due to the small size of the shaker used in the work, the table and moving coil are considered as a rigid assembly [5]. The corresponding DOF can be eliminated and the model can be reduced to a three DOFs model (two mechanical and one electrical).

Figure 2(a) and Figure 2(b) respectively represent the mechanical and electrical models.

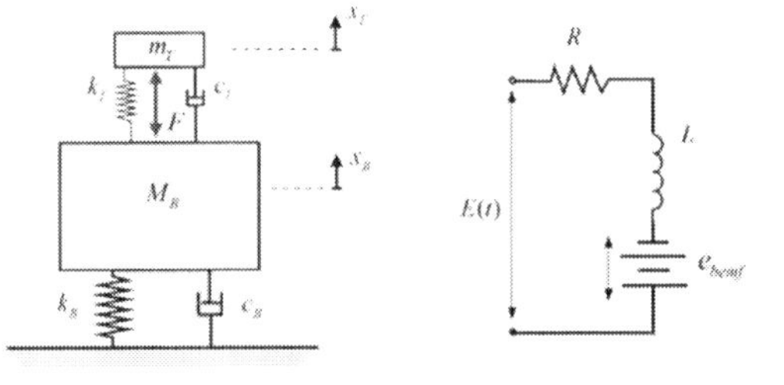

(a) (b)

Fig. 2. Shaker model as a mechanical 2DOFs system (a) added with the electrical 1DOF model of the moving coil (b).

The force F applied to the table is generated by the coil and, by reaction, also applies to the body of the shaker. The coil moves vertically and its relative speed in the magnetic field creates a back electromagnetic force (bemf).

The integration of the equations of motion with the expression of the drive voltage as a function of the current flowing through the coil leads to the following equations [2]:

$$\begin{bmatrix} m_T & 0 & 0 \\ 0 & M_B & 0 \\ 0 & 0 & 0 \end{bmatrix} \begin{bmatrix} \ddot{x}_T \\ \ddot{x}_B \\ 0 \end{bmatrix} + \begin{bmatrix} c_T & -c_T & 0 \\ -c_T & c_B + c_T & 0 \\ \mu_{bemf} & -\mu_{bemf} & L \end{bmatrix} \begin{bmatrix} \dot{x}_T \\ \dot{x}_B \\ di/dt \end{bmatrix} + \begin{bmatrix} k_T & -k_T & -\mu_f \\ -k_T & k_B + k_T & \mu_f \\ 0 & 0 & R \end{bmatrix} \begin{bmatrix} x_T \\ x_B \\ i \end{bmatrix} = \begin{bmatrix} 0 \\ 0 \\ E \end{bmatrix} \quad (1)$$

Where m_T, k_T, c_T, M_B, k_B and c_B are the mass, stiffness and damping of the table and shaker body, respectively, and R and L are the electrical resistance and inductance of the moving coil. μ_f is the force constant and μ_{bemf} the back electromagnetic force constant of the system.

The left hand side of (1) describes the shaker behavior while the right hand side represents the drive voltage applied to the terminals of the shaker.

3 Determination of the mechanical parameters

The six mechanical parameters are the mass, stiffness and damping of the two mechanical degrees of freedom of the system, and are determined experimentally.

The method used consists in alternatively blocking each of the degrees of freedom and letting only the studied degree oscillate. We obtain two configurations: the shaker without isolation studs (free table test) to characterize the suspension of the coil, and the shaker with blocking of the vibrating table (blocked table test) to characterize the isolation studs.

The system is mounted on a larger shaker which is used to perform a sine sweep. The response of the system for the corresponding configuration is measured and identified with respect to the classical response of an oscillator with one degree of freedom.

3.1 Experimental setup

The shaker used is a modal shaker of type B&K4809 whose mechanical characteristics supplied by the manufacturer are listed in Table 1.

In order to block the coil, we mounted a rigid structure between the body of the shaker and its vibrating table. It consists of a steel beam clamped at the top of the table and is depicted on Figure 3. We can notice the presence of an impedance head which will later be used in combination with a current sensor for the determination of the electrical parameters.

Fig. 3. The rigid structure limits the movement of the vibrating table. An impedance head is clamped between the vibrating table and the beam of the rigid structure.

We also have to eliminate the electrical degree of freedom by shorting the coil. During the free table test, the moving coil will have a relative speed with respect to the magnetic field. According the Lorentz force, a current tends to settle in the coil, but the latter can only circulate if the coil is short-circuited otherwise it creates a voltage at the terminals which would result in an electromagnetic damping on the coil.

3.2 Mechanical parameters identification

Figure 4 and Figure 5 show the plotted response curves as a least square estimation of the measurements for both blocked table and free table configurations.

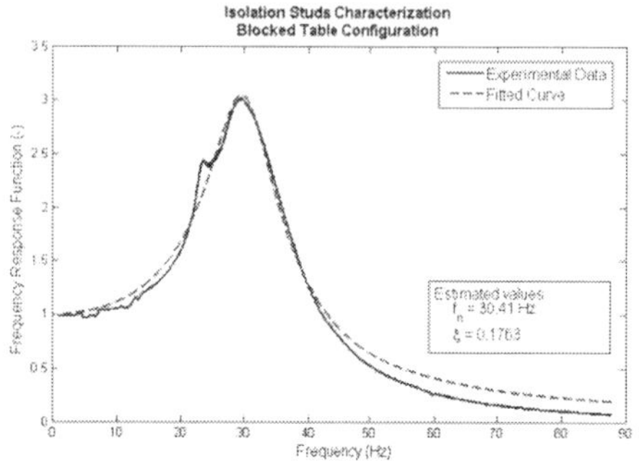

Fig. 4. Response curve of the isolation studs and the one DOF estimation in the blocked table configuration. With f_n the natural frequency and ξ the damping ratio.

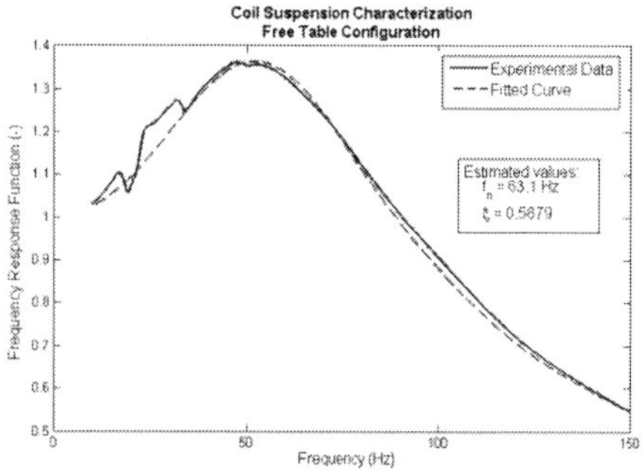

Fig. 5. Response curve of the vibrating table and the one DOF estimation in the free table configuration. With f_n the natural frequency and ξ the damping ratio.

The responses of the system correspond to 1DOF oscillators and using a least square estimation we obtained the values of the mechanical parameters. Those values are given in Table 1 with respect to the parameters given by the manufacturer.

3.3 Modal parameters

The developed method requires an external possibility of making the shaker vibrate and is quite difficult to apply to larger shakers. In this case, modal analysis is an alternative since the mechanical parameters and the modal ones are equivalent to characterize the system.

In this section, a short review of the modal theory will be presented to sufficiently master this equivalence as the relationship between movement, global coordinates, modal coordinates, and modal parameters.

Table 1. Mechanical parameters

Parameter	Symbol	Values	
		Given by the manufacturer[6]	Determined experimentally
Mass of the vibrating table	m_T	60 g	76 g[a]
Mass of the body	M_B	8.3 kg	8.375 kg
Damping of the suspension of the coil	c_T	-	35.43 Ns/m
Damping of the isolation studs	c_B	-	564.3 Ns/m
Stiffness of the suspension of the coil	k_T	$12 \cdot 10^3$ N/m	$11.95 \cdot 10^3$ N/m
Stiffness of the isolation studs	k_B	-	$3.06 \cdot 10^5$ N/m

[a]For testing purposes, the original vibrating table is equipped with an extra 16 g fastening bolt.

3.3.1 Theoretical aspects

If we only consider the purely mechanical system as depicted in Figure 2(a), the global equations (1) can be generalized by:

$$M\ddot{z} + C\dot{z} + Kz = F \tag{2}$$

Where M, C, and K are respectively the mass, damping, and stiffness matrices of the system.

Solving the eigenvalues problem for the free undamped system yields the transformation matrix P between the modal coordinates q and the global coordinates z according [7]:

$$z = Pq \tag{3}$$

The premultiplication by P^T allows to diagonalize M and K since both are symmetrical. Despite that C is also symmetrical, the damping matrix cannot be diagonalized and we can consider a linear combination of M and K [8]. Considering this, we now can express (2) as:

$$m_i\ddot{q}_i + c_i\dot{q}_i + k_i q_i = Q_i \ (i = 1{,}2) \tag{4}$$

The modal parameters m_i, c_i, and k_i are respectively the modal mass, damping and stiffness of the ith mode. Q_i is the modal force and q_i is the modal coordinate of the ith mode.

Whereas the system described by (2) is coupled, the system described by (4) is uncoupled since all the matrices have been diagonalized.

The solutions q_i to the decoupled equations are given by:

$$q_i = \frac{Q_i}{(k_i - m_i\omega^2 + jc_i\omega)} = \frac{Q_i}{Pk_i} \ (i = 1{,}2) \tag{5}$$

Pk_i is the characteristic polynomial of mode i and j is the complex constant.

Using the matrix formulation, the real coordinate solution z can be obtained from (3) and expressed by:

$$z = \frac{P^T P}{Pk} F = H_{ij} F \tag{6}$$

With F the global force vector representing the real forces applied in each DOF and H_{ij} the dynamic influence coefficient matrix.

3.3.2 Modal parameters calculation

Using both mechanical equations (1) and values from Table 1, we are able to compute the modal parameters and the dynamic influence coefficient matrix. These functions are represented in Figure 6 for the undamped system and in Figure 7 for the damped system.

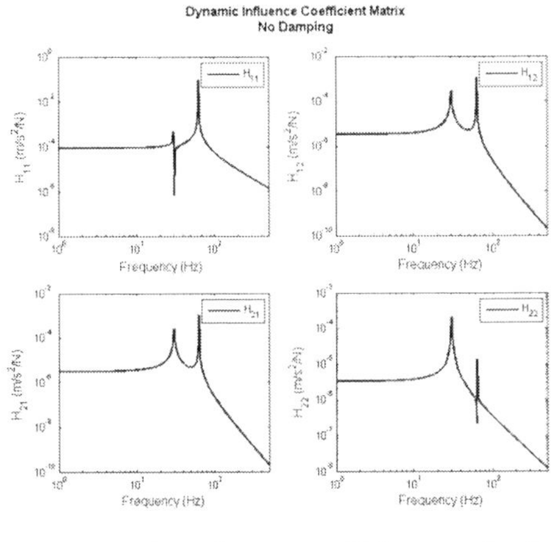

Fig. 6. Dynamic Influence Coefficient Matrix for the *undamped* system. The modal node 1 corresponds to the vibrating table whereas the modal node 2 corresponds to the body of the shaker.

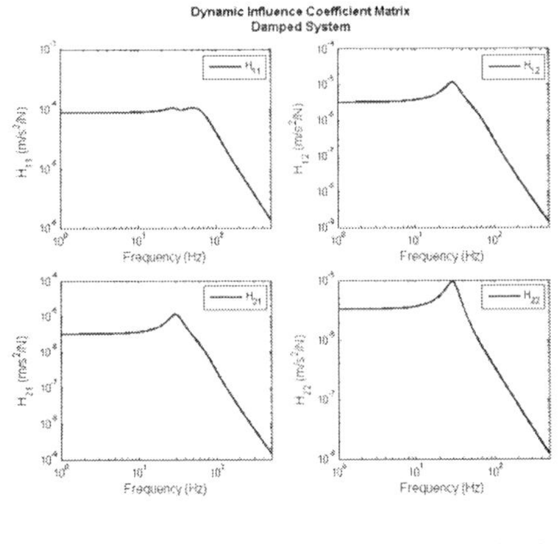

Fig. 7. Dynamic Influence Coefficient Matrix for the *damped* system. The modal node 1 corresponds to the vibrating table whereas the modal node 2 corresponds to the body of the shaker.

Figure 6 is clearer for mode detection than Figure 7 and shows two modes occurring at 30.2 Hz and 63.5 Hz. This is consistent with the observed response showed in Fig 4 and 5.

The calculated modal parameters are given in Table2.

3.3.3 Experimental modal analysis

As said before, the calculation is only possible if we possess all the mechanical parameters but for larger shaker it is often impossible to block all the DOFs alternatively and to make it vibrate in order to extract the parameters. However, as presented in the theoretical review, the global mechanical parameters are equivalent to the modal ones.

In [9], Guillaume presented a method for the spectral identification of modal parameters by Least Squares Complex Frequency-Domain estimation. The LMS PolyMax method is based on this method and is recognized to be very fast and accurate for modal estimation [10].

We introduced a force with an impact hammer in the two DOFs of the shaker (respectively the table and the body) and measured the acceleration of those DOFs. The LMS PolyMax method computed the transfer functions and estimated the modal parameters with the stabilization diagram presented in Figure 8.

The corresponding values are given in Table 2.

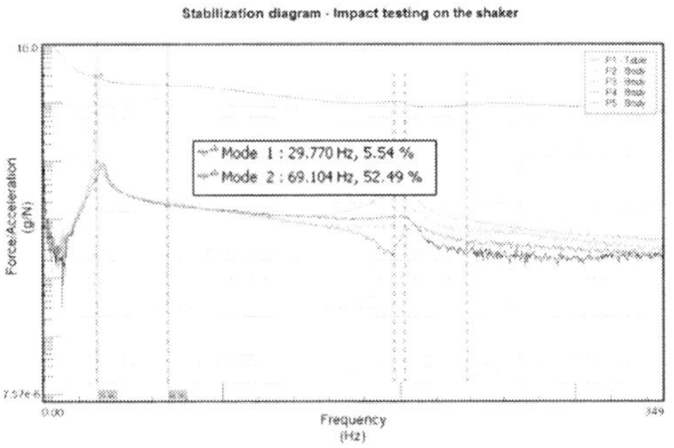

Fig. 8. Stabilization diagram of the impact testing on the shaker.

When looking at these values, we may observe that the experimental values match the calculated ones for the body-related parameters, especially for the modal mass and the modal stiffness, respectively m_2 and k_2, and in a less manner for the modal damping c_2. On the contrary, we have no match for the table-related parameters.

This difference is also noticeable in Figure 8. The response curves of the body (P2-P5) clearly show a resonance, making the body-related parameters identification possible, while the table resonance, identified in the stabilization diagram to be located at 69,104Hz, is not visible (P1). We may conclude that the experimentally determined table-related parameters are not valid.

This can be explained by the relative high damping in the suspension of the coil (see Figure 5) combined with the relative small force introduced into the shaker by the impact hammer, leading to a totally damped acceleration response of the table.

Table 2. Modal parameters

Modal parameter	Symbol	Values	
		Calculated	Determined experimentally
Modal mass of the vibrating table	m_1	77.2 g	18.2 g
Modal mass of the body	m_2	3.167 kg	3.33 kg
Modal damping of the suspension of the coil	c_1	15.93 kg/s	1.9 kg/s
Modal damping of the isolation studs	c_2	196.15 kg/s	76 kg/s
Modal stiffness of the suspension of the coil	k_1	$12.3 \cdot 10^3$ kg/s²	$28.02 \cdot 10^3$ kg/s²
Modal stiffness of the isolation studs	k_2	$1.143 \cdot 10^5$ kg/s²	$1.15 \cdot 10^5$ kg/s²

4 Determination of the electrical parameters

Looking at (1) gives us the needed electrical parameters to be defined. We find R and L, the resistance and the inductance of the coil, μ_f, the force constant, and μ_{bemf}, the bemf (back electromagnetic force) constant of the system.

R and L are the intrinsic electrical properties of the coil and are purely passive. A simple measurement on the RLC meter will allow us to get those values. These are given in Table 3.

The quantities μ_f and μ_{bemf} are representative of the same phenomenon – the expression of the Lorentz force – but have neither the same causes nor the same effects.

Lorentz expresses that the charge carriers of a conductor immersed in a (non-collinear) magnetic field are subjected to a force when moving and this according to the following relation:

$$\overline{F}_q = q\left(\overline{E} + \left(\overline{v} \times \overline{B}\right)\right) \tag{7}$$

With F_q the force exerted on a charge carrier of electrical charge q subjected to an electric field E, undergoing a speed v, and immersed in a magnetic field B.

The generalized expression of this force on a conductor of length l is:

$$F = Bli = \mu_f i \tag{8}$$

The current i flowing in the coil generates a force F on the latter with respect to the force constant μ_f.

The magnitude μ_{bemf} expresses the force undergone by the electrons in a conductor which moves in a magnetic field. The displacement of the coil generates an electric potential at the conductor terminals with respect to the bemf constant.

We study a system which on the one hand generates a *force* due to an applied *current* and on the other hand which creates an *electrical potential* with respect to the *speed of the coil*. This approach forms the basis of the development of a two-port network model and goes beyond the scope of this work. Different authors such as York, Smallwood or Guillaume develop this aspect deeper in detail [11–13].

4.1 Experimental determination of the force constant

First, we used the assembly already presented in Figure 3 which blocks the vibrating table. A voltage is applied to the terminals which generates a current, which creates a force. Due to the rigid structure, the table cannot move and no bemf can settle.

A measurement of the force with respect to the current yields the force constant. The latter is shown in Figure 9 for three different drive voltages.

We immediately notice a very strong structural response at 660 Hz due to the resonance of the structure originally intended to be rigid. By performing a modal analysis of this structure, we obtained its frequency response. The synthesized curve has been superimposed on the measured curves to emphasize the response due to the resonance of the beam structure.

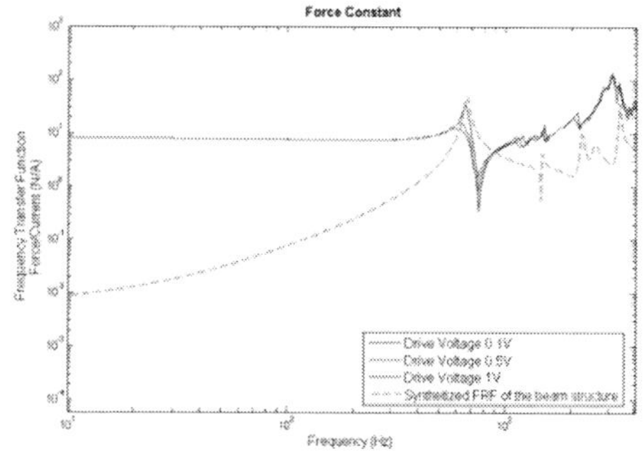

Fig. 9. Force Constant analysis for different drive voltages. The synthesized FRF of the beam structure is superimposed on the measured curves to emphasize the response due to the resonance of the beam structure.

Beyond this, the measurements tend to show an increase in damping with the drive voltage underlying a possible non-linearity of the system.

At this point any deduction is dangerous and should be avoided. Further analysis is required. Nevertheless, in the area below 100 Hz, the relative theoretical error due to the resonating structure with respect to the unity does not exceed 3%. Despite a relative variation in amplitude of 5% over the frequencies, the value of μ_f is considered to be constant. Its average value is 7.8493 N/A against an initial value form the manufacturer of 6.4 N/A. These data are shown in Table 3.

4.2 Experimental determination of the bemf constant

As we know the characteristics of the coil, we can build a test wherever the body of the shaker is fixed (to avoid measuring the influence of the isolation studs) but where the table is free to vibrate. The displacement speed of the table generates an electrical potential which counters the current and lets it decrease.

The electrical system is pure passive; the former current measurement allows us to determine the current loss. Taking as reference the current measured with the blocked table, the resonance of the beam structure will impose the same limitation.

Figure 10 illustrates the ratio between the bemf and the speed of the vibrating table. We conducted the test in two sets of added masses: 120 g and 502 g.

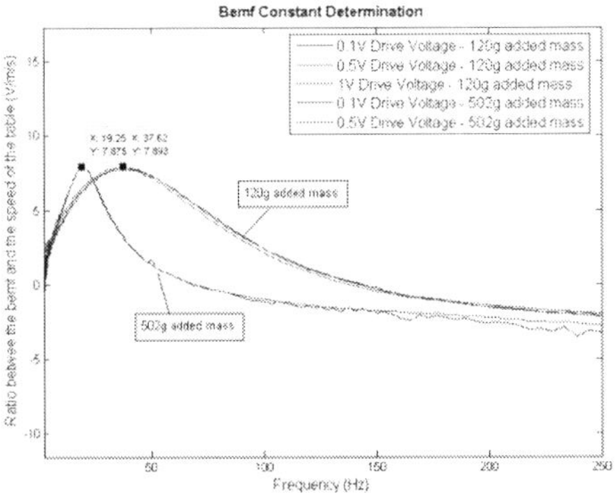

Fig. 10. Bemf Constant analysis for different drive voltages and different added masses.

As we could expect, the curves are gathered by set but with a constant top 'quality factor'. The bemf is related to the speed of the coil, itself dependent on the driven mass. Next to this, we must keep in mind that the shaker also has its own response. Here we can clearly observe the influence of the response of the shaker. As the frequency response of the shaker is not constant, it will also not be the case for the bemf constant which will be a function of the frequency.

Table 3. Electrical parameters

Electrical parameter	Symbol	Values	
		Given by the manufacturer	Determined experimentally
Resistance of the coil	R	~2 Ω	1.28 Ω
Inductance of the coil	L	–	76.4 μH
Force constant	μ_f	~6.4 N/A	7.8493 N/A
Back electromagnetic force constant	μ_{bemf}	–	Frequency varying

5 State-space modeling and virtual testing

5.1 State-space model

A State Space model allows a temporal approach of the simulation of a system. The equations governing the system must nevertheless be reformulated in order to correspond to the following formalism:

$$\begin{aligned} \dot{x} &= Ax + Bu \\ y &= Cx + Du \end{aligned} \tag{9}$$

x, y and u are respectively the states vector, the outputs vector and the inputs vector. The four matrixes A, B, C, and D characterize the system. It is in those matrixes that the mechanical en electrical parameters of the shaker will be introduced.

The shaker could be simulated based on a state space model derived from the general equations of motion (1) and computed using the 'lsim' function from *Matlab*.

On top of the simulation code, a test interface has been developed, managing the data (storing and generating) while the control algorithm manages the update process. A light routine splits the path into a physical acquisition or a simulated acquisition, sending and getting the signals to and from the shaker model instead of the real shaker. The data is then stored in an underlying layer of the interface and the process is totally transparent for the control algorithm.

This allows driving both physical testing and simulation over a same platform implementing and using the same control functions. This approach could offer a great correlation between physical and simulation results due to the fact that the difference mainly resides in the shaker modeling.

5.2 Experimental results

We will first retrieve the system's impulse response functions. A pulse of unit energy is introduced and the response is measured in each of the degrees of freedom. The responses of the body of the shaker and of the table are given in Figure 12 and Figure 13.

It can already be noted that the resonant frequency of the insulating pads is present. As for the table, it is not surprising to get this look. In fact, the input signal is a voltage applied to the terminals, which results in a force on the coil. The response on the table of a force introduced into the coil is the exact meaning of the transfer function H_{11} presented in Figure 7. These two curves are therefore very close.

In a second step, a 10s sine sweep up to 2000Hz is generated and introduced into the model. The Figure 11(a) and Figure 11(b) respectively show the temporal responses of the table and the body of the shaker.

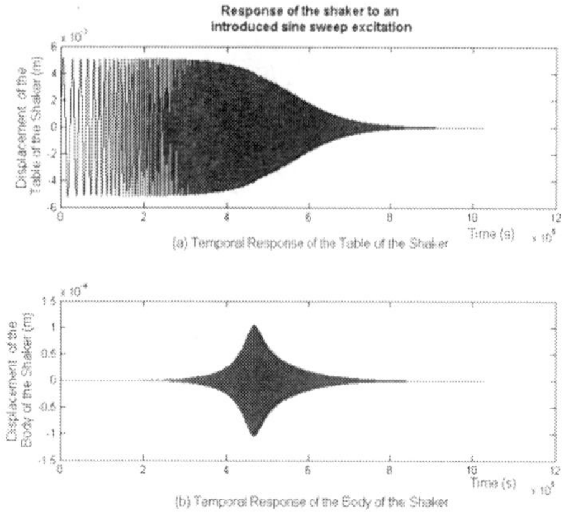

Fig. 11. Temporal response of the shaker to a sine sweep excitation. Response of the table (a) and response of the body (b).

We can see a very marked attenuation of the vibrations at the table, as well as a strong reaction of the body of the shaker at around 5s. We compute the frequency response function of those signals and superimpose them to the impulse response function of Figure 12 and Figure 13. We notice that the curves are very close to each other with a divergence above 2000Hz due to the limitation of the sine sweep.

The simulation carried out makes it possible to obtain a temporal response to a virtual test and for which the replies characterize the system well.

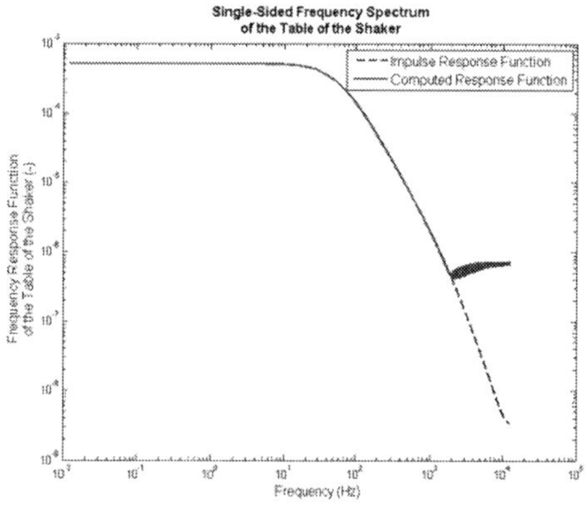

Fig. 12. Frequency Response Function and Impulse Response Function of the table of the shaker.

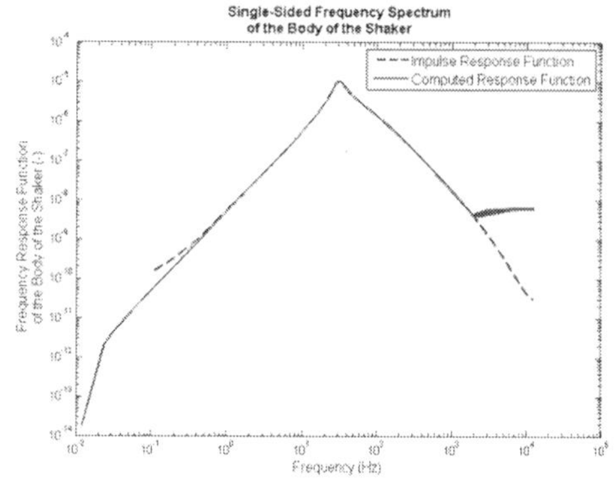

Fig. 13. Frequency Response Function and Impulse Response Function of the body of the shaker.

6 Conclusions

This article has highlighted an approach for determining the mechanical and electrical parameters of an electrodynamic shaker. Numerical and experimental modal analysis was also performed. The developed model has been successfully implemented in the linear state-space simulation environment for which the performed sine-sweeps illustrates the correlation with the experimentally observed behavior.

Nevertheless, many parameters presented in this work can be the object of a complementary research for their nonlinearity. Indeed, some parameters are expected constant what they are not at all. Some of these non-linear effects have not been presented in this paper and are beyond the scope of this work.

The mass-dependence of the bemf constant as well as a pronounced interaction of the amplifier opens the way to a complementary analysis to this work. Interesting tracks would be the determination of the impedance matrix of the system as well as the modeling following a 2 port network.

References

1. S. Ricci, *Proceedings of IMAC 27*, (2009)
2. J. Martino, *Proceedings of ISMA2016*, (2016)
3. N.L. Olsem, *Proceedings of the 4th International Modal Analysis Conference*, **2**, 1160-1167, (1986)
4. K.G. McConnell, *Vibration Testing: Theory and Practice*, (John Wiley & Sons, NY, 1995)
5. G.F. Lang, D. Snyder, J. Sound Vib., **35(10)**, 24-33, (2001)
6. Brüel & Kjaer, *Product Data: Vibration Exciter Type 4809*
7. D. Thorby, *Structural Dynamics and Vibration in Practice*, (Oxford: Elsevier, 127–129, 2008)

8. A.G. Piersol, T.L. Paez, *Harris' Shock and Vibration Handbook*, (6th ed., New-York: McGraw-Hill, 21.12, 2010)
9. P. Guillaume, *Proceedings of IMAC 21*, (2003)
10. B. Peeters, G. Lowet, H. Van der Auweraer, J. Leuridan, J. Sound Vib., **38(1)**, 24-29, (2004)
11. R. Yorke, The Radio and Electronic Engineer, **39(4)**, 2, (1970)
12. D. Smallwood, Journal of the IES, **40(5)**, 27-34, (1997)
13. P. Guillaume, R. Pintelon, J. Schoukens, *IMTC-91. Conference Record., 8th IEEE,* 263-271, (1991)

Dynamic responses of vehicle ballasted track interaction system for heavy haul trains

Yingjie Wang[1,*], *Zuzana* Dimitrovová[2,3], and *Jong-Dar* Yau[4]

[1]School of Civil Engineering, Beijing Jiaotong University, Beijing, China.
[2]Department of Civil Engineering, Faculdade de Ciências e Tecnologia, Universidade Nova de Lisboa, Caparica, Portugal
[3]IDMEC, Instituto Superior Técnico, Universidade de Lisboa, Lisboa, Portugal
[4]Department of Architecture, Tamkang University, New Taipei City, Taiwan.

Abstract. In this study, the dynamic responses of ballasted tracks subjected to heavy haul trains are investigated. The vehicle ballasted track interaction (VTI) model can be divided into three parts, which are vehicle model, ballasted track model and wheel/rail contact model. The vehicle is modelled as a multi-body system with 10 degrees of freedom. The ballasted track consisting of rail, sleeper, ballast, sub-ballast and sub-grade is modelled as four elastic components. The vehicle model and ballasted track model is linked by the wheel/rail contact model, which is determined by the nonlinear Hertzian theory. The VTI model is solved by an iterative procedure and validated with some field experiments. The dynamic responses of the vehicle/track system are compared with those from the moving load model. Moreover, the wheel/rail contact force, the force from rail to sleeper and the force from sub-ballast to subgrade are computed for different axle loads of 25t, 30t, 35t and 40t. It is deduced that maximum values of these forces increase in a linear form with the increasing of axle load.

1 Introduction

Railway has been playing a crucial role in the economic and social development during last decades[1]. Especially in China, several heavy haul railways such as Datong-Qinghuangdao railway, Shuozhou-Huanghua railway, and Watang-Rizhao railway were constructed and operated for raw coal transport from Inner Mongolia, Shanxi, and Shaanxi providences to east of China. Nowadays, the 30t per axle freight wagons have started operation in these three lines, which may cause much more damage for railway infrastructures.

To study the dynamic response of ballasted track under moving train, Oscarsson et al built a vertical train/track system model, where the ballast was modelled as rigid mass, and with this model the deflections, accelerations and forces in various track components are calculated[2]. Lei et al established a three layer beam model for ballasted track structures and studied the track vibration subjected to mixed passenger and freight cars by Fourier transform technique[3]. To model the ballast vibration, Zhai et al proposed a five-parameter model based on the assumption of pyramid distribution, which is validated using the experimental results[4]. Mosayebi et al built a locomotive car and ballasted track interaction model and the effect of some track parameters including sleeper type, track modulus on the displacements of track components was investigated[5]. From the available technical literatures, it is indicated that the sub-ballast layer is not considered in the railway track model. For this reason, this paper is focused on this topic.

In this paper, a vertical vehicle ballasted track interaction (VTI) model is developed, where the vehicle model is built using multi-body dynamics and the ballasted track model is developed with finite element method. The vehicle model and the ballasted track model are linked by the wheel/rail contact model, and the nonlinear Hertzian theory is used herein. In continuation, an iterative procedure is performed to compute the dynamic responses of the vehicle/track system, and the VTI model is validated by field experiments. Then the dynamic responses of ballasted track are computed through the numerical studies and some conclusions are drawn at the end.

2 Vehicle Ballasted Track Interaction Model

As shown in Figure 1, the heavy haul train is passing on the ballasted track, and each axle load of the freight vehicle is 30 tons. In this paper, the vehicle ballasted track interaction (VTI) model is developed, which consist of vehicle model, ballasted track model and wheel/rail contact model.

2.1 Vehicle model

For the vehicle, each freight vehicle is composed of one car body, two identical bogies and four identical wheel

* Corresponding author: wyj_bjtu@126.com

sets, which are regarded as rigid bodies as shown in Figure 2. The car body is supported at each side by a double-axle bogie through secondary suspension systems and each bogie is linked with wheel sets through the primary suspension systems. Both the primary and the secondary suspension systems are modelled as linear spring-dashpot units.

Fig. 1 Freight vehicle passing on ballasted track

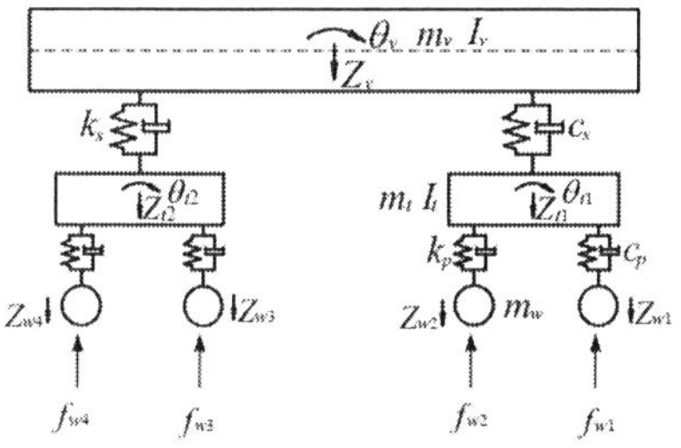

Fig. 2. Freight vehicle model

In the vehicle model, the vertical and pitching motion of the car body, front bogie and rear bogie are considered. For each wheel set, only the vertical displacement is considered. Therefore, the heavy haul vehicle is modelled as a multi-body system with 10 degrees of freedom.

Based on D'Alembert principle, the equations of motion for the vehicle system can be given as:

$$[M_V]\{\ddot{\mathbf{U}}_V\}+[C_V]\{\dot{\mathbf{U}}_V\}+[K_V]\{\mathbf{U}_V\}=\{\mathbf{F}_V\} \quad (1)$$

where $[M_V]$, $[C_V]$ and $[K_V]$ denote the mass, damping and stiffness matrices, respectively; $\{\ddot{\mathbf{U}}_V\}$, $\{\dot{\mathbf{U}}_V\}$ and $\{\mathbf{U}_V\}$ denote the accelerations, velocities and displacements, respectively; $\{\mathbf{F}_V\}$ is the external force vector acting on the vehicle system, which includes the gravities from the vehicle components and wheel/rail contact forces.

2.2 Ballasted track model

The ballasted track is modelled as four-layer structure consisting of rail, sleeper, ballast, sub-ballast and subgrade, which are connected one by one with spring-damping elements, as shown in Figure 3. For the ballasted track model, the rail is treated as an Euler-Bernoulli beam supported by discrete fasteners and sleepers, the ballast and sub-ballast layers are regarded as a series of lumped masses, and the spring-damping elements are used to link the fastening system, the elastic layer between ballast and sub-ballast, and the visco-elastic layer of the subgrade.

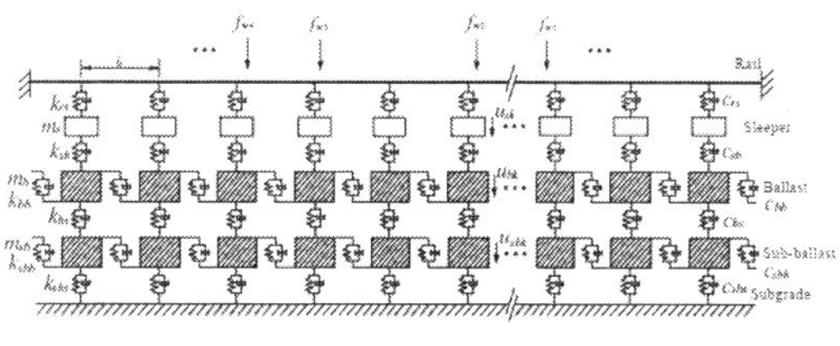

Fig. 3. Ballasted Track model

In this paper, the ballasted track is modelled by finite element method. To save the computing time and ensure the accuracy of the results at the same time, each rail element length is set as one sleeper space and supported by two adjacent sleepers[6]. For rail element, there are four degrees of freedom (DOFs): two vertical displacements u_j and u_{j+1} and two rotations θ_j and θ_{j+1} at two nodes. And each sleeper, ballast mass and sub-ballast mass only have one degree of freedom (DOF): vertical displacement u_{sk}, u_{bk}, u_{sbk}, respectively.

By assembling rail, sleeper, ballast mass and sub-ballast mass equations together, the equation of motion of the ballast track system under moving vehicle can be expressed as

$$[M_T]\{\ddot{\mathbf{U}}_T\}+[C_T]\{\dot{\mathbf{U}}_T\}+[K_T]\{\mathbf{U}_T\}=\{\mathbf{F}_T\} \quad (2)$$

where $[M_T]$, $[C_T]$ and $[K_T]$ denote the mass, damping and stiffness matrices of the track system, respectively, $\{\ddot{\mathbf{U}}_T\}$, $\{\dot{\mathbf{U}}_T\}$ and $\{\mathbf{U}_T\}$ denote the accelerations, velocities and displacements of the rail nodes, sleepers and ballast masses, respectively, $\{\mathbf{F}_T\}$ is the external force vector acting on the track system, that is the wheel/rail contact forces from the freight vehicle.

2.3 Wheel/rail contact model

In the vehicle ballasted track interaction (VTI) model, the upper vehicle and the substructure are linked by the assumed wheel/rail interaction relationship. In this study, the wheel/rail contact force $f_{wi}(t)$ is described by the nonlinear Hertzian theory as below,

$$f_{wi}(t)=\begin{cases} C_H\left[Z_{wi}(t)-u_{ri}(x,t)\right]^{3/2} & Z_{wi}(t)-u_{ri}(x,t)>0 \\ 0 & Z_{wi}(t)-u_{ri}(x,t)\le 0 \end{cases} \quad (3)$$

where $Z_{wi}(t)$ is the vertical displacement of the ith wheel set, $u_{ri}(x,t)$ is the vertical displacement of the rail at the point where the ith wheel set is located; C_H is the Hertzian constant and $C_H = 9.37\times10^{10}\text{N/m}^{3/2}$.

3 Solution and model verification

Due to the nonlinear feature of the wheel/rail contact force, as indicated in Section 2.3, the vehicle ballasted track interaction (VTI) model is solved by an iterative procedure, which proceeds as follows[7]:

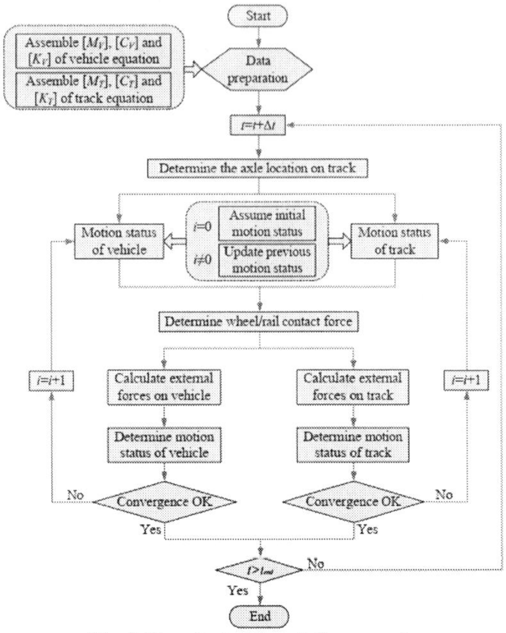

Fig. 4. Flowchart of the solution procedure

(a) Data preparation: Assemble $[M_V]$, $[C_V]$ and $[K_V]$ that are defined in Eq. (1) and $[M_T]$, $[C_T]$ and $[K_T]$ from Eq. (2).

(b) Determine the time duration t_{end} and the time interval Δt, required for the dynamic response time histories. In this study, the time interval Δt is set as 2×10^{-5} s.

(c) Determine the axle location on the track at time step $t + \Delta t$. For the first iteration step ($i = 0$), take the dynamic responses of the vehicle and track at the last time step t as the initial values for the iteration for the current time step $t + \Delta t$.

(d) Use the dynamic responses of the vehicle and the track at the last iteration step i to compute the wheel/rail contact forces in Eq. (3) for the current iteration step i + 1.

(e) Calculate the external forces acting on the vehicle and track, respectively, and solve the equation of motion of the vehicle and the track in Eqs. (1) and (2) using a step-by-step integration method such as the Newmark-β method, to find the vehicle and track responses at the iteration step $i + 1$.

(f) Use the vehicle and track responses determined under (e) for the next iteration step until the following convergence criterion is reached for both the vehicle and the track responses from Eqs. (1) and (2):

$$\left\| {}^{t+\Delta t}_{i+1}\mathbf{U}_T - {}^{t+\Delta t}_{i}\mathbf{U}_T \right\| \le CC, \quad \left\| {}^{t+\Delta t}_{i+1}\mathbf{U}_V - {}^{t+\Delta t}_{i}\mathbf{U}_V \right\| \le CC \quad (4)$$

The allowable tolerance for the convergence criterion is set as 10^{-6} m in this paper, which are found to be sufficient to determine the solution with acceptable accuracy from numerical comparison. If the convergence criterions in Eq. (4) are not satisfied, a new iteration step (i + 1) +1 can be started from (d). If the criterions are satisfied, a new time step can be started from (c).

To validate the vehicle ballasted track interaction (VTI) model, some experiment is performed. During the field testing, the freight train is composed of 100 wagons with 30 t per axle and the running speed of the train is 80 km/h. In Figure 5, the displacements of rail midpoint between two adjacent sleepers from the VTI model and field testing data are compared.

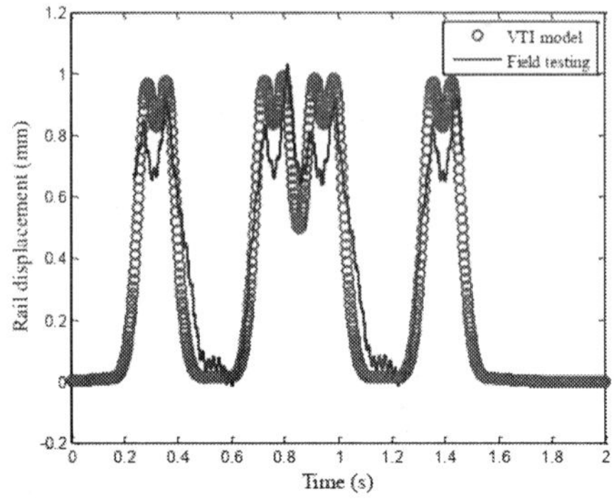

Fig. 5. Model validation

In the VTI model, there are only two vehicles considered due to the computation cost. Figure 5 indicates that the response obtained by the VTI model agrees reasonably well with the field testing data. Eight peaks, which are induced by the eight wheel sets from the two vehicles, are clearly shown in this figure.

4 Numerical studies

In these numerical investigations, the parameters of freight vehicle and ballasted track model are listed in Table 1 and Table 2, respectively[8]. In daily operation of the heavy haul train, the maximum running speed of the train is set as 80 km/h, which is also used in the following numerical examples.

Table 1. The vehicle model parameters

Notation	Parameters	Value	Unit
m_v	Mass of car body	55800	kg
m_t	Mass of bogie	900	kg
m_w	Mass of wheel set	600	kg
I_v	Mass moment of inertia of car body	2.7×10^6	kg m^2
I_t	Mass moment of inertia of bogie	855	kg m^2
k_p	Stiffness of primary suspension	1.3×10^7	N/m
k_s	Stiffness of secondary suspension	3.1×10^6	N/m
c_p	Damping of primary suspension	9.0×10^4	N s/m
c_s	Damping of secondary suspension	0.5×10^5	N s/m
l_v	Freight car length	14	m
l_t	Half of distance between two bogie centers	4.9	m
l_w	Half of distance between two axles in each bogie	0.89	m

Table 2. The ballasted track model parameters

Notation	Parameters	Value	Unit
m_r	Rail mass per unit length	74.414	kg/m
E_r	Elastic modulus of rail	2.06×10^{11}	N/m^2
I_r	Moment of inertia of rail	4.489×10^{-5}	m^4
m_s	Sleeper mass(half)	192.5	kg
m_b	Ballast mass	360	kg
m_{sb}	Sub-ballast mass	422	kg
l_s	Spacing of rail pad	0.6	m
k_{rs}	Rail pad stiffness	2.24×10^8	N/m
k_{sb}	Ballast stiffness	1.84×10^8	N/m
k_{bb}	Ballast shear stiffness	5.52×10^7	N/m
k_{bs}	Sub-ballast stiffness	9.66×10^8	N/m
k_{sbb}	Sub-ballast shear stiffness	2.90×10^8	N/m
k_{sbs}	Subgrade stiffness	1.18×10^8	N/m
c_{rs}	Rail pad damping	5×10^4	N s/m
c_{sb}	Ballast damping	3.61×10^4	N s/m
c_{bb}	Ballast shear damping	1.08×10^4	N s/m
c_{bs}	Sub-ballast damping	8.94×10^4	N s/m
c_{sbb}	Sub-ballast shear damping	2.68×10^4	N s/m
c_{sbs}	Subgrade damping	3.1×10^4	N s/m

The wheel/rail contact force is the connection of the vehicle model and the ballasted track model, and the contact forces for the four wheel sets of the first vehicle against rail location are computed, as shown in Figure 6.

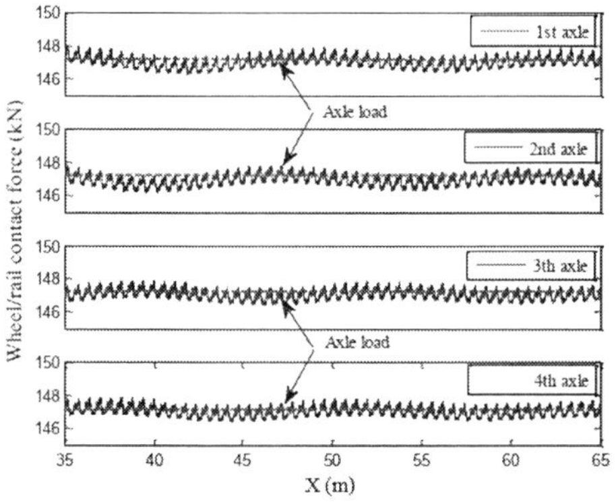

Fig. 6. Wheel/rail contact force

It can be concluded from Figure 6, that when the vehicle travels over the ballasted track, the wheel/rail contact force fluctuates around the vehicle axle load (147.15 kN), with the wavelength of 0.6 m, which equals to the distance of the two adjacent sleepers. It means that the wheel/rail contact force is a periodic force, which is dominated by the axle load and sleeper spacing.

The dynamic displacements of the ballasted track, including rail, sleeper, ballast mass and sub-ballast mass under the moving freight vehicle can be computed as a function of time, as shown in Figure 7.

Figure 7 indicates that the maximum displacements of ballasted track for rail, sleeper, ballast mass and sub-ballast mass are 1.00, 0.75, 0.43 and 0.37mm, respectively. It can be seen that the vibration amplitudes of the ballasted track decrease rapidly from rail to sub-ballast mass.

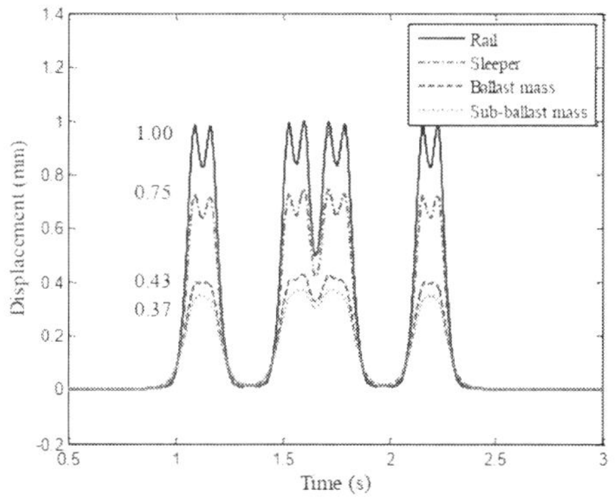

Fig. 7. Displacement of ballasted track

4.1. Effect of running speed

To investigate the effect of vehicle running speed on the dynamic responses of the vehicle ballasted interaction (VTI) model, the maximum values of wheel/rail contact force against different running speed of 20, 40, 60, 80, 100 km/h are calculated, as shown in Figure 8.

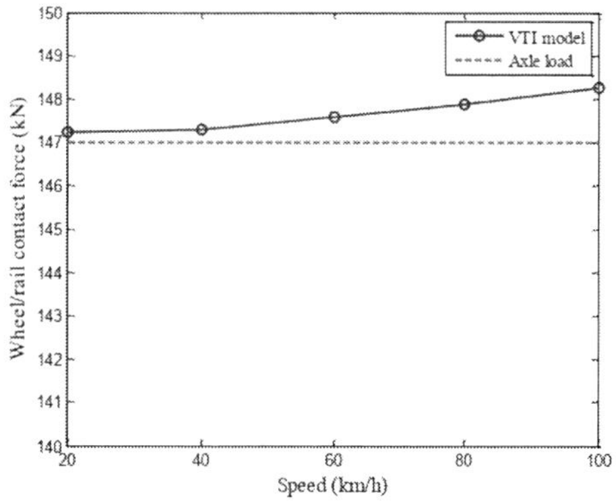

Fig. 8. Wheel/rail contact force against running speed

It is indicated that the maximum value of wheel/rail contact force for different running speed is a little higher than the value of axle load (147.15 kN), which is the dashed line in Figure 8.

The maximum displacements of ballasted track for rail, sleeper, ballast mass and sub-ballast mass are also computed against different running speed of 20, 40, 60, 80, 100 km/h, as shown in Figure 9. For comparison, the dynamic responses of ballasted track under moving load model are also computed.

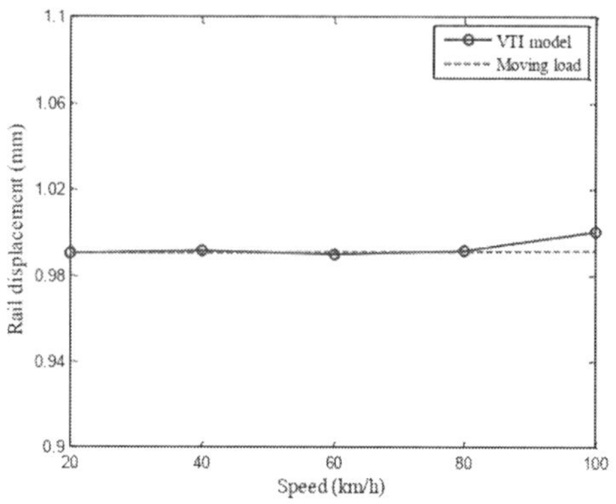

(a) Rail

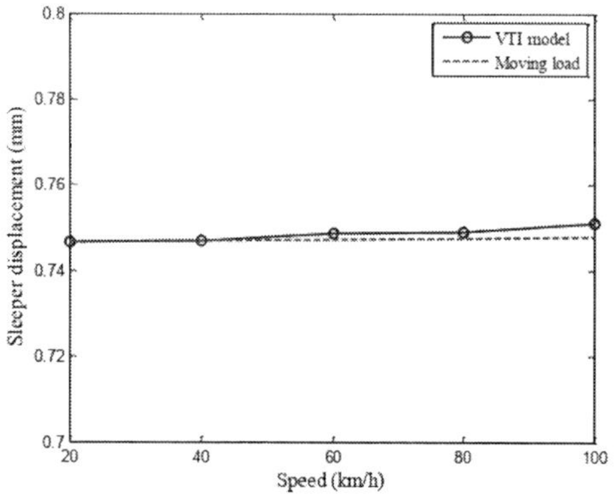

(b) Sleeper

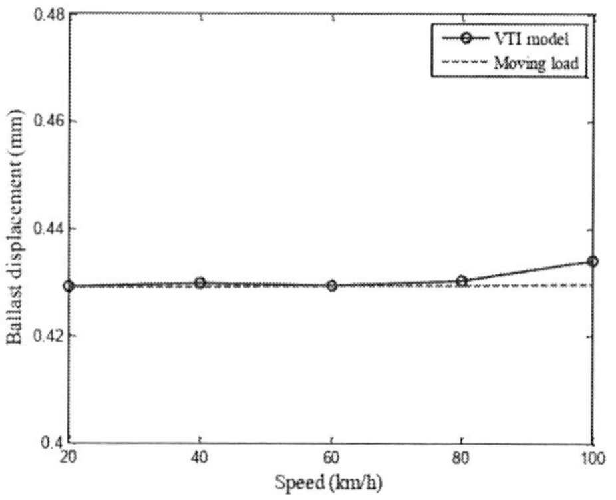

(c) Ballast

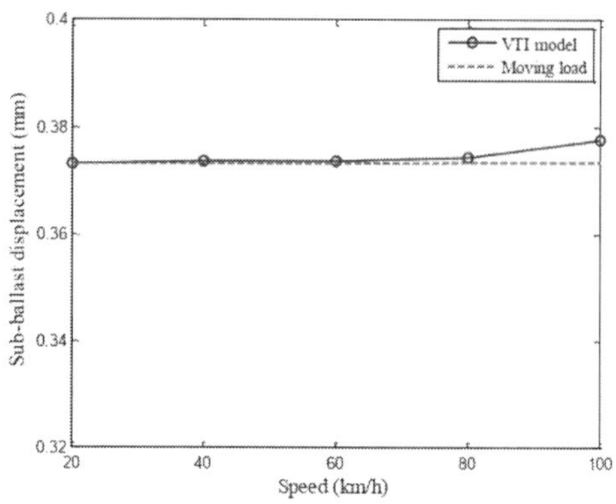

(d) Sub-ballast

Fig. 9. Displacement of ballasted track against running speed

As shown in Figure 9, for the lower running speed, especially less than 40 km/h, the maximum displacements of ballasted track computed from the VTI model are very close to these from the moving load model. However, when the running speed is higher, the results from VTI model are larger than those from the moving load model. In conclusion, the moving load model may underestimate the dynamic responses of ballasted tracks.

4.2 Effect of axle load

To investigate the effect of different axle load on the dynamic responses of the vehicle ballasted track interaction system, the maximum values of wheel/rail contact force against axle loads of 25, 30, 35 and 40 tones are computed, as shown in Figure 10.

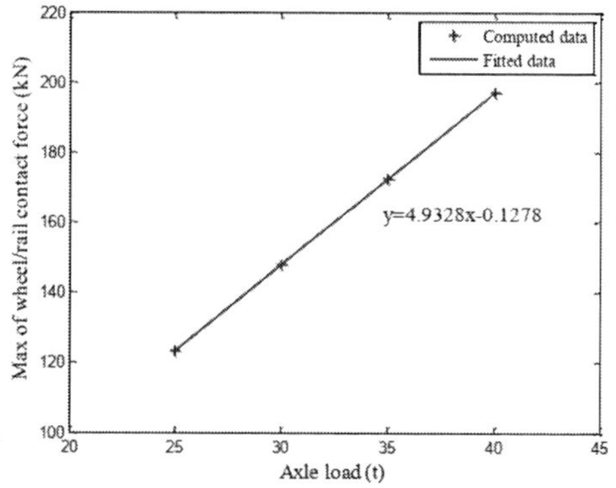

Fig. 10. Wheel/rail contact force against axle load

Figure 10 shows that with the increase of axle load, the maximum values of wheel/rail contact force increase. The regression equation of the wheel/rail contact force according to the axle load is presented in Table 3. It can be deduced that the wheel/rail contact force as a function of the axle load changes linearly according to y=a*x+b, where x stands for the axle load, y for the maximum value of wheel/rail contact force and a, b are fitted coefficients.

For the heavy haul railway, the rubber pad is used between the rail and sleeper to provide the elastic support. However, if the force passing from the rail to the sleeper is very high, the rubber may be worn out quickly. Another fact that should be carefully considered is the force transmitted from the sub-ballast to the subgrade, which may cause the failure of the subgrade. Therefore, the maximum values of these forces were extracted against axle loads of 25, 30, 35 and 40 tones and are shown in Figure 11.

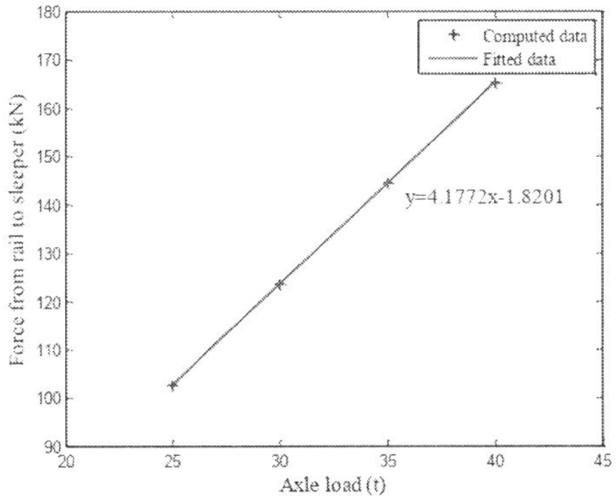

(a) Force from rail to sleeper

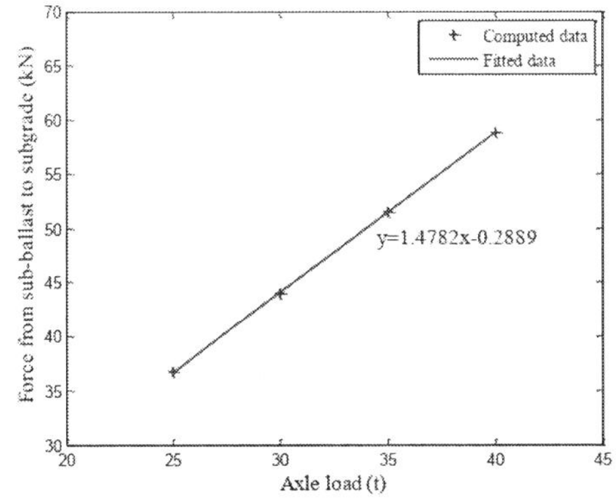

(b) Force from sub-ballast to subgrade

Fig. 11. Forces against axle load

Figure 11 indicates that when the axle load of the freight vehicle increases, the maximum values of the force from rail to sleeper and the force from sub-ballast to subgrade will also increase. The regression equation coefficients of the variation of these two forces with respect to the axle load are presented in Table 3.

Table 3. Regression equations

Items	Linear Eq 'y=a*x+b'		R^2 value
	a	b	
Wheel/rail contact force	4.9328	-0.1278	1
Force from rail to sleeper	4.1772	-1.8201	1
Force from sub-ballast to the subgrade	1.4782	-0.2889	1

From the results listed in Table 3, it can also be deduced that the dependences of the force from rail to sleeper and the force from sub-ballast to subgrade as a function of the axle load are linear (y=a*x+b). Moreover, is can be concluded that the coefficient "a" of these three regression equations, *i.e.* the slope of the corresponding lines, decreases from the top to the bottom of the ballasted track. Also, for the same axle load, the wheel/rail contact force is the largest one among these three, the force from rail to sleeper is next and the force from sub-ballast to subgrade is the lowest one.

5 Conclusions

In this paper, a vertical vehicle ballasted track interaction (VTI) model is developed and validated using the field test data. From the numerical studies, the following conclusions are drawn:
1) The wheel/rail contact force is a periodic force, which is dominated by the axle load and sleeper spacing.
2) For higher running speeds, the moving load model may underestimate the dynamic responses of ballasted tracks, because the wheel/rail contact force from the VTI model is a slightly larger than the axle load.
3) With an increase of the axle load, the maximum values of the wheel/rail contact force, of the force from rail to sleeper and of the force from sub-ballast to subgrade increase linearly. This point can be adopted for engineering analysis of wheel/track contact in railway design phase.

This work was supported in part by the National Natural Science Foundation of China under Grant No. 51408036 and Scientific Project of China Railway Group Limited under Grant No. 2014-重点-55.

References

1. S. Ponnuswamy. *Railway Transportation Engineering, Operation and Management* (Alpha Science International Ltd, Oxford, 2012).
2. J. Oscarsson, T. Dahlberg, Veh Syst Dyn, **29**, 73-84 (1998).
3. X. Lei, J.G. Rose, P I Mech Eng F-J Rai, **222**, 413-421 (2008).
4. W.M. Zhai, K.Y. Wang, J.H. Lin, J Sound Vib, **270**, 673-683 (2004).
5. S. Mosayebi, J. Zakeri, M. Esmaeili, J Mech Sci Technol, **30**, 1065-1069 (2016).
6. P. Lou, X.G. Zhong, J.F. Tang, Q.Y. Zeng, P I Mech Eng F-J Rai, **220**, 305-315 (2006).
7. Y.J. Wang, Y.X. Mao, A.Q. Hu, J Vibroeng, **17**, 1488-1498 (2015).
8. W.M. Zhai. *Vehicle-track coupled dynamics* (Science Press Ltd, Beijing, 2015).

Comparison of 1D and 2D solutions for a beam under transverse impact

Vítězslav Adámek[1,*]

[1]NTIS - New Technologies for the Information Society, Faculty of Applied Sciences, University of West Bohemia, Technicka 8, Pilsen, Czech Republic

Abstract. The problem of non-stationary vibration of an elastic beam caused by a transverse impact loading is studied in this work. In particular, two different approaches to the derivation of analytical solution of the problem are compared. The first one is based on the Timoshenko beam theory, the latter one follows the exact two-dimensional theory. Both mentioned methods are used for finding the response of an infinite homogeneous isotropic beam. The obtained analytical results are then compared and their agreement is discussed in relation to main factors, i.e. the beam geometry, the character of loading and times and points at which the beams responses are studied.

1 Introduction

Beams and beam-like structures are used in many engineering applications across the entire spectrum of industry. There exists a lot of simplified beam theories which have been developed in order to effectively investigate beams responses under different loading conditions. The basic assumptions under which these theories can be applied to static or dynamic beams problems are different. In the case of static problems, the ratio of beam length and its height plays the main role.

If we focus on dynamic problems of beams, there exist two main approaches: modal and wave. The modal method, making the use of eigen frequencies and eigen modes, is usually applied to vibration problems in which the maximal frequencies are of orders about ones or tens of kHz. This method is sometimes used also for problems of higher frequencies but it is usually associated with the problem of inaccurate determination of eigen values. On the other hand, the wave approach is applicable for nearly arbitrary range of frequencies but it becomes quite inefficient when used for lower frequency problems.

As can be seen from most of the published works dealing with problems of thin beams, the Bernoulli-Euler theory [1] is often used when the modal method is applied. In the case of wave approach, this elementary theory is inappropriate because it results in infinite phase velocity for increasing frequency. Another beam theory developed by Rayleigh in 1894 and taking into account the effect of rotary inertia [1] leads to finite phase velocity, but it is still limited to thin beams with negligible influence of shear forces. This effect has been added by Timoshenko in the early 20th century [2]. According to his theory, two characteristic wave velocities exist and they are very close to those predicted by the exact two-dimensional formulation. Unfortunately, the Timoshenko theory still neglects the warping of the beam cross-section so it can be applied to thick beams problems only under certain assumptions. This restriction is removed by so called higher-order beam theories developed after the mid-20th century, e.g. [3]-[6]. Such refined theories enable to solve more complicated problems of beams, e.g. beams made of functionally graded materials, laminated beams, nanocomposite beams etc.

In this study, the application of the Timoshenko beam theory to impact problems of beams is discussed from the wave propagation point of view. The analytical solution for an elastic infinite beam under transverse impact obtained using the Timoshenko theory is compared to that derived using the exact two-dimensional theory. The measure of compliance of both solutions is discussed in relation to the beam geometry, to the Fourier spectra of applied load and to spatio-temporal parameters which the both solutions are compared for.

2 Formulation and assumptions of the problem

We will assume an infinite thin beam with uniform rectangular cross-section of dimensions $b_0 \times d_0$ (width × height). The area of the beam cross-section will be denoted as $A = b_0 d_0$ in the following. The material of the beam will be isotropic and homogeneous and its elastic properties will be described by the modulus of elasticity E, shear modulus G and by the density ρ. The beam is in a rest at time $t = 0$ s. After that an external load perpendicular to the beam longitudinal axis x is applied to the upper edge of the beam. This load will be defined by the function $q(x, t)$

*e-mail: vadamek@kme.zcu.cz

which will be considered as an even function of the coordinate x. With respect to this assumption, the problem will be solved as a symmetric one on the spatio-temporal domain $\langle 0, +\infty) \times \langle 0, t_{max}\rangle$. Further, the bottom edge of the beam will be free of load and the beam deflection and the slope of the beam cross-section tend to zero for $x \to +\infty$. Finally, small displacements and rotations will be taken into account in the following sections.

3 Analytical solution - Timoshenko theory

3.1 Governing equations

As mentioned in Section 1, the Timoshenko beam theory takes into account the effect of shear forces and rotary inertia to the beam deformation. The equations of motion in this case can be formulated as [1]

$$\rho A \frac{\partial^2}{\partial t^2} v(x,t) - \frac{\partial}{\partial x} Q(x,t) = q(x,t),$$
$$Q(x,t) - \frac{\partial}{\partial x} M(x,t) - \rho J \frac{\partial^2}{\partial t^2} \alpha(x,t) = 0, \quad (1)$$

where $\alpha(x,t)$ and $v(x,t)$ denote the slope of the beam cross-section and the beam deflection, respectively, and J stands for the cross-sectional-area moment of inertia about the neutral axis. The functions $Q(x,t)$ and $M(x,t)$ denote the shear force and the bending moment present in the beam cross-section, respectively, and there hold

$$M(x,t) = -EJ \frac{\partial \alpha(x,t)}{\partial x},$$
$$Q(x,t) = \kappa AG \left(\frac{\partial v(x,t)}{\partial x} - \alpha(x,t) \right). \quad (2)$$

The constant κ is the so-called Timoshenko's shear coefficient introduced to quantify the variable distribution of shear force in the cross-section of the beam. Its value is $\kappa = 5/6 \doteq 0.833$ for the cross-section of a rectangular shape (see [1]).

3.2 Solution in Laplace domain

The solution of the system (1) can be found using the method of integral transforms, namely using the combination of Laplace transform in time and Fourier transform in space. The final formulas for the Laplace transforms of the unknown functions $\alpha(x,t)$ and $v(x,t)$ can be derived from the formulas presented in [7]. In the mentioned work, the analytical solution for the problem of a simply supported viscoelastic beam under transverse dynamic loading is presented for the case of special orthotropy. If we do that we can write the solution of 1D problem as

$$\bar{v}(x,p) = \frac{2}{\pi} \int_0^\infty V(\omega,p) \cos(\omega x)\, d\omega,$$
$$\bar{\alpha}(x,p) = \frac{2}{\pi} \int_0^\infty A(\omega,p) \sin(\omega x)\, d\omega, \quad (3)$$

where the new variables $p \in C$ and $\omega \in R$ have been introduced by the application of Laplace and Fourier transforms, respectively. The spectra of the Fourier integrals present in (3) can be expressed as complex rational functions in this case and there hold

$$V(\omega,p) = \frac{P_4(\omega,p)Q(\omega,p)}{P_6(\omega,p)},$$
$$A(\omega,p) = \frac{P_2(\omega,p)Q(\omega,p)}{P_6(\omega,p)}. \quad (4)$$

The complex polynomials $P_i(\omega,p)$ in (4) have the form

$$P_2(\omega,p) = -3\kappa\omega G,$$
$$P_4(\omega,p) = \frac{1}{4}\rho d_0^2 p^2 + 3\kappa G + \frac{1}{4} d_0^2 \omega^2 E,$$
$$P_6(\omega,p) = P_4(\omega,p)\left(\kappa\omega^2 G + \rho p^2\right) - \frac{1}{3}\left(P_2(\omega,p)\right)^2 \quad (5)$$

and the function $Q(\omega,p)$ represent the Fourier cosine integral of the Laplace transform of $q(x,t)$.

4 Analytical solution - exact 2D theory

4.1 Governing equations

When the exact 2D theory of continuum is used for solving beam bending problems, the state of plain stress is assumed. In this case, the beam is considered as a thin 2D strip and the beam deflection and the slope of the beam cross-section can be derived from the components of displacement. Let the functions $u(x,y,t)$ and $v(x,y,t)$ denote the horizontal and vertical displacement component, respectively. Both are the functions of the horizontal coordinate $x \in \langle 0, +\infty)$, the vertical coordinate $y \in \langle -d_0/2, d_0/2\rangle$ and time t. Introducing the function of volume dilatation $\Delta(x,y,t)$ and rotation $r(x,y,t)$ by relations

$$\Delta(x,y,t) = \frac{\partial u(x,y,t)}{\partial x} + \frac{\partial v(x,y,t)}{\partial y},$$
$$r(x,y,t) = \frac{1}{2}\left(\frac{\partial v(x,y,t)}{\partial x} - \frac{\partial u(x,y,t)}{\partial y} \right), \quad (6)$$

the equations of motion for a 2D element of the strip (beam) can be written as an uncoupled system of PDEs for $\Delta(x,y,t)$ and $r(x,y,t)$. This system of equations can be derived from the equations of motion presented in [8] for the case of an infinite viscoelastic strip. Doing so one obtains the following system:

$$\frac{\partial^2}{\partial t^2} \Delta(x,y,t) = c_3{}^2 (\nabla \Delta(x,y,t))^2,$$
$$\frac{\partial^2}{\partial t^2} r(x,y,t) = c_2{}^2 (\nabla r(x,y,t))^2, \quad (7)$$

where the constants c_2 and c_3 represent the phase velocities of shear and dilatational waves, respectively, and there hold

$$c_2 = \sqrt{\frac{G}{\rho}}, \quad c_3 = \sqrt{\frac{E}{\rho(1-\nu^2)}}. \quad (8)$$

The symbol ν stands in (8) for the Poisson's ratio.

4.2 Solution in Laplace domain

The solution of the system (7) can be simply derived in Laplace domain using the solution given in [8]. Neglecting the material viscosity, the Laplace transform of displacement components can be expressed as Fourier integrals having the following form:

$$\bar{u}(x, y, p) = \frac{1}{\pi} \int_0^\infty \Big\{ \left[A_1 \mathrm{ch}(y\lambda_1) - A_2 \mathrm{sh}(y\lambda_1) \right] k_6$$
$$+ \left[B_1 \mathrm{sh}(y\lambda_2) - B_2 \mathrm{ch}(y\lambda_2) \right] k_9 \Big\} \sin(\omega x)\, d\omega ,$$
$$\bar{v}(x, y, p) = \frac{1}{\pi} \int_0^\infty \Big\{ \left[A_1 \mathrm{sh}(y\lambda_1) - A_2 \mathrm{ch}(y\lambda_1) \right] k_8$$
$$+ \left[B_1 \mathrm{ch}(y\lambda_2) - B_2 \mathrm{sh}(y\lambda_2) \right] k_7 \Big\} \cos \omega)\, d\omega , \qquad (9)$$

where the positive sign of the vertical displacement transform $\bar{v}(x, y, p)$ corresponds to the positive value of the beam deflection transform $\bar{v}(x, p)$ from Section 3.2. The symbols sh and ch used in (9) represent for the function sinh and cosh, respectively.

The unknown complex functions $A_1(\omega, p)$, $A_2(\omega, p)$, $B_1(\omega, p)$ and $B_2(\omega, p)$ follows from the application of boundary conditions of the problem. Similarly to [8] one can write

$$A_1 = \frac{Q(\omega, p)\, k_5\, \mathrm{sh}(d\lambda_2)}{2k_{10}}, \quad A_2 = \frac{Q(\omega, p)\, k_5\, \mathrm{ch}(d\lambda_2)}{2k_{11}},$$
$$B_1 = \frac{Q(\omega, p)\, k_4\, \mathrm{ch}(d\lambda_1)}{2k_{11}}, \quad B_2 = \frac{Q(\omega, p)\, k_4\, \mathrm{sh}(d\lambda_1)}{2k_{10}}. \qquad (10)$$

The remaining complex functions $\lambda_i(\omega, p)$ and $k_i(\omega, p)$ present in (9) and (10) can be simply derived using the analogous functions given in [8].

5 Comparison of 1D and 2D solutions

5.1 Evaluation procedure

The evaluation process which leads to the results in spatio-temporal domain consists of two basic steps: (i) to evaluate the Fourier integrals contained in the derived formulas (3) and (9) and (ii) to perform the inverse Laplace transform (ILT). It is possible and advantageous to change the order of these steps and to perform the Laplace inversion of derived Fourier spectra first.

The inversion can be done analytically or numerically. The first approach, which is usually based on the residue theorem, is feasible and quite straightforward in the case of 1D solution. For the 2D problem, this procedure is much more complicated and so the application of a suitable numerical algorithm is more effective. To guarantee the same evaluation conditions for both solutions compared, the numerical approach to ILT is used in this work. In particular, an algorithm based on the combination of FFT and Wynn's epsilon accelerator was used for this purpose. This algorithm is described in detail in [9]. It is precise and quite robust for solving different problems of elastodynamics, as shown in the work mentioned. The evaluation of the Fourier integrals has been done also numerically. The classical Simpson rule with constant integration step $\Delta\omega$ has been used. The choice of $\Delta\omega$ and of the upper limit of integration ω_{max} depended on the value t_{max}. If $t_{max} < 200\,\mu$s, the parameters $\Delta\omega = 1$ and $\omega_{max} = 4000$ were used. For greater values of t_{max}, the choice was $\Delta\omega = 0.25$ and $\omega_{max} = 8000$ to guarantee high precision of obtained results.

In the following subsection, the results obtained for steel beams with material parameters $E = 2.1 \times 10^{11}$ Pa, $\nu = 0.3$ and $\rho = 7800\,\mathrm{kg/m^3}$ are presented. The effect of the beam height d_0, the character of the external loading $q(x, t)$ and the time and the position of observation are discussed in relation to the differences between both solutions.

5.2 Results comparison and discussion

The first comparison will be made for two beams with different heights $d_0 = 2$ mm and $d_0 = 10$ mm. Both beams are subjected to the load which is nonzero only for $x \in \langle 0, h \rangle$ and is defined as

$$q(x, t) = \sigma_a\, b_0 \cos\left(\frac{\pi}{2} \frac{x}{h}\right) \left[H(t) - H(t - t_0) \right], \qquad (11)$$

where $\sigma_a = 1$ MPa, $b_0 = 2$ mm, $h = 1$ mm and $t_0 = 2\,\mu$s. It means that the beam is excited by a pressure rectangular pulse lasting $2\,\mu$s and the amplitude of which is distributed along the x direction according to cosine function on a domain of total length 2 mm. If we compare the deflection histories $v(t)$ at points $x = \{10, 50, 100\}$ mm lying at beams longitudinal axes up to time $100\,\mu$s we find out that 1D and 2D approach lead for both values of d_0 to nearly identical results. To quantify the measure of this agreement, a relative error will be used. This error will be defined as

$$\delta(t) = \frac{v_{1D}(t) - v_{2D}(t)}{\max_t \left(|v_{2D}(t)| \right)}. \qquad (12)$$

The functions $v_{1D}(t)$ and $v_{2D}(t)$ in (12) represent the time histories of beam deflections determined by 1D and 2D theory, respectively, and the maximum in the denominator is looked for over the whole time interval $\langle 0, t_{max} \rangle$. Calculating $\delta(t)$ for all mentioned x one finds that it reaches its maximal value 5.8% in the case $d_0 = 2$ mm. When $d_0 = 10$ mm, the error is greater as expected. In this case the maximum error is about 19%. But it must be mentioned that this error is obtained only locally for time about $35\,\mu$s and for $x = 100$ mm.

The comparison of vertical beam velocities $dv(t)/dt$ is more important because it gives us deeper insight into the physics described by both approaches. This comparison is made in Fig. 1 and Fig. 2 for the same values of x as aforementioned. It is evident that the vertical beam velocity determined by the Timoshenko theory (thick solid curves) represents, let us say, a mean value around which the 2D solution oscillates (thin dashed curves). The amplitudes of oscillations for $d_0 = 2$ mm are quite small (see Fig. 1), so both theories give us very similar information about the vertical velocity. On the other hand, the greater value of d_0 leads to considerable differences between both approaches, see Fig. 2. It could be expected because also in static beam problems the higher height of the beam means worse compliance of 1D and 2D solution. But in the elastodynamics of beams this fact is primarily connected with the wavelength of the waves propagated through the beam, as will be discussed later.

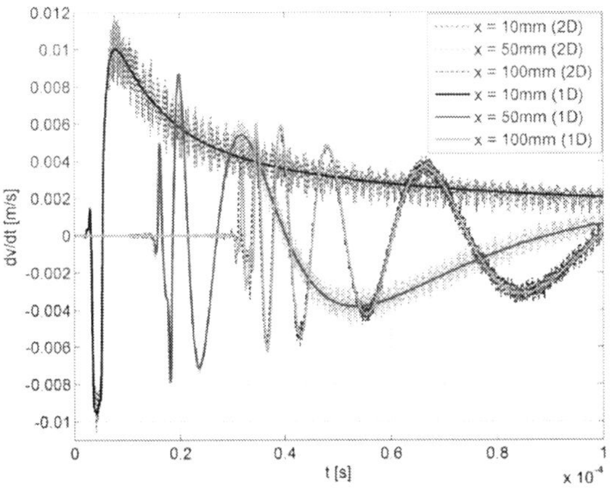

Figure 1. Comparison of $dv(t)/dt$ for different distances x and for $d_0 = 2\,\mathrm{mm}$

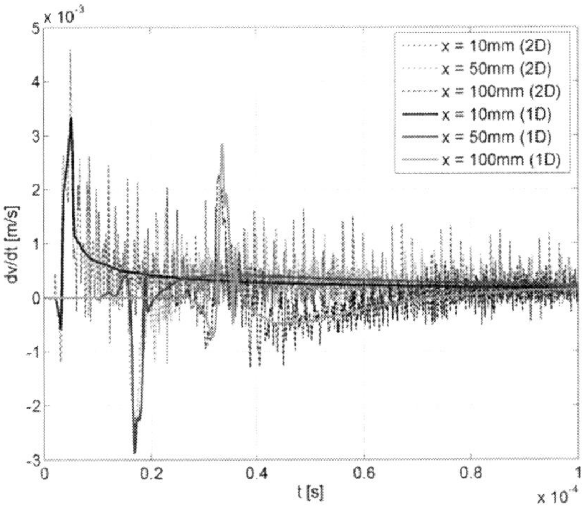

Figure 2. Comparison of $dv(t)/dt$ for different distances x and for $d_0 = 10\,\mathrm{mm}$

The significant oscillations of 2D solution in Fig. 2 correspond to real wave processes in the beam. For instance, the constant time-spacing between every two significant peaks corresponds to time during which the S-wave propagates from the beam axis to the beam edge and back. To make a better picture of the wave propagation in the beam, the spatial distributions of total velocity in 100 mm long part of one half of the beam (strip) are used in Fig. 3 to show the positions of wave fronts at different times. Firstly, the P-wave and S-wave are propagating from the place of loading (denoted by the arrow) through the strip. At time $t \doteq 1.6\,\mu s$, the front of P-wave is nearly reaching the lower strip boundary whereas the front of slower S-wave is located near the beam axis. At time $t \doteq 2.7\,\mu s$, i.e. approximately $0.7\,\mu s$ after the impact, the loading S-wave and reflected P-wave are interacting and the unloading P- and S-wave are propagated from the place of excitation. The process of waves interaction continues in time and the zone of transient state of stress moves farther from the vertical beam axis, see the state at $t \doteq 10.6\,\mu s$. When $t = 100\,\mu s$, the state in the observed part of the strip becomes quite homogeneous, the transient phenomena gradually disappear, so the amplitudes of oscillations of 2D solution around 1D results are decreasing, see Fig. 2.

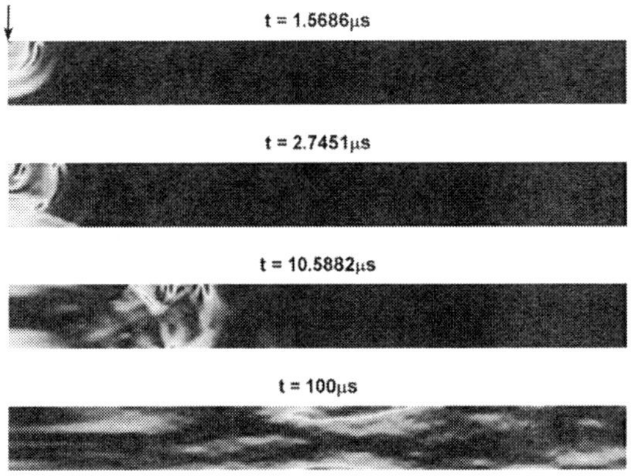

Figure 3. Wave phenomena in the beam (strip) at different times

Another fact which we meet if we compare both theories is that the 1D solution derived using the Timoshenko theory is delayed in comparison with that obtained using the exact 2D approach. This time lag is obvious from Fig. 4 where the history of $v(t)$ for $x = 300\,\mathrm{mm}$ is depicted for the case $d_0 = 2\,\mathrm{mm}$. This is not an error but this is the consequence of different dispersion properties of both types of continuum being compared. In the case of Timoshenko beam, there exist only two dispersion curves [1] with the asymptotes corresponding to two characteristic wave velocities $c_0 = \sqrt{E/\rho}$ and $c_{2T} = \sqrt{\kappa G/\rho}$. On the other hand, the beam modelled as a 2D strip has an infinite number of dispersion curves and two different characteristic velocities c_3 and c_2 given by the relations (8). Calculating all of these velocities for parameters mentioned in Section 3.1 and 5.1, one obtains: $c_0 = 5189\,\mathrm{m/s}$ vs. $c_3 = 5439\,\mathrm{m/s}$ and $c_{2T} = 2937\,\mathrm{m/s}$ vs. $c_2 = 3218\,\mathrm{m/s}$.

The next factor which significantly influences the compliance of both types of results is the maximal frequency contained in the amplitude spectrum of the time part of $q(x,t)$. From the elastodynamics point of view, a beam can be considered as thin only if the wavelength λ of the shortest wave present in the signal propagating through the beam is substantially larger than the beam height. For instance, when the Bernoulli-Euler beam theory is applied it must be true that $k_0/\lambda < 0.05$, where k_0 is the cross-

https://doi.org/10.1051/matecconf/201814805005

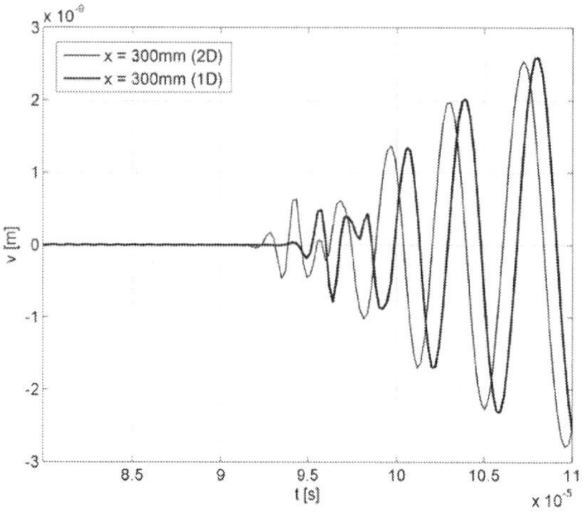

Figure 4. The time shift between both solutions for $x = 300$ mm and $d_0 = 2$ mm

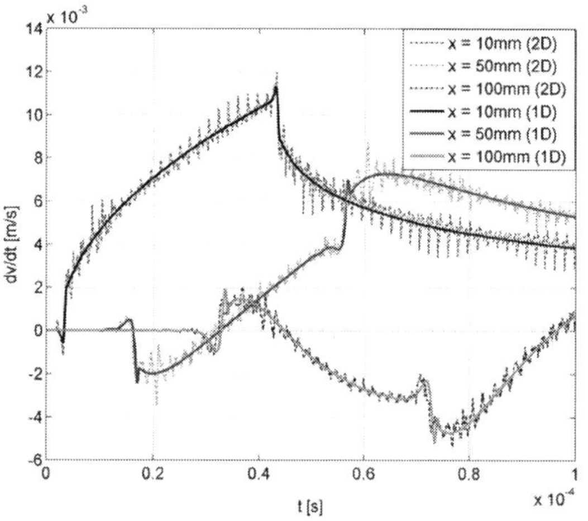

Figure 5. Comparison of $dv(t)/dt$ for different distances x and for $d_0 = 10$ mm, $t_0 = 40\,\mu$s

sectional inertia radius defined as $k_0 = \sqrt{I/A}$, see [1]. If we apply this criterion to the beam with rectangular cross-section $b_0 \times d_0$ we approximately obtain the inequality $\lambda < 5.77d_0$. It means that an Euler beam can be considered as thin only if its height is about six times smaller than the wavelength of the wave component with the highest frequency.

This fact will be demonstrated using two rectangular pulses of different duration. Let us assume that the external impact loading (11) takes $40\,\mu$s instead of $2\,\mu$s, as considered previously. It is clear that the high frequencies have much smaller contribution to the amplitude response than in the case of $t_0 = 2\,\mu$s. If we compare the histories of the beam deflection $v(t)$ determined for $t_0 = 40\,\mu$s by 1D and 2D approach, we find out that the maximal value of $\delta(t)$ for all $x = \{10, 50, 100\}$ mm and for $d_0 = 10$ mm is

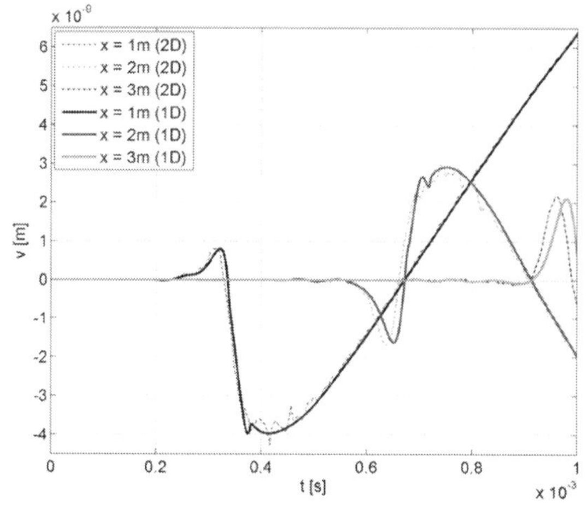

Figure 6. Comparison of $v(t)$ for different distances x and for $d_0 = 200$ mm, $t_0 = 40\,\mu$s

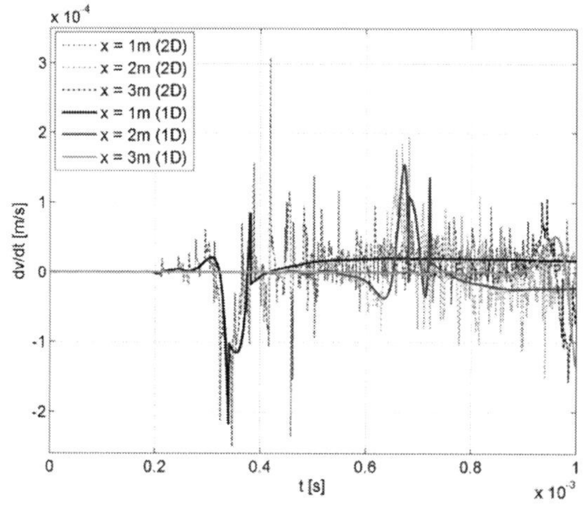

Figure 7. Comparison of $dv(t)/dt$ for different distances x and for $d_0 = 200$ mm, $t_0 = 40\,\mu$s

only 0.7% contrary to 19% in the case $t_0 = 2\,\mu$s, as mentioned above. Also the comparison of vertical velocities $dv(t)/dt$ shows better agreement between 1D and 2D solutions, as clear from Fig. 5. The time-spacing of peaks in the 2D solution is naturally same as in the case of short pulse but the amplitudes of its oscillations around 1D results are significantly smaller then in Fig. 2.

All of the comparisons presented previously have been done for beams of quite small dimensions and for very short times. To present the behaviour of both solutions for a more realistic case, the transient response of a beam with cross-sectional parameters $b_0 = 20$ mm and $d_0 = 200$ mm to the pulse (11) with $t_0 = 40\,\mu$s has been investigated as well. Fig. 6 and Fig. 7 show the comparison of 1D (thick solid curves) and 2D (thin dashed curves) solutions of $v(t)$ and $dv(t)/dt$, respectively, for $x = \{1, 2, 3\}$ m up to time t_{max} = 1 ms. It is clear from Fig. 6 that although the pulse takes 40 μs, the beam height is too large to satisfy the aforementioned criterion, so the agreement between the solutions $v(t)$ is not so good as in the previous cases. However, the Timoshenko beam theory gives us deflections the character of which is very similar to those followed from the exact 2D theory. Unfortunately, the same cannot be said in the case of $dv(t)/dt$, see Fig. 7. It is clear from this figure that the height of the beam is too high compared to the wavelength of the shortest wave to use the Timoshenko theory for the investigation of vertical beam velocity.

6 Conclusions

This work presents the comparison of two types of analytical solutions for the problem of transverse impact on an infinite homogeneous elastic beam. Based on the analyses of the results obtained for different geometric and loading parameters, the main factors influencing the agreement between 1D Timoshenko theory and 2D exact theory are discussed from the wave propagation point of view. It is shown that the Timoshenko theory is applicable also to impact problems of beams but the credibility of its results significantly depends on the frequencies of waves propagated through the beam.

This publication was supported by the project LO1506 of the Czech Ministry of Education, Youth and Sports.

References

[1] K.F. Graff, *Wave motion in elastic solids* (Dover Publications Inc., New York, 1975)
[2] S.P. Timoshenko, Philos. Mag. **41**, 744–746 (1921)
[3] J.N. Reddy, J. Appl. Mech.-T. ASME, **51**, 745–752 (1984)
[4] K.P. Soldatos, Acta Mech. **94**, 195–220 (1992)
[5] M. Touratier, Int. J. Eng. Sci. **29**, 901–916 (1991)
[6] M. Aydogdu, Compos. Struct. **89**, 94–101 (2009)
[7] V. Adámek, F. Valeš, B. Tikal, Nonlinear Anal.-Theor. **71**, e2569–e2576 (2009)
[8] V. Adámek, F. Valeš, J. Vib. Eng. Technol. **3**, 699–710 (2015)
[9] V. Adámek, F. Valeš, J. Červ, Adv. Eng. Softw. **113**, 120–129 (2017)

Online adaptive semi-active vibration damping of slender structures subject to moving loads

Dominik Pisarski[1], and *Andrzej* Myśliński[2,*]

[1]Institute of Fundamental Technological Research of Polish Academy of Sciences, 02-106 Warsaw, Poland
[2]Systems Research Institute of Polish Academy of Sciences, 01-447 Warsaw, Poland

Abstract. The paper deals with the adaptive optimal semi-active control of the slender vibrating structures subjected to the moving loads. The deflection of the structure is governed by Euler-Bernoulli beam equation approximated by the system of bilinear ordinary differential equations. The damping function of the structure support is chosen as the control function. The optimal control problem consists in finding such bang-bang control function to minimize the energy as well as the vibrations of the carrying structure. Although the switched optimal control is a very efficient tool in the reduction of structure vibrations it is very sensitive with respect to changes of the speed of the traveling load. This paper deals with the development of the adaptive descent type algorithm that enables the update of the optimal controls in real time based on the measured speed of the traveling load or structure's state. The proposed algorithm uses reference optimal controls computed for the constant speeds and the set of functions describing the sensitivity of the system dynamics with respect to the measured parameters. Numerical computations are carried out for different speed scenarios of the moving load. The obtained numerical results indicate that the proposed adaptive controller can significantly outperform the reference optimal solutions.

1 Introduction

The high speed traveling loads interacting with the slender structures generate their vibrations and/or reduce their stability. The primary aim of the structural control is to reduce these undesired oscillations and to enhance the structure's stability [1]. Many control methods of vibrating structures, including adaptive controllers (see references in [2]) are proposed in literature. Most of the existing optimal adaptive controllers (see [3]) are based on complex iterative procedures that often do not guarantee convergence to a solution in the required time.

The aim of this paper is to develop and test new method that enables online adaptation of the optimal damping of the structure's supports according to both the measured speed of the traveling load and the estimated state of the structure. The control aim consists in finding the bang-bang control function minimizing the objective functional related to the total energy of the structure. The design of the adaptive controller will be based on two optimal control subproblems. The first one will be the bilinear optimal control subproblem involving some given reference passage velocity. For this subproblem, we will formulate the necessary optimality conditions and identify the switched structure of the optimal controls. For the assumed reference velocity, the set of optimal switching times will be computed off line by solving the corresponding two point boundary value problem. In the second subproblem, intended for the sequential solution, we will consider the actual measured velocity. To find the set of optimal switching times for this subproblem, we will use the simple gradient-based iterative procedure involving the optimal solutions computed for the reference speed and the set of functions describing the sensitivity of the system dynamics with respect to the measured velocity. The sensitivity of reference solutions with respect to velocity is calculated based on Taylor series expansion rather than on the analysis of the second order optimality conditions as in [2, 4]. Since the sensitivity functions can be computed off line the solution to the second subproblem can be calculated very quickly. This feature allows the frequent update of the optimal controls. The performance of the designed adaptive method will be validated by means of numerical simulations for different speed scenarios of the moving load.

2 Structure model

Consider the carrying structure supported by the set of controlled dampers as illustrated in Fig. 1. The structure is governed by one-dimensional fourth order Euler–Bernoulli beam equation [5] and is characterized by the bending stiffness EI and density per unit length μ. The interaction between the mass and the beam is represented by contact point. The inertial forces associated to the mass are neglected [5]. The position and the velocity of the moving mass at time t are denoted by $\xi_M(t)$ and $v(t)$, respectively. The mass position is computed by formula $\xi_M(t) = \int_0^t v(\bar{t})\,\mathrm{d}\bar{t}$.

*Corresponding author: myslinsk@ibspan.waw.pl

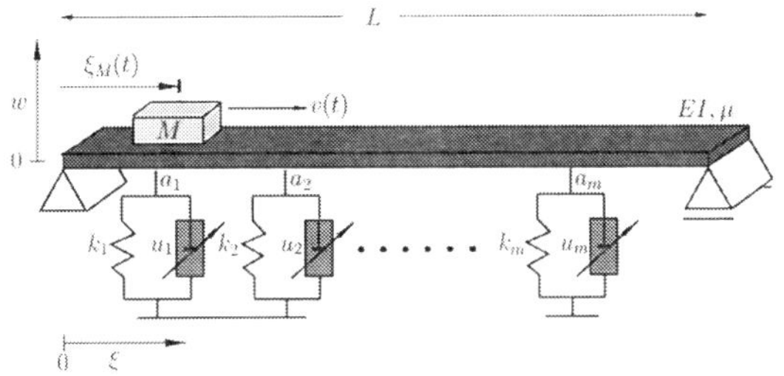

Fig. 1. A span supported by the set of controlled dampers and subjected to the traveling mass.

2.1 ODE representation

Using the separation technique the transverse deflection $w(\xi, t)$ of the vibrating beam at point ξ and time t governed by Euler–Bernoulli equation can be approximated [5] by the following Fourier series:

$$w(\xi, t) = \frac{2}{L}\sum_{\bar{j}=1}^{\infty} V_{\bar{j}}(t)\theta_{\bar{j}}(\xi), \quad \theta_{\bar{j}}(\xi) = \sin(\bar{j}\pi\xi/L). \tag{1}$$

We will rely on the approximated solutions by taking $\bar{j}$ from 1 to $n = 10$. The state vector $x(t) \in R^{2n}$ is defined as

$$x(t) = [V_1(t), \dot{V}_1(t), V_2(t), \dot{V}_2(t), ..., V_n(t), \dot{V}_n(t)]^T, \tag{2}$$

where $\dot{V}_j = dV_j/dt$, $j = 1, ..., n$. Let us also introduce the system matrices $A_{2n\times 2n}$, $B^i_{2n\times 2n}$, $i = 1, 2, .., m$ equal to: $(A)_{jk} = 1$ for $(j,k) = (2\bar{l}-1, 2\bar{l})$, $\bar{l} = 1, 2, ..., n$, $(A)_{jk} = -\frac{EI}{\mu}\frac{(j/2)^4\pi^4}{L^4}$ for $(j,k) = (2\bar{l}, 2\bar{l}-1)$, $\bar{l} = 1, 2, .., n$, $(A)_{jk} = -\frac{2}{\mu L}\sum_{i=1}^{m} \sin\frac{((j+1)/2)\pi a_i}{L} \sin\frac{(k/2)\pi a_i}{L}$ for $(j,k) = (2\bar{l}-1, 2\hat{l})\bar{l}, \hat{l} = 1, 2, ..., n$, $(A)_{jk} = 0$ in other cases. Moreover for $i = 1, ..., m$, $(B^i)_{jk} = -\frac{2}{\mu L}\sin\frac{(j/2)\pi a_i}{L}\sin\frac{(k/2)\pi a_i}{L}$ for $(j,k) = (2\bar{l}, 2\hat{l})$ $\bar{l}, \hat{l} = 1, 2, ..., n$ and $(B^i)_{jk} = 0$ else. The excitation vector $F_{2n\times 1}$ is given by: $(F)_j = -\frac{Mg}{\mu}\sin\frac{(j/2)\pi v t}{L}$ for $j = 2\bar{l}$, $\bar{l} = 1, 2, ..., n$ and $(F)_j = 0$ else. The beam dynamic equation can be written [2, 5] as the following system of the ordinary differential equations in time interval $t \in [0, T]$ with the fixed final time $T > 0$:

$$\dot{x}(t) = A\,x(t) + \sum_{i=1}^{m} u_i(t)\,B^i\,x(t) + F(v, t), \tag{3}$$

$$x(0) = x_0.$$

The time horizon T is equal to the time of the beam passage, i.e., $T = L/v$. For $\bar{j} = 1, 2, ..., n$ the function $x_0 \in R^{2n}$ in the initial condition is given and equal to

$$x_0 = [V_1(0), \dot{V}_1(0), V_2(0), \dot{V}_2(0), ..., V_n(0), \dot{V}_n(0)]^T. \tag{4}$$

Elements $V_{\bar{j}}(0)$ and $\dot{V}_{\bar{j}}(0)$ are computed based on the original boundary conditions imposed on vibrating beam [2]. The function $x_0 \in R^{2n}$ is assumed to be bounded. The matrices A and B^i, $i = 1, ..., m$, are bounded. Each control input $u_i(t)$, $i = 1, ..., m$, $t \in [0, T]$ is assumed bounded by two positive real values u_i^{min} and u_i^{max} corresponding to the minimum and maximum values of the admissible damping coefficients, i.e., $u = u(t) \stackrel{def}{=} \{u_i(t)\}_{i=1}^m$, $u \in \mathcal{U} \stackrel{def}{=} [u_i^{min}, u_i^{max}]^m \subset \mathbb{R}_+^m$. Moreover $0 < u_i^{min} < u_i^{max} \leq \tilde{\alpha}$, where for $i = 1, ..., m$ $\grave{\alpha} > u_i^{min}$ is a given real constant.

Note that the control variable u appears linearly in the state system (3). Due to the middle term in the right hand side of (3), the system is classified as the bilinear [2].

The excitation vector $F = F(v,t) = \{F_i(v,t)\}_{i=1}^{2n}$ is dependent on velocity $v = v(t) : [0,T] \to R$. The velocity v is assumed bounded, i.e., for all $t \in [0,T]$ $v^{min} < v(t) < v^{max}$ where v^{min} and v^{max} are a given real constants and continuously differentiable. Each component function $F_i(v,t) : R \times [0,T] \to R$, $i = 1, ..., 2n$, of the excitation vector F is assumed bounded and continuously differentiable with respect to time t and velocity v. In the paper, the bang-bang type of control $u \in U$ is considered. It generates the discontinuity of the right hand side of the system (3).

Let $v_{ref} \in R$ denote a given reference velocity of the moving load. For such a velocity and a positive definite $2n \times 2n$ matrix Q let us recall from [6–8] the optimal control problem considered for the system (3): For a given velocity v_{ref} find the pair of functions $(u(t), x(t)) \in \mathcal{U} \times R^{2n}$ minimizing the objective functional

$$J(u) = J(x(u), v_{ref}) = \int_0^T x^T(t)\,Q\,x(t)\,\mathrm{d}t, \tag{5}$$

subject to the constraints on the interval $[0, T]$

$$\dot{x}(t) = A\,x(t) + \sum_{i=1}^{m} u_i(t)\,B^i\,x(t) + F(v_{ref}, t), \tag{6}$$

$$x(0) = x_0^{v_{ref}}. \tag{7}$$

The solution to problem (5)-(6) will be later referred to as $(u^{\star v_{ref}}(t), x^{\star v_{ref}}(t))$. The objective functional $J : R^{2n} \times R \to R$ is associated with the energy of the structure. The aim of this optimal control problem is to find the control function, i.e., variable damping coefficients, so that for a given velocity, the energy of the beam subjected to the traveling load is minimized. As it is shown in [6], by minimizing the objective functional (5) we also provide the smooth trajectory of the moving load. The existence of an optimal solution $u^{\star v_{ref}}(t) \in U$ to problem (5)-(6) follows from Filippov theorem and the continuity of the objective functional (for details see [9]).

2.2 Necessary optimality condition for problem (5)-(6)

Let us introduce Hamiltonian function $H(t, x, u, p, v_{ref})$: $[0, T] \times R^{2n} \times \mathcal{U} \times R^{2n} \times R \to R$ defined as follows:

$$H(t, x, u, p, v_{ref}) = p^T(t)(A\,x(t) + \tag{8}$$

$$\sum_{i=1}^{m} u_i(t)\,B^i\,x(t) + F(v_{ref}, t)) - x^T(t)\,Q\,x(t),$$

where $p(t) \in R^{2n}$ is a column vector denoting the adjoint state. The part of Hamiltonian function associated with the control function u is called the switching function [4]

$$\sigma(t, x, p, v_{ref}) = (\sigma_1(t, x, p, v_{ref}), ..., \tag{9}$$

$$\sigma_m(t, x, p, v_{ref})),$$

$\sigma_i(t, x, p, v_{ref}) = p^T(t) B^i x(t)$ for $i = 1, 2, ..., m$. Therefore, the necessary optimality condition for (5)-(6) takes the following form [2, 5]: If $u^{\star v_{ref}}(t)$ is the optimal control for the problem (5)-(6) then there exists absolutely continuous function $p : [0, T] \to R^{2n}$ satisfying the following conditions:

$$\dot{p}(t) = -\frac{\partial H(t, x, u, p, v_{ref})}{\partial x}, \quad p(T) = 0, \qquad (10)$$

$$H(t, x, u^{\star v_{ref}}, p, v_{ref}) = \min_{u \in \mathcal{U}} H(t, x, u, p, v_{ref}). \qquad (11)$$

From the minimum condition (10) for the admissible control set $\mathcal{U}$ results the following control law for the ith optimal control component

$$u_i^{\star v_{ref}}(t) = \begin{cases} u_i^{min}, & \text{if } \sigma_i(t) < 0, \\ u_i^{max}, & \text{if } \sigma_i(t) > 0, \\ \text{undetermined}, & \text{if } \sigma_i(t) = 0. \end{cases} \qquad (12)$$

Throughout the rest of the paper, by the kth switching time we will mean the column vector $\tau_k = (\tau_{k1}, \tau_{k2}, ..., \tau_{km})$ where τ_{ki} is associated with the control component u_i.

3 Sensitivity of optimal bang-bang controls with respect to velocity

Assume that for a given velocity v_{ref} the optimal control structure τ_k for the problem (5)-(6) is known. As it was shown in [5], the solution to optimal control problem (5)-(6) exhibits significant sensitivity with respect to changes in the speed v of the traveling load. Denote by $t' \in [0, T)$ the measurement time instant of the traveling load and the state of the structure. Assuming that for the chosen time instant t' the real speed $v = v(t')$ of the traveling load and the real state $x(t')$ of the structure are known from the measurements our goal is to develop a method that enables the real-time adaptation of the optimal control functions according to both measured values. The method will use the reference optimal control functions computed for a given reference speed and the set of functions describing the sensitivity of the system dynamics to the measured parameters. Assume that the velocity perturbation

$$dv = v - v_{ref}, \qquad (13)$$

is suitably small. Let us now consider new finite-dimensional optimization problem involving the perturbed switching times as the optimization variables. First, we will introduce the following assumptions [2, 4]:

A1 The optimal control $u^{\star v_{ref}}$ (as well as the switching function $\sigma(t)$) has a finite number of switching times (zeroes) $\tau_k^{\star v_{ref}}$, $k = 1, ..., K$ satisfying

$$0 = \tau_0^{\star v_{ref}} < \tau_1^{\star v_{ref}} < \tau_k^{\star v_{ref}} < ... < \tau_{K+1}^{\star v_{ref}} = T. \qquad (14)$$

A2 There exists a set of twice continuously differentiable functions $u_k(t, x)$ such that the control $u^{\star v_{ref}}(t)$ is given in a state feedback form, i.e., for $k = 1, ..., K + 1$ it holds

$$u^{\star v_{ref}}(t) = u_k(t, x^{\star v_{ref}}(t)), \quad \forall t \in (\tau_{k-1}^{\star v_{ref}}, \tau_k^{\star v_{ref}}). \qquad (15)$$

A3 The time derivative $\dot{\sigma}$ of the switching function σ is assumed not to vanish, i.e., the strict bang-bang property [4] is assumed to be satisfied

$$\dot{\sigma}(\tau_k^{\star v_{ref}}) \neq 0, \quad k = 1, ...K. \qquad (16)$$

Based on the assumptions **A1**–**A3** we will reformulate [2, 4] the optimal control problem (5)-(6) as the finite-dimensional nonlinear induced optimization problem in terms of the switching times rather than solve it directly as the original problem (5)-(6). Let us introduce the matrix of switching times

$$\eta = (\tau_1, \tau_2, ..., \tau_K) \in R^{K \times m}, \qquad (17)$$

$$0 = \tau_0 < \tau_1 < \tau_k < ... < \tau_{K+1} = T,$$

taken from a neighborhood of the optimal reference vector $\eta^{\star v_{ref}} = (\tau_1^{\star v_{ref}}, \tau_2^{\star v_{ref}}, ..., \tau_K^{\star v_{ref}})$. For any η satisfying (17), we will write $u^\eta(t)$ for the nonsingular bang-bang control for $\tau_{k-1} < t < \tau_k$, $k = 1, ..., K + 1$ defined as follows:

$$u^\eta(t) = (u_1^\eta(t), u_2^\eta(t), ..., u_m^\eta(t)) = (u_{k1}, u_{k2}, ..., u_{km}). \quad (18)$$

Let us denote by $\mathcal{K}$ the following subset of the switching times indicators:

$$\mathcal{K} = \{k : k = 1, ..., K + 1, \quad \tau_k \geq t'\}. \qquad (19)$$

For the control $u^\eta(t)$, $\eta = ((\tau_k)_{k \in \mathcal{K}})$, let $x = x(t, x(t'), \eta, v)$ be the absolutely continuous function satisfying the initial value problem

$$\dot{x}(t) = A\,x(t) + \sum_{i=1}^{m} u_i^\eta(t)\,B^i\,x(t) + F(v, t), \qquad (20)$$

$$\text{for } \tau_{k-1} < t < \tau_k, \quad k \in \mathcal{K}, \quad x(t') = x_0^v.$$

Referring to problem (5)-(6), we have $x(t, x^{\star v_{ref}}(t'), \eta^{\star v_{ref}}, v_{ref}) = x^{\star v_{ref}}(t)$. Using the solution of the state equation (20), we can now formulate the parametric optimization problem in terms of η as the optimization variable: For a given velocity v find $\eta = ((\tau_k)_{k \in \mathcal{K}})$ minimizing the objective functional

$$J(\eta) = J(x(\eta), v) = \int_{t'}^{T} x^T(t)\,Q\,x(t)\,dt, \qquad (21)$$

subject to the constraints (20) on the interval $[t', T]$. The solution to problem (21) will be from now on referred to as $\eta^{\star v} = ((\tau_k^{\star v})_{k \in \mathcal{K}})$.

3.1 Directional derivatives with respect to switching time

To solve the optimization problem (21) with constraints (20) we develop an iterative algorithm based on the method of the steepest descent. For that purpose we need to estimate the relevant derivatives corresponding to the directions of the descent for the decision parameter η. It can be verified (see, for instance, [5]) that the derivative of the objective functional (21) with respect to the kth switching time of the ith control component is given by

$$\frac{dJ}{d\tau_{ki}} = -\int_{t'}^{T} \left(\frac{\partial H}{\partial \tau_{ki}}\right)_{|\tau_{ki}} dt. \tag{22}$$

From (11) it yields

$$\frac{dJ}{d\tau_{ki}} = (u_{(k+1)i} - u_{ki})\, p^T(\tau_{ki}, \tag{23}$$

$$x(t'), (\tau_k)_{k\in\mathcal{K}}, v)\, B^i\, x(\tau_{ki}, x(t'), (\tau_k)_{k\in\mathcal{K}}, v)\,.$$

Here, u_{ki} is the value of the ith control component in the time interval $t \in (\tau_{k-1}, \tau_k)$, $k \in \mathcal{K}$. In order to evaluate the derivatives (23), we employ approximations of the state x and the adjoint state p. The approximation of the state x, denoted later by x^{app}, will combine the optimal reference state $x^{\star v_{ref}}$ and the appropriate differentials due to the changes in the speed of the moving load, the initial condition, and the switching times. Similarly, for the approximate adjoint state p, later referred to as p^{app}, we will use $p^{\star v_{ref}}$ and the relevant differentials.

It follows from the assumption (13) on the velocity perturbation that the perturbation of the initial state $dx = x(t') - x^{\star ref}(t')$ can also be assumed to be suitably small. It implies small perturbations of the switching times, defined as follows:

$$d\tau_k = (d\tau_{k1}, d\tau_{k2}, ..., d\tau_{km})\,, \tag{24}$$

$$d\tau_{ki} = \tau_{ki} - \tau_{ki}^{\star ref}\,, \quad i = 1, 2, ..., m\,, \quad k \in \mathcal{K}\,.$$

By using the perturbations (13) and (24) the state $x(t, x(t'), (\tau_k)_{k\in\mathcal{K}}, v)$ can be evaluated as the function dependent on

$$x(t, x^{\star ref}(t') + dx, (\tau_k^{\star ref} + d\tau_k)_{k\in\mathcal{K}}, v_{ref} + dv)\,. \tag{25}$$

Since for $t \in [t', T]$ we have $x^{\star v_{ref}}(t) = x(t, x^{\star ref}(t'), (\tau_k^{\star ref})_{k\in\mathcal{K}}, v_{ref})$, by using Taylor series expansion formula and the standard arguments [2] we can employ the following approximation of the state function:

$$x^{app}(t) = x^{\star v_{ref}}(t) + \left(\frac{\partial x}{\partial v}\right)_{|t} dv + \frac{1}{2}\left(\frac{\partial^2 x}{\partial v^2}\right)_{|t} dv^2 + \tag{26}$$

$$\left(\frac{\partial x}{\partial x_0}\right)_{|t} dx + \sum_{i=1}^{m}\sum_{k\in\mathcal{K}}\left(\frac{\partial x}{\partial \tau_{ki}}\right)_{|t} d\tau_{ki}\,.$$

For $i = 1, 2, ..., m$ and $k \in \mathcal{K}$ let us introduce the following definitions of the sensitivity functions

$$S_v^x = \frac{\partial x}{\partial v}, \quad S_{vv}^x = \frac{\partial^2 x}{\partial v^2}, \quad S_{x_0}^x = \frac{\partial x}{\partial x_0}, \quad S_{\tau_{ki}}^x = \frac{\partial x}{\partial \tau_{ki}}. \tag{27}$$

Denote by f the right hand side of equation (6) for $t \in [t', T]$. Thus, $f = \dot{x}$ and due to (20) we have the following relations:

$$f = f(x(x_0, (\tau_k)_{k\in\mathcal{K}}, v), (\tau_k)_{k\in\mathcal{K}}, v)\,. \tag{28}$$

Thus, using the standard sensitivity approach [9] and differentiation with respect to the perturbed parameters we obtain ordinary differential equations satisfied by the sensitivity functions (27) in $[t', T]$:

$$\dot{S}_v^x = \frac{\partial f}{\partial x} S_v^x + \frac{\partial f}{\partial v}, \quad S_v^x(t') = 0, \tag{29}$$

$$\dot{S}_{vv}^x = \frac{\partial f}{\partial x} S_{vv}^x + \frac{\partial^2 f}{\partial v^2}, \quad S_{vv}^x(t') = 0, \tag{30}$$

$$\dot{S}_{x_0}^x = \frac{\partial f}{\partial x} S_{x_0}^x, \quad S_{x_0}^x(t') = 1, \tag{31}$$

and for $i = 1, 2, ..., m$ and $k \in \mathcal{K}$

$$\dot{S}_{\tau_{ki}}^x = \frac{\partial f}{\partial x} S_{\tau_{ki}}^x + \frac{\partial f}{\partial \tau_{ki}}, \quad S_{\tau_{ki}}^x(t') = 0\,. \tag{32}$$

From (6) we obtain

$$\frac{\partial f}{\partial x} = A + \sum_{i=1}^{m} u_i B^i, \tag{33}$$

$$\frac{\partial f}{\partial \tau_{ki}} = (u_{(k+1)i} - u_{ki})\, B^i\, x\, \delta(t - \tau_{ki}), \tag{34}$$

where $\frac{\partial f}{\partial v} = \frac{\partial F}{\partial v}$, $\frac{\partial^2 f}{\partial v^2} = \frac{\partial^2 F}{\partial v^2}$. The approximation p^{app} of the adjoint state p is based on the adjoint state $p^{\star v_{ref}}$ corresponding to the optimal state $x^{\star v_{ref}}$. It is computed similarly as the approximation x^{app} of the state function x, i.e., Taylor series expansion (26) is used with the adjoint function p rather than the state function x.

4 Optimal control algorithm

The developed adaptive control numerical algorithm is based on the concept of the Receding Horizon Control. This class of algorithms employs the solutions to the finite horizon optimal control problems where both the initial and the final times are repetitively being pushed forward. In this paper, a sequence of problems (21) with incrementally increasing time t', but fixed final time T, is considered. The final time is equal to the time in which a moving load is passing the structure with the constant velocity v_{ref}. The fixed final time enables us to employ the optimal reference solutions. Based on the measurements of the actual velocity $v(t')$ and the actual state $x(t')$ as well as using the formulas for the objective functional derivatives (23) combined with the formula (26) to calculate the approximation of the state and adjoint functions the algorithm is able to compute the descent directions to update the set of the optimally switched controls. Remark that all of the sensitivity state and adjoint functions (29)-(32) can be precomputed off line and stored in a controller's memory. Since the computation of the values of the objective functional derivatives (23) can be executed instantaneously, the adaptive method can be implemented in real-time. For detailed description of the algorithm see [2].

5 Numerical examples

The optimal control problem (21) with constraints (20) has been solved numerically for several passage scenarios including both constant and varying speed profiles. For each scenario the same reference optimal controls computed for given constant reference speed is used. Our primary goal is to estimate the range of velocity perturbation that guarantees high performance of the designed method. Matrix Q in the objective functional (21) is assumed as the identity matrix and the time horizon T is computed for $L = 4$ $[m]$. Here $v(t)$ is the actual velocity in the speed scenario. Other parameters are set as equal to $n = 10$, $m = 4$, $EI = 801$ $[Nm^2]$, $\mu = 2.3$ $[kg/m]$, $u^{min} = 500$ $[N/m]$, $u^{max} = 5000$ $[N/m]$, $a_i = 0.2L, 0.4L, 0.6L, 0.8L$, $M = 4$ $[kg]$. In the simulations, maximal number of iterations is set to $l_{max} = 20$ and stop criterion $\epsilon = 3 \cdot 10^{-10}$. Four control functions u_i, $i = 1,2,3,4$, associated with four controlled dampers are used. For each control function u_i two switching times τ_{1i} and τ_{2i} are assumed (see Fig. 2). The structure of the controls is defined as follows:

$$u_i(t) = \begin{cases} u^{min}, & \text{if } 0 \le t < \tau_{1i}, \quad i = 1,2,3,4, \\ u^{max}, & \text{if } \tau_{1i} \le t \le \tau_{2i}, \quad i = 1,2,3,4, \\ u^{min}, & \text{if } \tau_{2i} < t \le T, \quad i = 1,2,3,4. \end{cases} \qquad (35)$$

The solution to the reference optimal control problem (5)-(6) has been computed for the moving load speed $v_{ref} = 8$ [m/s] and zero initial condition $x_0^{v_{ref}} = 0$. The corresponding time horizon is equal to $T = 0.5$ $[s]$. The time interval $[0, T]$ is represented by the following set of time instants $\{t_j\}_{j=0,1,\ldots,1000}$, $t_0 = 0$, $t_{1000} = T$. For this representation we obtain the set of optimal switching times for problem (5)- (6):

$$\begin{aligned} &\text{damper no. 1} \quad \tau_{11}^{\star v_{ref}} = t_1, \quad \tau_{21}^{\star v_{ref}} = t_{366}, \\ &\text{damper no. 2} \quad \tau_{12}^{\star v_{ref}} = t_{152}, \quad \tau_{22}^{\star v_{ref}} = t_{596}, \\ &\text{damper no. 3} \quad \tau_{13}^{\star v_{ref}} = t_{322}, \quad \tau_{23}^{\star v_{ref}} = t_{732}, \\ &\text{damper no. 4} \quad \tau_{14}^{\star v_{ref}} = t_{80}, \quad \tau_{24}^{\star v_{ref}} = t_{972}. \end{aligned} \qquad (36)$$

The results for the exact optimal controls, later referred to as the optimal controls and denoted by $u^{\star\star}$ are computed as the solutions to the problem (5)-(6), where instead of the reference velocity v_{ref} and the initial state $x_0^{v_{ref}}$ we assume the actual speed profiles as equal either to constant velocity v_{const} or piecewise linear variable velocity v_{var}. For each case, the optimal control is assumed to preserve two switches structure as given by (35).

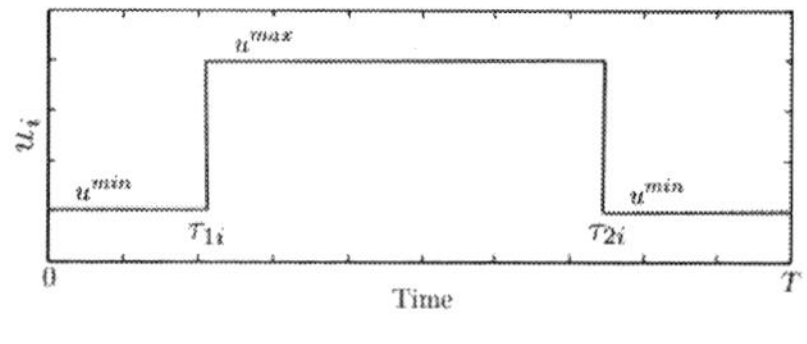

Fig. 2. The assumed structure of the switched controls u_i, $i = 1,2,3,4$.

The first set of simulations has been carried out for perturbed velocity scenarios. We have considered three scenarios of constant speed v_{const} and three scenarios of varying speed v_{var}. The assumed profile of the varying speed v_{var} is displayed in Fig. 3. For each of the constant speed

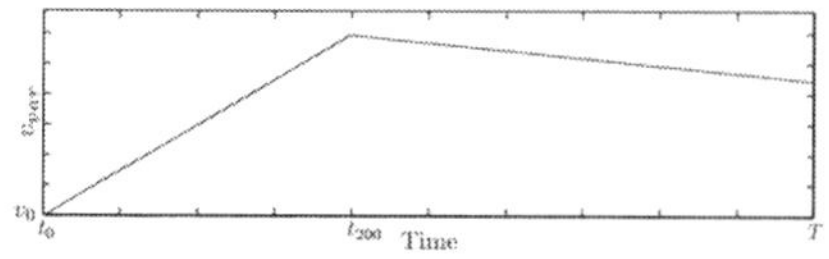

Fig. 3. The profile of the varying speed v_{var} used in the simulations.

Table 1. The values of the objective function J in the cases of the perturbed velocity scenarios.

Speed variant in [m/s]	Passive case ($u = u_{max}$)	Reference optimal control ($u = u^{\star v_{ref}}$)	Adaptive control ($u = u^{\star v}$)	Optimal control ($u = u^{\star\star}$)
$v_{const} = 8$	1.0000	0.4801	0.4801	0.4801
$v_{const} = 9$	1.0000	0.5295	0.4990	0.4881
$v_{const} = 12$	1.0000	0.9081	0.5448	0.5327
$v_{var}, v_0 = 8$	1.0000	0.5136	0.4939	0.4855
$v_{var}, v_0 = 9$	1.0000	0.5859	0.5202	0.5025
$v_{var}, v_0 = 12$	1.0000	1.1314	0.5696	0.5463

scenarios v_{const}, we assume that the traveling velocity is within the range of 8-12 [m/s]. The same range is assumed for the starting velocity v_0 in the cases of the varying speed v_{var}.

The comparison of the values of the objective function for the set of considered speed variants is presented in Table 1. For each speed scenario the objective values for the optimal, the reference and the adaptive control are normalized to the passive case. For moving load velocities higher than 8 [m/s], the adaptive strategy outperforms significantly (by 40.1 % or 43.1 % for velocity 12 [m/s]) the system driven by the open-loop reference controls. On the other hand this strategy results in only 2.2 % or 4.2 % loss compared to the optimal case.

Fig. 4 and 5 display the evolution of the adaptive switching times during the moving load passage with velocity $v_{const} = 12$ [m/s]. For the initial values, we assumed the set of the corresponding reference optimal solutions. Then, at each time instant, the switching times were repetitively updated until the time t_{667} corresponding to the end of the passage. The largest change in switching times values was observed at the beginning of the adaptive process. The convergence of the algorithm's gradient-based pro-

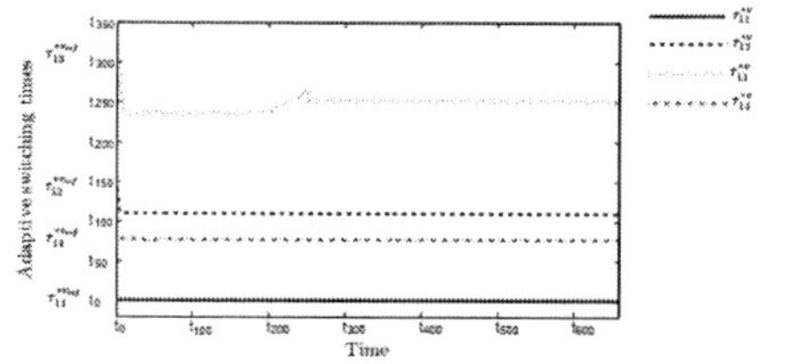

Fig. 4. Evolution of the optimal adaptive switching times $\{\tau_{1i}^{\star v}\}$ during the moving load passage for the case $v_{const} = 12$ [m/s]. The set of optimal reference switching times $\{\tau_{1i}^{\star v_{ref}}\}$ is assumed for the initial values.

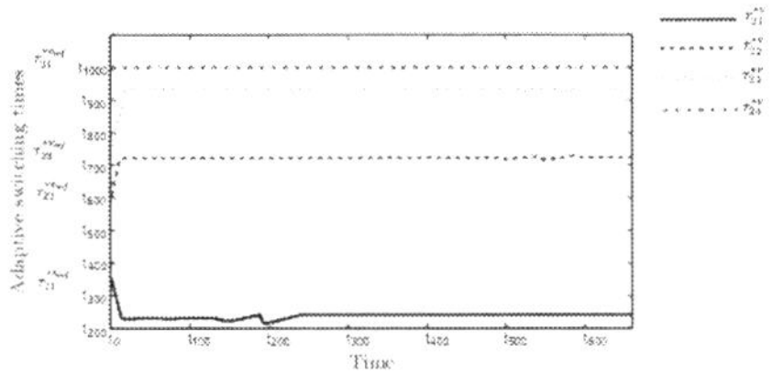

Fig. 5. Evolution of the optimal adaptive switching times $\{\tau_{2i}^{\star v}\}$ during the moving load passage for the case $v_{const} = 12$ [m/s]. The set of optimal reference switching times $\{\tau_{2i}^{\star v_{ref}}\}$ is assumed for the initial values.

cedure for different times instants is shown in Fig. 6. for the moving load passage with velocity $v_{const} = 12$ [m/s].

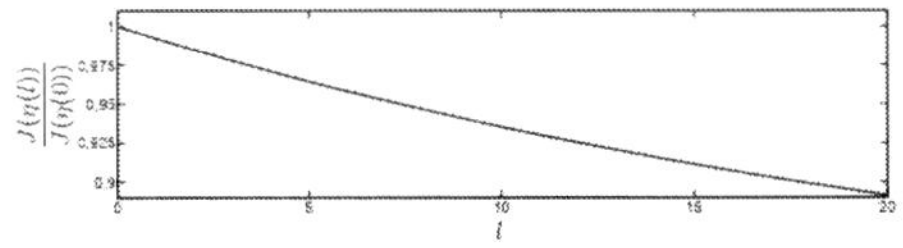

Fig. 6. Evolution of the objective functional during the iteration of the gradient-based procedure for the time instant t_0. The procedure is terminated by the condition $l = l_{max}$.

In Fig. 7 and 8 we demonstrate the comparison of the adaptive, reference and optimal control signals as well as the trajectories of the moving load for $v_{const} = 12$ [m/s]. One can notice that the switching times of the adaptive controls tend to approach their corresponding optimal control values.

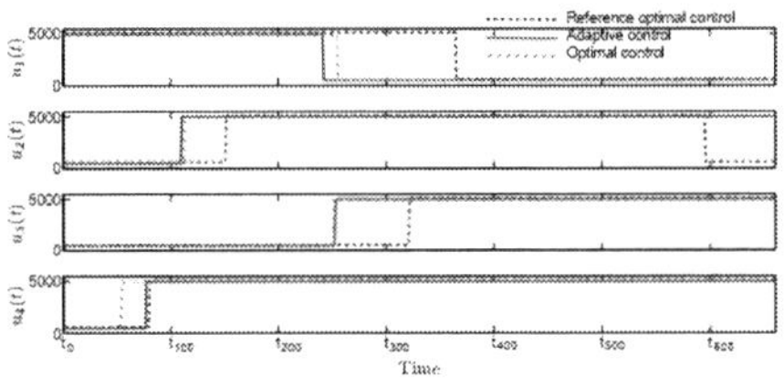

Fig. 7. Reference, adaptive, optimal controls. $v_{const} = 12$ [m/s].

6 Conclusion

Real time adaptive control method for structures subjected to a moving loads has been proposed and tested. The obtained results indicate that the proposed method can outperform the reference optimal solutions. Further numerical as well as experimental tests are being performed.

References

[1] C. Bajer, B. Dyniewicz, Arch. Appl. Mech. **79**, 499-508 (2009)

[2] D. Pisarski, A. Myśliński, Optim. Control Appl. Methods, Advance online publication, DOI:10.1002/oca.2321 (2017)

[3] M. Diehl, H.J. Ferreau, M. Abu-Khalaf, N. Haverbeke, *Efficient numerical methods for nonlinear MPC and moving horizon estimation in nonlinear Model Predictive Control: Towards new challenging applications* (Springer-Verlag, Berlin Heidelberg, 2009)

[4] G. Vossen, J. Optim. Theory Appl. **144**, 409-429 (2010)

[5] D. Pisarski, *Semi-active control system for trajectory optimization of a moving load on an elastic continuum* (PhD Thesis, Institute of Fundamental Technological Research of Polish Academy of Sciences, Warsaw, 2012)

[6] D. Pisarski, J. Vibr. Control, Advance online publication, DOI:10.1177/1077546316657244 (2016)

[7] D. Pisarski, Math. Problems Eng. **2015**, 1-12 (2015)

[8] D. Pisarski, C. Bajer, J. Sound Vibr. **329**, 140-149 (2010)

[9] D. Liberzon, *Calculus of variations and optimal control theory: A concise introduction* (Princeton University Press, New Jersey, 2012)

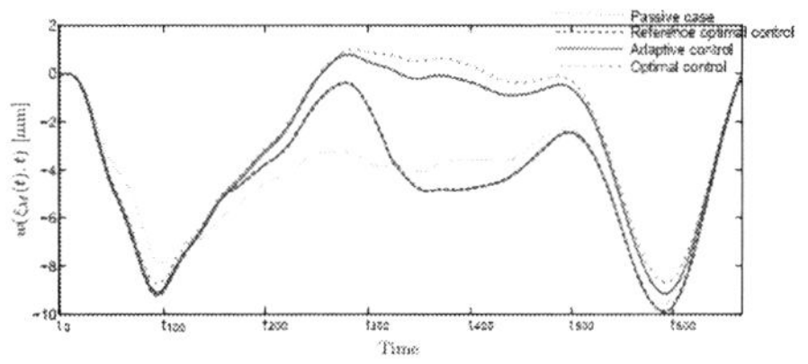

Fig. 8. Trajectories of the moving load for $v_{const} = 12$ [m/s].

Moving element analysis of high-speed rail system accounting for hanging sleepers

Jian Dai[1], *Kok Keng* Ang[1,*], and *Dongqi* Jiang[2,*]

[1]Department of Civil & Environmental Engineering, National University of Singapore, Singapore, 117576
[2]Department of Civil Engineering, School of Science, Nanjing University of Science and Technology, Nanjing, China, 210094

Abstract. It is very common in the ballasted track system that sleepers are not well supported by the ballast materials due to the uneven settlement of the ballast under repeated train passage. These unsupported track elements are often termed as hanging sleepers and they can lead to undesirable effects due to increased dynamic response of the train-track system, especially when the speed of the train is high. In this paper, we present a computation scheme in conjunction with the moving element method for the analysis of high-speed train-track dynamics accounting for hanging sleepers. The proposed computational scheme will be first verified by comparison with available analytical results. The dynamic response of a high-speed train traveling on a ballasted track considering unsupported sleepers is next investigated. Various factors affecting the response of the high-speed rail system including the speed of the train, the number of hanging sleepers and the pattern of the hanging sleepers will be examined and discussed.

1 Introduction

The high-speed train travel has become increasingly popular in continental people's daily life, for it does not only greatly shorten the inter-city travel time without causing traffic congestion but also being environmental friendlier as compared to the conventional car travel. The rapid increase in speed, however, also gives rise to concerns regarding the train-travel safety and increased demand on railway system maintenance. For example, the vibration of rails induced by a passing train would disturb the pavement of ballast materials [1]. According to [1], the ballast would undergo non-elastic deformation under vehicle passage which eventually leads to uneven track settlement and voids under sleepers. Literature has shown that it is very common that sleepers are not well supported by the ballast [2]. The gaps between sleepers and ballast put the safety of train operation in question, and the risk of disaster is believed to be amplified with regards to the increase in the running speed of the train.

The existence of unsupported sleepers, also known as hanging sleepers, has attracted railway engineers' attention, and quite a few research studies, both experimental and numerical, were carried out to investigate the effect of hanging sleepers on the dynamic response of a train-track system [3-5]. These works mainly assume that the hanging sleepers only appear in a consecutive arrangement. In reality, however, it is common that the sleepers are not aligned in a consecutive manner. In other words, well-bedded sleeper(s) may be surrounded by hanging sleepers. Unfortunately, such scenario has been rarely studied.

Due to the complexity of the hanging sleeper problem, most researchers resorted to numerical solutions such as finite element method (FEM) and finite difference method (FDM). Although these approaches are powerful and versatile, they encounter difficulties when the moving load approaches the boundary of the domain and even beyond. These difficulties can be overcome by employing a large enough domain size but at the cost of significantly increased computational effort. In addition, the moving load is more frequently located at an intermediate location within an element other than the nodal points. This has been shown to lead to a reduction in the accuracy of the solution [7]. On the other hand, the moving element method (MEM) adopts a moving frame-of-reference that travels with the moving load and thus the moving load is always located at a nodal point of one of the moving elements. Compared to the standard FEM, this numerical approach has been shown to be more accurate and efficient in dealing with various train-track dynamic problems [7-11].

In this paper, we present a computational study using the moving element method on the dynamic response of a high-speed train traveling on ballasted tracks accounting for hanging sleepers. The computational scheme presented in this paper will be first verified by comparison with available analytical results based on a simplified moving sprung mass model. Then a computational extended train-track model is put forward for the analysis of the dynamic response of the high-speed rail systems accounting for the effect of hanging sleepers. Parametric study will be conducted to investigate the effect of various factors affecting the dynamic response of the high-speed rail system. This

* Corresponding author: Kok Keng Ang: ceeangkk@nus.edu.sg
Dongqi Jiang: jiangdongqi@gmail.com

includes the speed of the train, the number of hanging sleepers, and the number of fully supported sleepers that are surrounded by the hanging sleepers.

2 Train-track model

The problem of concern comprises a high-speed train traveling at speed V over a ballasted track with or without hanging sleepers (see Fig. 1). The rail beam is resting over a number of equally spaced sleepers, which are again supported by the ballast. The ballast materials are idealized as discrete mass points that are interlocked with adjacent ballast points by their shear stiffness. These idealized ballast elements are supported by the subgrade above the rigid foundation. For sake of simplicity, only one rail beam of the railway track subjected to half weight of the train is considered.

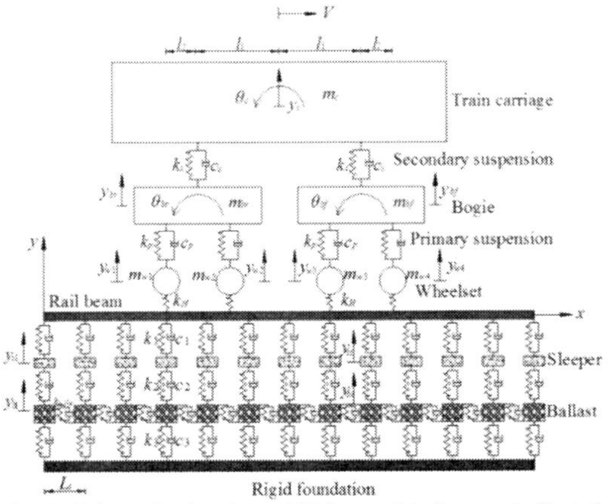

Fig. 1. Schematic drawing of high-speed train over ballasted track.

2.1 Ballasted track model

The ballasted track model comprises a rail beam that is supported by sleepers that have an equal spacing L_s. The rail beam is modelled according to Timoshenko beam theory with density ρ, cross-sectional area A, area moment of inertia I, Young's modulus E, shear modulus G and shear coefficient κ. The sleepers are modelled as rigid bodies with lumped mass m_s and they are supported by the ballast elements with lumped mass m_b. The rail pads between the rail beam and the sleepers have stiffness coefficient k_1 and damping coefficient c_1. The ballast pavement supports the sleepers through stiffness k_2 and damping coefficient c_2. The governing equations of motion for the track model can be written as

$$\rho A\frac{\partial^2 y_r(x,t)}{\partial t^2}-\kappa AG\frac{\partial^2 y_r(x,t)}{\partial x^2}+\kappa AG\frac{\partial \psi_r(x,t)}{\partial x} +\sum_{i=1}^{n}c_1\left(\frac{\partial y_r(x,t)}{\partial t}-\frac{\partial y_{si}(t)}{\partial t}\right)\delta(x-iL_s) +\sum_{i=1}^{n}k_1\left(y_r(x,t)-y_{si}(t)\right)\delta(x-iL_s)=-\sum_{j=1}^{m}F_{cj}\delta\left(x-X_j(t)\right) \quad (1a)$$

$$\rho I\frac{\partial^2 \psi_r(x,t)}{\partial t^2}-EI\frac{\partial^2 \psi_r(x,t)}{\partial x^2}-\kappa AG\frac{\partial y_r(x,t)}{\partial x} +\kappa AG\psi_r(x,t)=0 \quad (1b)$$

$$m_s\frac{\partial^2 y_{si}(t)}{\partial t^2}+c_1\left(\frac{\partial y_{si}(t)}{\partial t}-\frac{\partial y_r(x,t)}{\partial t}\bigg|_{x=(i-1)L_s}\right)+c_2\left(\frac{\partial y_{si}(x,t)}{\partial t}-\frac{\partial y_{bi}(x,t)}{\partial t}\right) +k_1\left(y_{si}(t)-y_r(x,t)\big|_{x=(i-1)L_s}\right)+k_2\left(y_{si}(t)-y_{bi}(t)\right)=0 \quad (1c)$$

$$m_b\frac{\partial^2 y_{bi}(t)}{\partial t^2}+c_2\left(\frac{\partial y_{bi}(t)}{\partial t}-\frac{\partial y_{si}(t)}{\partial t}\right)+c_3\frac{\partial y_{bi}(x,t)}{\partial t} +c_{bs}\left(2\frac{\partial y_{bi}(t)}{\partial t}-\frac{\partial y_{bi-1}(t)}{\partial t}-\frac{\partial y_{bi+1}(t)}{\partial t}\right)+k_2\left(y_{bi}(t)-y_{si}(t)\right) +k_3 y_{bi}(t)+k_{bs}\left(2y_{bi}(t)-y_{bi-1}(t)-y_{bi+1}(t)\right)=0 \quad (1d)$$

where y_r denotes the displacement of the rail beam; y_{si} the displacement of the ith sleeper; ψ_r the bending rotation of the rail; k_1 and c_1 the stiffness and damping coefficient of the rail pad; k_2 and c_2 the stiffness and damping coefficients of the ballast; k_3 and c_3 are the stiffness and damping coefficients of the subgrade. k_{bs} and c_{bs} are the shear stiffness and damping coefficient between two discrete ballast elements. F_{cj} is the contact force between the jth train wheel and the rail beam; X_j is the travel distance of the jth wheel; and δ the Dirac-delta function.

2.2 Train model

The train is modelled as a 10-DOFs (degrees of freedom) sprung-mass system comprising one train body, two bogies and four wheelsets. These train components are modelled as rigid bodies interconnected by suspension systems modelled as spring-dashpot units. The vehicle carriage has a lumped mass m_c and moment of inertia I_c. The two bogies have an identical mass m_{bg} and moment of inertia I_b. Each of the four wheelsets has a mass m_w. Based on Newton's second law of motion, the governing equations of the moving sprung-mass train model can be written in compact form as

$$\mathbf{M}_v\ddot{\mathbf{y}}_v+\mathbf{C}_v\dot{\mathbf{y}}_v+\mathbf{K}_v\mathbf{y}_v=\mathbf{P}_v \quad (2a)$$

where $\mathbf{M}_v$, $\mathbf{C}_v$, $\mathbf{K}_v$ and $\mathbf{P}_v$ denote the mass, damping, stiffness matrices and force vector of the vehicle model; $\mathbf{y}_v$ is the vector of train degrees of freedom and is given as

$$\mathbf{y}_v=\left\{y_c \quad \theta_c \quad y_{br} \quad \theta_{br} \quad y_{bf} \quad \theta_{bf} \quad y_{w1} \quad y_{w2} \quad y_{w3} \quad y_{w4}\right\}^{\mathrm{T}} \quad (2b)$$

herein y_c and θ_c denote the vertical displacement and pitching rotation of the carriage, respectively; y_{br}, θ_{br} and y_{bf}, θ_{br} are the vertical displacement and pitching rotation of the bogie at the rear and the one in the front of the vehicle, respectively; y_{wi} is the vertical displacement of the ith wheelset.

2.3 Wheel and rail interaction

The vehicle model and the track model are coupled together by means of wheel and rail interaction. Here in

 https://doi.org/10.1051/matecconf/201814805007

this paper, the nonlinear Hertz contact model is adopted by assuming that the contact area is always circular [6-8]. This contact model also allows the possible separation between the wheel and the rail known as the "jumping wheel phenomenon".

3 Moving element method

The moving element method (MEM) is employed here for the dynamic analysis of the train-track system in the time domain. This numerical method employs a moving r-axis whose origin is attached to the train and travels with it at the same speed V and requires the governing equations of motion to be reformulated in the moving frame-of-reference. For the analysis of discretely supported high-speed rail (HSR), the efficient computational scheme recently proposed by Dai *et al.* [11] which pre-prepares the global stiffness and damping matrices in one period for easy retrieval in the subsequent time domain computation can be adopted. In addition, the movement of the hanging sleeper(s) must be considered. The computation can thus be divided into three phases. In the first phase, the truncated track segment has not reached the location of the hanging sleeper, as shown in Fig. 2. The computation in this phase thus does not involve the unsupported sleeper. In the second phase, the truncated track has partially travelled across the unsupported sleeper. During this phase, the stiffness and damping matrices of the ballasted track system must be updated at each time step to account for the change due to the loss of ballast support under the incident sleeper. In the third phase, the truncated track has completely travelled past the unsupported sleeper and is now supported fully by well-bedded sleepers. The computation in this phase is identical to that in the first phase.

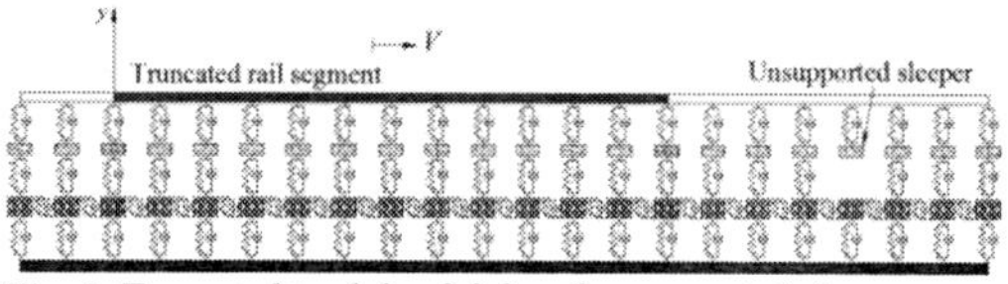

Fig. 2. Truncated track in vicinity of unsupported sleeper.

4 Numerical study

Two cases involving hanging sleepers are investigated in this paper. The first case involves a simplified discretely supported rail beam subject to a moving mass. The aim of this case is to examine the accuracy of the proposed numerical model through comparison with available analytical results in the literature. In the second case, the full model comprising the 10-DOF train model and 3-layer ballasted track model is employed to investigate the dynamic response of a HSR system accounting for one or more hanging sleepers with different patterns.

4.1 Comparison of results

In an attempt to examine the accuracy of the methodology presented in Section 3, the case of a simplified discretely supported track subjected to a moving sprung mass as studied by Sheng *et al.* [12] is first considered. The model of concern is shown in Fig. 3. The rail beam is represented by a Timoshenko beam resting on evenly spaced spring-dashpot units. The wheel load is modelled as a sprung-mass traveling at 80 m/s. Table 1 lists the parameters for the analysis model adopted in [12]. Note that the railhead roughness is represented by a sine function with amplitude $a_r = 10 \times 10^{-6}$ m and wavelength $\lambda_r = 0.04$ m. In view of the speed, the frequency of the excitation due to the railhead roughness is 2000 Hz.

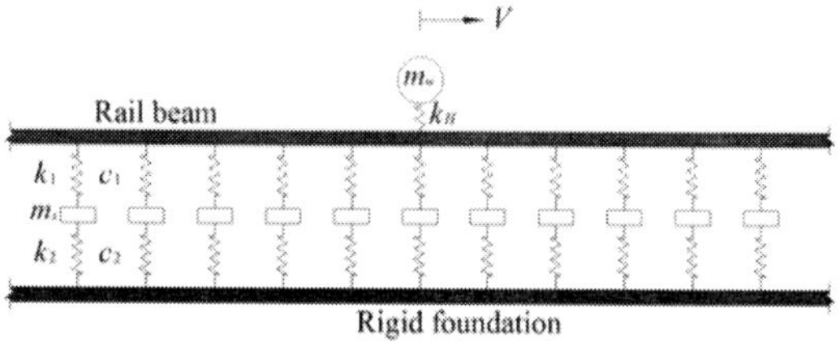

Fig. 3. Simplified model for numerical verification.

Table 1. Parameters for the simplified train-track model.

Parameter	Value	Parameter	Value
ρA	60.3665 kg/m	κ	0.4
EI	6.4155×10^6 Nm^2	m_w	1350 kg
GA	6.2289×10^9 N	m_s	162 kg
L_s	0.6 m	k_H	1.4×10^9 N/m
k_1	3.5×10^8 N/m	k_2	50×10^6 N/m

In the MEM model, a rail beam segment of length 18 m is truncated. This segment covers 30 sleeper bays and based on a convergence study, it is found that the length is long enough to avoid the boundary effect in simulating the behaviour of an infinitely long rail. The beam element size is set to 0.3 m. This means the track segment between two adjacent sleepers is discretized into two elements only. In order to capture the dynamic response due to the high-frequency excitation, the time step size is set to 7.5×10^{-6} s. This is equivalent to 100 steps for the sprung mass to travel across one entire sleeper bay.

Figure 4 shows the steady-state dynamic wheel-rail contact force predicted by the MEM for the case when the sprung mass travels from one sleeper to the next. Also presented in this figure are the results published by Sheng *et al.* [12]. As can be seen, both results match well with each other, thereby validating present computational approach for the analysis of discrete ballasted tracks under moving loads.

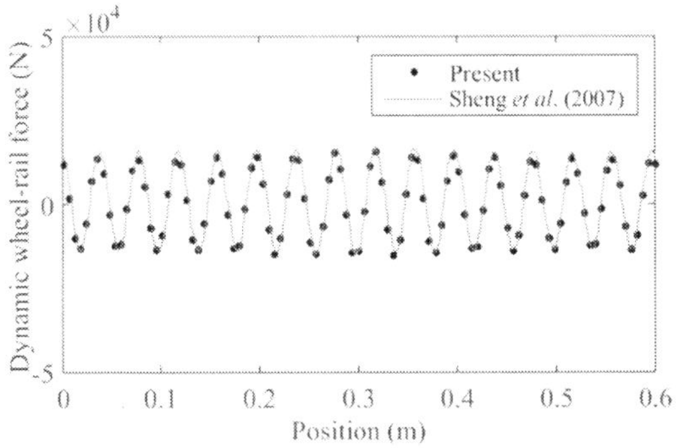

Fig. 4. Dynamic contact force over one sleeper bay.

4.2 Effect of hanging sleepers

The dynamic response of a high-speed rail system accounting for hanging sleepers is next investigated. In this study, the train-track coupled model described in Section 2 is employed for the analysis. The parameters for the train model are listed in Table 2. Table 3 lists the parameters for the track model where the rail beam model is based on the Chinese type 60 rail. The nonlinear wheel-rail contact constant is $1/k_H = 4.27 \times 10^{-8}$ m/N$^{2/3}$. Note that all the parameters listed above are reported by Zhu *et al.* [4]. Note that smooth railhead is purposely chosen such that the effect unsupported sleepers will not be overweighed by the existence of track irregularities.

Table 2. Parameters for the train model.

Parameter	Value	Parameter	Value
m_c	26000 kg	m_{bf}	2309 kg
m_{br}	2309 kg	m_w	750 kg
I_c	1.155×10^6 kgm^2	I_b	1970 kgm^2
k_p	1.87 MN/m	c_p	5×10^5 Ns/m
k_s	1.72 MN/m	c_s	1.96×10^5 Ns/m
l_1	9 m	l_2	1.25 m

In this study, various cases considering different patterns of hanging sleepers are investigated. Figure 5 shows 6 patterns. The first case involves only one hanging sleeper. The rest cases involve two hanging sleepers spaced apart by different numbers of fully supported sleepers. In view of this, the first case can also be regarded as the case involving two hanging sleepers that are spaced apart by an infinite number of fully supported sleepers.

Table 3. Parameters for the track model.

Parameter	Value	Parameter	Value
ρA	60.64 kg/m	κ	0.4
EI	6.7557×10^6 Nm^2	ν	0.3
m_s	130 kg	m_b	460 kg
k_1	120 MN/m	c_1	7.5×10^4 Ns/m
k_2	180 MN/m	c_2	5.88×10^4 Ns/m
k_3	120 MN/m	c_3	3.12×10^4 Ns/m
k_{bs}	78 MN/s	c_{bs}	8×10^4 Ns/m

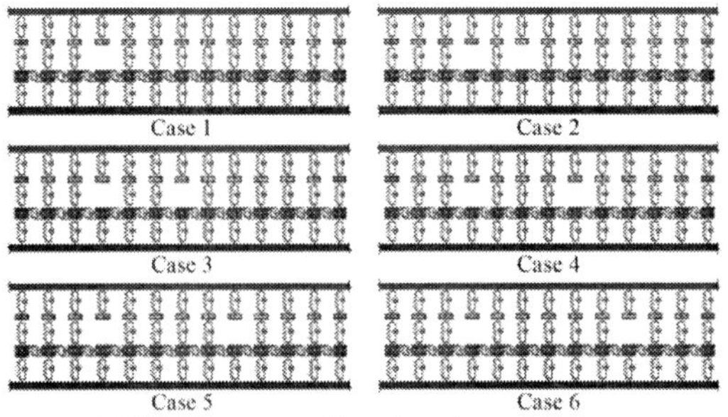

Fig. 5. 6 different patterns of hanging sleepers.

Figure 6 shows the wheel-rail contact forces in the vicinity of the hanging sleepers for different cases when the speed of the train is 320 km/h. As can be seen, the maximum contact force in the second case is smaller than the first case by about 9%, despite of the increase in the total number of hanging sleepers. This finding coincides with the results reported in [1]. However, as the number of fully supported sleepers between two hanging sleepers increases from case 3 to case 6, it is interesting to find that the maximum contact force increases to a peak of 122.9 kN in case 4 before decreases. As the number of fully supported sleepers increases to 5, the maximum contact force reduces to the same value as in the base case 1. Further study with more fully supported sleepers between two hanging sleepers revealed that the maximum contact force remains unchanged. In other words, the effect of the first hanging sleeper has been fully damped out before the wheelsets reach the second hanging sleeper for cases when the number of fully supported sleepers between the faulty exceeds 4.

Figure 7 presents the maximum DAF in wheel-rail contact force for case 1 to case 9. Note that cases 7, 8 and 9 refer to the patterns where two hanging sleepers are spaced apart by 6, 7 and 8 fully supported sleepers, respectively. As can be seen, case 2 tends to give the lowest maximum contact forces for all cruising speeds of the train considered in this study. The highest maximum contact forces are found to occur for cases 3-5, depending on the speed of the train. For smaller cruising speeds, the highest maximum contact force tends to occur for the case where the space between two hanging sleepers is narrower as compared to higher speeds. This may be explained as due to the phase difference in the dynamic response to the excitation, the maximum response is expected to occur slightly after the excitation. For higher speeds, the maximum contact force is expected to occur at longer distances after traveling across the unsupported sleeper. As the distance between the two hanging sleepers further increases, it is found that maximum contact forces reduce to those observed in case 1. This implies that these two hanging sleepers have independent effects on the dynamic response of the high-speed rail system when they are spaced far apart.

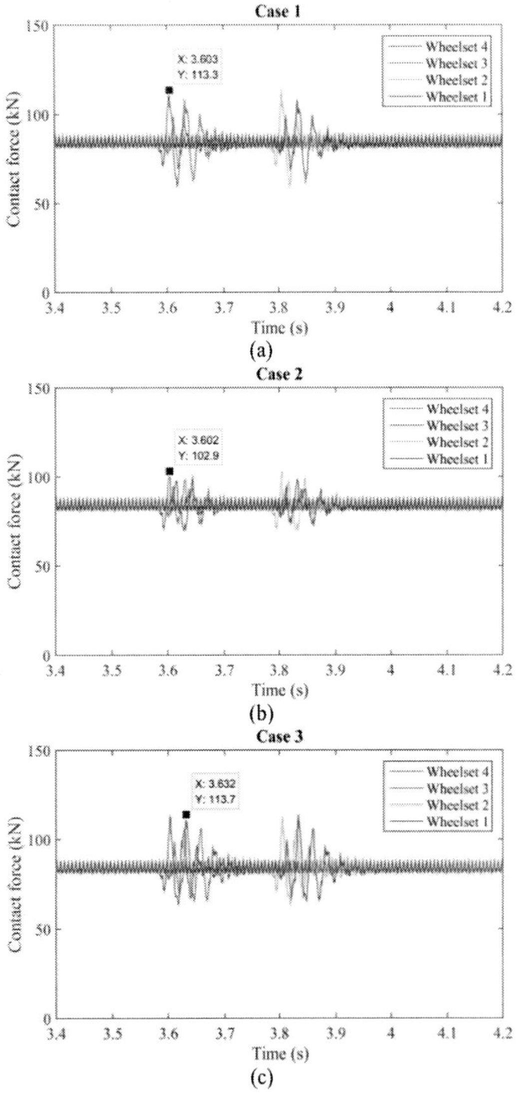

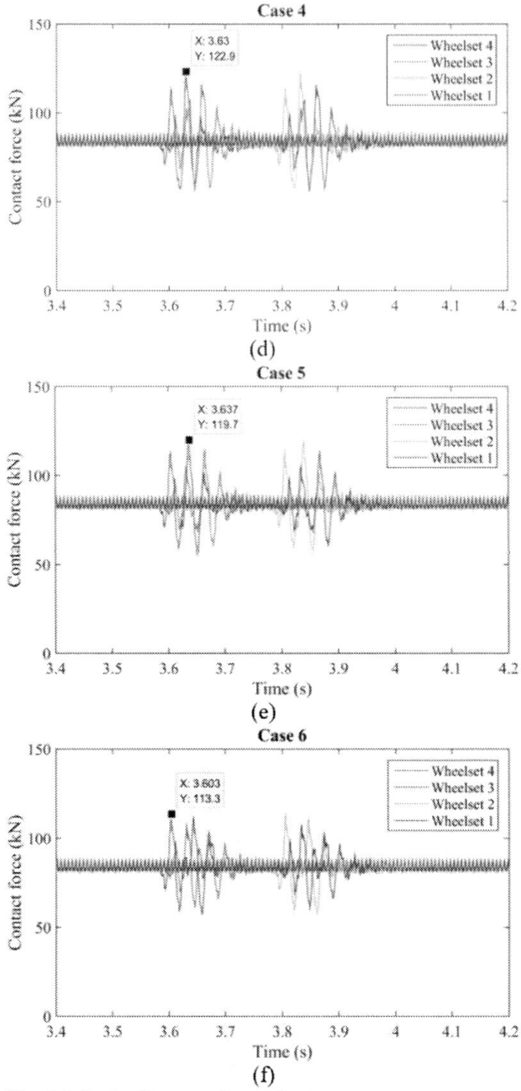

Fig. 6. Wheel-rail contact forces for 6 cases.

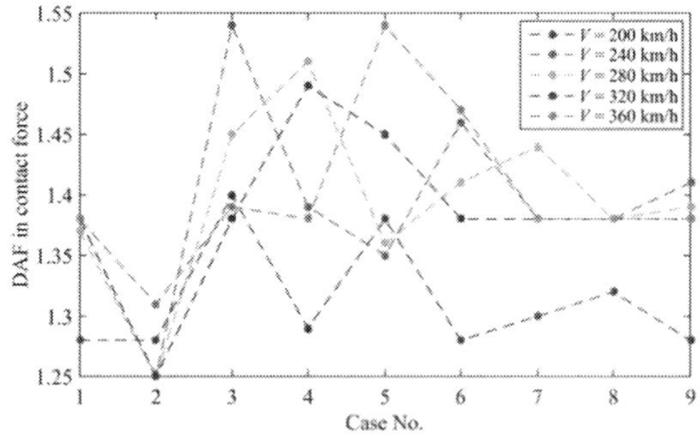

Fig. 7. DAF in contact force for different cases.

5 Conclusion

This paper is concerned with a computational study on the dynamic responses of a high-speed rail system accounting for hanging sleepers. A computational scheme in conjunction with the moving element method is proposed to account for the motion of the rail substructures and the unsupported sleepers with respect to the truncated rail segment. The accuracy of the proposed method is examined by comparison with available analytical results using a simplified moving sprung mass model. A good agreement is found between the present solution and the available results in the literature, thereby validating the proposed computational scheme for the analysis of ballasted track under moving loads.

In the analysis of high-speed rail systems accounting for hanging sleepers using an extended 10-degree-of-freedom moving sprung mass model, it is found that the speed of the train and the pattern of the hanging sleepers are two key factors affecting the dynamic response of the system. Generally, the lowest dynamic response in terms of wheel-rail contact force is expected to occur for the case of one fully supported sleeper surrounded by two hanging sleepers for all speeds considered in this study. The highest maximum contact force is observed to occur when the two hanging sleepers are spaced apart by 3 to 5 fully supported sleepers. This implies that the maintenance of the track must be conducted carefully to avoid the most undesirable situation. As the spacing between the two hanging sleepers further increases to 7 sleeper bays and beyond, the coupled effect of the two hanging sleepers becomes small and may be neglected.

References

1. A. Lundqvist, T. Dahlberg, Proc. IMechE. Part F: J. Rail Rapid Transit **219**, 67-77 (2005)
2. S. Augustin, G. Gudehus, G. Huber, A. Schunemann, *System dynamics and long-term behaviour of railway vehicles, track and subgrade*, **6**, 317-336 (Springer, Berlin, 2003)
3. J.J. Zhu, A.K.W. Ahmed, S. Rakheja, A. Khajepour, Vehicle Syst. Dyn., **48**, 1535-1552 (2010)
4. J.Y. Zhu, D.J. Thompson, C.J.C. Jones, Vehicle Syst. Dyn., **49**, 1389-1408 (2011)
5. J.A. Zakeri, M. Fattahi, M.M. Ghanimoghadam, J. Mech. Sci. Technol., **29**, 2289-2295 (2015)
6. C. Esveld, *Modern railway track* (MRT Productions, Duisburg, 2001)
7. K.K. Ang, J. Dai, J. Sound Vib., **332**, 2954-2970 (2013)
8. K.K. Ang, J. Dai, M.T. Tran, V.H. Luong, Int. J. Comp. Meth., **11**, 1-12 (2014)
9. C.G. Koh, J.S.Y. Ong, D.K.H. Chua, J. Feng, Int. J. Numer. Meth. Eng., **56**, 1549-1567 (2003)
10. M.T. Tran, K.K. Ang, V.H. Luong, J. Sound Vib., **333**, 5427-5442 (2014)
11. J. Dai, K.K. Ang, M.T. Tran, V.H. Luong, D. Jiang, Proc. IMechE. Part F: J. Rail Rapid Transit (2017)
12. X. Sheng, M. Li, C.J.C. Jones, D.J. Thompson, J. Sound Vib. **303**, 873-894 (2007)

A Reduced Model of a Thermo-Elastic Nonlinear Circular Plate

Anna Warminska[1,*], *Emil* Manoach[2,**], and *Jerzy* Warminski[3,***]

[1] *Department of Thermodynamics, Fluid Mechanics and Aviation Propulsion Systems, Lublin University of Technology, Lublin, Poland*
[2] *Institute of Mechanics, Bulgarian Academy of Sciences, Sofia, Bulgaria*
[3] *Department of Applied Mechanics, Lublin University of Technology, Lublin, Poland*

Abstract. Nonlinear vibrations of a circular plate subjected to mechanical and thermal loadings are presented in the paper. A model of the plate is based on the extended Mindlin theory, taking into account nonlinear geometrical terms and acting heat uniformly distributed along the plate span. The dynamics of a coupled thermo-mechanical problem is reduced from a set of partial differential equations to ordinary differential equations. Considering oscillations around the first natural frequency just one mode reduction is proposed. The analysis shows that elevated temperature shifts the resonance curve and new post-buckling oscillations arise. Depending on initial conditions for the post-buckling state various scenarios of bifurcations take place and transient irregular oscillations may occur. The proposed one degree of freedom model shows a good agreement with response of the model based on three or five-modes reduction.

1 Introduction

The ambient temperature which may vary in a large range may essentially influence dynamics of mechanical, aviation or aerospace structures, starting from low negative, up to high positive temperature values. The variation of temperature may be crucial while performing experimental tests. In some cases, close to critical points, even small temperature variation may give unexpected change in the system response. The fundamentals of thermo-elastic problems with aerospace engineering applications have been published in [1]. In publications [2] and [3] thermo-elastic vibrations of beam element have been studied in context of elevated abient temperature or heat flux acting directly on the beam. It has been shown that heat flux may change the transient as well as the steady state system's response. The coupled fields problem, represented by a set of partial differential equations (PDE) has been solved numerically by proposed by authors iterative procedure based on pseudo-normal modes approach. Reduced models of a nonlinear Timoshenko beam model have been presented in [4], [5]. The model has been obtained by just one mode reduction for the case of the elevated and uniformly distributed temperature. The influence of elevated temperature on the resonance curves and bifurcation points has been presented. The resonance curves have been shifted into lower frequencies if the temperature has been increased. Furthermore, above the certain threshold, the increased temperature has led to bucking phenomenon and small oscillations around a new post-buckled state or large chaotic oscillations around the neutral position. The thermo-elastic Timoshenko beam model has been extended for a nonuniform distribution of the temperature [6]. The reduced model allowed bifurcation analysis for nonuniform heat distribution.

Thermally induced oscillations of 2D structures e.g rectangular nonlinear plate models based on higher order shear deformation theory have been analysed in [7], [8]. The post-buckling phenomenon for isotropic or composite plates has been presented for different coefficient of thermal expansion and geometric imperfections. The theoretical results have been validated by experimental tests. Influence of temperature on moderately thick functionally graded plates is presented in [9]. Both material properies and the temperature have been assumed as varied nonlinearly through the plate thickness. Complex dynamics including periodic, quasi-periodic and chaotic motions have been found. The bifurcation analysis has been based on a large number of degrees of freedom which caused numerical complications, comparing with reduced low dimensional models. In recent publications [10], [11] a 2D continuum model of the nonlinear dynamic problem for a von Karman shear indeformable symmetric cross-ply laminated plate in a thermomechanical environment has been presented. The main features of nonlinear effects coupled with thermal field have been systematically analysed. A comparison of response by partially coupled and uncoupled mechanical reduced models has been demonstrated. The reduced model has consisted of one mechanical and two thermal ordinary differential equations. The reduced model preserved the main features of the full represented by PDEs model. The systematic numerical investigations

* e-mail: a.warminska@pollub.pl
** e-mail: e.manoach@imbm.bas.bg
*** e-mail: j.warminski@pollub.pl

of local and global dynamics of the plate for varying thermal and mechanical loadings have been presented.

The analysis performed for circular plate at elevated temperature and selected variants of external excitation has been presented in [6, 12]. The exact thermo-mechanical model of the circular plate described by PDEs presented has been reduced to one degree of freedom system. The preliminary results demonstrated thermally induced buckling phenomenon and then the post-buckling bifurcation scenario. In this paper we check correctness of the model proposed in paper [12] comparing one mode reduction with numerical simulations based on several pseudo-modes reduction.

2 A model of circular plate

A nonlinear model of the plate is based on extended Mindlin plate theory. The model proposed in this paper takes into account nonlinear displacement field, thermal and mechanical loadings. The circular plate taken under the study is presented in Fig. 1.

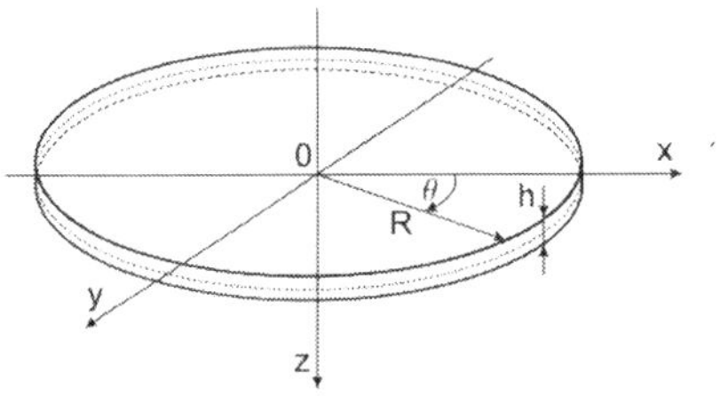

Figure 1. Circular plate model with indicated coordinates and dimensions.

The mathematical model of a fully coupled thermo-mechanical model is represented by a set of partial differential equations. The axis-symmetric vibrations of a nonlinear thermo-elastic plate has the form [12]

$$
\begin{aligned}
&Ah\left[\frac{\partial^2 u}{\partial r^2}+\frac{\partial w}{\partial r}\frac{\partial^2 w}{\partial r^2}+\frac{1}{r}\frac{\partial u}{\partial r}-\frac{u}{r^2}+\frac{1}{2r}(1-\nu)\left(\frac{\partial w}{\partial r}\right)^2\right]\\
&=A(1+\nu)\alpha_T\frac{\partial \gamma_T}{\partial r}\\
&D\left[\frac{\partial^2 \psi}{\partial r^2}+\frac{1}{r}\frac{\partial \psi}{\partial r}-\frac{\psi}{r^2}\right]-k^2Gh\left(\frac{\partial w}{\partial r}+\psi\right)-c_2\frac{\partial \psi}{\partial t}-\frac{\rho h^3}{12}\frac{\partial^2 \psi}{\partial t^2}\\
&=A(1+\nu)\alpha_T\frac{\partial \kappa^T}{\partial r}\\
&k^2Gh\left(\frac{\partial^2 w}{\partial r^2}+\frac{1}{r}\frac{\partial w}{\partial r}+\frac{\partial \psi}{\partial r}+\frac{\psi}{r}\right)\\
&+Ah\left[\frac{\partial u}{\partial r}+\frac{1}{2}\left(\frac{\partial w}{\partial r}\right)^2+\frac{\nu}{r}u-\frac{1+\nu}{h}\alpha_T\gamma_T\right]\left(\frac{\partial^2 w}{\partial r^2}+\frac{1}{r}\frac{\partial w}{\partial r}\right)\\
&+Ah\left(\frac{\partial^2 u}{\partial r^2}+\frac{\nu}{r}\frac{\partial u}{\partial r}-\frac{\nu}{r^2}u+\frac{\partial w}{\partial r}\frac{\partial^2 w}{\partial r^2}-\frac{1+\nu}{h}\alpha_T\frac{\partial \gamma_T}{\partial r}\right)\frac{\partial w}{\partial r}\\
&+c_1\frac{\partial w}{\partial t}-\rho h\frac{\partial^2 w}{\partial t^2}=-p(r,t)
\end{aligned}
\tag{1}
$$

where $u(r,t)$ is the in-plane displacement, $w(r,t)$ is the transverse displacement, $\psi(r,t)$ is the cross-section rotation angle and h and R are the plate thickness and radius. Material parameters are defined by the Young and shear modulus E, G, material density ρ, the coefficient of thermal expansion α_T. The rest of coefficients and functions is defined as: $A=\frac{E}{1-\nu^2}$, $D=\frac{Eh^3}{12(1-\nu^2)}$, $\gamma^T(r)=\int_{-h/2}^{h/2}\Delta T(r,z)dz$, $\kappa^T(r)=\int_{-h/2}^{h/2}\Delta T(r,z)zdz$, $\Delta T=T-T_0$, where T and T_0 are current and reference temperatures and $p(r,t)$ is external loading function. One can note that in Eq. $(1)_1$ the inertia term $\frac{\partial^2 u}{\partial t^2}$ is neglected. The above equations can be solved numerically by applying pseudo-modes method or reduced to ordinary differential equations taking a properly selected number of modes.

3 Circular plate reduced model

The full model of the plate represented by partial differential equations is reduced to the ordinary differential equations by the Galerkin method based on the modes projection. We assume that the temperature is uniformly distributed along the plate span and thickness and is equal to ambient temperature. In such a case Eq. $(1)_1$ is solved analytically and coordinate u is related to w. Assuming solution in the series

$$
w(r,t)=\sum_{n=1}^{N}q_n(t)W_n(r),\quad \psi(r,t)=\sum_{n=1}^{N}q_n(t)\Psi_n(r) \tag{2}
$$

where $W_n(r)$, $\Psi_n(r)$ are assume vibration modes and $q_n(t)$ generalised coordinates, the PDEs are transformed to ordinary differential equations (ODE) by the Galerkin method. The modes should satisfy geometrical boundary conditions. We take into consideration a case of the clamped plate. The mode shapes can be approximated by the Bessel functions which satisfy boundary conditions. The first there modes against radius normalised to one are presented in Fig. 2.

In this paper we take one mode reduction of the clamped plate, considering excitation distributed according to the first mode shape. Substituting solution Eq. (2) into PDEs, multiplying by the the selected mode function and integrating through the plate radius, we get a set of differential equations of motion. For a first mode reduction we obtain one degree of freedom nonlinear model

$$
\ddot{q}_1+2\xi_1\omega_1\dot{q}_1+\omega_1^2 q_1+F_{NL1}q_1^3+F_{NT1}\Delta T q_1=F_{NP1}\sin\omega t. \tag{3}
$$

The above equation is written in the form where time is given in seconds but displacements are represented in dimensionless form, $\bar{u}=u/R$, $\bar{w}=w/R$. The 'dot' denotes time derivative and q_1 dimensionless generalised coordinate, ω_1 means the first natural frequency, F_{NL1} is nonlinear geometrical term coming from the original model,F_{NT1} is a coefficient related to temperature variation, F_{NP1} means amplitude of mechanical loading. Modal damping is defined by coefficient ξ_1.

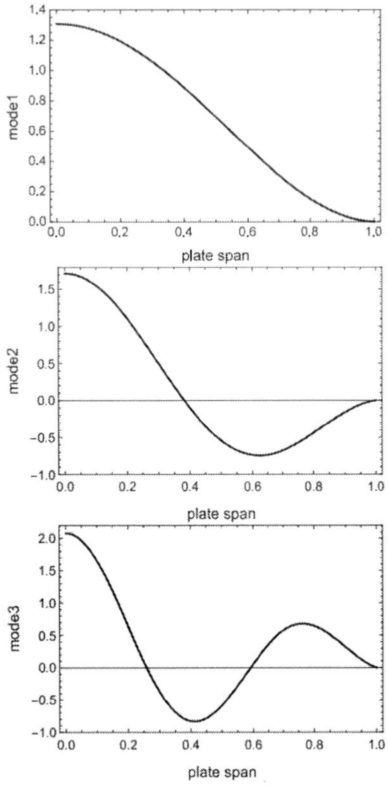

Figure 2. The first three vibration modes of the clamped circular plate described by the modified Bessel functions.

4 Numerical analysis of the reduced model

Example of numerical studies is performed for the clamped circular plate with diameter: $R = 0.1\,\mathrm{m}$, and thickness $h = 0.0025\,\mathrm{m}$,. The plate has homogenous properties and following material data: $E = 0.7 \times 10^{11}\,\mathrm{Pa}$, $\nu = 0.34$, $\rho = 2778\,\mathrm{kg/m^3}$. The reduced model coefficients computed on the basis of the physical parameters after the modal reduction procedure take values:

$$\omega_1 = 3928.1, \xi_1 = 0.01, F_{NL1} = 0.29497 \times 10^{11},$$

$$F_{NT1} = -0.358635 \times 10^7, F_{NP1} = 0.30849 \times 10^4$$

On the basis of the reduced analytical model Eq. (3) we computed a resonance curve for $\Delta T = 0$. It means that the temperature does not change with respect to the reference one. The plate is excited by harmonic loading with fixed amplitude and excitation frequency varied around the first natural frequency. In Fig. (3) the resonance curve demonstrates nonlinear behaviour with evident stiffening phenomenon. The solid and dashed lines represent stable and unstable solutions, respectively.

The influence of the elevated temperature on the plates response we can check by fixing mechanical excitation and then varying the temperature with respect to the reference one. Therefore we select from Fig. (3) frequency

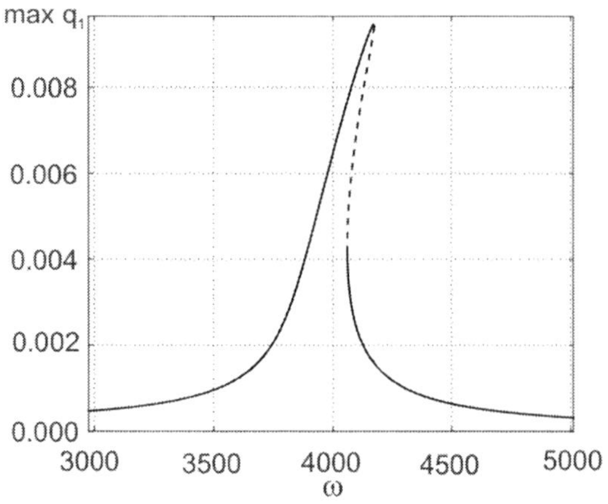

Figure 3. Resonance curve of the reduced model, $\Delta T = 0$, $F_{NP1} = 0.30849 \times 10^4$.

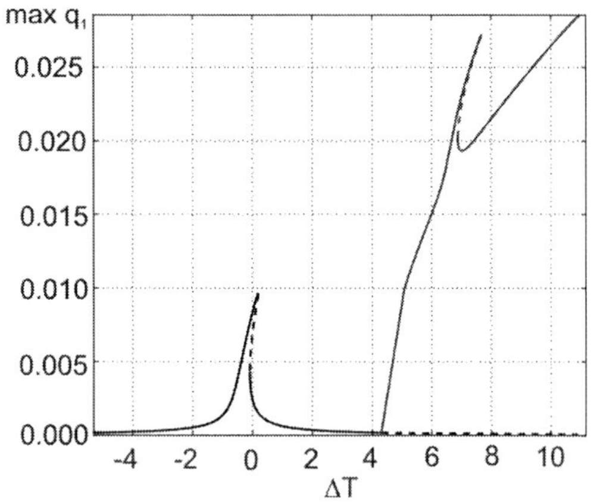

Figure 4. Bifurcation diagram of coordinate q_1 against temperature variation ΔT, $F_{NP1} = 0.30849 \times 10^4$, $\omega = 4100$.

$\omega = 4100$, which corresponds to a multi-solution case with stable, large or small, amplitudes and unstable solution between them. The bifurcation diagram for the fixed mechanical loading and varying temperature is presented in Fig. (4). In fact for $\Delta T = 0$ we observe three different solutions, corresponding to diagram Fig. (3). However, if temperature increases above the bifurcation point $DT = 4.3$ the buckling of the plate occurs, and the plate oscillates around post bucking state. The post-buckling oscillations are represented by the red curve in Fig. (4). The corresponding time history for post-buckling oscillations for $\Delta T = 6$, is presented in Fig. 5. The buckling phenomenon is observed by offset of the oscillation centre, and depending on initial conditions, it can buckle in a positive or negative direction.

In order to get knowledge how precise is just one mode reduction we compare the results with three and five modes models obtained from numerical simulation of exact PDEs and using the selected number of modes.

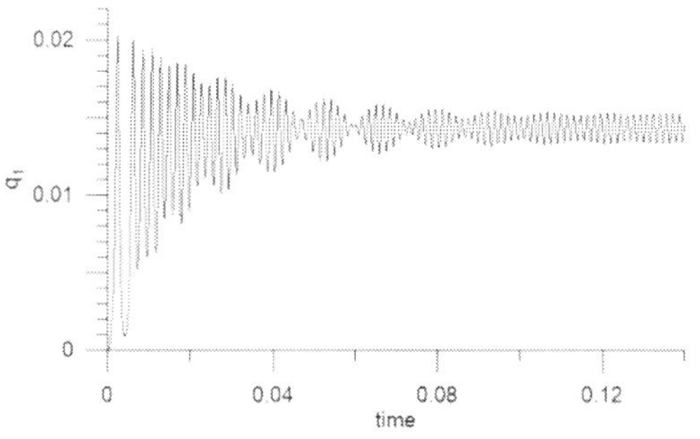

Figure 5. Time history for temperature $\Delta T = 6$, the solution obtained for 1 mode reduction; $\omega = 4100$.

The main assumption, however is that the plate is excited around the first natural frequency.

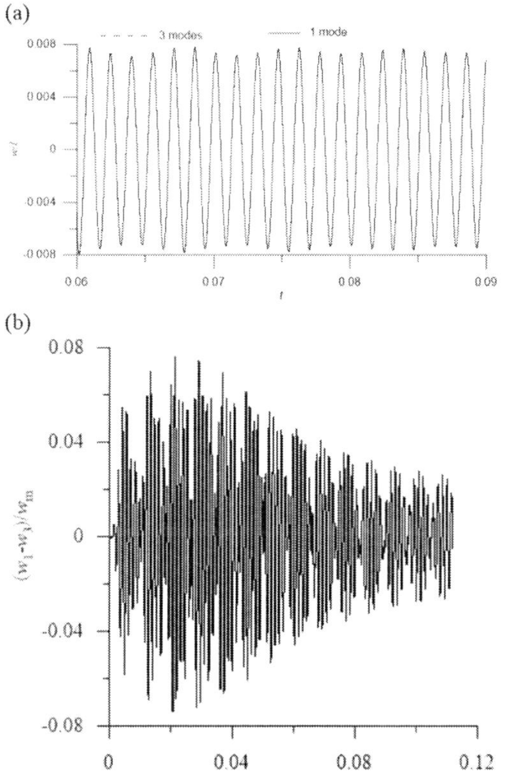

Figure 6. Time histories for temperature $\Delta T = 0$, (a) solution obtained for 1 mode (black) and 3 modes (blue) reduction; (b) time history of relative difference; $\omega = 4100$, $w_m = 0.0025$.

The example of the comparison of the model based on three and five modes with the one mode reduction is presented in Fig. 6 (a) and Fig. 7 (a) for $\omega = 4100$ and $\Delta T = 0$ and large oscillations corresponding to upper branch in Fig. 3. The relative difference between free(w_3) or five (w_5) modes reductions and one mode model (w_1), expressed versus mean value (w_m), is presented in Fig. 6 (b) and Fig. 7 (b), respectively. We note that for transient motion the difference is less than 8% and less then 4% for a steady state. In other cases for smaller amplitudes the difference is even smaller. Thus, the one degree of freedom model can be successfully implemented for detailed bifurcation analysis and only selected solutions verified by simulation of the exact model, which of course is more demanding and time consuming process.

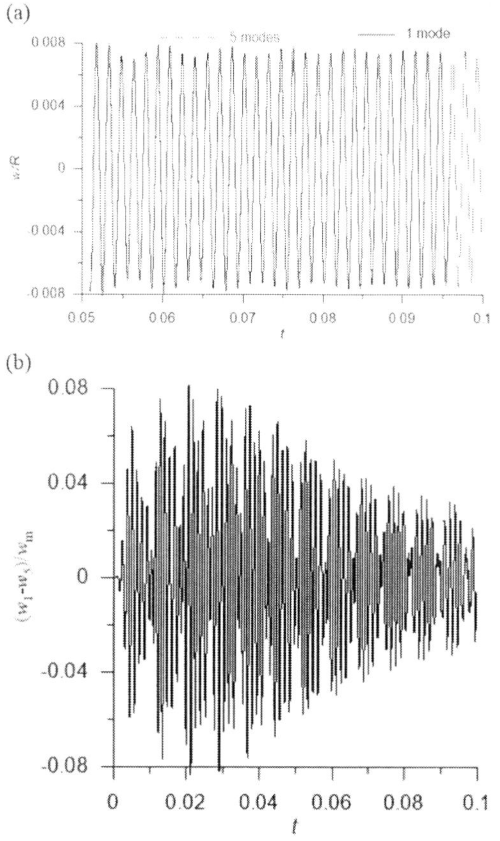

Figure 7. Time histories for temperature $\Delta T = 0$, (a) solution obtained for 1 mode (black) and 5 modes (red) reduction; (b) time history of relative difference; $\omega = 4100$, $w_m = 0.0025$.

The elevated temperature may lead to very complex response of the system. The bifurcation diagram in Fig. 8 demonstrates possible multi steady state solution (lines) or irregular oscillations (dark region) if temperature elevates up to $\Delta T = 5$. The detailed analysis of the Poincaré section confirmed that the dark region represents transient irregular oscillations which after long time motion become periodic.

5 Conclusions

The extended model of Mindlin plate with geometrical nonlinearities which occur due to large deformations is presented in the paper. The model takes into account

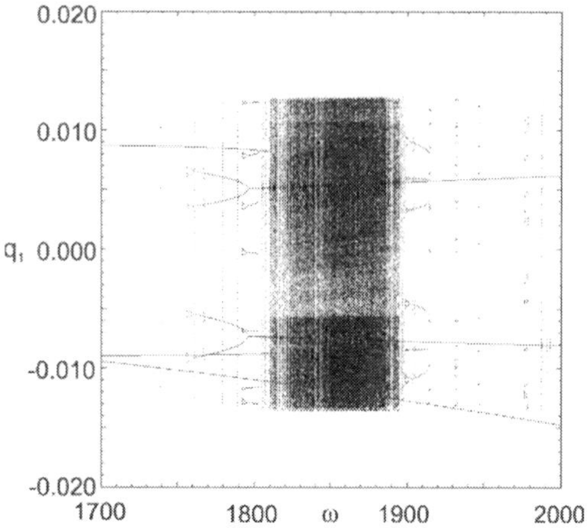

Figure 8. Bifurcation diagram of coordinate q_1 versus excitation frequency, $\Delta T = 5$, $F_{NP1} = 0.30849 \times 10^4$.

temperature which is assumed to be uniformly distributed along the plate spam. The exact model represented by PDEs is reduced to one degree of freedom system taking into account the nonlinear geometrical terms and temperature contribution. It is shown that one mode reduction can be successfully applied around the first resonance region, and the relative difference between one-mode and three or five-modes reduction is small. Thus, the proposed one degree of freedom nonlinear model can be used for effective bifurcation analysis.

The simulations based on the first mode reduction exhibit the stiffening effect of the resonance curve. The elevated temperature leads to bifurcation and buckling phenomenon and then post-buckling oscillations around a new steady state. The post-buckling oscillations are characterised by the positive or negative offset. The bifurcation analysis for elevated temperature shows multi-solutions, period doubling bifurcations and the transient irregular motion (transient chaos)which, after long time, goes to the periodic solution.

6 Acknowledgements

The first author would like to acknowledge the support received from the Polish NSF Grant S-38/M/2017, the second author the partial support from Bulgarian NSF Grant DN 02/7 2016 and H2020 Marie Skłodowska-Curie ITN under grant agreement No 643095, and the third author the support from Polish NSF grant S-3/M/2017.

References

[1] E.A. Thorton, *Thermal structures for aerospace applications* (AIAA Education Series, 1996) 1–479.

[2] E. Manoach, P. Ribeiro, Int. J. Mech. Sci., **46**, (2004) 1589–1606.

[3] P. Ribeiro, E. Manoach, J. Sound and Vibration, **285**, (2005) 1093–1107.

[4] A. Warminska, E. Manoach, J. Warminski, Meccanica, **49**, (2014) 1775–1793.

[5] A. Warminska, E. Manoach, J. Warminski, S. Samborski, Continuum Mechanics and Thermodynamics, (2014), DOI 10.1007/s00161-014-0381-6.

[6] A. Warminska, E. Manoach, J. Warminski, Proceedings of ICoEV 2015 Int. Conf. on Engineering Vibration, Ljubljana, Slovenia, (2015) 1057–1066.

[7] M. Amabili, S. Carra, J. Sound and Vibration, **321**, (2009) 936–954.

[8] M. Amabili, M. Reza, S. Tajahmadi, J Mechanical Engineering Science (2012) **226** 2393–2401.

[9] F. Alijani, F. Bakhtiari-Nejad, M. Amabili, Nonlinear Dynamics (2011) **66** 251–270.

[10] E. Saetta, G. Rega, Composite Structures, **174**, (2017) 420–441

[11] E. Saetta, G. Rega, Nonlinear Dynamics, **174**, (2017) DOI 10.1007/s11071-017-3648-1

[12] A. Warminska, E. Manoach, J. Warminski, Book Series: MATEC Web of Conferences, **83**, (2016) UNSP 05013

Synchronisation phenomenon in three blades rotor driven by regular or chaotic oscillations

Zofia Szmit[1,*] and *Jerzy* Warmiński[1,**]

[1]*Department of Applied Mechanics, Lublin University of Technology, Nadbystrzycka 36, 20-618 Lublin, Poland*

Abstract. The goal of the paper is to analysed the influence of the different types of excitation on the synchronisation phenomenon in case of the rotating system composed of a rigid hub and three flexible composite beams. In the model is assumed that two blades, due to structural differences, are de-tuned. Numerical calculation are divided on two parts, firstly the rotating system is exited by a torque given by regular harmonic function, than in the second part the torque is produced by chaotic Duffing oscillator. The synchronisation phenomenon between the beams is analysed both either for regular or chaotic motions. Partial differential equations of motion are solved numerically and resonance curves, time series and Poincaré maps are presented for selected excitation torques.

1 Introduction

The synchronisation phenomenon was discovered by Christiaan Huygens in the XVII centry. He hang two pendulums clocks on one beam and found out that after some transient time pendulums moved synchronously. Kapitaniak and et al. have recalled the same experiment in the paper [1]. Authors chose two pendulum clocks as identical as possible and observed the behavior of them, also they checked whether Huygens was able to observe all type of synchronisation in the XVII centry. Moreover, the computer simulations have been presented to check the influence of non identical clocks on synchronisation process. Although the synchronisation phenomenon is quite popular and well known issue, most of the scientist describe systems composed of different type of pendulums. The synchronisation of two self-excited double pendulums attached to the horizontal beam has been considered in [2]. Authors derived the synchronisation conditions for a model and explained types of synchronisation as well. The theoretical and experimental studies of synchronisation of two rotating in a vertical plane parametric pendulums attached to elastic support under harmonic excitation have been presented in [3]. In that model two types of synchronous states have been found by authors, complete and phase synchronisation. The experimental and numerical analysis of a set of double pendulums have been described in [4]. Authors was able to observe different type of synchronous configurations with a constant phase between pendulums. All previously mentioned papers, focused the vertical motion of pendulums. The synchronisation phenomenon as well as nonlinear dynamics for a model composed of two pendulums attached to a rotating hub in the horizontal plane have been studied in [5]. Authors found out that in case of non-symmetric pendulums the synchronisation with locked phase occurred. Furthermore, they proved that natural frequencies and vibration modes strongly depends on the mass of the hub. Similar conclusion about the influence of the mass hub on a rotating system has been presented in [6]. In that paper authors analysed beam with a heavy tip mass on one end fixed to a rotating beam. The nonlinear model included bending, tension and nonlinear curvature, authors proved that nonlinear terms in equation of motion cause softening effect on the system. While, dynamics of rotating structure with a thin-wall composite beams have been analysed in [7, 8], where authors derived equation of motion using Hamilton principle and Galerkin method to reduce the model to the ordinary differential equations. Moreover, the nonclassical effects as material anisotropy, transverse shear and rotary inertia have been considered in the hub-beam model. Most of studies on rotating structures have been motivated by mechanical and aerospace engineering; for instance helicopters rotors, jet engines and wind turbines are the typical examples of rotating structures in engineering. The equations of motion dedicated to helicopter blades have been presented in [9, 10].

The aim of this paper is to study synchronisation effect in the system composed of three composite beams attached to a hub rotating in the horizontal plane. The blades have been analysed as a Timoshenko beam. The first beam is considered as reference one while beam No. 2 is 10% thicker and beam No. 3 is 5% thicker compering to the reference one. Motivation for this study arises from observation of a production process of composite beams, in which it was not very easy to make three beams with the same geometric dimensions. Model with one de-tuned beam have been studied by the authors in their previous

*e-mail: z.szmit@pollub.pl
**e-mail: j.warminski@pollub.pl

paper [11]. The synchronisation phenomenon in the preset paper is studied when the model is exited by harmonic torque or torque given by a chaotic oscillator. The influence of aerodynamic loads is not taken into account in the analysed model.

2 Model and equations of motion

Let us consider a rotor composed of three slender, straight and elastic composite beams fixed to a rigid hub, of radius R_0 and mass moment of inertia J_h. The hub is able to oscillate or rotate about vertical axis CZ_0 as shown in Figure 1.

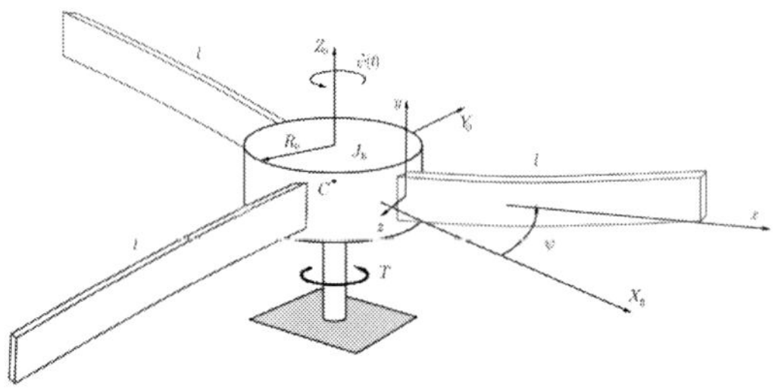

Figure 1. Model of the rotating hub with three elastic beams

The hub current position is represent by angle $\psi(t)$ with respect to the inertial reference frame (X_0, Y_0, Z_0). Angular velocity of the rotor $\dot{\psi}(t)$ is assumed to be not necessarily constant. The system is driven by external torque T, which is applied to the hub. Every single beam is described by the length l_i, width and thickness of the cross-section d_i and h_i $(i = 1, 2, 3)$, respectively. The model considers one nominal beam (beam No. 1) and two with 10% and 5% higher thickness, thus beam No.2 and No.3 are de-tuned. The system of partial differential equations (PDEs) for the hub and each beam has been derived by using the Hamilton's principle of least action as is presented in [7, 8]. Next, the Galerkin projection method has been applied to reduce system of PDEs to a set of ordinary equations of motions (ODEs). After conversion, the final set of governing equations of motion takes a form:

$$(J_h + 1 + \sum_{i=2}^{N} J_{bi} + \sum_{i=1}^{N} \alpha_{hi2} q_i^2)\ddot{\psi} + \sum_{i=1}^{N} \alpha_{hi1}\ddot{q}_i + \sum_{i=1}^{N} \alpha_{hi3} q_i \dot{q}_i \dot{\psi} + \zeta_h \dot{\psi} = \mu(t)$$
$$\ddot{q}_i + \alpha_{i2}\ddot{\psi} + \alpha_{i4} q_i \dot{q}_i \dot{\psi} + (\alpha_{i1} + \alpha_{i3}\dot{\psi}^2) q_i + \zeta_i \dot{q}_i = 0 \quad (1)$$

The first of the set of Equation 1 represent the dynamics of the hub, where term J_h and J_{bi} are dimensionless mass moment of inertia of the hub and dimensionless mass moment of inertia of each beam, respectively. All these are expressed as a magnitude of the inertia of the first beam. Whereas, the dimensionless angular velocity of the hub is denoted as $\dot{\psi} = \frac{d\psi}{dt}$. Term $\mu(t)$ is considered as a external torque (excitation of the rotor). The excitation is investigated in two variants, firstly as a driving torque, expressed as analytical function $\mu = \mu_0 + \rho \sin(\omega t)$ and secondly as a chaotic Duffing oscillator, where $\mu = \alpha_d x$. The variable x is calculated from the Duffing's equation $\ddot{x} + k\dot{x} + x^3 = \beta + \rho\cos(\omega t)$ as have been presented by Ueda in [12]. The second equation of the set 1 corresponds to a dynamics of each beam, where q_i $(i = 1, 2, 3)$ is the generalized coordinate of the complex flexural-torsional specimen deformation [8]. Furthermore, the viscous damping coefficients for a each beam ζ_i and the hub ζ_h are arbitrary included.

3 Results of numerical studies

Firstly, all coefficients that are presented in equation of motion as α_{ij} have been calculated on a basis of the real glass-epoxy laminate beams samples, the values are given in the Table 1.

Table 1. Dimensionless coefficients, based on the real physical model

Beam No.1	Beam No. 2	Beam No. 3
$\alpha_{11} = 12.0388$	$\alpha_{21} = 14.5664$	$\alpha_{31} = 13.2728$
$\alpha_{12} = 1.97656$	$\alpha_{22} = 1.97681$	$\alpha_{32} = 1.97669$
$\alpha_{13} = 0.49465$	$\alpha_{23} = 0.49462$	$\alpha_{33} = 0.49464$
$\alpha_{14} = -1.5517$	$\alpha_{24} = -1.5514$	$\alpha_{34} = -1.5515$
$\alpha_{h11} = 0.01120$	$\alpha_{h21} = 0.01232$	$\alpha_{h31} = 0.01176$
$\alpha_{h12} = -0.0084$	$\alpha_{h22} = -0.0093$	$\alpha_{h32} = -0.0088$
$\alpha_{h13} = -0.0169$	$\alpha_{h23} = -0.0186$	$\alpha_{h33} = -0.0177$

All the numerical calculation have been performed by direct numerical simulation of the system represented by Eqs. 1. As it has been mentioned before, the beam No. 2 and beam No. 3 are considered as de-tuned, 10% and 5% thicker then the remaining reference beam No. 1. The resonance curves and time histories plot are obtained for a hub exited by a harmonic torque ($\mu(t)$).

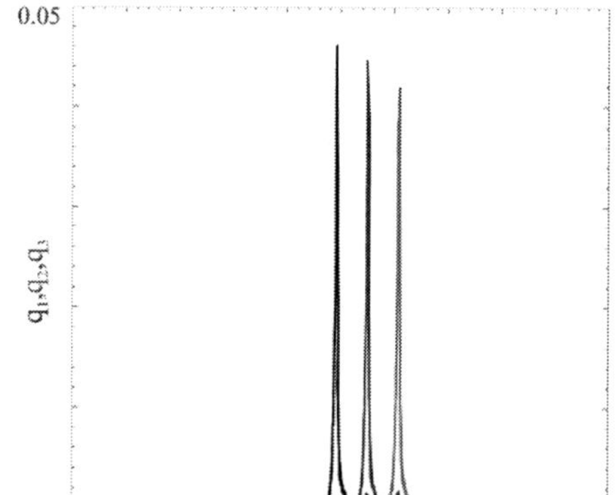

Figure 2. Resonance curves for amplitude of excitation $\rho = 0.01$ of displacements of each beam, beam No.1 - black line, beam No.2 - red line and beam No.3 - blue line

 https://doi.org/10.1051/matecconf/201814806002

The solutions for a system with a harmonic function are sought near the resonance zones for a selected amplitude of excitation ($\rho = 0.01$), also the mass moment of inertia of the hub is assumed as $J_h = 5.0$.

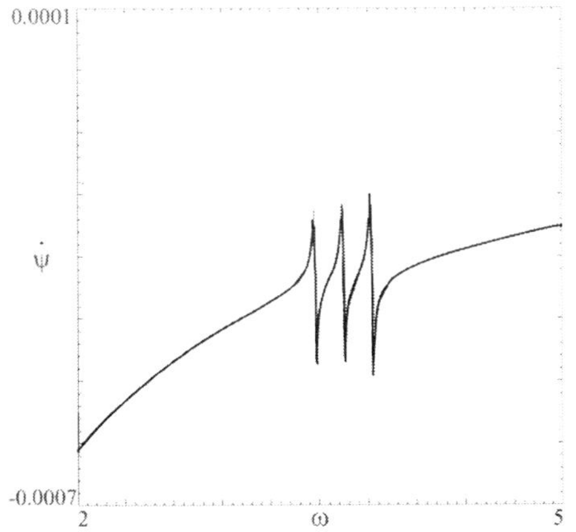

Figure 3. Resonance curves for amplitude of excitation $\rho = 0.01$ of angular velocity of the hub

The resonance curves for amplitude of displacement of each beam and angular velocity of the hub are shown in the Fig. 2 and 3, respectively. On both graphs three resonances can be noticed, first close to $\omega \approx 3.48$, second close to $\omega \approx 3.65$ and the third close to $\omega \approx 3.83$.

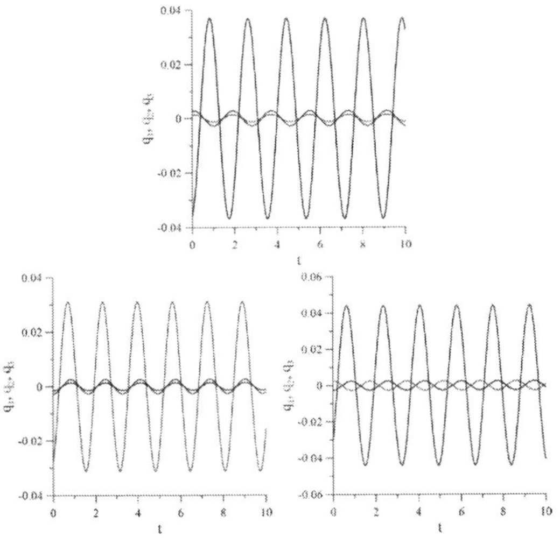

Figure 4. Time series of each beam (a) close to first resonance zone; (b) close to second resonance zone; (c) close to third resonance zone

The first resonance is observed for beam No. 1, with small oscillation of the second and third beam. In this resonance zone we observe anti-phase synchronisation with locked phase between beams No. 1 and No. 2 also between beams No. 1 and No. 3. Furthermore, second and third beam are completely synchronized close to the first resonance zone (see Fig.4a). Close to the second resonance zone large oscillations are localised in beam No.3, while motion of two other beams are very small. In this case the complete in-phase synchronisation can be observed for beam No. 1 and No. 3, while the motion of beam No. 2 compering to other beams is synchronized with locked phase (Fig. 4b). In the third resonance zone oscillations are found for a beam No. 2, also the synchronisation with locked phase can be observed between beam No. 2 and No. 3, as well as. between beam No. 1 and No. 3. While, motion of the first and the second beam is completely anti-phase synchronized (Fig. 4c).

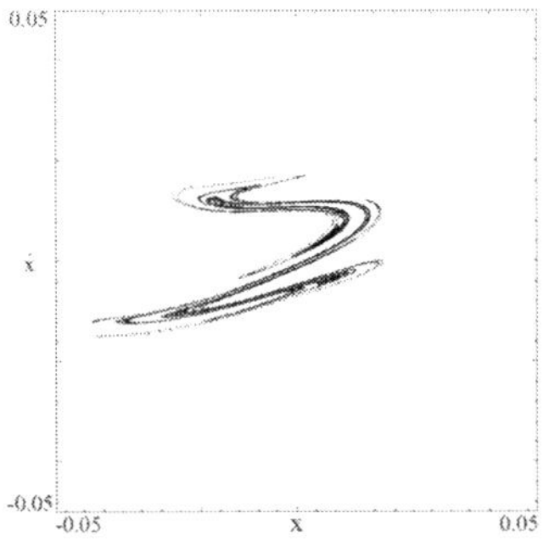

Figure 5. Portrait of a strange attractor obtained from the Duffing equation

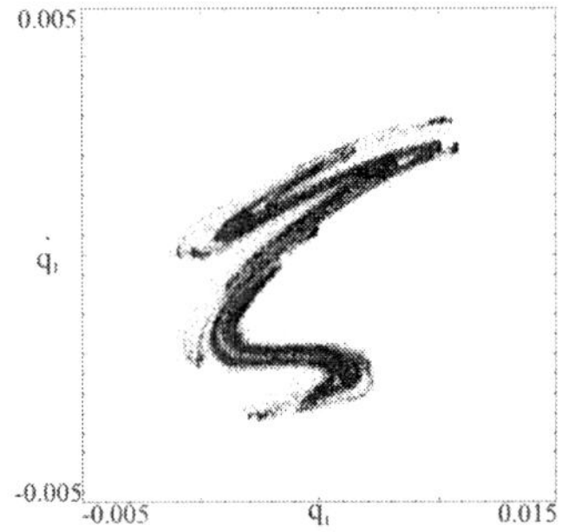

Figure 6. Portrait of a strange attractor for a beam No. 1

For the model excited by the chaotic Duffing oscillator, the Poincaré maps have been obtained. The parameters for a Duffing equation have been taken from [12], and $k = 0.05$, $\beta = 0.03$, $\rho = 0.16$. In Figure 5, the chaotic strange attractor for Duffing equation have been shown. The strange chaotic attractor have been observed for each beam, see Fig. 6, 7, 8. Furthermore, the time series are studied to analyse the synchronisation phenomenon. The complete in-phase synchronisation can be observed between every beam, when the rotating de-tuned structure is exited by Duffing oscillator (see Fig. 9).

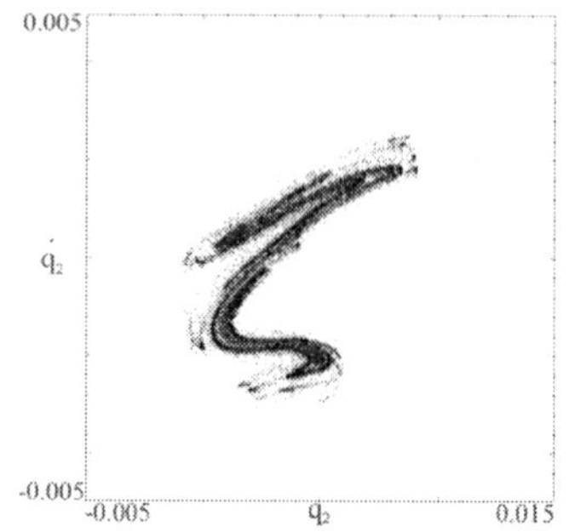

Figure 7. Portrait of a strange attractor for a beam No. 2

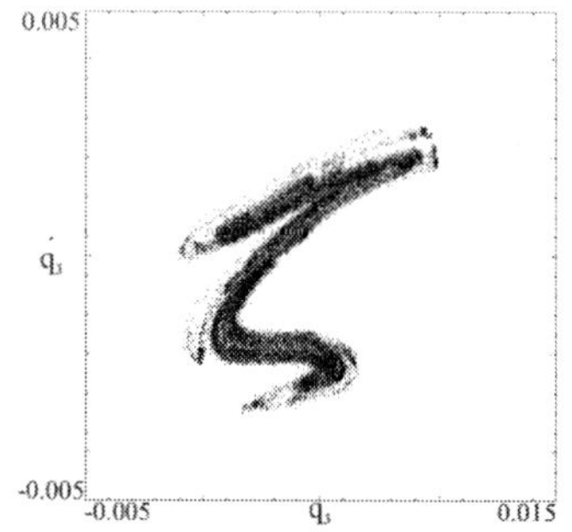

Figure 8. Portrait of a strange attractor for a beam No. 3

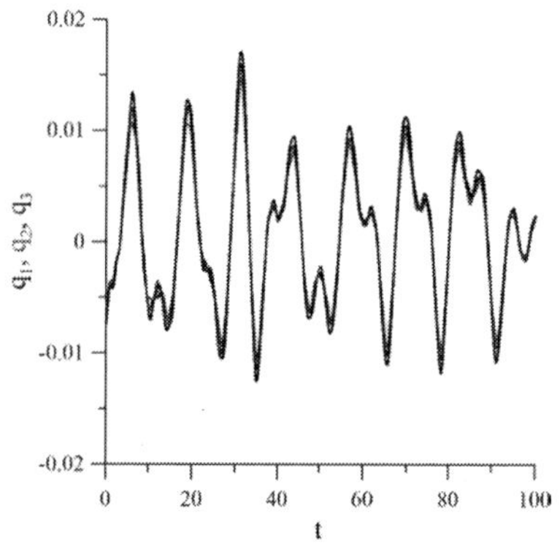

Figure 9. Time series of each beam in case of excitation by chaotic oscillator

4 Conclusions

In case of the rotating de-tuned structure excited by the harmonic torque three resonances have been observed, first close to $\omega \approx 3.48$, second around $\omega \approx 3.65$ and the third one close to $\omega \approx 3.83$. Resonance curves for amplitudes for every beam are different, due to complete de-tuning of the rotor. Moreover, different synchronisation scenario can be observed in each resonance zone, at the first resonance zone the synchronisation with locked phase and complete in-phase synchronisation have been found. Around the second resonance zone similar synchronisation scenario is noticed but for different beams. While close to third resonance the anti-phase synchronisation occurred between beams No. 1 and No. 2. Moreover, the portrait of strange chaotic attractor have been obtained for every single beam, in case of excitation by Duffing oscillator. The complete in-phase synchronisation of motion can be observed between every beam, when the rotor is excited by chaotic oscillations. To sum up the type of excitation completely change the process of synchronisation in the analysed de-tuned rotating structure.

5 Acknowledgments

This work was supported by grant 2015/19/N/ST8/03906 from Polish National Science Centre.

References

[1] M. Kapitaniak, K. Czolczynski, P. Perlikowski, A. Stefanski, T. Kapitaniak, Physics Reports **517**, 1 (2012)
[2] P. Koluda, P. Perlikowski, K. Czolczynski, T. Kapitaniak, Communications in Nonlinear Science and Numerical Simulation **19**, 977 (2014)
[3] A. Najdecka, T. Kapitaniak, M. Wiercigroch, International Journal of Non-Linear Mechanics **70**, 84 (2015)
[4] J. Strzalko, J. Grabski, J. Wojewoda, M. Wiercigroch, T. Kapitaniak, Chaos: An Interdisciplinary Journal of Nonlinear Science **22**, 047503 (2012)
[5] J. Warminski, Z. Szmit, J. Latalski, The European Physical Journal Special Topics **223**, 827 (2014)
[6] J. Warminski, J.M. Balthazar, 7th Conference on Dynamical Systems - Theory and Applications: proceedings **1** (2003)
[7] F. Georgiades, J. Latalski, J. Warminski, Meccanica **49**, 1833 (2014)
[8] J. Latalski, J. Warminski, G. Rega, Mathematics and Mechanics of Solids (2016)
[9] M.R.M. Crespo da Silva, C.C. Glynn, Journal of Structural Mechanics **6**, 437 (2007)
[10] M.R.M. Crespo da Silva, C.C. Glynn, Journal of Structural Mechanics **6**, 449 (2007)
[11] Z. Szmit, J. Warminski, **after reviews** (2017)
[12] Y. Ueda, Annals of the New York Academy of Sciences **357**, 422 (1980)

Vibration analysis of an axially moving multiscale composite plate subjected to thermal loading

Krzysztof Marynowski*

Department of Dynamics, Lodz University of Technology, Poland

Abstract. This study investigated the effects of temperature on free vibrations and critical transport speeds of an axially moving multiscale composite plate. On the basis of the frequency-temperature equivalence principle, a linear mathematical model of the moving multiscale composite plate is derived in the complex frequency domain. Fractional standard rheological model of the plate material as the function of reduced frequency depended on the temperature is determined. In numerical investigations carbon nanotubes- and graphene-reinforced copper plate was taken into account.. To describe thermomechanical properties of the plate material, the investigation results obtained from the molecular dynamics studies and the experimental characteristic of beryllium copper in low temperatures presented in literature is taken into account.. The effects of temperature, transport speed, and internal damping on natural frequencies and critical transport speed are analyzed. The critical transport speeds of the graphene-reinforced multiscale composite are higher than both carbon nanotubes-reinforced composite as well as the comparable copper alloy.

1 Introduction

Axially moving composite objects can be found in many different technical applications. These include for example paper and textile webs during production and processing, conveyor belts, and flat objects moving at high speeds in space. Many factors influence the dynamic behavior of the moving object. The most important are: transport speed, tension, material properties and influence of external environment. Among the external factors affecting the dynamics, temperature plays an important role. Accurate identification of the temperature effect on the dynamics of the axially moving plant is important to properly control these devices. A literature review on dynamics of axially moving systems can be found, for example, in the study [1].

Studies in the field of dynamics of stationary composites are carried out for a long time. At the turn of the centuries composite materials began to appear particularly in the aerospace, maritime and space industries because they offer a number of advantageous mechanical properties., These properties include resistance to electrochemical corrosion, high strength and rigidity with less weight than conventional materials. The state of researches in the dynamics of laminated composite and sandwich plates in the paper [2] is presented.

In recent years, considerable research effort has been devoted to multiscale composites: carbon nanotubes (CNTs) and graphene-reinforced materials. Carbon nanotubes are very good reinforcement because of their unique mechanical properties and large surface area per unit volume. Experimental studies show that CNTs have Young's modulus greater than the graphite fibers, and strengths at least an order of magnitude higher than typical graphite fibers. Graphene attracted considerable interest because of its extraordinary mechanical, electronic and thermal properties. Experimental studies have shown that the single-layer graphene can present exceptional thermal conductivity which outperforms all other known materials.

For recent several years, graphene and CNTs have been used as reinforcement for polymeric composites because of their relative ease of processing. However, in recent times there has been increased interest in using matrices such as metals, semiconductors, and ceramics in addition to that of polymers. This study deals with metal matrix multiscale composites (MMMCs). Copper has been used as the metal matrix. MMMCs have less weight and higher strength in comparison to the conventional metals and alloys. MMMCs also have lower coefficient of friction and better resistance to wear and corrosion. Therefore these can be used in a variety of engineering applications such as automobile and aerospace industries. The state of research on mechanical and tribological properties of MMMCs in the paper [3] is presented.

The objective of this study is vibration analysis of the axially moving multiscale composite plate taking into account thermal effects. To describe thermo-mechanical properties of the plate material, properties of multiscale reinforced stationary composites presented in literature are taken into account. Fractional standard rheological model of the plate material as the function of reduced frequency depended on the temperature is taken into considerations. The effects of temperature, transport speed, and internal damping on natural frequencies and critical transport speed are analyzed.

* Corresponding author: krzysztof.marynowski@p.lodz.pl

2 Factional standard rheological model

The basis dependence of linear viscoelasticity, that the stress function is linearly dependent on the past history of the strain function, is expressed by Boltzmann's superposition principle in the form

$$\sigma(t) = \int_0^t r(t-\tau)\frac{\partial \varepsilon(\tau)}{\partial \tau}d\tau, \qquad (1)$$

where: σ(t) – stress function, ε(t) – strain function, r(t) – relaxation function. From the mathematical point of view Eq. (1) is a convolution of functions, thus using Laplace transformation one can find equivalent of Eq. (1) in the complex frequency domain

$$\sigma(\omega) = R(\omega)\,\varepsilon(\omega), \qquad (2)$$

where: $R(\omega)$ – transform of relaxation function in the frequency domain. On the base of Eq. (2) one can solve viscoelasticity problems as elasticity problems using a complex modules of relaxation function which depends on frequency. Eq. (2) defines the elastic – viscoelastic equivalence principle.

Most often in studies of dynamics of viscoelastic systems to determine the complex moduli the fractional model is used [4]. To model the behavior of the moving plate under thermal loading the fractional standard rheological model was used in this paper, whereas the hydroelastic behavior was considered as elastic

The constitutive equation of the fractional standard rheological model

$$\frac{\gamma}{E_2}\frac{d^\beta \sigma}{dt^\beta} + \sigma = \frac{\gamma(E_1+E_2)}{E_2}\frac{d^\beta \varepsilon}{dt^\beta} + E_1\varepsilon \qquad (3)$$

where: $0 < \beta < 1$. Taking into account the relaxation time constants

$$\tau_1 = \frac{\gamma}{E_1}; \quad \tau_2 = \frac{\gamma}{E_2}; \quad \tau_0 = \tau_1 + \tau_2, \qquad (4)$$

the constitutive equation has the following form:

$$\tau_2\frac{d^\beta \sigma}{dt^\beta} + \sigma = E_1\left(\tau_0\frac{d^\beta \varepsilon}{dt^\beta} + \varepsilon\right). \qquad (5)$$

In the free vibration problem the stress and strain functions can be expressed as

$$\sigma = \sigma_0\exp(i\omega t); \qquad \varepsilon = \varepsilon_0\exp(i\omega t). \qquad (6)$$

In accordance with the frequency-temperature equivalence principle, the frequency ω can be replaced by the reduced frequency $\omega\alpha(T)$ depended on the temperature. Substituting Eqs (6) into the constitutive equation (5) one receives the Young's modulus in the frequency domain

$$E(\omega,T) = \frac{\sigma_0}{\varepsilon_0} = E_1\frac{1+\tau_0(i\omega\alpha(T))^\beta}{1+\tau_2(i\omega\alpha(T))^\beta}. \qquad (7)$$

On the base of eq. (7) Fig.1 shows the shift of the storage Young modulus for different four temperatures in the frequency domain. The highest curve corresponds to the lowest temperature T_1 and vice versa the lowest curve corresponds to the highest temperature T_4.

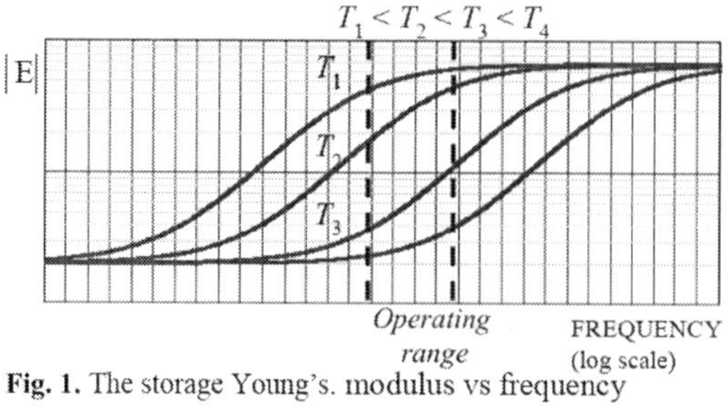

Fig. 1. The storage Young's. modulus vs frequency

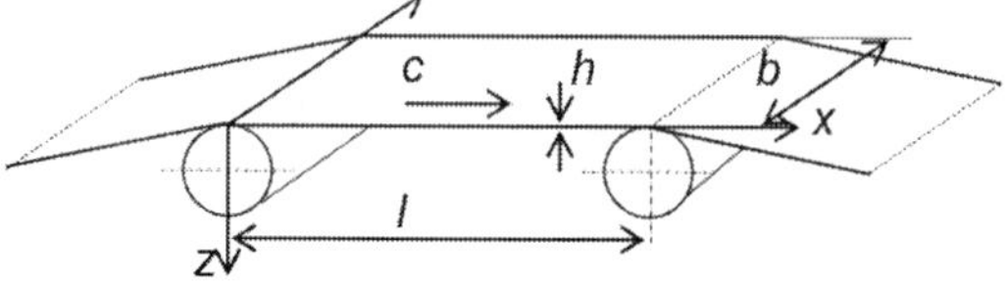

Fig. 2. Axially moving plate

3 Mathematical model of the plate

A thin multiscale composite plate of the length l, the width b and the thickness h is considered. The plate moves along the longitudinal direction x at the constant velocity c. The co-ordinate system and the geometry are shown in Fig. 2.

The partial differential equation resulting from the Hamilton's principle for a transverse motion of the two-dimensional axially moving plate was derived in the book [5] and has the following form:

$$\rho h\left(\frac{\partial^2 \overline{w}}{\partial t^2} + 2c\frac{\partial^2 \overline{w}}{\partial x\partial t} + c^2\frac{\partial^2 \overline{w}}{\partial x^2}\right) - \frac{\partial(N_x\overline{w}_{,x})}{\partial x} - \frac{\partial(N_y\overline{w}_{,y})}{\partial y} -$$
$$-\frac{\partial(N_{xy}\overline{w}_{,x})}{\partial y} - \frac{\partial(N_{xy}\overline{w}_{,y})}{\partial x} - \frac{\partial^2 M_x}{\partial x^2} - 2\frac{\partial^2 M_{xy}}{\partial x\partial y} - \frac{\partial^2 M_y}{\partial y^2} = 0 \qquad (8)$$

where: ρ – mass density; h – thickness; w– transverse displacement of the plate; M_x, M_y, M_{xy} – bending moments per unit length; N_x, N_y, N_{xy} – in-plane forces per unit length.

The equilibrium state of the moving plate is defined as the non-trivial equilibrium position when time-dependent forces do not interact on the plate and the form of deflection is dependent only on the transport speed. The equation of equilibrium states of the axially moving plate with a constant speed which is tensioned with the constant longitudinal load N_{x0} can be defined on the basis of governing Eq. (8) in the form:

$$\rho h c^2\frac{\partial^2 w}{\partial x^2} - N_{x0}\frac{\partial^2 w}{\partial x^2} - \frac{\partial^2 M_x}{\partial x^2} - 2\frac{\partial^2 M_{xy}}{\partial x\partial y} - \frac{\partial^2 M_y}{\partial y^2} = 0 \qquad (9)$$

Bending moments can be expressed as:

$$M_x = \int_{-h/2}^{h/2}\sigma_x z\,dz, \quad M_y = \int_{-h/2}^{h/2}\sigma_y z\,dz,$$
$$M_{xy} = \int_{-h/2}^{h/2}\sigma_{xy} z\,dz. \qquad (10)$$

Taking into account the thin-plate theory, Eqs (10) have the forms:

$$M_x = -D(\omega,T)\left(\frac{\partial^2 w}{\partial x^2} + \eta\nu(\omega,T)\frac{\partial^2 w}{\partial y^2}\right),$$

$$M_y = -\eta D(\omega,T)\left(\frac{\partial^2 w}{\partial y^2} + \nu(\omega,T)\frac{\partial^2 w}{\partial x^2}\right), \quad (11)$$

$$M_{xy} = -D_1(\omega,T)\frac{\partial^2 w}{\partial x\partial y},$$

where: η - orthotropy coefficient,

$$D(\omega,T) = \frac{h^3}{12}\frac{E(\omega,T)}{\left(1-\eta\nu(\omega,T)^2\right)}, \quad (12)$$

$$D_1(\omega,T) = \frac{G(\omega,T)h^3}{6}.$$

In the free vibration problem, the transverse displacement can be defined as:

$$w(x,y,t) = w(x,y,\omega)\exp(i\omega t) \quad (13)$$

Substituting Eqs (11) into governing Eq. (9), one receives a differential equation of equilibrium states of the moving plate in the frequency domain:

$$\rho h c^2\frac{\partial^2 w}{\partial x^2} - N_{x0}\frac{\partial^2 w}{\partial x^2} + D(\omega,T)\times$$

$$\left[\frac{\partial^4 w}{\partial x^4} + 2\left(\eta\nu(\omega,T) + \frac{D_1(\omega,T)}{D(\omega,T)}\right)\frac{\partial^4 w}{\partial x^2\partial y^2} + \eta\frac{\partial^4 w}{\partial y^4}\right] = 0. \quad (14)$$

Eq. (14) can be transposed to a dimensionless form using the following terms:

$$z = \frac{w}{h};\ \xi = \frac{x}{l};\ \varphi = \frac{y}{l};\ r = \frac{b}{l};\ c_w = \sqrt{\frac{N_{x0}}{h\rho}};\ s = \frac{c}{c_w} = c\sqrt{\frac{h\rho}{N_{x0}}};$$

$$\tau = \frac{tc_w}{l};\ \omega = \frac{\omega l}{c_w};\ \Delta(\omega,T) = \frac{D(\omega,T)}{N_{x0}l^2};\ \Delta_1(\omega,T) = \frac{D_1(\omega,T)}{N_{x0}l^2}. \quad (15)$$

The dimensionless equation of the equilibrium state in the frequency domain has the form:

$$(s^2-1)\frac{\partial^2 z}{\partial\xi^2} + \Delta(\omega,T)\times \quad (16)$$

$$\left[\frac{\partial^4 z}{\partial\xi^4} + 2\left(\eta\nu(\omega,T) + \frac{\Delta_1(\omega,T)}{\Delta(\omega,T)}\right)\frac{\partial^4 z}{\partial\xi^2\partial\varphi^2} + \eta\frac{\partial^4 z}{\partial\varphi^4}\right] = 0$$

The dimensionless boundary conditions referring to simple supports at transverse ends and longitudinal free ends are as follows:

$$z\Big|_{\substack{\xi=0\\ \xi=1}} = 0;\ -\Delta(\omega,T)\left(\frac{\partial^2 z}{\partial\xi^2} + \nu(\omega,T)\frac{\partial^2 z}{\partial\varphi^2}\right)\Bigg|_{\substack{\xi=0\\ \xi=1}} = 0,$$

$$-\Delta(\omega,T)\left(\frac{\partial^2 z}{\partial\varphi^2} + \nu(\omega,T)\frac{\partial^2 z}{\partial\xi^2}\right)\Bigg|_{\substack{\varphi=0\\ \varphi=r}} = 0;. \quad (17)$$

$$-\Delta(\omega,T)\left(\frac{\partial^3 z}{\partial\varphi^3} + (2-\nu(\omega,T))\frac{\partial^3 z}{\partial\xi^2\partial\varphi}\right)\Bigg|_{\substack{\varphi=0\\ \varphi=r}} = 0.$$

Comparing Eq.(14) with the equation of motion for a stationary plate subjected to in-plane tension along the longitudinal direction, a necessary condition for the existence of non-trivial equilibrium positions of the plate can be obtained in the following form:

$$s^2 - 1 \geq 0 \quad \text{or} \quad c \geq \sqrt{\frac{N_{x0}}{\rho h}}. \quad (18)$$

Assuming the notation:

$$\psi(\omega,T) = \frac{s^2-1}{\Delta(\omega,T)}, \quad (19)$$

the equilibrium state equation has the form:

$$\psi(\omega,T)\frac{\partial^2 z}{\partial\xi^2} + \frac{\partial^4 z}{\partial\xi^4} + \quad (20)$$

$$2\left(\eta\nu(\omega,T) + \frac{\Delta_1(\omega,T)}{\Delta(\omega,T)}\right)\frac{\partial^4 z}{\partial\xi^2\partial\varphi^2} + \eta\frac{\partial^4 z}{\partial\varphi^4} = 0.$$

The solution to Eq. (20) enables us to determine non-trivial equilibrium positions of the plate z_{nk} (n, k = 1, 2, ...) complying with multiple values $\psi(\omega,T)_{nk}$. The solution to Eq.(20) is forecast in the form:

$$z_{nk}(\xi,\varphi,\omega) = A_k\exp(k(\omega)\varphi)\sin(n\pi\xi). \quad (21)$$

Substituting Eq. (21) into equation (20), the fourth-order characteristic equation is obtained:

$$k^4 - 2n^2\pi^2\left(\nu(\omega,T) + \frac{\Delta_1(\omega,T)}{\eta\Delta(\omega,T)}\right)k^2 + \frac{n^4\pi^4}{\eta} - \frac{n^2\pi^2}{\eta}\psi(\omega,T) = 0. \quad (22)$$

Having introduced the notations:

$$\Delta_2(\omega,T) = \nu(\omega,T) + \frac{\Delta_1(\omega,T)}{\eta\Delta(\omega,T)};\ \Delta_3(\omega,T) = \Delta_2^2(\omega,T) - \frac{1}{\eta}, \quad (23)$$

the solutions of characteristic Eq. (23) have the forms:

$$\alpha_{1n} = \sqrt{\Delta_2(n\pi)^2 + n\pi\sqrt{\Delta_3(n\pi)^2 + \psi/\eta}};$$

$$\alpha_{2n} = \sqrt{\Delta_2(n\pi)^2 - n\pi\sqrt{\Delta_3(n\pi)^2 + \psi/\eta}}; \quad (24)$$

$$\alpha_{3n} = \sqrt{-\Delta_2(n\pi)^2 + n\pi\sqrt{\Delta_3(n\pi)^2 + \psi/\eta}}$$

Solutions (24) comply with three sets of non-trivial equilibrium positions:

- for $0 \leq \psi < (n\pi)^2$:

$$z_{n1} = [C_{11}\sinh(\alpha_1\varphi) + C_{12}\cosh(\alpha_1\varphi) + C_{13}\sinh(\alpha_2\varphi) + C_{14}\cosh(\alpha_2\varphi)]\sin(n\pi\xi), \quad (25)$$

- for $\psi > (n\pi)^2$:

$$z_{n2} = [C_{21}\sinh(\alpha_1\varphi) + C_{22}\cosh(\alpha_1\varphi) + C_{23}\sin(\alpha_3\varphi) + C_{24}\cos(\alpha_3\varphi)]\sin(n\pi\xi), \quad (26)$$

- for $\psi = (n\pi)^2$:

$$z_{n3} = [C_{31}\sinh(\alpha_1\varphi) + C_{32}\cosh(\alpha_1\varphi) + C_{33} + C_{34}\varphi]\sin(n\pi\xi), \quad (27)$$

where: C_{11}, C_{12}, ... C_{34} - constants, which are determined from boundary conditions (17). Fig. 3 shows the plots of z_{11}, z_{12}, z_{21} and z_{22} equilibrium states of the moving composite plate. To determine coefficients ψ_n which describe non-trivial equilibrium positions, Eqs (25) – (27) are substituted into boundary conditions (17). The critical transport speed is the lowest transport speed at which the non-trivial equilibrium positions exist.

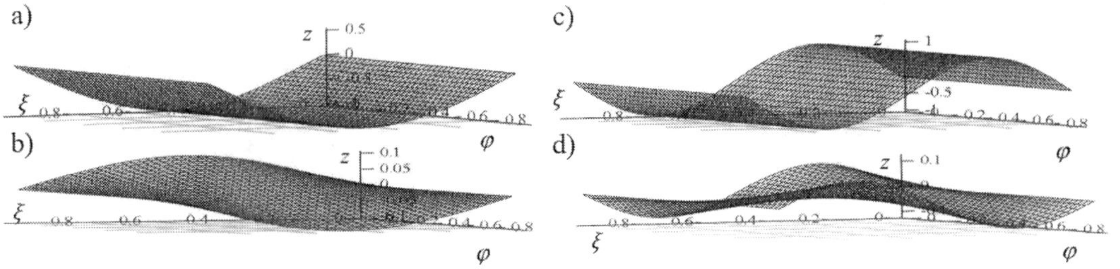

Fig. 3. Deflection forms at equilibrium states: z_{11} (a), z_{12} (b), z_{21} (c), z_{22} (d).

4 Investigation results

To describe thermomechanical properties of the plate material, the investigation results obtained from the molecular dynamics studies presented in literature are taken into account. Thermomechanical properties of the multiscale materials presented in the study [6] are shown in Tab.1 and 2. The single-layer graphene sheets (SLGS) and carbon nanotubes were appled as nanofillers in copper matrix in this studies. The Young's and Kirchhoff's moduli are summarized in Table 1 along with the analogical results of pure Cu. Table 2 lists the Young's modulus in the axial direction of these multiscale composites at different temperatures along with the results of pure Cu for comparison.

Table 1. Comparison of mech. properties of Cu based nano-composite reinforced with 9% vol. fraction of nanofiller [6]

	E_{11} [GPa]	E_{22} [GPa]	G_{23} [GPa]	G_{12} [GPa]
Pure Cu	56	56	22	22
SLGS–Cu	164.8	93.5	26.3	23.7
CNT–Cu	140.7	67.9	24.2	22.6

Table 2. Young's modulus (E_{11} [GPa]) of Cu-based nanocomposite at different temperatures [6]

	Temperature [°K]		
	1 K	300 K	500 K
Pure Cu	86	56	45
SLGS–Cu	175.4	164.8	156.4
CNT–Cu	150.4	140.7	131.2

Numerical investigations of free vibrations were carried out the axially moving multiscale copper plate shown in Fig.2. The plate is tensioned with the constant longitudinal load N_{x0}. Parameters of the moving plate are presented in Table 3.

Table 3. Parameters of the plate

Parameter	Units	Notation	Value
thickness	mm	h	0.8
length	m	l	1.0
width	m	b	0.1
axial load	N/m	N_{x0}	100
Cu mass density	kg/m^3	ρ	8930
wave velocity	m/s	c_w	3.74

4.1 Identification of the fractional model

To identify the parameters of the fractional rheological model of the plate material, analysis of dynamics of the beam model of this structure was carried out [5]. Fig.4 shows the plot of the lowest natural frequencies of the beam model of the pure copper plate vs transport speed in the ambient temperature. On the base of these results the first natural frequency of the stationary beam model ω_{10} was treated as the frequency coordinate of the operating point on the storage Young's modulus plot in the ambient temperature. The low- and high-frequency asymptotes of this plot constitute the values of the Young's moduli from the molecular dynamic study for 500 K and 1 K, respectively. The frequency characteristic in the ambient temperature with different slop was carried out by the operating point. Fig.5 shows the characteristic of the pure copper plate for the slop $\beta = 0.03$.

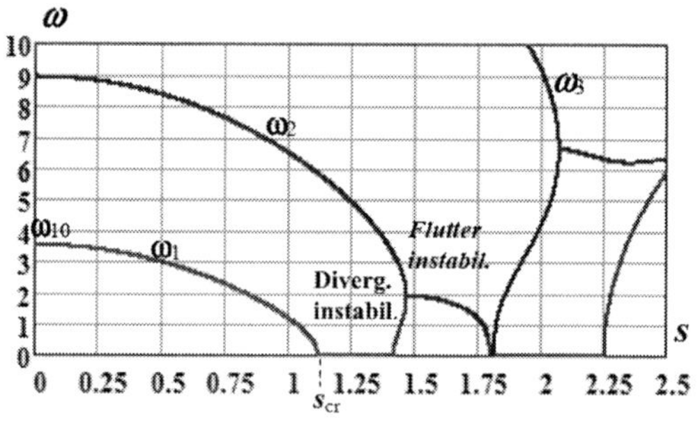

Fig. 4. The lowest natural frequencies of the beam model of the pure copper plate vs transport speed

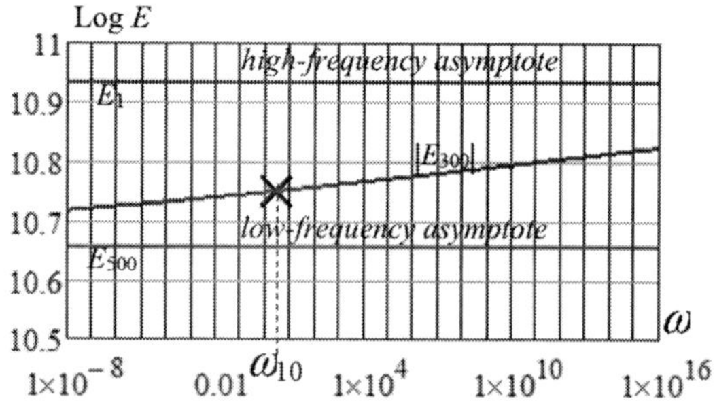

Fig. 5. The storage Young's modulus vs frequency (pure Cu, $T = 300$ K, $\beta = 0.03$)

The temperature dependent the reduced frequency factor in the fractional model of the plate material was determined by the Willims-Ferre-Lendel method using the formula

$$\log[\alpha(T)] = \frac{c_1 (T - T_0)}{c_2 + T - T_0} \qquad (28)$$

where: c_1 and c_2- experimental constants, T – measuring temperature, T_0 - reference temperature.

However, in the literature very limited information on the temperature characteristics of the multiscale composites can be found. To determine the values of the experimental constants C_1 and C_2 in equations (28) the experimental characteristic of beryllium copper from the references [7] was used (Fig.6.).

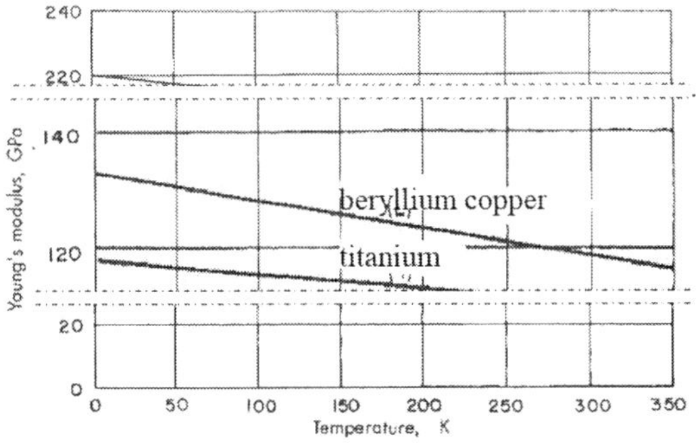

Fig. 6. The plot of the Young's modulus vs temperature for beryllium copper [7]

To determine the values of experimental constants for multiscale copper composites the same experimental characteristic of the copper alloy in low temperatures was taken into consideration.

4.2 Results of numerical investigations

Fig.7 shows the graphs of the decimal logarithm of the determinant of the matrix Q received from boundary conditions versus the frequency value for the three lowest modes.

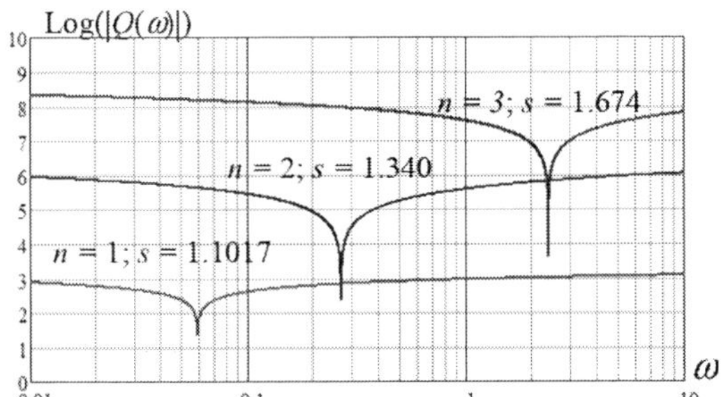

Fig.7. Non-trivial equilibrium positions (pure Cu, material properties from molecular dynamic study, T = 350 K, τ = 0.1487, τ_0 = 0.2842, β = 0.03)

The local minima on the curves in Fig.7 determine the natural frequencies of vibrations. The critical transport speed of the plate corresponds to the speed at which the lowest natural frequency vanishes. In numerical investigations the transport speed at which the lowest natural frequency equals 0.01 has been treated as the critical speed. Fig. 8 illustrates the procedure of determining the critical transport speed of the copper plate.

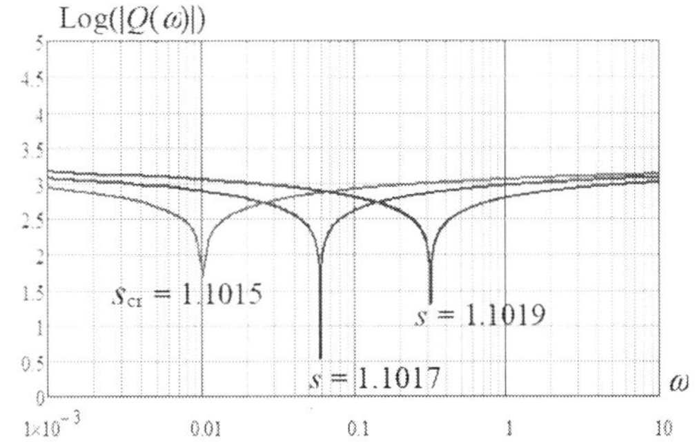

Fig. 8. Determining the critical transport speed of the moving copper plate (material properties from molecular dynamic study, T = 350 K, n = 1, τ = 0.1487, τ_0 = 0.2842, β = 0.03)

Results of numerical investigation in Fig.9 show effect of temperature on the lowest non-trivial equilibrium positions of three successive modes for the pure copper plate. The lowest solid curve presents the plot of the critical transport speed versus temperature.

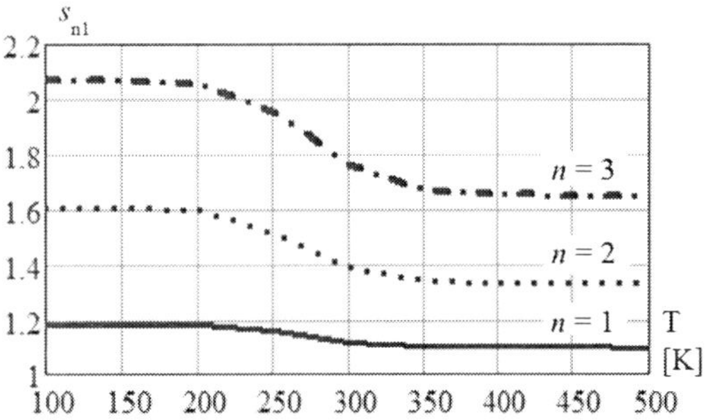

Fig. 9. Effect of temperature on the lowest non-trivial equilibrium position (pure Cu - material properties from molecular dynamic study, β = 0.03)

The plot of the critical transport speed for the pure copper plate is also visible in Fig.10. This is comparison of investigation results of critical speeds for both multiscale plates, beryllium copper and copper plates.

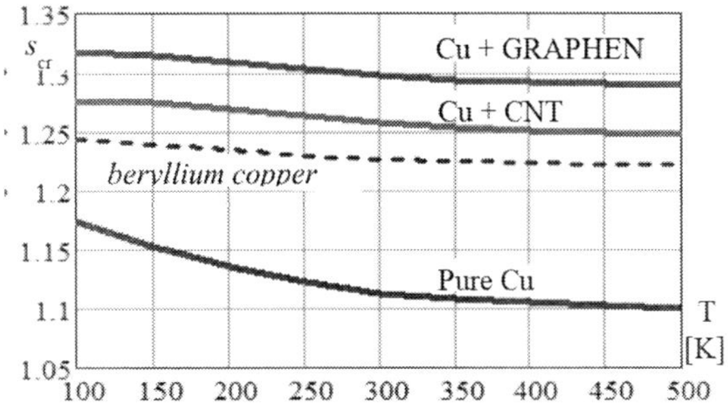

Fig.10. Effect of temperature on the critical transport speed (material properties from molecular dynamic study – solid line, and experiment – dashed line, β = 0.01)

The plots in Fig. 10 show that critical transport speeds of all investigated plates decrease with increasing temperature. In the considered temperature range, the critical speeds of the graphene-reinforced composite are higher than the others. Figures 11, 12 and 13 show effect of temperature and the slope of the frequency characteristics on the critical speeds.

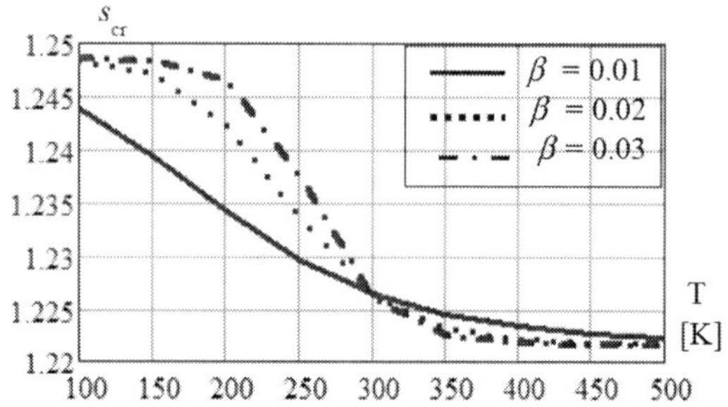

Fig.11. Effect of temperature and the rheological parameter β on the critical transport speed for beryllium copper

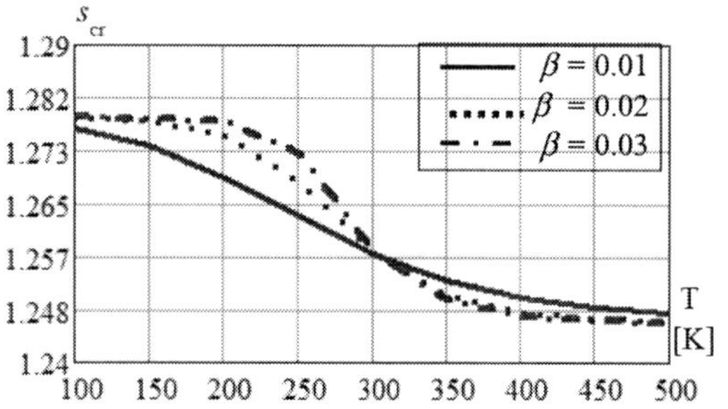

Fig.12. Effect of temperature and the rheological parameter β on the critical transport speed for the CNT/Cu composite

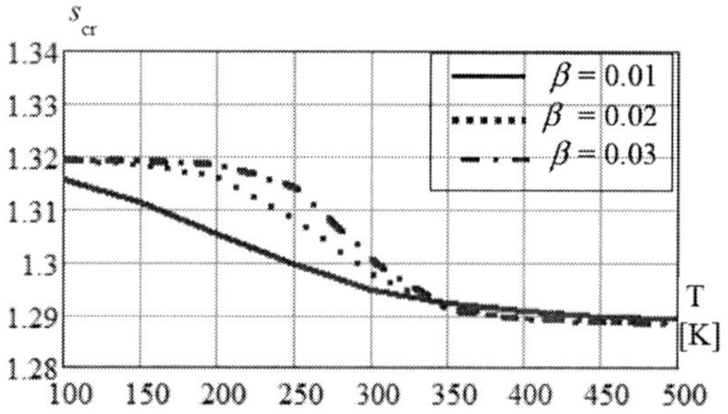

Fig.13. Effect of temperature and the rheological parameter β on the critical transport speed for the graphene/Cu composite

Figures 11, 12 and 13 show increasing the slope of the frequency characteristics results in an increase the value of the critical speeds at lower temperatures and reducing this value at the higher temperatures.

5 Conclusions

Although recently big research effort has been devoted to multiscale composites, there is still a need to know the dynamics of these structures. Dynamics of the axially moving multiscale composite plate subjected to thermal load is analyzed in this study. To describe thermomechanical properties of the plate material, properties of multiscale stationary composites presented in the literature are taken into account. Fractional standard rheological model of the plate material as the function of reduced frequency depended on the temperature is determined. On the base of the Hamilton's principle for the transverse motion the equation of equilibrium states of the plate in the frequency domain is derived.

Research plants were CNT-reinforced and graphene-reinforced metal composites with copper matrix. In the literature very limited information on the temperature characteristics of these composites can be found. Therefore, to determine the temperature-depended reduced frequency in the fractional rheological model, the experimental characteristic of beryllium copper in low temperatures was used.

The free vibrations of multiscale composite plates are analyzed. Critical transport speeds of both multiscale composites decrease with increasing temperature. In the examined temperature range, the critical transport speeds of the graphene-reinforced composite are higher than both the CNT-reinforced composite as well as the comparable copper alloy. Increasing the slope of the frequency characteristics of all composites results in an increase of the critical transport speeds in low temperatures and a lower reducing the critical transport speed in the high temperature range.

References

1. K. Marynowski, T. Kapitaniak. Int. J. Mech. Sc. **81**, 26 (2014)
2. A.S. Sayyad, Y.M. Ghugal, Compos. Struct. **129**, 177 (2015)
3. A.D. Moghadam, E. Omrani, P.L Menezes, P.K Rohatgi. Comp Part B, **77**, 402, (2015)
4. T .Pritz. J, Sound and Vibr. **195**(1), 103 (1996)
5. K. Marynowski..*Dynamics of the axially moving orthotropic web*. (Springer-Verlag: 2008)
6. S, Sharma, P. Kumar, R. Chandrai. J Comp. Mater, DOI: IO.1177/0021998316682363 (2016),
7. S. Van Sciver, J. Pfotenhauer USPAS Cryogenics Short Course, Boston, MA 6/14/2010 – 6/18/2010 http://uspas.fnal.gov/materials/10MIT/Lecture_1.2.pdf

Nonlinear dynamics of a planar hinged-supported beam with one end spring system

Łukasz Kłoda[1,2,*], *Stefano* Lenci[1] and *Jerzy* Warmiński[2]

[1]Department of Civil and Buildings Engineering, and Architecture, Polytechnic University of Marche, Ancona, Italy
[2]Department of Applied Mechanics, Lublin University of Technology, Lublin, Poland

Abstract. Forced nonlinear oscillations of a planar, initially straight, Timoshenko beam are studied. The goal of this paper is to examine large amplitude vibrations of a hinged-simply supported beam with a linear axial spring on one end, allowing to consider the influence of varying axial boundary conditions. Influence of geometrical nonlinearity coming from very large deformations in axial, transversal and rotational directions on frequency response curves are presented for a wide spectrum of the spring stiffness. The complete frequency response curve is computed by a special pseudo continuation method of explicit numerical simulations. For selected cases, a jump phenomenon between branches and super/sub harmonic resonances have been observed.

1 Introduction

Large beam vibrations of a planar initially straight beam, where shear deformation and rotational bending effects are taken into account are presented e.g. in [1]. The nonplanar, nonlinear and inextensional beam which is undergoing flexure about two principal axes and torsion around longitudinal axis is studied in [2,3]. The beam is modeled as a set of mathematically consistent partial differential equations in [2], and its resonant forced oscillations are studied in [3]. However, those papers discussed only transversal and shear vibrations. In [4] the model of hinged beam is extended to take into account also longitudinal and rotatory inertia. Using the Galerkin method of modal expansion with arbitrary boundary conditions, a system of coupled nonlinear ordinary differential equations are first obtained and then investigated.

Lenci, Rega and Clementi [5-8] developed the exact equations of motion of a planar, initially straight beam, by considering axial, bending and shear deformations, and axial, transversal and rotatory inertia. It allowed to study the influence of longitudinal vibrations on dynamics of hinged-movable supported beam with axial linear spring on its end. The works [5-8] are focused on free nonlinear oscillations considering various spring stiffnesses. Using the Poincarè-Lindstedt method [5], the effect of variable slenderness [6] and axial-transversal coupling vibrations [7] has been obtained analytically and then validated numerically in [8].

The goal of this paper is to study forced and damped oscillations numerically by the Finite Element Method (FEM), using to commercial Abaqus_CAE® software. The FEM is applied by a special technique, similar to continuation method. In particular, for fixed values of the harmonic excitation amplitude, the frequency response curves are obtained by continuously varying the excitation frequency.

2 The beam model

Let us consider the Timoshenko beam presented in Fig. 1. It has length L, is straight in the initial configuration and has prismatic cross section A with homogeneous material properties: Young modulus E, shear modulus G and density ρ. The system is hinged-simply supported with one end axial spring with stiffness κ in the moving end, see Fig. 1. Displacements in the axial and the transversal directions are denoted by $W(Z,T)$ and $U(Z,T)$, respectively, where Z is a spatial coordinate and T is time.

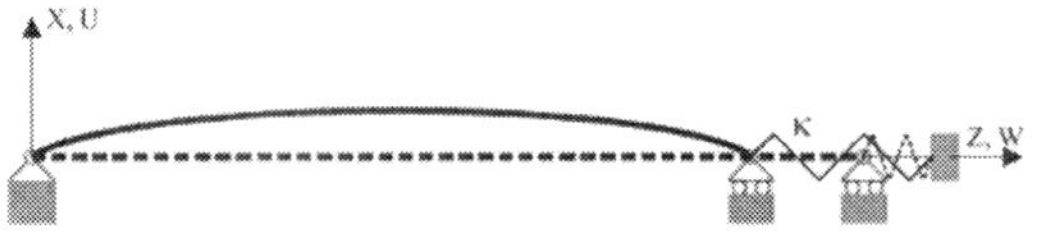

Fig. 1. Deformed configuration of the initially straight beam with one end linear spring.

Detailed description of kinematics, constitutive elastic behavior, boundary conditions and the derivation of Partial Differential Equations (PDEs) of motion are given in [5]. In this paper we extend the mathematical model presented in [5] adding harmonic excitation and linear viscous damping, as shown in Fig. 2. The equations of forces and moments balance for horizontal (1), vertical (2) motions and rotation (3) are:

---
* Corresponding author: l.kloda@pollub.pl

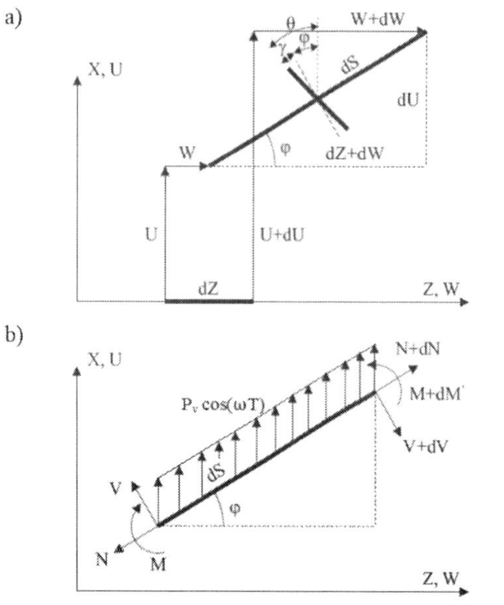

Fig. 2. The beam element deformations and rotation (a): dZ – infinitesimal element in its initial configuration, dS – deformed element, dU, dW – incremental displacements, Θ – rotation of the beam cross-section, φ – slope angle of beam axis, γ – shear strain and (b) balance of forces and moments: N – axial force, V – shear force, M – bending moment, P_v – X-directional uniformly distributed external harmonic excitation of frequency ω.

$$(N\cos\varphi + V\sin\varphi)' = \rho A\ddot{W} + 2C_h\dot{W}, \tag{1}$$

$$(N\sin\varphi - V\cos\varphi)' = \rho A\ddot{U} + 2C_v\dot{U} + P_v\cos(\omega T), \tag{2}$$

$$M' - VS' = \rho J\ddot{\theta} + 2C_r\dot{\theta}. \tag{3}$$

In Eqs.(1)-(3) the dot corresponds to the derivative with respect to time T and the prime denotes the spatial derivative with respect to axial coordinate Z. C_h, C_v and C_r are linear viscous damping factors in horizontal, vertical and rotational directions, respectively. Formulas (1), (2), (3) lead to the exact forced-damped PDEs of motion without approximations or condensation.

The boundary conditions are:

$$U(0,T) = U(L,T) = 0, \tag{4}$$

$$M(0,T) = M(L,T) = 0, \tag{5}$$

$$W(0,T) = 0, \tag{6}$$

while the boundary conditions provided by the axial spring in horizontal direction has the form:

$$[N(L,T)\cos\varphi + V(L,T)\sin\varphi] + k_s W(L,T) = 0. \tag{7}$$

Note that for $\kappa = \frac{k_S L}{EA} = 0$ we get simply supported beam. When κ tends to ∞ it implies $W(L,T) = 0$ and the beam becomes hinged-hinged. Other cases of κ represent hinged-simply supported-spring systems.

3 Finite Element Model

The problem is addressed by the FEM, where we consider classic Timoshenko beam with length to width ratio equal to 10. Beam has been discretized by using 100 finite elements of B31 type. This element type has linear shape between nodes, in three-dimensional space. Beam element is linear elastic and allows transverse shear deformation [9]. Planarity was guaranteed by constraining the out of plane nodes displacements. Mechanical properties are collected in Table 1. Boundary conditions are assumed as in (4-7).

Table 1. Mechanical characteristics of the FEM beam model

Initial length [L]		500 mm
Cross section [A]		50 mm × 50 mm
Density [ρ]		7850 kg/m^3
Young modulus [E]		210 Gpa
Poisson's ratio [ν]		0.3
Beams shear factor [Sk]		0.85
Viscous damping factor [ζ]		6%
Spring stiffness $\kappa = \frac{k_S L}{EA}$	Hinged- simply supported beam	$\kappa = 0$
	Hinged-simply supported-axial spring beam system	$0 < \kappa < \infty$
	Hinged-hinged	$\kappa = \infty$
External excitation amplitude [P_v]		40.7992 kN

Using Lanczos method, the first natural frequency and corresponding free vibration mode shape has been determined. The results show, as expected, that in first

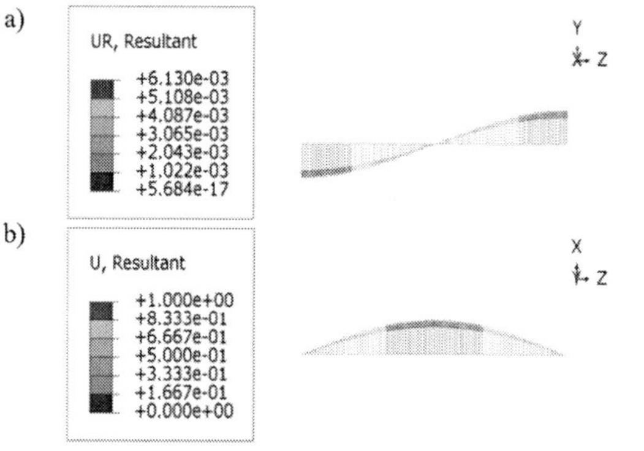

Fig. 3. First natural frequency mode shape, dimensionless vectors of rotation and displacement.

bending mode we have $W(Z,T) = 0$, namely that longitudinal vibrations have no effects on first linear bending shape and natural frequency (Fig. 3). In other words, small amplitudes of transversal vibrations are independent of axial spring stiffness. For all inspected types of axial boundary conditions the first bending linear frequency is $\omega_0 = 461.47$ Hz (2899.50 rad/s), period T=0.002167 sec.

Dynamic explicit calculations have been performed to study nonlinear vibrations. External excitation has been modeled as concentrated force in $Z = L/2$ of the beam in transversal direction, mathematically described by the Dirac delta function $P_v = p_v \delta(Z - \frac{L}{2})$. Amplitude p_v is constant (see Tab. 1 for its value) and corresponds to 1 mm static deflection of middle point of the hinged-simply supported beam, which is equals 2% its thickness.

In order to get the complete frequency response curve, i.e. how the amplitude of the response varies by varying the excitation frequency, forward and backward frequency sweeping are performed. Each excitation frequency ω has been applied individually, step by step, first with increasing frequency from 0 Hz to 922.94 Hz (double of the natural frequency ω_0). Next, the simulation is reseted and then repeated with decreasing frequency from 922.94 Hz to 0 Hz. To minimize the computational time a special methodology, similar to a continuation method has been applied, see Tab. 2. This reduced the unwanted jump between the upper and lower branches (when they coexist). The initial conditions to each step have been taken from upshot of the previous step, consistent with the excitation phase.

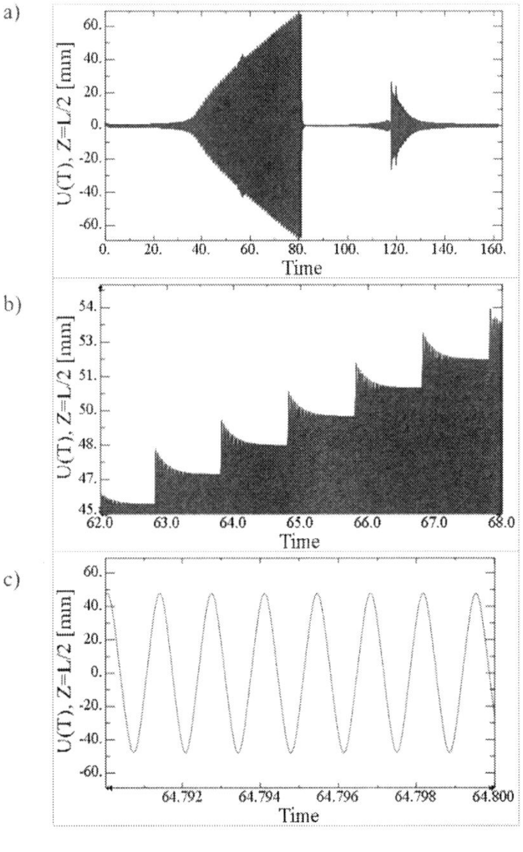

Fig. 4. Time histories of vibrations of hinged-supported-spring $\kappa = 4$, reported in three different time scales: (a) global sweep forward from $\frac{\omega}{\omega_0} = 0$ to $\frac{\omega}{\omega_0} = 2$, reset $\frac{\omega}{\omega_0} = 0$ and backward from $\frac{\omega}{\omega_0} = 2$ to $\frac{\omega}{\omega_0} = 0$, (b) stabilization of the each frequency from $\frac{\omega}{\omega_0} = 1.55$ to $\frac{\omega}{\omega_0} = 1.625$ and (c) time history of a steady-state solution for $\frac{\omega}{\omega_0} = 1.625$.

Table 2. Sampling of excitation frequencies

→ → → → Increment → → → → →											
$\frac{\omega}{\omega_0}$	Static Load	...	0.5	0.6	...	0.97813	0.98125	0.98438	...	2	↓ Reset ↓
← ← ← ← Decrement ← ← ← ← ←											

The displacement of the middle point of the beam has been recorded in transversal direction. The time step was constant for each frequency and equal to $\frac{1}{40}$ of excitation period. After a frequency increment/decrement, the simulations run for a sufficient number of periods during which the solution ends its transient and approaches the new attractor (Fig. 4b). The response amplitude is measured at the end of the transient, just before changing again the excitation frequency. An example of a time history obtained with this approach is reported in Fig. 4.

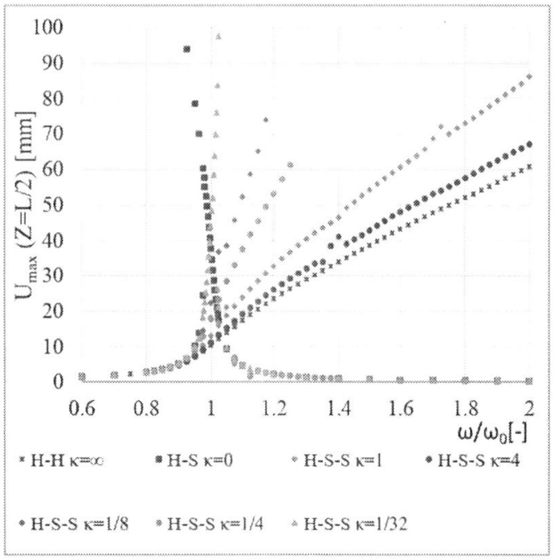

Fig. 5. Frequency response curves of hinged-supported-spring systems, obtained by FEM model, for selected spring stiffness from 0 to +∞. Curves demonstrate effect from softening to hardening, respectively.

Computations are demanding and very time consuming. Notwithstanding, such an approach allows to study nonlinear systems. Post-processing the many time histories obtained for different values of ω, the final result are the frequency response curves, which are reported in Fig. 5 for different values of κ.

Jumps from upper branches to lower branches occur for the beam axial stiffness κ equal $\frac{1}{8}$ and $\frac{1}{4}$. Moreover, unexpected super or sub harmonic resonances appeared for $\kappa = 4$ and $\kappa = 1$ for frequencies $\omega/\omega_0 = 1.725$ and $\omega/\omega_0 = 1.40$, respectively. Particular attention has been paid to these zones, and the corresponding time histories are plotted in Fig. 6.

a)
b)

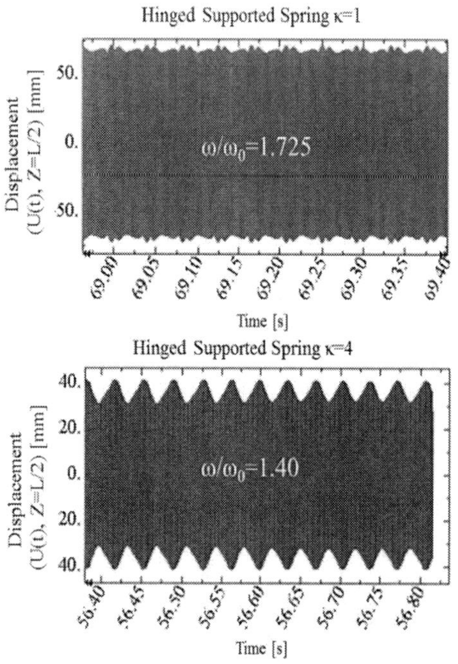

Fig. 6. Time history of superharmonic resonances of hinged-supported-spring (a) $\kappa = 1$ and (b) $\kappa = 4$.

The Fourier Transform shows relationship between longitudinal and transversal vibrations. In case $\kappa = 1$ peaks corresponds to $\omega_{11} \cong 1.725\,\omega/\omega_0$, $\omega_{12} \cong 5.2\,\omega/\omega_0$, $\omega_{13} \cong 6.3\,\omega/\omega_0$ and for $\kappa = 4$ peaks are at $\omega_{41} \cong 1.4\,\omega/\omega_0$ and $\omega_{42} \cong 4.2\,\omega/\omega_0$. The ratio between harmonics in both cases are $\omega_{12} = 3\omega_{11}$ and $\omega_{42} = 3\omega_{41}$ (Fig. 7). These cases correspond to internal resonance between the first and the second bending modes.

4 Conclusions

The paper presents the results of explicit simulation of damped and forced motion of hinged-supported beam with one end axial spring system. Using the FEM large amplitude vibrations under harmonic excitation are considered. The influence of axial boundary conditions as varying spring stiffness on softening or hardening effect on the frequency response curves is presented. In numerical FEM analyses it has been shown that a 1:3 internal resonances for spring supported cases may occur.

a)
b)

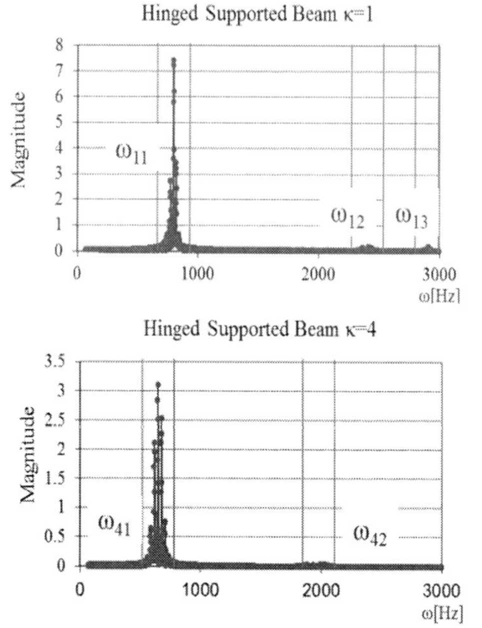

Fig. 7. Fast Fourier transforms of time series of hinged-supported-spring systems (a) $\kappa = 1$ and (b) $\kappa = 4$.

In further developments the Multiple Time Scales Method will be applied to the mathematical model, and the frequency response curves and their backbone curves will be determined analytically. The proposed analysis can be adapted to any system i.e. nonhomogeneous and with varied cross-section beams, higher modes, different boundary conditions, etc. Other internal resonances will be investigated as well, and for selected cases, the results will be confirmed experimentally.

References

1. J.N. Reddy, I.R. Singht. Int. J. Num. Mech. Eng. **17(6)**, p.829-852 (1981)
2. M.R.M Crespo da Silva, C.C. Glynn, J. of Struc. Mech., **6(4)**, 437-448 (1978)
3. M.R.M Crespo da Silva, C.C. Glynn, J. of Struc. Mech., **6(4)**, 449-461 (1978)
4. S. Atluri, J. Appl. Mech. **40**, 121-126 (1973)
5. G.Rega S. Lenci, Procedia IUTAM **19**, 35-42 (2016)
6. S. Lenci, F. Clementi, G. Rega, Meccanina **51(11)**, 2589-2606, (2016)
7. S. Lenci, G. Rega, Proc. Royal Soc. A, (2016)
8. F. Clementi, S. Lenci, G. Rega, Archive of Appl. Mech. **(5)87**, 865-880 (2017)
9. Abaqus 6.9 Documentation

Sophisticated finite strip for vibration analysis of a rotating cylindrical shell

Ivo Senjanović, *Ivan* Ćatipović, *Neven* Alujević[1], *Nikola* Vladimir, *Damjan* Čakmak[1,*]

[1]Fakultet strojarstva i brodogradnje, Sveučilište u Zagrebu, Ivana Lučića 5, 10 000 Zagreb, Croatia

Abstract. In this article a two-node finite strip with eight degrees of freedom for free vibration analysis of pre-stressed rotating cylindrical shells is formulated. The circumferential mode shape profiles are described exactly using trigonometric functions. The axial mode shape profiles are approximated by bar and beam shape functions for membrane and bending displacements, respectively. In this way a semi-analytical formulation is facilitated so that the discretization is required only in the axial direction. The developed finite strip is validated by comparisons with analytical results. An excellent agreement is observed both for stationary and rotating shells.

1 Introduction

The theory of shells and plates has been covered in a systematic manner in a number of books [1-8]. For example, an instructive approach to thin shell theory, written in a relatively simple way and adapted to the engineering level for practical usage, is presented in [9].

The general theory and specific discussions regarding shells of revolution exposed to uniform load can be found in [10]. Such problems are more particular for submarine or aircraft pressure hull designs with pronounced axial symmetry [11-13].

It is often the case that axisymmetric shells rotate around the axis of symmetry [14-27]. For example, rotating shells of revolution are found in engineering practice in rotor systems of gas turbine engines, high-speed centrifugal separators, rotating satellite structures, and automotive tires. Rotation makes their dynamic behaviour significantly more complex. One of the first investigations into vibration of rotating structures has been carried out by Bryan [14]. He studied vibrations of a rotating ring and described the travelling modes phenomenon. These phenomena result from the Coriolis effect, as shown on the example of infinitely long rotating cylindrical shells, [15, 16] as well as on finite rotating cylinders [17, 18]. An experimental study on flexural vibrations of a thin rotating ring is given in [19]. Furthermore, the influence of a pre-stress on free vibrations of rotating cylinders has been studied in [20].

Huang and Soedel [21] used the nonlinear strain-displacement relationships [13], to formulate the corresponding set of differential equations of motion of a rotating cylindrical shell. They solved the free and forced vibration problem of a simply supported cylindrical shell exactly, by assuming simple sine and cosine displacement functions of the circumferential and axial variables. In this case, postulating the eigenvalue - eigenvector problem results in a characteristic polynomial of sixth order. Its solution gives three positive and three negative natural frequencies.

If the shell does not rotate, the polynomial is bi-cubic. There are three pairs of positive/negative frequencies characterised by the same absolute value. This is physically explained through pairs of backward and forward rotating modes. The two modes of a pair rotate with the same speed in opposite directions and thus superimpose into a stationary mode. The reason why there are three pairs of modes and natural frequencies is that there are three types of dominant modes: bending (radial), longitudinal (axial), and shear (circumferential).

If the shell rotates around its axis with a constant speed, then the polynomial is no longer bi-cubic. The odd polynomial coefficients occur. The positive and negative natural frequencies have distinct absolute values and the so-called frequency veering (bifurcation) occurs. This means that the forward and backward rotating modes no longer rotate with the same rotation speed and thus cannot superimpose into the stationary modes (standing waves). As a result, with spinning shells the modes rotate independently. For example, this phenomenon comes about with tires rolling over the road surface [22-26]. This significantly influences the overall NVH (Noise and Vibration Harshness) characteristics of the vehicle.

It is very common in the literature on vibration of cylindrical shells, either rotating or stationary, to assume that the two ends of the shell are simply supported [21, 27, 28]. This type of boundary support is sometimes referred to as shear diaphragm set of boundary conditions [8, 9]. The reason why this set of boundary conditions is often used is that it is probably the only one for which a relatively simple solution can be obtained analytically. In other words, mode shapes assumed as

* Corresponding author: damjan.cakmak@fsb.hr

appropriate products of trigonometric functions of the circumferential and axial variables, usually satisfy both the differential equations and the boundary conditions. This results in a mathematically convenient model.

However, such a model is not necessarily suitable to describe a particular engineering problem. For example, although it is very instructive to investigate dynamics of rotating tires assuming simply supported edges of a tire tread-band [21, 27, 28], it is difficult to accept, from an engineering point of view, that the tire sidewall is infinitely stiff in radial or tangential directions [25, 26].

In case of cylindrical shells with boundary conditions other than simply supported, the mathematics significantly complicates. Closed form solutions become difficult to obtain. A number of investigations have been undertaken to tackle this problem [29-37]. One of the solutions has been obtained by assuming shell displacement field as a product of Fourier series in the axial direction and trigonometric functions in the circumferential direction [32]. This procedure has been recently extended to rotating cylindrical shells [33]. The problem of free vibration of rotating cylindrical shell, having arbitrary boundary conditions, can be also solved by employing the Rayleigh-Ritz method. Such a solution, using characteristic orthogonal polynomials for displacements variations along the axial direction, can be found in [34].

A completely analytical solution for free vibrations of a rotating cylindrical shell with arbitrary boundary conditions has been recently offered in [35]. The equations of motion in [35] are based on the strain-displacement relationships of Hermann and Armenakas [13]. However, the solution, although exact, requires a rather complicated procedure of the undetermined shell length to perform the analysis.

In this paper, a semi-analytical approach is used which is based on the theory of finite strips [38]. A new, sophisticated finite strip for modelling pre-stressed rotating cylindrical shells is formulated. This methodology benefits from describing the shell displacement field in the circumferential direction exactly so that the discretization is only needed in the axial direction. The energy approach is used with the strain-displacement relationships given in [13, 21] to develop and validate such a special finite strip. The strip element is deliberately made quite simple including two nodes and eight degrees of freedom (DOF). Its reliability is checked in a number of numerical examples by comparison of the numerical with analytical results. An excellent agreement is observed for all boundary conditions considered.

2 The finite strip

2.1 Strain and kinetic energies

A thin cylindrical shell rotating around its axis of symmetry with constant angular speed Ω, is shown schematically in Fig. 1 a). The shell dimensions are the following: L is the length, a is the radius, h is the thickness. The shell mid-surface is defined in the cylindrical coordinate system, where x and φ are the axial and angular coordinates, respectively. The displacement of a point P on the mid-surface, whose position is defined by x and φ, is specified by axial, tangential and radial displacement components, u, v and w, respectively.

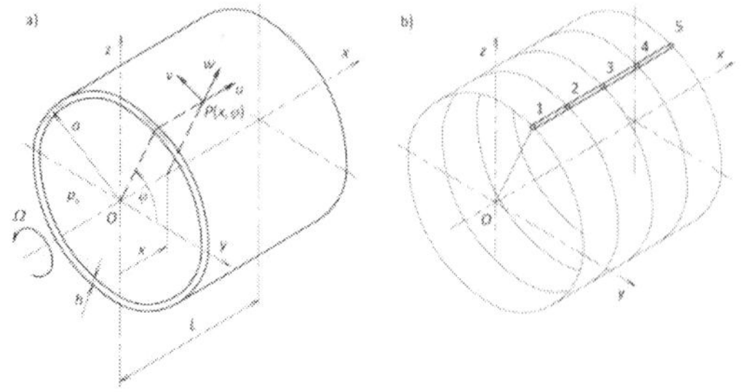

Fig. 1. Rotating cylindrical shell with: a) the main dimensions and the displacement components. b) finite strips with nodes

The strain energy stored in an infinitesimal shell element of thickness h is

$$dE_s = F_s(x,\phi,t)\,dx\,d\phi, \tag{1}$$

where F_s is the strain energy density function. The kinetic energy stored in an infinitesimal shell element is given by

$$dE_k = F_k(x,\phi,t)\,dx\,d\phi, \tag{2}$$

where the kinetic energy density function F_k is a product of the specific mass per unit area and the total velocity squared. The total velocity consists of the velocity components in the axial, circumferential and radial directions. It also includes contributions due to the shell rotation.

The total potential and kinetic energy of a general cylindrical shell is given by

$$E = \int_0^{2\pi}\int_0^{L}\left[F_s(x,\phi,t)+F_k(x,\phi,t)\right]dx\,d\phi, \tag{3}$$

where the energy density functions F_s and F_k are denoted in Eq. (1) and (2), respectively.

For a cylindrical shell with a closed cross-section and arbitrary boundary conditions at the two ends (at $x = 0$ and $x = L$, where L is the length of the shell), the three displacement components can be assumed in the form

$$\begin{aligned} u(x,\phi,t) &= U(x)\cos(n\phi+\omega t),\\ v(x,\phi,t) &= V(x)\sin(n\phi+\omega t),\\ w(x,\phi,t) &= W(x)\cos(n\phi+\omega t), \end{aligned} \tag{4}$$

where ω is a natural frequency.

2.2 Stiffness matrix

In this section the properties of the finite strip of a rotating cylindrical shell are developed by employing the energy approach. The rotating cylindrical shell is modelled by a number of finite strips, as shown in Fig. 1 b). Due to the trigonometric variation of the displacement field in the circumferential direction, the dynamic behaviour of the finite strip becomes representative by its axial generatrix which behaves like a sophisticated beam. The stiffness matrix is obtained using the element strain energy from Eq. (1)

$$E_s = \int_0^{2\pi}\int_0^{l} F_s(x,\phi,t)\,\mathrm{d}x\,\mathrm{d}\phi, \tag{5}$$

where the expression for F_s can be found in [39], and l is the strip width. Substituting the assumed displacement functions (4) into the expression for F_s and integrating Eq. (5) over the circumference ϕ, the strain energy is obtained in the form

$$E_s = \int_0^{l}\begin{bmatrix} \frac{1}{2}a_1(U')^2 + \frac{1}{2}a_2U^2 + \frac{1}{2}a_3(V')^2 + \\ \frac{1}{2}a_4V^2 + a_5U'V + a_6UV' + \\ +\frac{1}{2}b_1(W'')^2 + \frac{1}{2}b_2(W')^2 + \frac{1}{2}b_3W^2 + \\ +b_4W''W + b_5WU' + b_6W'V' + b_7WV + b_8W''V \end{bmatrix}\mathrm{d}x, \tag{6}$$

where the coefficients $a_1 \ldots a_6$ and $b_1 \ldots b_8$ can be found in [39]. The strain energy (6) is time-invariant. This is due to the fact that the rotating modes are characterised by fixed profiles which rotate along the circumference.

A relatively simple finite strip with two-nodal lines is introduced next. The amplitudes of the displacement components, Eqs. (4), are approximated by the following interpolation functions:

$$\begin{aligned} U(x) &= \sum_{j=1}^{2} U_j g_j(x), \\ V(x) &= \sum_{j=1}^{2} V_j g_j(x), \\ W(x) &= \sum_{j=1}^{4} \Delta_j f_j(x), \end{aligned} \tag{7}$$

where

$$\{U\} = \begin{Bmatrix} U_1 \\ U_2 \end{Bmatrix}, \quad \{V\} = \begin{Bmatrix} V_1 \\ V_2 \end{Bmatrix}, \quad \{\Delta\} = \begin{Bmatrix} W_1 \\ \Theta_1 \\ W_2 \\ \Theta_2 \end{Bmatrix}, \tag{8}$$

are vectors of the amplitudes of the nodal line displacements, *i.e.* nodal displacements for short, Fig. 2.

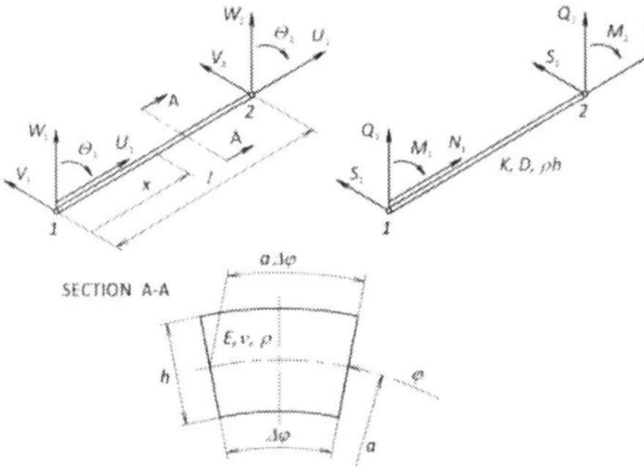

Fig. 2. Nodal displacements and nodal forces of the finite strip and its cross-section.

For membrane displacements, ordinary polynomial bar shape functions are used. For bending displacements, beam shape functions (Hermitian polynomials) are used:

$$\begin{aligned} &g_1 = 1-\xi, \quad g_2 = \xi \\ &f_1 = 1-\xi^2(3-2\xi), \quad f_2 = l\xi(1-\xi)^2, \\ &f_3 = \xi^2(3-2\xi), \quad f_4 = -l\xi^2(1-\xi), \end{aligned} \tag{9}$$

where $\xi = x/l$ is the dimensionless axial coordinate.

Substituting expressions (7) into (6) and differentiating the strain energy with respect to the nodal displacements, a system of eight algebraic equations is obtained, which can be written in the matrix notation:

$$\frac{\partial E_s}{\partial\{\delta\}} = [K]\{\delta\} \tag{10}$$

where

$$\langle\delta\rangle = \{\delta\}^T = \langle U_1 \;\; U_2 \;\; V_1 \;\; V_2 \;\; W_1 \;\; \Theta_1 \;\; W_2 \;\; \Theta_2\rangle \tag{11}$$

is the vector of the nodal displacements, Fig. 2, and $[K]$ is the strip stiffness matrix. The stiffness matrix consists of eleven submatrices as follows:

$$[K] = \begin{bmatrix} a_1[K]_1 + a_2[K]_2 & a_5[K]_3 + a_6[K]_3^T & b_5[K]_8 \\ a_5[K]_3^T + a_6[K]_3 & a_3[K]_1 + a_4[K]_2 & b_6[K]_9 + b_7[K]_{10} + b_8[K]_{11} \\ b_5[K]_8^T & b_6[K]_9^T + b_7[K]_{10}^T + b_8[K]_{11}^T & b_1[K]_4 + b_2[K]_5 + b_3[K]_6 + b_4[K]_7 \end{bmatrix} \tag{12}$$

Submatrices $[K]_i, i = 1,2\ldots11$ represent integrals of a different combination of the shape function products. They are calculated in the closed form and given in [39] with some comments on their physical meaning. The stiffness matrix in fact incorporates a geometric stiffness matrix, which is due to the initial tension forces, and which tend to increase the natural frequencies [39].

2.3 Mass matrices

Mass matrices are determined from the shell kinetic energy from Eq. (2)

$$E_k = \int_0^{2\pi}\int_0^{l} F_k(x,\phi,t)\,\mathrm{d}x\,\mathrm{d}\phi. \tag{13}$$

The kinetic energy density function F_k is given by

$$F_k(x,\phi,t) = \frac{1}{2}a\rho h\left\{\left(\frac{\partial u}{\partial t}\right)^2 + \left[\frac{\partial v}{\partial t} + (a+w)\Omega\right]^2 + \left(\frac{\partial w}{\partial t} - v\Omega\right)^2\right\} \tag{14}$$

Substituting (14) and (4) into (13), one obtains after integration over the circumference angle

$$E_k = \pi a\rho h\int_0^{l}\begin{bmatrix} \frac{1}{2}\omega^2U^2 + \frac{1}{2}(\omega^2+\Omega^2)V^2 + \\ +\frac{1}{2}(\omega^2+\Omega^2)W^2 + 2\omega\Omega WV + a^2\Omega^2 \end{bmatrix}\mathrm{d}x \tag{15}$$

The kinetic energy (15) is also time invariant for the same reason of rotating fixed axial mode profiles.

Furthermore, substituting expressions (7) into (15) and differentiating the kinetic energy with respect to the nodal displacements yields

$$\begin{aligned} \left\{\frac{\partial E_k}{\partial U_i}\right\}_{i=1,2} &= \pi a\rho h\omega^2[K]_2\{U\}, \\ \left\{\frac{\partial E_k}{\partial V_i}\right\}_{i=1,2} &= \pi a\rho h(\omega^2+\Omega^2)[K]_2\{V\} + 2\pi a\rho h\Omega\omega[K]_{10}\{\delta\}, \\ \left\{\frac{\partial E_k}{\partial \Delta_i}\right\}_{i=1,2,3,4} &= \pi a\rho h(\omega^2+\Omega^2)[K]_6\{\delta\} + 2\pi a\rho h\Omega\omega[K]_{10}^T\{V\}. \end{aligned} \tag{16}$$

Submatrices $[K]_i$, i = 2,6,10, have been calculated and can be found in [39]. The system of algebraic equations (16) can be written in the matrix form

$$\frac{\partial E_k}{\partial\{\delta\}} = [A]\{\delta\}, \tag{17}$$

where $\{\delta\}$ is a vector of nodal displacements, Eq. (11). Matrix $[A]$ can be separated into three matrices with respect to the power of frequency ω

$$[A]=[B]+\omega[C]+\omega^2[M], \tag{18}$$

where matrices $[B]$, $[C]$ and $[M]$ are given in [39].

2.4 The finite strip equation

By employing the minimum total energy principle, one obtains the finite strip equation in the following form

$$\{F\}=([K]-[B]-\omega[C]-\omega^2[M])\{\delta\}. \tag{19}$$

Matrix $[B]$, which depends on Ω^2, reduces the element stiffness due to the rotating mass, whereas the geometric stiffness matrix contained in $[K]$ has the opposite effect. However, the net effect of terms depending on Ω^2 is an increase of natural frequencies. Details on how to re-arrange the rows and columns of all matrices in (19) in order to facilitate a chain-like assembly of the strip equations are given in [39].

3 Numerical examples

3.1 Simply supported shell

In order to illustrate the application of the developed finite strip and to evaluate its accuracy, a number of numerical examples considering various boundary conditions are presented. First, a free vibration analysis a of simply supported cylindrical shell with the following geometric and material properties of rubber is carried out using the example from [35]: $L = 0.2$ m, $a = 0.1$ m, $h = 0.002$ m, $E = 0.45$ GPa, $\nu = 0.45$, $\rho = 1\,452$ kg/m^3. The developed finite strip is validated against an analytical solution [35]. The bifurcation of natural frequencies due to the angular velocity Ω, for a few $m\times n$ natural modes, has been calculated using 20 strip elements, and shown in Fig. 3 for two example modes.

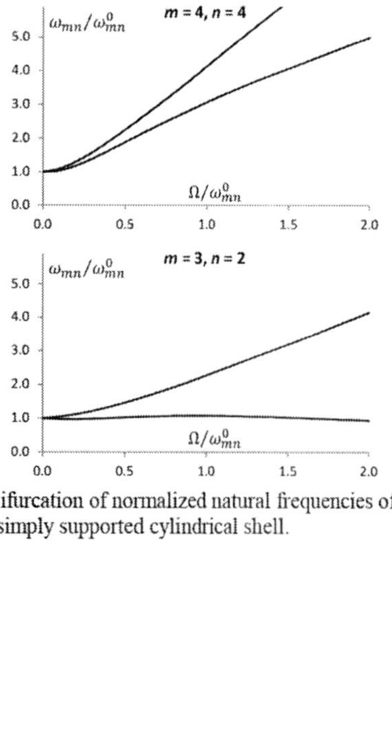

Fig. 3. Bifurcation of normalized natural frequencies of a rotating simply supported cylindrical shell.

Parameter m denotes the number of vibration nodes of the axial mode profile and the parameter n denotes the number of the nodes of the circumferential mode profile. The same results are obtained analytically so that the curves perfectly overlap.

Parametric analysis considering the internal pressure p_0 is performed next. Fig. 4 shows the increase of natural frequencies due to increased internal pressure. Also here the curves obtained analytically perfectly overlap with those calculated using the present finite strip method.

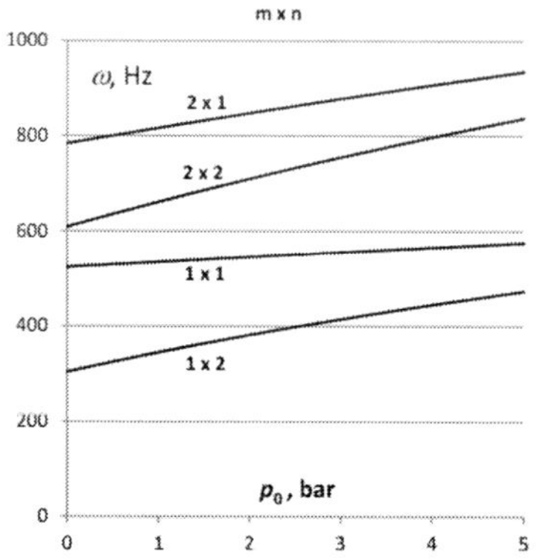

Fig. 4. Increase of natural frequencies of a simply supported cylindrical shell due to internal pressure.

3.2 Free-Free shell

Free vibration analysis of the cylindrical shell having the same geometric and material properties as in the previous example is now carried out for the case of free boundary conditions at the two ends of the shell. Two natural modes are shown in Fig. 5.

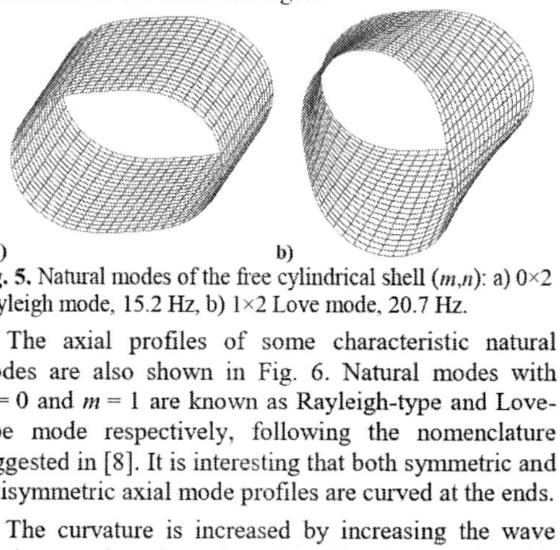

a) b)

Fig. 5. Natural modes of the free cylindrical shell (m,n): a) 0×2 Rayleigh mode, 15.2 Hz, b) 1×2 Love mode, 20.7 Hz.

The axial profiles of some characteristic natural modes are also shown in Fig. 6. Natural modes with $m = 0$ and $m = 1$ are known as Rayleigh-type and Love-type mode respectively, following the nomenclature suggested in [8]. It is interesting that both symmetric and antisymmetric axial mode profiles are curved at the ends.

The curvature is increased by increasing the wave number n in circumferential direction [35]. This phenomenon manifests the effect of the boundary conditions. A more detailed analysis of the mode shapes of free-free shells can be found in [31], [35-37].

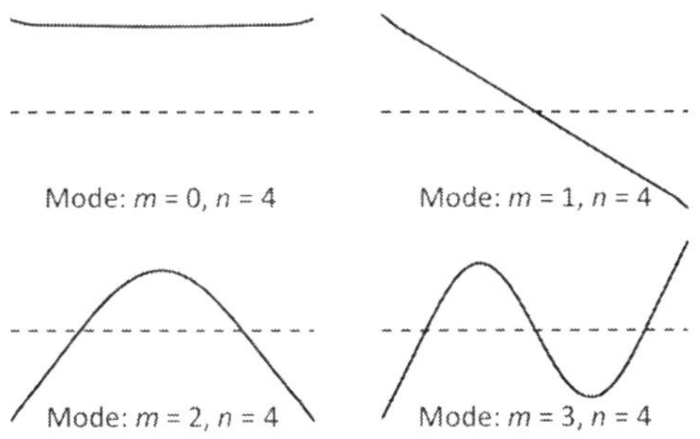

Fig. 6. Natural modes of the free cylindrical shell: axial mode profiles, $\Omega = 0$, $p_0 = 0$

4 Conclusions

In this paper, a new two-node finite strip with eight degrees of freedom for free vibration analysis of rotating and pre-stressed cylindrical shells is formulated. In order to develop the stiffness and mass matrices of the finite strip, simple bar and beam shape functions are used. The element stiffness matrix is a sum of an ordinary stiffness matrix and a geometric stiffness matrix. The geometric stiffness matrix is due to the initial tension forces. On the other hand, the inertia matrix is constituted of three mass matrices. The first one is the ordinary mass matrix related to squared natural frequencies (inertia load), the second one is related to squared angular velocity (centrifugal load), and the third mass matrix is related to the multiple of a natural frequency and angular velocity (Coriolis load). The stiffness and mass matrices are constituted of eleven submatrices, which are effectively spatial integrals of products of different combinations of the shape functions and their derivatives. These integrals have been calculated exactly such that the proposed finite strip method can be used without further numerical burden.

The application of the developed finite strip is illustrated by numerical examples considering simply supported and free-free boundary conditions at the two ends of the shell. The reliability of the finite strip is validated by a comparison with a benchmark analytical solution and an excellent agreement is observed. Also, a parametric analysis is performed in order to appreciate the effects of the angular velocity and pre-stress on natural frequencies. It is shown that veering of natural frequencies due to Coriolis effect, as well as their upward shift due to the pressurisation and the centrifugal effect, can be very accurately simulated.

Nowadays, vibrations of rotating tires are extensively investigated primarily due to their importance regarding the Noise and Vibration Harshness (NVH) of a road vehicle. The cylindrical finite strip developed in this paper can be used to model the belt of a rotating tyre. In order to model a tyre sidewall, a conical of toroidal finite strip would be necessary. In order to reduce the number of the finite strips necessary to accurately model a shell structure, a sophisticated finite strip with increased number of nodes and possibly faster convergence may also be considered.

This project has received funding from the European Union's Horizon 2020 research and innovation programme under the Marie Sklodowska-Curie grant agreement no. 657539. This investigation also received funding within the international collaborative project Global Core Research Center for Ships and Offshore Plants (GCRC SOP), established by South Korean Government (MSIP) through the National Research Foundation of South Korea (NRF).

References

1. S. Timoshenko, S. Woinowsky-Krieger, *Theory of Plates and Shells* (McGraw-Hill Book Company, New York, 1959)
2. V.V. Novozhilov, *The Theory of Thin Elastic Shells* (P. Noordhoff, Groningen, The Netherlands, 1964)
3. W. Flügge, *"Statik and Dynamik der Schalen"* (*Statics and dynamics of shells*) (Springer-Verlag, Berlin, 1934)
4. A.L. Goldenveizer, *Theory of Thin Shells* (Pergamon Press, Elmsford, New York, 1961)
5. V.Z. Vlasov, *General Theory of Shells and Its Applications in Engineering* (translation from Russian) (NASA TTF-99, U.S. Governmental Printing Office, Washington, D.C., 1964)
6. J.E. Gibson, *Linear Elastic Theory of Thin Shells* (Pergamon Press, London, 1965)
7. L.H. Donnell, *Beams, Plates, and Shells* (McGraw-Hill, New York, 1976)
8. A.W. Leissa, *Vibrations of Shells* (NASA SP-288, U.S. Government Printing Office, Washington, D.C., 1973)
9. W. Soedel, *Vibrations of Shells and Plates, 3rd edition, revised and expanded* (Marcel Dekker Inc., New York, 2004)
10. I. Senjanović, *Theory of Shells of Revolution* (Ship Research Institute, Zagreb, 1972)
11. C.T.F. Ross, *Pressure Vessels Under External Pressure: Statics and Dynamics* (Elsevier Applied Science, London, 1990)
12. Group of Authors, in E.E. Allmenidinger, (Ed.), *Submersible Vehicle System Design* (SNAME, Jersey City, 1990)
13. G. Herrmann. A.E. Armenakas, Dynamic Behavior of Cylindrical Shells under Initial Stress, Proc. 4th U.S. Nat. Congr. Appl. Mech, ASME, 203-213, (1962)
14. G.H. Bryan, On the beats in the vibrations of a revolving cylinder or bell. In: Proc. of the Camb. Philos. Soc., 101–111 (1880)
15. R.A. Di Taranto, M. Lessen, Coriolis Acceleration Effect on the Vibration of a Rotating Thin-Walled Circular Cylinder, J. Appl. Mech. **31**, 700-701 (1964)
16. A.V. Srinivasan, G.F. Lauterbach, Traveling Waves in Rotating Cylindrical Shells, J. Eng. Ind. ASME **93**, 1229-1232 (1971)

17. A. Zohar, J. Aboudi, The free vibrations of a thin circular finite rotating cylinder, Int. J. Mech. Sci. **15**, 269–278 (1973)
18. T. Saito, M. Endo, Vibration of finite length, rotating cylindrical shells, J. Sound Vib. **107**, 17–28 (1986)
19. M. Endo, K. Hatamura, M. Sakata, O. Taniguchi, Flexural vibration of a thin rotating ring, J. Sound Vib. **92**, 261–272 (1984)
20. J. Padovan, Natural frequencies of rotating prestressed cylinders. J. Sound Vib, **31**, 469–482 (1973)
21. S.C. Huang, W. Soedel, On the forced vibration of simply supported rotating cylindrical shells, J. Acoust. Soc. Am. **84**(1), 275-285 (1988)
22. C. Gonzalez Diaz, P. Kindt, J. Middelberg, S. Vercammen, C. Thiry, R. Close, J. Leyssens, Dynamic behaviour of a rolling tyre: Experimental and numerical analyses, J. Sound Vib. **364**, 147–164 (2016)
23. J. Lee, S. Wang, P. Kindt, B. Pluymers, W. Desmet, Identification of the direction and value of the wave length of each mode for a rotating tire using the phase difference method, Mech. Sys. and Signal Process. **68-69**, 292–301 (2016)
24. P. Kindt, C. G. Diaz, S. Vercammen, C. Thiry, J. Middelberg, B. Kimble, J. Leyssens, Effects of rotation on the tire dynamic behavior: Experimental and numerical analyses, Tire Sci. and Tec. **41**(4), 248-261 (2013)
25. W.R. Graham, Modelling the vibration of tyre sidewalls, J. Sound Vib. **332**(21), 5345–5374 (2013)
26. C. Lecomte, W.R. Graham, M. Dale, A shell model for tyre belt vibrations, J. Sound Vib. **329**(10), 1717–1742 (2010)
27. Y.-J. Kim, JS. Bolton, Effects of rotation on the dynamics of a circular cylindrical shell with application to tire vibration. J. Sound Vib. **275**, 605–621 (2004)
28. L.R, Molisani, R.A. Burdisso, D. Tsihlas, A coupled tire structure/acoustic cavity model. Int. J. Solids Struct. **40**, 5125–5138 (2003)
29. K. Forsberg, Influence of boundary conditions on the modal characteristics of thin cylindrical shells, AIAA J. **2**(12), 2150-2157 (1964)
30. V.I. Weingarten, *On the free vibration of thin cylindrical shells* (Aerospace corporation, Systems research and planning division, Report No. TDR.169(3560.30)TN-3, El Segundo, California, 1962)
31. G.B. Warburton, Vibration of thin cylindrical shells, J. Mech. Eng. Sci. **7**(4), 399-407 (1965)
32. H. Chung, Free vibration analysis of circular cylindrical shells, J. Sound Vib. **74**, 331–350 (1981)
33. S. Sun, S. Chu, D. Cao, Vibration characteristics of thin rotating cylindrical shells with various boundary conditions, J. Sound Vib. **331**, 4170–4186 (2012)
34. S. Sun, D. Cao, Q. Han, Vibration studies of rotating cylindrical shells with arbitrary edges using characteristic orthogonal polynomials in the Rayleigh–Ritz method, Int. J. of Mec. Sci. **68**, 180–189 (2013)
35. N. Alujević, N. Campillo-Davo, P. Kindt, W. Desmet, B. Pluymers, S. Vercammen, Analytical solution for free vibrations of rotating cylindrical shells having free boundary conditions. Eng. Str. **132**, 152-171 (2017)
36. N. Alujević, N. Campillo-Davo, P. Kindt, W. Desmet, B. Pluymers, S. Vercammen, A simplified tire model based on a rotating shell, *Proceedings of the 4th International Tyre Colloquium, University of Surrey, Surrey* (2015)
37. N. Alujević, N. Campillo-Davo, P. Kindt, W. Desmet, B. Pluymers, S. Vercammen, A simplified model of a rotating tire using cylindrical shells with free ends supported by an elastic foundation, *Proceedings of ISMA2014, Katholieke Universiteit Leuven, Leuven* (2014)
38. Y.K. Cheung, *Finite Strip Method in Structural Analysis* (Pergamon Press, Oxford, 1976)
39. I. Senjanović, I. Ćatipović, N. Alujević, N. Vladimir, D. Čakmak, A finite strip for the vibration analysis of rotating cylindrical shells, Thin-Wall. Struct. (accepted).

Experimental and numerical investigation of eigenfrequencies of rectangular plates, interacting with a fluid

Sergey Bochkarev[1,*], *Alexander* Kamenskikh[2], and *Sergey* Lekomtsev[1]

[1] *Institute of Continuous Media Mechanics RAS, 1 Acad. Korolev str., Perm, 614013, Russian Federation*
[2] *Perm National Research Polytechnic University, 29 Komsomolsky ave., Perm, 614990, Russian Federation*

Abstract. The method of laser vibrometry was applied to investigate the vibrations of elastic plates interacting with a liquid for different variants of plate clamping. The obtained results are used to verify the developed finite-element algorithm designed to solve the problem of spatial vibrations of one or two parallel rectangular plates interacting with a quiescent viscous fluid. The results of calculations of the natural frequencies and damping decrement obtained under different boundary conditions and the height of the fluid layer are discussed. It has been demonstrated that for the configurations considered, viscosity of the liquid has a significant effect on the damping decrement, in contrast to frequencies of vibrations.

1 Introduction

Interest in the problems of natural vibrations of plates interacting with a fluid is due to the availability of various technical applications with strict requirements for safety and performance characteristics of structures. Typical examples include trash racks of hydroelectric plants [1] or parallel-plate fuel assemblies of power plants. Recent literature includes numerous papers, in which the influence of various factors on the natural vibrational frequencies of plates interacting with a fluid is analyzed. Detailed chronology and review of relevant papers are presented in articles [2, 3].

Experimental studies on this topic are represented by a limited number of works. In particular, horizontal circular plates completely immersed in a liquid are studied in [2], and rectangular ones — in [3, 4]. One of the main results of these studies is the conclusion that natural frequencies of vibrations tend to an asymptotic value when the characteristic size of the structure exceeds the height of the fluid layer. The setup developed in [3] allows one to change the position of the plate and to examine partially submerged vertical plates. In the recently published paper [5], the natural frequencies and mode shapes of a horizontal plate, overlaid with a fluid layer have been investigated theoretically and experimentally. Here, measurements were made in two ways. In the first case, a sensor based on polyvinylidene difluoride (PVDF) was used, and in the second case, measurements were taken using the amplitude-fluctuation electronic speckle pattern interferometry (AF-ESPI) method.

In the dynamics of thin-walled structures, along with the frequencies and modes of vibrations, it is of practical interest to study the damping characteristics of the system. In the case of hydroelastic interaction, they can also be determined by viscosity properties of the liquid. In the majority of works, the authors used model of an ideal fluid, in which damping was not taken into account. In this connection, the development of a numerical algorithm for solving the problem of spatial vibrations of plates interacting with a quiescent viscous fluid, and its verification with the help of experiment, are the problems of direct interest.

2 Mathematical formulation

We consider a prismatic tank of length a, width b, and height H, containing fluid. It has perfectly rigid walls and is bounded from below and above (or only from above) by elastic rectangular plates of thickness h. Let the system be set in motion under the action of some perturbations. First, we formulate the constitutive relations for the case of one plate.

The vortex-free perturbed motion of a viscous fluid is described in the framework of the acoustic approximation in terms of the perturbed velocity potential ϕ [6]

$$\frac{\partial^2\phi}{\partial t^2} = \left(c^2 + \frac{4\nu}{3}\frac{\partial}{\partial t}\right)\nabla^2\phi, \qquad (1)$$

where c, ν are the sound velocity and kinematic viscosity of the fluid.

The components of the stress tensor of the fluid are determined by relation [6]

$$\sigma_{ij}^f = -\left(-\rho_f\frac{\partial\phi}{\partial t} + 2\mu\nabla^2\phi\right)\delta_{ij} + 2\mu\frac{\partial^2\phi}{\partial x_i\partial x_j}, \qquad (2)$$

where μ and ρ_f are the coefficients of viscosity and fluid density, δ_{ij} is the Kronecker symbol, i, j take the values x, y, z.

* e-mail: bochkarev@icmm.ru

On the boundary of the elastic plate and fluid S_σ the condition of equality of velocities and surface forces are prescribed

$$\dot{\mathbf{u}} = \mathbf{v}, \tag{3}$$

$$\sigma^f_{ij} n_j = -\sigma^s_{ij} n_j, \tag{4}$$

where $\mathbf{u} = \{u, v, w\}$, $\mathbf{v} = \{v_x, v_y, v_z\}$ are the vectors of the plate displacement and fluid velocity, u, v, w are the displacements of points at the midsurface of the plate in the direction of the corresponding axes of the coordinate system (x, y, z), σ^s_{ij}, σ^f_{ij} are the components of stress tensors of the plate and the fluid, n_j are the vector components of the unit normal $\mathbf{n} = \{n_x, n_y, n_z\}$ to the surface S_σ.

In the case of small displacements, the vector of the normal $\mathbf{n}$ to the surface S_σ slightly deviates from the Oz-axis of the Cartesian coordinate system. Then, the boundary conditions (3), (4), with account of (2) take the following form:

$$\frac{\partial \phi}{\partial x} = \dot{u}, \quad \frac{\partial \phi}{\partial y} = \dot{v}, \quad \frac{\partial \phi}{\partial z} = \dot{w}. \tag{5}$$

$$t_x = -2\mu \frac{\partial^2 \phi}{\partial x \partial z}, \quad t_y = -2\mu \frac{\partial^2 \phi}{\partial y \partial z}, \tag{6}$$

$$t_z = -\rho_f \frac{\partial \phi}{\partial t} + 2\mu \left(\nabla^2 \phi - \frac{\partial^2 \phi}{\partial z^2} \right).$$

Here $\mathbf{t} = \{t_x, t_y, t_z\}$ is the vector of surface forces of the plate.

At the interface between the fluid and the rigid wall S_w we set the following condition

$$v_n = \partial \phi / \partial n = 0. \tag{7}$$

The small deformations of the elastic plate resulting from the hydroelastic interaction are determined by the first-order shear deformation plate theory [7]

$$\boldsymbol{\varepsilon} = \{\varepsilon_{xx}, \varepsilon_{yy}, \gamma_{xy}, \gamma_{xz}, \gamma_{yz}\}^{\mathrm{T}} = \tilde{\boldsymbol{\varepsilon}} + z\tilde{\mathbf{k}}, \tag{8}$$

$$\tilde{\boldsymbol{\varepsilon}} = \left\{ \frac{\partial u}{\partial x}, \frac{\partial v}{\partial y}, \frac{\partial u}{\partial y} + \frac{\partial v}{\partial x}, \frac{\partial w}{\partial x} + \theta_x, \frac{\partial w}{\partial y} + \theta_y \right\}^{\mathrm{T}},$$

$$\tilde{\mathbf{k}} = \left\{ \frac{\partial \theta_x}{\partial x}, \frac{\partial \theta_y}{\partial y}, \frac{\partial \theta_x}{\partial y} + \frac{\partial \theta_y}{\partial x}, 0, 0 \right\}^{\mathrm{T}},$$

where θ_x and θ_y are the angles of rotation about the y and x -axes.

The physical equation that establishes the relation between the force and moment vector $\mathbf{T}$ and the strain vector $\boldsymbol{\varepsilon}$, is written in the matrix form as

$$\mathbf{T} = \{N_{xx}, N_{yy}, N_{xy}, M_{xx}, M_{yy}, M_{xy}, Q_x, Q_y\}^{\mathrm{T}} = \mathbf{D}\boldsymbol{\varepsilon}. \tag{9}$$

The elements of the matrix $\mathbf{D}$ for isotropic material are determined in terms of the elastic modulus E, the Poisson's ratio ν and the shear modulus in the known manner.

A mathematical formulation of the problem of plate dynamics is based on the variational principle of virtual displacements, which includes terms that take into account the work of hydrodynamic pressure and inertia forces. It can be written in the matrix form as

$$\int_{S_s} \delta \boldsymbol{\varepsilon}^{\mathrm{T}} \mathbf{D} \boldsymbol{\varepsilon} dS + \int_{V_s} \rho_s \delta \mathbf{d}^{\mathrm{T}} \ddot{\mathbf{d}} dV - \int_{S_\sigma} \delta \mathbf{d}^{\mathrm{T}} \mathbf{p} dS = 0, \tag{10}$$

$$\mathbf{p} = \{t_x, t_y, t_z, 0, 0\}^{\mathrm{T}},$$

where ρ_s is the density of the plate, $\mathbf{d}$ and $\mathbf{p}$ are the vectors of the generalized displacements and surface loads, S_s is the surface that bounds the volume of the plate V_s. Dots signify the first and second time derivatives.

3 Numerical formulation

When solving the problem of fluid dynamics by the Bubnov–Galerkin method, the formulation involving the wave equation (1) and boundary conditions (5), (7), is transformed to a weak form. The subsequent application of the standard procedures of the finite element method (FEM) leads to a matrix expression

$$\mathbf{M}_f \ddot{\mathbf{f}} + \mathbf{C}^v_f \dot{\mathbf{f}} + \mathbf{K}_f \mathbf{f} + \mathbf{M}^v_{fs} \ddot{\mathbf{w}} + \mathbf{C}_{fs} \dot{\mathbf{w}} = 0, \tag{11}$$

$$\mathbf{M}_f = \sum_{m_f} \int_{V_f} \frac{\mathbf{F}^{\mathrm{T}} \mathbf{F}}{c^2} dV, \; \mathbf{C}^v_f = \sum_{m_f} \int_{V_f} \frac{4\nu \Phi}{3c^2} dV,$$

$$\mathbf{K}_f = \sum_{m_f} \int_{V_f} \Phi dV, \; \mathbf{M}^v_{fs} = -\sum_{m_\sigma} \int_{S_\sigma} \frac{4\nu \mathbf{F}^{\mathrm{T}} \overline{\mathbf{N}}}{3c^2} dS,$$

$$\mathbf{C}_{fs} = -\sum_{m_\sigma} \int_{S_\sigma} \mathbf{F}^{\mathrm{T}} \overline{\mathbf{N}} dS,$$

$$\Phi = \frac{\partial \mathbf{F}^{\mathrm{T}}}{\partial x} \frac{\partial \mathbf{F}}{\partial x} + \frac{\partial \mathbf{F}^{\mathrm{T}}}{\partial y} \frac{\partial \mathbf{F}}{\partial y} + \frac{\partial \mathbf{F}^{\mathrm{T}}}{\partial z} \frac{\partial \mathbf{F}}{\partial z}.$$

Here m_f, m_σ is the number of finite elements, which are used to decompose the fluid domain V_f and the boundary S_σ; $\mathbf{f}$ is the vector of nodal values of the perturbation velocity potential ϕ; $\mathbf{F}, \overline{\mathbf{N}}$ are the matrices of the shape functions of the perturbation velocity potential and the normal component w of the displacement vector of the plate.

Applying the known procedures of the finite element method to the variational equation (10) taking into account (6), (8), (9), we obtain the following matrix relation

$$\mathbf{M}_s \ddot{\mathbf{d}} + \mathbf{K}_s \mathbf{d} + \mathbf{C}_{sf} \dot{\mathbf{f}} + \mathbf{K}^v_{sf} \mathbf{f} = 0, \tag{12}$$

$$\mathbf{M}_s = \sum_{m_s} \int_{V_s} \rho_s \mathbf{N}^{\mathrm{T}} \mathbf{N} dV, \mathbf{K}_s = \sum_{m_s} \int_{S_s} \mathbf{B}^{\mathrm{T}} \mathbf{D} \mathbf{B} dS,$$

$$\mathbf{C}_{sf} = \sum_{m_\sigma} \int_{S_\sigma} \rho_f \mathbf{F}^{\mathrm{T}} \overline{\mathbf{N}} dS, \mathbf{K}^v_{sf} = \sum_{m_\sigma} \int_{S_\sigma} 2\mu \mathbf{H} dS.$$

Here, m_s is the number of finite elements used to decompose the surface of the plate; $\mathbf{N}$ is the matrix of the shape function of the finite element, $\mathbf{B}$ is the matrix of the strain-nodal displacement relationship, $\mathbf{H}$ is the matrix of the nodal displacement-velocity potential relationship.

The problem of studying the dynamic behavior of elastic plate interacting with a compressible viscous fluid is reduced to a simultaneous solution of the systems of equations (11) and (12). We consider the perturbed motion of the structure and fluid written as $(\mathbf{d}, \mathbf{f}) = (\bar{\mathbf{d}}, \bar{\mathbf{f}}) \exp(\lambda t)$,

 https://doi.org/10.1051/matecconf/201814807002

where $\bar{\mathbf{d}}$ and $\bar{\mathbf{f}}$ are the vector functions of the coordinates, and $\lambda = \delta + i\omega$ the characteristic exponent. Here, ω and δ have the following physical meaning: ω is the natural frequency, and δ is damping. As a result, we obtain a generalized eigenvalue problem defined as

$$\begin{bmatrix} \mathbf{C} & \mathbf{K}+\mathbf{A} \\ -\mathbf{I} & 0 \end{bmatrix} \begin{Bmatrix} \lambda\mathbf{x} \\ \mathbf{x} \end{Bmatrix} + \lambda \begin{bmatrix} \mathbf{M} & 0 \\ 0 & \mathbf{I} \end{bmatrix} \begin{Bmatrix} \lambda\mathbf{x} \\ \mathbf{x} \end{Bmatrix} = 0, \quad (13)$$

where $\mathbf{I}$ is the unit matrix, $\mathbf{x} = \{\bar{\mathbf{d}}, \bar{\mathbf{f}}\}^T$.

The complex eigenvalues of the coupled system of two equations (13) are calculated using an algorithm, which is based on the implicitly restarted Arnoldi method [8].

The model described above and the finite-element algorithm of its numerical implementation can easily be generalized to a configuration that is the structure comprising several parallel plates. Equation (10) is written for each elastic body, and expressions (1), (5), (7) — for each volume of the fluid. In the case of two plates, we obtain

$$\mathbf{M}\begin{Bmatrix} \ddot{\mathbf{d}}_1 \\ \ddot{\mathbf{f}} \\ \ddot{\mathbf{d}}_2 \end{Bmatrix} + \mathbf{C}\begin{Bmatrix} \dot{\mathbf{d}}_1 \\ \dot{\mathbf{f}} \\ \dot{\mathbf{d}}_2 \end{Bmatrix} + \mathbf{K}\begin{Bmatrix} \mathbf{d}_1 \\ \mathbf{f} \\ \mathbf{d}_2 \end{Bmatrix} = 0, \quad (14)$$

$$\mathbf{M} = \begin{bmatrix} \mathbf{M}_{s_1} & 0 & 0 \\ \mathbf{M}^v_{fs_1} & \mathbf{M}_f & \mathbf{M}^v_{fs_2} \\ 0 & 0 & \mathbf{M}_{s_2} \end{bmatrix}, \quad \mathbf{C} = \begin{bmatrix} 0 & \mathbf{C}_{sf_1} & 0 \\ \mathbf{C}_{fs_1} & \mathbf{C}^v_f & \mathbf{C}_{fs_2} \\ 0 & \mathbf{C}_{sf_2} & 0 \end{bmatrix},$$

$$\mathbf{K} = \begin{bmatrix} \mathbf{K}_{s_1} & \mathbf{K}^v_{sf_1} & 0 \\ 0 & \mathbf{K}_f & 0 \\ 0 & \mathbf{K}^v_{sf_2} & \mathbf{K}_{s_2} \end{bmatrix}.$$

Here the subindices «1» and «2» are used to identify variables belonging to the lower and upper plate, respectively. The eigenvalue problem will take the form of (13) up to the notation used.

4 Experiment

4.1 Experiment setup

The experimental setup is a demountable prismatic tank, the top cover (or part of it) of which is a plate of length a and width b. Its thickness h is an order of magnitude smaller than the thickness of the walls and the bottom of the container. The structure of the tank is shown in figure 1.

In the process of assembly, the tank is first filled with a fluid such that its level is slightly higher than the plate location. Then, a plate is dropped in special grooves cut in the side walls of the tank and additionally fixed by means of guide pins and demountable clamping frame. Under the weight of the plate, excess fluid is squeezed out through the technological holes or pumped out. This experimental setup provides the possibility for clamping or leaving free any of the four edges of the plate. This allows one to analyze various options for clamping, but requires the manufacture of test specimens, the dimensions of which correspond to the distance between the inner walls of the tank.

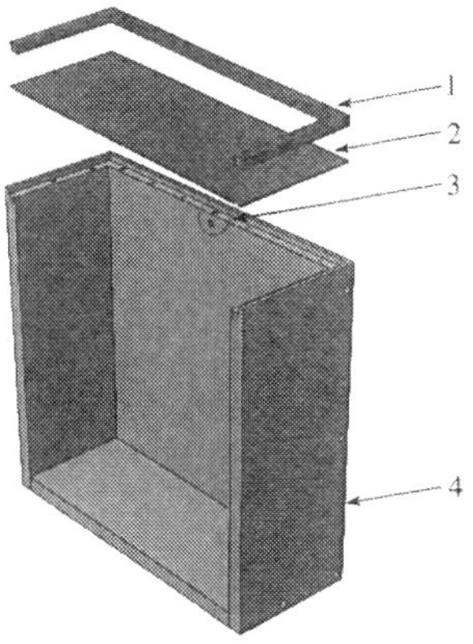

Figure 1. Experimental setup (in section): 1 — clamping frame; 2 — plate; 3 — technological hole; 4 — demountable tank

Table 1. Geometrical dimensions and boundary conditions for the plates.

N	h, m	a, m	b, m	BC
1	1.00×10^{-3}	20.00×10^{-2}	14.90×10^{-2}	CFCF
2	7.08×10^{-4}	20.00×10^{-2}	28.15×10^{-3}	CFCF
3	7.50×10^{-4}	14.10×10^{-2}	19.15×10^{-3}	CFFF

Natural frequencies of the structure were investigated using a laser Doppler vibrometry method. As a source of excitation we used a speaker, which received a signal from computer's sound card. The duration of excitation was 12 seconds, during which the frequency varied from 1 to 2000 Hz. The readings were taken by the Polytec PDV-100 digital laser vibrometer. The control point, at which the measurements were taken, was marked on the surface of the plate. The device was installed at a distance of 510 mm above the experimental setup and was oriented so that the angle between the laser ray and the surface of the plate was 90°. Such an arrangement made it possible to measure the normal component of the velocity vector $\dot{w}$. The natural vibration frequencies were determined from the Fourier transform analysis of the signal (corresponding to significant peaks). All test specimens were examined 30 times. Before each new measurement, the fluid was pumped out and the plate was demounted.

4.2 Experiment results

Geometrical parameters of the test plates are given in table 1. Here, a and b should be understood to mean dimensions that are not involved in clamping. In fact, the length and width of the samples is slightly larger. For the kinematic boundary conditions specified at the plate edges (BC) we introduce the following notation: F corresponds to the free edge and C — to rigid clamping. In designation of the specific boundary condition letter symbols are given in the clockwise direction starting from the left edge. For example, the variant CFFF corresponds to a cantilever plate clamped only at the left edge.

Table 2. Experimentally obtained values of natural frequency (Hz) of the plate interacting with air and fluid.

Sample		ω_1	ω_2	ω_3	ω_4	ω_5
1	Air	230.4 ±0.5	257.1 ±0.5	359.3 ±0.4	591.5 ±0.3	667.0 ±0.8
	Fluid	102.2 ±0.5	184.1 ±0.3	267.8 ±0.1	312.5 ±0.4	333.0 ±0.3
2	Air	92.5 ±1.3	255.0 ±2.2	417.8 ±0.5	502.4 ±3.5	833.1 ±5.4
	Fluid	52.9 ±10.9	150.7 ±11.2	304.5 ±11.9	310.0 ±11.3	520.7 ±13.3
3	Air	32.3 ±0.2	202.7 ±1.3	463.1 ±0.7	568.7 ±3.5	1115.9 ±6.6
	Fluid	22.4 ±0.7	138.0 ±0.8	374.0 ±1.5	393.3 ±2.8	786.8 ±5.2

The average values of the five lowest natural frequencies of sample vibrations are given in table 2. In each cell of the table the numbers in the second line show the value of the mean square deviation s, obtained from 30 measurements:

$$s = \sqrt{\frac{\sum_{i=1}^{30} (\omega_n - \omega)^2}{n-1}}, \tag{15}$$

where ω is the arithmetic mean value.

The results demonstrate that when the plate interacts with a fluid, the frequency spectrum is significantly reduced. The frequency ω_1 of each of the test specimens decreases by 55.6%, 42.8% and 30.6%, respectively. Note that the maximum difference corresponds to the plate with the largest contact area (specimen 1).

5 Numerical results

Verification of the developed finite-element algorithm was carried out by comparing the calculated natural oscillation frequencies with the experimental values given in Section 4.2. The calculations were made using the following characteristics: E = 201 GPa, ρ_s = 7850 kg/m^3, ν = 0.30, ρ_f = 1000 kg/m^3, $\mu = 8.94\times10^{-4}$ Pa s, c = 1500 m/s. The obtained results are presented in table 3. The numbers in the second line in each cell of the table indicate the relative error compared to the experimental data.

The computational scheme applied for samples 2 and 3 did not take into account the gravitational waves excited at the free surface of the fluid under vibration. The corresponding boundary condition used in the model is represent as $\phi = 0$. Despite such an assumption, the obtained numerical results agree with the experimental data within the relative accuracy of 6%.

To reveal new relationships determining the dynamic characteristics of plates interacting with a viscous fluid, a series of computational experiments was carried out. The calculations were performed for the following geometric dimensions and physicomechanical characteristics: a = 0.24 m, b = 0.32 m, $h = 2\times10^{-3}$ m, $H = 4\times10^{-2}$ m, E = 69 GPa, ν = 0.30, ρ_s = 2700 kg/m^3, ρ_f = 1000 kg/m^3, $\mu = 8.94\times10^{-4}$ Pa s, c = 1500 m/s.

Table 3. Comparison of natural frequencies (Hz) of plates interacting with air and fluid.

Sample		ω_1	ω_2	ω_3	ω_4	ω_5
1	Air	243.7 -5.8%	270.5 -5.2%	373.9 -4.1%	587.0 0.8%	672.1 -0.8%
	Fluid	97.8 4.3%	174.3 5.3%	288.8 -8.6%	313.5 -0.3%	327.7 1.6%
2	Air	93.6 -1.2%	257.9 -1.1%	422.6 -1.2%	506.0 -0.7%	837.9 -0.6%
	Fluid	55.0 -3.9%	155.7 -3.3%	313.4 -2.9%	325.5 -5.0%	533.4 -2.4%
3	Air	31.6 2.2%	197.6 2.5%	448.5 3.1%	553.8 2.6%	1087.3 2.6%
	Fluid	22.4 0.1%	140.9 -2.1%	383.2 -2.4%	400.2 -1.7%	788.3 -0.2%

The investigation of the natural vibrations of a plate interacting with an ideal fluid was carried out in works [9, 10]. The results obtained with a new model confirm the well-known fact that viscosity of a liquid has practically no effect on the vibrational frequency. In this regard, the curves shown in figure 2a, a completely coincide with those published in [9]. The dependence of the lowest natural frequency of a single plate on the height of the fluid layer is asymptotic if the ratio $H/b > 0.75$. However, for a parameter δ, characterizing damping of the system, such a tendency manifests itself to a lesser extent and is determined by clamping conditions (see figure 2b).

The analysis of two parallel plates showed that the dependence of any fixed frequency in the spectrum and its damping on the ratio H/b is strongly nonmonotonic in nature and may have kinks due to a change in the vibration modes [10]. To illustrate this fact, we present figure 3a, which shows that the two lowest vibration modes $(1, 0)$ and $(0, 0)$ correspond to the lowest natural frequency ω_1 in the examined interval H/b. Here, (n, m) denotes the number of nodal lines in the direction of the $Ox-$ and $Oy-$ axes. Figure 3b demonstrates the response of the parameter δ to a change of the vibration mode in the vicinity of point $H/b \approx 0.265$. It is seen that at this instant of time an abrupt change of damping occurs.

6 Conclusions

Using the developed test setup, we carried out an experimental study of the vibration frequencies of a plate in contact with a fluid for different types of fixing.

For the developed mathematical formulation of the problem of spatial vibrations of plates interacting with a viscous fluid, we constructed a finite-element algorithm. Its reliability was verified by comparing the obtained results with experimental data. The calculations made it possible to conclude for two parallel plates that the dependence of any fixed frequency in the spectrum and its damping on the height of the liquid layer is of strongly pronounced nonmonotonic character and may have kinks due to a change in vibration mode.

The work is supported by the Russian Foundation for Basic Research grant (project N 16-41-590646).

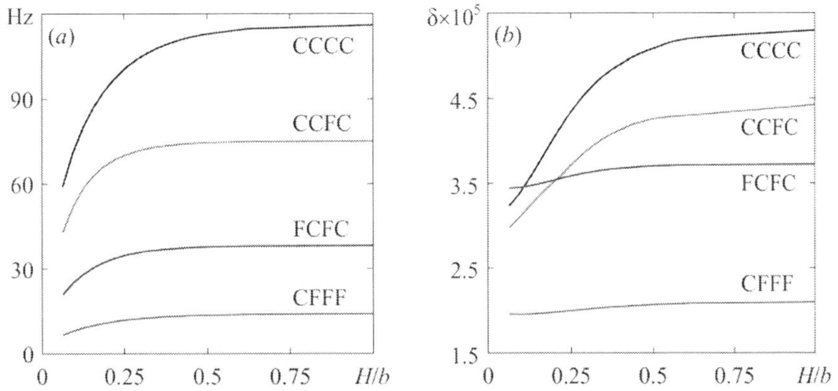

Figure 2. Plots of lowest natural frequencies of the vibration ω (a) and damping δ (b) versus the height of the fluid layer H for the plate at different variants of boundary conditions

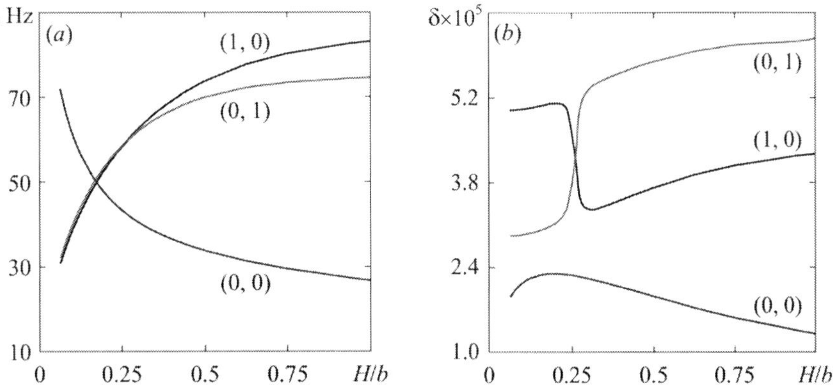

Figure 3. Natural frequencies of vibration ω (a) and damping δ (b) versus the height of the fluid layer (CCFC), for two plates

References

[1] X. Huang, C. Valero, E. Egusquiza, A. Presas, A. Guardo. Computers & Fluids. **71**, 54–64 (2013)

[2] E. Askari, K.-H. Jeong, M. Amabili. J. Sound Vib. **332**, 3064–3085 (2013)

[3] U.S. Lindholm, D.D. Kana, W.H. Chu et al. J. Ship Res. **9**, 11–22 (1965)

[4] M.R. Haddara, S. Cao. Mar. Struct. **9**, 913–933 (1996)

[5] C.-Y. Liao, Y.-C. Wu, C.-Y. Chang, C.-C. Ma. J. Sound Vib. **394**, 545–574 (2017)

[6] D.D. Joseph. J. Fluid Mech. **479**, 191–197 (2003)

[7] J.N. Reddy. *An introduction to nonlinear finite element analysis*, 2nd edn. (Oxford University Press, Oxford, UK, 2015)

[8] R.B. Lehoucq, D.C. Sorensen. SIAM J. Matrix Anal. Appl. **17**, 789–821 (1996)

[9] S.A. Bochkarev, S.V. Lekomtsev, V.P. Matveenko. Fluid Dyn. **51**, 821–833 (2016)

[10] S.A. Bochkarev, S.V. Lekomtsev. Solid State Phenomena. **243**, 51–58 (2016)

Size effects on free vibration of heterogeneous beams

Bahman Hassanati[1,*], *Marcus Wheel*[1]

[1]Mechanical & Aerospace Engineering Department, University of Strathclyde, 75 Montrose Street, Glasgow G1 1XJ, Scotland, UK

Abstract. In this paper the influence of microstructure on the free vibration of geometrically similar heterogeneous beams with free-free boundary conditions was numerically investigated by detailed finite element analysis (FEA) to identify and quantify any effect of beam size on transverse modal frequencies when the microstructural scale is comparable to the overall size. ANSYS Mechanical APDL was used to generate specific unit cells at the microstructural scale comprised of two isotropic materials with different material properties. Unit cell variants containing voids and inclusions were considered. At the macroscopic scale, four beam sizes consisting of one, two, three or four layers of defined unit cells were represented by repeatedly regenerating the unit cell as necessary. In all four beam sizes the aspect ratio was kept constant. Changes to the volume fractions of each material were introduced while keeping the homogenized properties of the beam fixed. The influence of the beam surface morphology on the results was also investigated. The ANSYS results were compared with the analytical results from solution to Timoshenko beam and nonlocal Timoshenko beam as well as numerical results for a Micropolar beam. In nonlocal Timoshenko beams the Eringen's small length scale coefficients were estimated for some of the studied models. Numerical analyses based on Micropolar theory were carried out to study the modal frequencies and a method was suggested to estimate characteristic length in bending and coupling number via transverse vibration which verifies the use of Micropolar elasticity theory in dynamic analysis.

1 Introduction

In recent years the progress in technologies such as aerospace, biomedical, nanotechnology etc. have demanded the need for the application of small scale structures and that has created a whole new era for researchers to investigate the dynamic behaviour of structures where the classical theories of elasticity become increasingly invalid to use in cases such as small scale heterogeneous beams. Heterogeneity in general is known as a discontinuity of physical properties of material in either a specific direction or multidirectionally. In the literature generally the homogenisation methods are sought to represent the properties of materials with a selected unit cell. Rabboh et al used the rule of mixture to calculate the elastic constants and Poisson's ratio for functionally graded material sandwich beam and investigated the effect of functionally graded material (FGM) on the beam's dynamic performance[1]. Della and Shu used Eshelby's equivalent inclusion method to investigate the vibration of piezoelectric beams and their analytically obtained results also indicate that the size effect due to the size of piezoelectric inclusions, their location in the beam and their volume fraction influence their dynamic behaviour[2].

Homogenisation methods become severely problematic where the size of constituent materials such as inclusions and/or voids becomes comparable to the overall size of the specimen. Modifications to classical elasticity theories are only useful as long as the internal length scale parameters are considered very small. The size dependent characteristics and behaviour of materials have been reported by many researches such as in the work by Groh & Weaver[3], Gherlone[4], and Schulze et al in laminated beams[5]. Alghamdi and Dasgupta's results in modelling active damping of adaptive structures mostly shows how the beam's time to decay changes by changing the device aspect ratio, inclusion shape, location, and volume fraction, and also how changes in host stiffness results in changes in time to decay and electrical field[6]. Timoshenko beam theory is widely used by researches as it is able to include rotary inertia and shear deformation and therefore it is considered nonlocal if Eringen's small scale effect is incorporated in the governing equations[7]. C M Wang et al used finite segments method to calibrate Eringen's small length scale coefficient for initially stressed vibrating nonlocal beams and stated that 'e_0 does not depend on buckling or vibration modes' [8] . In our FE results, however, we observed that in such cases that $\alpha = (e_0 a)$ is not size independent and therefore it cannot be accepted as a unique property of the material. Available results on the influence of size effect on the behaviour of heterogeneous materials reported by researches show deviation from elastic theories in static loading cases when the beam or plate L/h ratio reduces [9]–[15]. Nakamura and Lakes used a two dimensional FE method

* Corresponding author: bahman.hassanati@strath.ac.uk

and studied the localised end loads on a strip sample and concluded that as the characteristic length increases, the rate of decay of stress and strain energy reduces. Micropolar theory, however, takes into account couple stresses and thus requires a characteristic length constitutive parameter and an additional degree of freedom, a micro rotation. Micropolar theory by nature incorporates the size effect in the material. A 2D Micropolar strip loaded at one end was investigated by Nakamura & Lake [16] and the influences of elastic constants especially coupling number and characteristic length are investigated. They concluded that for a very small characteristic length (in comparison with the strip's width), the rate of stress/or strain energy decreases as the characteristic length increases. In the dynamic case this may be shown by wave dispersion. Their work predominantly includes studying the models for various characteristic length and coupling number, N, and provides no method to determine them. However in another work by Lakes, a comprehensive comparison has been made between various theories such as Eringen's nonlocal theory and Micropolar theory[17]. This indicates that the elastic constants can be determined via a dynamic wave propagation method. Nevertheless, in any method such as a size effect based method, field method and/or wave method, there are limitations with regard to the smallest characteristic length. Also, special attention must be given to the coupling number, N, near the limits of zero and one while performing numerical analysis otherwise numerical errors in computation may result.

Wheel et al studied size effects in heterogeneous beams in static cases when loaded in 3 point bending [16]. They investigated size effects in beams with voids and showed that sample stiffness relates to the sample size, as measured by the reciprocal of its depth squared, in a linear manner. They also reported that there are both negative and positive effects of beam size depending on the beam boundary morphology [15]. Waseem et al also investigated the influence of void size on the constitutive properties of a perforated ring model and by relating the diametrical loads, displacement and strain energy derived the final equation relating the stiffness to the specimen size[14]. In the case of smooth specimen surface (rings circumference), the stiffness changes linearly with sample size measure. McGregor provides the same conclusions[18]. Beveridge et al investigated the Micropolar behaviour of perforated beams in 3 point bending and by using a control volume finite element method and iteration to model static 3 point bending test results determined the coupling number for the models[10].

2 Finite Element Analyses

2.1. FE modelling

Three types of 2D beams were modelled: perforated beams (beams with voids), beams with compliant inclusions and beams with a compliant matrix. Modal analysis was carried out using ANSYS APDL version 16.2. Unit cell variants containing voids and inclusions were considered. The height and length of unit cells are 0.866 and 1mm. Parametric APDL codes were written to regenerate models for various voids/inclusions volume fractions.

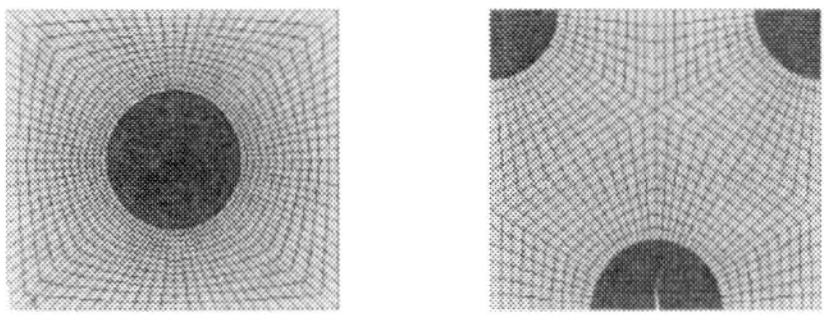

Figure 1: Unit cell consisting of two isotropic materials r=0.2 mm, cell on the right is for when inclusions intercept the beams surface.

At the macroscopic scale, unit cells were repeatedly regenerated to produce four beam sizes consisting of one, two, three or four layers of cells while keeping the aspect ratio constant at 10.4:1. Changes to the volume fraction of each material were introduced while keeping the homogenized properties of the beam fixed. This method is meant to account for studying the size effect of free vibration for various volume fractions and beam sizes while keeping the mass of the unit cell and beams overall properties constant. The aim is to study frequency changes for various beam sizes and void/or inclusions volume fractions for lateral vibration modes.

2.2. Size effect

In figure 2, the normalised non-dimensional frequencies for the first 10 transverse modes based on the reciprocal measure of 4 beam sizes are shown. The blue lines (dotted lines with squared markers) represent the homogeneous case and the results show that in this case the modal frequency λ is size independent. The size effect is greatest for beams which are smaller. It is also evident that the size effect is mode dependent as there is a shift in the nature of size effect, after mode 3 changes in void radius causes a decrease rather than an increase in normalised frequencies. In figure 3, it is shown that beams with compliant inclusions show similar behaviour.

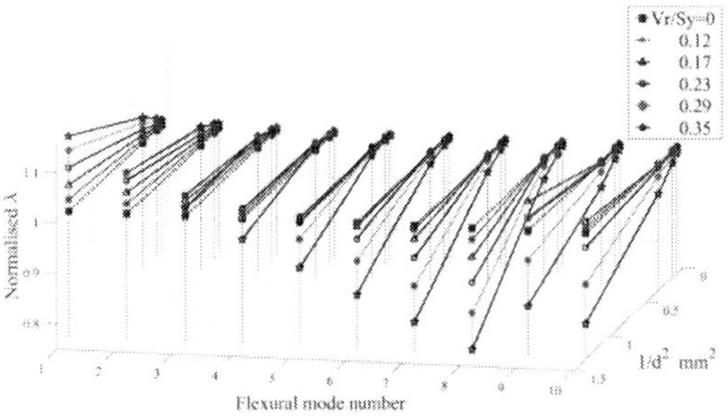

Figure 2: Normalised flexural modal frequencies of the first 10 modes, four beam sizes for beams with voids and smooth morphology

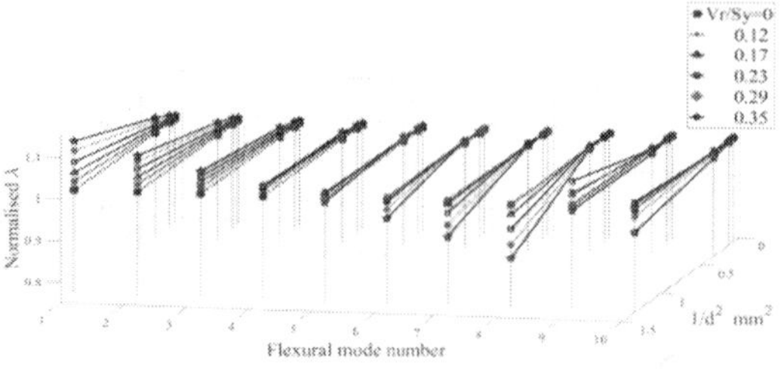

Figure 3: Normalised flexural modal frequencies of the first 10 modes, four beam sizes for smooth surfaced beams with compliant inclusions

From the results we can see that the nature of the size effect changes and NN-λ (Normalised non-dimensional frequency parameter) decreases rather than increases. It has been observed that after some cut-off frequency the FE results increasingly become even more nonlinear and need more research which is beyond the scope of this paper. The results for specimens with stiff inclusions show a different dynamic behaviour, figure 4, and that it displays a totally different size effect which indicates that increasing volume fraction causes a decrease in NN-λ and this rate of change decreases at higher mode numbers. However, for beams with voids and compliant inclusions for which the voids and/or inclusions intercept the surfaces, this rate of change increases although the resulting NN-λ remains below the homogenised case.

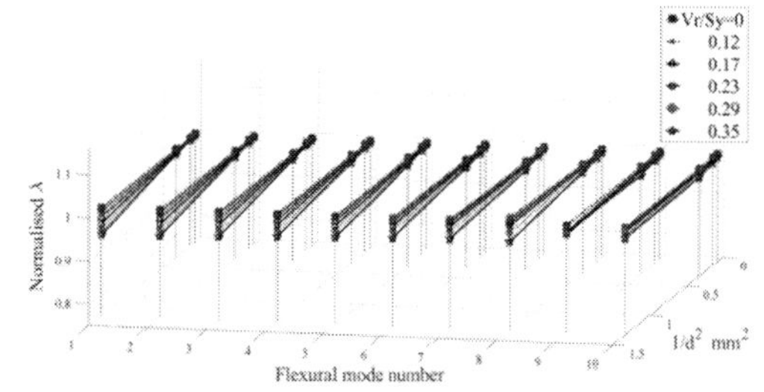

Figure 4: Normalised flexural modal frequencies of the first 10 modes, four beam sizes for smooth surfaced beams with compliant Matrix

3 Nonlocal Timoshenko Beam

The nonlocal Timoshenko governing equations can be obtained by applying Hamilton's principle and incorporated the Eringen's small scale coefficient $\alpha = e_0 a$ into the Timoshenko beam model[7]:

$$EI\frac{d^2\phi}{dx^2} - \kappa GA\left(\phi + \frac{dw}{dx}\right) + \rho I\omega^2\phi - (e_0 a)^2\left(\rho A\omega^2\frac{dw}{dx} + \rho I\omega^2\frac{d^2\phi}{dx^2}\right) = 0 \quad (1)$$

$$\kappa GA\left(\frac{d\phi}{dx} + \frac{d^2w}{dx^2}\right) + \rho A\omega^2 w = 0 \quad (2)$$

where ϕ is the rotation, w is the transverse displacement and ω is the circular frequency. After decoupling equations (1) and (2) and applying free-free boundary conditions, the following equation may be derived:

$$\cosh\beta\cos\gamma + \left(\frac{[(H_2H_3)^2-(H_1H_4)^2]}{2H_1H_2H_3H_4}\right)\sinh\beta\sin\gamma = 1 \quad (3)$$

where:

$$H_1 = \alpha^2\lambda^2 + \left(\frac{\alpha^2\lambda^2}{\xi^2} - 1\right)\beta\Psi_\beta,\ H_2 = \alpha^2\lambda^2 + \left(\frac{\alpha^2\lambda^2}{\xi^2} - 1\right)\gamma\Psi_\gamma$$

$$H_3 = \Psi_\beta + \beta,\ H_4 = \Psi_\gamma + \gamma$$

$$a = \left(1 - \frac{\alpha^2\lambda^2}{\xi^2}\right),\ b = \lambda^2\left(\Omega + - \frac{1-\Omega\alpha^2\lambda^2}{\xi^2} + \alpha^2\right)$$

$$c = \lambda^2\left(\frac{\lambda^2\Omega}{\xi^2} - 1\right)$$

$$\begin{pmatrix}\beta\\ \gamma\end{pmatrix} = \left(\frac{\pm b + \sqrt{b^2 - 4ac}}{2a}\right)^{1/2}$$

$$\Psi_\beta = -\frac{\beta^2+\lambda^2\Omega}{\beta}\ ,\quad \Psi_\gamma = \frac{\gamma^2+\lambda^2\Omega}{\gamma}$$

$$\bar{x} = \frac{x}{L},\ \bar{w} = \frac{w}{L},\ \lambda^2 = \omega^2\frac{\rho AL^4}{EI},$$

in which $\lambda = \omega^{1/2}(\rho AL^4/EI)^{1/4}$ is the frequency parameter, $\Omega = EI/(K_s GAL^2)$ the shear deformation parameter, $\alpha = e_0 a/L$ a scaling effect parameter and finally $\xi = L(A/I)^{1/2}$ is the slenderness ratio.

Equation (3) has been solved by the bisection method and the normalised transverse modal frequencies for the first 38 modes of a beam with aspect ratio L/d=10.4 are provided in figure 5. By comparing these results and the modal frequencies obtained in FE analysis using ANSYS APDL it is possible to obtain Eringen's scale parameter for our heterogeneous models. A direct comparison between the results in figure 5 and the FE results in section 2.2, confirms that it is not always possible to easily determine α for cases similar to our models. However, by applying a constant $(\lambda/\lambda_0)_2 = (\lambda/\lambda_0)_1 - [C1 \times (1 - (\lambda/\lambda_0)_{Mode\_1})]$ and shifting the results below the homogeneous line, it is possible to obtain α for beams with voids and inclusions.

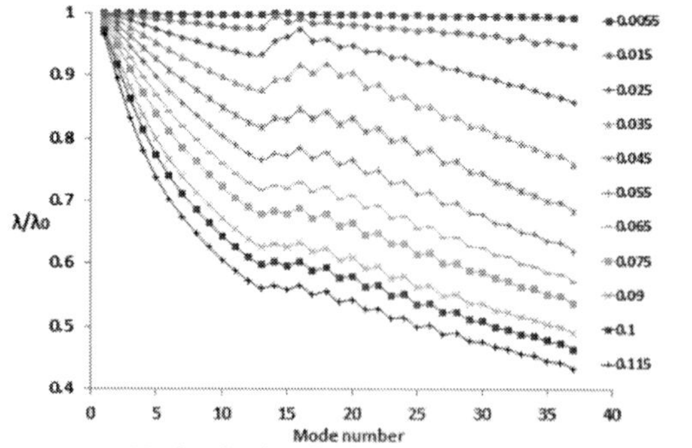

Figure 5: Nonlocal Timoshenko Frequency Parameters for Various α's, F-F BC

The values for α via this curve fitting method show 'α' is not size independent as illustrated in figure 6 therefore it cannot be regarded as a unique property of the material. When voids and inclusions intercept the surfaces, there is no need to consider any constant but the small scale coefficient still showed size dependency.

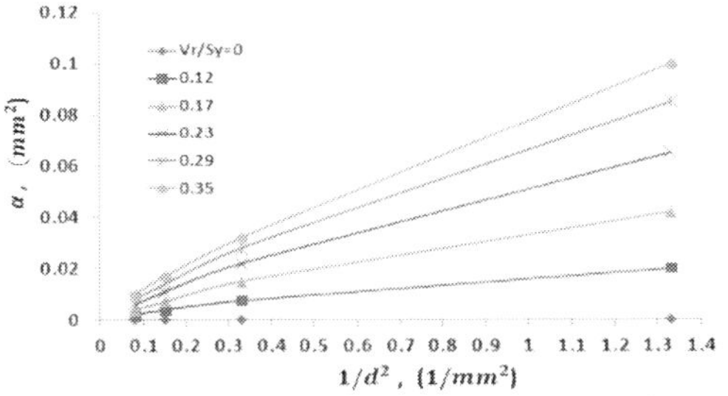

Figure 6: Smooth surfaced perforated beam. Scale Coefficient 'α' (Alpha), Obtained by curve fitting FE results with NLTB

4 Micropolar Beam

4.1 2D Micropolar elastic materials

The stress-strain relations in classical elasticity may be stated as $\tau_{ij}=\lambda\varepsilon_{kk}\,\delta_{ij}+2\mu\varepsilon_{ij}$ where λ and μ are the Lamé constants but in Micropolar elasticity there are four extra elastic constants. In linear, three dimensional, micropolar elasticity the force stresses, τ_{ij}, and couple stresses, m_{ij}, are related to the deformations by [17]:

$$\tau_{ij}=\lambda\varepsilon_{kk}\,\delta_{ij}+(2\mu^*+\kappa)\,\varepsilon_{ij}+\kappa e_{ijk}\,(\theta_k-\emptyset_k) \quad (4)$$

$$m_{ij}=\alpha\emptyset_{k,k}\,\delta_{ij}+\beta\emptyset_{i,j}+\gamma\emptyset_{j,i} \quad (5)$$

where τ is the force stress tensor, ∅ is the micro rotation, θ is the macro rotation and *m* is a length scale dependent couple stress. The strain components, ε, are given in terms of the displacements, *u*, and micro rotations, ∅, by:

$$\varepsilon_{ij}=u_{j,i}+e_{jik}\,\emptyset_k$$

Macro rotation and strain tensors are:

$$\theta_k=(e_{ijk}\,u_{k,j})/2 \quad , \quad \varepsilon_{ij}=(u_{i,j}+u_{j,i})/2$$

where i,j,k=1,2,3, δ is the Kronecker delta and is equal to 1 if i=j otherwise it is zero, e_{ijk} the permutation tensor. For even permutation of of ijk, $e_{ijk}=+1$, for odd permutation of ijk, $e_{ijk}=-1$ and otherwise zero.

There are four additional elastic constants α, β, γ and κ. However, in 2D Micropolar elasticity the number of constants reduces to 4 independent engineering constants:

$$E_m=\frac{(2\mu^*+\kappa)(3\lambda+2\mu^*+\kappa)}{(2\lambda+2\mu^*+\kappa)} \quad , \quad v_m=\frac{\lambda}{(2\lambda+2\mu^*+\kappa)}$$

$$l_b^2=\frac{\gamma}{2(2\mu^*+\kappa)} \quad , \quad N^2=\frac{\kappa}{2(\mu^*+\kappa)}$$

E_m is Micropolar modulus v_m is Micropolar Poisson ratio l_b is a length scale parameter and it should reflect the microstructural scale. N is the coupling number and characterises the shear stress asymmetry. In the dynamic case micro inertia also needs to be included. In 2D finite element methods a mass matrix needed to be introduced in to the equations. The asymmetric components of the shear stress:

$$G_{11}=\mu^*+\kappa \quad , \quad G_{12}=\mu^* \quad , \quad G_{11}=G_{22} \text{ and } G_{12}=G_{21}$$

$$G_{sy}=(G_{11}+G_{12})=2\mu^*+\kappa \; , \; G_{asy}=G_{11}-G_{12}=\kappa$$

are controlled by N as noted which can therefore be identified at higher order modes involving increased shear deformation using an iteration method that will be described.

4.2 Micropolar elastic constants

4.2.1 Characteristic length in bending, l_b

E_{fm} and v_m can be determined from static tensile tests. l_b can be obtained from mode 1 behaviour and equation 1 for $\lambda_{mode=1}$. In a slender beam where $d/L \ll 1$
$Microrotation\ \emptyset_z \cong Macrorotation\ \theta_z$
Moment, M, is driven by internal force stress, τ_{xx}, and couple stress, m_{xz}, Thus:

$$M=\int_A (y\tau_{xx}+m_{xz})dA \quad (6)$$

$$\frac{1}{R}=\frac{d\theta}{dx}=\frac{d\emptyset_z}{dx}=-\frac{d^2W}{dx^2} \text{ and: } m_{xz}=\gamma\frac{d\emptyset_z}{dx} \; , \; \tau_{xx}=\frac{E_{fm}y}{R}$$

$$If\ \emptyset_z \cong \theta_z \rightarrow m_{xz}=\frac{\gamma}{R} \; , \; I=\int_A y^2 dA \; , \; A=\int_A dA$$

Substituting for $\frac{1}{R}=-\frac{d^2W}{dx^2}$ in the moment curvature relationship:

$$\frac{d^2W}{dx^2}=-\frac{M}{E_{mf}I+\gamma A}=-\frac{M}{D_{mf}} \quad (7)$$

where $D_{mf}=E_{mf}I+\gamma A$ =Micropolar flexural rigidity (8)

From an unloaded Euler-Bernoulli beam in the dynamic case:

$$D_{mf}\frac{d^4W}{dx^4}-\mu\omega^2W=0 \; , \; \mu=\rho A \quad (9)$$

$$\omega=\lambda^2\sqrt{\frac{D_{mf}}{\rho AL^4}} \longrightarrow \omega=\lambda^2\sqrt{\frac{E_{mf}I+\gamma A}{\rho AL^4}} \quad (10)$$

$$I=\frac{bd^3}{12} \; , \; A=bd \; , \; \gamma=\frac{E_{mf}l_b^2}{12}$$

$$\omega^2=\frac{E_{fm}\lambda^4}{12\rho L^4}(d^2+l_b^2) \quad (11)$$

$D=EI$ In absence of couple stress

$$m.\omega^2=\frac{E_{fm}\lambda^4 b}{12}\left(\frac{d}{L}\right)^3\left(1+\left(\frac{l_b}{d}\right)^2\right) \quad \frac{\text{Kg}}{\text{s}^2} \quad (12)$$

https://doi.org/10.1051/matecconf/201814807003

Equation (12) can provide valuable information. It relates characteristic length, l_b, non-dimensional frequency, λ, flexural Micropolar modulus, E_{fm}, on the right hand side to the beam's mass multiplied by squared frequency on the left hand side. In fact, if the mass times frequency squared can be calculated for beams with various sizes and plotted against reciprocal size measure, $(1/d^2)$, then it is possible to obtain the E_{fm} or λ from the line intercept and the characteristic length from the line slope. As equation (12) is obtained based on the assumptions of slender beam, therefore, the equation is valid for a slender beam and hence mode one only.

4.2.2 Coupling number, N

If α, β, γ and $\kappa = 0$, then the solid should behave in a classical manner while if N=1 and therefore, microrotation and macrorotation are exactly equal which means they are not kinematically distinct, $\phi_z \cong \theta_z$.

Having obtained the characteristic length in bending and based on the results from ANSYS FE analysis and iteration process the coupling number, N, may be estimated. The iteration is based on linear regression[10] and fits the curves in the graphs for $m.\omega^2$ vs mode numbers (or wavelength). In the iteration process the first three transverse modes were used to iterate for coupling number, N; the reason is to obtain a coupling number which can satisfy all modal frequencies and model depths. It was found that N changes with void and/or inclusions radius linearly. But this change is not significant for our specimens 0.1mm < V_r or 0.1155 < V_r/S_y.

A control volume finite element method (CVFEM) Matlab code has been developed by us which incorporates Micropolar theory and is able to model, mesh and perform both static and modal analysis. A second code is also developed which automatically estimates N.

5 Results

Using the CVFEM code and linear regression as described in section 4.2.2, a unique value for N was identified as N=0.0528 for beams with compliant inclusions and N=0.0538 for beams with voids. Characteristic length does not vary with beam size and only depends on volume fraction. See table 1 below:

Table 1. Characteristic length changes with volume fraction

Void or Inclusions radius' volume fraction	lc for beams with Voids, mm	lc for beams with compliant inclusions, mm
4%	0.2717	0.2555
8%	0.4139	0.3612
15%	0.5432	0.4681
23%	0.6522	0.5621
33%	0.7334	0.6367

In figures 7 and 8, the finite element results from ANSYS and Micropolar CVFEM codes for mode 1 and 2 are compared after convergence of the iteration process; $m\omega^2$ is obtained within two limits of N=0 and N=0.9 (N value at higher bound must not equal 1 due to numerical errors it causes.)

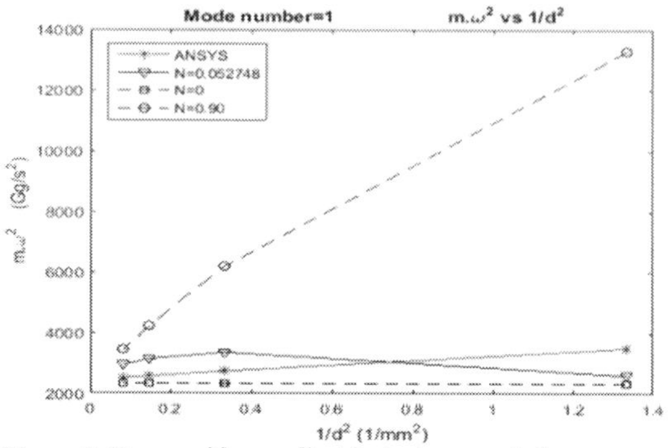

Figure 7: Beams with compliant inclusions, mode 1

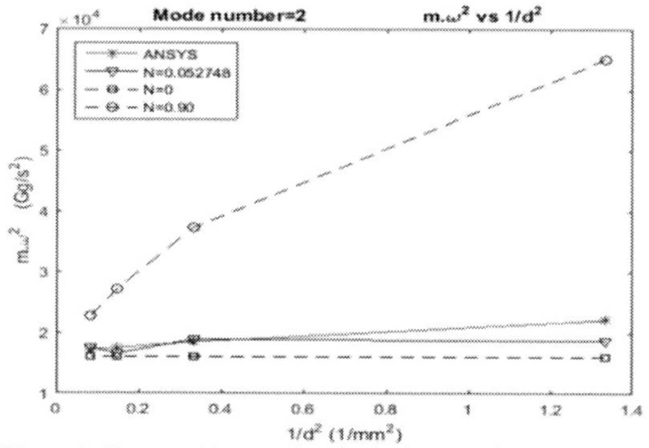

Figure 8: Beams with compliant inclusions, mode 2

In figure 9 the size effect shown in a smooth surfaced beam with voids which is represented by the Micropolar CVFEM code. Beams specifications are:
Model: Beam with voids (Smooth surface), Size: 2, AR=10.4, d=2*0.866mm, DOF=1107, ANSYS DOF =257596, BC=FF, N=0.05, lc=0.652 mm, Void radius=0.25mm

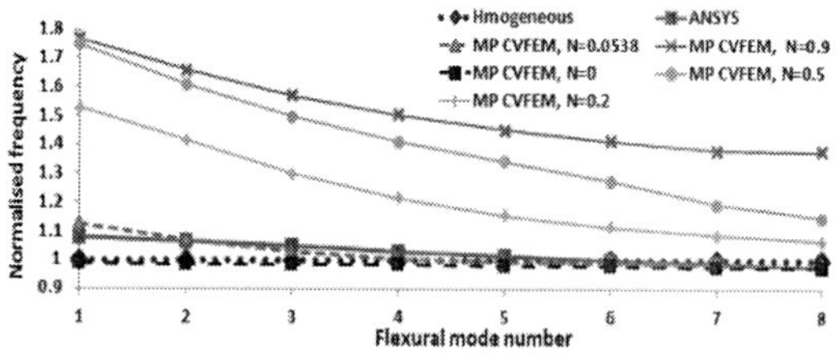

Figure 9: Normalised flexural modal frequencies produced by ANSYS squared markers and CVFEM code shown in dotted line and triangular markers. $lc = 0.6522\ mm$, $N = 0.0538$

Micropolar CVFEM code models homogeneous beams with 6 node triangular elements and characterises the heterogeneity through incorporating the characteristic length and coupling number. Comparing the total degrees of freedom, it is clear that CVFEM code uses significantly less DOF's and therefore less computation is needed. Figure 9 also illustrates the sensitivity of frequencies with changes in N e.g. N is more sensitive at higher modes, see N=0.5.

5 Conclusions

A Micropolar CVFEM code for modal analysis was developed which is able to model, mesh and perform both static and modal analysis using Micropolar theory to account for micro rotation and couple stress. In Micropolar beam vibration, characteristic length in bending is obtained through dynamic analysis. Obtained coupling numbers and Characteristic lengths for models with various void or inclusions volume fraction used in Modal Analysis are fairly accurate and may be used to predict modal frequencies in free vibration with the exceptions of compliant inclusions and/or Voids intersect the beam surface and beams with compliant matrices. The dynamic behaviour of the heterogeneous beams strongly depends to the beam's aspect ratio, mode numbers for which frequencies are obtained, material discontinuity across the height and length of the structure as well as surface morphology. Although Eringen's Nonlocal Timoshenko beam (NTB) could explain the dynamic behaviour of our models to some extent the FE results for our specific models showed that even NTB has shortcomings and is not constant for all model sizes with the same aspect ratio e.g. multiplying the number of constitutive layers require new scaling effect parameter as the layers add up for the models with the same aspect ratios.

References

1. S. A. Rabboh, N. E. Bondok, T. S. Mahmoud, and H. I. El Kholy, "The Effect of Functionally Graded Materials into the Sandwich Beam Dynamic Performance," vol. **2013**, no. November, pp. 751–760, 2013.
2. C. N. Della and D. Shu, "Natural frequency of beams with embedded piezoelectric sensors and actuators," *Aerospace*, vol. **137**, pp. 137–138, 2007.
3. R. M. J. Groh and P. M. Weaver, "On displacement-based and mixed-variational equivalent single layer theories for modelling highly heterogeneous laminated beams," *Int. J. Solids Struct.*, vol. **59**, pp. 147–170, 2015.
4. M. Gherlone, "On the Use of Zigzag Functions in Equivalent Single Layer Theories for Laminated Composite and Sandwich Beams: A Comparative Study and Some Observations on External Weak Layers," *J. Appl. Mech.*, vol. **80**, no. 6, p. 61004, 2013.
5. S.-H. Schulze, M. Pander, K. Naumenko, and H. Altenbach, "Analysis of laminated glass beams for photovoltaic applications," *Int. J. Solids Struct.*, vol. **49**, no. 15–16, pp. 2027–2036, 2012.
6. A. Dasgupta, "Eigenstrain Techniques for Modeling Adaptive Structures: II. Active Damping," *J. Intell. Mater. Syst. Struct.*, vol. **11**, no. 8, pp. 631–641, 2000.
7. C. M. Wang, Y. Y. Zhang, and X. Q. He, "Vibration of nonlocal Timoshenko beams," *Nanotechnology*, vol. **18**, no. 10, p. 105401, 2007.
8. Z. Zhang, N. Challamel, and C. M. Wang, "Eringen's small length scale coefficient for buckling of nonlocal Timoshenko beam based on microstructured beam model," *J. Appl. Phys.*, vol. **114**, no. 11, 2013.
9. A. J. Beveridge, M. A. Wheel, and D. H. Nash, "A higher order control volume based finite element method to predict the deformation of heterogeneous materials," *Comput. Struct.*, vol. **129**, no. 0, pp. 54–62, 2013.
10. A. J. Beveridge, M. A. Wheel, and D. H. Nash, "The micropolar elastic behaviour of model macroscopically heterogeneous materials," *Int. J. Solids Struct.*, vol. **50**, no. 1, pp. 246–255, 2013.
11. A. R. Dehkordi, "3D Finite Element Cosserat Continuum Simulation of Layered Geomaterials," *PhD Thesis*, p. 263, 2008.
12. J. C. Frame, "A computational and experimental investigation into the micropolar elastic behaviour of cortical bone," *PhD Thesis_University Strat.*, pp. 1–199, 2013.
13. A. Riahi and J. H. Curran, "Full 3D finite element Cosserat formulation with application in layered structures," *Appl. Math. Model.*, vol. **33**, no. 8, pp. 3450–3464, 2009.
14. A. Waseem, A. J. Beveridge, M. A. Wheel, and D. H. Nash, "The influence of void size on the micropolar constitutive properties of model heterogeneous materials," *Eur. J. Mech. A/Solids*, vol. **40**, pp. 148–157, 2013.
15. M. a. Wheel, J. C. Frame, and P. E. Riches, "Is smaller always stiffer? On size effects in supposedly generalised continua," *Int. J. Solids Struct.*, vol. **67–68**, pp. 84–92, 2015.
16. S. . R. S. L. Nakamura, "Finite element analysis of Saint-Venant end effects in micropolar elastic solids," *Adapt. from Eng. Comput. 12, 571-587*, pp. 1–10, 1995.

17. R. Lakes, "Experimental methods for study of Cosserat elastic solids and other generalized elastic continua," *Contin. Model. Mater. with Micro-structure*, no. 1, pp. 1–22, 1996.
18. M. McGregor and M. A. Wheel, "On the coupling number and characteristic length of micropolar media of differing topology," *Proc. R. Soc. A Math. Phys. Eng. Sci.*, vol. **470**, no. 2169, pp. 20140150–20140150, 2014.

On displacement based non-local models for non-linear vibrations of thin nano plates

Tomás R. C. Chuaqui[1], *Pedro* Ribeiro[1,*]

[1]DEMec/INEGI, Faculdade de Engenharia da Universidade do Porto, Portugal

Abstract. This paper addresses the formulation of displacement based, non-linear, plate models adopting Eringen's non-local elasticity, to study the modes of vibration of thin, nano plates. Plate models governed by ordinary differential equations of motion with generalized displacements as unknowns have some advantages over mixed type formulations, but difficulties arise in the development of such non-linear models when non-local effects are taken into account. To circumvent those difficulties, approximations of debatable justification can be imposed. Different approximations are discussed here and the accuracy of the best non-local, non-linear displacement based model achieved is put to test, by carrying out comparisons with a model based on Airy's stress function.

1 Introduction

Micro and nano systems are rapidly evolving and increasingly attracting the interest of researchers and engineers. Vibrating plates frequently appear as components in these systems. In particular, due to their promising properties and possible applications, substantial attention has been given to graphene sheets [1, 2].

Atomistic simulations have been extensively used in nano mechanics [3], but they require high computational resources [4]. Continuum modelling techniques do not suffer from this handicap and classical elasticity theories have been applied in attempts to replicate the mechanical behaviour of components used in nanotechnology. However, some analyses indicate that classical elasticity theories are not always able to correctly predict the behaviour of nanostructures, because they miss small scale effects [5, 6]. In order to solve this problem, non-local continuum theories that take into account the interatomic forces and the internal length scales have been applied. In particular, Eringen's non-local elasticity theory [7, 8] has been used as the basis of many studies [5, 9-13].

Nano plates can achieve large displacements, in comparison to their dimensions, inducing geometric non-linearities, which lead to an effective stiffening of the plate, consequently altering the natural frequencies and mode shapes of vibration. In the geometrical non-linear regime, the non-local differential constitutive equations cause difficulties in plate models, because of the coupling between the diverse stress resultants [9]. A way to circumvent this issue and arrive at displacement based workable models, is to neglect some non-linear non-local terms of the equilibrium equations, but the rationale for neglecting those terms and the ensuing errors are issues of concern. Using Airy's stress function – hence not following a totally displacement based approach – offers another solution to sidestep the aforementioned coupling [12-15].

In this paper, the implementation of displacement based non-linear and non-local models of nano plates is discussed. The models are based on Kirchhoff's thin plate theory and include Von Kármán's geometrical non-linear terms [16]. To investigate their performance, the backbone curves of a nano plate computed using them are compared with the curves obtained using a model based on Airy's stress function [15].

2 Displacement based models

2.1 Non-local equations of motion

2.1.1 Equilibrium equations

The displacement components in directions x, y and z - represented by u, v and w, respectively - are given by

$$u = u^0 - z\frac{\partial w^0}{\partial x},\quad v = v^0 - z\frac{\partial w^0}{\partial x},\quad w = w^0 \tag{1}$$

where u^0, v^0 and w^0 represent the displacement components of points on the middle plane (plane Oxy). The strain-displacement relations are

$$\begin{Bmatrix}\varepsilon_x\\ \varepsilon_y\\ \gamma_{xy}\end{Bmatrix} = \begin{Bmatrix}\dfrac{\partial u^0}{\partial x}\\ \dfrac{\partial v^0}{\partial y}\\ \dfrac{\partial u^0}{\partial y}+\dfrac{\partial v^0}{\partial x}\end{Bmatrix} + \begin{Bmatrix}\dfrac{1}{2}\left(\dfrac{\partial w^0}{\partial x}\right)^2\\ \dfrac{1}{2}\left(\dfrac{\partial w^0}{\partial y}\right)^2\\ \dfrac{\partial w^0}{\partial x}\dfrac{\partial w^0}{\partial y}\end{Bmatrix} - z\begin{Bmatrix}\dfrac{\partial^2 w^0}{\partial x^2}\\ \dfrac{\partial^2 w^0}{\partial y^2}\\ 2\dfrac{\partial^2 w^0}{\partial x\partial y}\end{Bmatrix} \tag{2}$$

* Corresponding author: pmleal@fe.up.pt

https://doi.org/10.1051/matecconf/201814807004

with the remaining strains equal to zero.

The following equations of motion correspond to the free vibration case and can be derived, for example, applying Hamilton's principle

$$\frac{\partial N_{xx}}{\partial x}+\frac{\partial N_{xy}}{\partial y}=\rho h\frac{\partial^2 u^0}{\partial t^2} \quad (3)$$

$$\frac{\partial N_{xy}}{\partial x}+\frac{\partial N_{yy}}{\partial y}=\rho h\frac{\partial^2 v^0}{\partial t^2} \quad (4)$$

$$\frac{\partial^2 M_{xx}}{\partial x^2}+2\frac{\partial^2 M_{xy}}{\partial y\partial x}+\frac{\partial^2 M_{yy}}{\partial y^2}+N_{nlinear}= \rho h\frac{\partial^2 w^0}{\partial t^2}-\rho h^3\left(\frac{\partial^4 w^0}{\partial t^2\partial x^2}+\frac{\partial^4 w^0}{\partial t^2\partial y^2}\right) \quad (5)$$

where

$$N_{nlinear}=\frac{\partial}{\partial x}\left(N_{xx}\frac{\partial w^0}{\partial x}+N_{xy}\frac{\partial w^0}{\partial y}\right)+ +\frac{\partial}{\partial y}\left(N_{xy}\frac{\partial w^0}{\partial x}+N_{yy}\frac{\partial w^0}{\partial y}\right) \quad (6)$$

The effect of rotary inertia will be neglected. In-plane inertia can also be neglected, because its importance in periodic oscillations of thin plates is, in general, small [17]. Due to the latter simplification, equation (6) leads to

$$N_{nlinear}=N_{xx}\frac{\partial^2 w^0}{\partial x^2}+2N_{xy}\frac{\partial^2 w^0}{\partial y\partial x}+N_{yy}\frac{\partial^2 w^0}{\partial y^2} \quad (7)$$

2.1.2 Non-local elasticity

So far, non-local effects were not introduced. According to [8, 18], the equations of motion for a generic non-local linear elastic solid, are:

$$t_{ij,j}+f_i=\rho\ddot{u}_i \quad (8)$$

with ρ, f_i, $\ddot{u}_i$ designating the mass density, applied forces and the acceleration vector respectively. The stress tensor t_{ij} in Eringen's non-local elasticity is defined by the following volume integral over the region V occupied by the body

$$t_{ij}(\mathbf{x})=\int_V \alpha(|\mathbf{x'}-\mathbf{x}|)\sigma_{ij}(\mathbf{x'})\mathrm{d}v(\mathbf{x'}) \quad (9)$$

where $\mathbf{x}$ is a reference point in the body, $\alpha(|\mathbf{x'}-\mathbf{x}|)$ is the non-local kernel function [8] and $\sigma_{ij}(\mathbf{x'})$ is the local, or stress tensor of the classical theory of elasticity at any point $\mathbf{x'}$. Based on this theory, the stress at an arbitrary point is not only a function of the components of strain at that point but also a function of strain at all other points of the structure. It is a size-dependent theory; as the dimensions increase - and, consequently, the internal characteristic length becomes much smaller than the external or macroscopic length of the structure - it approaches the traditional, local, theory of elasticity.

Instead of writing equation (8) as an integro-partial differential equation using (9), it is possible to write it in partial differential form making [8]

$$\left[1-(e_0a_i)^2\nabla^2\right]t_{ij}=\sigma_{ij} \quad (10)$$

where ∇^2 is the Laplacian, a_i is an internal characteristic length (lattice parameter, granular size or molecular diameter), and e_0 is an empirical material constant. Employing the local constitutive equations, (10) leads to

$$\left[1-(e_0a_i)^2\nabla^2\right]t_{ij}=C_{ijkl}\varepsilon_{kl} \quad (11)$$

where C_{ijkl} are the components of the fourth order stiffness or elasticity tensor, and ε_{kl} is the strain tensor. In the rest of this section, a non-local parameter $\mu=(e_0 a_i)^2$ is employed, for the sake of simplicity.

Now, we extend the non-local elasticity theory to the Kirchhoff plate, and write the constitutive equations in terms of stress resultants as [9]

$$\begin{Bmatrix}N_{xx}\\N_{yy}\\N_{xy}\end{Bmatrix}-\mu\nabla^2\begin{Bmatrix}N_{xx}\\N_{yy}\\N_{xy}\end{Bmatrix}= =\begin{bmatrix}A_{11}&A_{12}&0\\A_{12}&A_{22}&0\\0&0&A_{66}\end{bmatrix}\begin{Bmatrix}\frac{\partial u^0}{\partial x}+\frac{1}{2}\left(\frac{\partial w^0}{\partial x}\right)^2\\ \frac{\partial v^0}{\partial y}+\frac{1}{2}\left(\frac{\partial w^0}{\partial y}\right)^2\\ \frac{\partial u^0}{\partial y}+\frac{\partial v^0}{\partial x}+\frac{\partial w^0}{\partial x}\frac{\partial w^0}{\partial y}\end{Bmatrix}, \quad (12)$$

$$\begin{Bmatrix}M_{xx}\\M_{yy}\\M_{xy}\end{Bmatrix}-\mu\nabla^2\begin{Bmatrix}M_{xx}\\M_{yy}\\M_{xy}\end{Bmatrix}= =-\begin{bmatrix}D_{11}&D_{12}&0\\D_{12}&D_{22}&0\\0&0&D_{66}\end{bmatrix}\begin{Bmatrix}\frac{\partial^2 w^0}{\partial x^2}\\ \frac{\partial^2 w^0}{\partial y^2}\\ 2\frac{\partial^2 w^0}{\partial x\partial y}\end{Bmatrix}, \quad (13)$$

where coefficients A_{ij} and D_{ij} are

$$A_{ij}=hQ_{ij},\; D_{ij}=\frac{h^3}{12}Q_{ij} \quad (14)$$

with the plane-stress reduced stiffnesses Q_{ij} as given, for example, in [19].

Applying the differential operator $\mathcal{L}=1-\mu\nabla^2$ to equations (3)-(5), and the constitutive relations (12) and (13), we arrive at the equations of motion for non-local Kirchhoff plates with Von Kármán's type non-linearity

$$(1-\mu\nabla^2)\left(\frac{\partial^2 M_{xx}}{\partial x^2}+2\frac{\partial^2 M_{xy}}{\partial y\partial x}+\frac{\partial^2 M_{yy}}{\partial y^2}\right)+ +(1-\mu\nabla^2)N_{nlinear}=(1-\mu\nabla^2)\left(\rho h\frac{\partial^2 w^0}{\partial t^2}\right) \quad (15)$$

2.2 Simplifications of displacement based non-linear, non-local equations of motion

The difficulty in displacement based non-linear, non-local plate models resides in dealing with term

$$\left(1-\mu\nabla^2\right)N_{nlinear}= \left(1-\mu\nabla^2\right)\left(N_{xx}\frac{\partial^2 w^0}{\partial x^2}+2N_{xy}\frac{\partial^2 w^0}{\partial y\partial x}+N_{yy}\frac{\partial^2 w^0}{\partial y^2}\right) \quad (16)$$

due to the couplings introduced by the stress resultants. This issue was already mentioned in [9], where the static case is addressed. The following sections provide a detailed description of simplification options available in order to achieve displacement based formulations.

2.2.1 Neglecting all non-linear terms

An oversimplifying way to deal with the issue, stated here for the sake of completeness, is to linearise equation (15), which would become

$$\left(1-\mu\nabla^2\right)\left(\frac{\partial^2 M_{xx}}{\partial x^2}+2\frac{\partial^2 M_{xy}}{\partial y\partial x}+\frac{\partial^2 M_{yy}}{\partial y^2}\right)= \left(1-\mu\nabla^2\right)\left(\rho h\frac{\partial^2 w^0}{\partial t^2}\right) \quad (17)$$

Now, using equation (13), ones gets

$$-D_{11}\frac{\partial^4 w^0}{\partial x^4}-\left(2D_{12}+4D_{66}\right)\frac{\partial^4 w^0}{\partial x^2\partial y^2}-D_{22}\frac{\partial^4 w^0}{\partial y^4}= =\left(1-\mu\nabla^2\right)\left(\rho h\frac{\partial^2 w^0}{\partial t^2}\right) \quad (18)$$

So, if non-linearity is disregarded in a free vibration problem, then non-local effects only change the inertia forces. This approach has been widely implemented, namely to analyse the importance of non-local effects on the linear modes of vibration [20].

2.2.2 Neglecting all non-local non-linear terms

Another oversimplifying option at hand consists in directly neglecting terms that are simultaneously non-local and non-linear, so that the partial differential equation of motion reads

$$\left(1-\mu\nabla^2\right)\left(\frac{\partial^2 M_{xx}}{\partial x^2}+2\frac{\partial^2 M_{xy}}{\partial y\partial x}+\frac{\partial^2 M_{yy}}{\partial y^2}\right)+N_{nlinear}^{local} =\left(1-\mu\nabla^2\right)\left(\rho h\frac{\partial^2 w^0(x,y,t)}{\partial t^2}\right) \quad (19)$$

with $N_{nlinear}$ as in local problems, i.e., given by Eq. (12) with μ=0. As occurred in the former section, in this approach the non-local effects do not influence the stiffness, they would only be considered in the external excitation – which is absent in this paper - and in the inertia forces.

One might also be tempted to implement the approximation described in the following lines. Let's start by writing equation (16) in the ensuing form

$$\left(1-\mu\nabla^2\right)\left(N_{xx}\frac{\partial^2 w^0}{\partial x^2}+2N_{xy}\frac{\partial^2 w^0}{\partial y\partial x}+N_{yy}\frac{\partial^2 w^0}{\partial y^2}\right)= \left(\left(1-\mu\nabla^2\right)N_{xx}\right)\frac{\partial^2 w^0}{\partial x^2}+2\left(\left(1-\mu\nabla^2\right)N_{xy}\right)\frac{\partial^2 w^0}{\partial y\partial x}+ +\left(\left(1-\mu\nabla^2\right)N_{yy}\right)\frac{\partial^2 w^0}{\partial y^2}+o.t. \quad (20)$$

where $o.t.$ stands by "other terms", which have in common the facts that all contain non-local parameter μ and either a third or fourth order derivative of w.

Using the non-local constitutive relation (12), we have

$$\left(\left(1-\mu\nabla^2\right)N_{xx}\right)\frac{\partial^2 w^0}{\partial x^2}+2\left(\left(1-\mu\nabla^2\right)N_{xy}\right)\frac{\partial^2 w^0}{\partial y\partial x}+ +\left(\left(1-\mu\nabla^2\right)N_{yy}\right)\frac{\partial^2 w^0}{\partial y^2}= =\left[A_{11}\left(\frac{\partial u^0}{\partial x}+\frac{1}{2}\left(\frac{\partial w^0}{\partial x}\right)^2\right)+A_{12}\left(\frac{\partial v^0}{\partial y}+\frac{1}{2}\left(\frac{\partial w^0}{\partial y}\right)^2\right)\right]\frac{\partial^2 w^0}{\partial x^2} +2A_{66}\left(\frac{\partial u^0}{\partial y}+\frac{\partial v^0}{\partial x}+\frac{\partial w^0}{\partial x}\frac{\partial w^0}{\partial y}\right)\frac{\partial^2 w^0}{\partial x\partial y}+ \left[A_{12}\left(\frac{\partial u^0}{\partial x}+\frac{1}{2}\left(\frac{\partial w^0}{\partial x}\right)^2\right)+A_{22}\left(\frac{\partial v^0}{\partial y}+\frac{1}{2}\left(\frac{\partial w^0}{\partial y}\right)^2\right)\right]\frac{\partial^2 w^0}{\partial y^2} \quad (21)$$

Neglecting the "other terms" in (20), which allows replacing the right hand side of that equation by the right side of equation (21), one arrives at equations of motion that would read as

$$-D_{11}\frac{\partial^4 w^0}{\partial x^4}-\left(2D_{12}+4D_{66}\right)\frac{\partial^4 w^0}{\partial x^2\partial y^2}-D_{22}\frac{\partial^4 w^0}{\partial y^4}+ +\left[A_{11}\left(\frac{\partial u^0}{\partial x}+\frac{1}{2}\left(\frac{\partial w^0}{\partial x}\right)^2\right)+A_{12}\left(\frac{\partial v^0}{\partial y}+\frac{1}{2}\left(\frac{\partial w^0}{\partial y}\right)^2\right)\right]\frac{\partial^2 w^0}{\partial x^2} +2A_{66}\left(\frac{\partial u^0}{\partial y}+\frac{\partial v^0}{\partial x}+\frac{\partial w^0}{\partial x}\frac{\partial w^0}{\partial y}\right)\frac{\partial^2 w^0}{\partial x\partial y}+ \left[A_{12}\left(\frac{\partial u^0}{\partial x}+\frac{1}{2}\left(\frac{\partial w^0}{\partial x}\right)^2\right)+A_{22}\left(\frac{\partial v^0}{\partial y}+\frac{1}{2}\left(\frac{\partial w^0}{\partial y}\right)^2\right)\right]\frac{\partial^2 w^0}{\partial y^2}= =\left(1-\mu\nabla^2\right)\left(\rho h\frac{\partial^2 w^0(x,y,t)}{\partial t^2}\right) \quad (22)$$

So, neglecting terms $o.t.$ in equation (20) – which is essentially justified by mathematical convenience -, leads to the removal of all non-local terms related with stiffness, that is, it leads again to equation (19).

2.2.3 Neglecting some non-local non-linear terms

In order to consider non-local effects on the stiffness, whilst still arriving at a manageable displacement based formulation, the ensuing reasoning will be followed.

Consider, by way of example, term $(1-\mu\nabla^2)\left(N_{xx}\frac{\partial^2 w^0}{\partial x^2}\right)$ of $(1-\mu\nabla^2)N_{nlinear}$. Employing the first row of equation (12), the following equality can be written

$$(1-\mu\nabla^2)\left(N_{xx}\frac{\partial^2 w^0}{\partial x^2}\right)=(1-\mu\nabla^2)\left\{\left[A_{11}\left(\frac{\partial u^0}{\partial x}+\frac{1}{2}\left(\frac{\partial w^0}{\partial x}\right)^2\right)+ A_{12}\left(\frac{\partial v^0}{\partial y}+\frac{1}{2}\left(\frac{\partial w^0}{\partial y}\right)^2\right)\right]\frac{\partial^2 w^0}{\partial x^2}\right\}+\underline{(1-\mu\nabla^2)\left((\mu\nabla^2 N_{xx})\frac{\partial^2 w^0}{\partial x^2}\right)} \quad (23)$$

In the term underlined in equation (23), the non-local effects appear in operator $(1-\mu\nabla^2)$ and multiplying by the Laplacian of the stress resultant. We will neglect this term and later analyse the consequences.

Following this line of thought, the equations of motion become

$$\begin{aligned}&-D_{11}\frac{\partial^4 w^0}{\partial x^4}-(2D_{12}+4D_{66})\frac{\partial^4 w^0}{\partial x^2\partial y^2}-D_{22}\frac{\partial^4 w^0}{\partial y^4}\\&+(1-\mu\nabla^2)\left\{\left[A_{11}\left(\frac{\partial u^0}{\partial x}+\frac{1}{2}\left(\frac{\partial w^0}{\partial x}\right)^2\right)+\right.\right.\\&\left.A_{12}\left(\frac{\partial v^0}{\partial y}+\frac{1}{2}\left(\frac{\partial w^0}{\partial y}\right)^2\right)\right]\frac{\partial^2 w^0}{\partial x^2}+\\&2A_{66}\left(\frac{\partial u^0}{\partial y}+\frac{\partial v^0}{\partial x}+\frac{\partial w^0}{\partial x}\frac{\partial w^0}{\partial y}\right)\frac{\partial^2 w^0}{\partial x\partial y}+\\&\left.\left[A_{12}\left(\frac{\partial u^0}{\partial x}+\frac{1}{2}\left(\frac{\partial w^0}{\partial x}\right)^2\right)+A_{22}\left(\frac{\partial v^0}{\partial y}+\frac{1}{2}\left(\frac{\partial w^0}{\partial y}\right)^2\right)\right]\frac{\partial^2 w^0}{\partial y^2}\right\}\\&=(1-\mu\nabla^2)\left(\rho h\frac{\partial^2 w^0(x,y,t)}{\partial t^2}\right)\end{aligned} \quad (24)$$

3 The ordinary differential equations of motion and their solution

The partial differential equations of motion (24) were transformed into ordinary differential equations of motion employing a p-version finite element type approach; the details do not fit in this paper, but are given in [21]. The resulting ODEs constitute a system of coupled Duffing type equations, which is, for the sake of simplicity, written as

$$\begin{aligned}&\left[\mathbf{M}+\mathbf{M}_\mu\right]\ddot{\mathbf{q}}_w(t)+\mathbf{K}^0\mathbf{q}_w(t)+\\&+\left[\mathbf{K}_\mu^2(\mathbf{q}_w(t))+\mathbf{K}^2(\mathbf{q}_w(t))\right]\mathbf{q}_w(t)=\mathbf{0}\end{aligned} \quad (25)$$

where $\mathbf{M}$ is used to represent a mass matrix and $\mathbf{K}$ to represent a stiffness matrix. Superscript ^{0} indicates a constant stiffness matrix and superscript ^{2} a matrix that is a quadratic function of the generalized transverse displacements, $\mathbf{q}_w(t)$. Subscript μ indicates that the matrix is due to non-local effects.

Provided that modal coupling does not exist and discarding the study of bifurcations [22, 23], a one harmonic approach often provides rather accurate backbone curves. In this case, the vector of generalized displacements is written as:

$$\mathbf{q}_w(t)=\mathbf{u}\cos(\omega t)). \quad (26)$$

The algebraic system that results from substituting (26) in (25) can be solved in a number of ways. We employ the arc-length continuation method presented by R. Lewandowski in [24].

4 Numerical results

4.1 Local non-linear model

In order to verify the local part of the computational code, tests were carried out on macroscopic plates, making μ=0 [21]. An example of the tests performed is shown in Table 1, which contains the ratio between the first non-linear natural frequency and the linear one, ω/ω_l, computed by the present approach, and ratios presented in [25], [26], obtained by other procedures. The plate is "simply supported" with immovable edges, boundaries we designate as "pinned". Furthermore, it is isotropic and rectangular, with the following properties: E= 205 800 MPa, a=20 mm, b=10 mm, h=0.1 mm, ν=0.3, ρ=7800 kg/m^3. The results of the present model were obtained using 9 symmetric out-of-plane bi-dimensional shape functions, and either 25 or 49, both symmetric and anti-symmetric, in-plane shape functions. Ratio ω/ω_l is calculated for different values of w_{max}/h, which designates the ratio between the amplitude of vibration at x=y=0 and the thickness of the plate.

Table 1. Ratio ω/ω_l, isotropic rectangular plate

$\frac{w_{max}}{h}$	Present model (9,25)	Present model (9,49)	Harmonic balance* [26], [27]	Elliptic function [26], [27]	Sarma et al. [25]
0.2	1.026	1.026	1.0241	1.0241	1.0254
0.4	1.095	1.096	1.0933	1.0927	1.0975
0.6	1.206	1.206	1.1998	1.1975	1.2072
0.8	1.349	1.349	1.3347	1.3292	1.3447
1.0	1.518	1.518	1.4903	1.4808	1.5023

* These results are given in [26] under label "perturbation method"; they are based on [27], where, apparently, the "perturbation method" is the harmonic balance method with one harmonic.

The results presented in Table 1 are very close to each other, suggesting that the local part of the computational code is correct; maybe we ought to point out that other approaches – not shown - give results that diverge more from the ones presented in Table 1 (see [26]), due to differences in hypothesis assumed or solution methods employed.

4.2 Non-local non-linear model

To analyse the consequences of the approximation suggested in section 2.2.3, results from this model were compared with data given by A.Farajpour *et al.* in [15]. A square nanoplate, assumed to be on $BiTiO_3$-$CoFe_2O_4$ composites with the following properties was considered: a=b=60 nm, h=4 nm, c_{11}=c_{22}=226 GPa, c_{12}=125 GPa, c_{13}=124 GPa, c_{33}=216 GPa, c_{44}= c_{55}=44.2 GPa and c_{66}=50.5 GPa, the remaining values of c_{ij} are zero. a and b represent the width and length of the plate, respectively. Details on the application of the values of elastic stifnesses c_{ij} to arrive at the reduced stiffnesses - which appear on the constitutive relation between the non-null stresses and strains in a Kirchhoff plate, where plane stress is assumed - are given, for example, in [15, 16]. The plate is considered to be pinned.

We are specifically comparing with [15] because this work stands on the following accurate assumptions:

- a formulation based on Airy's stress function is adopted, allowing one to derive equations of motion of non-local, geometrically non-linear, plates, without neglecting any of the non-linear terms that stem from a Von Kármán's approach;

- the non-local compatibility equation is used (this is not always the case in the literature, some authors employ a local compatibility equation).

Nonetheless, there are upsetting differences between the way boundary conditions are imposed on Airy's stress function based formulations and on displacement based formulations. In the specific case at hand, whilst we impose that the boundaries are immovable in-plane by enforcing precisely that (note that the origin of the coordinate system coincides with the geometric centre of the plate)

$$u(\pm a/2, y, t) = 0 \quad v(x, \pm b/2, t) = 0, \tag{27}$$

in [15], immovable boundary conditions of the geometric type are written as

$$x = \pm a/2 \Rightarrow \int_{-b/2}^{b/2}\int_{-a/2}^{a/2} \frac{\partial u(x,y,t)}{\partial x} \mathrm{d}x\mathrm{d}y = 0, \tag{28}$$

$$y = \pm b/2 \Rightarrow \int_{-b/2}^{b/2}\int_{-a/2}^{a/2} \frac{\partial v(x,y,t)}{\partial y} \mathrm{d}x\mathrm{d}y = 0, \tag{29}$$

This averaging procedure is characteristic of formulations based on stress or force functions [11], but equations (28) and (29) are not strictly the same as equations (27).

In Fig. 1 the non-linear natural frequencies computed by the present approach are compared with the ones of [15]. Different values of the non-local parameter, which is represented by its dimensionless value χ=e_0a_i/a, were considered, from χ=0 for the results on the top, to χ=0.3 for the results on the bottom.

In the local case (χ=0), the present model was verified (see Table 1). Nevertheless, there are some dissimilarities between the results of the approach proposed in this paper and the results of ref. [15].These specific differences are reasonably small and have as possible reasons the diverse ways adopted to impose boundary conditions, the different methods of solution of the equations of motion and the fact that the values were read from a graph published in [15].

The simplified displacement based approach of section 2.2.3 is able to predict that increasing the non-local effects, increases the hardening spring effect. However, as the non-local parameter and the vibration amplitude rises, the simplified displacement based model progressively deviates from the one based on Airy's stress function. Hence, the simplified displacement based formulation presented in 2.2.3 should not be employed in situations where non-local effects are more prevalent and vibration amplitudes larger.

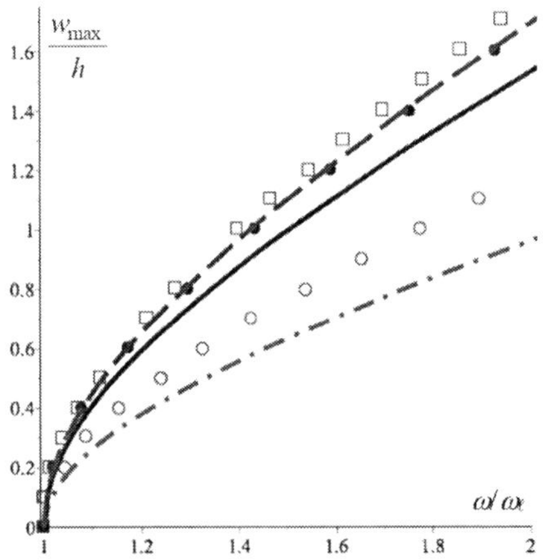

Fig. 1. Backbone curves of the present displacement based approach (lines) and of Airy's stress function based approach [15] (points; read from graph); non-local parameters from top-to-bottom (colour online): χ=0 red, χ=0.1 blue, χ=0.3 black.

5 Final comments

The implementation of displacement based, non-local formulations, for non-linear vibrations of plates was analysed. The models were based on Eringen's non-local elasticity, Kirchhoff hypothesis and von Kármán type non-linear strain-displacement relations. A displacement based formulation was not achieved without recurring to simplifications. Some of the simplifications that can be employed, and which appear to have been suggested in the literature, lead to a complete removal of non-local terms from the stiffness and were therefore disregarded. An alternative simplification was implemented, where only some non-local terms were neglected.

Natural frequencies of vibration computed by a method based on the latter approach were compared with the frequencies computed via an – apparently more accurate – model where Airy's stress function is applied. The simplified displacement based approach correctly predicted an increasing of the hardening spring effect with the non-local parameter. However, the disagreement between the displacement based model and the one that employs Airy's stress function increases as the vibration amplitude and the non-local parameter increase. Therefore, the domain of application of the simplified displacement based procedure is restricted, as it can only be applied if the non-local parameter and the non-linear effects are not large.

The second author gratefully acknowledges the funding of Project NORTE-01-0145-FEDER-000022 - SciTech - Science and Technology for Competitive and Sustainable Industries, cofinanced by Programa Operacional Regional do Norte (NORTE2020), through Fundo Europeu de Desenvolvimento Regional (FEDER).

References

1. Randviir, E.P., D.A.C. Brownson, and C.E. Banks, Mater. Today **17**(9): p. 426-432 (2014).
2. Wang, Q. and B. Arash, Comput. Mater. Sci. **82**(Supplement C): p. 350-360 (2014).
3. Eom, K., et al., Phys. Rep. **503**: p. 115-163 (2011).
4. Narendar, S., D.R. Mahapatra, and S. Gopalakrishnan, Int. J. Eng. Sci. **49**: p. 509–522 (2011).
5. Arash, B. and Q. Wang, Comput. Mater. Sci. **51**: p. 303-313 (2012).
6. Wang, Q., J. Appl. Phys. **98**(12) (2005).
7. Eringen, A.C., Int. J. Eng. Sci. **10**: p. 425-435 (1972).
8. Eringen, A., J. Appl. Phys. **54**: p. 8 (1983).
9. Reddy, J.N., Int. J. Eng. Sci. **48**: p. 1507-1518 (2010).
10. Ansari, R., S. Sahmani, and B. Arash, Phys. Lett. A **375**(1): p. 53-62 (2010).
11. Shen, L., H.-S. Shen, and C.-L. Zhang, Comput. Mater. Sci. **48**(3): p. 680-685 (2010).
12. Jomehzadeh, E. and A.R. Saidi, Comput. Mater. Sci. **50**(3): p. 1043-1051 (2011).
13. Jomehzadeh, E., A.R. Saidi, and N.M. Pugno, Phys. E: Low-dimensional Systems and Nanostructures **44**(10): p. 1973-1982 (2012).
14. Li, H.B., et al., J. Sound Vib. **358**: p. 285-300 (2015).
15. Farajpour, A., et al., Compos. Struct. **140**: p. 323-336 (2016).
16. Chia, C.Y., *Nonlinear Analysis of Plates* (McGraw-Hill Int. Book Company, 1980).
17. Ribeiro, P., Compos. Sci. Technology **69**(2): p. 176-185 (2009).
18. Lu, P., et al., Proc. R. Soc. A **463**: p. 3225–3240 (2007).
19. Reddy, J.N., *Mechanics of Laminated Composite Plates and Shells: Theory and Analysis* (CRC Press, Boca Raton, 2004).
20. Thai, H.-T., et al.,. Compos. Struct. **177**(Supplement C): p. 196-219 (2017).
21. Chuaqui, T.R.C., *Linear and non-linear vibrations of single-layer graphene sheets*, (DEMec, Faculty of Engineering, University of Porto, Porto, 2016).
22. Ribeiro, P., J. Sound Vib. **322**(1-2): p. 8-14 (2009).
23. Ribeiro, P. and M. Petyt, Int. J. Non-Linear Mech. **35**(2): p. 263-278 (2000).
24. Lewandowski, R., J. Sound Vib. **170**(5): p. 577-593 (1994).
25. Sarma, M.S., et al., J. Sound Vib. **159**(3): p. 540-545 (1992).
26. Mei, C. and K. Decha-Umphai, AIAA J. **23**(7): p. 1104-1110 (1985).
27. Eisley, J.G., J. Appl. Math. and Phys. (ZAMP) **15**(2): p. 167-175 (1964).

https://doi.org/10.1051/matecconf/201814807005

Static stability and load capacity of pyramidal trusses

Murillo V. B. Santana[1,*], *Paulo B.* Gonçalves[1], and *Ricardo A. M.* Silveira[2]

[1] *Departamento de Engenharia Civil - Pontifícia Universidade Católica do Rio de Janeiro*
[2] *Departamento de Engenharia Civil - Escola de Minas - Universidade Federal de Ouro Preto*

Abstract. The analysis of pyramidal trusses has an immediate practical interest since these structures are currently used in many present-day civil constructions, either as main parts or a constitutive element. They can be used to represent tripod-like structures, cap of masts, tower cranes, big span roofs, and even a portion of a single-layer geodesic dome or of a generic-shaped reticulated shell. This paper examines the nonlinear static stability and load capacity for a simple class of space trusses in the shape of a regular pyramid. Joints located at the vertices of the base polygon are fixed while the joint at the apex is subjected to static loads acting in either the vertical direction, in the horizontal plane, or along a generic oblique direction. Despite their apparent simplicity, these structural systems exhibit a wide variety of post-critical responses, not exhausted by the classical snapping and bifurcation phenomena. In addition to regular primary and secondary branches, the equilibrium paths may include neutral branches, namely branches entirely composed of bifurcation or limit points. The analysis is conducted using the Finite Element Method together with a corotational formulation for the bars. The numerical results are validated in the elastic domain using the closed-form solutions found in literature.

1 Introduction

In this work the large displacement response of a class of space trusses in the shape of a regular pyramid is studied using a finite element formulation. Pyramidal trusses possess an immediate practical interest since they are currently used in many present-day civil constructions, either as main parts or as minor elements. For instance, if the number n of the composing bars is 3, we find a common tripod-like structure loaded on its upper joint. For n = 4 and a large height to base ratio, the truss may represent the cap of a mast or, if the load acts transversally, the extremity of a three-dimensional cantilever beam or of the jib of a tower crane. Finally, for n = 5 or n = 6 and a small height to base ratio, the truss may represent a molecule of a single-layer geodesic dome or of a generic-shaped reticulated shell [6].

The paper is organized as follows: in Section 2, the system equilibrium equations are posed considering an elastic response and thick elements. Section 3 presents the nonlinear effects in the structural response: snap-through and bifurcation. Finally, some concluding remarks are given in Section 4.

2 Analytic Model

The pyramidal truss considered in this study is shown in Fig. 1. It consist of $n \geq 3$ bars pinned at the joints and equally spaced so that the base nodes form a regular polygon of n sides simply supported and equally separated at the regular base Γ and connected at the apex node. The pyramid has height H, base radius B and bar's length $L = \sqrt{B^2 + H^2}$.

In the initial configuration, the apex and i^{th} base nodes are located at $\begin{bmatrix}0 & 0 & H\end{bmatrix}^T$ and $\begin{bmatrix}B\cos(\theta_i) & B\sin(\theta_i) & 0\end{bmatrix}^T$, respectively, with $\theta_i = 2\pi i/n$. The apex node is free to move and it's position is given by $\mathbf{x} = \begin{bmatrix}x & y & z\end{bmatrix}^T$ or, in a cylindrical coordinates system, by $\mathbf{x} = \begin{bmatrix}\theta & r & z\end{bmatrix}^T$. The structure is subjected to a load $\mathbf{p} = \lambda\begin{bmatrix}0 & 0 & p_z\end{bmatrix}^T$. Here, it is assumed that all joints behave as ideal frictionless hinges, so that no moments are generated in them by nodal loads.

Following [6], considering no imperfections, the current length of the bar connecting the apex and i^{th} nodes is:

$$l_i^2 = r^2 - 2rB\cos(\theta - \theta_i) + B^2 + z^2 \tag{1}$$

As in the bar element, the Green-Lagrange strain is given by $\varepsilon_i = (l_i^2 - L^2)/(2L^2)$ and thus, considering an elastic material response, the axial force is $N_i = (l_i/L)\, EA\varepsilon_i$. The system internal $W(\mathbf{x})$ and external $V(\mathbf{x}, \lambda)$ potential energies can be written as:

$$W(\mathbf{x}) = \frac{nEA}{8L^3}\left[\left(r^2 + z^2 - H^2\right)^2 + 2B^2r^2\right] \tag{2}$$

$$V(\mathbf{x}, \lambda) = \lambda p_z (z - H) \tag{3}$$

*e-mail: murilvbs@aluno.puc-rio.br

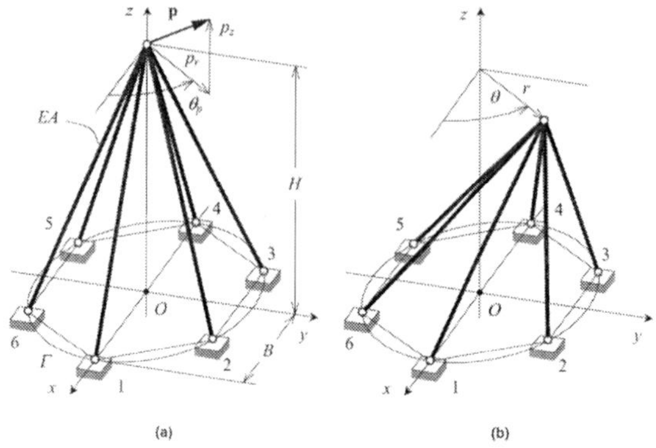

Figure 1. Pyramidal truss: (a) Initial and (b) Current equilibrium configuration. Source [6]

The system's total potential energy is given by $\Pi(\mathbf{x}, \lambda) = W(\mathbf{x}) - V(\mathbf{x}, \lambda)$. Applying the minimum potential energy principle, the nonlinear equilibrium equations are obtained as:

$$\mathbf{r}(\mathbf{x}, \lambda) = \frac{\partial \Pi}{\partial \mathbf{x}} = \mathbf{F}_i(\mathbf{x}) - \mathbf{F}_e(\lambda) = \mathbf{0} \qquad (4)$$

where the system internal $\mathbf{F}_i(\mathbf{x})$ and external $\mathbf{F}_e(\lambda)$ force vectors are given by:

$$\mathbf{F}_i(\mathbf{x}) = \frac{nEA}{2L^3}\begin{bmatrix} x\mu_1^2 & y\mu_1^2 & z\mu_2^2 \end{bmatrix}^T \qquad (5)$$

$$\mathbf{F}_e(\lambda) = \lambda\begin{bmatrix} 0 & 0 & p_z \end{bmatrix}^T \qquad (6)$$

In the above equations we have $\mu_1^2 = r^2 + z^2 + B^2 - H^2$ and $\mu_2^2 = r^2 + z^2 - H^2$. The system stiffness matrix is obtained by taking the variation of the internal force vector in Eq. 5:

$$\mathbf{K}(\mathbf{x}) = \frac{\partial \mathbf{F}_i}{\partial \mathbf{x}} = \frac{nEA}{2L^3}\begin{bmatrix} 2x^2 + \mu_1^2 & 2xy & 2xz \\ 2xy & 2y^2 + \mu_1^2 & 2yz \\ 2xz & 2yz & 2z^2 + \mu_2^2 \end{bmatrix} \qquad (7)$$

3 Analysis

In this section, the static stability and load capacity of the system are studied. The effects of two types of nonlinear behavior are considered: snap-through and bifurcation. The snap-through behavior is characterized by two limit points along the nonlinear equilibrium path, where a dynamic jump to an inverted position occurs. This buckling behavior is typical of very shallow systems, where jump can occur in the elastic range.

For deeper structures, however, it's possible that the primary equilibrium path becomes unstable before the structure reaches the limit points. In such cases, the truss looses stability through an unstable symmetric bifurcation, where the secondary path connects this critical point to its mirror inverted position. This bifurcation is associated to an asymmetric buckling mode. Again after reaching the bifurcation point the structure jumps to an inverted configuration.

Along both paths and during the dynamic jump the bars are subjected to compressive axial forces, which can become greater than the Euler's buckling load [2] or the stresses may reach the material elastic limits [5], leading to an inelastic nonlinear response. In such cases, the local instability of the bars and the plastic behavior of the material may influence the load carrying capacity of the structure. In addition, when subjected to unexpected loads, it's possible that the structure suffers members loss. The number and the way that the that the system losses it's elements may have great influence in it's stability and load capacity.

In all the numerical models, the pyramidal truss has $n = 6$ bars and base radius $B = 7.00$m. The structure height H is chosen depending on the nonlinear effect studied. Also, the bars have a hollow circle cross section with diameter $d = 508.0$mm and thickness $t = 12.7$mm. The material has an elastic modulus $E = 200$GPa and yield stress $\sigma_y = 400$MPa. The reference load is taken as $p_z = -(H/L)^3(nEA/2)$. The solution of the nonlinear equations is obtained through an incremental - iterative Newton - Raphson process [3] associated with an arc-length method [4]. In addition, the following non-dimensional parameters are used in the analysis $\bar{z} = z/H$, $\bar{r} = r/H, \alpha = B/H, \beta = L/H = \sqrt{1+\alpha^2}$ and $\bar{\omega}_i = \omega_i/\omega_{1,0}$.

The static global behavior of the pyramidal lattice truss is mainly governed by the topology of the total potential energy. In terms of non-dimensional quantities, substituting the parameters defined before in Eq. (2), the static potential energy of the structure ($\overline{\Pi}$) is given by:

$$\overline{\Pi} = \Pi/(p_z H) \qquad (8)$$

$$\overline{\Pi} = \left[\left(\bar{r}^2 + \bar{z}^2 - 1\right)^2 + 2\alpha^2\bar{r}^2\right]/4 - \lambda(\bar{z} - 1) \qquad (9)$$

The static unloaded equipotential curves for the three selected values of α are shown in Fig. 2. For $\alpha = 0.80 < 1.00$, Fig. 2(a), the potential energy displays five equilibrium positions, the origin, a local maximum (unstable), two additional equilibrium positions along the $\bar{z}$ axis at $\bar{z} = \pm 1.00$, corresponding to two local minima (stable), and two additional equilibrium positions along the $\bar{r}$ axis at $\bar{r} = \pm\sqrt{1-\alpha^2}$, corresponding to two symmetric saddles (unstable).

At $\alpha = 1.00$, Fig. 2(b), the two saddles coalesce with the maximum, leading to a degenerated saddle, with the only eigenvector in the direction of the axis, reducing the number of equilibrium positions to three. For $\alpha = 1.20 > 1.00$, Fig. 2(c), there are only three equilibrium positions, a saddle (unstable) at the origin and two symmetric minimum (stable) along the $\bar{z}$ axis. So, $\alpha = 1.00$ corresponds to a critical value separating two types of global behavior.

The static loaded equipotential curves with $\alpha = 0.80$ and $\lambda = 0.25$ are shown in Fig. 3. The presence of a load changes the topology of the total potential energy, modifying the number of solutions, the position of the equilibrium points on the $(\bar{r}, \bar{z})$ plane and their stability. The load breaks the symmetry of potential with respect to the $\bar{r}$ axis, shifting the equilibrium points away from it.

One should have in mind that Figs. 2 and 3 represents the cross sections of the four dimensional phase-space of the conservative system and that the stable and unstable manifolds of the saddles separate the initial conditions that lead to bounded solutions surrounding the stable equilibrium positions of the unloaded structure. The knowledge of these frontiers helps the designer to separate the phase space into safe and unsafe domains and evaluate the degree of safety of the system. In the following, a parametric analysis of the truss under vertical and horizontal loads is presented.

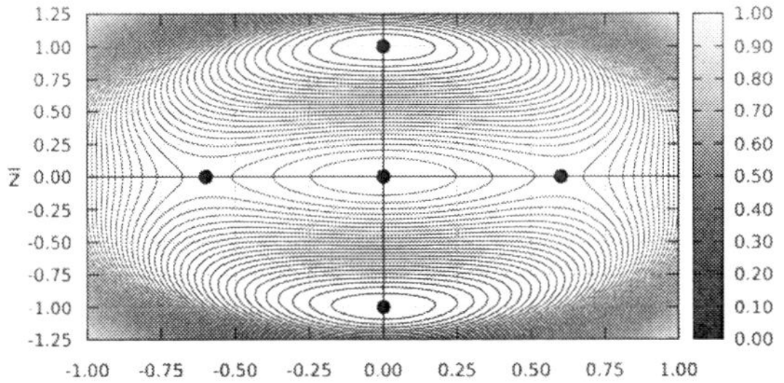

(a) $\alpha = 0.80$

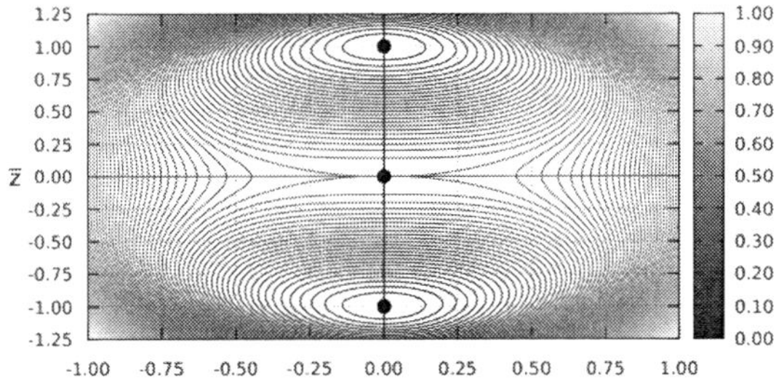

(b) $\alpha = 1.00$

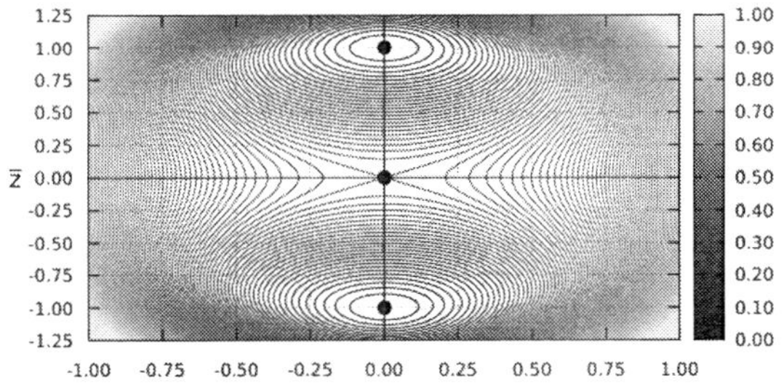

(c) $\alpha = 1.20$

Figure 2. Unloaded equipotential energy curves.

3.1 Snap-through

Initially the elastic symmetric response of the system (nonlinear primary path) subjected to a vertical load at the apex node is considered, for which case we have $r = 0$ and, from Eqs. 4-6:

$$\lambda = \bar{z}\left(1 - \bar{z}^2\right) \tag{10}$$

The differentiation of the previous equation give the limit points $(\bar{z}_l^v, \lambda_l^v)$ as:

$$\bar{z}_l^v = \pm\sqrt{3}/3 \quad \lambda_l^v = \pm 2\sqrt{3}/9 \tag{11}$$

From the stiffness (Eq. 7), the instability conditions are $|\bar{z}| > \sqrt{3}/3$ and $|\bar{z}| > \sqrt{1-\alpha^2}$. The first condition is associated with the limit points of the snap-through behavior while the second one only appears in deep systems and is associated with the bifurcation behavior (Sec. 3.2), for shallow systems this condition is automatically satisfied. The equilibrium path and limit points with $\alpha = 1.20$ are shown in Fig. 4.

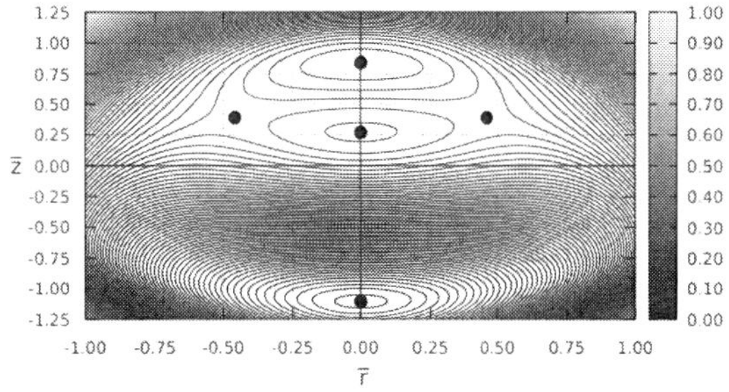

Figure 3. Loaded equipotential energy curves.

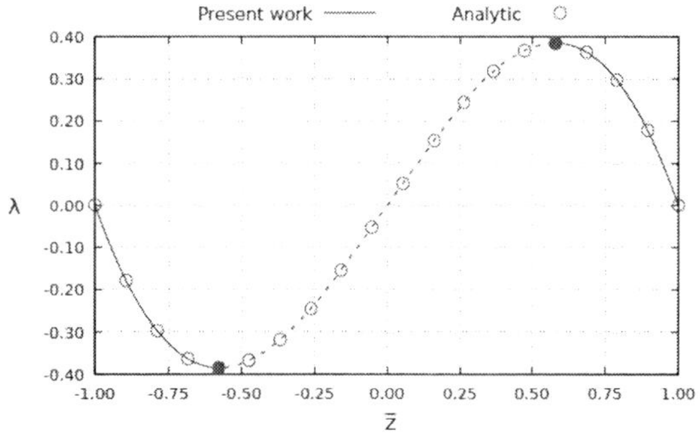

Figure 4. Equilibrium path: snap - through ($\alpha = 1.20$).

3.2 Bifurcation

Now we study the bifurcation (nonlinear secondary path) response of the system. For the elastic structure we consider a $0.1\%H$ imperfection in the x direction of the apex node for the numerical analysis. Considering $r \neq 0$ and combining the x and y components in Eqs. ??-6 we have $r = \sqrt{H^2 - B^2 - z^2}$ and:

$$\lambda = \alpha^2\bar{z} \tag{12}$$

The bifurcation points $(\bar{z}_b^v, \lambda_b^v)$ are the intersections of the equilibrium paths in Eqs. 10 and 12:

$$\bar{z}_b^v = \pm\sqrt{1-\alpha^2} \quad \lambda_b^v = \pm\alpha^2\sqrt{1-\alpha^2} \tag{13}$$

The previous equations provides the bifurcation condition as $H \geq H_b = B$. In addition, from Eqs. (11) and (13), the limit point $(\bar{z}_l^v, \lambda_l^v)$ and bifurcation point $(\bar{z}_b^v, \lambda_b^v)$ under vertical load coincide if $\alpha = \sqrt{6}/3$. For $\alpha > \sqrt{6}/3$, the bifurcation point appears on the unstable part of the nonlinear equilibrium path, without influence on the load capacity but changing the static behavior. The bifurcation point disappears for $\alpha > 1$.

The apex node movement is illustrated in Fig. 5. The equilibrium path is shown in Fig. 6 for $\alpha = 0.7$. We can notice that, in the secondary branch, the system loses stability, but later recovers it, returning to the primary branch, as the compressive axial force in the bars is reduced.

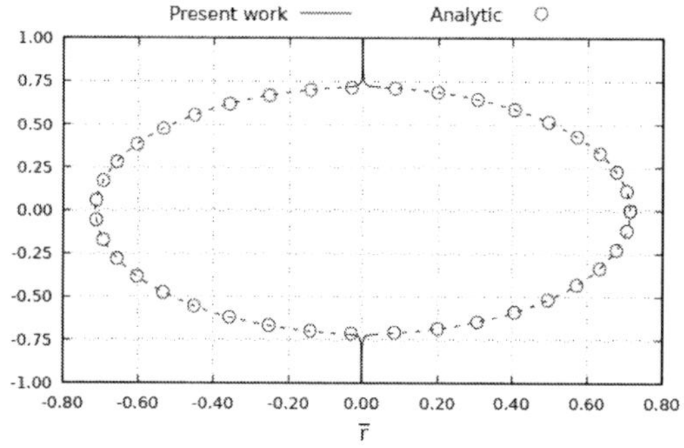

Figure 5. Apex movement: bifurcation ($\alpha = 0.7$).

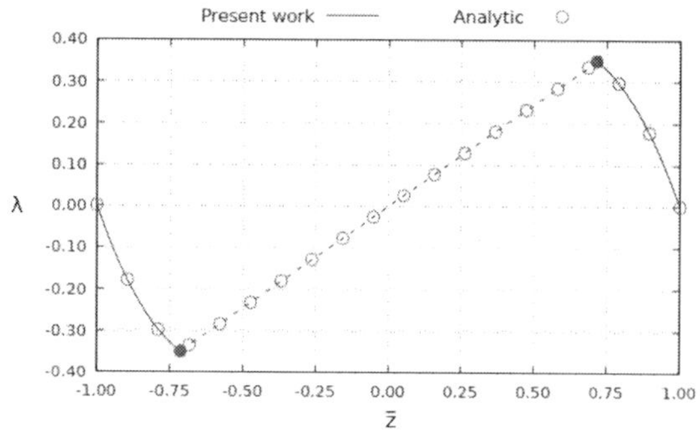

Figure 6. Equilibrium path: bifurcation ($\alpha = 0.7$).

4 Conclusions

In this work, the stability and load capacity for space trusses in the shape of a regular pyramid was studied. Using a Finite Element formulation, the numeric response was first validated with the analytical one proposed by [6] in the elastic range for the system snap-through and bifurcation. The snap-through effect is present regardless of the system parameters, creating limit load points and instability. For deep structures, global instability may appear as a bifurcation in the equilibrium path.

References

[1] Battini, J. M., Pacoste, C., Comput. Methods Appl. Mech. Engrg, 191, 1755–1789 (2002)

[2] Bazzucchi, F., Manuello, A., Carpinteri, A., International Journal of Non-Linear Mechanics, 88, 11-20 (2017)

[3] Crisfield, M. A., Wiley, Chichester (1991)

[4] Krenk, S., Cambridge University Press, New York (2009)

[5] Jiang, X. M., Chen, H., Liew, J. Y. R., Journal of Constructional Steel Research, 58, 193–212 (2002)

[6] Ligarò, S. S., Valvo, P. S., International Journal of Solids and Structures, 43, 4867–4887 (2006)

Chaos prediction in nano-resonators based on nonlocal elasticity theory

Hassan Nahvi [1*] and *Mehdi Maleki* [1]

[1]Department of Mechanical Engineering, Isfahan University of Technology, Isfahan 8415683111, Iran

Abstract. By decreasing the thickness of micro- and nano- beams, classical continuum theory is not accurate to predict the static and dynamic response due to the absence of length scale parameter. In this paper, nonlocal elasticity theory is used to detect chaos in nano-resonators. In this way, first mode shape of the nano-beam is found and Galerkin method is used to convert the governing partial differential equation to an ordinary differential equation. Melnikov method is used to determine the critical value of AC actuation voltage resulting chaotic motion. Effects of nonlocal parameter and beam thickness on the stability region of the resonator are investigated. It will be shown that increasing the nonlocal parameter and decreasing the beam thickness increases the difference between stability regions obtained by classical and nonlocal theories. Moreover, increasing the nonlocal parameter decreases the nonlinear stiffness and increases the critical actuation voltage which may lead to chaotic motion.

1 Introduction

Several decades have passed by since the discovery and development of micro-electro-mechanical systems (MEMS). One important type of MEMS is resonator which has various applications [1]. Due to nonlinear behavior of this device caused by nonlinear electrical force, squeezing film damping, mid-plane stretching and other geometrical nonlinearities, some complicated phenomena have been seen in the dynamic behavior of resonators. Nonlinearities also may lead to chaotic behavior [2]. Haghighi and Markazi [3] used Melnikov analysis method to predict chaos in MEMS resonators. They proposed an analytical criterion for homoclinic chaos in the form of an inequality which is written in terms of the system parameters.

Experimental tests have shown that static and dynamic behavior of micro- and nano- beams cannot be predicted by classical continuum theory with the absence of an intrinsic length scale in the constitutive relations [4]. Nonlocal elasticity theory is one of the various higher-order elasticity theories which have been used to develop microstructure-dependent beam models in order to study the static and dynamic response of micro- and nano-beams [5, 6].

Many researches on the local and nonlocal theories have been carried out not only considering the beam model but also considering the thin shell model, in the linear and nonlinear fields, with particular application to carbon nanotubes [7, 8].

In this paper, based on Melnikov analysis method, chaotic region for a nano-resonator obtained with nonlocal elasticity theory will be compared with that of classical one and the effect of nonlocal parameter on the safe region of the resonator in which the system has harmonic response will be investigated.

2 Mathematical modeling

Figure 1 shows layout of a conventional resonator under electrical actuation.

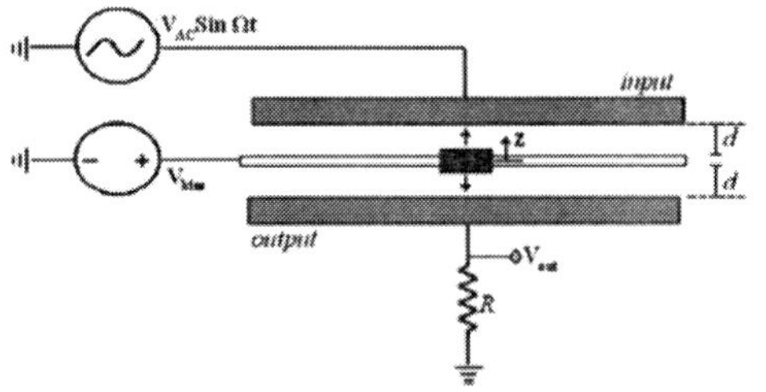

Figure 1. Layout of a nano-resonator under electrical actuation [3].

Dynamic equation of the nano-resonator based on nonlocal theory can be written as [6]:

$$-2EI(e_0a)^2\frac{\partial^6 w}{\partial x^6}+(EA+(e_0a)^2P)\frac{\partial^4 w}{\partial x^4}+\rho A\frac{\partial^2 w}{\partial t^2}+C\frac{\partial w}{\partial t}-(P+\frac{EA}{2L}\int_0^1(\frac{\partial w}{\partial x})^2dx)\frac{\partial^2 w}{\partial x^2}-q(x,t)=0 \qquad (1)$$

where w is beam deflection, EI is flexural rigidity, A is beam cross-sectional area, ρ is material mass density, P is axial force, C is damping coefficient,

* Corresponding author: hnahvi@cc.iut.ac.ir

q is external load and e_0a is the size effect parameter which can be obtained by fitting against experiments and molecular dynamic simulations. The terms appear in Eq. (1) correspond to the size effect (which is ignored in the classical Euler-Bernoulli beam model), micro-beam elastic force, micro-beam acceleration, damping, axial force and mid-plane stretching, and external load, respectively.

Using the following non-dimensional parameters, Eq. (1) reduces to Eq. (3) as:

$$\bar{x} = \frac{x}{L}, \quad \bar{w} = \frac{w}{g}, \quad \bar{t} = \frac{t}{T} \tag{2}$$

$$\Re(\bar{w}) = -2\tau_n^2 \frac{\partial^6 \bar{w}}{\partial \bar{x}^6} + (1+\bar{P}\tau_n^2)\frac{\partial^4 \bar{w}}{\partial \bar{x}^4} + \frac{\partial^2 \bar{w}}{\partial \bar{t}^2} + c\frac{\partial \bar{w}}{\partial \bar{t}}$$
$$-(\bar{P} + \alpha_1 \int_0^1 (\frac{\partial \bar{w}}{\partial \bar{x}})^2 d\bar{x})\frac{\partial^2 \bar{w}}{\partial \bar{x}^2} - \bar{q} = 0 \tag{3}$$

where L is the micro-beam length and g is the gap between two electrodes and

$$\alpha_1 = \frac{EAg^2}{2EI}, \quad \bar{P} = \frac{L^2}{EI}P, \quad \tau_n = \frac{e_0 a}{L},$$
$$c = \frac{CL^4}{EIT}, T = \sqrt{\frac{\rho bhL^4}{EI}} \tag{4}$$

In order to use Galekin method, linear mode shapes of beam based on nonlocal theory should be found. The solution of Eq. (3) can be assumed as [9]:

$$\bar{w}_n(\bar{x},\bar{t}) = \psi_n(\bar{x})\sin(\omega_n \bar{t}) \tag{5}$$

where $\psi_n(\bar{x})$ is the nth mode function and ω_n is the dimensionless natural frequency. The natural frequencies and mode shapes of the beam are found by setting nonlinear terms and external excitation in Eq. (3) equal to zero. Substituting Eq. (5) into Eq. (3) yields:

$$-2\tau_n^2 \frac{d^6\psi_n}{d\bar{x}^6} + (1+\bar{P}\tau_n^2)\frac{d^4\psi_n}{d\bar{x}^4} - \omega_n^2 \psi_n = 0 \tag{6}$$

Micro-beam boundary conditions for clamped ends based on the nonlocal theory are [7]:

$$\psi_n(0) = \psi_n(1) = 0$$
$$\frac{d\psi_n(0)}{d\bar{x}} = \frac{d\psi_n(1)}{d\bar{x}} = 0$$
$$\frac{d^3\psi_n(0)}{d\bar{x}^3} + 3\tau_n^2 \frac{d^5\psi_n(0)}{d\bar{x}^5} = \frac{d^3\psi_n(1)}{d\bar{x}^3} + 3\tau_n^2 \frac{d^5\psi_n(1)}{d\bar{x}^5} = 0 \tag{7}$$

General solution of Eq. (6) can be written as:

$$\psi_n(\bar{x}) = C_1 e^{\beta_1 \bar{x}} + C_2 e^{\beta_2 \bar{x}} + C_3 e^{\beta_3 \bar{x}} + C_4 e^{\beta_4 \bar{x}}$$
$$+ C_5 e^{\beta_5 \bar{x}} + C_6 e^{\beta_6 \bar{x}} = C_k \sum_{k=1}^{6} e^{\beta_k \bar{x}} \tag{8}$$

In the above equation, C_1 to C_6 are integration constants of the sixth-order differential equation, Eq. (6), and β_1 to β_6 are the roots of the following equation:

$$-2\tau_n^2 \beta^6 + (1+\bar{P}\tau_n^2)\beta^4 - \omega_n^2 = 0 \tag{9}$$

Substituting Eq. (8) into Eq. (7) gives the following matrix equation:

$$\mathbf{A}(\beta_1,\beta_2,\beta_3,\beta_4,\beta_5,\beta_6)[C_1,C_2,C_3,C_4,C_5,C_6]^T = \mathbf{0} \tag{10}$$

To obtain non-trivial solution for the constants C_i, the determinant of matrix A should be zero, thus:

$$|\mathbf{A}| = 0 \tag{11}$$

By using Eqs. (9) and (11), the seven unknown quantities $(\beta_1,\beta_2,\beta_3,\beta_4,\beta_5,\beta_6,\omega_n)$ can be found. Substituting the results into Eq. (10) and setting C_1 as an arbitrary constant gives the constants C_2 to C_6. The one-parameter Galerkin's solution may be determined using the following relation:

$$\int_0^1 \Re(\bar{w})\psi_n(\bar{x})d\bar{x} = 0 \tag{12}$$

where $\bar{w} = \psi_n(\bar{x})z(\bar{t})$ based on the variable separation and z is the beam mid-point deflection.

The governing differential equation of the beam mid-point, $z(\bar{t})$, is found to be:

$$mz'' + k_1 z + k_3 z^3 + bz' = F_{act} \tag{13}$$

where ($'$) represents derivative with respect to $\bar{t}$. The parameters in Eq. (13) are:

$$m = \int_0^1 \psi_n(\bar{x})^2 d\bar{x}$$

$$k_1 = \int_0^1 \psi_n(\bar{x})(-2\tau_n^2 \frac{d^6\psi_n(\bar{x})}{d\bar{x}^6} + (1+\bar{P}\tau_n^2)\frac{d^4\psi_n(\bar{x})}{d\bar{x}^4} - \bar{P}\frac{d^2\psi_n(\bar{x})}{d\bar{x}^2})d\bar{x}$$

$$k_3 = -\int_0^1 \alpha_1 \psi_n(\bar{x})\frac{d^2\psi_n(\bar{x})}{d\bar{x}^2}(\int_0^1 (\frac{d\psi_n(\bar{x})}{d\bar{x}})^2 d\bar{x})d\bar{x}$$

$$b = c\int_0^1 \psi_n(\bar{x})^2 d\bar{x} \tag{14}$$

In Eq. (13), F_{act} is the electrostatic force which is the combination of DC and AC actuations. By introducing non-dimensional time as $\tau = \omega_0 \bar{t}$, Eq. (13) can be rewritten as:

$$\ddot{z} + z + \beta z^3 + \mu \dot{z} = \gamma\left(\frac{1}{(1-z)^2} - \frac{1}{(1+z)^2}\right) - \frac{A\cos(\omega\tau)}{(1-z)^2} \quad (15)$$

where $\dot{(\ )}$ denotes derivation with respect to τ. The parameters in Eq. (15) are defined as:

$$\omega_0 = \sqrt{\frac{k_1}{m}};\ \gamma = \frac{\varepsilon b l^4 V_b^2}{2\omega_0^2 E I g^3},\quad A = 2\gamma\frac{V_{AC}}{V_b},$$
$$\beta = \frac{k_3}{m\omega_0^2},\quad \mu = \frac{b}{m\omega_0},\ \omega = \frac{T\Omega}{\omega_0} \quad (16)$$

In Eq. (16), ε is the permittivity of vacuum, V_b is the bias DC voltage, V_{AC} is the amplitude of applied AC voltage and as shown in Figure (1), Ω is the frequency of AC voltage.

3 Melnikov analysis

In order to apply Melnikov method, Taylor expansion is applied to Eq. (15) and then is written in state space form as [3]:

$$\dot{z} = y$$
$$\dot{y} = (1-4\gamma)z + (\beta - 8\gamma)z^3 + \varepsilon\left(-\bar{\mu}\dot{z} + \frac{\bar{A}}{(1-z)^2}Sin\,\omega\tau\right) \quad (17)$$

where $\mu \equiv \varepsilon\bar{\mu}$ and $A \equiv \varepsilon\bar{A}$ (i.e., $\mu = O(\varepsilon)$, $A = O(\varepsilon)$) and ε is a small parameter. These assumptions are valid for a MEMS resonator with high quality factor and AC diving voltage much smaller than the bias voltage. The Melnikov function is proportional to the first variation of the distance function between stable and unstable manifolds. The Melnikov function of Eq. (17) is:

$$M(\tau_0) = \int_{-\infty}^{\infty}\left[y_0\left(-\bar{\mu}y_0 + \frac{\bar{A}}{(1-z_0)^2}Sin\,\omega(\tau+\tau_0)\right)\right]d\tau \quad (18)$$

where (z_0, y_0) is the solution corresponding to the unperturbed homoclinic trajectory, as

$$\dot{z} = y$$
$$\dot{y} = (1-4\gamma)z + (\beta-8\gamma)z^3 \quad (19)$$

The solution of Eq. (19) is analytically obtained as:

$$(z_0, y_0) = \mp z_e(\text{sech}(\sqrt{1-4\gamma}\tau), \text{sech}(\sqrt{1-4\gamma}\tau)\tanh(\sqrt{1-4\gamma}\tau)) \quad (20)$$

where z_e is found by setting potential function equal to zero.

The integral in Eq. (18) is solved using residue theorem and the following inequality was found for the chaos region [3]:

$$\frac{\bar{A}}{\bar{\mu}} \geq \left|\frac{z_e\sqrt{4\gamma-1}\sqrt{1-z_e^2}\,Sinh[\bar{\omega}\pi]}{3\pi\bar{\omega}\,Sinh[-\bar{\omega}(\text{ArcCos}[z_e])]}\right| \quad (21)$$

where $\bar{\omega} = \frac{\omega}{\sqrt{4\gamma-1}}$. For more details the reader is referred to the reference [3].

4 Simulations and results

In this section, the Melnikov method will be used to analyse the effect of the size on the chaotic motion of nano-resonators. In the following, a nano-beam made of silicon will be considered as an illustrative example. From experimental and numerical atomic analyses, the relevant bulk material properties for silicon are: $E = 165\text{GPa}$, $\nu = 0.24$ and $\rho = 2332\text{kg/m}^3$. The nano-resonator geometrical parameters used here are listed in Table 1.

Table 1. Nano-resonator parameters.

Parameter	Value
Length	$L = 50\mu m$
Width	$b = 5\mu m$
Thickness	$h = 1\mu m$
Initial gap	$g = 3\mu m$

Figure 2 shows the effect of nonlocal parameter on the chaotic motion of the nano-resonator using Eq. (21). It should be noted that the region below the curve corresponds to the harmonic motion and chaotic motion will not occur in this region. As shown, for small values of the nonlocal parameter, both theories lead to the same result but increasing the nonlocal parameter increases the difference between the results obtained by the two theories.

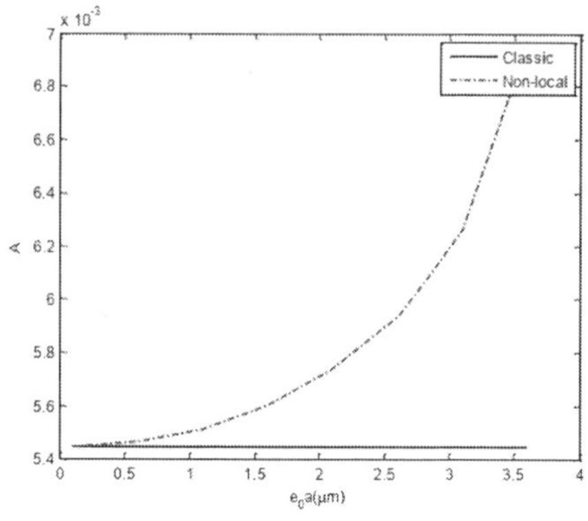

Figure 2. Critical values for AC as a function of nonlocal parameter.

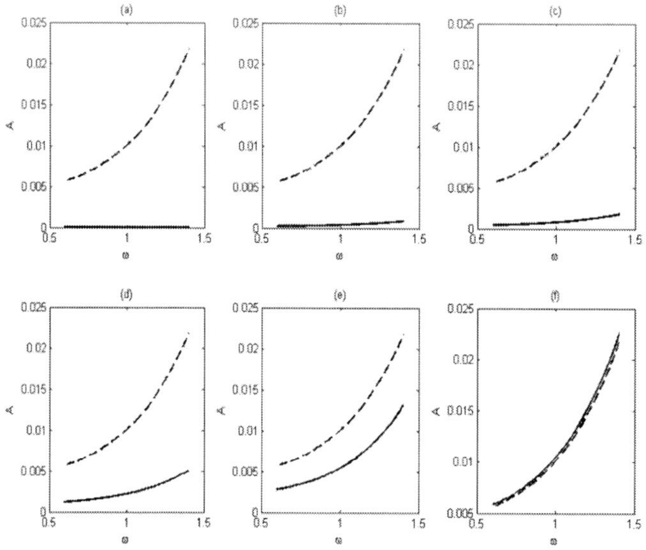

Figure 3. Critical values for AC actuation for (a) $h = 10\text{nm}$, (b) $h = 20\text{nm}$, (c) $h = 100\text{nm}$, (d) $h = 500\text{nm}$, (e) $h = 1\mu\text{m}$, (f) $h = 10\mu\text{m}$; classical theory -------- , nonlocal theory - -

The predicted region for the chaotic motion based on Melnikov method, Eq. (21), is shown in **Грешка! Източникът на препратката не е намерен.**3 for different beam thicknesses. The nonlocal parameter for this simulation is set as $e_0 a = 6\mu\text{m}$.

It can be seen that, for beam thickness values higher than $10\,\mu\text{m}$, the predicted regions for chaos based on classical continuum and nonlocal theories emerge and size effect is not present. By decreasing the beam thickness, the size effect becomes dominant and the harmonic region based on the nonlocal theory becomes different to the one obtained by classical continuum theory.

The nano-resonator response for $\omega = 1$, $h = 10\text{nm}$ and $A = 0.003$ is simulated using classical and nonlocal theories. Figure (a) and 4(b) show the time response and phase plane of the nano-resonator based on classical theory, respectively. As can be seen, the response of the system is chaotic, which is in agreement with Melnikov analysis.

Figure (5) shows the Poincare map of the system response based on classical theory which indicates chaotic motion.

The corresponding results for the nano-resonator based on the nonlocal theory are shown in Figure 6. It can be seen that the response of the system is not chaotic and the vibration of the nano-resonator is harmonic.

Figure (7) shows the Poincare map of the system based on the nonlocal theory which indicates harmonic motion.

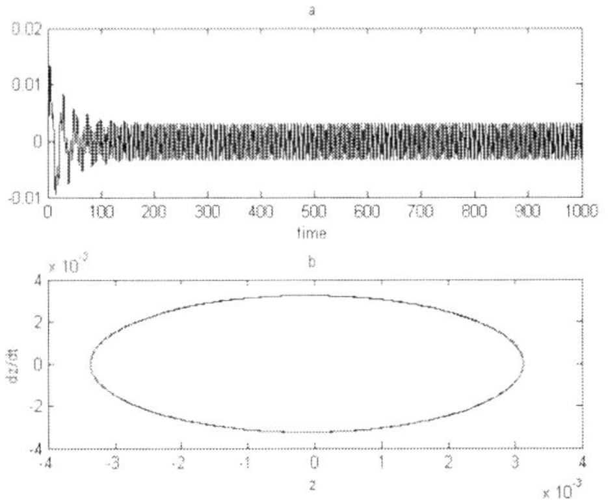

Figure 4. (a) Time response and (b) phase plane of the nano-resonator using classical continuum theory.

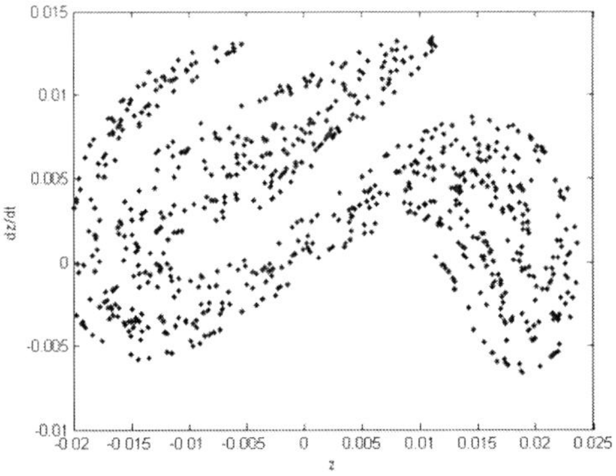

Figure 5. Poincare map of the nano-resonator based on classical theory.

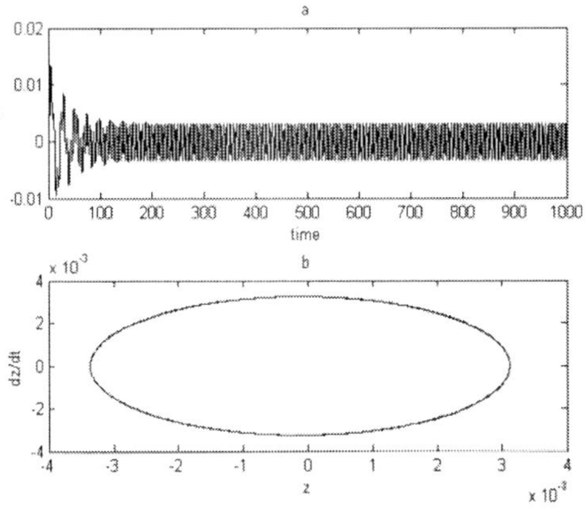

Figure 6. (a) Time response and (b) phase plane of the nano-resonator using nonlocal continuum theory.

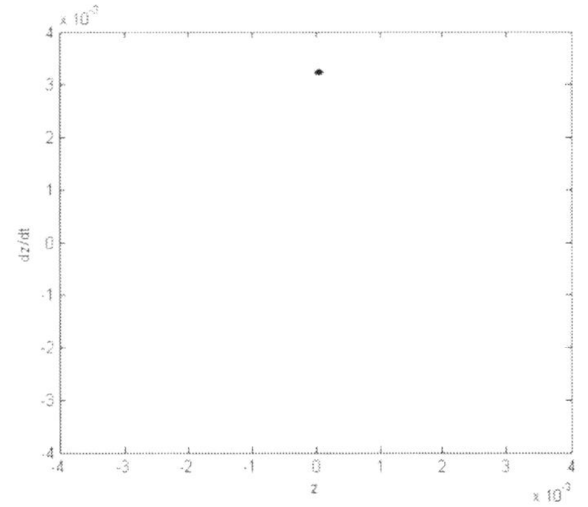

Figure 7. Poincare map of the nano-resonator system based on nonlocal theory.

5 Conclusions

In this paper, nonlocal elasticity theory is used to predict chaos in MEMS/NEMS resonators. In order to use Galerkin decomposition method, the mode shape of the beam based on nonlocal theory is found and used to convert the governing differential equation to ordinary one. Melnikov method is used to obtain the critical actuation amplitude leading to chaotic motion. Effects of the nonlocal parameter and the beam thickness are investigated and the following results are obtained:

- Increasing the nonlocal parameter decreases the nonlinear stiffness and increases the critical actuation voltage which may lead to chaotic motion.
- For the beams thicknesses more than 10 micrometer, size effect can be ignored.
- Decreasing the beam thickness increases the difference between the results of the two theories.

References

1. A. Tocchio, A. Caspani, G. Langfelder, *Sens. J. IEEE*, **12**, 1719-25, (2012)
2. A.C. Luo, F.-Y. Wang, *Commun. Nonl. Sci. Num. Simul.*, **7**, 31-49, (2002)
3. H.S. Haghighi, A.H. Markazi, *Commun. Nonl. Sci. Num. Simul.*, **15**, 3091-9, (2010)
4. C. Tang, G. Alici, *J. Phys. D: Appl. Phys.*, **44**, 33550, (2011)
5. A.C. Eringen, *Nonlocal continuum field theories*: Springer, (2002)
6. C. Lim, C. Li, J.-L. Yu, *Acta Mech. Sin.*, **26**, 755-765, (2010)
7. Y.-G. Hu, K.M. Liew , Q. Wang, X. He, B. Yakobson, *J. Mech. Phys. Sol.*, **56**, 3475-3485, (2008)

8. M. Strozzi, L.I. Manvitch, F. Pellicano, V.V. Smirnov, D.S. Shepelev, *J. Sou. Vib.*, **233**, 2936-2957, (2014)
9 C.W. Lim, C. Li, J.L. Yu, *Interact. Multis. Mech.*, **2**, 223-233, (2009)

Stochastic Bifurcation Analysis of an Elastically Mounted Flapping Airfoil

Chandan Bose[1,*], *Sunetra* Sarkar[2,**], and *Sayan* Gupta[1,***]

[1] *Department of Applied Mechanics, Indian Institute of Technology Madras, Chennai 600036, India*
[2] *Department of Aerospace Engineering, Indian Institute of Technology Madras, Chennai 600036, India*

Abstract. The present paper investigates the effects of noisy flow fluctuations on the fluid-structure interaction (FSI) behaviour of a span-wise flexible wing modelled as a two degree-of-freedom elastically mounted flapping airfoil. In the sterile flow conditions, the system undergoes a Hopf bifurcation as the free-stream velocity exceeds a critical limit resulting in a stable limit-cycle oscillation (LCO) from a fixed point response. On the other hand, the qualitative dynamics changes from a stochastic fixed point to a random LCO through an intermittent state in the presence of irregular flow fluctuations. The probability density function depicts the most probable system state in the phase space. A phenomenological bifurcation (P-bifurcation) analysis based on the transition in the topology associated with the structure of the joint probability density function (pdf) of the response variables has been carried out. The joint pdf corresponding to the stochastic fixed point possesses a Dirac delta function like structure with a sharp single peak around zero. As the mean flow speed crosses the critical value, the joint pdf bifurcates to a crater-like structure indicating the occurrence of a P-bifurcation. The intermittent state is characterized by the co-existence of the unimodal as well as the crater like structure.

1 Introduction

Fluid-structure interaction (FSI) is known to lead to undesirable phenomena like aero-elastic flutter which is usually detrimental for engineering systems. On the contrary, biological flyers take the advantage of FSI to augment their propulsive efficiency by exploiting the coupling between the flexible wings and the surrounding unsteady flow-field. A proper understanding of the role of FSI in natural flights may directly benefit biologically-inspired design of Micro Aerial Vehicles (MAVs) [1]. Moreover, the performance of very light weight flapping wing MAVs is significantly altered in the presence of gusty flow-field.

Flapping wing MAVs have received significant research attention in recent years due to their potential application in surveillance and environmental monitoring [2, 3]. Extensive experimental and computational studies have been carried out to understand the unsteady aerodynamics of flapping wings [4–6]. However, in most of these studies, the effects of flow fluctuations on the flapping flight are ignored by considering a uniform flow condition. However, MAVs often need to operate in severe weather conditions. Therefore, the consideration of realistic flow fluctuations is crucial to investigate the performance of flapping wing MAVs.

There are substantial literature available regarding the effect of wind gust on the dynamical behaviour and structural safety of aircrafts at high Reynolds number regime [7–9]. However, MAVs are more susceptible to gusty flows due to their low inertia and low flight speed. The studies investigating the effect of realistic flow fluctuations in flapping wing aerodynamics are limited. Lian and Shyy [10] observed that a flapping airfoil can significantly alleviate the impact of wind gust as compared to fixed airfoil thus potentially benefiting the MAV design. Shyy *et al.* [11] compared the aerodynamic performance of rigid and flexible airfoils subjected to gusty flows by carrying out numerical simulations as well as experiments and observed that a flexible airfoil is able to maintain high lift to drag ratio as compared to a rigid airfoil. Thus, flexibility can make the flapping flight gust tolerant. Watkins *et al.* [12] reported that the wind gust made up of small scale eddies produce uneven lift distribution over a flapping wing, causing a rolling motion. On the other hand, gusts having large scale eddy motion cause pitch motion. Lian and Shyy [13] observed that both the lift and drag coefficients are altered significantly under gust environment. Wind tunnel tests are also carried out to study the impact of gusty flows on MAVs [14, 15].

However, it is to be noted that most of these studies have been carried out based on rigid wing assumption, where wing flexibility is ignored. The inclusion of flow fluctuations in the flapping wing aerodynamic models poses several computational challenges as the difference in the gust frequency and the flapping frequency leads to a multi-scale problem [10]. The different time scales present in the problem increases the computational cost. The fluid-elastic coupling adds to the difficulty level. Although, wind gusts encompass a broad range of frequencies and amplitudes, majority of the available studies con-

*e-mail: cb.ju.1991@gmail.com
**e-mail: sunetra.sarkar@gmail.com
***e-mail: gupta.sayan@gmail.com

sider a simplified model of harmonic gust with a single frequency [10, 16]. The aerodynamic performance of flexible airfoils under irregular flow-fluctuations is still not well understood. Therefore, a stochastic bifurcation analysis is essential to carry out to understand the effects of irregular fluctuations in the flow on the dynamical stability characteristics of the coupled system. The present study is focussed to investigate the behaviour of flexible wing section under irregular gust. A two degrees of freedom model with pitch-plunge flexibility has been considered to investigate the dynamical behaviour. The irregular gust is modelled as stochastic input and the wing output behaviour is examined qualitatively through a Phenomenological bifurcation (P-bifurcation) analysis [17]. P-bifurcation behaviour is indicative of stochastic bifurcation.

In this present study, the large deformation of a spanwise flexible wing has been modelled by incorporating cubic nonlinearity in the rotational spring along the pitch (torsion) degree of freedom. The nonlinear structural model has been coupled with an inviscid flow solver using a weak coupling strategy to build the FSI framework. The equations related to flow are solved by unsteady vortex lattice method (UVLM). A stochastic bifurcation analysis has been carried out for the two degree-of-freedom pitch-plunge airfoil subjected to noisy flow fluctuations in inviscid condition, with the mean speed of the flow taken to be the bifurcation parameter.

The remainder of the paper consists of the following sections: Section 2 describes the FSI framework and the gust model; P-bifurcation results are presented in Section 3. The paper ends with the concluding remarks at Section 4.

2 Computational methodology

In the present FSI framework, the 2-DOF pitch-plunge flexible flapping system has been coupled (using a partitioned approach based 'weak coupling' strategy) with a potential flow solver where both flow and structure solvers exchange information explicitly at each time step.

2.1 Pitch-plunge flexible flapping model

The nonlinear structural model consists of a rigid NACA 0012 airfoil elastically supported by a cubic nonlinear rotational spring and a translational spring along pitch and plunge degrees of freedom (see Fig. 1). The structural model takes limited mode spanwise flexibility into account along the translational plunge (y) and rotational pitch (α) direction. The elastic center where the axis system is located is at a distance $0.25b$ towards the foil leading edge from the mid-chord with the chord length being $c = 2b$. The center of mass (c.m.) is located at a distance $x_\alpha b$ from the elastic axis toward the foil trailing edge. The direction of positive lift (L) and moment (M) is indicated in Fig. 1. The nondimensionalized structural equations of motion are given as [18],

$$h'' + x_\alpha \alpha'' + 2\,\zeta_h\,(\frac{\bar{\omega}}{U^*})\,h' + (\frac{\bar{\omega}}{U^*})^2\,h = \frac{1}{\pi\mu}\,C_l, \quad (1)$$

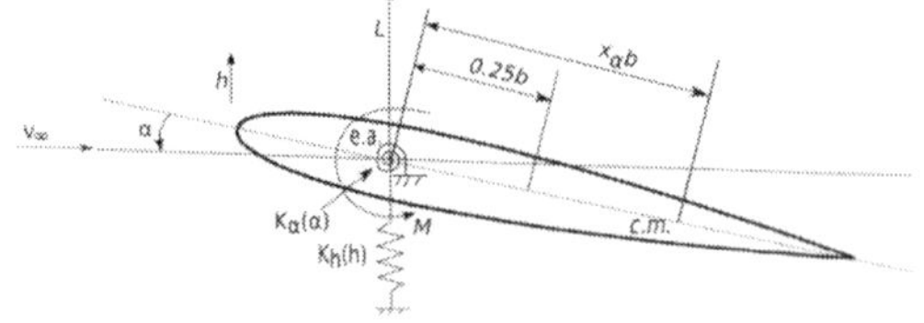

Figure 1. Schematic of airfoil in pitch and plunge degrees of freedom.

$$\frac{x_\alpha}{r_\alpha^{\,2}}\,h'' + \alpha'' + 2\,(\frac{\zeta_\alpha}{U^*})\,\alpha' + (\frac{1}{U^*})^2\,(\alpha + \beta_\alpha \alpha^3) = \frac{2}{\pi\mu r_\alpha^{\,2}}\,C_m. \quad (2)$$

The non-dimensional variables are defined as follows: $h = y/b$; $x_\alpha = S/mb$; $r_\alpha = \sqrt{I/mb^2}$; $U^* = v_\infty/b\omega_\alpha$; $\tau = v_\infty t/b$; $\bar{\omega} = \omega_\xi/\omega_\alpha$; $\mu = m/\pi\rho b^2$. Here, b is the semi-chord, $r_\alpha b$ is the radius of gyration, $x_\alpha b$ is the distance between the center of mass and elastic center; ω_ξ and ω_α are the uncoupled natural frequencies of plunge and pitch respectively; cubic nonlinear co-efficient β_α determines the extent of non-linearity in the spring stiffness along the pitch degree of freedom; v_∞ is the free-stream; C_l and C_m are the unsteady lift and moment coefficients about the aerodynamic center of the airfoil and are computed at each time step using an unsteady vortex lattice method (UVLM) based potential flow solver. The unsteady potential solver details are given in the next subsection. The structural responses are obtained by numerically integrating Eq. 1-2 using an explicit fourth-order Runge-Kutta method. The time step for integrating structural equations of motion is taken to be equal to that of the flow solver.

2.2 Flow solver

An 2D unsteady vortex lattice method (UVLM) based potential flow solver is implemented following the Hess and Smith approach to estimate the aerodynamic loads in the coupled system governing equations [19, 20]. The flow solver assumes the flow to be inviscid, incompressible and irrotational and is governed by the Euler equation,

$$\frac{\partial \vec{u}}{\partial t} + \vec{u}.\nabla\vec{u} = \frac{-\Delta p}{\rho}, \quad (3)$$

where $\vec{u}$ is the flow velocity, p is the pressure and ρ is the fluid density. A schematic of the UVLM approach has been presented in Fig. 2.

Unlike the classical analytical approach (Theodorsen's Method [21]), in this method, the unsteady wake is not assumed to be rigid and is free to evolve with its own local velocities. Hence, the actual shape of the wake is captured. The wake consists of discretized wake elements. Furthermore, the actual shape of the airfoil is considered to compute the flow-field variables. The airfoil surface is divided into 'N' number of small segments called panels. In the present study, the number of panels (N) is chosen to be 400 based on a panel convergence study. Each panel is represented using two types of singularity elements, sources and vortices. The source singularity strength is considered

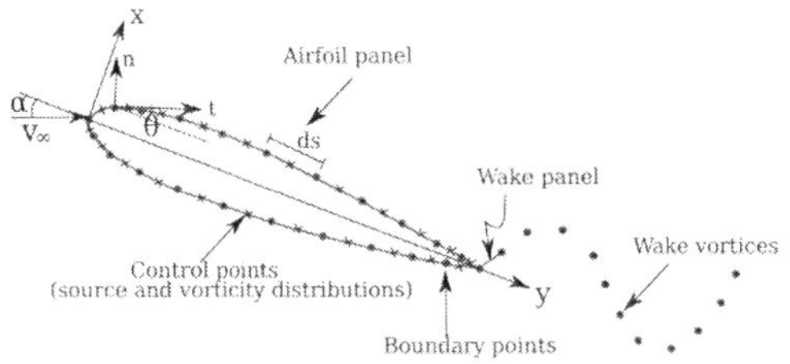

Figure 2. Schematic representation of UVLM.

to be constant over a particular panel (q_j, $j = 1, 2, ..., N$) and the vorticity strength is considered to be constant over all the panels (τ). For unsteady flows, the time dependent source strengths and the time dependent vorticity strength are denoted as $(q_j)_k$ and τ_k respectively for the time step index k. The velocity at any point (x,y) in the flow-field is the vector sum of velocities of the undisturbed flow (free-stream) and the disturbance field due to the presence of the oscillating body and also the wake behind the body. The boundary condition that the surface of the body is a streamline is satisfied by taking the summation of velocities induced by body bound singularities, free-stream and wake vortices to be equal to zero in the direction normal to the surface at each discretized body element or panel as,

$$\sum_{j=1}^{N} A^n_{ij}(q_j)_k + \tau_k \sum_{j=1}^{N} B^n_{ij} + (\vec{v}_{str}.n_i)_k + (\tau_w)_k (B^n_{i,n+1})_k + \sum_{m=1}^{k-1} (C^n_{im})_k(\Gamma_{m-1} - \Gamma_m) = 0, i = 1, 2, ..., N. \tag{4}$$

Here, A^n_{ij} and B^n_{ij} are the influence co-efficients at the ith panel control point by a unit strength source distribution and unit strength vorticity distribution on the jth panel respectively. $\vec{v}_{str}$ denotes the unsteady stream velocity at body fixed co-ordinate system including airfoil motion. $(\tau_w)_k$ is the uniform vorticity distribution over the wake panel attached to the trailing edge at t_k. $(B^n_{i,n+1})_k$ and $(C^n_{im})_k$ are the normal velocity induced at the ith panel control point by unit strength vorticity distribution on the shed wake panel and unit strength m^{th} wake vortex at time-step t_k respectively. Γ_m and Γ_{m-1} are the overall circulation on the airfoil surface at time-steps t_m and t_{m-1} respectively. The control point is situated at the mid point of each panel where the zero normal flow boundary condition is satisfied. The unknown body bound vortex strength, unknown source strengths and the immediately shed wake vortex at the current time step are computed using the no-normal boundary condition along with the additional condition of Kelvin's circulation theorem given by,

$$\Gamma_k + \Delta_k(\tau_w)_k = \Gamma_{k-1}, \tag{5}$$

where Δ_k is the length of the wake panel attached to the trailing edge at t_k. Kelvin's theorem states that the total circulation in the flow-field must be preserved and that any changes in the circulation about the body are balanced by an equal and opposite circulation added in the wake in the form of new vortices. These vortices influence the local velocity field significantly, and as a result the forces on the airfoil at any instant become a function of the past motion of the airfoil. All the wake vortices, shed from the trailing edge of the airfoil, are carried downstream by their local velocities induced by the free-stream, body bound vortices and the other shed vortices. In unsteady potential flow, the calculation of pressure at any point on the body is done by using unsteady Bernoulli's equation [20]. The pressure coefficients at the ith panel control point at time-step t_k can be written as,

$$[(C_p)_i]_k = \frac{[(v_{str})_i]^2_k}{v^2_\infty} - \frac{[(v_t)_i]^2_k}{v^2_\infty} - \frac{2}{v^2_\infty}\frac{(\phi_i)_k - (\phi_i)_{k-1}}{t_k - t_{k-1}}, \tag{6}$$

where v_t is the tangential velocity induced at the ith panel; $(\phi_i)_k$ and $(\phi_i)_{k-1}$ are the velocity potential of the ith panel at time-steps t_k and t_{k-1} respectively.

The small effective angle of attack of the system justifies the use of the potential flow solver in the present study. The present UVLM based potential flow solver has been quantitatively validated by comparing the peak lift coefficients with the Navier-Stokes (N-S) results reported by Young [22] as well as with lumped vortex method (LVM) results computed by an inhouse code (see Fig. 3). LVM is a similar panel method based potential flow solver [23]. In LVM, each panel has a lumped vortex at the control point situated at one-fourth of its length. The zero normal flow boundary condition is satisfied at the collocation point situated at three-fourth of its length. It can be seen that the present results show an excellent match with the LVM results and are also in close agreement with the N-S results especially at low κh values.

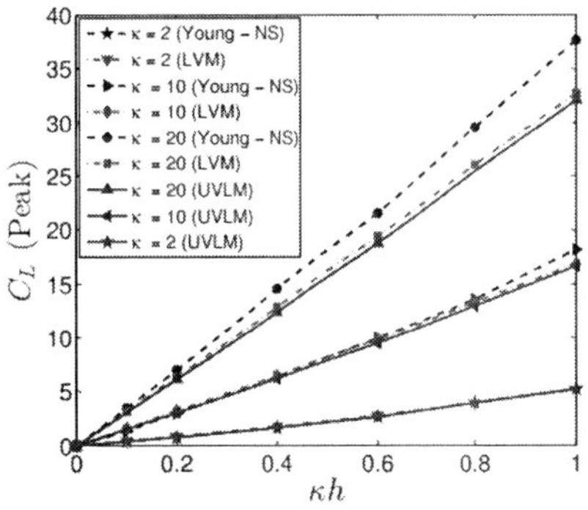

Figure 3. Comparison of peak lift coefficients, estimated using UVLM (present method), LVM (inhouse code) and from Navier-Stokes simulation presented in [22].

2.3 Noisy flow fluctuations

The fluctuating flow field is defined in the present study as,

$$U^*(\tau) = U^*_{mean} + u(\tau); \tag{7}$$

where $u(\tau)$ is the irregular flow fluctuation modelled as Ornstein-Uhlenbeck (O-U) process [24, 25]. The O-U is a stationary Gauss-Markov process and is defined by the following stochastic differential equation:

$$du(\tau) = -\Delta\omega u(\tau)\, d\tau + \sqrt{\Delta\omega q}\, dW(\tau); \quad (8)$$

where $W(\tau)$ is a Wiener process, $\Delta\omega$ and q are noise parameters. The correlation function of the O-U process is given by,

$$\rho(\tau) = < u(\tau)u(0) > = q\exp(-\Delta\omega|\tau|). \quad (9)$$

The O-U process is simulated by numerically integrating the stochastic differential equation using the Euler-Maruyama method [26]. It is to be noted that $\Delta\omega$ is independently varied keeping q fixed to vary the correlation of the O-U noise. The variation of the correlation function of the O-U noise with different values of $\Delta\omega$ for $q = 1$ is presented in Fig. 4. It can be seen that as the $\Delta\omega$ is increased to 5 from 1, the correlation length of the process is decreased significantly. Thus, this noise model allows us to control the time-scale present in the fluctuations by changing $\Delta\omega$ of the process. The time histories of the noisy inflow with different noise correlation ($\Delta\omega = 1$ and 5) have been presented in Fig. 5. It is observed that the time-scales present in the noisy inflow drastically change as the correlation length of the noise is varied. A long time-scale gusty inflow is obtained at $\Delta\omega = 1$; it becomes a short time-scale gusty inflow at $\Delta\omega = 5$ as the correlation of the O-U noise is decreased.

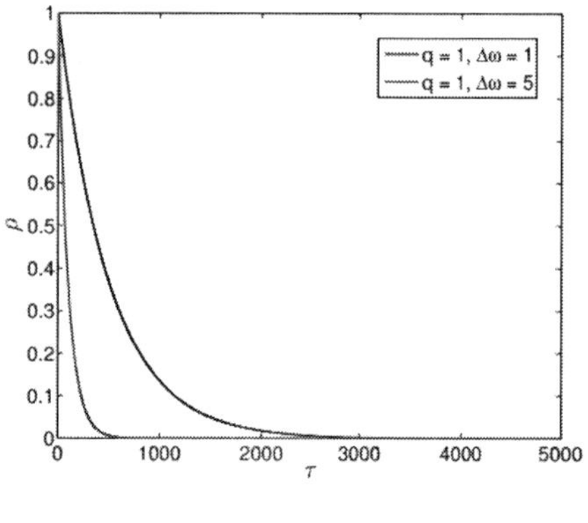

Figure 4. Variation of the correlation function of the O-U noise for (a) $\Delta\omega = 1$, (b) $\Delta\omega = 5$.

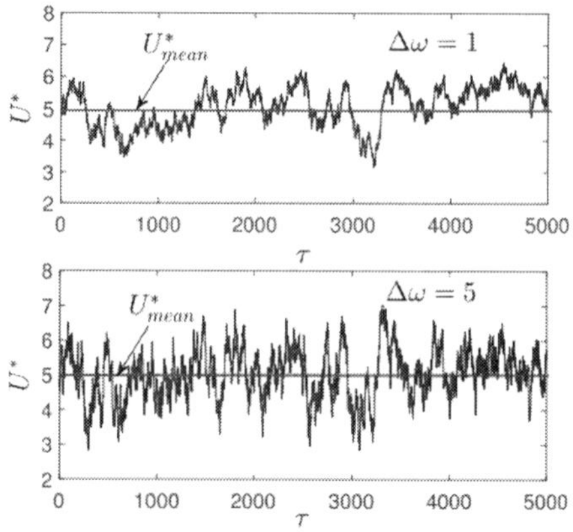

Figure 5. Time histories of the noisy inflow with different noise correlation: (a) $\Delta\omega = 1$, (b) $\Delta\omega = 5$.

coupled system response damps down to a fixed equilibrium point ($h = 0$) below a critical value of the control parameter U^*. The system undergoes a supercritical Hopf bifurcation at $U^* = U^*_{cr}$ and stable limit-cycle oscillations (LCO) emerge at $U^* > U^*_{cr}$. The Hopf bifurcation point (U^*_{cr}) is observed to be 3.75 from Fig. 6. For $U^* > U^*_{cr}$ LCOs persists and the amplitude of LCOs increase with U. The pitch response at $U^* = 5$ has been presented in Fig. 7 which shows a typical LCO in the post Hopf-bifurcation regime.

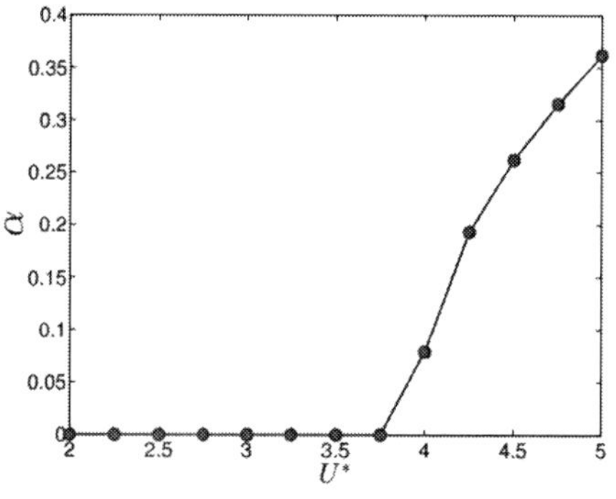

Figure 6. Bifurcation diagram of pitch response for uniform flow condition.

3 Results and discussions

3.1 Bifurcation in uniform flow

The bifurcation behaviour of a pitch-plunge flexible flapping system in the presence of uniform flow is well investigated in the literature. The bifurcation diagram of the pitch response of the coupled system is presented in Fig. 6 considering the non-dimensional free-stream velocity (U^*) to be the control parameter. It is seen that the

3.2 Bifurcation in fluctuating flow

This section presents the results of the bifurcation analysis of the flexible flapping system in the presence of irregular flow fluctuations. Typical atmospheric wind-spectra modelled using Dryden or von Kármán spectra are known to capture the long time scales. Hence, a long time-scale gust (O-U noise with $\Delta\omega = 1, q = 1$) has been imposed on the present system. The mean of the fluctuating flow

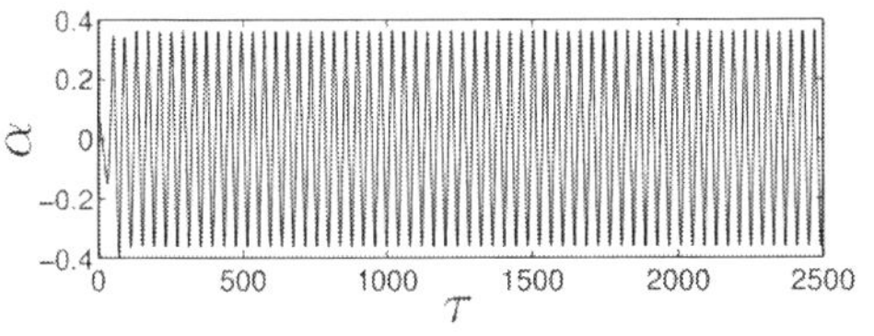

Figure 7. Pitch response at $U^* = 5$: LCO in the post Hopf bifurcation regime.

speed (U^*_{mean}) has been taken to be the bifurcation parameter. The pitch time histories are presented at different U^*_{mean} values in Fig. 8. It can be seen that the coupled system response does not transition from a damped oscillatory state to well developed limit cycle oscillations abruptly beyond a particular critical mean flow velocity. Instead, the system response gradually transforms from a stochastic fixed point (at U^*_{mean} =2) to a randomly modulated LCO (at U^*_{mean} =6) through irregular alternations between two qualitatively different dynamics as the bifurcation parameter is increased. In the dynamical systems theories, this kind of transitional dynamics is known as intermittency. In the presence of long time-scale fluctuations, an 'on-off' type intermittency phenomenon [27, 28] has been observed where the system response is seen to irregularly alter between an 'off' state (very low amplitude aperiodic fluctuations) and an 'on' state (high amplitude periodic bursts) at U^*_{mean} = 4.0 and 4.5. As the mean flow speed is gradually increased, the segments of periodic bursts are seen to be more pronounced. Eventually the periodic oscillations completely dominate the time history as random LCOs on further increasing the mean flow speed to U^*_{mean} = 6.

To obtain a better insight into the system dynamics, a P-bifurcation analysis is carried out considering U^*_{mean} as the control parameter. Conventionally, P-bifurcation is defined by the qualitative changes in the structure of the stationary joint-pdf of the response variables and their instantaneous time derivatives, as the bifurcation parameter is varied [17, 29, 30]. Figure 9 shows the typical joint pdfs $p(\alpha, \dot{\alpha})$ for the pitch response and its instantaneous time derivatives for some representative values of U^*_{mean}. It can be clearly seen that at U^*_{mean} = 2.0, the joint $p(\alpha, \dot{\alpha})$ appears to have a unimodal structure centered about the origin; see Fig. 9(a). Figure 9(b) shows the j-pdf structure for U^*_{mean} = 4.5. It is observed that while a peak at the origin still exists, a ring like structure appears to form around the peak at the origin. It is observed that the amplitude of the peak at the origin has decreased from what is observed in Fig. 9(a). It is quite obvious that the strength of the attractor associated with the annular ring is significantly weaker than the one at the origin. As U^*_{mean} is increased to 5, it can be observed from Fig. 9(c) that the ring like structure has become more prominent and the peak at the center has become smaller. Finally, at U^*_{mean} = 6.0, it can be seen from Fig. 9(d) that the joint pdf resembles a crater like structure with the peak at the origin having completely disappeared. Clearly, one can see that there have been topolog-

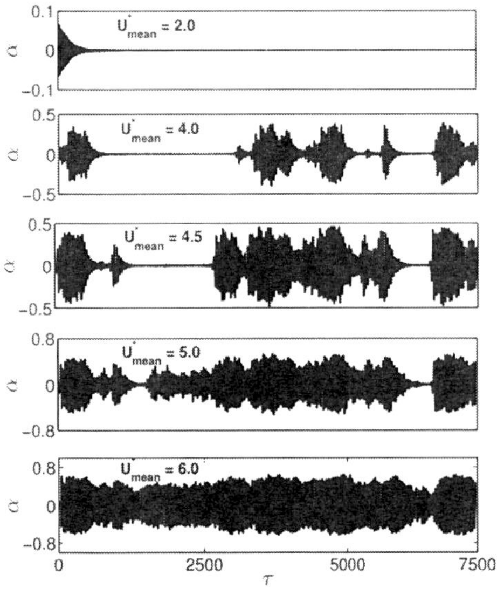

Figure 8. Pitch time histories at different U^*_{mean} values: 'On-Off' intermittent transition

ical changes associated with the joint pdf $p(\alpha, \dot{\alpha})$ as U^*_{mean} is increased. Similar qualitative changes can be observed with the joint pdf associated with the plunging response and its instantaneous time derivatives; these figures have not been included in the paper for the sake of brevity.

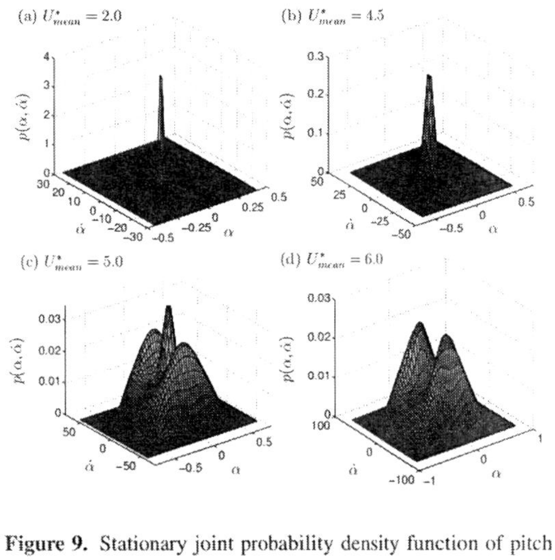

Figure 9. Stationary joint probability density function of pitch response and its instantaneous derivative $p(\alpha, \dot{\alpha})$ for (a) U^*_{mean} = 2, (b) U^*_{mean} = 4.5, (c) U^*_{mean} = 5, (d) U^*_{mean} = 6.

The underlying physics associated with these joint pdfs can be better appreciated from an inspection of the contour plots of the joint pdfs and shown in Fig. 10. Note that these pdf contours provide a measure of the time spent by the trajectories in a volume element of the state space which is indicative of different stochastic attractors. The contour plots of $p(\alpha, \dot{\alpha})$ for $U^*_{mean} = 2.0$ reveal a circular structure about the origin with very small radius (Figure 10(a)) as the trajectories spend most of the time around the zero state. At $U^*_{mean} = 4.5$, the contours shown in Fig. 10(b) reveal that the radius of the attractor at the origin has increased due to the presence of periodic bursts denoting the wider base of the corresponding jpdf. As U^*_{mean} is increased to 5, two attractors- one at the origin and another the annular ring are seen to co-exist characterizing the 'off' and the 'on' states respectively in the intermittent phase; see Fig. 10(c). However, as the annular ring like attractor emerges reflecting prominent periodic bursts, the contour level values of the existing attractor at the origin gets reduced. It signifies the fact that the probability of the trajectories spending time in the zero state becomes lower. Finally, Fig. 10(d) shows that at $U^*_{mean} = 6.0$, the attractor at the origin has disappeared and only the annular ring like attractor exists characterizing the random limit-cycle oscillations.

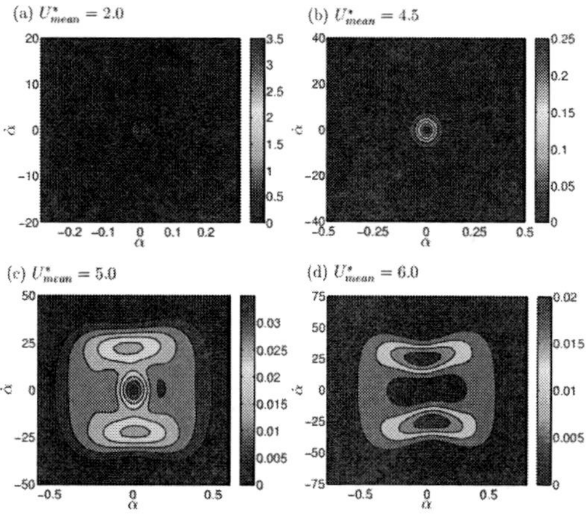

Figure 10. Contour plots of $p(\alpha, \dot{\alpha})$ representing the stochastic attractors in the phase space for (a) $U^*_{mean} = 2$, (b) $U^*_{mean} = 4.5$, (c) $U^*_{mean} = 5$, (d) $U^*_{mean} = 6$.

4 Concluding remarks

A stochastic bifurcation analysis has been carried out for a two degree-of-freedom pitch-plunge airfoil subjected to long time-scale noisy flow fluctuations, with the mean speed of the flow taken to be the bifurcation parameter. A P-bifurcation analysis based on studying the changes in the topology associated with the structure of the joint pdf of the response variables has been presented. The presence of long time-scale fluctuations in the flow result in an 'on-off' type intermittency phenomenon in flow regimes prior to Hopf bifurcation. The system exhibits intermittency at flow regimes where the phase space is characterized by two co-existing attractors. This is evidently reflected in the joint pdf of response variables. The effect of correlation on the noise induced intermittency is currently being investigated by the authors.

References

[1] K.D. Jones, M.F. Platzer, AIAA paper **37** (2006)
[2] W. Shyy, H. Aono, S.K. Chimakurthi, Trizila, C.K. Kang, C.E.S. Cesnik, H. Liu, Progress in Aerospace Sciences **46**, 284 (2010)
[3] D. Mackenzie, Science **335**, 1430 (2012)
[4] Z.J. Wang, Journal of Fluid Mechanics **410**, 323 (2000)
[5] J. Young, J. Lai, AIAA journal **42**, 2042 (2004)
[6] H. Liu, K. Kawachi, Journal of Computational Physics **146**, 124 (1998)
[7] G. Yang, S. Obayashi, Journal of Aircraft **41**, 1353 (2004)
[8] V. Golubev, R. Mankbadi, M. Visbal, J. Scott, R. Hixon, AIAA Paper **2426** (2006)
[9] M.J. Patil, D. Taylor, AIAA Paper **1638** (2006)
[10] Y. Lian, AIAA Paper **539** (2009)
[11] W. Shyy, D.A. Jenkins, R.W. Smith, AIAA journal **35**, 1545 (1997)
[12] S. Watkins, J. Milbank, B.J. Loxton, W.H. Melbourne, AIAA journal **44**, 2591 (2006)
[13] Y. Lian, W. Shyy, AIAA journal **45**, 1501 (2007)
[14] Y. Lian, W. Shyy, AIAA paper **71** (2007)
[15] E. Johnson, J. Jacob, AIAA Paper **64** (2009)
[16] J. Zhu, L. Jiang, H. Zhao, B. Tao, B. Lei, Journal of Mechanical Science and Technology **29**, 4681 (2015)
[17] K.R. Schenk-Hoppé, Nonlinear dynamics **11**, 255 (1996)
[18] B.H.K. Lee, L. Jiang, Journal of Fluids and Structures **13**, 75 (1999)
[19] J.L. Hess, A.O. Smith, Progress in Aerospace Sciences **8**, 1 (1967)
[20] T. Cebeci, M. Platzer, H. Chen, K.C. Chang, J.P. Shao, *Analysis of low-speed unsteady airfoil flows* (Springer Berlin Heidelberg, 2005)
[21] Y.C. Fung, *An introduction to the theory of aeroelasticity* (Courier Dover Publications, 2008)
[22] J. Young, *Numerical simulation of the unsteady aerodynamics of flapping airfoils* (University of New South Wales, Australian Defence Force Academy, School of Aerospace, Civil and Mechanical Engineering., 2005)
[23] J. Katz, A. Plotkin, *Low-speed aerodynamics*, Vol. 13 (Cambridge university press, 2001)
[24] R. Graham, A. Schenzle, Physical Review A **26**, 1676 (1982)
[25] D.T. Gillespie, Physical review E **54**, 2084 (1996)
[26] D.J. Higham, SIAM Review **43**, 525 (2001)
[27] D.C. Poirel, S.J. Price, Journal of Aircraft **34**, 619 (1997)
[28] D.C. Poirel, S.J. Price, AIAA journal **39**, 1960 (2001)
[29] L. Arnold, *Random dynamical systems* (Springer Science & Business Media, 2013)
[30] S. Ariaratnam, Nonlinearity and chaos in engineering dynamics pp. 267–274 (1994)

The behavior of bilinear impact oscillators subjected to random forcings

Aasifa Rounak[1,*] and *Sayan* Gupta[1,**]

[1] *Department of Applied Mechanics, Indian Institute of Technology Madras, Chennai 600036 India*

Abstract.
A multistable regime of a bilinear vibro-impact oscillator in the presence of noise has been studied, for which relatively lesser attractors coexist with a well-defined basin boundary structure in the phase space. The effect of adding parametric noise in this model has been investigated and the switching that the system undergoes between the basins of its different attractors has been examined.

1 Introduction

Strongly non-linear systems could exhibit multiple stable solutions for the same set of parameters. A consequence of this is that the stable solution to which the system converges at $t \to \infty$ is dictated by the basins of attraction on which the initial conditions of the system lie. The mathematical modelling of a physical system involves simplifying assumptions and approximations that introduce uncertainties into the model. These uncertainties are responsible for the deviations in the behaviour of the physical system with the predictions obtained from the mathematical model. This, in turn, can lead to erroneous predictions about the long term behaviour of a dynamical system. As these uncertainties are unknowns which can be characterized only in the probabilistic sense, it is important to carry out an analysis that enables predicting the long term behaviour in a probabilistic sense. This will enable identifying the most likely long term dynamical state of the system. For such multistable systems, a better understanding of their dynamics would help to demarcate the optimum operational conditions. As an example, for the classical tuned mass absorber, a device that is widely used to absorb energy and mitigate unwanted vibrations, optimal damping ability is achieved in the neighbourhood of its multistability zone [1]. However, among all the closely spaced co-existing solutions only one mitigates oscillations effectively. A slight disturbance would lead to other solutions that might amplify the amplitude of the base system, making its use counterproductive. Similarly, in systems with impacts, one of the stable solutions can ensure the desirable operation of the system while others may lead to vibration amplification and damage. Multistability in impacting systems has been studied in [2–4]. Vibro-impacting scenarios appear in a wide range of practical problems such as percussive drilling tools [5], print hammers [6] and vibro-impact moling systems [7]; most of these studies report multistability in deterministic scenarios. The present study aims at analysing the co-existing stable solutions of the bilinear impacting systems in the presence of slight disturbances.

On adding parametric noise with a high signal to noise ratio over the actual input forcing signal, the phenomenon of basin hopping as described in [2] has been observed. However unlike the studies discussed in the literature where the system gets stabilized indefinitely to the attractors that were previously present, the responses were seen to be flipping between the pre-existing attractors. Typically, the system showed such transition in behaviour for parameter regimes in the neighbourhood of its stability boundaries. Well defined basin boundaries cannot be defined in such scenarios. This is because, the presence of noise leads to alteration in the regimes of different attractors[8]. Thus, a detailed analysis of such scenarios is necessitated.

This paper is organized as follows. In Section 2 the model and the equations of motion for the impacting system have been presented. Section 3 investigates the coexistence of different regimes and their basins of attraction has been investigated. Section 4 summarizes the salient features arising from this study.

2 Problem Statement

A soft impacting system with bilinear stiffness has been considered; see Fig. 1 for a schematic diagram. It comprises of a mass-spring-damper system with mass M, spring stiffness k_1, and damping factor R_1. The cushioned impacting surface is modelled by a spring with stiffness k_2 and has been assumed to have negligible mass as compared to M. The equilibrium position of the left end of the spring k_1 is at zero and k_2 is at d. The mass M is subjected to a harmonic excitation of the form $u = A\sin(\Omega t)$ where Ω is the angular frequency. The state variable y is the elongation of the spring from the unstretched position. Initially when both the springs are relaxed, the mass is at a

*e-mail: aasifa.rounak@gmail.com
**e-mail: sayan@iitm.ac.in

distance of d from the impacting surface. Let $\mathbf{Y} = [y; \dot{y}]^T$ be the state vector with $\dot{y}$ representing the velocity (in m/s) of the mass. The dynamics of the mass M are given by the Eq. 1.

$$M\ddot{y} = \begin{cases} -R_1\dot{y} - k_1 y + MA\Omega^2 \sin(\Omega t), & y \leq d \\ -R_1\dot{y} - (k_1 + k_2)y + MA\Omega^2 \sin(\Omega t) + k_2 d, & y > d \end{cases} \tag{1}$$

The non-dimensionalised equations of motion for the bilinear oscillator is given by Eq. 2

$$\ddot{x} + 2\zeta\dot{x} + x + \beta(x - e)H(x - e) = F_0 \sin(\omega\tau), \tag{2}$$

where $x = y/y_0$ is the non-dimensionalised displacement, $\dot{x}$ is the non-dimensionalised velocity, $\tau = \omega_n t$ is the non-dimensional time, $\beta = k_2/k_1$ is the stiffness ratio, $e = d/x_0$ is the non-dimensional gap, F_0 is the non-dimensional forcing amplitude, $\zeta = c/2m\omega_n$ is the damping ratio and the over-dot represents the derivative with respect to τ. $H(.)$ represents the Heaviside step function that demarcates the region starting where, the mass M and secondary stiffness K_2 are in contact. A simplifying assumption that the discontinuity surface is not in motion and is located at $x - e = 0$, has been made. The swiching between the two state spaces, thus always happens at $x = e$, making the analyses easier.

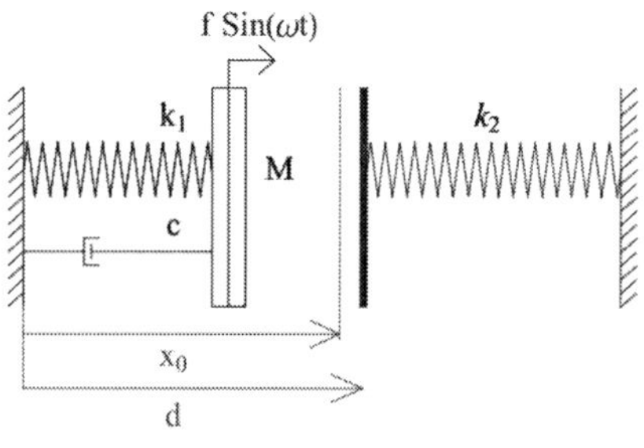

Figure 1: Schematic of the impact oscillator

The system of equations were solved using point-wise mapping technique using 4th order Runge-Kutta method. For the value of parameters defined as $\zeta = 0.01, \beta = 29, e = 1.26, m = 2.1, \omega = 2/m, a = 0.75, f = a\omega^2$, two co-existing stable solutions of periodicity one and three were noted as seen in Fig. 2.

The group of fundamental periodic motions were characterized by the different number of impacts in one motion period, which equals the excitation force period. The Poincaré sections were taken stroboscopically *i.e.*, each at a time period of $2\pi/\omega$. These are shown as blue circles in Figs 2(a), 2(b). It can also be confirmed with the corresponding bifurcation plot and basin of attraction that no other stable attractors exist for the regime being taken into study.

The explanation for other parameters that were chosen is as follows. It is known from [9] that in impacting mechanical systems large-amplitude chaotic oscillations develop where collisions with the secondary stiffness are

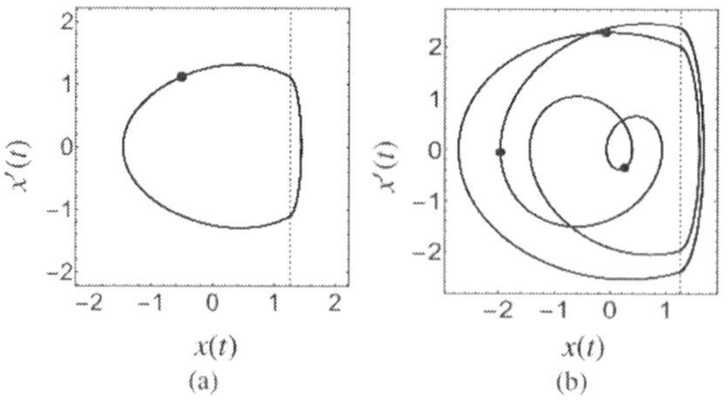

Figure 2: (a)Phase portrait and Poincaré section for period 1 oscillations, (b) period 3 oscillations; the blue points correspond to the Poincaré points.

with near-zero velocity, *i.e.* when grazing is encountered. These oscillations vanish when the forcing frequency is an integer multiple or is nearly equal to an integer multiple of the natural frequency of the system. Here $\omega_n = 1$, hence our choice of $\omega = 0.952$ (obtained as the ratio of frequencies $m = 2\omega_n/\omega$ is taken to be 2.1) happens to evade the condition where chaotic orbits would be encountered, thus simplifying the analyses of the effect of noise perturbations. The parameter of choice avoids the condition of grazing, hence chaotic orbits are not created for the domain being studied.

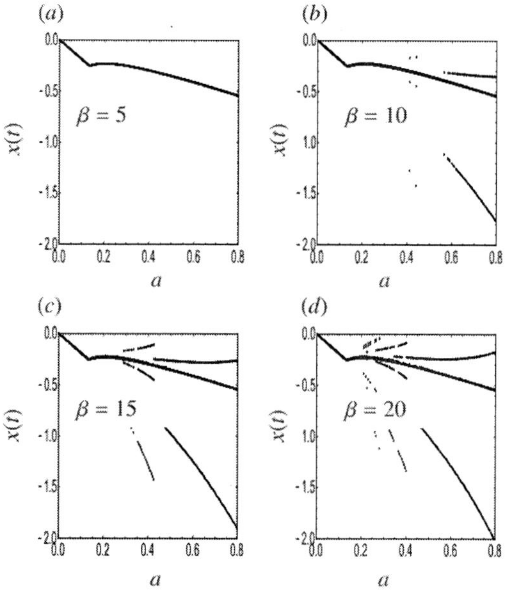

Figure 3: Figure showing how co-existing attractors come to originate as the secondary subspace gets stiffer.

Fig. 3 shows the bifurcation diagram with a for various values of β. Fig 3 (a) shows a single curve illustrating that only one stable solution exists for $\beta = 5$. As β is increased, one can see the emergence of multiple lines indicating the presence of coexisting attractors. These other attractors appear as a consequence of unstable periodic orbits born due to hitting at the secondary boundary, which become stable as a increases. Further increase of β ensures that the unstable periodic orbits gain stability at smaller values of a, thereby not creating unstable periodic or chaotic saddles near the regime being studied. Thus, a significantly higher value of β, *i.e.*,29 was chosen. The results of the brute force bifurcation plots showing stable sets are available in [9].

3 Introduction of noise

A random perturbation over the sinusoidally varying forcing was input to the system. The forcing F(t) now consists of a harmonic function superimposed by small amplitude noise modelled as an Ornstein Uhlenbeck process and is mathematically expressed as:

$$F(t) = F_0 \sin(\omega\tau) + n\lambda, \tag{3}$$

where n is the noise level of the parametric fluctuations and $\lambda(t)$ is a sample realisation of the Ornstein-Uhlenbeck(OU) process with process parameters mean, volatility, and mean reversion speed as 0, 0.1 and 0.3 respectively. The noise was added at each step of the carried out fixed-step Runge-Kutta integration. The input λ to the forcing was made to take in a new value at every new step of integration performed. This led to a deviation from the deterministic behaviour of the system. The OU process has been chosen as it is the only Gauss–Markov process with a bounded variance that admits a stationary probability distribution. It thus facilitates the study of the effect of correlation lengths of the noise input on the behaviour of the system. The OU process has been deployed to study the effect of over-damped linear springs in the presence of random fluctuations and thus very well suits the model being studied [10]. Over time, the process tends to drift towards its long-term mean. Such a process is called mean-reverting. Hence, the mean reversion speed is an important parameter of the Ornstein-Uhlenbeck process. The autocorrelation for the process is given by $\exp[\theta|\tau|]\sigma^2/2\theta$. When $n = 0.385f$, the percentage of noise, with respect to the deterministic signal, that is input to the system is $0.107 \pm 3.45\%$ and the forcing input is as shown in Fig. 4. The forcing thus becomes amplitude modulated, the modulation happening at random, as it still retains its sinusoidal character.

The governing equation was integrated for 10^4s. The phase plot diagram obtained from the numerically integrated trajectories of the field equations are shown in Fig 5. These phase diagrams were plotted with trajectories from $(0.45 - 0.75) \times 10^4$ s and $(0.8 - 1) \times 10^4$ s. These figures are superimposed with the phase plot trajectories corresponding to period 1 (red line) and period 3 (green line) oscillations for the corresponding deterministicially excited system. Additionally, the Poincare points obtained by sampling at period of $2\pi/\omega$ are also shown in these figures as blue points. It is seen that these blue points form a cluster in the vicinity where the Poincare points were

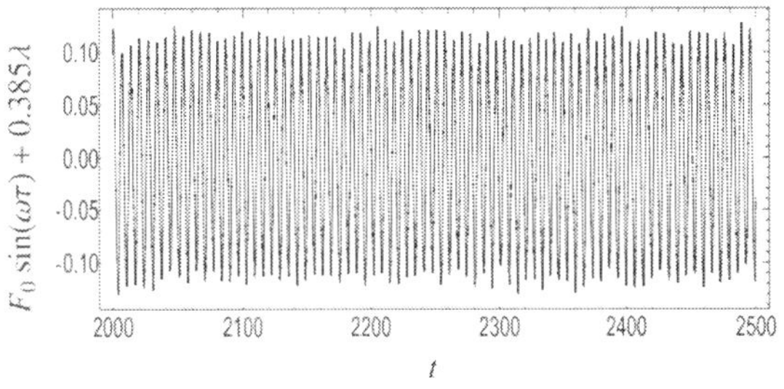

Figure 4: Noisy sinusoidal forcing to the vibroimpact system.

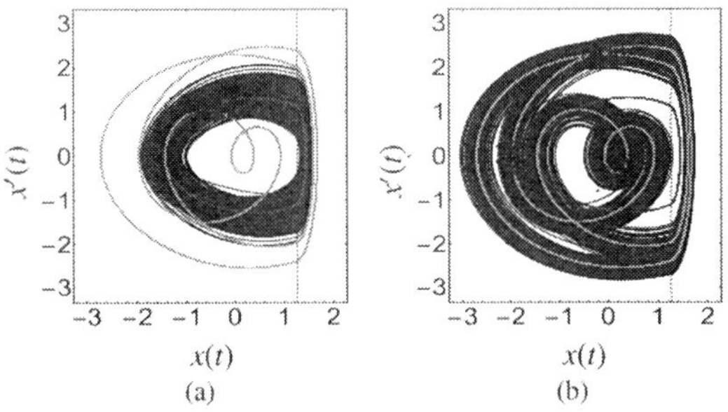

Figure 5: (a) Phase portrait and Poincaré section for time $(0.8 - 1)10^4$. The response of random input seen to follow modulated period 1 oscillations. (b) Phase portrait and Poincaré section for time $(0.45 - 0.75)10^4$. The response of random input seen to follow modulated period 3 oscillations.

observed in the phase space diagram corresponding to the deterministic system; see Fig 5.

Thus, it is observed that the system's long term behavior fluctuates from a modulated period 3 motion to a modulated period 1 motion. The corresponding time histories presented in Fig. 7 confirm that the response of the system flips between two qualitatively distinct dynamics. Such irregular alternation between qualitatively distinct dynamics in the presence of noise is known as noise-induced intermittency in nonlinear dynamical systems literature, see [8]. This explanation for why such a flipping behavior is observed can be provided with the help of bifurcation diagrams constructed in the presence of noise. Such bifurcation diagrams constructed give an idea of the spread of the modulated periodic attractor [11].

When $n \leq 0.385f$, in the bifurcation diagram at $a = 0.75$, the red regions, corresponding to the bifurcation points in the presence of noise, following period 3 and period 1 branches do not overlap each other. However, as the intensity of noise is increased, the regions begin to overlap, causing the sytem to exhibit such switching back and forth between two different dynamics as is seen in Fig. 6. The deterministic bifurcation diagram, in black, shows four arms at $a = 0.75$ corresponding to three arms of a period 3 cycle at $x = -1.962, -0.083, 0.25$ and one arm

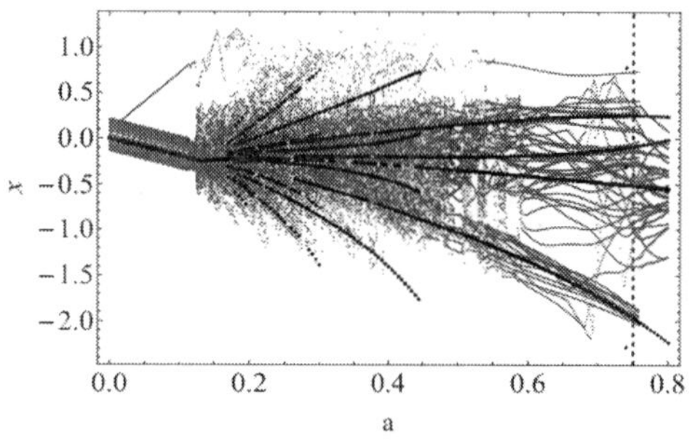

Figure 6: Bifurcation diagram of the bilinear oscillator system in the presence of noise (in red) of intensity $n = 0.385f$ superimposed over the deterministic bifurcation plot (in black).

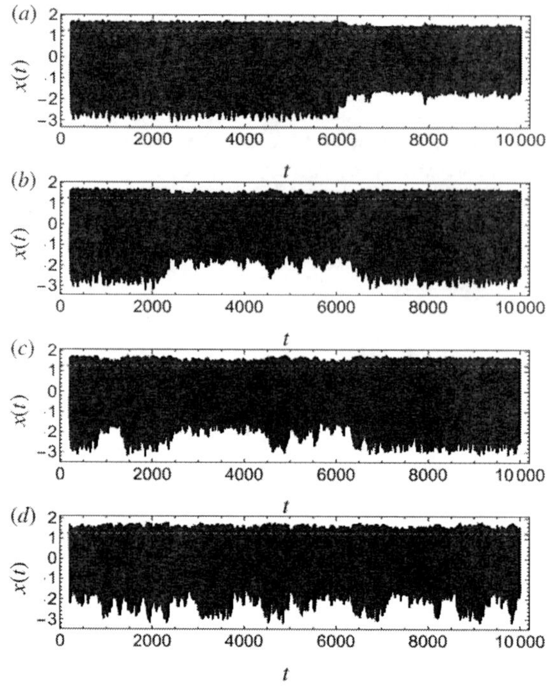

Figure 7: Time histories showing flipping behavior when (a) $n = 0.385f$, (b) $n = 0.6f$, (c) $n = 0.7f$, (d) $n = 0.9f$. The red line indicates the presence of the discontinuity boundary.

of a co-existing period 1 invariant set at $x = -0.511$. The red regions in the bifurcation diagram correspond to the stroboscopically recorded bifurcation points in the noisy case. The corresponding time histories for different noise intensities are as seen in Fig. 7. It has been observed that increasing the noise levels leads to more increase in the basin hopping phenomenon. The higher amplitude period 3 and lower amplitude period 1 oscillations are observed to persist for significant amounts of time until the external forcing makes the system change its behavior. The trajectory takes turns to visit, by sudden bursts, to the system's two coexisting attractors.

To further investigate the effects of a parametric noise perturbation on the final structure of the basins of the co-existing attractors, a two-dimensional section of the phase space $(x_0; x'_0)$ with $(x_0; x'_0)$ as initial conditions were taken and evolved for a time of $10^4 s$. The first 300 cycles were dropped off and the responses of the system in the next 1400 cycles were analyzed in a grid of 160×160 pixels and the final basin structures were drawn.

In order to draw the basins, the spread of the Poincaré points were studied. A small elliptical region with major and minor axes taken along confidence lines of σ around x and $\dot{x}$ about the four deterministic Poincaré points. These ellipses were then considered as thresholds for the recorded modulated period 1 and period 3 responses. Thus, if > 66.67% of the points happened to congregate around the period 1 Poincaré point and almost none fell into the regions of period 3 points, say, then the behaviour was defined to be modulated period 1 throughout. For a case where all the regions around period 1 and period 3 were visited and the duration the system spent exhibiting transitory behaviour, given by the number of points outside the thresholds, was less in comparision to what it spent in the basins of the co-existing invariant sets, the behavior was considered to be intermittent. Thus, if the outliers from both the regions were within a certain number, which here is taken as $1/4^{th}$ of the total number of points, then the system was defined to be undergoing intermittent oscillations. Otherwise the system was said to undergo random oscillations. The following color scheme has been adopted in Fig 8: (a) if the behavior is modulated period 1, it is marked as red, (b) if the behavior is modulated period 3, it is marked as blue, (c) if the system is understood to be visiting both the attractors and the system has quick transition times *i.e.* if less points fall outside the designated domains, it is marked magenta, (d) if the system is understood to be visiting both the attractors and more points fall outside the designated domains, it is understood that the system exhibits random oscillations and is marked in yellow.

It is seen in Fig. 8 that when the amplitude of noise is $n = 0.385f$, the intermittent regions are barely existent. As the forcing input signal gets noisier, the regions of modulated throughout-homogeneous periodic structures shrink until they are completely lost to irregular or random oscillations. The basin structures are however seen to have been retained but the basin boundaries get smeared out as the noise intensifies.

4 Conclusion

In this paper it has been shown that the parametric noise in the forcing input leads the system to hop between the stable co-existing periodic attractors of the bilinear oscillator system, even when the basins of attraction have well defined and significant basin volumes for each of the present attractors. It has been shown that the long term behavior of the system might alternate between the possible distinct stable states. An explanation for basin hopping

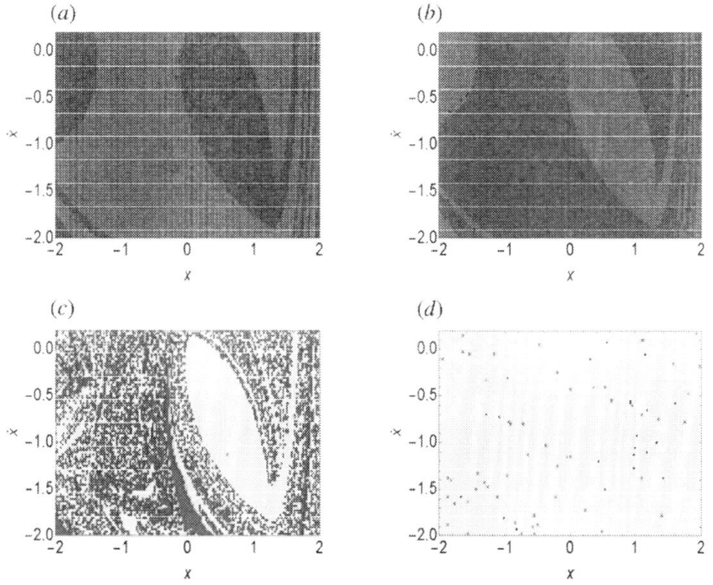

Figure 8: Basins of attraction drawn for 1400 cycles beyond transience for noise levels (a) $n = 0.385f$, (b) $n = 0.5f$, (c) $n = 0.6f$, (d) $n = 0.8f$. The colours red, blue, magenta and yellow correspond to initial conditions that display modulated period 1 motion, modulated period 3 motion, intermittent flipping and random motions respectively.

was described by [12] to be due to the properties of the non attracting stable states, also known as saddles, lying in between the basins. The trajectory, while wandering in the vicinity of such saddles, experience a transient irregular motion. The noise term pushes the trajectory out of an open neighborhood of some attractor into the basin boundary. The trajectory spends there a certain amount of time until it reaches again the neighborhood of the same or another attractor[4]. This study can be very complex for structures where many stable and chaotic states co-exist. More studies are needed to be carried out to study such scenarios.

References

[1] Brzeski, P., Lazarek, M., Kapitaniak, T., Kurths, J., & Perlikowski, P. (2016). Basin stability approach for quantifying responses of multistable systems with parameters mismatch. Meccanica, 51(11), 2713-2726.

[2] Blażejczyk-Okolewska, Barbara, and Tomasz Kapitaniak. "Co-existing attractors of impact oscillator." Chaos, Solitons & Fractals 9.8 (1998): 1439-1443.

[3] Ing, James, et al. "Bifurcation analysis of an impact oscillator with a one-sided elastic constraint near grazing." Physica D: Nonlinear Phenomena 239.6 (2010): 312-321.

[4] de Souza, Silvio LT, Iberê L. Caldas, and Ricardo L. Viana. "Multistability and self-similarity in the parameter-space of a vibro-impact system." Mathematical Problems in Engineering 2009 (2009).

[5] Wiercigroch, Marian, Anton M. Krivtsov, and Jerzy Wojewoda. "Vibrational energy transfer via modulated impacts for percussive drilling." Journal of Theoretical and Applied Mechanics 46.3 (2008): 715-726.

[6] Jerrelind, Jenny, and Harry Dankowicz. "A global control strategy for efficient control of a Braille impact hammer." Journal of vibration and acoustics 128.2 (2006): 184-189.

[7] K.-C. Woo, A. A. Rodger, R. D. Neilson, and M. Wiercigroch, "Application of the harmonic balance method to ground moling machines operating in periodic regimes," Chaos, Solitons & Fractals, vol. 11, no. 13, pp. 2515–2525, 2000.

[8] Venkatramani, J., et al. "Precursors to flutter instability by an intermittency route: A model free approach." Journal of Fluids and Structures 61 (2016): 376-391.

[9] Suda, Narasimha, and Soumitro Banerjee. "Why does narrow band chaos in impact oscillators disappear over a range of frequencies?." Nonlinear Dynamics 86.3 (2016): 2017-2022.

[10] Figueroa, Leonardo E., and Endre Süli. "Greedy approximation of high-dimensional Ornstein–Uhlenbeck operators." Foundations of Computational Mathematics 12.5 (2012): 573-623.

[11] Simpson, David JW, Stephen John Hogan, and Rachel Kuske. "Stochastic regular grazing bifurcations." SIAM Journal on Applied Dynamical Systems 12.2 (2013): 533-559.

[12] de Souza, Silvio LT, et al. "Noise-induced basin hopping in a vibro-impact system." Chaos, Solitons & Fractals 32.2 (2007): 758-767.

Manifestation of additional frequencies in vortex induced vibrations in the presence of noise

M.S. Aswathy[1,*] and *Sunetra* Sarkar[1,**]

[1] *Department of Aerospace Engineering, Indian Institute of Technology Madras*

Abstract. In this study, we examine the effect of random input fluctuations in the mean flow to a circular cylinder undergoing transverse oscillations. A Duffing-Van der pol combined system has been used to model the structure and wake oscillators in the VIV system. We observe that the addition of noise brings in major qualitative and quantitative changes on the structural response of the system compared to the deterministic cases. It has been observed that the stochastic system is always influenced by the presence of structural frequency. In contrast, the system under mean flow condition aligns with the structural frequency, only in the lock-in range. This feature is seen as noise exciting multiple frequencies in the response of the cylinder in the pre lock-in and post lock-in regimes.

1 Introduction

Vortex induced vibrations (VIV) of circular cylinders is an important area of research in many real life situations including heat exchangers, bridges, tall buildings, riser tubes, towing pipes, power transmission lines etc. VIV, considered and studied as a Fluid Structure Interaction (FSI) problem, demands a crucial attention because of the high amplitude vibrations undergone by the structures which can lead to large scale damages. Flow around circular cylinders stands as the first step in understanding such flows. An important phenomenon which has to be discussed while discussing VIV of cylinders is lock-in. When a cylinder is allowed to oscillate freely in a flow, for lower velocities, it oscillates with lower amplitudes whose magnitude varies as per the system parameters. During this stage, the vortex shedding frequency of the cylinder, will be based on the non dimensional quantity, Strouhal number (St). Strouhal number follows the Strouhal relationship, which is, $St=\frac{f_v D}{U}$, where f_v is the vortex shedding frequency at the particular flow velocity, D is the diameter of the cylinder, U is the flow velocity. But as velocity is increased, the response undergoes major qualitative changes. The vortex shedding frequency gets closer to the cylinder's natural frequency and f_v gets locked onto f_n, resulting in a resonance like condition, reflected as high amplitude oscillations for a range of velocities. During lock in, the frequency of vortex shedding is not in accordance with the Strouhal relationship mentioned earlier, but becomes the structure's natural frequency itself. As the velocity is further increased, lock-out happens again, with the vortices shedding as per the Strouhal relationship again. There have been numerous amounts of experimental works with focus onto the lock-in phenomenon, structural dynamics and the corresponding flow patterns [1], [2], [3], [4]. Experimental studies explain the hydrodynamics part in VIV through explaining the vortex shedding modes and synchronisation regimes in terms of the flow fields [5], [6], [7]. Later, analytical formulations based on 'wake oscillators' gained prominence due to the reason that they are effective in understanding the physics and overall dynamics of the problem through lesser computational efforts [8], [9], [10].

The focus of the present work is to investigate the changes in the dynamics in the presence of noise. A real time system can be subjected to a lot of system uncertainties, which can alter the behaviour of the system as a whole. Studies emphasising the importance of modelling physical systems with the inclusion of noise have been prevalent [11], [12]. Modelling of noise in engineering systems have also gathered attention recently. Triggering mechanism and the role of stochastic noise has been studied for thermoacoustic systems by Waugh *et al.* [13] and Noiray *et al.* [14]. Waugh *et al.* [13] have reported that the intensity of noise has a major role in deciding the practical stability limits. Poirel *et al.* [15], [16] and Zhao et al. [17] have addressed that aerofoils undergoing flutter with cubic and freeplay nonlinearities are subjected to major qualitative changes when subjected to turbulence in the environment. The number of studies pertaining to discussions on stochastic uncertainties which can be inherently present or externally induced on VIV system are limited. So it is important that we have an understanding on the impact that the fluctuations will have on the system dynamics by studying how big or small changes they can bring onto the response.

In short, we make efforts to capture the role of noise superimposed on an uniform incoming flow in affecting

*e-mail: aswathymek@gmail.com
**e-mail: sunetra.sarkar@gmail.com

the dynamics of an elastically mounted rigid cylinder undergoing transverse oscillations. In the base deterministic model, at each time step, we add a parametric noise modelled through a uniform distribution to the mean flow. Then the system response and their changes are studied with change in the bifurcation parameter. Section 2 explains the formulation of the base deterministic model used for the present study. The bifurcation behaviour of the deterministic system captured using the continuation techniques and direct numerical integration (DNI) are given in section 3. Later, section 4 discusses the stochastic responses obtained using DNI and their qualitative features. In the present work, a low value of noise intensity has been chosen, which is σ=0.1.

2 Methodology

The model consists of a single degree of freedom elastically supported circular cylinder subjected to incompressible constant flow with uniformly distributed noise superimposed on it, as in Figure 1. As described earlier, the physical model is captured using an analytical model where a mechanical spring mass system of the Duffing type models the transverse vibrations of an infinitely long cylinder and a van der Pol oscillator to describe the wake dynamics.

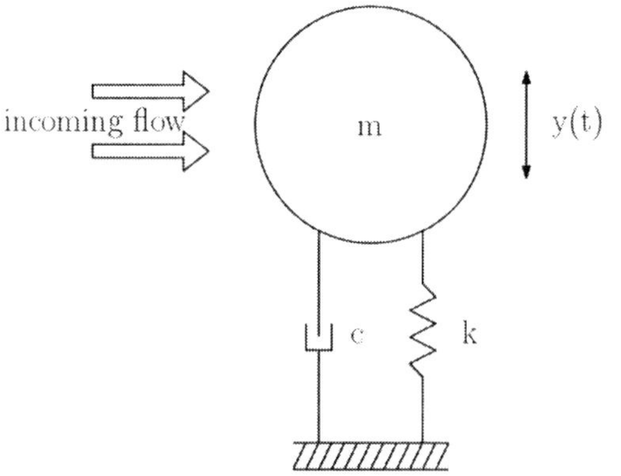

Figure 1: Schematic of cylinder undergoing vortex induced vibrations

The formulation of the model as done by Fachinetti *et al.* [10] is briefly given as follows. A cubic term has been additionally incorporated in the final structural equation in order to account for the nonlinearities present. The structure oscillator is considered to be a spring mass system oscillating with a single degree of freedom as

$$m\ddot{Y} + r\dot{Y} + hY = S \quad (1)$$

The above equation is given in the dimensional form. Here, the dots denote differentiation with respect to the dimensional time T, $m = m_s + m_f$, $r = r_s + r_f$, where $m_s, m_f, r_s, r_f, h, Y, S$ are the structural mass, fluid added mass, structural damping, fluid added damping, spring stiffness, structural displacement, forcing due to the effect of wake, respectively. Here, $m_f = \frac{\pi \rho D^2 C_M}{4}$ and $r_f = \gamma\Omega_f \rho D^2$, where C_M is the added mass coefficient taken as 1 in case of circular cylinder, γ is a stall parameter assumed to be 0.8, $\Omega_f = \frac{2\pi StU}{D}$ is the angular vortex shedding frequency of the cylinder with St the Strouhal number, U as the incoming flow velocity, D as the diameter of the cylinder. Dividing equation (1) by m gives

$$\ddot{Y} + ((2\zeta\Omega_s) + \frac{\gamma\mu}{\Omega_f}) + \Omega_s{}^2 Y = S/m \quad (2)$$

with μ being a non dimensional mass ratio as $\frac{m_s+m_f}{\rho D^2}$, $\Omega_s = \sqrt{\frac{h}{m}}$, being the angular structural natural frequency.

The wake oscillator is modelled by a nonlinear Van der pol equation as

$$\ddot{q} + \epsilon\Omega_f(q^2 - 1)\dot{q} + \Omega_f^2 q = F \quad (3)$$

with F being the forcing term on the wake due to the cylinder motion.

Nondimensionalising equations (2) and (3) using $t = T\Omega_s$ and $y = Y/D$ and incorporating a cubic term accounting for the geometrical non linearities, as done by Srinil *et al.* [18] , The final forms of non dimensional equations of motion are

$$\ddot{y} + (2\zeta + \frac{\gamma\Omega}{\mu})\dot{y} + y + \alpha y^3 = M\Omega^2 q \quad (4)$$

$$\ddot{q} + \epsilon\Omega(q^2 - 1)\dot{q} + \Omega^2 q = A\ddot{y} \quad (5)$$

In the above equations, overdots mean derivatives with respect to the non dimensional time t. Here, Ω is the ratio of the vortex shedding frequency to the cylinder natural frequency in the medium. It is also defined as $\Omega = StU_r$, where U_r is the reduced velocity which is the bifurcation parameter in the present case. The Strouhal number for the present study is assumed to be 0.2, in the subcritical range of Reynolds numbers [19]. Here, μ is the mass ratio related to conventional mass ratio as defined by Williamson *et al.* [7] as $m^* = \frac{4\mu}{pi} - C_m$; ϵ and A are empirical coefficients chosen as by [10] as 0.3 and 12 respectively and α is taken as 0.7. M is a mass number which is a function of the mass ratio defined as $M = \frac{Cl_0}{16\pi^2 St^2}$ [10], with Cl_0 being the associated lift coefficient of a stationary cylinder, taken as 0.3 [19]. The coupling model used for the wake in the present work is an acceleration coupling model [10]. This has been used due to the reason that acceleration coupling model which considers that the structure has a linear inertial effect on the fluid. This coupling model was successful in explaining numerous qualitative aspects of the VIV phenomenon like the lift magnification at lock-in, range of lock-in etc in contrast to the previous coupling models that existed. In the current case, we have used a m^*=248 and $m^*\zeta$ = 0.013. Further details of the parameters chosen can be referred from [10].

However, a numerical integration of the differential equations alone would not give an exact picture of the type of bifurcations involved, the stability of the trajectories, detection of any unstable branches present etc. In order to understand these, we do a continuation study of the set of differential equations. This has been achieved by using the tool XPPAUT [20], which contains codes and algorithms for the continuation and bifurcation software AUTO [21]. Numerical continuation techniques [22] help in finding solutions to a set of parameterised nonlinear equations, from an initial guess for starting the problems. Initially, iterative schemes such as the Newton method are used for solving the system of equations and convergence towards the end solution. The stability of the solutions is determined by calculating eigenvalues and eigenvectors of the fixed points or the Floquet multipliers in case of periodic orbits. Subsequently, this solution is continued to get a curve of solutions of the system for different values of the parameter. Continuation can also be done by starting from a periodic solution. This initial starting orbit is determined by integration and the period of this orbit is used as an approximate period to do continuation by XPPAUT.

3 Analysis of the deterministic system

Before studying the stochastic dynamics, it is important to have a clear understanding regarding the dynamics of the deterministic VIV system. We first discuss the bifurcation behaviours and dynamics of the deterministic VIV system obtained using continuation. Later, (DNI) is done to obtain the time histories and frequency spectrums of the structural responses.

Figure 2 shows the bifurcation plot of the VIV system captured using the continuation method. Closed circles show the amplitudes of the stable limit cycles and the open circles show the amplitudes of the unstable limit cycle. Arrows represent the direction of movement of the responses at points where bifurcation of the orbits happen. It is seen that the system is characterised by regions of bistability (two stable solutions for a single parameter) and the steady state solutions of the given system are dependent on the initial conditions given. For example, consider a region consisting of an unstable branch, as in the region between U_r = 4.29927 to U_r = 4.77048. Any initial condition above the unstable branch would take the system to the high amplitude branch, whereas any initial condition below the unstable branch would take it to the low amplitude branch. The unstable solutions become the steady state solution only when our initial condition is also exactly that value. The region between U_r = 4.29927 to U_r = 4.77048 and U_r = 5.31756 to U_r = 6.1896 are regions of hysterisis. The orbit chosen by the system on these regions depends on whether we move in the forward or reverse direction in the bifurcation diagram. It has been understood that the type of bifurcation undergone by the system while transiting from the low amplitude to high amplitude limit cycles is a saddle node bifurcation. These have been confirmed from a Floquet analysis near the bifurcation points and are not shown here for the sake of brevity.

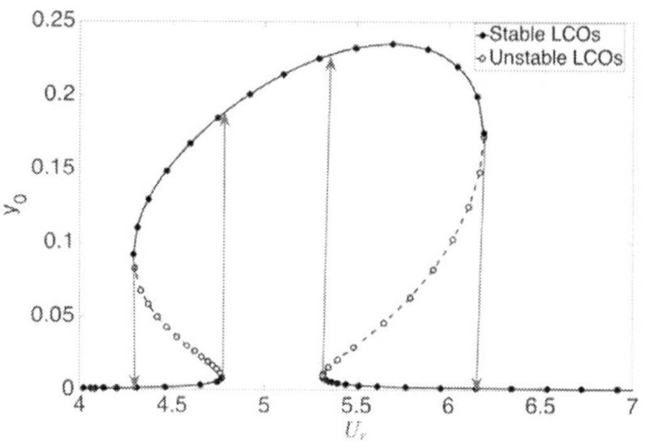

Figure 2: Bifurcation diagram of deterministic system

In the basic formulation of the deterministic model itself, the response y and q are considered to be structural response and wake response respectively with a phase difference, oscillating with a common angular frequency [10]. This common frequency comes out to be the vortex shedding frequency outside lock-in and the structural natural frequency during lock-in. The differential equations have been numerically integrated using a fourth order Runge Kutta method with the bifurcation parameter U_r being incremented in steps of 0.1. As the parameter is increased from U_r = 4, the low amplitude response jumps to a higher branch of response at U_r = 4.8. From U_r = 4.8 to U_r = 5.3, system exhibits high amplitude oscillations. Figure 3 shows the bifurcation diagram obtained by plotting the maximum non dimensional amplitude versus reduced velocity. Figure 4 (a-L) and (a-R) shows time histories and FFTs respectively for U_r=4.7, figure 4 (b-L) and (b-R) for U_r=4.8, figure 4 (c-L) and (c-R) for U_r=5.4, where L and R represents left and right sides respectively. The X-axis of the power spectrum plots shows f, which is the ratio of oscillation frequency of the response (f_o) to the structural natural frequency (f_n).

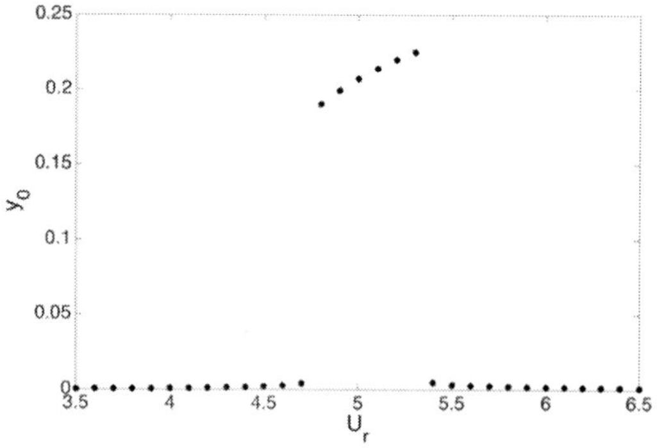

Figure 3: Bifurcation diagram of deterministic system using direct numerical integration

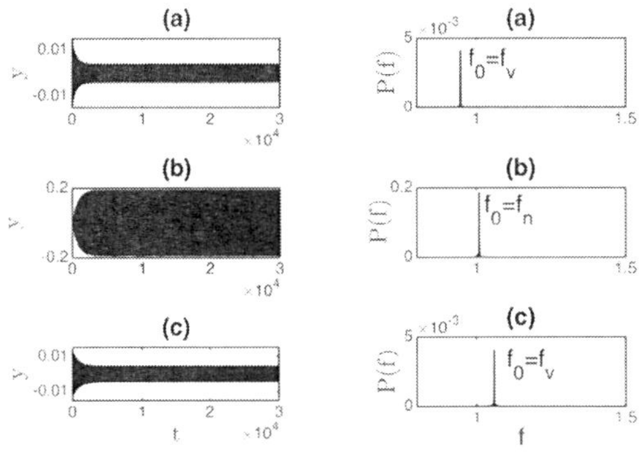

Figure 4: Time histories and FFTs corresponding to the deterministic case for U_r= 4.7(a-L) and (a-R), U_r= 4.8(b-L) and (b-R), U_r= 5.4(c-L) and (c-R)

From Figure 3, it is evident that lock-in occurs from U_r=4.8 to U_r=5.3 in the deterministic case. As is seen from the FFTs of Figure 4, all these responses are characterised by single frequencies. At U_r=4.7, we have small amplitude oscillations of the order of 10^{-3}, characterised by a single frequency, which is the vortex shedding frequency at this particular velocity. At U_r=4.8, the oscillations suddenly jump to a high amplitude of order of 10^{-1}(2 orders higher), again characterised by a single frequency. But this time, the vortex shedding frequency has locked onto the cylinder's frequency and in turn from U_r=4.8 to U_r=5.3, the cylinder oscillates with its own natural frequency of oscillation in the fluid. Again at U_r=5.4, the system has moved out of lock-in, oscillating with very low amplitudes, with the vortex shedding frequency at that particular velocity. As we further increase velocities, it is seen that the oscillation amplitudes get lower though the frequency of oscillation would be the vortex shedding frequencies at the particular velocities.

4 Noise induced dynamics of the VIV system

In this section, we investigate the effect of noise on the system dynamics and compare the results with the deterministic cases. The stochastic modelling of velocity is done as follows: The reduced velocity is considered to be a fluctuating parameter, since it is one of the best choices to account for the incoming flow uncertainties. That is, the total reduced velocity at every time considered becomes, $U_{rtotal} = U_r + \sigma u'(t)$ where U_{rtotal} = total reduced velocity , σ is the intensity of noise and $u'(t)$ is the fluctuating parameter changing at every time step. The input noise fluctuations have been mathematically modelled through a uniform distribution. A distribution like the normal distribution can lead to very large net values and even negative velocity values, which is an unrealistic situation. On the contrary, uniform distribution bounds or limits the maximum value of the net velocity. The values of the fluctuations are determined over a time scale which is very less compared to the system time scale, through a linear interpolation. That means, the input noise is a rapidly changing one compared to the system responses. Fluid particles will have such high frequency random fluctuations superimposed on their mean velocity generally while operating in the transition regimes to turbulence. These fluctuations can be present not only in the incoming stream, but also in the associated wake of the bluff body, which is encountering vortex shedding. Slight local disturbances around the body, along with the combined effects of the above-mentioned effects, might change the dynamics of the FSI problem subjected to transient fluctuations [23].

The output VIV response is analysed for a noise intensity (σ) of 0.1. Figure 5 and Figure 6 show the time histories and FFTs at different values of U_r. Figure 5 (a) corresponds to the response prior to lock-in, at U_r = 4.6. Comparing these and the deterministic responses prior to lock in discussed previously, it is noticeable that the response is no longer periodic characterised by a single frequency. A Fourier analysis of the structural response shows the existence of two frequencies in the spectrum. One of these frequencies correspond to the vortex shedding frequency itself at the particular velocity. The other frequency seen was found to be corresponding to the structural natural frequency in the oscillating system. These can be seen in Figure 6 (a). That is, the system no longer oscillates with a single frequency, but oscillates as a combination of the inherent system frequencies. The lock-in region is still influenced by the structural dynamics alone, as in the deterministic case, as seen in Figure 6 (b), which is for U_r = 5. Outside lock-in, the system again tries to follow the Strouhal law, following the vortex shedding frequency, but the presence of noise induces a part of structural component in this also, which is seen through the presence of the extra frequency as in Figure 6 (c), for U_r = 5.4. The effects of noise are prominent in the non lock-in states.

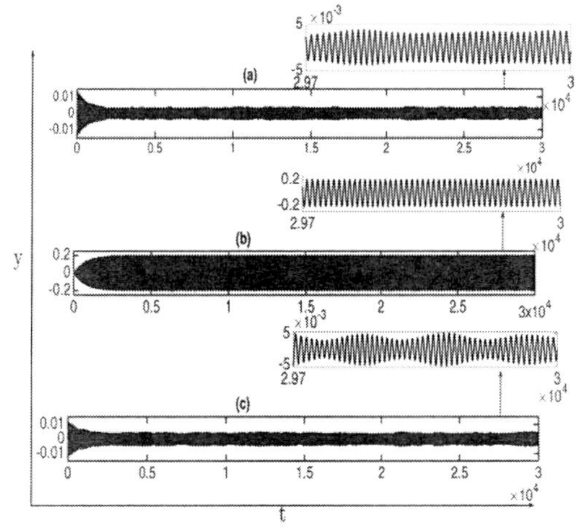

Figure 5: Responses in the presence of noise at U_r (a)=4.6 [pre lock-in] (b)=5 [lock-in] (c)=5.4 [just after lock-in]

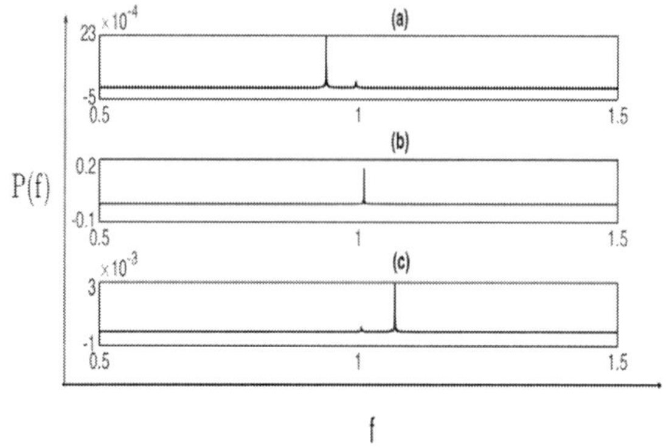

Figure 6: Frequency spectrums of the responses in the presence of noise at U_r (a)=4.6 [pre-lock in] (b)=5 [lock in] (c)=5.4 [just after lock-in]

The existence of two frequencies under the effect of free stream turbulence has been discussed through experiments by So *et al.* [24], where a circular cylinder was allowed to undergo cross flow vibrations. They report that the two frequencies correspond to vortex shedding frequency and structural frequency and observe that the two frequencies coalesce at the lock-in point. However, free stream turbulence has been reported to change the Strouhal number post lock-in (unlike in the current work), due to which the domination of the flow or the structural component also changes. In any case, their experiments are consistent with our observations that external noise play a major role in altering the frequency characteristics of the system. Noise seems to act a mechanism through which energy is supplied to the cylinder motion and hence the cylinder's tendency to be in locked-in state remains high. However, the present study has been based on a noise intensity which is very low. The manifestation of frequencies with a higher level of noise intensity might be different from the current scenario. Investigation on these features, along with the identification of the transitional states by a probabilistic approach and through time series analysis are being carried out by the authors presently.

5 Conclusion

In the present study, we investigated the effect of stochastic uncertainties in altering the response characteristics of a circular cylinder undergoing Vortex Induced Vibrations. We have observed that uncertainties in the incoming flow can alter the response behaviour considerably both in quantitative and qualitative ways. The responses become fluctuating signals owing to the presence of noise. Noise is seen to invoke additional frequencies in the response oscillations compared to the deterministic cases. The non lock-in states which were characterised by the vortex shedding frequency only in the deterministic cases become subjected to the presence of structural frequency also in the presence of noise. Thus, noise plays a role in driving a part of the system dynamics always controlled by the structure as well. An investigation on whether similar behaviour occurs under a higher noise intensity and studies based on stochastic bifurcations are being pursued by the authors.

References

[1] P.W. Bearman, Annual review of fluid mechanics **16**, 195 (1984)
[2] C.C. Feng, Ph.D. thesis, University of British Columbia (2011)
[3] A. Khalak, C.H.K. Williamson, Journal of Fluids and Structures **11**, 973 (1997)
[4] A. Khalak, C.H.K. Williamson, Journal of fluids and Structures **13**, 813 (1999)
[5] R. Govardhan, C.H.K. Williamson, Journal of Fluid Mechanics **420**, 85 (2000)
[6] A. Khalak, C.H.K. Williamson, Journal of Wind Engineering and Industrial Aerodynamics **69**, 341 (1997)
[7] C.H.K. Williamson, Journal of Wind Engineering and Industrial Aerodynamics **69**, 3 (1997)
[8] R. Bishop, A. Hassan, *The lift and drag forces on a circular cylinder oscillating in a flowing fluid*, in *Proceedings of the Royal Society of London A: Mathematical, Physical and Engineering Sciences* (The Royal Society, 1964), Vol. 277, pp. 51–75
[9] R.T. Hartlen, I.G. Currie, Journal of the Engineering Mechanics Division **96**, 577 (1970)
[10] M.L. Facchinetti, E. De Langre, F. Biolley, Journal of Fluids and structures **19**, 123 (2004)
[11] L. Arnold, *Random dynamical systems* (Springer Science & Business Media, 2013)
[12] W. Horsthemke, in *Non-Equilibrium Dynamics in Chemical Systems* (Springer, 1984), pp. 150–160
[13] I.C. Waugh, M.P. Juniper, International journal of spray and combustion dynamics **3**, 225 (2011)
[14] N. Noiray, A. Denisov (2016)
[15] D. Poirel, S.J. Price, Nonlinear dynamics **48**, 423 (2007)
[16] D.C. Poirel, S.J. Price, AIAA journal **39**, 1960 (2001)
[17] D. Zhao, Q. Zhang, Y. Tan, Nonlinear Dynamics **58**, 643 (2009)
[18] N. Srinil, H. Zanganeh, Ocean Engineering **53**, 83 (2012)
[19] R.D. Blevins, *Flow-induced vibration* (New York, NY (USA); Van Nostrand Reinhold Co., Inc., 1990)
[20] B. Ermentrout, *Simulating, analyzing, and animating dynamical systems: a guide to XPPAUT for researchers and students* (SIAM, 2002)
[21] E.J. Doedel, A.R. Champneys, T.F. Fairgrieve, Y.A. Kuznetsov, B. Sandstede, X. Wang et al., AUTO97, Concordia University, Canada (1997)
[22] E.L. Allgower, K. Georg, *Numerical continuation methods: an introduction*, Vol. 13 (Springer Science & Business Media, 2012)

https://doi.org/10.1051/matecconf/201814808003

[23] G. Buresti, Lecture Notes, International Advanced School on Wind-Excited and Aeroelastic Vibrations, Genoa, Italy (2000)

[24] R.M. So, X. Wang, W.C. Xie, J. Zhu, Journal of Fluids and Structures **24**, 481 (2008)

https://doi.org/10.1051/matecconf/201814808004

A Stochastic source model for the 2015 M_W 7.9 Gorkha, Nepal, Earthquake using Multi-Dimensional Ensemble Empirical Mode Decomposition technique

S Sangeetha[1], and *S.T.G* Raghukanth[2]

[1]Research scholar, Department of Civil Engineering, Indian Institute of Technology Madras, Chennai 600036, India.
[2]Associate professor, Department of Civil Engineering, Indian Institute of Technology Madras, Chennai 600036, India.

Abstract. The present study aims at developing a new strategy to model the spatial variability of slip on the rupture plane using multi-dimensional ensemble empirical mode decomposition (MEEMD) technique. Here, the earthquake slip distribution is split into finite number of empirical modes of oscillation called the intrinsic mode functions (IMFs). This help in identifying the fluctuation component and trend in the slip data. The trend is positive and characterizes the nonstationary mean of the slip distribution. The fluctuation component can be modelled as a stationary random field using an exponential power spectral density function. The trend can be modeled as an elliptic patch. This new technique is demonstrated for the slip distribution of the recent Nepal Earthquake, 2015. It is observed that the new model can be used to simulate the spatial complexity of slip distribution of any earthquake.

1 Introduction

Finite-fault inversion provides a key to understand the earthquake source complexities which is essential to predict realistic ground motion for a scenario earthquake. The derived finite-fault rupture models from inversion contain information on fluctuations of slip on the rupture plane. In addition, these earthquake slip distributions are inherently nonstationary due to the fact that slip vanishes towards the buried edges of the fault. This clearly states the level of complexity which cannot be modelled by simple deterministic functions. In such situations, stochastic approaches provide an alternate for characterizing the spatial complexity of slip on the rupture plane. Research has been carried along this direction, over the last two decade (Somerville *et al.*, Mai and Beroza, Lavallée *et al.*, Raghukanth and Iyengar, Raghukanth, Raghukanth and Sangeetha [1-6]). The method for investigating earthquake slip distribution in these works are based on stationary random processes models with Gaussian properties. Limitation with these studies are, the developed models are based on Fourier analysis. But, it is well known that the PSD exists if and only if the signal is a wide-sense stationary process. For nonstationary process PSD does not exist. In such situations, Fourier-based spectral analysis which uses sine and cosine functions as the basis, in analyzing nonlinear and nonstationary data can yield distorted and incomplete information on the nature of the nonstationary slip distribution.

Huang *et al.* [7] pointed out that the only way to represent the physics of nonlinear and nonstationary processes is through adaptive basis functions. They have demonstrated the superiority of Empirical mode decomposition technique (EMD) technique to others by analysing wind speed, tidal velocity and earthquake data. The technique has been previously applied to understand strong motion data of past earthquakes (Huang *et al.*, Loh *et al.*, Zhang *et al.*, Raghukanth and Sangeetha [8-11]). Mallikarjuna *et al.* [12] has used the EMD approach to forecast the air traffic by modelling the fluctuation part using artificial neural network (ANN) technique and the trend through simple regression analysis. Raghukanth [5] has proposed a new data-driven technique, namely Bi-dimensional empirical mode decomposition (BEMD) to characterize earthquake slip distribution of past earthquakes. Here the slip distribution is modelled as a sum of fluctuation and trend component. The fluctuation component is simulated as an exponential random field. The shortcoming with this work is that modelling of the trend component has not been explored in detail. The trend component is deterministic for a particular earthquake, hence it has to be modelled as 2D nonstationary function and cannot be taken as mean value of trend part as considered in Raghukanth [5]. In this study, an attempt has been made to model the trend of the slip data.

* Corresponding author: ssangee.civil@gmail.com

2 Source model of Nepal earthquake, 2015

The 25th April 2015 (M_w 7.9) earthquake triggered many broadband instruments across the globe. Researchers have reported this event as the largest to have occurred in the Nepal region in the last eight decade. A multitude of inversion studies have been carried out (Hayes *et al.*, Yagi and Okuwaki, Lay *et al.*, Tung and Masterlark, Yue *et al.*, Zhang *et al.*, McNamara *et al.*, [13-19]) using seismological (teleseismic and strong motion) and geodetic (high-rate GPS, static GPS, InSAR,ALOS-2) observations of both the mainshock and its largest aftershock. Nevertheless, we restrict our study to the modelling and analysis of finite fault slip model proposed by Yagi and Okuwaki [14] (see Figure 1). The source model indicates a north dipping low-angled thrust fault in MCT with strike/dip/rake angle of 285/10/58.6. It can also be observed from the figure that the rupture extends about 160km along the strike and 88km along the dip direction with an average rupture velocity of 3km/s. A maximum slip of about 7.5m is concentrated at a radial distance of 50km to the east of epicentre (28.147 °N lat, 84.708°E long). The total seismic moment is 9.1 $\times 10^{20}$ Nm corresponding to M_w 7.9. The focal depth and depth to the top of fault are 15km and 4.58km respectively. The hypocentre of the event is located at a distance of 140 and 60km along the strike and dip directions, respectively. Rise time, which is defined as the time taken by each subfault to rupture completely is 0.8s while the source duration is 50s. The entire rectangular fault in divided into 220 subfaults of size 8km x 8km.

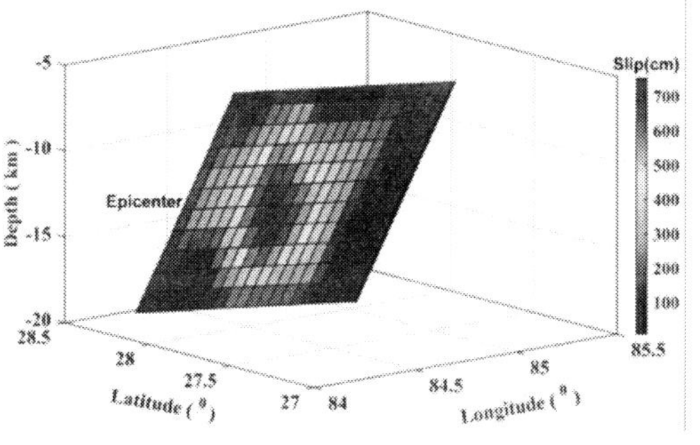

Fig.1. Finite fault slip distribution of the 25th April 2015 Gorkha, Nepal Earthquake

3 MEEMD

The present study introduces an advanced adaptive data analysis technique, namely the multi-dimensional ensemble empirical mode decomposition (Wu *et al.*, [20]) technique, to analysis the spatial variability of the earthquake slips. This MEEMD method has been formulated based on the EMD (Huang [7]) and its more robust version EEMD (Huang and Wu, Wu and Huang [21-22]). In the EMD approach 1-D data, $X(t)$ is decomposed into finite (n) modes of oscillation with certain frequencies called the intrinsic mode functions, IMFs such that

$$X(t) = \sum_{j=1}^{n} IMF_j + R_n(t) \tag{1}$$

where $R_n(t)$ represents the residue of data which can either be a trend (monotonic function with one maxima) or a constant value. The EMD/EEMD approach finds its application in the recently developed MEEMD method. The theory behind MEEMD is to visualize a multi-dimensional dataset as a combination of data in each dimension, which is followed by the application of EEMD to slices of data in each and every dimension and finally reconstructed by combining components of similar scale along each dimension.

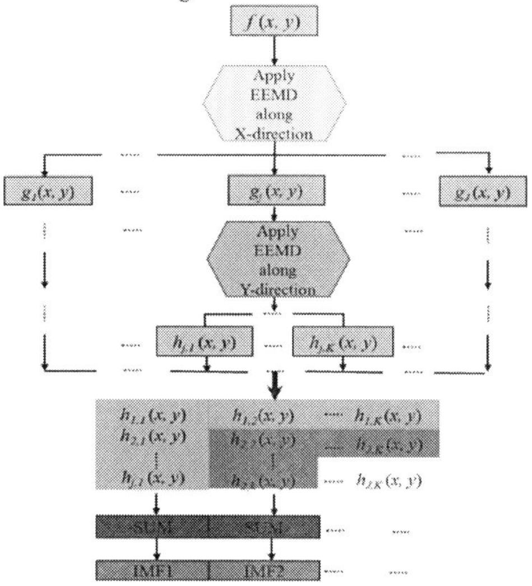

Fig.2. Schematics of MEEMD

A schematic flowchart explaining the MEEMD method for a two-dimensional spatial data such as earthquake slip distribution $f(x, y)$ is shown in figure 2. To start with, we assume the 2-D slip data of Nepal earthquake, $f(x, y)$ as a set of 1-D slices in x-direction (i.e. along the strike). Now, each 1-D slice is decomposed using EEMD resulting in a 2-D pseudo IMF like component $g(x, y)$. This new 2-D data is further decomposed, but this time, the data is considered as a set of 1-D slices in y-direction (i.e. down dip direction) at each location along the x-direction. These 1-D slices of $g(x, y)$ upon EEMD decomposition results in collection of 2-D components of similar scales namely $h(x, y)$. The derived 2-D components are further combined into a reduced set of final IMF components based on the minimal scale combination principle.

3.1 Intrinsic Mode Function

The slip distribution of Nepal earthquake has been decomposed into three dynamic IMFs each evolving around a specific frequency. With the help of these extracted IMFs, one can reconstruct the original slip data as

$$f(x,y)=\sum_{j=1}^{n} IMF_j(x,y)+R(x,y) \qquad (2)$$

$R(x, y)$ in equation (2) refers to the residue of the slip data which is similar to $R(t)$ in equation (1). Figure 3 depicts the extracted IMFs along with the residue and slip data. A closer look at the IMFs in figure shows that the number of extremas decreases with increasing hierarchy of IMFs. The statistics of these IMFs are presented in Table 1. The 5x5 correlation matrix clearly shows that there exists a strong correlation between data and IMFs.

Table1. Correlation matrix of IMFs of slip data.

	Data	IMF1	IMF2	IMF3	Trend
Data	1.000	0.200	0.635	0.880	0.900
IMF1		1.000	0.261	0.025	-0.116
IMF2			1.000	0.699	0.294
IMF3				1.000	0.768
Trend					1.000

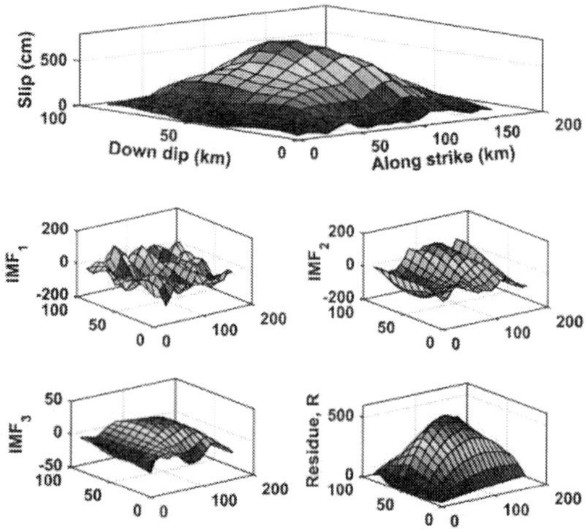

Fig. 3. IMFs and residue of 25th April 2015 Gorkha, Nepal Earthquake

From figure 3, it can also be observed that the residue of the data is positive and slowly varies about the long term average. One may interpret this as a nonstationary component about which the spatial variability of the slip oscillates. This residual component which represents trend of the slip data (as in equation 3) has the highest percentage variance of 74.5% and exhibits the highest correlation of 0.9. On the other hand the fluctuation component of slip which is defined as the summation of all IMFs (as in equation 4) exhibits a percent variance of 25.5%.

$$\text{Trend: } D_T(x,y)=R(x,y) \qquad (3)$$

$$\text{Fluctuation: } D_F(x,y)=\sum_{j=1}^{n} IMF_j(x,y) \qquad (4)$$

This statistical measure indicates the relative contribution of each component (fluctuation and trend) to the data. It is clearly observed that the fluctuation component characterizes the small scale variability in the data. Contour maps of both these components for the Nepal earthquake are shown in figure 4.

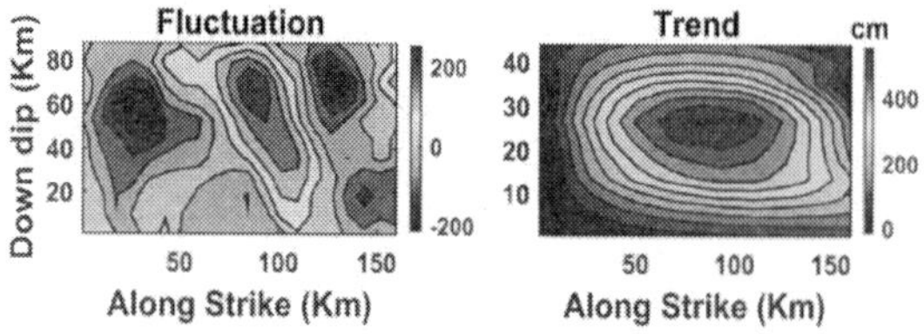

Fig.4. Fluctuation and trend of the slip field shown in figure 1.

The fluctuation component is oscillatory and can be modelled as a random filed. In the second-order analysis, variation of random field at two different locations is characterized either in space by an auto-correlation function (ACF) or in the wave number domain by a power spectral density (PSD). Power spectral density of the slip field and fluctuation are shown in figure 5. Here, the wavenumber ranges between $-\pi/dx$ to π/dx. The spatial frequency content of the fluctuation component is found to be 0-0.2 km^{-1}

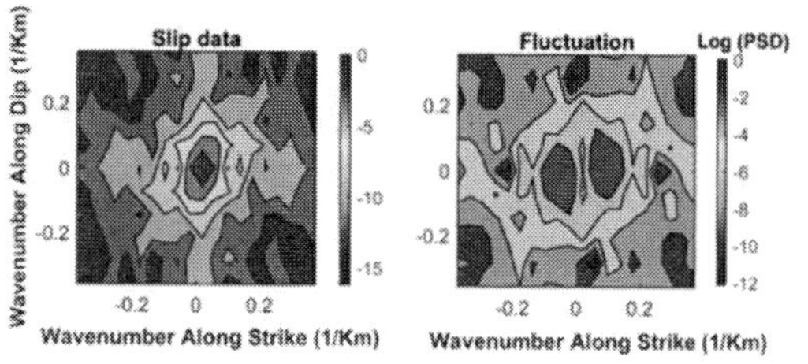

Fig.5. PSD of the slip field and fluctuation component

The article presents a new strategy wherein, one can attempt modelling the slip distribution as a function of fluctuation and trend component instead of modelling the entire slip which is highly erratic.

4 Modelling the fluctuation component

The fluctuation component of the Nepal earthquake as seen in figure 4 is highly complex or random. This erratic nature of the fluctuation component (D_F) can be associated with the randomness observed in ground motion records. The spatial variability of this component can only be modelled as a stationary random field. Assuming the fluctuation component as ergodic, two-dimensional PSD [$D_F(\kappa_x, \kappa_y)$] is obtained from the slip distribution as

$$D_F(\kappa_x,\kappa_y)=\left|\int_{-\infty}^{\infty}\int_{-\infty}^{\infty} D_F(x,y)\ e^{-i\kappa_x x-i\kappa_y y}dxdy\right|^2 \qquad (5)$$

where κ_x and κ_y are the spatial wave numbers along the along the strike and dip direction respectively. In the present study, we adopt an exponential random field model to characterize the fluctuation component and is defined as

$$S(\kappa_x, \kappa_y) = \frac{a_x a_y}{(1+k^2)^{3/2}} \quad (6)$$

where $k = \left(a_x^2 k_x^2 + a_y^2 k_y^2\right)^{0.5}$, a_x and a_y denote the correlation length along strike and dip direction respectively. The correlation lengths, a_x and a_y are estimated by minimizing mean square error between the PSD of slip data and PSD computed for the random field model in Equations (6). The deduced correlation length for the Nepal earthquake is estimated as a_x=10 and a_y=15. In Figure 6, the fitted 1D exponential PSD at cross-section k_y =0 and k_x =0 is shown along with the PSD of the data. The best-fit correlation measures are also shown in the same figure for Nepal earthquake, 2015. It can be observed from the figure that the misfit associated with exponential function is negligible. Hence, it can be concluded that the preferred model that is most consistent with the data and which characterizes the complexities of the fluctuation component follows an exponential PSD.

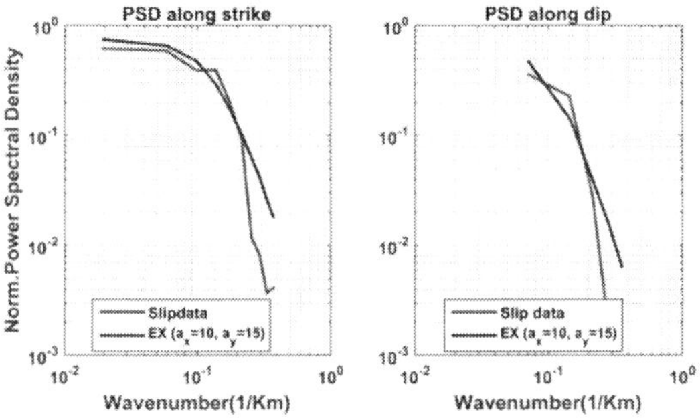

Fig. 6. PSD at the cross sections k_y = 0 and k_x = 0 for the fluctuation component of Nepal event shown in figure 5.

5 Modelling the trend component

The trend represents a positive mean value of the slip field. This nonstationary characteristic of the trend can be modelled by fitting a simple deterministic function. In this study a single elliptic patch model is used to model the trend component. This method has been previously used by Vallée and Bouchon [23] to model entire slip data. In the present, we apply the same technique but, instead of modelling the entire slip data, we try to model only the trend component.

5.1 Single elliptic patch model

As a first step, we estimate the effective rupture area of the trend based on 5%-95% Husid plot which turns out to be 128 x 68 sq. km. This effective rupture patch is shown in figure 7 as a black rectangle. Now, we model this effective rupture area as a single elliptic patch with a constant rise time constant, d_k and rupture velocity, V_r. The precise parameterization of trend component of slip is detailed in Figure 8. A sum of 8 parameters are required to describe this rupture patch. This includes

D	-Spatially varying slip distribution of trend
V_r	- Rupture velocity
d_k	- Rise time
a, b	- Geometry of the elliptic patch
D_x,D_z	-Shift of the patch centre from the hypocentre of the event
α	-Orientation of the ellipse

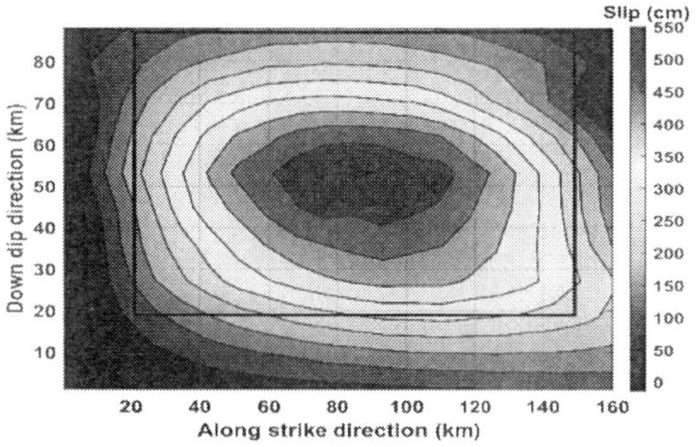

Fig.7. Rupture patch of trend component

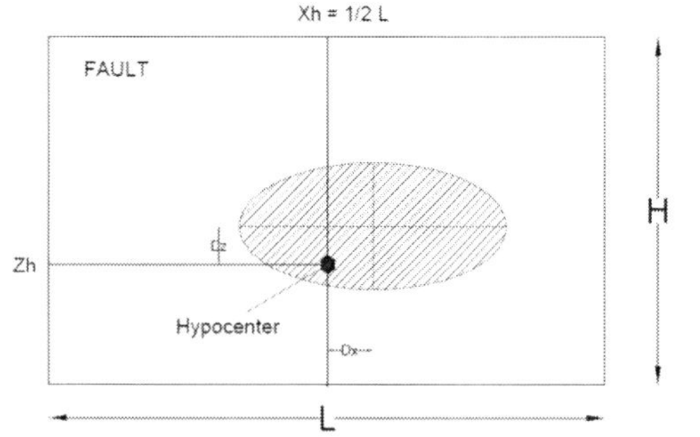

Fig.8. Parameterization for trend modelling of the Nepal earthquake

The parameters a and b are estimated to be 64 km and 34 km respectively. Values of V_r, d_k, D_x and D_z are mentioned in section 2. The orientation of the ellipse, α in taken as 0° as seen in figure 8 for Nepal event. The elliptical perimeter of ellipse is then defined as a collection of points $P_{e,i}$ ($i=1,2,3...m$) and the nucleation point P_n (x_n,y_n) is given in fault-plane coordinates with origin at the centre of the rupture patch. Now, the slip distribution of the trend, $D(r)$ along line i from P_n to each point $P_{e,i}$ ($x_{e,i}$, $y_{e,i}$) on the perimeter of the ellipse can be expressed as

$$D(r) = \frac{D_p}{1-D_{\min}}(e^{-k_i r^2 - D_{\min}}), \qquad 0 \le r \le R_i \quad (7)$$

where,

D_p	– Max slip
D_{min}	– 0.1 D_p
k_i	– Exponential decay coefficient
R_i	– Distance from r=0 to r=R_i

Equation (7) yields continuous, spatially smooth pseudo-Gaussian slip distribution of the trend on the elliptical rupture patch. Rupture initiates at P_n and propagates radially at rupture velocity V_r. Boundary conditions (equation 8) are satisfied to estimate slip, $D(r)$. We finally superimpose $D(r)$ on a Cartesian co-ordinate system as D_T (x, y). The estimated slip distribution of trend is shown in figure 9.

$$BC1: \quad D(r=0) = D \ and \frac{d}{dr} D(r=0) = 0, \qquad (8)$$
$$BC2: \quad D(r=R) = 0$$

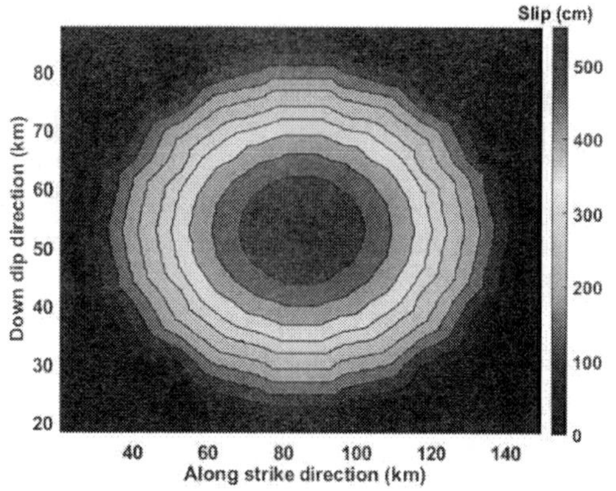

Fig. 9. Estimated trend component

6 Simulated slip field

The final slip field can be obtained as a weighted sum of the fluctuation and trend components. In this case, explained variance of each component is taken as weightage. Figure 10 shows the final simulated slip field for the 25th April 2015 Nepal earthquake. The simulated slip field is in good agreement with original slip data shown in figure 1. The total moment of the simulated slip is comparable to total moment of the data.

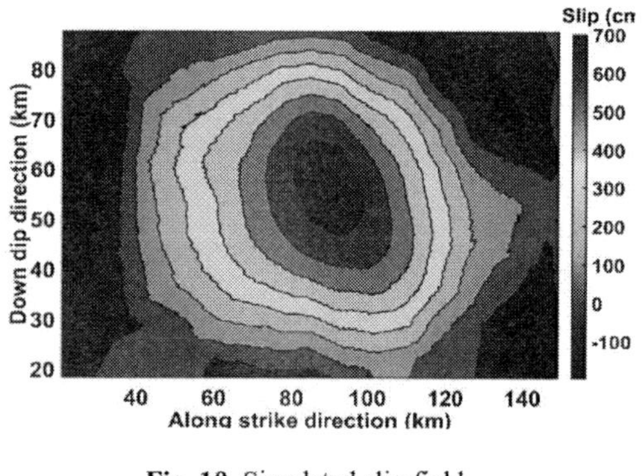

Fig. 10. Simulated slip field

7 Simulation procedure for a hypothetical future event

1. Fix the focal mechanism (strike, dip, rake)
2. Assume faulting-style (Strike-slip, Reverse, Normal)
3. Choose the Moment magnitude, M_w of the potential earthquake
4. Determine the moment magnitude, M_0 from M_w=2/3 $\log_{10}(M_0)$-10.7
5. Determine the fault length L, fault width W, mean slip d and σ_d from scaling relations
6. Estimate correlation lengths a_x, a_y from scaling relations for an exponential random field.
7. Simulate fluctuation component as an exponential random field from the spectral representation method of Shinozuka and Deodatis [24] or Fourier integral method of Pardo-Igu´zquiza and Chica-Olmo [25] using the estimated correlation lengths
8. Simulate tend component as a single patch rupture model using the parameters a, b, α, D_p for a given event.
9. Obtain the final random field as a weighted sum of fluctuation and trend component.
10. Fix hypocentre of the event.

An ensemble of slip fields can be simulated using the proposed methodology. The simulated slip field from this approach finds its application in the synthesis of several realizations of accelerograms using any of the analytical approaches.

7 Discussion and conclusion

A novel approach to model spatial complexity of slip on the fault plane using MEEMD technique has been explored in this article. The decomposition of the fault slip into IMFs brings out certain interesting features of the slip distribution of Nepal earthquake. The most interesting feature that has been observed is that, MEEMD sorts the spatial frequency content of the slip distribution into a finite set of IMFs with the highest and lowest frequency corresponding to the first IMF and trend, respectively. One can try modelling the randomness associated with each IMF separately, however, this may lead to an increase in the number of model parameters whereby reducing the efficiency of the developed model. To overcome this limitation, in the present study, it is proposed that the slip field can be modelled as a combination of stationary random field (fluctuation) and a nonstationary trend. Given a particular fault and magnitude, the proposed model can handle uncertainties in the slip distribution. The developed stochastic source model together with seismological method or Spectral Finite Element Method, can be used to synthesize an ensemble of ground motion time histories for any future earthquake and for regions lacking strong motion data. Further, randomness in the simulated time histories can be attributed to the fluctuation component while peak ground displacement and peak ground residual displacement being associated with the trend term. Another advantage with the developed model is that the estimated slip field from this approach finds its real time application in generation of seismic or tsunami hazard maps.

8 Reference

1. Somerville, P., Irikura, K., Graves, R., Sawada, S., Wald, D., Abrahamson, N., Iwasaki, Y., Kagawa, T., Smith, N., Kowada, A. (1999). Characterizing Crustal Earthquake Slip Models for the Prediction of Strong Ground Motion. *Seismological Research Letters* **70**(1), 59-80.
2. Mai, P. M. and Beroza, G. C. (2002). A spatial random field model to characterize complexity in earthquake slip. *Journal of Geophysical Research* **187**(11), 1-28.
3. Lavallée, D., Liu, P. and Archuleta, R. J. (2006). Stochastic model of heterogeneity in earthquake spatial distribution. *Geophysical Journal International* **165**, 622–640.
4. RaghuKanth, S. T. G. and Iyengar, R. N. (2009). Engineering source model for strong ground motion. *Soil Dynamics and Earthquake Engineering* **29**, 483-503.
5. RaghuKanth, S. T. G. (2010). Intrinsic mode functions of earthquake slip distribution. *Advances in Adaptive Data Analysis* **2**(2), 193–215
6. Raghukanth, S.T.G. and Sangeetha, S. (2014), A stochastic model for earthquake slip distribution of large events, *Geomatics, Natural Hazards and Risk*, DOI:10.1080/19475705.2014.941418
7. Huang, N. E., Shen, Z., Long, S. R., Wu, C. M., Shih, H. H., Zheng, Q., Yen, N. C., Tung, C. C., and Liu, H. H. (1998). The empirical mode decomposition and the Hilbert spectrum for nonlinear and non-stationary time series analysis. *Proceedings of the Royal Society London* **454**, 903–995.
8. Huang, N. E., Chern, C. C., Huang, K., Salvino, L. W., Long, S. R., and Fan, K. L. (2001). A new spectral representation of earthquake data: Hilbert spectral analysis of station TCU129, Chi-Chi, Taiwan, 21 September 1999. *Bulletin of the Seismological Society of America* **91**(5), 1310–1338.
9. Loh, C. H., Wu, T. C. and Huang, N. E. (2001). Application of the empirical mode decomposition-Hilbert spectrum method to identify near-fault ground-motion characteristics and structural responses. *Bulletin of the Seismological Society of America* **91**, 1339–1357.
10. Zhang, R. R., Ma, S., Safak, E. and Hartzell, S. (2003). Hilbert-Huang Transform Analysis of Dynamic and Earthquake Motion Recordings. *Journal of Engineering Mechanics* **129**(8), 861– 875.
11. Raghukanth, S.T.G. and Sangeetha, S. (2013), Empirical mode decomposition of earthquake accelerograms, *Advances in Adaptive Data Analysis* **4**(3). DOI: 10.1142/S1793536912500239
12. Mallikarjuna, C. and Raghukanth, S. T. G. (2010). Forecasting the air traffic for north-east indian cities, *Advances in Adaptive Data Analysis* **2**(1),1-16
13. Hayes, G.P., Briggs, R.W., Barnhart, W.D., Yeck, W.L., McNamara, D.E., Wald, D.J., Nealy, J.L., Benz, H.M., Gold, R.D., Jaiswal, K.S., Marano, K., Earle, P.S., Hearne, M.G., Smoczyk, G.M., Wald, L.A., Samsonov, S.V., 2015. Characterization of the 2015 M_w 7.8 Nepal (Gorkha) earthquake sequence and its seismotectonic context. *Seismological Research Letters* **86**(6), 1557–1567.
14. Yagi, Y., and Okuwaki, R. (2015). Integrated seismic source model of the 2015 Gorkha, Nepal, earthquake, *Geophysical Research Letters* 42, 6229–6235.
15. Lay, T., Ye, L., Koper, K.D., Kanamori, H., (2016). Assessment of teleseismically determined source parameters for the April 25, 2015 M_w7.9 Gorkha, Nepal earthquake and the May 12, 2015 M_w 7.2 aftershock. *Tectonophysics* 714-715, 4-20.
16. Tung, S., and T. Masterlark (2016), Coseismic slip distribution of the 2015 M_w 7.8 Gorkha, Nepal, earthquake from joint inversion of GPS and InSAR data for slip within a 3-D heterogeneous Domain, *Journal of Geophysical Research Solid Earth* 121, 3479–3503.
17. Yue, H., Simon, M., Duputel, Z., Jiang, J., Fielding, E., Liang, C., Owen, S., Moore, A., Riel, B., Ampuero, J.P., Samsonov, S.V. (2016). Depth varying rupture properties during the 2015 M_w 7.8 Gorkha (Nepal) earthquake. *Tectonophysics* 714–715 (2017) 44–54
18. Zhang, L., Lia, J., Liao, W., Wang, Q. (2016). Source rupture process of the (2015) Gorkha, Nepal M_w7.9 earthquake and its tectonic implications. *Geodesy and geodynamics* 7(2) , 124 -131
19. McNamara, D.E., Yeck, W.L., Barnhart, W.D., Schulte-Pelkum,V., Bergmane.,L., Adhikarid, L.B., Dixit, A., Hough, S.E., Benza, H.M., .Earlea, P.S. (2016).Source modeling of the 2015 M_w 7.8 Nepal (Gorkha) earthquake sequence: Implications for geodynamics and earthquake hazards. *Tectonophysics* 714–715, 21–30
20. Wu, Z., Huang, N. E., and Chen, X. (2009). The multi-dimensional ensemble empirical mode decomposition method. *Advances in Adaptive Data Analysis* 1, 339–372.
21. Huang, N. E., and Wu, Z. 2008. A review on Hilbert-Huang transform: Method and its applications on geophysical studies. *Reviews of Geophysics* 46, RG2006, doi: 10.1029/2007RG000228.
22. Wu, Z., and Huang, N. E. (2009). Ensemble empirical mode decomposition: A noise-assisted data analysis method. *Advances in Adaptive Data Analysis* 1, 1–41
23. Vallee, M. and Bouchan, M. (2004) .Imaging coseismic rupture in far Þeld by slip patches. *Geophysical Journal International* **156**, 615-630.
24. Shinozuka, M., and Deodatis, G. (1998). Stochastic process models for Earthquake source. *Probablistic Engineering Mechanics* **3**(3), 114-123.
25. Pardo-Igu′ zquiza, E., Chica-Olmo, M., (1993). The Fourier integral method—an efficient spectral method for simulation of random fields. *Mathematical Geology* **25** (2), 177–217.

Dynamics of SMA micro-actuator in biomechanical system

Rafal Rusinek[1,*], *Andrzej* Weremczuk[1,**], *Marcin* Szymanski[2,***], and *Jerzy* Warminski[1,****]

[1] *Lublin University of Technology, Nadbystrzycka 36, 20-618 Lublin, Poland*
[2] *Medical University of Lublin, Jaczewskiego 8, 20-854 Lublin, Poland*

Abstract. In this paper the polynomial model of shape memory alloy is used to characterise properties of a micro-actuator which is applied as a new middle ear prosthesis. A two degrees of freedom model of the reconstructed middle ear is solve by means of multiple time scales method. The system has various behaviours near the primary resonance depending on ambient temperature. The special case when relative temperature $\theta = 1.0$ characterises untypical resonance curve. Increasing temperature to the normal human body one the resonance curves are typical. Then the system has only one periodic solution if the excitation is not too strong.

1 Introduction

Shape memory alloys (SMAs) are a family of metals with the ability to change shape depending on their temperature [1]. SMA may exhibit one or two-way shape memory effect which is attractive in various science and engineering applications starting from biomedical and ending to aerospace industry. SMA is used to built thermal actuators, stabilizing mechanisms and so on. SMAs undergo thermo–elastic phase transformations between martensite and austenite induced by temperature or stress. Two variants of the martensitic phase are possible: twinned or detwinned. The conversion from twinned to detwinned martensite takes place by means of loading process. When the loading - unloading process is finished, some amount of residual strain remains, meaning that the reverse transformation from detwinned to twinned martensite is not completed. The shape memory effect appears by heating which activates the transformation from detwinned martensite to austenite. When SMA is stressed at temperatures that are above the austenite transformation temperature an inelastic deformation (hysteresis loop) is observed. This phenomenon is known as pseudoelastic effect [2, 3].

These thermomechanical properties of SMAs are studied in literature by microscopic or macroscopic models. The microscopic models treat phenomena on a molecular level, therefore the macroscopic phenomenological models are more attractive from practical point of view. In the literature one can find a lot of interesting and thorough description of SMAs [4–10] and their applications [11–14]. Different models of SMA give various results depending on nonlinear phenomenon which are taken into account. A comparison of popular SMA models is presented in [15].

In this paper the model with polynomial free-energy based on Devonshire theory is used to model SMA micro actuator. Temperature induced phase transition, combined with hysteresis was proposed by Falk [16] and Falk and Konopka [17]. In that model free energy depends on the temperature and the one-dimensional strain. On the other hand, the SMA micro actuator in the form of prosthesis is applied to reconstruct a damaged human middle ear. The prosthesis made of SMA is able to change its shape and length while being heated and remain unchanged after cooling. Therefore, the SMA micro actuator could provide better sound transfer conditions from an outer to an inner ear. The idea of using SMA to build middle ear prosthesis is presented in [18–20]. However nowadays, a typical middle ear prosthesis is made of titanium. Its size and shape is widely discussed in literature e.g. [21–24].

Here, in this contribution an analysis of the reconstructed middle ear (RME) by means of the SMA micro actuator (shape memory prosthesis - SMP) is presented. A two degrees of freedom (2dof) model of the RME is solved analytically with the help of the multiple times scale method (MTSM). Finding the resonance curves at different temperatures is the main goal of the paper.

2 Modelling

The human middle ear composed of three parts: the outer, the middle and the inner ear. Sound is transmitted from the outer to the inner through the middle ear which is goal of the present analysis. The middle ear consists of three small bones: the malleus, the incus and the stapes connected each other. When the ossicular chain structure is destroyed by inflammatory diseases such as chronic suppurative otitis media or cholesteatoma, the middle ear should be reconstructed. In the medical practice it is called ossiculoplasty (the reconstruction of the middle ear ossicles). For over 50 years, middle ear surgery techniques

* e-mail: r.rusinek@pollub.pl
** e-mail: a.weremczuk@pollub.pl
*** e-mail: marcinszym@onet.pl
**** e-mail: j.warminski@pollub.pl

have improved hearing that had been destroyed by various diseases. Numerous procedures are currently used in clinical practice, and a variety of middle ear prostheses made of titanium or titanium alloy are available. Most prostheses can be modified only once they have been prepared. If a prosthesis is too short, it cannot be remodelled. However, prostheses of different sizes can be chosen or tailored by a surgeon to the desired length. Their length can be adjusted by cutting the prosthesis leg. To improve and fasten implementation process we propose a shape memory prosthesis (SMP) that can adjust its size and angulation to the requirements of a particular medical case. The prosthesis can be made of the Nitinol with one–way shape memory effect. It must change its shape while being heated and remain unchanged after cooling. In medical practice, when the middle ear is damaged usually the incus has to be removed to implement a prosthesis between the malleus and the stapes. Therefore 2dof model of reconstructed middle ear is proposed and characterised in Sect. 2.2. The mathematical model of the SMA element is presented in Sect. 2.1.

In the present work, we focus on analytical solutions of the nonlinear system with SME which allows wide range parametric studies for any mechanical and biomechanical systems. However, the detailed investigations are performed for the parameters corresponding to the model of a reconstructed human middle ear.

2.1 SMA actuator

The SMA actuator (prosthesis) is modelled by a polynomial dependence developed by Falk [16]. The spring force of the prosthesis can be written as

$$F_{SMA} = \sigma A \tag{1}$$

where A is a cross-section of the prosthesis. The stress σ is described by a polynomial proposed by Falk [16] and used to describe the pseudoelasticity and shape memory effect as well [25]. The stress of the SMA is given by

$$\sigma = a(T - T_M)\varepsilon - b\varepsilon^3 + \frac{b^2}{4a(T_A - T_M)}\varepsilon^5 \tag{2}$$

where, ε is a strain, T means temperature, a and b are positive material constants. T_M and T_A mean the temperature when martensite (subscript M) and austenite (subscript A) is stable. Finally, after calculations the force of SMA prosthesis is defined as

$$F_{SMA} = a_1(T - T_M)(x_P - x_S) - a_2(x_P - x_S)^3 + a_3(x_P - x_S)^5 \tag{3}$$

where, x_P and x_S mean displacements of the masses in the model presented in Fig.1.

2.2 Middle ear with SMA actuator

The 2dof model of RME consists of two lumped masses, springs and dampers as shown in (Fig.1). In the model the stapes (m_S) and the malleus (m_M) are connected each other by the SME (in this case called also as the shape memory prosthesis). The tympanic membrane, the cochlea and the annular ligament stiffness and damping are defined as k_{TM}, k_C, k_{AL}, c_{TM}, c_C and c_{AL}, respectively. The micro-actuator

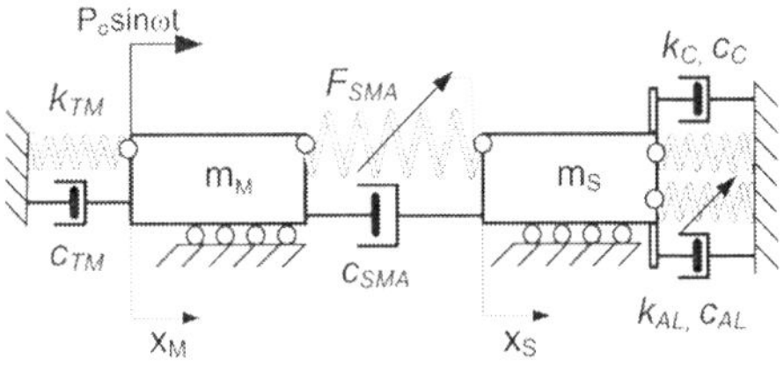

Figure 1. Model of a middle ear with SMA prothesis

as a SMA spring (F_{SMA}) is described by Eq. 3. The differential equations of motion of the system with the SMA micro-actuator take the form

$$m_M\ddot{x}_P + k_{TM}x_M + F_{SMA} + c_{TM}\dot{x}_M = P_0 \sin\omega\tau$$
$$m_S\ddot{x}_S + k_C x_S + k_{AL}x_S + k_{2AL}x_S{}^2 + k_{3AL}x_S{}^3 +$$
$$c_C\dot{x}_S + c_{AL}\dot{x}_S - c_{SMA}(\dot{x}_M - \dot{x}_S) - F_{SMA} = 0 \tag{4}$$

The model is solved analytically by means of the multiple time scales method in Sect. 3

3 Analytical solution

The dimensional equations of motion (Eq.4) is transformed to the nondimensional, and next the quasi-normal coordinates and linear eigenvalue problem is solved. Then, an approximate analytical solution is determined by means of the multiple time scale method [26]. According to this method the solution is expressed by series of a small parameter ε

$$\eta_j(\tau,\varepsilon) = \eta_{j0}(T_0,T_1) + \varepsilon\eta_{j1}(T_0,T_1) + \ldots j = 1,2 \tag{5}$$

Finally, the equation of the resonance curve near the first natural frequency has the form

$$\frac{9}{16}\alpha_2^2 a_1^6 + \frac{3}{2}\left(\omega_{01}^2 - \omega^2\right)\alpha_2 a_1^4 +$$
$$\left[\zeta_1^2\omega^2 - \left(\omega_{01}^2 - \omega^2\right)^2\right]a_1^2 - p_1^2 = 0 \tag{6}$$

The more detailed description of the methodology and transformation from dimensional parameters shown in Eqs. (3)-(4) and the dimensionless presented in Eq.(6) can be found in [27]. Eq. (6) is used to draw the primary resonance curves (Fig.2) for two variants of relative temperature $\theta = 1$ and $\theta = 1.0787$ which relates to the normal temperature of the human body. Moreover, the excitation amplitude varies from P_0 to $10P_0$. Black colour depicts stable while red - unstable periodic solutions. For the calculations the following set of parameters is used [27]:

$$m_P = 2.5525 \times 10^{-5}kg,\ m_S = 1.78 \times 10^{-6}kg,$$

$$k_{TM} = 300N/m,\ k_C = 200N/m,\ k_{AL} = 623N/m,$$

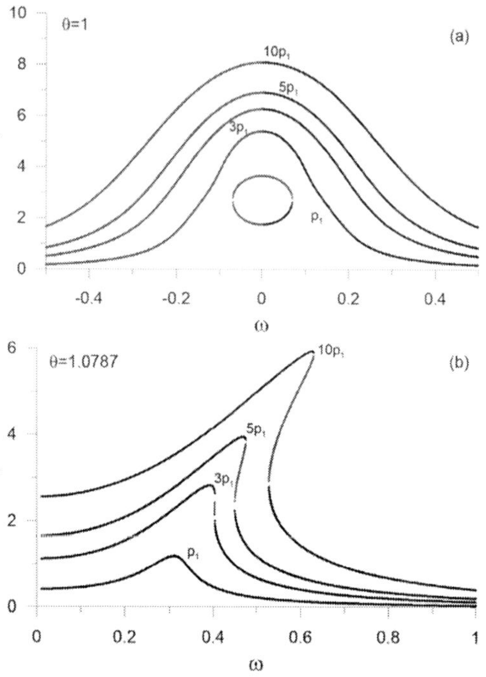

Figure 2. Primary resonance curves, $\theta = 1$ (a) and $\theta = 1.0787$ (b)

$$k_{AL2} = 3 \times 10^8 N/m^2,\ k_{AL3} = 2.25 \times 10^{15} N/m^3,$$
$$a_1 = 157080 kg/(Km),\ a_2 = 2.5233 \times 10^{14} kg/(ms)^2,$$
$$a_3 = 3.8666 \times 10^{21} kg/(m^2 s)^2,\ T_M = 287.0K,$$
$$P_0 = 1.2 * 10^{-4} N,\ c_{TM} = 0.06 kg/s,$$
$$c_{SMA} = 0.055 kg/s,\ c_C = 0.0017 kg/s,\ c_{AL} = 0.00004 kg/s.$$

When the relative temperature is low $\theta = 1.0$ the system resonance curve is untypical. This is because at $\theta = 1.0$, $T - T_M = 0$. For excitation P_0 a circle appears in the middle. Then two stable solutions are possible. For larger excitation amplitude only one stable solution exists. In case of $\theta = 1.0787$ the system behaves typically. For excitation P_0 the system is fully predictable because it has only one solution, that is important from practical point of view. Larger excitation can lead to two stable solution. This situation is not wanted in the middle ear.

4 Conclusions

The micro-actuator proposed here as the prosthesis of the human middle ear has various behaviour depending on temperature. The special case when $\theta = 1.0$ characterises untypical resonance curve. Increasing temperature to the normal human body one the resonance curves are typical. Then the system has only one periodic solution if the excitation is not too strong. That is a positive aspect from practical point of view. More examples and information about primary resonance can be found in [27] where the problem of the reconstructed middle ear by means of the shape memory prosthesis is discussed.

References

[1] M.A. Savi, P.M. Pacheco, A.M. Braga, International Journal of Non-Linear Mechanics **37**, 1387 (2002)
[2] M.A. Savi, A.M.B. Braga, Journal of the Brazilian Society of Mechanical Sciences and Engineering **15**, 1 (1993)
[3] M.A. Savi, M.A. Sá, A. Paiva, P.M. Pacheco, Chaos, Solitons & Fractals **36**, 828 (2008)
[4] L.C. Brinson, Journal of Intelligent Material Systems and Structures **4**, 229 (1993)
[5] F. Auricchio, A. Reali, Materials Science Forum **583**, 257 (2008)
[6] V. Piccirillo, J.M. Balthazar, A.M. Tusset, D. Bernardini, G. Rega, Proceedings of the Institution of Mechanical Engineers, Part C: Journal of Mechanical Engineering Science **230**, 120 (2015)
[7] de Aguiar, Ricardo Alexandre Amar, de Castro Leão Neto, Waldyr Cardoso, M.A. Savi, Calas Lopes Pacheco, Pedro Manuel, Materials Science Forum **758**, 147 (2013)
[8] A. Rao, A. Ruimi, A.R. Srinivasa, International Journal of Solids and Structures **51**, 4554 (2014)
[9] J.H. Roh, Mathematical Problems in Engineering **2014**, 1 (2014)
[10] M. Fabrizio, M. Pecoraro, V. Tibullo, Mechanics Research Communications **74**, 20 (2016)
[11] E. Nikbakht, K. Rashid, F. Hejazi, S.A. Osman, Structure and Infrastructure Engineering **11**, 297 (2014)
[12] Z.W. Zhu, Y.N. Liu, J. Xu, H.L. Wang, *Modeling and application of Shape Memory Alloy based on hysteretic nonlinear theory and data mining*, in *2008 3rd International Conference on Intelligent System and Knowledge Engineering* (IEEE, 2008), pp. 329–334, ISBN 978-1-4244-2196-1
[13] L.G. Machado, M.A. Savi, Pacheco, Pedro Manuel C.L., International Journal of Solids and Structures **40**, 5139 (2003)
[14] D. Bernardini, G. Rega, Mathematical and Computer Modelling of Dynamical Systems **11**, 291 (2005)
[15] D. Bernardini, G. Rega, International Journal of Mechanical Sciences **130**, 458 (2017)
[16] F. Falk, Acta Metallurgica **28**, 1773 (1980)
[17] F. Falk, P. Konopka, Journal of Physics: Condensed Matter **2**, 61 (1990)
[18] R. Rusinek, J. Warminski, M. Szymanski, K. Kecik, K. Kozik, International Journal of Mechanical Sciences **127**, 163 (2017)
[19] R. Rusinek, M. Szymanski, J. Warminski, Journal of Computational and Nonlinear Dynamics (2017)
[20] J. Latalski, R. Rusinek, The European Physical Journal Plus **132**, 645 (2017)
[21] K.D. Shah, R.A. Bradoo, A.A. Joshi, D.D. Sapkale, Indian Journal of Otolaryngology and Head & Neck Surgery **65**, 298 (2013)
[22] S. Şevik Eliçora, D. Erdem, A.E. Dinç, M. Damar, S. Bişkin, European Archives of Oto-Rhino-

Laryngology **274**, 773 (2017)
[23] A. Coordes, L. Jahreiss, U. Schönfeld, M. Lenarz, The Laryngoscope **127**, 500 (2017)
[24] O. Fawole, S.E. Mowry, American Journal of Otolaryngology **37**, 136 (2016)
[25] A. Paiva, M.A. Savi, Mathematical Problems in Engineering **2006**, 1 (2006)
[26] A.H. Nayfeh, *Perturbation methods* (Wiley, New York, 1973), ISBN 9780471630593
[27] R. Rusinek, J. Warminski, A. Weremczuk, M. Szymanski, International Journal of Non-Linear Mechanics **(under review)** (2018)

Experimental studies of a single flexibly-mounted rod in a triangular rod bundle in cross-flow

Sabine Upnere[1,2,*], *Normunds* Jekabsons[3,2], *Sergejs* Dementjevs[4], and *Michael* Wohlmuther[4]

[1] *Institute of Mechanics, Riga Technical University, Latvia*
[2] *Ventspils University College, Latvia*
[3] *Faculty of Physics and Mathematics, University of Latvia, Latvia*
[4] *Paul Scherrer Institute, Switzerland*

Abstract. Experiments on flow-induced vibrations using a closely-packed triangular rod array with a pitch-to-diameter ratio of 1.1 in water cross-flow was carried out at Paul Scherrer Institute. The bundle consists of 21 row of five rods in each one. Single flexibly-mounted test rod (TR) is in the fourth row in an otherwise fixed array. The test rod can freely move in the transverse and in-line direction. Two accelerometer sensors were attached at both ends of the TR to measure the rod response on the fluid flow. The effect of flow rate on the stability of the flexibly-mounted TR has been analysed. During experiments, it reveals a set of conditions and tendencies for the flow-induced vibration in the closely-packed multi-rod system.

In mechanical, civil or nuclear engineering communities for the structural health monitoring vibration analysis is commonly used. This work is a contribution to investigations directed on development of SINQ (Swiss Spallation Neutron Source) target accident protection system. SINQ target is located in Paul Scherrer Institute (PSI).

The mock-up of SINQ target with single flexibly-mounted test rod (TR) in an otherwise rigid triangular rod array, with a pitch-to-diameter ratio of 1.1 has been used to experimentally study the flow-induced vibrations (FIV) due to the water cross-flow (flow is perpendicular to the rod axis). Cross-flow-induced vibrations are responsible for the vast majority of failures [1] in the multi-tube systems. Due to the mounting of the TR, the oscillating system has relatively small stiffness features.

Although an experimental investigation of the system could be expensive compared with analytical or numerical solutions, it is necessary to get an exact knowledge of the excitation mechanism of FIV in the particular rod array. The behaviour of the vibrations in the rod bundle is tightly related to the geometrical or mechanical variation of the array layout [2]. In the case of liquid as a surrounding substance, there are several basic principles of FIV excitation, for example, vortex shedding, turbulence buffeting, fluid-elastic instability [3], [4]. Weaver and Grover [5], [6] experimentally found all mentioned excitation mechanisms in a tube bank with 19 flexibly mounted tubes surrounded by 116 rigid. Fluid excitation forces originated from vortex shedding and turbulence are experimentally measured in tightly packed rigid tube bundle by Liu et al. [7]. They conclude that a rod position is the most critical point regarding excitation forces acting on the system. Lin and Yu [8] have done experiments to analyse the influence of surrounding cylinders on critical velocity or vibration amplitude. Experimental results showed that the cylinder induced by the cross-flow vibrates initially only at its natural frequency around the onset of fluid-elastic instability but increasing the fluid velocity the cylinder vibrations include the frequency components of the surrounding cylinders too.

In the present study, non-contact measurements are used to analyse the TR response to environmental changes. Experiments were done at PSI. The amplitudes, oscillation orbits and frequency spectra of the TR vibration are obtained by measurements of accelerometers and laser sensors. A set of conditions and tendencies for the FIV in the very closely-packed multi-rod system is revealed using experimental results.

1 Experimental set-up

The experimental installation was developed at PSI. It consists of rectangular water channel with 21 row of rigid, cylindrical stainless steel rods. Rods are rigidly fixed in hollows in the plexiglass cover. All rods are identical, 30 mm long with 8 mm diameter. Each row contains five rods or four rods and two half-rods, see figure 1. The rods are arranged in a normal triangular array with a small pitch-to-diameter ratio ($P/d = 1.1$).

Based on the conclusion by Weaver and El-Kashlan [9] that the critical tubes for instability are in the third and fourth rows the single flexibly-mounted test rod (TR) is located in the middle of the fourth row. Two cylindrical elastic beams allow free movement of the TR in the transverse

*e-mail: upnere@gmail.com

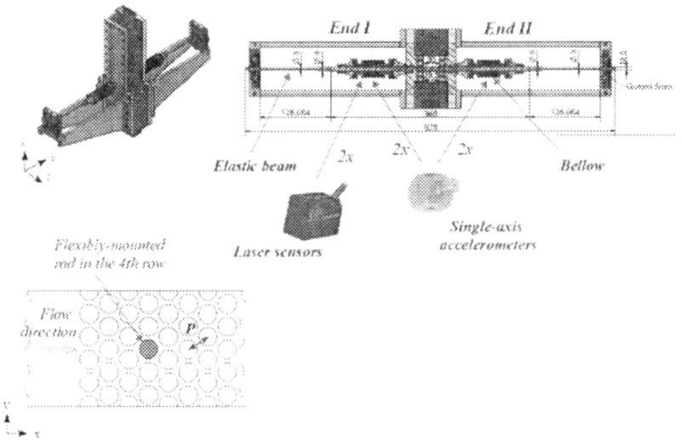

Figure 1. Experimental set-up

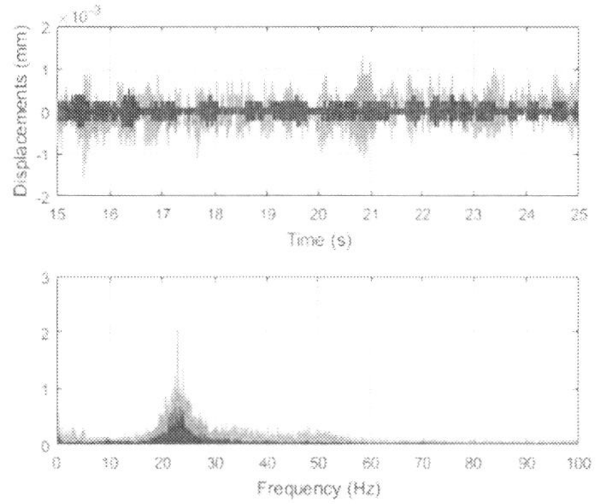

Figure 2. Measured displacements in mm when flow rate $Q = 0$ l/s, red line and $Q = 0.53$ l/s, green line

and in-line direction. Maximum oscillation amplitude is 1 mm. Bellows at the rod ends provide waterproofing.

Maximum flow rate, Q_{max}, in the channel inlet is 2.9 l/s or 2.06 m/s. The Reynolds number based on the maximum inlet velocity U_∞ and cylinder diameter d is 16454. The mean gap flow velocity (1) at maximum inlet velocity corresponds to Reynolds number of 181280. Therefore in the water flow in the rod bundle is fully developed turbulence.

$$U_p = \frac{U_\infty P/d}{P/d - 1} \tag{1}$$

The coordinate system (Cartesian) is defined such that the x-axis points in the direction of the flow, the y-axis in the perpendicular direction and the z-axis is the axis of the rod and is pointing out of the plane in the longitudinal direction of the rod. The origin of the coordinate system is in the centre of the oscillating rod.

1.1 Sensors

Non-contact measurements have been done. The instrumentation includes four single-axis accelerometers and two laser triangulation sensors.

Kistler's 8640A5 accelerometers with sensitivity 1000 mV/g and frequency response of 0.5-3000 Hz are used for acceleration measurements. On the TR End I has been applied two sensors, one for perpendicular (Acc1) and one for parallel (Acc2) direction concerning to the flow. In the same manner, the second end (End II) of the TR is instrumented with two accelerometers Acc3 and Acc4.

Displacements of the flexibly-mounted TR are measured using optoNCDT 1320 laser triangulation sensor with 1 μm repeatability. Two laser sensors are installed on the TR End I, one parallel (L1) and one perpendicular (L2) to the flow. Locations of the sensors can see in the figure 1.

Measurements show that analysed system has a high level of environmental noise and used sensors are not quite sensitive in cases when the flow rate is low ($Q < 1$ l/s), see figure 2.

Due to the laser sensor lower accuracy, if compared with an accelerometer in given system, the analysis mostly is based on TR acceleration measurements.

2 Mathematical formulations

To obtain displacements from the accelerometer time history the following algorithm are used:

- $\ddot{X}(\omega) = FFT(\ddot{x}(t))$
- $\dot{X}(\omega) = \frac{\ddot{X}(\omega)}{\omega i}$
- $X(\omega) = -\frac{\ddot{X}(\omega)}{\omega^2}$
- $\dot{x}(t) = IFFT(\dot{X}(\omega))$
- $x(t) = IFFT(X(\omega))$

$\ddot{x}$ is measured acceleration, FFT is Discrete Fourier transform, $IFFT$ is Inverse Fourier transform, and ω is a frequency. Validation of the applied algorithm is done using measurements of laser sensors.

The Power Spectral Density (PSD) where the power is normalized by the frequency resolution is used in the frequency domain analysis.

$$PSD = \frac{1}{F_s \cdot N}|X(\omega)|^2, \tag{2}$$

where F_s is sampling frequency and N is a number of samples.

For displacement calculations and spectral analysis of the behaviour of the TR in cross-flow, the MATLAB scripts have been developed.

2.1 Comparison of transitional and rotational movement

For modelling and simulation purposes it would be useful to reduce the complexity of the analysed oscillation system. One way would be to move from 3-dimensional to 2-dimensional cases. To check if this is acceptable in a given case, transitional and rotational movements of the TR are compared. The mass-damper-spring system (see figure 3) has been used for evaluation.

Lagrangian equations (3) have been drawn up for two generalised coordinates: a displacement of the TR midpoint x and angle θ of rotation around the centre of inertia.

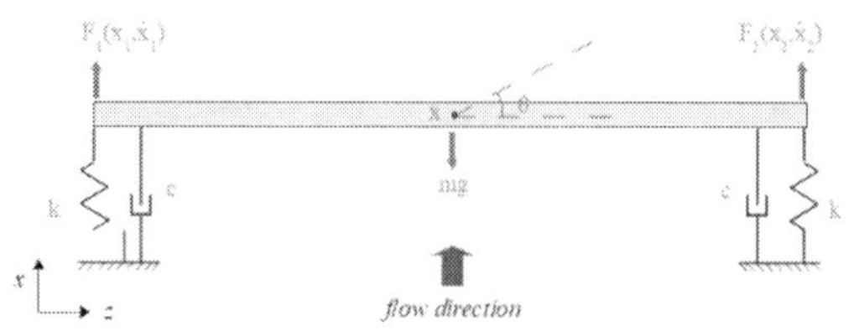

Figure 3. Mass-damper-spring system

Both generalized coordinates are used for perpendicular and parallel axis.

$$\frac{d}{dt}\left(\frac{\partial L}{\partial \dot{q}_i}\right) - \frac{\partial L}{\partial q_i} = F_i, \quad i = 1, 2, 3, 4, \tag{3}$$

where $L = T - V$, T is kinetic energy, V is potential energy, q_i is generalized coordinates and F_i is generalized forces.

Since it is not possible to compare coordinate and angle in a direct way, the kinetic energy is used. The ratio of kinetic energies of the transitional and rotational movements (4) are shown in figure 4.

$$\frac{T_x}{T_\theta} = \frac{1/2m\dot{x}^2}{1/2I\dot{\theta}^2}, \tag{4}$$

where m is mass and $I = mL^2/12$.

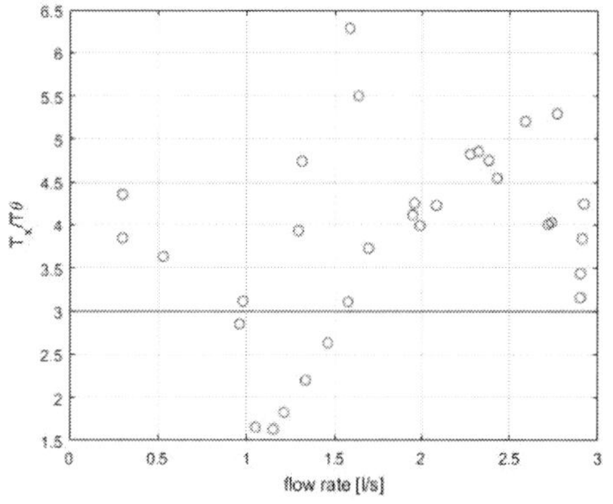

Figure 4. Ratio of the kinetic energies of x and θ

As can see from figure 4, in most cases the kinetic energy of the displacement of the midpoint is at least 3 times larger than the kinetic energy of the rotation, therefore, can assume that in these cases the 2D approximations are permissible.

3 Experimental results

To find sensitive or hard-to-calculate system constants for the numerical and analytical models the force hammer tests are done. Different flow rates have been used to analyse the relationship between the flow and oscillation frequencies as well between flow and oscillation amplitudes.

3.1 Error estimation

Random features are observed in virtually all measurements [10]. To estimate random error, the standard deviation σ of acceleration and displacement measurements at maximum flow rate has been calculated. Average standard deviation and standard error (5) of four experiments are summarized in table 1. Measurements of the rod End I response is chosen (Acc1, Acc2 and L1, L2 sensors).

$$e_\sigma = \frac{\sigma}{\sqrt{n_{exp}}}, \tag{5}$$

where e_σ is a standard error, n_{exp} is the number of experiments.

Table 1. Standard deviation and error of measurements of Acc1, Acc2, L1 and L2 sensors

	Acc1	Acc2	L1	L2
σ	0.1984	0.2014	0.0089	0.0079
e_σ	0.0992	0.1007	0.0047	0.0037

From table 1 follows that accelerometer measurements are more dissipated. Therefore laser sensors have the smaller standard error comparing to accelerometers.

3.2 Damping ratio of the system

The maximum amplitudes of the TR oscillations during the hammer tests correspond to the amplitude when the flow rate is higher than 2.5 l/s. The system damping ratio is found using a logarithmic decrement. The logarithmic decrement is calculated by the following equation:

$$\lambda = \frac{1}{N} ln \frac{x(t)}{x(t + NT)}, \tag{6}$$

where λ is a logarithmic decrement, N is the integer of successive, positive peaks, $x(t)$ is the amplitude at time t, T is the oscillation period. The damping ratio, $\zeta_x = 0.020$ and $\zeta_y = 0.019$ with standard deviations $\sigma_x = 0.003$ and $\sigma_y = 0.0017$, respectively, has been obtained.

The impulse of the unity force is used as excitation in hammer test. The transfer function H from force F to acceleration a is:

$$H(s) = \frac{a(s)}{F(s)} = \frac{1}{m} \frac{s^2}{s^2 + 2\zeta\omega_n s + \omega_n^2}, \tag{7}$$

where $\omega_n = \sqrt{k/m}$ and s is Laplace operator. Example of Bode diagram with frequency response and phase shift is shown in figure 5.

3.3 The influence of the rod mass

The damage of the rod may result in a reduction in the mass of the rod. To check is the rod mass reduction observable in the frequency or space domain in the experimental installation is attached two mass elements at the TR ends increasing the total TR mass, see figure 6.

The larger mass of the TR causes oscillation amplitude increase and frequency decrease. In the figure 7 is shown

https://doi.org/10.1051/matecconf/201814809002

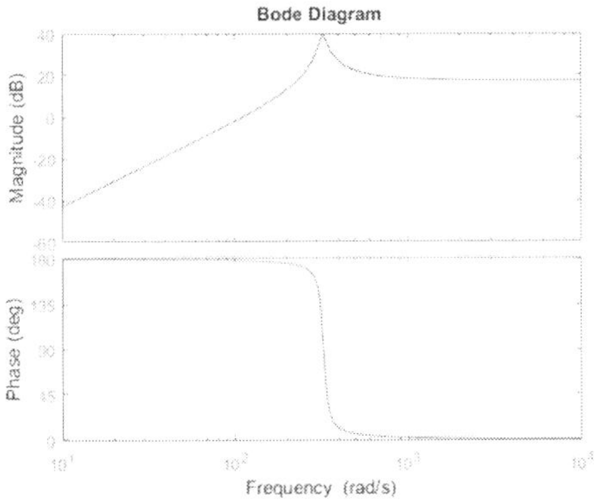

Figure 5. Bode diagram

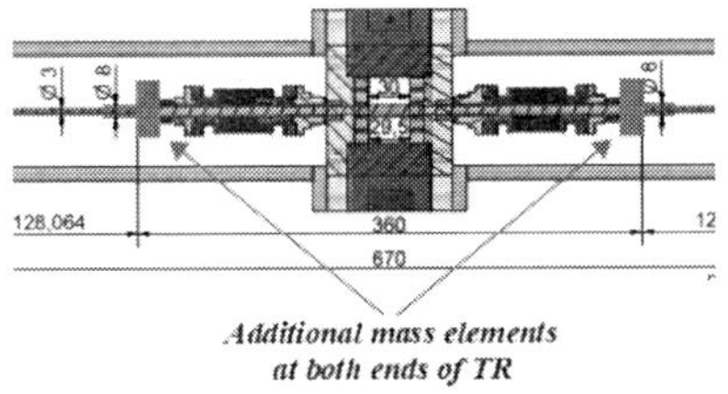

Figure 6. Experimental installation with increased mass of the TR

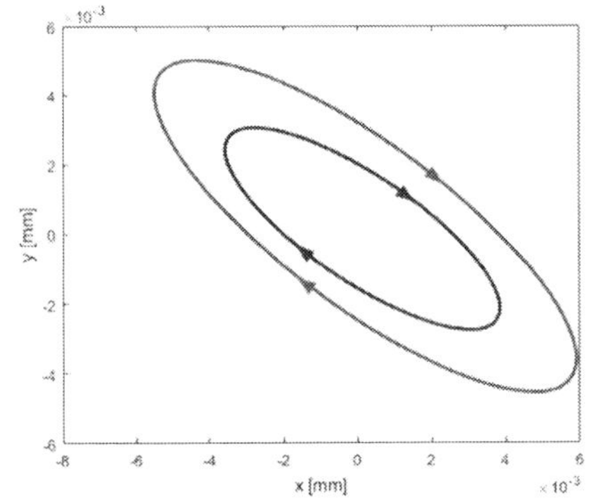

Figure 7. Left: TR movement in x-y plane (mm); m_0 - blue line, m_1 - red line.

TR movement in the x-y plane with original mass, m_0 = 469 g (red line) and with larger mass, m_1 = 581 g of the rod (blue line).

Measurements show that increasing the mass by 1.24 times causes frequency decrease by 1.13 times. The original mass gives the stiffness coefficient k of 44283 N/m. Using calculated stiffness k from the original case the frequency ratio is 1% larger. It can conclude that the fluid added mass has the minor influence in a particular case.

The TR mass changes are detectable from the hammer test in both in the frequency spectra (especially in parallel direction) and in the space domain.

3.4 Maximum frequencies depending on flow rate

PSD is used to analyse obtained accelerometer data in the frequency domain. In figure 8 is shown three maximum frequencies of each accelerometer measured in one experimental case.

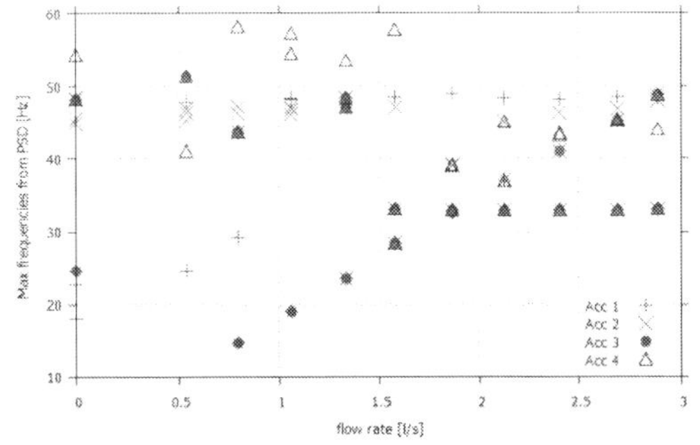

Figure 8. Maximum frequencies (Hz) from the accelerometer measurements

In general, all registered frequencies can divide into three groups: a) frequencies around 50 Hz; b) frequencies around 32 Hz starting after 1.5 l/s; c) gradually growing frequencies when a flow rate is higher than 0.8 l/s.

First two groups are related to the system and environmental noises. The last one should describe the characteristic frequency of the TR due to the flow. The PSD spectra for large flow rates is given in figure 9. As mentioned by Ziada [11], a narrow band peak indicates that it is basically a periodic phenomenon. As illustrated in figure 9, the linear relationship between the frequency, f and the flow rate, Q has been observed.

3.5 Oscillation mechanisms

The probability P_i, (8) is used to find areas in the x-y plane where the end of the TR is the most common during the time. This allows to better understand causes of the rod excitation.

$$P_i = \frac{n_i(x)}{N}, \tag{8}$$

where $n_i(x)$ is number of samples in given area i, N is number of total samples.

The interesting part of the x-y plane is split into small squares and for each of these areas P_i is calculated, see figure 10.

An analyse of the measured TR movement in the x-y plane depending on the water flow velocity shows two possible oscillation scenarios. In one case, the turbulent fluctuations are dominated by the system. If the behaviour of the oscillation system is dictated by the turbulence, with a higher probability the rod will be around the zero point, see figure 11. In this case, cannot detect changes of the

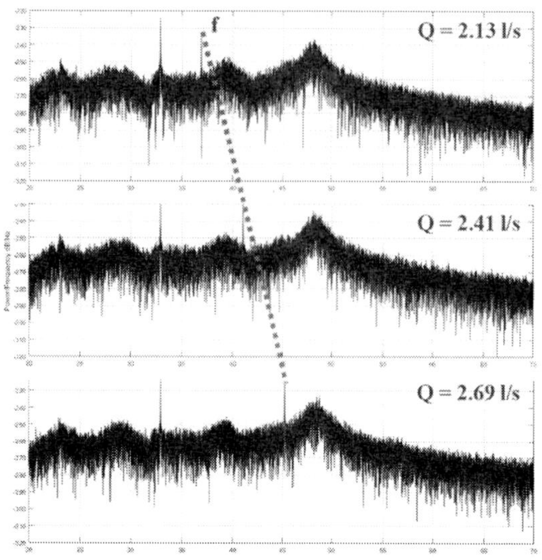

Figure 9. Typical spectra of PSD. With dotted line is shown a frequency, f depending on the flow rate, Q. On x-axis is frequencies in Hz and on y-axis is power/frequency in dB/Hz

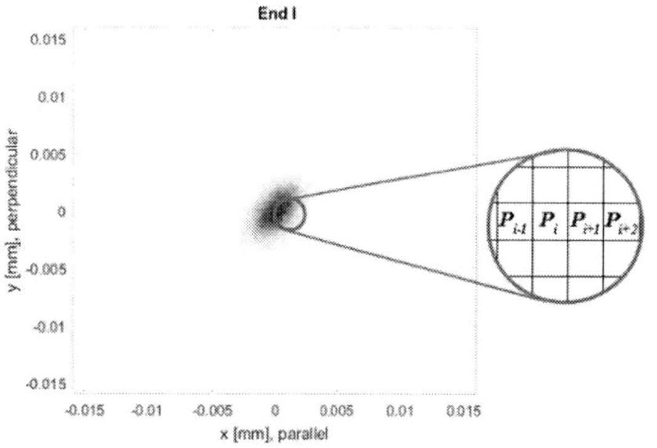

Figure 10. The plane splitting for the calculations of the probability function

characteristic frequencies of the system. The TR response on fluid flow is random and no periodic motion has been observed.

When in the probability graph is observed two maximum points, the rod has two unstable equilibrium states and its oscillation behaviour can describe as the pendulum system (see figure 12). In this case, can analyse processes related with theses system frequencies.

The boundary when the analysed system starts work as a pendulum system can be found investigating TR oscillation amplitude depending on the flow rate. Figure 13 shows that after 2.2 l/s the vibration amplitude of the TR starts to increase. When the flow rate is between 2.7 and 2.8 l/s, the oscillation amplitude reaches its maximum. After maximum, the TR amplitude decrease although the flow velocity continues to grow. Thus, at high flow rates,

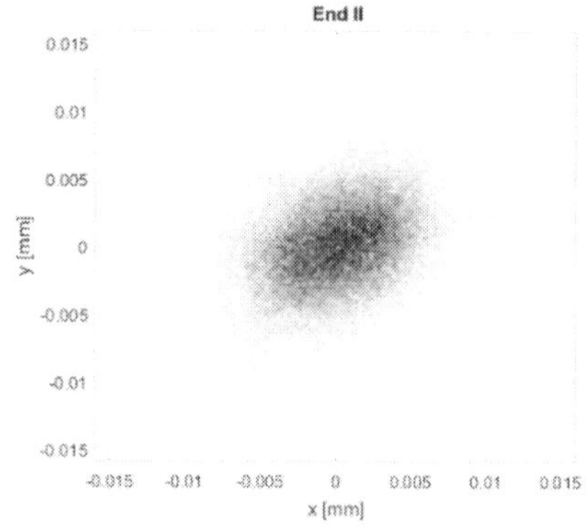

Figure 11. Probability of the TR location if turbulent fluctuations are dominant

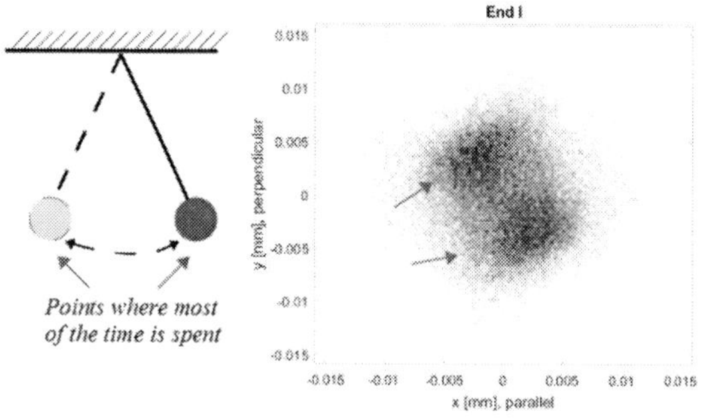

Figure 12. Probability of the TR location if two unstable equilibrium states are observable

a hump in vibrations amplitudes is created. It looks like a typical pattern of the vortex-induced vibrations but considering a small gap between rods and TR location in the 4th row more investigations are needed to get a definitive conclusion.

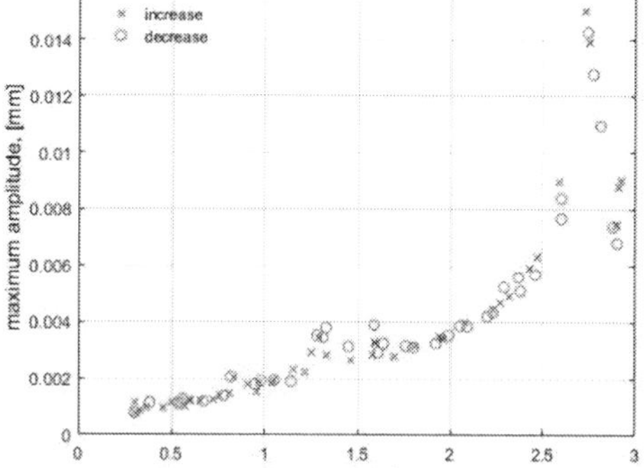

Figure 13. Magnitude of maximum amplitudes of TR depending on flow rates

Examples of distribution of average displacement amplitudes calculated from accelerometers (Acc1, Acc2 at End I and Acc3, Acc4 at End II) at two flow rates, Q = 2.32 l/s and Q = 2.74 l/s is shown in figure 14. In most cases the random distribution of average amplitudes is observable. Although, at the flow rate 2.74 l/s in a parallel direction the distribution of amplitudes is similar to the normal distribution (magenta line in Acc2 and Acc4 graph in figure 14). Similar trends are found in other cases when the maximum amplitude of oscillations is reached (between 2.7 and 2.8 l/s).

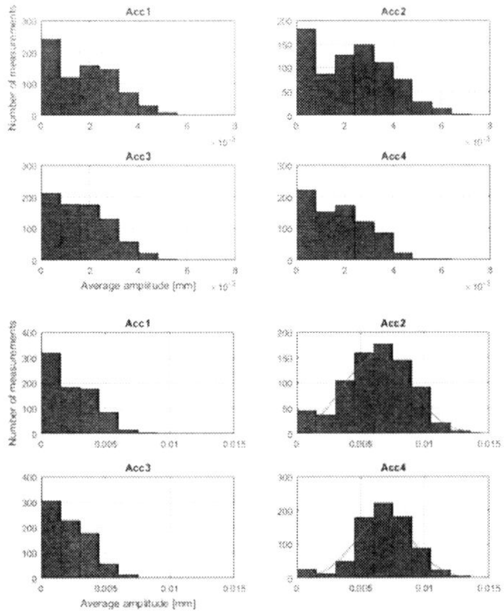

Figure 14. Distribution of average amplitudes of displacements at Q = 2.32 l/s (top) and Q = 2.74 l/s (bottom). With magenta line is presented Gaussian distribution function

4 Conclusions

A single flexibly-mounted rod is investigated as a model system for the flow-induced vibration evaluation in the closely-packed (P/d = 1.1) triangular rod bundle. Characteristic features of the system with relatively small stiffness was analysed. The inlet flow velocity is in a range from 0 m/s till 2.06 m/s which at maximum gives Reynolds number of 16454 concerning to the rod diameter. Maximum reached oscillation amplitude of the rod is 0.016 mm.

In the analysed system observed two excitation mechanisms. At low flow rates, primary oscillations of the rod come from turbulent fluctuations. When the flow rate reaches a critical value, the rod occurs two unstable equilibrium states and starts large amplitude oscillations. A linear relationship between characteristic frequencies of the system and a flow rate was found at high flow velocity.

The test rod mass impact on the detected frequency and amplitude has been investigated. Correlation between mass increasing and a vibration amplitude increase revealed in the hammer tests. Results show that the fluid added mass effect on the oscillation frequency is minor.

Lagrangian equations for two generalised coordinates: the movement of the test rod midpoint and the rotational angle around the rod centre of inertia, are obtained for perpendicular and parallel direction related to the flow. Comparison of kinetic energies of generalised coordinates shown that in the most cases 2D approximation is allowed.

Based on results of the hammer tests, system parameters for numerical simulations and analytical investigation have been found.

References

[1] Paidoussis, M.P., *Practical experiences with flow-induced vibrations* (Springer-Verlag, Berlin, 1980) 1-81

[2] Andjelić, M., Austermann, R., Popp, K. Journal of Pressure Vessel Technology **114**, 336-343 (1992)

[3] Jafari, H.H., Dehkordi, B.G. Journal of Fluids Engineering **135**, 031102-1-031102-13 (2013)

[4] El Bouzidi, S., Hassan, M. Journal of Fluids and Structures **57**, 264-276 (2015)

[5] Weaver, D.S., Grover, L.K. Journal of Sound and Vibration **59**, 263-276 (1978)

[6] Weaver, D.S., Grover, L.K. Journal of Sound and Vibration **59**, 277-294 (1978)

[7] Liu, L., Feng, J., Wu, H., Xu, W., Tan, W. Journal of Pressure Vessel Technology **139**, 031307-1-031307-8 (2017)

[8] Lin, T., Yu, M. Experimental Thermal and Fluid Science **29**, 523-536 (2005)

[9] Weaver, D.S., El-Kashlan, M. Journal of Sound and Vibration **75**, 265-273 (1981)

[10] Wheeler, A.J. and Ganji, A.R. *Introduction to engineering experimentation, 3rd. ed.*, (Pearson, 2010) 480

[11] Ziada, S. Journal of the Braz. Soc. of Mech. Sci. & Eng. **28**, 186-199 (2006)

Chatter identification in milling of Inconel 625 based on recurrence plot technique and Hilbert vibration decomposition

Paweł Lajmert[1,*], *Rafał* Rusinek[2], and *Bogdan* Kruszyński[1]

[1]Lodz University of Technology, Institute of Machine Tools and Production Engineering, Stefanowskiego 1/15 Street, Lodz, Poland
[2]Lublin University of Technology, Department of Applied Mechanics, Nadbystrzycka 36 Street, Lublin, Poland

Abstract. In the paper a cutting stability in the milling process of nickel based alloy Inconel 625 is analysed. This problem is often considered theoretically, but the theoretical finding do not always agree with experimental results. For this reason, the paper presents different methods for instability identification during real machining process. A stability lobe diagram is created based on data obtained in impact test of an end mill. Next, the cutting tests were conducted in which the axial cutting depth of cut was gradually increased in order to find a stability limit. Finally, based on the cutting force measurements the stability estimation problem is investigated using the recurrence plot technique and Hilbert vibration decomposition method.

1 Introduction

The need to produce responsible machine parts in aerospace, chemical process or seawater applications requires the use of special metal alloys, such as nickel-chromium-molybdenum alloys, e.g. Inconel 625. Inconel 625 belongs to category of superalloys, which have high tensile strength in high temperatures, good fatigue strength and outstanding corrosion and oxidation resistance [1]. The properties of this material that make it an excellent choice in many demanding applications causes however, many difficulties in its machining. Additionally, in industry, because of competitiveness reasons, there is a constant need for growth in productivity and product quality that requires the material removal rate to be increased [2]. A higher productivity may be achieved using high speed machining (HSM) which however can induce so called self-excited vibrations caused by a regenerative mechanism [3, 4]. To avoid regenerative chatter, the working region corresponding to stable machining must be identified. For this purpose, a stability lobe diagrams (SLDs) are usually created, where rotational speed and depth of cut determine the boundary between stable and unstable cutting [3]. However, due to non-stationary nature of cutting processes the exact boundary position is difficult to predict in advance and may change for different technological cases. This in turn forces the machine tool operators to use undervalued cutting parameters. Although, there exist methods for eliminating self-excited vibrations they are quite difficult to apply in industrial practice. From these reasons there is a need to use automatic supervision systems which could diagnose the process online and change the cutting parameters when some symptoms of process instabilities appear [5-9]. Moreover, some papers focus on different methods of chatter identification directly from workpiece displacement or cutting force signals [3, 10-14].

This paper deals with the problem of stability limit prediction of chatter vibrations in milling process of Inconel 625 on the basis of experimental measurements of cutting force components. In order to find the initial point of chatter vibrations, the recurrence quantification analysis and Hilbert vibration decomposition are employed [15, 16].

2 Experimental tests

The experimental tests on Inconel 625 were performed using the Haas MiniMill CNC milling machine. The experimental setup, presented schematically in Fig. 1, is composed of two parts: a modal analysis system and a dynamometer for measuring cutting forces. The former was used to measure viscoelastic properties of the machine-tool system. The modal system consists of the PCB-086C03 modal hammer, the PCB-352B10 accelerometer and the NI-9234 data acquisition card. The rotating, piezoelectric dynamometer Kistler 9123C can measure three components F_x, F_y and F_z of the resultant cutting force and torque M_z. The dynamometer is connected to the Kistler 5223 signal conditioner and the 2855A4 data acquisition card. Both measurement systems are integrated with a PC computer and controlled by the DynoWare software.

The measurements were carried out in two steps. First, a single point impact test is performed to determine modal parameters of the cutting system, that is: stiffness, natural vibration frequency and a damping ratio. The modal hammer is used to excite the tool, and then the output signal is measured by the low mass accelerometer

* Corresponding author: pawel.lajmert@p.lodz.pl

mounted at the tool tip. The test is repeated for the two perpendicular directions x and y.

Next, the modal parameters are implemented to the CutPro9 software as a frequency response function (FRF) in order to determine stability lobes diagram (SLD). The SLD, generated for the up-milling of Inconel 625 using a 12 mm diameter end mill cutter with four flutes is shown in Fig. 2. The radial depth of cut, a_e, is set to 4 mm and the feed per tooth, f_z, is set to 0.05 mm. Stable cutting occurs in the region below the stability lobes, while unstable machining should occur above the lobes. The unconditionally stable region is below the critical cutting depth a_{pcr}=0.65 mm. Then, the cutting process is stable regardless the spindle speed.

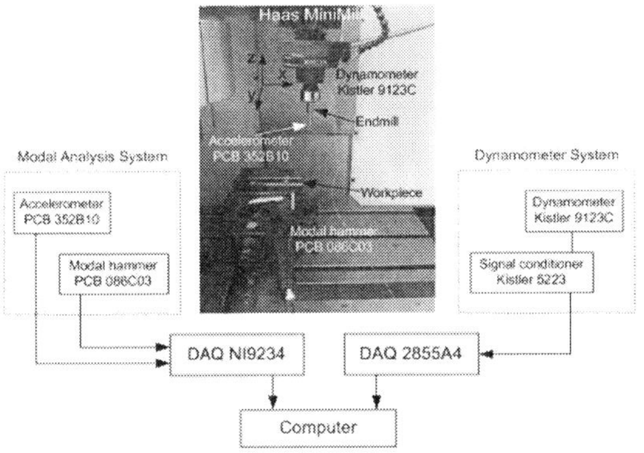

Fig. 1. Experimental setup scheme.

The second stage of experiment is performed on the milling machine in order to verify the SLD and to explain chatter vibrations nature. The cutting parameters are set as follows: spindle rotational speed n=1900 rpm, radial depth of cut a_e=4 mm, feed per tooth f_z=0.05 mm, while the axial depth of cut a_p is being increased from 0 to 1 mm during the test. The cutting force components, F_x, F_z, and the torque, M_z, are measured and recorded with a sampling rate of 2 kHz.

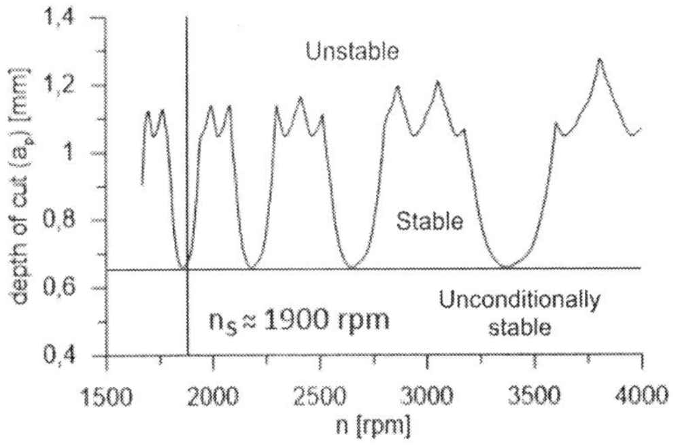

Fig. 2. Stability lobes diagram for Inconel 625 milling.

The analyses in the following section are undertaken to verify whether the critical depth of cut, presented in Fig.2, is found properly as well as to identify chatter vibrations component in the measured signal.

3 Time series analysis

In this section the analysis of the force time series is done for the application in stability and chatter identification of milling process. To achieve this aim, the recurrence plot (RP) technique and Hilbert vibration decomposition (HVD) methods are applied.

3.1. Recurrence plot technique

Based on the experiment and CutPro analysis the SLD is achieved. However, industry practice suggests that the stability boundary is not quite proper. Therefore, the obtained stability limit from the SLD is verified here experimentally using the recurrence plot (RP) technique. The RP approach provides a qualitative interpretation of hidden patterns of dynamical systems based on phase space reconstruction. This technique was originally introduced by Eckmann [15]. According to the RP theory any time series can be presented as a delayed vector in an m-dimensional space called "the reconstructed phase space". The embedding parameters, i.e. a dimension m and a time delay d are estimated by applying the average mutual information function (AMI) and the false nearest neighbours method (FNN). More information about the AMI and FNN functions can be found in the papers [3, 10, 17].

In this study, we applied the Tisean software [18] to determine the embedding parameters. A recurrence plot is a graph which shows all time instants at which a state of a dynamical system recurs. If the trajectory in the reconstructed phase space returns at time i into the neighbourhood of ε where it was before at time j, then the results are plotted as black point in RP. Black points create a specific pattern which can be analysed to achieve some information about system dynamics. From a practical point of view, RPs can be presented also by means of the recurrence quantification analysis (RQA) which is not considered in this contribution. The RP for the cutting force components F_x and F_z are presented in Fig.3 and 4. Note that during the milling test depth of cut a_p was being constantly increasing from 0 to 1 mm.

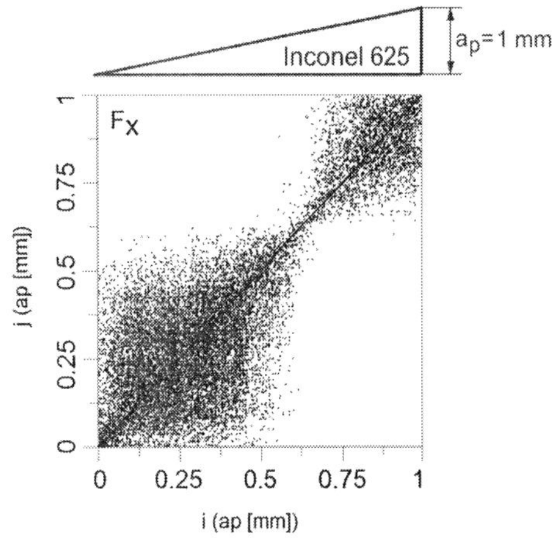

Fig. 3. Recurrence plots for cutting forces F_x.

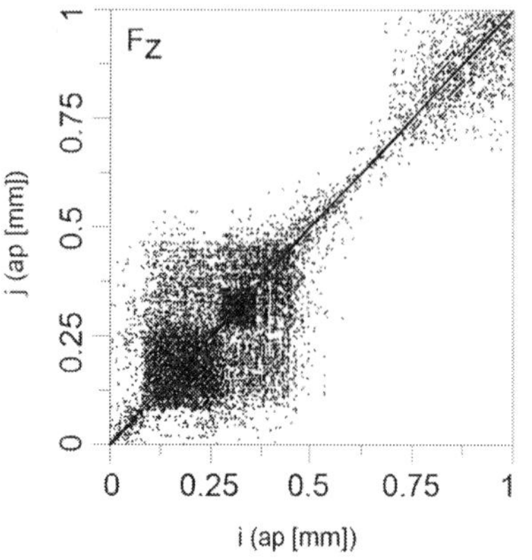

Fig. 4. Recurrence plots for cutting forces F_z.

The transition from chatter free milling to chatter is visible at a_p=0.5 mm, that is just before the value obtained from CutPro. In case of transition from chatter free to chatter region the change of RP pattern is observed both for F_x and F_z force.

3.2. Hilbert vibration decomposition

The Hilbert vibration decomposition was developed by Feldman for separation of non-stationary and multicomponent dynamical signals [16, 19]. In contrast to Fourier transform, in Hilbert transform it is assumed that the signal may be represented by a single, but amplitude and frequency modulated sinusoidal waveform [19]. As a consequence, the HVD method, which is based on Hilbert transform, allows to decompose the original signal into a sum of quasi-harmonics with slow-varying instantaneous amplitudes and frequencies:

$$x(t)=\sum_{i=1}^{N} A_i(t)\cos\left(\int_0^t \omega_i(t)dt\right) \tag{1}$$

where $A_i(t)$ and $\omega_i(t)$ are the instantaneous amplitude and frequency of the *i-th* component. The decomposition of the initial signal $x(t)$ starts with the estimation of its instantaneous frequency $\omega(t)$ with the use of Hilbert transform. For nonstationary signals with multiple components the instantaneous frequency $\omega(t)$ is composed of two parts. The first is slow varying frequency of the largest energy component and the other is a rapid varying part which oscillate around the first one [16]. Thanks to this property, the mean value of instantaneous frequency $\omega(t)$ is equal to the instantaneous frequency $\omega_1(t)$ of the largest energy component. Having the instantaneous frequency $\omega_1(t)$ it is possible to extract from the signal its instantaneous amplitude $A_1(t)$ using a synchronous demodulation algorithm [16, 19]. Finally, based on the instantaneous frequency and amplitude, the first isolated component $x_1(t)$ with the highest energy is generated. The isolated component is subtracted from the original signal and the whole procedure is repeated to isolate the subsequent components. The flowchart of the HVD method is shown in Fig. 5.

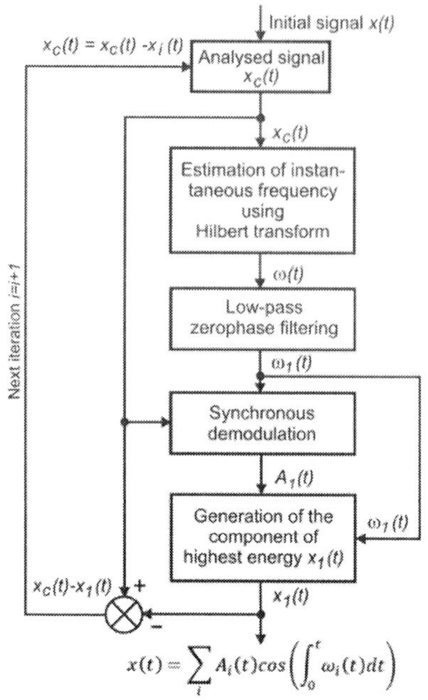

Fig. 5. Flowchart of HVD method.

The presented method was used to extract components from real force measurements. Based on a preliminary experiments the number of desired components to be isolated was established. In Fig. 6 the first three dominant components, isolated from F_z force time series are shown.

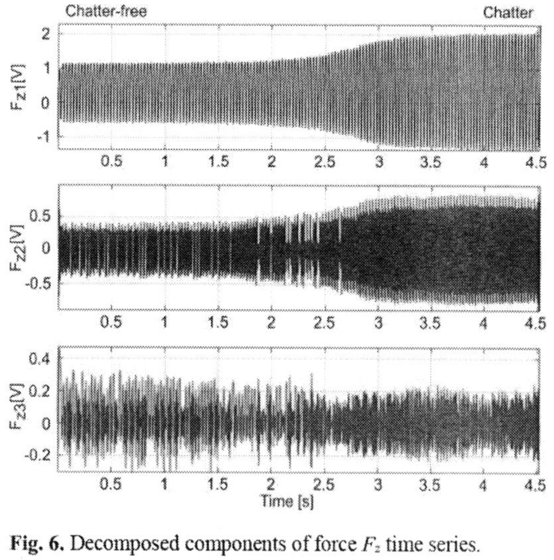

Fig. 6. Decomposed components of force F_z time series.

The first component is responsible for the force changes caused by the material removal by the individual wings of the end mill cutter (tool rotational speed). The second component may be responsible for the vibrations of the machine tool structure, while the third one corresponds to a milling cutter vibrations. In Fig. 7 the course of this component (red line) for chatter-free and chatter milling is presented. It may be seen that for stable machining, the milling cutter vibrations are dampened after each cutter engagement. On the other hand for chatter milling the cutter vibrations are not dampened.

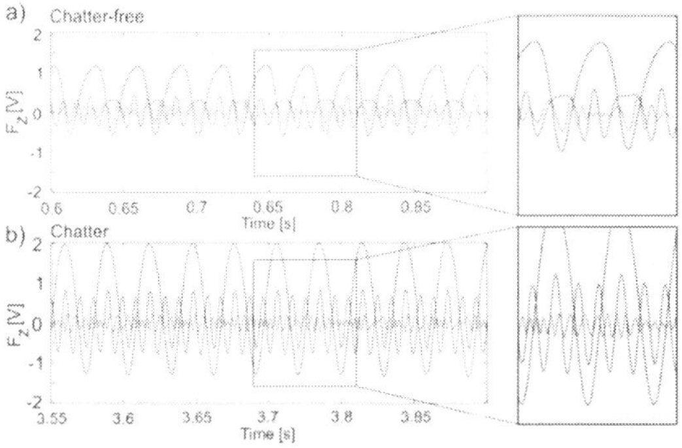

Fig. 7. The course of the three components for: a) chatter-free and b) chatter milling.

In Fig. 8 changes of instantaneous amplitudes for the three extracted components are presented. Particularly for the first and second component (black and blue line) a growth of the amplitude may be noticed, being the evidence of chatter vibrations.

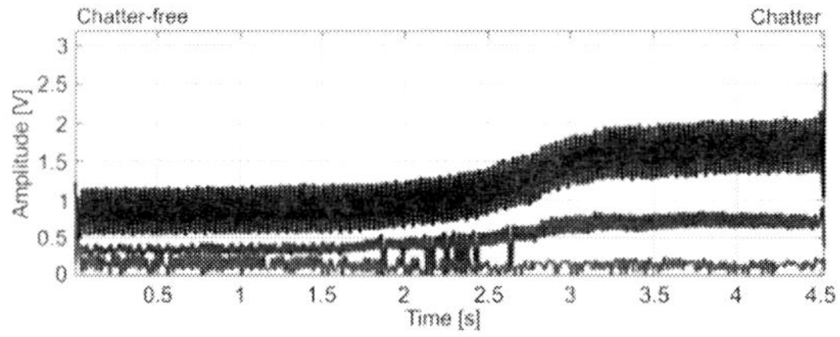

Fig. 8. Changes of the instantaneous amplitude for the three decomposed signal components.

In Fig. 9 Hilbert spectrum for three decomposed components are presented. As may be seen the increase of variance of instantaneous frequency for third component (red one) is clearly visible.

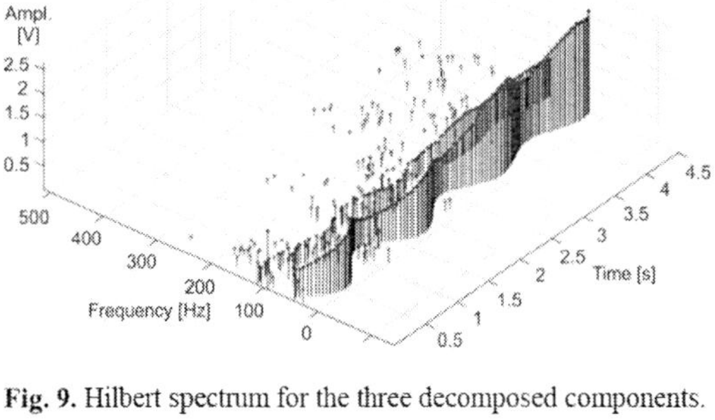

Fig. 9. Hilbert spectrum for the three decomposed components.

5 Conclusions

The paper focused on assessment of milling process stability by the recurrence plot technique and Hilbert vibration decomposition. The recurrence plots drawn for the milling force components show exactly the moment when chatter vibrations appear. The critical depth of cut, determined by means of RP is a bit different compared to the one obtained by means of modal test and CutPro software. It was shown that dynamical features of individual force components can be revealed by means of advanced signal analysis methods, e.g. the HVD method. The use of this method enables determination of the force signal components which are related to individual cutting processes or machine tool vibration modes. Therefore, these components have a greater potential to distinguish between chatter-free and chatter states. It must be noted that major disadvantage of HVD method is the difficulty in proper selection of algorithm parameters, mainly the parameters of low-pass filter.

References

1. E. O. Ezugwu, Z. M. Wang and A. R. Machado, J. of Materials Processing Technology **86** (1999) 1-16.
2. Gdula M., Burek J. Zylka L., Plodzen M., Aircraft Enginering and Aerospace Technology, (2015).
3. Kecik K., Rusinek R., Warminski J., Int. J. of Bifurcation and Chaos **21/10**, 2943-2954 (2011).
4. R. Rusinek, Maintenance and Reliability **3**, 48–55 (2010).
5. B. Kruszynski, P. Lajmert, Proc. Inst. Mech. Eng., Part B: J. Eng. Manuf. **220**, 355-363 (2006).
6. B. Krusznski, P. Lajmert, CIRP Ann. – Manuf. Technol. **54/1**, 305-308 (2005).
7. P. Lajmert, M. Sikora, B. Kruszynski, D. Ostrowski, Lecture Notes in Mech. Eng, Adv. in Manufacturing, 707-716 (2018).
8. Zylka L., Burek J., Mazur D., Adv. in Production Engineerign & Management **12**, 221-232 (2017).
9. Wolszczak P., Lygas K., Litak G., Adv. In Science and Technology Research Journal **11/1**, 96-103 (2017).
10. Rusinek R., J. of Engineering Manufacture, **226/12**, 1976-1985 (2012).
11. Litak G., Kecik K., Rusinek R., Latin Americal J. of Solids and Structures **10/1**, 133-138 (2013).
12. Litak G., Rusinek R., Kecik K., Rysak A., Syta A., Discontinuity and complexity in nonlinear physical systems, Springer 359-367 (2014).
13. R. Rusinek, P. Lajmert, K. Kęcik, B. Kruszynski, J. Warmiński, Int. J. of Mechanical Sciences **99**, 196-207 (2015).
14. Kecik K., Borowiec M., Rusinek R., European Physical Journal Plus 131/1, (2016).
15. J.-P. Eckmann, S.O. Kamphorst, D. Ruelle, Europhys. Lett. **4**, 973–977 (1987).

16. M. Feldman, J. of Sound and Vibration **295**, 518-530 (2006).
17. R. Rusinek, Int, J. of Non-Linear Mechanics **45**, 458–462 (2010).
18. R. Hegger, H. Kantz, T. Schreiber, The TISEAN package, Chaos **9**, 413–435 (1999).
19. M. Feldman, Mechanical Systems and Signal Processing **25**, 735-802 (2011).

A dynamic model of cylindrical plunge grinding process for chatter phenomena investigation

Paweł Lajmert[1,*], *Małgorzata* Sikora[1], and *Dariusz* Ostrowski[1]

[1]Lodz University of Technology, Institute of Machine Tools and Production Engineering, Lodz, Poland

Abstract. In the paper, chatter vibrations in the cylindrical plunge grinding process are investigated. An improved model of the grinding process was developed which is able to simulate self-excited vibrations due to a regenerative effect on the workpiece and the grinding wheel surface. The model includes a finite-element model of the workpiece, two degrees of freedom model of the grinding wheel headstock and a model of wheel-workpiece geometrical interferences. The model allows to studying the influence of different factors, i.e. workpiece and machine parameters as well as grinding conditions on the stability limit and a chatter vibration growth rate. At the end, simulation results are shown and compared with exemplified real grinding results.

1 Introduction

In grinding, all vibration processes are interrelated by physical and dynamic parameters of the machine tool, the workpiece and the grinding tool system [1]. Since grinding is a final operation, aimed at achieving high dimensional and shape accuracy of machined parts, vibrations should be minimised. It may be done through the selection of optimal grinding conditions at the design stage and next by monitoring of these vibrations to correct the grinding conditions or to decide about the grinding wheel redressing [2-6].

A chatter physical mechanism may be divided into two main categories, i.e. forced vibrations and self-excited vibrations [7-11]. Forced vibrations originate from rotating machine tool elements, mainly unbalance of a rotating grinding wheel. In industrial practice, this kind of vibrations is eliminated by means of automatic wheel balancing systems. However, during the grinding, even a small unbalance of any machine tool element or eccentricity of the grinding wheel may induce other machine tool vibration modes, which are selectively amplified due to a regenerative effect [6-7]. The regenerative effect appears as waviness is worn into the grinding wheel or the workpiece. This waviness produces oscillatory forces which are transferred back to the process causing a gradual increase of the waviness amplitude until a vibration level is no longer acceptable from the point of view of the assumed part quality criteria. To recognize the regenerative mechanism, many models of plunge grinding process were created [12-16]. Basically, these models may be divided into analytical and time-domain models [14]. In literature, the most frequently used are the analytical models. These models are based on a model proposed by Snoeys at all [12], which contains blocks, representing two main closed loops, corresponding to the wheel and the workpiece regenerative effect. However, in these models, the machine tool dynamical structure is modelled as a simple transfer function block. Additionally, the wheel-workpiece geometrical interactions are treated in a simplified manner and modelled using analytical relationships. On the contrary, the time-domain models usually take these geometrical interactions into account [13, 15]. Li and Shin [15] presented a time-domain model of the plunge grinding process which includes geometrical interactions between the wheel and workpiece to identify grinding chatter boundaries due to the regenerative effect on the workpiece. However, still structure dynamics is modelled using a modal system which must be first identified using modal analysis for each technological case. In [16] a model of the plunge grinding process was introduced to investigate the self-interrupted grinding chatter. The workpiece is modelled as an Euler-Bernoulli beam and the grinding wheel as a spring-mass system. However, real wheel-workpiece geometrical interactions are not taken into consideration.

This paper presents an improved simulation model which takes into consideration a multi-degree of freedom model of grinding wheel headstock, a finite element model of the workpiece and a model of the wheel-workpiece geometrical interactions, including the wear mechanism of the grinding wheel. The model is useful in a simulation of grinding chatter development and prediction of stable grinding conditions. The proposed model does not require identification of modal parameters each time for different technological cases.

2 Model description

The model of the cylindrical plunge grinding process is composed of three components: a model of grinding

* Corresponding author: pawel.lajmert@p.lodz.pl

wheel headstock, a model of workpiece transverse vibrations and a model of wheel-workpiece geometrical interactions.

2.1. Model of grinding machine

The model of the grinding process is illustrated in Figure 1. The grinding wheel headstock was modelled using a damped spring-mass system, representing grinding wheel having mass m_s, grinding wheel headstock having mass m_w and a massless element, representing an infeed drive, which is responsible for a kinematic force input to the model. The position of this driving element is denoted by x_f. From the other side, on the grinding wheel two components of the grinding force act in normal F_n and tangential F_t directions.

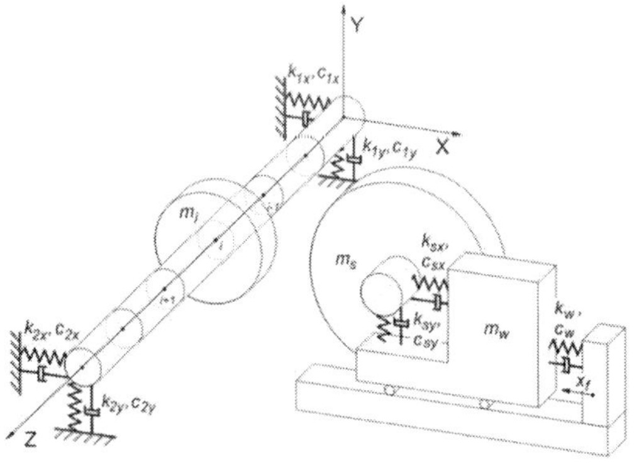

Fig. 1. Model of grinding wheel headstock and workpiece.

In the model, the mass representing the grinding wheel may displace in X and Y directions, while the wheel headstock only in X direction. The matrix equation of motion of this system may be expressed in the following form:

$$\begin{bmatrix} m_w & 0 & 0 \\ 0 & m_s & 0 \\ 0 & 0 & m_s \end{bmatrix}\begin{bmatrix} \ddot{x}_w \\ \ddot{x}_s \\ \ddot{y}_s \end{bmatrix}+\begin{bmatrix} c_{sx} & -c_{sx} & 0 \\ -c_{sx} & c_{sx} & 0 \\ 0 & 0 & c_{sy} \end{bmatrix}\begin{bmatrix} \dot{x}_w \\ \dot{x}_s \\ \dot{y}_s \end{bmatrix}+$$
$$\begin{bmatrix} k_{sx} & -k_{sx} & 0 \\ -k_{sx} & k_{sx} & 0 \\ 0 & 0 & k_{sx} \end{bmatrix}\begin{bmatrix} x_w \\ x_s \\ y_s \end{bmatrix}=\begin{bmatrix} c_w(\dot{x}_f-\dot{x}_w)+k_w(x_f-x_w) \\ -F_n \\ -F_t \end{bmatrix} \quad (1)$$

where x_f, x_w, x_s, y_s are displacements of the infeed element, wheel headstock and the grinding wheel in X and Y direction.

The ground workpiece was discretized using the finite element method. A two-node beam elements with four degrees of freedom in a node (two translational and two rotational) were used. Far-right and far-left nodes of the workpiece model were supported using springs and dampers, which substitute the centres of workpiece headstock and a tailstock. A matrix equation of the workpiece may be expressed in the following form:

$$[M_g]\ddot{q}+[C_g]\dot{q}+[K_g]q=F \quad (2)$$

where M_g, C_g, K_g are mass, damping and stiffness matrixes respectively, q is a vector of generalized displacements and F is a force vector containing normal and tangential force components spread between two nodes where the force is applied. In equation (2) the damping matrix is a linear combination of the mass and stiffness matrixes. For both models. adequate coefficients were chosen based on modal experiments and the peak picking method.

2.2. Model of geometrical interactions

In the model of wheel-workpiece geometrical interactions the surface of the grinding wheel and the workpiece is discretized using equally spaced supporting points as shown in Figure 2.

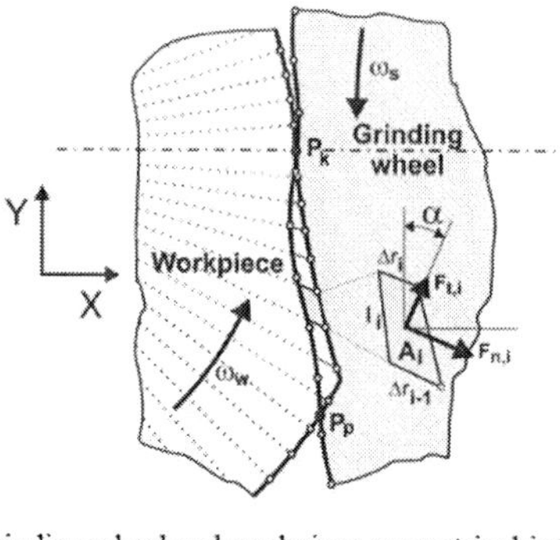

Fig. 2. Grinding wheel and workpiece geometrical interactions.

Force components are calculated based on discretized cross-section area of workpiece and the grinding wheel. For each elementary cross-section element A_i normal and tangential force components are calculated. The resultant force components may be given in the following form:

$$\begin{cases} F_n = \sum_i (F_{n,i}\cos\alpha + F_{t,i}\sin\alpha) \\ F_t = \sum_i (-F_{n,i}\sin\alpha + F_{t,i}\cos\alpha) \end{cases} \quad (3)$$

The simulation algorithm includes the following steps:

- displacement and rotation of the workpiece and the grinding wheel resulting from their rotational velocities and infeed velocity of the wheel,
- determination of the force components based on cross-section area of the wheel and the workpiece,
- displacement of the grinding wheel and the workpiece resulting from the dynamical models and estimated force components,
- movement of the supporting points on the grinding wheel due to the wear of abrasive grains,
- movement of the supporting points of the workpiece based on current contact conditions of the wheel and workpiece.

3 Measurement test stand

To verify the usefulness of the proposed model, experimental research has been carried out for different grinding conditions on cylindrical grinding machines [17-18]. These machines were equipped with sensors for the measurement of grinding force components and vibrations. Force components were measured using Kistler piezoelectric sensors mounted between the body of the wheel headstock and the housing of the wheel spindle. Vibration signals were acquired using piezoelectric sensors mounted on the grinding wheel headstock and the tailstock centre. Signals from these sensors were used to detect unstable grinding conditions. During the experiments, the rings were ground made of 34CrAl6C steel hardened to 52 HRC of width 10-20 mm mounted on a shafts of different lengths (a rigid and flexible system). A 38A60K5V aluminum oxide grinding wheel was used. After grinding tests, the workpiece and grinding wheel shape were measured using a specially designed device, equipped with a laser triangulation sensor, see Figure 3.

Fig. 3. General view of grinding wheel shape measurement device.

3 Simulation results

Based on the described grinding process model, a simulation software was created which allows determining a static and dynamic course of normal and tangential force components, displacement and acceleration of any element of the system and the generated shape of the workpiece and the grinding wheel due to the wheel wear.

First, a simulation of workpiece regenerative effect was conducted. During the exemplified simulation, workpiece peripheral speed was changed at time 20sek from n_w=1 rev/s (stable conditions) to 9 rev/s which corresponds to the unstable region of grinding. A course of the normal force component is shown in Figure 4. After the speed change, the amplitude of normal force component increases gradually and after a certain time, the workpiece losses the contact with the grinding wheel which bears the evidence of process instability. It is observable that this kind of chatter develops relatively fast.

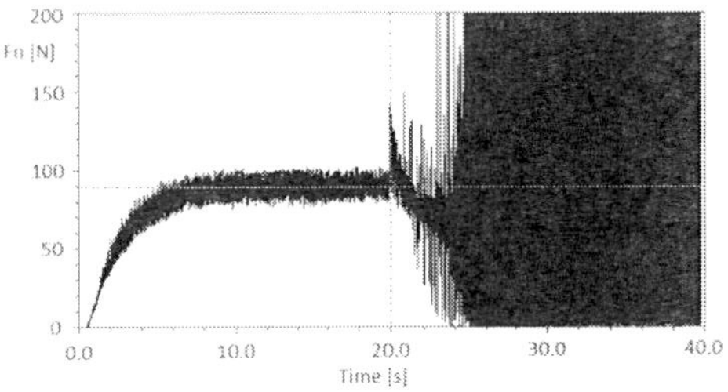

Fig. 4. Course of normal force component during the change in workpiece peripheral speed from n_w=1 to 9 rev/s (v_f=10μm/s, b=15 mm).

In Figure 5 a corresponding spectrum of normal force component is shown. The dominant frequency amplitude is about 710 Hz. This frequency multiplied by workpiece rotational speed should produce about 78 waves on the workpiece periphery. In Figure 6, the resultant workpiece shape and waviness distribution are presented. As it may be observed, the number of obtained waves on the workpiece periphery closely corresponds to the anticipated number of waves.

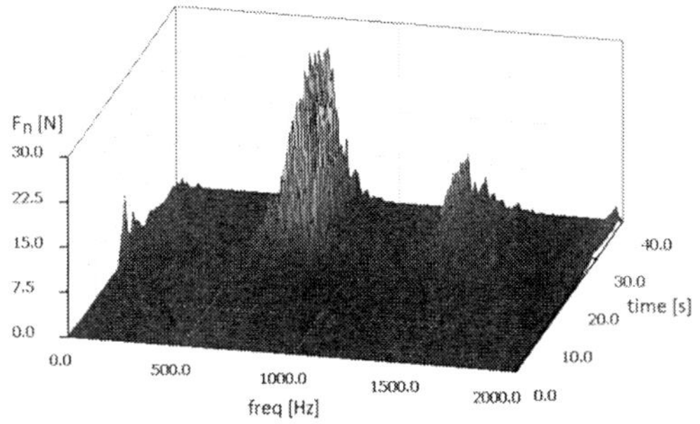

Fig. 5. Specturm of normal force component during the simulation.

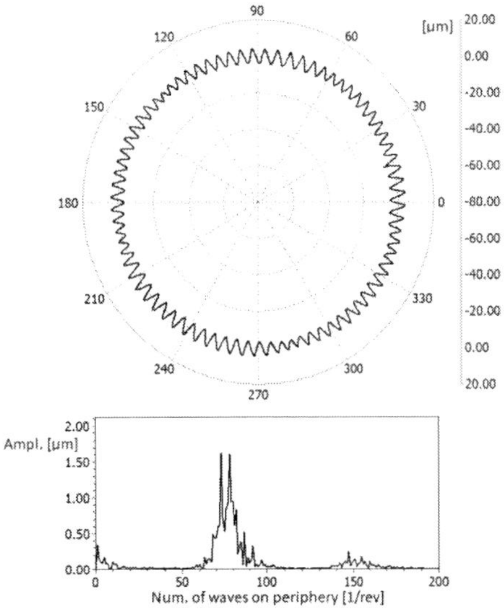

Fig. 6. Workpiece shape and wavines distribution after grinding in unstable conditions.

The model was also used to predict the stable grinding region and to compare with experimental results. Simulation and experimental tests were done for different specific material removal rates Q'_w and workpiece peripheral speeds v_w. The identified stable region obtained during the simulation and experiment (red line) is shown in Figure 7. Additionally, in Figure 7, restrictions due to the workpiece thermal damage are shown. As it may be seen, simulation results well match the experimental ones.

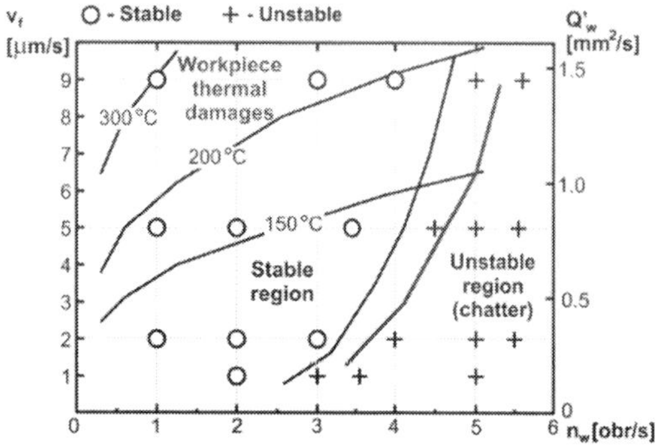

Fig. 7. The identified stable grinding region obtained during simulation and experiment (red line).

The developed simulation program allows also to investigate the regenerative effect on the grinding wheel surface and predict the resultant waviness on the ground workpiece. In Figure 8, a comparison of grinding wheel shapes at the end of wheel life obtained during the experiment and simulation are presented. The obtained shapes are quite similar but the shape obtained during the experiment is more irregular.

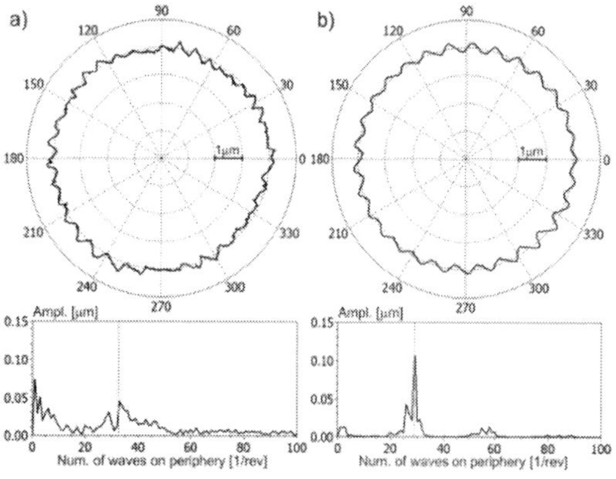

Fig. 8. Grinding wheel shapes and waviness amplitude distribution for a worn girnding wheel obtained during: a) experiment and b) simulation.

3 Conclusions

In this paper, a time domain model of cylindrical plunge grinding process is presented, which combines finite element model of workpiece deflection and two degrees of freedom model of grinding wheel headstock. The nonlinear effect caused by the wheel workpiece geometrical interactions has been taken into account. Using the developed model it is possible to find the optimal working conditions in respect to waviness development on the workpiece. Based on diagrams of stable grinding conditions, the correct operating kinematic parameters can be established for the following grinding phases what should advantageously affect the final part quality parameters and increase machining efficiency. Furthermore, the model allows studying the influence of grinding conditions, a machine tool and workpiece physical and dynamical parameters on grinding force and vibrations growth due to the wheel regenerative effect. Using the proposed model, it will be also possible to develop more conducive, diagnostic systems, necessary to identify process instability and to develop effective methods of chatter suppression.

References

1. M. Weck, K. H. Schiefer, Ann. of the CIRP **38/1**, 281-285 (1979)
2. B. Kruszynski, P. Lajmert, Ann. of the CIRP **54/1**, 305-308 (2005)
3. B. Kruszynski, P. Lajmert, *Proc. Inst. Mech. Eng., Part B: J. Eng. Manuf.* **220**, 355-363 (2006)
4. L. Zylka, Babiarz R., J. of Mechanical Science and Technolgy **31/9**, 4411-4417 (2017)
5. L. Zylka, J. Burek, D. Mazur, Adv. in Production Engineerign & Management **12**, 221-232 (2017)
6. P. Lajmert, M. Sikora, B. Kruszynski, D. Ostrowski, *Lecture Notes in Mech. Eng, Adv. in Manufacturing*, 707-716 (2018)
7. I. Inasaki, B. Karpuschewski, H.-S. Lee, Ann. of the CIRP **50/2**, 515-534 (2001)
8. P. Wolszczak, K. Lygas, G. Litak, Adv. In Sci. and Tech. Research Journal **11/1**, 96-103 (2017)
9. Y. Altintas, M. Weck, Ann. of the CIRP **53/2**, 619-642 (2004)
10. G. Litak, R. Rusinek, K. Kecik, A. Rysak, A. Syta, Discontinuity and complexity in nonlinear physical systems, Springer 359-367 (2014)
11. K. Kęcik, R. Rusinek, J. Warmiński, Int. J. of Bifurcation and Chaos **21/10**, 2943-2954 (2011)
12. R. Snoyes, D. Brown, *Proc. 10th Int. Machine Tool Des. and Res. Conf.*, Manchester, 325-348 (1969)
13. M. Weck, N. Hennes, A. Schulz, Ann. of the CIRP **50/1**, 213-216 (2001)
14. J. Biera, J. Vinolas, F.J. Nieto, Int. J. Mach. Tools and Manufact. **37/11**, 1555-1572 (1997)
15. H. Li, Y.C. Shin, Int. J. of Machine Tools & Manufacture **47**, 1563-1572 (2007)
16. Y. Yan, J. Xu, M. Wiercigroch, Meccanica **51**, 3185-3202 (2016)
17. P. Lajmert, P. Leżański, Archives of Mechanical Technology and Automation **33/3**, 3-15 (2013)
18. P. Lajmert, D. Ostrowski, M. Sikora, B. Kruszyński, Mechanic **8/9**, 195-202 (2014)

Particular aspects regarding the effects of whole body vibration exposure

Mihaela Picu[1,*], *Laurentiu* Picu[1]

[1] "Dunarea de Jos" Galati University, Romania

Abstract. This paper analyses the influence of whole-body vibrations on human performance; for this it was investigated how a group of men (20-29 years of age) and a group of woman (21-31 years of age) answered to specific requirements after being subjected to vertical vibrations under controlled laboratory conditions for 10-25 min. The vibrations were generated by a vibrant system with known amplitudes and frequencies. Accelerations were measured with NetdB - complex system for measuring and analysing human vibration and they were found in the range 0.4 - 3.1m/s^2. The subjects' performances were determined for each vibration level using specific tests. It can be concluded that exposure to vibrations higher than those recommended by ISO 2631 significantly disrupts how subjects responded to tests requirements.

1 Introduction

Our body is exposed to vibrations in the occupational environment because of several mechanisms such as construction equipment, heavy equipment (grinders, pick-hammers) and electric hand tools. It has been shown that vibrations lead to the disruption of the muscular and bone systems of both arms, of the neck and of the back [1, 2].

Occupational WBV exposure, especially when chronic, is suspected to cause adverse health effects such as fatigue, lower back pain, vision problems, interference with or irritation to the lungs, abdomen, or bladder, and adverse effects to the digestive, genital/urinary, and female reproductive systems [3]. In addition to the problems of muscular and bone systems, the exposure of the whole body to occupational vibrations also creates psychomotor, physiological and psychological problems [4].

Contact with a vibrating machine transfers the vibration energy to the human body. Depending on the type of exposure, vibrations can affect much of the worker's body or only a certain organ. The effect of vibration exposure also depends on its frequency. Each body organ has its own resonance frequency. If the exposure occurs at that resonance frequency or near one, the resulting effect is increased [5].

The negative effect of vibrations transmitted by moving equipment on operators have been studied by many researchers [6÷17]. Newell and Mansfield [18] found out on a group of 21 subjects exposed to vibrations of 1÷20Hz and placed in various positions, that their work performances dropped significantly, the reaction times increased and the workload decreased. Including the armrest improved the working conditions, thus the performance.

The frequency dependence of discomfort caused by vertical mechanical shocks has been investigated with 20 seated males exposed to upward and downward shocks at 13 fundamental frequencies and 18 magnitudes. The rate of growth of discomfort with increasing shock magnitude depended on the fundamental frequency of the shocks, so the frequency dependence of equivalent comfort contours varied with shock magnitude. The rate of growth of discomfort was similar for acceleration and force, upward and downward shocks, and lower and higher magnitude shocks [19].

Liu, Qiu and Griffin [20] analyzed fourteen male subjects who stayed on a rigid seat with feet hanging while exposed to random vertical vibration (0.5 to 20 Hz) at three vibration magnitudes (0.25, 0.5, and 1.0 ms^{-2} r.m.s.). The frequency of the principal resonance in the vertical apparent mass was higher at the front thighs (8 to 10 Hz) than at the ischial tuberosities (around 5 Hz). There was greater nonlinearity in the vertical apparent mass measured at the front thighs than at the ischial tuberosities.

Apparent mass frequency response functions of the seated human body have been measured with random vibration in the vertical direction at frequencies up to 20 Hz.

Relative movement between the feet and the seat was found to affect the apparent mass at frequencies below resonance, particularly near zero-frequency.

The magnitude of the vibration had a consistent effect: the resonance frequency decreased from about 6 to 4 Hz when the magnitude of the vibration was increased from 0.25 to 2.0 ms^{-2} r.m.s. The apparent masses of all the subjects were remarkably similar when normalized with respect to sitting weight [21].

Matsumoto and Griffin [22] analyzed the influence of the posture of the legs and the vibration magnitude on

* Corresponding author: mihaelapicu@yahoo.com

the dynamic response of the standing human body exposed to vertical whole-body vibration has been investigated. Motions were measured on the body surface at the first and eighth thoracic and fourth lumbar vertebrae (T1, T8 and L4), at the right and left iliac crests and at the knee. Twelve subjects took part in the experiment with three leg postures (normal, legs bent and one leg), and five magnitudes of random vibration (0.125–2.0 ms^{-2}r.m.s.) in the frequency range till 30 Hz. The main resonance frequencies of the apparent masses at 1.0 ms^{-2}r.m.s. differed between postures: 5.5 Hz in the normal posture, 2.75 Hz in the legs bent posture and 3.75 Hz in the one leg posture.

Multi-directional vibration was also investigated. In the biodynamic research, subjects were exposed to 0.55, 1.75 and 2.25 ms^{-2} r.m.s. vibration levels in the 0.5–40 Hz frequency domain. The influence of sitting position on human body behaviour under two axial vibrations was also examined. Data analysis showed that the human body behaviour under two-directional random vibrations could not be approximated by superposition of one-directional random vibrations [23].

2. Materials and methods

Subjects: The experiment involved 6 women and 6 men (to investigate variability between genders) aged between 20-31 years, in good health. Each participant was asked to complete a health screening questionnaire, to give written consent for tests and was instructed in writing concerning the experiment. Participants were informed that they may abandon the experiment at any time for any reason.

Experimental conditions: Experiments were made inside the laboratory, the temperature and humidity have been set to 22^{0}C, respectively 60%.

The vibrations were transmitted by Modal Exciter Brüel & Kjær type 4827 and were analysed with NetdB - Complex system to measure human vibration using PCB Piezotronics 356A16 - Triaxial Accelerometers. Each experiment lasts 10-25 minutes.

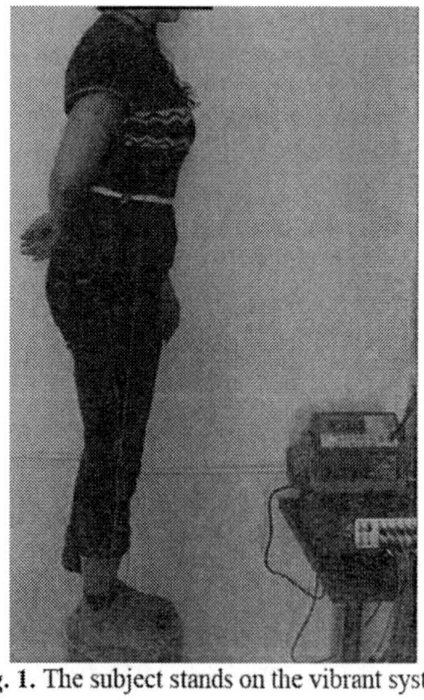

Fig. 1. The subject stands on the vibrant system

The values of vibrations level transmitted to the subject were chosen as close as possible to those in real conditions: frequency v = 2.65; 4.55; 8.25 Hz; acceleration $a_{r.m.s}$ = 0.425; 2.755; 3.100 m/s^2; vibration time t =10; 20; 25 min.

There was also a control case: $a_{r.m.s}$ = 0 (a subject placed on the vibrant system - but it does not work without the subject being informed of this situation).

Two cases were studied: subject standing on the vibrant system and subject placed on the vibrating system in sitting position (Fig. 1 and 2).

Discomfort determination using Likert scale: The first part of the experiment was to determine the subjects' discomfort exposed to stressors, using the Likert scale: Very strongly: 4÷5; Strongly: 3÷4; Slightly: 2÷3; A little: 1÷2; Not at all: 0÷1. Each subject was given a score for what he felt after the experiment ended [24].

Tests of human performance: Subjects answered 3 types of tests: Logical and mausing (locate a point with the mouse pointer) tests, Psychological Test, Stress tests.

3 Results and discussions

In order to determine how to quantify the action of these external factors on human performance, it will be considered that each of the stressors equally concurs to subject's disturbance: $S = 1/3 \cdot a_{r.m.s.} + 1/3 \cdot v + 1/3 \cdot t$ [A.U.] , where S is the stress corresponding to the considered case [24].

The performance (P) of a subject was determined as follows: $P = 0.5 \cdot [0.5 \cdot CP + LM]$ [A.U.], where CP is cognitive performance measurement: $CP = 0.5 \cdot (Psy + ST)$; LM is Logical and mausing tests; Psy is Psychological Test and ST is Stress tests [24].

In order to analyze how subjects responded to the three stressors, different combinations were made between them. In all experiments sets were randomly introduced "placebo" situations (where v=0 and $a_{r.m.s.}$=0 and S=3.333, 6.666 and 8.333 respectively).

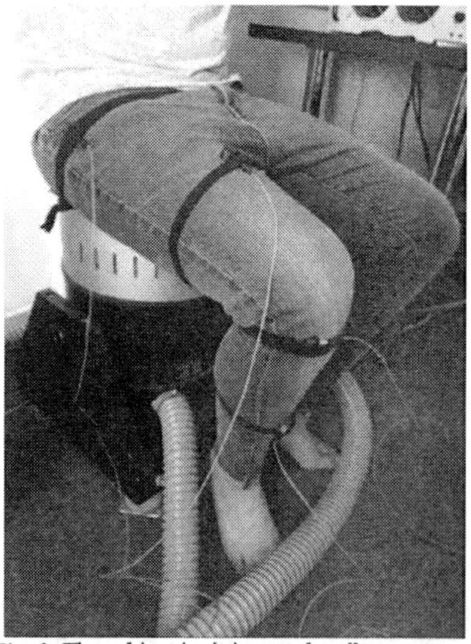

Fig. 2. The subject is sitting on the vibrant system

The following symbols are used in all graphical representations: Female (□); Male (■); ▒ placebo - female; ‖ placebo - male.

Set 1 Figures 3-5 show the effect of acceleration r.m.s. and vibration time on human performance (for constant frequencies).

Figure 3 shows that as the acceleration and vibration time increases, the stress value increases. Women proved to be more resistant than men for lower stressors (performance values are higher). In cases 1, 5 and 8 the "placebo" cases appear. Even if in these situations the subjects are not subject to vibration, the test results were not very good (with higher values performance for males).

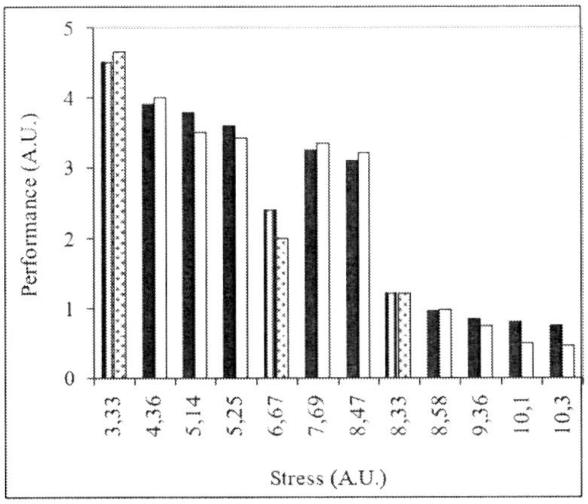

Fig. 3. Effect of acceleration r.m.s. and time on human performance (ν = 2.65 Hz) (Cases 1÷9)

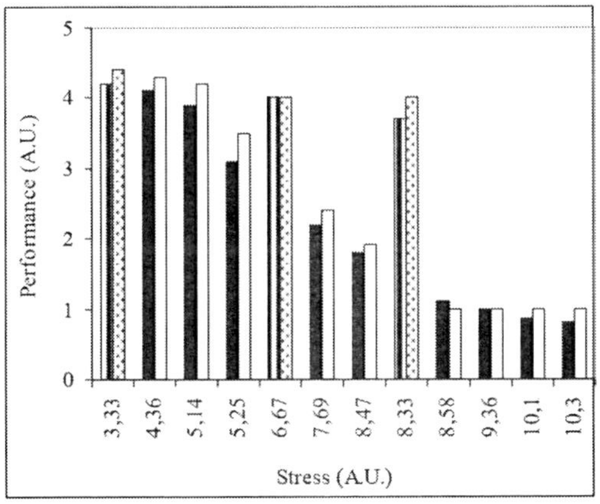

Fig. 4. Effect of acceleration r.m.s. and time on human performance (ν = 4.55 Hz) (Cases 10÷18)

Figure 4 shows the "placebo" cases in situations 1, 5 and 7. Performance values for case 7 (S = 8.33333) are lower than those similar in Figure 3. Women also had better results than men for lower stressors.

Both for women and men there is a sudden decrease in performance at high stressors.

Figure 5 shows that subjects have high performance values in almost all situations. The test values for "placebo" situations are unexpectedly small, especially for situations 3 and 6. It may be concluded that fatigue is beginning to develop.

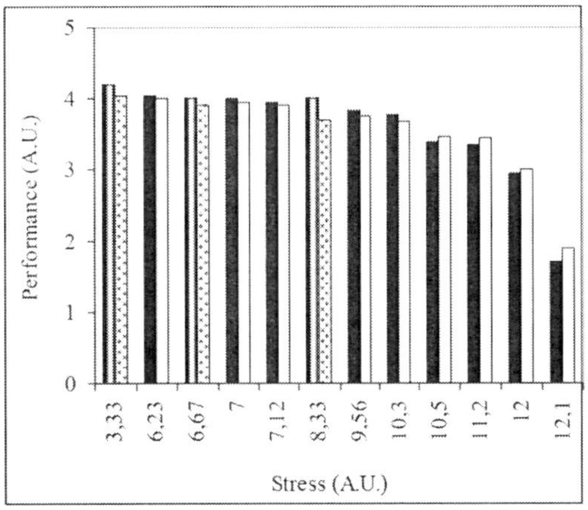

Fig. 5. Effect of acceleration r.m.s. and time on human performance (ν = 8.25 Hz) (Cases 19÷27)

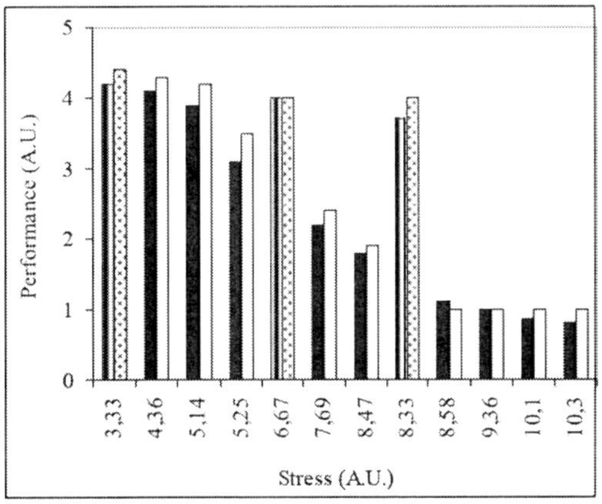

Fig. 6. Effect of frequency and time on human performance ($a_{r.m.s.}$ = 0.425 m/s^2) (Cases 28÷36)

Set 2 In sets 2 and 3 we found the following: each combination represents - from a mathematical point of view - the same situation as one of the previous one. However, the reaction of the subjects is different, namely, the higher the stress values, the subjects considered that they "feel" the "placebo" more intense.

Figures 6-8 show the effect of frequency and vibration time on human performance (for constant acceleration r.m.s).

Figure 6 shows that - with the increase of stressors values - women achieved better results in tests than men, even in "placebo" situations (1, 5 and 8).

Figure 7 shows a slightly different situation from the precedents: for low stress levels, test results are good and considerably equal both for women and men.

Also for "placebo" situations, the same thing is found. From the middle of the experiment to the end it is seen that men get 2 values of unexpectedly good performances; afterwards these values drop sharply. Instead, women maintain some constancy.

Figure 8 shows that the first dropouts occurred: women - in the last 3 experiments and men in the last experiment.

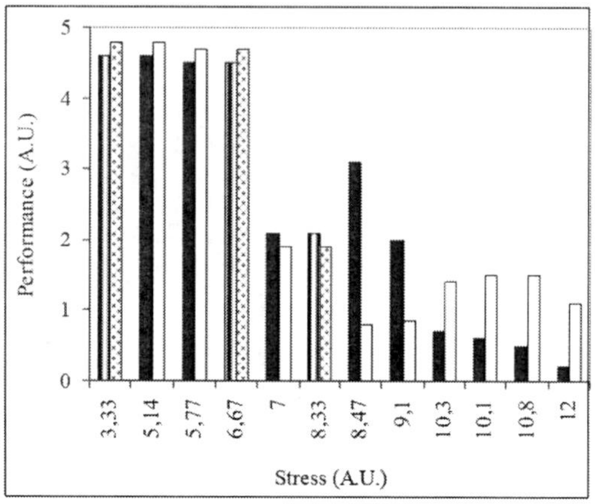

Fig. 7. Effect of frequency and time on human performance ($a_{r.m.s.}$ = 2.755 m/s^2) (Cases 37÷45)

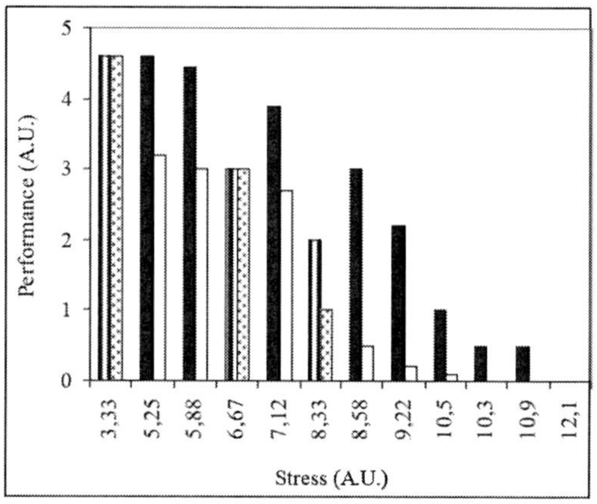

Fig. 8. Effect of frequency and time on human performance ($a_{r.m.s.}$ = 3.100 m/s^2) (Cases 46÷54)

Even the results for the third "placebo" are extremely modest. The men had much better results than women.

It is clear that women's fatigue has occurred and, implicitly, the power of concentration needed to solve the tests.

Set 3 Figure 9 shows the effect of frequency and acceleration r.m.s. on human performance. Figure 9 shows that in the first case men had a higher test score; in the following cases, women had a higher score, but in the latter case, women gave up.

There is no logical correlation between the level of stress and the performance of subjects. It can be seen that at a higher stress (9.1) better performance has been achieved than at lower stress value (7.69). In the latter case, women gave up.

In these last experiments, women gave up on the first attempt and the men gave up the last experiment. There have been no "placebo" situations here. The test results were modest and had equal values in both situations.

The effects of frequency and acceleration on human performance (t=20min and 25 min) (Cases 64÷72 and Cases 73÷81,) were no longer represented due to the fact that there were many abandonments.

4. Conclusions

After analysing the results obtained in the tests it is ascertained that, as the intensity of the stressors increases, the performance decreases. There is, however, a situation (cases 10-18) where performance has increased. It could be an explanation that subjects already had the experience of the previous 9 cases, becoming more careful in solving the tests.

For Cases 1-36 (Figures 3 – 6), performance (P) can be considered as parabolic dependence of stress (S): $P = a{\cdot}S^2 + b{\cdot}S + c$ ($R^2 \approx 0,8$), where a, b and c are constants that depend on each individual situation.

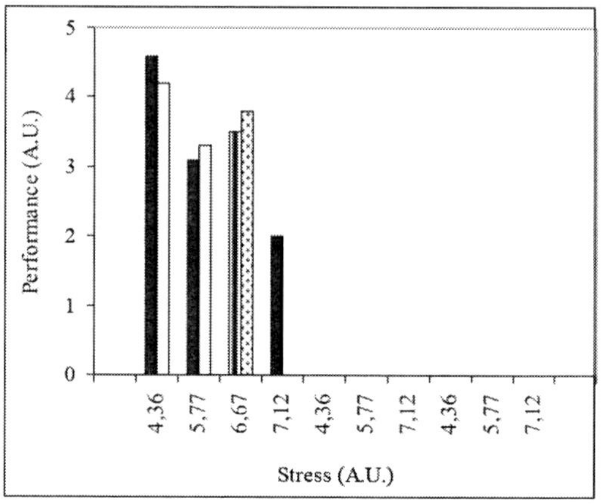

Fig. 9. Effect of frequency and acceleration on human performance (t = 10 min) (Cases 55÷63)

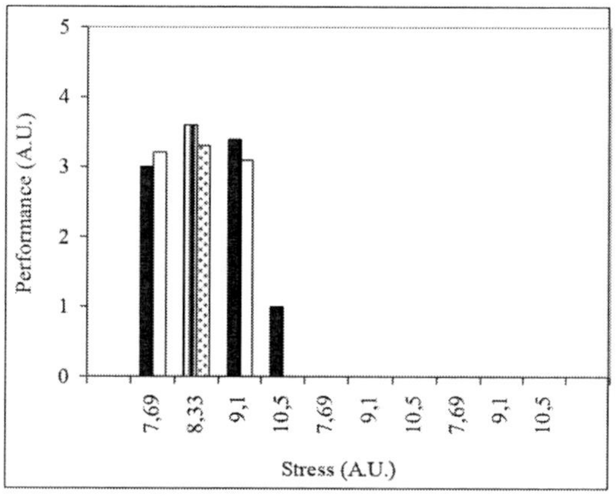

Fig. 10. Effect of frequency and acceleration on human performance (t = 20 min) (Cases 64÷72)

For cases 37 – 45 (Figure 7), there was no dependence between performance (P) and stress (S). For cases 46-54 (Figure 8), there was also a parabolic dependence, with the observation that (R2 ≈ 0.93). Starting Case 55 to the end (Case 81), the performances of the subjects were becoming more modest, the results obtained being not proportional to the induced stress (Figures 9 – 11). It should be emphasized that all these cases represent a repetition of previous cases, so the lack of concentration of the subjects can be attributed to physical and psychological fatigue.

In most situations, women and men have obtained similar results in tests. In the latter cases, part of the subjects abandoned the experiments: for Cases 46-54 (Figure 8) and 73-81 (Figure 11), women abandoned 3 experiments and men 1 experiment; for Cases 55-72, women abandoned 1 experiment.

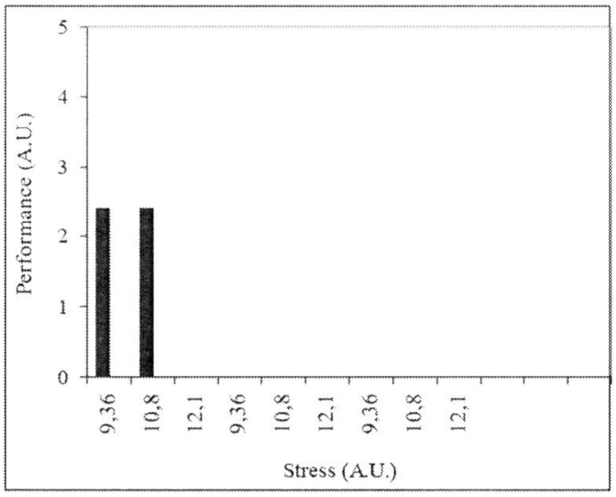

Fig. 11. Effect of frequency and acceleration on human performance (t = 25 min) (Cases 73÷81)

We must emphasize that the subjects who abandoned were those who were younger.

Smokers are among those who have abandoned, but there were also non-smokers who have given up.

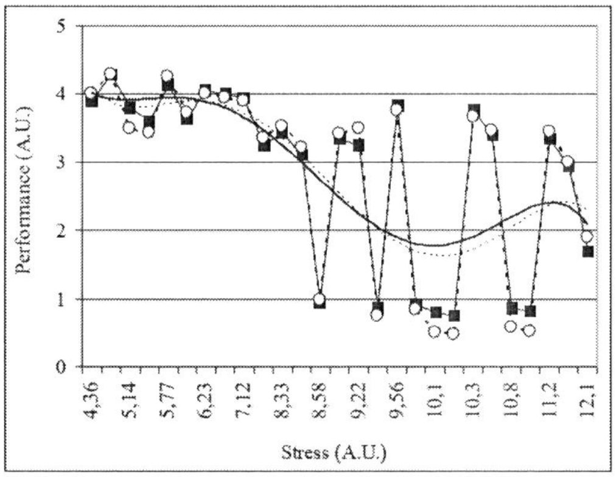

Fig. 12. Human performance variation depending on vibration parameters, for all sets of experiments (—■—) = male; (– –o– –) = female; (——) fitting for men; (– –) fitting for female

In conclusion, there were no significant differences between men and women (Figure 12). What really influences people's performance is personality, as well as personal problems, fatigue. It is very important to know the problems of each worker in order to know what professional tasks he / she can do.

References

1. C.H. Lewis, M.J. Griffin, J. Sound Vib. **215**: 915-926 (2005)
2. M. Cardinale, M.H. Pope, Acta Physiol Hung. **90**(3):195-206 (2003)
3. H.W. Paschold, A.G. Mayton, Professional Safety **56**: 30–35 (2011)
4. N.J. Mansfield, M.J. Griffin, J. Sound Vib. **215**, 813-825 (2005)
5. R. Lundström, P. Holmlund, J. Sound Vib. **215**: 801-812 (2007)
6. M. Bovenzi, F. Ronchese, M. Maurom, Int. Arch. Occup. Environ. Health **84**, 325-334 (2011)
7. H. Dupuis, G. Zerlett, *The Effects of Whole-Body Vibration*, Ed. Springer-Verlag Berlin Heidelberg (2012)
8. L. Gerhardsson, L. Burstrom, R. Lundstrom, T. Nilsson, J. Occup. Med. Toxicol. **8**(1):8 (2013)
9. G.S. Paddan, N.J. Mansfield, C.I. Arrowsmith, A.N. Rimell, S.K. King, S.R. Holmes, Ergonomics **55**(8): 923-936 (2012)
10. J.J. Pilcher, E. Nadler, C. Busch, Ergonomics **45**(10): 682-698 (2002)
11. M. Picu, J. Mater. Sci. Eng. A, **4**(9): 315-320 (2014)
12. M. Picu, Adv. Mater. Res. **837**: 452-457 (2014)
13. M. Picu, *24th International Conference of Noise and Vibration, Nis, Serbia*: 209-216 (2014)
14. M. Picu, Appl. Mech. Mater. **325-326**: 152-157 (2013)
15. M. Picu, Int. J. Mod. Manuf. Tech. **V**(1): 87-90 (2013)
16. M. Picu, Int. J. Mod. Manuf. Tech. **IV**(2): 61-66 (2012)
17. J. Village, C. Trask, Y. Chow, J.B. Morrison, M. Koehoorn, K. Teschke, Ergonomics **55**(4): 415-424 (2012)
18. G.S. Newell, N.J. Mansfield, Int. J. Ind. Erg. **38**(5-6): 499– 508 (2008)
19. Z. Zhou, M.J. Griffin, Ergonomics **60**(3) (2017)
20. C. Liu, Y. Qiu, M.J. Griffin, *50th UK Conference on Human Responses to Vibration, University of Southampton, England* (2015)
21. T.E. Fairley, M.J. Griffin, J. Biomech. **22**(2): 81-94 (1989)
22. Y. Matsumoto, M.J. Griffin, J. Sound Vib. **212**(1): 85-107 (1998)
23. M. Demić, J. Lukić, Ž. Milić, J. Sound Vib. **253**(1): 109-128 (2002)
24. M. Picu, JMEST **2**(9) (2015)

State Switching of a Multistable Impacting System using PD-Like Control

Boying Liu[1], *Wai-keung* Fung[1], *Yang* Liu[2]

[1] School of Engineering, Robert Gordon University, Aberdeen, United Kingdom, AB10 7GJ
Email: {b.liu4, w.k.fung}@rgu.ac.uk
[2] College of Engineering, Mathematics and Physical Sciences, University of Exeter, Exeter, UK, EX4 4QF
Email: y.liu2@exeter.ac.uk

Abstract. A new PD-like control law is proposed in this paper to control a multistable impacting system. This control law can switch the system from a current, undesired state to a desired one by using the differences of the displacement and velocity between the current and desired states. The control law can control the multistable system without affecting its original dynamics, and its stability is proved by using the Lyapunov direct method. Numerical results are compared with the results obtained by using the intermittent control studied in [1]. The proposed PD-like control shows a better performance in terms of the smoothness of its control signal, which is easier to be implemented in practical applications.

1. Introduction

Multistability exists in nature and engineering systems in various fields, such as biology [2], electronics [3] and mechanics [4]. In recent years, many researchers have been paying attention on how control of multistability can be achieved with regard to its high sensitivity to system parameters and external perturbation [5]. Control of multistability can enhance system efficiency by switching its state from an undesired state with large amplitude periodic motion to a desired one with small amplitude periodic motion. For example, maintaining small amplitude periodic oscillations in drill-strings can improve the efficiency of drilling, and therefore, extending lifespan of equipment and reducing maintenance cost. Maintaining large amplitude motion for energy harvester can generate more energy and improve its efficiency. In order to control multistable systems, several control methods have been proposed, including steering the system by a feedforward control strategy [6], applying a short pulse to multistable systems [7], and using a pseudo-periodic force to destroy the undesired attractor [8]. These proposed methods are capable of altering an undesired state to a desired state in multistable systems. However, these control laws have limitations. For example, when co-existing attractors in a multistable system are destroyed, its parameters may be affected, and the structure of basins of attraction of the system may become complex. This makes control the system difficult. If the system is under the influence of noise or perturbation, the system may be switched back to undesired state easily. The control law is then switched on and off continuously in high speed and the system will exhibit chattering behaviour. Therefore, it is important to keep the original basin of attraction intact, and thus, guarantee that the system stays in desired state in the presence of noise.

Liu *et al.* proposed an intermittent control law for state switching in multistable systems [1]. Intermittent control applies an impulse force to the system in order to switch the system to the desired state. However, this control law requires the information of the nonlinearity of the plant model, which is sometimes difficult to acquire accurately. In this paper, a new PD-like control that does not depend on the nonlinearity of the plant is proposed. The proposed control law is simple to be implemented and the basin of attraction of the system will not be changed.

In order to evaluate the performance of this proposed control law, an impact oscillator, which is a typical multistable system, is selected for investigation in this work. It is a simple one degree of freedom system that models the basic mechanism for a variety of engineering systems, such as rotor, gearbox and percussive drilling system [9]. The study of basins of attraction of this system has been extensively reported in literature through numerical analysis and experiment [10-12].

The paper is organised as follows. In Section 2, the mathematical model of the selected multistable system is presented. The proposed PD-like control law and its stability are studied using the Lyapunov direct method in Section 3. In Section 4, the performance of the proposed control law is evaluated using the selected multistable system through simulation. Finally, some concluding remarks are given in Section 5.

2. Impact oscillator

The mathematical model of an impact oscillator is presented as follows. The physical model of the impact oscillator is shown in Figure 1. It includes a mass which is connected to a linear spring and a viscous damper, and the impact oscillator is excited by harmonic force. When the displacement of the mass exceeds the gap between the mass and the secondary spring, impact will occur. Nondimensional model of the impact oscillator is shown below.

$$\ddot{x} = \Gamma sin(\omega\tau) - 2\xi\dot{x} - x - \beta(x - e)H(x - e) \quad (1)$$

Equation 2 shows the process of changing dimensional parameters of the impact oscillator in figure 1 to the nondimensional parameters.

$$x = \frac{y}{x_0}, \quad \tau = \omega_n t, \quad \beta = \frac{k_2}{k_1}, \quad \omega = \frac{\Omega}{\omega_n},$$
$$\xi = \frac{c}{2m\omega_n}, \quad \Gamma = \frac{a}{x_0}, \quad e = \frac{g}{x_0} \quad (2)$$

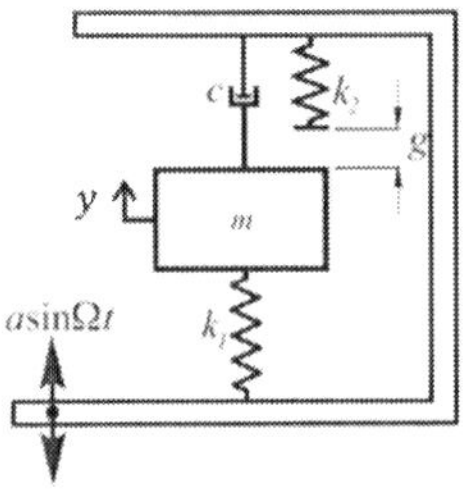

Fig. 1. Schematics of impact oscillator [10]

Two reference values are used for the nondimensional formula. One is the reference distance $x_0 > 0$ and the other is the natural frequency $\omega_n = \sqrt{k_1/m}$. Moreover, x and τ are nondimensional displacements of the mass and time separately. β is the stiffness ratio, ω is the frequency ratio and ξ is the damping ratio, whilst Γ and e denote the nondimensional amplitude and the gap between mass and the second spring respectively. $H(\,.\,)$ is the Heaviside step function.

3. The Proposed Control Law

In this section, we present the proposed PD-like contro law and its stability proof on an impact oscillator.

3.1 The Governing equations

In general, when a multistable system is controlled by a state switching control law $u(t)$, its dynamic model is written as follow.

$$\ddot{x} = f\,(x, \dot{x}) + p(t) + u(t) \quad (3)$$

After the system is switched to its desired state, its dynamic model becomes.

$$\ddot{x}_d = f\,(x_d, \dot{x}_d) + p(t) \quad (4)$$

where x, $\dot{x}$ and x_d, $\dot{x}_d$ are the displacement and velocity of current state and desired state of the system respectively, $f(x, \dot{x})$ and $f(x_d, \dot{x}_d)$ model the nonlinearity of the system at current and desired states respectively, $p(t)$ is the external harmonic excitation and $u(t)$ is the control law applied to the system.

Define

$$e_1 = x - x_d$$
$$e_2 = \dot{x} - \dot{x}_d \quad (5)$$

which are the difference in displacement and velocity between the current and desired trajectories respectively. The proposed PD-like control law is shown as follows:

$$u(t) = -\,k_p e_1 - k_d e_2 \quad (6)$$

where k_p and k_d are the positive gains of the control law.

When the control law is switched on, the system will be driven to the desired state $(x_d, \dot{x}_d)$ within a certain duration of time i.e. e_1 and e_2 are zeros and the control input $u(t)$ becomes zero. Thus, the system is switched from the current undesired state to the desired state. This control law can steer dynamical systems without changing the parameters of the system and thus the structure of basin of attraction is not affected. Furthermore, when the system is affected by sensor noise and external disturbance, this control law forces the system to remain in the desired state, or move back to the desired state. Indeed, the control law guides the system by exploiting only the difference between the current state and desired state under the same parameters. Exact knowledge of the nonlinearity of the plants is not required. In order to prevent damage to the actuator within the system, the control input $u(t)$ is bounded by actuator limit u_{max} and the control law is modified as follows:

$$u(t) = \begin{cases} -k_p e_1 - k_d e_2 & |u| < u_{max} \\ sign(u) \cdot u_{max} & |u| \geq u_{max} \end{cases} \quad (7)$$

3.2 Stability Analysis

The stability of the proposed PD-like control law is studied on the selected multistable system using Lyapunov Stability Theorem. The error dynamics after the control law is applied to the system is defined as:

$$\dot{e}_2 = f^* - k_p e_1 - k_d e_2 \quad (8)$$

where $f^* = f(x, \dot{x}) - f(x_d, \dot{x}_d)$. Note that the effect of the external harmonic excitation $p(t)$ is cancelled in the error dynamics and it is an autonomous system.

A Lyapunov function is proposed for the error dynamics of the system.

$$V(e_1, e_2) = \frac{1}{2}(e_2 + \lambda e_1)^2 + \frac{1}{2}(k_p + \lambda k_d - \lambda^2)e_1{}^2 \quad (9)$$

where $0 < \lambda < k_d$ so that the Lyapunov function is positive definite.

Time derivative of the Lyapunov function is given as:

$$\dot{V}(e_1, e_2) = (e_2 + \lambda e_1)f^* + (\lambda - k_d)e_2^2 - \lambda k_p e_1^2 \quad (10)$$

For the impactor oscillator, f^* is defined as,

$$f^* = -2\xi e_2 - e_1 - \beta[(x - e)H(x - e) \quad (11)$$
$$-(x_d - e)H(x_d - e)]$$

if

$$k_d > -\xi + \sqrt{\xi^2 + (1 + \beta)} \quad (12)$$

$\dot{V}(e_1, e_2)$ is negative definite and the system converges to the desired trajectory asymptotically. Detailed stability proof is presented in the Appendix.

4. Numerical Simulation

In order to evaluate the performance of the proposed control law, MATLAB simulation was conducted using the impact oscillator.

4.1 Impact oscillator

Figure 2(a) shows the basin of attraction of the impact oscillator with parameters (ξ =0.01, β =29, e =0.01, Γ=1.0385 and ω =0.686). There are two co-existing attractors. Figure 2(b) shows the trajectory of the two co-existing attractors, one is a period-2 response with one impact (purple region in figure 2(a) and red line in figure 2(b)) and the other one is a period-2 response with two impacts (blue region in figure 2(a) and black line in figure 2(b)). The process of state switching is shown in blue line in figure 2(b). Figures 2(c) and (d) show respectively the displacement and velocity responses of the system when the proposed control law is applied to the system at around 45 seconds under the control parameters (k_p= 1 and k_d= 5.5). k_d is equal to the minimum value according to equation 12. The system switches to the desired state within 15 seconds. Moreover, when the system converges to the other state, the control input $u(t)$ becomes 0, as shown in Figure 2(e).

In order to study the sensibility to the control parameters, more sets of the control parameters are used. Different performances of the control signal are observed. Figure 3(a) shows the maximum peak of nondimensional control force $u(t)$ with different sets of k_p and k_d. In figure 3(a), the maximum peak of control force $u(t)$ varies from 8.89 to 29.47. The maximum peak of control force $u(t)$ increases with the increasing k_d. Moreover, the maximum peak of control force $u(t)$ decreases with increasing k_p. Moreover, the switching duration of different k_p and k_d are shown in figure 3(b). The switching duration varies from 2.51 seconds to 13.31 seconds. Both of the maximum peak of the control force $u(t)$ and the switching duration are sensitive to the gains in control law.

(a)
(b)
(c)
(d)

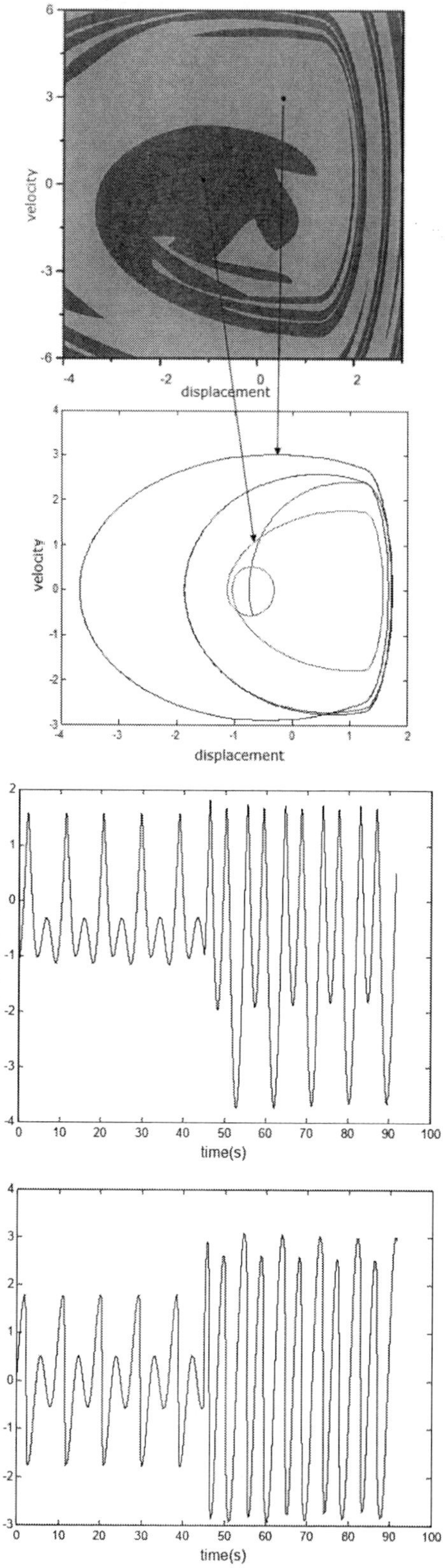

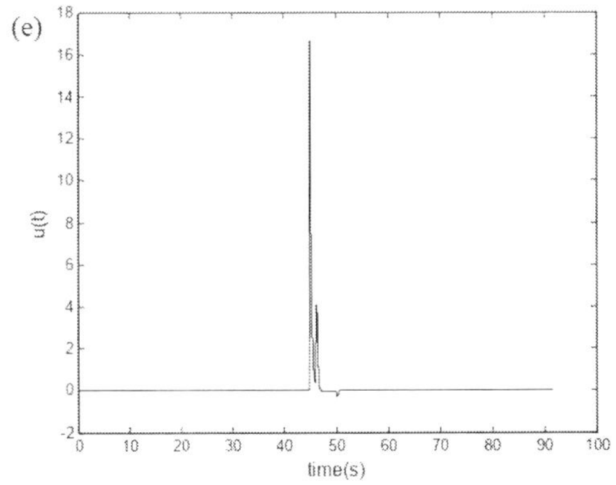

Fig. 2. The simulation result of applying the PD-like control law on an impact oscillator. Parameters of the system: ξ =0.01, β=29, e=0.01, Γ=1.0385 and ω=0.686. Parameter of control law: k_p =1 and k_d =5.5. (a) Basin of attraction. (b) State trajectories of the two co-existing attractors on phase plane. (c), (d) Displacement and velocity of mass respectively. (e) The applied control force

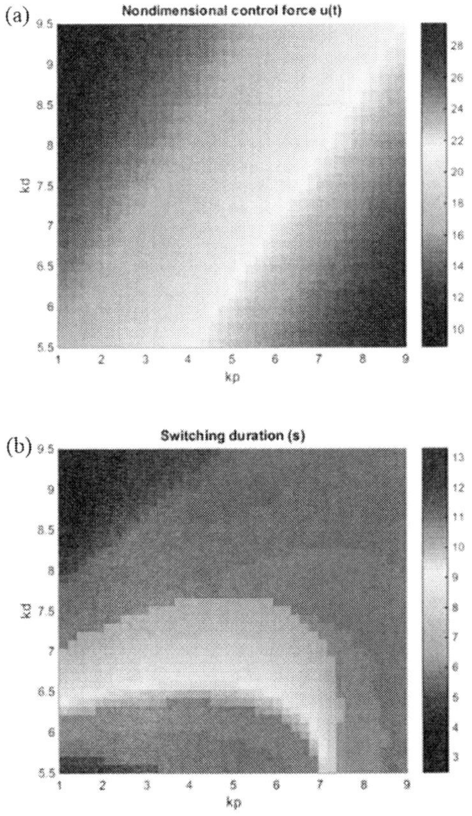

Fig. 3. The simulation result of apply the PD-like control law on impact oscillator with different k_p and k_d. The colour bars show the magnitude of nondimensional control force $u(t)$ and switching duration (a) The nondimensional control force $u(t)$ with different k_p and k_d. (b) The switching duration of the control law with different k_p and k_d

4.2 Comparison with Intermittent Control

The intermittent control law proposed in [8] is shown as follows:

$$\ddot{x} = f\,(x, \dot{x}) + p(t) + u(t)$$

$$u(t) = \begin{cases} 0 & |e_1| > \delta \\ -f^* - k_p e_1 - k_d e_2 & |e_1| \le \delta \end{cases} \quad (13)$$

$$f^* = f\,(x, \dot{x}) - f\,(x_d, \dot{x}_d)$$

$$k_d \ge \frac{ln[|\dot{e}_1(t_b)|/\epsilon]}{\tau} \quad (14)$$

where δ and ϵ define the error of the displacement and velocity between the current undesired and desired trajectories respectively and they are adequately small. When $|e_1| \le \delta$ and $|e_2| \le \epsilon$, the intermittent control switches on. t_b is the first time the system satisfies this condition and τ is the duration of time when the intermittent control law is on. Moreover, the control parameter k_d can be chosen according to equation (14). The proposed PD-like control law looks like the intermittent control law except the nonlinearity term f^* is absent.

The performance of the proposed PD-like control law and the intermittent control law are compared by applying them to the impact oscillator, discussed in Section 4.1, with the following parameters:

δ = 1.2, $\epsilon = 10^{-5}$, k_p = 1 and k_d = 5.5

both control laws are applied with the same k_p and k_d gains. Figure 4 shows the control input $u(t)$ to the impact oscillator when the proposed PD-like control and intermittent control are applied to the impact oscillator under the same system parameters and control parameters. There is no time limit for both control laws or $\tau = +\infty$. In Figure 4, both control laws can switch the system to the desired state. Moreover, the trough magnitude of the control input, generated by the PD-like control law, is smaller than that of the intermittent control law. The control input generated by the PD-like control law has fewer troughs. Furthermore, the PD-like control law takes less time to switch the system from the present state to the desired state than the intermittent control law.

Figure 5 depicts simulation results when the external control input is bounded in both control laws due to actuator limit u_{max}. Equation (15) shows the constrained PD-like control, and equation (16) shows the constrained intermittent control.

$$u(t) = \begin{cases} -k_p e_1 - k_d e_2 & |u| < u_{max} \\ sign(u) \cdot u_{max} & |u| \ge u_{max} \end{cases} \quad (15)$$

$$\mathrm{u}(t) = \begin{cases} 0 & |e_1| > \delta \\ -f^* - k_p e_1 - k_d e_2 & |e_1| \le \delta\,, |u| < u_{max} \\ sign(u) \cdot u_{max} & |e_1| \le \delta\,, |u| \ge u_{max} \end{cases} \quad (16)$$

where u_{max} is the bound of the absolute magnitude of control input $u(t)$.

Both control laws have the same control parameters and they are applied at the same time to the impact oscillator. Both of them are capable of switching the system into the desired state with various u_{max}. For example, when u_{max}= 5, both control laws exhibit very similar behaviours in $u(t)$ shortly after application of the control action. Overall, the control input generated by the PD-like control law has fewer peaks and troughs than those generated by intermittent control, and it takes less time to complete state switching using the PD-like control laws.

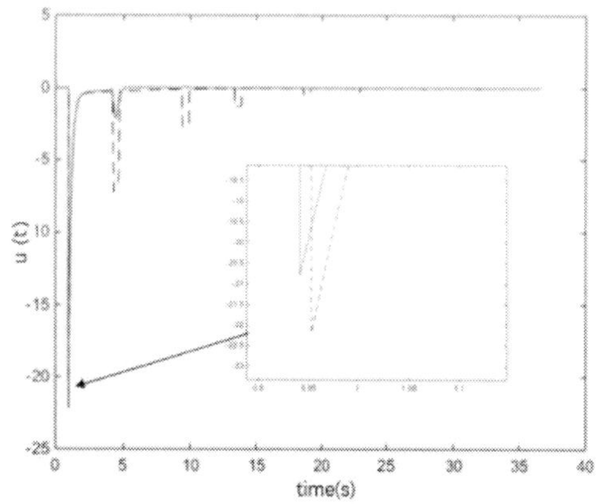

Fig. 4. Control input $u(t)$ to the impact oscillator using intermittent control law (blue dashed line) and PD-like control law (red solid line). System parameters: ξ=0.01, β=29, e=0.01, Γ=1.0385 and ω=0.686. Control parameters: δ=1.2, $\epsilon=10^{-5}$, k_p=1 and k_d=5.5. The inset shows the minimum of the control input generated by two control laws.

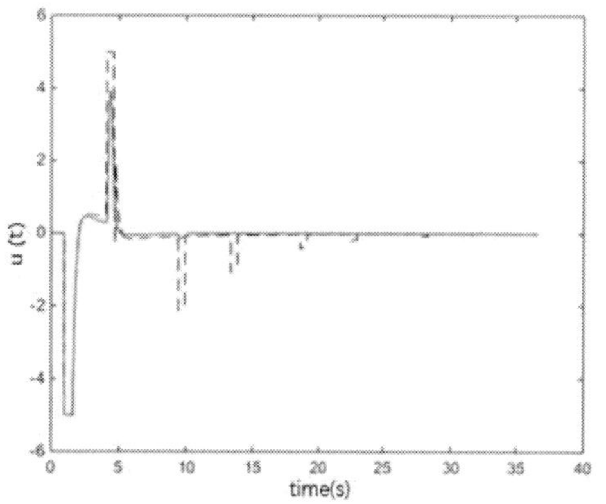

Fig. 5. Simulation results of the impact oscillator with intermittent control law (blue dashed line) and PD-like control law (red solid line). System parameters: ξ=0.01, β=29, e=0.01, Γ=1.0385 and ω=0.686. Control parameters: u_{max}=5, δ=1.2, $\epsilon=10^{-5}$, k_p=1 and k_d=5.5.

Figure 6 shows the simulation results of the control input $u(t)$ when the constrained value of the external control input is very small (u_{max}= 0.5). The intermittent control law fails to switch the impact oscillator to the desired state. Figure 6(a) shows that the control input $u(t)$ oscillates with high frequency and does not stop. On the other hand, the suggested PD-like control law can switch the system to the desired state in 35 seconds. Obviously, the process of state switching takes a longer time when small actuator limits compared to Figures 4 (u_{max}= ∞) and 5 (u_{max}= 5).

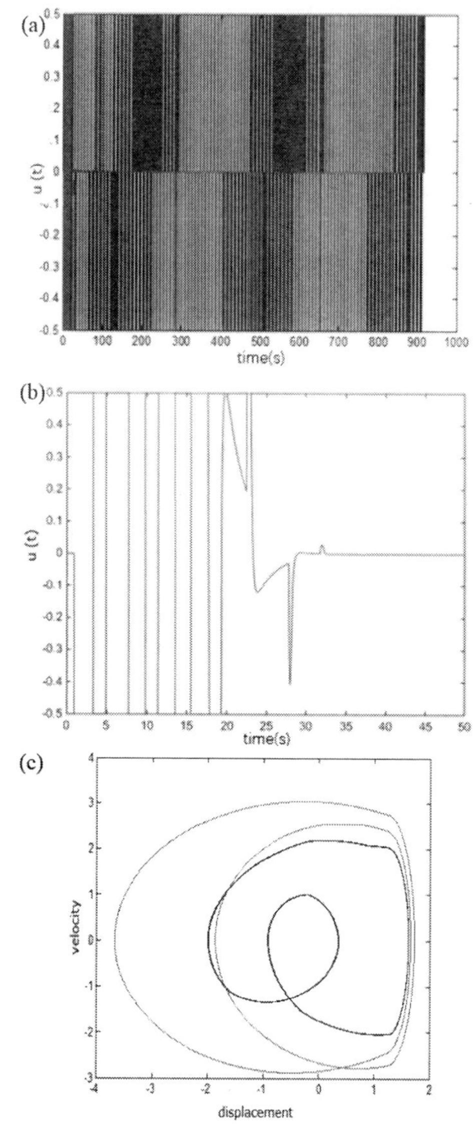

Fig. 5. Simulation results of the impact oscillator with the intermittent control law (blue line) and the PD-like control law (red line). System parameters: ξ=0.01, β=29, e=0.01, Γ=1.0385 and ω=0.686. Control parameters: u_{max}=0.5, δ=1.2, $\epsilon=10^{-5}$, k_p=1 and k_d=5.5. (a) Control input generated by the two control laws. (b) Detail of control input generated by the PD-like control. (c) State trajectories of response on the phase plane after applying the two control laws.

5. Conclusion and Future Work

A PD-like control law was proposed to switch multistable systems to their desired states, without knowing exact knowledge of their nonlinearity, and without destroying the structure of the basin of attraction of the system. Numerical simulations using an impact oscillator demonstrate that the proposed control law can switch impact oscillator from its current, undesired state to the desired state successfully. Moreover, the control input generated by the proposed PD-like control is smoother than that generated by the intermittent control. Constrained control force is considered to avoid exceeding actuator limit. The smaller the control force limit, the longer the switching duration.

The peak value of control force $u(t)$ and switching duration of control law are sensitive to the control gains. Smoothness of the control force $u(t)$ can be optimised by minimising the number of peaks, the peak value in $u(t)$ and the switching duration through finding the optimal k_p and k_d gains and switching time. Some of these objectives are conflicting and multi-objective optimisation of the smoothness of $u(t)$ will be investigated.

References

[1] Y. Liu, M. Wiercigroah, J. Ing, E. Pavlovskaia, Intermittent control of coexisting attractors, Philos. Trans. R. Soc. (2012)

[2] J. Foss, A, Longtin, B. Mensour, J. Milton, Multistability and Delayed Recurrent Loops, Phys. Rev. Lett, **76**, 708-711 (1996)

[3] J. Maurer, A.Libchaber, Effect of the Prandtl number on the onset of turbulence in liquid-He-4, J. de Phys. Lett., **41**, 515-518 (1980)

[4] J.M.T Thompson, H.B. Stewart, Nonlinear Dynamics and Chaos, (1986)

[5] A.N. Pisarchik, U. Feudel, Control of Multistability, Phys. R, **540,** 167-218 (2014)

[6] D.W.M Veldman, R.H.B. Fey, H. Zwart, Impulsive steering between coexisting stable periodic solutions with an application to vibrating plates, J. Comput. Nonlin. Dyn. (2016)

[7] K. Kaneko, Clustering, coding, switching, hierarchical ordering, and control in a network of chaotic elements, Phys D, **41**, 137-172 (1990)

[8] J. Guckenheimer, P. Holms, *Nonlinear Oscillations, Dynamical Systems and Bifurcation of Vector Fields*, Springer-Verlag, (1983)

[9] Y. Liu, J.P. Chavez, Controlling coexisting attractor of an impacting system via linear augmentation, Phys. D, **348**, 1-11 (2017)

[10] J. Ing, E. Pavlovskaia, M. Wiercigroah, S. Banerjee, Experimental study of impact oscillator with one-side elastic constraint, Philos. Trans. R. Soc. (2007)

[11] J. Ing, E. Pavlovskaia, M. Wiercigroah, S. Banerjee, Bifurcation analysis of an impact oscillator with a one-sided elastic constraint near grazing, Phys. D, **239** (2010)

[12] J. Ing, Near Grazing Dynamics of Piecewise Linear Oscillators, Ph.D. dissertation, School of Engineering, University of Aberdeen, (2008)

Appendix

The stability proof of the proposed PD-like control law on impact oscillator is presented.

Given a Lyapunov function

$$V(e_1, e_2) = \frac{1}{2}(e_2 + \lambda e_1)^2 + \frac{1}{2}\left(k_p + \lambda k_d - \lambda^2\right)e_1^{\,2} \quad (17)$$

Where $0 < \lambda < k_d$ so that the Lyapunov function is a positive definite.

In order to prove the $\dot{V}(e_1, e_2)$ is negative definite. Four case are considered

1. $x < e$ and $x_d < e$

$$\dot{V}(e_1, e_2) = (\lambda - k_d)e_2^{\,2} - \lambda k_p e_1^{\,2} \quad (18)$$
$$-(e_2 + \lambda e_1)(2\xi e_2 + e_1)$$

a. When e_1 and e_2 have the same sign, $\dot{V}(e_1, e_2)$ is negative definite.

b. When e_1 and e_2 have different signs and $|e_1| > |e_2|$. There exists a λ which makes $(e_2 + \lambda e_1)(2\xi e_2 + e_1) > 0$ and thus $\dot{V}(e_1, e_2)$ is negative definite.

c. When e_1 and e_2 are different signs and $|e_1| < |e_2|$, There is a bound for k_d : $k_d > 1 - 2\xi$ to guarantee $\dot{V}(e_1, e_2)$ is negative definite.

2. $x > e$ and $x_d > e$

$$\dot{V}(e_1, e_2) = (\lambda - k_d)e_2^{\,2} - \lambda k_p e_1^{\,2} \quad (19)$$
$$-(e_2 + \lambda e_1)[2\xi e_2 + (1 + \beta)e_1]$$

a. When e_1 and e_2 have the same sign, $\dot{V}(e_1, e_2)$ is negative definite.

b. When e_1 and e_2 have different signs and $|e_1| > |e_2|$. There exists a λ which make sure that $(e_2 + \lambda e_1)[2\xi e_2 + (1 + \beta)e_1] > 0$ and thus $\dot{V}(e_1, e_2)$ is negative definite.

c. When e_1 and e_2 are different signs and $|e_1| < |e_2|$, There is a bound for k_d: $k_d \geq -\xi + \sqrt{\xi^2 + (1 + \beta)}$ to ensure that $\dot{V}(e_1, e_2)$ is negative definite.

3. $x > e$ and $x_d < e$

$$\dot{V}(e_1, e_2) = (\lambda - k_d)e_2^{\,2} - \lambda k_p e_1^{\,2} \quad (20)$$
$$-(e_2 + \lambda e_1)[2\xi e_2 + e_1 + \beta(x - e)]$$

Since $x > e$ and $x_d < e$, $0 < x - e < e_1$

a. When $e_2 > 0$, $\dot{V}(e_1, e_2)$ is negative definite.

b. When $e_2 < 0$ and $|e_1| > |e_2|$, There exists a λ to guarantee that $(e_2 + \lambda e_1)[2\xi e_2 + e_1 + \beta(x - e)] > 0$ and thus $\dot{V}(e_1, e_2)$ is negative definite.

c. When $e_2 < 0$ and $|e_1| < |e_2|$. There is a bound for k_d : $k_d \geq -\xi + \sqrt{\xi^2 + (1 + \beta)}$ to make sure that $\dot{V}(e_1, e_2)$ is negative definite.

4. $x < e$ and $x_d > e$

$$\dot{V}(e_1, e_2) = (\lambda - k_d)e_2^{\,2} - \lambda k_p e_1^{\,2} \quad (21)$$
$$-(e_2 + \lambda e_1)[2\xi e_2 + e_1 - \beta(x_d - e)]$$

Since $x < e$ and $x_d > e$, $e_1 < e - x_d < 0$

a. When $e_2 < 0$, $\dot{V}(e_1, e_2)$ is negative definite.

b. When $e_2 > 0$ and $|e_1| > |e_2|$, There exists a λ to ensure that $(e_2 + \lambda e_1)[2\xi e_2 + e_1 - \beta(x_d - e)] > 0$ and thus $\dot{V}(e_1, e_2)$ are negative definite.

c. When $e_2 > 0$ and $|e_1| < |e_2|$. There is a bound for k_d : $k_d \geq -\xi + \sqrt{\xi^2 + (1 + \beta)}$ to make sure that $\dot{V}(e_1, e_2)$ is negative definite.

Optimization of the Vibro-Impact Capsule System for Promoting Progression Speed

Liao Maolin[1,*], *Yan* Yao[2], and *Liu* Yang[3]

[1] *CNPC Drilling Research Institute, Beijing, 102206, China*
[2] *University of Electronic Science and Technology of China, Chengdu, 611731, China*
[3] *College of Engineering, Mathematics and Physical Sciences, University of Exeter, Rennes Drive, Exeter, EX4 4RN, UK*

Abstract. This paper studies the dynamics of vibro-impact capsule systems with one-sided and double-sided constraints under variations of control parameters, including frequency of excitation, mass ratio, and stiffness ratio. The aim of this study is to optimize the progression speed of the capsule system. Extensive comparisons reveal that the capsule system with one-sided constraint is better than the one with double-sided constraints in terms of progression speed. Moreover, the system's period-one one-right-impact motion is proved as the ideal vibration condition due to its lowest energy consumption for impacts. According to the dynamic analyses of control parameters, an inner mass with its weight similar as the weight of capsule and a right constraint with a relative weak stiffness are beneficial for further accelerating the capsule system forwards.

1 Introduction

Pipelines play an important role in a large number of modern industries, which are essential for oil and gas transport, water supplies, and so on. With many pipelines being located in remote and harsh locations, such as underground or seabed, access for inspection, maintenance and repair work could be extremely difficult. Moreover, it becomes a costly issue if the pipeline has to be drained and production stopped while repair work takes place. Therefore, pipeline inspection devices capable of moving independently, with or against product flow would yield significant advantages over traditional pressure driven inspection tool in certain situations. In recent years, investigation of such a self-propelled mechanism moving rectilinearly under internal vibration force when overcoming medium resistance has attracted great attention from researchers, e.g. [1–5]. The principle of such mechanism is that the rectilinear motion can be obtained by overcoming external resistance described as dry friction using an additional internal mass interacting with the main body of the system.

The dynamics of the vibro-impact capsule system, which consists of a capsule main body interacting with a harmonically driven internal mass, has been studied extensively by Liu et al. [5–11]. In [5], the model of the vibro-impact capsule system was firstly studied to provide a fundamental understanding of its dynamics. Dynamics of the system in various environments was investigated in [6], and numerical results show that the behaviour of the system becomes very complex when the capsule is moving in a fluid, but the nature of the friction mechanism becomes less significant when the weight of the internal mass is smaller than the weight of the capsule. In [7], the nonlinear dynamics analysis has been conducted to identify the optimal amplitude and frequency of the applied force to achieve the required motion and the maximal speed. In [10], Páez Chávez et al. studied two practical problems for the capsule system, which were maximizing the rate of progression and directional control of the system by following a typical period-1 trajectory by means of path-following techniques. However, the above studies were based on the dynamics of the capsule system with one-sided constraint, and the performance of the system with double-sided constraints has not been investigated. Thus, it is reasonable to carry out a comparative study of vibro-impact capsule systems with one-sided and double-sided constraints in this paper, which can provide a better insight for design of such system with consideration of the optimum rate of progression.

The rest of this paper is organized as follows. In Section 2, mathematical modelling of the vibro-impact capsule systems with one-sided and double-sided constraints is studied. In Section 3, a comparative study of both the capsule systems through varying control parameters to explore the optimum rate of progression are conducted. Finally, some concluding remarks are drawn in Section 4.

2 Mathematical Modelling

Consider a two degrees-of-freedom system as shown in Fig. 1(a), which is composed of a movable internal mass m_1 interacting with a rigid capsule m_2 via a primary linear spring with stiffness k and a viscous damper with damping coefficient c. The internal mass is driven by an external harmonic force with amplitude P_d and frequency Ω. On

*e-mail: liaomldr@cnpc.com.cn

the right of the internal mass, a weightless plate is connected to the capsule by a linear spring with stiffness k_1, and a secondary weightless plate is connected to the capsule by a linear spring with stiffness k_2 on the left of the internal mass. Here, X_1 and X_2 represent the absolute displacements of the internal mass and the capsule, respectively. The internal mass will contact with the right plate when the relative displacement $X_1 - X_2$ is larger or equals to the gap G_1, or contact with the left plate when the relative displacement $X_2 - X_1$ is larger or equals to the gap G_2. When the left spring is removed (i.e. $k_2 = 0$) as shown in Fig. 1(b).

Introduce nondimensional parameters

$$\begin{aligned} &\Omega_0 = \sqrt{\frac{k}{m_1}}, \quad \omega = \frac{\Omega}{\Omega_0}, \quad \alpha = \frac{P_d}{P_f}, \\ &\zeta = \frac{c}{2m_1\Omega_0}, \quad \gamma = \frac{m_2}{m_1}, g_1 = \frac{k}{P_f}G_1, \\ &g_2 = \frac{k}{P_f}G_2, \quad \beta_1 = \frac{k_1}{k}, \quad \beta_2 = \frac{k_2}{k}, \end{aligned} \tag{1}$$

and nondimensional variables

$$\begin{aligned} &\tau = \Omega_0 t, \ x_1 = \frac{k}{P_f}X_1, \ x_2 = \frac{k}{P_f}X_2, \\ &v_1 = \frac{dx_1}{d\tau} = \frac{k}{\Omega_0 P_f}\dot{X}_1, \ v_2 = \frac{dx_2}{d\tau} = \frac{k}{\Omega_0 P_f}\dot{X}_2. \end{aligned} \tag{2}$$

The nondimensional equations of motion can be written as [12]

$$\begin{aligned} &\dot{x}_1(\tau) = v_1(\tau), \\ &\dot{v}_1(\tau) = \alpha\cos(\omega\tau) + f_i, \\ &\dot{x}_2(\tau) = v_2(\tau), \\ &\dot{v}_2(\tau) = \frac{1}{\gamma}\left(f_i + (1-\delta_v)S_v + \delta_v H_f S_f + \delta_v(1-H_f)f_i\right), \end{aligned} \tag{3}$$

where

$$\begin{aligned} f_i &= -2\zeta(v_1(\tau) - v_2(\tau)) - (x_1(\tau) - x_2(\tau)) \\ &\quad -H_1\beta_1(x_1(\tau) - x_2(\tau) - g_1) \\ &\quad -H_2\beta_2(x_1(\tau) - x_2(\tau) + g_2), \end{aligned} \tag{4}$$

$$\begin{aligned} H_1 &= H(x_1(\tau) - x_2(\tau) - g_1), \\ H_2 &= H(x_2(\tau) - x_1(\tau) - g_2), \\ H_f &= H(|f_i| - 1), \\ \delta_v &= \delta(v_2(\tau)), \\ S_v &= \text{sign}(v_2(\tau)), \\ S_f &= \text{sign}(f_i). \end{aligned}$$

3 Bifurcation analysis

In order to compare the system dynamics and optimize the progression speed of capsule, bifurcation analyses were carried out for both the capsule systems with one-sided and double-sided constraints by sweeping control parameters, including frequency of excitation, mass ratio, and stiffness ratio. The simulations were run for 200 cycles of the external excitation, and the data for the first 180 cycles were omitted to ensure the steady state responses, where the last 20 cycles were used to plot bifurcation diagrams, phase portraits, and time series.

3.1 Frequency of excitation

To investigate the influence of excitation frequency ω on the capsule progression, numerical simulations are carried out for $\omega \in [0.5, 2.0]$, and the simulation results are presented in Fig. 2. According to the comparison of subplots (a) and (b), the dynamic responses of the capsule system with one-sided constraint are mainly periodic; while the chaotic motion is observed for the capsule system with two-sided constraints when the excitation frequency is low. Subplots (c), (d) show the progressions per unit time and per unit power for the capsule system with one-sided constraint, respectively. Both of their optimum progressions are achieved at $\omega = 1.005$ where a period-one one-right-impact motion is detected. Subplots (e), (f) show the corresponding results for the capsule system with double-sided constraint, whose maximal progression speed is about 1/3 of that for the capsule system with one-sided constraint.

3.2 Mass ratio

The effects of mass ratio γ on the dynamic responses of both the capsule systems are studied via sweeping $\gamma \in [1, 10]$, and the obtained numerical results are shown in Fig. 3. One can observe from subplots (a) and (b) that, both the capsule systems experience a period-doubling and a reverse period doubling bifurcations as the increase of mass ratio. For the capsule system with double-sided constraints, the negative capsule progression lasts, see subplots (e) and (f); namely, the backward impacts dominate the capsule movement. As a comparison, the capsule system with one-sided constraint maintains forward progression with the progression speed deceasing when the mass ratio increases, see subplots (c) and (d). This observation indicates that a tiny inner mass is hard to vibrate a huge capsule; therefore, it is better to design the weight of the inner mass closing to that of the capsule.

3.3 Stiffness ratio

The dynamic responses of the capsule systems under varying the right stiffness ratio, $\beta_1 \in [2, 20]$, are investigated, and the obtained numerical results are presented in Fig. 4. As can be seen from subplots (c) and (d), the fastest progression speed of the capsule system with one-sided constraint is achieved at $\beta_1 = 6.05$ where a period-doubling bifurcation is observed. Similarly, the fastest progression speed of the capsule system with double-sided constraint is achieved at $\beta_1 = 4.10$ where a grazing bifurcation is detected, see subplots (e) and (f). Before appearance of the bifurcations, period-one motions are observed, and the progression speed keeps increasing as the increase of stiffness ratio, and then it drops quickly when passing the bifurcations. In addition, compare the two capsule systems with the same right stiffness ratio, the capsule system with one-sided constraint moves forwards faster.

4 Concluding Remarks

Optimization of vibro-impact capsule system to promote progression speed was carried out in this paper. Different capsule structures were compared and a series of control parameters were analyzed. Extensive comparisons revealed that the capsule system with one-sided constraint was better than that with double-sided constraints in terms of progression speed; moreover, the system's period-one one-right-impact motion was proved as the ideal vibration condition due to its lowest energy consumption for impacts. Meanwhile, according to the dynamic analyses of control parameters, an inner mass with its weight closing to the weight of capsule and a right constraint with the relative weak stiffness were benefit for further accelerating the capsule system forwards.

Acknowledgements

This research is funded by National Basic Research Program of China (2015CB251206), CNPC Scientific Research and Technology Development Program, and China Postdoctoral Science Foundation.

References

[1] F.L. Chernousko, J. Appl. Maths Mechs. **66**, 1 (2002)
[2] N.N. Bolotnik, I.M. Zeidis, K. Zimmermann, S.F. Yatsun, J. Computer and Systems Sciences International **45**, 157 (2006)
[3] H.B. Fang, J. Xu, J. Vibration and Control **17**, 19 (2011)
[4] Z. Nagy, R.I. Leine, D.R. Frutiger, C. Glocker, B.J. Nelson, IEEE Trans. Robotics **28**, 1058 (2012)
[5] Y. Liu, M. Wiercigroch, E. Pavlovskaia, H. Yu, Int J Mechanical Sciences **66**, 2 (2013)
[6] Y. Liu, E. Pavlovskaia, M. Wiercigroch, Int J Mechanical Sciences **72**, 39 (2013)
[7] Y. Liu, M. Wiercigroch, E. Pavlovskaia, Z.K. Peng, Int. J. Non-Linear Mech. **70**, 30 (2015)
[8] Y. Liu, E. Pavlovskaia, M. Wiercigroch, Nonlinear Dyn. **83**, 1029 (2016)
[9] Y. Liu, S. Islam, E. Pavlovskaia, M. Wiercigroch, Strojniški vestnik - Journal of Mechanical Engineering **62**, 430 (2016)
[10] J. Páez Cháez, Y. Liu, E. Pavlovskaia, M. Wiercigroch, Communications in Nonlinear Science and Numerical Simulation **37**, 102 (2016)
[11] Y. Liu, J. Páez Cháez, Nonlinear Dyn. **88**, 1289 (2017)
[12] Y. Yan, Y. Liu, M. Liao, Nonlinear Dyn. **89**, 1063 (2017)

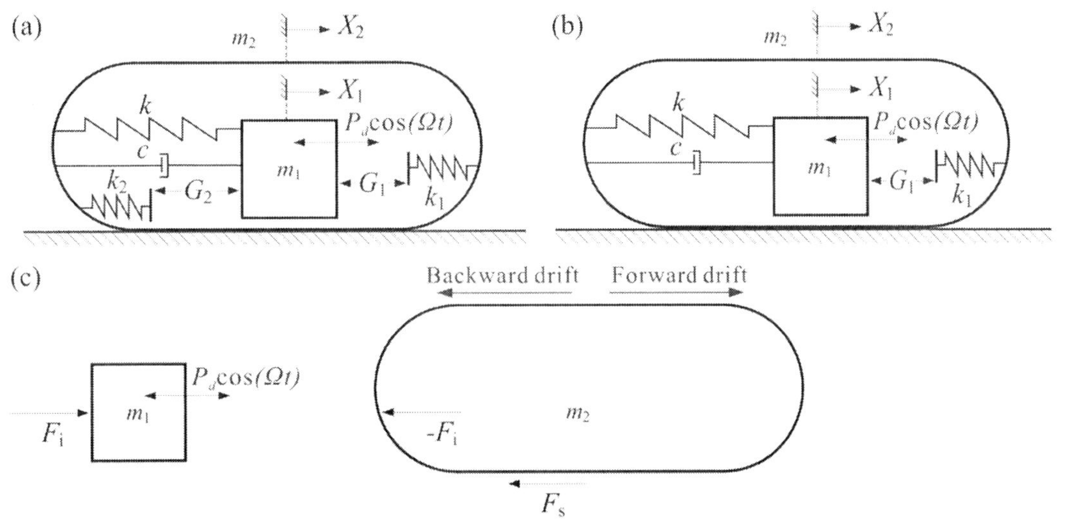

Figure 1. Physical models of the vibro-impact capsule systems, with (a) double-sided constraint, and (b) one-sided constraint. (c) Free body diagrams of the inner mass m_1 and the capsule m_2 [12].

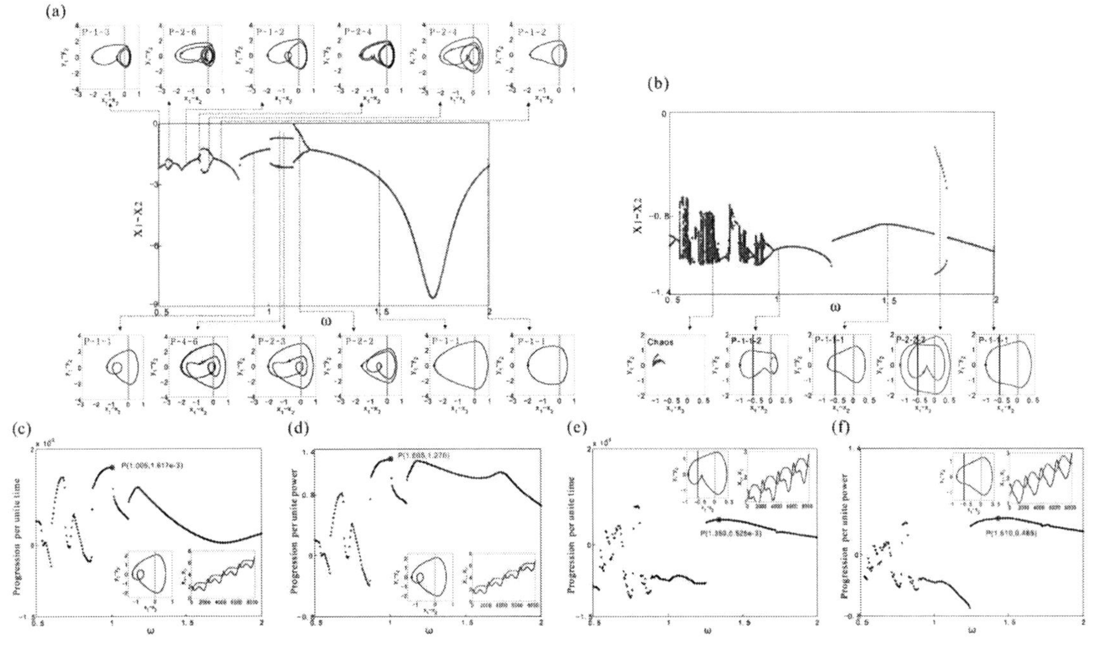

Figure 2. Bifurcation analyses of excitation frequency for the capsule systems with one-sided (a) and double-sided (b) constraints. The nondimensional parameters are $\omega \in [0.5, 2.0]$, $\xi = 0.05$, $\gamma = 4$, $a = 1.6$, $\beta_1 = 12$, $g_1 = 0.02$, $\beta_2 = 5$, $g_2 = -0.06$. The labels shown as P-n-m-l indicate that the trajectory is a period-n response with m left impacts and l right impacts. The vertical green and orange lines in panels represent the impact boundaries $x_1 - x_2 = g_2$ and $x_1 - x_2 = g_1$, respectively. Subplots (c), (d) show the progressions per unit time and per unit power for the capsule system with one-sided constraint, respectively. Subplots (e), (f) show the progressions for the capsule system with double-sided constraint. The position P marked by a red circle indicates the obtained optimum progression, its corresponding phase portrait and time series of displacements for the inner mass (black) and the capsule (red) are displayed.

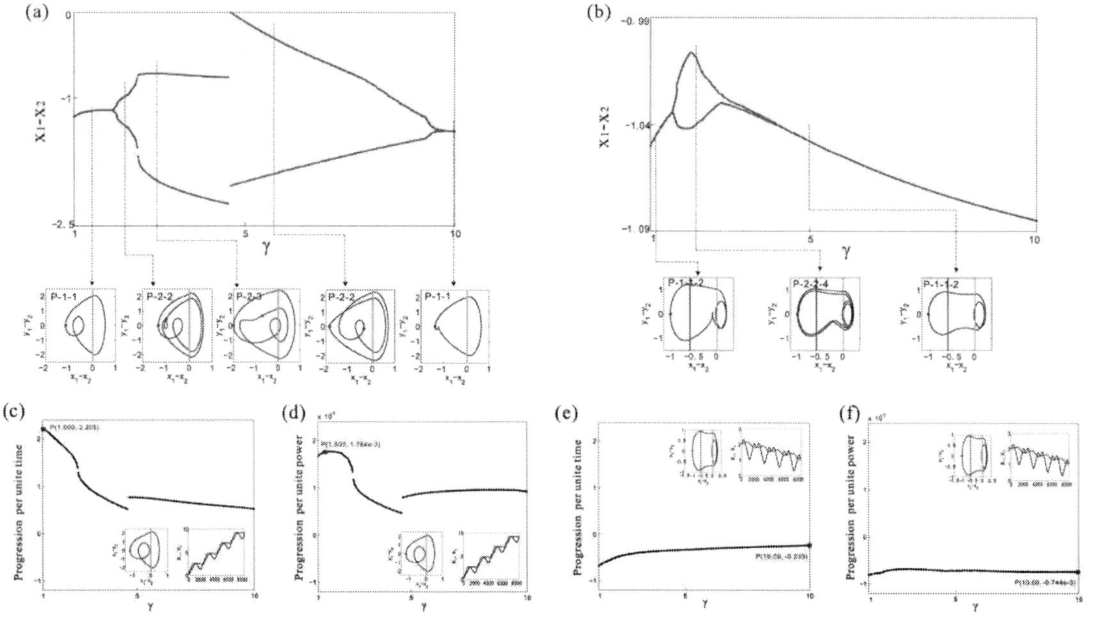

Figure 3. Bifurcation analyses of mass ratio for the capsule systems with one-sided (a) and double-sided (b) constraints. The nondimensional parameters are $\gamma \in [1, 10]$, $a = 1.6$, $\omega = 1.1$, $\xi = 0.05$, $\beta_1 = 12$, $g_1 = 0.02$, $\beta_2 = 5$, $g_2 = -0.06$. Subplots (c), (d) show the progressions per unit time and per unit power for the capsule system with one-sided constraint. Subplots (e), (f) show the progressions for the capsule system with double-sided constraint.

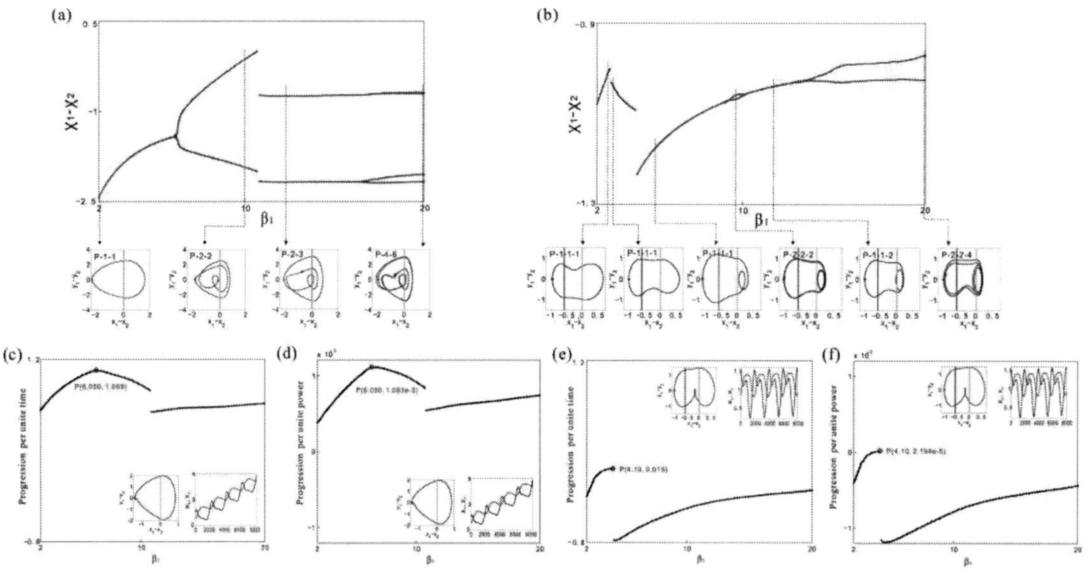

Figure 4. Bifurcation analyses of stiffness ratio for the capsule systems with one-sided (a) and double-sided (b) constraints. The nondimensional parameters are $\beta_1 \in [2, 20]$, $\gamma = 4$, $a = 1.6$, $\omega = 1.1$, $\xi = 0.05$, $g_1 = 0.02$, $\beta_2 = 5$, $g_2 = -0.06$. Subplots (c), (d) show the progressions per unit time and per unit power for the capsule system with one-sided constraint. Subplots (e), (f) show the progressions for the capsule system with double-sided constraint.

Spectrum of Lyapunov exponents in non-smooth systems evaluated using orthogonal perturbation vectors

Marek Balcerzak[1], *Artur* Dąbrowski[1], *Andrzej* Stefański[1*], and *Jerzy* Wojewoda[1]

[1]Division of Dynamics, Lodz University of Technology, Stefanowskiego 1/15, 90-924 Lodz, Poland

Abstract. This paper covers application of the novel method of Lyapunov exponents (LEs) spectrum estimation in non smooth mechanical systems. In the presented method, LEs are obtained from a Poincaré map. By analysing the map instead of the full trajectory, problems with transition of perturbations through discontinuities can be avoided. However, the explicit formula of the map is usually not known. Therefore, the Jacobi matrix of the map is estimated using small perturbations of the initial point. In such a manner, direct calculation of the Jacobi matrix can be avoided. The article provides a detailed description of the method accompanied by clear schemes. The algorithm of Jacobi matrix estimation is elaborated and an example is given. Efficiency of the method is confirmed by a numerical experiment. The mechanical oscillator with impact has been simulated. Bifurcation diagrams and Lyapunov exponents graphs have been generated. It has been shown that the method provides values of the whole Lyapunov exponents spectrum with high accuracy.

1 Introduction

Lyapunov exponents (LEs) are one of the most useful criteria for determining the stability of solutions of dynamical systems in local [1] and also global sense [2]. They are extremely useful instruments for identification of the motion character in these systems. Determination of these numbers is one of the fundamental tasks in nonlinear dynamics. Definition of LEs was established by V. I. Oseledec [3], in a form suited for the theory of dynamical systems. These numbers were named after A. M. Lyapunov, due to the fact that their values are qualitative and quantitative illustration of his criterion of dynamical systems stability [1]. LEs can be treated as the exponential measure of divergence or convergence of orbits in the phase space that start close to each other. Mathematically, LEs are eigenvalues of the state transition matrix that determine evolution of an infinitesimal perturbation in the dynamical system. Hence, they can be treated as a measure of the sensitivity to initial conditions in the phase space.

For preliminary assessment of the system dynamics, it is enough to know the largest Lyapunov exponent. Positive value of the largest LE indicates high sensitivity to initial conditions of the system and is a sign of an irregular response, i.e. chaotic or even hyperchaotic. The largest LE equal to zero means regular, periodic or quasi-periodic system dynamics. If all the LEs are negative, then the limit set is a stable fixed point. However, for more detailed and accurate analysis of the system, the knowledge of the complete spectrum of LEs is essential. For any dynamical system, such spectrum is a set of real numbers sorted in the non-increasing order. Number of LEs is equal to the dimension of the phase space. Knowing all the LE values in the spectrum enables to estimate the fractal dimension of the system attractor (according to the Kaplan-Yorke formula [4]), determine the dimension of the system quasiperiodicity (defined by the number of LEs of zero value) or the scale of hyperchaos (defined by number of positive LEs). On the other hand, Lyapunov exponents also provide a criterion for global stability of the system [2]. From properties of the LEs one can conclude that the sum of all of them is equal to the divergence of the phase flow [5]. If the sum of all components in the spectrum is negative, then the dynamical system under consideration is dissipative and its attractor is globally stable. This sum is equal to zero for any conservative system. In the case of positive sum of all the LEs, an exponential divergence of the phase flow is observed, so the system is globally unstable.

The analytical determination of Lyapunov exponents is possible only in the considerations of simple linear dynamical systems. Owing to this fact, numerical methods are used to calculate LEs in most cases. The first numerical characteristics of chaotic behaviour of the dynamical system, which presented a divergence of neighbouring trajectories, was conducted by Henon and Heiles [6]. An efficient algorithm for calculating the complete spectrum of the Lyapunov exponents based on the Oseledec theorem was independently formulated by both Benettin et al. [7] and Shimada, Nagashima [8]. Then, it was improved by Benettin et al. [9-10], Wolf [11], Parker, Chua [12] and Nusse, Yorke [13]. Recently, another method for the LE spectrum calculation, based

* Corresponding author: Andrzej.Stefanski@p.lodz.pl

on the scalar product of vectors, was proposed [14-20]. These techniques allow to calculate the complete spectrum of the LEs. They can be successfully used for many dynamical systems, both autonomous and non-autonomous. But the range of their use is limited by the condition of continuity of the system equations in the analysed time interval. Discontinuity in equations usually cause significant numerical problems, because derivative of the trajectory with respect to the initial conditions (Jacobi matrix) becomes ill conditioned or cannot be calculated at all. Therefore, when typical methods are applied, error of LEs estimation becomes very large or calculation of the LEs is not possible.

In the real world there are many dynamical systems with discontinuities. Among them mechanical systems with impacts, dry friction or piecewise linear stiffness characteristics are especially important. Two decades ago, a method of the LEs calculation based on the Oseledec theorem adapted to dynamical systems with discontinuities was proposed [21]. In order to apply this method, linearization of the equations of motion must be accompanied by a clear statement of the transition conditions while the trajectory is passing through the discontinuity. Another class of methods for the LEs calculation employ reduction of the dynamics of the phase flow in the k-dimensional phase space to a discrete map of a lower dimension, such as a Poincaré map [12], an impact map [22], a local map [23] or a transcendental map [24]. Then, the LEs of such mapping are determined using classical approaches mentioned above. First examples of such an approach can be found in the works by Hilborn [25] and Oestreich [26-27]. The main problem in application of these algorithms is determination of the Jacobi matrix of the mapping, where consecutive iterations are not explicitly defined by a known difference equation, but they are reconstructed from the flow. The example of such a map-based algorithm is the method by Galvanetto [28], who applied implicitly defined maps for calculation of the two largest LEs of the 2-DoF stick–slip system. Other interesting application of such approach is the algorithm for impact oscillators formulated by Souza & Caldas [24], which exploits the so-called transcendental maps that take into account the solution of integrable differential equations between impacts, supplemented by transition conditions at the instants of the impact. On the other hand, Jin et al. [23] focused on the local mappings of non-smooth systems and found a general calculation method of the spectrum of LEs for k-dimensional non-smooth dynamical systems, which could be generally applied to the systems with rigid or flexible constraints. The largest LE can be estimated also from a one-dimensional mapping of the dynamics of the system – the so-called H-map, without defining the Jacobi matrix [26]. The idea of such an attempt is calculation of the value of the largest LE on the basis of the averaged value of the slope (directional coefficient) of a line tangent do the H-map in the point of its contact with the phase plane trajectory [29]. The next practical method for estimation of the largest LE, elaborated by authors of this paper, is based on the synchronization phenomenon. Previously, we successfully employed this approach for discontinuous oscillators with impacts and friction [30-31], discrete-time maps [32] and systems with time delay [33].

Other technique, which can be also applied in the case of systems with discontinuities, is an algorithm of estimation of the LEs from the scalar time signal. This method is based on a procedure of reconstruction of the system attractor from the time dependence of one of the coordinates that was introduced by Takens [34]. The first numerical algorithm for estimation of the largest LE, derived from it, was formulated by Wolf [35]. In the following years, this algorithm has been improved by other researchers in order to estimate the entire spectrum of the LEs and to recognize the so-called spurious Lyapunov exponents [36].

In this paper we demonstrate a method for calculating the spectrum of LEs which is based on the idea of estimation of the Jacobian matrix using small perturbations of the initial orthogonal vectors. This technique is especially useful for non-smooth oscillators, because such estimation procedure allows us to avoid direct calculation of the Jacobi matrix. We have successfully applied this procedure in order to test the hyperbolic properties of the attractor of stick-slip oscillator [37]. Recently, an analogous approach of determining the LEs has been presented by Soriano et al. [38]. They named it cloned dynamics approach. We have developed and applied this technique to a typical mechanical system with discontinuity: the impact oscillator.

The paper is organized as follows. The section 2 contains a detailed description of the applied method of the LEs spectrum calculation. In the section 3 examples of the method application are demonstrated. Finally, in the section 4, a discussion of our results and conclusions are presented.

2 The method

The method under consideration is based on the classical algorithm by Parker & Chua [12]. However, calculation of linearized equations is replaced with the estimation of Jacobi matrices using small orthogonal perturbation vectors. Scheme of the Jacobi matrix estimation method is presented in the Fig. 1.

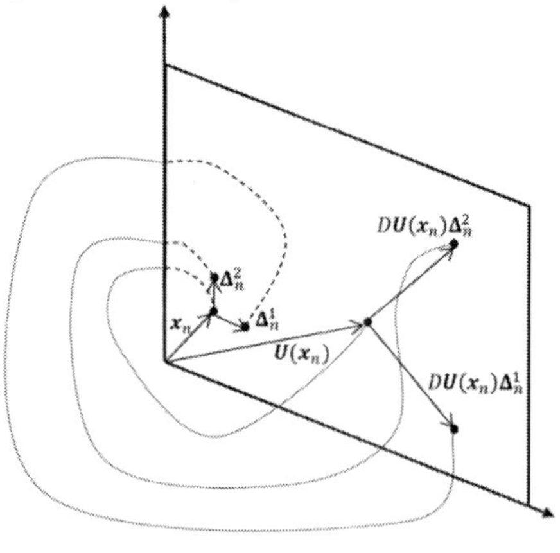

Fig. 1. Scheme of the Jacobi matrix estimation method

The idea of the estimating procedure is as follows. For any N-dimensional time-continuous system:

$$\dot{\mathbf{y}} = \mathbf{F}(\mathbf{y}) \quad (1)$$

where $\mathbf{y}=[y_1, y_2, \ldots, y_N]^T \in R^N$ and $\mathbf{F}=[f_1, f_2, \ldots, f_N]^T$, a ($N$-1)-dimensional map (e.g. Poincaré cross-section) can be defined in the following general form:

$$\mathbf{x}_{n+1} = \mathbf{U}(\mathbf{x}_n) \quad (2)$$

where $\mathbf{x}=[x_1, x_2, \ldots, x_{N-1}]^T \in R^{N-1}$. For the discrete map presented in the Eq. (2), direct calculation of Jacobi matrices is impossible, because the right side of the Eq. (2) is not known. In this case, each iteration of the map is reconstructed numerically from the solution of the investigated system, described by the Eq. (1). Therefore, the Jacobi matrix of the map (2) cannot be determined analytically. However, an approximate value of the Jacobi matrix can be evaluated by means of the method of small perturbations vectors. This idea works as follows. After introducing the perturbation vector $\mathbf{\Delta}_n = [\delta_1, \delta_2, \ldots, \delta_{N-1}]^T$, where δ_i (i = 1, 2, … N-1) is a small magnitude value, into the Eq. (2), we have

$$\mathbf{x}_{n+1} + \mathbf{\Delta}_{n+1} = \mathbf{U}(\mathbf{x}_n + \mathbf{\Delta}_n) \approx D\mathbf{U}(\mathbf{x}_n)\mathbf{\Delta}_n + \mathbf{U}(\mathbf{x}_n) \quad (3)$$

where $D\mathbf{U}(\mathbf{x}_n)$ is a Jacobi matrix composed of (N-1) column vectors $D\mathbf{U}_i(\mathbf{x}_n)$ = $[\partial f_i(x_1, \ldots, x_{N-1})/\partial x_1, \ldots, \partial f_i(x_1, \ldots, x_{N-1})/\partial x_{N-1}]^T$, i.e.:

$$D\mathbf{U}(\mathbf{x}_n) = \left[D\mathbf{u}_1(\mathbf{x}_n), D\mathbf{u}_2(\mathbf{x}_n), \ldots, D\mathbf{u}_{N-1}(\mathbf{x}_n)\right] \quad (4)$$

From Eq.(3) the Jacobi matrix can be approximated from the formula:

$$D\mathbf{U}(\mathbf{x}_n)\mathbf{\Delta}_n \approx \mathbf{U}(\mathbf{x}_n + \mathbf{\Delta}_n) - \mathbf{U}(\mathbf{x}_n) \quad (5)$$

Consequently, each column vector can be estimated as:

$$D\mathbf{u}_i(\mathbf{x}_n) \approx [\mathbf{U}(\mathbf{x}_n + \mathbf{\Delta}_n^i) - \mathbf{U}(\mathbf{x}_n)]/\delta \quad (6)$$

where $\mathbf{\Delta}_n^1=[\delta, 0, \ldots, 0]^T$, $\mathbf{\Delta}_n^2=[0, \delta, \ldots, 0]^T \ldots$, $\mathbf{\Delta}_n^{N-1}=[0, 0, \ldots, \delta]^T$ are orthogonal vectors of small perturbation δ. Columns of the Jacobi matrix of the map $D\mathbf{u}_i(\mathbf{x}_n)$ can be estimated numerically from the solution of the system defined by the Eq. (1). Procedure is as follows. For each iteration of the map, trajectory has to be calculated once for the reference path, starting from the original initial conditions $\mathbf{x}_n$, and N-1 times for the initial conditions disturbed by each perturbation vector: $\mathbf{x}_n+\mathbf{\Delta}_n^i$. In such manner, after the iteration of the map, the reference point $\mathbf{x}_{n+1}=\mathbf{U}(\mathbf{x}_n)$ and N-1 perturbed points $\mathbf{U}(\mathbf{x}_n+\mathbf{\Delta}_n^i)\approx \approx\mathbf{x}_{n+1}+D\mathbf{U}(\mathbf{x}_n)\mathbf{\Delta}_{n+1}^i$ are obtained (see the Fig. 1). Then, columns of the Jacobi matrix described in the Eq. (4) can be obtained from the Eq. (6). Obviously, this way of Jacobian matrix estimation can be also applied for the maps given by known difference equations.

Let's follow how this procedure works on the example of classical Henon map, i.e.:

$$\mathbf{U}(\mathbf{x}_n) = \begin{cases} x_{n+1} = 1 - ax_n^2 + y_n \\ y_{n+1} = bx_n \end{cases} \quad (7)$$

with the Jacobi matrix :

$$D\mathbf{U}(\mathbf{x}_n) = \begin{bmatrix} -2ax_n & 1 \\ b & 0 \end{bmatrix} \quad (8)$$

Orthogonal perturbation vectors are $\mathbf{\Delta}_x=[\delta, 0]^T$ and $\mathbf{\Delta}_y=[0, \delta]^T$ ($\delta_x = \delta_x = \delta$). Substituting Henon map into the Eq.(6), we obtain following column vectors:

$$D\mathbf{u}_x(\mathbf{x}_n) = \left[\mathbf{U}(\mathbf{x}_n + \mathbf{\Delta}_x) - \mathbf{U}(\mathbf{x}_n)\right]/\delta = \\ = \begin{bmatrix} \dfrac{1 - a(x_n+\delta)^2 + y_n - (1 - ax_n^2 + y_n)}{\delta} \\ \dfrac{b(x_n+\delta) - bx_n}{\delta} \end{bmatrix} = \begin{bmatrix} -2a(x_n+\delta) \\ b \end{bmatrix} \quad (9)$$

Finally, in accordance with the Eq.(4), estimated Jacobi matrix of the Henon map is obtained in the following form:

$$D\mathbf{U}(\mathbf{x}_n) = \left[D\mathbf{u}_x(\mathbf{x}_n) \quad D\mathbf{u}_y(\mathbf{x}_n)\right] = \begin{bmatrix} -2a(x_n+\delta) & 1 \\ b & 0 \end{bmatrix} \quad (10)$$

The example above implies that the precision of the Jacobi matrix estimation depends on the magnitude of the small parameter δ. Note that, in general, this procedure requires high accuracy of the trajectory simulation, especially in the vicinity of the discontinuity. Then, such numerically generated Jacobi matrix can be implemented in a classical algorithm for calculating Lyapunov exponents spectrum, including the Gramm-Schmidt orthonormalization [12].

3 Numerical example

In this paper, the method presented in the previous section has been applied to a simple mechanical system with impacts. Scheme of the system is presented in the Fig. 2. The system is a typical linear mass-spring-damper system. However, motion of the mass is constrained by the wall placed in the position X_w. The equation of motion of the system is as follows:

$$m\ddot{X} = -c\dot{X} - kX + F\cos(\Omega t) \quad (11)$$

where X is the coordinate of the oscillator, m is the mass of the oscillator, k is the spring stiffness, c is the damping coefficient, F is the amplitude of forcing, Ω is the angular frequency of forcing and X_w is the position of the wall. Obviously, this equation is not correct when a contact with the wall occurs. If t_c is the moment of time in which collision with the wall takes place, then transition of the oscillator's velocity is defined by the following equation:

$$\dot{X}(t_c^+) = -e\dot{X}(t_c^-) \qquad (12)$$

where e is the coefficient of restitution (CoR).

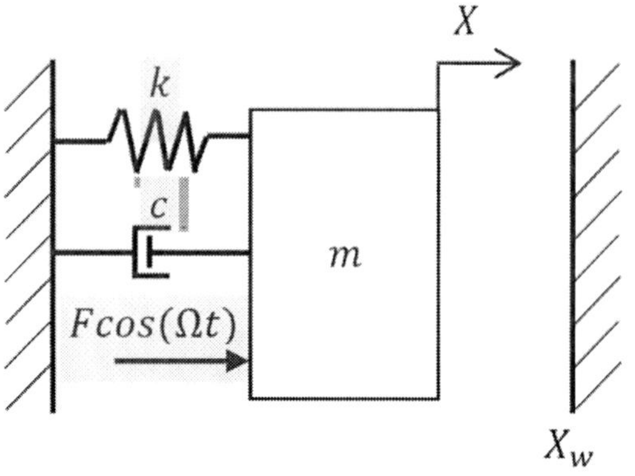

Fig. 2. Scheme of the analysed system with impacts

By simple rearrangements, Eq. (11-12) can be transformed to the form:

$$X'' = -2\zeta X' - X + f\cos(\eta\tau) \qquad (13)$$

$$X'(\tau_c^+) = -eX'(\tau_c^-) \qquad (14)$$

where ζ is the damping ratio, $\omega^2=k/m$ is the squared natural frequency of the system, $f=F/(m\omega^2)$, $\tau=\omega t$ is the dimensionless time, $\eta=\Omega/\omega$ is the dimensionless frequency of forcing and τ_c is the dimensionless time of collision. Equations (13-14) have been simulated numerically. The values of parameters were chosen as follows: $\zeta=0.05$, $f=1.0$, $X_w=2.0$, $e=1.0$ and the dimensionless forcing frequency η was used as the control parameter. To simulate the system a simulation program has been written in C++ by means of CodeBlocks environment. Throughout the simulation, Runge-Kutta method of the fourth order has been used to integrate the equation (13). The selected step length was dependent on the dimensionless frequency in such a way, that 1800 integration steps were performed for each forcing period. In order to detect the collision time τ_c with a proper accuracy, the linear interpolation method has been used when the coordinate X approached X_w. The spectrum of Lyapunov exponents has been calculated by means of the method described in the previous section. The obtained bifurcations diagrams and Lyapunov exponents graphs are presented in the Fig. 3.

4 Observations and conclusions

It can be noticed that the bifurcation diagram presented in the Fig. 3 is consistent with the Lyapunov exponents graph. Moreover, it can be shown that the divergence of the system (13-14) is equal to $2\zeta=-0.1$. As one can notice in the graph, the sum of both Lyapunov exponents from the spectrum is also equal to -0.1 independently of the value of the control parameter η, which confirms validity of calculations.

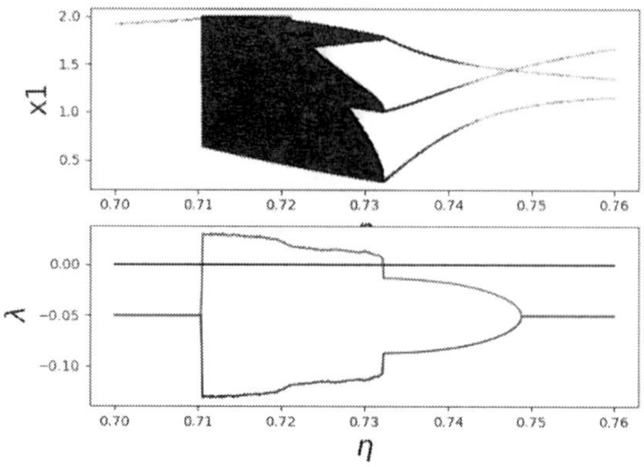

Fig. 3. Bifurcation diagram and Lyapunov exponents graphs of the system defined by the Eq. (21-22)

This paper provides an introduction to the new method of Lyapunov exponents spectrum estimation for non continuous dynamical systems. Foundations of the method have been explained and detailed description of the algorithm has been given. The numerical example proves validity and efficiency of the method. It has been shown that this method of LEs spectrum estimation can be successfully applied to mechanical systems with discontinuities. In the future, authors plan to apply the method to other kinds of non continuous mechanical systems, for example as stick-slip oscillator or a network of such oscillators.

This study has been supported by Polish Ministry of Science and Higher Education under the program "Diamond Grant", project no. D/2013 019743.
This study has been supported by Polish National Centre of Science (NCN) under the program MAESTRO: "Multi-scale modelling of hysteretic and synchronous effects in dry friction process", project no. 2012/06/A/ST8/00356.

References

1. A. M. Lyapunov: *Probleme General de la Stabilité du Mouvment.* Annales Mathematical Study, **17**, Princeton University Press, Princeton, New Jersey (1947)

2. G. D. Birkhoff: *Dynamical Systems,* AMS Colloquium Publications, Providence, (1927)

3. V. I. Oseledec: *A multiplicative ergodic theorem: Lyapunov characteristic numbers for dynamical systems.* Trans. Mosc. Math. Soc. **19**, 197–231 (1968)

4. J. L. Kaplan, J. A. Yorke: *Chaotic behavior of multidimensional difference equations*, in Lecture Notes in Mathematics, vol. **730**, edited by H. O. Peitgen and H. O. Walther, pp. 228 -- 237, Springer, Berlin (1978)

5. V. C. Anishchenko: *Complex oscillations in simple systems*, **312**, Nauka Publisher, Moscow (1990)

6. M. Henon, C. Heiles: *The applicability of the third integral of the motion: some numerical results*, Astron. J. **69**, 77 (1964)

7. G. Benettin, L. Galgani, J.M. Strelcyn: *Kolmogorov entropy and numerical experiment*, Phys. Rev. A **14**, 2338– 2345 (1976)

8. I. Shimada, T. Nagashima: *A numerical approach to ergodic problem of dissipative dynamical systems*, Prog. Theor. Phys. **61(6)**, 1605–1616 (1979)

9. G. Benettin, L. Galgani, A. Giorgilli, J.M. Strelcyn: *Lyapunov exponents for smooth dynamical systems and Hamiltonian systems; a method for computing all of them, part I: theory*, Meccanica **15**, 9–20 (1980)

10. G. Benettin, L. Galgani, A. Giorgilli, J.M. Strelcyn: *Lyapunov exponents for smooth dynamical systems and Hamiltonian systems; a method for computing all of them, part II: numerical application*. Meccanica **15**, 21–30 (1980)

11. A. Wolf: *Quantifying chaos with Lyapunov exponents*, Chaos, Manchester University Press, Manchester, 273-290 (1986)

12. T.S. Parker, L.O. Chua: *Practical Numerical Algorithms for Chaotic Systems*, Springer-Verlag, Berlin (1989)

13. H. Nusse, J. Yorke: *Dynamics: Numerical Explorations*, Springer-Verlag (1994)

14. A. Dąbrowski: *Estimation of the largest Lyapunov exponent from the perturbation vector and its derivative dot product*, Nonlinear Dynamics **67(1)**, 283-291 (2012)

15. M. Balcerzak, A. Dąbrowski, D. Pikunov: *The fastest, simplified method of Lyapunov exponents spectrum estimation for continuous dynamical systems*, Submitted to Nonlinear Dynamics

16. A. Dąbrowski: *The largest transversal Lyapunov exponent and master stability function from the perturbation vector and its derivative dot product (TLEVDP)*, Nonlinear Dynamics **69(3)**, 1225-1235 (2012)

17. M. Balcerzak, A. Dąbrowski, T. Kapitaniak, A. Jach: *Optimization of the control system parameters with use of the new simple method of the largest Lyapunov exponent estimation*, Mechanics and Mechanical Engineering **17(3)**, 225-239 (2013)

18. K. Pijanowski, A. Dąbrowski, M. Balcerzak: *New method of multidimensional control simplification and control system optimization*, Mechanics and Mechanical Engineering **19(2)**, 127-139 (2015)

19. A. Dąbrowski: *Estimation of the the Largest Lyapunov exponent-like (LLEL) stability measure parameter from the perturbation vector and its derivative dot product (part 2) experiment simulation*, Nonlinear Dynamics **78(3)**, 1601-1608 (2014)

20. M. Balcerzak, D. Pikunov: *The fastest, simplified method of estimation of the Largest Lyapunov exponent for continuous dynamical systems with time delay*, Mechanics and Mechanical Engineering (accepted)

21. P. Müller: *Calculation of Lyapunov exponents for dynamical systems with discontinuities*, Chaos, Solitons Fractals **5(9)**, 1671–1681 (1995)

22. A. Stefański, T. Kapitaniak: *Estimation of the dominant Lyapunov exponent of non-smooth systems on the basis of maps synchronization*, Chaos, Solitons & Fractals, **15(2)**, pp. 233–244 (2003)

23. L. Jin, Q.-S. Lu, E.H. Twizell: *A method for calculating the spectrum of Lyapunov exponents by local maps in non-smooth impact-vibrating systems*, Journal of Sound and Vibration, **298(4–5)**, pp. 1019-1033 (2006)

24. S. De Souza, I. Caldas: *Controlling chaotic orbits in mechanical systems with impacts*, Chaos, Solitons & Fractals, **19**, 171-178 (2004)

25. R. C. Hilborn: *Chaos and Nonlinear Dynamics: An Introduction for Scientists and Engineers*, Oxford University Press (1994)

26. M. Oestreich, N. Hinrichs, K. Pop: *Bifurcation and stability analysis for a non-smooth frictional oscillator*, Arch. Appl Mech. **66**, 301–314 (1996)

27. M. Oestreich: *Untersuchung von Schwingern mit nichtglatten Kennlinien*, Fortsrchritt-Berichte VDI, Reiche 11: Schwingungstechnik, Nr. **258** (in German) (1998)

28. U. Galvanetto: *Numerical computation of Lyapunov exponents in discontinuous maps implicitly defined*, Computer Physics Communications, **131(1-2)**, 1-9 (2000)

29. N. Hinrichs, M. Oestreich, K. Popp: *Dynamics of oscillators with impact and friction*, Chaos, Solitons & Fractals, **4(8)**, 535–58 (1997)

30. A. Stefański, T. Kapitaniak: *Using chaos synchronization to estimate the largest Lyapunov exponent of non-smooth systems*, Discrete Dyn. Nat. Soc. **4**, 207–215 (2000)

https://doi.org/10.1051/matecconf/201814810003

31. A. Stefański: *Estimation of the largest Lyapunov exponent in systems with impacts*, Chaos, Solitons Fractals **11(15)**, 2443–2451 (2000)

32. A. Stefański, T. Kapitaniak: *Estimation of the dominant Lyapunov exponent of non-smooth systems on the basis of maps synchronization*, Chaos, Solitons Fractals **15**, 233–244 (2003b)

33. A. Stefański, A. Dąbrowski, T. Kapitaniak: *Evaluation of the largest Lyapunov exponent in dynamical systems with time delay*, Chaos Solitons Fractals **23**, 1651–1659 (2005)

34. F. Takens : *Detecting strange attractors in turbulence,* Lecture Notes in Mathematics, **898(1)**, 366-381 (1981)

35. A. Wolf, J.B. Swift, H.L. Swinney, J.A. Vastano: *Determining Lyapunov exponents from a time series*, Physica D: Nonlinear Phenomena, **16(3)**, 285-317 (1985)

36. J.A. Tempkin, J.A. Yorke : *Spurious Lyapunov exponents computed from data*, SIAM Journal on Applied Dynamical Systems, **6(2)**, 457-474 (2007)

37. A. Stefański, J. Wojewoda, A. Chudzik, T. Kapitaniak: *Hyperbolic-like properties of Popp's attractor*, International Journal of Bifurcation and Chaos, **22(08)**, 1250187 (2012)

38. D.C. Soriano, F.I. Fazanaro, R. Suyama, J.R. de Oliveira, R. Attux, M.K. Madrid: *A method for Lyapunov spectrum estimation using cloned dynamics and its application to the discontinuously-excited FitzHugh–Nagumo model,* Nonlinear Dynamics, **67(1)**, 413-424 (2012)

Cluster synchronization of dry friction oscillators

Michał Marszal[1,*] and *Andrzej* Stefański[1,**]

[1] *Division of Dynamics, Lodz University of Technology, ul. Stefanowskiego 1/15, 90-924 Lodz, Poland*

Abstract. Synchronization is a well known phenomenon in non-linear dynamics and is treated as correlation in time of at least two different processes. In scope of this article, we focus on complete and cluster synchronization in the systems of coupled dry friction oscillators, coupled by linear springs. The building block of the system is the classic stick-slip oscillator, which consists of mass, spring and belt-mass friction interface. The Stribeck friction itself is modelled using Stribeck friction model with exponential non-linearity. The oscillators in the systems are connected in nearest neighbour fashion, both in open and closed ring topology. We perform a numerical study of the properties of the dynamics of the systems in question, in two-parameter space (coupling coefficient vs. angular excitation frequency) and explore the possible configurations of cluster synchronization.

1 Introduction

Synchronization is well known phenomenon in the non-linear dynamics. One of the first scientific paper on that topic dates back to the 17th century when Christian Huygens observed mutual synchronization of two pendulum clocks hanging on the common support [1]. Synchronization can be encountered in the dynamical systems in various fields of science including, mechanics, electric engineering, biology [2, 3]. It is defined as adjustment of rhythms of oscillating objects due to their weak interaction [4].

There are many types of synchronization. In scope of this article we examine complete and cluster synchronization. Complete synchronization means that the trajectories of all oscillators in the system converge. It can happen only in the case of identical oscillators. Cluster synchronization is similar, however here the system is not in complete synchronization but respective oscillators form a subsets of oscillators, which are in sync with each other. More formal definition is presented in Section 2.

Synchronous state stability is a feature of synchronization, which enable us to analyse the synchronization thresholds. Master stability function (MSF) proposed by Pecora and Caroll [5] is a powerful mathematical tool for for that purpose. In this method one can separate the network topology from the local dynamics of the system. The synchronous state stability is estimated by means of Lyapunov exponents of variational equation representing local dynamics of the system, for parameters, which correspond to eigenvalues of the connectivity matrix (see Section 2).

Here, a more direct estimation of the master stability function is presented, which is based on two-oscillator probe [6]. Two-oscillator probe is suitable for system, possessing only real eigenvalues of the connectivity matrix. There is no need to calculate the Lyapunov exponents. MSF is estimated by synchronization error for reference probe of two coupled oscillators (see Section 2). Such a procedure is particularly convenient for non-smooth systems (e.g. stick-slip oscillator), where it is numerically difficult to calculate Lyapunov exponents.

Friction is an ubiquitous force in physics and is the main source of the energy dissipation. The phenomenon of friction was investigated by famous scientists including: da Vinci [7], Euler [8], Coulomb [9]. In many engineering applications the classic Coulomb model is sufficient. Nowadays various mathematical models of friction have been developed, such as: [10–12]. More advanced models include differential equation representing the internal states of the model to fully describe the friction phenomenon.

In this paper, we presents findings of our research on synchronization properties of systems consisting of coupled dry friction oscillators. Such kind of oscillators belong to the class of non-smooth systems, whose dynamics is influenced by the discontinuity plane, corresponding to sticking phase. We also show the application of master stability function based on two-oscillator probe in systems with discontinuity. This article is based on authors' previous works [13, 14] and consists of following parts. Section 1 gives brief introduction to the topic of synchronization and friction. Section 2 discusses the methodology used in the research in more details. In Section 3 the mathematical model of the system in question is explained. Section 4 presents the results of numerical investigation. Finally, the conclusions are drawn in Section 5.

*e-mail: michal.marszal@p.lodz.pl
**e-mail: andrzej.stefanski@p.lodz.pl

2 Methodology

In this Section we present the methodology used in the research. Subsection 2.1 gives basic notion of synchronization, while the 2.2 introduces the concept of master stability function (MSF).

2.1 Synchronization

Let us consider a a following dynamical system:

$$\dot{x} = F(x) + \sigma(G \otimes H)x, \tag{1}$$

where $x = (x_1, \ldots, x_N) \in \Re^N$ is a state vector and $F(x) = (f_1(x_1), ..., f_N(x_N))$. The second term in (1) describes the coupling properties. G is a connectivity matrix, $H : \Re^N \to \Re^N$ linking function, H a linking matrix, σ coupling coefficient, $\otimes$ denotes the Kronecker product of two matrices. For the general case the properties of G and H matrices can be arbitrary.

Let us narrow our investigations to only two types of synchronization, namely complete (global) synchronization and cluster synchronization. We assume that the oscillators in the network are identical and so are the linking functions between them. Complete synchronization can be observed when two trajectories converge to the same value and later hold that condition [15]. In [16] complete synchronization is defined for two dynamics systems represented with their phase plane trajectories $x(t)$ and $y(t)$, respectively, when for all $t > 0$, the following relation is fulfilled:

$$\lim_{t\to\infty} \|x(t) - y(t)\| = 0. \tag{2}$$

Complete synchronization requires the synchronization of oscillators both in phase and in amplitude. However, there may be a situation when the whole system is not in complete synchronization, but one can distinguish a subset of oscillators which are in sync with each other and out of sync with the members of the other subset. Subsets of synchronized oscillators are called clusters. It is important to mention that we can talk about cluster synchronization when whole system is not in complete synchronization. The motion of different clusters may be uncorrelated or one can observe a shift phase between them. The topic of clusters can be found in literature in [17–19].

2.2 Master stability function

Master stability function is a power mathematical tool used in determining the stability of the synchronous state [5]. The classic algorithm to estimate the MSF is to calculate transversal Lyapunov exponents (TLE) of the Eq. (6) derived below. Let us start with deriving a variational equation of Eq. (1):

$$\dot{\xi} = [1_N \otimes DF + \sigma G \otimes DH]\xi, \tag{3}$$

where ξ_i is the variation of the i-th node, $\xi = (\xi_1, \xi_2, ...\xi_N)$ is variation vector, DF is the Jacobian of any node, DH is the Jacobian of the linking function. Diagonalization of Eq. (3) leads to uncoupling the variational Eq.(3) into N block having a form of:

$$\dot{\xi}_k = [DF + \sigma\gamma_k DH]\xi_k, \tag{4}$$

where γ_k is the k-th eigenvalue of the G, $i = 0, 1, 2, ..., N-1$, ξ_k is transverse mode of perturbation from the synchronous state.

For $k = 0$, eigenvalue is $\gamma_0 = 0$, and consequently Eq. (4) can be reduced to

$$\dot{\xi}_0 = Df\xi_0, \tag{5}$$

associated with the longitudinal direction located within the synchronization manifold. The other k-th eigenvalues correspond to transverse eigenvectors [5]. The tendency to synchronization is a function of eigenvalues γ_k. Let us substitute $\sigma\gamma = \alpha + i\beta$ in Eq. (4), where α and β are respective real and imaginary part of eigenvalues.

$$\dot{\xi} = [DF + (\alpha + i\beta) DH]\xi, \tag{6}$$

where ξ is an arbitrary transverse mode.

Invariant synchronization manifold exists when the connectivity matrix G has zero sum rows [5]. All the real parts of eigenvalues, which correspond to transversal modes, are negative ($\mathrm{Re}(\gamma_{k\neq0}) < 0$). The spectrum of eigenvalues has the descending form, i.e., $\gamma_0 \geq \gamma_1 \geq ... \geq \gamma_{N-1}$. In general case, [5] defines MSF as the largest transversal Lyapunov exponent λ_T surface, computed basing on Eq. (6), on a complex numbers plane (α,β). If the interaction between the nodes be mutual (e.g. mechanical systems), then the eigenvalues have only real part and then MSF is represented only by a curve describing the largest TLE as a function of real number α, defined as

$$\alpha = \sigma\gamma. \tag{7}$$

The synchronous state of dynamical system is stable when all eigenmodes of the discrete eigenvalue spectrum $\sigma\gamma_k$ lay in ranges of the largest negative TLE. Even if only one eigenvalue is in the range, where $\lambda_T > 0$, the global synchronization is unstable, however, cluster synchronization is still possible.

In case of non-smooth dynamical systems, the computation of TLE requires is numerically troublesome and requires algorithms. Solutions to that problem are techniques called *three-oscillator probe* and *two-oscillator probe* [6, 20]. Oscillator probe is based on estimating the MSF on the complex plane by direct detection of the complete synchronization by means of numerical integration or by experimental measurement. The methodology is simple and efficient. When calculating MSF in three-oscillator probe for system containing N oscillators, one initially investigates the reference probe of three oscillators. The area on the complex plane (α,β), where the complete synchronization or imperfect complete synchronization occurs, is the equivalent of the area of negative transversal Lyapunov exponents. The two-oscillator probe can be applied for mechanical systems, where due to mutual interaction between the nodes, the eigenvalues of the connectivity matrix G are real and the MSF is reduced to a

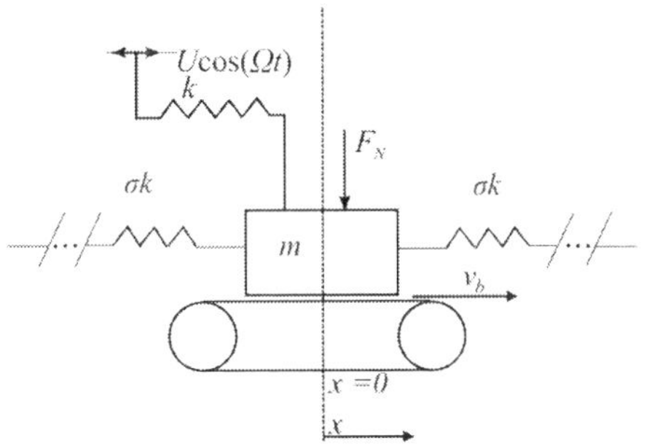

Figure 1. Building block of the system - single stick-slip oscillator.

curve. Should all non-zero eigenvalues of the connectivity matrix of the system of N oscillators be in the surface or range of complete (imperfect complete) synchronization for the reference probe, then the complete synchronization (imperfect complete

3 Mathematical model

Let us consider a system of coupled dry friction, self-induced oscillators with additional kinematic excitation (see Fig. 1).

The equation of motion in a non-dimensional form can be as follows:

$$\begin{Bmatrix} \ddot{\chi}_1 \\ \vdots \\ \ddot{\chi}_N \end{Bmatrix} = -\begin{Bmatrix} \chi_1 \\ \vdots \\ \chi_N \end{Bmatrix} + \sigma \boldsymbol{G}_N \begin{Bmatrix} \chi_1 \\ \vdots \\ \chi_N \end{Bmatrix} + \begin{Bmatrix} \epsilon f(\vartheta_{r_1}) + u\cos\omega\tau \\ \vdots \\ \epsilon f(\vartheta_{r_N}) + u\cos\omega\tau \end{Bmatrix}. \quad (8)$$

where χ is a non-dimensional displacement, σ is a coupling coefficient, corresponding the strength of the coupling, G_N is the connectivity matrix, ϵ normal load coefficient, ϑ_r is a relative tangential velocity between contacting interfaces, u amplitude of excitation, ω angular excitation frequency, τ non-dimensional time.

Stribeck friction model with the exponential nonlinearity is applied as the basis for friction modelling in this work. The other models of friction may also be applied, however, they do not make significant difference from the point of view of the synchronization properties. The Stribeck friction model with the exponential nonlinearity is formulated as follows:

$$f(\vartheta_r) = \left(\mu_k + (\mu_s - \mu_k)\, e^{-a|\vartheta_r|}\right)\mathrm{sgn}\vartheta_r, \quad (9)$$

where μ_s - static friction coefficient, μ_k - kinetic friction coefficient, a is constant defining the shape of the friction - relative velocity curve.

4 Results

In this Section let us show the result of two parameter study in (σ, ω) space, as well an example of MSF with two-oscillator probe.

The numerical simulations are based on author's own program written in C++. Following non-dimensional parameters are used in simulations: $\vartheta = 0.1$, $\mu_s = 0.35$, $\mu_k = 0.2$, $a = 2.5$, $u = 0.1$. If different values are used, information is placed in figure caption or legend respectively. A transient time equal to 1000 excitation periods ($\Delta\tau_t = 1000 \cdot 2\pi/\omega$) is applied, which is followed by measurement of average synchronization error for time interval corresponding to 200 excitation periods. The investigated systems are started from the initial conditions, when the complete synchronization is slightly perturbed. Should the synchronous state be stable, the trajectories return to synchronous state after time $\Delta\tau_t$.

Fig.s 2 and 3 depict different behaviour of the investigated system for particular set of parameters. Black colour marks the complete synchronization areas. Red colour in Fig. 2 denotes the cluster synchronization zone. The cluster synchronization layout is presented on the right hand side of each sub-figure. In Fig. 3, which shows result for the closed ring topology, the variety of cluster synchronization configurations is richer. The cluster layouts are marked with respective colour. Both for open and close ring topology the complete synchronization region is located in the weak coupling zone. For stronger coupling the complete synchronization transforms to cluster synchronization eventually to disappear.

More detailed analysis of global (CS) and cluster synchronization thresholds is performed for certain selection of ω, including verification of the obtained results by means of MSF (see Fig.s 4, 5). The MSF is defined as average synchronization error for two-oscillator probe $\langle e_{II}\rangle(\alpha)$ as a function of real number α (see Equation (7)). Next, MSF $\langle e_{II}\rangle(\alpha)$ is projected via eigenvalues of connectivity matrix $\vec{G}_N$ onto bifurcation diagrams of average synchronization error for networks consisting of N oscillators with σ as a bifurcation parameter. The complete synchronization for network of N oscillators occurs, provided all eigenvalues spectra of connectivity matrix $\vec{G}_N$ lie within zero $\langle e_{II}(\alpha)\rangle$ function. The areas of zero MSF within the eigenvalues spectrum, as well as complete synchronization regions in networks of N oscillators are marked with grey colour respectively.

The method described above is robust for predicting the global CS thresholds and along with the analysis of eigenvectors can be used to explain the cluster synchronizability. However, due to additional coupling factors (i.e., excitation, friction or coexistence of the system attractors) the MSF method indicates only the tendencies of the oscillators to synchronize and might be not always verified in given network configuration.

for the networks consisting from three to six oscillators. For the case of three oscillators $\epsilon = 1.5, N = 3$, $\omega = 1.6$ (Fig. 4) the CS region is in range $\sigma \in [0.12, 0.27]$. It is worth mentioning that the first and last oscillators in the network are in cluster synchronization for almost all investigated range of σ. Similar behaviour can be seen in Fig. 2a, marked as red region, wherein cluster synchronization occupies large area of the investigated parameter space. For the case of four oscillators (see Fig. 5) with $\omega = 1.4$, $\epsilon = 1$ CS region is located for $\sigma \in [0, 0.21]$.

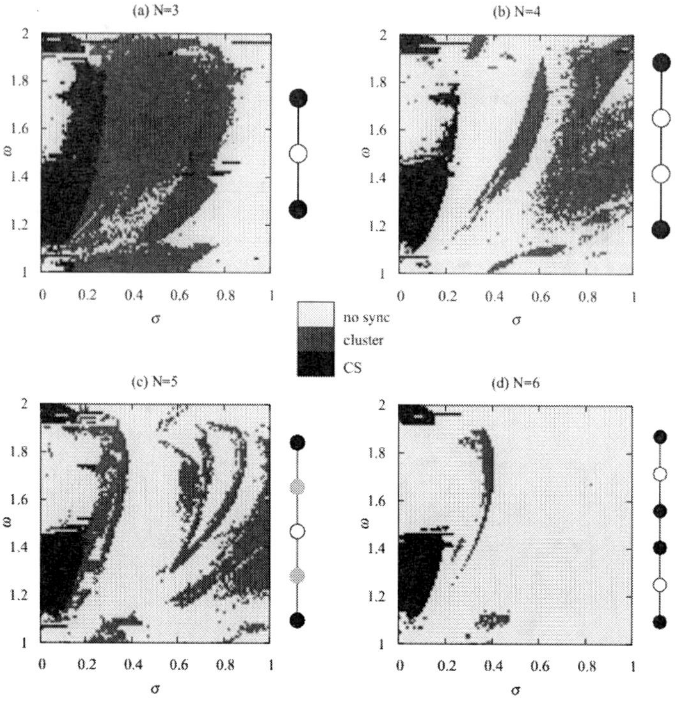

Figure 2. Synchronization regions for open ring of N oscillators in two-parameter space (σ vs. ω), ϵ = 1.5, on sides of scheme of observed cluster layouts: (a) N = 3, (b) N = 4, (c) N = 5, (d) N = 6 (colours online).

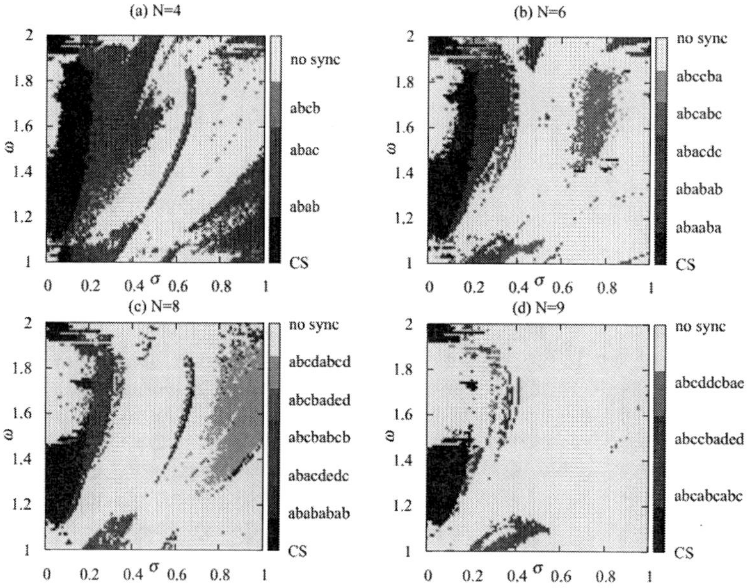

Figure 3. Synchronization regions for open ring of N oscillators in two-parameter space (σ vs. ω), ϵ = 1.5, on sides of scheme of observed cluster layouts: (a) N = 3, (b) N = 4, (c) N = 5, (d) N = 6 (colours online).

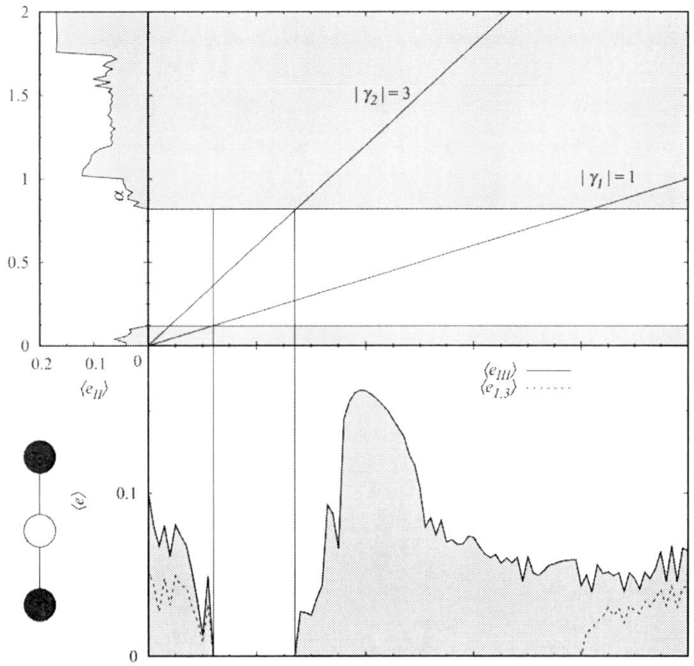

Figure 4. MSF $\langle e_{II}\rangle(\alpha)$ projected onto respective average synchronization errors $\langle e_3\rangle$, $\langle e_{1,3}\rangle$ for N coupled oscillators via eigenvalues of connectivity matrix $\vec{G}_3$. $N = 3$, $\omega = 1.6$, $\epsilon = 1.5$ initial conditions $\vec{\chi} = [0.2913, 0, 0.2945, 0, 0.2922, 0]^T$

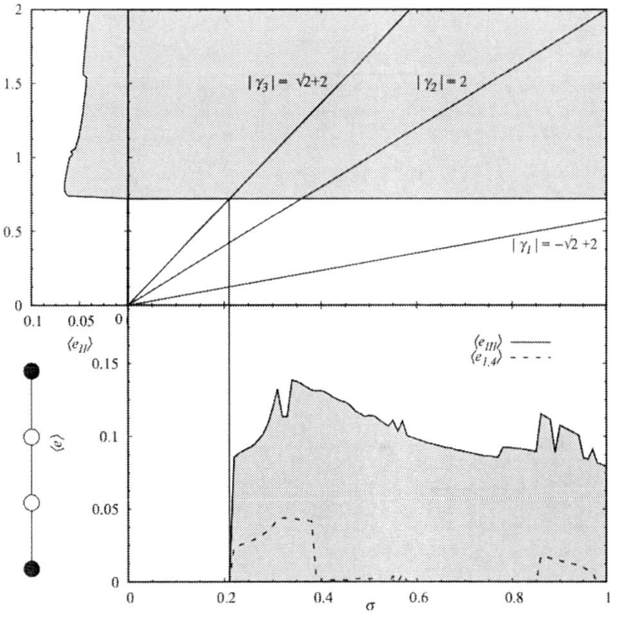

Figure 5. MSF $\langle e_{II}\rangle(\alpha)$ projected onto average synchronization error between 1st and 4th oscillator $\langle e_{1,4}\rangle$ (dashed line) and average global synchronization error $\langle e_{IV}\rangle$ (solid line) for three coupled oscillators via eigenvalues of connectivity matrix $\mathbf{G_4}$ for excitation angular frequency $\omega = 1.4$. Initial conditions for each value of σ: $\chi_0 = [0.2913, 0, 0.2945, 0, 0.2922, 0]^T$.

In the presented systems it is possible to observe the so called ragged synchronizability phenomenon [21]. In Fig.s 4 one can observe global ragged synchronizability, where all oscillators in the network are in synchronous or de-synchronous state.

5 Conclusions

To sum up, we have performed a study of synchronization properties in arrays of coupled dry friction oscillators. Oscillators are connected in nearest neighbour fashion. The goal was to find synchronization thresholds in various networks of oscillators. The used methodology is based on master stability function. However, MSF is not estimated in a traditional way by TLE but by means of more direct approach, namely, two-oscillator probe.

One needs to bear in mind that MSF describes tendencies of the system to synchronize. It cannot be treated as the final condition for the synchronization. The oscillators in question are coupled also by common excitation and the friction force itself, which may contribute also to the synchronizability of the system. Velocity of the conveyor belt, which is equal for all oscillators as well as the same friction model, provide identity of the parameters, which is necessary condition for the occurrence of CS. Moreover, the common harmonic excitation correlates in time with the driving components of all oscillators and as a consequence facilitates the synchronization.

On the contrary, important factor leading to the desynchronization or appearance of cluster is the coexistence of attractors, which is characteristic and often encountered for the systems with friction and impact oscillators. Coexistence of attractors is a property of non-linear systems, which can occur also in smooth, time-continuous dynamical systems. Therefore, the MSF concept and eigenvectors analysis can be treated only as a tool for estimating the overall, global predisposition of the system of coupled oscillators to synchronize or to cluster. This may explain fact, that for some configuration in closed ring topology, some of the cluster layouts cannot be explained by the eigenvectors or eigenvalues interpretation. The results presented in this paper also show ragged synchronization phenomenon (i.e., complete synchronization windows).

In general, the synchronization stability criterion given by the MSF does not provide for proper detection of global network synchronization state even in the case of smooth systems described by continuous differential equations. The more this problem occurs in non-smooth systems where the structure of attractors coexistence and their basins of attraction is usually more complex than in smooth systems. Hence, on the basis of our research, we can conclude that for non-smooth dynamical systems the MSF estimated with two oscillators probe can be even more effective than one calculated with use of the TLE, because then we can be sure that the synchronous region was really detected and it is not only a projection of an interval of the negative TLE.

This work has been supported by National Science Centre - Poland (NCN) in frame of Project no. DEC - 2012/06/A/ST8/00356. This research was supported in part by PL-Grid Infrastructure, as well as, PLATON project's infrastructure at the Lodz University of Technology Computer Centre.

References

[1] C. Huygens, *Horologium oscillatorium* (Muget, Paris, 1673)
[2] B. van der Pol, Lond. Edinb. Dubl. Phil. Mag. **3**, 65 (1927)
[3] J. Buck, E. Buck, Science **159**, 1319 (1968)
[4] A. Pikovsky, M. Rosenblum, J. Kurths, *Synchronization* (Cambridge University Press, Cambridge, 2003)
[5] L.M. Pecora, T.L. Carroll, Phys. Rev. Lett. **80**, 2109 (1998)
[6] K.S. Fink, G. Johnson, T. Carroll, D. Mar, L. Pecora, Phys. Rev. E **61**, 5080 (2000)
[7] L. da Vinci, *Codex Atlanticus* (Biblioteca Ambrosiana, Milan, 1478-1518)
[8] L. Euler, Novi Comment. Acad. Sci. Imp. Petropol. **6**, 223 (1761)
[9] C.A. Coulomb, *Theorie des machines simples* (Bachelier, Paris, 1821)
[10] R. Stribeck, Z. Ver. Dtsch. Ing. **46**, 39 (1902)
[11] C. Canudas de Wit, H. Olsson, K.J. Astrom, P. Lischinsky, IEEE Trans. Autom. Control **40**, 419 (1995)
[12] F. Al-Bender, V. Lampaert, J. Swevers, Tribol. Lett. **16**, 81 (2004)
[13] M. Marszal, A. Saha, K. Jankowski, A. Stefański, Eur. Phys. J. Spec. Top. **225**, 2669 (2016)
[14] M. Marszal, A. Stefański, Phys. Lett. A **381**, 1286 (2017)
[15] L.M. Pecora, T.L. Carroll, Phys. Rev. Lett. **64**, 821 (1990)
[16] A. Stefański, *Determining Thresholds of Complete Sychronization and Application*, Vol. 67 (World Scientific, 2009)
[17] K. Kaneko, Physica D **41**, 137 (1990)
[18] V.N. Belykh, I.V. Belykh, E. Mosekilde, Phys. Rev. E **63**, 036216 (2001)
[19] W. Wu, W. Zhou, T. Chen, IEEE Trans. Circuits Syst. I Regul. Pap. **56**, 829 (2009)
[20] C.W. Wu, IEEE Int. Symp. Circuits Syst. **3**, 261 (2001)
[21] A. Stefański, P. Perlikowski, T. Kapitaniak, Phys. Rev. E **75**, 016210 (2007)

An inerter-based active vibration isolation system

Neven Alujević[1], *Damjan* Čakmak[1,*], *Hinko* Wolf[1], and *Marko* Jokić[1]

[1]Fakultet strojarstva i brodogradnje, Sveučilište u Zagrebu, Ivana Lučića 5, 10 000 Zagreb, Croatia

Abstract. This paper presents a theoretical study on passive and active vibration isolation schemes using inerter elements in a two degree of freedom (DOF) mechanical system. The aim of the work is to discuss basic capabilities and limitations of the vibration control systems at hand using simple and physically transparent models. Broad frequency band dynamic excitation of the source DOF is assumed. The purpose of the isolator system is to prevent vibration transmission to the receiving DOF. The frequency averaged kinetic energy of the receiving mass is used as the metric for vibration isolation quality. It is shown that the use of inerter in the passive and active vibration isolation schemes considered enhances the isolation effect.

1 Introduction

Inerter is a one port element in mechanical networks which resists relative acceleration across its two terminals [1, 2]. The coefficient of this resistance is called inertance and is measured in kilograms. An appealing property of inerters is that they can be designed and realized in practice having their inertance significantly larger than their mass [1, 2]. This opens many interesting possibilities so that many authors reported on how to design and use inerters to suppress mechanical vibrations [1-20].

The concept of "relative mass" has been considered by Schönfeld [21] in connection with mechanical–electrical analogies. He mentioned the possibility of a two-terminal mechanical inertance and gave a rudimentary scheme of a physical realization of the concept. Smith [1], and Smith and Wang [2] developed this idea by investigating how to design such a device in practice and pointed out a number of peculiarities that the new element brings into a mechanical network. The authors instilled that inerter is the analogue of the capacitor element in electrical networks [2]. Therefore, adding the inerter to classical dampers and springs fills an empty niche enabling a complete synthesis of passive mechanical networks [2, 21].

Smith and Wang designed their inerter using a plunger sliding in a cylinder which drives a flywheel through a rack, pinion and gears [2]. Alternatively, an electromagnetic transducer (voice coil, linear motor) can be shunted with electrical impedance. If the total shunt impedance is properly tuned, then the whole electromechanical network theoretically behaves exactly as if it incorporated an ideal inerter mounted in series with a parallel spring damper-pair [3]. In this context, self-powered configurations employing a simultaneous active control and energy harvesting have been considered to synthesize mechatronic inerters [3]. Another type of mechatronic inerter utilizes a rotary DC motor shunted with an appropriate electrical circuit [4]. An inertance-like behaviour can also be accomplished through a scheme in which hydraulic fluid is accelerated [5, 6] with a piston which pushes the fluid through a helical channel [6].

Inerters can be very useful in vibration absorber systems. Performance of vibration absorbers, especially Tuned Mass Dampers (TMDs) is known to very much depend on the proof mass added to a primary structure to reduce its vibration. As this mass is added to structures exclusively to control their vibrations, it is penalized in lightweight automotive and aerospace applications [13, 14]. In this context the use of inerter elements can be interesting given the fact that their inertance can be significantly larger than their mass. Consequently a number of new concepts have arisen. These include tuned inerter damper (TID), tuned mass–damper–inerter (TMDI), and inerter–based dynamic vibration absorber (IDVA) [15-19]. Various applications have been considered using tuned inerter dampers including vibration reduction of cables in cable-stayed bridges [15, 16].

Dynamic vibration absorbers can be made active by using inertial actuators with a velocity or velocity + displacement feedback control scheme. Inertial actuators are typically designed with a low mounted natural frequency in order to widen the range of frequencies where they can efficiently actuate. However, the low natural frequency is usually associated with increased static sags. This limits the applicability of inertial actuators in presence of large accelerations (*i.e.* vehicle manoeuvring or centrifugal accelerations in rotating structures [22-24]).

Zilletti investigated an active vibration absorber system in which the inerter is attached in parallel with the suspension spring, damper, and the actuator [20]. The author has shown that with such a design it is possible to reduce the natural frequency of the actuator without increasing the proof mass or reducing the

* Corresponding author: damjan.cakmak@fsb.hr

suspension stiffness. He considered only an idealized inerter element, which neglects the inertial, stiffness and damping of the gearing mechanism. However, Kras and Gardonio studied the effective weight and dynamic effects of an inerter element composed by a single flywheel which is either pinned or hinged to the base mass or to the proof mass of the actuator [25].

Inerters can also be very useful in vibration isolation systems. In this sense, many authors focused their efforts on improving vehicle suspension systems using inerters [2], [7-10]. Further applications of inerters include vibration isolation in civil engineering structures, such as multi-storey buildings under earthquake base excitation [11]. In vibration isolation problems it is often necessary to tune the impedance of the isolator elements based on some optimization criteria. This can be done by either minimizing maxima of the response (minimax or H_∞ optimization), or by minimizing the energy in the response signals (H_2 optimization) [12].

In this paper an active vibration isolation problem is considered. It is shown that the use of inerter can significantly improve the stability and performance of the active vibration isolation system in certain situations. In particular, it is shown analytically on a simplified model problem that the use of inerter enables successful active vibration isolation in a family of mechanical systems that are otherwise difficult to control. This family of systems has been referred to as subcritical 2 DOF systems [26]. Subcritical systems are those characterized by the natural frequency of the receiving body larger than that of the source body. In such vibration isolation problems the use of inerter is shown to stabilize the feedback loop and therefore to enable a remarkable active vibration isolation effect. In addition to the active vibration isolation system, an inerter-based and an inerter-free passive isolator scheme are proposed and analysed, with the aim of establishing fair benchmarks for the evaluation of the performance of the active isolators studied later in the paper.

In each system, either active of passive, tuneable parameters are adjusted in order to minimize the kinetic energy of the receiving body per unit, spectrally white, dynamic excitation of the source body.

2 Mathematical model

As shown in Fig. 1., the problem studied is represented by a lumped parameter two degree of freedom (DOF) mechanical system. The system consists of two masses m_1 and m_2 coupled by a spring k_2, a viscous damper c_2 and an inerter of inertance b_2. The inerter produces a force proportional to the relative acceleration between masses m_1 and m_2. The two masses are attached to fixed reference bases via the two mounting springs k_1 and k_3. The lower mass m_1 is excited by the disturbance force F_1. It is assumed that the force F_1 has characteristics of an ideal white noise and that the power spectral density (PSD) of the force equals one over all frequencies. The effects of dampers between the source mass m_1 and the ground and between the receiving mass m_2 and the ground are neglected *i.e.* $c_1 \approx c_3 \approx 0$.

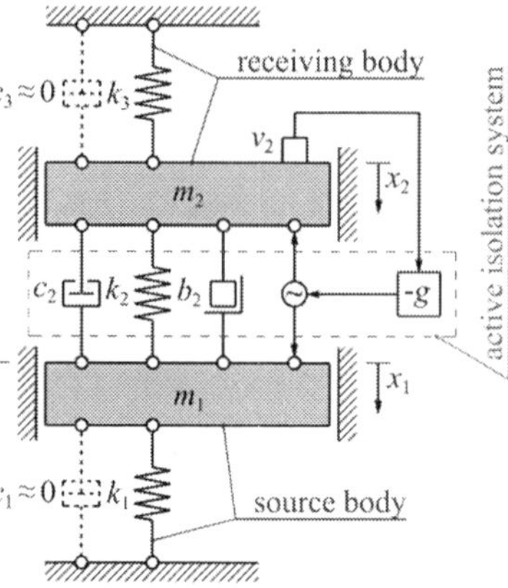

Fig. 1. The two degree of freedom lumped mass active vibration isolation system.

More complex systems may also be representable by the general configuration shown in Fig. 1. [27-30].

The active part of the vibration isolation system is realized through a skyhook damping unit [30, 31]. The velocity sensor is mounted onto mass m_2 in order to realize a disturbance rejection control scheme. In this scheme the actuator is driven with a signal proportional to the negative absolute velocity of the receiving body amplified by a constant control gain g. The equations of motion can be written in the matrix form as [32]

$$\mathbf{M}\ddot{\mathbf{x}} + \mathbf{C}\dot{\mathbf{x}} + \mathbf{K}\mathbf{x} = \mathbf{F}, \quad (1)$$

where $\mathbf{M}$ is the mass matrix, $\mathbf{K}$ is the stiffness matrix, $\mathbf{C}$ is the damping matrix, $\mathbf{x}(t)$, $\dot{\mathbf{x}}(t)$ and $\ddot{\mathbf{x}}(t)$ are the displacement, velocity and acceleration column vectors respectively, and $\mathbf{F}(t)$ is excitation column vector. Assuming a simple harmonic excitation and expressing the excitation and the steady-state response in the exponential form $\mathbf{F}(t) = \hat{\mathbf{F}}e^{j\omega t}$ and $\mathbf{x} = \hat{\mathbf{x}}e^{j\omega t}$, Eq. (1) can be differentiated with respect to time t and written as

$$\dot{\mathbf{x}}(j\omega) = \mathbf{Y}(j\omega)\mathbf{F}(j\omega), \quad (2)$$

where $\dot{\mathbf{x}}(j\omega) = j\omega\,\mathbf{x}(j\omega)$ is the velocity vector, $\mathbf{S}(j\omega)$ is the dynamic stiffness matrix and $\mathbf{Y}(j\omega) = j\omega\,\mathbf{S}^{-1}(j\omega)$ is the mobility matrix containing four frequency response functions (FRFs) between velocities and forces. With the aim of more general approach, the element, Y_{21}, of the mobility matrix in Eq. (2) can be expressed in the following dimensionless form

$$\Upsilon_{21}(j\Omega) = \frac{B_0 + (j\Omega)B_1 + (j\Omega)^2 B_2 + (j\Omega)^3 B_3}{A_0 + (j\Omega)A_1 + (j\Omega)^2 A_2 + (j\Omega)^3 A_3 + (j\Omega)^4 A_4}, \quad (3)$$

where coefficients $A_0 \ldots A_4$ and $B_0 \ldots B_3$ are given by

$$\begin{aligned} A_0 &= \mu_1(\alpha\beta\mu_1 + \alpha + \beta) & B_0 &= 0 \\ A_1 &= 2\eta_2(\beta\mu_1 + \lambda + 1) & B_1 &= \alpha\mu_1, \\ A_2 &= \mu_1(\alpha\mu_1 + \beta\mu_1\mu_2 + \alpha + \beta + \mu_2 + 1) & B_2 &= 2\eta_2 \\ A_3 &= 2\eta_2(\mu_1 + \lambda + 1) & B_3 &= \mu_1\mu_2 \\ A_4 &= \mu_1(\mu_1\mu_2 + \mu_2 + 1) \end{aligned} \quad (4a\text{-}i)$$

where

$$\alpha = \left(\frac{\Omega_2}{\Omega_1}\right)^2, \beta = \left(\frac{\Omega_3}{\Omega_1}\right)^2, \eta_2 = \frac{c_2}{2\sqrt{m_1 k_1}}, \lambda = \frac{g}{c_2}, \mu_1 = \frac{m_2}{m_1}, \mu_2 = \frac{b_2}{m_2}, \Omega = \frac{\omega}{\Omega_1}, \quad (5a\text{-}g)$$

and $\Upsilon_{21} = m_1\Omega_1 Y_{21}$ is now the dimensionless transfer mobility. In Eqs. (5a-g), α and β are squared natural frequency ratios, η_2 is the damping ratio, λ is the feedback gain normalized with respect to the passive damping coefficient, and μ_1 and μ_2 are the mass and inertance ratios respectively. Furthermore, Ω is dimensionless circular frequency normalized with respect to the natural frequency of the uncoupled source body Ω_1 (as if the source body was uncoupled by removing spring k_2), Ω_3 is the natural frequency of the uncoupled receiving body (as if the receiving body was uncoupled by removing spring k_2), and Ω_2 is the natural frequency of the receiving body as if it was attached to a fixed reference base through the spring of stiffness k_2 only. The three natural frequencies $\Omega_1 \ldots \Omega_3$ are thus

$$\Omega_1 = \sqrt{\frac{k_1}{m_1}},\quad \Omega_2 = \sqrt{\frac{k_2}{m_2}},\quad \Omega_3 = \sqrt{\frac{k_3}{m_2}}. \tag{6}$$

Given that the excitation force F_1 with unit power spectral density (PSD) has been assumed, the specific kinetic energy of the receiving body (per unit mass, per unit excitation force) can now be calculated as

$$I_k = \int_{-\infty}^{\infty} \left|Y_{21}(\mathrm{j}\Omega)\right|^2 \mathrm{d}\Omega, \tag{7}$$

according to the Parseval's identity. The specific kinetic energy index I_k is used throughout this study as the measure of the performance of broad frequency band vibration isolation. Formulae for integrating ratios of polynomial functions of a complex variable can be found in [33]. Throughout the paper it is investigated how can the kinetic energy of the receiving body be minimised by minimising the specific kinetic energy index in Eq. (7).

3 Passive control

In this section, two benchmark passive vibration isolation schemes are considered. In the first scheme, an inerter is not employed and in the second, an inerter of dimensionless inertance μ_2 is considered.

3.1 Without inerter

The modulus of the transfer mobility $|Y_{21}|$ is shown in Fig. 2. for three values of the passive damping ratio η_2: a relatively small one (solid line), a medium (dashed line) and a large damping ratio (dash-dotted line).

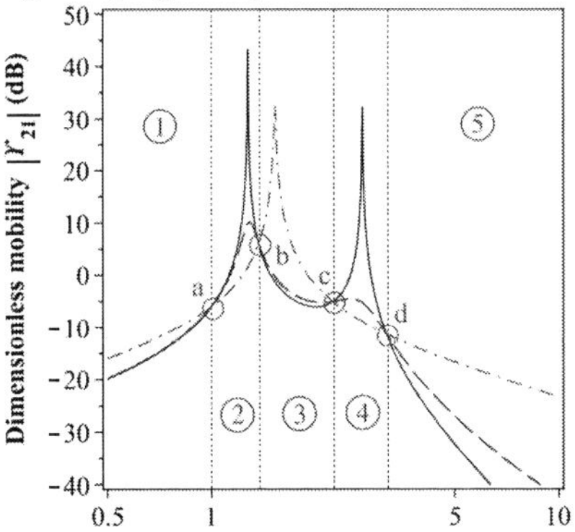

Fig. 2. Isolation system performance without inerter b_2 ($\mu_2 = 0$): Transfer mobility function $|Y_{21}(\mathrm{j}\Omega)|$, $\eta_2 = \eta_{2\mathrm{opt1}}/100$ (solid line), $\eta_2 = \eta_{2\mathrm{opt1}}$ (dashed line), $\eta_2 = 100\eta_{2\mathrm{opt1}}$ (dash-dotted line).

Parameters that characterize this example system from Fig. 1. are $\alpha = 2$, $\beta = 5$ and $\mu_1 = 1/2$. The variation of the passive damping ratio shows that it is possible to optimise the system so that the receiving body vibration measured through the metrics defined by Eq. (7) is minimised. This is because: i) for a small damping ratio the response is very large at resonances (solid line); ii) for a very large damping ratio the damper locks the source and receiving bodies and a new lightly damped resonance is generated (dash-dotted line). It is therefore best to use: iii) the optimal damping ratio which minimises the kinetic energy of the receiving body. The optimal damping ratio corresponds to the dashed line in Fig. 2. Also shown in the figure are five frequency ranges, 1-5, in which the variation of the damping ratio causes either an increase or a reduction of the amplitude of the transfer mobility function. The corresponding bordering frequencies are designated by letters a-d. In particular, around resonances the increase in damping decreases the vibration transmission, but elsewhere it increases the vibration transmission. Thus the optimal damping ratio exists.

3.2 With inerter

Fig. 3. shows the amplitude of the dimensionless transfer mobility $|Y_{21}|$ for the same parameters as in Fig. 2., except that an inerter is now employed. A minimisation of the specific kinetic energy index has been carried out with respect to the inertance and the damping. The minimisation indicates that there is an optimal inertance-damping pair which minimises the receiving body kinetic energy. The results in Fig. 3 are shown for the optimised inertance, however the damping ratio has been varied from suboptimal towards the optimal one and above it. Anti-resonance effect at the dimensionless frequency of $\Omega_A = \sqrt{\alpha/\mu_2}$ (corresponding to the dimensional frequency $\omega_A = \sqrt{k_2/b_2}$) is observed (solid line) which is a result of using the inerter.

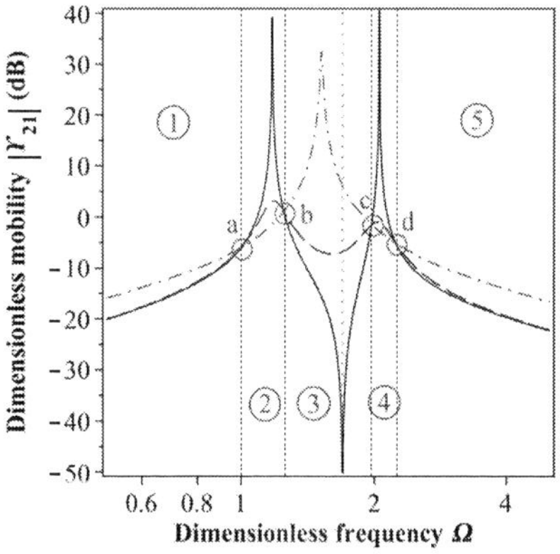

Fig. 3. Isolation system performance with inerter b_2 ($\mu_2 = \mu_{2\mathrm{opt}}$): Transfer mobility function $|Y_{21}(\mathrm{j}\Omega)|$, $\eta_2 = 0$ (solid line), $\eta_2 = \eta_{2\mathrm{opt2}}$ (dashed line), $\eta_2 = 100\eta_{2\mathrm{opt2}}$ (dash-dotted line).

The use of inerter improves the broadband passive vibration isolation performance by ~ 3 dB for the case considered where $\eta_2 = \eta_{2\mathrm{opt2}}$ and $\mu_2 = \mu_{2\mathrm{opt}}$ in comparison to the system studied in the previous subsection. This improvement certainly depends on the parameters that characterise the system without inerter and the particular vibration isolation problem at hand.

4 Active control

4.1 Stability analysis

In this section, two active vibration isolation schemes are analysed with respect to the stability of the feedback control. In the first scheme the inerter is not used, and in the second it is. The stability study is carried out using the Routh-Hurwitz criterion [34, 35].

If the inerter is not used then the analysis of the stability indicates that there are two basic families of systems. The first family can be referred to as *supercritical* and it is characterized by $\beta < 1$ (the natural frequency of the source body is higher than that of the receiving body). The systems belonging to this group allow for the implementation of unconditionally stable active vibration isolation scheme based on the direct feedback of the absolute velocity of the receiving body even if the inerter is not used. The second family is characterized by $\beta > 1$ and it can be referred to as *subcritical*. The systems belonging to this group do not allow for the implementation of unconditionally stable absolute velocity feedback scheme, as the feedback loop is only conditionally stable with a limited maximum feedback gain g, *i.e.* λ.

Fig. 4. shows the Hurwitz principal diagonal minors plotted as a function of the dimensionless feedback gain λ when no inerter is used. The system is subcritical with characteristic parameters $\alpha = 2$, $\mu_1 = 1/2$, $\eta_2 = 1$ and $\beta = 5$.

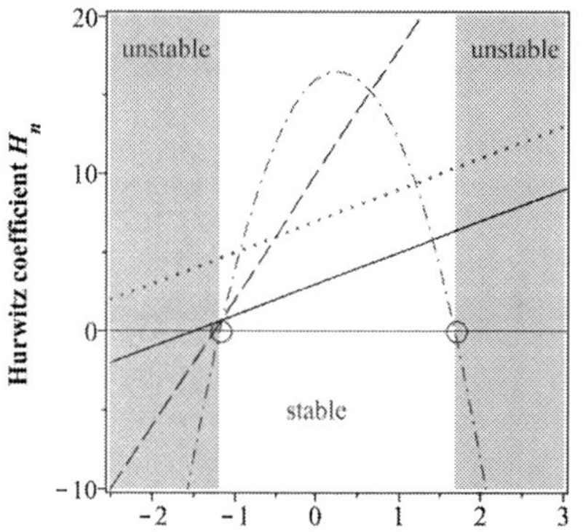

Fig. 4. Dependence of Hurwitz coefficients H_1 (solid line), H_2 (dashed line), H_3 (dash-dotted line) and A_1 (dotted line) magnitude on active damping ratio λ without inerter b_2 ($\mu_2 = 0$): $\beta_{\text{II}} > 1$.

A limited stability range as a function of λ is observed. By a detailed analysis of the sign of the Hurwitz principal diagonal minors it can be shown that by including the inerter with inertance $\mu_2 > \alpha$, a subcritical system becomes unconditionally stable for any chosen positive λ. This is illustrated in Fig. 5. on the same passive system but this time employing an inerter with dimensionless inertance of $\mu_2 = 4$.

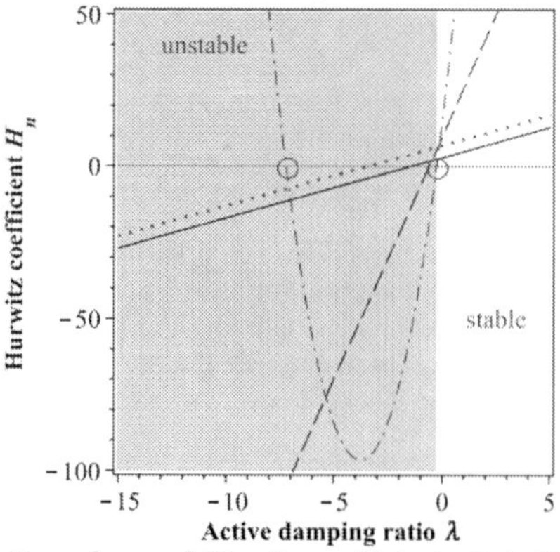

Fig. 5. Dependence of Hurwitz coefficients H_1 (solid line), H_2 (dashed line), H_3 (dash-dotted line) and A_1 (dotted line) magnitude on active damping ratio λ with inerter b_2 ($\mu_2 \neq 0$) and $\beta_{\text{II}} > 1$: $\mu_2 > \alpha$.

4.2 Performance without inerter

Fig. 6. shows the amplitude of the dimensionless transfer mobility $|Y_{21}|$ of a subcritical system characterised by the parameters $\alpha = 1/2$, $\beta = 5$ and $\mu_1 = 1/2$ and $\eta_2 = 0.02$. The dimensionless feedback gain λ is varied from zero to 1.5.

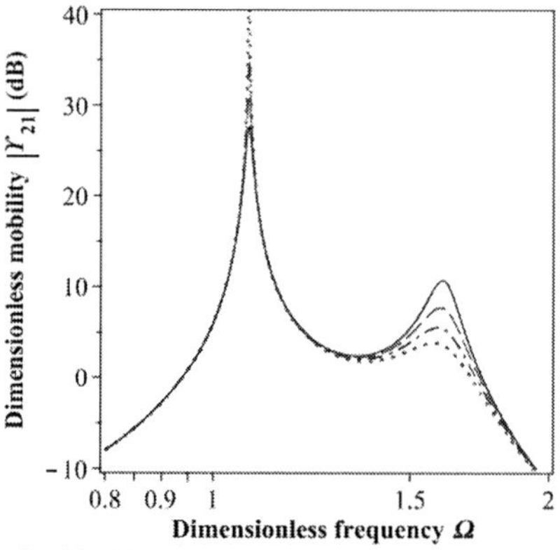

Fig. 6. Subcritical isolation system performance without inerter b_2 ($\mu_2 = 0$) and $\beta_{\text{II}} > 1$: Transfer mobility function $|Y_{21}(j\Omega)|$, $\lambda = 0$ (solid line), $\lambda = 0.5$ (dashed line), $\lambda = 1$ (dash-dotted line), $\lambda = 1.5$ (dotted line).

A significant overshoot can be observed in the vicinity of the first dimensionless natural frequency Ω_{n1} for rising λ, which also results with system instability if the feedback gain is further increased. Thus the performance of the active control scheme considered is jeopardised by the limited stability of the feedback loop.

4.3 Performance with inerter

Fig. 7. shows the subcritical system transfer mobility amplitude $|Y_{21}|$ with $\alpha = 1/2$, $\beta = 2$, $\mu_1 = 1/2$ and $\eta_2 = 0.02$, equipped with an inerter of inertance $\mu_2 = 2$. Therefore, an inertance large enough to stabilize the feedback loop is used ($\mu_2 > \alpha$).

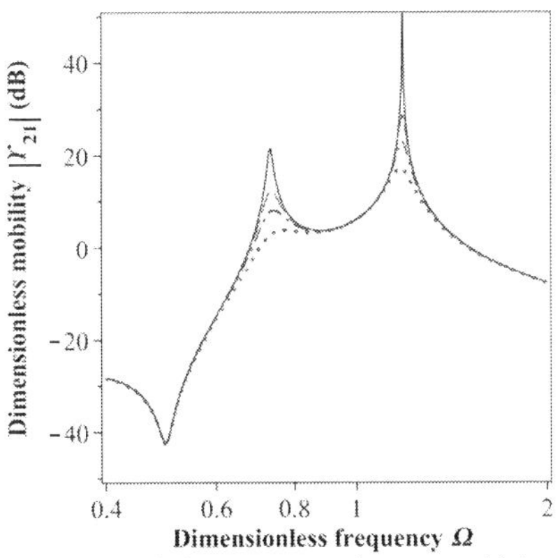

Fig. 7. Subcritical isolation system performance with inerter b_2 ($\mu_2 \neq 0$), $\beta_{II} > 1$ and $\mu_2 > \alpha$: Transfer mobility function $|\Gamma_{21}(j\Omega)|$, $\lambda = 0$ (solid line), $\lambda = 5$ (dashed line), $\lambda = 10$ (dash-dotted line), $\lambda = 20$ (dotted line).

It can be seen that with an increase in the dimensionless gain λ, the amplitude of the dimensionless mobility $|\Gamma_{21}|$ decreases at all frequencies, which indicates that the desired vibration isolation effect is now accomplished. Anti-resonance is observed below the first natural frequency which is due to the use of the inerter in the isolator. The amplitude at this frequency does not increase with the increase in the dimensionless gain.

It should be noted that adding the inerter into the isolator effectively generates a sort of relative acceleration feedback that stabilizes the control loop. It is important to mention that direct acceleration feedback would probably not be possible in practice due to very pronounced stability problems [36]; therefore the passive element which mimics such feedback is very useful.

5 Conclusions

In this paper, an inerter-based active vibration isolation system is presented. Two fundamental passive benchmark isolators are also investigated, one not employing the inerter and the other employing the inerter. The methodology is studied on a simple two degree of freedom system so that many conclusions can be drawn based on analytically derived expressions. Such a simplified system can be seen as a reduced order model of a potentially more complex structure. It is shown in the paper that the vibration isolation performance of the fundamental passive isolator not employing the inerter can be improved by adding the inerter in parallel with the isolator spring and damper. By investigating the stability of the active control when no inerter is used, it is found that there are two fundamental families of vibration isolation problems. With the first family (supercritical systems), which is characterized by the natural frequency of the uncoupled source body larger than the natural frequency of the uncoupled receiving body, large feedback gains can be used without compromising the stability of the feedback control system. This results in a convincing broadband vibration isolation effect. With the second family of systems (subcritical systems), the natural frequency of the uncoupled source body is below the natural frequency of the uncoupled receiving body. The range of stable feedback gains is limited which results in poor vibration isolation performance. However with the inclusion of the inerter, broadband active vibration isolation can also be achieved in the subcritical family of systems. In fact, it is shown that the minimum inertance to stabilize the loop is proportional to the stiffness of the isolator spring and inversely proportional to the squared natural frequency of the source body.

This project has received funding from the European Union's Horizon 2020 research and innovation programme under the Marie Sklodowska-Curie grant agreement no. 657539 STARMAS.

References

1. M.C. Smith, IEEE T. Autom. Control **47**, 1648-1662 (2002)
2. M.C. Smith, F.C. Wang, Veh. Syst. Dyn. **42**, 235–257 (2004)
3. A. Gonzalez-Buelga, L.R. Clare, S.A. Neild, J.Z. Jiang, D.J. Inman, Smart Mater. Struct. **24**, 1-10 (2015)
4. D. Li, Y.-Z. Liu, App. Mech. Mater. **620**, 28-32 (2014)
5. F.-C. Wang, M.-F. Hong, T.-C. Lin, *Proceedings of the Institution of Mechanical Engineers, Part C: J. Mech, Eng. Sci.* **225**, 66-72 (2011)
6. S.J. Swift, M.C. Smith, A.R. Glover, C. Papageorgiou, B. Gartner, N.E. Houghton, Int. J. Control, Systems, Modelling and Feedback Control: A Special Issue Dedicated to Sir Alistair G.J. MacFarlane **86**, 2035-2051 (2013)
7. M.Z.Q. Chen, Y. Hu, C. Li, G. Chen, IEEE T. Contr. Syst. T. **23**, 1571-1577 (2015)
8. M.Z.Q. Chen, K. Wang, Y. Zou, J. Lam, IEEE T. Autom. Control **58**, 1841-1846 (2013)
9. M.Z.Q. Chen, Y. Hu, C. Li, G. Chen, J. Vib. Acoust. **138**, 1-11 (2016)
10. A. Kuznetsov, M. Mammadov, I. Sultan, E. Hajilarov, Arch. Appl. Mech. **81**, 1427–1437 (2011)
11. I.F. Lazar, S.A. Neild, D.J. Wagg, Earthq. Eng. Struct. D. **43**, 1129-1147 (2014)
12. Y. Hu, M.Z.Q. Chen, Z. Shu, L. Huang, J. Sound Vib. **346**, 17-36 (2015)
13. T. Bein, D. Mayer, J. Bös, *INTER-NOISE 2012* **5**, 3759-3771 (2012)
14. T. Bein, J. Bös, S. Herold, D. Mayer, T. Melz, M. Thomaier, Aerosp. Sci. Technol. **12**, 62-73 (2008)
15. I.F. Lazar, S.A. Neild, D.J. Wagg, *Conference Proceedings of the Society for Experimental Mechanics Ser.* **2**, 433-441 (2015)
16. I.F. Lazar, S.A. Neild, D.J. Wagg, Eng. Struct. **122**, 62-71 (2016)

17. P. Brzeski, T. Kapitaniak, P. Perlikowski, J. Sound Vib. **349**, 56-66 (2015)
18. P. Brzeski, E. Pavlovskaia, T. Kapitaniak, P. Perlikowski, Int. J. Non-Lin. Mech. **70**, 20-29 (2015)
19. Y. Hu, M.Z.Q. Chen, Int. J. Mech. Sci. **99**, 297–307 (2015)
20. M. Zilletti, J. Sound Vib. **369**, 16-28 (2016)
21. J.C. Schönfeld, Appl. Sci. Res. **3**, 417–450 (1954)
22. G. Zhao, N. Alujević, B. Depraetere, G. Pinte, J. Swevers, P. Sas, J. Sound Vib. **348**, 15-30 (2015)
23. G. Zhao, N. Alujević, B. Depraetere, G. Pinte, P. Sas, J. Intel. Mat. Syst. Str. **27**, 1584-1599 (2016)
24. G. Zhao, N. Alujević, B. Depraetere, P. Sas, J. Intel. Mat. Syst. Str. **26**, 1995-2010 (2015)
25. A. Kras, P. Gardonio, J. Sound Vib. **402**, 31-50 (2017)
26. N. Alujević, H. Wolf, P. Gardonio, I. Tomac, J. Sound Vib. **330**(21), 4981–4997 (2011)
27. R.M. Chalasani, *ASME Symposium on Simulation and Control of Ground Vehicles and Transportation Systems, AMD-vol.* **80**, *DSC-vol.* **2**, 187–204 (1986)
28. A. Caiazzo, N. Alujević, B. Pluymers, W. Desmet, *J. Phys. Conference series* **744**, 012026 (2016).
29. A. Preumont, *Vibration Control of Active Structures, 2nd Edition* (Kluwer Academic, London, 2002, Chapter 6, "Active vibration isolation")
30. C.E. Kaplow, J.R. Velman, AIAA J. Guid. Control Dynam. **3**(3), 227–233 (1980)
31. D.C. Karnopp, A.K. Trikha, Trans. ASME, J. Eng. Ind., Ser. B **91**, 1128–1132 (1969)
32. W.T. Thomson, *Theory of vibration with applications* (Prentice Hall, London, 1997)
33. D.E. Newland, *An introduction to Random Vibrations, Spectral & Wavelet Analysis, Third Edition* (Dover Publications Inc, New York, 2005)
34. E.J. Routh, *A Treatise on the Stability of a Given State of Motion: Particularly Steady Motion* (Macmillan and Co., London, 1877)
35. A. Hurwitz, Math. Ann. **46**, 273–284 (1895)
36. N. Alujević, I. Tomac, P. Gardonio, J. Sound Vib. **331**, 2731–2728 (2012)

Real-Time Order Tracking of Gear Mesh Vibration in High Speed Planetary Gearboxes

Daniel Fritz Plöger[1,*], *Philipp* Zech[1], and *Stephan* Rinderknecht[1]

[1] *Mechatronic Systems in Mechanical Engineering, Technische Universität Darmstadt*

Abstract. Possible approaches to real-time order tracking are discussed. Two methods for real-time order tracking are developed and validated experimentally for the entire audible spectrum. An adaptive heterodyne filter bank is compared to a direct integral transform. The performance of both methods is adequate for usage in an active vibration control (AVC) algorithm. Vold-Kalman filters are not suitable for AVC. The vibration data of three different planetary gearboxes is analyzed using order tracking. While some of the existing research could be reproduced, the data contradicts statements made by several authors. Lastly, the architecture of a novel AVC algorithm is sketched out.

1 Introduction

The vibration of planetary gearboxes is highly modulated. An explanation for this has been reported first by McFadden and Smith [1]. For an accurate analysis of the vibration a technique called *order tracking* has to be employed. Order tracking decomposes the vibration signal into its individual *spectral components*, often referred to as orders. It is mostly performed off-line using recorded vibration data. Borghesani et al. [2] explain how this can be performed using state of the art methods. They categorize the known approaches into three classes: "the resampling methods, the Kalman filter based methods and the transform based methods". For off-line analysis, resampling remains a popular choice because of its accuracy and robustness.

The goal of this paper is to extend order tracking to real-time and to fast running planetary gearboxes. The spectral components of the vibration signal are extracted at the same time they occur. While real-time order tracking can still be used for off-line analysis, it may also be used for on-line condition monitoring or active vibration control (AVC). Here, the focus will be on AVC.

The class of resampling methods will not be investigated in this paper, because these inherently rely on recorded data. Instead, a transform based concept will be developed, leading to two different methods. The Vold-Kalman filter (VKF) will also be discussed briefly. The obtained methods will be implemented on a test rig and a brief analysis of the experimentally acquired gearbox vibration will be given. Lastly, an AVC algorithm using real-time order tracking will be sketched out. However, an implementation of AVC is not within the scope of this paper.

*e-mail: ploeger@ims.tu-darmstadt.de

2 Experimental Set-Up

The experimental set-up is depicted in Figure 1 and consists of an asynchronous motor (1) that drives the sun gear of the planetary gearbox (3) and an eddy current brake (5) that is connected to the planet carrier. Metal bellow couplings (2) and (4) connect the shafts of motor, eddy current brake and gearbox. Using this setup the gearbox can be operated at input speeds up to $10\,000\,\mathrm{min}^{-1}$ and output torques up to $40\,\mathrm{N\,m}$. The gearbox mount is supported by two triaxial force sensors (6). Figure 1 shows the coordinate systems for both sides of the mount. Furthermore the gearbox mount is equipped with eight accelerometers (7). The experimental setup allows measuring the gear mesh forces that are transmitted through the gearbox mount (8) into the surrounding structure. This work focuses on the data of the accelerometer on the right side of the gearbox mount in x_1 direction to investigate the online order-tracking approaches. The experimental setup enables measurement of run-ups as well as stationary operation. A speedgoat performance real-time target machine featuring an Intel Core i7 3770K 3.5 GHz quadcore CPU allows for a frequency of 80 kHz in closed loop operation. The piezoelectric accelerometers are of type MMF KS78B.10 and their signal is fed into a MMF M208B measurement amplifier featuring a low-pass filter with cut-off frequency of 30 kHz. Three commercially available planetary gearboxes are selected for the investigation as presented in Table 1.

3 Modeling Gear Mesh Vibration

It is known from existing research [3] that the vibration of planetary gearboxes may be modeled accurately as a Fourier series. Using this approach, a vibration model has been presented by the authors [4]. This paper will extend its application to real time. On behalf of briefness, only a

Table 1. Selected planetary gearboxes.

Gearbox	A	B	C
Gear	spur	spur	helical
Z	84	84	108
Ratio	3	3	3
Outer diameter	60 mm	40 mm	68 mm
Mass	0.9 kg	0.35 kg	1.9 kg

Figure 1. Set-up: Motor shaft (1), metal bellow couplings (2, 4), planetary gearbox (3) within gearbox mount (8), eddy current brake (5), triaxial force sensors (6) and accelerometers (7).

rough outline on the vibration model will be discussed. For a detailed derivation refer to [4]. In this paper planetary gearboxes with a fixed ring gear will be discussed.

The vibration of a planetary gearbox can be expressed as a modulated signal. This has been first recognized by McFadden and Smith [1]. The fundamental frequency corresponds to the nominal gear mesh frequency ω_0, which is defined by the angular velocity of the planet carrier $\dot{\varphi}$ and the number of teeth on the ring gear Z.

$$\omega_0 = Z\dot{\varphi} \tag{1}$$

If the angular velocity is constant, the modulation will be periodic w. r. t. time. In transient operation the modulation A can be modeled periodic w. r. t. the carrier angle φ, if the angular acceleration is not too large. In this model, a complex modulation A will be used. Two counter rotating phasors are introduced in order to constrain the signal υ to real values.

$$\upsilon = \frac{1}{2}A(\varphi)e^{i\omega_0 t} + \frac{1}{2}\overline{A}(\varphi)e^{-i\omega_0 t} \tag{2}$$

The signal υ may represent any observable vibration, such as a force, acceleration, displacement or similar. Because the modulation A is periodic w. r. t. the angle φ of the planet carrier it may be expressed as a Fourier series. In anticipation of the next step the index of the series has already been shifted by the number of teeth Z.

$$A(\varphi) = \sum_{k=-\infty}^{\infty} c_{k-Z} e^{ik\varphi} \tag{3}$$

The shift is only a change in notation from a mathematical standpoint. However, the shifted index k is closely connected to the properties of the vibration signal. This will be validated later in Section 7. In summary it means that the most significant vibration occurs at frequencies close to the nominal gear mesh frequency ω_0. It also means that the spectral components are spaced evenly in the frequency domain at a distance corresponding to the frequency $\dot{\varphi}$ at which the planet carrier rotates. Thus, the index k uniquely identifies a spectral component. Now (3) is inserted into (2) and (1) is applied to the result.

$$\upsilon = \frac{1}{2}\left[\sum_{k=-\infty}^{\infty} c_k e^{ik\varphi} + \sum_{k=-\infty}^{\infty} \overline{c}_k e^{-ik\varphi}\right] \tag{4}$$

Without loss of generality it may be assumed that the coefficients c are complex conjugate pairs.

$$c_k = \begin{cases} a_0 & \text{for } k = 0 \\ \frac{1}{2}(a_k - ib_k) & \text{for } k > 0 \\ \frac{1}{2}(a_{-k} + ib_{-k}) & \text{for } k < 0 \end{cases} . \tag{5}$$

Therefore (4) reduces to

$$\upsilon = \sum_{k=-\infty}^{\infty} c_k e^{ik\varphi} = \frac{a_0}{2} + \sum_{k=1}^{\infty} (a_k \cos k\varphi + b_k \sin k\varphi) . \tag{6}$$

The expressions $\cos k\varphi$ and $\sin k\varphi$ may be interpreted as *reference oscillators*.

This very basic model of the vibration signals may serve different purposes. It is a useful analytical tool because it reduces substantial amounts of data to a set of coefficients. It compares favorably to the use of the FFT, which does not reduce the amount of data without averaging of the spectrum. In contrast to FFT based approaches it also utilizes the sparsity of the vibration signal in the frequency domain. The FFT of steady-state gearbox vibration is zero at many frequencies. In addition, approximately 80 % of the nonzero spectral lines are only nonzero because of spectral leakage. The model (6) presumes this structure inherently and does not need to express it explicitly by yielding vanishing spectral components at certain frequencies. Also it performs acceptably in the case of non-constant angular velocity. Here FFT based approaches need to re-sample the measured data as explained in [2].

If used for analysis, the task is to compute the coefficients c_k from measured data. However, the model is not limited to analysis. Because it describes the vibration of planetary gearboxes precisely and efficiently, it is suited as the foundation from which active vibration control can be derived. This will be sketched out in Section 8. The intermediate goal is to arrive at a real-time capable estimation of c_k, where the estimated coefficients are updated on-line at a rate of 80 kHz, observing the entire audible

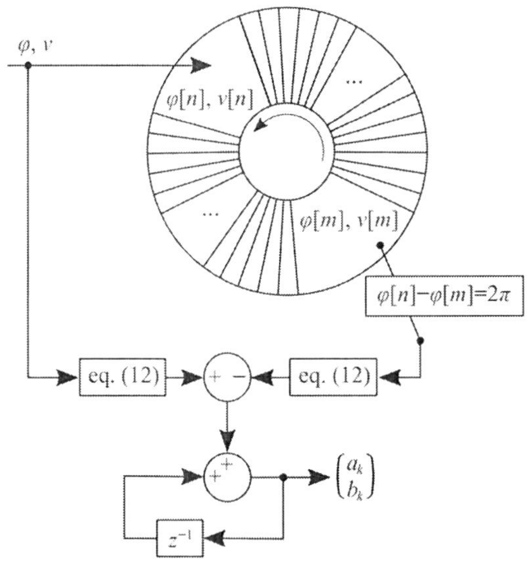

Figure 2. Ring buffer structure.

range. In the following sections, three possible approaches will be discussed: First, the coefficients may be calculated by integral transform. Second, a heterodyne filter bank may be employed. Third, a Vold-Kalman filter can be used.

4 Integral Transform

From the infinite series of (6) any coefficient c_k may be computed by solving

$$a_k = \frac{1}{\pi}\int_0^{2\pi} v(\varphi)\cos k\varphi \, d\varphi \quad (7)$$

and the analogous Equation for b_k. Because the only difference between a_k and b_k is the use of sin instead of cos, the derivation of b_k is mostly omitted from this study. More details may be found in [4]. For a numerical computation the continuous time signals have to be substituted by their discrete time approximations

$$a_k \approx \frac{1}{2\pi}\sum_{l=m}^{n} v[l]\,(\varphi[l+1]-\varphi[l-1])\cos k\varphi[l] \quad (8)$$

$$b_k \approx \frac{1}{2\pi}\sum_{l=m}^{n} v[l]\,(\varphi[l+1]-\varphi[l-1])\sin k\varphi[l] \quad (9)$$

where the start index m and the stop index n have to satisfy the limits of the original integral, such that

$$\varphi[n]-\varphi[m] \approx 2\pi \quad . \quad (10)$$

The method was originally used for off-line analysis in [4]. This approach may be adapted to real-time. In this real-time context the index n denotes the current step. From (10) it follows that the index $m[n]$ is a function of n, because at any time n there needs to be one full turn of the planet carrier between m and n.

The sum in Eq. (8) may be split up in its individual summands.

$$a_k[n] = \sum_{l=m[n]}^{n} \Delta a_k[l] \quad (11)$$

$$\Delta a_k[l] = \frac{1}{2\pi} v[l]\,(\varphi[l+1]-\varphi[l-1])\cos k\varphi[l] \quad (12)$$

$$a_k[n] = a_k[n-1] + \Delta a_k[n] - \sum_{l=m[n-1]}^{m[n]-1} \Delta a_k[l] \quad (13)$$

During stationary operation in every time step there will be one summand added and one will be subtracted. If the gearbox speeds up, more than one summand needs to be removed. Conversely at time steps where the gearbox slows down there it is possible that no summand needs to be subtracted.

In a practical implementation this will be realized using a circular buffer memory as visualized in Figure 2. This buffer is able to store a certain amount of past data. Its length is adjusted to the lowest angular velocity of the planet carrier at which the algorithm is to be executed.

Numerical stability has to be considered. In fixed point arithmetic, the summands added can be removed exactly by subtraction. In a floating point arithmetic this is not always true, as the exponent might change after addition. If single precision floating point arithmetic is used, the coefficients may have to be reset in regular intervals, which can be done without loss of information. Using double precision, the round-off error can safely be ignored. Here, double precision arithmetic will be used.

5 Heterodyne filtering

The heterodyne principle moves the contents of a time domain signal in the frequency domain. This technique has been known for more than a century. For the historical context see [5] or [6]. The principle has been used in radio applications since. It also may be employed to create digital adaptive band pass filters as discussed in [7].

The heterodyne principle has been applied to planetary gearboxes before. [8]. However, it has been employed to shift high frequency acoustic emission signals below the Nyquist frequency of the DAQ hardware. This paper uses the heterodyne principle for an entirely different purpose. Here, instead of a single heterodyne unit, a filter bank consisting of several units is used to decompose the signal in its spectral components. The filter bank needs to be adaptive because the frequencies at which the components are located depend on the angular velocity of the planet carrier.

Figure 3 shows how an adaptive band pass filter may be realized. From the signal v the k-th spectral component will be extracted. The instantaneous angle φ of the planet carrier is fed into the reference oscillator. Here, the filter is split into a real path (cosine) and an imaginary path (sine). All operations are performed equally on both paths. The mixer multiplies the signal v by the output of the reference oscillator.

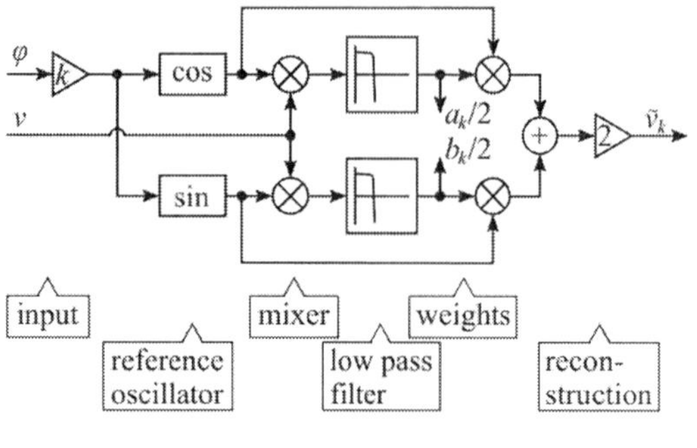

Figure 3. Adaptive heterodyne band pass filter.

This mixing is the key step in the heterodyne principle. It shifts the frequency of the k-th spectral component to the base band. Because of this, an ordinary low pass filter can be used to remove all other spectral components. The design of the filter determines the actual performance of the heterodyne unit. It will be discussed in detail later. Because the frequency of the spectral component of interest has been shifted to DC, the two low pass filters extract real and imaginary part of the Fourier coefficient c_k. Multiplication of these weights by the reference oscillation yields a reconstruction of the time domain signal.

Of course, this arrangement only extracts one spectral component $\tilde{v}_k$ from v.

$$\tilde{v}_k = a_k \cos k\varphi + b_k \sin k\varphi \tag{14}$$

In order to extract multiple components, a parallel array of identical units has to be employed. This is known as a *filter bank*. It is able to extract any number of Fourier coefficients concurrently. In turn, using the extracted coefficients and the reference oscillators, the original signal v may be reconstructed. The quality of a filter design may be judged by the reconstruction error

$$\varepsilon = v - \sum_{k=0}^{N} \tilde{v}_k \tag{15}$$

where N is the Number of spectral components present in the signal.

However if the filter exhibits a phase of $\pm\pi/2$ in the stop band, the reconstruction error will be a superficial measure of the performance of the filter design. This can be most easily explained using the experimental data of Section 7. Consider the spectral component of the highest magnitude. For gearbox A this is $k = 83$ at most operating conditions. If the planet carrier turns at $\dot{\varphi}/2\pi = 3000\,\text{min}^{-1}$, the distance between the spectral components will amount to 50 Hz. Heterodyne unit number 82 will shift spectral component to 50 Hz with a phase of $-\pi/2$ and an attenuation specific to the filter design. Unit 84 will also shift it to 50 Hz, but with a phase of $\pi/2$ and the same amount of attenuation. In Equation (15) this will cancel out. While the reconstruction is successful, the decomposition is not.

Any low pass filter may be used in the heterodyne unit, but the performance relies on a good choice. Generally FIR filters are not investigated in this paper. Compared to IIR filters they require a much higher order to achieve equal performance. At a desired cutoff frequency of less than 1/1000 of the Nyquist frequency FIR filters are computationally too expensive. A FIR design would need a length of more than 5000 to achieve acceptable performance.

A Chebyshev type II design has been chosen for the low pass filter. If the order of this filter is odd, the phase will be $-\pi/2$ in the stop band. Only for even orders numerical optimization of the reconstruction error is sensible. Because the filter will be implemented as a biquad structure, an order of 3 would offer no advantage over an order of 4 regarding the computational effort. Orders 1 or 2 cannot yield valid filter designs. Therefore, an order of 4 has been chosen. Generally, no improvements could be achieved by higher filter orders.

The filter may be characterized by the parameters stopband frequency f_{stop} and stopband attenuation. They determine the selectivity of the filter. This means the ability to extract the spectral component of interest and to reject any other component. As explained in Section 3, the distance between the spectral components corresponds to the rate at which the planet carrier rotates. Therefore the stopband frequency f_{stop} must never exceed this.

$$0 < f_{stop} < \frac{\dot{\varphi}}{2\pi} \tag{16}$$

This means that there is a minimal speed of the gearbox the filter bank cannot operate below. Therefore it would seem that the stopband frequency should be as low as possible.

However, a high selectivity comes at a cost. In transient operation, e. g. a run-up of the gearbox the weights are not constant, but change slowly. If the selectivity of the filter is chosen too high, it will also reject fluctuation of the weights. This will degrade performance significantly. Therefore, at given minimum and maximum speed of the gearbox there is an optimal stopband frequency. On the measured data of gearbox A a stop band frequency of $f_{stop} = 22\,\text{Hz}$ and an attenuation of 84 dB yield optimal performance for a Chebyshev type II filter of order 4.

6 Vold-Kalman filtering

The Vold-Kalman filter serves the same purpose of decomposing a signal into spectral components using a reference oscillator. It has initially been proposed in [9]. There are two distinct formulations of the algorithm. The first type calculates the decomposition of the signal v into the spectral components $\tilde{v}_k$ in the time domain. The second type calculates the complex magnitude of the spectral components. This is equivalent to the determination of the Fourier coefficients c_k. It is generally agreed upon that the accuracy of Vold Kalman filters is very good. Therefore, researchers have tried to adapt the filter to real-time [10] or to make it on-line capable [11].

Because the Vold-Kalman filter calculates a globally optimal decomposition of the signal into its spectral components, the computational effort is higher than that of comparable methods. While [11] reports real-time capability on similar computer hardware as used in this paper, the

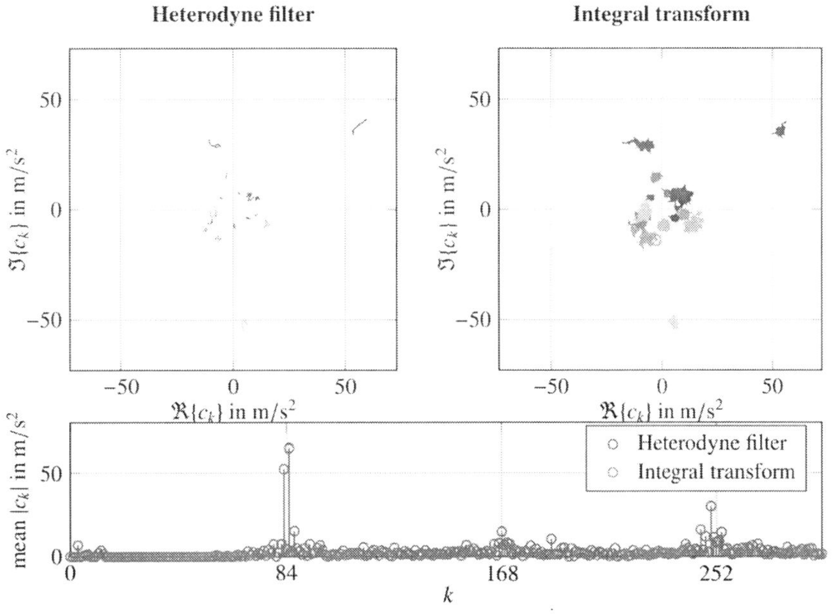

Figure 4. Computed c_k from recorded data. Top: The 20 most significant; Bottom: Mean 0 to 3.5Z.

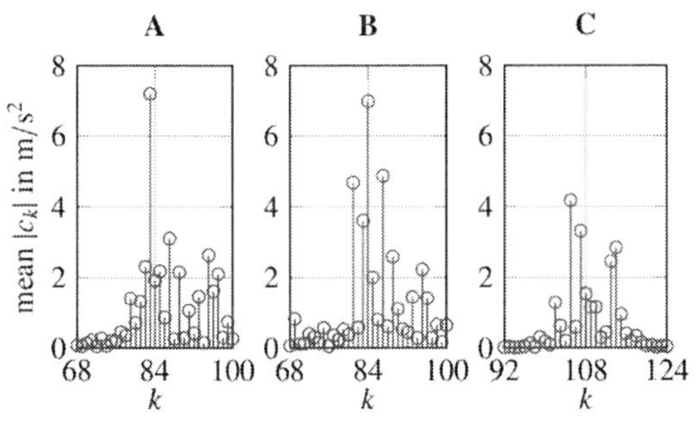

Figure 5. Input speed: 7000 min^{-1}, load: 10 N m.

conditions are less demanding than in the case of high speed planetary gearboxes. There, a sampling rate of 2048 Hz is used while in this study a sampling frequency of 80 kHz is required. Also, only five orders are tracked in [11]. In addition the algorithm is executed in a block-wise fashion leading to a delay of up to 146 ms. An other investigation [10] reports that it takes at least 21.7 s to track four orders in a 5 s signal at a sampling rate of 2 kHz.

These computational properties may be acceptable for condition monitoring, but are very problematic for active vibration control. Incorporation of the VKF into an AVC algorithm seems unlikely in this high speed application. Therefore, it will be not investigated further.

7 Results

There are two performance measures for a real-time order tracking algorithm: Accuracy and execution time. Figure 4 depicts the results of both the heterodyne filter and the integral transform. The vibration of gearbox A running at an input speed of 10 000 min^{-1} and an output load of 30 N m has been recorded for a time of 0.8 s. From this data, the Fourier coefficients c_k are computed. The most significant coefficients are in the vicinity of the number of teeth on the ring gear Z. This is expected from Section 3 and Equation (3). Also, the nonlinear dynamics of the gear meshing excite higher gear mesh order around 2Z and 3Z.

The results of both approaches are very similar. The integral transform yields a greater fluctuation of the coefficients than the heterodyne filter bank. The mean values of the computed coefficients are nearly identical. Therefore it has to be concluded that there is no significant difference in accuracy between the two approaches. Both approaches are then tested in terms of execution time on the test rig. The number of coefficients to be extracted is increased until CPU overload occurs. The heterodyne filter bank is able to extract the 100 most significant spectral components in real-time. The integral transform can track 20 orders. However, the program code has not been optimized and an improvement to the level of the heterodyne filter bank seems possible. In practice, it may be advisable to leave out some spectral components. If a certain component does not contribute to the total power of the signal, unit dedicated to its extraction will only yield noise. In this application the best reconstruction of the signal is achieved using 34 orders. Therefore, neither approach has a significant advantage in terms of execution speed.

Figure 5 compares the results of all three gearboxes during equal operating conditions. The load has been set to the maximum of gearbox B while the speed is limited by C. Only the vicinity of the gear mesh order is depicted because

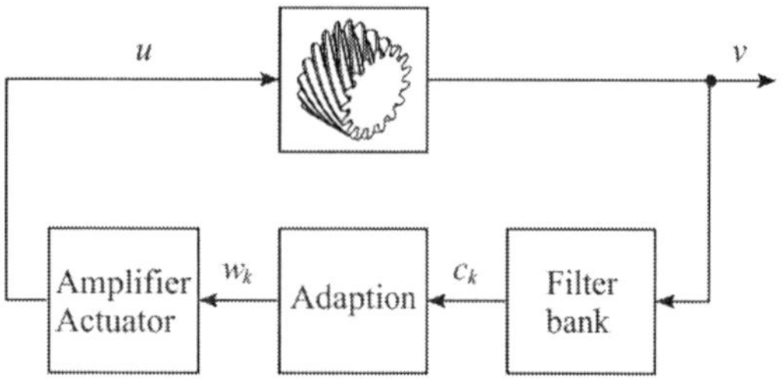

Figure 6. Architecture of the proposed AVC method.

here the most significant spectral component are located. The difference between spur (A,B) and helical (C) gears is to be expected. However, gearboxes A and B differ more than it would be expected from existing research. These two gearboxes share most constructive parameters (see Table 1). While the base frequency is mostly suppressed in A, it is most significant in B. This clearly contradicts existing research such as [1] or [3].

8 Active Vibration Control

Active vibration control of the gear meshing needs to utilize the specific structure of the excitation. Only this way sufficient performance may be achieved. One approach is to adapt the FxLMS algorithm to the planetary gearbox. It uses reference oscillators similar to those of the heterodyne filter bank. While it offers good performance, there is one drawback: The FxLMS algorithm relies on an accurate model of the secondary path. At frequencies of more than 4 kHz small disturbances, even variations in temperature may change the frequency response of the secondary path considerably. Therefore, the performance of the FxLMS algorithm may degrade quickly. There have been remedies proposed for this problem. For example, Li et al [12] use on-line identification of secondary path. But this solution introduces new problems: The computational requirements will increase. Also the stability will depend on an accurate estimation of the secondary path, which is difficult to do outside of a laboratory.

An algorithm which does not rely on a model of the dynamics of the actuator or the structure will avoid this issue. Therefore, a new algorithm using real-time order tracking is proposed in Figure 6. From the measured vibration v the Fourier coefficients c_k are estimated using real-time order tracking. An optimization algorithm then chooses the output weights w_k. The output u of the controller is determined by the complex weights w_k and the reference oscillators.

$$u = \sum_{k \in \mathcal{K}} w_k e^{ik\varphi} \tag{17}$$

Only a relevant subset $\mathcal{K}$ is used for reasons discussed in Section 7. Because the performance measure is defined in terms of the Fourier coefficients, the problem will decompose into independent sub-problems. The objective is to minimize the magnitude of each coefficient $|c_k|$ individually by adjusting the corresponding output weight $\Re\{w_k\}$ and $\Im\{w_k\}$. This leads to a much easier estimation of the gradient, because the dimension is reduced to 2. Also, the step size may be adjusted more efficiently because an estimation of the Hessian matrix is feasible. A discussion of an actual implementation would go beyond the scope of this paper.

9 Conclusion

Two methods of real-time order tracking have been developed: The heterodyne filter bank and the integral transform. Their respective accuracy and computational efficiency are both sufficient for the proposed high-speed real-time use. This has been validated using a test rig. The methods utilize the characteristic modulation of the vibration excited by a planetary gearbox. From existing research it it concluded that the Vold-Kalman filter does not meet the requirements on the computational effort. The architecture for an active vibration control algorithm using real-time order tracking has been proposed. In the next step, an actual implementation of AVC should be performed. Also the experimental data shows that the modulation in planetary gearboxes is more complex than commonly assumed. Therefore, a comparative study of different commercial gearboxes is planned.

References

[1] P. Mcfadden, J. Smith, Proceedings of the Institution of Mechanical Engineers, Part C: Journal of Mechanical Engineering Science **199**, 65 (1985)
[2] P. Borghesani, P. Pennacchi, S. Chatterton, R. Ricci, Mechanical Systems and Signal Processing **44**, 118 (2014)
[3] J. McNames, Journal of Vibration and Acoustics **124**, 150 (2001)
[4] D.F. Plöger, P. Zech, S. Rinderknecht, *Experimental identification of high-frequency gear mesh vibrations in a planetary gearbox*, in *Proceedings of ISMA2016 International Conference on Noise and Vibration Engineering* (Leuven, Belgium, 2016), pp. 911–924
[5] J.L. Hogan, Proceedings of the Institute of Radio Engineers **1**, 75 (1913)
[6] L.C. Godara, Proceedings of the IEEE **87**, 1975 (1999)
[7] M.A. Soderstrand, in *Adaptive Filtering*, edited by L. Garcia (InTech, 2011), pp. 359–398, ISBN 978-953-307-158-9
[8] Y. Qu, D. He, J. Yoon, B. Van Hecke, E. Bechhoefer, J. Zhu, Sensors **14**, 1372 (2014)
[9] H. Vold, J. Leuridan, SAE Technical Paper 931288, SAE International, Warrendale, PA (1993)
[10] M.C. Pan, C.X. Wu, Mechanical Systems and Signal Processing **21**, 2957 (2007)
[11] M. Čala, P. Beneš, *Implementation of the vold-kalman order tracking filters for online analysis*, in *23rd International Congress on Sound and Vibration 2016 (ICSV 23)* (2016), Vol. 1, pp. 367–374
[12] M. Li, T.C. Lim, W.S.S. Jr, Y.H. Guan, Smart Materials and Structures **14**, 917 (2005)

Dynamic Characteristics of Rotors on Passive and Active Thrust Fluid-film Bearings with Fixed Pads

Alexander Babin[1,*], *Leonid* Savin[1], and *Sergey* Majorov[1]

[1] Dept. of Mechatronics, Mechanics and Robotics, Orel State University, Naugorskoe sh. 29, 302020, Orel, Russia

Abstract. Application of fluid-film bearings in rotor machines in many cases could have no alternative due to obvious advantages when compared to roller element bearings. Integration of information technology in mechanical engineering resulting in emergence of a new field of research – mechatronic bearings which allowed tracking condition of the most important parts of a machine and adjusting operational parameters of the system. Application of servo valves to control the flow rate through a fluid-film bearing is the most universal and simple way of rotor's position control due to relative simplicity of modelling and absence of need to radically change the design of conventional hydrodynamic bearings. In the present paper numerical simulations of passive (conventional as opposed to mechatronic) and active hybrid thrust fluid-film bearings with a central feeding chamber are presented, that are parts of a mechatronic rotor-bearing node. Numerical model of an active thrust bearing is based on solution of equations of hydrodynamics, rotor dynamics and an additional model of a servo valve. Various types of control have been investigated: P, PI and PID control, and the dynamic behaviour of a system has been estimated under various loads, namely static, periodic and impulse. A design of a test rig has been proposed to study passive and active thrust fluid-film bearings aimed at, among other, validation of numerical results of active bearings simulation.

1 Introduction

Investigation in the field of rotor's dynamic behaviour control in fluid-film bearings is carried out in the following fields: study of the possibility of profile adjustment of a gap between a rotor and a sleeve or a thrust disk of a bearing, e.g. tilting-pad and some foil bearings [1, 2]; study of the possibility of application of lubricants with adjustable parameters such as viscosity or density, e.g. ferromagnetic fluids [3, 4]; and study of hybrid fluid-film bearings [5, 6], where load capacity is generated as a combination of hydrodynamic forces in the film and hydrostatic force from the lubricant supplied into a bearing under pressure, where rotor's position can be adjusted by means of electrohydraulic devises like servo valves.

Application of servo valves to control the flow rate through a bearing is the most universal and simple way of rotor's position control due to relative simplicity of modeling and absence of need to radically change the design of conventional hydrodynamic bearings. However, the field of active hybrid bearings is not as investigated, as, for instance, magnetic bearings. Despite the fact, that fluid-film bearings with additional elements to form a control system are significantly cheaper than the magnetic bearings with complex structure and maintenance, the modelling of them requires a lot of numerical tests to solve the design problems. Both thrust and journal fluid-film bearings are used in industrial applications, such as centrifugal pumps, that extrude crude oil, and bearings in them are lubricated with the same medium they extrude. The fields of application also include turbo generators, rockets, etc. In every case, a unique set of operational parameters dictates the design of a bearing.

In the present paper numerical simulations of passive (conventional as opposed to mechatronic) and active hybrid thrust fluid-film bearings with a central feeding chamber are presented, that are parts of a mechatronic rotor-bearing node.

2 Numerical simulations of passive hybrid thrust fluid-film bearings

In the present paper, a fixed-pad type of a hybrid thrust fluid-film bearing is considered. This type is the easiest to manufacture and requires no complications of a model, as opposed to, for instance, tilting pad thrust bearings. For further clarification, in this case 'hybrid' stands for a combination of hydrodynamic lifting force, generated due to the lubricant being dragged into the converging gap between a fixed pad and the rotor's disk, and hydrostatic lifting force, generated by the lubricant being supplied to the axial gap under pressure. Numerical simulations of passive hybrid thrust bearings (PHTB) are a base for further modelling of active hybrid thrust bearings (AHTB), the conceptual model of which is described below.

* Corresponding author: alex.mech.osu@gmail.com

 https://doi.org/10.1051/matecconf/201814811003

2.1. Conceptual model of an ATHB

The operational principle of an ATHB (Figure 1) is as follows: the pump (1), located in the tank with the fluid (water, oil, etc.), supplies the lubricant into the housing (2) of the bearing (3), which supports the rotor (4) in the axial direction. The pressure data is acquired with the sensor (8). The data on the present position of the rotor is acquired with the proximity sensor (7). Data from the sensors is analogue and is acquired with the analogue input modules of a DAQ system (5). This system also controls the servovalve (6) by means of adjusting the opening rate ξ generating an analogue voltage signal, and the pump (descrete output module).

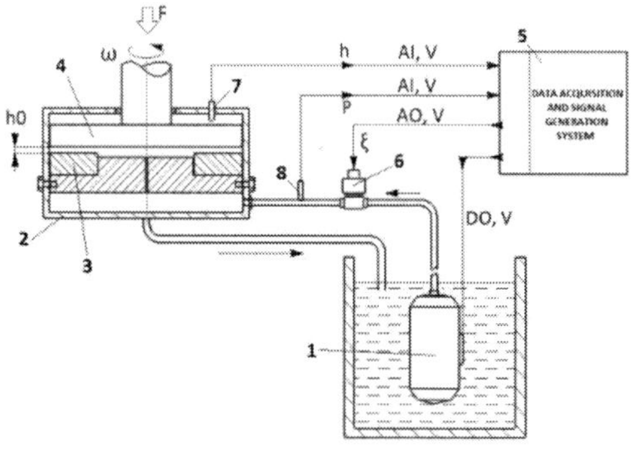

Fig. 1. Conceptual model of an AHTB.

2.2 Basic equations of PHTBs modelling

To simulate the operation of a rotor on a PHTB, one has to determine the reaction forces of the lubricant film, thus being able to obtain dynamic coefficients, i.e. stiffness and damping, to simulate the rotor's dynamic behaviour.

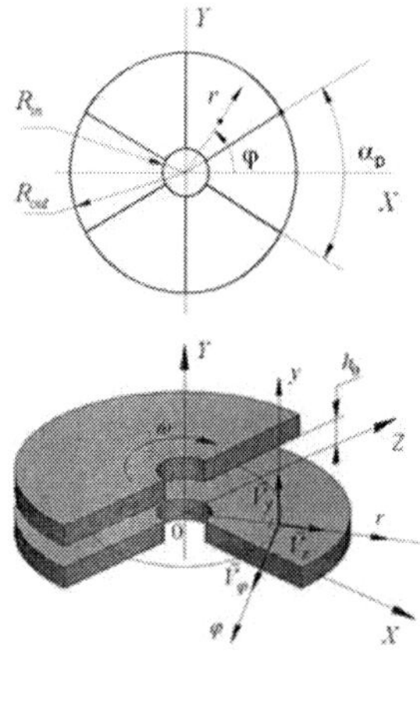

Fig. 2. Calculation diagram of a PHTB.

The first step in this modelling is determination of pressure distribution over the surface of the rotor's disk, which is governed by the Reynolds equation in polar coordinates in accordance to the diagram in the Figure 2:

$$\frac{\partial}{\partial \varphi}\left[\frac{\Phi_\varphi h^3}{rK_\varphi \mu}\frac{\partial p}{\partial \varphi}\right]+\frac{\partial}{\partial r}\left[\frac{\Phi_r h^3}{K_r \mu}\frac{\partial p}{\partial r}\right]= \\ =6\omega r\frac{\partial h}{\partial \varphi}+6\omega r\sigma\frac{\partial \Phi_s}{\partial \varphi}+12V_y, \quad (1)$$

where Φ_r, Φ_φ and Φ_s – radial, circumferential pressure and shear flow rate coefficients accordingly, which are determined as described in [7,8], h – axial gap function, K_φ and K_r – viscosity coefficients determined according to [9], p – pressure, μ - dynamic viscosity of the lubricant and is a function of temperature: $\mu=\mu_{in}e^{-\lambda(T-T_s)}$, λ – temperature/viscosity coefficient, μ_{in} – viscosity of the supplied medium, T_s – temperature of the supplied lubricant, ω – angular speed of the shaft, σ – RMS roughness of the surfaces of the rotor's, V_y – velocity of the shaft in the direction of film thickness.

The solution of (1) in the present case requires the following boundary conditions for a single pad:

$$p\big|_{r=Rout}=p_a,\ p\big|_{r=Rin}=p_s,$$

$$p\big|_{\varphi=0}=p\big|_{\varphi=\alpha_p},\ \frac{\partial p}{\partial \varphi}\bigg|_{\varphi=0}=\frac{\partial p}{\partial \varphi}\bigg|_{\varphi=\alpha_p},$$

where p_a – ambient pressure, p_s – supplied pressure, i.e. pressure in the feeding chamber, α_p – angular extent of a single pad.

The pressure distribution in a fluid is greatly influenced by the temperature of the fluid and the surrounding surfaces. The temperature distribution is governed by the energy equation, which in many cases has to be solved for a three-dimensional flow. However, for relatively low rotational speeds as shown in [10] a two-dimensional energy equation gives a quite accurate results. The energy equation for the present case is as follows:

$$\rho c_p\left(-\frac{h^2}{12\mu}\frac{\partial p}{\partial r}\frac{\partial T}{\partial r}+\left(-\frac{h^2}{12\mu r}\frac{\partial p}{\partial \varphi}+\frac{\omega r}{2}\right)\frac{\partial T}{r\partial \varphi}\right)= \\ =\mu\left(\left(\frac{1}{12}\left(\frac{h}{\mu}\frac{\partial p}{\partial r}\right)^2\right)+\left(\frac{1}{12}\left(\frac{h}{\mu}\frac{\partial p}{r\partial \varphi}\right)^2+\left(\frac{\omega r}{h}\right)^2\right)\right), \quad (2)$$

where ρ – fluid's density, c_p – specific heat of the fluid, T – temperature.

The solution of (2) requires the following boundary conditions:

$$T\big|_{r=Rin}=T_s,$$

$$T\big|_{\varphi=0}=T\big|_{\varphi=\alpha_p},$$

$$\frac{\partial T}{\partial \varphi}\bigg|_{\varphi=0}=\frac{\partial T}{\partial \varphi}\bigg|_{\varphi=\alpha_p},$$

where T_s – temperature of the supplied lubricant.

Numerical solution of (1) and (2) yields pressure and temperature distribution over the surface of the rotor's disk. The procedure is implemented in a series of iterations until a load capacity convergence accuracy is achieved. Equations (1) and (2) are solved using the finite difference method.

2.3 Static characteristics of PTHBs

Determination of static characteristics, namely load capacity R_Y, friction torque M_{FR} and power loss due to friction N_{FR} being of the most significance in this case, is a further step in modelling the dynamic behaviour of the rotor. These characteristics are determined by means of integration of the pressure distribution:

$$R_Y = z \int_{Rin}^{Rout} \int_{0}^{\alpha_p} prd\varphi dr,$$

$$M_{FR} = r \int_{Rin}^{Rout} \int_{0}^{\alpha_p} \left[\frac{h}{2} \frac{\partial p}{r \partial \varphi} + \frac{\omega r \mu K_\varphi}{h} \right] rd\varphi dr,$$

$$N_{FR} = M_{TP}\omega,$$

where z – number of pads.

In the Figure 3 and the Figure 4 the results of calculation of axial force R_Y and power loss due to friction accordingly depending on the minimum film thickness for various pad inclination angles are presented. For the Figure 3 the following modelling parameters were used : $Rout$ = 0.06 m, Rin = 0.03 m, lubricant – H_2O @ 23°C , ω = 900 rad/s, $\sigma_{pad,\ rotor}$ = 5μm. For the Figure 4 the following modelling parameters were used : $Rout$ = 0.06 m, Rin = 0.03 m, lubricant – H_2O @ 23°C, ω = 900 rad/s $\sigma_{pad.\ rotor}$ = 5μm, α = 0.25°.

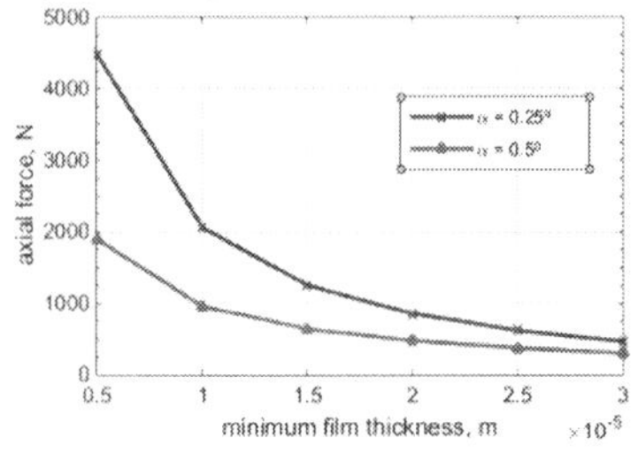

Fig. 3. Axial force of a PHTB versus minimum film thickness.

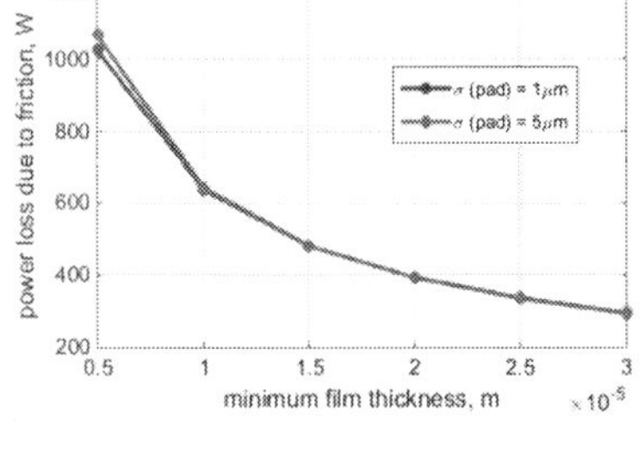

Fig. 4. Power loss due to friction versus minimum film thickness.

From these results one could tell, that geometry plays a big role in providing energy efficient operation with sufficient load capacity, which is crucial for most of the rotor machines. Additionally, when the fluid film is very thin, if, for instance, the load is high or the speed is low, one could notice the influence of the pads' roughness on the power loss due to friction. This leads to the necessity of high precision during the mechanical treatment of the parts of various machines and additional simulations in every particular case of operational conditions.

2.3 Static characteristics of PTHBs

Determination of dynamic characteristics requires representation of the rotor-bearing node as shown in the Figure 5.

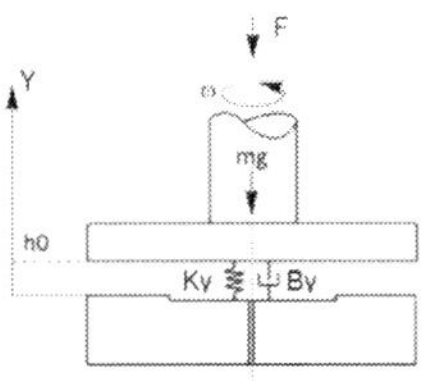

Fig. 5. Rotor-bearing dynamic system.

The present step of modelling aims at determination of stiffness and damping coefficients of the bearings, K_Y and B_Y accordingly.

Under steady conditions the rotor takes the equilibrium position, when the hydrodynamic forces counterbalance the inertia forces, i.e. $R_{Y0} = mg$, where m – rotor's mass, g – freefall acceleration. When the rotor's position changes under the influence of some impulse, some additional forces start acting on the rotor, so, assuming these displacements are relatively small, the reaction force of a bearing could be linearized using Taylor series. The reaction force of the bearing could be determined as follows:

$$R_y = R_{Y0} - K_Y \Delta Y - B_Y \Delta \dot{Y},$$

and, consequently, the motion law of the system is as follows:

$$m\ddot{Y} - B_Y \dot{Y} - K_Y Y = F(t),$$

where $F(t)$ is some external force as a function of time t.

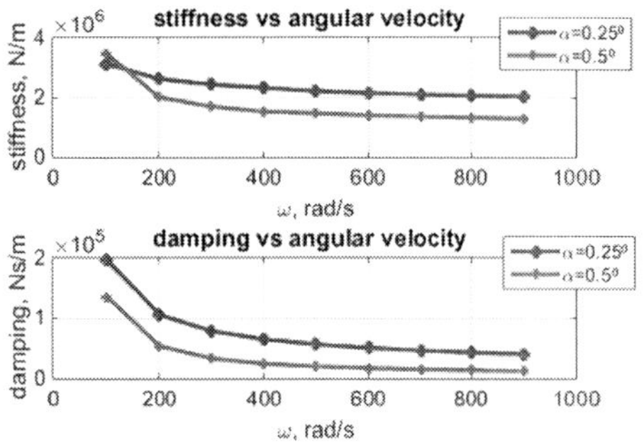

Fig. 6. Stiffness and damping of a PHTB versus angular velocity.

In the Figure 6, the results of the simulations show, how stiffness and damping of the PHTB are influenced by the angular velocity for various inclination angles of a single pad. The following modelling parameters were used: *Rout* = 0.06 m, *Rin* = 0.03 m, lubricant – H_2O @ 23°C, σ_{pad}= 2.4 µm, σ_{rotor}≈ 0 µm, mass of the rotor – 5.5 kg.

It could be seen, that for most of the operational conditions both stiffness and damping are higher for lower inclination angles of the pad, which could be explained by the fact, that with lower inclination angles the area of the pad where pressure is generating the load capacity is bigger, thus increasing the dynamic characteristics.

For further simulations the following types of the external loading are modelled: zero external load, periodic external load of 30 N at approx. 8 Hz, static load of 30 N and impulse load of 30 N applied in the middle of modelling time. The resulting rotor trajectories are presented in the Figure 7.

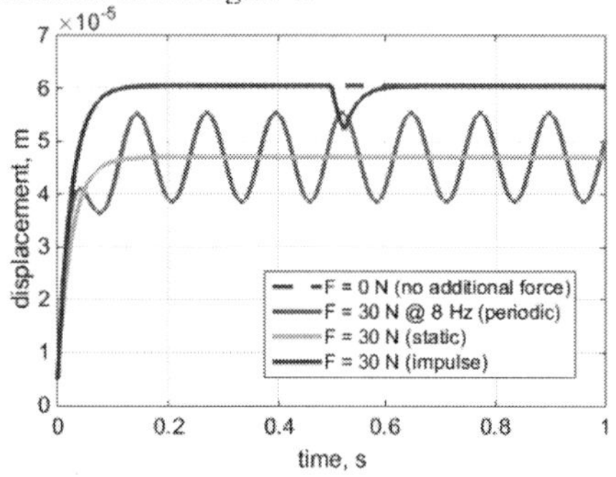

Fig. 7. Trajectories of the rotor with different loads.

3 Numerical modelling of AHTBs

According to the Figure 1, in the present case the active control of rotor's axial displacement is implemented by means of application of a servo valve. This kind of electrohydraulic actuators is simulated using the known relation between the pressure it supplies the lubricant with and the voltage on the input of the valve. The most important term here is the time constant of the valve T_{SV}, i.e. the time it takes for a fully closed valve to fully open. This is a crucial characteristic for systems subjected to high-frequency periodic loads.

The expression that allows simulation of a servo valve is as follows:

$$T_{SV}\frac{dp(t)}{dt}+p(t)=K_{SV}u(t),$$

where K_{SV} – gain coefficient of a servo valve, $u(t)$ – control signal, i.e. voltage on the servo valve's input.

Operation of the control system introduced to a PHTB significantly influences the stability of the system. It is possible to see how rotor's displacement is influenced depending on the simulated servo valve arbitrarily setting the time constant to various settings. Such comparison is shown in the Figure 8.

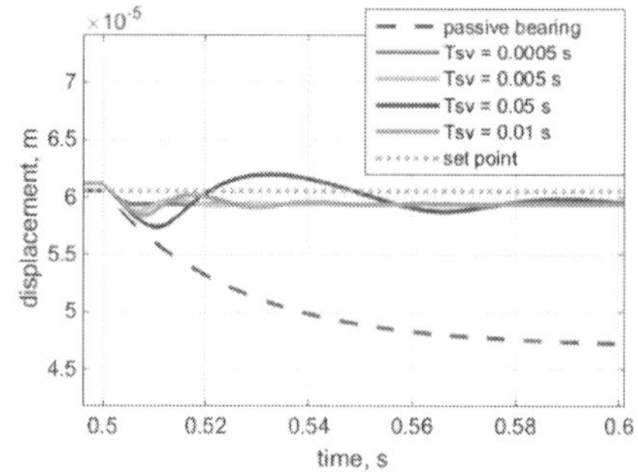

Fig. 8. Trajectories of the rotor depending on the time constant of a servo valve.

The set point value is in this case chosen to be the equilibrium position of the rotor on a PHTB. The simulated load was applied in the middle of the calculated time, and it could be seen from the Figure 8, that depending on the value of T_{SV} the control system affects the rotor's position differently in terms of induced oscillations. At this stage, it becomes obvious, that the choice of an actuator for specific application is indeed a very important task, especially if a system is subjected to high frequency vibrations or chaotic impacts of external nature.

In the light of this fact, it is also necessary to see how the controller setting influence the dynamic coefficients of stiffness and damping of a bearing. In the Figure 8 the simulation results are presented for a case of an AHTB under static load applied in the middle of the simulation time.

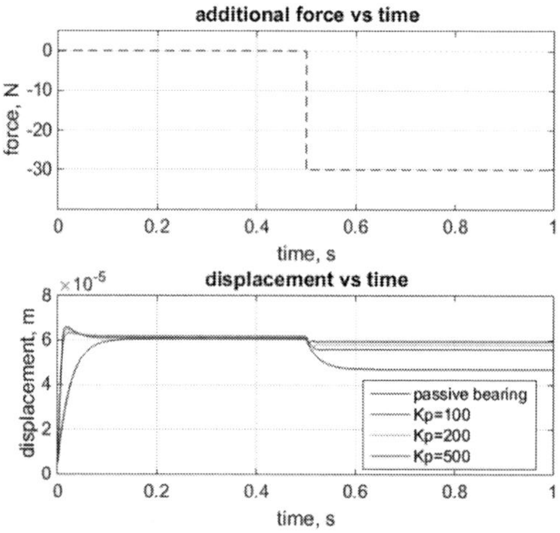

Fig. 8. Trajectories of the rotor on an AHTB under static load

The modelling parameters are as follows: ω = 500 rad/s, *Rout* = 0.06 m, *Rin* = 0.03 m, lubricant – H_2O @ 23°C, σ_{pad}= 2.4 µm, σ_{rotor}≈ 0 µm, mass of the rotor – 5.5 kg, additional force = 30 N @ 0.5 s, T_{SV} = 0.005 s.

Various arbitrarily set values of the proportional coefficient were chosen in order to see the influence of the *Kp* on the rotor's dynamic behaviour. Primary objective here was to compare the stiffness coefficients.

For a passive bearing, the stiffness coefficient is approx. $2.2 \cdot 10^6$ N/m, while for an active bearing, stiffness increases accordingly with the increase of proportional coefficient *Kp* of a P-controller (Table 1).

Table 1. Comparison of the stiffness coefficients.

Proportional coefficient *Kp*	**Stiffness, N/m**
0 (passive bearing)	$2.2 \cdot 10^6$
100	$4.9 \cdot 10^6$
200	$8.1 \cdot 10^6$
500	$16.7 \cdot 10^6$

Application of active control based on P-, PI- and PID control could significantly reduce the amplitude of rotor's oscillations. The simulation results in the Figure 9 below show the behaviour of the modelled system under the influence of a periodic external force with various settings of a control system.

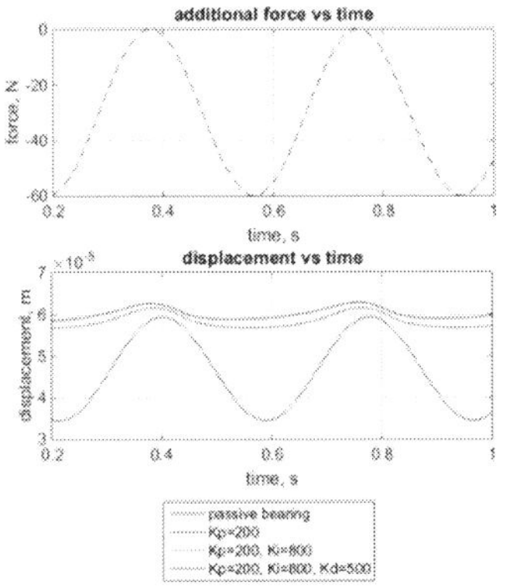

Fig. 9. Trajectories of the rotor on an AHTB under periodic load

Increase of the *Kp* of a P-controller allows decreasing of the oscillation amplitude, however a constant control error remains. Introduction of the *Ki* in a PI-controller appears to not have any significant influence. Addition of the *Kd* in a PID-controller allows this error to be decreased quite rapidly, the higher the setting, the faster the amplitude reaches values close to zero.

4 Results and discussion

Presented mathematical model allows evaluation of the influence of the control system based on an electrohydraulic device on the dynamic behavior of a rotor on an active thrust fluid-film bearing with fixed pads. Presented results show the importance of a careful choice of a servo valve, characteristics of which must meet the requirements of a particular application. Peculiarity of the presented type of control has been discovered: 'single-side' control only allows adjustment of the rotor's position below the set point, while decrease of the supplied pressure does not allow momentary adjustment of the position due to the properties of the lubricant and inertia properties of the rotor. This could be possibly avoidable by means of development of specific control algorithms or introduction of an additional hydraulic supply tract. Insignificance of *Ki* has been discovered, probably, due to small values of displacements (10^{-6} order of magnitude).

A test rig is being manufactured to verify the presented results and possibly to identify effects which occur and need to be taken into account. The schematic design of the rig is presented in the Figure 10.

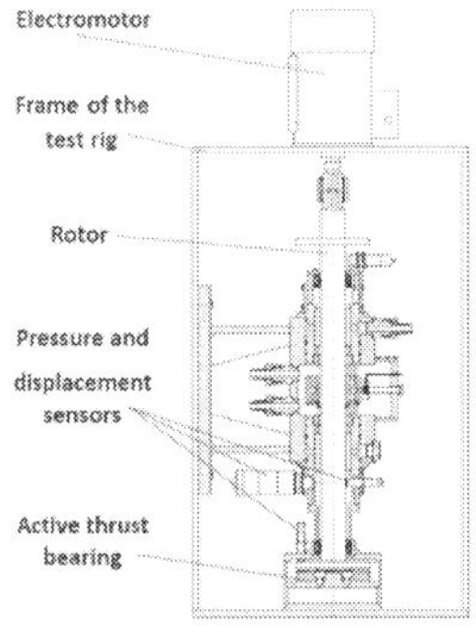

Fig. 10. Design of the test rig to study the AHTBs

The present research has been implemented as a part of the Russian Science Foundation project No. 16-19-00186 and Ministry of Education and Science of the Russian Federation project No. 9.2952.2017/4.6.

References

1. A. M. Haugaard, I. F. Santos, *Proceedings of the STLE/ASME International Joint Tribology Conference*, pp. 335–337 (2008)
2. G. Aguirre, F. Al-Bender, H. Van Brussel *PROCEEDINGS OF ISMA*, pp. 105-118 (2008)
3. H. Urreta, Z. Leicht, *Journal of Physics Conference Series*, **149** (2009)
4. A. Yu. Albagachiev, V.D. Danilov, *Bulletin of the Bryansk State Technical University*, **3**. pp. 90-93 (2016)
5. P. Kytka, B. Riemann, R. Nordmann, *Proceedings of the 9th International Conference on Motion and Vibration Control*, (2000)
6. L.A. Savin, S.V. Mayorov, D.V. Shutin, A.Y. Babin, *Proceedings of ICMMR MATEC Web of Conferences* **77**, 5 (2016)
7. N. Patir, H. S. Cheng, *ASME J. Lubr. Technol.*, **100**, pp. 12–17 (1978)

8. N. Patir, H. S. Cheng, *ASME J. Lubr. Technol.*, **101**, pp. 220–229 (1979)
9. N. Tepey, V.N. Constantinescu, *Publishing House of the Academy of Sciences of the RNR,* 458 p. (1964)
10. I. Chatzisavvas, A. Boyaci, A. Lehn, M. Mahner, B. Schweizer, P. Koutsovasilis, *Proceedings of ASME Turbo Expo 2016: Turbomachinery Technical Conference and Exposition,* 10 p. (2016)

Development of Approaches to Creation of Active Vibration Control System in Problems of the Dynamics for Granular Media

Andrei P. Khomenko[1], *Sergey K.* Kargapoltsev[2] and *Andrey V.* Eliseev[3,*]

[1]Irkutsk State Transport University, 15 Chernishevskiy str., 664074, Irkutsk, Russia

Abstract. The article deals with the development of mathematical models and evaluation criteria of the vibration field in the dynamic interactions of the elements of the vibrational technological machines for the processes of vibrational strengthening of long-length parts with help of a steel balls working medium. The study forms a theoretical understanding of the modes of motions of material particles in interaction with a vibrating surface of the working body of the vibration machine. The generalized approach to the assessment of the dynamic quality of the work of vibrating machines in multiple modes of tossing, when the period of free flight of particles is a multiple of the period of the surface oscillations of the working body, is developed in the article. For the correction of vibration field of the working body, the characteristics of dynamic interactions of granular elements of the medium are taken into account using original sensors. The sensors that can detect different particularities of interaction of the granular medium elements at different points of the working body are proposed to evaluate the deviation from a homogeneous and one-dimensional mode of vibration field. Specially developed sensors are able to register interactions between a single granule, a system of granules in filamentous structures, and multipoint interactions of the elements in a close-spaced cylindrical structure. The system of regularization of the structure of vibration fields based on the introduction of motion translation devices is proposed using the multi-point sensor locations on the working body. The article refers to analytical approaches of the theory of vibration displacements. For the experimental data assessment, the methods of statistical analysis are applied. It is shown that the peculiar features of the motion of granular medium registered by the sensors can be used to build active control systems of field vibration.

1 Introduction

Modern technological machines work in intense dynamic modes of interaction of working bodies with the processed parts and working media [1-3]. Vibrating machines implement a variety of processes of transportation, classification and separation of granular mixtures, strengthening etc. [4-6]. Special aspects of joint motions of the elements of the system with unilateral constraints must be considered for the stable operation of these technological systems [7-8]. Fundamentals of the theory of vibrational technological processes and the vibratory displacements of the working media are reflected in the works [9-12]. This gives insight into the variety of motion forms and indicates the necessity of a comprehensive consideration of interactions between system elements in a spatial structure. Some of the issues of possible regular forms of working media are discussed in [8]. However, generalized approaches to integrated perception of the problems of interaction between the elements have not yet been given proper attention to the level of detail.

The aim of this work is to develop of a comprehensive approach to the forms of the trajectories of the medium elements in interaction with the vibrating support surface and methods for controlling integral parameters of dynamic state of the technical system.

2 Formulation of a problem. Mathematical model of continuous tossing of material particles

Generalized approach to the estimation of trajectories of particles with the formation of a free phase of flying followed by interaction with the vibrating surface is developed in the article. It is assumed that a horizontal surface makes the vertical harmonic vibrations. The interaction of material particles with the surface is inelastic in nature and is implemented taking into account the additional forces and viscous friction. The main interest is the continuous tossing of material particles. Analytical approach is developed to solve the problem of determining the trajectory of the working medium element. Concepts of a gap function, the conditions and order of separation from the surface are used in this approach.

* Corresponding author: eavsh@ya.ru

The method of gap functions is the inverse problem relative to the direct problem of determination of a particle trajectory after detachment from the surface in a known point in time. The gap function allows carrying out detailed study of modes of tossing a material particle on the vibrating surface subject to unilateral constraints relations. In the case of harmonic oscillations of the surface, the gap function takes the form [8]:

$$R_H(t,t_0) = A\sin(\omega t_0) - A\sin(\omega t) + \\ + A\omega(t-t_0)\cos(\omega t_0) - \frac{1}{2}g(t-t_0)^2, t \geq t_0, \quad (1)$$

where A, ω is the amplitude and frequency of oscillation of the surface, t_0 is the moment of separation. The gap function is based on the consideration of all possible trajectories. A set of possible trajectories in the phase of flight is shown in Figure 1.

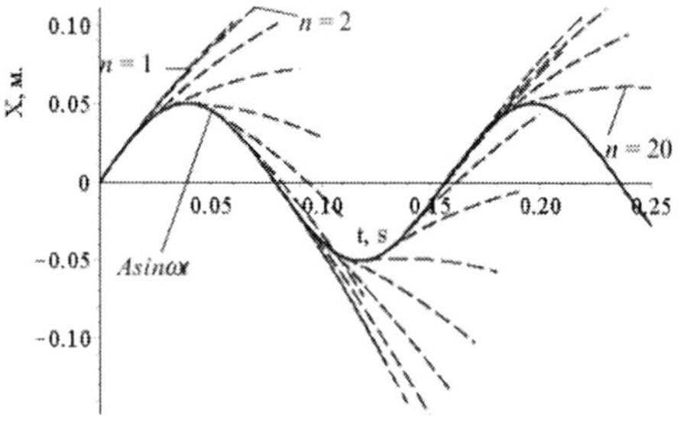

Fig. 1. A set of possible trajectories in the phase of flight

A number of sufficient conditions in terms of derivatives of the gap function are formulated to determine the moment of separation. The notion of the order of the separation for a multiple mode of tossing is used to differentiate different types of trajectories. In that case, trajectories with a multiple ratio of the approach time to the period of the surface oscillations are preferred in a certain way. A number of key characteristics of the trajectories are obtained using the gap functions for multiple modes of tossing. These characteristics are: the conditions of separation of material particles from the oscillation surface, the flight duration, the dependence on the constant additional forces and the forces of resistance from the working medium. In particular, the relationship between the trajectories of different multiplicity modes with the continuous tossing can be defined. A complex graph of the approach height of the material particles with different types of separation for a one-touch tossing mode is shown in Fig. 2. The curve S is graph of height of throw of the particle from the area of separation of the third order depending on the frequency at a fixed oscillation amplitude A = 5 mm.; curves k_1,... k_5 are the graphs of heights of throw of particles in case of the separation from the second-order separation area with multiplicity 1,2,...5 accordingly. The variety of modes is determined by the detailed characteristics of the tossing modes of material particles. Therefore, these modes can be viewed as an effect of unilateral constraints when a material particle interacts with a vibrating surface. Concepts on the motions of the working media are based on concepts on the trajectory of the individual granules.

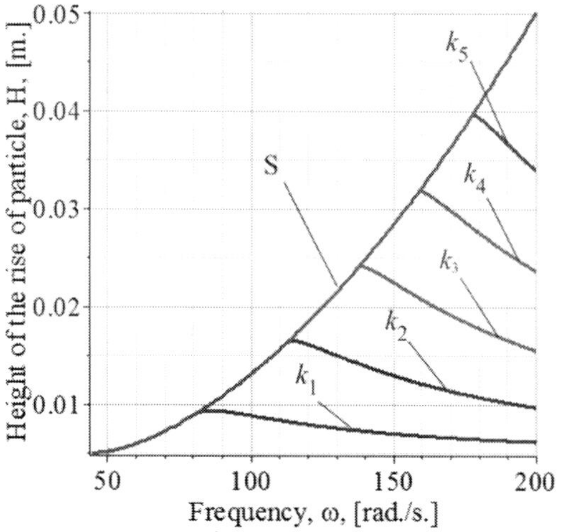

Fig. 2. The heights of throw. S – is graph of height of throw of the particle from point separation of the third order. Curves k_1,... k_5 - are the graphs of heights of throw of the particle from point separation of the second order

Thus, the dynamical interactions of the elements of technological complexes are studied due to effects of unilateral constraints. Analytical approaches are identified and developed during the evaluation of vibration technology processes with continuous tossing of the working medium. The generalized approach to problems of dynamic synthesis of vibration technologies during the continuous tossing is developed. The introduction of the concept of the gap function is the basis for the approach. Thus, the trajectories of motions of material particles are constructed with help of gap function in accordance with the properties of the interaction of the material particles with a vibrating surface. Mathematical models of formation of vibrational interactions are developed taking into account the factors having technological values, or criteria which include the conditions for the implementation of multiple modes, application of additional forces and elastic constraints [13-14].

3 A mathematical model of vibration technology machines

A mathematical model of the vibration stand which implements the process of vibration strengthening, with some assumptions, is reduced to a linear model with two degrees of freedom of small oscillations of a solid body placed on the elastic elements. The requirement of homogeneity and one-dimensionality of the vibrational field generated by one or more vibrators is a peculiar feature of the process. In the mathematical model, the influence of vibrators placed on the vibration table is reflected by the application of harmonic forces to specific points of the reference solid body. Adjustment of the vibration field is implemented by introducing a motion translation device with a predetermined rigidity and mass and inertia characteristics into the structure of the oscillating system. However, the vibrational field control system is also possible, when the mass and inertia elements are moved along the guides.

Solid body on elastic foundations under inertial excitation of the working body at a certain point (in this case, at the point of A) is shown in Fig. 3 in the form of a schematic diagram of the vibration machine. A model of the vibrating working body is formed by the solid body (1, Fig.3) of the mass M with a moment of inertia J in relation to the center of gravity O, elastic elements k_1, k_2 fixed at the points A_1 and A_2 at the distance l_1 and l_2 from the center of gravity O. It is assumed that the center of gravity is equal to the center of the swing for the solid body 1. The harmonic force $Q_{l_0} = A_f \sin(\omega_f t)$ applied at the point A_{l_0} at a distance l_0 from the point O is a source of excitation. The device of motion transformation is inserted in the oscillation circuit to control the structure of vibration field. The device is characterized by stiffness k_0 and mass-inertia coefficient $L>0$. The device is installed vertically between the static surface 3 (Fig.3) and the center of gravity of the solid body 1 (Fig.3).

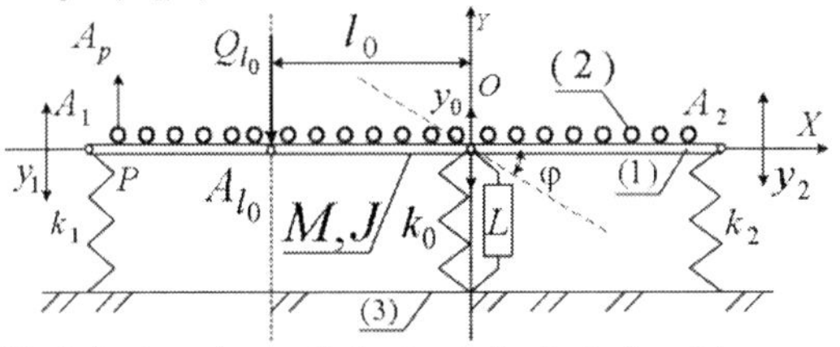

Fig.3. Design scheme of vibration technological machine. y_1, y_2 - the coordinates of the points A_1, A_2, M – *mass*, J - moment of inertia, L- the device of motion transformation, y_0 -coordinate of the center of mass, φ - the angle of inclination

Configuring the vibration table in some cases is carried out by varying parameters of vibration table (weight of vibrators, stiffness of elastic elements, etc.) so that the amplitude and frequency of oscillations of the working body are in accordance with the technological requirements. In the present model problem, the technological requirements for the vibrating field points of the working body are homogeneity, frequency, amplitude and multiplicity of oscillations of the continuous tossing of the working media. In particular, the frequency for the fluctuation of the characteristic points A_1, A_2 of the working surface with equal amplitudes is given by:

$$\omega_0^2 = \frac{1}{M+L}\left(k_0 + k_1 + k_2 + \frac{l_2 k_2 - l_1 k_1}{l_0}\right). \quad (2)$$

The frequency range of vibrations in which the vibration table of the considered structure provides a homogeneous vibratory field of the working surface points, is extended if the “adjustment item” is included into the contour of the vibration table [15, 16].

4 Approaches to the construction of the measuring system

The formation of the experimental base of physical effects identified with help of a number of special sensors is implemented consequently with the development of the theoretical foundations of the technology of controlling the vibration field of the working body. Following the generalized principles, specialized sensors are developed that reflect the characteristics of the motion of the working mixture of the vibrational technological machines on the basis of indirect signs. Namely, the proposed measurement tools are based on the registering of dynamic effects of contact violation [17,18]. A variant of the schematic diagram of the sensor for evaluation the different dynamic states of the vibrating media is shown in Fig.4.

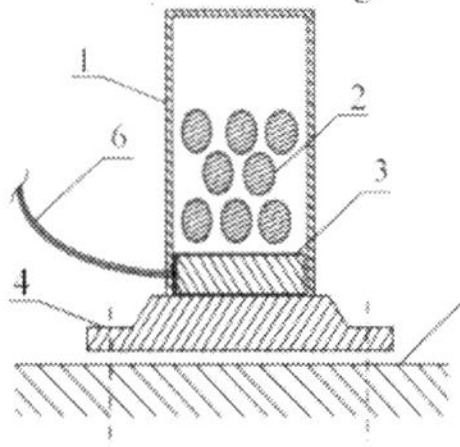

Fig. 4. A sensor for estimating the dynamic conditions of the working medium

Experimental sensor 5 is installed on the surface of vibration table and is comprised of a cylindrical body 1; inertial elements 2 of the working medium; a piezoelectric element 3; a fastening element 4; conductors of current 6. The housing of the dynamic states evaluation sensor contains a few "representatives" of the working media of the vibration technological machine. It is assumed that resolution ability of sensor is sufficient for registration of the physical effects occurring inside the granular layer.

The assumption about the similarity or relationship between of the dynamic characteristics of the "working medium" of vibration technological machine and the "modeled medium" placed inside the sensor is the main design idea of the proposed sensors. Inertial elements of the "modeled medium" of sensor are equal to inertial elements of "working medium" of vibration technological machine. However, the problem of identity of dynamics properties of "modeled medium" and "working medium" has not been solved yet.

The method of four sensors to be simultaneously installed on the working body of the vibration table is proposed in the approach to control the vibration field by changing the mass and inertia characteristics of the working body.

The peculiar features of vibration of the working body as a solid body performing complex oscillatory motion at the harmonic excitation are taken into account by location of sensors. Method for analyzing the vibration field structure is focused on the measurement of the vertical component of the velocities, displacements and accelerations at the corresponding information carrier to determine the distribution of amplitudes of displacements, velocities and accelerations.

It is assumed that the homogeneity of the vertical motion of the working body of the vibration table is provided by the respective limiters of motion and by the ability to perform adjustments by keeping certain ratios of the parameters of motion in other directions except vertical.

Schematic diagram of the vibratory table in the form of the solid body undergoing vertical oscillations is shown in Fig.5. The working body of the vibration table 2 is mounted on the elastic elements 5. Sensors 1 are installed at the corners of the working surface of vibratory table. Rails 4 with mass and inertia elements 3 that can be displaced by means of incremental principle of operation are mounted along the side surfaces of the working body of the vibratory machine.

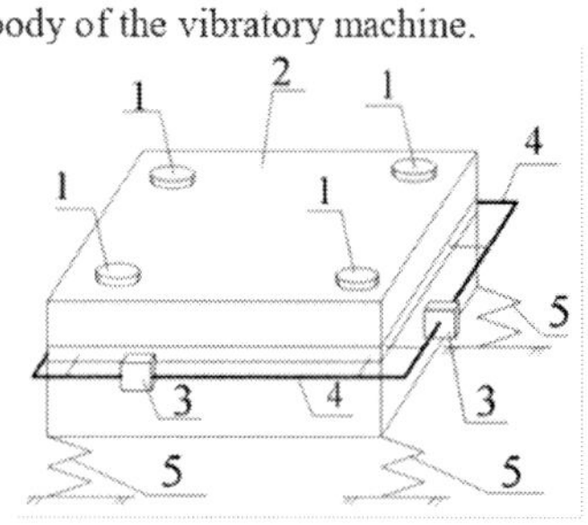

Fig. 5. Schematic diagram of the vibratory table in the form of one solid body which makes a vertical oscillation

The control capabilities are based on the assumption that there is a constructive position of the mass inertial elements, after the transition to which the vibrational field acquires the characteristics of homogeneity with four control points of the unmatched signals.

5 Discussion of results

Experimental testing efficiency of the proposed approach has been carried out in a production environment [19-20]. The performance of the sensors has been verified by special experiments in which statistical certainty of the signals is evaluated, and the dependences of forms of the vibrational modes on the "modeled medium" contained inside the casing of the sensors have been obtained.

A number of experiments on the measurement of signals with different number of "modeled medium" elements have been carried out to establish the dependence of the waveforms on the "modeled medium". The statistical characteristics of the signal, including the standard deviation, are determined by the number of elements of the "modeled medium" and by shape of the sensor casing. Peculiar features of functional dependencies between the basic parameters of the signal are of special interest. Graphics plots of the model sensor signals with 2 and 22 elements of a "modeled medium" which are represented by the vertical chains of steel balls inside the cylindrical casing are shown in Fig.6 and Fig.7 respectively. The signals are represented by the process of "quasi pulsation" which is characterized by a periodic increase of the oscillation amplitude in the control points of a model vibratory table. The same processes are plotted by various graphs of signals from sensors by the same design, but with a different number of elements of "modeled medium".

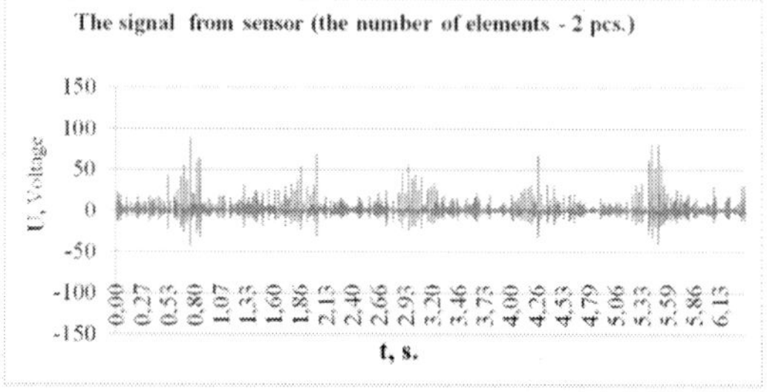

Fig. 6. The signal from sensor with 2 elements.

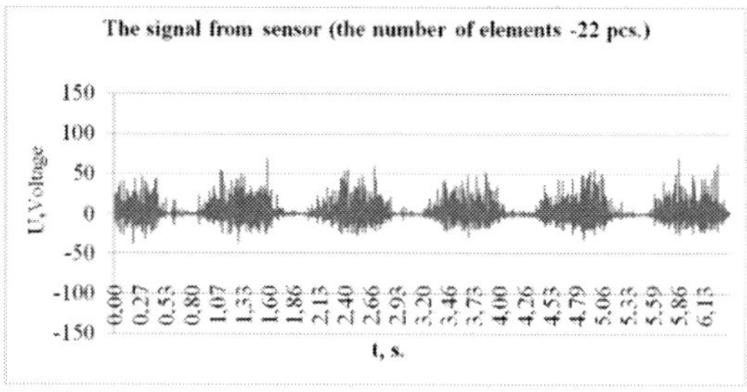

Fig. 7. The signal from sensor with 22 elements.

In addition, methods of mathematical statistics are used to justify the possibility of obtaining a technical result. Thus, the statistical significance of differences in signals from sensors which design includes the introduction of additional elements is checked on the basis of the homogeneity criterion. The registration of the useful signal is confirmed by the statistical significance of the differences between the signal I recorded by a sensor with the inertial elements and signal II recorded by a sensor without the inertial elements, on the basis of the Smirnov homogeneity criterion with a significance level of α=0.05. Fig. 8 shows the empirical distribution functions of signals I and II.

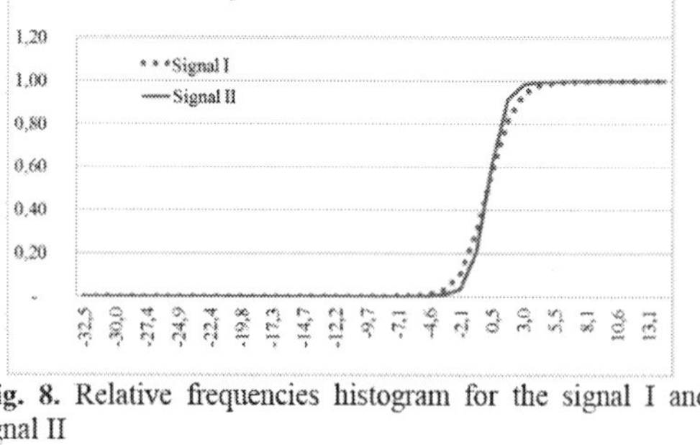

Fig. 8. Relative frequencies histogram for the signal I and signal II

The hypothesis H0 is rejected (H0 is a hypothesis of the homogeneity of the signal I and signal II) on the basis of the Smirnov (Kolmogorov-Smirnov) homogeneity criterion. The rejection of the homogeneity hypothesis indicates the statistical significance of differences between signal I and signal II. This is interpreted as the ability to obtain useful signal of the "modeled medium" of the sensor characterizing the dynamic qualities of the "working medium" of the vibration technological machine.

6 Conclusion

On the basis of studies related to the review and evaluation of measurement methods, it can be stated that fixation and control of current parameters of the vibration process – the displacement, velocity and acceleration – are the object of attention of the developers in the first place. However, for specific technical applications, such as vibration strengthening, compliance with a continuous tossing is of most interest. Despite the perfection of mathematical models, the transition from consideration of the interactions of one material particle with a vibrating surface to the concepts of the processes of interaction of the granular medium layer with the surface is quite complicated. Using sets of sensors to realistically reproduce the process of vibrational interaction can be considered as a solution. The controlling system of the vibration field is designed to configure the dynamic quality of the process of vibration strengthening of long-length parts using the working medium made of steel balls. It has been proposed to change the characteristics of the motion translation device (which imposes additional constraints) and mass and inertia characteristics of the working body by redistributing the locations of the inertial elements to control the vibration field. Measuring devices which construction is based on the assumption that the motion of the "working medium" is similar to that of the "modeled medium" elements of the sensor have been developed to evaluate the integral properties of the "working medium" motion for the vibration technological machine. The article presents an approach to the construction of the control system of the vibratory field of technological machines with the large working body formed by the solid body.

References

1. G.Ja. Panovko, *Dynamics of vibrational technological processes* (Moscow, Izhevsk: RHD Publ., 176 p., 2006) (In Russian)
2. Ju.R. Kopylov, *Dynamics of vibro-impact hardening* (Voronezh: Nauchnaja kniga Publ., 568 p., 2011) (In Russian)
3. A.P. Babichev, I.A. Babichev, *Basics of vibration technology* (Rostov-na-Donu: DGTU Publ., 620 p., 1999) (In Russian)
4. Ju.R. Kopylov, Spline estimation of the distribution parameters of surface layer quality in vibro-impact hardening. Journ. of Mech. Eng., No.7, pp. 53-60 (2016) (In Russian)
5. A.Ju. Marchenko, N.N. Kuznecova, G.V. Serga, New construction of equipment for surface hardening of parts in engineering production. *Proceedings of the VII International scientific-practical conference (InMash-2015)*, pp. 95-99 (2015) (In Russian)
6. M.A. Tamarkin, D.V. Kazakov, A.S. Shvedova, R.V. Grebenkin, Improving the reliability of technological processes of machining by the dynamic methods of surface plastic deformation. Strength. Techn. and Coat., No.8 (128), pp. 23-27 (2015) (In Russian)
7. A.V. Eliseev, Features of mathematical modeling in problems of vibrating interactions between elements of technological systems subject to unilateral constraints relations. *Proceedings of the DYVIS-2015. XVIII International Symposium "Dynamics of vibroimpact (strongly nonlinear) systems"*. Moscow: IMASh RAN Publ., pp. 107-114 (2015) (In Russian)
8. A.V. Eliseev, V.V. Sel'vinsky, S.V. Eliseev, *The dynamics of a vibratory interaction of elements of technological systems subject to unilateral constraints relations* (Novosibirsk: Nauka Publ., 332 p., 2015) (In Russian)
9. I.I. Blehman, *Vibrational mechanics* (Moscow: Fizmatlit Publ., 400 p., 1994) (In Russian)
10. Clarence W. de Silva. Vibration. Fundamentals and Practice. Boca Raton, London, New York, Washington, D.C.: CRC Press, 2000. 957 p.
11. S. V. Eliseev, A. I. Artyunin, *Applied theory of oscillations in problems of dynamics of linear mechanical systems* (Novosibirsk: Nauka Publ., p. 459., 2016) (In Russian)
12. S.V. Eliseev, A.V. Lukyanov , Yu.N. Reznik , A.P. Khomenko, *Dynamics of mechanical Systems with Additional Ties* (Irkutsk: Irkutsk State University Publ., 315 p.)(In Russian).
13. S. V. Eliseev, K. K. Markov, Some problems of dynamics of the oscillatory process at not-holding ties. *Mechanics and control processes* (Irkutsk: Institute of Informatics Problems Publ., pp. 71-83., 1971) (In Russian)
14. V. I. Babitsky, *Theory of vibro-impact systems (approximate methods)*(Moscow:Nauka Publ., 1978, 352 p. (In Russian)
15. A.V. Eliseev, I.S. Sitov, D.X. Nguen, Some questions of formation of structure of vibrating fields of vibration table: features of the measurement system, configuration tools. Syst. Meth. Techn. (Bratsk: BrSTU Publ.), Bratsk, No.2 (30), pp. 17-26 (2016) (In Russian)
16. K. V. Frolov, *Vibrations in engineering: Handbook in 6 volumes, Vol. 6. Protection of equipment from vibration* (Moscow: Mechanical Engineering Publ. 1981, 452 p.) (In Russian)
17. S. V. Eliseev, A. P. Khomenko, Artyunin A. I., V. B. Kashuba, *Sensor to monitor the dynamic state of vibration field (patent)*. The patent for useful model RF no. 166422 U1. Published: 27.11.2016, Bulletin No. 33 (2016) (In Russian)

18. A.V. Eliseev, S.K. Kargapolcev, I.S. Sitov, A.V. Nikolaev, The integral estimation of the state parameters of granular medium under vibration interactions. *7th International Conference on Science and Technology. London, SCIEURO*, 23-29 October 2016. pp. 53-68, (2016) (In Russian)
19. S.V. Eliseev, V.B. Kashuba, A.G. Pnjov, A.V. Eliseev, I.S. Sitov, Unilateral constraints in the dynamic interactions of the granular medium and the vibrating surface: the scientific and methodological substantiation of the technology of vibration hardening. Syst. Meth. Techn. (Bratsk: BrSTU Publ.), No.3 (23), Bratsk, pp. 17-31 (2014) (In Russian)
20. A.V. Eliseev, S.V. Eliseev, Approaches in the development of motion sensors in the problems of dynamics of granular media. *Proceedings of Irkutsk State Technical University*. No. 8 (115), pp. 20–35 (2016). DOI: 10.21285/1814-3520-2016-8-20-35. (In Russian)

Active vibration control of a flexible link robot with the use of piezoelectric actuators

Darren Williams[1,*], *Hamed* Haddad Khodoparast[1], and *Chenyuang* Yang[1]

[1]College of Engineering, Bay Campus, Swansea University, Wales, UK.

Abstract. Within robot systems the use of flexible links could solve many issues raised by their rigid counterparts. However, when these flexible links are integrated within systems which include moving parts their main issue lies in the vibrations experienced along their length due to disturbances. Much research effort has been made to solve this issue, with particular attention being paid to the application of piezoelectric patches as actuators within active vibration control (AVC). The study will consist of accurate models of a flexible link and two surface bonded piezoelectric patches, where the link and the piezoelectric patches will be modelled through the use of Euler-Bernoulli beam theory (EBT). The link will be subject to an initial displacement at its free end, and the resulting displacement of this end of the beam is to be controlled using a classic proportional-differential (PD) controller. The voltages to be applied across each of the actuators is to be controlled in accordance with the displacement of the free end of the beam, the actuators will then induce a strain upon the link opposing the movement of the tip. This research outlines this general method, obtains the best location of the piezoelectric patches and the control gains to be used, and proves that the method can be used to attenuate the vibrations experienced by a flexible link.

1 Introduction

The attenuation of unwanted vibrations is a popular topic within engineering research as multiple fields suffer from the issue when designing, manufacturing and implementing a multitude of designs. The issue affects the majority of the fields of engineering wherein designs include moving parts or are subject to uncontrollable external forces.

Within robotics active vibration control (AVC) is a prevalent area of research as most robot systems will contain moving parts which will be subject to unwanted vibrations. Flexible links within said systems are especially susceptible to unwanted vibrations, even more so than their rigid counterparts. However, the advantages of using flexible links opposed to rigid links greatly outweigh this disadvantage. Flexible links are easier to incorporate into a design, less expensive, and most importantly lighter. Having a system or design which is lighter overall is a highly sought property especially within the fields of robotics and aerospace where payload is a priority.

This leads to research becoming focussed on the AVC of these flexible links. The use of piezoelectric transducers as actuators and sensors has multiple advantages. The application of piezoelectric patches is non-intrusive to a design in the sense that the weight and functionality of said design will not be affected significantly. These lightweight components have a relatively high force output in comparison to their size, and can either be bonded to the surface or embedded within a component.

Crawley and Luis [1] were one of the first to suggest the use of piezoelectric actuators as elements of intelligent structures. They presented analytical models based on the Euler-Bernoulli theory (EBT) for long, slender beams, as well as physical experimentation. *Rathi and Khan* [2] compared the use of surface bonded and embedded piezoelectric actuators, as well as single input single output (SISO) and multiple input multiple output (MIMO) systems. *Nor et al.* [3] employed a proportional integral derivative (PID) controller to alter the voltage across a piezoelectric actuator in accordance with the transverse velocity at the location of the actuator on a plate. With the PID controller being designed within MATLAB® and implemented within COMSOL Multiphysics® via LiveLink™ for MATLAB®. *Wang and Wang* [4] presented their findings on the optimal placements and sizes of a collocated actuator pair surface bonded onto a beam. *Salmasi et al.* [5] produced an analytical model of a flexible link with surface bonded actuators and used simulation software to validate the analytical results through the comparison of the natural frequencies.

Within this research an analytical model of a flexible link with collocated piezoelectric actuators has been created using EBT. A model of the link has then been created within COMSOL Multiphysics® with the same dimensions and properties for the validation of the analytical results. A proportional-differential (PD) controller has been designed and implemented within the

* Corresponding author: 708199@swansea.ac.uk

analytical model. A parametric study has been conducted to obtain the best locations of the piezoelectric actuators and the optimal control gains, and these values have been implemented within the analytical model.

2 Modelling the flexible link

2.1 Description and analytical model

The flexible link and bonded piezoelectric actuator patches have been modelled mathematically as a beam based on the Euler-Bernoulli beam theory, as the link has dimensions which conform to those for which the theory holds (a long slender beam). The link is of length L, width W, thickness T and cross-sectional area A, and has Young's Modulus E, density ρ and a second moment of area I_x. The bonded piezoelectric patch has the same variables, but those corresponding to the patch are denoted through the subscript p. Fig.1 is a visual representation of the model, where $w(x,t)$ represents the transverse motion of the link, and x_1 and x_2 represent the start and end locations, respectively, of the piezoelectric patches along the x axis. The bond between the piezoelectric patch and the link has been considered as perfect for this analytical model.

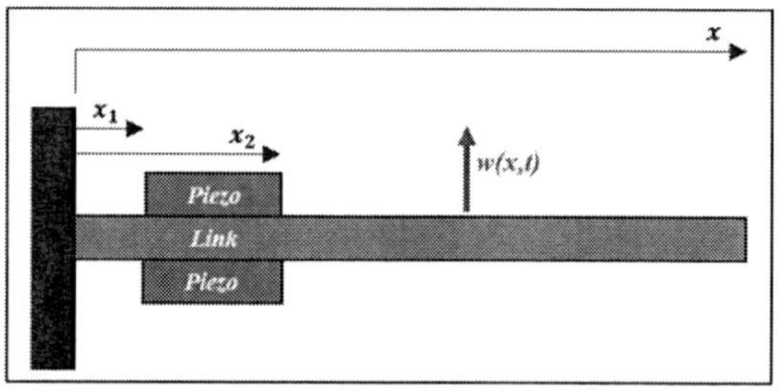

Fig. 1 Visualisation of the model

The equation of motion for the flexible link with a bonded piezoelectric actuator can be written as [6]:

$$\frac{\partial^2}{\partial x^2}\left(EI(x)\frac{\partial^2 w(x,t)}{\partial x^2}\right)+\rho A(x)\frac{\partial^2 w(x,t)}{\partial t^2} +c_a\frac{\partial w(x,t)}{\partial t}=F_0\sin(\Omega t)\,\delta(x-L) +\vartheta V sin(\Omega_V t)\left(\frac{d\delta(x-x_1)}{dx}-\frac{d\delta(x-x_2)}{dx}\right) \quad (1)$$

where, c_a is the viscous air damping coefficient, F_0 is the force applied to the structure and Ω and Ω_V are the frequencies of the force and the voltage applied within the model, respectively. As the density and Young's Modulus of the structure varies over its length, $EI(x)$ and $\rho A(x)$, are functions which vary with respect to x (see equations 2 and 3). The Dirac delta function is denoted as $\delta(x)$, and ϑ is the piezoelectric coupling term and can be calculated using equation 4 [7].

$$\rho A(x)=\rho A+\rho_p A_p[H(x-x_1)-H(x-x_2)] \quad (2)$$

$$EI(x)=EI_x+E_p I_{x_p}[H(x-x_1)-H(x-x_2)] \quad (3)$$

$$\vartheta=\frac{\bar{e}_{31}W}{2OT_p}\left[\left(T_p+\frac{T}{2}\right)^2-\frac{T^2}{4}\right] \quad (4)$$

Within equations 2, 3 and 4 the Heaviside function has been denoted as $H(x)$ and $\bar{e}_{31}$ is the effective transverse piezoelectric coefficient. The model is subject to the boundary conditions of a clamped-free beam.

Substituting equations 2 and 3 into equation 1, and employing the product rule yields:

$$\frac{\partial^2 EI(x)}{\partial x^2}\frac{\partial^2 w(x,t)}{\partial x^2}+2\frac{\partial EI(x)}{\partial x}\frac{\partial^3 w(x,t)}{\partial x^3} +EI(x)\frac{\partial^4 w(x,t)}{\partial x^4}+\rho A(x)\frac{\partial^2 w(x,t)}{\partial t^2} +c_a\frac{\partial w(x,t)}{\partial t}=F_0\sin(\Omega t)\,\delta(x-L) +\vartheta V sin(\Omega_V t)\left(\frac{d\delta(x-x_1)}{dx}-\frac{d\delta(x-x_2)}{dx}\right) \quad (5)$$

2.2 Reduced order model

The reduction of the order of the model requires the employment of the Galerkin Decomposition method [8]. Within this method the spatial dependence is eliminated though the substitution of the transverse motion of the link for a series expansion in terms of the eigenfunctions of the model. For the nth mode the transverse motion can be expressed as:

$$w(x,t)=\sum_{n=1}^{N}T_n(t)X_n(x) \quad (6)$$

where, $T_n(t)$ is the nth generalised coordinate, and $X_n(x)$ is the nth mode shape of the link. Assuming single mode excitation (n^{th} mode) and using the Galerkin method yields:

$$M\ddot{T}_n(t)+C\dot{T}_n(t)+KT_n(t) = F_b\sin(\Omega t)+\theta_p V sin(\Omega_V t) \quad (7)$$

where,

$$M=\rho A(x)\int_0^L X_n^2(x)dx \quad (8)$$

$$C=c_a\int_0^L X_n^2(x)dx \quad (9)$$

$$K=\int_0^L\left[\frac{\partial^2 EI(x)}{\partial x^2}X_n''(x)+2\frac{\partial EI(x)}{\partial x}X_n'''(x) +EI(x)X_n^{iv}(x)\right]X_n(x)dx \quad (10)$$

$$F_b=F_0\int_0^L X_n(x)\delta(x-L)dx \quad (11)$$

$$\theta_p=\vartheta\left(\frac{dX_n(x_2)}{dx}-\frac{dX_n(x_1)}{dx}\right) \quad (12)$$

To solve for the generalised coordinate $T_n(t)$ Equation 7 can be rewritten as:

$$\ddot{T}_n(t) + 2\zeta\omega_n\dot{T}_n(t) + \omega_n^2 T_n(t) = F\sin(\Omega t) + \chi V sin(\Omega_V t) \quad (13)$$

where,

$$\zeta = \frac{c}{2M\omega_n}, \quad \omega_n = \sqrt{\frac{K}{M}}, \quad F = \frac{F_b}{M}, \quad \chi = \frac{\theta_p}{M}. \quad (14)$$

2.3 Validation

The simulation software COMSOL Multiphysics® was employed along with its MEMS module and the LiveLink™ for MATLAB® interface to create a finite element (FE) model of the link and the two piezoelectric patches for comparison with the analytical counterpart. The LiveLink™ for MATLAB® interface was used to control input variables and the computation of the COMSOL® model from the MATLAB® script. Thus, the results were returned from the simulation within the MATLAB® desktop, which ensured easier comparison. Table 1 shows the details of the model used for both methods of testing, and Fig.2 shows the set-up of the model within the software.

Table 1 Details of the link and piezoelectric patch

Parameters	Link (Aluminium)	Piezoelectric Patch (PZT-5H)
Length (m)	0.4	0.056
Width (m)	0.035	0.028
Thickness (m)	0.003	0.0003
Density (kg/m^3)	2700	7500
Young's Modulus (GPa)	70	61.05
Piezoelectric coefficient, $\bar{e}_{31}$ (C/m^2)	N/A	-11

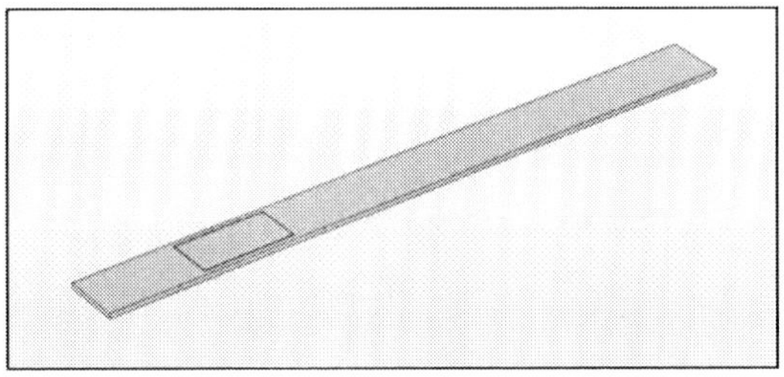

Fig. 2 COMSOL Multiphysics® model of link and bonded piezoelectric patches.

A free tetrahedral mesh was used on the COMSOL Multiphysics® model, and the maximum element size of this mesh was determined through a mesh convergence study. As can be seen in Fig.3 there was convergence observed when the maximum element size was around $0.02m$, this is very close to the predefined mesh size within the software 'Finer', was employed.

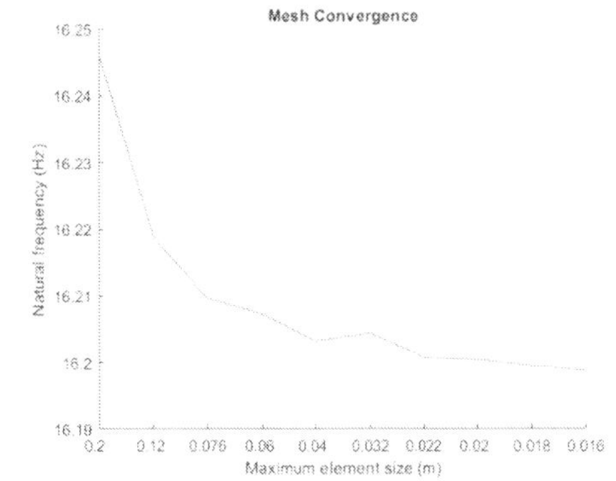

Fig. 3 Mesh convergence study on the first natural frequency of the beam.

The natural frequencies (ω_n) were calculated both analytically and through the COMSOL Multiphysics® model, with the first comparison not including the piezoelectric patches. The first three modes were tested, and the results are shown in Table 2.

Table 2 Natural frequency results for the first three modes

Mode	Analytical (H_z)	COMSOL® (H_z)	Error (%)
1	15.438	15.422	0.10
2	96.747	96.633	0.12
3	270.894	270.510	0.14

From the results, it is clear to see that the two modelling methods produce very similar natural frequencies. The error between the analytical and the COMSOL® results is expected due to two reasons; the link modelled analytically in two dimensions and modelled within COMSOL® in three dimensions, and that COMSOL® uses approximate numerical methods (such as FE) and the analytical method is more accurate.

For the second comparison, the analytical and simulation models were edited to include the two piezoelectric patches. The piezoelectric patches were set to a location along the beam (x_1) of 0.06m within both models. The natural frequencies retrieved from the two models are presented in Table 3 below.

Table 3 Natural frequency results for the first three modes (models include piezoelectric patches)

Mode	Analytical (H_z)	COMSOL® (H_z)	Error (%)
1	15.954	16.2	1.58
2	94.738	95.393	0.70
3	266.606	266.250	0.13

As can be seen from the results in Table 3 above, the inclusion of the piezoelectric patches produced slightly larger errors between the two different models for the first three modes. As the errors are all below three percent, the analytical model can be deemed as validated. The slight errors can, again, be explained by the reasons mentioned previously.

3 Active Vibration Control

3.1 Design of control system

A Proportional-Differential (PD) controller was implemented for the attenuation of the transverse motion of the link as a result of an initial displacement at the free end of the link. The control system took advantage of the indirect piezoelectric effect, so that the output of the system was the voltage to be applied across the piezo patches. Fig.4 below shows the model of the control system used.

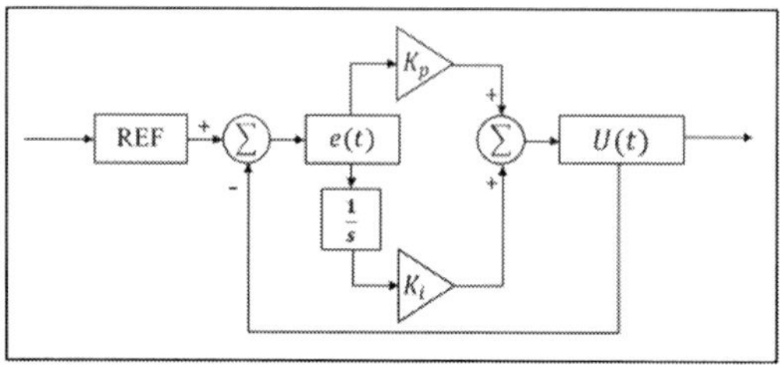

Fig. 4 Model of the PD Control system

The control model can be expressed as the following equation for use within the analytical model:

$$U(t) = K_p e(t) + K_d \frac{de(t)}{dt} \tag{15}$$

where, REF is the desired displacement of the tip of the link ($0\ m$), and $e(t)$ is the error, which is the difference between REF and the measured displacement. $U(t)$ is the controlled output of the system, which, in this instance is the displacement of the tip of the beam, and K_p and K_d are the proportional and differential gains, respectively. As the output of the control system is the displacement required for the attenuation of the vibration of the tip, and the voltages required to be applied across the two piezoelectric patches are to be obtained, the following equation is to be used.

$$V(t) = \frac{1}{2\chi_2}\left(M\frac{d^2U(t)}{dt^2}\right) \tag{16}$$

where, $V(t)$ is the total voltage required, M is the mass of the system, $U(t)$ is the controlled output of the system, and χ_2 is as follows:

$$\chi_2 = \frac{\bar{e}_{31}W}{2T_p}\left[\left(T_p + \frac{T}{2}\right)^2 - \frac{T^2}{4}\right] \times \left(\frac{dX_n(x_2)}{dx} - \frac{dX_n(x_1)}{dx}\right) \tag{17}$$

Although the operating voltage range of piezoelectric patches generally include negative voltages, the magnitude of the negative voltages is a lot smaller than the positive voltages i.e. a piezoelectric patch may have an operating voltage of -60V to 360V. Therefore, within this paper the operating voltage range will be considered to be positive only. The magnitude of the positive voltages obtained within equation 16 will be applied to the piezoelectric patch on the upper surface of the link, and the magnitude of the negative voltages will be applied to the piezoelectric patch on the lower surface of the link.

3.2 Analytical model of control system

To incorporate the control system into the analytical model, equation 13 is to be re-written so that it includes the control law $u(t)$ (which in this case represents force) as follows [9].

$$\ddot{T}_n(t) + 2\zeta\omega_n\dot{T}_n(t) + \omega_n^2 T_n(t) = u(t) \tag{18}$$

Writing the above equation in terms of the velocity and acceleration, so that the equations to be solved become second order differential equations, yields:

$$\begin{aligned} \dot{y}_1 &= y_2 \\ \dot{y}_2 &= -2\zeta\omega_n y_2 - \omega_n^2 y_1 + u \end{aligned} \tag{19}$$

where, y_1, $\dot{y}_1$ and $\dot{y}_2$ are the displacement, velocity and acceleration of the system, respectively. M is the mass of the system, and the control law u may now be defined as:

$$u = -ky = -\begin{bmatrix} k_p & k_d \end{bmatrix}\begin{bmatrix} y_1 \\ y_2 \end{bmatrix} \tag{20}$$

Using the above relation and writing equation 20 into the matrix form:

$$\dot{y} = \boldsymbol{F}y + \boldsymbol{G}u \tag{21}$$

yields:

$$\begin{bmatrix} \dot{y}_1 \\ \dot{y}_2 \end{bmatrix} = \begin{bmatrix} 0 & 1 \\ -\omega_n^2 & -2\zeta\omega_n \end{bmatrix}\begin{bmatrix} y_1 \\ y_2 \end{bmatrix} + \begin{bmatrix} 0 \\ 1 \end{bmatrix} u \tag{22}$$

The above equation (23) can then be used to produce the following equations to be solved to find the displacement and the velocity of the system (y_1 and y_2).

$$\begin{aligned} \dot{y}_1 &= y_2 \\ \dot{y}_2 &= -(\omega_n^2 + k_p)y_1 - (2\zeta\omega_n + k_d)y_2 \end{aligned} \tag{23}$$

To find the voltage required for the control the value of $U(t)$ can be calculated using the following equation.

$$U(t) = k_d y_2(t) + k_p y_1(t) \tag{24}$$

4 Parametric Study

4.1 Objective function

To obtain the best of both the location of the piezoelectric patches and the control gains an objective function was required. In this case this has been chosen to be the amount of time necessary to attenuate the displacement of the tip of the link to a value below an arbitrarily chosen displacement value (t_{Atten}). For the purpose of the optimisation this chosen displacement value will be referred to as the Effective Zero Displacement (EZD) value, any amplitudes of the displacement below this value will be considered to be zero. The EZD value is chosen based on when the tip displacement becomes acceptably attenuated (barely noticeable to the naked eye); to allow the vibrations to stop completely would require a significantly greater amount of time.

The parametric study will focus on reducing t_{Atten} for the harmonic motion of the link, when subject only to an initial displacement at its free end. This initial displacement has been arbitrarily chosen as a value that a link within a robot system of the physical parameters within Table 1 is likely to be subject to. The viscous air damping coefficient (c_a) is taken to be $4.4\ kg/s$.

4.2 Location of the piezoelectric patches

The location of the piezoelectric patches along the link (x_1) affects both the natural frequency of the model, and the moment produced by the actuators on the beam. These two variables, in turn, effect the time taken to attenuate the amplitude of the vibrations below the EZD. For a range of actuator locations $(0.02{:}\,0.001{:}\,0.12)$, set values of the gains ($K_p = 2$ and $K_d = 1s$), and an initial tip displacement of $0.01m$, t_{Atten} was obtained. The locations were limited to this range as in previous research the ideal location was proven to be nearest to the fixed end of the beam when the length of the piezoelectric patches was less than the length of the link [4]. The results are presented in Fig.5 below.

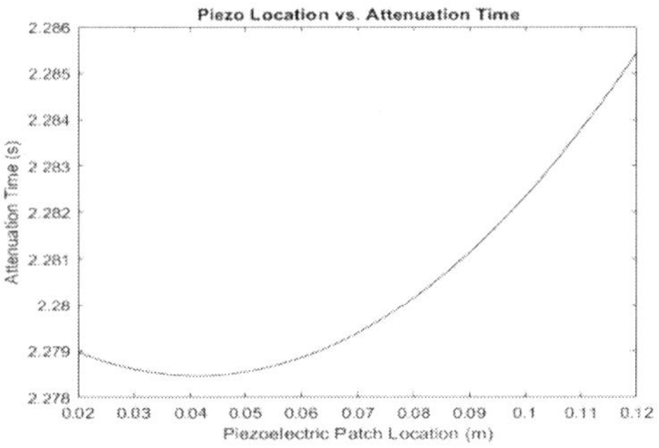

Fig. 5 Attenuation time for each location of the piezoelectric patches

The results show that for the dimensions and properties of the link and piezoelectric patches (see Table 1) the best location (x_1) lies around 0.04m. This is to be expected as the maximum strain energy is at the clamped end of the link.

4.3 Control gains

The proportional and differential control gains affect the voltage to be applied across the piezoelectric patches, thus affecting the moment the actuators produce upon the link. As both types of gains are used within the controller it was important to obtain the best values of the gains as a combination of K_p and K_d, so that for each value of K_p within a set range, a range of K_d values were used. This meant that for the proportional gain range of $0{:}\,0.1{:}\,60$, and a differential gain range of $0{:}\,0.01{:}\,4.4s$ there were 601x441 combinations of values tested.

An issue with acquiring the best combination of the gains is that most piezoelectric actuators will have a range for the operating voltage. In this case the operating voltage of the actuators was assumed to be 0 to 360V, and the maximum voltage produced for each combination of gains was recorded. Therefore, any gains which produced a voltage exceeding this operating voltage were eliminated. Although, it should be noted that the range of the gains tested were chosen to eliminate voltages above the maximum operating voltage of the piezoelectric patches to a certain extent.

The locations of the actuators were kept at a constant value throughout the optimisation of the gains ($x_1 = 0.04m$), and the tip of the link was subject to an initial displacement of $0.01m$. A small selection of the results is presented in Fig.6, note that the differential gain range has been reduced to improve the presentation.

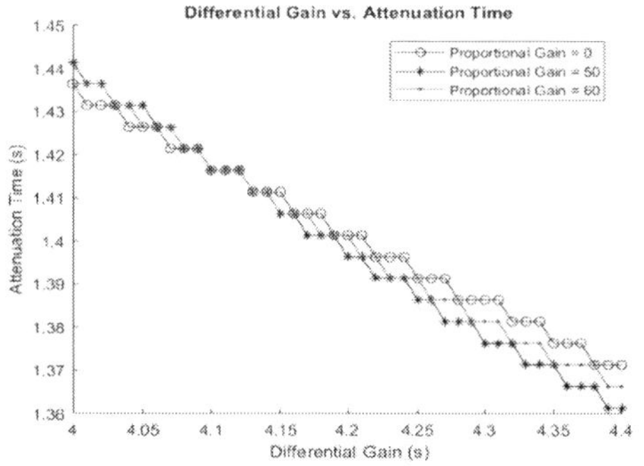

Fig. 6 Attenuation time against the proportional gain for three different proportional gain values.

The results show that as the differential gain is increased, the time required to attenuate the tip displacement below the EZD is reduced. The alteration of the proportional control value did not have as a significant effect on the attenuation time. However, as can be seen in the above figure, the best combination of gains (which still allow the maximum voltage required to lie within the operating voltage) is $K_p = 50$ and $K_d = 4.4s$.

5 Results

Once the best location of the piezoelectric patches and the best combination of control gains were found, the open and closed loop response of the system could be predicted through the analytical model. Fig.6 shows the open and closed loop response of the system when the dimensions and properties of the link and piezoelectric patches are those within Table 1, for the first mode of the model. The initial displacement at the tip of the link was set to 0.01m, and the free vibration of the tip of the clamped-free link as a result of the initial displacement was estimated. The location of the piezo patch and the control gains were the ideal values found through the parametric study ($x_1 = 0.04m$, $K_p = 50$ and $K_d = 4.4s$), and the viscous air damping coefficient (c_a) was set to $4.4\ kg/s$ ($\zeta = 0.02$).

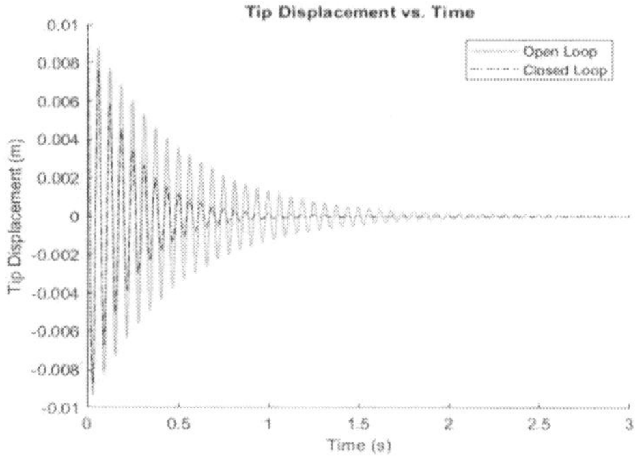

Fig. 7 The open and closed loop responses of the system over time

Fig.7 clearly shows that the closed loop response has a faster attenuation time compared to that of the open loop. Although some damping can be observed in the open loop response, which is due to the viscous air damping coefficient (c_a), the closed loop response can be seen to have a much greater damping.

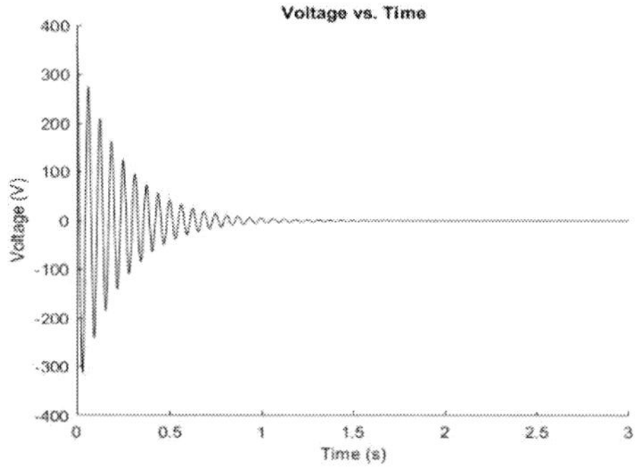

Fig. 8 The voltage applied across the upper and lower piezoelectric patches over time

The voltages that are to be applied across the piezoelectric patches can be seen within Fig.8. Here it can be seen that the maximum operating voltage of the piezoelectric patches is not exceeded. The magnitudes of the positive voltages are to be applied to the piezoelectric patch attached to the upper surface of the link, and negative to the patch attached to the lower surface.

6 Conclusion

This research has proven theoretically that the use of piezoelectric patches as actuators within AVC is an effective method of attenuated unwanted vibrations within a flexible link for the first mode. The modelling method of the link (EBT) has been validated through the use of a finite element model created using COMSOL Multiphysics®. A PD controller has been modelled analytically and has been effective at controlling the voltages to be applied across the actuators as a result of the free end displacement of the link. A parametric study was used to find the ideal variables, before the open and closed loop responses of the theoretical system were compared.

Although the research has proven theoretically that the proposed method for AVC is effective, the results have not been fully validated. Both the analytical models of the link with and without the inclusion of the piezoelectric patches have been proven to be accurate, through the comparison with the finite element model. However, it would then be necessary to verify the results produced through the use of AVC within the analytical model with results produced from a finite element model which also includes AVC.

The results of the research can then be validated further through the conduction of a physical experiment, which would be set up as the model in Fig.1. This research would include attempting to use low voltages (up to 50V) to power the piezoelectric patches for situations in which higher voltages are unobtainable, taking advantage of the negative operational voltages of piezoelectric patches. In order to fully comprehend and predict the vibrations experienced by a flexible link as part of a robot system future research would include a simulation of a robot arm with a flexible link attached to the end effector. It would then be possible to analyse the vibrations within the flexible link as a result of the motions of the robot arm, and eventually employ the AVC method used within this paper to attenuate said vibrations.

References

1 E.F.Crawley and J.Luis. AIAA Journal. *Use of Piezoelectric Actuators as Elements of Intelligent Structures*. **25**. 1373-1385. (1987).
2 V.Rathi and A.H.Khan. LAJSS. *Vibration attenuation and shape control of surface mounted, embedded smart beam*. **9, No. 3**. (2012).
3 K.A.M.Nor et al. IEEE Conference Publications. *Optimization in Active Vibration Control: Virtual Experimentation Using COMSOL Multiphysics – MATLAB Integration*. **5**. 385-389. (2014).
4 Q.Wang and C.M.Wang. IOPSCIENCE Journal. *Optimal placement and size of piezoelectric patches on beams from the controllability perspective*. **9**. 558-567. (2000).
5 H.Salmasi et al. CAODIE. *Active Vibration Suppression of a Flexible Link Manipulator Using a Piezoelectric Actuator*. **X**. 199-208 (2007).
6 Inman, D. J. (2014). *Engineering vibration*.
7 Erturk, A., & Inman, D. J. (2011). *Piezoelectric energy harvesting*. John Wiley & Sons.
8 H.Madinei et al. MSSP Journal. *Design of MEMS piezoelectric harvesters with electrostatically adjustable resonance frequency*. **81**. 360-374. (2016).
9 Bolton, W. (2002). *Control systems*. Newnes.

Synthesis of control algorithms for robotic platform with electro-hydraulic drive

Larisa Rybak[1,*], *Elena* Gaponenko[1], and *Dmitry* Malyshev[1]

[1]BSTU after V. G. Shoukhov, Dep. of machine building technology, 308012 Kostyukov str., Russia

Abstract. The article deals with a robotic vibroprotection platform with electro-hydraulic drive. Built mathematical model systems from the account for characteristics of the drive. Synthesized invariant controller by system state that uses reverse ties coefficients. Controller synthesized at discrete area with a restriction in the magnitude of the relative displacement of the object relative to the base. To ensure the stability of the controller are selected in regulation law amendment providing the desired value of the relative displacement. The results of mathematical modeling are presented.

1 Introduction

Vibration accompanies us everywhere and in most cases it is undesirable. First of all, one can call vibration and fluctuations in road and rail transport, in the mining industry, vibration of equipment, motors and machines, oil and gas platforms, buildings and structures in the zone of high seismic hazard, undesirable vibrations of laboratory tables (especially optical ones) etc. In all these cases, the task is to isolate the person or the payload from the source of vibration or, conversely, isolate the source of vibration. Particularly important today is the issue of human vibroprotection (including operator-driver of mobile machines and equipment).

General vibration, as a result of the mechanical interaction of a person with the external environment, is an important ecological factor throughout biological evolution and has great biological activity. The formation of physiological and pathological shifts on the part of various systems of the body, the frequency and severity of these disorders depends on the ergonomic data of the workplace and human characteristics.

The nature and depth of vibration disorders are caused by changes in the vestibular analyzer, musculoskeletal system, cardiovascular system, hemodynamics of blood circulation, neurohumoral sphere. Sensitive to vibration is the opto-vestibulospinal system with the development of the main clinical manifestations, such as a headache, non-systemic dizziness, impaired coordination, lability of arterial pressure, changes in the vessels of the fundus, signs of the neurosis-like syndrome and autonomic dysfunction, pain in the back muscles, lumbar spine. Vibration causes a decrease in visual acuity, a violation of color perception, a narrowing of the field of view, a decrease in the stability of clear vision, a decrease in functional mobility, a disorder in fixing objects with the eye, a violation of the sharpness of the perception of objects, a difficulty in reading instrumental information.

The level and frequency spectrum of vibration in the driver's workplace depends on the road profile, driving speed, engine vibration level, chassis and cabin suspension and, finally, the driver's seat. For example, the vibration levels of the driver's seats of excavators ECG-8I, ECG-4,6B in the quarry of the Lebedinsky ore mining and processing combine exceed permissible by 2 ... 10 dB in the frequency range 2 ... 16 Hz. Spectral characteristics of input excitations of operators' workplaces show that the vibrational energy of quarry machines is concentrated at the lower end of the frequency range and has characteristic peaks.

In many respects, the effect of vibration on a person depends on its spectral composition. Areas of frequencies are known in which vibration levels can affect the performance of production operations. So the frequency of 4 ... 5 Hz is a resonant area for a person working standing, 11 Hz - sitting.

Reducing the level of vibration of workplaces and improving their ergonomic characteristics will solve one of the most important problems of improving working conditions and preserving the health of workers.

Methods of protection from vibration are divided into two groups: methods that reduce the parameters of vibration by influencing the source of excitation, and methods that reduce the parameters of vibration along the path of its propagation from the source of excitation.

Vibration isolation system is divided into active and passive. Despite all the structural differences, the essence of the vibration damping system is the same. The passive system consists of a spring and a damper. The spring is designed to soften vibrations and shocks, and the damper - to quench the fluctuations that have arisen in the system. The active system also uses an additional pair consisting of an accelerometer and an electromagnetic drive, which makes it possible to achieve an exceptionally high degree of vibration isolation.

* Corresponding author: rl_bgtu@intbel.ru

It is proposed to develop an active system that provides effective vibration isolation from 0.1 Hz, the absence of resonances and the necessary rigidity under the action of constant accelerations.

2 Description of schemes and statement of the problem

Consider a system with an electrohydraulic actuator (Figure 1), which includes a hydraulic cylinder located between the object and the base, an electrohydraulic transducer, as well as the actual vibration protection object, accelerometers on the object and the base, a relative displacement sensor and a controller.

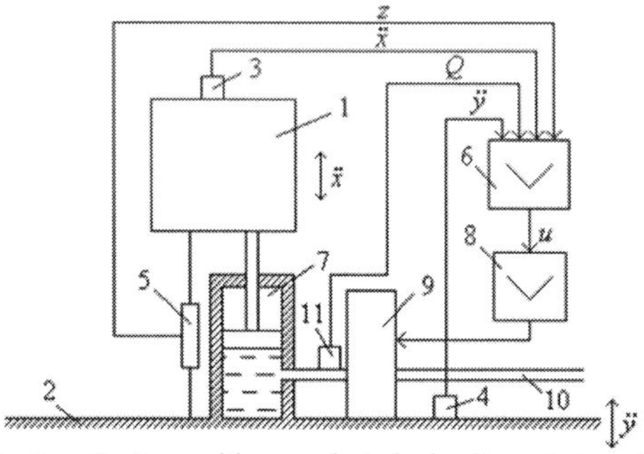

Fig. 1. System with an electrohydraulic actuator 1 - vibroprotection object; 2 - base; 3, 4 - accelerometers on the object and the base; 5 - a sensor of the relative displacement; 6 - a controller; 7 – hydraulic cylinder; 8 - power amplifier; 9 - electro-hydraulic converter; 10 - hydraulic line; 11 - working fluid flow sensor

The electrohydraulic conversion unit consists of a throttling distributor with a stepping motor which, through a screw pair, moves the four-spool valve in the axial direction. The stepper motor supplied voltage pulses, and each pulse corresponds to the rotation of its shaft at a certain angle-pitch (angular increments), which often is 1,5°. Thus, the angle of rotation is determined by the number of pulses fed, and the rotational frequency by the frequency of their following. The throttle valve is essentially a hydraulic power amplifier that converts the input mechanical action into a corresponding displacement of the output link, which allows the flow and direction of the oil flow to be varied in several lines simultaneously, depending on the external control action.

Consider the synthesis of invariant control [1, 2], which will create such a system operating modes which provide full stabilization vibration isolation object [3].

The problem of the synthesis of invariant control is essentially a task of synthesizing an algorithm for controlling the inverse dynamic problem. The basis of the method of inverse problems of dynamics is the following: the algorithm for the normalization of the control function is constructed by the principle of symmetry of the structure and inversion of operations with respect to the structure and group of operations corresponding to the mathematical model of the controlled process. Consequently, the symmetry properties inherent to the object and a control system that uniquely define the structure and parameters of the algorithm of forming control functions assigned to the path of movement $\mathbf{X}^*(t)$ the structure and parameters of the control algorithm be determined by the structure and parameters of a mathematical model of the object.

This provision is the methodological basis for constructing motion control algorithms for dynamic systems. In this case, in fact, the task of designing the control algorithm is presented as the inverse problem of dynamics: it is required to find a control function (power) u^*, which implements the prescribed trajectory of motion $\mathbf{X}^*$ [4].

Let us consider the procedure for the synthesis of an invariant control perturbation applied to the system under consideration with an electrohydraulic actuator. We describe the entire system, including the actuator, in the state space. In developing the mathematical model of this executive mechanism, a system of equations

$$\begin{cases} \ddot{z} = \dfrac{Ak_{\text{тр}}}{m} p, \\ \dot{p} = -\dfrac{4AE}{V}\dot{z} - \dfrac{4Ek_{\text{ут}}}{V} p + \dfrac{4E}{V} Q \end{cases} \quad (1)$$

We introduce 2 state variables: $x_1 = \dot{z}$, $x_2 = p$ The system of equations (1) becomes

$$\begin{cases} \dot{x}_1 = \dfrac{Ak_{\text{тр}}}{m} x_2, \\ \dot{x}_2 = -\dfrac{4AE}{V} x_1 - \dfrac{4Ek_{\text{ут}}}{V} x_2 + \dfrac{4E}{V} Q \end{cases} \quad (2)$$

or in vector-matrix form

$$\dot{\mathbf{X}} = \mathbf{AX} + \mathbf{B}Q, \quad (3)$$

Synthesis optimal controller problem will be solved in the discrete area, to which the system of equations (3) translate into discrete form

$$\mathbf{X}[i+1] = \mathbf{A}_\Delta \mathbf{X}[i] + \mathbf{B}_\Delta Q[i], \quad (4)$$

where $\mathbf{A}_\Delta = \exp(\mathbf{A}T) = \mathbf{I} + \sum_{i=1}^{\infty} \dfrac{\mathbf{A}^i T^i}{i!}$,

$$\mathbf{B}_\Delta = \left(\int_0^T \exp(\mathbf{A}T)\, dt \right) \mathbf{B} = \left(\mathbf{I}T + \sum_{i=1}^{\infty} \frac{\mathbf{A}^i T^{i+1}}{(i+1)!} \right) \mathbf{B}.$$

In the equations (4), the symbols $\mathbf{A}_\Delta$ and $\mathbf{B}_\Delta$ designate analogs of matrices $\mathbf{A}$ and $\mathbf{B}$ for a discrete system, and the symbol T is entered to designate the sampling period. Because of the high complexity of mathematical calculations is not appropriate to display the general form and the matrices $\mathbf{A}_\Delta$ and $\mathbf{B}_\Delta$. Preferred to calculate the matrix $\mathbf{A}_\Delta$ and $\mathbf{B}_\Delta$ in numerical form for each specific pair of matrices $\mathbf{A}$ and $\mathbf{B}$. Next, to simplify the entries, we return to the notation without indices: $\mathbf{A} = \mathbf{A}_\Delta$; $\mathbf{B} = \mathbf{B}_\Delta$;.

From the system equation I (1) can obtain the equation

$$\ddot{x}[i] - \ddot{y}[i] = \frac{Ak_{\text{тр}}}{m} p[i] \quad (5)$$

Having put $\ddot{x}[i]=0$. You can obtain an expression for the desired differential etc. The pressure in the hydraulic cylinder::

$$p[i]=-\frac{m}{Ak_{\text{тр}}}\ddot{y}[i] \quad (6)$$

Let $\mathbf{A}=\begin{bmatrix} a_{11} & a_{12} \\ a_{21} & a_{22} \end{bmatrix}$, $\mathbf{B}=\begin{bmatrix} b_1 \\ b_2 \end{bmatrix}$, where a_{ij}, b_k – specific numerical values obtained for a particular system after its reduction to form (4). Then from II system equation (4) we obtain

$$Q[i]=\frac{1}{b_2}p[i+1]-\frac{a_{22}}{b_2}p[i]-\frac{a_{21}}{b_2}\dot{z}[i] \quad (7)$$

In view of (6), (7) we obtain the value of the manipulated variable flow

$$u_Q[i]=-\frac{m}{Ak_{\text{тр}}b_2}\ddot{y}[i+1]+\frac{ma_{22}}{Ak_{\text{тр}}b_2}\ddot{y}[i]-\frac{a_{21}}{b_2}\dot{z}[i] \quad (8)$$

For formation of the algorithm of the control law (8) must be the acceleration value based on the following clock cycle $\ddot{y}[i+1]$. It is possible to use a method of predicting a given value, as proposed in [3, 6], which is to use extrapolation Lagrange formula

$$\hat{\ddot{y}}[i+1]=3\ddot{y}[i]-3\ddot{y}[i-1]+\ddot{y}[i-2] \quad (9)$$

Using Lagrange estimation, considering the formulas (8) and (9), to form the expression control law

$$u_Q[i]=\frac{m(a_{22}-3)}{Ak_{\text{тр}}b_2}\ddot{y}[i]+\frac{3m}{Ak_{\text{тр}}b_2}\ddot{y}[i-1]-$$
$$-\frac{m}{Ak_{\text{тр}}b_2}\ddot{y}[i-2]-\frac{a_{21}}{b_2}\dot{z}[i] \quad (10)$$

We synthesize the control system for the electrohydraulic system.

The matrices **A** and **B** of the equation (3) of equation (3) for such a following system:

$$\mathbf{A}=\begin{bmatrix} 0 & 3,656\times10^{-5} \\ -2,014\times10^{10} & -9,680\times10^{3} \end{bmatrix}$$

$$\mathbf{B}=\begin{bmatrix} 0 \\ 5,232\times10^{12} \end{bmatrix}.$$

Going to a discrete task, taking a sampling period $T=200\,\text{ms}$, we get the following matrix $\mathbf{A}_\Delta$ and $\mathbf{B}_\Delta$ equation (4):

$$\mathbf{A}_\Delta=\begin{bmatrix} 0,992 & 3,217\times10^{-9} \\ -1,772\times10^{6} & 0,140 \end{bmatrix},$$

$$\mathbf{B}_\Delta=\begin{bmatrix} 2,201 \\ 4,604\times10^{8} \end{bmatrix}.$$

After performing the appropriate calculations, we define the control law for the invariant controller

$$u_Q[i]=(-1,699\times10^{-4})\ddot{y}[i]+(1,782\times10^{-4})\ddot{y}[i-1]+$$
$$(-5,941\times10^{-5})\ddot{y}[i-2]+(3,848\times10^{-3})\dot{z}[i] \quad (11)$$

To control algorithm (11), the mathematical modeling of the system when harmonic distortion. Graphs of acceleration at the object and displacement of the object relative to the base for the frequency of the disturbing effect $f=2\,\text{Hz}$ are shown in Fig. 2. The amplitude of the disturbance is assumed equal to $A_y=g$. In this case, the disturbance is almost completely suppressed, if we do not take into account the noise caused by the stepping nature of the electric motor.

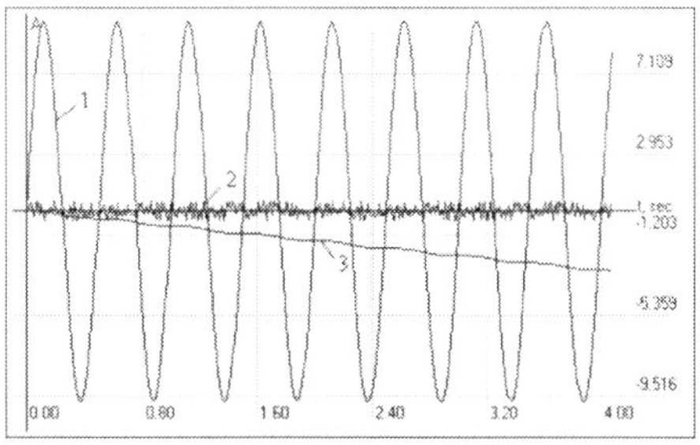

Fig. 2. Modeling the system operation at the disturbance frequency $f=2$ Hz: *1* – basis acceleration; *2* – object acceleration; *3* – moving the object relative to the base.

According to the expression (10), a control law there is only one value being a coordinate system state - is the relative velocity of the object $\dot{z}$. Can be introduced into the control channel of the relative speed of the object defining a small correction Δ [5], whereas the control law takes the form

$$u_Q[i]=\frac{m(a_{22}-3)}{Ak_{\text{тр}}b_2}\ddot{y}[i]+\frac{3m}{Ak_{\text{тр}}b_2}\ddot{y}[i-1]-$$
$$-\frac{m}{Ak_{\text{тр}}b_2}\ddot{y}[i-2]-(1-\Delta)\frac{a_{21}}{b_2}\dot{z}[i] \quad (12)$$

The introduction of an amendment to the nature of the impact on the system is equivalent to introducing restrictive feedback into the control channel at the relative velocity of the object. When $\Delta=0$ expression (12) gives a control law for a perfect invariant controller.

In addition, the control signal must be limited from above and from below in accordance with expression

$$u'_Q[i]=\begin{cases} u_Q[i], & |u_Q[i]|\le Q_{\max}, \\ \operatorname{sgn}(u_Q[i])Q_{\max}, & |u_Q[i]|>Q_{\max}, \end{cases} \quad (13)$$

where $Q_{\max}$ is maximum allowable working fluid flow, defined by parameters of the throttle control valve and hydraulic lines. As before, in the control loop for the flow of working fluid as a controller, we select a three-position relay controller.

In Fig. 3 is a structural diagram of a vibroprotection system invariant controller taking into account the control law (12), and the constraints on the manipulated variable.

 https://doi.org/10.1051/matecconf/201814811006

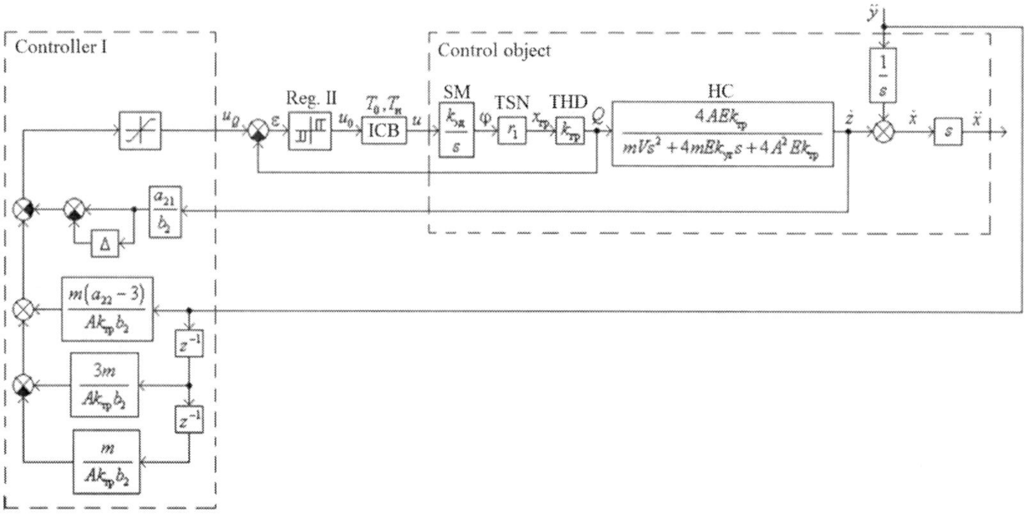

Fig. 3. Active vibroprotection system structural diagram with the electro-hydraulic actuating mechanism and invariant controller: SM - stepper motor, TSN - transmission "screw-nut", THD - throttling hydraulic distributor, HC - hydraulic cylinder.

For the control algorithm (1.131) (the magnitude of the correction $\Delta = 0{,}01$) a simulation of the operation of the system was carried out under perturbations of various types. Graphs of acceleration at the object and displacement of the object relative to the base for the frequency of the disturbing effect $f = 1\,\text{Hz}$ are shown in Fig. 4.

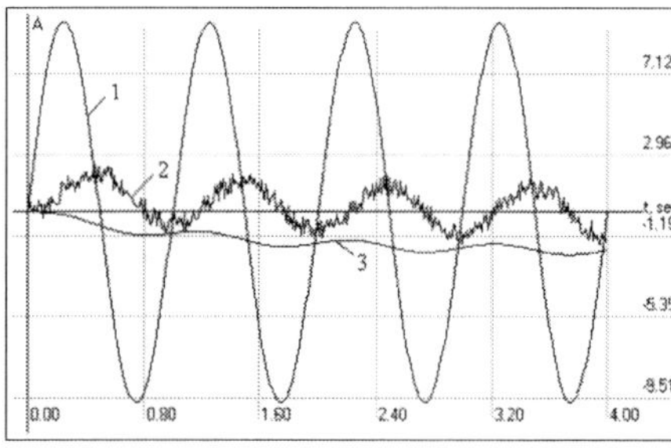

Fig. 4. Simulation of the system at a frequency of 1 Hz disturbance.

The amplitude of the disturbing effect was assumed equal to $A_{\ddot{y}} = g$. In this case, the coefficient of suppression of the disturbing vibrational action in amplitude is $\frac{A_{\ddot{x}}}{A_{\ddot{y}}} = 0{,}2$. It should also be noted that the relative movement of the object is limited and can be kept at an acceptable level. Level steady relative movement can be reduced by increasing the correction Δ a control law (12). However, it should be borne in mind that in this case, the acceleration amplitude on the object will increase. In any case, the decision will have a compromised character.

In the course of mathematical modeling, it was found that disturbances in the range 1 ... 20 Hz are suppressed by a system with an invariant controller better than a system with an optimal controller. This means that with the same level of relative movement of the object and comparable control costs, it is possible to achieve a somewhat lower level of acceleration at the site. However, in general, the frequency range of operation of the system in the case of using the invariant controller narrows.

This work was supported by the Russian Science Foundation, the agreement number 16-19-00148.

References

1. F.A. Aliyev. V.B. Larin, K.I. Naumenko, V.N.Suntsev, *Optimization of time-invariant control systems*, 327 (1978)
2. A.V. Gradetsky, A.I. Pashkov, L.A. Rybak, A.V. Sinev, V.S. Solovyov, Probl. of mach. build. and mach. reliab., **4**, 78-86 (1995)
3. L.A. Rybak, A.V. Sinev, A.I. Pashkov, *The synthesis of the active vibration isolation systems in space objects*, 160 (1997)
4. E.A. Fedosov, A.A. Krasovsky, E.P. Popov et al., *Machine building*, **1-4**, 688 (2000)
5. L.A. Rybak, V.K. Arshinov, R.V. Erzhukov, XIII Int. Symp. "Dyn. of vib-imp. (hi. nonl.) sys.", 82 (2001)

* Corresponding author: rl_bgtu@intbel.ru

Middle ear vibrations with SMA prosthesis. Experimental research

Rafal Rusinek[1,*], *Marcin* Szymanski[2,**], and *Grzegorz* Teresinski[2,***]

[1] *Lublin University of Technology, Nadbystrzycka 36, 20-618 Lublin, Poland*
[2] *Medical University of Lublin, Jaczewskiego 8, 20-854 Lublin, Poland*

Abstract. The paper focuses on experimental research of a middle ear prosthesis made of shape memory alloy. The prosthesis provides better adjustment to individual patient than classical prosthesis. The shape memory prosthesis is implemented to a fresh temporal bone and vibrations of the round window are recorded by means of the Laser Doppler Vibrometer. Finally, the results are presented in the form of transfer function and compared to the intact and damaged middle ear.

1 Introduction

The human middle ear forms a sound conduction system that transmits sound from the external ear to the fluids of the inner ear through three ossicles: the malleus, the incus, and the stapes. The ossicles are connected to each other by the incudo-mallear and incudo-stapedial joints. The whole ossicular chain is supported by the tensor tympani muscle and the stapedius muscle to the temporal bone. The malleus is also firmly connected to the tympanic membrane, whereas the stapes is attached to the bony walls of the oval window by an annular ligament forming a stapediovestibular junction.

The ossicular chain can be destroyed by inflammatory diseases such as chronic suppurative otitis media or cholesteatoma. For patients with middle ear dysfunctions, the medical practice called ossiculoplasty (reconstruction of the middle ear ossicles) can improve the hearing process. This technique has been used in clinical practice for over 50 years. Various procedures are currently used, and a variety of middle ear prostheses are available. The classical TORP (Total Ossicular Replacement Prosthesis) and PORP (Partial Ossicular Replacement Prosthesis) are made of various materials with titanium or titanium alloy being most frequently used. They connect the tympanic membrane or the malleus to the stapes head or stapes footplate. Results of the middle ear reconstruction are presented in many medical papers [1–12]. Meulemans et.al. [13] have tested titanium Kurz variac PORP. All patients have underwent a pure-tone audiometry hearing test that has determined pure-tone thresholds in the frequency range from 250 to 4000 Hz by air and bone conduction. Pure tone audiograms have been also performed after surgery reconstruction made by classical titanium prosthesis in [8], to estimate effects of different ossiculoplasty materials in [14] and to test the active middle ear implant [15]. Bony ankylosis hydroxyapatite prostheses in the middle ear have been presented in [11] but without any experimental verification.

Audiometry is relatively simple and therefore quite common technique, however in most cases more precisely the Laser Doppler Vibrometer (LDV) is used to measure ossicular vibrations of the middle ear. LDV is applied to measure motion of the intact ossicular chain [16], tympanic membrane [17] and the reconstructed middle ear [18, 19]. The reconstruction by means of adjustable-length titanium prosthesis is presented in [20], optimal tension for PORP is discussed in [7]. The effects of vibration transmission to the footplate of inserting cartilage of varying sizes, and materials of varying rigidities is examined in [12].

Most of the presented prostheses cannot be modified after they have been prepared. If a prosthesis is too short, it must be lost. Prostheses length can be chosen only once before surgery. Sometimes, a reduction of the prosthesis length [20] and head plate area is possible by breaking some rings of the head plate. Therefore, here a new idea of the prosthesis made of Nitinol or Flexinol is presented. In the literature, the prostheses made of Nitinol are known as the piston–stapes prosthesis (NiTiBOND), and self-crimping prostheses [21, 22]. However, to the authors' knowledge, no study has considered the problems of shape memory prostheses (SMP) which are able to adjust their length to individual patient's need.

The main goal of this research is to test experimentally the handmade shape memory prosthesis (SMP) in the human middle ear. The idea of the SMP is reported in previous papers where the static model of C-shape SMP [23] and dynamical behaviour of the middle ear with SMP [24] is analysed. The paper is organized as follows: the section two presents an experimental procedure, bone prepa-

* e-mail: r.rusinek@pollub.pl
** e-mail: marcinszym@poczta.onet.pl
*** e-mail: grzegorz.teresinski@umlub.pl

ration and testing setup. The section three shows results of measurements in the form of transfer function. Finally, discussion and closing remarks are collected in the section four.

2 Experimental procedure

Vibrations of the round window (RW) are measured on the fresh human temporal bone specimen (figure 1) by means of the Laser Doppler Vibrometer (LDV) in case of the intact, damaged and reconstructed middle ear (ME). Firstly, the human temporal bone and the shape memory prosthesis (SMP) have been prepared as it is described in sections 2.1 and 2.2. Next, the middle ear is stimulated by sound in the range of 0.2 to 8 kHz and motion of RW has been recorded in the following cases:

- intact middle ear (IME)
- middle ear without the incus (damaged middle ear - DME)
- reconstructed middle ear with the help of the SMP

First, measurements have been taken from the intact middle ear to obtain a baseline for the RW vibration. Next, the incus has been removed to simulate a damaged ear and to be able to reconstruct ME with the help of the shape memory prosthesis. The results of experiment are presented as the transfer function between the RW velocity and the sound pressure applied to the ear.

2.1 Bone preparation

The fresh human temporal bone has been prepared to the experiment (figure 1). Firstly, the soft tissue has been removed and a standard antromastoidectomy with posterior tympanotomy performed. The mastoid facial nerve has been removed to visualize the stapes arch and the footplate. An artificial external ear canal of length 25 mm and diameter 9 mm was then attached to the bone with epoxy resin. The artificial canal is equipped with two ports, one for an ear microphone (ER-7C Etymotic Research) and one for a sound source (ER2 Etymotic Research). The artificial canal was closed with the glass plate to create a sound sealed chamber. Pieces of reflective tape (weighing less than 0.05 mg) were placed over the round window membrane. The temporal bone specimen was then embedded in dental cement and put in the temporal bone holder (Storz).

2.2 Shape memory prosthesis

The shape memory prosthesis consists of two parts. The main part is a stick made of Flexinol and a holder made of titanium. The stick diameter is 0.2mm. An activation temperature of the Flexinol is 70°C. The prototype of the SMP is shown in figure 2. Initially, the SMP (stick) curved to be shorter in order to implement easier in the middle ear (figure 2a), next after heating the SMP gets longer and takes the shape presented in figure 2b. More details cannot be published because the SMP project is applied to Polish Patent Office and it is still under consideration.

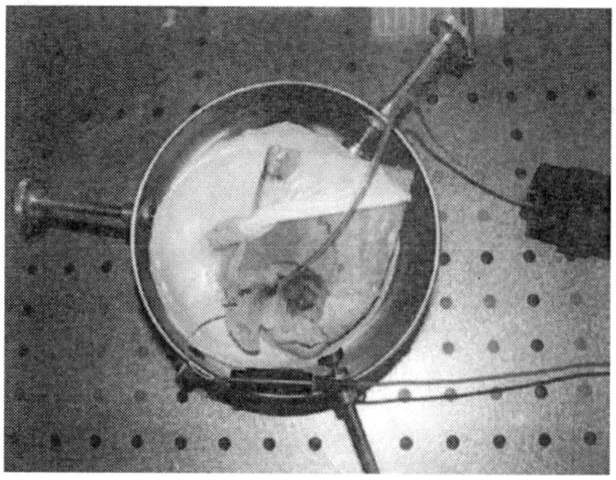

Figure 1. Temporal bone prepared for measurements with the microphone and the sound source

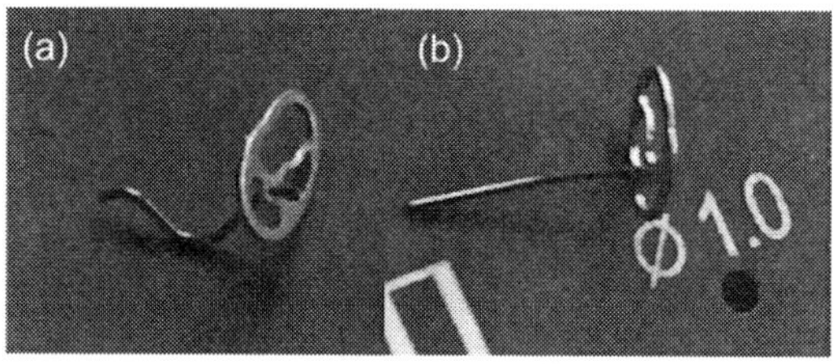

Figure 2. Handmade shape memory prosthesis, (a) before heating, (b) after heating

2.3 Testing setup

The measurements of the RW motion have been performed on the antivibration table inside the sound booth (figure 3a). The sound stimuli has been frequency sweeps from 0.2 - 8kHz. The frequency range is important for speech recognition, therefore this range is analysed in the paper. Indeed, a one - point modal analysis is performed here because the sweep signal is an excitation and the output signal (velocity) is measured. The sound source has been connected to a power amplifier (AP-12) to produce an adequate signal output. The round window velocity has been measured with the Polytec Laser Doppler Vibrometer system (figure 4) composed of an OFV-5000 controller with VD-06 velocity decoder and an OFV-534 laser head. The head has been connected to a joystick-operated micromanipulator, which is mounted on the microscope. Using the micromanipulator, the helium-neon laser beam has been directed onto retroreflective targets on the RW through the artificial ear canal. The real view and the scheme of the experimental setup is presented in figure 3 and 4. Velocity of the RW has been analysed and acquired by means of The LMS Scadas Mobile operated by TestXpress software.

3 Middle ear vibration

Vibration of the round window is presented as the transfer function in decybel scale in figure 5. The intact middle ear

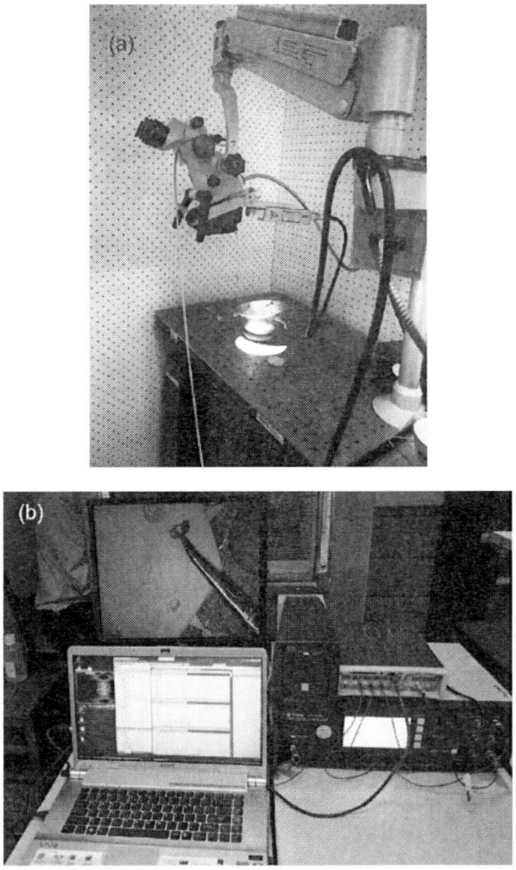

Figure 3. Real view of experimental standing for middle ear vibrations, (a) the sound booth with the laser head, the microscope and the antivibration table, (b) the measuring setup

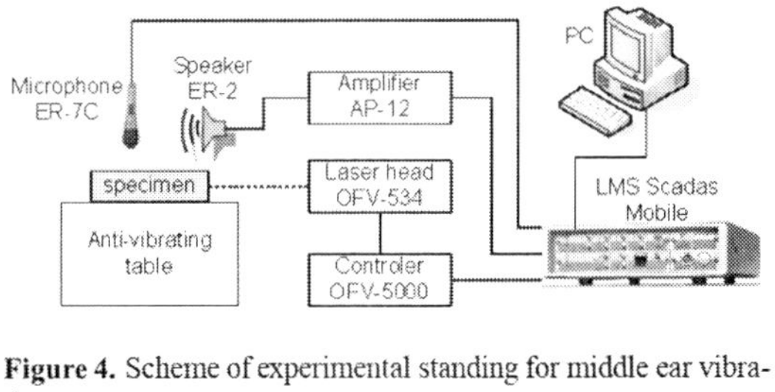

Figure 4. Scheme of experimental standing for middle ear vibrations

(IME) is the base line with two characteristic picks at 1kHz and 5kHz. These are the natural frequencies of a human middle ear. When the incus is removed the sound transmission process is perturbed and consequently the damaged middle ear (DME) line is below the base line (IME), according to our expectation. Hearing loss is about 20dB. Moreover, the natural frequencies are shifted left. After middle ear reconstruction with the help of the shape memory prosthesis natural frequencies are almost the same as in the intact middle ear (IME) and hearing loss is only

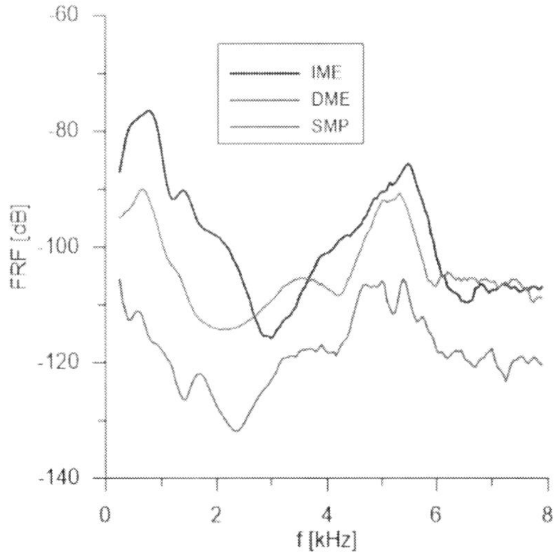

Figure 5. Vibration of round window for intact middle ear (IMR), damaged middle ear (DME) and reconstructed by means of shape memory prosthesis (SMP)

about 5dB at 5kHz and 15dB at 1kHz. Of course, any prostheses (reconstructed middle ear) cannot be as good as the intact ear. Note, that possibility of easier implementation is the most important advantage of the shape memory prosthesis.

4 Conclusions

From theoretical point of view the idea of shape memory prosthesis is presented in [23, 24]. The successful theoretical consideration are confronted here with the experimental test. The shape memory prosthesis of the human middle ear improves hearing compared to the damaged ear but we should remember that possibility of easier implementation is the most important advantage of the shape memory prosthesis. In the next step, the results of the sound transmission through reconstructed middle ear with the use of the shape memory prosthesis will be compared to classical prostheses available commercially.

References

[1] A.M. Huber, F. Ma, H. Felix, T. Linder, The Laryngoscope **113**, 853 (2003)

[2] M. Bance, A. Campos, L. Wong, D.P. Morris, R.G. van Wijhe, Otolaryngology-Head and Neck Surgery **137**, 70 (2007)

[3] F. Gentil, M. Marques, M. Parente, P. Martins, C. Santos, R.N. Jorge, Journal of Mechanics in Medicine and Biology **15**, 1540006 (2015)

[4] D.J. Kelly, P.J. Prendergast, A.W. Blayney, Otology & Neurotology pp. 11–19 (2003)

[5] G.W. Knox, H. Reitan, The Laryngoscope **115**, 1340 (2005)

[6] R.M. Lord, R.P. Mills, E.W. Abel, Hearing research **145**, 141 (2000)
[7] D.P. Morris, M. Bance, R.G. van Wijhe, M. Kiefte, R. Smith, The Laryngoscope **114**, 305 (2004)
[8] K.D. Shah, R.A. Bradoo, A.A. Joshi, D.D. Sapkale, Indian Journal of Otolaryngology and Head & Neck Surgery **65**, 298 (2013)
[9] A.M. Huber, D. Veraguth, S. Schmid, T. Roth, A. Eiber, Otology & Neurotology **29**, 893 (2008)
[10] A. Eiber, H.G. Freitag, C. Burkhardt, W. Hemmert, M. Maassen, J. Rodriguez Jorge, H.P. Zenner, Audiol Neurootol **4**, 178 (1999)
[11] O. Fawole, S.E. Mowry, American Journal of Otolaryngology **37**, 136 (2016)
[12] D.P. Morris, M. Bance, R.G. van Wijhe, Otolaryngology-Head and Neck Surgery **131**, 423 (2004)
[13] J. Meulemans, F.L. Wuyts, Forton, Glen E. J., JAMA Otolaryngology–Head & Neck Surgery **139**, 1017 (2013)
[14] S. Şevik Eliçora, D. Erdem, A.E. Dinç, M. Damar, S. Bişkin, European Archives of Oto-Rhino-Laryngology **274**, 773 (2017)
[15] A. Coordes, L. Jahreiss, U. Schönfeld, M. Lenarz, The Laryngoscope **127**, 500 (2017)
[16] S.E. Voss, J.J. Rosowski, S.N. Merchant, W.T. Peake, Hearing research **150**, 43 (2000)
[17] J.K.R. Whittemore, S.N. Merchant, B.B. Poon, J.J. Rosowski, Hearing research **187**, 85 (2004)
[18] M. Asai, A.M. Huber, R.L. Goode, Acta Otolaryngol **119**, 355 (1999)
[19] M. Asai, K.E. Heiland, A.M. Huber, R.L. Goode, Acta Otolaryngol (Stockh) **119**, 573 (1999)
[20] S. Zhao, N. Hato, R.L. Goode, Acta Oto-Laryngologica **125**, 33 (2005)
[21] A.M. Huber, T. Schrepfer, A. Eiber, Otology & Neurotology **33**, 132 (2012)
[22] K.D. Brown, B.J. Gantz, Archives of otolaryngology–head & neck surgery **133**, 758 (2007)
[23] J. Latalski, R. Rusinek, The European Physical Journal Plus **132**, 645 (2017)
[24] R. Rusinek, J. Warminski, M. Szymanski, K. Kecik, K. Kozik, International Journal of Mechanical Sciences **127**, 163 (2017)

Modelling and Investigation of a Hybrid Thermal Energy Harvester

Todor Todorov[1,*], *Nikolay* Nikolov[1], *Georgi* Todorov[2] *Yanko* Ralev[1]

[1]Technical University - Sofia, Faculty of Industrial Technology, Department of Theory of Mechanisms and Machines, 1797, Sofia, Bulgaria
[2]Technical University - Sofia, Faculty of Industrial Technology, Dept. of Technology of Machine Tools and Manufacturing, 1797, Sofia, Bulgaria

Abstract. The presented paper deals with dynamical and experimental investigations of a hybrid energy harvester containing shape memory alloy (SMA) wire and elastic cantilever with piezoelectric layer. The SMA wire periodically changes its temperature under the influence of a heated plate that approaches and moves away from the SMA wire. The change of SMA wire length causes rotation of the hot plate. The plate is heated by a heater with constant temperature. The repeated SMA wire extensions and contractions bend the piezoelectric cantilever which generates electric charges. The shape memory effect is presented as a temperature approximation of the Young's modulus. A dynamical model of the energy harvester is created and some analytical investigations are presented. With the help of an experimental setup the acceleration, the force, the temperature, and the output voltage have been measured. The theoretical results are validated experimentally. Some conclusions are made about the best performance of the energy harvester.

1 Introduction

Energy harvesting devices convert otherwise unused ambient energy into small but useful levels of electrical energy. The most utilized energy sources are solar energy, mechanical vibration energy, and thermal energy. The utilization of waste heat is one of the most important fields of the energy harvesting. The currently mostly used methods for harvesting a mid to low grade thermal energy can be divided into thermal to direct electrical, and thermal-to-mechanical-to-electrical ones.

The thermal to electrical energy converters include thermionic, pyroelectric, and Seebeck generators, alkali metal thermal-to-electric converters, Johnson thermo-electrochemical converter, Johnson electrochemical heat pipe, and infrared photovoltaic harvesters [1-4]. Pyroelectric converters rely on the property of certain crystalline materials to generate an electrical potential when subjected to a temperature-time gradient [5, 6]. Seebeck thermoelectric generators produce electricity by two different types of semiconductor materials when their junction is exposed to a temperature difference [7].

Thermomechanical heat engines are based on Stirling, Brayton, and Rankie thermodynamic cycles [8-12]. Thermoacoustic heat engines, thermomagnetic engines, hydride heat generators can be divided into two groups: traveling wave, and standing wave devices [13-15]. Thermomagnetic engines rely on the change of magnetization of certain materials that occurs with change in temperature [16-18].

Shape memory alloy (SMA) are materials that allow relatively large strain about 8% when deformed and they completely recover their original shape when heated [19].

SMA engines transform thermal energy by SMAs using change of the temperature of water or air flow. Depending of type of the motion SMA engines can be divided into SMA engines with continuous and oscillating motion. SMA engines with continuous motion convert the reciprocating linear motion of an SMA actuator into continuous rotary motion [20-22]. Oscillating SMA engines consist of SMA elastic element (cantilever, bridge, wire) which moves between two end positions corresponding to heated and cooled state [23].

All above mentioned engines convert thermal-to-mechanical energy and they could be the first part of thermal-to-mechanical-to-electrical converter. For the second part mechanical-to-electrical the well-known piezoelectric, magnetic, and electrostatic converters could be applied. Such energy harvesters with two different type converters are called hybrid or combined energy harvesters. Namli end Taya present Piezo-SMA composite for thermal energy harvesting under fluctuating temperature. Strain induced in the SMA material causes stress in the piezoelectric material, inducing electric charge by direct piezoelectric effect [24]. A remote light energy harvesting and actuation hybrid system, comprising SMA wire, attached to a piezoelectric bimorph cantilever, generates electric energy when wire is heated by a pulse laser [25]. A

* Corresponding author: tst@tu-sofia.bg

thermal energy harvester with SMA/piezoelectric laminated composite consisting of 30 μm thick $Ti_{50}Ni_{25}Cu_{25}$ ribbons and piezoelectric layer of Micro Fiber Composite (MFC) induces electric charge under temperature fluctuations [26].

The aim of the present paper is to model and investigate a new SMA-piezoelectric energy harvester (SMA-PEH) consisting of SMA wire and piezoelectric cantilever. The energy source of the SMA-PEH is a heater with constant temperature. The constant temperature of the heater is delivered to the SMA wire as fluctuating temperature with the help of a simple vibrating mechanical system, driven by the SMA wire.

2 Design concept of SMA-PEH

In Figure 1(a) a schematic description of the proposed SMA-PEH in the beginning of SMA wire heating is presented. The SMA wire 1 is strained by cantilever 3 with piezoelectric bimorph 2. A brass plate 4, which is heated by heater 5, touches the SMA wire 1, which starts to heat as well. Figure 1 (b) illustrates the SMA-PEH with hot SMA wire. When the SMA wire gets hot, it changes its crystallographic structure from martensite to austenite, and it contracts. The free end of the cantilever moves. The deformation of the cantilever induces electric charge in the piezoelectric bimorph. At the same time by the limiters 6 the hot SMA wire rotates the brass plate 4 about the heat source 5 and heating of the SMA wire is off. Meanwhile, SMA wire cooling leads to austenite-to-martensite transformation, which decreases the SMA stiffness. The SMA wire strains again under the force action of the cantilever and the brass plate rotates counterclockwise to its position shown in Figure 1 (a). The motion of the free end of the cantilever deforms piezoelectric layer and a second electric charge is induced. Then all the processes described above are repeated.

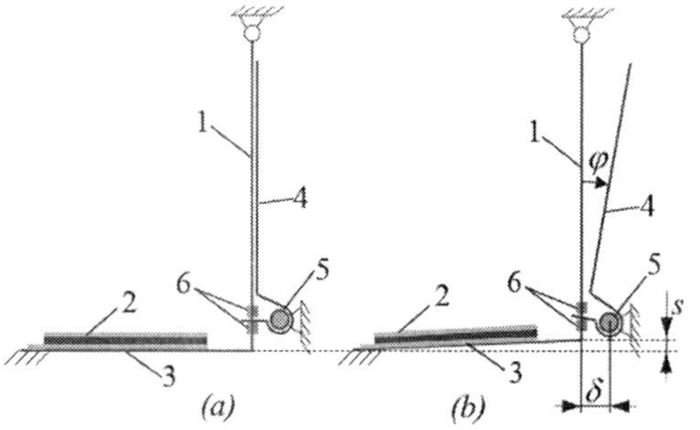

Fig. 1. Schematic representation of the thermal energy system: (a) SMA wire is cold (b) SMA wire is heated.

3 Dynamical modelling of SMA-PEH

Let consider the deformations of the piezoelectric cantilever and of the SMA wire shown in Figure 2.

The segments CD and AE are the initial undeformed states of the SPF wire and the cantilever when not connected (Figure 2a). At the beginning the cantilever and the SMA wire have a common point A. In this

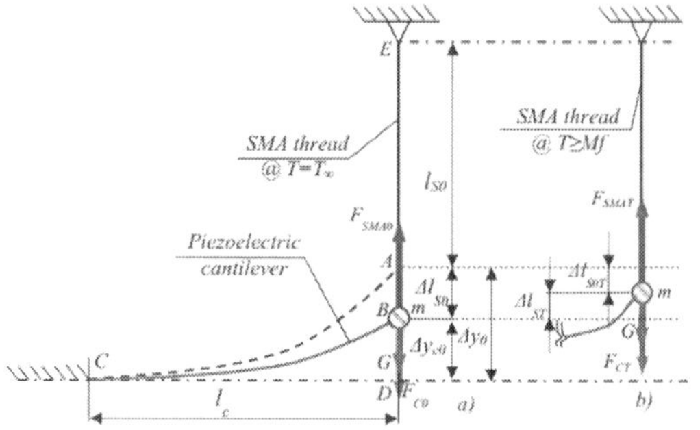

Fig. 2. Deformations of the SMA wire and of the piezoelectric cantilever.

state, the free cantilever end deformation is Δy_0, and the SMA wire length is l_{S0}. At the equilibrium position, corresponding to point B, the deformation of the free cantilever end decreases to Δy_{c0} and the SMA wire deformation is Δl_{S0}. The force F_{c0} at the free cantilever end can be expressed as

$$F_{c0} = k_c \Delta y_{c0}, \tag{1}$$

where

$$k_c = \frac{3E_c I}{l_c^3} \tag{2}$$

is the stiffness of the cantilever, E_c is the Young modulus of cantilever material, l_c is the length of the cantilever. The cross-sectional moment of the cantilever is $I = \frac{wh^3}{12}$, where w is the width and h is the thickness of the cantilever. The equilibrium force F_{S0} in the SMA wire is presented in the form

$$F_{S0} = \frac{E_M A}{l_{S0}} \Delta l_{S0} \tag{3}$$

where E_M is the Young's modulus for starting martensite temperature of nitinol, and A is the cross-section area of the SMA wire, d is the wire diameter. We assume that at point B a generalized mass m with weight G also acts as

$$G = mg \tag{4}$$

From the equilibrium of the forces described by (1), (3), and (4) follows the equation

$$k_c \Delta y_e + mg = \frac{E_M A}{l_{S0}} \Delta l_{S0}. \tag{5}$$

Taking into account that the deformation Δy_0 is initially given according to Figure 2 (a) we can write

$$\Delta y_0 = \Delta l_{S0} + \Delta y_{c0}. \tag{6}$$

Combining two above equations into a system, we can express

$$\Delta y_{c0} = \frac{\Delta y_0 E_M A - l_{S0} mg}{E_M A + k_c l_{S0}}, \tag{7}$$

$$\Delta l_{S0} = \frac{(\Delta y_0 k_c + mg) l_{S0}}{E_M A + k_c l_{S0}}. \tag{8}$$

In Figure 2 b) it is described the static load of the considered mechanical system, when the wire is heated to an arbitrary temperature over the final martesite temperature Mf and less than the final austenite temperature Af. The wire heating leads to structure changes that increase the Young's modulus. The wire hardens, and it decreases it's elastic deformation. The distance Δl_S, that is the reduction of the elastic deformation due to the heating, we call temperature deformation. Expressing the equilibrium of the vertical forces in this case, we can write

$$F_{CT} + G = F_{SMAT}, \tag{9}$$

where the force of the cantilever at temperature deformation is

$$F_{CT} = k_c\left(\Delta y_C + \Delta l_S\right) \tag{10}$$

and the force in the SMA wire is

$$F_{SMAT} = \frac{E(T)A}{l_{S0}}\Delta l_{S0T} \tag{11}$$

Taking in account that

$$\Delta y_0 = \Delta l_{S0T} + \Delta y_{c0} + \Delta ls \tag{12}$$

substituting (4), (10), and (11) into (9) and combining with (12) we can express

$$\Delta l_S = \frac{Al_{S0}\left(k_c\Delta y_0 + gm\right)\left[E(T) - E_M\right]}{\left(AE_M + k_c l_{S0}\right)\left[AE(T) + k_c l_{S0}\right]}, \tag{13}$$

$$\Delta l_{S0T} = \frac{\left(k_c\Delta y_0 + mg\right)l_{S0}}{E(T)A + k_c l_{S0}}. \tag{14}$$

We assume that the Young's modulus $E(T)$ is a temperature function presented in the form

$$E = 2E_M + \frac{E_A - E_M}{2}\tanh\frac{T - As}{k_E} \tag{15}$$

where

$$k_E = \frac{E_A - E_M}{Af - Mf} \tag{16}$$

E_A is SMA Young's module for austenite structure, Af is the final austenite temperature, Mf is the final martensite temperature, and As is the start austenite temperature. We neglected hysteresis because the strain of the SMA wire is smaller than 0,3%.

The graph of the approximated Young's modulus is shown in Figure 3.

With the help of the Young's modulus approximation (15) we can express the strain-temperature graph shown in Figure 4.

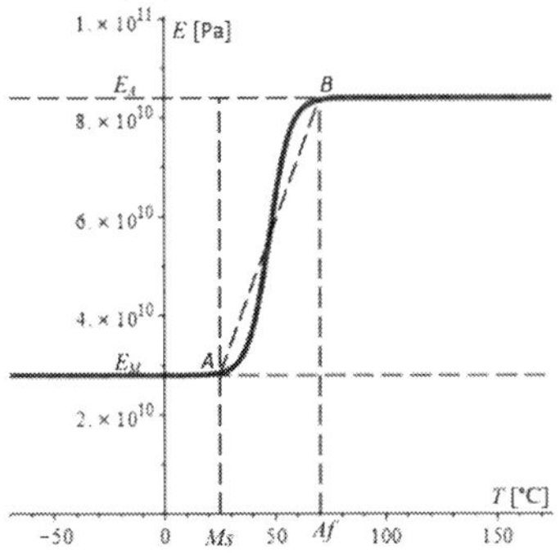

Fig. 3. Young's modulus approximation.

The analysis shows that the maximum temperature deformation does not recover the total elastic strain of the SMA wire. This is so because at the austenite phase there is an elastic strain, which can vanish only when there is no stress in the SMA wire.

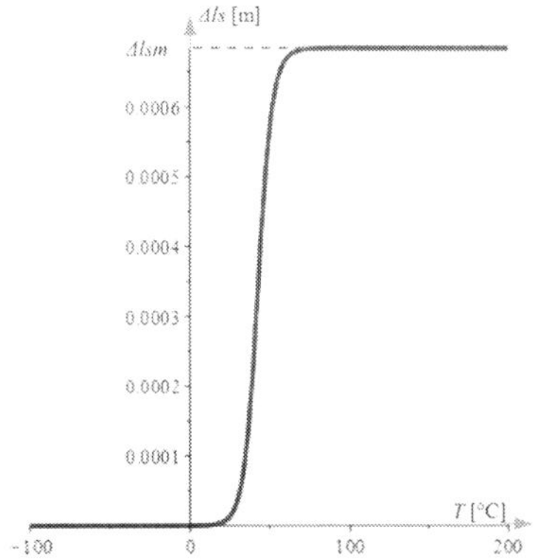

Fig. 4. Temperature deformation vs. time.

For creating a dynamical model we assume that there is a lumped mass at the free end of the cantilever. We denote the position of this mass as y. The force F_{SMA} of the SMA wire that exerts on the mass is

$$F_{SMA} = \begin{cases} \dfrac{E(T)A\left(\Delta l_{S0} - y + \Delta y_{c0}\right)}{l_{S0}} & y < \Delta y_{c0} + \Delta l_{S0} \\ 0 & y \geq \Delta y_{c0} + \Delta l_{S0} \end{cases} \tag{17}$$

In the above expression we take into account that the force F_{SMA} is unidirectional and it can only strain the SMA wire.

From Figure 1 we see that the angle φ is constrained to be positive only, because the negative value is possible if the plate penetrates into the SMA wire. This consideration leads to the formula

$$\varphi = \begin{cases} \arctan\dfrac{y - \Delta y_e}{b} & y - \Delta y_e > 0 \\ 0 & y - \Delta y_e \leq 0 \end{cases} \tag{14}$$

For the heat transfer between hot plate and SMA wire the following relationship is assumed:

$$T_{hp} = T_r + \frac{T_h}{2}\left(1 + \cos\frac{\pi\varphi}{\varphi_m}\right), \tag{15}$$

where T_h is the temperature of the hot plate, T_{hp} is the temperature which is transferred to the SMA wire, φ_m is the maximum angle of the hot plate, and T_r is an experimentally determined temperature.

The unidirectional action of the SMA wire can change the damping coefficient of the system in the form

$$\beta = \begin{cases} \beta_{SMA} + \beta_c & y < \Delta y_{c0} + \Delta l_{S0} \\ \beta_c & y \geq \Delta y_{c0} + \Delta l_{S0} \end{cases} \tag{16}$$

where β_{SMA} is the damping coefficient of the SMA wire and β_c is the damping coefficient of the cantilever.

On the basis of the above assumptions a dynamical model was created, presented in the view

$$\begin{cases} m\dfrac{d^2y}{dt^2}+\beta\dfrac{dy}{dt}+k_c y=F_{SMA} \\ \rho A c_p\dfrac{dT}{dt}-h_c A_c\left\{T_\infty - T + T_{hp}\left[\varphi(y)\right]\right\}=0 \end{cases} \quad (17)$$

where m is the lumped mass, ρ is the density of the SMA wire material, c_p is the specific heat, h_c is the heat convective transfer coefficient, A_c is the circumferential area per unit length of the wire, and T_∞ is the room temperature.

In the above system the first equation describes the force balance, and the second equation describes the temperature processes of the energy harvester [28, 29].

The system (21) is solved numerically by Euler's implicit method of 5th order.

The solution of the system (16) shows that the amplitude of the displacement is modulated and it increases in the transition period. The same is the behavior of the acceleration (Figure 5).

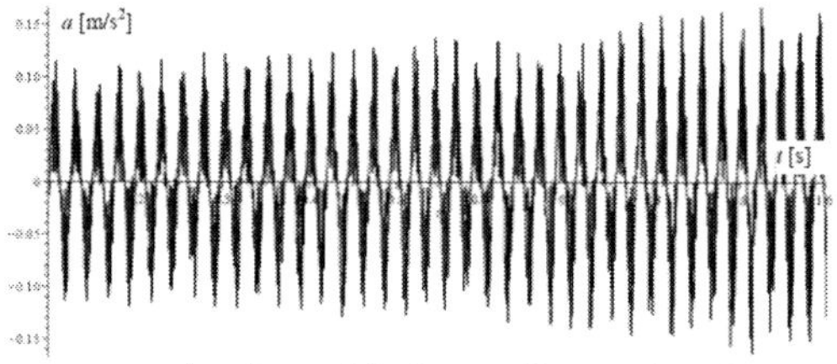

Fig. 5. Acceleration a of the free cantilever end vs. time t.

The temperature of the SMA increases modulated as it is presented in Figure 6. There is an equilibrium temperature at 62°C corresponding to the temperature of the heater in a middle position.

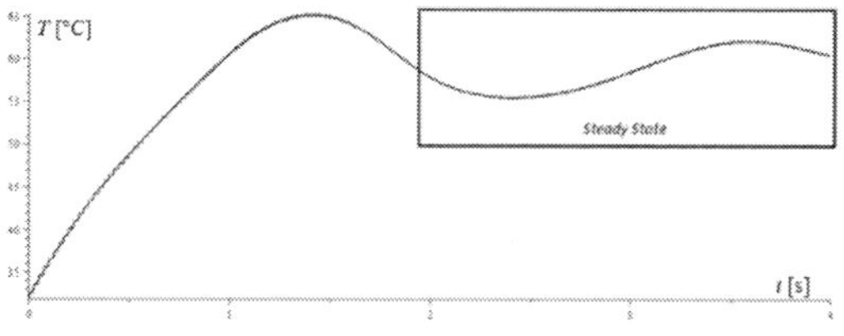

Fig. 6. Temperature of the SMA wire vs. time.

Form the obtained above solution we can conclude that in the behavior of the mechanical and of the thermal parameters two types of vibrations are observed: The first one – proportional to the temperature time constant, and the second one – equal to the natural frequency of the mechanical part. According to Donald J. L. [29] the temperature time constant τ is determined by the relation

$$\tau=\frac{\rho A c_p}{h_c A_c}. \quad (22)$$

Neglecting the damping, the mechanical natural frequency can be calculated by the formula

$$\Omega_\Sigma=\sqrt{\frac{k_c+k_{sma}}{m}}, \quad (23)$$

where

$$k_{sma}=\frac{E(T)A}{l_{S0}} \quad (24)$$

is the stiffness of the SMA wire.

Formula (23) shows that besides the amplitude modulation there is an additional frequency modulation. The amplitude modulation is due to the thermal shortening of the SMA wire. The frequency modulation is due to the stiffness change of the wire when heated. Both modulations are due to the thermal change but they reflect two different properties of the SMA material: shortening and stiffening.

4 Design concept of the experimental setup

The evaluation setup for investigation of SMA-PEH is shown in Figure 7. The sensor signals of the SMA-PEH 1 are processed by a Data Acquisition System (DAQ) 2. The sensors and heater are powered up by two rectifiers 3, 4 and a thermal regulator 5. The DAQ output enters into a laptop 6, where the experiment is controlled by a LabVIEW program.

In Figure 8 the SMA-PEH with the sensors is presented. A piezorestive MEMS force sensor 1 measures the longitudinal force in SMA wire. The hot plate 2 rotates about the fixed axis of a heater 3, which is a part of soldering iron. The acceleration of the cantilever beam 4 along vertical axis is measured by a three-axis MEMS accelerometer 5.

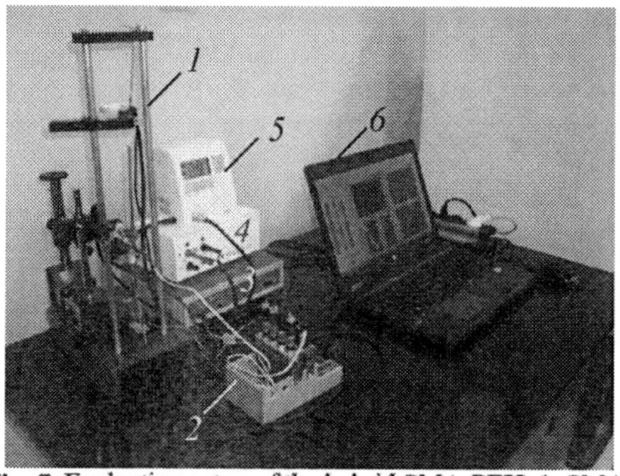

Fig. 7. Evaluation setup of the hybrid SMA-PEH: 1. SMA-PEH; 2. DAQ; 3.4. Rectifiers; 5. Thermal regulator. 6. Laptop.

For obtaining the necessary data of the temperature three thermocouples are used. The thermocouple 6 measures the heater temperature. Thermocouple 7 is used to measure the room temperature. In order to increase the frequency response of the thermocouple for the SMA wire, a thermocouple type-T with a small diameter of 100 μm is used, as shown in Figure 9.

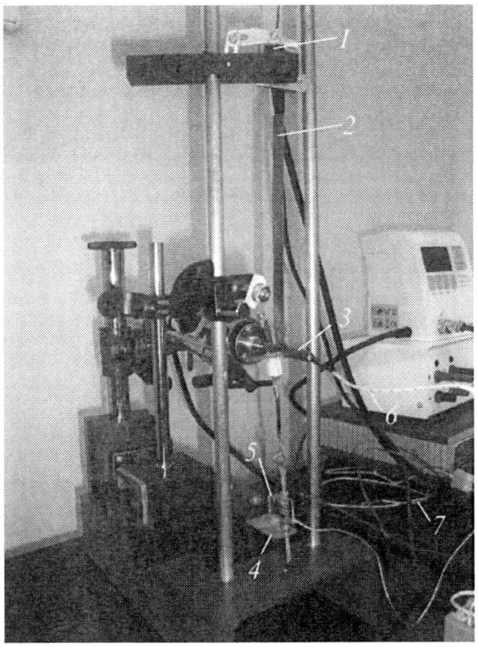

Fig. 8. Closeup of SMA-PEH: 1. Force sensor; 2. Hot plate; 3. Heater; 4. Piezoelectric cantilever; 5. Accelerometer. 6. Thermocouple for the heater, 7 Thermocouple for the room temperature.

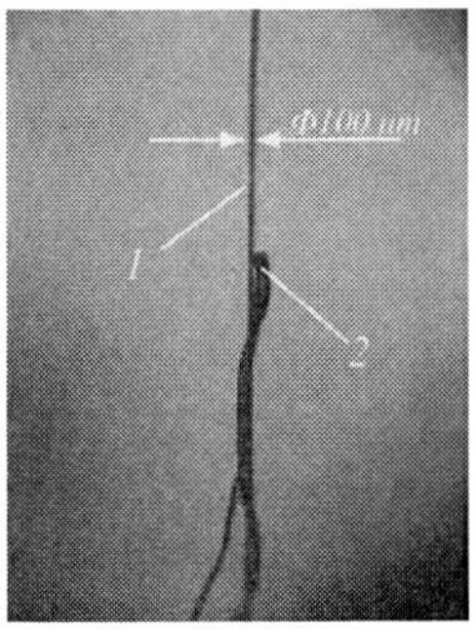

Fig. 9. The SMA wire and its thermocouple: 1.SMA wire, 2. Thermocouple.

5 An experimental validation and interpretation of the model

By the evaluation setup, described above, consequently series of experiments were performed. Results of one such an experiment are shown in Figures 13-16.

In Figure 10 the experimental results for the acceleration of the free end of the cantilever are presented. We see that the output of the accelerometer is a periodic oscillation with stable behavior. Compared to the model acceleration (Figure 5), we can register a relative good fit with respect to the amplitude, shape and frequency.

The experimentally obtained temperature change of the SMA wire is shown in Figure 10. Compared to the theoretical result from Figure 6, for the steady state period a good fit is demonstrated.

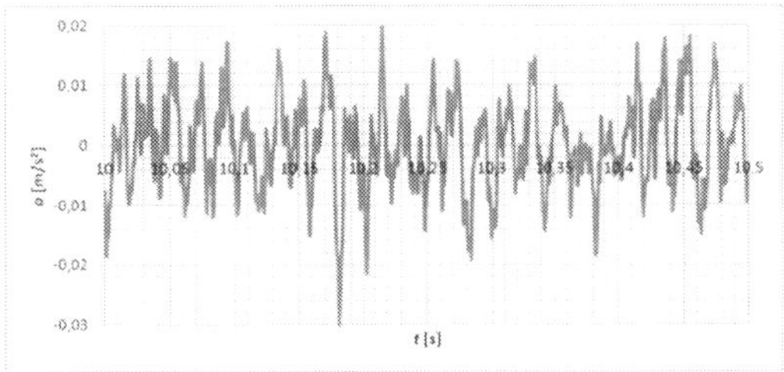

Fig. 10. Experimental acceleration a of the free cantilever end vs. time t.

The measured output voltage is presented in Figure 12. Maximum amplitude of about 50 mV is registered at load resistance of 10 kΩ which corresponds to 0,5 μW peak-to-peak power.

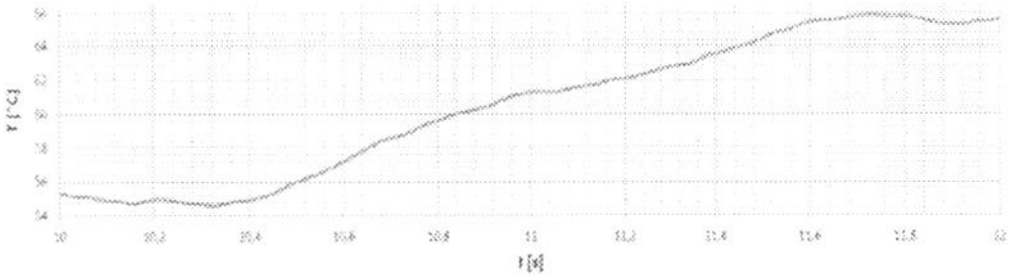

Fig. 11. Experimentally determined temperature T of the SMA wire for the steady state of 2 seconds.

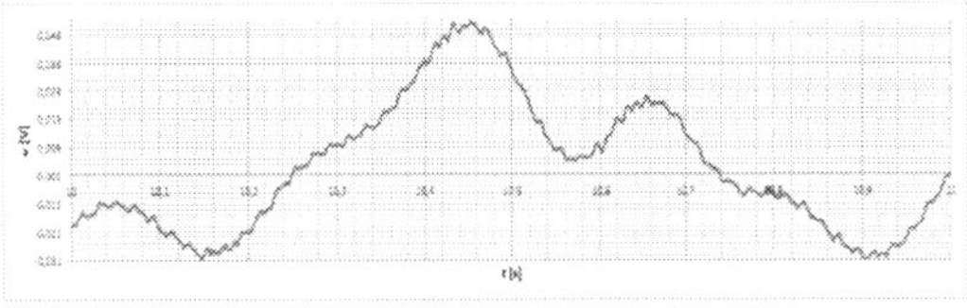

Fig. 12. Measured output u voltage of the energy harvester

The best efficiency of the energy harvester we expect to have when the frequency of the mechanical and of the thermal oscillations coincide. As it was already clarified, the energy harvester has a variable frequency. Therefore, the condition for frequency coincidence is possible to be satisfied only for one value of the temperature of the SMA wire. The frequency equalizing in the considered device requires a wire with a very small diameter and the smallest possible lumped mass.

Conclusions

A new design of a combined thermoelectric energy harvester, containing SMA wire and piezoelectric flexible member for generation of electric energy from constant heat source is presented.

A dynamical model of the energy harvester is built and investigated.

With the help of an experimental setup model, an experimental validation is performed.

The comparison between model results and experiments indicates a good validation of temperature during the steady state period, and in terms of acceleration satisfactory coincidences of amplitudes and vibration periods are obtained.

The condition for the best performance of the energy harvester according to parametric resonant systems is the coincidence of twice frequency and the system's natural mechanical frequency. Because the two frequencies vary with temperature change, this condition can only be achieved for a certain temperature value of the SMA wire.

The presented work is carried out and funded under contract DUNK-01/3 between National Science Fund, MES of RB and R&D Sector of TU-Sofia.

References

1. Percy S. Knight C. McGarry S. Post A. Moore T. Cavanagh K., *Thermal Energy Harvesting for Application at MEMS Scale*, Springer, (2014)
2. Hernqvist, K.G. Kanefsky M. Norman F.H., Thermionic Energy Converter. RCA Rev. **19**, 2 (1958)
3. Sievers R.K. Ivanenok J.F. Hunt T.K. Alkali metal thermal to electric conversion, Mechanical Engineering; **117**, 10 (1995)
4. Johnson L.G. Muller J.R. Johnson reversible engine, US Patent, US7160639 B2, (2007)
5. Hossain A. Rashid M.H. Pyroelectric detectors and their applications. IEEE Trans Ind Appl. **27** (1991)
6. Whatmore R.W. Pyroelectric devices and materials. Rep Prog Phys. **49** (1986)
7. Liang G. Zhou J. Huang X. Analytical model of parallel thermoelectric generator, Applied Energy **88,** 12 (2011)
8. Patterson D.Ee Jamison K,D, Durrett M. *CVD diamond based miniature stirling cooler* International cryocooler conference (2007)
9. Nakajima N. Ogawa K. Fujimasa I. Study on micro engines-miniaturizing Stirlingengines for actuators and heatpumps. Sens Actuators **20**, 75–82, (1989)
10. Rosa R.J. *Characteristics of a closed Brayton cycle piston engine*, IEEE Western Canada conference on computer, power and communications systems in a rural environment, pp 153–159, (1991)
11. Johnston J.R. Evaluation of expanders for use in a solar-powered Rankine Cycle Heat Engine. Doctoral dissertation, Ohio State University, pp 1–138, (2001)
12. Cho J. Lin C. Richards C *Demonstration of an external combustion micro-heat engine.* Proc Combust Inst **32**, 3099–3105 (2009)
13. Pierens M. Thermeau J.P. Le Pollès T. Duthil P. *Experimental characterization of a thermoacoustic travelling-wave refrigerator.* International conference on fluid mechanics, heat transfer and thermodynamics (2011)
14. Ceperley P.H. A Pistonless Stirling Engine—the Traveling Wave Heat Engine.J Acoust Soc Am **66**, 1508–1513, (1979)
15. Swift G.W. Thermoacoustic Engines, J Acoust Soc Am **84** 1145 (1988)
16. Solomon D (1991) Design of a thermomagnetic generator. Energy Convers Manage 31:157–173.
17. Hsu C-J, Sandoval SM, Wetzlar KP, Carman GP (2011) Thermomagnetic conversion efficiencies for ferromagnetic materials. J Appl Phys 110:123923–123927.
18. Bulgrin KE, Ju YS, Carman GP, Lavine AS (2009) A coupled thermal and mechanical model of a thermal energy harvesting device. ASME 2009 International Mechanical Engineering Congress & Exposition. Lake Buena Vista, Fl, pp 327–335
19. Otsuka K. Wayman C.M. *Shape memory materials*, 1st edn. Cambridge University Press, Cambridge (1998)
20. Schiller E.H. Heat Engine Driven by Shape Memory Alloys:Prototyping and Design, Master thesis, Virginia Polytechnic Institute (2002)
21. Hochstein P.A. *Some Engineering Parameters for a Nitinol Engine Design*, Proc. of the Nitinol Heat Engine Conf., Silver Spring, Maryland, (1978)
22. Sandoval D.J. *Nitinol Belt Engine*, Proc. of the Nitinol Heat Engine Conf., Silver Spring, Maryland, (1978)
23. Kahn H, Huff MA, Heuer AH The TiNi shape-memory alloy and its applications for MEMS. J Micromech Microeng **8** 213–221 (1998)
24. Namli O.C. Taya M. Design of piezo-SMA composite for thermal energy harvester under fluctuating temperature, J. Appl. Mech., **78**, 8 (2011)
25. Avirovik D. Kumar A. Bodnar R.J. Priya S. Remote light energy harvesting and actuation using shape memory alloy-piezoelectric hybrid transducer, Smart Mater. Struct., **22**, 6 (2013)
26. Lebedev G.A. Gusarov B.V. Viala B. Delamare J. Cugat O. Lafont T. Zakharov D.I. *Thermal energy harvesting using shape memory alloy piezoelectric composite*, in Proc. 16th Int. Solid-State Sens. Actuators Microsyst. Conf. (Transducers'11), Beijing, China, pp. 669–670 (2011)
27. Ralev Y. Todorov T. Modelling of Shape Memory alloys and experimental setup for their investigation, CAx Technologies, **4**, (2016) 46-51,
28. Bhargaw H. N. Ahmed M., Sinha P. Thermo-electric behaviour of NiTi shape memory alloy, Trans. Nonferrous Met. Soc. China, **23**, (2013) 2329–2335
29. Donald J. L. *Engineering analysis of smart material systems*, Hoboken, New Jersay: John Wiley & Sons, Inc, (2007) 333-350

Broadband frequency response of a nonlinear resonator with clearance for energy harvesting

Krystian Łygas[1,*], *Piotr* Wolszczak[1,**], and *Grzegorz* Litak[1,2,***]

[1] *Department of Automation, Lublin University of Technology, Nadbystrzycka 36, PL-20-618 Lublin, Poland*
[2] *Department of Process Control, AGH University of Science and Technology, Mickiewicza 30, PL-30-059 Krakow, Poland*

Abstract. We examine the response of an inverted pendulum excited harmonically with amplitude limiters. Such a nonlinear resonator is suggested to work in the transduction of ambient vibration energy into the electrical power through bending of the moving beam. Using a multibody model we follow rotation and bending of the elastic beam and we provide the solutions of various frequencies. We found that the corresponding bending is the largest for subharmonic solutions of beam oscillations.

1 Introduction

Energy harvesting based on nonlinear resonators can provide more efficient ambient energy converters for various frequencies [1–4] comparing to linear devices [5]. The role of nonlinearities [6] is important not only in broadening of frequency in the main resonance region but also in involving the subharmonic and superharmonic frequencies in multiple solutions. For instance, possible solutions in double well system resonators [7] are related to the small amplitude single well solution and the large amplitude cross barrier solution. Especially, subharmonic solutions with a large amplitude are very useful for energy harvesting [7, 8]. Following the concept of the inverted elastic pendulum which undergoes a bifurcation of single to double equilibria under load of a tip mass [9], we propose a system with the effect of clearance in the clamping.

In this note, we examine effect of clearance as a typical nonlinearity which can appear during exploitation. Furthermore, we show how the response of the system is changing with frequency sweep and discuss particular solutions.

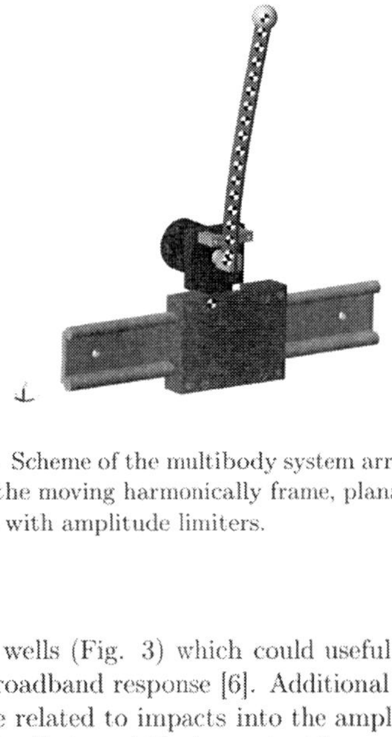

Figure 1. Scheme of the multibody system arrangements including the moving harmonically frame, planar inverted pendulum with amplitude limiters.

2 Dynamical Model

Our model (Fig. 1) is based on a vertical flexible beam with a tip mas excited horizontally (see [9]). Instead of the usual clamping we consider the hinged connection with additional clearance. Because of contacts with limiters and motion around the hinged point the pendulum has two different possibilities to move: rotation and bending (see in Fig. 2). It possesses two potential wells (Fig. 3) which could useful for a frequency broadband response [6]. Additional nonlinear effects are related to impacts into the amplitude limiters. This effect would be important for energy transduction [12, 13] as the final design is expected to be equipped in the piezoelectric patch along the bending beam surface (above the impacting region).

Flexible nonlinear beam systems have been already discussed in the previous works [14–16]. In particular, Tang and Ren [14] and Xu and Yu [15] examined a control algorithm to stabilize the pendulum in the upright position. On the other hand, Semenov et al. [16] discussed of a concept of elastic inverted Kapica like pendulum with the vertical excitation. Interestingly, their pendulum was also exposed on backlash following the hysteretic switching coupling. Later, Semenov at al. [17, 18] considered

*e-mail: k.lygas@pollub.pl
**e-mail: p.wolszczak@pollub.pl
***e-mail: g.litak@pollub.pl

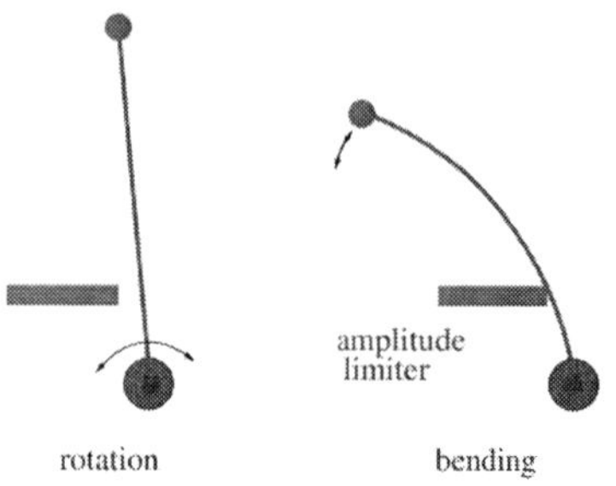

Figure 2. Schematics of two different possible motions of the pendulum (marked by arrows).

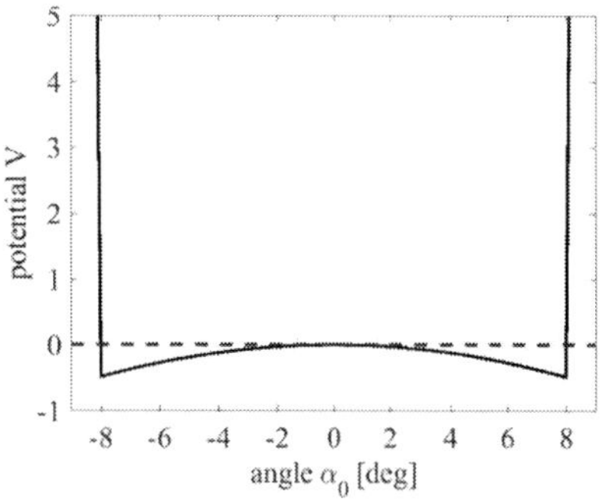

Figure 3. Normalized potential V of the simplified rigid pendulum versus bearing angle α_0.

also the horizontal excitation in the proposed system with a hysteretic coupling between the excitation frame and the inverted pendulum. We continue these studies on an inverted elastic pendulum by investigating capability of impacts for energy conversion. In our dynamical system the contact between amplitude limiters and the pendulum is defined in a different way (comparing to earlier proposals [17, 18]).

The computational model is based on multistage inverted pendulum with 17 shorter and rigid beams and a tip mas. Kinetic, T, and potential, V, energies in the time intervals between impacts can be written as follows:

$$T = \sum_{i=1}^{N} \frac{\Delta m_i(\mathbf{v}_i + \mathbf{v}_{i-1})^2}{8} + \sum_{i=1} \frac{\Delta I_i \dot{\alpha}_i{}^2}{2} + \frac{M\mathbf{v}_N^2}{2}, \tag{1}$$

$$V = \sum_{i=1}^{N} \Delta m_i g \frac{(y_i + y_{i-1})}{2} + \sum_{i=1}^{N} \frac{k(\alpha_i - \alpha_{i-1})^2}{2} + Mgy_N, \tag{2}$$

where $g = 9.81\mathrm{m/s}^2$ is the gravity acceleration, M is a tip mass attached to the end node of beam element No. N, $\mathbf{v}_i = [x_i, y_i]$ are vectors denoting local connecting node velocities, α are angles between the neighbour short beam elements. All the elements have the same shape and sizes. Note that α_0 is the bearing rotation and the assumed $v_0 = \dot{x}_0$ of harmonic motion in the horizontal direction:

$$x_0 = A\sin(\omega t), \quad y_0 = 0, \tag{3}$$

where A is the amplitude while ω is the corresponding frequency of kinematic excitation. Furthermore, $\Delta m_i = \Delta m$ is mass of a short beam element while $\Delta I_i = \Delta I$ is the moment of inertia for defined elements, y_N is the height on the tip mas M. Finally, k is the local spring coefficient.

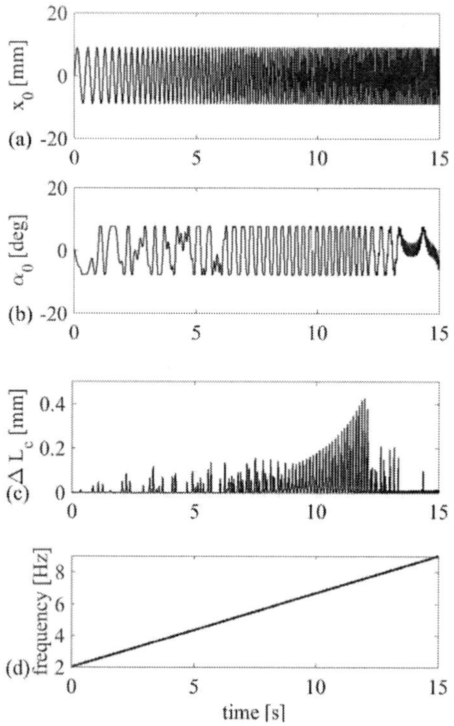

Figure 4. Frequency sweep is defined in the interval from 2 to 9 Hz (d). (a) shows the frame excitation x, (b) presents the α_0 response, (c) is showing the time dependence ΔL_c as estimated bending in terms of the chord (connecting the beam ends) reduction.

The Lagrangian of the beam without further constraints (limiters) has the following form:

$$\begin{aligned} & L(q_1, ..., q_i, ..., q_N, \dot{q}_1, ..., \dot{q}_i, ..., \dot{q}_N, t) \\ = \; & T(q_1, ..., q_i, ..., q_N, \dot{q}_1, ..., \dot{q}_i, ..., \dot{q}_N, t) \\ - \; & V(q_1, ..., q_i, ..., q_N, \dot{q}_1, ..., \dot{q}_i, ..., \dot{q}_N, t), \end{aligned} \tag{4}$$

where the generalized coordinates this multistage inverted pendulum $q_i \in \{\alpha_1, ..., \alpha_n\}$ and their corresponding velocities $\dot{q}_i \in \{\dot{\alpha}_1, ..., \dot{\alpha}_n\}$. Consequently, the general equation of motions for such a kinetically forced inverted pendulum can be written as:

$$\frac{d}{dt}\left(\frac{\partial L}{\partial \dot{q}_j}\right) - \frac{\partial L}{\partial q_j} = Q_j + Q_n^{imp}. \quad (5)$$

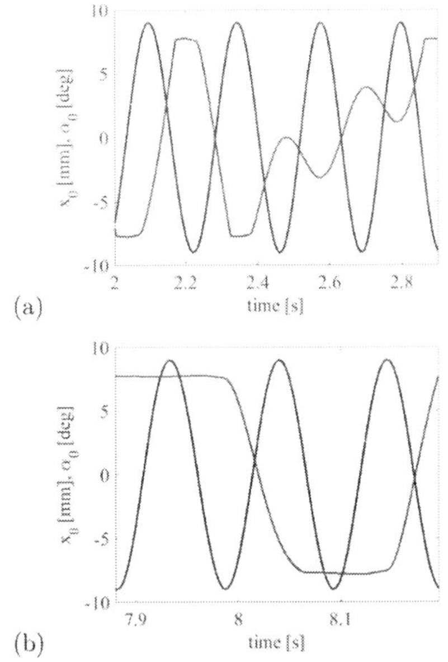

Figure 5. Waveform responses based on the frequency sweep results (Fig. 4), where blue line denotes a frame displacement, x_0 and red line the angular α_0 displacement of bearings, α_0. Responses are ploted for different excitation frequencies $f = 3.1, 5.7, 6.7, 7.8, 8.8$ Hz, for cases (a)-(e), respectively.

The generalized forces Q_j appearing in the above equation are defined:

$$Q_j = -\beta(\dot{\alpha}_j - \dot{\alpha}_{j-1}) - \beta(\dot{\alpha}_j - \dot{\alpha}_{j+1}), \quad (6)$$

while the additional generalized force Q_n^{imp} is caused by an inelastic contact to one of amplitude limiters with local contact elasticity and damping properties (see Tab. 1 for parameter values and Figs. 1 and 2 for view). They are attached to the beam elements and cause the whole beam to bounce on the amplitude limiters.

3 Simulation results

Using the equations of motion (Eqs. 5-6) and parameters of the system are presented in Tab. 1 we performed simulations. The calculations of the dynamical response of the introduced pendulum-impact model were performed by using *Maltab-Simulink* and *Simscape Multibody* with *Flexible Beam* and *Contact Forces* libraries.

The frequency sweep results for of calculation are presented in Fig. 4. Here, in the time interval of 15 s we changed steadily the excitation frequency in the range of 2-9 Hz. Interestingly, the system behavior shows a number of bifurcations. In the low frequency the response is nonperiodic while in the range $f \in [5-7]$ subharmonic 3 dominate. For larger frequency we found subharmonic 5 and transition to nonperiodic behaviours.

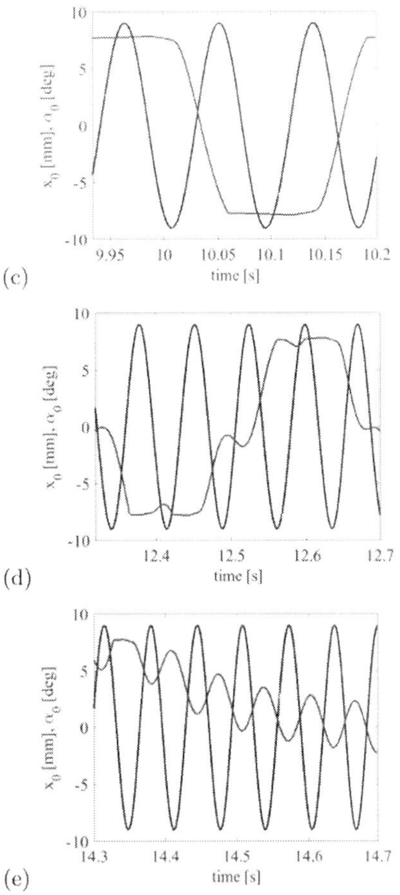

Figure 5. Continuation.

For better clarity it the characteristic waveforms are ploted also in Fig. 5. Note that Figs. 5a and e shows clearly intervals of nonperiodic responses in the small ($f = 3.1$ Hz) and high (f=8.8 Hz) frequencies from the interval $f \in [2$ Hz, 9 Hz$]$ (Fig. 4), while in the middle frequencies (f=5.7 and 6.7 Hz) correspond to a subharmonic solution of period 3 with respect to the excitation period (see blue and red lines in Figs. 5b and c). Finally, at the excitation frequency $f = 7.8$ Hz (Fig. 5d) a subharmonic solution of period 5 is present. The periodic waveforms (shown in Figs. 5b-d) are characterized by the large amplitude oscillations with passing through the potential barrier (Fig. 3). Interestingly, the maximum of bending response (Fig. 4c) is reached for subharmonic solution of period 3.

Table 1. Material and geometrical properties of beam and tip mass.

ultimate tensile strength (ASTM D638)	7.5 MPa
Young Modulus (ASTM D638)	350 MPa
elongation at break (ASTM D638)	3 - 5%
typical parts density	0.85 g/cm^3
beam dimensions (width x thickness x length)	17 × 3 × 195 mm
natural vibration of elastic beam	first mode above 9 Hz
tip mass	1.129 g
distance between limiters	12 mm
contact stiffness	10^4 N/m
contact damping	10^2 Ns/m

4 Conclusions

The obtained results show that the system response of elastic inverted pendulum with amplitude limiters of fixed distance has a nonlinear features of double well oscillator with soft impacts. For the fixed distance between the amplitude limiters excitation, excitation frequency was changed quasi-statically during simulation. Examining the response wave forms one can notice that there are both periodic and nonperiodic responses. Periodic responses of the large amplitude appeared to be subharmonic solutions of the period 3 and 5. Especially, the subharmonic solution of excitation period 3 was interesting for energy harvesting showing simultaneously the capability of fairly high bending (Fig. 4c). This solution is consistent with the recent results by Huguet et al. [8] and Syta et al. [7] where the subharmonic solutions were present. Here, the beam deflection response is formed in a sequence of peaks depending on impacts in the chosen system geometry and dynamical conditions. Such peaks will produce an electrical impulses [19] on the system once it is adapted to energy harvesting by the piezoelectric patches. For energy harvesting, both maximal values and higher frequency are important, Therefore harmonic and subharmonic solution of period 3 and those of period 5 with smaller bending, or noperiodic solutions can be of importance. Finally, we advocate the application of the proposed mechanical system as a resonator in a broadband frequency energy harvester including the subharmonic and also nonperiodic solutions.

References

[1] P.D. Mitcheson, E.M. Yeatman, G.K. Rao, A.S. Holmes, T.C. Green, Proceedings of the IEEE **96**, 1457 (2008)

[2] R.L. Harne, K.W. Wang, Smart Mat. Struct. **22**, 023001 (2013)

[3] S.P. Pellegrini, N. Tolou, M. Schenk, J.L. Herder, J. Intell. Mater. Syst. Struct. **24**, 1303 (2013)

[4] J. Twiefel, H. Westermann, J. Intell. Mater. Syst. Struct. **24**, 1291 (2013)

[5] S.P. Beeby, M.J. Tudor, N.M. White, Meas. Sci. Technol. **17**, R175 (2006)

[6] M.F. Daqaq, R. Masana, A. Erturk, D.D. Quinn, Applied Mechanics Reviews **66**, 040801 (2014)

[7] A. Syta, G. Litak, M.I. Friswell, S. Adhikari, Eur. Phys. J. B **89**, 99 (2016)

[8] T. Huguet, A. Badel, A., M. Lallart, Appl. Phys. Lett. **111**, 173905 (2017)

[9] M.I. Friswell, S.F. Ali, S. Adhikari, A.W. Lees, O. Bilgen, G. Litak, J. Intell. Mater. Syst. Struct. **23**, 1505 (2012)

[10] A. Erturk, J. Hoffmann, D.J. Inman, Appl. Phys. Lett. **94**, 254102 (2009)

[11] F. Cottone, H. Vocca, L. Gammaitoni, Phys. Rev. Lett. **102**, 080601 (2009)

[12] S. Stoykov, E. Manoach, G. Litak, Eur. Phys. J. Special Topics **224**, 2755 (2015)

[13] M. Borowiec, G. Litak, S. Lenci, Inter. J. Struct. Stab. Dynamics **14**, 1440020 (2014)

[14] J. Tang, G. Ren, Tsinghua Science and Technology **14**, 22 (2009)

[15] C. Xu, X. Yu, J. Contr. Theory Applic. **3**, 281 (2004)

[16] M.E. Semenov, P.A. Meleshenko, A.M. Solovyov, A.M. Semenov, Hysteretic nonlinearity in inverted pendulum problem, in *Structural Nonlinear Dynamics and Diagnosis*, edited by M. Belhaq (Springer, Cham, 2012), pp. 463–506

[17] M.E. Semenov, A.M. Solovyev, Stabilization of elastic pendulum with hysteresis, in *International Conference Information Technology and Nanotechnology*, CEUR Workshop Proceedings DOI: 10.18287/1613-0073-2016-1638-650-657 (Samara, 2016), pp. 650–657

[18] M.E. Semenov, A.M. Solovyov, P.A. Meleshenko, Nonlin. Dynamics **82**, 677 (2015)

[19] S. Adhikari, M.I. Friswell, G. Litak, H.H. Khodaparast, Smart Mater. Struc. **25**, 065009 (2016)

Electromagnetic Vibration Energy Harvesting for Railway Applications

S. Bradai[1,2*], S. Naifar[1,2], C. Viehweger[1], O. Kanoun[1]

[1]Dept. of Electrical Engineering and Information Technology, Technische Universität Chemnitz, Germany
[2]Dept. of Mechanical Engineering, National School of Engineers of Sfax, University of Sfax, Tunisia

Abstract. Safe localization of trains via GPS and wireless sensors is essential for railway traffic supervision. Especially for freight trains and because normally no power source is available on the wagons, special solutions for energy supply have to be developed based on energy harvesting techniques. Since vibration is available in this case, it provides an interesting source of energy. Nevertheless, in order to have an efficient design of the harvesting system, the existing vibration needs to be investigated. In this paper, we focus on the characterization of vibration parameters in railway application. We propose an electromagnetic vibration converter especially developed to this application. Vibration profiles from a train traveling between two German cities were measured using a data acquisition system installed on the train's wagon. Results show that the measured profiles present multiple frequency signals in the range of 10 to 50 Hz and an acceleration of up to 2 g. A prototype for a vibration converter is designed taking into account the real vibration parameters, robustness and integrability requirements. It is based on a moving coil attached to a mechanical spring. For the experimental emulation of the train vibrations, a shaker is used as an external artificial vibration source controlled by a laser sensor in feedback. A maximum voltage of 1.7 V peak to peak which corresponds to a maximum of 10 mW output power where the applied excitation frequency is close to the resonant frequency of the converter which corresponds to 27 Hz.

1 Introduction

Contrarily to passenger's train, electricity is not available in all carriages of freight trains. However, there is a growing need to install several sensors in different wagons of the train in order to measure different parameters such as temperature and position, etc.... In such application, vibration is one of the promising source in this case to power sensors. Nevertheless, to harvest energy form such source is not easy because of its irregularity in term of characteristic frequency and amplitude.

Currently, several energy harvestings for railway are developed. Mainly two mechanisms are used to convert this mechanical energy to electrical one which are piezoelectric [1-3] and electromagnetic principles [4, 5].

[6] developed a piezoelectric energy harvesting system with magnetic pendulum movement able to generate 40.245 $\mu W/cm^3$ in passenger's train. Using similar principle, [7] proposed a piezoelectric converter based cantilever beam, it generates 19.64 Mw when 8 permanent magnets were added as tip mass on the cantilever for an operating commercial high-speed Korean train with a speed greater than 300 km/s. In [8], a piezoelectric converter using patch-type and stack-type inducing energy by passing trains is developed. Results prove, the converter harvests energy in the range of mJ from track vibration. Another type of piezoelectric converter placed on the bottom of rail track is realized by [9]. The system is based on the use of piezoelectric films and harvests 1 mW of power.

Other research focused on the development of vibration converter based on electromagnetic principle. [10] developed an electromagnetic converter which is placed near the rail. Once the train passes, energy will be generated through the coil fixed near the rail. [11] Another method is used in rail to harvest energy, it uses a gear and rack mechanism, and maintaining the rotational motion of the generator with roller clutches and a flywheel. An energy outcome in the range of W is reached. Another architecture based on the use of spherical magnet and a spiral coil placed around a central disc magnet is realized in [12] and implemented in train. It generates an average power of 0.1 mW in the case of attracting field (discs and spherical magnets have same polarization) and 0.65 mW for opposing field (discs and spherical magnets have opposite polarization).

These solutions are useful to profit from train vibration to power systems placed on the rail but not adequate to

* Corresponding author: sonia.bradai@s2013.tu-chemnitz.de

be placed on the train wagons. The aim of this work is to develop a solution that can be implemented in train wagons to harvest energy from existing vibration.

In this work, an electromagnetic converter able to be placed on train wagons is proposed. This paper is organized in 3 sections. As first, trains vibration is measured and characterized in real time in term of acceleration and frequency. Section 2 is devoted to the design of an electromagnetic converter adequate for implementation in train. At the end, the converter is realized and tested under random vibration profiles.

2 Railways vibration characteristics

Vibration profiles from a freight train travelling from Leipzig to Hamburg and back were recorded. For this purpose, a data logger was designed and was attached to one of the train carriages as shown in Fig. 1. It processes the signals from an acceleration sensor and saves them in the form of profiles. The vibration measurement was carried out with two acceleration sensors, which operate with a sampling frequency of 100 Hz. According to sampling theorems, frequencies up to 50 Hz can thus be measured.

The measurement was done for the three axes; X, axis in the direction of travel, Y horizontal axis orthogonal to the direction of travel and the Z axis in the height. However, vibration along the Z axis have showed the greatest acceleration amplitudes and then only data along this axis will be considered.

After collecting of the measured data from the field test, the profiles were analyzed and investigated and promising ones were identified.

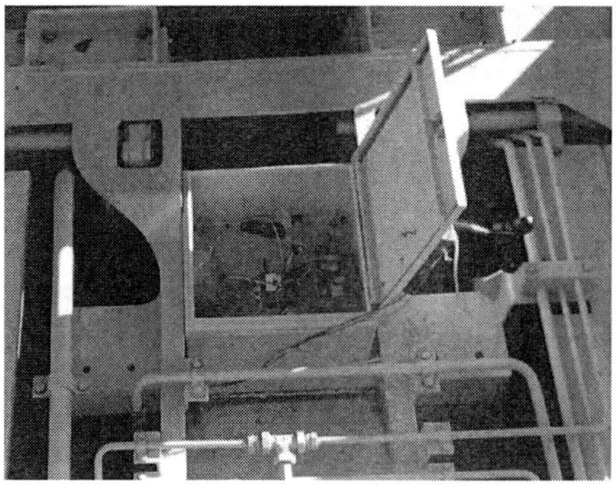

Fig. 1. Data logger attached to one of the train carriages for vibration measurement.

2.1. Measured acceleration

Fig. 2 presents an example of measured acceleration profiles for 1 hour of travel. It can be seen that the maximum measured acceleration is in the range of 2 g, which corresponds to an acceleration of about 20 m/s^2.

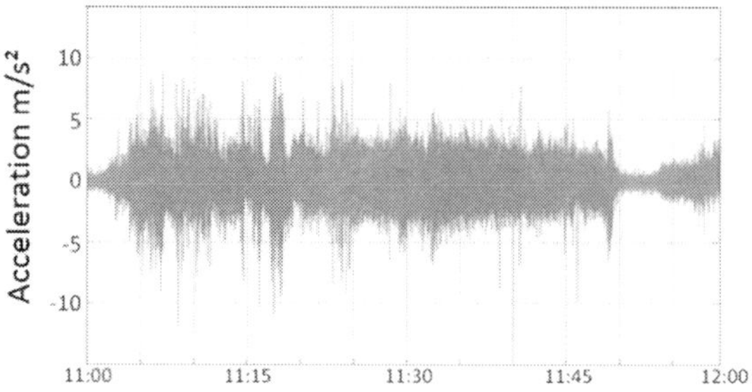

Fig. 2. Example of measured vibration profile for 1 hour of travel.

On average, the amplitudes are about 3 m/s^2.

2.1. Frequency analysis

Two selected acceleration profiles were processed and converted to amplitude responses in order to be able to regenerate the measured excitations using an artificial vibration source. In order to investigate the presented frequencies in the measured profiles, Fourier spectrum for one-minute vibration is presented in Fig.3.

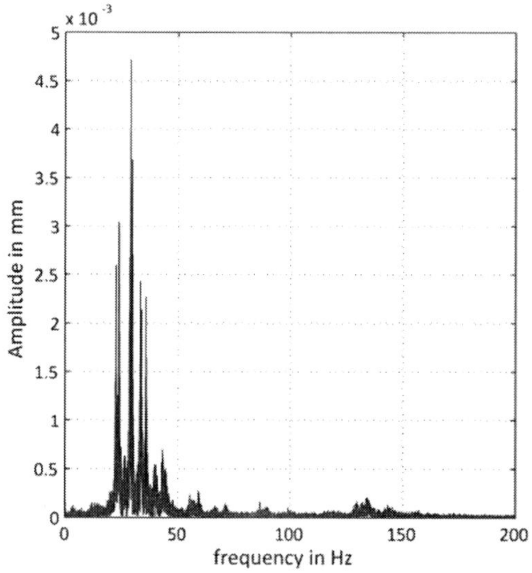

Fig. 3. Fourier transforms performed on an example for one-minute vibration profile.

Results indicates that most of the presented frequencies in the examined profile are in the range of 20 Hz to 40 Hz. Furthermore, the main measured peak frequency is about 29 Hz. Based on these results and the state of the art of vibration converter [13], electromagnetic principle is one of the most promising solution for this frequency range. Therefore, in the following design of an electromagnetic converter is proposed to harvest energy from such vibration source.

3 Design

This section is devoted to the design of an electromagnetic vibration converter suitable for the presented source in section 2. In general, an electromagnetic converter is basically formed with magnets and coil moving relative to each other in order to generate a certain energy. To ensure the system functionality, two main concepts are used for the converter design. The first one consists of the use of mechanical spring where a coil or a magnet is attached and moving relative to the applied excitation (Fig.4. (a)). The second concept is based on the use of magnetic spring principle to transmit the applied vibration. In this case, the coil is fixed surrounding the moving magnet from the magnetic spring (Fig.4. (b)).

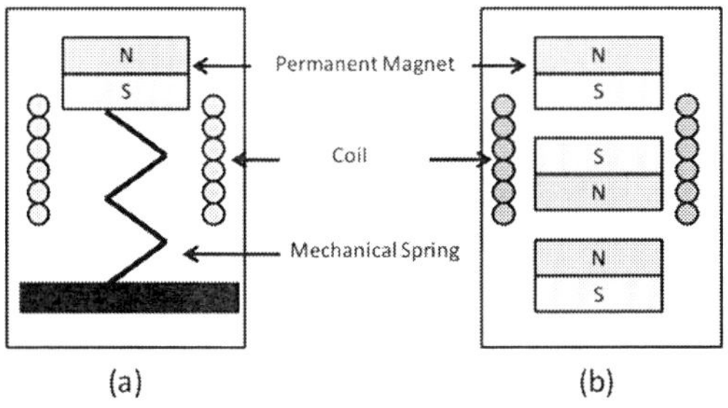

Fig. 4. Electromagnetic converter architecture based (a) mechanical spring, (b) magnetic spring.

For the magnetic spring architecture, the main advantage is to have a system working for larger frequency bandwidth due to the nonlinearity of the system. However, this architecture has a functional limitation that such solution cannot be implemented in a metallic environment because of the use a moving magnet. Instead of that, the mechanical spring architecture can be used for such environment by attaching the coil to the mechanical spring using a non-magnetic mechanical spring. Taking into account the application, the mechanical spring using moving coil will be adopted in order to be implemented in a magnetic environment.

In this case, the frequency bandwidth for the converter will be limited. Therefore, for the design, the resonant frequency of the converter should be well chosen in order to harvest maximum of the energy relative to the applied random excitation measured in railway. The proposed design is presented in Fig.5. It is formed with a mechanical spring where a coil is attached. Magnets are placed inside the coil. The main advantages of this proposed architecture are the possibility to be placed in magnetic environment, its compactness and the possibility to tune the resonant frequency by changing the stiffness spring or by using an added mass. At the end, the design overcomes one of the main limitation of the moving coil by avoiding the contact between the coil wire with the mechanical spring which leads to a better robustness for the converter.

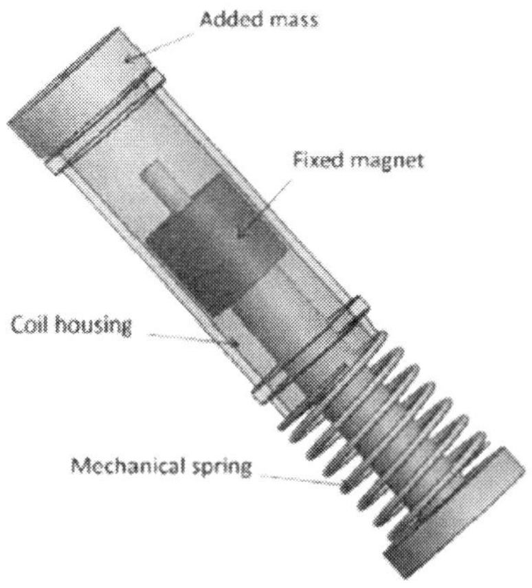

Fig. 5. Proposed mechanical design for the electromagnetic converter.

Table 1 presents the different parameters characteristics selected for the proposed converter design robustness for the converter.

Table 1. Specification of the proposed converter.

Parameter	Value
Housing size	Diameter x length: 14 x 80 mm^2
Mechanical spring stiffness	0.06 N/mm^2
Magnet size	6 x 2 x 2 mm^3
Magnets number	10
Coil length	20 mm
Copper wire thickness	0.1 mm
Number of turns	600
Coil resistance	35 Ω
Total mass acting on the spring	2 g

4 Experimental results

In the following the proposed converter was realized and tested under some real profiles measured in train. The profiles are reproduced by an electrodynamic shaker. The resonant frequency of the harvester was determined experimentally using a harmonic excitation for 1 mm excitation amplitude. Results indicates that the designed converter which has an inner resistance of 35 Ω can generate 0.72 V as open circuit peak to peak voltage at 27 Hz.

Experiment investigations were carried out using two vibration amplitude profiles of 30 seconds length each as shown in Fig. 6 (a) and (c). The harvester was fixed on the top of the shaker and the amplitude of vibration was measured by a displacement laser sensor. Fig. 6 (b) and (c) present the measured open circuit voltages corresponding to profiles given in Fig. 6 (a) and (c) respectively.

For the profile (A), the maximum amplitude is around 0.5 mm while the average amplitude is around 0.3 mm. In this case, the maximum output voltage is around 1.7 V peak to peak voltage (Fig.6. (b)). It can be seen that the maximum voltage is reached not at the maximum applied amplitude displacement. This can be explained by the effect of the corresponding frequency for each cycle.

For the profile (B), the maximum amplitude is around 0.45 mm while the average amplitude is around 0.2 mm. 1.2 V peak to peak voltage is the maximum reached output voltage (Fig.6. (d)). This voltage is achieved for the maximum applied amplitude. This means that for this cycle, the applied frequency is close to the resonant frequency of the converter.

It can be seen that the maximum voltage is reached at the maximum applied amplitude displacement. This can be explained by the effect of the corresponding frequency for each cycle.

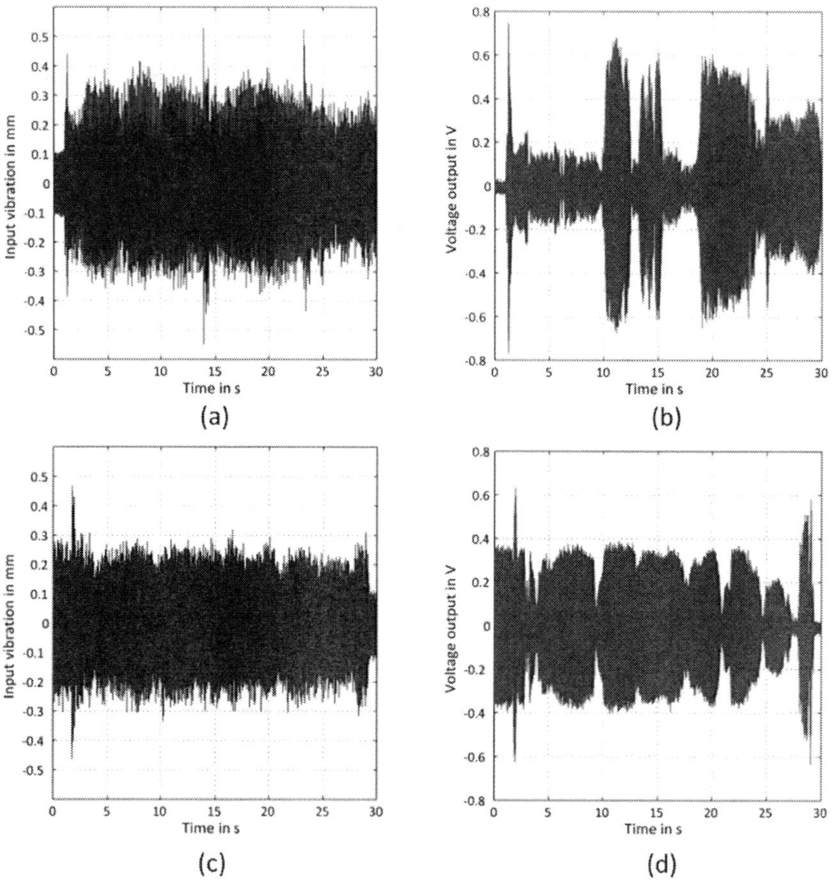

Fig. 6. (a), (c) applied random excitation for the converter, (b), (d) generated voltage from the converter. Profile A (a), (b), profile B (c), (d)

5 Conclusion

In this work, an electromagnetic vibration converter is investigated in order to convert mechanical vibration from railway to useful energy. Vibration profiles were measured in train travelling between two German cities. Promising acceleration profiles were identified and characterized. The measured profiles are characterized by an acceleration amplitude up to 10 m/s^2 and a frequency range varying from 20 to 40 Hz. Measured profiles are regenerated using an electrodynamic shaker and tested on the proposed converter. Results indicate that a maximum of 1.7 V peak to peak voltage which corresponds to a maximum of 10 mW output power can be reached for specific cycle where the frequency is close to the resonant frequency of the converter. An average voltage of 1.2 V peak to peak can be generated for an applied random excitation up to 1 mm of amplitude and for a frequency range from 20 to 40 Hz.

References

1. Mouapi, A., Hakem, N., Kandil, N., & Kamani, G. V., Energy harvesting design for autonomous Wireless Sensors Network applied to trains. In 2016 IEEE International Ultrasonics Symposium (IUS), (pp. 1-4), 2016.
2. Wischke, M., Masur, M., Kröner, M., & Woias, P., Vibration harvesting in traffic tunnels to power wireless sensor nodes. Smart Materials and Structures, 20(8), 085014, 2011.
3. Li, J., Jang, S., & Tang, J., Optimization of piezoelectric energy harvester for wireless smart sensors in railway health monitoring. In Proc. of SPIE, Vol. 8692, pp. 86924L-1, 2013.
4. Gao, M., Wang, P., Cao, Y., Chen, R., & Cai, D., Design and Verification of a Rail-Borne Energy Harvester for Powering Wireless Sensor Networks in the Railway Industry. IEEE Transactions on Intelligent Transportation Systems, 18(6), 1596-1609, 2017.
5. Bradai, S., Naifar, S., Keutel, T., & Kanoun, O., Electrodynamic resonant energy harvester for low frequencies and amplitudes. In 2014 IEEE International Instrumentation and Measurement Technology Conference (I2MTC) Proceedings, pp. 1152-1156, 2014.
6. Cho, J. Y., Jeong, S., Jabbar, H., Song, Y., Ahn, J. H., Kim, J. H., ... & Sung,. Piezoelectric energy harvesting system with magnetic pendulum movement for self-powered safety sensor of trains. Sensors and Actuators A: Physical, 250, 210-218, (2016).
7. Song, D., Yang, C. H., Hong, S. K., Kim, S. B., Woo, M. S., & Sung, Feasibility study on application of piezoelectricity to convert vibrations of Korea Train eXpress. In Applications of Ferroelectrics held jointly with 2012 European Conference on the Applications of Polar Dielectrics and 2012 International Symp Piezoresponse Force Microscopy and Nanoscale Phenomena in Polar Materials (ISAF/ECAPD/PFM), (2012, July).
8. Wang, J., Shi, Z., Xiang, H., & Song, G, Modeling on energy harvesting from a railway system using piezoelectric transducers. Smart Materials and Structures, 24(10), 105017, (2015).
9. Nelson, C. A., Platt, S. R., Albrecht, D., Kamarajugadda, V., & Fateh, M, Power harvesting for railroad track health monitoring using piezoelectric and inductive devices. In Proc. SPIE Vol. 6928, pp. 69280R-69280R, 2008.
10. Z. F. Mian, 2009, Wireless railroad monitoring US patent 12/493,789.
11. Wang J J, Penamalli G and Zuo L, Electromagnetic energy harvesting from train induced railway track vibrations 2012 IEEE/ASME Int. Conf. on Mechatronics and Embedded Systems and Applications (MESA) pp 29–34, 2012.
12. Hart, G., Moss, S., Nagle, D., Jung, G., Wilson, A., Ung, C., ... & Crew, G., Vibration Energy Harvesting for Aircraft, Trains and Boats. In Proceedings of Acoustics, 2013.
13. Naifar, S., Bradai, S., Viehweger, C., & Kanoun, O., Survey of electromagnetic and magnetoelectric vibration energy harvesters for low frequency excitation. Measurement, 106, 251-263, 2017.

Nonlinear Vibrations of a Mechanism for the Miller-Atkinson Cycle

Ionuţ Dragomir[1], *Bogdan* Mănescu[1], *Nicolae–Doru* Stănescu[2,*]

[1]AKKA ROMSERV, Bucharest, Romania
[2]University of Piteşti, Piteşti, Romania

Abstract. The distribution mechanism for the Miller-Atkinson is presented based on some previous papers of the authors. The cam is obtained in an analytical way and the optimization procedure of its shape based on Jarvis March algorithm is also described. The non-linear vibrations of the mechanism are analytically obtained used the Fourier transformation and they are numerically validated for different types of cams.

1 Introduction

The references were discussed in our previous papers [9-11] and will not repeat here. The most of the scientists usually study the phenomena in an engine working after the Miller-Atkinson cycle. The study of the distribution mechanism is a very poor one because of its complexity and of the fact that the obtained cam is not a convex one. In our paper we will use the Jarvis March in order to obtain a convex cam and we will compare the results obtained in classical way with those deduced for the convex cam.

The distribution mechanism considered in our paper is presented in previous works of the authors [9-11]. It consists in a valve moving on the vertical direction (Figs. 1-4), a lever OB rotating around the revolute joint situated at the point O, at the end of the lever being situated a roll of radius R_2. The lever is rigidly linked to the bar OD, at the end D of which being situated another roll of radius R_3 (Fig. 5). The cam of center O (Fig. 5) rotates around its center, rotating the element DO_2B and acting the valve.

2 Synthesis of cam

There exist four different variants to be considered:

i) the valve is asymmetrically acted, the head of valve being a spherical one;

ii) the valve is symmetrically acted, the head of valve being a spherical one;

iii) the valve is asymmetrically acted, the head of valve being a planar one;

iv) the valve is symmetrically acted, the head of valve being a planar one.

In the case i) one knows the maximum displacement of the valve s_{max}, its law of motion $s = s(\varphi)$, the initial angle β_0 for which it results $\theta = 0$. The radius of the spherical head of valve is equal to R_1. The lever OD (Fig. 1) has a length equal to l_1. A displacement s of the valve leads to a rotation of angle θ for the lever.

a)

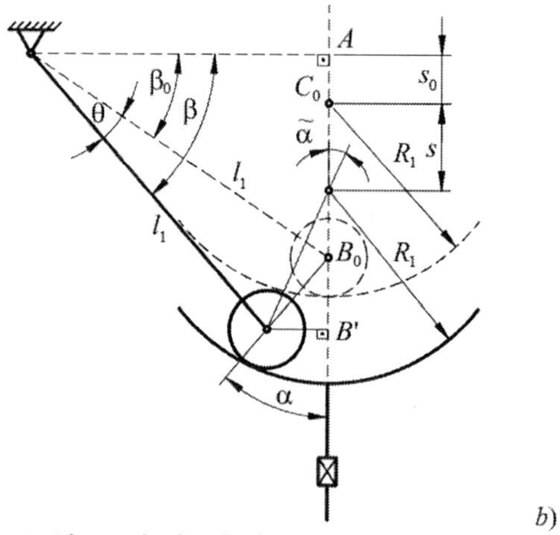

b)

Fig. 1. The mechanism in the asymmetric case: *a*) analytic approach; *b*) geometric approach.

An analytic approach (Fig. 1, *a*)) offers [9-11] the implicit equation from which one may obtain the rotational angle β

$$\begin{aligned}
&\{2x_C(l_1 \sin\beta - y_C)^2 + (l_1\cos\beta - x_C) \\
&\quad \times [C_1 - 2y_C(l_1\sin\beta - y_C)]\} \\
&- (l_1^2 + x_C^2 + y_C^2 - 2l_1x_C\cos\beta - 2l_1y_C\sin\beta) \\
&\quad \times \{4x_C^2(l_1\sin\beta - y_C)^2 \\
&\quad + [C_1 - 2y_C(l_1\sin\beta - y_C)]^2 \\
&\quad - 4R_1^2(l_1\sin\beta - y_C)\} = 0,
\end{aligned} \tag{1}$$

* Corresponding author: s_doru@yahoo.com, doru.stanescu@upit.ro

where the parameters C_1, x_C and y_C have the following expressions $C_1 = R_1^2 - R_2^2 + l_1^2 - \left(x_C^2 + y_C^2\right)$, $x_C = d$, $y_C = h + s - R_1$.

Following a geometric approach (Fig. 1, *b*)) one obtain the relation

$$\begin{gathered} -s + l_1\left[\sin(\theta + \beta_0) - \sin\beta_0\right] \\ + (R_1 - R_2)(1 - \cos\tilde{\alpha}) = 0, \end{gathered} \quad (2)$$

where

$$\tilde{\alpha} = \arcsin\left(\frac{l_1\left[\cos\beta_0 - \cos(\beta_0 + \theta)\right]}{R_1 - R_2}\right). \quad (3)$$

For the situation described in ii) (Fig. 2) one may write [11] the geometric relations

$$\beta_0 = \arccos\left(\frac{d_1 + d_2}{l_1}\right),\ \tilde{\alpha}_0 = \arcsin\left(\frac{d_2}{R_1}\right),$$

$$\alpha = \beta_0 + \frac{\theta}{2},\ B_0B = \tilde{d} = 2l_1\sin\left(\frac{\theta}{2}\right),$$

$$\sin\tilde{\alpha} = \frac{d_2 - \tilde{d}\sin\alpha}{R_1},$$

$$\tilde{\alpha} = \arcsin\left(\frac{d_2 - 2l_1\sin\alpha\sin\left(\frac{\theta}{2}\right)}{R_1}\right), \quad (4)$$

$$CB' = R_1\cos\tilde{\alpha},\ C_0B' = R_1\cos\tilde{\alpha}_0 + \tilde{d}\cos\alpha,$$
$$s = C_0B' - CB',$$

$$s = R_1\cos\tilde{\alpha}_0 - R_1\cos\tilde{\alpha} + 2l_1\cos\alpha\sin\left(\frac{\theta}{2}\right),$$

$$\tilde{\alpha}_{final} = \tilde{\alpha}_f = -\alpha_0$$

wherefrom

$$\sin\tilde{\alpha}_f = \frac{d_2 - 2l_1\sin\alpha\sin\left(\frac{\theta}{2}\right)}{R_1} = -\sin\alpha_0,$$

$$d_2 - 2l_1\sin\left(\beta_0 + \frac{\theta}{2}\right)\sin\left(\frac{\theta}{2}\right) = -d_2,$$

$$2d_2 = l_1\left[\cos\beta_0 - \cos(\beta_0 + \theta)\right], \quad (5)$$

$$\theta_f = \arccos\left(\frac{l_1\cos\beta_0 - 2d_2}{l_1}\right) - \beta_0,$$

$$\theta_f = \arccos\left(\frac{d_1 - d_2}{l_1}\right) - \beta_0.$$

An iterative algorithm may be developed in order to determine the distance d. If the maximum displacement of the valve $s_{\max}$ is known, then we follow the steps:

1. consider $d_2 = (d_2)_0$ as initial value;

2. calculate in order β_0, $\tilde{\alpha}_0$, θ_f, α, s, the deviation Δ with the formula $\Delta = s - s_{\max}$;

3. if $|\Delta| > \varepsilon$, where ε is the a priori imposed error, we consider a new value for d_2, $d_2 \mapsto d_2 + increment$, and repeat the step 2; otherwise the distance d_2 was determined.

Finally, $\beta_f = \beta_0 + 2\arcsin\left(\frac{\sqrt{4d_2^2 + s_{\max}^2}}{2l_1}\right)$.

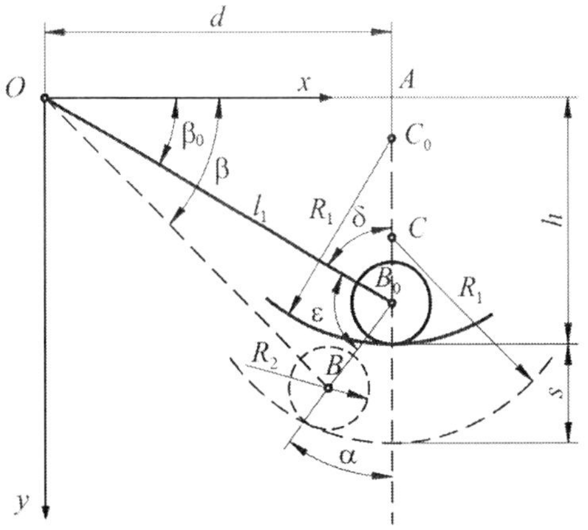

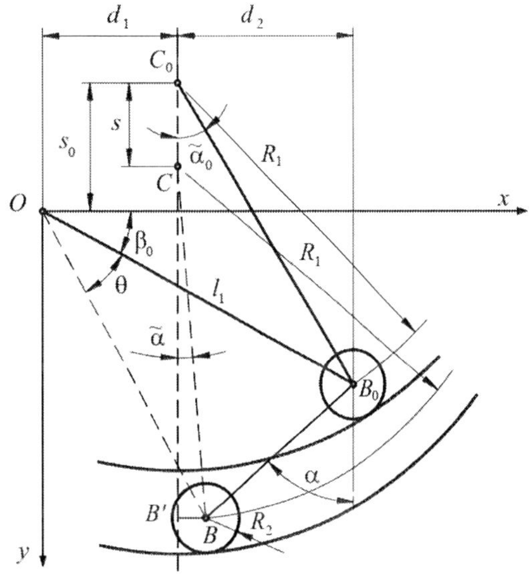

Fig. 2. The mechanism in the symmetric case.

The cases iii) and iv) are characterized by simpler formulae (Figs. 3 and 4)

$$\beta = \arcsin\left(\frac{s + l_1\sin\beta_0}{l_1}\right) \quad (6)$$

$$\theta = \beta - \beta_0 = \arcsin\left(\frac{s + l_1\sin\beta_0}{l_1}\right) - \beta_0 \quad (7)$$

and

$$\beta = \arcsin\left(\sin\beta_0 + \frac{s}{l_1}\right) \quad (8)$$

respectively.

At the repose position (Fig. 5) one may write the following relations [11]

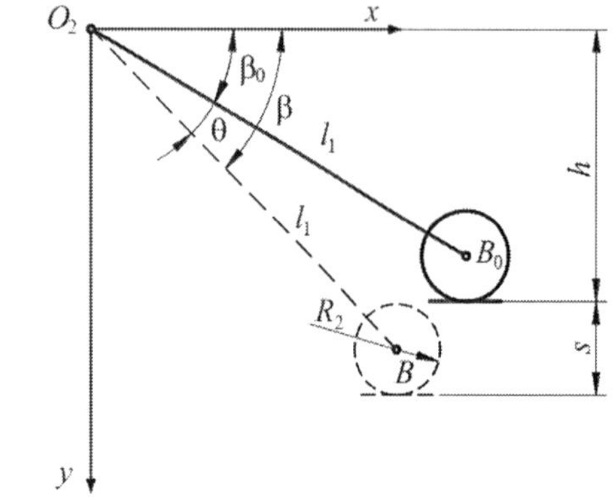

Fig. 3. The asymmetric case of planar head of valve.

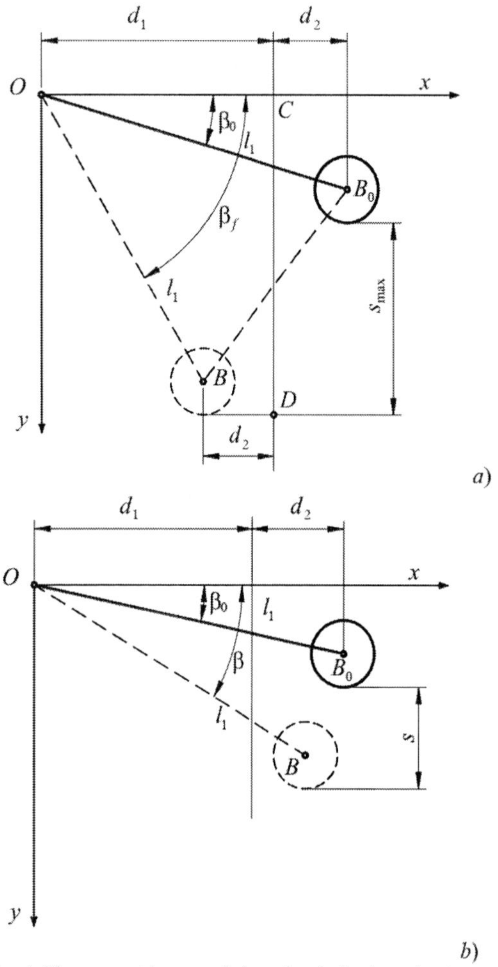

Fig. 4. The symmetric case of planar head of valve: *a*) extreme position; *b*) intermediate position.

$$\tilde{\beta}_0 = \pi - \beta_0 - \delta_0,$$
$$\gamma = \arccos\left(\frac{l_2^2 + d^2 - (R_0 + R_3)^2}{2 l_2 d}\right), \tag{9}$$

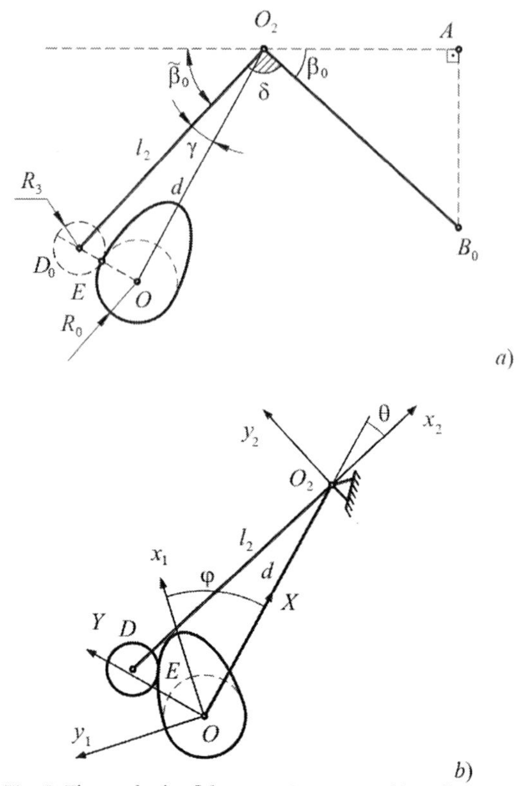

Fig. 5. The synthesis of the cam: *a*) repose position; *b*) current position.

The parametric equations of the cam are [10, 11]

$$\begin{aligned} x_1 &= d\cos\varphi - l_2\cos(\theta + \gamma + \varphi) \\ &\quad - R_3\sin(\theta + \gamma + \varphi + \lambda), \\ y_1 &= -d\sin\varphi + l_2\sin(\theta + \gamma + \varphi) \\ &\quad - R_3\cos(\theta + \gamma + \varphi + \lambda), \end{aligned} \tag{10}$$

where

$$\tan\lambda = \frac{d\cos(\theta + \gamma) - l_2\left(1 + \dfrac{d\theta}{d\varphi}\right)}{d\sin(\theta + \gamma)}. \tag{11}$$

The resulted cam is not always a convex one. Considering an incremental step $\Delta\varphi$ for the rotational angle of the cam, one obtains a table of values $x_1(\varphi_j)$ and $y_1(\varphi_j)$, where $\varphi_j = \varphi_0 + j\Delta\varphi$. From these values, the Jarvis March selects only those that define the convex covering of the cam. In this way, only a part of the characteristic points are selected and the cam obtained by Jarvis March contains segments of straight lines corresponding to more than one step $\Delta\varphi$.

3 Analysis of cam

From the equations

$$x_2 = -l_2 - R_3\sin\lambda,\ y_2 = -R_3\cos\lambda \tag{12}$$

it results

$$\frac{dy_2}{dx_2} = -\tan\lambda \tag{13}$$

The tangency condition leads to

$$\tan(\varphi+\theta+\gamma) = \frac{-\frac{dy_1}{dx_1}+\frac{dy_2}{dx_2}}{1+\frac{dy_1}{dx_1}\frac{dy_2}{dx_2}} = \frac{-\frac{dy_1}{dx_1}-\tan\lambda}{1--\frac{dy_1}{dx_1}\tan\lambda}, \tag{14}$$

wherefrom we obtain the expression

$$f_1(\varphi,\theta,\lambda) = \sin(\varphi+\theta+\lambda) \times\left(\cos\lambda-\frac{dy_1}{dx_1}\sin\lambda\right)+\cos(\varphi+\theta+\lambda) \times\left(\cos\lambda\frac{dy_1}{dx_1}+\sin\lambda\right)=0. \tag{15}$$

Another two functions are obtained from the contact point's coordinates, that is,

$$f_2(\varphi,\theta,\lambda) = -x_1\cos\varphi+y_1\sin\varphi+d - l_2\cos(\theta+\gamma)-R_3\sin(\theta+\gamma+\lambda)=0, \tag{16}$$

$$f_3(\varphi.\theta,\lambda) = -x_1\sin\varphi-y_1\cos\varphi + l_2\sin(\theta+\gamma)-R_3\cos(\theta+\gamma+\lambda)=0. \tag{17}$$

One has to solve a nonlinear system of three equations $f_i = 0$, with three unknowns φ, θ, and λ. This system is solved using the Newton method, starting with the initial values $\varphi_0 = 0$, $\theta_0 = 0$, and $\lambda = \lambda_0$, where $\lambda_0 = \arctan\left(\frac{d\cos\gamma - l_2}{d\sin\gamma}\right)$.

4 Fourier analysis

We denote by $\overline{\omega}$ the angular velocity of the cam-shaft. In these conditions, one may write

$$s(\varphi) = s(\overline{\omega}t) = \sum_{i=1}^{\infty} a_i\cos(i\overline{\omega}t)+\sum_{i=1}^{\infty} b_i\sin(i\overline{\omega}t), \tag{18}$$

where s is the displacement of the valve, while a_i and b_i are the Fourier coefficients.

The acceleration of the valve has the expression

$$a(\varphi) = a(\overline{\omega}t) = -\sum_{i=1}^{\infty} a_i(i\overline{\omega})^2\cos(i\overline{\omega}t) - \sum_{i=1}^{\infty} b_i(i\overline{\omega})^2\sin(i\overline{\omega}t), \tag{19}$$

while the magnitude of the inertial force is

$$F(\varphi) = F(\overline{\omega}t) = m_{eq}a(\varphi), \tag{20}$$

m_{eq} being the equivalent mass of the distribution mechanism.

The force induced by the j th valve reads

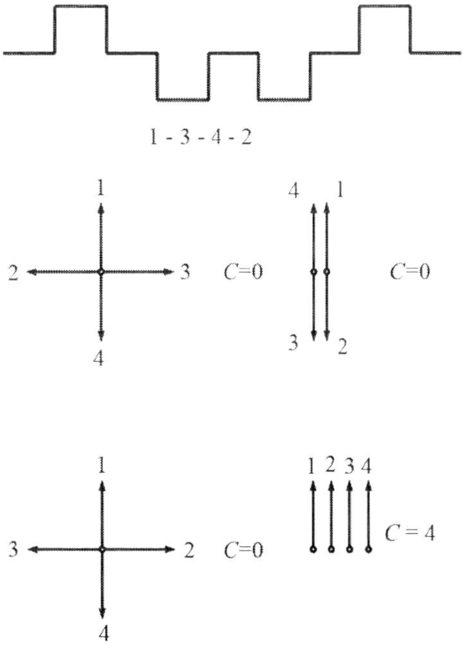

Fig. 6. The cam-star for the Fourier analysis.

$$F_{ij}(\overline{\omega}t) = -m_{eq}A_i(i\overline{\omega})^2\cos(i(\overline{\omega}t+\gamma_j+\theta)); \tag{21}$$

the force induced by all four valves (we assumed a four-stroke engine with 4 cylinders, the firing order being 1-4-3-2 (Fig. 6)) is

$$F_i(\overline{\omega}t) = -m_{eq}A_i(i\overline{\omega})^2\sum_{j=1}^{4}\cos(i(\overline{\omega}t+\gamma_j+\theta)), \tag{22}$$

and the net inertial force is given by

$$\sum_{j=1}^{4}\cos(i(\overline{\omega}t+\gamma_j+\theta)) = C_i\cos(i(\overline{\omega}t+\gamma_j+\theta)). \tag{23}$$

Analyzing Fig. 6, we may conclude the most important coefficients are $C_{4k} = 4$, k integer, the rest of coefficients being zero.

It results the magnitude of the inertial force and the amplitude of oscillations as

$$\hat{F}_i = 4A_i m_{eq}(4k\overline{\omega})^2,\ \hat{A}_i = \frac{\hat{F}_i}{M_e(i\overline{\omega})^2}, \tag{24}$$

M_e being the mass of the engine.

One may write the following relations

$$\{\varphi_i\} = \{0,\pi,\pi,0\} \tag{25}$$

$$\{\gamma_j\} = \left\{0,\frac{\pi}{2},\frac{3\pi}{2},\pi\right\} \tag{26}$$

$$i=1 \Rightarrow \{i\gamma_j\} = \left\{0,\frac{\pi}{2},\frac{3\pi}{2},\pi\right\} \tag{27}$$

$$i=2 \Rightarrow \{i\gamma_j\} = \{0,\pi,3\pi,2\pi=0\} \tag{28}$$

$$i=3 \Rightarrow \{i\gamma_j\} = \left\{0,\frac{3\pi}{2},\frac{9\pi}{2}=\frac{\pi}{2},3\pi=\pi\right\}, \tag{29}$$

$$i = 4 \Rightarrow \{i\gamma_j\} = \{0{,}2\pi = 0{,}6\pi = 0{,}4\pi = 0\}. \quad (30)$$

5 Numerical simulation

The following values were selected for the numerical simulation: $l_1 = 0.01612$ m, $R_1 = 0.025$ m, $R_2 = 0.010$ m, $\beta_0 = 15^0$, $s_{max} = 0.00935$ m, $R_0 = 0.018$ m, $d = 0.042$ m, $l_2 = 0.060$ m, $R_3 = 0.01867$ m, $d_1 = 0.010$ m, the angles $\varphi_1 = 92^0$ (valve starts to open), $\varphi_2 = 161^0$ (valve is completely open), $\varphi_3 = 180^0$ (valve starts to close), and $\varphi_4 = 244^0$ (valve is completely closed). Two laws of motion are considered for the valve displacement [9-11]

$$s_1(\varphi) = \begin{cases} 0, \varphi \in [0, \varphi_1), \\ s_{max} \sin^2\left(\dfrac{90(\varphi - \varphi_1)}{\varphi_2 - \varphi_1}\right), \varphi \in [\varphi_1, \varphi_2], \\ s_{max}, \varphi \in (\varphi_2, \varphi_3), \\ s_{max} \sin^2\left(\dfrac{90(\varphi + \varphi_4 - 2\varphi_3)}{\varphi_4 - \varphi_3}\right), \varphi \in [\varphi_3, \varphi_4], \\ 0, \varphi \in (\varphi_4, 360] \end{cases} \quad (25)$$

$$s_2(\varphi) = \begin{cases} 0, \varphi \in [0, \varphi_1), \\ s_{max} \dfrac{(\varphi - \varphi_1)^2 (\varphi - 2\varphi_2 + \varphi_1)^2}{(\varphi_2 - \varphi_1)^4}, \varphi \in [\varphi_1, \varphi_2], \\ s_{max}, \varphi \in (\varphi_2, \varphi_3), \\ s_{max} \dfrac{(\varphi - \varphi_4)^2 (\varphi - 2\varphi_3 + \varphi_4)^2}{(\varphi_3 - \varphi_4)^4}, \varphi \in [\varphi_3, \varphi_4], \\ 0, \varphi \in (\varphi_4, 360]. \end{cases} \quad (26)$$

The resulted cams are given in Fig. 7.

In Fig. 8 are drawn the real laws of motion for the normal cams, and the laws of motion for the convex cams in some cases.

Some Fourier coefficients a_i are given in Table 1.

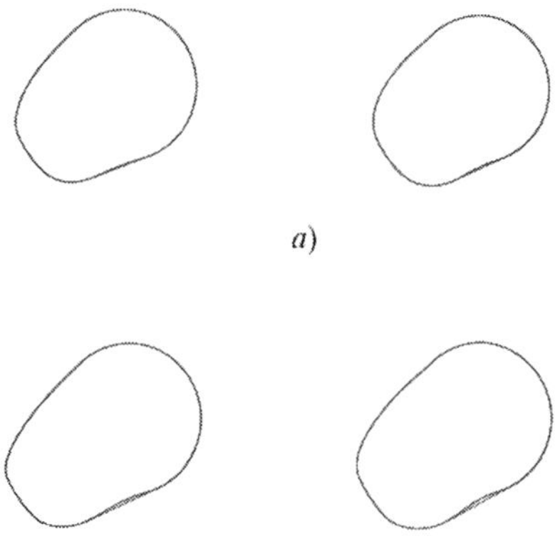

a) *b*) *c*) *d*)

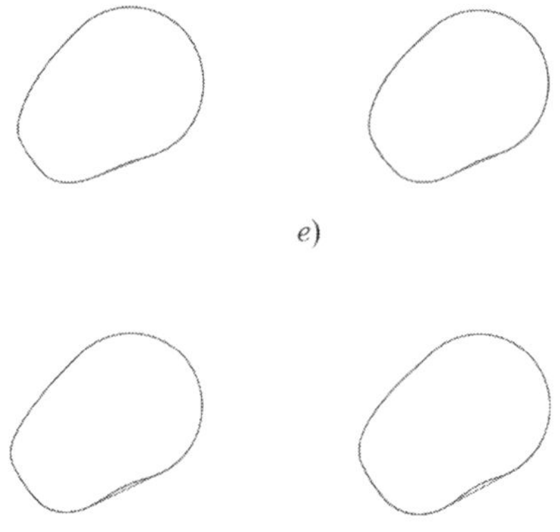

e) *f*) *g*) *h*)

Fig. 7. The obtained cam for: *a*) case i) first law; *b*) case i) second law; *c*) case ii) first law; *d*) case ii) second law; *e*) case iii) first law; *f*) case iii) second law; *g*) case iv) first law; *h*) case iv) second law. The normal cam in drawn in blue, the convex one obtained with the aid of Jarvis March in red.

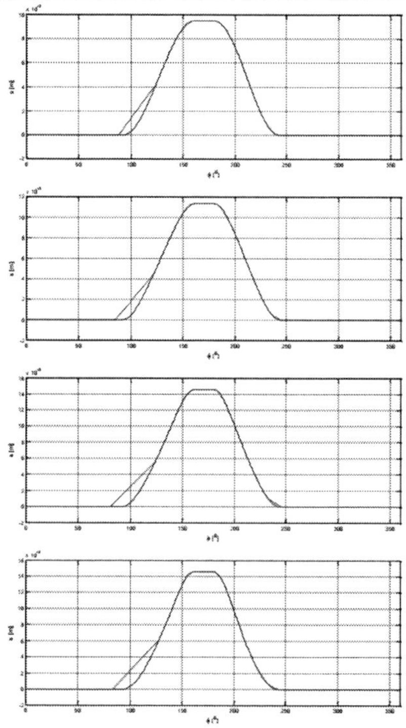

a) *b*) *c*) *d*)

Fig. 8. The laws of motion for the real cams (blue) and convex cams (red) for: *a*) case i) first law; *b*) case iii) second law; *c*) case ii) second law; *d*) case iv) first law.

One may write the following observations:

– symmetric case leads to bigger cams;

– the two laws of motion lead to similar cams;

– the obtained cams are not convex and the Jarvis March must be applied;

– the coefficient a_4 is the most important one.

The Fourier coefficients b_i are not presented in this paper, but for non-symmetric angles of opening and closing of valves these coefficients do not vanish.

Table 1. Fourier coefficients in some cases.

Law	First			
Case	i)	i) Jarvis	ii)	ii) Jarvis
a_0	0.0044413	0.0046178	0.0044413	0.0064828
a_1	0.0038423	0.0038930	0.0038423	0.0054888
a_2	0.0024231	0.0022955	0.0024231	0.0033474
a_3	0.0009880	0.0008999	0.0009880	0.0015438
a_4	0.0001240	0.0002080	0.0001240	0.0006347
a_8	-0.0000165	-0.0000294	-0.0000165	-0.0000505
a_{12}	0.0000015	-0.0000083	0.0000015	0.0000335
a_{16}	0.0000059	0.0000190	0.0000059	0.0000222
a_{20}	0.0000009	0.0000060	0.0000009	-0.0000187
Law	Second			
Case	iii)	iii) Jarvis	iv)	iv) Jarvis
a_0	0.0046717	0.0054925	0.0046717	0.0069123
a_1	0.0039965	0.0045939	0.0039965	0.0057659
a_2	0.0024178	0.0026665	0.0024178	0.0033530
a_3	0.0008729	0.0010631	0.0008729	0.0014174
a_4	0.0000108	0.0002919	0.0000108	0.0005069
a_8	-0.0000149	-0.0000006	-0.0000149	-0.0000112
a_{12}	-0.0000039	0.0000073	-0.0000039	0.0000093
a_{16}	0.0000066	0.0000168	0.0000066	-0.0000002
a_{20}	0.0000019	-0.0000061	0.0000019	-0.0000234

6 Conclusions

Some conclusions can be drawn:

– the Miller–Atkinson cycle does not lead to convex cams and the Jarvis March must be applied. Some simulations for larger radii for the base circle were performed, but, for acceptable radii the cams are not convex;

– the Fourier coefficients a_i are greater for the convex cam;

– the non-convexity of the cam is mainly caused by the opening of the valve.

The influence of different parameters on the dynamics of the mechanism will be the goal of our future work.

References

1. G. Gonca, B. Sahin, A. Parlak, Y. Ust, V. Ayhan, I. Cesur, B. Boru, Applied Energy **138**, 11 (2015)
2. M.E.S. Martins, T.D.M. Lanzanova, Applied Thermal Engineering **90** 274 (2015)
3. R. Ebrahimi, Mathematical and Computer Modelling **53**, 1289 (2011)
4. J.P. Zammit, M.J. McGhee, P.J. Shayler, T. Lawa, I. Pegg, Energy **79**, 100 (2015)
5. Y. Murata, J. Kusaka, M. Odaka, Y. Daisho, D. Kawano, H. Suzuki, H. Ishii, Y. Goto, SAE paper 2006-01-0203 (2006)
6. W. De Ojeda, SAE paper 2010-01-1124 (2010)
7. A. Kéromnès, B. Delaporte, G. Schmitz, L. Le Moyne, Energy Conversion and Management **82**, 259 (2014)
8. S. Zhu, K. Deng, S. Liu, S. Qu, Energy **93**, 75 (2015)
9. I. Dragomir, B. Mănescu, N.-D. Stănescu, *AVMS 2017*, Timişoara (2017)
10. I. Dragomir, B. Mănescu, N.-D. Stănescu, N. Pandrea, *ICMSAV XLI*, Cluj-Napoca (2017)
11. I. Dragomir, B. Mănescu, N.-D. Stănescu, N. Pandrea, A. Clenci, D. Popa, *CAR 2017*, Piteşti (2017)
12. B. Yan, H. Wang, Z. Zheng, Y. Qin, M. Yao, Applied Thermal Engineering **122**, 439 (2017)
13. B. Wu, Q. Zhan, X. Yu, G. Lv, X. Nie, S. Liu, Applied Thermal Engineering **118**, 621 (2017)
14. J. Zhao, Applied Energy **185**, 300 (2017)
15. T. Li, B. Wang, B. Zheng, Energy Conversion and Management **123**, 140 (2016)
16. Q. Luo, B. Sun, Energy Conversion and Management **123**, 209 (2016)
17. G. Gonca, Energy **109**, 152 (2016)
18. Y. Wang, B. Zu, Y. Xu, Z. Wang, J. Liu, Energy Conversion and Management **119**, 316 (2016)
19. E. Dobrucali, Energy **103**, 119 (2016)
20. G. Gonca, B. Sahin, Applied Mathematical Modelling **40**, 3764 (2016)

Influences of the Control on the Nonlinear Vibrations of a Variable Compression Ratio Mechanism

Bogdan Mănescu[1], *Ionuț* Dragomir[1], *Nicolae–Doru* Stănescu[2,*]

[1]AKKA ROMSERV, Bucharest, Romania
[2]University of Pitești, Pitești, Romania

Abstract. For the mechanism described in references the study of the nonlinear vibrations is performed considering a multibody approach for the elements of the mechanism and different laws of motion for the control element. A great attention is paid to the equilibrium of the motion. The influence of different parameters of control is highlighted in the paper. The results are numerically validated.

1 Introduction

The references were presented in our previous papers [12-14] and we will not repeat here.

The mechanism (Fig. 1) consists in the shaft OA which rotates uniformly around the point O, the triangular shell ABC, the levers CE and BD, and the piston situated at the point D.

The distance between the point E and the OY-axis is denoted by d, while the distance between the piston and the same axis is equal to e.

The triangular shell ABC and the levers BD and CE have plane-parallel motions.

The end E of the lever CE can move in vertical direction during the functioning of the engine, resulting different compression ratios and different points of extreme for the piston.

We denote by C_i, $i=\overline{1,5}$, the centers of mass for different elements of the mechanism, and by φ_i, $i=\overline{1,5}$, the rotational angles for the same elements.

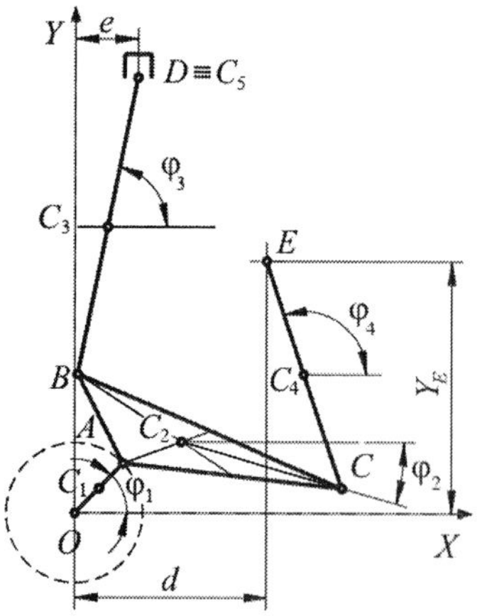

Fig. 1. The mechanism.

2 Multibody approach

The position of the mechanism is completely known if we know the position of the center of mass for each element (that is, the coordinates X_{C_i} and Y_{C_i}) and the corresponding rotational angles φ_i. It results a number of 15 unknowns.

Let us denote by $[\mathbf{A}_i]$, $i=\overline{1,5}$, the rotation matrix of an element,

$$[\mathbf{A}_i]=\begin{bmatrix}\cos\varphi_i & -\sin\varphi_i\\ \sin\varphi_i & \cos\varphi_i\end{bmatrix}, \tag{1}$$

and let us consider a generic point P that belongs to the elements i_1 and i_2.

If $x_P^{(i_1)}$ and $y_P^{(i_1)}$ are the coordinates of the point P relative to the mobile reference system $C_{i_1}x_{i_1}y_{i_1}$, while X_P and Y_P are the coordinates of the same point relative to the fixed reference system OXY, then one may write the relation

$$\begin{bmatrix}X_P\\ Y_P\end{bmatrix}=\begin{bmatrix}X_{C_{i_1}}\\ Y_{C_{i_1}}\end{bmatrix}+[\mathbf{A}_{i_1}]\begin{bmatrix}x_P^{(i_1)}\\ y_P^{(i_1)}\end{bmatrix}=\begin{bmatrix}X_{C_{i_1}}+x_P^{(i_1)}\cos\varphi_{i_1}-y_P^{(i_1)}\sin\varphi_{i_1}\\ Y_{C_{i_1}}+x_P^{(i_1)}\sin\varphi_{i_1}+y_P^{(i_1)}\cos\varphi_{i_1}\end{bmatrix}, \tag{2}$$

in which $X_{C_{i_1}}$ and $Y_{C_{i_1}}$ are the coordinates of the center of mass C_{i_1} relative to the fixed reference system.

Considering now the element i_2 for which the coordinates of the center of mass C_{i_2} relative to the fixed reference system are $X_{C_{i_2}}$ and $Y_{C_{i_2}}$, while the coordinates of the point P in the local reference system are $x_P^{(i_2)}$ and $y_P^{(i_2)}$ one may write a similar expression

* Corresponding author: s_doru@yahoo.com, doru.stanescu@upit.ro

$$\begin{bmatrix} X_P \\ Y_P \end{bmatrix} = \begin{bmatrix} X_{C_{i_2}} \\ Y_{C_{i_2}} \end{bmatrix} + \left[\mathbf{A}_{i_2}\right] \begin{bmatrix} x_P^{(i_2)} \\ y_P^{(i_2)} \end{bmatrix} = \begin{bmatrix} X_{C_{i_2}} + x_P^{(i_2)} \cos\varphi_{i_2} - y_P^{(i_2)} \sin\varphi_{i_2} \\ Y_{C_{i_2}} + x_P^{(i_2)} \sin\varphi_{i_2} + y_P^{(i_2)} \cos\varphi_{i_2} \end{bmatrix}. \quad (3)$$

Equating the relations (2) and (3), it results

$$\begin{bmatrix} X_{C_{i_1}} + x_P^{(i_1)} \cos\varphi_{i_1} - y_P^{(i_1)} \sin\varphi_{i_1} \\ Y_{C_{i_1}} + x_P^{(i_1)} \sin\varphi_{i_1} + y_P^{(i_1)} \cos\varphi_{i_1} \end{bmatrix} = \begin{bmatrix} X_{C_{i_2}} + x_P^{(i_2)} \cos\varphi_{i_2} - y_P^{(i_2)} \sin\varphi_{i_2} \\ Y_{C_{i_2}} + x_P^{(i_2)} \sin\varphi_{i_2} + y_P^{(i_2)} \cos\varphi_{i_2} \end{bmatrix}, \quad (4)$$

wherefrom one gets two constraint functions

$$X_{C_{i_1}} + x_P^{(i_1)} \cos\varphi_{i_1} - y_P^{(i_1)} \sin\varphi_{i_1} = X_{C_{i_2}} + x_P^{(i_2)} \cos\varphi_{i_2} - y_P^{(i_2)} \sin\varphi_{i_2}, \quad (5)$$

$$Y_{C_{i_1}} + x_P^{(i_1)} \sin\varphi_{i_1} + y_P^{(i_1)} \cos\varphi_{i_1} = Y_{C_{i_2}} + x_P^{(i_2)} \sin\varphi_{i_2} + y_P^{(i_2)} \cos\varphi_{i_2}. \quad (6)$$

We denote by $\{\mathbf{q}\}$ the column matrix

$$\{\mathbf{q}\} = \left[X_{C_1} \; Y_{C_1} \; \varphi_1 \; \dots \; X_{C_5} \; Y_{C_5} \; \varphi_5\right]^{\mathrm{T}}. \quad (7)$$

The coordinates of the point C_1 offer the first two constraint functions

$$f_1(\{\mathbf{q}\}) = X_{C_1} - OC_1 \cos\varphi_1 = 0, \quad f_2(\{\mathbf{q}\}) = Y_{C_1} - OC_1 \sin\varphi_1 = 0. \quad (8)$$

Assuming a homogeneous triangular shell, denote by θ_1 and θ_2 the angles BC_2C and AC_2C, and by m_a, m_b, m_c the lengths of the medians of the triangle ABC, one may write

$$m_a^2 = \frac{2\left[(AC)^2 + (AB)^2\right] - (BC)^2}{4}, \quad m_b^2 = \frac{2\left[(AB)^2 + (BC)^2\right] - (AC)^2}{4}, \quad m_c^2 = \frac{2\left[(BC)^2 + (AC)^2\right] - (AB)^2}{4}, \quad (9)$$

$$\theta_1 = \arccos\left(\frac{\left(\frac{2}{3}m_c\right)^2 + \left(\frac{2}{3}m_b\right)^2 - (BC)^2}{2 \times \frac{2}{3}m_c \times \frac{2}{3}m_b}\right), \quad \theta_2 = \arccos\left(\frac{\left(\frac{2}{3}m_a\right)^2 + \left(\frac{2}{3}m_c\right)^2 - (AC)^2}{2 \times \frac{2}{3}m_a \times \frac{2}{3}m_c}\right), \quad (10)$$

$$x_A^{(2)} = \frac{2}{3}m_a \cos\theta_2, \; y_A^{(2)} = \frac{2}{3}m_a \sin\theta_2, \quad x_B^{(2)} = \frac{2}{3}m_b \cos\theta_1, \; y_B^{(2)} = \frac{2}{3}m_b \sin\theta_1, \quad x_C^{(2)} = \frac{2}{3}m_c, \; y_C^{(2)} = 0. \quad (11)$$

Since point A belongs to the elements 1 and 2, point B belongs to the elements 2 and 3, point C belongs to the elements 2 and 4, and the point D belongs to the elements 3 and 5, one may write the relations (5), resulting the following constraint functions

$$f_3(\{\mathbf{q}\}) = X_{C_1} + x_A^{(1)} \cos\varphi_1 - y_A^{(1)} \sin\varphi_1 - X_{C_2} - x_A^{(2)} \cos\varphi_2 + y_A^{(2)} \sin\varphi_2 = 0, \quad (12)$$

$$f_4(\{\mathbf{q}\}) = Y_{C_1} + x_A^{(1)} \sin\varphi_1 + y_A^{(1)} \cos\varphi_1 - Y_{C_1} - x_A^{(2)} \sin\varphi_2 - y_A^{(2)} \cos\varphi_2 = 0, \quad (13)$$

$$f_5(\{\mathbf{q}\}) = X_{C_2} + x_B^{(2)} \cos\varphi_2 - y_B^{(2)} \sin\varphi_2 - X_{C_3} - x_B^{(3)} \cos\varphi_3 + y_B^{(3)} \sin\varphi_3 = 0, \quad (14)$$

$$f_6(\{\mathbf{q}\}) = Y_{C_2} + x_B^{(2)} \sin\varphi_2 + y_B^{(2)} \cos\varphi_2 - Y_{C_3} - x_B^{(3)} \sin\varphi_3 - y_B^{(3)} \cos\varphi_3 = 0, \quad (15)$$

$$f_7(\{\mathbf{q}\}) = X_{C_2} + x_C^{(2)} \cos\varphi_2 - y_C^{(2)} \sin\varphi_2 - X_{C_4} - x_C^{(4)} \cos\varphi_4 + y_C^{(4)} \sin\varphi_4 = 0, \quad (16)$$

$$f_8(\{\mathbf{q}\}) = Y_{C_2} + x_C^{(2)} \sin\varphi_2 + y_C^{(2)} \cos\varphi_2 - Y_{C_4} - x_C^{(4)} \sin\varphi_4 - y_C^{(4)} \cos\varphi_4 = 0, \quad (17)$$

$$f_9(\{\mathbf{q}\}) = X_{C_3} + x_D^{(3)} \cos\varphi_3 - y_d^{(3)} \sin\varphi_3 - X_{C_5} - x_D^{(5)} \cos\varphi_5 + y_D^{(5)} \sin\varphi_5 = 0, \quad (18)$$

$$f_{10}(\{\mathbf{q}\}) = Y_{C_3} + x_D^{(3)} \sin\varphi_3 + y_D^{(3)} \cos\varphi_3 - Y_{C_5} - x_D^{(5)} \sin\varphi_5 - y_D^{(5)} \cos\varphi_5 = 0. \quad (19)$$

Keeping into account that the point C_5 moves only in the vertical direction, and the element 5 (the piston) does not rotate, one gets another two constraint functions

$$f_{11}(\{\mathbf{q}\}) = X_{C_5} - e = 0, \quad (20)$$

$$f_{12}(\{\mathbf{q}\}) = \varphi_5 = 0. \quad (21)$$

In this paper we assume that the point E has a fixed position, that is, another two constraint functions are obtained

$$f_{13}(\{\mathbf{q}\}) = X_{C_4} + C_4E \cos\varphi_4 - d = 0, \quad (22)$$

$$f_{14}(\{\mathbf{q}\}) = Y_{C_4} + C_4E \sin\varphi_4 - (Y_E)_0 = 0, \quad (23)$$

where $(Y_E)_0$ is the fixed coordinate of the point E in the vertical direction.

Since we obtained 14 constraint functions $f_i(\{\mathbf{q}\}) = 0$, $i = \overline{1,14}$, it results that the mechanism has one degree of freedom.

We apply the theory of the extremes with constraints resulting the Lagrange function

$$F(\{\mathbf{q}\}, \lambda_1, \dots, \lambda_{14}) = Y_{C_5} + \sum_{i=1}^{14} \lambda_i f_i(\{\mathbf{q}\}). \quad (24)$$

One has to solve the following nonlinear system of 29 equations with 29 unknowns (X_{C_1}, Y_{C_1}, φ_1, ..., X_{C_5}, Y_{C_5}, φ_5, λ_1, ..., λ_{14})

$$\frac{\partial F}{\partial q_j} = 0, \quad \frac{\partial F}{\partial \lambda_i} = 0, \quad i = \overline{1,14}, \quad j = \overline{1,15}. \quad (25)$$

If the coordinate Y_E is not a fixed one, that is, the end E of the lever may move in the vertical direction, then the mechanism may have one degree of freedom (if the law of motion $Y_E = Y_E(t)$ is known), or two degrees of freedom (if the law is not known). The discussion is made in [12].

3 Numerical simulation

We selected the following standard values for different parameters of the mechanism: $(AB)_0 = 0.043$ m, $(BC)_0 = 0.128$ m, $(CA)_0 = 0.099$ m, $(CE)_0 = 0.103$ m, $(OA)_0 = 0.030$ m, $(BD)_0 = 0.130$ m, $(d)_0 = 0.086$ m, $(e)_0 = 0$ m, $(Y_E)_0 = 0.108$ m. For each parameter a maximum deviation of ± 0.002 m is used during the numerical simulation.

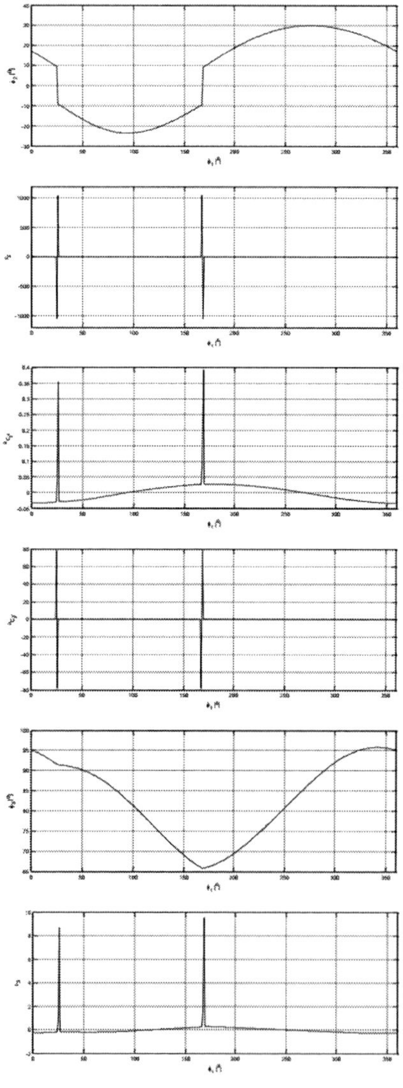

a) *b)* *c)* *d)* *e)* *f)*

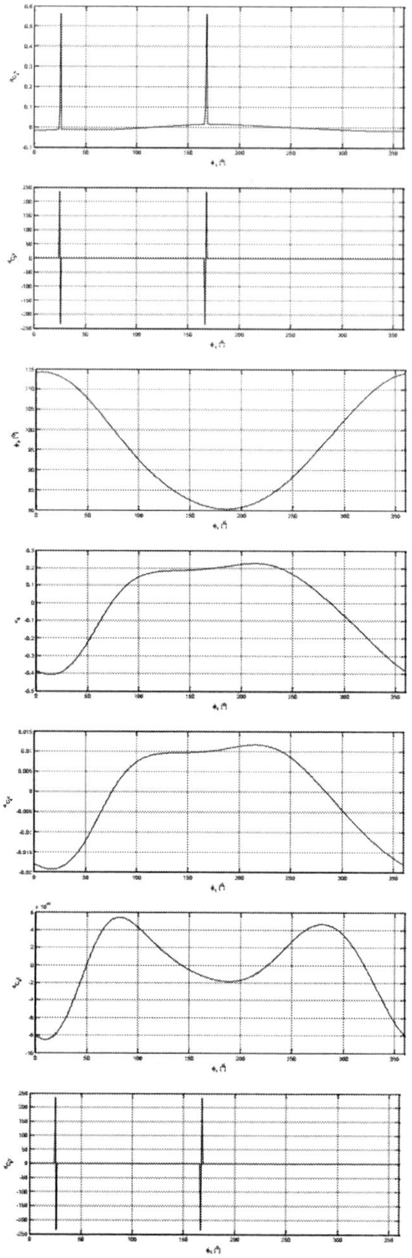

g) *h)* *i)* *j)* *k)* *l)* *m)*

Fig. 2. Different variations in the standard case: *a)* $\varphi_2 = \varphi_2(\varphi_1)$; *b)* $\varepsilon_2 = \varepsilon_2(\varphi_1)$; *c)* $a_{C_2x} = a_{C_2x}(\varphi_1)$; *d)* $a_{C_2y} = a_{C_2y}(\varphi_1)$; *e)* $\varphi_3 = \varphi_3(\varphi_1)$; *f)* $\varepsilon_3 = \varepsilon_3(\varphi_1)$; *g)* $a_{C_3x} = a_{C_3x}(\varphi_1)$; *h)* $a_{C_3y} = a_{C_3y}(\varphi_1)$; *i)* $\varphi_4 = \varphi_4(\varphi_1)$; *j)* $\varepsilon_4 = \varepsilon_4(\varphi_1)$; *k)* $a_{C_4x} = a_{C_4x}(\varphi_1)$; *l)* $a_{C_4y} = a_{C_4y}(\varphi_1)$; *m)* $a_{C_5y} = a_{C_5y}(\varphi_1)$.

We are interesting in the reduced linear and angular accelerations of different elements. Some diagrams are captured in the Figs. 2 (for the standard values), 3 (by varying Y_E), and 4 (by varying CE).

One may observe that when Y_E increases the linear acceleration decrease and angular accelerations increase. This statement holds true for all elements; consequently, when the inertial forces decrease, the inertial torques increase.

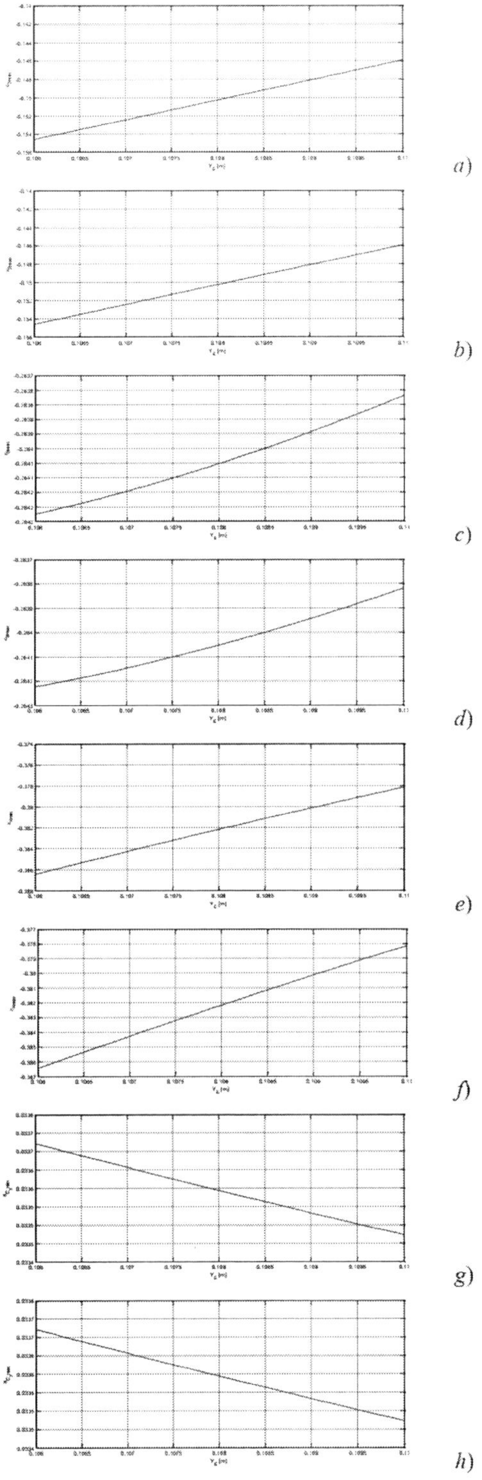

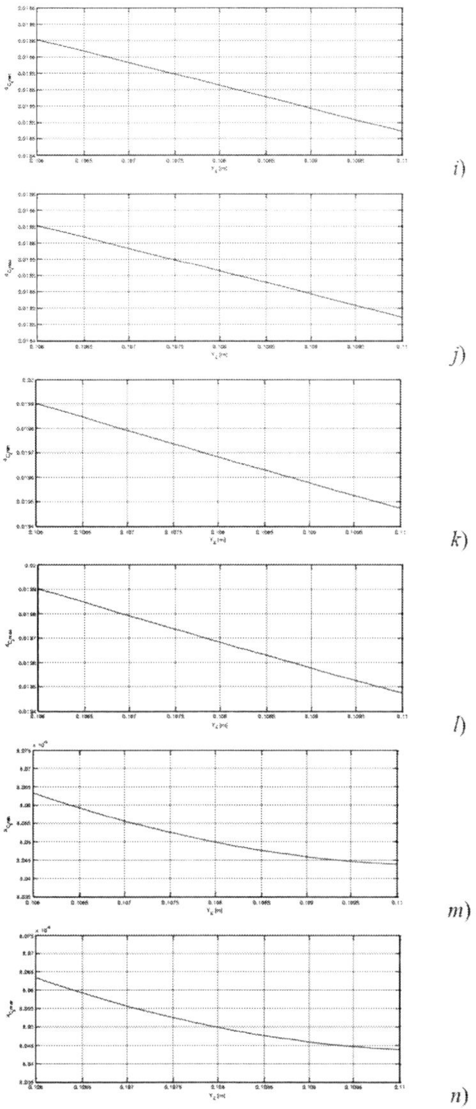

Fig. 3. Different variations of parameters when Y_E varies: *a)* $\varepsilon_{2\,\min} = \varepsilon_{2\,\min}(\varphi_1)$; *b)* $\varepsilon_{2\,\max} = \varepsilon_{2\,\max}(\varphi_1)$; *c)* $\varepsilon_{3\,\min} = \varepsilon_{3\,\min}(\varphi_1)$; *d)* $\varepsilon_{3\,\max} = \varepsilon_{3\,\max}(\varphi_1)$; *e)* $\varepsilon_{4\,\min} = \varepsilon_{4\,\min}(\varphi_1)$; *f)* $\varepsilon_{4\,\max} = \varepsilon_{4\,\max}(\varphi_1)$; *g)* $a_{C_2\,\min} = a_{C_2\,\min}(\varphi_1)$; *h)* $a_{C_2\,\max} = a_{C_2\,\max}(\varphi_1)$; *i)* $a_{C_3\,\min} = a_{C_3\,\min}(\varphi_1)$; *j)* $a_{C_3\,\max} = a_{C_3\,\max}(\varphi_1)$; *k)* $a_{C_4\,\min} = a_{C_4\,\min}(\varphi_1)$; *l)* $a_{C_4\,\max} = a_{C_4\,\max}(\varphi_1)$; *m)* $a_{C_5\,\min} = a_{C_5\,\min}(\varphi_1)$; *n)* $a_{C_5\,\max} = a_{C_5\,\max}(\varphi_1)$.

We may state that the contribution of the variation of CE for the variations of the extreme values of linear and angular accelerations does not exist, as it results from Fig. 4. Moreover, the contribution of the variation of CE for the variations of the instantaneous values of linear and angular accelerations is not significant.

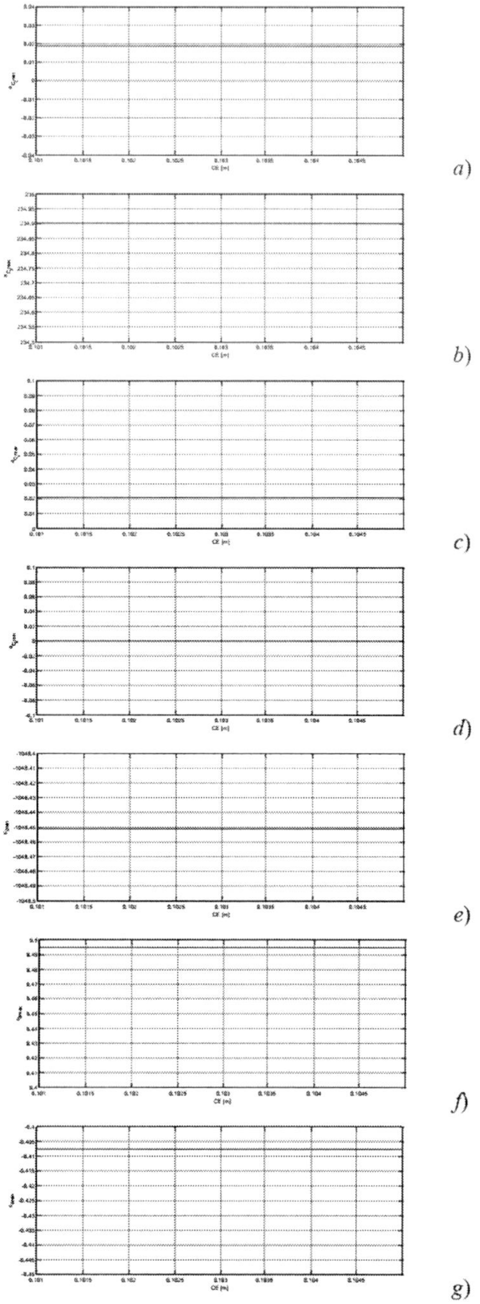

Fig. 4. Different variations of parameters when CE varies: *a*) $a_{C_2\,\min} = a_{C_2\,\min}(\varphi_1)$; *b*) $a_{C_3\,\max} = a_{C_3\,\max}(\varphi_1)$; *c*) $a_{C_4\,\max} = a_{C_4\,\max}(\varphi_1)$; *d*) $a_{C_5\,\min} = a_{C_5\,\min}(\varphi_1)$; *e*) $\varepsilon_{2\,\min} = \varepsilon_{2\,\min}(\varphi_1)$; *f*) $\varepsilon_{3\,\max} = \varepsilon_{3\,\max}(\varphi_1)$; *g*) $\varepsilon_{4\,\min} = \varepsilon_{4\,\min}(\varphi_1)$.

It results that the only element with a significant contribution is the coordinate Y_E of the control lever.

One may easily observe that the motion is always stable. The stability is a consequence of the previous diagrams which show that the variations of different parameters are small when the parameters Y_E and CE vary. It results the stability with respect these two parameters. Because the variations of the extreme values of the linear and angular accelerations are bounded, it also results that the variations of the instantaneous linear and angular accelerations are bounded and they can be made as small as we want, resulting the stability of the motion.

4 Conclusions

From the previous discussion one may conclude:

– the existence of jumps for linear and angular accelerations, which lead to greater inertial forces and torques;

– a very complicate synthesis of mechanism.

Particular values for the dimensions of particular elements [25] may lead to an useless mechanism, that is, the mechanism works like a simple crank-shaft one, no matter which is the position of the point E.

The previous discussion implies to involve more elements in the study. The classical synthesis of the mechanisms, that is, the establishing of certain dimensions of the elements, must be continued with the study of the resulted linear and angular accelerations in order to determine the inertial forces and torques, and the study of angular velocities in order to determine the wear of the joints.

Keeping into account the previous remark [25], it is possible to determine a variable compression ratio mechanism which satisfies the constructive and dynamical requirement.

References

1. F. Freudenstein, E.R. Maki, Journal of Mechanisms, Transmissions and Automation in Design **105**, 259 (1983)
2. T. Hoeltgebaum, R. Simoni, D. Martins, Mechanism and Machine Theory **96**, 308 (2016)
3. V. Hara, N. Pandrea, D. Popa, M. Stan, S. Boncea, 1995, *Adaptive thermal engines* (Publishing House of the University of Pitești, 1995)
4. A. Clenci, *Automobile equipped with engine with variable compression ratio* (Research report 2003)
5. I. Miller, www.me.udel.edu/eeg425 / SaabVarComp.doc, 2001.
6. P.J. Mahlesh, V.K. Aparna, International Journal of Mechanical Engineering Research and Development (IJMERD) **2**, 9 (2012)
7. A. Teodorczyk, published on Institute of Heat Engineering Warsaw (1995)
8. http://freerepublic.com/ focus/f-news/1818771/posts by Red Badger (2007)
9. SAE article http://articles.sae.org/6043/ published

by John Kendall (2009)

10. SAE article http://articles.sae.org/15040/ published by John Kendall (2016)
11. S. Nagarajaa, K. Sooryaprakashb, R. Sudhakaran, *Proceedia Earth and Planetary Science* (2015)
12. B. Mănescu, I. Dragomir, N.-D. Stănescu, *AVMS 2017*, Timişoara (2017)
13. B. Mănescu, I. Dragomir, N.-D. Stănescu, N. Pandrea, *ICMSAV XLI*, Cluj-Napoca (2017)
14. B. Mănescu, I. Dragomir, N.-D. Stănescu, N. Pandrea, A. Clenci, D. Popa, Aspects in the Synthesis of a Variable Compression Ratio Mechanism, *Car 2017*, Piteşti (2017)
15. K. Sivaramakrishnan, 2017, Egyptian Journal of Petroleum in press (2017)
16. K. Satyanarayana, A.K. Viswanath, T.V. Hanumanta Rao, S.V. Uma Maheswara Rao, Procedia Engineering **173**, 1672 (2017)
17. B. Wu, Q. Zhan, X. Yu, G. Lv, X. Nie, S. Liu, Applied Thermal Engineering **118**, 621 (2017)
18. J.A.A. Yamin, M.H. Dado, Applied Energy **77**, 447 (2004)
19. L. Eriksson, I. Nielsen 2014, *Modeling and control of engines and drivelines* (John Wiley & Sons, London 2014)
20. L.W. Tsai, *Mechanism design: Enumeration of kinematic structures according to function* (CRC Press, London 2000)
21. H. S. Yan, *Creative design of mechanical devices* (Springer, Berlin 1998)
22. Y. Nilsson, *Modelling for fuel optimal control of a variable compression engine* (Linköping University 2007)
23. P. Podevin, A. Clenci, 2008, *Moteurs à taux de compression variable* (BM 2525, Editions Techniques de l'Ingénieur, Paris 2008)
24. A. Shaik, N. Shenbaga Vinayaga Moorthi, R Rudramoorthy, Proc. IMechE Part D: J. Automobile Engineering **221**, 1159 (2007)
25. Pandrea N, *Private communications* (2017)

Optimal homotopy asymptotic method to large post-buckling deformation of MEMS

Nicolae Herisanu[1,2,*], and *Vasile* Marinca[1,2]

[1]University Politehnica Timisoara, 300222 Bd. Mihai Viteazu 1, Timisoara, Romania
[2]Center for Advanced Technical Research, Romanian Academy Branch of Timisoara, Bd. Mihai Viteazu 24, Timisoara, Romania

Abstract. In the present paper, the post-buckling response of an axially stressed clamped-clamped actuator, modeled as a beam and subjected to a symmetric electrostatic field is analyzed. An analytical approximate method, namely the Optimal Homotopy Asymptotic Method (OHAM) is applied to the governing nonlinear integro-differential equation. The analytical results obtained through the proposed procedure show excellent agreement with numerical solution, proving the validity of the proposed procedure, which is simple and easy to use.

1 Introduction

Micro electromechanical systems (MEMS) have high popularity in industry and engineering, being an extension of microelectronics and integrated circuits technology. MEMS are intensively used in various fields such as wireless communication, automotive, aviation, navigation, biomedicine, consumer electronics, and so on [1]. There exists a large variety of MEMS, which may be classified as pneumatic, thermal, electrical or piezoelectric.

Several experimental or numerical studies have been conducted on dynamic behavior of MEMS devices. The mathematical study of the MEMS started by the pioneering work of Nathanson et al [2]. They considered a model of a parallel-plates capacitor of which one plate is fixed and their other is a movable mass attached to a spring. MEMS resonators have been used as transducers in mechanical microsensors [3] or as a flexible alternative to conventional large-size resonators [4]. An analytical approach and a reduced-order model to investigate the behavior of electrically actuated microbeam-based MEMS are presented in [5] by Younis et al. Nayfeh et al. [6] studied the pull-in instability in MEMS resonators and established that characteristics of the pull-in phenomenon in the presence of alternate current (AC) loads differ from those under purely direct current (DC) loads. A new test structure and method for measuring residual stresses was presented by Abu-Salih and Elata [7]. They showed that a single test structure of the proposed design may be used to measure compressive and tensile residual stress in a continuous wide range. Hu applied three analytical models, namely the full-order, the fourth-order and the third-order models and the corresponding closed form solutions for the pull-in voltages of micro beams subjected to electrostatic loads [8]. The electromechanical post-buckling response of an axial stressed clamped-clamped beam actuator subjected to a symmetric electrostatic field is analyzed by Yu et al. [9]. Rezazadeh et al [10] presented the static and dynamic responses of a fixed-fixed and cantilever microbeam, using the lumped and the distributed models to a DC and a step DC voltage. Squeeze-film characteristics of electrostatically actuated microbeams under large DC load coupled with small AC component are presented for the first three flexural modes of vibration of the resonator operating in different ambient pressure conditions, by Chaterjee and Pokit [11]. Zhu and Espinoza presented in [12] a methodology for a coupled-field finite element analysis (FEM) to examine the performance of radio frequency MEMS switches as a function of temperature. Also, the structure of out-of-plane profile was investigated in detail. Beni and Heidari investigated the pull-in instability of micro-beams with a curved grown electrode under action of electric force within the framework of von-Karman nonlinearity and the Euler-Bernoulli beam theory [15]. The effects of geometric parameters such as beam lengths, width, thickness, gaps, and size effect are discussed in detail through a numerical study. Mojahedi et al. [14] studied static pull-in instability of electrostatically-actuated microbridges and microcantilevers considering different nonlinear effects. The nonlinear differential equations are converted by means of Galerkin's decomposition method into nonlinear integro-differential equations. In [15], Singh et al. studied the linear and nonlinear frequency characteristics of non-uniform beam with tip mass using the method of multiple scales to obtain the approximate solution. The amplitude-voltage response of MEMS resonator cantilevers under soft electrostatic actuation is studied by Caruntu et al [16]. The electrostatic actuation is of the voltage of frequency near half natural frequency of the system. It is showed that the instabilities for high

* Corresponding author: nicolae.herisanu@upt.ro

amplitudes of the resonator are not captured by the method of multiple scales. Daneshpajooh and Moghimi Zand used the homotopy analysis method to investigate the nonlinear oscillatory behavior of clamped-clamped initially curved micro/nano beam, under electrostatic force actuation. The influence of initial elevation, input voltage, midplane stretching are also examined. The effect of squeeze air-film and AC actuation voltage on dynamic stability of microswitch actuated electrostatically has been investigated in squeeze film domain by Harish et al. [19]. Using trajectories in phase plane and time history, characteristics of the pull-in phenomena have been studied in the presence of DC voltage combined with AC component.

Taking into account these considerations, the main objective of the present paper is to propose an accurate procedure to investigate the electromechanical post-buckling response of an axially stressed clamped-clamped actuation, subject to a symmetric electrostatic field. The analytical approximate solutions are obtained through a version of OHAM applied to a nonlinear integral-differential equation established in [7] and [9]. Our procedure, which is independent of the presence of small or large parameters, is based on the construction and determination of the linear operator and of the auxiliary functions, combined with a convenient way to optimally control the convergence of the solution. The efficiency of the present procedure is proved, while an accurate solution is explicitly analytically obtained after only one iteration.

2 Equation of motion

In this section, we consider the post-buckling response of an axially stressed clamped-clamped actuator, modeled as a beam and subjected to a symmetric electrostatic field (fig.1).

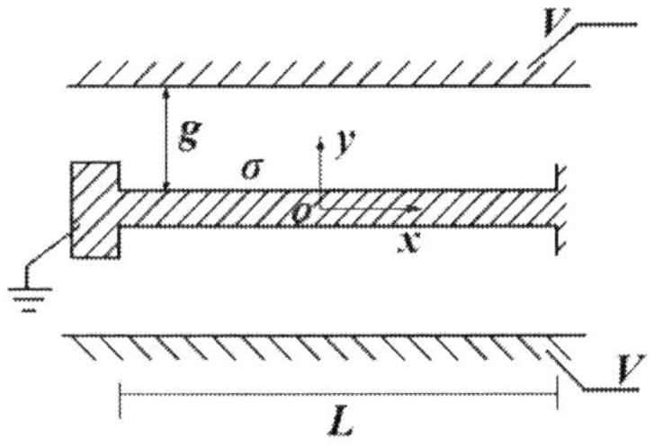

Fig.1. Clamped-clamped beam subject to an axially residual stress and a symmetric electrostatic field.

The effect of residual stress gradients is neglected but the residual stress is considered.

The electromechanical system is governed by the following nonlinear integral-differential equation [7], [9]:

$$E^* I\frac{d^4y}{dx^4} - \sigma A\frac{d^2y}{dx^2} - EA\left[\frac{1}{L}\int_{-0.5L}^{0.5L}\left(\frac{dy}{dx}\right)^2 dx\right]\frac{d^2y}{dx^2} = \\ = \frac{1}{2}\varepsilon_0\lambda_0 V^2\left[\frac{1}{(g-y)^2} - \frac{1}{(g+y)^2}\right] \quad (1)$$

with the boundary conditions

$$y(-0.5L) = y(0.5L) = \frac{dy}{dx}(-0.5L) = \frac{dy}{dx}(0.5L) = 0 \quad (2)$$

Here $E^* = \dfrac{E}{(1-\nu^2)}$ is the effective modulus in bending, E is the Young's modulus and ν is the Poisson ratio, I is the second moment of the beam cross-section, b_0 is the beam width and h is the beam thickness, L is the length, σ is the axial residual stress, ε_0 is the permittivity of the free space, $A=b_0h$ is the cross-sectional area, g is the nominal gap and V is the potential of the two fixed electrodes. It is assumed that b>>h and $x \in [-L/2, L/2]$ is the horizontal coordinate, while y is the transverse coordinate and denotes the post-buckling deflection of the beam. The three terms on the left hand side of Eq.(1) are the distributed mechanical forces associated with bending, axial stress and membrane stretching of the beam. The right side of Eq.(1) is the distributed electrostatic force.

Only symmetric modes of the deformed beam are considered and any possibility of symmetry and microbeam is disregarded.

The normalized form of the equilibrium equation can be expressed as

$$Y^{(IV)}(s) - \tilde{\sigma}Y''(s) - 12\tilde{E}r^2\left[\int_0^{0.5}\left(Y'(s)\right)^2 ds\right]Y''(s) = \\ = U^2 Y(s)\left[1 - Y^2(s)\right]^2 \quad (3)$$

where prime denotes the derivative with respect to s and

$$s = \frac{x}{L}, \quad Y = \frac{y}{g}, \quad r = \frac{g}{h}, \quad \frac{\tilde{\sigma}}{(2\pi)^2} = \frac{\sigma}{|\sigma_E|}, \\ \tilde{E} = \frac{E}{E^*}, \quad U^2 = \frac{24\varepsilon_0\lambda_0 L^4 V^2}{E^* g^3 h^3} \quad (4)$$

The Euler buckling stress is [7], [9]

$$\sigma_E = -4\pi^2 E^* I/(AL^2) \quad (5)$$

The boundary conditions of the beam are

$$Y(-0.5) = Y(0.5) = Y'(-0.5) = Y'(0.5) = 0 \quad (6)$$

Now, by introducing a new independent variable

$$\tau = \frac{2\pi x}{L} + \pi \quad (7)$$

we transform Eqs. (3) and (6) into the following dimensionless form:

$$\left(1 - Y^2\right)^2\left[Y^{(IV)} - \lambda Y'' - \beta Y''\int_0^{2\pi} Y'^2(\tau)d\tau\right] - \mu Y = 0 \quad (8)$$

$$Y(0) = Y(2\pi) = Y'(0) = Y'(2\pi) = 0 \quad (9)$$

where

$$\lambda = \frac{\widetilde{\sigma}}{(2\pi)^2}, \quad \beta = \frac{3\widetilde{E}r^2}{\pi}, \quad \mu = \frac{u^2}{2\pi^4} \tag{10}$$

3 Basic ideas of the optimal homotopy asymptotic method (OHAM)

The OHAM was proposed by Vasile Marinca and Nicolae Herisanu [19-24].

To apply OHAM, we consider the nonlinear differential equation in the general form

$$L[Y(\tau)] + N[Y(\tau)] = 0, \quad \tau \in D \tag{11}$$

subject to the boundary/initial conditions

$$B\left(Y(\tau), \frac{dY}{d\tau}\right) = 0 \tag{12}$$

where L is a linear operator, τ denotes independent variable, $Y(\tau)$ is an unknown function, $N[Y(\tau)]$ is a nonlinear operator, D is the domain of interest and B is a boundary operator.

By means of OHAM, one constructs a family of equations

$$\begin{aligned}&(1-p)L(\widetilde{Y}(\tau, C_i)] = \\ &= pH(\tau, C_i)[L(\widetilde{Y}(\tau, C_i) + N(\widetilde{Y}(\tau, p, C_i)\end{aligned} \tag{13}$$

where $p \in [0,1]$ is an embedding parameter, $H(\tau, C_i)$ is a nonzero auxiliary function for $p \neq 0$, $\widetilde{Y}(\tau, C_i)$ is an unknown function, C_i, i=1,2,…,s are s unknown parameters which will be optimally determined later. For p=0 and p=1 it holds

$$\widetilde{Y}(\tau, 0, C_i) = Y_0(\tau) \tag{14}$$

$$\widetilde{Y}(\tau, 1, C_i) = Y(\tau) \tag{15}$$

Therefore, as p increases from 0 to 1, the solution $\widetilde{Y}(\tau, p, C_i)$ varies from $Y_0(\tau)$ to $Y(\tau)$, where the initial approximation $Y_0(\tau)$ is obtained from Eq.(1) for p=0:

$$L[Y_0(\tau)] = 0 \tag{16}$$

and

$$B\left(Y_0, \frac{dY_0}{d\tau}\right) = 0 \tag{17}$$

Let us consider the solution of Eq.(13) in the form

$$\widetilde{Y}(\tau, p, C_i) = Y_0(\tau + pY_1(\tau, C_i)), \quad i = 1,2,\ldots,s \tag{18}$$

Substituting Eq. (18) into Eq. (15) and equating the coefficients of p^o and p^1, we obtain the governing equation of the initial approximation $Y_0(\tau)$ given by Eq.(16) with the boundary/initial conditions (17) and then the governing equation of the first (and last) approximation Y_1, i.e.

$$L[Y_1(\tau, C_i)] = H(\tau, C_i)N_0[Y_0(\tau)] \tag{19}$$

with

$$B\left(Y_1, \frac{dY_1}{d\tau}\right) = 0 \tag{20}$$

where $N_m(Y_0(\tau))$ is the coefficient of p^m obtained expanding $N\left[\widetilde{Y}(\tau, p, C_i)\right]$ in series with respect to the embedding parameter p:

$$\begin{aligned}&N\left[\widetilde{Y}(\tau, p, C_i)\right] = N_0[Y_0(\tau)] + pN_1(Y_0(\tau), Y_1(\tau, C_i) + \ldots \\ &+ p^m N_m[Y_0(\tau), Y_1(\tau, C_i), \ldots, Y_m(\tau, C_i)] + \ldots)\end{aligned} \tag{21}$$

In our case, we consider only a single iteration and therefore we need to choose m=0.

It should be emphasized that Y_0 and Y_1 are governed by the linear Eqs. (16) and (19) with the boundary/initial conditions given by Eqs.(17) and (20), respectively. The convergence of the solution (18) depends upon the auxiliary function $H(\tau, C_i)$. There are some possibilities to choose the function H and it is important to mention that $H(\tau, C_i)$ must follow the terms appearing in Eq.(19). Therefore, we try to choose $H(\tau, C_i)$ so that in Eq. (19) the product $H(\tau, C_i)N_0[Y_0(\tau)]$ be of the same shape with the terms which appear into $N_0[Y_0(\tau)]$.

If the sum (18) is convergent at p=1, one has

$$\widetilde{Y}(\tau, C_i) = Y_0(\tau) + Y_1(\tau, C_i) \tag{22}$$

The Eq.(22) is the approximate solution of the first order.

Inserting Eq. (22) into Eq.(11), we obtain the residual

$$\begin{aligned}&R(\tau, C_i) = L[\widetilde{Y}(\tau, C_i)] + N[\widetilde{Y}(\tau, C_i)], \\ &i = 1,2,\ldots,s, \quad \tau \in D\end{aligned} \tag{23}$$

If $R(\tau, C_i)=0$, then $\widetilde{Y}(\tau, C_i)$ will be the exact solution, but this rarely happens to nonlinear problems. The convergence-control parameters C_i, i=1,2,…s, may be optimally identified via various methods, like minimizing the square residual error, using:

$$J(C_1, C_2, \ldots, C_s) = \int_D R^2(\tau, C_1, C_2, \ldots C_s)d\tau \tag{24}$$

The values of the unknown parameters will be determined from the conditions:

$$\frac{\partial J}{\partial C_1} = \frac{\partial J}{\partial C_2} = \ldots = \frac{\partial J}{\partial C_s} = 0 \tag{25}$$

Alternatively, the system of algebraic equations necessary to identify the unknown convergence-control parameters may be obtained imposing the conditions

$$R(\tau_1, C_i) = R(\tau_2, C_i) = \ldots = R(\tau_s, C_i) = 0 \tag{25'}$$

or by using the Galerkin method, the collocation method, the Ritz method, the Kantorovich method and so on. For more details see [20], [21].

With these optimal values of the convergence-control parameters known, the first-order approximate solution (22) is well determined.

4 Application of OHAM to post-buckling deformation of MEMS

We use the basic ideas of the OHAM by considering Eqs. (8) and (9). We may choose the linear operator in the form

$$L[Y(\tau)] = Y^{IV}(\tau) + Y''(\tau) \tag{26}$$

We mention that, generally, the choosing of the linear operator is not unique. Taking into account (26), the Eqs. (16) and (17) become, respectively:

$$Y_0^{IV}(\tau) + Y_0''(\tau) = 0 \tag{27}$$

$$Y_0(0) = Y_0(2\pi) = Y_0'(0) = Y_0'(2\pi) = 0 \tag{28}$$

which have the following solution:

$$Y_0(\tau, C_1) = C_1(1 - \cos\tau) \tag{29}$$

with C_1 unknown for the moment.

The nonlinear operator $N[Y(\tau)]$ is obtained from Eqs. (8), (11) and (26):

$$N[Y(\tau)] = -Y'' + Y^{IV}(Y^4 - 2Y^2) - -(1-Y^2)^2[\lambda + \beta\left(\int_0^{2\pi} Y'^2 d\tau\right)Y'' - \mu Y \tag{30}$$

such that substituting Eq. (29) into Eq. (30), we obtain

$$N[Y(\tau, C_i)] = -C_1(1 + \lambda + \pi\beta C_1^2)[M_0 + M_1\cos\tau + + M_2\cos 2\tau + M_3\cos 3\tau + M_4\cos 4\tau + M_5\cos 5\tau] \tag{31}$$

where

$$M_0 = \frac{\mu}{1+\lambda+\pi\beta C_1^2} + 2C_1^2 - \frac{7}{2}C_1^4,$$
$$M_1 = -\frac{\mu}{1+\lambda+\pi\beta C_1^2} + 1 - \frac{7}{2}C_1^2 + \frac{49}{8}C_1^4$$
$$M_2 = 2C_1^2 - 4C_1^4, \quad M_3 = \frac{29}{16}C_1^4 - \frac{C_1^2}{2}, \tag{32}$$
$$M_4 = -\frac{C_1^4}{2}, \quad M_5 = \frac{C_1^4}{16}$$

Having in view that within Eq. (31) we can found a trigonometric function and that the auxiliary function $H(\tau, C_i)$ must follow the terms appearing in Eq. (31), then we may choose the function H in the following forms:

$$H(\tau, C_i) = -\frac{(C_2 + 2C_3\cos\tau + 2C_4\cos 2\tau)}{C_1(1+\lambda+\pi\beta C_1^2)} \tag{33}$$

or

$$H(\tau, C_i) = -\frac{1}{C_1(1+\lambda+\pi\beta C_1^2)}(C_2' + 2C_3'\cos\tau) \tag{34}$$

or

$$H = -\frac{C_2^* + 2C_3^*\cos\tau + 2C_4^*\cos 2\tau + 2C_5^*\cos 3\tau}{C_1(1+\lambda+\pi\beta C_1^2)} \tag{35}$$

or yet

$$H(\tau, C_i) = -\frac{(C_2'' + 2C_3''\cos\tau + 2C_4''\cos 4\tau)}{C_1(1+\lambda+\pi\beta C_1^2)} \tag{36}$$

and so on, where C_i, C_i', C_i'' and C_i^* are unknown parameters at this moment.

If we choose only the expression (33) for $H(\tau, C_i)$, then using (31), (33) and (19) we can obtain the equation in the first approximation Y_1:

$$\begin{aligned} &Y_1^{IV} + Y_1'' = M_0C_2 + M_1C_3 + M_2C_4 + [M_1(C_2 + C_4) \\ &+ 2M_0C_3 + M_2C_3 + M_3C_4]\cos\tau + [M_2C_2 + (M_1 + \\ &+ M_3)C_3 + (2M_0 + M_4)C_4]\cos 2\tau + [M_3C_2 + (M_2 + \\ &+ M_4)C_3 + (M_1 + M_5)C_4]\cos 3\tau + [M_4C_2 + (M_3 + \\ &+ M_5)C_3 + M_2C_4]\cos 4\tau + (M_5C_2 + M_4C_3 + \\ &+ M_3C_4)\cos 5\tau + (M_5C_3 + M_4C_4)\cos 6\tau + \\ &+ M_5C_4\cos 7\tau, \\ &Y_1(0) = Y_1(2\pi) = Y_1'(0) = Y_1'(2\pi) = 0 \end{aligned} \tag{37}$$

No secular terms in Y_1 requires

$$M_1(C_2 + C_4) + 2M_0C_3 + M_2C_3 + M_3C_4 = 0 \tag{38}$$

The solution of Eq. (37) can be written as

$$\begin{aligned} &Y(\tau, C_1, C_2, C_3, C_4) = \\ &\frac{M_2C_2 + (M_1 + M_3)C_3 + (2M_0 + M_4)C_4}{12}(\cos 2\tau - \cos\tau) + \\ &+ \frac{M_3C_2 + (M_2 + M_4)C_3 + (M_1 + M_5)C_4}{72}(\cos 3\tau - \cos\tau) + \\ &+ \frac{M_4C_2 + (M_3 + M_5)C_3 + M_2C_4}{240}(\cos 4\tau - \cos\tau) + \\ &+ \frac{M_5C_2 + M_3C_4}{600}(\cos 5\tau - \cos\tau) + \frac{M_5C_3 + M_4C_4}{1260}(\cos 6\tau - \\ &- \cos\tau) + \frac{M_5C_4}{2352}(\cos 7\tau - \cos\tau) \end{aligned} \tag{39}$$

The first-order approximate solution of Eqs. (8) and (9) is obtained from Eqs. (29), (39) and (22):

$$\tilde{Y}(\tau, C_1, C_2, C_3, C_4) = Y_0(\tau, C_1) + Y_1(\tau, C_1, C_2, C_3, C_4) \tag{40}$$

4 Numerical example

In order to show the validity of the proposed procedure, we consider the following values of the parameters which are involved in Eq.(8):

https://doi.org/10.1051/matecconf/201814813003

$$\lambda = -0.96, \quad \beta = \frac{4.32}{\pi}, \quad \mu = 0.025 \tag{41}$$

Applying the described procedure for minimizing the residual, the optimal values of the convergence-control parameters are obtained as:

$$\begin{aligned} C_1 &= 0.08387148\ 7738863\ , C_2 = 0.00100704\ 014882 \\ C_3 &= -0.0006214\ 8786228,\ C_4 = 0.00100759\ 265364 \end{aligned} \tag{42}$$

so that the approximate solution of Eqs.(8) and (9) can be obtained in the form:

$$\begin{aligned} Y(\tau) &= 0.083871487738 - 0.083911246033\cos\tau + \\ &+ 0.000031180822\cos 2\tau + 8.516244\cdot 10^{-6}\cos 3\tau + \\ &+ 6.6998\cdot 10^{-8}\cos 4\tau - 5.7507\cdot 10^{-9}\cos 5\tau - \\ &- 2.131\cdot 10^{-11}\cos 6\tau + 1.32\cdot 10^{-12}\cos 7\tau \end{aligned} \tag{43}$$

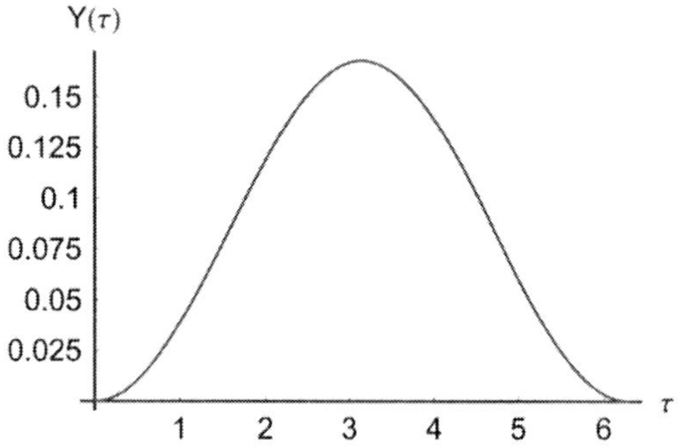

Fig.2. Approximate solution (43) of Eqs. (8) and (9)

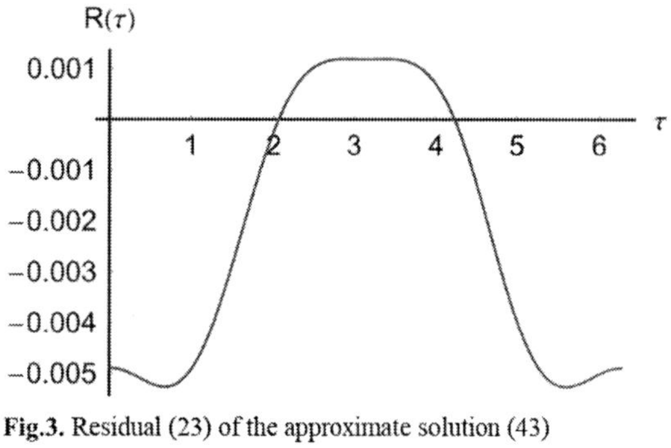

Fig.3. Residual (23) of the approximate solution (43)

In fig.2 we present the approximate solution (43) of Eqs. (8) and (9) obtained through OHAM while in fig.3 it is presented the residual (13) obtained for the approximate solution (43). It can be observed the accuracy of the solution, since the obtained residual is less than 0.005.

5 Conclusions

In this study we proposed an analytical approximate solution to the electromechanical post-buckling response of an axially stressed clamped-clamped actuator, modeled as a beam, subjected to a symmetric electrostatic field.

Our procedure is valid even if the nonlinear differential equation does not contain any small or large parameters. In the construction of the proposed homotopy appear some distinctive concepts such as the auxiliary function $H(\tau, C_i)$ and several convergence-control parameters C_i whose optimal values ensure a fast convergence of the solution.

The example presented in this work lead to the conclusion that the obtained results are quite accurate after the first iteration. This technique is effective, explicit and accurate for a nonlinear integral-differential equation. Our construction of homotopy is different from other approaches, especially in what concerns the linear operator L and the auxiliary convergence-control function H.

As it was demonstrated, the proposed method is straightforward, concise, and can be applied to other nonlinear problems.

References

1. S.B. Yamgoue, A.J. Tchiegang, J. Nonlinear Dynamics, ID 735712 (2014)
2. H.C. Nathanson, W.E, Newell, R.A. Wickstrom, T.R. Davis Jr., IEEE Transactions on Electron Devices, **14**, 117 (1996)
3. M. Elwespoek, R. Wiegerink, *Mechanical microsensors* (Springer, Berlin, 2001)
4. V.M Varadan, K.J. Visroy, K.A. Jose, *RF MEMES and their applications* (Wiley, NY, 2003)
5. M.I. Younis, E.M. Abdel Rehman, A. Nayfeh, J. Microelectromechanical Syst., **12**, 672 (2003)
6. A.H. Nayfeh, M.I. Younis, E.M. Abdel Rehman, Nonlinear Dynamics, **48**, 153 (2007)
7. S. Abu-Salih, D. Elata, Using electromechanical buckling for measuring residual stress, EuroSime 2006 - 7th International Conference on Thermal, Mechanical and Multiphysics Simulation and Experiments in Micro-Electronics and Micro-Systems, **DOI:** 10.1109/ESIME.2006.1643991
8. Y-C Hu, J. Micromech. Microeng., 16, 648 (2006)
9. Y Yu, B.S. Wu, C.W. Lim, Int. J Mech. Sci., **55**, 95 (2012)
10. G. Rezazadeh, M. Fathalilou, M. Sadeghi, Imaging, **12**, 117 (2011)
11. S. Chaterjee, G. Pohit, Int. J. Eng. Sci. Techn., **2**, 187 (2010)
12. Y. Zhu, H.D. Espinoza, J. Micromech. Microeng., **14**, 1270 (2004)
13. Y.T. Beni, M. Heidari, Int. J. Eng. Appl. Sci., **4**, 41 (2012)
14. M. Mojahedi, M. Moghimi Zand, M.T. Ahmadian, Appl. Math. Modelling, **34**, 1032 (2010)
15. S.S. Singh, P. Pal, A.K. Pandey, J. of Vibration and Acoustics, **138**, 064502 (2016)
16. D.I Caruntu, I. Martinez, K.N. Taylor, Mech, Research Commun., **52**, 25 (2013)

17. H. Daneshpajooh, M. Moghimi Zand, J. of Mech. Sci. and Technology, **29**, 3855 (2015)
18. C.S. Harsha, C.S.R. Prasanth, B. Pratiher, Procedia Engineering, **14**, 891 (2016)
19. N. Herisanu, V. Marinca, T. Dordea, Gh. Madescu, Proc. Rom. Acad, Ser.A, **9**, 229 (2008)
20. V. Marinca, N. Herisanu, J. Sound Vibr. **329**, 1450 (2010)
21. V. Marinca, N. Herisanu, *Nonlinear dynamical systems in engineering. Some approximate approaches* (Springer, Heidelberg, 2011)
22. V. Marinca, N. Herisanu, *The optimal homotopy asymptotic method. Engineering applications* (Springer, Cham, 2015)
23. N. Herisanu, V. Marinca, Gh. Madescu, Wind Energy, **18**, 1657 (2015)
24. V. Marinca, N. Herisanu, Int. J. Heat and Mass Transf **79,** 146 (2014)

The nonlinear thermomechanical vibration of a functionally graded beam on Winkler-Pasternak foundation

Vasile Marinca[1,2], and *Nicolae* Herisanu[1,2,*]

[1]University Politehnica Timisoara, 300222 Bd. Mihai Viteazu 1, Timisoara, Romania
[2]Center for Advanced Technical Research, Romanian Academy Branch of Timisoara, Bd. Mihai Viteazu 24, Timisoara, Romania

Abstract. By using the Optimal Auxiliary Functions Method (OAFM), nonlinear free thermomechanical vibration of functionally graded beam (FGB) on Winkler-Pasternak elastic foundation is studied. Based on von Karman geometric nonlinearity, on Euler-Bernoulli beam theory and also on Galerkin procedure we obtain a second-order nonlinear differential equation with quadratic and cubic nonlinear terms. The results obtained by means of OAFM are compared and shown to be in an excellent agreement with available solutions known in the literature.

1 Introduction

The concept of functionally graded materials (FGM) appeared in 1984 in Sendai area of Japan. These are heterogeneous, anisotropic materials and are made from a mixture of ceramics and metals. The mechanical properties of them varies smoothly and continuously from the surface to other surface of the material. Initially FGM were designed as thermal barrier materials for aerospace application and fusion reactors. These materials had found application in various fields of engineering like automotive, semiconductor industry, manufacturing industry, biomedical science, aerospace, defense industry and general structural element in thermal environments.

Many researchers have investigated different aspects of FGM. The effect of damage on free and forced vibration of a FG cantilever beam is studied by Birman and Byrd [1]. The modes of damage include a region with degraded stiffness adjacent to the root of the beam, a single delamination crack and a single crack at the root cross section of the beam propagating in the thickness direction. Dokmeci presented a system of 1-D equations so that to analyze the thermoviscoelastic behavior of an axially FGB of rectangular cross section at high-frequency vibration [2]. The system of 1-D equation governs the extensional, flexural, torsional and thickness shear and also the coupled vibrations of the beam at high frequency. Khorrambadi [3] analyzed the free vibration of FGB with piezoelectric actuators subjected to axial compressive loads. The elasticity modulus of beam is assumed to vary as a power form of the thickness coordinate variable. The effect of the applied voltages, axial compressive loads and FG index of the vibration frequency are discussed. The static Green's functions for FG Euler-Bernoulli and Timoshenko beams are presented by Carl et al [4]. All material properties are arbitrary functions along the beam thickness direction. For symmetrical material properties along the beam thickness directions and symmetric cross-sections, the resulting stress distribution is also symmetric. Alshabatat and Naghshineh presented in [5] a design method to optimize the material distribution of FGM with respect to some vibration and acoustic properties. Two novel volume fraction laws are used to describe the material volume distributions through the length of the FGB. Ke et al [6] discussed the effects of material property distribution and end supports on the nonlinear dynamic behavior of FGB. The direct numerical integration method and Runge-Kutta method are employed to find the nonlinear vibration of FGB with different end supports. Yaghobi and Torabi [7] investigated an analytic solution using variational iteration method for nonlinear vibration and post-buckling of beams made of FGM resting on a nonlinear elastic foundation, subject to an axial force. Mohammadi [8] considered the non-linear terms in von-Karaman's strain-displacement relation with the help of Hamilton's principle. The equation of potential and kinetic energy of the beam are derived and as a result, the nonlinear motion equation could be reached. Fu et al [9] used the finite difference method or dynamic equations of FGB with piezoelectric patches. The Eshelby-Mori-Tanaka approach based on an equivalent fiber is used by Thomas et al [10] to investigate free vibration of functionally graded nanocomposite beam reinforced by randomly oriented straight single-walled carbon nanotubes. The first five normalized mode shapes for this type of beam with different boundary conditions and different carbon nanotubes orientation are presented. The large-amplitude free vibration of clamped immovable thin beams made of FGM is investigated by Elmaguiri et al [11] using the

* Corresponding author: nicolae.herisanu@upt.ro

energy method and a multimode approach. By means of harmonic balance method, the equations of motion are converted into a nonlinear algebraic form and are solved by an iterative numerical method.

The linear and nonlinear vibration behavior of monomorph and bimorph beam made from a mixture of PZT4 and PZT-5H with material composition are investigated by Yang et al [12]. Maganti et Nalluri [13] considered the deformation variables to determine flapwise bending of rotating functionally graded double tapered beam attached to a rigid hub. The equations of motions are derived using hybrid deformation variables employing Lagrange's approach and Rayliegh-Ritz method is used to evaluate the frequencies of the beam. The effect of temperature field on the natural frequencies of FGB with different conditions is studied by Kashyzadeh and Asforjani [14]. Modal analysis has been performed for a FGB with clamped-clamped and clamped-free supports. The vibration of Euler-Bernoulli beam with FGM which is modeled by fourth-order partial differential equations with variable coefficients are examined by Yigit et al [15] by using the Adomian Decomposition Method. Su et al [16] presented a unified solution for free and transient analyses of a functionally graded piezoelectric curved beam with general boundary conditions within the framework of Timoshenko beam theory. The formulation is derived by means of the variational principle in conjunction with a modified Fourier series. Fundamental frequency of sandwich beams with functionally graded face sheet and homogenous core is studied by Mhu and Zhao [17].

The classical plate theory is used to analyze the face sheet and a higher-order theory is used to analyze the core of sandwich beams in which both the transverse normal and shear strains of the core are considered. Shwartsman and Majak [18] studied free vibration of axially functionally graded Euler-Bernoulli beams with elastically restrained ends. The method of initial parameters in differential form is treated for the numerical solution of the problem. Numerical method proposed has fourth order of accuracy and the Richardson extrapolation of results with different step sizes gives solutions of the sixth order of accuracy.

In the present paper are investigated the nonlinear free thermomechanical vibrations of FGB on Winkler-Pasternak elastic foundation. Considering von Karaman geometric nonlinearity, the Euler-Bernoulli beam theory and Galerkin procedure, we obtain a second-order nonlinear differential equation with quadratic and cubic nonlinear terms. This equation is solved by using a new analytical approximate approach – OAFM. The accuracy of the analytical results obtained through the proposed approach is proved by numerical simulations developed in order to validate analytical results. Several numerical examples show that the proposed procedure is simple and easy to use.

2 The governing equations

In what follows, we consider a FGB of length L, width b, and thickness h resting on an elastic foundation of Winkler-Pasternak type and subjected to an axial force P (fig.1).

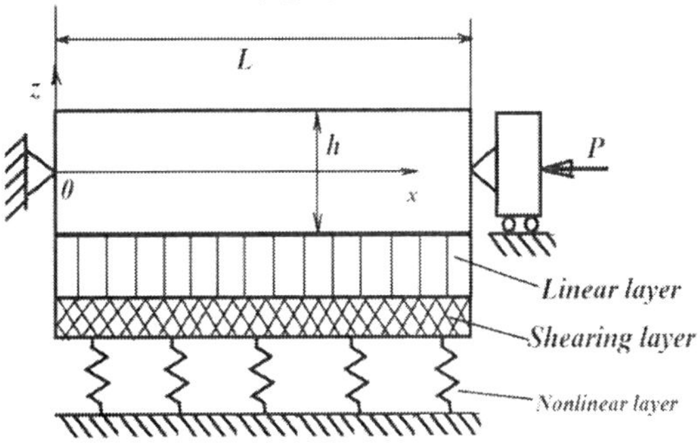

Fig.1. Schematic of the FGB with nonlinear foundation

The mechanical properties of the FGB can be varied as a function along thickness, based on the rule of mixtures. Taking into account the rule of mixtures, we have

$$E(z) = (E_2 - E_1)\left(\frac{2z+h}{2h}\right)^k + E_1 \tag{1}$$

$$\rho(z) = (\rho_2 - \rho_1)\left(\frac{2z+h}{2h}\right)^k + \rho_1 \tag{2}$$

$$\nu(z) = (\nu_2 - \nu_1)\left(\frac{2z+h}{2h}\right)^k + \nu_1 \tag{3}$$

where subscript 1 and 2 denote the top surface (z=-h/2) and bottom surface (z=h/2) respectively. The constant k characterizes the distributions of material properties. The case k=0 corresponds to an isotropic homogenous beam.

For a small strain, moderate deformation and rotation, the axial strain of the midplane of the beam accounting for the midplane stretching is given by [19]:

$$\varepsilon_x = \frac{\partial \overline{U}}{\partial \overline{x}} + z\frac{\partial^2 \overline{W}}{\partial \overline{x}^2} + \frac{1}{2}\left(\frac{\partial \overline{W}}{\partial \overline{x}}\right)^2 \tag{4}$$

where, based on Euler-Bernoulli beam theory, the displacement of an arbitrary point along the x and z axes are $\overline{U}(x,z,t)$ and $\overline{W}(x,z,t)$ respectively. If $U(x,t)$ and $W(x,t)$ are displacement components in the midplane, then we have

$$\overline{U}(x,z,t) = U(x,t) + z\frac{\partial W}{\partial x} \tag{5}$$

$$\overline{W}(x,z,t) = W(x,t) \tag{6}$$

where t is the time. The normal stress for the von Karaman type of geometry is given by the law

$$\sigma_{xx} = \frac{E(z)}{1-\nu^2(z)}\left[\frac{\partial \overline{U}}{\partial \overline{x}} + z\frac{\partial^2 \overline{W}}{\partial \overline{x}^2} + \frac{1}{2}\left(\frac{\partial \overline{W}}{\partial \overline{x}}\right)^2\right] \tag{7}$$

The curvature of the beam is given by

$$\frac{1}{R}=\frac{\partial^2\overline{W}}{\partial\overline{x}^2} \tag{8}$$

The axial, coupling and bending stiffness are defined by

$$A_{11}=\int_{-h/2}^{h/2}\frac{E(z)}{1-\nu^2(z)}dz \tag{9}$$

$$B_{11}=\int_{-h/2}^{h/2}\frac{E(z)}{1-\nu^2(z)}zdz \tag{10}$$

$$D_{11}=\int_{-h/2}^{h/2}\frac{E(z)}{1-\nu^2(z)}z^2dz \tag{11}$$

Assuming that the initial temperature of the beam is zero, resultant force and thermal momentum are defined as

$$N_{T\overline{x}}=-\int_{-h/2}^{h/2}\frac{E(z)}{1-\nu^2(z)}\alpha(z)Tdz \tag{12}$$

$$M_{T\overline{x}}=-\int_{-h/2}^{h/2}\frac{E(z)}{1-\nu^2(z)}\alpha(z)Tzdz \tag{13}$$

The total induced axial force N_x and bending moment M_x are related to the stress resultants as

$$N_x=A_{11}\left[\frac{\partial\overline{U}}{\partial\overline{x}}+\frac{1}{2}\left(\frac{\partial\overline{W}}{\partial\overline{x}}\right)^2\right]+B_{11}\frac{\partial^2\overline{W}}{\partial\overline{x}^2}+N_{T\overline{x}} \tag{14}$$

$$M_x=B_{11}\left[\frac{\partial\overline{U}}{\partial\overline{x}}+\frac{1}{2}\left(\frac{\partial\overline{W}}{\partial\overline{x}}\right)^2\right]+D_{11}\frac{\partial^2\overline{W}}{\partial\overline{x}^2}+M_{T\overline{x}} \tag{15}$$

The equations of motion for axial and transverse vibration of FGB based on Euler-Bernoulli beam theory and von Karaman geometric nonlinearity are

$$I_1\frac{\partial\overline{U}}{\partial t}-\frac{\partial N_x}{\partial\overline{x}}=0 \tag{16}$$

$$\frac{\partial^2 M_x}{\partial\overline{x}^2}-\frac{\partial}{\partial\overline{x}}\left(N_x\frac{\partial\overline{W}}{\partial\overline{x}}\right)+I_1\frac{\partial^2\overline{W}}{\partial t^2}=F_w \tag{17}$$

where

$$I_1=\int_{-h/2}^{h/2}\rho(z)dz \tag{18}$$

and F_w is reaction of the elastic Winkler-Pasternak foundation.

Now, if the axial inertia is neglected, then from Eq.(14) gives

$$N_x=N_{x0}=A_{11}\left[\frac{\partial\overline{U}}{\partial\overline{x}}+\frac{1}{2}\left(\frac{\partial\overline{W}}{\partial\overline{x}}\right)^2\right]+B_{11}\frac{\partial^2\overline{W}}{\partial\overline{x}^2}+N_{T\overline{x}} \tag{19}$$

and therefore N_x is independent of $\overline{x}$. From Eq.(17) we obtain

$$A_{11}^{-1}\left(N_x-B_{11}\frac{\partial^2\overline{W}}{\partial\overline{x}^2}-N_{T\overline{x}}\right)-\frac{1}{2}\left(\frac{\partial\overline{W}}{\partial\overline{x}}\right)^2=\frac{\partial\overline{U}}{\partial\overline{x}} \tag{20}$$

Supposing that the beam has immovable ends,

$$\overline{U}(0,t)=\overline{U}(L,t)=0 \tag{21}$$

and integrating Eq.(20) with respect to $\overline{x}$ and having in attention that N_x is independent of $\overline{x}$, we obtain

$$N_x=\frac{A_{11}}{2L}\int_0^L\overline{W}'^2dx+\frac{B_{11}}{L}[\overline{W}'(L,t)-\overline{W}'(0,t)]+N_{T\overline{x}}-P \tag{22}$$

From Eqs.(14) and (22) results

$$\frac{\partial\overline{U}}{\partial\overline{x}}=A_{11}^{-1}(N_x-N_{T\overline{x}})-\frac{1}{2}\overline{W}'^2-B_{11}\overline{W}'' \tag{23}$$

or

$$\frac{\partial\overline{U}}{\partial\overline{x}}=\frac{1}{2L}\int_0^L\overline{W}'^2dx+\frac{B_{11}}{A_{11}L}[\overline{W}'(L,t)-\overline{W}'(0,t)]-\frac{1}{2}\overline{W}'^2-\frac{B_{11}}{A_{11}}\overline{W}'' \tag{24}$$

Differentiating Eq.(24) with respect to $\overline{x}$ yields

$$\frac{\partial^2\overline{U}}{\partial\overline{x}^2}=-\overline{W}'\overline{W}''-\frac{B_{11}}{A_{11}}\overline{W}''' \tag{25}$$

where prime denotes derivative with respect to $\overline{x}$.

Substituting Eq.(24) into Eq.(15), we obtain

$$M_x=\frac{B_{11}}{2L}\int_0^L\overline{W}'^2dx+\frac{B_{11}^2}{A_{11}L}[\overline{W}'(L,t)-\overline{W}'(0,t)]-\frac{B_{11}}{A_{11}}\overline{W}''+D_{11}\overline{W} \tag{26}$$

Furthermore, using Eqs.(15) and (26) we have

$$\frac{\partial^2 M}{\partial\overline{x}^2}=\left(D_{11}-\frac{B_{11}^2}{A_{11}L}\right)\overline{W}^{(IV)} \tag{27}$$

By replacing the Eqs.(22) and (27) into Eq.(17), the governing nonlinear thermomechanical vibration equation of FGM is obtained as follows

$$I\ddot{\overline{W}}'' + \left(D_{11} - \frac{B_{11}^2}{A_{11}L}\right)\overline{W}^{(IV)} - \overline{W}''\left[\frac{A_{11}}{2L}\int_0^L \overline{W}'^2 dx + \right.$$
$$\left. + \frac{B_{11}}{L}\left(\overline{W}'(L,t) - \overline{W}'(0,t)\right) + N_{T\overline{x}} - P\right] = F_W \tag{28}$$

where the dot denotes derivative with respect to time.

Due to the nonlinear elastic Winkler-Pasternak foundation, the second side of the Eq.(8) is defined as follows:

$$F_W = -k_L\overline{W} - k_{NL}\overline{W}^3 + k_s\frac{\partial^2\overline{W}}{\partial\overline{x}^2} \tag{29}$$

where k_L and k_{NL} are the linear and nonlinear coefficients respectively and k_s is the shear coefficient of elastic foundation. In order to derive general results independent of dimensions and specific sizes, we introduce the dimensionless parameters:

$$x = \frac{\overline{x}}{L},\quad W = \frac{\overline{W}}{h},\quad r = \sqrt{\frac{I_1}{A}},\quad 2\alpha = \frac{A_{11}r^2}{D_{11} - \frac{B_{11}^2}{A_{11}}}$$
$$t = \frac{\bar{t}}{L}\sqrt{\frac{1}{D_{11}}\left(D_{11} - \frac{B_{11}^2}{A_{11}L}\right)}, N_T = \frac{N_{T\overline{x}}L^2}{D_{11} - \frac{B_{11}^2}{A_{11}}},$$
$$P = \frac{\overline{P}L^2}{D_{11} - \frac{B_{11}^2}{A_{11}}},\quad \lambda = \frac{B_{11}r^2}{D_{11} - \frac{B_{11}^2}{A_{11}}}, k_1 = \frac{k_L L^4}{D_{11} - \frac{B_{11}^2}{A_{11}}} \tag{30}$$
$$, k_2 = \frac{k_{NL}r^2L^4}{D_{11} - \frac{B_{11}^2}{A_{11}}},\quad k_3 = \frac{k_s L^2}{D_{11} - \frac{B_{11}^2}{A_{11}}} + \alpha(P - N_T)$$

By replacing Eq.(30) into Eq.(28) it results

$$W'' + W^{(IV)} + k_1W + k_2W^3 - k_3W'' -$$
$$-\alpha W''\left[\int_0^1 W'^2 dx + \lambda\left(W'(1,t) - W'(0,t)\right)\right] = 0 \tag{31}$$

The Eq.(31) is a partial differential equation in two dimension, displacement x and time t. Using the Galerkin method, Eq.(31) becomes an ordinary differential equation. Assuming that the transverse displacement is expressed as

$$W(x,t) = X(x)T(t) \tag{32}$$

where X(x) is the linear fundamental vibration mode and T(t) is the time dependent function to be determined. Substituting Eq.(32) into Eq.(31) and applying Galerkin's method, in which the orthogonality property of the mode shapes is used, yields:

$$\ddot{T} + aT + bT^2 + cT^3 = 0 \tag{33}$$

where

$$a = \int_0^1 X(x)X^{(IV)}(x)dx + k_1\int_0^1 X^2(x)dx - k_3\int_0^1 X(x)X''(x)dx$$
$$b = -\alpha\lambda\left[X'(1) - X'(0)\right]\int_0^1 X(x)X''(x)dx$$
$$c = k_2\int_0^1 X^{(IV)}(x)dx - \alpha\left(\int_0^1 X'^2(x)dx\right)\left(\int_0^1 X(x)X''(x)dx\right) \tag{34}$$

For the case of simply supported beam, the fundamental vibration mode is

$$X(x) = \sqrt{2}\sin\pi x \tag{35}$$

The Eq.(33) contains a quadratic nonlinear term due to the presence of bending-extension coupling effect in FGB and cubic nonlinear term due to the Winkler-Pasternak foundation. This equation does not have an exact solution, but by means of OAFM may be obtained an approximate solution using a set of auxiliary functions. We remark that in Eq. (33) there exists no small or large parameter.

The beam centroid is subjected to the following initial conditions

$$T(0) = A,\quad \dot{T}(0) = 0 \tag{36}$$

3 The Optimal Auxiliary Functions Method

In the following, we consider the general form of a nonlinear differential equation

$$L[T(t)] + N[T(t)] = 0,\quad t \in D \tag{37}$$

with the initial/boundary conditions

$$B\left(T(t), \frac{dT(t)}{dt}\right) = 0 \tag{38}$$

where L is a linear operator and N is a nonlinear operator, t denotes the independent variable and T(t) is an unknown function.

It is known that an exact solution for strongly nonlinear differential equations of type (37) and (38) is frequently hard to be found. To find an approximate solution, we suppose that this solution can be expressed in the form

$$\overline{T}(t, C_i) = T_0(t) + T_1(t, C_i),\quad i = 1,2,...,n \tag{39}$$

where the initial and the first approximation will be determined as described in what follows.

Substituting Eq.(39) into (37), it results in

$$L[T_0(t)] + L[T_1(t, C_i)] + N[T_0(t) + T_1(t, C_i)] = 0 \tag{40}$$

The initial approximation $T_0(t)$ can be identified from the linear equation

$$L[T_0(t)] = 0 \tag{41}$$

with the initial/boundary conditions

$$B\left(T_0, \frac{dT_0}{dt}\right) = 0 \tag{42}$$

and the first approximation from the equation

$$L[T_1(t, C_i)] + N[T_0(t) + T_1(t, C_i)] = 0, \quad i = 1, 2, ..., n \tag{43}$$

$$B\left(T_1, \frac{dT_1}{dt}\right) = 0 \tag{44}$$

The nonlinear term in the last equation is expanded in the form

$$N[T_0(t) + T_1(t, C_i)] = N[T_0(t)] + \sum_{k=1}^{\infty} \frac{T_1^k}{k!} N^{(k)}[T_0(t)] \tag{45}$$

To avoid the difficulties that appear in solving the nonlinear differential equation (43), and to accelerate the rapid convergence of the first approximation T_1, instead of the last term arising in (43), we propose another expression such that Eq.(43) can be written in a new form

$$\begin{aligned} &L[T_1(t, C_i)] + A_1\left(T_0(t), \dot{T}_o(t), C_j\right) P[N(T_0(t))] + \\ &+ A_2(T_0, \dot{T}_0, C_k) = 0 \\ &B\left(T_1(t, C_i), \frac{dT_1(t, C_i)}{dt}\right) = 0 \end{aligned} \tag{46}$$

where A_1 and A_2 are two arbitrary auxiliary functions depending on the initial approximation $T_0(t)$ and several unknown parameters C_j and C_k i=1,2,…,n, j=1,2,…,s, k=s+1,s+2,…,n and $P[N(T_0(t)]$ is a part of $N[T_0(t)]$. The auxiliary function A_1 and A_2 (namely auxiliary functions) are not unique and are of the same form as $T_0(t)$. If $N[T_0(t)]=0$ then it is clear that $T_0(t)$ is an exact solution of Eq.(37).

The unknown parameters C_i can be optimally identified via different methods, such as Galerkin method, Ritz method, the least square method, colocation method and so on [19-23]. If

$$R(t, C_i) = L[\bar{T}(t, C_i)] + N[\bar{T}(t, C_i)] \tag{47}$$

where $\bar{T}(t, C_i)$ is given by Eq.(39), then we construct the functional

$$J(C_1, C_2, ..., C_n) = \int_{(D)} R^2(t, C_i) dt \tag{48}$$

and the values of the parameters C_i are obtained from the equations

$$\frac{\partial J}{\partial C_1} = \frac{\partial J}{\partial C_2} = ... = \frac{\partial J}{\partial C_n} = 0 \tag{49}$$

Alternatively, an equivalent system of algebraic equations useful for identifying the values of the parameters C_i would be given by

$$R(t_1, C_i) = R(t_2, C_i) = ...R(t_n, C_i) = 0 \tag{49'}$$

By means of this novel method, the first-order approximate solution given by Eq.(39) is well determined after the identification of the optimal values of the initial unknown convergence-control parameters C_i, i=1,2,…,n.

It will be proved that our procedure is a powerful tool for solving nonlinear problems without small or large parameters.

4 Application of OAFM to Eqs. (33) and (36)

If we introduce the independent variable $\tau = \omega t$ and the dependent variable T=Ay, then Eqs. (33) and (36) become, respectively

$$y'' + \frac{a}{\omega^2} y + \frac{bA}{\omega^2} y^2 + \frac{cA^2}{\omega^2} y^3 = 0 \tag{50}$$

$$y(0) = 1, \quad y'(0) = 0 \tag{51}$$

where prime denotes derivative with respect to τ and ω is the frequency of the system.

For Eq.(50) the linear operator may be written

$$L[y(\tau)] = y'' + y \tag{52}$$

and the corresponding nonlinear operator becomes

$$N[y(\tau)] = \left(\frac{a}{\omega^2} - 1\right) y + \frac{bA}{\omega^2} y^2 + \frac{cA^2}{\omega^2} y^3 \tag{53}$$

The Eqs.(41) and (42) become

$$y_0'' + y_0 = 0, \quad y_0(0) = 1, y'(0) = 0 \tag{54}$$

and has the solution

$$y_0(\tau) = \cos\tau \tag{55}$$

Substituting Eqs.(55) into (53), we obtain

$$\begin{aligned} N[y_0(\tau)] &= \frac{bA}{2\omega^2} + \left(\frac{a}{\omega^2} - 1 + \frac{3cA^2}{4\omega^2}\right)\cos\tau + \\ &+ \frac{bA}{\omega^2}\cos 2\tau + \frac{cA^2}{4\omega^2}\cos 3\tau \end{aligned} \tag{56}$$

The Eq.(46) can be written as

$$\begin{aligned} &y_1'' + y_1 + A_1(\cos\tau, C_j) P[N(y_0(\tau))] + \\ &+ A_2(\cos\tau, C_j) = 0, \quad y_1(0) = y_1'(0) = 0 \end{aligned} \tag{57}$$

where

$$P[N(y_0(\tau))] = \frac{bA}{2\omega^2} + \left(\frac{a}{\omega^2} - 1 + \frac{3cA^2}{4\omega^2}\right)\cos\tau + \frac{bA}{\omega^2}\cos 2\tau \quad (58)$$

$$A_1(\cos\tau, C_j) = -(C_1 + 2C_2\cos\tau + 2C_3\cos 2\tau) \quad (59)$$

$$A_2(\cos\tau, C_k) = -(C_4\cos 3\tau + C_4\cos 4\tau) \quad (60)$$

where C_1, C_2, C_3 and C_4 are unknown parameters at this moment. However, the function A_1, A_2 and P are not unique. Alternatively, we may choose these functions in the form:

$$P[N(y_0(\tau))] = \frac{bA}{2\omega^2} + \left(\frac{a}{\omega^2} - 1 + \frac{3cA^2}{4\omega^2}\right)\cos\tau \quad (61)$$

$$A_1(\cos\tau, C_j) = -(C_1 + 2C_2\cos\tau) \quad (62)$$

$$A_2(\cos\tau, C_k) = -(C_3\cos 3\tau + C_4\cos 5\tau) \quad (63)$$

or yet

$$P[N(y_0(\tau))] = \frac{bA}{2\omega^2} + \frac{bA}{\omega^2}\cos 2\tau + \frac{cA}{4\omega^2}\cos 3\tau \quad (64)$$

$$A_1(\cos\tau, C_j) = -(C_1 + 2C_2\cos\tau + 2C_3\cos 2\tau + 2C_4\cos 3\tau) \quad (65)$$

$$A_2(\cos\tau, C_k) = 0 \quad (66)$$

and so on. Using only Eqs. (58), (59) and (60), Eq.(57) becomes

$$y_1'' + y_1 = \frac{bA}{2\omega^2}C_1 + \left(\frac{4a+3cA^2}{4\omega^2} - 1\right)C_2 + \frac{bAC_3}{2\omega^2} + \left[\frac{3bA}{2\omega^2}C_2 + \left(\frac{4a+3cA^2}{4\omega^2} - 1\right)(C_1 + C_3)\right]\cos\tau + \left[\frac{bA}{\omega^2}(C_3 + \frac{1}{2}C_1) + \left(\frac{4a+3cA^2}{4\omega^2} - 1\right)C_2\right]\cos 2\tau + \left[\left(\frac{4a+3cA^2}{4\omega^2} - 1\right)C_3 + \frac{bA}{\omega^2}C_2 + C_4\right]\cos 3\tau + \left(\frac{bA}{\omega^2}C_3 + C_5\right)\cos 4\tau \qquad y_1(0) = y_1'(0) = 0 \quad (67)$$

The solution of Eq.(67) is chosen so that to contain no secular terms, which lead to the condition

$$\omega^2 = a + \frac{3}{4}cA^2 + \frac{3}{2}bA\frac{C_2}{C_1 + C_3} \quad (68)$$

The solution of Eq.(67) is given by

$$y_1(\tau, C_i) = \frac{bA}{2\omega^2}C_1 + \left(\frac{4a+3cA^2}{4\omega^2} - 1\right)C_2 + \frac{bA}{2\omega^2}C_3 + \frac{1}{3}\left[\frac{bA}{2\omega^2}(C_1 + 2C_3) + \left(\frac{4a+3cA^2}{4\omega^2} - 1\right)C_2\right](\cos\tau - \cos 2\tau) + \frac{1}{8}\left[\frac{bA}{2\omega^2}C_2 + \left(\frac{4a+3cA^2}{4\omega^2} - 1\right)C_3 + C_4\right](\cos\tau - \cos 3\tau) + \frac{1}{15}\left(\frac{bA}{2\omega^2}C_3 + C_4\right)(\cos\tau - \cos 4\tau) \quad (69)$$

From Eqs. (55), (69) and (39) and taking into account that $\tau = \omega t$ and T=Ay, we can get the first-order approximate solution of Eqs.(33) and (36) in the form

$$T(t) = A\cos\omega t + \left[\frac{bA^2}{2\omega^2}C_1 + A\left(\frac{4a+3cA^2}{4\omega^2} - 1\right)C_2 + \frac{bA^2C_3}{2\omega^2}\right](1 - \cos\omega t) + \frac{1}{3}\left[\frac{bA^2(C_1 + 2C_3)}{2\omega^2} + \left(\frac{4aA+3cA^3}{4\omega^2} - A\right)C_2\right](\cos\omega t - \cos 2\omega t) + \frac{1}{8}\left[\frac{bA^2C_2}{2\omega^2} + \left(\frac{4aA+3cA^2}{4\omega^2} - A\right)C_3 + AC_4\right](\cos\omega t - \cos 3\omega t) + \frac{1}{15}\left(\frac{bA^2C_3}{2\omega^2} + AC_5\right)(\cos\omega t - \cos 4\omega t) \quad (70)$$

where ω is given by Eq.(68).

5 Numerical examples

We consider the following three cases, in which the parameters a, b and c are obtained from Eqs. (34) and (35):

$$a = \pi^4 + k_1 + \pi^2 k_3, \quad b = -2\sqrt{2}\pi^3\alpha\lambda, \quad c = \frac{3k_2}{2} + \alpha\pi^4 \quad (71)$$

5.1 Case 1

For k_1=k_3=50 and k_3=25, α=1, λ=0.1, A=0.6, the parameters C_i and ω are obtained using the described procedure as

$$C_1 = -0.803588048507, C_2 = -0.286807831152, C_3 = 0.165036386496, C_4 = -0.024155608542, C_5 = 0.000054512958, \omega = 25.9571 \quad (72)$$

The first-order approximate solution of Eqs.(33) and (36) is

$$\bar{T}(t) = 0.000590623 + 0.598237\cos\omega t - \\ - 0.0000679857\cos 2\omega t + 0.00121644\cos 3\omega t + \\ + 0.0000235972\cos 4\omega t \quad (73)$$

To emphasize the accuracy of the obtained solution, we compare the approximate result obtained through OAFM with the numerical integration results. Fig.2 presents a comparison of the present solution (72) and numerical results obtained by a fourth-order Runge-Kutta method.

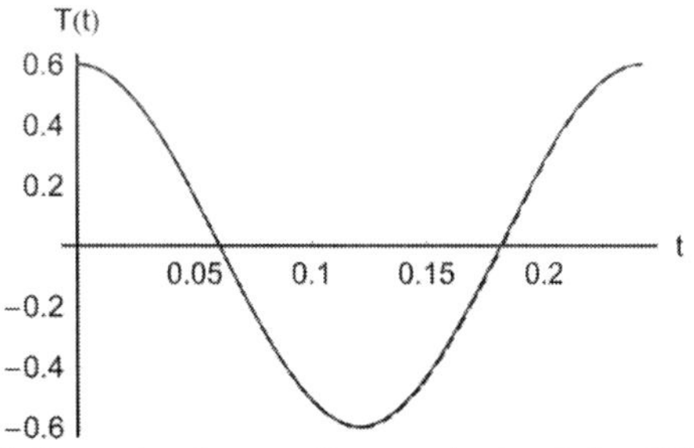

Fig.2. Comparison between the approximate solution (73) and numerical integration results
________ numerical; _ _ _ _ analytical

5.2 Case 2

In the second case, for k_1=50, k_2=40 and k_3=20, we have

$$C_1 = -0.8086487353162, C_2 = -0.2991968209976, \\ C_3 = 0.1707921303513, C_4 = -0.0271212510421, \\ C_5 = 0.00005828000552259946, \omega = 23.9352 \quad (74)$$

and therefore

$$\bar{T}(t) = 0.00059746 + 0.59808\cos\omega t - \\ - 0.0000422829\cos 2\omega t + 0.00133548\cos 3\omega t + \\ + 0.0000290429\cos 4\omega t \quad (75)$$

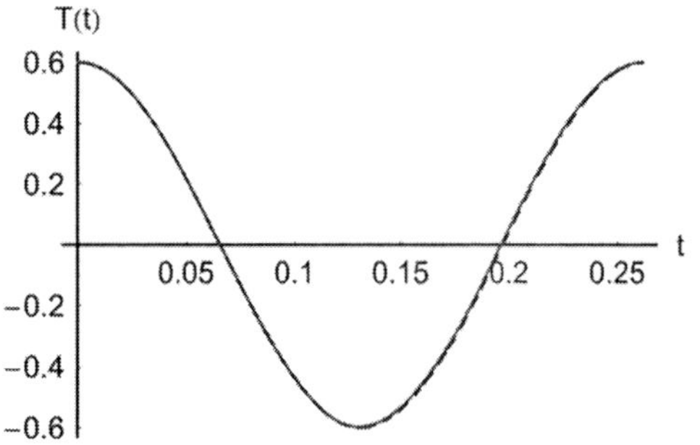

Fig.3. Comparison between the approximate solution (75) and numerical integration results
________ numerical; _ _ _ _ analytical

5.3 Case 3

In the last considered case, k_1=40, k_2=30 and k_3=50, such that

$$C_1 = -0.8652502750014, C_2 = -0.3458418183991, \\ C_3 = 0.18086458531247, C_4 = -0.0266366276983, \\ C_5 = 0.000032366254671716754, \omega = 25.7944 \quad (76)$$

and therefore

$$\bar{T}(t) = 0.000379825 + 0.598285\cos\omega t + \\ + 0.0000164287\cos 2\omega t + 0.00129126\cos 3\omega t + \\ + 0.0000273128\cos 4\omega t \quad (77)$$

In figs.3 and 4 are compared the solutions (75) and (77) with the corresponding numerical integration results.

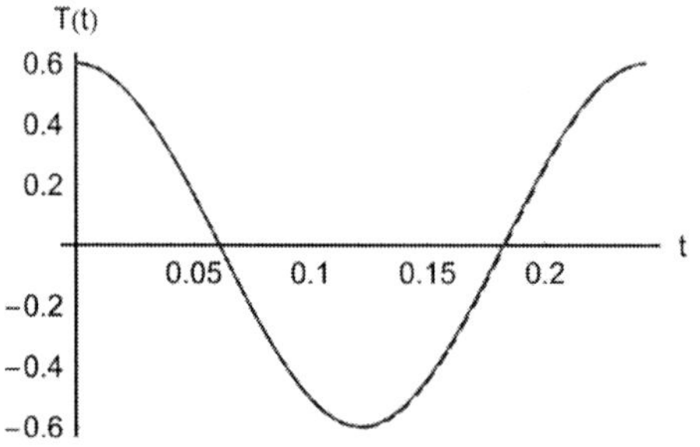

Fig.4. Comparison between the approximate solution (77) and numerical integration results
________ numerical; _ _ _ _ analytical

From Figs.2-4 one can be observed that the first-order approximate analytical results obtained by means of OAFM are almost identical with the numerical simulation results in all considered cases for various values of the parameters k_i, i=1,2,3.

6 Conclusions

In this work, we proposed a reliable new technique, namely the Optimal Auxiliary Functions Method (OAFM), which accelerates the rapid convergence of the approximate analytical solution of nonlinear thermomechanical vibration of a FGB on Winkler-Pasternak foundation.

The proposed procedure is valid even if the nonlinear equation does not contain any small or large parameters. Our construction of the first iteration is different from other traditional approaches especially concerning the optimal auxiliary functions which depend on some initially unknown parameters C_i.

The initially unknown parameters, called convergence-control parameters, after identifying their optimal values, provide a fast convergence of the approximate analytical solutions using only one iteration.

The main advantage of the proposed procedure is the possibility to optimally control and adjust the convergence of the solutions through the auxiliary functions.

The obtained approximate analytical solutions are in excellent agreement with the numerical integration results. This proves the validity of our approach, and on the other hand, proves that this method is very efficient in practice.

References

1. V. Birman, L.W. Byrd, AIAA Journal, **45**, 2747 (2007)
2. M.C Dokmeci, J. of the Acoustical Society of America, **119**, 3336 (2006)
3. M.K. Khorrambadi, J. of Solid Mech. **1**, 22 (2009)
4. O. Carl, P. Villamil, C. Zhang, Proc. Appl. Math. Mech., **11**, 199 (2011)
5. N.T. Alshabatat, K. Naghshineh, Advances in Acoustics and Vibration, ID 752361, 2014
6. L-L. Ke, J. Yong, S. Kitipornchai, Meccanica **45**, 743 (2010)
7. H. Yaghoobi, M. Torabi, J. Theor. Appl. Mech., **51**, 39 (2013)
8. N. Mohammadi, Int. J. of Eng. and Tech. Sci., **3**, 144 (2015)
9. Y Fu, J. Wang, Y Mao, J. of Intelligent Material Syst. and Struct., **22**, 2093 (2011)
10. B. Thomas, P. Inamdar, T. Roy, B.K. Nanda, Int. J. on Theor. and Appl. Res. in Mech. Eng., **2**, 97 (2013)
11. M.N. Elmaguri, M. Haterbouch, A. Bouayad, O. Oussouaddi, J. Mater. Environ. Sci., **6**, 3620 (2015)
12. J. Yang, S. Kitipornchai, C. Feng, Int. J. of Structural Stability and Dynamics, 15, 1540015 (2015)
13. N.V.R. Maganti, M.R. Nalluri, Int. J. Mech. and Material Eng., 10, 21 (2015)
14. K.R. Kashizadeh, A.A. Asfarjani, Advances in Materials, **5**, 57 (2016)
15. G. Yigit, A. Sahim, M. Bayram, Open Math., **14**, 661 (2016)
16. Z. Su, G. Jin, T Ye, Smart Materials and Struct., 25, 065003 (2016)
17. L Mu. G. Zhao, Shock and Vibration, ID 3287645, 2016
18. B.S. Shvartsman, J. Majak, Adv. in Theoretical and Appl. Mech., **9**, 31 (2016)
19. S. Emam, A.H. Nayfeh, Composite Structures, **88**, 636 (2009)
20. N. Herisanu, V. Marinca, Springer Proc. Math. and Statistics, **182**, 169 (2016)
21. V. Marinca, N. Herisanu, *Nonlinear dynamical systems in engineering. Some approximate approaches* (Springer, Heidelberg, 2011)
22. V. Marinca, N. Herisanu, *The optimal homotopy asymptotic method. Engineering applications,* (Springer, Cham, 2015)
23. V. Marinca, N. Herisanu, Int. J. Heat and Mass Transf **79**, 146 (2014)

The effect of nylon nanofibers on the dynamic behaviour and the delamination resistance of GFRP composites

Cristobal Garcia [1, 2, *], *Irina* Trendafilova [1], *Andrea* Zucchelli [3], and *Justin* Contreras [1]

[1] University of Strathclyde, Department of Mechanical and Aerospace Engineering, 75 Montrose Street, G1 1XJ, Glasgow, UK
[2] IMDEA Materials Institute, C/ Eric Kandel 2, 28906, Getafe, Spain
[3] University of Bologna, Department of Industrial Engineering, Viale Risorgimento 2, 40125, Bologna, Italy

Abstract. Vibrations are responsible for a considerable number of accidents in aircrafts, bridges and other civil engineering structures. Therefore, there is a need to reduce the vibrations on structures made of composite materials. Delamination is a particularly dangerous failure mode for composite materials because delaminated composites can lose up to 60% of their strength and stiffness and still remain unchanged. One of the methods to suppress vibrations and preventing delamination is to incorporate nanofibers into the composite laminates. The aim of the present work is to investigate how nylon nanofibers affect the dynamic behaviour and delamination resistance of glass fibre reinforced polymer (GFRP) composites. Experiments and numerical simulations using finite element modelling (FEM) analysis are used to estimate the natural frequencies, the damping ratio and inter-laminar strength in GFRP composites with and without nylon nanofibers. It is found that the natural frequencies of the nylon nano-modified composites do not change significantly as compared to the traditional composites. However, nano-modified composites demonstrated a considerable increase in damping ratio and inter-laminar shear strength due to the incorporation of nylon nanofibers. This work contributes to the knowledge about the mechanical and dynamic properties of glass fibre reinforced polymer (GFRP) composites with nylon nanofibers.

1. Introduction

Composite laminates interleaved with nanofibers has become rather popular recently because the inclusion of nanofibers can enhance some material properties drastically. One of the most frequent failure modes in composite materials is delamination. Delamination is difficult to detect because it takes place and grows under the surface without being visible from outside and it may cause catastrophic failure in the whole structure. One of the methods to prevent delamination is the introduction of nanofibers into epoxy-based composite laminate materials [1]. Composites interleaved with nanofibers are supposed to increase their delamination resistance as compared to the virgin ones [2]. The interleaving with nano-fibers can be also used to increase the stiffness and other macro-properties of epoxy-based composite laminates.

Vibrations are one of the most common and important motions of most mechanical, civil and aerospace structures. Vibrations can directly and adversely affect the working condition and the health of machines and structures. In such a sense vibration analysis is of critical importance for purposes of equipment maintenance and environmental safety. Vibrations are responsible for a considerable number of failures in structures made of composite materials including aircrafts, buildings and bridges failures. The interleaving with nano-fibres of composite laminates can be also used for purposes of vibration amplitude reduction. Thus several studies suggest the inclusion of nylon nanofibers in composite laminates with the aim of improving its properties [3, 4]. It should be said that on the other hand very little is known about the effect of nylon interleaving on the macro properties of the resulting composite including its dynamics and vibratory behaviour. There are only a couple of studies that consider the dynamic behaviour of nano-interleaved composite laminates [5].

Delamination is known to be one of the most dangerous and in the same time widespread damage and failure mode. Delamination affects the vibratory behaviour of laminates and thus vibrations can be useful delamination diagnosis [6]. From such a view point this study also investigated the delamination resistance of nylon interleaved GFRP laminates.

Nylon nanofibers fabricated by electrospinning are particularly suitable to improve the properties of composites because of their excellent properties which include high heat resistance, good adhesion with epoxy resin and excellent mechanical characteristics such as high tensile and flexural strength, stiffness and superior wear resistance compared to other polymers. The high melting temperature allows the nanofibers to maintain their morphology during the curing process of the

* Corresponding author: cristobal.garcia@strath.ac.uk

laminate. However, these enhancements which nanofiber interleaving is expected to bring in composite materials still have to be tested and proven desirably using both experiments and FEM numerical analysis.

This work aims to investigate some dynamic properties of nylon nano-modified GFRP composites and their delamination resistance. The dynamical properties are evaluated experimentally using free vibration test and numerically using finite element modelling (FEM). In terms of delamination resistance, the interlaminar strength is determined using ASTM D 2344 international standard test for determining short beam strength of Polymer Matrix Composite Materials and Their Laminates [7]. The results of the latter experiments are also simulated in ANSYS Workbench. The paper offers experimental results which come to verify our FEM models and simulation results. Eventually a discussion on the experimental and the simulation results is proposed. The paper ends with some conclusions and trends for future research in terms of potential applications for the nano-modified composites.

2. Dynamic behaviour of nylon nano interleaved GFRP laminates

This section discusses the dynamic behaviour of nylon interleaved composite beams as compared to the traditional composite laminates. The experimental analysis for the virgin and the nano-modified composites was carried out using free vibration tests which are described in subsection 2.1. Consequently the finite element model used to simulate the experiments is introduced. Eventually, the experimental and FE model results are compared and discussed.

2.1 Experimental investigation

The vibration properties of composite laminates with and without nylon nanofibers are analysed in this paragraph. The composite specimens tested have the following dimensions 168 x 32 x 3 mm and are fabricated with a symmetric stacking sequence of $[0/90/0/90/0]_s$. The specimens are composed of 10 GFRP layers. The nylon modified specimens are interleaved with networked membranes of nylon nanofibers into the ply interfaces. The membranes of nylon nanofibers are interleaved at each ply composite interface excluding the central one.

The incorporation of the membrane of nanofibers did not affect the thickness of the composite laminates. Ten specimens made of glass fibre epoxy composite identical in size, number of layers and stacking sequence are tested. Five of these are virgin (non-interleaved) and the other five are nylon modified.

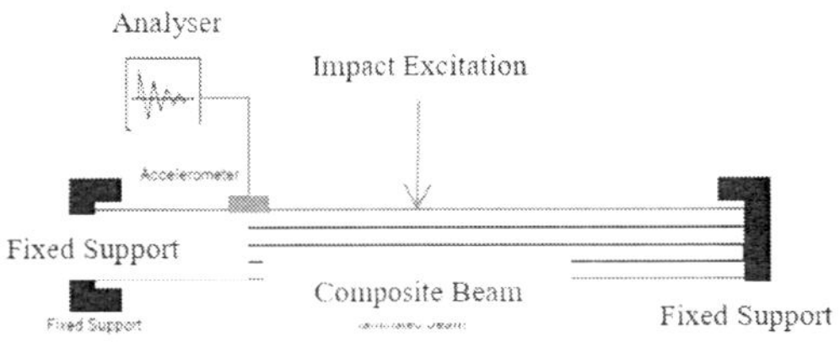

Figure 1: Dynamic free experiment for measuring the vibratory responses of the composites laminated beams

Free vibration tests were carried out with specimens fixed at both ends as indicated in figure 1. An impact excitation was applied at the middle of the composite laminate using a modal hammer. After that, the free decay vibration response of the composites was recorded using a RT-440 portable analyser. The experiments were repeated 10 times for each specimen. The natural frequencies and the damping ratios for each specimen were determined applying the Fast Fourier Transform (FFT) and the Logarithmic decrement method [8, 9] respectively, on the recorded free decay acceleration signals. The presented results are averaged over the tests and the specimens.

2.2 Numerical modelling

The finite element model (FEM) used to simulate the dynamic behaviour of the virgin and nylon nano-modified composite laminates was developed using ANSYS Workbench. The free decay time domain signal obtained after the hammer impact was simulated using transient vibration analysis. The number of elements used for each lamina was 936 elements and the criteria employed to select the number of elements was convergence [10]. The element type used was Shell 181 and it was chosen among several elements on the basis of having high accuracy and minimum computations time. Then the simulated free decay response obtained was used to calculate the natural frequencies and the damping ratio. The simulations for the traditional and nano-modified composites beams used the same dimensions, number of layers and ply orientations as those of the experimentally manufactured composite beams (see &2.1). The glass fibre reinforced polymer (GFRP) composite layers are simulated using the materials constants given in table 1, where E is the young modulus, G is the shear modulus, v is the poison's ratio and ρ is the density of the material [11]. For the nano-modified composites, as the nylon nanofibers are interleaved at the ply interfaces are embedded into epoxy resin, the material constants are calculated according to the rule of mixture as indicated in reference [12]. The material parameters for the interfaces composed of nylon nanofibers and epoxy resin are also given in table 1.

 https://doi.org/10.1051/matecconf/201814814001

Table 1: Material properties for the layers of GFRP composite material and ply interfaces of nylon nanofibers and epoxy resin

Magnitude	*GFRP layers*	*Nano interfaces*	*Units*
E_1	44000	3000	MPa
$E_2 = E_3$	11000	750	MPa
$G_{12} = G_{13}$	4500	800	MPa
G_{23}	500	400	MPa
v_{12}	0.28	0.35	dimensionless
$v_{13} = v_{12}$	0.10	0.2	dimensionless
ρ	1085.9	1140.0	Kg m^{-3}

2.3 Results and comparison

Table 2 and 3 shows the first five experimentally measured and numerically calculated natural frequencies for the virgin and the nano-modified composite specimens. In tables 2 and 3 the differences in % between the frequencies for the virgin and the nano-modified specimens are also indicated. The results indicate that the natural frequencies are not affected by the interleaving with nylon nanofibers. The miniscule changes in the natural frequencies which are below 5 % are within the range of the experimental and measurement error. These very little changes in the frequencies can be explained by the fact that nylon nano-fibres have rather low stiffness which is much lower than the stiffness of the glass fibres, and the interleaving did not change the mass of the specimens.

Table 2: Variation of natural frequencies due to nylon nanofibers measured experimentally

Frequencies	*Virgin (Hz)*	*Nano-Modified (Hz)*	*Variation (%)*
f_1	484.45	498.80	2.88
f_2	930.93	965.90	3.62
f_3	1373.75	1434.80	4.25
f_4	1857.50	1929.00	3.71
f_5	2303.75	2399.30	3.98

Table 3: Variation of natural frequencies due to nylon nanofibers calculated using FEM

Frequencies	*Virgin (Hz)*	*Nano-Modified (Hz)*	*Variation (%)*
f_1	515.00	535.00	3.74
f_2	940.00	945.00	0.53
f_3	1330.00	1343.00	0.97
f_4	1770.00	1775.00	0.28
f_5	2420.00	2423.00	0.12

The damping ratio of the composite laminates with and without nylon nanofibers is also investigated. Table 4 shows the values measured experimentally and calculated numerically using the FE model.

Table 4: Variation in damping ratio due to the interleaving of nylon nanofibers measured experimentally and using FEM

Damping	*Virgin (dimensionless)*	*Nano-Modified (dimensionless)*	*Variation (%)*
Experiments	0.0121	0.0165	36
FEM	0.0160	0.0233	45

Both, the experiments and FEM simulations showed that the interleaving with nylon nanofibers increased the damping significantly with 36 % and 45 % respectively. The higher the damping value the more energy is dissipated; and thus it can be concluded that GFRP composite laminates modified with nylon nanofibers dissipate more energy than virgin ones under the same mechanical conditions. This can be used to reduce the amplitude of vibrations in structures made of such composites. The significant increase in the damping of the nano-modified composites in comparison with traditional ones can be explained by the larger interface area produced due to the incorporation of nanofibers.

3. Delamination resistance of nylon nano interleaved GFRP laminates

This paragraph examines the delamination resistance of the nylon-nano modified specimens as compared to the virgin ones. The international standard ASTM D 2344 [13] was performed to experimentally determine the inter-laminar shear strength of the composites. It should be noted that these experiments were simulated using cohesive zone FE modelling via ANSYS Workbench 17.2. The following paragraphs present the experiment performed, the FE modelling and eventually the comparison between the experimental and the numerically obtained results.

3.1 Experimental study

Fourteen specimens made of nylon modified composites and virgin laminates were prepared in accordance with the ASTM D 2344 standard. The dimensions of the composite beams were 18 mm long, 6 mm wide and 3.21 mm thick with a stacking sequence of $[0/90/0/90/0]_s$ for a total of 10 layers.

The analysis of the inter-laminar shear stress was performed in accordance with the international standard ASTM D-2344 and the experimental test is illustrated in figure 2. The specimens were located on the two 3 mm diameter supports. During the test, the specimens are loaded in the centre of the beam at a rate of 1.0 mm/min. The load was increased until each specimen fractured and the fracture load was recorded for each trial.

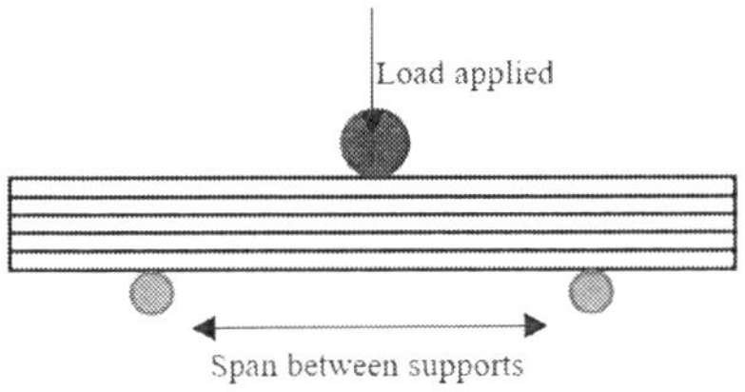

Figure 2: Standard test method to determinate the inter-laminar shear strength in composite laminates with and without nylon nanofibers.

After that, the interlaminar shear strength was calculated for each specimen using equation 1 [13], where F^{sbs} is the interlaminar strength, P_m is the fracture load, b is the specimen width and h is the specimen thickness.

$$F^{sbs} = 0.75 \times \frac{P_m}{b \times h} \qquad (1)$$

3.2 Numerical study

The finite element model used to simulate the experiments was developed using cohesive zone modelling (CZM) in ANSYS Workbench. The CZM used the bi-linear traction separation law. The material properties for the layers of GFRP composite material are given in table 1. The parameters used to simulate the interfaces of the virgin and the nano-modified composites are given in table 5, where σ_{max} and τ_{max} are the maximum normal stress and the tangential traction stress respectively, u_n^c and u_t^c denote the normal and the tangential displacement jump at the completion of the debonding, and α is the ratio between the normal displacement jump at maximum traction and at the completion of debonding. The virgin and the nano-modified composite beams were modelled with the same dimensions, number of layers and ply orientations as the experimentally manufactured composite beams.

Table 5: Cohesive zone parameters identified for the interfaces of traditional and nano-modified composite materials

Parameter	*Virgin*	*Nano-Modified*	*Units*
σ_{max}	70	73	MPa
τ_{max}	38	49	MPa
u_n^c	0.025	0.04	mm
u_t^c	0.08	0.09	mm
α	0.1	0.085	dimensionless

3.3 Results and comparison

A summary of the experimental results obtained following the ASTM D-2344 standard experiment is presented in table 6. The experiments showed that the inter-laminar strength for the seven virgin composite specimens was 50.21 ± 4.05 MPa. However, this value increased to 63.81 ± 2.48 MPa for the nylon nano-modified composite specimens. Thus it can be concluded that the inter-laminar shear strength increased with 27.07% due to the incorporation of nylon nanofibers.

Table 6: Variation in interlaminar shear strength due to the interleaving of nylon nanofibers measured experimentally by ASTM D-2344 and numerically using FEM

Interlaminar Strength	*Virgin (MPa)*	*Nano-Modified (MPa)*	*Variation (%)*
Experiments	50.21	63.81	27.07
FEM	54.15	65.03	20.10

The numerical inter-laminar strengths as per table 6 are 54.15 MPa and 65.03 MPa for the virgin and the nano-modified laminates respectively. The FEM results given show that the presence of nylon nanofibers in the composites increases the inter-laminar strength with 20.10 %. According to these results, it can be concluded that the incorporation of nanofibers has a significant impact on the interlaminar strength. According to [14], the increase in interlaminar strength can be explained with the presence of significant bridging phenomena due to the interaction between the epoxy resin and the nylon nanofibers

4. Conclusions and future research

In this work, membranes of networked nylon nanofibers were interleaved into the ply-interfaces of GFRP composite laminates with the aim to suppress vibrations and to prevent delamination. For this purpose, experiments and FE simulations analysed the effect of nylon nanofibers on the vibratory behaviour and delamination resistance in terms of the natural frequencies, damping and on the inter-laminar shear strength. On the basis of the experimental and the numerical results, the following conclusion can be drawn:

Nylon nanofibers did not affect the natural frequencies of the nano-modified specimens. The frequencies underwent very small changes which can be considered within the range of the noise and the numerical errors. Thus in terms of frequencies the vibratory behaviour of the nano-modified specimens is the same as the behaviour of the virgin ones. On the other hand the damping ratio considerably increased due to the inclusion of nanofibers according to both the experiments and the numerical modelling. The experimental results showed an increment of 36% in the damping ratio of the nano-modified specimens while the numerical simulations reported a 45% increase. From these results it was concluded that nano-modified composite materials dissipate much more energy than virgin ones. This property can be used for purposes of reduction of the vibration amplitude which can be achieved by the interleaving with nylon nano-fibres. On the other hand the vibratory behaviour of the nano-modified laminates is not supposed to change in terms of frequencies. As the natural frequencies have not changed and the mass remained unchanged as a result of the nylon nano modification it can be assumed that the stiffness of the nano-modified composites has not undergone significant changes.

In terms of delamination resistance, it can be concluded that GFRP composites interleaved with nylon nanofibers are less prone to delamination than the virgin ones. Both the experimental and FEM results showed a significant increase in the inter-laminar strength. An increment of 27% in the inter-laminar strength was observed experimentally using the ASTM D 2344 standard as compared to a 20% increment observed from the numerical results. The increase in the inter-laminar strength can be explained by the presence of significant bridging phenomena between the nanofibers and the composite laminates.

This study suggests a finite element model for the dynamic behaviour of the nano-modified composite laminates. The results of this model can be verified from the corresponding experimental tests with freely vibrating composite specimens. The modelling was also done for the purposes of understanding the vibratory behaviour of nano-modified GFRP laminates as compared to the behaviour of virgin ones. The findings from the modelling regarding the stiffness and the damping are being checked with other experimental methods.

The delamination resistance of the virgin and the nano-modified laminates was tested numerically as well as experimentally using the ASTM D 2344 standard. The goal of this test is to determine the inter-laminar strength which was done according to equation (1). Both the experiments and the FE modelling confirmed the increase of the inter-laminar strength as a result of the modification with nylon nano-fibres. The increase obtained numerically was 20% which agrees well with the 27% increase from the experimental study. Thus one can conclude that nylon nano-fibres can be used for improving the delamination resistance of GFRP laminates.

For the future, the authors plan to continue the investigation of interleaving with nylon nanofibers for the purposes of delamination prevention and restriction. In this line an investigation of the effect of nylon nanofibers on the impact resistance of the composite laminates to damage. This will be done experimentally and numerically while in the same time the FEM investigation will be done to bring more understanding about the behaviour of nylon nano-modified GFRP laminates.

This study contributes to the knowledge about the effect of nylon nanofibers in GFRP composite laminates. It gives more insight to the dynamic and vibratory behaviour of such composites. And it draws important conclusions regarding the influence of nylon nano-fibrous interleaving on some mechanical and dynamic macro-properties of GFRP laminates.

References

1. Y.A. Dzenis, D.H. Reneker, Patent, *Delamination resistant composites prepared by small fiber reinforcement at ply interfaces,* University of Nebraska Lincolm (1999)
2. L. Daelemans, S. Heijden, I. Baere, H. Rahier, W. Paepegem, K. Clerck, 17th European Conference on Composite materials, *Increasing the damage resistance of composites by interleaving them with electrospun nanofibrous veils*, 1-8 (2016)
3. A. Zucchelli, M.L. Foccarete, C. Gualandi, C. Ramakrishna, J Polym Adv Technol, *Electrospun nanofibers for enhancing structural performance of composite materials*, **22**, 339-349 (2011)
4. H. Saghafi, S.R.Ghaffarian, T.M. Brugo, G. Minak, A. Zucchelli, H.A. Saghafi, J Compos Part B, *The effect of nanofibrous membrane thickness on fracture behaviour of modified composite laminates-A numerical and experimental study*, **101**, 116-123 (2016)
5. C. Garcia, J. Wilson, I. Trendafilova, L. Yang, J Compos Struct, *Vibratory behaviour of glass fibre reinforced polymer (GFRP) interleaved with nylon nanofibers,* **176**, 923-932 (2017)
6. R. Sultan, S. Guirguis, M. Younes, E. El-Soaly, Int J Mech Eng, *Delamination identification on composite material by free vibration test*, **1**, 227-236 (2012)
7. C.A. Berg, J. Tirosh, M. Israeli, Conference of Composite Materials: Testing and Design, *Analysis of Short Beam Bending of Fiber Reinforced Composites*, ASTM, 206-218 (1972)
8. E. Barkanov, W. Hufenbach, L. Kroll, J Comput Method Appl M, *Transient response analysis of systems with different damping models*, **192**, 33-46 (2003)
9. H. Mevada, P. Dipal, 12th International conference on vibration problems (ICOVP), *Experimental determination of structural damping of different materials* (2015)
10. K. Alnefaie, J Compos Struct, *Finite element modelling of composite plates with internal delamination*, **90**, 21-27 (2009)
11. H. Saghafi, PhD thesis, *Mechanical behaviour of flat curved laminates interleaved by electrospun nanofibers*, University of Bologne (2013)
12. P. Zhu, Z.X. Lei, K.M. Liew, J Compos Struc, *Static and free vibration analyses of carbon nanotube-reinforced composite plates using finite element method with first order shear deformation plate theory*, **94**, 1450-1460 (2012)

13. ASTM Standard, D 2344/D 2344 M, *Standard test method for short-beam strength of polymer matrix composite materials and their laminates* (2000)
14. R. Palazzetti, I. Trendafilova, A. Zucchelli, J Compos Struc, *The self-reinforcing effect of nylon 6,6 nano-fibres on CFRP composite laminates subjected to low velocity impact*, **106**, 661-671 (2013)

A new methodology for fault detection in rolling element bearings using singular spectrum analysis

Hussein Al Bugharbee[1], and *Irina Trendafilova*[2]

[1] Department of Mechanical engineering, the University of Wasit, Iraq
[2] Department of Mechanical and Aerospace Engineering, the University of Strathclyde, Glasgow G1 1XJ

Abstract. *This paper proposes a vibration-based methodology for fault detection in rolling element bearings, which is based on pure data analysis via singular spectrum method. The method suggests building a baseline space from feature vectors made of the signals measured in the healthy/baseline bearing condition. The feature vectors are made using the Euclidean norms of the first three PC's found for the signals measured. Then, the lagged version of any new signal corresponding to a new (possibly faulty) condition is projected onto this baseline feature space in order to assess its similarity to the baseline condition. The category of a new signal vector is determined based on the Mahalanobis distance (MD) of its feature vector to the baseline space. A validation of the methodology is suggested based on the results from an experimental test rig. The results obtained confirm the effective performance of the suggested methodology. It is made of simple steps and is easy to apply with a perspective to make it automatic and suitable for commercial applications.*

1 Introduction

Rolling element bearings (REB) constitute a major part of most rotating machinery. No matter how well and precisely the bearings are designed, how normally loaded, correctly assembled and properly lubricated they are, they fail due to material fatigue after a certain number of revolutions or operational time. The defects in REBs are considered as one of the main reasons for rotating machinery failures. The usual monitoring philosophy for REB's is vibration-based monitoring and it is based on their measured vibration response signals. During recent decades, a number of studies suggested different approaches for purposes of REB fault detection. Several of these focus on the review and the comparison of some of the approaches for REB's vibration-based health monitoring.

Most fault detection methods based on bearing vibration signal analysis usually include the extraction of certain features representative of the bearing state, which are used as means of comparison against the healthy state. There are many features suggested using a number of methods that are based on analysing the vibration signal in the different domains (*i.e* time, frequency and time-frequency) [1, 2]. Several challenges accompany the feature selection and extraction which include the sensitivity of the features to the change of the bearing condition and their noise perceptiveness.

The methodology proposed in this study is based on singular spectrum analysis (SSA) and is simple in structure and easy to apply. SSA is a time series analysis method that is primarily used to uncover the trend and the periodic components of the original time series [3]. It has two main stages: decomposition and reconstruction. Only the decomposition stage is used in this study, whereas bearing vibration signals are decomposed into a new space made of a number of their principal components (PCs), which correspond to the directions of the biggest variability in the original data [4]. Thus most of the variance of the original data is preserved as the original signal is projected onto the space of some of its first PC's. Accordingly a certain amount of the information contained in the original signal is preserved in each PC in terms of its variance. In this study the norms of the first several PC's are used to build a reference space corresponding to the healthy bearing condition. Thus, the reference space preserves a certain amount of the information for the healthy state of the bearings depending on the number of PC's used to build it. From this perspective any new signal is projected onto the reference space in order to assess its similarity to the healthy/ baseline condition.

As the main goal of the current study targets only the detection of faults and, due to the limitation of the manuscript length, the capabilities of the method regarding location identification and fault severity estimation will not be discussed in details here. We'll merely mention that the same method can be used for purposes of further fault identification [5].

This paper offers a validation of the methodology using data measured on an experimental test rig, which was built specifically for the purpose at Strathclyde University.

* Corresponding author: hrazzaq@uowasit.edu.iq, husseinmec@yahoo.com

The rest of the paper is organised into the following sections. The next section 2 presents the methodology for fault detection. Section 3 covers an experimental case study where the application of the developed methodology is demonstrated. Section 4 discusses the results obtained in the previous section 3. The last section 5 gives some general conclusions and directions for future research.

2. Methodology for fault detection in rolling element bearings.

The methodology suggested in this paper is based mainly on the decomposition stage of the singular spectrum analysis. It has two main steps: building a baseline/reference space and fault diagnosis. These two steps are detailed below.

2.1 Building the reference space

To build a reference space, a number L of discretized vibration signals measured on healthy bearing (i.e x_i=[$x(1)$, $x(2)$,...,$x(n)$]) are arranged into a matrix $\mathbf{X}_{ref}$ as shown in equation (1):

$$\mathbf{X}_{ref} = \begin{bmatrix} x_1(1) & x_1(2) & x_1(3) & . & . & x_1(n) \\ x_2(1) & x_2(2) & x_2(3) & . & . & x_2(n) \\ x_3(1) & x_3(2) & x_3(3) & . & . & x_3(n) \\ \vdots & \vdots & \vdots & \vdots & & \vdots \\ x_L(1) & x_L(2) & x_L(3) & . & . & x_L(n) \end{bmatrix} \qquad (1)$$

Where L is the window size of the signal decomposition. Then the covariance matrix of the matrix $\mathbf{X}_{ref}$is obtained as shown in equation (2).

$$\mathbf{C_X} = \frac{\mathbf{X}_{ref}X_{ref}^T}{L} \qquad (2)$$

The letter ''T'' denotes the transpose. Then the covariance matrix is subjected to singular value decomposition to extract its eigenvectors $\mathbf{U}_i$ and eigen values λ_i according to the equation (3).

$$\mathbf{C_X U}_i = \lambda_i \, \mathbf{U}_i \qquad (3)$$

Each eigenvalue λ_i represents the percentage of the variance of the original signal contained in the direction of $\mathbf{U}_i$. In this study the first three eigenvectors are used to build the reference space where all the lagged versions of others signals are projected as it will be described in the next section.

2.2. Fault assessment methodology

The fault assessment methodology has two main stages: feature extraction and a fault detection process. The fault detection process involves the setting of a threshold to distinguish between features corresponding to the healthy and the faulty states and the classification process.

2.2.1 Feature extraction

The signals measured on the healthy bearings are initially divided into a training and a testing sample. While all other signals obtained from the faulty bearing are used as a testing sample. The signals from the training sample are transformed into their lagged version (see equation (4)) and then multiplied (i.e projected) by the eigenvectors $\mathbf{U}_i$,obtained in equation (3). This is presented in equation (5).

$$\mathbf{X} = \begin{bmatrix} x(1) & x(2) & x(3) & . & . & x(K) \\ x(2) & x(3) & x(4) & . & . & x(K+1) \\ x(3) & x(4) & x(5) & . & . & x(K+2) \\ \vdots & \vdots & \vdots & & \vdots & \vdots \\ x(L) & x(L+1) & x(L+2) & . & . & x(n) \end{bmatrix} \qquad (4)$$

where K=n-L+1 and L is the pre-selected window size of the signal decomposition.

Projecting the jth signal onto a number of eigen vectors results in a set of PC's, i=1,2,...., L as detailed in equation (5) below.

$$PC_{ij}(m) = \sum_{k=1}^{L} X^T(m+k-1) * U_i(m) \qquad (5)$$

where PC_{ij} is the ith projection corresponding to j^{th} observation, i=1,2,..,L, j=1,2, *[please complete,]*, and m=1,2,,...,K are the components of the PC. Then the Euclidean norms of the first Q PCs are calculated according to equation (6).

$$f_{ij} = \sum_{m=1}^{K} \left(PC_{ij}(m)\right)^2 \qquad (6)$$

Where $i = 1,2, \dots, Q$ and $j = 1,2,3 \dots k$. The notation Q refers to the number of PC's used in the analysis and in this case Q=3. The notation k refers to the number of the signals used in the analysis. The above features are arranged to form k feature vectors as shown in equation (7). Each FV has Q =3. components.

$$\mathrm{fv_j} = [\mathrm{f_{1j}} \;\; \mathrm{f_{2j}} \;\; \mathrm{f_{3j}}\,] \qquad (7)$$

Accordingly the feature vector space is defined. In the present study, only the first three features have been used in the analysis. Thus, the FVs dimension is three and it can be represented as a point into the three dimensional space.

2.2..2 Fault detection

The fault detection in this case is regarded as a classification problem where two classes are introduced – the class of signals from healthy bearings and the class of signals from faulty bearings. Thus each signal from the testing sample can be assigned to one of these classes. The classification process has two steps: setting a threshold and comparison to of the features calculated for a signal under test to the threshold.

For setting a threshold, a feature matrix ($\mathbf{F}_{baseline}$) is first made by arranging all the training FVs in rows. Then the Mahalonobis distance (D) of each of the training FVs to the $\mathbf{F}_{baseline}$ is calculated as in equation (8).

$$D_i = \sqrt{(\mathbf{fv}_i - \mathbf{E}_{baseline}).\mathbf{S}^{-1}.(\mathbf{fv}_i - \mathbf{E}_{baseline})^T} \quad (8)$$

Where $\mathbf{E}_{baseline}$ is the mean of the rows of $\mathbf{F}_{baseline}$ and $\mathbf{S}^{-1}$ is the inverse of the covariance matrix of $\mathbf{F}_{baseline}$ which can be calculated as per equations (2) and (3) applied for $\mathbf{F}_{baseline}$.
The threshold (Thr_{ref}) is eventually selected as the maximum value of the distances D_i calculated for all the training FV's (9)

$$\mathrm{Thr}_{baseline} = \max(D_i) \quad (9)$$

where $i=1,2,..N$ and N is the number of training FV's used.
Once threshold is set, the distance of the testing FVs to $\mathbf{F}_{baseline}$ is calculated as per equation (8). Then each new testing FV is assigned to one of the classes-healthy or faulty -by comparing it to $\mathrm{Thr}_{baseline}$ as detailed in equation (10).

$$\left.\begin{array}{ll} D_i > \mathrm{Thr}_{baseline} & \text{a } FV_i \text{ is assigned to F class} \\ D_i \le \mathrm{Thr}_{baseline} & \text{a } FV_i \text{ is assigned to H class} \end{array}\right\} \quad (10)$$

3. Methodology verification

3.1 Case study and experimental setup

The experimental setup for this case study, which is built at the Department of Mechanical and Aerospace Engineering at Strathclyde University consists of 1 HP shunt DC motor, bearing assembly and a mechanical loading system, and is presented in Figure 1.

Fig. 1. Bearing test rig at Strathclyde university.

The vibration acceleration data is collected from the test bearing housing (in this case a SKF deep groove 6308 ball bearing) is mounted inside. A pinion-toothed belt mechanism is used to transmit the torque from the motor to the bearing assembly. A fault diameter of 0.05 inch was introduced by an electrical discharge machine on the inner race, on a ball and on the outer race for different bearings. The bearing vibration acceleration datasets is collected at a sampling rate of 12 kHz for different speeds (*ie* 250 r/min, 750 r/min and 1250 r/min) using a magnetic-based accelerometer.

3.2 Data collection and signal analysis

The data used in the analysis is shown in Table 1,where (H) refers to a healthy bearing condition, IRF refers to a bearing with an inner race fault, BF refers to a bearing with a ball fault and ORF refers to a bearing with an outer race fault.

Table 1. Signals used in the verification

Case No	Motor speed	Signal category	No of signals
1	250	Healthy,IRF,BF,ORF	320 (4X80)
2	750	Healthy,IRF,BF,ORF	320 (4X80)
3	1250	Healthy,IRF,BF,ORF	320 (4X80)

The signals are segmented into 320 equal non-overlapping signals (*ie* 4 signal classes × 80 signals per class, each one containing 2048 data points). Forty of the healthy signals are used as training sample and the remaining 280 (*i.e* 40 healthy and 240 faulty) signals are used as a testing sample.

Figure 2 presents a signal measured on a healthy bearing at a speed of 1250 r/min. The x-axis represents the number of data points while the vertical y-axis is the acceleration of the signal vibration.

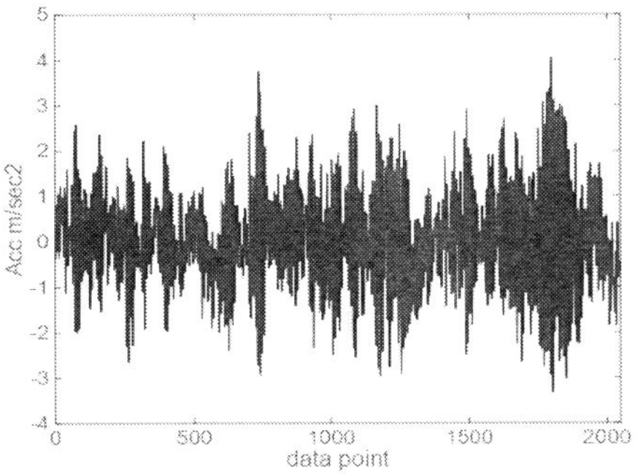

Fig. 2. An example of healthy bearing vibration signal

4. Results and discussion

As an illustration Figure 3 shows a 3D visualisation of the 40 FVs corresponding to the baseline-training sample, which are used to form the baseline features.

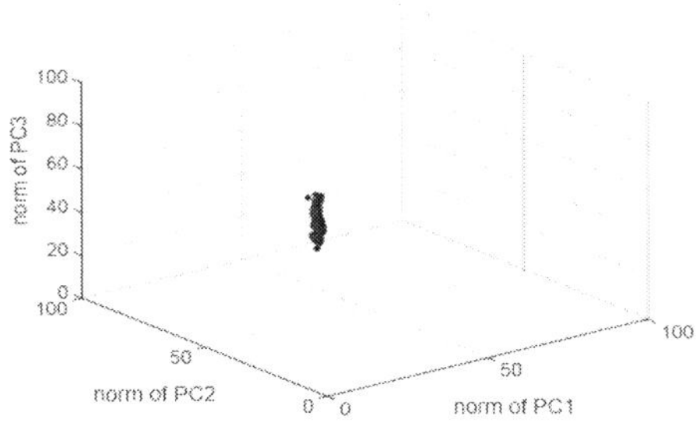

Fig. 3. 3D visualization of the training sample of signals measured on healthy bearings

In this study, only the first three PCs are used to form the feature vectors. The feature vectors are formed using the norms of these first three PC's.

The next Figure 4 shows a 3D visualization of the features corresponding to all the signals from the different bearing conditions.

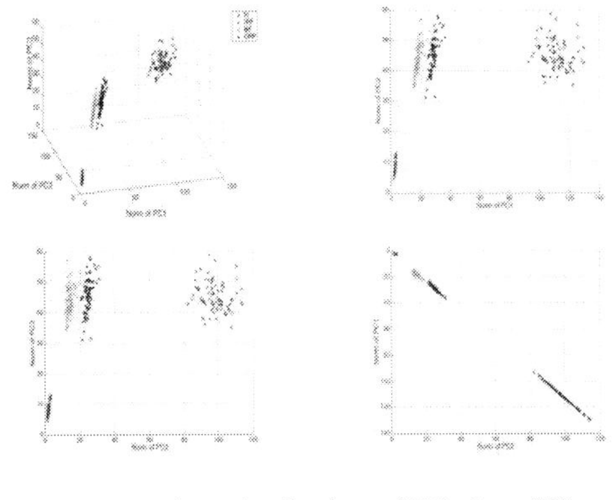

Fig. 4. 3D and 2D visualizations of FV's from different bearing conditions.

The blue dotes refer to healthy FVs, red dotes refer to IRF FVs, green dotes refer to BF FVs and the black dots refer to ORF FVs. It can be seen that the performed signal analysis has resulted in clustering the signals form the different fault categories and in terms of the resulting features the fault categories clearly distinguishable. From the figures on Figure 4 it can be seen that there is separability not only between the baseline and non-baseline FVs but also among the different fault categories.

The next Figure 5 shows the Mahalanobis distances (MDs) of both the training and the testing FVs measured to the baseline testing FVs category.

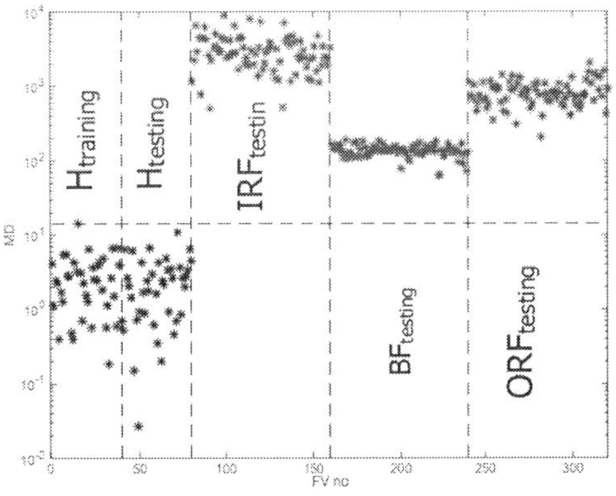

Fig. 5. MD's for the training and the testing samples corresponding to different fault states

As the MD is used as a fault indicator in this study, it is compared to the predetermined threshold constructed as detailed in equation (9), &2.2.2 .

Figure 5 is divided into 5 zones corresponding to the different bearing conditions and also separating the testing and the training MD's, using the vertical dashed lines. The $H_{training}$ Zone corresponds to training sample of healthy bearings signals (H), the $H_{testing}$ zone is made of the testing sample for healthy signals. The other three zones ($IRF_{testing}$, $BF_{testing}$ & $ORF_{testing}$) represent the testing FVs corresponding to IRF, BF and ORF classes respectively. The horizontal dashed line represents the value of the threshold calculated. The separation between the baseline and non-baseline FVs is very clear and it results in very accurate classification and fault detection. As can be seen from figure 5 all the testing feature vectors (100%) are correctly classified.

The performance of the methodology using the datasets of this case study is shown in Table 2. The first column shows the type of dataset as corresponding to the revolutions of the motor at which the data is collected. The second, the fourth and the fifth columns show the correct classification rate of the training and testing FVs for the healthy and the faulty signals, respectively.

Table 2. Percentages of correct classification rates of signals used in the analysis

Speed (rev/min)	% correct classification		
	H training	H testing	F testing
250	100%	100%	100%
750	100%	100%	100%
1250	100%	100%	100%

From Table 2, it can be seen the all the 320 FVs are correctly classified to their correct real classes.

5. Conclusions and future work.

The present study suggests a rather simple and easy to apply but accurate method for fault detection in rolling element bearings based on singular spectrum analysis. The suggested method is simple in application as it solely uses the signals from the healthy bearing state to build the reference space. As such, it does not require any previous measurements corresponding to faulty/ anomalous conditions. A number of signals measured in the baseline/healthy condition can be used to build the reference space. SSA is used only for the purposes of building this reference space. The new signals, which are to be classified, are not subjected to SSA; their lagged versions are simply projected onto the first principal directions of the reference space. As such, the transformations applied to the measured signals are minimal and very simple. The classification rule, which is based on a threshold of the Mahalanobis distance, is also a robust and simple one.

As a result of this, the method holds considerable potential for automatization and practical implementation.

It should also be noted that the method is rather general and can be applied to any measured signals regardless of their stationarity. The 3D visualisation shows that not only baseline and non-baseline signal categories can be distinguished but in some cases the method can be used to separate different faulty categories. This remains a subject of further research and as such the interpretation of the PCs/the principal directions will be very helpful. Based on the results obtained, it has been shown for the case studies considered that SSA is capable of extracting essential information regarding the presence of faults in signals where different fault locations are detected at different motor speeds. As the primary goal of the current study is limited to the detection of faults, further investigations are targeted in terms of expanding the method for fault qualification and quantification purposes and also for online monitoring of the bearing condition.

References

1 A. Sharma, M. Amarnath, and P. Kankar, "Feature extraction and fault severity classification in ball bearings," *Journal of Vibration and Control,* vol. **22**, pp. 176-192, (2016).

2 B. Nayana and P. Geethanjali, "Analysis of Statistical Time Domain Features Effectiveness in Identification of Bearing Faults from Vibration Signal," *IEEE Sensors Journal,* (2017).

3 H. Hassani, "Singular spectrum analysis: methodology and comparison," (2007).

4 H. Al-Bugharbee and I. Trendafilova, "A fault diagnosis methodology for rolling element bearings based on advanced signal pretreatment and autoregressive modelling," *Journal of Sound and Vibration,* vol. **369**, pp. 246-265, (2016).

5 H. R. Al-Bugharbee, "Data-driven methodologies for bearing vibration analysis and vibration based fault diagnosis," *PhD thesis,* University of Strathclyde, 2016.

Study on Singular Spectrum Analysis as a data-driven technique for damage diagnosis. Comparison between time and frequency domain.

David Garcia[1,*], and *Irina* Trendafilova[1]

[1]Mechanical and Aerospace Engineering, University of Strathclyde, 75 Montrose Street - Glasgow G1 1XJ – United Kingdom

Abstract. Vibration-based Structural Health Monitoring methodologies have been developed in many different applications with the aim of damage diagnosis. Recently, purely data-driven methods have been gained popularity because these methods do not assume any linearity or model in their analysis. Data-driven methods use the measured vibration signals as data-input to extract features that can conclude obtain useful information for the damage diagnosis. In this work a methodology based on Singular Spectrum Analysis (SSA) technique is presented which decomposes the measured vibration responses in a certain number of principal components which reveal the rotational patterns at any frequency in the motion. One of the steps of the methodology is to create a reference state where the observations can be compared for damage assessment. The data used to create the reference state determines how the information is represented in the reference state and therefore how meaningful and informative are the feature vectors for damage assessment. This study presents of the effect of the data representation considered on the creation of the reference state when the data is introduced in the time or frequency domain. The results obtained are different depending on the signal representation and hence they should have different interpretation when the state is created based on vibratory signals represented in the time or frequency domain.

1 Introduction

Structural health Monitoring (SHM) is a strategy that Engineers and Scientifics have been developing to understand and monitor the performance of structures for their design life expectation. SHM is considered a multidisciplinary technology with the goal of identifying and objectively quantify any occurrence regarding to the structural integrity. For decades monitoring the vibration responses on a structure have been used as an indicator of its condition based on assumption of any change due to a damage results in a more or less pronounced change of the dynamic behaviour [1], this is known as vibration-based SHM (VSHM). Recently, purely data-driven methods have gained popularity because these methods do not assume any linearity or model in their analysis. Data-driven methods use the measured vibration signals as a data-input to subtract features that can conclude in useful information for damage diagnosis [2].

Over the last years, the philosophy of SHM has been defined in different manners. One of the most common one was introduced by Farrar et al. in [3]. The authors defined SHM procedure in four steps: (i) operation evaluation, (ii) data acquisition, (iii) feature extraction and (iv) decision making and classification. Each of these steps can devote a separate research to find useful and reliable techniques for developing successful SHM methods and address some of the fundamental axioms presented in [4].

The data-driven VSHM methodology based on singular spectrum analysis (SSA) presented by the authors in [5] is divided in four steps; data collection, creation of the reference state, feature extraction and damage assessment. The data used to create the reference state determines how much information is contained in the reference state and therefore how meaningful and informative are the feature vectors for damage assessment.

This manuscript presents the effects of the data representation (time or frequency domain) on the creation of the reference state. Although the idea is similar and the creation of the reference state follows the same steps for both cases of the data representation, the results obtained are different and therefore a different interpretation should be applied when the reference state is created based on vibration signals represented in the

* Corresponding author: david.garcia@strath.ac.uk

time or frequency domain. In order to compare the effects on the creation of the reference state, the same vibration responses were considered in both analysis. The results were different for each case and hence they should have different interpretation.

2 Damage assessment methodology – a short introduction.

The methodology is based on singular spectrum analysis (SSA) [6] a technique which is a nonparametric method for data compression and information extraction. The technique finds linear combinations of the initial variables that describe and smooth the major trends and oscillations patterns in the vibration responses measured on the structure/system in consideration.

The steps considered in the methodology presented from the authors in [5] are data collection, creation of the reference state, feature extraction and damage assessment. The aim of this study is to find the differences when the reference state is constructed in the time or frequency domain. Therefore this will be the only step to be analysed in this manuscript.

Creation of the reference state: The main purpose of the methodology is to detect the differences between healthy and damaged structures. The reference state is generally constructed by signal vectors measured in the healthy/reference structure. This is the state where the newly observed signals, measured on the structure, will be compared to the reference state for making the decision whether the structure is healthy or damaged.

However the aim of this study is not to identify if a structure is healthy or damaged but how informative the reference state will be when the vector signals used to build the reference state, are considered/represented in the time or in the frequency domain. For this reason the reference state will be created first by signal vectors in the time domain and secondly by signals in the frequency domain. The steps will be the same in both exercises but the length of the vectors will be different such as N in time domain and N/2 frequency domain. The steps considered in the construction of the reference state are defined below.

(i) Data collection: The vibration responses measured on the structure to be monitored are discretised into a vector. More than one measurement is considered having a total M vector signal realisations.

(ii) Representation of the signal vector: If the reference state wants to be computed in the time domain, the signal vectors are made of the discretised signal vectors of the measured signal with length N and ordered in columns within the data matrix $\mathbf{X}$ as shown in Equation 1. However, when the signal vectors used to construct the reference state are in the frequency domain representation, the signal vectors with length N/2 contain the amplitudes of the frequency spectra of measured vibration signals and ordered in columns within the matrix $\mathbf{Y}$ and shown in Equation 2.

$$\mathbf{X} = (\mathbf{x}_1\, \mathbf{x}_2\, \dots\, \mathbf{x}_m \dots\, \mathbf{x}_M) \qquad (1)$$

$$\mathbf{Y} = (\mathbf{y}_1\, \mathbf{y}_2\, \dots \mathbf{y}_m \dots\, \mathbf{y}_M) \qquad (2)$$

(iii) Whether the reference estate is to be constructed in the time or in the frequency domain, this step will be applied in Equation 1 or Equation 2 respectively. This step is called embedding process. Each signal vector realisation is embedded into a matrix $\widetilde{\mathbf{X}}_m$ or $\widetilde{\mathbf{Y}}_m$ by W-lagged copies of itself where W is the sliding window size and m=1…M is the number of signal realisations. As the embedding process is applied to each signal vector realisation $\mathbf{x}_m$ or $\mathbf{y}_m$, the full embedded matrix will have a dimension $\mathrm{N} \times (\mathrm{M} \times \mathrm{W})$ or $\mathrm{N}/2 \times (\mathrm{M} \times \mathrm{W})$ whether the signal vectors are in the time (see Equation 3) or in the frequency domain (see Equation 4).

$$\widetilde{\mathbf{X}} = (\widetilde{\mathbf{X}}_1\, \widetilde{\mathbf{X}}_2\, \dots \widetilde{\mathbf{X}}_m \dots\, \widetilde{\mathbf{X}}_M) \qquad (3)$$

$$\widetilde{\mathbf{Y}} = (\widetilde{\mathbf{Y}}_1\, \widetilde{\mathbf{Y}}_2\, \dots \widetilde{\mathbf{Y}}_m \dots\, \widetilde{\mathbf{Y}}_M) \qquad (4)$$

(iv) The eigen-decomposition of the covariance matrix of the full embedded matrix is computed to obtain the eigenvalues and their corresponding eigenvectors ordered in decreasing order. The original signal vectors are then projected onto the eigenvectors to obtain the Principal Components (PCs).

(v) The last step is to define the Reference state. Finally, the Reconstructed Components (RCs) are obtained by convolving the Principal components with their associated eigenvector. Therefore, the reference state is constructed and represented by the matrix $\mathbf{R}$ that contains all Reconstructed Component in columns as shown in Equation 5.

$$\mathbf{R} = (\mathbf{R}_1\, \mathbf{R}_2\, \dots\, \mathbf{R}_m \dots\, \mathbf{R}_M) \qquad (5)$$

Therefore, the matrix **R** defines the reference state of the structure/system where the new observation signal vectors will be compared. Full details about the methodology are given in [5]. This manuscript only gives a short summary of the creation of the reference state step.

3 Creation of the reference state in the time and in the frequency domain.

In order to compare the effects on the creation of the reference state based on the time and frequency domain the same vibratory responses were considered in both analysis. The vibration signals utilised as input-data are represented in Fig. 1. Fig. 1(a) represents a free-decay vibration responses and Fig. 1(b) is its frequency spectrum representation.

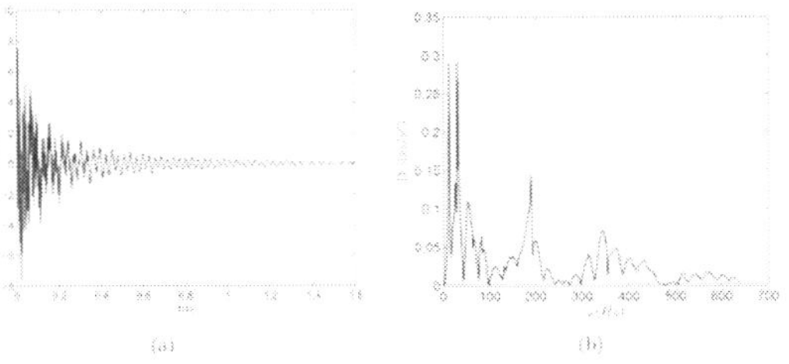

Fig. 1. Vibration signal. a) Free-decay acceleration signal and b) Frequency spectrum of the free-decay acceleration signal.

3.1. Reference state based on vibration responses represented in the time domain

This section studies the effect on the creation of the reference state based on the vibratory responses in the time domain representation. Seven signal realisations were considered (M = 7) for this analysis. Each signal realisation was discretised into a vector $\mathbf{x}_m$ with a dimension of N = 2048 where the index m labels each realisation. The seven signal vectors were arranged in columns into the matrix **X** to define the data set considered for the construction of the reference state. The data set **X** is then processed by the steps described in the section 2. Two sliding window sizes were considered in the creation of the embedding matrix. One case uses a W=7 and other case W=50 to create the embedded matrix shown in Equation 3. The eigen-decomposition of the covariance matrix of the full embedded matrix $\tilde{\mathbf{X}}$ yields to a (M x W) total number of eigenvalues with their corresponding eigenvectors, which more specifically are 49 for the case of W = 7 and 350 for the case of W = 50. As explained in section 2, the eigenvalues and eigenvectors are ordered in decreasing order by means of their variance content in the vibratory signal. Each eigenvalue defines the percentage of variance contained in the direction of its corresponding eigenvector. The percent of the partial variance of each eigenvalue in the total decomposition is calculated by Equation 6 where λ_k are the eigenvalues and k is the eigenvalue index which varies from k=1,..., K being K the total number of eigenvalues equal to (M x W).

$$\%\,variance = 100 \times \frac{\lambda_k}{\sum_{k=1}^{K} \lambda_k} \qquad (6)$$

Fig. 2 (a) and 2(b) represent the partial variance percent contained in each eigenvalue for the decomposition by using W = 7 and W = 50, respectively. These graphs represent the percentage variance only for 20 eigenvalues since the information contained in the rest of the eigenvalues is insignificant. Comparing the two graphs can be observed that for the decomposition with W = 50 the variance information contained in the first eigenvalues is smaller than for the case of W = 7. This behaviour stands in the fact that the information is distributed along the eigenvectors. As observed, for the case of W = 50 the number of eigenvalues increases and hence the variance percent contained in the first eigenvalues is reduced and distributed over all the components. However, for the case of W = 7 it can be observed that the variance percentage contained in the first eigenvalue is considerably larger than the rest of the eigenvalues.

The decomposition of the covariance matrix of the data set in eigenvalues and eigenvectors is to distribute the oscillatory components contained in the original signal vectors by means of their variance content. In this case, when the value of W is small, the variance contained in the first eigenvalues is large because the eigenvector corresponding to the first eigenvalue contains more than one oscillatory component and hence more variance of the original signal vector. On the other hand, when W is relatively large, the variance in the first eigenvalue reduces and hence less oscillatory components are contained in its corresponding eigenvector. In this case, it can be stated that large values of W will give better separable components in comparison with small values of W [7].

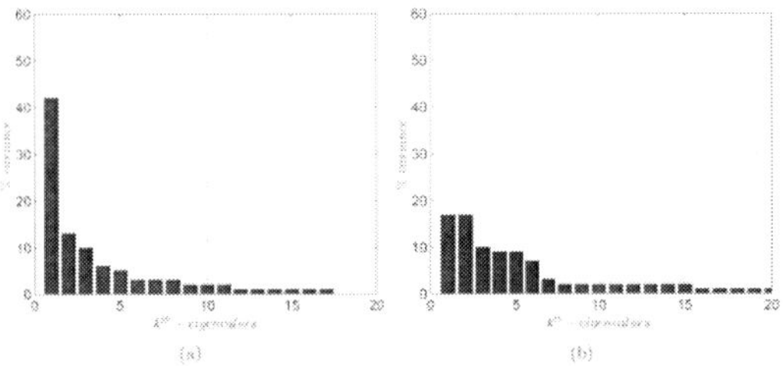

Fig. 2. Partial variance percent of each eigenvalue for a) W = 7 and b) W = 50 when the reference space is based in the time domain

Each RC is obtained based on the information contained in its corresponding PC weighted by its corresponding eigenvector. Each RC contains oscillatory components with the same content of variance in the vibratory response. The oscillatory components are less or more separated depending on the complexity of the vibratory system and also on the number of PCs obtained in the decomposition. In order to observe the oscillatory components contained in the RCs, the Fourier transform was applied to each RC separately of the matrix **R**. Fig. 3(a) and 3(b) represent the reconstructed signal using only the first two RCs (RC1 and RC2) for the case of W = 7 and W = 50, respectively. Figures show the comparison of the reconstructed signal with the original signal to identify what oscillatory components are present in the reconstruction. In Fig. 3(a) can be observed that the two first RCs contain the oscillatory components for low frequencies up to 200 Hz. It can be observed that the oscillatory components belonging to higher frequencies are not contained in these RCs. In a similar way, Fig. 3(b) shows the frequency spectrum of the reconstructed signal using the first two RCs for a W = 50. It can be observed that only the oscillatory components belonging to the low frequencies up to 35 Hz are contained in the reconstructed signal. Also it can be observed that for the decomposition of the vibratory signal by using W = 50, the oscillatory components contained in the first RCs are very well separated in comparison with the decomposition by W = 7 where more oscillatory components are contained in the reconstructed signal.

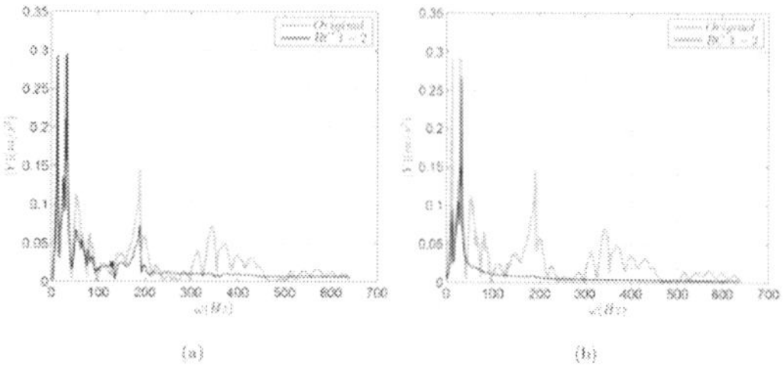

Fig. 3. Comparison between the original vibratory signal and the reconstructed signal using 2-RCs. The graphs show the effect of the methodology performed in the time domain. a) Comparison when W = 7 and b) W = 50

As explained above, when the decomposition is done by large values of W the oscillatory components are better separated. Therefore, the RCs have interpretable oscillatory components which can be used to identify particular modes. This can be beneficial when the interest is to extract particular modes of vibration.

3.2. Reference state based on vibration responses represented in the frequency domain

This section studies the effect on the creation of the reference state based on the vibratory responses in the frequency domain representation. Seven signal realisations were considered (M = 7) for this analysis like in the previous section 3.1. Each signal realisation was discretised into a vector $\mathbf{y}_m$ with a dimension of N/2 = 1024 where the index m labels each realisation. As explained above each signal vector represents the frequency domain representation of its corresponding $\mathbf{x}_m$ vector.

The data set $\mathbf{Y}$ is then processed by the steps described in the section 2. Two sliding window sizes were also considered in the creation of the embedded matrix $\tilde{\mathbf{Y}}$. One case uses a W = 7, the other case W = 50 to create the embedded matrix shown in Equation 4. The same sliding window sizes that in section 3.1 were considered in this analysis. The eigen-decomposition of the covariance matrix of the full embedded matrix $\tilde{\mathbf{Y}}$ yields to 49 eigenvalues and eigenvectors for the case of W = 7 and 350 for the case of W = 50. In this way, the number of RCs obtained in the signal vectors decomposition will be the same that in section 3.1. The partial variance percent of each eigenvalue over the total eigenvalues was calculated by Equation 6.

Fig 4(a) and 4(b) represent the partial variance percent contained in each eigenvalue for the decomposition by using W = 7 and W = 50, respectively. These graphs represent the percentage variance only for 20 eigenvalues since the information contained in the rest of the eigenvalues is insignificant. Comparing the two Figures, it can be observed that for the decomposition with W = 50, the variance information contained in the first eigenvalue is smaller than for W = 7. This behaviour stands in the fact that the information is distributed along the eigenvectors and when large values of W are considered, the number of eigenvectors increases, and hence the information is then distributed over all of them. As observed, for the case of W = 50 the number of eigenvalues increases with respect to the case of W = 7 and hence the variance percent contained in the first eigenvalues is reduced and distributed over all the rest of the eigenvalues. However, the variance percent contained in the first eigenvalue is the largest in both cases (W = 7 and W = 50) with a significant difference in comparison with the other eigenvalues. Therefore, the first eigenvector contains the majority of the variance and it will have much more global contribution in the reconstruction of the signal vector. However, the rest of the eigenvectors have less variance content and their contribution in the reconstruction will be more local.

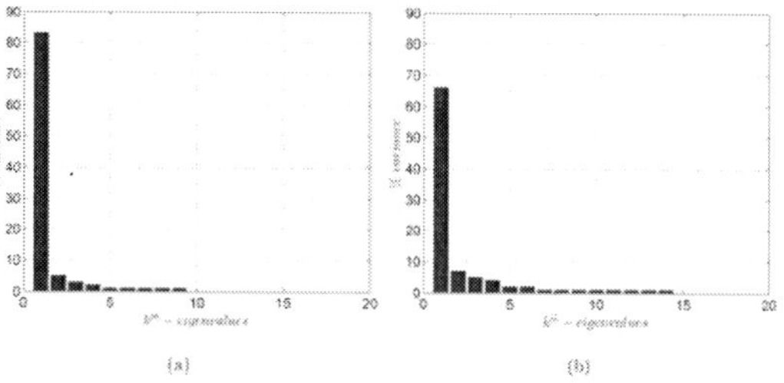

Fig. 4. Partial variance percent of each eigenvalue for a) W = 7 and b) W = 50 when the reference space is based in the frequency domain

Fig. 5(a) and 5(b) represent the reconstruction of the original signal vector by using only the first two RCs (RC1 and RC2) for the case of W = 7 and W = 50, respectively. In both graphs it can be observed that the reconstructed signal describes approximately the general trend of the original frequency spectrum. The estimated reconstructed signal contains information along the entire spectrum but it is more significant in the frequencies where the higher peaks are presented, in other words, at frequencies where the maximum of energy contained in the spectrum is concentrated. For the case of W = 7 (see Fig. 5(a)), the reconstructed signal is very good approximated and it contains well depicted peaks at low frequencies. However, for the case of W = 50 the reconstructed signal is smoother than in Figure 5(a), although the reconstruction still has a contribution to the entire spectrum (see Figure 5(b)). As mentioned above, the first eigenvectors contain the majority of the variability in terms of variance content and this is clearly represented in the reconstructed spectrum as observed in

Figure 5(a) and 5(b) where a smooth trend is achieved by using only two RCs in the reconstruction. The selection of the sliding window size W is very important in the dimension of the decomposition and it depends on the goal of the study. Depending on what analysis is going to be addressed, the selection of W will help to obtain more understandable information [8].

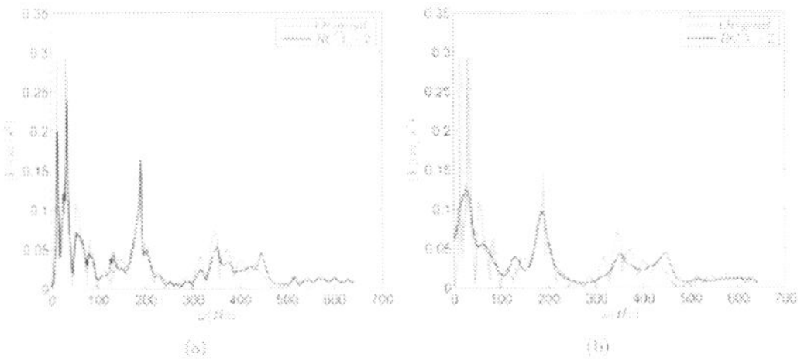

Fig. 5. Comparison between the original vibratory signal and the reconstructed signal using 2-RCs. The graphs show the effect of the methodology performed in the frequency domain. a) Comparison when W = 7, b) W = 50

4 Discussions

The analysis in sections 3.1 and 3.2 was to explain the effects when the reference state is constructed in the time or frequency domain. The comparison in the eigen-decomposition and in the reconstruction was analysed. For both cases, it was observed that the first eigenvalues contain the majority of the variance and it decreases for the rest of the eigenvalues. In order to compare the eigen-decomposition for both, time and frequency domain, the eigenvalue spectra was calculated where the normalised eigenvalue and the normalised index are defined in Equation 7 respectively

$$\log\left(\frac{\lambda_k}{\sum_k^K \lambda_k}\right), k = \frac{k}{K} \qquad (7)$$

where λ_k is the k-th eigenvalue. The results are represented in Fig. 6(a) and 6(b) when the decomposition, based on the time and frequency domain, is compared for the case of W = 7 and W = 50, respectively. For both sliding window size cases, it is observed that the first eigenvalues are smaller for the time domain case in comparison with the frequency domain case. The magnitude of eigenvalues decreases for both cases even though in the time domain case, the eigenvalues are larger than in frequency domain. This can explain that for the time domain case the difference in variance content between the first and the rest of the eigenvalues is less than for the frequency domain case. It is observed that the variance content is constantly distributed as shown in Fig. 6(a) for the time domain case, while a big jump can be observed between the first and the second eigenvalue for the case of the frequency domain. After this jump, the magnitude of the eigenvalues goes constantly down. In Fig. 6(b), it is observed that groups of eigenvalues are created in the time domain decomposition which means that they have similar amount of variance content. This proves that large values of W leads to more separable oscillatory components in the time domain decomposition. For the case of the frequency domain, the separation is clearly achieved between the first and the rest of the eigenvalues. In this case the effect on the selection of W will change the variance content in the first eigenvalue.

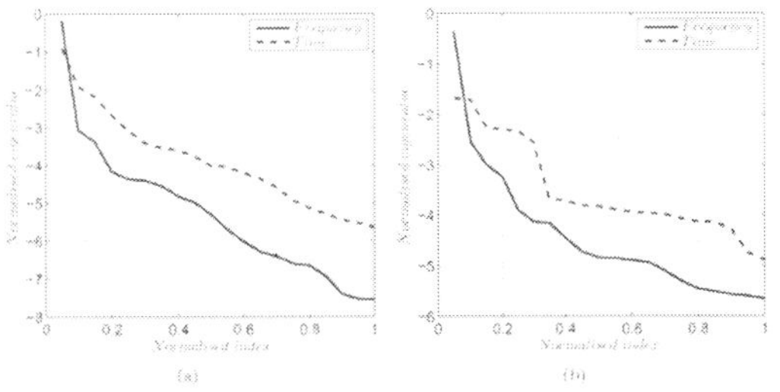

Fig. 6. Comparison of the eigenvalue spectra between time and frequency domain decomposition for a) W = 7 and b) W = 50

This means that when the frequency domain is used, the first eigenvector will contain the general information in the spectrum which leads to a smooth trend in the reconstruction as shown in Fig. 5. Then, all frequency modes captured in the original frequency domain are also included in the reconstructed spectrum. The rest of the eigenvectors will add the fluctuations on the spectral line described by the first RC. It is true that the reconstructed spectrum does not have the same amplitude as the original signal vector but it still conserves the general weight of distribution in the entire spectrum.

In the case of the time domain, the behaviour observed in the reconstruction has a different interpretation than in the case of the frequency domain. The decomposition in the time domain leads to a number of oscillatory components which can be interpreted separately. As shown in Fig. 3 only a few frequency bands are included in the reconstruction signal. The separability of these oscillatory components depends on the value of sliding window size. As a general rule, large values of W will give more separated oscillatory components. The benefit of obtaining separated oscillatory components is to identify the particular modes which can contain relevant information.

With the comparison of the creation of the reference state based on time or frequency domain developed in the above sections, it can be concluded that the selection of one of these analysis depends on the case to study. When a time domain representation is used in the methodology, separated oscillatory components are obtained. The information contained in the reconstructed signal will depend on which oscillatory components are considered in the reconstruction. The separability of

these oscillatory components will depend on the complexity of the vibratory system and also on the dimension of decomposition defined by W. This analysis can be beneficial when a predominant mode of vibration is to be analysed and therefore the methodology can be useful for mode extraction or identification. However, when the frequency domain representation is used, it can be observed that a smoother version of the original spectrum is obtained by the reconstructed spectrum. The reconstructed spectrum is more approximated to the original one, when more RCs are considered in the reconstruction. This analysis can be beneficial when more than one vibration mode is considered in the analysis.
As the reconstruction describes the general trend of the spectral line, all rotational patterns are considered in the reconstructed spectrum with small number of RCs.

5 Conclusions

In this work is studied a stage of a methodology based on Singular Spectrum Analysis which decomposes the vibration responses in a certain number of principal components having in consideration all rotational patterns at any frequency. One of the steps of the methodology is to create a reference state where the observations can be compared for damage assessment. The data used to create the reference state determines how the information is represented in the reference state and therefore how meaningful and informative are the feature vectors for damage assessment. The study presents the effects on the creation of the reference state when the data is introduced in the time or frequency domain and the main conclusions are summarised below.

When the reference state is constructed in the time domain representation, the reference state provides separated and interpretable oscillatory components distributed in the reconstructed components. If the aim is to have information about all the oscillatory components presented in the vibratory response more reconstructive components should be retained in order to reconstruct an approximated original vibration response. As the reconstructed components are separated, the selection of particular oscillatory components can depend on the application.
On the other hand, when the reference state is created in the frequency domain representation, the first reconstructed component describes the general trend of the original spectral line of the frequency spectrum. The rest of the reconstructed components represent the fluctuations along the spectral line. Therefore, the reconstructed spectrum with only the first RC gives more weight to the frequencies bands where the maximum energy is concentrated by means of the amplitudes/peaks at different frequencies. All the rotational patterns are considered with only the first reconstructive component.

References

1. C. P. Fritzen, Key Eng. Mat. **293**, 3-20 (2005)
2. G. Gui, H. Pan, Z. Lin, Y. Li, and Z. Yuan, KSCE J. Civil Eng. **21**, 523-534 (2017)
3. C. R. Farrar, k. Worden, Phil. Trans. R. Soc. A **365**, 303–315 (2007)
4. K. Worden, C.R Farrar, G. Manson and G. Park, Phil. Trans. R. Soc. A **463**, 1639–1664 (2007).
5. D. Garcia, and I. Trendafilova, J. Sound and Vib. **333**, 7036-7050 (2014).
6. N. Golyandina, and A. Zhigljavsky, *Singular Spectrum Analysis for Time Series* (Springer briefs in statistics, 2013)
7. M. C. R. Leles, A. S. V. Cardoso, M. G. Moreira, H. N. Guimares, C. M. Silva, and A. Pitsillides, *IEEE International Symposium on Signal Processing and Information Technology* (2016).
8. J. B. Rafert, J. Zabalza, S. Marshall, and J. Ren, App. Spectrosc. **70** 1528-1588 (2016).

Use of damping identification technique for damage detection

Vikas Arora*

Department of Technology and Innovation, University of Southern Denmark, 5230 Odense, Denmark

Abstract. Stiffness-based structural health monitoring methods are widely used for detecting the damage in a structure. These stiffness-based structural health monitoring methods uses change in natural frequencies and modeshapes for damage detection. These methods are based on identifying the change in stiffness of the healthy and damage structure to predict the damage in the structure. These stiffness-based methods are not efficient for detecting a small damage in a structure as there is a negligible change in natural frequencies and modeshapes due to a small damage in a structure, however the damping characteristics of the structure are highly sensitive to the damage in a structure. In this paper, new damping-based damage detection procedure has been proposed. In the proposed procedure, the changes in damping matrix of the structure has been used to detect the damage in the structure. The proposed procedure is able (or can) to detect both the location of the damage and the extend of the damage in the structure. The proposed procedure of damping-based damage detection is a 2-step procedure. In the first step, damping matrices of both the healthy and damage structure are identified and in the second step, the identified damping matrices are used for damage detection. Numerical and experimental case studies are presented to demonstrate the effectiveness of the proposed procedure. The results have shown that the proposed damping-based damage detection procedure can be used for detecting damage in a structure with confidence.

1 Introduction

Stiffness-based methods have been developed and applied for detecting the damage in the structures. The basic idea behind these methods is that modal properties (natural frequencies and modeshapes) of the structure depends upon the stiffness of the structure. Therefore, the variation in the stiffness of the structure cause changes in the natural frequencies and modeshapes of the structure. Damage in a structure may be due to cracks, loose bolts, broken welds, corrosion and fatigue. Cawley and Adams [1] and Kim et al. [2] used natural frequencies of the structure to detect the damage in the structure. These methods are very simple as natural frequencies can be measured with high accuracy. Pandey and Biswas [3] used natural frequencies and modeshapes for detecting the damage in the structure. This method predicts changes in flexibility, which is inverse of stiffness, of the structure to detect damage. These methods are non-parametric and direct. However, these methods can't be used for predicting the exact location, extend of the damage and multiple damages in the structure.

Finite element method (FEM) is the most widely used for predicting the dynamic characteristics of any structure. It is well known that finite element predictions will be erroneous due to inevitable difficulties in modelling of joints, boundary conditions and damping. The experimental data are generally considered to be more accurate. This has led to the development of model updating which aims at reducing the inaccuracies present in the analytical model in the light of measured dynamic test data. Many model updating methods have been proposed in recent years. Model updating methods can be broadly classified into direct methods, (These methods are essentially non-iterative) and the iterative methods. A significant number of methods, (Baruch and Berman and Nagy [4]; Baruch [5]), which were first to emerge belonged to the direct category. Although, these methods are computationally cheaper and reproduce the measured modal data exactly, they violate structural connectivity and updated structural matrices are difficult to interpret. On the other hand, by using iterative methods, structural connectivity can be easily maintained and corrections suggested in the selected parameters can be physically interpreted. These methods also provide a wide choice of parameters for updating. Iterative methods either use eigendata or frequency response functions (FRF). Collins et al. [6] used the eigendata sensitivity for analytical model updating in an iterative framework and Lin and Ewins [7] used measured FRF data to update an analytical model. The finite element method along with model updating (Jaishi and Ren [8]; Wu and Li [9]) has been used to detect and locate the damage elements in a structure. Most of the model updating methods are iterative and parametric undamped methods and uses changes in the stiffness for predicting the damage in the structure. These methods can predict multiple damages in the structure however,

* Corresponding author: viar@iti.sdu.dk

these methods can't be used to detect small damages in the structure due to low sensitivity of natural frequencies and the modeshapes to damage.

Some research has been carried out to detect the damage using damping instead of stiffness. Zhou and Yang [10] demonstrated that visually undetectable cracks have very little change in natural frequencies and modeshapes, while there is a significant change in damping characteristics of the structure. Kawiecki [11] studied the changes in modal damping, obtained using half-power bandwidth method, to detect damage in the structure and concluded that changes in damping characteristics of the structure can be a useful indicator of damage in the structure.

In this paper, new damping-based damage detection procedure has been proposed. In the proposed procedure, the changes in damping matrix due to damage are used for detecting the damage in a structure. The proposed procedure of damping-based damage detection is a 2-step procedure. The proposed damping-based damage detection procedure not only predict the location of the damage but also the extent of damage in the structure. In the first step, damping matrices of both the healthy and damage structure are identified using finite element model updating approach and in the second step, the identified damping matrices are used for damage detection. Numerical and experimental case studies are presented to demonstrate the effectiveness of the proposed procedure. The results have shown that the proposed damping-based damage detection procedure can be used for detecting damage in a structure with confidence.

2 Simulated numerical case study

In this section, a simulated numerical case study of damped lumped mass system is presented to demonstrate that the changes in damping is a better indicator of damage in a dynamic system in comparison to stiffness. 4 degree of freedom lumped mass system shown in the Fig. 1. In the first case, the system is considered undamped and stiffness of one spring (K_4) is varied to demonstrate the damage in the system. In the second case, spring stiffnesses are considered constant and damping (C_4) is varied to simulate damage in the system. The lumped masses M_1, M_2, M_3, M_4 have a constant value of 5 Kg each. In the first case, stiffness of springs K_1, K_2 and K_3 are 2×10^6 N/m each. For simulated healthy structure, stiffness of spring K_4 is 2×10^6 N/m considered. To simulate damage in the system, stiffness of spring 4 (K_4) is decreased by 10 to 50%. The natural frequencies are calculated for different level of damage in the system. The percentage change in natural frequencies at various modes due to percentage change in stiffness value of K_4 is shown in Fig. 2. It can be observed from Fig. 2 that there is maximum change of 15.53% in natural frequency with respect to 50% change in stiffness value of K_4 for second mode.

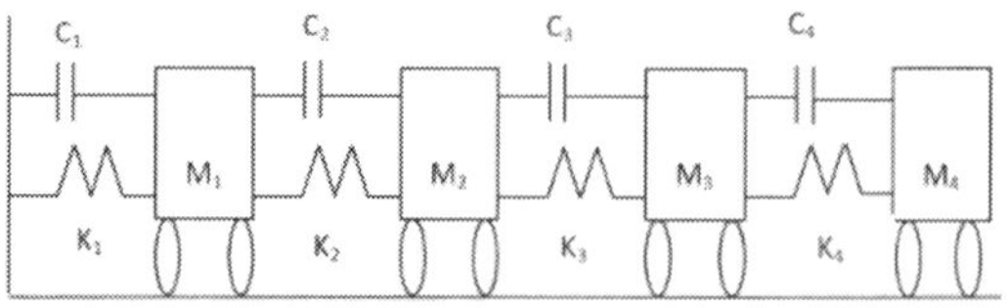

Fig. 1. Lumped mass system

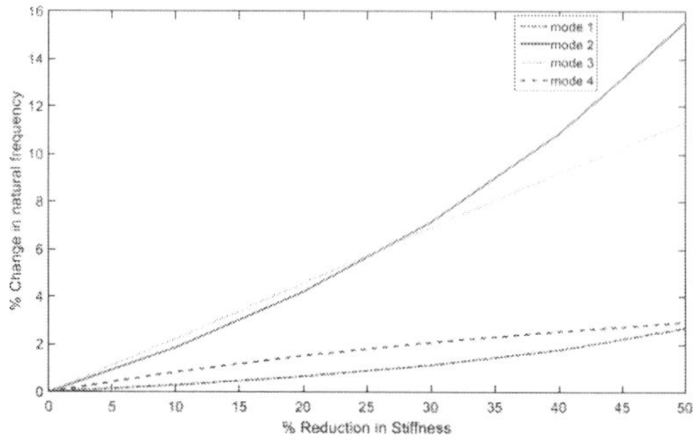

Fig. 2. % Change in natural frequencies with change in stiffness (K_4)

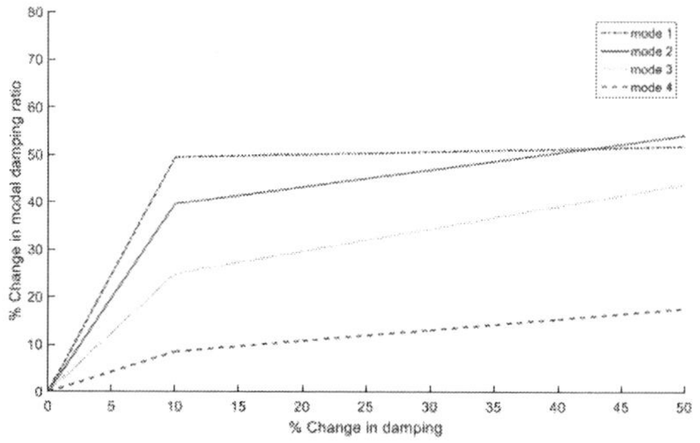

Fig. 3. % Change in modal damping values with change in damping

A similar study has been carried out in which the damage is represented by change in damping. In this case, lumped masses M_1, M_2, M_3 and M_4 have a constant value of 5 Kg each and stiffness of springs K_1, K_2, K_3 and K_4 have constant values of 2×10^6 N/m each. Viscous damping values for C_1 is 400 Ns/m, C_2 and C_3 are 300 Ns/m each. For simulated healthy structure, C_4 is 400 Ns/m. The damping value increase with the damage, therefore to simulate damage, the value of damping (C_4) has been increased by 10 to 50%. The change in modal damping values due to change in damping has been calculated. The modal damping is calculated using the formulae given below.

$$\zeta_i = \frac{-\alpha_i}{\sqrt{\alpha_i^2 + \beta_i^2}} \tag{1}$$

The desired damping matrix is also provided to create a better understanding of the process. Because this is a contrived example, the described result is known ahead of time. The percentage change in modal damping values due to change in damping values C_4 is shown in Fig. 3. It can be observed from the Fig. 3 that maximum change in second modal damping due to 50% change in damping value. Fig 4. shows change in natural frequency and modal damping for the first mode with respect to change in stiffness and damping values. It can be observed from the Fig. 4 that modal damping is very sensitive to change in damping as compare to the natural frequency with respect to change in stiffness.

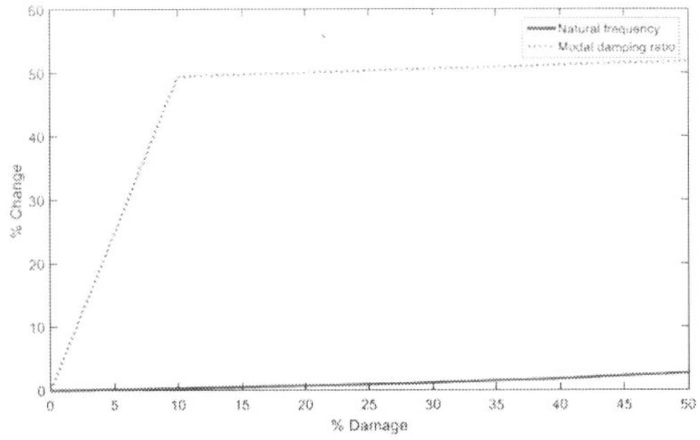

Fig. 4. Change in modal damping and natural frequencies for first mode with respect to damage.

It can be concluded from this simulated experimental study that modal damping is very sensitive with respect to damage and therefore can be used for detecting damage in the structure.

3 Theory

In this section, damping identification method [12] is presented. The damping identification method, this damping identification method requires prior knowledge of accurate stiffness and mass matrices. The identified damping matrix (C) is both symmetric and positive definite. This method is based on Lancaster's formulation [13] in which damping matrix is identified from the measured complex modal eigen data and damping matrix can be computed from the formula given as:

$$C = -M(\Phi\Lambda^2\Phi^T + \overline{\Phi\Lambda^2\Phi^T})M \qquad (2)$$

where the overall bar represents complex conjugate. Λ is the diagonal matrix of complex eigenvalues λ_i and Φ are complex eigenvectors. This formulation requires normalized complex eigenvectors (Φ). The complex eigenvectors are normalized by direct method given by Arora et al. [12] as:

$$\Phi_i^T(M\lambda_i^2 - K)\Phi_i = \lambda_i \qquad (3)$$

where M and K are accurate mass and stiffness matrices. The accurate mass and stiffness matrices can be obtained by undamped finite model updating method [14].

4 Experimental case study

An experimental study on an aluminium cantilever beam is also conducted to evaluate the effectiveness of the proposed method. The dimensions of the beam are 600×50×20 mm as shown in Fig. 5. To evaluate the proposed damping based damage detection method, a small damage has been induced in the cantilever structure as shown in Fig 6. The structure is excited by an impact hammer and responses are measured by accelerometer. The structure is excited at 5 different locations, so 5 frequency response functions (FRFs) are acquired both for healthy and damaged structures.

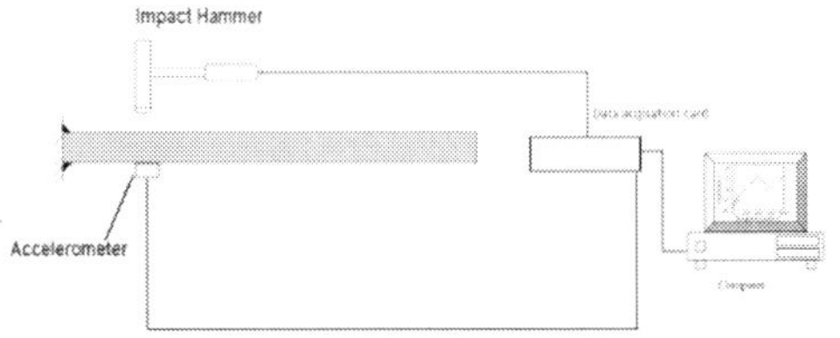

Fig. 5. Experimental setup for modal testing of cantilever beam.

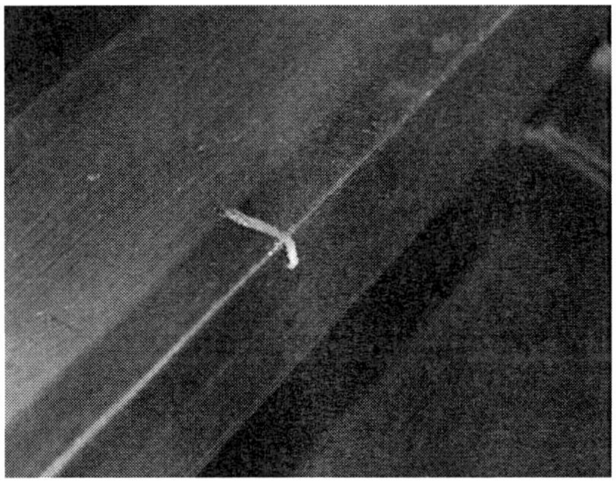

Fig. 6. Pictorial view of induced damage in the cantilever beam

The FRFs of the healthy and damaged cantilever beam are plotted in Fig.7. It can be observed from the Fig. 7 that there is no change in natural frequencies due to damage in the cantilever beam. It can be concluded from the healthy and damaged FRFs that existing stiffness-based damage detection methods cannot be applied to detect small damage. The proposed damping-based damage detection method has been applied to detect both location and extend of the damage in the cantilever beam structure.

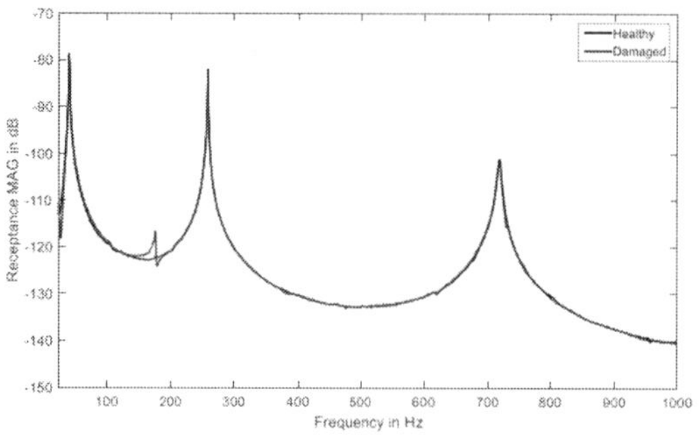

Fig. 7. Overlay of FRFs obtained from healthy and damaged cantilever beam structure.

The beam is modelled using five, two-dimensional frame elements (Two translational degrees of freedom in x and y direction and one rotational degree of freedom) and the fixed end is modelled by taking coincident nodes. Thus, two nodes that are geometrically coincident are taken as fixed end instead of one node. A horizontal, a vertical and a torsional spring couples two nodes at each of such coincident pair of nodes and the stiffness of these springs is K_x, K_y and K_t respectively. An overlay of the measured FRF from healthy cantilever beam and the corresponding undamped FE model FRF is shown in Fig. 8. It can be observed from Fig. 8 that the FRF predicted by FE model doesn't match with the corresponding experimental FRF, which is due error in the stiffness matrix of the FE model. For better estimation of damping, the stiffnesses at the end of the cantilever beam are updated and subsequently damping is identified using mass and updated (accurate) stiffness matrices.

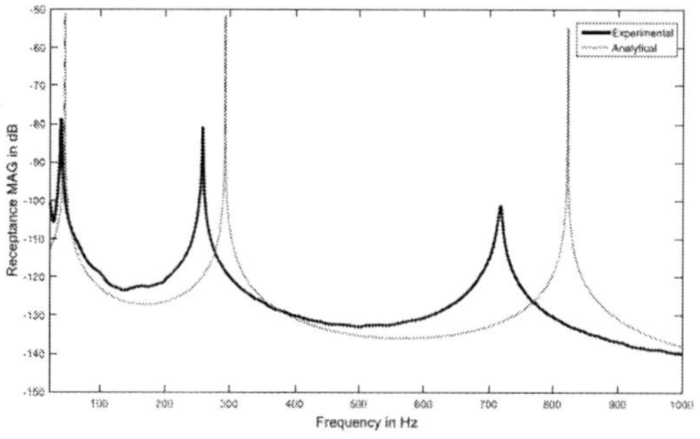

Fig. 8. Overlay of experimental FRF of the healthy structure and analytical FRF.

Undamped FE model is updated using resonance and anti-resonance frequencies [14]. Choice of updating parameters based on engineering judgment about the possible locations of modelling errors in a structure is one of the strategies to ensure that only physical meaningful corrections are made. In case of cantilever beam structure, modelling of stiffness at the end is expected to be the dominant source of inaccuracy in the FE model. The three spring stiffnesses at the end are chosen as updating variables. Natural frequencies and anti-resonance frequencies in the range of 0-1000 Hz are selected for updating the finite element model. The initial and final values of three springs at the end are given in the Table 1. It is observed that the values of stiffness of all the springs at the end are reduced. Fig. 8 shows the overlay of measured healthy and undamped updated FRF. It can be observed from Fig. 9 that the shape of the updated FRFs is same as that of measured FRFs. Since the undamped updating of FE model is performed, experimental FRF and updated FRF do not match near resonance and anti-resonance frequency points as the effect of damping is maximum near the resonance and anti-resonance regions.

Table 1. Stiffness values of springs before and after updating.

Updating variable	Initial Value	Updated Value
K_x (N/m)	3.28×10^6	$1.41\ \times10^5$
K_y (N/m)	3.28×10^6	1.53×10^5
K_t (N m/rad)	3.28×10^6	1.89×10^5

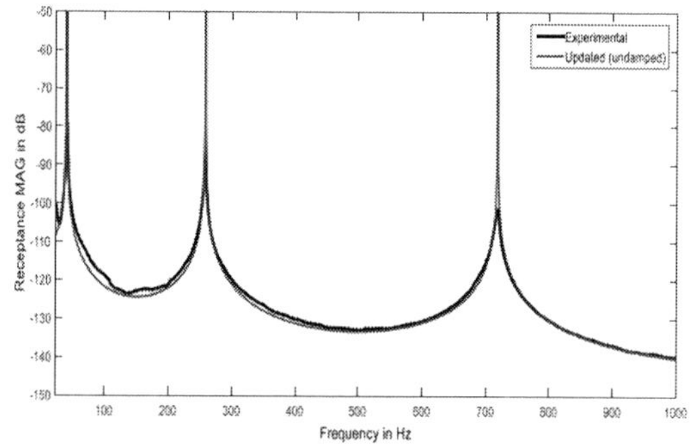

Fig. 9. Overlay of experimental FRF of the healthy structure and updated FRF

The updated FE model obtained in the previous step is used to identify the viscous damping matrix of the cantilever beam structure. The size of complex eigendata matrices is [5×5] whereas, the size of mass and stiffness matrices obtained in previous step is [18×18]. To apply the damping identification method, the size of mass and stiffness matrices are reduced to [5×5] according to the measured degrees of freedom using iterated IRS method [15] given by Friswell et al. Damping of the healthy cantilever beam is identified using damping identification method using accurate mass and stiffness matrices. Fig. 10 shows the overlay of experimental and corresponding damped identified FRF for healthy cantilever beam structure. It can be observed from Fig. 10 that the damping identification method is able to predict FRF accurately up to considered frequency range 0-1000 Hz. Similarly, same procedure is applied to identify the damping matrix for the damage cantilever beam structure.

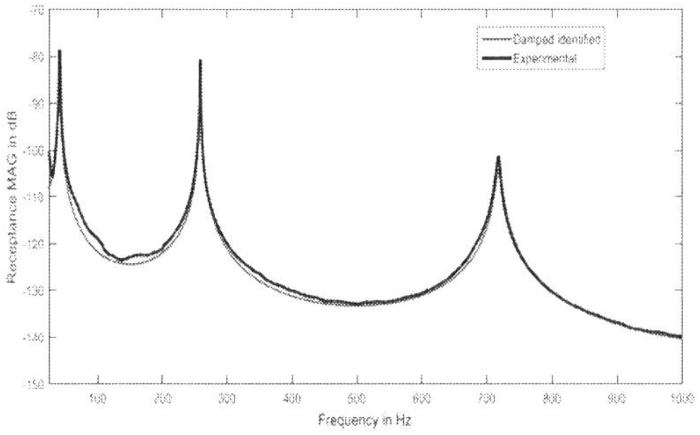

Fig. 10. Overlay of experimental FRF of the healthy structure and damped updated FRF

It can be observed from the identified damping matrices of the healthy and damaged cantilever beam that there is a 3.06% damage in the element 3 of the cantilever beam structure. It can be concluded from the experimental study that the proposed damping-based damage detection method is able to predict small damage in a structure.

5 Conclusions

In this paper, a damping-based damage detection method method is proposed. The proposed method can detect the location as well as the extent of the damage in a structure. The proposed damping-based damage detection method can detect a small damage in a structure as modal damping is very sensitive to small changes in the damping matrix due damage in the structure. Whereas, stiffness-based damage detection cannot able to detect small damages in the structure as natural frequencies and modeshapes are not very sensitive to small changes in stiffness matrix due to damage. Numerical and experimental case studies based on a lumped mass system and cantilever beam structure have been carried out to demonstrate that the changes in modal damping is better indicator of damage in a structure than the natural frequencies and modeshapes. The proposed method is subsequently applied to real experimental data of the cantilever beam structure to detect small damage in the structure. The proposed method able to detect successfully small damage in the cantilever beam structure. So, it can be concluded that damping-based damage detection method can be applied with confidence.

References

1. P. Cawley, R. Adams, J. Strain Anal. Eng. **14**, (1979)
2. J. T. Kim, H.M. Cho, N. Stubbs, Eng. Struct. **25** (2003)
3. A.K. Pandey, M. Biswas, J. Sound Vib. **169** (1994)
4. A. Berman, E.J. Nagy, AIAA J. **21** (1983)
5. M. Baruch, AIAA J. **22** (1984)
6. J.D. Collins, G.C. Hart, T.K. Hasselman, B. Kennedy, AIAA J. **12** (1974)
7. R.M. Lin, D.J. Ewins, Mech. Syst. Sig. Pro. **8** (1994)
8. B. Jaishi, W.X. Ren, J. Sound Vib. **290** (2006)
9. J.R. Wu, Q.S. Li, J. Constr. Steel Res. **62** (2006)
10. L. Zhou and G. Yang, Smart Struct. Syst. **2** (2006)
11. G. Kawiecki, *European COST F3 conference on system identification and structural health monitoring* (Madrid, Spain, 2000)
12. V. Arora, S.P. Singh, T.K. Kundra, J. Sound Vib. **324** (2009)
13. P. Lancaster, J. Aero. Sc. **28** (1961)
14. V. Arora, Inv. Prob. Sc. Eng. **22** (2014)
15. M.I. Friswell, S.D. Garvey, J.E.T. Penny, J. Sound Vib. **1211** (1998)

Triboelectric nanogenerator as self-powered impact sensor

Cristobal Garcia [1] *, *Irina* Trendafilova [1], *Roberto* Guzman de Villoria [2], and *Jose* Sánchez del Río [3]

[1] University of Strathclyde, Department of Mechanical and Aerospace Engineering, 75 Montrose Street, G1 1XJ, Glasgow, UK
[2] FIDAMC, Avda. Rita Levi Montalcini 29, 28906, Getafe, Spain
[3] IMDEA Materials Institute, C/ Eric Kandel 2, 28906, Getafe, Spain

Abstract. In recent years, triboelectric nanogenerators (TENGs) are used to harvest mechanical energy from ambient environment. These devices convert ambient energies (e.g. vibrations, breathing-driven, impacts or human body motions) into electricity based on the triboelectric effect. Furthermore, some TENGs can be successfully employed as self-power active sensors because the electric response from the TENG is proportional to the magnitude of the mechanical motion. This study report on the design and development of a novel triboelectric nanogenerator, and its potential application as self-powered impact sensor. To prepare the TENG device, membranes of polyvinylidene fluoride (PVDF) and polyvinylpyrrolidone (PVP) nanofibers are sandwiched between copper electrode films and wrapped on PET films. The TENG works based on the triboelectric interaction between the membranes of nanofibers. After the preparation, the TENGs are subjected to several impacts by the drop-ball impact test. The purpose of the experiment is to analyse if the electric response of TENG is dependent on the energy of the impact. The results of the experiment are presented and discussed. The main contributions of this work are the preparation of a novel nanogenerator (TENG) based on the triboelectric interaction between polyvinylidene fluoride and polyvinylpyrrolidone sub-micron polymer fibers and the investigation of its potential use as a self-powered impact sensor.

1. Introduction

A few years ago, impact sensors are required for a number of applications including structural health monitoring [1, 2]. Such sensors should be designed so that they can detect the impact but also measure some of its characteristics. Recently TENGs as self-powered active sensors are gaining a lot of popularity as they save energy but in the same time some of them can detect touch, impacts, vibrations, pressures and so on. These devices demonstrate excellent sensitivity to certain mechanical motions. Triboelectric nano-generators (TENG) as self-powered active sensors work on the basis of the triboelectric effect [3, 4 and 5]. This type of sensing has shown a number of desirable properties including extremely high output and efficiency in comparison to other sensing devices, low cost production technology, as it does not require highly sophisticated instruments for the manufacturing process, outstanding stability to high temperatures in comparison to piezoelectric devices, as well as environmental friendliness.

Traditional sensors are usually powered by batteries; however, TENGs do not need a power supply unit. The number of sensors used for monitoring different structures and machinery is increasing continuously. In a lot of cases the sensors are distributed across a wide range of area and integrated within the structure, and as such they need to operate wirelessly. For these systems, it is very important for the sensor to have the capability of operating independently, sustainably and maintenance-free. Therefore, the ability of the sensors to perform self-powered operations in a wireless sensor networks is critically important.

Vibrations are one of the most important forms of mechanical motions that can be found in ambient environment. Some examples of these are the motions generated for a washing machine/engine during its operational state or the vibrations produced on buildings and bridges due to an earthquake. Vibrations can directly reflect the working state and the health of the machineries and infrastructures, therefore the sensing of vibrations is of critical importance in equipment maintenance and environment monitoring. Since TENGs can generate electricity from any type of mechanical motions in the natural environment including vibrations

* Corresponding author: cristobal.garcia@strath.ac.uk

[6, 7], human body motions [8, 9], impacts [10, 11], breathing-driven [12] and so on, they can be used to develop self-powered active sensors for all these types of mechanical motions.

In this work we build a novel TENG and investigate its applicability as impact sensor. This device is constructed with the aim to develop a vibration sensor. But at this stage we only investigate its capabilities to detect and measure impact. The developed TENG is based on the interaction between PVDF and PVP fibres. The nano-structured materials are sandwiched between copper electrodes and wrapped in PET films.

To the best of our knowledge, this is the first attempt to produce a TENG using this selection of nano-structured materials. This work investigates the potential application of the newly developed TENG as a self-powered impact sensor. The impact testing is supposed to check and verify the sensor sensitivity to impact. In this respect the output voltage for different impact heights, velocities and energies is measured. There are other studies that test the application of different sensors to impact [13]. As the sensor developed is a novel one this testing and all the results are new. Therefore, this work presents solid progress toward the practical applications of the developed TENG.

The authors plan to use this TENG for purposes of development of an active vibration sensor and investigate its capabilities for vibration-based structural health monitoring in their future work.

The rest of the paper is organised as follows: The second section gives some insight on the fabrication of the necessary polymer nanofibers. &3 is dedicated to the sensor assembly using these nano-structured mats. Section 4 explains the experiments carried out to demonstrate the potential application of the TENG as self-powered impact sensor. The experimental results are introduced and discussed briefly in section 5. The last section offers some conclusions and ideas for extending the research.

2. Nano-material preparation

Electrospinning is a versatile process to produce sub-micron polymer fibers having high surface to volume ratio. Membranes of polyvinylidene fluoride and polyvinylpyrrolidone submicron fibers are prepared via electrospinning. The surface morphology of the fibers was observed using scanning electron microscopy (SEM) from Carl Zeiss (EVO MA15). An SEM image of PVDF nanofibers is given in figure 1 (a). The image shows fibers with different sizes distributed randomly in the membrane. The average diameter of the beaded fibers is about 700 nm. PVP fibers are shown in figure 1 (b). The SEM image shows microfibers with approximately the same diameter which can be measured to be about 1.4 micras.

PVDF in form of pellets is dissolved in DMF/acetone (40/60) mixture solvent to prepare a 20% w/v solution. The PVDF solution was loaded in a 5 mL plastic syringe with 21G steel needle. The electrospinning was carried out under the following conditions: voltage = 15 kV, spinning distance = 15 cm, flow rate = 0.6 ml/h. Membranes of PVDF nanofibers with a thickness of 0.25 mm were deposited on copper foil. PVP powder was dissolved in ethanol to prepare a 10% w/v solution. The PVP solution was transferred to a plastic syringe to be electrospun in Nanon-01 A. The electrospinning was carried out under the following conditions: voltage 18 kV, spinning distance 12 cm, 21 gauge needle and feed rate 0.5 ml/h. Membranes of PVP nanofibers with thickness of 0.12 mm were electrospun on copper foil.

(a)

(b)

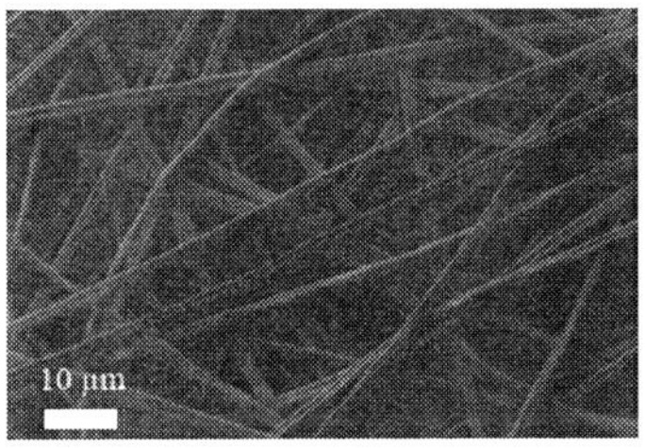

Figure 1: (a) SEM image of the electrospun PVDF fibers and (b) SEM image of the electrospun PVP fibers.

3. Sensor assembly

The sensor is fabricated by using electrospun nanofiber mats. The electrospun nanofiber mats are deposited at 0.05 mm thick square copper electrodes with a size of 4 x 4 cm^2. The sensor is made of four layers. The first and fourth layers are the bottom and the top copper electrodes respectively. The electrospun nanofiber mats are located as layers in between the electrodes, which face with each other. Eventually, the device is wrapped with polyethylene terephthalate film (PET) using a standard laminator (Inspire A3). The schematic structure of the sensor is shown in Figure 2.

The selection of triboelectric materials was done in accordance to the triboelectric series [14]. PVDF is one of the electronegative materials from the series. Thus, it tends to attract electrons and charge negatively. PVP is an electropositive material in the series, therefore, it tends to give electrons and charge positively. It is known that the contact area between the triboelectric materials affects the performance of the TENG. This is why the authors chose to use nanofibers instead of films. The larger surface increases the generation of triboelectric charges and thus improves the performance of the sensor.

The working mechanism of the sensor is the vertical contact-separation mode described elsewhere [15]. It is out of the scope of this work to assets it, however, we are going to show the sensing mechanism of a TENG based in the schemes of other authors [16, 17]. The basic working principle of the device is schematically depicted in figure 2. In the original state of the device, there is no charge generation and no electric potential difference. Under pressure from an external motion, the two triboelectric layers are brought into physical contact with each other. The contact between the triboelectric layers generates opposite triboelectric charges onto both nano-structured surfaces. Thus, the membranes of PVDF and PVP nanofibers are oppositely charged. When the pressing force is released, the two oppositely charged nanostructured surfaces are separated. The gap between the opposite triboelectric charges creates a dipolar moment with a potential difference across the two electrodes. Last, the material reverts to its original state because the triboelectric charges created disappear. Thus the voltage is created by the contact and the consequent separation of the two PVDF and PVP membranes of nanofibers. The contact and separation result in changes of the measured voltage.

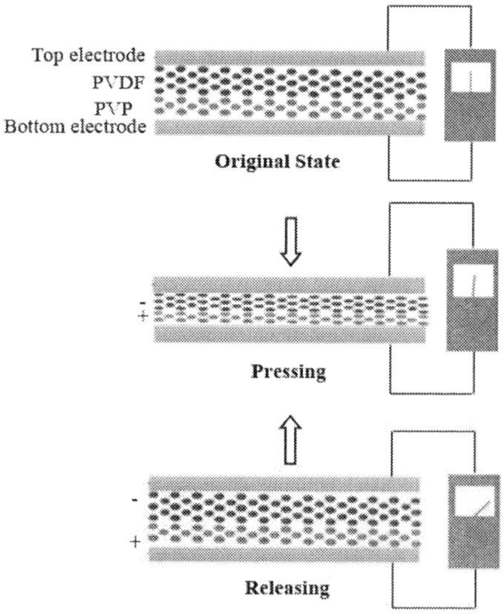

Figure 2: A schematically description of the structure and mechanism of generation of electricity in the TENG

4. Application of the device for impact detection and measurement.

This section describes the experiment used to measure the sensitivity of the manufactured sensor to impact. The prepared self-powered TENG was subjected to impacts with different energies applied via drop ball test. In this test, a free-falling glass ball of 21.22 g is dropped from four different heights: 10, 35, 65 and 100 cm. A schematic description of the experimental set-up is shown in figure 3. The diameter and the density of the ball is 2 cm and 5,06 g/cm^3 respectively. The ball is dropped within a plastic tube so that the impacts were applied perpendicularly to the sensor. The TENG is connected to a commercial oscilloscope to measure the electric output in response to impacts in the form of voltage using Tektronix 2012B commercial oscilloscope.

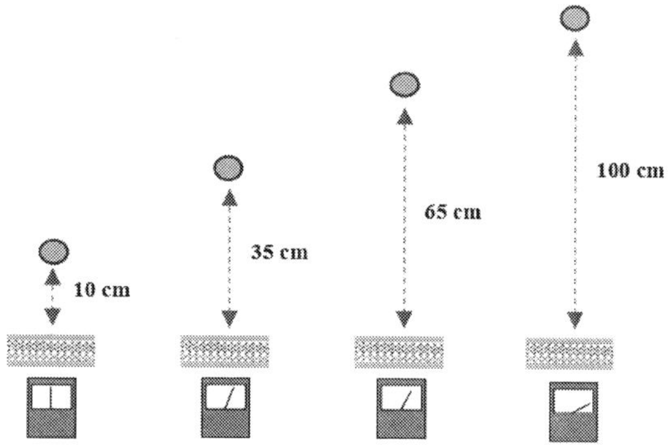

Figure 3: A schematic diagram of the experimental set-up to measure the electrical response to impacts

When the ball hits the sensor, both membranes of sub-micron fibres are brought into physical contact. After the impact, the ball rebounds and the nano-structured membranes revert back to their initial position. Thus, the nano-structured membranes are brought into contact and separation which generates a potential difference between the electrodes.

In the case of a drop ball test, according to the law of conservation of energy the potential energy before the collision equals the kinetic energy after the collision, which can be written as in equation 1, where m is the mass of the ball, g is the acceleration due to gravity, h is the drop ball height and v is the impact velocity. In the experiment, the drag forces caused by the air resistance and the side walls of the plastic tubes during the ball drop were neglected to simplify the analysis. Solving from the law of conservation of energy, the energy of the impact and impact velocity can be calculated using equations 1 and 2 respectively.

$$U = mgh = \frac{1}{2} mv^2 \qquad (1)$$

$$v_{impact} = \sqrt{2gh} \qquad (2)$$

5. Results and discussion

This section presents and discusses the results obtained in the drop-ball test. The TENG fabricated as indicated in section 3 was tested by drop-ball impact test using a free-falling ball. The ball was dropped from four different heights of 10, 35, 65 and 100 cm respectively. Table 1 presents the velocity and energy of the impact calculated theoretically for the particular drop height. The table indicates that when the balls were dropped from greater heights, the velocity and energy of the impact increase.

Table 1: The velocity and energy of the impact calculated theoretically as a function of the drop height.

Drop Height (cm)	*Velocity (m/s)*	*Impact Energy (mJ)*
10	1.40	21
35	2.62	73
65	3.57	135
100	4.43	208

The output voltage responses obtained for the impacts on the sensor are given in figure 4. The graph shows the electric response as a function of the time that result in response to the different impacts. The graph shows the variations of voltage caused by the ball impacts at 10, 35, 65 and 100 cm drop height respectively. The results show that increasing the impact drop height leads to increase of the voltage output. A number of peaks are observed for each impact due to multiple collisions, with the first peak corresponding to the first impact of the ball falling on the sensor. The following peaks correspond to the subsequent rebounds. The graph indicates the drop height for the first impact. The time interval between the bounces is also indicated in figure 4. The time interval between the first impact and the first bounce is the biggest one. However, the time interval between the subsequent bounces decreased gradually.

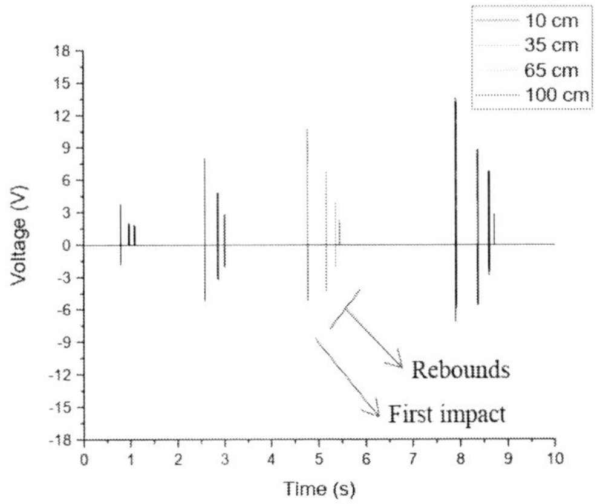

Figure 4: Voltage output response in time domain for the different ball impacts

Figure 4 also shows the peak-to-peak voltage values for the first impact and the next three rebounds. The highest voltage amplitude corresponds to the first ball impact. The peak-to-peak voltage values decrease gradually for the subsequent rebounds. The voltage generated at all the impacts from the different drop heights follow the same trend.

Figure 5 (a) and (b) shows the peak-to-peak voltage amplitude (V) for the first impact as a function of the velocity and the energy of the impact respectively. Five impacts tests were done for each drop height. The results given are the average voltage response and the standard deviation for the five tests.

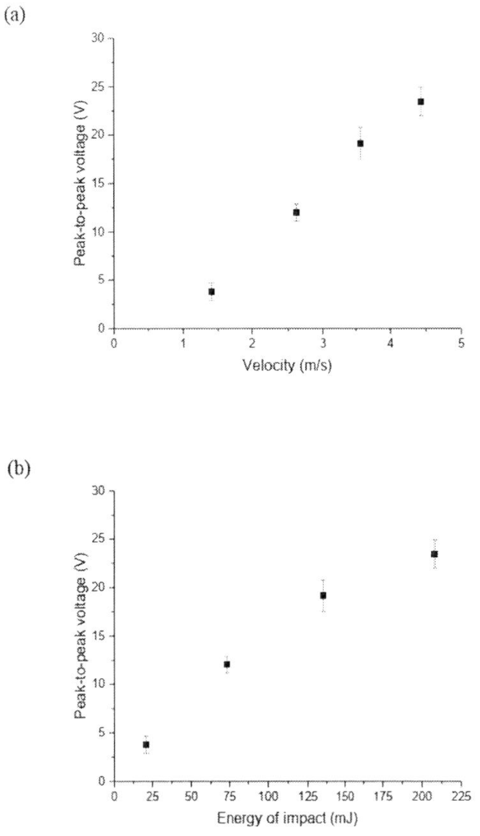

Figure 5: Peak-to-peak voltage response as a function of the velocity (a) and impact energy (b)

Impacts at higher velocity and energy lead to increase in the voltage response in the range from 1 to 5 m/s and 20 to 225 mJ respectively. When the ball hits the sensor at a small energy and velocity, the nano-structured rough membranes prevents completely close contact. Therefore, large areas of PVDF nanofibers are not in contact with PVP nanofibers which decreases the generation of electricity. When a higher energy and velocity is applied, the sensor deformation increases leading to more contact area between the membranes and this increases the voltage response.

The relation between the velocity of the impact and the voltage as per figure5 (a) is approximately linear. While the relation between the energy of the impact and the voltage does not seem to be linear and it looks as if for the higher energies the relative changes in the voltage are smaller as compared to the changes at lower energy.

6. Conclusions and future research

In this work, we demonstrated the potential applications as self-powered impact sensor of a TENG based on triboelectric interaction between submicron fibres of PVDF and PVP. An electrospinning method was utilized to fabricate nanofiber-networked membranes of both polymers, which was beneficial to enhance the performance of the TENG. The sensor is composed by four layers. The first and the four layers are the bottom and top copper electrodes respectively and the middle layers are the membranes of polymer fibres. The sensitivity of the manufactured TENG to impacts was analysed using the drop ball test. The experimental results showed that the TENG was very sensitive to impacts and it can clearly discriminate between impacts at different energies and velocities in the ranges between 0 and 225 mJ, and between 1 and 5 m/s, respectively.

For the future, authors plan to analyse the sensitivity of this TENG to mechanical compressions. For such analysis, the dynamic mechanic analyser (DMA; Q800; TA Instruments) equipped with the compression clamp used. The DMA equipment can apply mechanical compressions with forces in the range from 0.1 to 18 N. Moreover, DMA equipment can also apply mechanical compressions with different frequencies in the range between 0.10 to 20 Hz. Authors will study the electric output response of the TENG in response to different compressive forces and frequencies.

References

1. D. Garcia, I. Trendafilova, J. Sound Vib., *A multivariate data analysis approach towards vibration analysis and vibration-based damage assessment. Application for delamination detection in composite materials*, **25**, 7036-7050 (2014)

2. D. Garcia, I. Trendafilova, D. Inman, Smart Mater. Struct, *A study on the vibration-based self-monitoring capabilities of nano-enriched composite laminated beams*, **25**, 1-19 (2016)

3. Y. Zheng, L. Cheng, M. Yuan, Z. Whang, L. Zhang, Y. Qin, T. Jing, Nanoscale, *An electrospun nanowire-based triboelectric nanogenerator and its application in a fully self-powered UV detector,* **6**, 7842-7846 (2014)

4. S. Wang, L. Lin, Z.L. Wang, Nano Energy, *Triboelectric nanogenerators as self-powered active sensors,* **11**, 436-462 (2015)

5. B. Zhang, L. Zhang, W. Deng, L. Jin, F. Chun, H. Pan, B. Gu, H. Zhang, Z. Lv, W. Yang, Z.L. Wang, ACS Nano, *Self-powered acceleration sensor based on liquid metal triboelectric nanogenerator for vibration monitoring*, **11**, 7440-7446 (2017)

6. H. Zhang, Y. Yang, Y. Su, J. Chen, K. Adams, S. Lee, C. Hu and Z.L. Wang, Adv. Funct. Mater., *Triboelectric nanogenerator for harvesting vibration energy in full space and as self-powered acceleration sensor,* **24**, 1401-1407 (2014)

7. W. Yang, J. Chen, Q. Jing, J. Yang, X. Wen, Y. Su, G. Zhu, P. Bai, and Z.L. Wang, Adv. Funct. Mater., *3D stack integrated nanogenerator for harvesting vibration energy,* **24,** 4090-4096 (2014)

8. G. Zhu, W.Q. Yang, T.J. Zhang, Q.S. Jing, J. Chen, Y.S. Zhou, P. Bai, Z.L. Wang, Nano Lett., *Self-powered ultrasensitive, flexible tactile sensors based on contact electrification,* **14**, 3208-3213 (2014)

9. Y. Yang, H.L. Zhang, Z.H. Lin, Y.S. Zhou, Q.S. Jing, Y.J. Su, J. Su, J. Yang, J. Chen, C.G. Hu, Z.L. Wang, ACS Nano, *Human skin based triboelectric nanognerators for harvesting biomechanical energy as self-powered active tactile sensor system,* **7**, 9213-9222 (2013)

10. Y. Su, X. Wen, G. Zhu, J. Yang, J. Chen, P. Bai, Z. Wu, Y. Jiang, Z.L. Wang, Nano Energy, *Hybrid triboelectric nanogenerator for harvesting water wave energy and as a self-powered distress signal emitter,* **9**, 186-195 (2014)

11. G. Cheng, Z. H. Lin, Z.L. Du, Z.L. Wang, ACS Nano, *Simultaneously harvesting electrostatic and mechanical energies from flowing water by a hybridized triboelectric nanogenerator,* **8**, 1932-1939 (2014)

12. Q. Zheng, B. Shi, F. Fan, X. Wang, L. Yan, W. Yuan, S. Wang, H. Liu, Z. Li, Z.L. Wang, Adv. Mater., *In vivo powering of pacemaker by breathing-driven implanted triboelectric nanogenerator*, **26**, 5851-5856 (2014

13. S. Joshi, G.M. Hegde, M.M. Nayak, K. Rajanna, Sens Actuators A Phys, *A novel piezoelectric thin film impact sensor: Application in non-destructive material discrimination,* **199**, 272-282 (2013)

14. Z.L. Wang, L. Lin, J. Chen, S. Niu, Y. Zi, Green energy and technology, *Triboelectric nanogenerators*, 10-11 (2016)

15. Z.L. Wang, L. Lin, J. Cheng, S. Niu, Y. Zi, Green energy and technology, *Triboelectric nanogenerators*, 23-46 (2016)

16. W. Li, J. Sun, M. Chen, Nano energy, *Triboelectric nanogenerator using nano-Ag ink as electrode material*, **3**, 95-101 (2014)

17. S.Y. Shin, B. Saravanakumar, A. Ramadoss, S.J. Kim, Int J Energy Res, *Fabrication of PDMS-based triboelectric nanogenerator for self-sustained power source application*, **40**, 288-297 (2015)

Vibration-based damage detection of structural joints in presence of uncertainty

Hussein Al-Bugharbee[1,*], *Ali* Abolfathi[2], and *Irina* Trendafilova[3]

[1] Dept. of Mechanical Engineering, University of Wasit, Iraq.
[2] Dept. of Mechanical Engineering, University College London, London, UK.
[3] Mechanical and Aerospace Engineering, University of Strathclyde, Glasgow, UK

Abstract. Early damage detection of structure's joints is essential in order to ensure the integrity of structures. Vibration-based methods are the most popular way of diagnosing damage in machinery joints. Any technique that is used for such a purpose requires dealing with the variability inherent to the system due to manufacturing tolerances, environmental conditions or aging. The level of variability in vibrational response can be very high for mass-produced complex structures that possess a large number of components. In this study, a simple and efficient time frequency method is proposed for detection of damage in connecting joints. The method suggests using singular spectrum analysis for building a reference space from the signals measured on a healthy structure and then compares all other signals to that reference space in order to detect the presence of faults. A model of two plates connected by a series of mounts is used to examine the effectiveness of the method where the uncertainty in the mount properties is taken into account to model the variability in the built-up structure. The motivation behind the simplified model is to identify the faulty mounts in trim-structure joints of an automotive vehicle where a large number of simple plastic clips are used to connect the trims to the vehicle structure.

1 Introduction

Joints such as clips and bolts are commonly used in the built-up structures to connect different components. The damage in these joints can adversely affect the integrity of the whole structure and may lead to catastrophic incidents. Thus, an early detection of damage in the structure with faulty joints is crucial in order to maintain the safety and to extend the service life span of structures.

Damage detection in structural joints has attracted a lot of attention and many techniques were developed during the past few decades. Vibration-based analysis is one of the most popular strategies for the damage detection as it is non-destructive and repeatable. The deviation of natural frequencies and damping ratios from the baseline values due to bolt looseness is investigated in [1] where a couple of Euler beams with a single bolted lap joint is used in the analysis. The results illustrate that the bolts looseness affect the structure's natural frequencies and damping ratios. However, the change is more significant at the higher frequency range.

In reference [2], an experimental investigation is conducted for identifying the looseness in cargo bolts under random excitation. The experiment is conducted on twelve bolt groups and seven different severity of damages (i.e. looseness). For each simulated damage type, vibration signals are acquired using accelerometers and time series are used for detection. Two kinds of autoregressive models were constructed. The residual errors of the models are used as damage index for different levels of damage. The results showed that the suggested methodology has the possibility of early detection of bolt looseness severity.

A damage detection method based on the analysis of the subharmonic resonance is presented in [3]. The study proposed the structure bolted joint as a two-degree of freedom nonlinear model and uses a multiple timescale method for illustrating the generation of subharmonic resonance. Experiments were conducted on a single bolt-joint aluminum beam and the damage in the joint is simulated by bolts looseness. The excitation of the beams and acquisition of the corresponding response signals are conducted by piezoelectric transducers. The results showed that the subharmonic frequencies appear in the structure response spectrum when it is excited by a double of its natural frequency.

The study in [4] presents a technique for the looseness detection of bolted structure. The technique based on the frequency response function (FRF) data. The experimental results of the study are obtained using both accelerometer and strain gauge and their measurements are compared to assess the bolt looseness. The results show that presence of bolt looseness causes and abrupt change in the orthogonal modes.

* Corresponding author: hrazzaq@uowasit.edu.iq, husseinmec@yahoo.com

The detection of undesired structural changes in space vehicle due to bolt looseness is investigated in [5]. Special kind of piezoelectric sensors are used on a real satellite panel with 49 bolts. The damage in the satellite panel is simulated as looseness in a bolt. The capability of two different methods, namely Acousto-Elastic and Electro-Mechanical Impedance methods, of detecting damage is investigated. The experimental results of the study show that the present techniques have a potential possibility for both damage detection and localization.

A combined methodology based on vibration and electro-mechanical impedance techniques for the purpose of structural joint damage detection is presented in [6]. A number of structure joint damage scenarios are designed in the study for the purpose of damage detection. The experimental results such as modal parameters are analyzed and showed that the present methodology is useful in extracting damage information of the joint.

The study in Ref. [7] proposes an artificial neural network (ANN) based method for the estimation of damage severities in truss bridge's joint. The mode shapes and natural frequencies are used as input features to the ANN in order to assess the damage. A numerical analysis is presented in order to assess the method's efficiency and accuracy.

In this study, the effectiveness of a new method to detect the damage in joints of two connected plates is investigated. The uncertainty in joint property are modelled using the measured variability of plastic mount that are used to connect the trim to the structure of an automotive vehicle [8]. A large number of these clips are used in modern vehicles. These clips should be firmly connected and any rattling can be a source of unwanted noise in a vehicle's cabin. It is shown that the variability in the effective stiffness and damping of such clips can affect the vibration response of the vehicle [8]. Such variability makes the damage detection process more difficult and uncertain.

A simple and easy methodology based on Singular Spectrum Analysis (SSA) is suggested here for damage detection. In this method, the time domain vibration acceleration signals are subjected to the SSA for the decomposition purposes. From the response of structure with healthy joints a reference space is made. Other signals of the healthy structure will be projected onto this space and allow an estimation of a threshold value. Any new signal will be projected on that baseline space and their distance to the cluster of healthy signal will be compared to the threshold value to classify them as healthy or damaged.

In the following section, the mathematical formulation of the problem is described. In section 3, the fundamentals of the SSA are described briefly. Results and discussion are presented in section 4 and the section 5 focuses on conclusions made in this paper.

2 Mathematical formulation of the problem

A simple model of two connected steel plates with multiple mounts is used here in order to simulate a built-up structure. For the purposes of modeling, the steel plates are simply supported and thus an analytical solution for their vibration response exists according to Kirchhoff–Love plate theory [9]. The two plates are connected by eleven mounts which are distributed randomly between the plates. This allows obtaining an analytical solution using the impedance-mobility technique (FRF coupling) [10,11] The plates are made of steel and have the following dimensions: plate 1 has a thickness of 1.5 mm, width of 350mm and length of 500mm; plate 2 has a thickness of 1.3mm, width of 400mm and length of 600mm. A schematic view of two plates is shown in figure 1. The mobility functions between point 1 on the first plate and point 2 on the second plate are obtained.

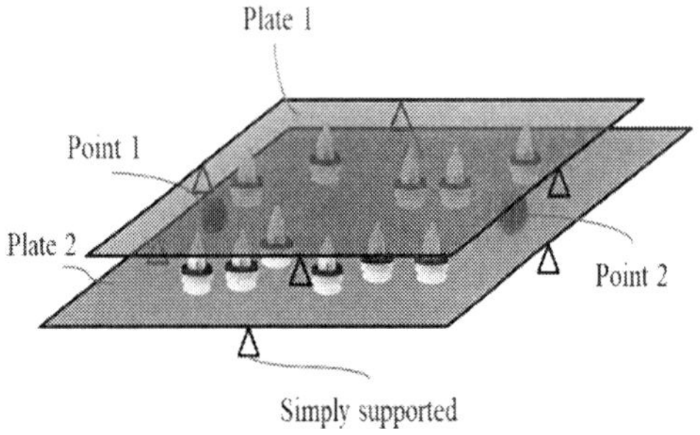

Fig. 1. A schematic view of two plates connected by eleven clips.

The uncertainty in the clips properties are considered based on the measurement conducted in [8]. A Monte Carlo simulation is used to obtain the point and the transfer mobilities while the clips properties are varied. The mobility Y_{12} for point 1 of the first plate and point 2 of the second plate is shown in figure 2 for 250 realisations where all clips are considered connected. It can be seen that at lower frequencies there are distinct modes which are not affected by the variability in clips properties. However, the effect of variability in the clips properties becomes more prevalent with the increase of the frequency.

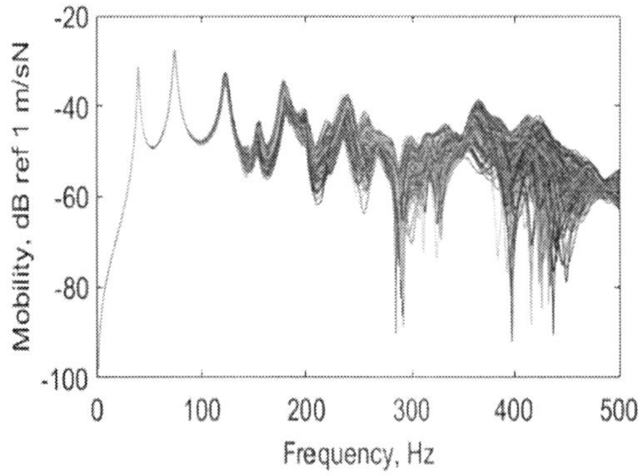

Fig. 2. Monte Carlo simulation results of mobility Y_{12} of two connected plate with 11 clips.

Damage is modelled by removing one clip at a time. Then again, a Monte Carlo simulation is used to simulate the effect of uncertainty in the clip properties. The mobility Y_{12} for the two connected plates, where one of the clips is removed is shown in figure 3. Here, again 250 realisations for each case are produced. Overall 11 cases of damaged joint and one case with no damage are presented in figure 3. The variability in the clips properties make it impossible to distinguish between each case and there are overlap between healthy and damaged connector for all the cases.

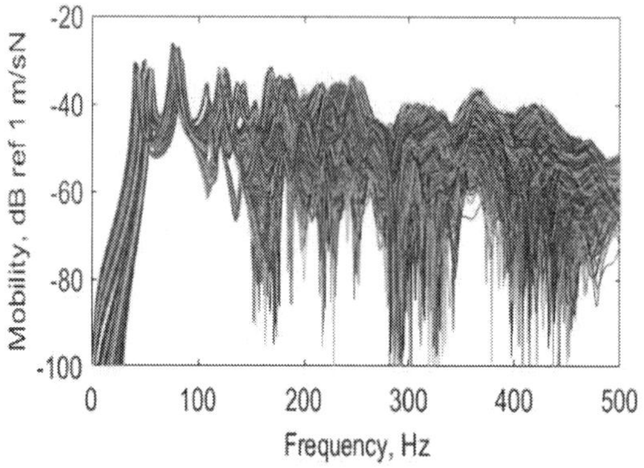

Fig. 3. Monte Carlo simulation of the mobility Y_{12} of two connected plate for healthy and damaged connections where one of the clips is removed each time.

3 Methodology

SSA is a time series analysis method that popularly used in biomedical and meteorological sciences [12-14]. Recently, it was used for the purposes of engineering application such as fault diagnosis of rolling element bearings [15-19], tool wear health monitoring [20, 21] and delamination in composite materials [22].

The SSA has two main stages; decomposition and reconstruction. In the first stage, a signal which is discretized as a time series is decomposed into a number of independent components, the principal components (PC's). Each component contains a certain percentage of the original signal variance. The reconstruction stage, which is not used in this methodology, uses all or some of the principal components to reconstruct the original signal. Further details of the SSA can be found in [23, 24].

The methodology suggested has two key steps: building baseline space and damage detection methodology. The later contains extraction of feature vectors (FVs), setting a threshold and a classification process.

3.1 Building reference/baseline space.

A baseline space is made from subjecting a signal measured on a healthy structure to the decomposition stage of the SSA. First a trajectory matrix **X** of dimension L by K is made from the time-lapped signal x of a length n (i.e $x = [x(1), x(2), x(3) \dots x(n)]$) as shown in Eq. (1):

$$\mathbf{X} = \begin{bmatrix} x(1) & x(2) & x(3) & \dots & x(K) \\ x(2) & x(3) & x(4) & \dots & x(K+1) \\ x(3) & x(4) & x(5) & \ddots & x(K+2) \\ \vdots & \vdots & \vdots & & \vdots \\ x(L) & x(L+1) & x(L+2) & \cdots & x(K+L-1) \end{bmatrix} \tag{1}$$

where $x(n)$ is a nth element of the measured signal. The L is the window of decomposition (i.e the number of PCs resulted from the decomposition stage) and $K=n-L+1$. The covariance matrix of the trajectory matrix can be obtained using to Eq. 2,

$$\mathbf{C}_\mathrm{x} = \frac{\mathbf{XX}'}{L} \tag{2}$$

Eigenvalues and eigenvectors of the covariance matrix $\mathbf{C}_\mathbf{x}$ can be obtained by subjecting it to singular value decomposition,

$$\mathbf{C}_\mathbf{x}\mathbf{U}_i = \lambda_i \mathbf{U}_i \tag{3}$$

Each eigenvalue (i.e. λ_i) represents a fraction of the original signal's variance in the direction of the corresponding eigenvector $\mathbf{U}_i$. The eigenvalues are usually arranged in a decreasing order and the corresponding variances are represented in the so-called scree plot [25].

For building the baseline space, all or only a part of eigenvectors obtained above, which correspond to the healthy condition can be used.

3.2 Damage detection methodology

After building the baseline/reference space from a healthy signal, any new signal will be projected on the baseline space and the fault detection process is conducted. The detection process has two main steps: feature vector extraction and classification.

3.2.1 Feature vector extraction

Supposing signals from healthy and faulty condition are available, the healthy signals are divided equally into training and testing samples, while all the faulty signals are used as testing samples.

For the training samples, the trajectory matrix of every signal is projected onto the baseline space. The projection means multiplying the transpose of the trajectory matrix **X** by each of the baseline eigenvector $\mathbf{U}_i$. This projection will provide the corresponding principal components $\mathbf{PC}_i$,

$$\mathbf{PC}_i = \frac{\mathbf{X}'\mathbf{U}_i}{\lambda_i} \tag{4}$$

The symbol ($'$) denotes the transpose. Then, the Euclidean norm of each of the three principal components is calculated as in Eq.5,

$$f_{ij} = \sum_{m=1}^{K} \left(\mathbf{PC}_{ij}(m)\right)^2 \tag{5}$$

where f_{ij} is the feature, i is the number of principle component and j is the number of the signal that is projected.

In the present study, the baseline space is basically made from the first three eigenvectors. Hence, all feature vectors will be of three dimensions. More eigenvectors can also be used but in this study the first three vectors were sufficient to achieve a very good classification . Then the feature vector (FV) obtained from j^{th} signal will have the form,

$$\mathbf{f}_j = [f_{1j}\ f_{2j}\ f_{3j}]' \quad (6)$$

The reason for choosing three feature components only is that these can be visualized in a 3D space.

3.2.2 Classification

When a baseline space is made and the training samples are projected, the resultant FVs are arranged in rows to form the baseline feature matrix $\mathbf{F}_{baseline}$. Then, the Mahalanobis distance (D_i)of each feature vectors to the $\mathbf{F}_{baseline}$ is calculated,

$$D_i = \sqrt{(fv_i - E_{baseline}).S^{-1}.(fv_i - E_{baseline})^T} \quad (7)$$

Where $\mathbf{E}_{baseline}$ is the mean of the rows of the $\mathbf{F}_{baseline}$ and S^{-1} is the inverse of the covariance matrix of $\mathbf{F}_{baseline}$.

From the values of D_i corresponding to the baseline training sample a suitable probability distribution is fitted. In the present study, a lognormal probability distribution if fitted. From this probability distribution, a suitable threshold T_r is found. The threshold here is selected in a way the cumulative probability distribution become equal to 0.99 [26]. A new signal will be classified based on its D_i to the class of healthy or faulty signals, according to Eq. (8).

$$\begin{matrix} D_i > T_r & singal\ is\ assigned\ to\ healthy\ class \\ D_i \leq T_r & signal\ is\ assigned\ to\ faulty\ class \end{matrix} \quad (8)$$

where D_i is the Mahalanobis distance of ith signal measured from the healthy space and T_r is the calculated threshold.

4 Results and discussion

As was mentioned in Section 3, 250 realisations were obtained for the case when all clips are connected that is for the case of healthy clips. Similarly, 250 signals are simulated for each case when one clip is removed. In total, there will be 3000 realisations (250*12 (1 Healthy case +11 removed clip cases)). The signals from the healthy structure are divided equally (i.e. each is 125 realisations) into training and testing samples, each one containing 125 signals. All the other 2750 signals are used as testing samples.

Figure 4 represents the 3D visualization of training FVs which were obtained according to Eq.6.

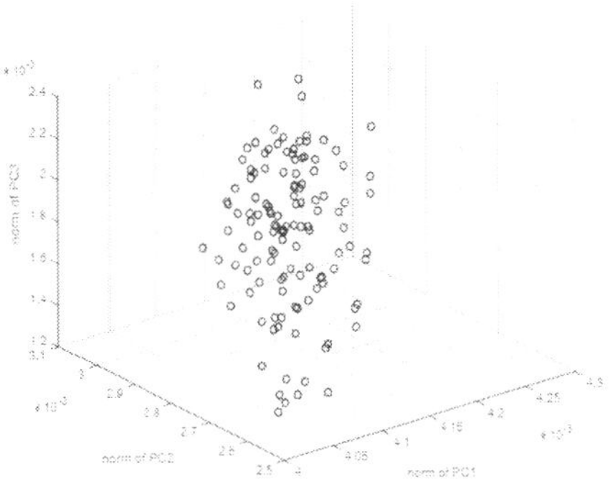

Fig. 4. 3D visualisation of baseline FVs.

Figure 5 shows the projection of the testing FVs on the baseline feature space. The healthy training and testing FVs are represented in blue color. The FVs corresponding to faulty class are represented in red color. It can be seen that the testing FVs corresponding to the faulty cases can be visually recognized from the baseline FVs.

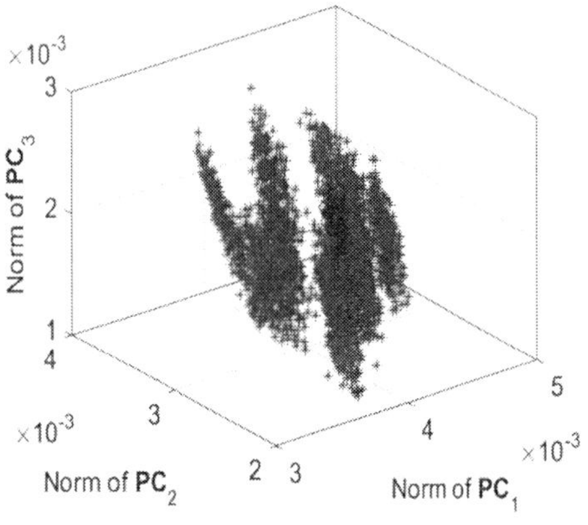

Fig. 5. Clustering of 3D visualisation of baseline FVs.

Figure 6 represents the Mahalanobis distances of all the feature vectors that were measured to the training FVs. The horizontal dashed line represents the threshold *Tr*. The figure clearly shows that the *D* level of the faulty FVs is higher than the *D* level of the training FVs. In this sense, the FVs made from faulty class are dissimilar to those made from training FVs. The FV's from the faulty class will be recognized as such according to the rule defined in Eq. (8).

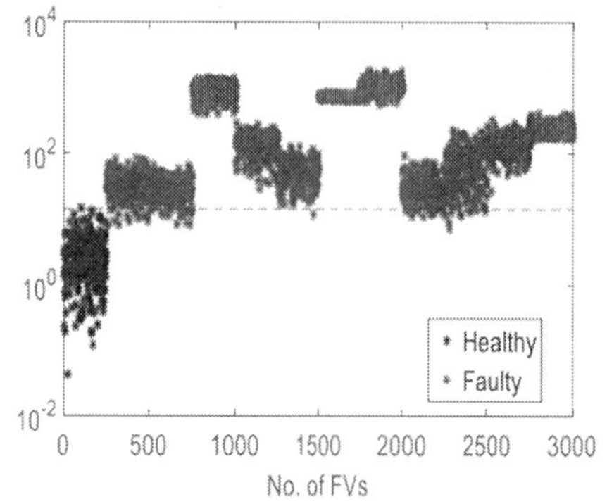

Fig. 6. The levels of *D* for both training and testing sample.

Table 1 shows the confusion matrix that represents the rates of correct and incorrect classification in percentages for the healthy and the faulty signals. The first column denotes the real class and the first row corresponds to the recognized class. The numbers on the main diagonal shows the correct classification rates while the off-diagonal numbers show the percentage of misclassified signals. All the 125 healthy FVs from the testing samples are correctly assigned to the healthy class. Only 31 out of the 2750 (1.13 %) testing FVs corresponding to faulty conditions were misclassified as healthy while all the rest 2719 FVs (i.e. 98.87%) were correctly classified as faulty.

Table 1. Confusion matrix

Real class/recognized class	**H**	**F**
H	100%	0%
F	1.13%	98.87%

5 Conclusions

This study suggests a vibration-based monitoring method to identify damage in joints in presence of variability. The proposed method is used to detect damage in a faulty joint of two connected plates. Two plates are assumed to be connected by small plastic clips that are used to connect the vehicle's trim to its structure. The variability in clips' effective stiffness and damping are modelled here using a Monte Carlo simulation.

The methodology proposed here does not require previous measurements of signals corresponding to faulty cases. The results obtained in the paper demonstrate the accuracy of the method, which supports the claim of using it for purposes of automatic damage assessment.

References

1 A. Dekal, A. Raol, S. Kamath, A. Gaurav, K. Gangadharan, presented at the 2nd Mechanical Engineering Graduate Research Symposium (MEGRES), IIT Bombay, Mumbai, India (2015).
2 G. Dong, F. Zhao, and X. Zhang, Adv. Mech. Eng., **9**(3), 1-12 (2017).
3 M. Zhang, Y. Shen, L. Xiao, and W. Qu, Nonli. Dyn., **88**, 1643-1653 (2017).
4 H.C. Eun, Y.J. Ahn and S.G. Lee, "Experimental study on damage detection at bolted joint using frequency response function." in ACEM14, Busan, Korea, (2014).
5 D. Doyle, A. Zagrai, and B. Arritt, "Bolted joint integrity for structural health monitoring of responsive space satellites," in *the Proceeding of 50th AIAA/ASME/ASCE/AHS/ASC Structures, Structural Dynamics, and Materials Conference,* Palm Springs, CA (2009).
6 W. Na, J. Lee, J. Kim, and D. Hong, "Hybrid Health Monitoring for Damage Detection in Structural Joints," (2006).
7 M. Mehrjoo, N. Khaji, H. Moharrami, and A. Bahreininejad, Exp. Sys. Appl., **35**, 1122-113 (2008).
8 A. Abolfathi, D. J. O'Boy, S. J. Walsh, A. M. Dowsett, and S. A. Fisher, Proc. IMechE, C: J Mech. Eng. Sci., 0954406217721724 (2017).
9 S. S. Rao, *Vibration of continuous systems,* (John Wiley & Sons, 2007).
10 P. Gardonio and M. Brennan, Mobility and impedance methods in structural dynamics, in *Advanced Applications in Acoustics, Noise and Vibration* (CRC Press, 2004).
11 E. L. Hixson, *Shock and vibration handbook,* **10**, 1-46, (1988).
12 S. Sanei and H. Hassani, *Singular spectrum analysis of biomedical signals* (CRC Press, 2015).
13 S. M. Mohammadi, S. Enshaeifar, M. Ghavami, and S. Sanei, "Classification of awake, REM, and NREM from EEG via singular spectrum analysis," in *2015 37th Annual International Conference of the IEEE Engineering in Medicine and Biology Society (EMBC)*, 4769-4772 (2015).
14 A. Hossein Vahabie, M. M. R. Yousefi, B. N. Araabi, C. Lucas, and S. Barghinia, "Combination of singular spectrum analysis and autoregressive model for short term load forecasting," in *Power Tech, IEEE,* Lausanne, 1090-1093 (2007).
15 B. Kilundu, X. Chiementin, and P. Dehombreux, J Vib. Acous., **133**, 051007 (2011).
16 T. Liu, J. Chen, and G. Dong, J Vib Cont., 1506-1521 (2013).
17 B. Muruganatham, M. Sanjith, B. Krishnakumar, and S. Satya Murty, Mech. Sys. Sig. Proc., **35**, 150-166 (2013).
18 H. Al-Bugharbee and I. Trendafilova, JSV, **369**, 246-265 (2016).
19 H. Al-Bugharbee and I. Trendafilova, Int. J. Cond. Mont., 7, 26-35 (2017).

 https://doi.org/10.1051/matecconf/201814814006

20 D. Salgado and F. Alonso, J Mat. Proc. Tech., **171**, 451-458 (2006).
21 B. Kilundu, P. Dehombreux, and X. Chiementin, Mech. Sys. Sign. Proc., **25**, 400-415 (2011).
22 D. Garcia, I. Trendafilova, and H. Al-Bugharbee, "Vibration-based health monitoring approach for composite structures using multivariate statistical analysis," in *EWSHM-7th European workshop on structural health monitoring* (2014).
23 N. Golyandina and A. Zhigljavsky, *Singular Spectrum Analysis for time series.* (Springer Science & Business Media, 2013).
24 H. Hassani, "Singular spectrum analysis: methodology and comparison," (2007).
25 L. Zhang, J. Marron, H. Shen, and Z. Zhu, J Comp. Grap. Stat., **16**, 833-854 (2007).
26 H. R. Al-Bugharbee, *Data-driven methodologies for bearing vibration analysis and vibration based fault diagnosis*, (PhD thesis, University of Strathclyde, 2016).

Vibro-acoustic and nonlinear analysis of cadavric femoral bone impaction in cavity preparations

Sebastian Oberst[1,2,3,6,⋆], *Johanna* Baetz[4], *Graeme* Campbell[4], *Frank* Lampe[5], *Joseph C.S.* Lai[6], *Norbert* Hoffmann[2,3], and *Michael* Morlock[4]

[1] *University of Technology Sydney, Centre for Audio, Acoustics and Vibration, Faculty of Engineering and Information Technology, Sydney, Australia*
[2] *Hamburg University of Technology, Dynamics Group, Mechanical Engineering, Hamburg, Germany*
[3] *Imperial College London, Department of Mechanical Engineering, London, United Kingdom*
[4] *Hamburg University of Technology, Institute of Biomechanics, Mechanical Engineering, Hamburg, Germany*
[5] *Hamburg University of Applied Sciences, Department of Life Sciences, Hamburg, Germany*
[6] *University of New South Wales Canberra, School of Engineering and IT, Canberra, Australia*

Abstract. Owing to an ageing population, the impact of unhealthy lifestyle, or simply congenital or gender specific issues (dysplasia), degenerative bone and joint disease (osteoarthritis) at the hip pose an increasing problem in many countries. Osteoarthritis is painful and causes mobility restrictions; amelioration is often only achieved by replacing the complete hip joint in a total hip arthroplasty (THA). Despite significant orthopaedic progress related to THA, the success of the surgical process relies heavily on the judgement, experience, skills and techniques used of the surgeon. One common way of implanting the stem into the femur is press fitting uncemented stem designs into a prepared cavity. By using a range of compaction broaches, which are impacted into the femur, the cavity for the implant is formed. However, the surgeon decides whether to change the size of the broach, how hard and fast it is impacted or when to stop the excavation process, merely based on acoustic, haptic or visual cues which are subjective. It is known that non-ideal cavity preparations increase the risk of peri-prosthetic fractures especially in elderly people.

This study reports on a simulated hip replacement surgery on a cadaver and the analysis of impaction forces and the microphone signals during compaction. The recorded transient signals of impaction forces and acoustic pressures ($\approx$ 80 μs - 2 ms) are statistically analysed for their trend, which shows increasing heteroscedasticity in the force-pressure relationship between broach sizes.

TIKHONOV regularisation, as inverse deconvolution technique, is applied to calculate the acoustic transfer functions from the acoustic responses and their mechanical impacts. The extracted spectra highlight that system characteristics altered during the cavity preparation process: in the high-frequency range the number of resonances increased with impacts and broach size. By applying nonlinear time series analysis the system dynamics increase in complexity and demand for a larger minimum embedding dimension. The growing number of resonances with similar level of the transfer function indicates a higher propensity to dissipate energy over sound; the change in embedding dimension indicates a decrease in linearity. The spectral changes as well as the altered dimension requirements indicate either an improved coupling between the bone and the broach or the onset of micro-fractures caused by growing stress levels within the bone.

1 Introduction

Inflammatory bone degeneration (osteoarthritis) often requires the replacement of the whole synovial joint, i.e. total arthroplasty [1] with total hip (THA) and knee replacements (TKA) become more common as reported by the Australian National Joint Replacement Registry [2]. Metrics such as the ASA score (American Society of Anaesthesiologists - Physical Status Classification) or the Body Mass Index (BMI) [2] indicate that an longer life expectancy, increasingly unhealthy lifestyles and the growing risk of overweight result in more problems associated with joints such as the knee, the shoulder or the hip [1]. In 2015 about 498,660 hip, and 592,577 knee joint replacement procedures (1,123,643 in total) were registered, an increase of 2.6% and 5.9%, respectively over those in 2014 [2]. Hip prosthesis systems consist of a ball joint which is tapered to a stem component. The stem is press fitted to the femur and the ball joint runs in the liner of the acetabular cup which is inserted into the acetabulum [3]. Implant materials are *hard* (H) metal alloys (steel, titanium-based or more often made of cobalt-chromium), alumina/composite ceramics or *soft* (S) plastics e.g. the ultra-high molecular weight polyethylene [4]. Hard-on-hard (HoH) bearing articulations, i.e. ceramic-on-ceramic or hard-on-soft combinations can be installed [4].

⋆e-mail: sebastian.oberst@uts.edu.au

The cavity for the stem is prepared using the impaction of broaches and a surgical hammer; here in particular the primary stability of the artificial joint is important. Primary stability is achieved through an interference-fit which strongly depends on the bone-implant contact [5]. Preparation of the proximal femur cavity prior to implantation is achieved via compaction (bone is crushed to form the cavity) or extraction (bone is removed to form the cavity) using incremental broaching. As a result of the cavity preparation generally a higher bone densification close to the cavity region in the bone-implant envelope can be found [6]. During THA the surgeon decides when and whether to change the size of the broach, how hard and fast to impact it and when to stop the cavity preparation [7]. This decision is subjective, generally based on changes in pitch, haptic or visual cues [7]. However, owing to non-ideal cavity preparation, lysis (implant loosening) and post-operative prosthesis dislocation due to insufficient stability at the time of implantation may occur which is a major cause for revision surgeries [2, 4]. On the other hand, by increasing the primary stability too far the risk of peri-prosthetic fractures actually grows [4, 8].

Yagi et al. [9] used an ultrasonic system within an instrumented hammer to measure the acoustic signal's attenuation time to decide on the most suitable broach and showed that the attenuation time reduced for broaches with a larger femoral canal, presumably due to increased compaction of the cancellous bone and an enhanced connection between the implant and the tissue [9]. Mathieu et al. [10] studied the variation of the impact duration during in-vitro insertions of acetabular cup implants which decreased at first as a function of the impact number, then reached a stationary value equal to about 4.2 ± 0.7 ms after an average number of about 4 impacts. It was hypothesised that the impact duration is related to the bone-implant contact variations.

Sakai et al. [7] studied the relation between the hammering frequency during insertion by measuring both force and sound on a clamped femur. The internal stress during implantation is estimated using the dynamic explicit finite element solver LSDYNA. A reduction in frequency correlated well with the increased stress levels σ which suggests that once the maximum σ is reached the hammering force can be reduced [7].

Whitwell et al. [11] studied the spectral components of the acoustic signal during impaction for nine plant insertion surgeries which were independently conducted by two surgeons (six and three samples). For the last broach a low-frequency component at around 0.4-1.2 kHz was detected, which they attributed to the resonant frequency of a standing sound wave within the femoral canal (estimated at about 894 Hz); other peaks were assumed to be transverse vibration modes of the implant components. Whitwell et al. [11] suggested the low frequency component as an indicator of the optimal broach size being reached. Morohashi et al. [12] studied the sound pressure levels of the impacting hammer of the first and the second dominant resonances in 71 total hip arthroplasties; a pattern with a smaller number accentuated resonances in the spectrum showed significantly less intraoperative fractures and subsidence.

Especially within the last five years some progress has been made in identifying acoustic features which help surgeons when to stop impacting the insertion broach. However, this progress is very limited and the findings do not have any practical use yet. From past studies important information, such as the sampling rate is missing and never the change in the system itself has been studied but only its acoustic response alone. Here we measure the radiated sound and the impaction forces during cavity preparation using compaction broaches in a cadaver THA study in the theatre. We analyse the sequence of impactions for each broach and between broaches and calculate acoustic transfer functions as unique descriptors of the system to radiate sound due to impulsive excitation, cf. [13]. Nonlinear dynamics tools and complexity measures, in particular nonlinear time series analyses (NTSA) methods have rarely been applied in biomechanics [14] although these methods could provide insights for complex nonlinear data which linear methods could not [15]. We therefore apply NTSA tools to analyse the impulse response signal in order to determine objectively the change of complexity i.e. the change of phase space dimensions required in course of the impaction process.

2 Materials and Methods

2.1 Experiments

The hip of a female 54-year-old cadaver was scanned with quantitative computer tomography (Philips Brilliance 16, 1 mm slice thickness) including a measurement sphere for templating. An experienced surgeon determined the adequate implant size with templates and resected the femoral head using an anterolateral approach. A chisel was used to excise the residual femoral neck and a blunt reamer demonstrated the canal axis. Six stainless steel compaction broaches of increasing size were introduced to prepare the cancellous bone bed. Impaction forces from

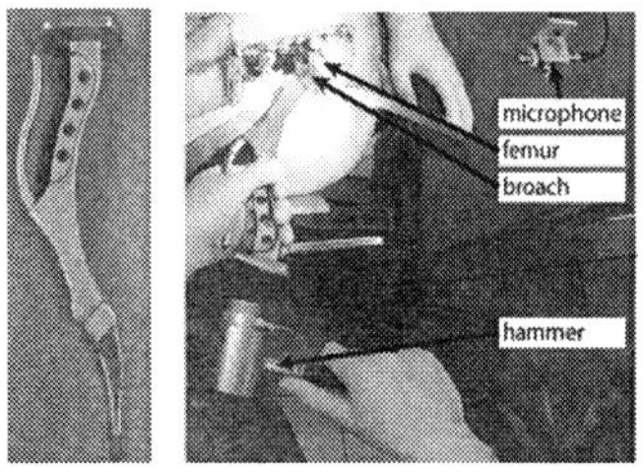

(a) (b)

Figure 1. (a) Broach attached to the broach handle and (b) experimental setup in the operation theatre

the surgeon were recorded at the tip of an instrumented surgical hammer (1.4 kg) that was modified with a dynamic piezoelectric load cell (capacity 89 kN, 200C20, PCB Piezotronics, Depew, New York, USA). The acoustic signal was recorded with an ICP microphone system

(378B02, PCB Piezotronics, Depew, New York, USA)) within 28 cm distance to the calcar. For data acquisition a 16-bit module was used (NI 9222, National Instruments, Austin, TX, USA). Both signals were triggered by hammer impaction and were recorded for 3 ms with a sampling frequency of 500 kHz via a Labview (National Instrument, Austin TX, USA) interface.

2.2 Statistical analysis

Before the data was analysed, outliers (near zero impaction force and or near zero sound pressure level, as identified visually and by box plots) were discarded. Histograms of all recorded impaction and all acoustic signals were generated. The theoretical continuous density is estimated using the EPANECHNIKOV kernel method and the probabilities of occurrence were calculated. Then the maximum value in acoustic pressure and impact force for each hit and broach were identified and plotted in a scatter plot in order to identify obvious changes between single hits or between different broaches. The maximum values of sound pressure (Pa) are plotted against their maximum impaction force (kN).

2.3 Acoustic transfer function

To determine the system characteristics, i.e. the ability of the tissue/bone-broach interface of the cadaver-implant-system to radiate sound owing to impaction, acoustic transfer functions are determined. We therefore recall the principle of an open-loop controller, i.e. the output (z) does not have any effect on the control of the input values (x), bounded by the system (g) considering only single hits [16]. To calculate the transfer function the spectrum of the impact as source signal and that of the microphone as receiver signal is used, similar to [17] only that now the transfer function is unknown (cf. water level method, Gubbins [18]). In Eq. (1) this relation between the input (source), the system and output (receiver) is expressed in the time domain using convolution integrals; by using the FOURIER domain ($\mathcal{F}$) the convolution changes to a multiplication; and considering a spectrum, a vector notation is used (**bold** letters).

$$x(t) * g(t) = z(t) \xrightarrow{\mathcal{F}} X(\omega)G(\omega) = Z(\omega) \xrightarrow{\vec{(\cdot)}} \mathbf{XG} = \mathbf{Z} \quad (1)$$

Calculating backwards, simplifying $X(\omega) = X$ and conducting an inverse FOURIER transform provides the system's impulse response function $g(t)$, of which we take the real part $\Re$

$$\mathbf{XG} = \mathbf{Z} \xrightarrow{X^{-1}(\cdot)} \mathbf{G} = \mathbf{X}^{-1}\mathbf{Z} \xrightarrow{\mathcal{F}^{-1}} \Re\,(g(t))\,. \quad (2)$$

However, a real-life system is often ill-conditioned with noise in the signal and even for simulation data, the application of an inverse FOURIER transformation generally is problematic [17]. Attempts of signal enhancement with deconvolution can be found in image processing with a point-spread function using wavelet-filtering with soft/hard BAYESIAN thresholding (regularisation) [19]. Another well established method is the water level method [18] or the TIKHONOV regularisation [20], which we use here by adding uncorrelated noise to the denominator to improve the matrix condition [20].

Let $\mathbf{X}$ be the Discrete Fourier Transform of the impaction signal (complex vector), $\mathbf{G}$ be the transfer function of the broach and the cadaver, and $\mathbf{Z}$ be the measured acoustic response. Minimisation of the EUCLIDEAN norm of the difference [18] $\epsilon^2 = \|\mathbf{X}\cdot\mathbf{G}-\mathbf{Z}\|_2^2$ gives an estimate of $\mathbf{G}$, which is generally overdetermined. Due to overdetermination the so-called TIKHONOV or water-level matrix for regularisation is incorporated which leads to the following expression of the excitation signal.

$$\mathbf{G} = \frac{\mathbf{Z}\bar{\mathbf{X}}}{\mathbf{X}\bar{\mathbf{X}} + \bar{\mathbf{\Gamma}}} \quad (3)$$

with $\bar{\mathbf{G}}$ being the HERMITIAN of matrix $\mathbf{G}$ and $\mathbf{\Gamma} = \sqrt{\lambda \mathbf{1}}$ being the regularisation matrix equal to the square-root of the water-level control parameter λ times a vector, which includes only entries of unity [20].

To calculate the FOURIER spectra, a sampling rate of 0.5 MHz and a time series length of $L = 1,500$ samples were selected. The choice of λ is based on comparing spectra within a defined frequency range, here zero to 250 kHz [17] (for $\lambda = 1.2 \times 10^2$). For each hammer hit a transfer function was attracted and plotted for each broach separately. Then, the fundamental frequencies and their peaks (i.e. the fundamental resonances) of the transfer functions were extracted using MATLAB's peak search algorithm.

2.4 Nonlinear time series analysis

In order to confirm a change in the characteristics of the system from being mono-frequent to multi-frequent with greater broach size, (i.e. an increase in complexity), a nonlinear time series analysis is conducted next. The transfer function is converted into its real-valued impulse response $g(t)$ (as a linear time-invariant system's response valid for all frequencies) by conducting an inverse FOURIER transformation (Eq. 2).

The changes of the system characteristics with progressing impaction were estimated by estimating a time delay and an embedding dimension of the impulse response function $g(t)$ (Eq. 2). As complexity measures, the averaged auto-mutual information [21, 22] (10 bins) and the global false nearest neighbour algorithm (maximum norm, falseness of 5 neighbours) were employed [23–25]. To study significant differences, one-way analysis of variance (ANOVA) tests were conducted and combined with a BONFERRONI correction in a *multi-compare* study to account for group size effects, both at a 5% significance level; impacts belonging to one broach were assigned to one group. All analyses were conducted using MATLAB 2016b.

3 Results

3.1 Experiments

Broaches one to six were hit 27, 22, 18, 25, 25 and 35 times. Normalised time traces of the impaction and the microphone signal of the second broach are exemplified in Fig. 2. No obvious difference between the different hits could be observed. After the maximum impaction force is reached, the impulse attenuates quickly and it takes about 1 ms until the maximum sound pressure is reached due to travel time of the sound wave to the microphone.

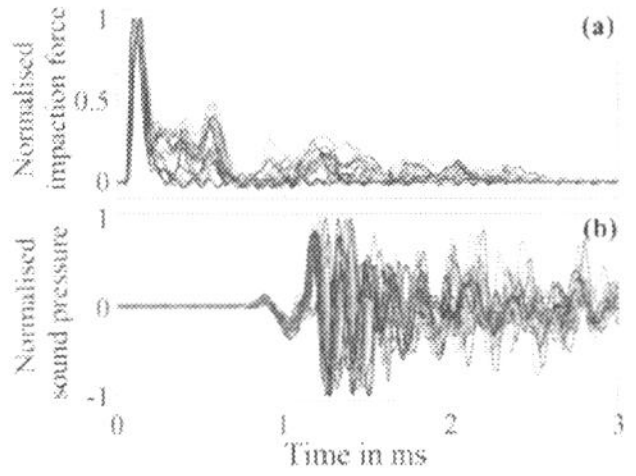

Figure 2. Normalised (a) impaction forces; and (b) sound pressures for impacting the second broach (22 measurements each in total)

3.2 Statistical analysis

For broaches 2, 4 and 6, one data point each with either near zero force or frequency is discarded. Distributions of all maximum impaction amplitudes and all maximum sound pressure in Fig. 3 reveal that they are positively skewed. While the impact force appears uni-modal in the range between 1 kN and 19 kN, the sound pressure signal (1 Pa to 37 Pa) appears to be multimodal consisting of at least two superposed GAUSSIANS. Plotting maximum

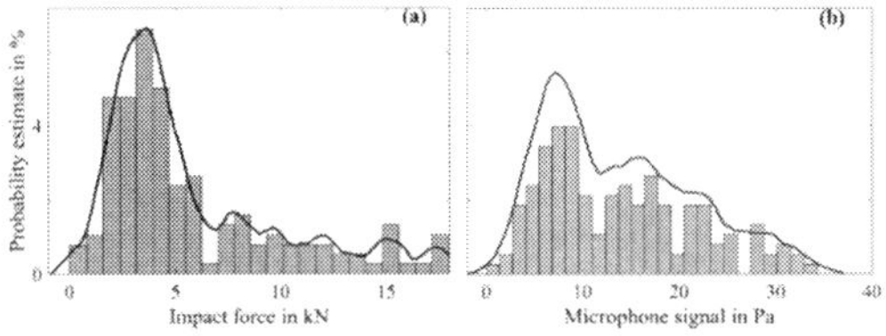

Figure 3. Statistical distribution of maximum amplitudes: (a) impaction signal; and (b) sound pressure signal

sound pressure (dependent variable) against the maximum impaction force (independent variable) for each impact and each broach in a scatter plot (Figure 4) shows that for the first three broaches, maximum impaction forces were less than 5 kN, but that for the last three broaches, they were up to 19 kN. Overall, the data appears linearly correlated but heteroscedastic in character, i.e. an increased variability with larger broaches used. Further, for a given impaction force, the sound pressure produced for broach 4 seems to be significantly higher than all the other broaches except for low impaction force.

3.3 Acoustic transfer function

After varying λ in a parametric study from 10^{-6} to 10^{6} in 101 steps it was set to 1.2×10^{2}. If λ was chosen too

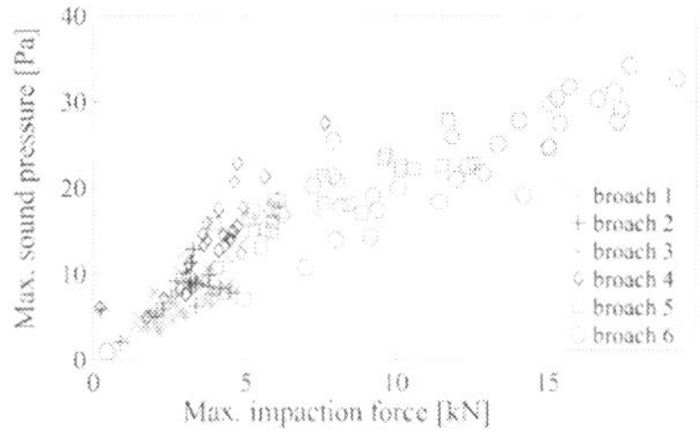

Figure 4. Scatter plot data plotted in Fig. 3 of all six broaches.

small, the transfer function (TF) experienced a low frequency distortion; for a λ too large the amplitude of the TF decreased too quickly after having converged to the same level [17]. Families of transfer functions for broaches 1 to 6 are depicted in Figure 6. Diamond markers indicate the fundamental frequency for each hit. Dashed vertical lines indicate fundamental resonance frequencies overlapping for two hits within ± 1 kHz; bold dashed vertical lines indicate fundamental frequencies overlapping for more than 2 hits within ± 1 kHz. The change in the fundamental fre-

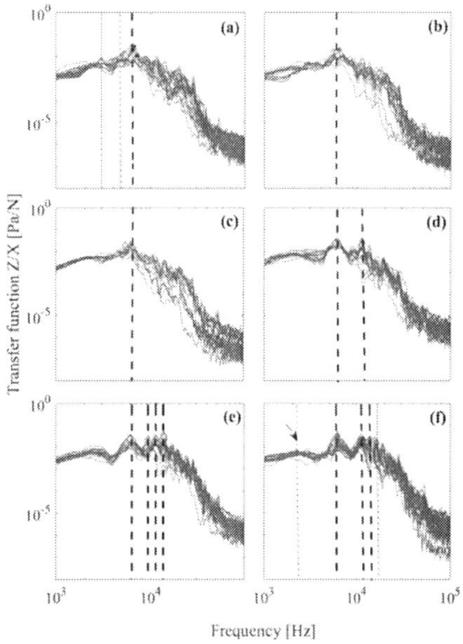

Figure 5. Vibroacoustic transfer functions of all six broaches: For broaches 1 to 3 compared to broaches 4 to 6 the transfer function does not change its fundamental frequency very much as indicated by the smaller number dashed lines.

quency with each impact does not show a trend. However, with increasing broach size, the fundamental frequency initially found at about 6.1 kHz changes its position more, with a tendency to move towards higher frequencies with the spectra becoming more broadband. Similar to [11], also a low frequency resonance is picked up (indicated by arrow in Fig. 5(f)): whether this is due to a standing wave

https://doi.org/10.1051/matecconf/201814814007

within the bone [11] or due to other effects cannot be determined.

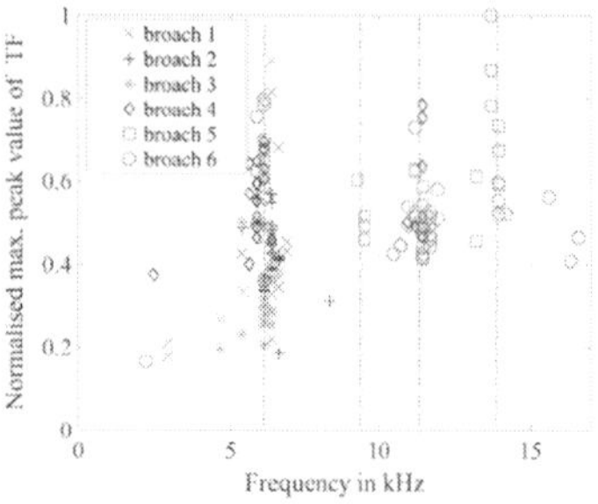

Figure 6. Normalised maximum peak values of the fundamental frequency of the transfer function over resonance frequencies

Next, the peak values as picked from the transfer function are normalised relative to their maximum absolute value and plotted against their frequencies in Fig. 6. Four clusters become visible: around 6.10 kHz (the most often occurring), 9.28 kHz (only for broach 5), 11.47 kHz and 13.92 kHz (indicated by vertical dashed lines).

3.4 Nonlinear time series analysis

Fig. 7(a) shows the real-valued impulse response $g(t)$ and Figs. 7(b) and (c) exemplify the extraction of the delay and a suitable embedding dimension[23, 25].

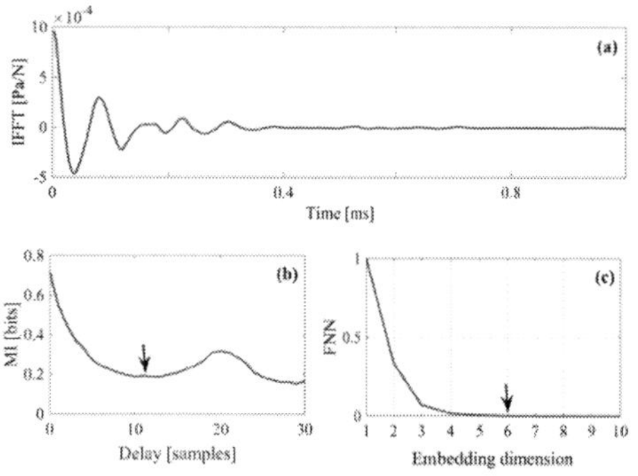

Figure 7. Steps of estimating the delay and the dimension (indicated by arrows) using the (a) impulse response: (b) mutual information, and (c) false nearest neighbour algorithm.

The delay and the embedding dimension for each broach are grouped and analysed using one-way ANOVA tests. The results for the delay and the embedding dimension showed significant differences between the broaches (SS = 845.83 , df = 148, MS = 78.15, F = 24.56, $p << 0.001$) and (SS = 169.05 , df = 148, MS = 8.71, F = 9.92, $p << 0.001$). Results of the posthoc analysis with BONFERRONI correction are depicted in Fig. 8 together with its 95 % confidence intervals (lines). The larger the broach, the smaller the embedding delay and the larger the embedding dimension. Two groups are formed with significantly different delays and dimensions since their 95 % confidence intervals do not overlap; only broach 4 is not significantly different, to either group.

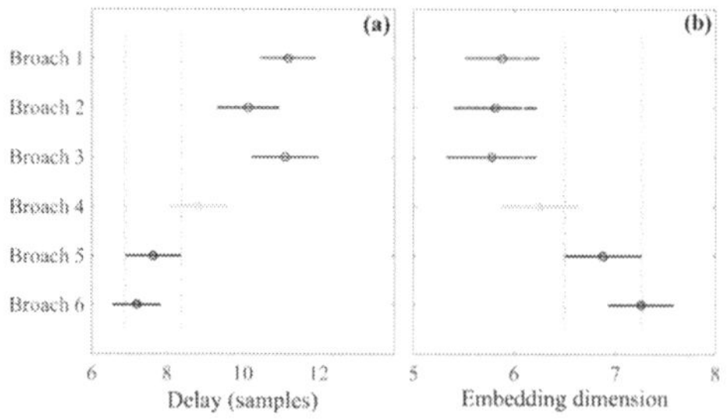

Figure 8. Results of the statistical analysis multi-compare post-hoc test: average (a) delay, and (b) embedding dimension with 95 % confidence interval.

4 Discussion

In this cadaver study we simulated the procedure of a total hip arthroplasty using compaction broaches. By analysing qualitatively the relationship between impaction forces and their resulting sound pressure, we identified at least two groups of broaches with different behaviours as indicated by a superposed GAUSSIAN distribution (multi-modal) and an increase of heteroscedasticity which cannot be explained with the statistical analysis alone. In particular the fourth broach showed relative loud impaction sounds.

The acoustic transfer functions obtained here show that the acoustic radiation of the tissue/bone-broach interface of the cadaver-implant-system depends on the size of the broaches and the working progress during impaction. With increasing broach size the magnitude at the fundamental frequency seems to show a tendency to increase; especially the last three broaches show an up to 5 times higher ability to radiate sound per unit impaction force; however considering their magnitudes the difference would be generally only below 2 dB. With increasing working progress the data also seems to scatter more (heteroscedasticity).

Whitwell et al. [11] observed a shift of the fundamental frequency to a lower frequency and assumed that this is due to a standing wave within the bone once the ideal cavity is prepared. However, we found that as a result of the cavity preparation impacting increasing broach sizes, the system's fundamental frequency changed its location and started fluctuating around the system's initial fundamental frequency of around 6.1 kHz with a tendency towards higher frequencies and also to become more broadband.

This qualitative change in the complexity of the system was consequently quantified by calculating *complexity measures* [25] such as the averaged auto-mutual information and the false nearest neighbour algorithm from the impulse response function to obtain the delay and the global embedding dimension respectively [25]. As the broach size increased, the delay time decreased but the embedding dimension increased, indicating a decreased predictability of the system and an increased complexity of the impulse response function. More complex dynamics require a greater phase space dimension in order to describe the system's dynamics [23]. Whether this increased complexity is due to the broach positioning (better coupling between bone and broach) or micro-fractures (hence an indication to stop the impaction process [12]) needs to be studied further e.g. by ultrasound for damage detection.

Results of this study encourage the development of an instrumented hammer to use nonlinear analysis tools based on complexity measures in a data stream. This hammer could serve the surgeon alone or in conjunction with a robotic arm as a diagnostic tool during the implantation procedure and enhance implant positioning. Similar devices could be applied for knee and shoulder prostheses. The development of such a technology would reduce the likelihood of revision surgeries (currently at about 9.6 % [2]), would have a major impact on patients recovery times and well-being, and reduce the overall treatment costs of total arthroplasties.

The determination of a proper time lag and embedding dimension, however, can be problematic. Here the ideal numbers of neighbours to be searched for is difficult to determine. While the mutual information and the false nearest neighbourhood algorithm are well-established and mature methods, more sophisticated e.g. adaptive neighbourhood search algorithms [26] could enhance the performance in cases of varying measurement quality. Owing to the rather good quality of our data, we did not encounter problems during the nonlinear time series analysis step and the TIKHONOV regularisation performed well. However, during total hip arthroplasties operations, employing an instrumented hammer, signal processing in a data stream and optimised algorithms for real-time processing capability would be desirable.

Declarations and Acknowledgements

The authors declare that no conflict of interest exists and the study was approved by the Ethics Commission of the Medical Association Hamburg. SO acknowledges the receipt of the Australia Award/ Endeavour Research Fellowship (ERF PDR 5563 2016) awarded by the Australian Government.

References

[1] A. Carr, O. Robertsson, S. Graves, A. Price, N. Arden, A. Judge, B. D.J., The Lancet **379**, 1331 (2012)

[2] N.J.R. Registry, Tech. rep., Australian Orthopaedic Association (2016)

[3] S. Oberst, Z. Zhang, G. Campbell, M. Morlock, J. Lai, N. Hoffmann, *Towards the understanding of hip squeak in total hip arthroplasty using analytical contact models with uncertainty*, in *Proceedings of the INTER-NOISE 2016 - 45th International Congress and Exposition on Noise Control Engineering: Towards a Quieter Future 21-24 August 2016, Pages 5539-5549* (2016)

[4] E. Askari, P. Flores, D. Dabirrahmani, R. Appleyard, Tribology International **93**, 239 (2016)

[5] K. Sø balle, E.S. Hansen, H. B.-Rasmussen, P.H. Jø rgensen, C. Bü nger, Journal of Orthopaedic Research **10**, 285 (1992)

[6] J. Bätz, K. Püschel, A. Klein, M. Morlock, G. Campbell, *Determination of bone-cavity interface densification after femoral cavity broaching using QCT*, in *22nd Congress of the European Society of Biomechanics, July 10 - 13, 2016, Lyon, France* (1997), p. 1

[7] R. Sakai, A. Kikuchi, T. Morita, N. Takahira, K. Uchiyama, T. Yamamoto, M. Moriya, K. Uchida, K. Fukushima, K. Tanaka et al., Hip International **21**, 718 (2011)

[8] D. Romness, D. Lewallen, Bone & Joint Journal **72-B**, 761 (1990)

[9] N. Yagi, Y. Oshiro, O. Ishikawa, Y. Hata, N. Shibanuma, 2011 IEEE Workshop on Robotic Intelligence In Informationally Structured Space **12133871**, 32 (2011)

[10] V. Mathieu, A. Michel, C. Flouzat Lachaniette, A. Poignard, P. Hernigou, J. Allain, G. Ha iat, Medical Engineering & Physics **35**, 1558 (2013)

[11] G. Whitwell, C. Brockett, S. Young, M. Stone, T. Stewart, Proceedings of the Institution of Mechanical Engineers. Part H, Journal of Engineering in Medicine **227**, 175 (2012)

[12] I. Morohashi, H. Iwase, A. Kanda, T. Sato, Y. Homma, A. Mogami, O. Obayashi, K. Kaneko, SICOT J 2017, 3, 13 **3**, 1 (2017)

[13] N.H. Fletcher, T.D. Rossing, *The physics of musical instruments* (Springer-Verlag New York, 1998)

[14] E. Askari, P. Flores, D. Dabirrahmani, R. Appleyard, Nonlinear Dynamics **82**, 1039 (2015)

[15] S. Oberst, J.C.S. Lai, Journal of Sound and Vibration **330**, 2978 (2011)

[16] J. DiSteffano, A. Stubberud, I. Williams, *Schaum's Outline of Feedback and Control Systems, Second Edition* (McGraw-Hill Education: New York, Chicago, San Francisco, Athens, London, Madrid, Mexico City, Milan, New Delhi, Singapore, Sydney, Toronto, 2014)

[17] E. Oberst, S. Nava-Baro, J.C.S. Lai, T.A. Evans, Acoustics Australia **43(1)**, 87 (2015)

[18] D. Gubbins, *Time Series Analysis and Inverse Theory for Geophysicists* (Cambridge University Press, 2004)

[19] S. Grace Chang, B. Yu, M. Vetterli, IEEE Transactions on image processing **9**, 1532 (2000)

[20] A. Tikhonov, V.Y. Arsenin, F. John, *Solutions of ill-posed problems* (John Wiley & Sons, New York, Toronto, London, Sydney, 1977)

[21] A.M. Fraser, H.L. Swinney, Physical Review A **33**, 1134 (1986)

[22] D. Abásolo, J. Escudero, R. Hornero, C. Gómez, P. Espino, Medical & Biological Engineering & Computing **46**, 1019 (2008)

[23] H.D.I. Abarbanel, *Analysis of observed chaotic data* (Springer: New York, 1996)

[24] N. Marwan, M. Carmen Romano, M. Thiel, J. Kurths, Physics Reports **438**, 237 (2007)

[25] S. Oberst, J. Lai, Journal of Sound and Vibration **334**, 120 (2015)

[26] N. Garcia-Pedrajas, J. A. Romero del Castillo, G. Gonzalo Cerruela-Garcia, IEEE Transactions on Neural Networks and Learning Systems pp. 470–475 (2015)

Vibration Based Damage Detection of Rotating Beams

Stanislav Stoykov[1,*], *Emil Manoach*[2] *and Maosen Cao*[3]

[1]Institute of Information and Communication Technologies, Bulgarian Academy of Sciences, Bulgaria
[2] Institute of Mechanics, Bulgarian Academy of Sciences, Bulgaria
[3] Department of Engineering Mechanics, Hohai University, Nanjing, China

Abstract. The early detection and localization of damages is essential for operation, maintenance and cost of the structures. Because the frequency of vibration cannot be controlled in real-life structures, the methods for damage detection should work for wide range of frequencies. In the current work, the equation of motion of rotating beam is derived and presented and the damage is modelled by reduced thickness. Vibration based methods which use Poincaré maps are implemented for damage localization. It is shown that for clamped-free boundary conditions these methods are not always reliable and their success depends on the excitation frequency. The shapes of vibration of damaged and undamaged beams are shown and it is concluded that appropriate selection criteria should be defined for successful detection and localization of damages.

1 Introduction

Modelling of the dynamic behavior of rotating beams is very important for many engineering applications like wind power generators, helicopter blades, etc. The early detection and localization of damages of these structures is essential for their operation and maintenance. Vibration based methods for damage detection are among the most popular methods for health monitoring of structures. The easiest ones are based on the modal properties of the structures. The lowest natural frequencies, however, are not very sensitive to small damages and they are not indicative to predict the presence of damage and even less its location. The modes of vibrations are more sensitive to damages. The difficulty for their use comes from the fact that many sensors have to be used. This difficult is overcame by using the scanning laser Doppler vibrometer (SLDV) techniques. Even by using SLDV the modal based features could be difficult to extract when the objects of monitoring are large and/or complex structures and when some parts of these structures are either inaccessible or very difficult for taking measurements. In the last decades the vibration based methods for damage detection are not limited to monitoring the changes in modal properties of the structures but are extended to analysis of the response of the structures in the time domain. The researches using such methods try to extract some features from the signals in the forced response of structures. These methods are more appropriate for in-operation structural health monitoring (SHM) of structures subjected to dynamic loading and they poses a generality. A review of the time domain method for SHM with accent on the ones using the nonlinear response of the structure could be found in [1]. A big and very promising group of methods, which use measured time series are the one, used the statistical approaches [1- 3]. One of the time-series methods is the method based on the Poincaré maps of the forced response of the structures [3, 4]. It was tested and verified experimentally for composite beams with delamination in [5]. This approach was not tested, however, for the important from the practical point of view case for rotating beam. In the present work the Poincaré map based method for damage assessment was tested for rotating beams and conclusions about its applicability and further development was drawn.

2 Mathematical model

It is assumed that the beam can vibrate in longitudinal and transverse directions and rotate in a plane perpendicular to the plane of transverse vibrations. The Timoshenko's beam theory is considered and the rotation of the beam is introduced into the equation of motion through the inertia forces. Geometrical type of nonlinearity is included in the model. The equation of motion is derived by the principle of virtual work and it is discretized by the finite element method.

Considering Timoshenko's beam hypothesis, the displacements in longitudinal and transverse directions are expressed by the displacements on the middle line and the rotation of the cross section:

$$u(x,y,z,t) = u_0(x,t) + z\phi_y(x,t),$$
$$w(x,y,z,t) = w_0(x,t), \quad (1)$$

*Corresponding author: stoykov@parallel.bas.bg

where $u_0(x,t)$ and $w_0(x,t)$ represent the displacements in longitudinal and transverse directions and $\phi_y(x,t)$ represent the rotations of the cross section about y axis.

Nonlinear strain-displacement relations are considered in the model, which are known as geometrical type of nonlinearity. Hence the strains are expressed as:

$$\varepsilon_x = \frac{\partial u}{\partial x} + \frac{1}{2}\left(\frac{\partial w}{\partial x}\right)^2, \qquad (2)$$
$$\gamma_{xz} = \frac{\partial w}{\partial x} + \frac{\partial u}{\partial z} + \frac{\partial w}{\partial x}\frac{\partial w}{\partial z}.$$

The material of the beam is assumed to be homogeneous and isotropic. The following constitutive relations are used:

$$\begin{Bmatrix} \sigma_x \\ \tau_{xz} \end{Bmatrix} = \begin{bmatrix} E & 0 \\ 0 & \lambda G \end{bmatrix} \begin{Bmatrix} \varepsilon_x \\ \gamma_{xz} \end{Bmatrix}, \qquad (3)$$

where σ_x and τ_{xz} are bending and shear stresses, E is the Young's modulus of the material, G is the shear modulus and λ is shear correction factor.

The rotation of the beam is included in the equation of motion through the inertia forces. For that purpose, two coordinate systems are considered: one fixed, denoted by S_0, and another one rotating about the fixed coordinate system denoted by S_1. The displacement fields (1) are written in the rotating coordinate system S_1, which is assumed to rotate with constant speed $\dot{\vartheta}$ (rad/s). The absolute acceleration of an arbitrary point P(x,y,z) from the beam is expressed with respect to the fixed coordinate system S_0 [6]:

$$\boldsymbol{a}_{P_{S_0}} = \boldsymbol{a}_{P_{S_1}} + \boldsymbol{\omega} \times (\boldsymbol{\omega} \times \boldsymbol{r}), \qquad (4)$$

where $\boldsymbol{a}_{P_{S_0}}$ represents the absolute acceleration of point P, $\boldsymbol{a}_{P_{S_1}}$ represents the relative acceleration of point P, ω represents the angular velocity vector of the rotating coordinate system, and $\boldsymbol{r}$ represents the position vector of point P with respect to the origin of S_1.

The equation of motion is derived by the principle of virtual work:

$$\delta W_V + \delta W_{in} + \delta W_E = 0 \qquad (5)$$

where δW_V represents the virtual work of internal forces, δW_{in} represents the virtual work of inertia forces, δW_E represents the virtual work of external forces.

The resulting partial differential equations are discretized by the finite element method. Quadratic elements with three nodes are used for the discretization. The following system of nonlinear ordinary differential equations is obtained:

$$\mathbf{M}\ddot{\mathbf{q}} + \beta\mathbf{K}_l\dot{\mathbf{q}} + \mathbf{K}_l\mathbf{q} + \mathbf{K}_{nl}(\mathbf{q})\mathbf{q} + \mathbf{T}(\dot{\vartheta})\,\mathbf{q} = \mathbf{f} - \mathbf{r}(\dot{\vartheta}) \qquad (6)$$

where $\mathbf{M}$ represents the mass matrix, $\mathbf{K}_l$ represents the stiffness matrix of constant terms, $\mathbf{K}_{nl}(\mathbf{q})$ represents the stiffness matrix that depends on the vector of generalized coordinates $\mathbf{q}$, it results from the geometrical nonlinearity of the model and introduces quadratic and cubic terms at the equation of motion. $\mathbf{f}$ represents the vector of generalised external forces. $\mathbf{T}(\dot{\vartheta})$ and $\mathbf{r}(\dot{\vartheta})$ are matrix and vector that arise from the rotation of the beam and depend on the speed of rotation. Stiffness proportional damping is introduced in the model.

3. Damage detection formulas

The Poincaré map based method introduced in [3, 4] is based on the differences which arise in the phase space of the responses of the healthy and damaged structures. The Poincaré maps preserves many properties of periodic and quasiperiodic orbits of the system and has a lower-dimensional state space then the time-history series.

The method requires calculation of a damage index constructed in the following way:

$$I_i^d = C\left(S_i^u - S_i^d\right) \qquad (7)$$

where

$$S_i^u = \sum_{j=1}^{N_p-1} \sqrt{\left(w_{i,j+1}^u - w_{i,j}^u\right)^2 + \left(\dot{w}_{i,j+1}^u - \dot{w}_{i,j}^u\right)^2}$$
$$S_i^d = \sum_{j=1}^{N_p-1} \sqrt{\left(w_{i,j+1}^d - w_{i,j}^d\right)^2 + \left(\dot{w}_{i,j+1}^d - \dot{w}_{i,j}^d\right)^2} \qquad (8\text{ a,b})$$

$i=1,2\ldots N_{\text{nodes}}$

In Eqs. (7) and (8) , N_{nodes} is the number of nodes, N_p is the number of points on the Poincaré map $(w_{ij}^u, \dot{w}_{ij}^u)$ and $(w_{ij}^d, \dot{w}_{ij}^d)$ denote the jth point on the Poincaré maps of the undamaged and the damaged states of ith node , respectively.

C is a constant which could be connected with the number of the nodes in the Poincaré map (for example $C=1/N_p$.). In our previous studies, C was chosen to be a function of x, as $C = S_i^u$. In the most cases this approach gives very good results. In some cases, however S_i^u could be very small at the areas close to the clamped ends of the structure and this could lead to wrong results.

This difference given with Eq. (7) is taken as a measure for the global change (during the full period of vibration) in the dynamic behaviour of the damaged structure comparing with the undamaged one. If the function $I^D(x)$ has a maximum and it is strong concave in the vicinity of the maximum it means that the damage is located close to the nodes where is located the maximum. Thus for the structure with damage the nodes close to the maximal value of the function will represent the damaged area.

In [7], by using the analogy of the Poincaré map based method with the modal displacement method and the fact that the modal curvature method gives better results then the modal displacement method, the damage index (eq. (7)) was improved .

The following damage index based again on the Poincaré maps was suggested there:

$$DI2_i = C\left(\left|\frac{\partial^2(S_i^u - S_i^d)}{\partial x^2}\right|\right) \quad i = 1, 2, \ldots N_{nodes}$$

This expression, of course, is far more complicated than the modal curvature damage index but the second derivatives of functions of the displacements and the velocity of the response are included in it.

If one wants to enlarge the differences in the second derivatives of these functions the following formula can be used:

$$DI3_i = C^2\left(\left|\left(\frac{\partial^2 S_i^u}{\partial x^2}\right)^2 - \left(\frac{\partial^2 S_i^d}{\partial x^2}\right)^2\right|\right) \quad i = 1, 2, \ldots, N_{nodes}$$

4 Numerical experiments

A beam with homogeneous and elastic material is considered for the numerical experiments. The damage is modelled by reducing the thickness of the damaged area of the beam. The material properties of the beam are: Young modulus$E = 70$ GPa, density$\rho = 2778$ kg/m^3, Poisson ratio $\nu = 0.3$ and the geometrical properties are: length $l = 0.58$ m, width $b = 0.02$ m, thickness $h = 0.002$ m. The reduced thickness of the damaged beam is assumed to be $h^d = 0.0015$ m. The equation of motion is discretized by quadratic finite elements. Selective and reduced integration is used for the shear strain energy, in order to avoid shear locking. 32 finite elements are used for discretization which leads to model with 65 nodes. The damage is considered to be in elements 21 and 22 which is between nodes 41 and 45. The total number of DOF of the system is 195.

Numerical experiments for detection and localization of the damage are performed on beams with clamped-clamped and clamped-free boundary conditions. Newmark's method is used for solving the ODE in time domain and Newton's method is used for solving the nonlinear system. Different speeds of rotation are assumed for the clamed-free boundary conditions.

First, a beam with clamped-clamped boundary conditions is analysed. An excitation frequency of 209 rad/s is applied in all examples. Very often in real-world applications, like damage detection in wind turbine blades, the excitation frequency of the external conditions cannot be controlled, thus the methods for damage detection should work for wide variety of excitation frequencies. The selected excitation frequency is close to the fundamental frequency of the clamped-clamped beam. It is close to the second linear frequency of clamped-free beam, and when the beam rotates with speed of 100 rad/s, the excitation frequency is between the first and the second natural frequencies of the rotating beam.

The results for damage indexes of a beam with clamped-clamed boundary conditions calculated according to formulas (7) and (9) are presented in Fig. 1 and 2, respectively. DI2 has similar behavior as DI3. It can be seen that the location of the damage is successfully detected.

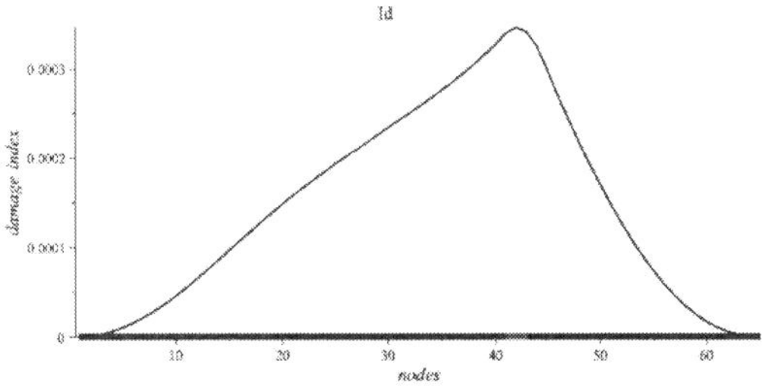

Fig. 1. Damage index I^d applied to beam with clamped-clamped boundary conditions. The red segment presents damaged zone.

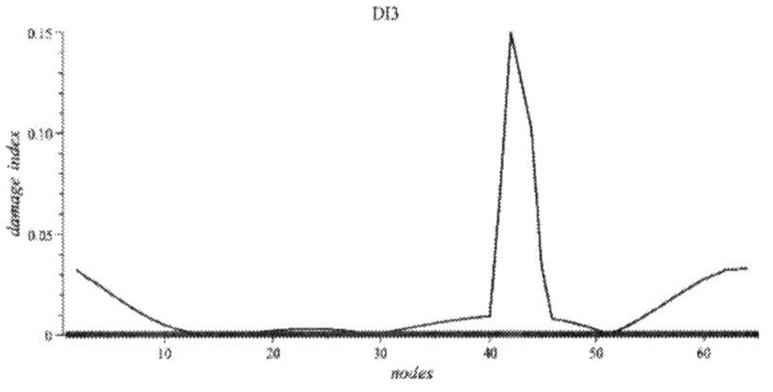

Fig. 2. Damage index DI3 applied to beam with clamped-clamped boundary conditions. The red segment presents damaged zone.

When the boundary conditions of the beam are clamped-free, the presented formulas do not always localize the damage. The formulas have higher amplitudes on different location of the beam, depending on the shape of vibration. Fig. 3 and 4 present damage index DI3 for non-rotating and rotating beams.

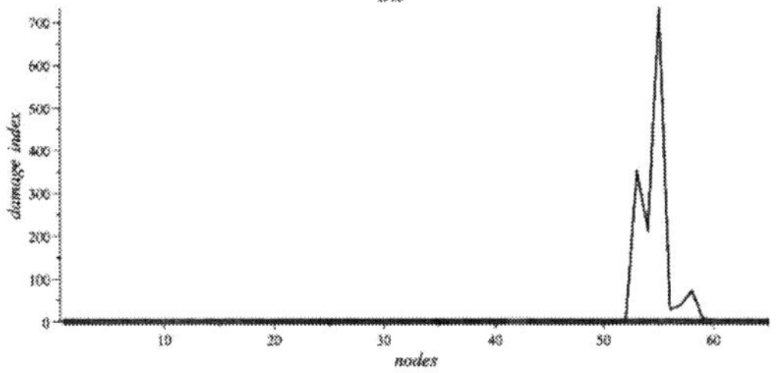

Fig. 3. Damage index DI3 applied to non-rotating beam with clamped-free boundary conditions.

In both cases, the damage is not appropriately localized. The reason for not localizing the damage is the free boundary condition. This boundary condition allows displacements of the beam with higher amplitude not only near the damaged area, but also near the free end and in other areas along the length, depending on the excitation frequency and the shape of vibration. The shapes of vibration of damaged and undamaged beams and their curvatures are presented in Figs. 5-8.

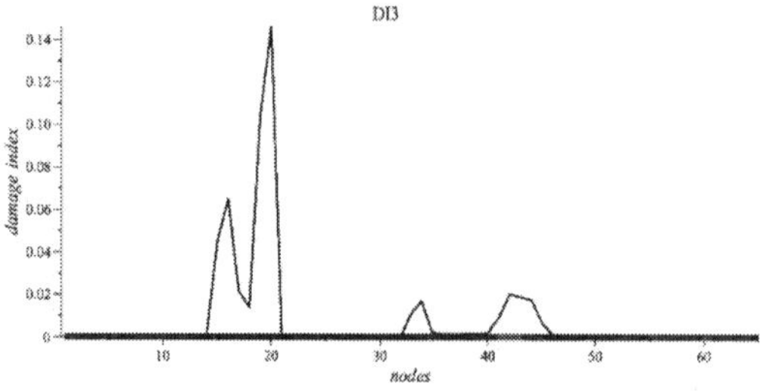

Fig. 4. Damage index DI3 applied to rotating beam with speed 100 rad/s with clamped-free boundary conditions.

The first two figures present the transverse displacements and the curvature of non-rotating clamped-free beam and the other two figures present the displacements and curvatures of rotating beams. It can be seen that the differences between damaged and undamaged beams are significant on locations different from the damaged area.

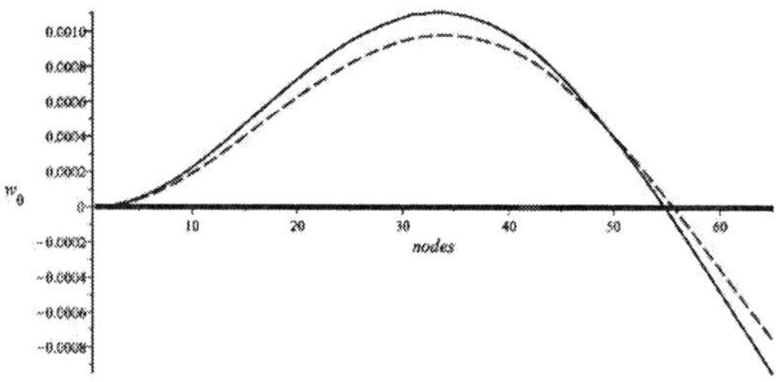

Fig. 5. Comparison of transverse displacement w_0 of damaged and undamaged beam at time t = 4.962 s. Non-rotating beam with clamped-free boundary conditions. Blue dash line presents damaged beam.

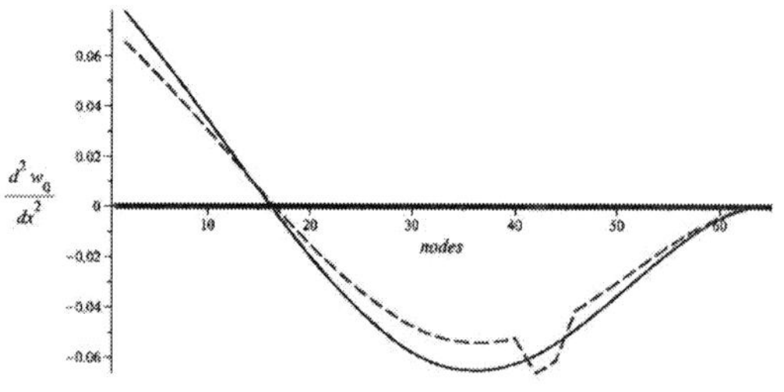

Fig. 6. Comparison of second derivative of transverse displacement w_0 of damaged and undamaged beam at time t = 4.962 s. Non-rotating beam with clamped-free boundary conditions. Blue dash line presents damaged beam.

There are several possibilities for implementation of the selection criteria. For example, the selection criteria could be applied to the nodes which have higher amplitude in the curvature of the damaged beam. This criterion gives very good results in the numerical experiments, shown in Fig. 9. Further numerical experiments and improvements of the selection criteria will be developed in the future work of the authors

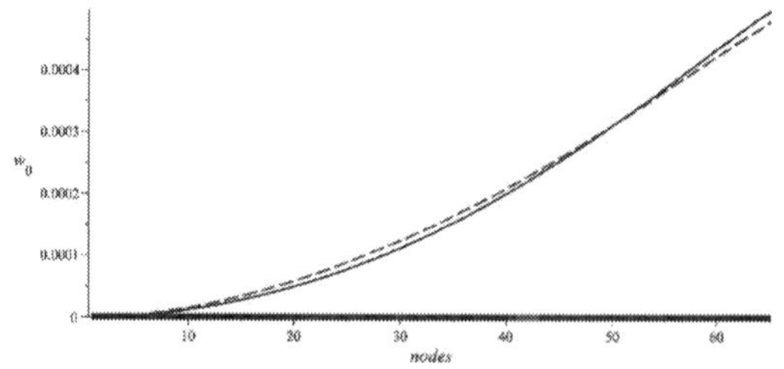

Fig. 7. Comparison of transverse displacement w_0 of damaged and undamaged beam at time t = 1.2 s. Rotating beam with speed 100 rad/s with clamped-free boundary conditions. Blue dash line presents damaged beam.

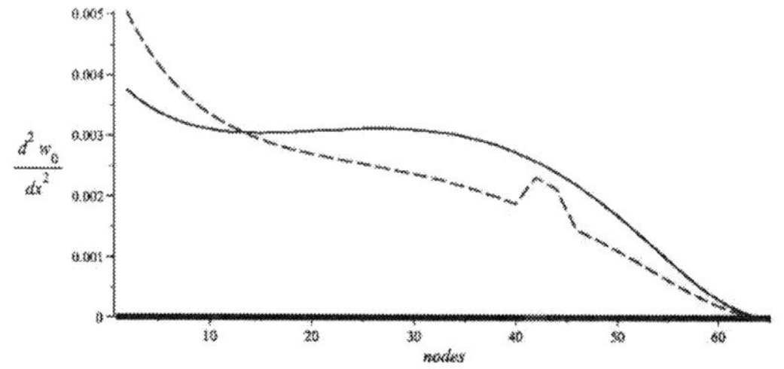

Fig. 8. Comparison of second derivative of transverse displacement w_0 of damaged and undamaged beam at time t = 1.2 s. Rotating beam with speed 100 rad/s with clamped-free boundary conditions. Blue dash line presents damaged beam.

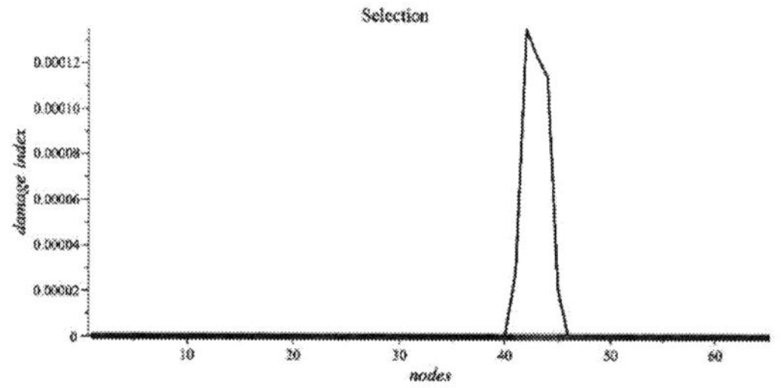

Fig. 9. Damage index obtained by selection of second order derivatives applied to rotating beam with speed 100 rad/s with clamped-free boundary conditions.

5 Conclusions

The equation of motion for rotating beams that vibrate in one plane was derived and presented. Numerical experiments for localization of damages were performed by using vibration based methods and Poincaré maps. The influence of the speed of rotation on the shapes of vibration was outlined. It was shown that in the case of clamped-free boundary conditions, the methods not always localize correctly the damage. Additional methods which used appropriate selection of nodes were developed and it is shown that they perform much better. The future work will continue with the development and improvement of criteria for selection of the nodes and analysis of clamped-free beams with damages.

Acknowledgements

The support of this work through the Bulgarian NSF Grant DNT-SO14 is gratefully acknowledged.

References

1. Trendafilova, I., Manoach, E., Nonlinear dynamics time series methods for fault detection and identification in structural and robotic systems. Book chapter in a book "Fault diagnosis in robotic and industrial systems", (G. G. Rigatos –Editor) iConceptPress, Queensland, Australia, I, 111-138, 2012.
2. García D., Cherniak, T., Trendafilova I. J. Phys.: Conf. Ser., 2015 , 628, 012086
3. Trendafilova I., Manoach E. Mech. Syst. Signal Pr., **22**, 1092-1106 (2008).
4. Manoach E., Trendafilova I.. J.Sound Vib., **315** (3), 591-606 (2008)
5. Manoach, E., Samborski, S., Mitura, A., Warminski J.. . Int. J. Mech. Sci.; 62, 120–132 (2012).
6. Shabana, A. *Dynamics of multibody systems*, Cambridge University Press, (2013).
7. E. Manoach, J. Warminski, L. Kloda, A. Teter, , Compos. Struct. **170,** 26-39 (2017).

Reflection and Transmission of Acoustic Waves through the Layer of Multifractional Bubbly Liquid

Damir Anvarovich Gubaidullin[1], and *Ramil Nakipovich* Gafiyatov[1,*]

[1]Institute of Mechanics and Engineering KazSC RAS, Department of Continuum Mechanics, 420111, Kazan, Russian Federation

Abstract. The mathematical model that determines reflection and transmission of acoustic wave through a medium containing multifractioanl bubbly liquid is presented. For the water-water with bubbles-water model the wave reflection and transmission coefficients are calculated. The influence of the bubble layer thickness on the investigated coefficients is shown. The theory compared with the experiment. It is shown that the theoretical results describe and explain well the available experimental data. It is revealed that the special dispersion and dissipative properties of the layer of bubbly liquid can significantly influence on the reflection and transmission of acoustic waves in multilayer medium.

1 Introduction

The investigations of wave dynamics and acoustics of disperse media are of significant interest. Currently, the basics of mechanics and thermophysics of bubbly liquids, as well as the most significant results in the study of wave processes in these environments are presented in monographs [1-5]. Hao and Prosperetti [6] obtained the results of study of the vapor bubble dynamics in liquid. Commander and Prosperetti [7] presented a model of the propagation of plane waves with small amplitude in a mixture of liquid and of gas bubbles. This model is adequate for the frequencies far from resonance at the volume content of a disperse phase of 1-2%. Nigmatulin et al. [8] showed the need of allowing for the effect of liquid compressibility for problems of acoustics of bubbly liquids. Shagapov [9] considered a problem of propagation of small disturbances in liquids with polydisperse bubbles. Nigmatulin et al. [10] studied the small radial oscillations of vapor-gas bubbles in liquid under the influence of an acoustic field. They showed that capillary effects and phase transitions lead to the new resonant frequency of small bubbles. This frequency is different from the Minnaert frequency.

The propagation of acoustic waves in a two-fractional mixture of liquid with bubbles is investigated in [11-13]. A mathematical model of the propagation of sound waves in the two-fractional mixture of liquid with polydisperse vapor-gas and gas bubbles with phase transitions is shown in [14-16]. The propagation of acoustic waves in multifractional mixtures of liquid with gas and vapor-gas bubbles of different sizes and different compositions is studied in [17-19]. The problem of reflection of an acoustic wave from a two-layer medium containing a layer of bubble liquid is considered in [20-22].

In this study the acoustic wave reflection and transmission through a multilayer object containing a multifractional bubbly fluid layer is investigated.

2 Dispersion relation

Linearized equations for one-dimensional disturbances in a multifractional mixture of liquid with bubbles are obtained from the general equations of motion for bubbly mixtures [1]. The dispersed phase consists of N+M fractions having various gases in bubbles and different in the bubbles radii. Phase transitions accounted for N fractions. The total bubble volume concentration is small (less than 1%). Solving this system the dispersion relation is obtained:

$$\left(\frac{K_*}{\omega}\right)^2 = \frac{1}{C_f^2} + \frac{\rho_{10}}{p_0}\left(\frac{(S_{1i}+S_{1j})(S_{3i}+S_{3j})}{S_{2i}+S_{2j}-i\omega} + S_{Ri} + S_{Rj}\right) \quad (1)$$

where C_f is the frozen speed of sound ($C_f = C_1/\alpha_{10}$) and also the following notations are adopted:

$$S_{1i} = \sum_{i=1}^{M} H_{1i}\,,\ S_{1j} = \sum_{j=1}^{N} H_{1j}\,,\ S_{3i} = \sum_{i=1}^{M} H_{3i}\,,$$

$$S_{3j} = \sum_{j=1}^{N} H_{3j}\,,\ S_{2i} = \sum_{i=1}^{M}\left(\frac{m_i}{\tau_{T1i}} + H_{2i}\right),$$

$$S_{2j} = \sum_{j=1}^{N}\left(\frac{m_j}{\tau_{T1j}} + H_{2j}\right),\ S_{Ri} = \sum_{i=1}^{M}\left(\frac{\alpha_{20i}}{N_{Ri}}\left(1-\frac{M_{4i}}{M_{3i}}\right)\right),$$

* Corresponding author: gafiyatov@mail.ru

$$S_{Rj}=\sum_{j=1}^{N}\left(\frac{\alpha_{20j}}{N_{Rj}}\left(1-\frac{M_{4j}}{M_{3j}}\right)\right),$$

$$H_{1j}=\frac{m_j}{\tau_{T1j}}\left(\frac{M_{1j}M_{4j}}{M_{3j}}+M_{2j}\right),$$

$$H_{1i}=\frac{m_i}{\tau_{T1i}}\left(\frac{M_{1i}M_{4i}}{M_{3i}}+M_{2i}\right),\ H_{2j}=\frac{m_j}{\tau_{T1j}}\frac{M_{1j}b_j}{M_{3j}},$$

$$H_{2i}=\frac{m_i}{\tau_{T1i}}\frac{M_{1i}b_i}{M_{3i}},\ H_{3j}=\frac{\alpha_{j0}}{N_{Rj}}\frac{b_j}{M_{3j}},\ H_{3i}=\frac{\alpha_{i0}}{N_{Ri}}\frac{b_i}{M_{3i}},$$

$$M_{1j}=G_j-M_{2j}-\frac{L_{1j}N_{3j}}{L_{4j}-\delta N_{2j}},\ M_{2j}=\frac{N_{2j}L_{1j}}{N_{Rj}(L_{4j}-\delta N_{2j})},$$

$$M_{3j}=\frac{N_{3j}\delta-L_{2j}N_{3j}}{L_{4j}-\delta N_{2j}}+L_{3j}+M_{4j},$$

$$M_{4j}=\frac{L_{4j}-L_{2j}N_{2j}}{N_{Rj}(L_{4j}-\delta N_{2j})},\ M_{1i}=-\left(N_{3i}+M_{2i}\right),$$

$$M_{2i}=\frac{N_{2i}}{N_{Ri}},\ M_{3i}=1+N_{3i}\left(1+b_i\right)+M_{4i},$$

$$M_{4i}=\frac{1}{N_{Ri}}\left(1+N_{2i}\left(1+b_i\right)\right),\ N_{1j}=\frac{i\omega\tau_{T1j}}{m_j}-1,\ N_{1i}=1,$$

$$N_{2j}=i\omega\tau_{Tj}-1,\ N_{2i}=i\omega\tau_{Ti}-1,$$

$$N_{3j}=k_{2j}\left(c_j-R_j\right)-1+G_j,\ N_{3i}=k_{2i}\left(c_i-R_i\right)-1,$$

$$L_{1j}=E_j\left(1-i\omega\tau_j\right),\ E_j=\frac{R_V}{R_j}\frac{p_0}{l_0\rho^{\circ}_{V0}}\left(1-k_{Vj0}\right),$$

$$L_{2j}=-\frac{l_0k_{2j}}{\left(1-k_{Vj0}\right)T_0}+\Delta R_j-L_{1j}\left(1+b_j\right),$$

$$L_{3j}=1-G_j\left(1+b_j\right),\ L_{4j}=L_{1j}+\Delta R_j N_{2j},\ k_{2j}=\frac{i\omega\tau_{Tj}}{c_j},$$

$$k_{2i}=\frac{i\omega\tau_{Ti}}{c_i},\ b_j=\frac{c_1\tau_{Tj}}{c_j\tau_{T1j}},\ b_i=\frac{c_1\tau_{Ti}}{c_i\tau_{T1i}},$$

$$N_{Rj}=\frac{-(i\omega)(a_{j0})^2G_{Rj}\rho^{\circ}_{10}}{3(t_{Aj}G_{Rj}+1)p_0},\ N_{Ri}=\frac{-(i\omega)(a_{i0})^2G_{Ri}\rho^{\circ}_{10}}{3(t_{Ai}G_{Ri}+1)p_0},$$

$$G_{Rj}=\frac{1}{t_{Rj}}-i\omega,\ G_{Ri}=\frac{1}{t_{Ri}}-i\omega,\ t_{Rj}=\frac{(a_{j0})^2}{4\nu_1},$$

$$t_{Ri}=\frac{(a_{i0})^2}{4\nu_1},\ t_{Aj}=\frac{a_{j0}}{C_1(\alpha_{j0})^{1/6}},\ t_{Ai}=\frac{a_{i0}}{C_1(\alpha_{i0})^{1/6}},$$

$$m_{j0}=\frac{\rho^{\circ}_{j0}}{\rho^{\circ}_{10}},\ m_{i0}=\frac{\rho^{\circ}_{i0}}{\rho^{\circ}_{10}},\ m_j=\frac{\rho_{j0}}{\rho_{10}},\ m_i=\frac{\rho_{i0}}{\rho_{10}},$$

$$i=1,2,\ldots M,\ j=1,2,\ldots N.$$

Here, a – the radius of the bubbles, α – the volume concentration, ρ°, ρ – the true and average densities, p – the pressure, c – the specific heat capacity, ν – the kinematic viscosity, R – the gas constant, T – the temperature, l – the specific heat of evaporation, λ – the thermal conductivity, k – the weight concentration, τ – the relaxation time. Subscript 1 refers to the parameters of the liquid phase, 2 – disperse phase, C_1 – the velocity of sound in pure liquid.

The dispersion relation (1) considered as the dependence of the complex wavenumber K_* on the vibration frequency ω determines propagation of acoustic perturbations in multifractional mixtures of a liquid with vapor-gas and gas bubbles (of various initial radii, various initial volume contents, and various thermophysical properties of gases in the bubbles).

3 Mathematical model

In analysing the interaction between an acoustic signal and a multilayer object (Fig. 1), the following method of calculations is used. According to [23], the result of the reflection and transmission of a plane monochromatic wave $\exp(iK_*x-\omega t)$ from a multilayer object is the plane waves $R\exp(iK_*x+\omega t)$ and $T\exp(iK_*x-\omega t)$, where R and T are the wave reflection and transmission coefficients, respectively, determined in terms of the layer impedances Z_i and the entry impedances of the layer boundaries Z_i^{in}. For a three-layer medium the coefficients R and T are as follows:

$$R=\frac{Z_2(Z_1-Z_3)-i(Z_2^2-Z_1Z_3)\mathrm{tg}(K_2d_2)}{Z_2(Z_1+Z_3)-i(Z_2^2+Z_1Z_3)\mathrm{tg}(K_2d_2)}$$

$$T=\prod_{j=1}^{2}\left(\frac{Z_j^{in}+Z_j}{Z_j^{in}+Z_{j+1}}\exp[iK_jd_j]\right) \qquad (2)$$

$$Z_j=\rho_j\frac{\omega}{K_j},\ Z_2^{in}=\frac{Z_1-iZ_2\mathrm{tg}(K_2d_2)}{Z_2-iZ_1\mathrm{tg}(K_2d_2)}Z_2,\ \ j=1,2,3.$$

Here, d_2 is the thickness of the layer with bubbles, K is the wave number, ω is the perturbation frequency, and ρ is the layer density. For a homogeneous layer, the wave number is determined as $K_j=\omega/C_j$, where C_j is the speed of sound in the j-th layer.

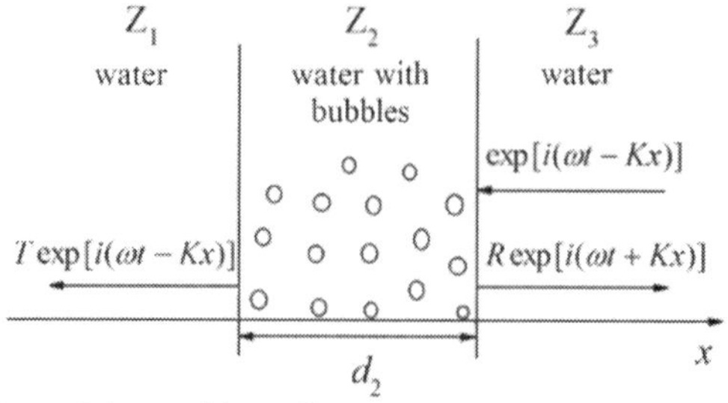

Fig. 1. Scheme of the medium.

4 Results

We will consider the acoustic signal transmission through the three-layer medium, namely, water-water with bubbles-water. Disperse phase of bubbly liquid contain vapor-air bubbles, bubbles of carbon dioxide with water vapor and bubbles of helium. Let the bubble layer thickness be d_2=4 mm, the bubble radii – 2mm (vapor-air bubbles), 1mm (bubbles of carbon dioxide with water vapor), 1.5mm (bubbles of helium), f_0=1630Hz (resonance frequency of 2mm bubbles) and the volume content α_{20}=0.01. The mixture pressure p_0=0.1 MPa and the temperature T_0=288 K. The calculations were performed according equations (1), (2). The dependences of the coefficients of the wave reflection and transmission through the given bubble layer on the dimensionless disturbance frequency are presented in the figure 2. Clearly that a minimum of the transmission coefficient and a maximum of the reflection coefficient are observable in the region of the resonance frequency of the bubbles. This means that at the given frequency the bubble layer almost completely reflects the incident acoustic wave. In addition, three fractions with different initial radii of the disperse phase leads to the appearance of three local minima and maxima on this coefficients. This is due to the difference in the values of the resonance frequencies of the intrinsic vibrations of bubbles of each fraction.

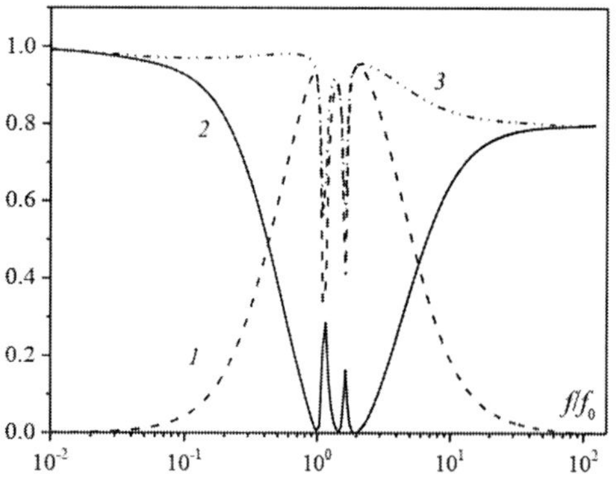

Fig. 2. Coefficients of sound reflection and transmission through a bubbly screen, 1 – R^2, 2 – T^2, 3 – R^2+T^2.

On the figures 3, 4 the influence of the bubble layer thickness on the acoustic wave transmission and reflection is illustrated. Parameters of bubbly liquid are the same as on figure 2, curves *1* – d_2=4mm, curves *2* – d_2=8mm, curves *3* – d_2=12mm, curves *4* – d_2=16mm. An increase in the bubble layer thickness leads to a decrease in the transmission coefficient and, correspondingly, to an increase in the reflection coefficient on the entire frequency range, except from a vicinity of the resonance frequency of the bubbles. With increase in the bubble layer thickness the range near the resonance frequency ($f \approx f_0$), where the wave transmission coefficient takes near-zero values (opposite effect on the reflection coefficient), also expands.

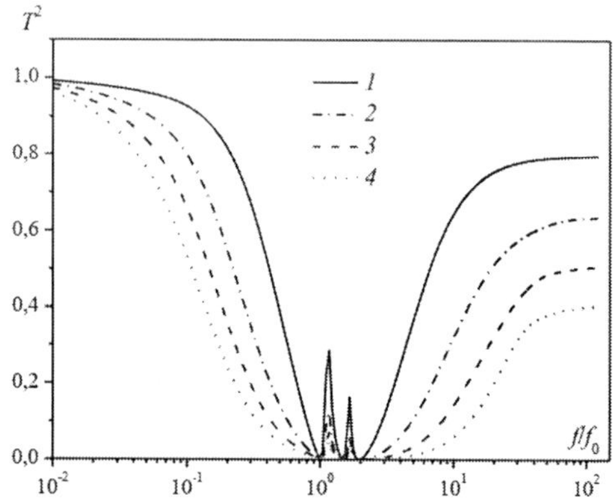

Fig. 3. Coefficient of sound transmission through a bubbly screen at different layer thickness.

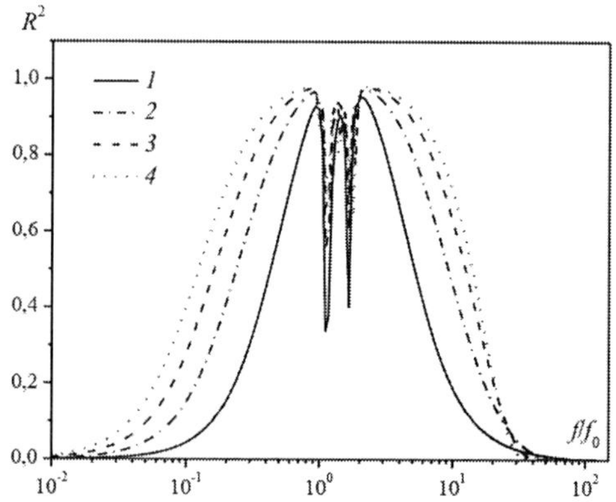

Fig. 4. Coefficient of sound reflection through a bubbly screen at different layer thickness.

On the figure 5 we have compared the theory with the experimental data [24] for coefficients of sound reflection and transmission through a bubbly screen. In this paper, heat and mass transfer are taken into account in our model, in contrast to [24], where only heat exchange is taken into account. Parameters of dispersed medium: α_{20}=0.0146131, a_0=2mm, d_2=1.5sm, f_0=1630Hz. Curves *R* and *T* for the mixture of water with air bubbles are plotted on the basis of the dispersion equation [25], which, in the absence of phase transitions ($k_V \to 0$), is consistent with Eq. (1). We note that for the transmission coefficient agreement with the experiment is obtained. Our model, in a particular case, describes the experimental data well.

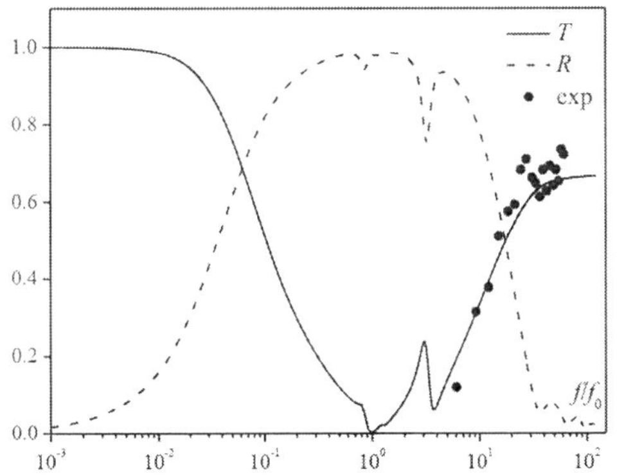

Fig. 5. Comparison of the coefficients of sound reflection and transmission through a bubbly screen with the experimental data [24].

This study was supported by the Russian Science Foundation (project No. 15-11-20022).

References

1. R.I. Nigmatulin, *Dynamics of multiphase media* (New York: Hemisphere, 1991)
2. S. Temkin, *Suspension acoustics: an introduction to the physics of suspensions* (Cambridge: Cambridge University Press, 2005)
3. T.G. Leighton, *The acoustic bubble* (London: Academic Press, 1994)
4. V.E. Nakoryakov, B.G. Pokusaev, I.R. Shreiber, *Wave propagation in gas-liquid media* (New York: CRC Press, 1993)
5. D.A. Gubaidullin, *Dynamics of two-phase gas-vapor-droplet media* (Kazan: Kazan mathematical society, 1998)
6. Y. Hao, A. Prosperetti, Physics of Fluids **11**, 2008 (1999)
7. K.W. Commander, A. Prosperetti, *J. Acoustical Society of America* **85**, 732 (1989)
8. R.I. Nigmatulin, V.Sh. Shagapov, N.K. Vakhitova, USSR Academy of Sciences Doklady **304**, 1077 (1989)
9. V.Sh. Shagapov, *J. of Applied Mechanics and Technical Physics* **18**, 77 (1977)
10. R.I. Nigmatulin, N.S. Khabeev, F.B. Nagiev, *International Journal Heat and Mass Transfer* **24**, 1033 (1981)
11. D.A. Gubaidullin, A.A. Nikiforov, R.N. Gafiyatov, High Temperature **50**, 250 (2012)
12. R.N. Gafiyatov, D.A. Gubaidullin, A.A. Nikiforov, Fluid Dynamics **48**, 366 (2013)
13. D.A. Gubaidullin, R.N. Gafiyatov, J. Phys.: Conf. Ser. **567**, 012020 (2014)
14. D.A. Gubaidullin, D.D. Gubaidullina, Yu.V. Fedorov, J. Phys.: Conf. Ser. **567**, 012011 (2014)
15. D.A. Gubaidullin, Yu.V. Fedorov, Acoustical Physics **62**, 179 (2016)
16. D.A. Gubaidullin, D.D. Gubaidullina, Yu.V. Fedorov, J. Phys.: Conf. Ser. **669**, 012011 (2016)
17. D.A. Gubaidullin, A.A. Nikiforov, R.N. Gafiyatov, High Temperature **53**, 240 (2015)
18. D.A. Gubaidullin, A.A. Nikiforov, R.N. Gafiyatov, J. Phys.: Conf. Ser. **669**, 012019 (2016)
19. D.A. Gubaidullin, R.N. Gafiyatov, J. Phys.: Conf. Ser. **789**, 012019 (2017)
20. D.A. Gubaidullin, Yu.V. Fedorov, Doklady Physics **61**, 499 (2016)
21. D.A. Gubaidullin, Yu.V. Fedorov, Fluid Dynamics **52**, 107 (2017)
22. D.A. Gubaidullin, D.D. Gubaidullina, Yu.V. Fedorov, J. Phys.: Conf. Ser. **789**, 012016 (2017)
23. L.M. Brekhovskikh, O.A. Godin, *Acoustics of layered media* (Moscow: Nauka, 1989)
24. T.M. Tien, *Sound propagation through a bubble screen at finite gas-volume fraction* (Taiwan: National Cheng Kung University, 2001)
25. D.A. Gubaidullin, A.A. Nikiforov, R.N. Gafiyatov, Izv. Vuzov. Problemy Energetiki **3-4**, 3 (2009)

Axial Wave Reflection and Transmission in Stepped Nanorods Using Doublet Mechanics Theory

*Metin Aydogdu**, *Ufuk Gul*

Department of Mechanical Engineering, Trakya University, Turkey

Abstract. A numerical investigation of the reflection and transmission of axial waves at stepped nanorods is presented. The scale dependent doublet mechanics theory is used in the analysis. The main difference of the doublet mechanics from other scale dependent models (stress gradient, strain gradient and couple stress theories) is its direct dependence to the micro/nano structure of the solid. Scale parameter is directly related to atomic structure of the material in doublet mechanics theory and it is assumed as carbon-carbon bond length in the present study. However, identification of scale parameters in other scale dependent theories is difficult compared to doublet mechanics theory. Governing equations of stepped nanorods are derived in the framework of doublet mechanics using the Hamilton Principle. The numerical results predicted by doublet mechanics are shown and compared with the classical elasticity.

1 Introduction

In the last two decades, carbon nanotubes have been attractive due to their excellent mechanical, chemical and thermal properties so they have been used for designing of new nano-electronics devices, nanocomposites, nano electro-mechanical systems (NEMS). The determination of transmission and reflection coefficients in stepped nanorods is also a crucial problem while studying the structure-borne sound, propagation and radiation in nanostructures. Cramer and Heckl [1] have studied the reflection and transmission coefficients at the discontinuity cross section using the one dimensional wave equation. For larger wavelengths, these coefficients have been computed by Hagedorn and Seemann [2] in rods and beams considering the classical rod and beam theory. Love's theory which takes into account the kinetic energy due to lateral motion was used in order to calculate the transmission and reflection coefficients of a stepped rod by Seemann [3]. He compared the obtained results with the FEM solution and 3-dimensional theory of elasticity for validation. Mace [4] derived the transmission and reflection matrices for the cases of a point support and a sudden change in cross-section of the Euler-Bernoulli beams. The results were given for the free and forced vibration cases. Wave propagation of tunable fluid-filled beams has been investigated considering the wave reflection and transmission at a change in cross-section by Harland et al. [5]. Also, some studies which are related to wave reflection and transmission in Timoshenko beams have been examined in [6-8]. Recently, wave reflection and transmission of carbon nanotubes have been analysed by using finite difference time domain method [9]. The main aim of these papers is to derive the reflection and transmission matrices under the discontinuities in structures. This provides a concise and systematic approach for analysis of the structures.

Classical continuum models are not suitable for modelling of nano scale structures due to their size effects and molecular dynamics simulation has some difficulty while computing the nano-scaled structures. So, some different size dependent continuum models have been proposed in the open literature such as strain and stress type gradient models, modified couple stress theory, peridynamics. They have been used in static and dynamic analysis of nanostructures [10-12]. The main purpose of these theories was the consideration of the internal length scale parameter of the material while modelling their mechanical responses. An alternative micro mechanics scale dependent theory, DM was proposed by Granik [13]. Granik and Ferrari [14] and Ferrari et al. [15] have used DM theory in granular materials modelling. DM theory is not a phenomenological theory in contrast to other size dependent continuum models. In DM theory, atomic distance of elastic material is directly used as an intrinsic length scale parameter. This makes DM model more physical compared to other size-dependent continuum models. Also the approximation of the micro structure of the solid can be controlled by using different number of terms in the Taylor series expansion. Vajari and Imam [16,17] investigated the axial and torsional vibration of single-walled carbon nanotubes (CNTs) using DM. Axial vibration and wave propagation of a nanorod embedded in an elastic medium using DM has been discussed in [18]. Recently, dynamic and static analyses of CNTs have been examined via DM theory in [19,20]. The best agreement was obtained between DM model

* Corresponding author: metina@trakya.edu.tr

and experiment for graphite in Ref. [20]. Considering previous studies according to the best knowledge of the authors, there is no study related to wave reflection and transmission in stepped nanorods by using DM. In this paper, expressions for the propagation, reflection and transmission matrices are derived for waves in stepped nanorods by using DM and elementary rod theories. These reflection and transmission matrices can provide a concise and systematic approach to dynamic analysis of nanorods. The advantages and physical basis of DM model are discussed in detail.

2 DM Model of Nanotubes

Analysis in DM theory starts with arraying each atom of elastic material as a node. These atoms (nodes) are located at certain finite distances (bond length) which are taken as 0.1421 nm for CNTS. Any two nodes are called as doublet and distance between them is called "doublet distance" of the considered material. Each node in the material can be elongate and/or rotate with respect to reference node. In this theory, elongation micro-stress p_α, shear micro-stress t_α and torsional micro-stress m_α (Fig. 1) are considered in the elastic deformation. It is noted that these micro stresses are vector quantities. A symbolic deformation model for the DM theory is presented in Fig. 2. Each node α and their neighboring nodes are separated by doublet distance $\eta = |\xi_\alpha|$.

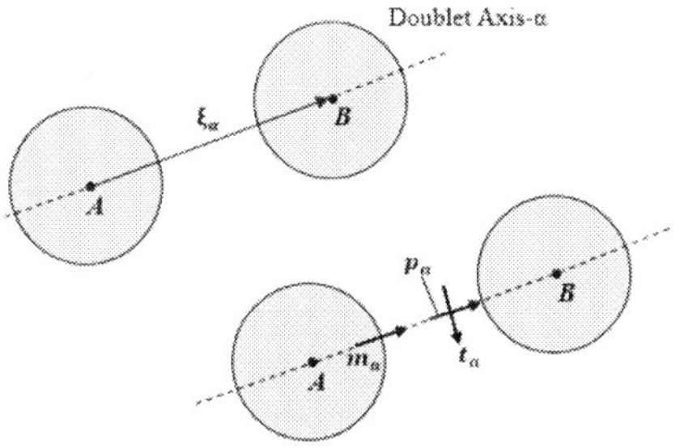

Fig. 1. Micro stresses in a doublet geometry [15]

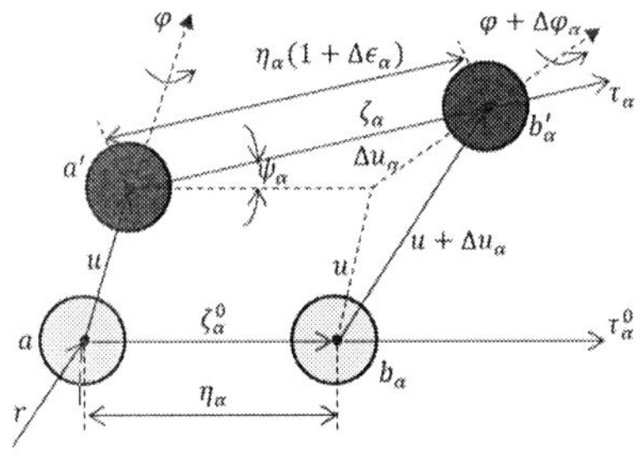

Fig. 2. Deformation of the doublet nodes [15]

After deformation of the nodes in a solid shown as a'_α and b'_α. The increment of displacement is [15]:

$$\Delta u_\alpha = \vec{u}(\vec{r} + \vec{\xi}^0_\alpha, t) - \vec{u}(\vec{r}, t) \tag{1}$$

Here, $\vec{r}$ is the position vector of node α, t is the time and α=1,2,...m, m is the number of doublets. Elongation micro strain can be defined as [15]:

$$\epsilon_\alpha = \frac{\vec{\tau}^0_\alpha \Delta\vec{u}_\alpha}{\eta_\alpha} \tag{2}$$

where η_α is the doublet separation distance and $\vec{\tau}^0_\alpha$ is the unit vector in the α-direction. Then, micro strain ϵ_α can be expanded in a Taylor series as follows:

$$\epsilon_\alpha = \sum_{\chi=1}^{M} \frac{(-\eta_\alpha)^{\chi-1}}{\chi!} \vec{\tau}^0_\alpha . (\vec{\tau}^0_\alpha . \vec{\nabla})^\chi \vec{u} \tag{3}$$

where Del operator and the number of the terms in the Taylor series expansion are denoted by $\vec{\nabla}$ and M, respectively. It should be noted that if M equals to 1 it gives the classical elasticity theory equations. For the present study we considered the M=3 in DM theory. The elongation micro stress is defined as [15]:

$$p_\alpha = \sum_\beta B_{\alpha\beta}\, \epsilon_\beta \tag{4}$$

where p_α is the elongation micro stress in the α doublet and ϵ_β is the axial micro strain in β doublet and $B_{\alpha\beta}$ is the tension micro modulus between nodes α and β. The micro stress is linked to macro stress as below [15]:

$$\sigma = \sum_{\alpha=1}^{n} \vec{\tau}^0_\alpha \vec{\tau}^0_\alpha \sum_{\chi=1}^{M} \frac{(-\eta_\alpha)^{\chi-1}}{\chi!} (\vec{\tau}^0_\alpha \cdot \vec{\nabla})^{\chi-1} p_\alpha \tag{5}$$

The expressions presented above is given for 3-D formulations. In the remaining part of the study plane stress assumption will be considered. The macro stress-micro stress relation can be written with using transformation matrix [H] in the following form:

$$\{\sigma\} = [H]\{p\} \tag{6}$$

By assumption of plane stress condition σ, ε and H are

$$\{\sigma\} = \begin{bmatrix} \sigma_{xx} \\ \sigma_{yy} \\ \sigma_{xy} \end{bmatrix}, [H] = \begin{bmatrix} (\tau^0_{11})^2 & (\tau^0_{21})^2 & (\tau^0_{31})^2 \\ (\tau^0_{12})^2 & (\tau^0_{22})^2 & (\tau^0_{32})^2 \\ \tau^0_{11}\tau^0_{12} & \tau^0_{21}\tau^0_{22} & \tau^0_{31}\tau^0_{32} \end{bmatrix}, \{p\} = \begin{bmatrix} p_1 \\ p_2 \\ p_3 \end{bmatrix} \tag{7}$$

where $\vec{\tau}^0_{ij}$ are the cosines of the angles between the micro-stresses and Cartesian coordinates. Relation between micro strain and macro strain is:

$$\{\epsilon\} = [H]\{\varepsilon\} \tag{8}$$

where

$$\{\varepsilon\} = \begin{bmatrix} \varepsilon_{xx} \\ \varepsilon_{yy} \\ \varepsilon_{xy} \end{bmatrix}, \quad \{\epsilon\} = \begin{bmatrix} \epsilon_1 \\ \epsilon_2 \\ \epsilon_3 \end{bmatrix} \tag{9}$$

The Hooke's law for the macro level is defined as

$$\{\sigma\} = [C]\{\varepsilon\} \tag{10}$$

where C denotes the matrix of rigidities which incorporates the Lamé's constants (λ, μ):

$$C = \begin{bmatrix} 4\mu\left(\frac{\lambda+\mu}{\lambda+2\mu}\right) & \frac{2\mu\lambda}{\lambda+2\mu} & 0 \\ \frac{2\mu\lambda}{\lambda+2\mu} & 4\mu\left(\frac{\lambda+\mu}{\lambda+2\mu}\right) & 0 \\ 0 & 0 & \mu \end{bmatrix} \tag{11}$$

Using Eqs. (6), (8) and (10), following expression can be written as

$$[C] = [H][B][H]^T \tag{12}$$

where B is

 https://doi.org/10.1051/matecconf/201814815002

$$B = \begin{bmatrix} \frac{4}{9}\mu\frac{7\lambda+10\mu}{\lambda+2\mu} & \frac{4}{9}\mu\frac{\lambda-2\mu}{\lambda+2\mu} & \frac{4}{9}\mu\frac{\lambda-2\mu}{\lambda+2\mu} \\ \frac{4}{9}\mu\frac{\lambda-2\mu}{\lambda+2\mu} & \frac{4}{9}\mu\frac{7\lambda+10\mu}{\lambda+2\mu} & \frac{4}{9}\mu\frac{\lambda-2\mu}{\lambda+2\mu} \\ \frac{4}{9}\mu\frac{\lambda-2\mu}{\lambda+2\mu} & \frac{4}{9}\mu\frac{\lambda-2\mu}{\lambda+2\mu} & \frac{4}{9}\mu\frac{7\lambda+10\mu}{\lambda+2\mu} \end{bmatrix} \quad (13)$$

For simplicity $\lambda = 2\mu$ may be chosen and from Eq. (13), it yields $B_{12} = 0$ which leads to

$$B_0 = \frac{8\mu}{3} = E. \quad (14)$$

A zigzag model of nanotube is shown in Fig. 3. For the present problem a node of α has three neighboring nodes. Their orientation can be given in terms of direction cosines: τ1(1,0,0), τ2 (-1/2, $\sqrt{3}$/2,0), τ3 (-1/2, $\sqrt{3}$/2,0) (Fig. 3). It is noted that by changing chiral angle of the doublet axis, the other models of nanotube can be obtained such as chiral and armchair models.

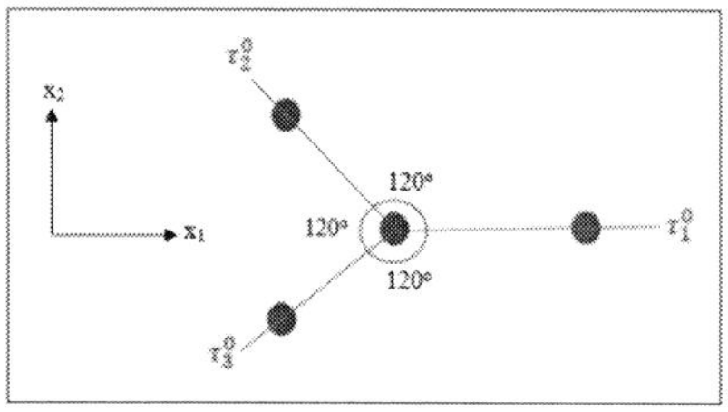

Fig. 3. Arrangement of atoms in a zigzag nanotube

The elongational micro strains (Eq. (3)) can be expanded in a Taylor series for the first three terms (M=3) by

$$\epsilon_\alpha = \vec{\tau}^0_\alpha.(\nabla\vec{u})\vec{\tau}^0_\alpha + \frac{1}{2}\eta_\alpha[\vec{\tau}^0_\alpha.(\vec{\tau}^0_\alpha.\nabla).(\vec{\tau}^0_\alpha.\nabla\vec{u})] + \frac{1}{6}\eta_\alpha^2[\vec{\tau}^0_\alpha.(\vec{\tau}^0_\alpha.\nabla)(\vec{\tau}^0_\alpha.\nabla)(\vec{\tau}^0_\alpha.\nabla\vec{u})] \quad (15)$$

in Cartesian coordinate $\vec{\nabla} = \frac{\partial}{\partial x_i}\vec{e}_i$ and expanding $\vec{\tau}^0_\alpha$

$$\dot{o}_\alpha = \tau^0_{\alpha m}\tau^0_{\alpha n}\left(\varepsilon_{mn} + \frac{1}{2}\eta_\alpha\tau^0_{\alpha s}\frac{\partial\varepsilon_{mn}}{\partial x_s} + \frac{1}{6}\eta_\alpha^2\tau^0_{\alpha t}\tau^0_{\alpha s}\frac{\partial^2\varepsilon_{mn}}{\partial x_t\partial x_s}\right) \quad (16)$$

Similarly, the relation between macro stress and micro stress can be written as:

$$\sigma_{ij}^{(M)} = \sum_{a=1}^{n}\tau^0_{\alpha i}\tau^0_{\alpha j}\left\{\left(p_\alpha - \frac{1}{2}\eta_\alpha\overset{-0}{\tau}_\alpha.(\nabla p_\alpha)\right) + \frac{1}{6}\eta_\alpha^2\left[\left(\overset{-0}{\tau}_\alpha.\nabla\right)\left(\overset{-0}{\tau}_\alpha.\nabla p_\alpha\right)\right]\right\} \quad (17)$$

The elongational micro stress can be defined as

$$p_\alpha = B_0\tau^0_{\alpha m}\tau^0_{\alpha n}\left(\varepsilon_{mn} + \frac{1}{2}\eta_\alpha\tau^0_{\alpha s}\frac{\partial\varepsilon_{mn}}{\partial x_s} + \frac{1}{6}\eta_\alpha^2\tau^0_{\alpha t}\tau^0_{\alpha s}\frac{\partial^2\varepsilon_{mn}}{\partial x_t\partial x_s}\right) \quad (18)$$

Then, using Eqs. (17) and (18), the relation between macro stress and macro strain is:

$$\sigma_{ij}^{(M)} = \sum_{\alpha=1}^{n}B_0\tau^0_{\alpha i}\tau^0_{\alpha j}\tau^0_{\alpha m}\tau^0_{\alpha n}\left(\varepsilon_{mn} + \frac{1}{12}\eta_\alpha^2\tau^0_{\alpha t}\tau^0_{\alpha s}\frac{\partial^2\varepsilon_{mn}}{\partial x_t x_s}\right) \quad (19)$$

Under the plane stress condition $-\frac{\varepsilon_{33}}{\varepsilon_{11}} = -\frac{\varepsilon_{22}}{\varepsilon_{11}} = \nu = \frac{1}{3}$ and considering only the axial strain, stress equation for DM theory can be obtained as

$$\sigma_{xx}^{M} = B_0\left(\varepsilon_{11} + \frac{1}{12}\eta^2\frac{\partial^2\varepsilon_{11}}{\partial x^2}\right) \quad (20)$$

For plane stress condition, $B_0 = E$ in Eq. (14) in the nanotube. The first term in Eq. (20) is classical Cauchy Stress and the second term is doublet stress with a scale effect. It should be noted that doublet term depends on the number of Taylor series expansion in the solid.

3 Governing Equations of a Nanorod

Equations of motion and possible boundary conditions for a nanorod can be obtained by using Hamilton principle:

$$\delta\left[\int_{t_2}^{t_1}(T - W - U)dt\right] = 0 \quad (21)$$

where δ is the variational symbol, U is the strain energy of the nanorod, T is the kinetic energy and W is the total work of external forces. The first variation of strain energy is

$$\delta U = -\int[AE(u'' + s^2u^{IV})] + [AE(u' + s^2u''')\delta u]_0^L + [AE(s^2u'')\delta u']_0^L \quad (22)$$

where $s^2 = \frac{\eta^2}{12}$.

where A is the cross sectional area of the rod, u is the axial displacement and E is the elasticity modulus of the nanorod. The first variation of the kinetic energy is

$$\delta T = \delta\left[\frac{1}{2}\int_0^L\rho\dot{u}^2dx\right] \quad (23)$$

where ρ denotes the mass density of the nanorod and dot () defines the time derivative. The first variation of total work done by external forces (concentrated force P, distributed force f and double forces R) can be written as:

$$\int_{t_0}^{t_1}\delta W dt = \int_{t_0}^{t_1}\int_0^L f(x,t)\,\delta u dx dt + \int_{t_0}^{t_1}\{[P\delta u]_0^L + [R\delta u']_0^L\}dt \quad (24)$$

The variational equation (Eq. 21) takes the following form:

$$\int_{t_0}^{t_1}\int_0^L[AE(u'' + s^2u^{IV}) + f - \rho\ddot{u}]\delta u dx dt + \int_{t_0}^{t_1}[\{P - AE(u' + s^2u''')\}\delta u]_0^L dt$$

$$+\int_{t_0}^{t_1} [\{R - AE(-s^2u'')\}\delta u']_0^L dt$$
$$+\int_0^L [\rho\dot{u}\delta u]_{t_0}^{t_1} dx = 0 \quad (25)$$

Each term of the variational equation (25) must be zero. Thus, the equation of motion of the nanorod is obtained as

$$AE[(u''(x,t) + s^2u^{IV}(x,t)] + f = \rho\ddot{u}(x,t) \quad (26)$$

The initial conditions satisfy the equation:

$$\dot{u}(x,t_1)\delta u(x,t_1) - \dot{u}(x,t_0)\delta u(x,t_0) = 0 \quad (27)$$

and the boundary conditions are:

$$[P(L,t) - AE[u'(L,t) + s^2u'''(L,t)]]\delta u(L,t)$$
$$-[P(0,t) - AE[u'(0,t) + s^2u'''(0,t)]]\delta u(0,t) = 0,$$
$$[R(L,t) - AE[-s^2u''(L,t)]]\delta u'(L,t)$$
$$-[R(0,t) - AE[-s^2u''(0,t)]]\delta u'(0,t) = 0 \quad (28)$$

where the classical boundary conditions are displacement u and the axial force $P=AE(u' + s^2u''')$. The boundary strain u' or the boundary double force AEs^2u'' are defined as the non-classical boundary conditions of the present DM theory. If by setting $s^2=0$ in Eq. (26), classical elasticity equation of motion is obtained. s^2 denotes the length scale parameter which is defined in terms of doublet separation distance.

4 Wave Reflection and Transmission in Stepped Nanorods

This section considers the wave reflection and transmission in a stepped nanorod (Fig. 4). It is seen that the neutral axes of the nanorods coincide.

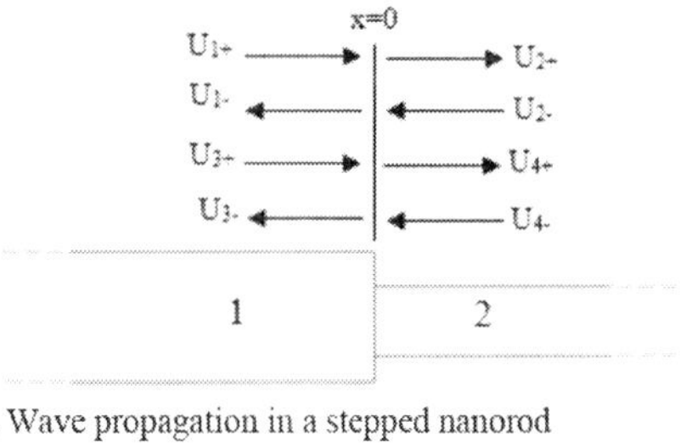

Fig. 4. Wave propagation in a stepped nanorod

Based on Eq. (26), equations of motions of a stepped nanorod can be expressed as:

$$EA_1u_1''(x,t) + s^2EA_1u_1^{IV}(x,t) - \rho A_1\ddot{u}_1(x,t) = 0,$$
$$EA_2u_2''(x,t) + s^2EA_2u_2^{IV}(x,t) - \rho A_2\ddot{u}_2(x,t) = 0. \quad (29)$$

Denoting the axial displacements of the nanorod on the right and left sides of the neutral axis (x=0), the solution can be written by assuming the time-harmonic motion in the following form:

$$u_1(x,t) = U_{1+}e^{i(\omega t-k_1x)} + U_{1-}e^{i(\omega t+k_1x)}$$
$$+U_{3+}e^{i(\omega t-k_2x)} + U_{3-}e^{i(\omega t+k_2x)},$$
$$u_2(x,t) = U_{2+}e^{i(\omega t-k_1x)} + U_{2-}e^{i(\omega t+k_1x)}$$
$$+U_{4+}e^{i(\omega t-k_2x)} + U_{4-}e^{i(\omega t+k_2x)}. \quad (30)$$

where the eight amplitudes are related to reflection $(U_{1-},U_{3-},U_{2-},U_{4-})$ and transmission $(U_{1+},U_{3+},U_{2+},U_{4+})$ waves. It is seen that depending on the frequency there are four propagating waves (two positive going and two negative going) for each cross-section of the stepped nanorod. Here, the wavenumbers are

$$k_1 = \pm\sqrt{E - \sqrt{E(E-4\rho\omega^2s^2)}/2Es^2},$$
$$k_2 = \pm\sqrt{E + \sqrt{E(E-4\rho\omega^2s^2)}/2Es^2}. \quad (31)$$

These wavenumbers can be obtained by assuming harmonic motion in Eq. (29). Eq. (29) gives a polynomial in k^2. So the waves are dispersive in Eq. (31). The relation between the amplitudes and reflection and transmission coefficients can be explicitly defined as

$$U_{1-} = r_1U_{1+} + t_2U_{2-},$$
$$U_{2+} = t_1U_{1+} + r_2U_{2-},$$
$$U_{3-} = r_3U_{3+} + t_4U_{4-},$$
$$U_{4+} = t_3U_{3+} + r_4U_{4-}. \quad (32)$$

where r_1,r_2,r_3,r_4 are reflection coefficients and t_1,t_2,t_3,t_4 are transmission coefficients. When an incident wave with amplitudes of U_{1+} , U_{3+} comes into rod-1, they produce reflected waves in rod-1 with amplitudes r_1U_{1+}, r_3U_{3+} and transmitted into rod-2 with amplitudes t_1U_{1+}, t_3U_{3+}. If there is an incident wave with amplitudes U_{2-}, U_{4-}, in rod-2, reflected waves become r_2U_{2-}, r_4U_{4-} and transmitted waves become t_2U_{2-}, t_4U_{4-}. To compute the reflection and transmission coefficients, idealized transition conditions can be defined as

$$u_1(0,t) = u_2(0,t),$$
$$A_1E\left[u_1'(0,t)+s^2u_1'''(0,t)\right]$$
$$= A_2E\left[u_2'(0,t)+s^2u_2'''(0,t)\right],$$
$$A_1Es^2u_1''(0,t) = A_2Es^2u_2''(0,t),$$
$$u_1'(0,t) = u_2'(0,t). \quad (33)$$

By using Eqs. (32) and (33), values of the coefficients $r_1,r_2,r_3,r_4,t_1,t_2,t_3,t_4$ can be computed. Two conditions can be considered to compute the coefficients. Firstly, for $U_{1+} = U_{3+} = 1$ and $U_{2-} = U_{4-} = 0$, the amplitudes become $U_{1-} = r_1$, $U_{3-} = r_3$, $U_{2+} = t_1$, $U_{4+} = t_3$. Thus, the reflection and transmission matrices can be obtained as

$$\begin{bmatrix} 1 & 1 & -1 & -1 \\ k_1 & k_2 & k_1 & k_2 \\ -A_1Es^2k_1^2 & -A_1Es^2k_2^2 & A_2Es^2k_1^2 & A_2Es^2k_2^2 \\ S_{41} & S_{42} & S_{43} & S_{44} \end{bmatrix} \begin{bmatrix} r_1 \\ r_3 \\ t_1 \\ t_3 \end{bmatrix} =$$
$$\begin{bmatrix} -2 \\ k_1 + k_2 \\ A_1Es^2(k_1^2 + k_2^2) \\ A_1E(k_1 + k_2) - A_1Es^2(k_1^3 + k_2^3) \end{bmatrix} \quad (34)$$

where

$$S_{41} = A_1Ek_1 - A_1Es^2k_1^3,$$
$$S_{42} = A_1Ek_2 - A_1Es^2k_2^3,$$
$$S_{43} = A_2Ek_1 - A_2Es^2k_1^3,$$
$$S_{44} = A_2Ek_2 - A_2Es^2k_2^3. \quad (35)$$

By solving the linear equation system (34), coefficients r_1,r_3,t_1,t_3 can be easily obtained.

Secondly, for $U_{1+} = U_{3+} = 0$ and $U_{2-} = U_{4-} = 1$, the amplitudes become $U_{1-} = t_2$, $U_{3-} = t_4$, $U_{2+} = r_2$, $U_{4+} = r_4$. Thus, the reflection and transmission matrices are

$$\begin{bmatrix} 1 & 1 & -1 & -1 \\ k_1 & k_2 & k_1 & k_2 \\ -A_1Es^2k_1^2 & -A_1Es^2k_2^2 & A_2Es^2k_1^2 & A_2Es^2k_2^2 \\ S_{41} & S_{42} & S_{43} & S_{44} \end{bmatrix} \begin{bmatrix} t_2 \\ t_4 \\ r_2 \\ r_4 \end{bmatrix} = \begin{bmatrix} -2 \\ k_1 + k_2 \\ A_2Es^2(-k_1^2 - k_2^2) \\ A_2E(k_1 + k_2) - A_2Es^2(k_1^3 + k_2^3) \end{bmatrix} \quad (36)$$

Similar to previous case, other four coefficients r_2, r_4, t_2, t_4 can be obtained by solving the linear system of equation (36).

5 Numerical Results

In this section, some typical reflection and transmission coefficients are presented for the classical elasticity and DM models in Fig. (5-7). The material properties of the nanorod are taken as ρ=2300 kg/m^3 and E=1 TPa. The reflection and transmission coefficients are real for both theories. This corresponds to a single pair of propagating waves in a stepped nanorod. It is clearly seen that the reflection and transmission coefficients are independent from the frequency ω for the elementary rod theory (Fig. 5). Unlike the elementary rod theory, the reflection and transmission coefficients change with the frequency for the DM model (Figs. 6-7). The coefficients obtained from the two theories agree well with each other especially for low frequencies. The magnitude of the reflection and transmission coefficients increases with increasing the wave frequency for DM (especially when $\omega > 10^{10}$) in Figs. (6-7). It should be noted that the difference of the coefficients are more apparent for higher frequencies especially for $\omega > 10^{10}$. This is due to softening effect of the DM theory.

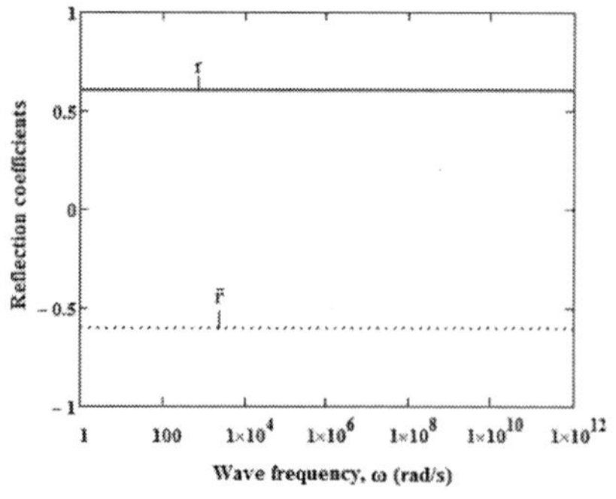

a)

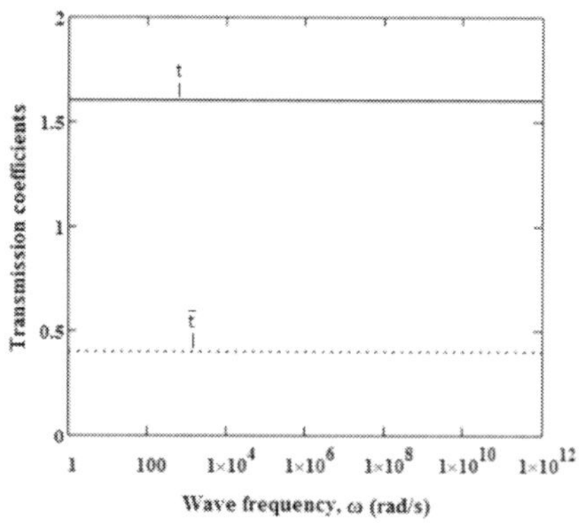

b)

Fig. 5. Reflection and transmission coefficients for classical elasticity theory (A_2/A_1=0.25)

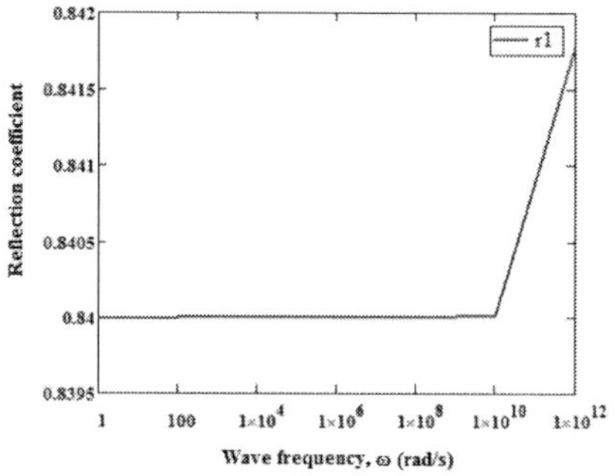

a)

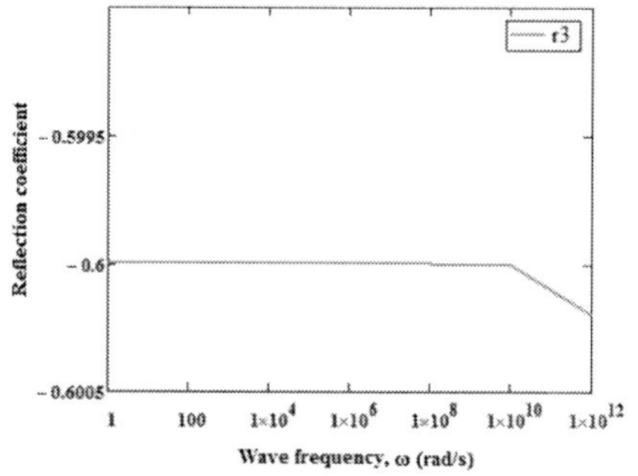

b)

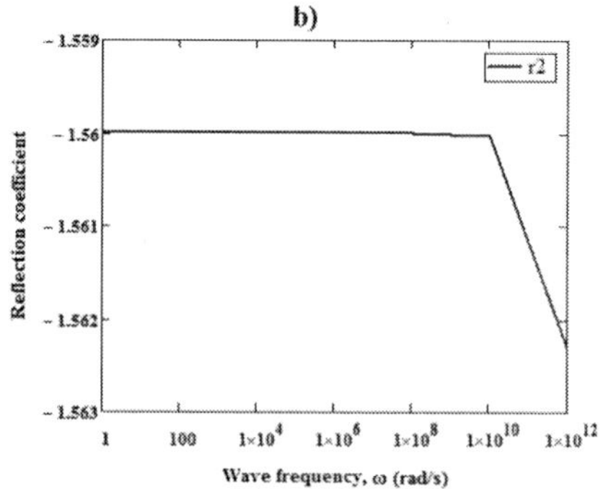

c)

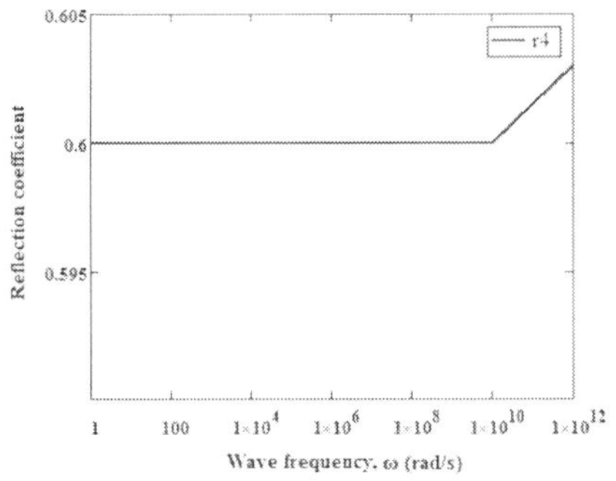

d)

Fig. 6. Reflection coefficients for DM theory (A_2/A_1=0.25, η=0.1421 nm)

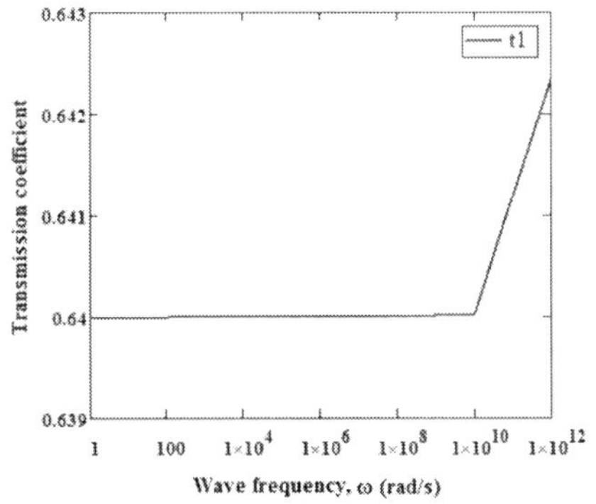

a)

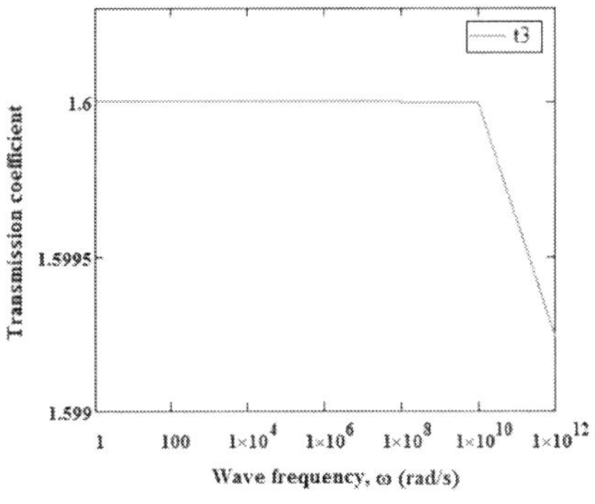

b)

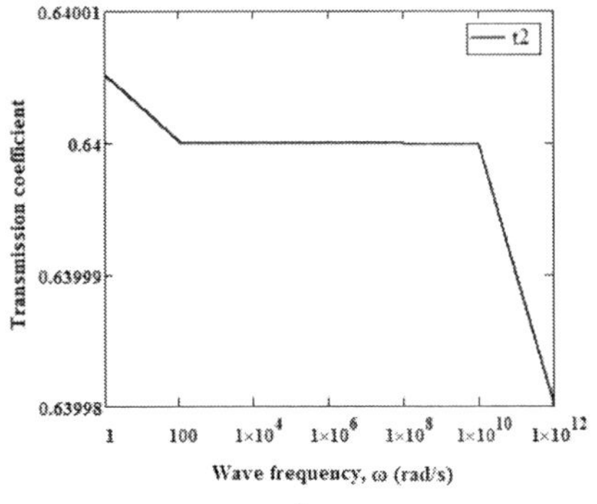

c)

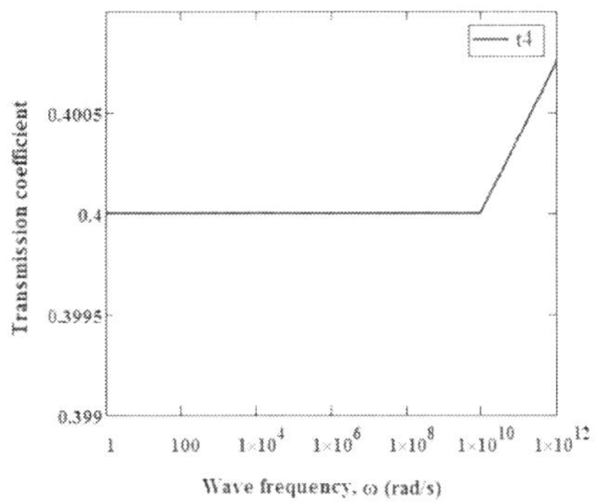

d)

Fig. 7. Transmission coefficients for DM theory (A_2/A_1=0.25, η=0.1421 nm)

6 Conclusions

In this study, DM theory was used to determine the wave propagation, reflection and transmission relations in the stepped nanorods. The vibration analysis of the nanorods become systematic and concise with the availability of propagation, reflection and transmission matrices. It was obtained that magnitudes of the coefficients are almost same for the low frequencies however there is an important difference for high frequencies for DM and classical theory. These results may be given for such discontinuities caused by general point supports in the future studies.

References

1. L. Cremer, M. Heckl, *Structure-borne Sound* (Berlin: Springer-Verlag, 1988)
2. P. Hagedorn, W. Seemann, (VPI-Symposium, Blacksburg, 1995)
3. W. Seemann, J. Sound Vib. **197**, 5 (1996)
4. B.R. Mace, J. Sound Vib. **97**, (1984)
5. N.R. Harland, B.R. Mace, R.W. Jones, J. Sound Vib. **241**, 5 (2001)
6. C. A. Tan, B. Kang, J. Sound Vib. **213**, 3 (1998)
7. C. A. Tan, B. Kang, Int. J. Solids Struct. **36**, (1999)
8. J.F. Doyle, *Wave Propagation in Structures* (Spring-Verlag, New York, 1989)
9. M.Y. Wang, H.L. Li, Y.L. Dong, G.P. Li, C.L. Zhong, J. Xu, ACES J. **31**, (2016)
10. E. Cosserat, F. Cosserat, *Sur la théorie des corps déformables* (Herman,Paris, 1909)
11. R.D. Mindlin, Arch. Ration Mech. Anal. **16**, (1964)
12. S. Kong, S. Zhou, Z. Nie, K. Wang, Int. J. Eng. Sci. **47**, (2009)

13. V.T. Granik, *Technique report IM/MGU* (Inst. of Mech. of Moscow State Univ., 1978)
14. V.T. Granik, M. Ferrari, Mech. Mater. **15**, (1993)
15. M. Ferrari, V.T. Granik, A. Imam, J. Nadeau, *Advances in Doublet Mechanics* (Springer, 1997)
16. A.F. Vajari, A. Imam, Indian J. Phys. **90**, 4 (2016)
17. A.F. Vajari, A. Imam, ZAMMP (to be published)
18. U. Gul, M. Aydogdu, G. Gaygusuzoglu, Comp. Struct. **160**, (2017)
19. U. Gul, M. Aydogdu, J. Engng. Math. (to be published)
20. U. Gul, M. Aydogdu, Physica E **93**, (2017)

Plasticity Detection and Quantification in Monopile Support Structures Due to Axial Impact Loading

P.C. Meijers[1,*], A. Tsouvalas[1], and A.V. Metrikine[1]

[1] Dept. of Hydraulic Engineering, Delft University of Technology, the Netherlands

Abstract. Recent developments in the construction of offshore wind turbines have created the need for a method to detect whether a monopile foundation is plastically deformed during the installation procedure. Since measurements at the pile head are difficult to perform, a method based on measurements at a certain distance below the pile head is proposed in this work for quantification of the amount of plasticity. By considering a one-dimensional rod model with an elastic-perfectly plastic constitutive relation, it is shown that the occurrence of plastic deformation caused by an impact load can be detected from these measurements. Furthermore, this plastic deformation can be quantified by the same measurement with the help of an energy balance. The effectiveness of the proposed method is demonstrated via a numerical example.

1 Introduction

In recent years, the number of offshore wind farms has considerably increased to meet the demand for energy from renewable resources. The most commonly used foundation concept in offshore wind is a steel monopile [1]. Steel monopiles are thin-walled cylindrical structures that are normally driven into the seabed with a hydraulic impact hammer. Each hammer blow induces stress waves in the pile which help the latter to gradually progress into the soil.

As soon as the pile reaches the desired penetration depth, the wind turbine is installed on top of the support structure. Until recently, grouted connections were used between the foundation pile and the superstructure, however, bolted connections have become increasingly popular in recent years [1]. Compared to a grouted connection, a bolted connection is more sensitive to material damage inflicted by the pile driving process since the wind turbine is directly connected to the pile head. In addition, material damage in the pile reduces the service life of the support structure. A method to detect and quantify the damage caused by the pile driving process is therefore essential in order to assess the structural health of a monopile.

Measuring forces and deformations directly at the pile head is a cumbersome task not only due to the high stress levels expected at this critical location, but, additionally, due to the practical implementation of the sensors very close to the pile head. Furthermore, sensors at this location need careful installation to withstand high strains, which is difficult given the hard conditions at sea and the tight working schedules. Thus, a plasticity detection method based on acceleration or strain measurements a few meters below the pile head seems more feasible. Currently, sensors at this location are routinely used for the analysis of the driving process [2]. Given that, the proposed method of quantification of plasticity fits well within the current practice of the offshore industry. This work focuses on using the measurement data from these sensors to detect and quantify plastic deformations inflicted by the pile driving process.

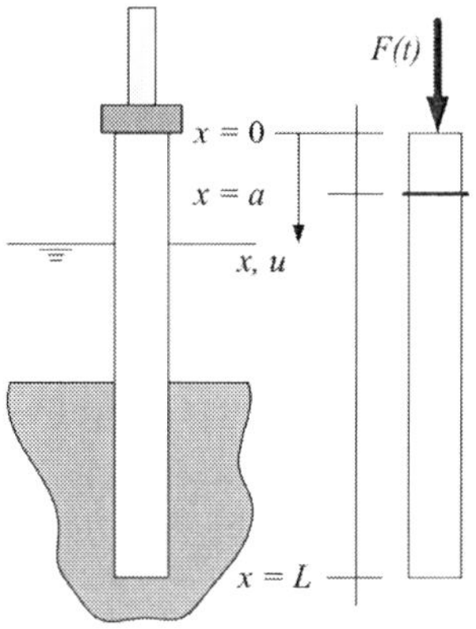

Figure 1: Sketch of the pile driving process (left) and the mathematical one-dimensional model (right).

2 Model

A sketch of the pile driving process is shown in Figure 1. The monopile is modelled by the one-dimensional rod theory [4], which includes the Rayleigh-Love correction term [5] to account for stress wave dispersion expected for

*e-mail: p.c.meijers@tudelft.nl

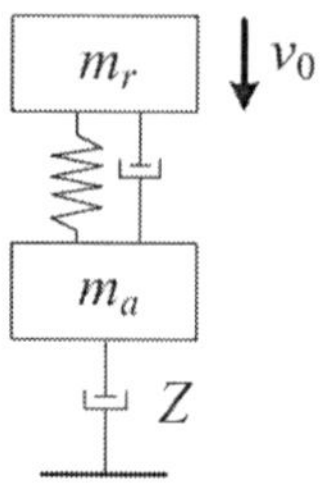

Figure 2: The hammer model of Deeks and Randolph. [3]

piles of large diameter. The model relates the axial displacement $u(x,t)$ and the axial stress $\sigma(x,t)$ as follows:

$$\rho\frac{\partial^2 u}{\partial t^2} = \frac{\partial\sigma}{\partial x}, \tag{1a}$$

$$\varepsilon = \frac{\partial u}{\partial x}, \tag{1b}$$

$$\sigma = E\left(\varepsilon - \varepsilon_p\right) + \underbrace{\nu^2 r_g^2 \frac{\partial^2\sigma}{\partial x^2}}_{\text{RL-correction}}, \tag{1c}$$

in which ρ, E and ν are the density, Young's modulus and Poisson's ratio, respectively. The radius of gyration of the cross-section is denoted by r_g and the total axial strain by ε. To account for plastic deformation in the equations above, auxiliary relations for the plastic axial strain, $\varepsilon_p(x,t)$, are needed. Assuming perfect plasticity, these relations read [6]:

$$\frac{\partial\varepsilon_p}{\partial t} = \gamma\,\text{sign}(\sigma), \tag{2a}$$

$$f(\sigma) = |\sigma| - \sigma_y \le 0, \tag{2b}$$

$$\gamma \ge 0, \quad f(\sigma) \le 0, \quad \gamma f(\sigma) = 0, \tag{2c}$$

$$\gamma\frac{\partial f(\sigma)}{\partial t} = 0 \quad \text{if } f(\sigma) = 0, \tag{2d}$$

where γ is the magnitude of the plastic flow rate, $f(\sigma)$ the yield function and σ_y the yield stress of the material.

The hammer force is described in this model by a time signal, which can be generated by a hammer model as shown in Figure 2 [3]. The time signal and the corresponding amplitude spectrum of the exerted hammer force obtain by a linear pile model are presented in Figure 3 with the dashed-dotted line. The input energy of the hammer is governed by the impact velocity of the ram v_0.

3 Detection

To show the effect of the non-linear material behaviour on the shape of the stress wave, simulated time signals of strain measurements at $x = a$ are considered. The measurement location is indicated in Figure 1 and it is assumed that no plastic deformation occurs at this cross-section. Figure 3 shows the strain signals of an elastic and an elasto-plastic model of the pile. Both cases are simulated with the same force signal as input. In the elastic case, the time signal and the amplitude spectrum have the same shape as the input signal. However, a plateau and an elongation of the peak value in the strain time signal are seen for the elasto-plastic case. Furthermore, a shift in frequency content compared to the elastic case is observed. This shift is caused by the non-linear material behaviour. Thus, by comparing the measured strain signal with the expected strain computed on the basis of the elastic model of the structure, one can reach some preliminary conclusions regarding plastic deformation at the pile head.

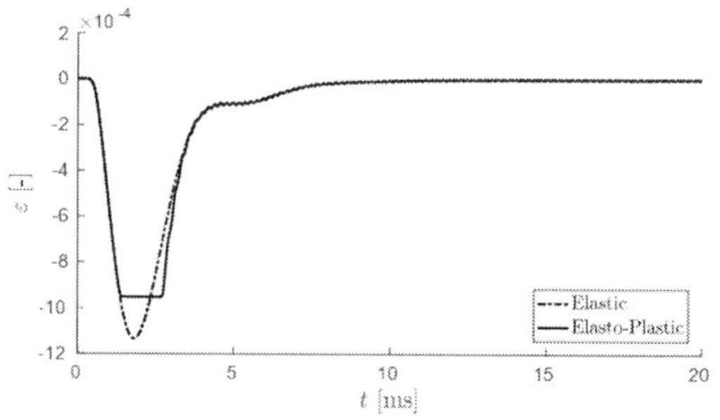

(a) Time signals.

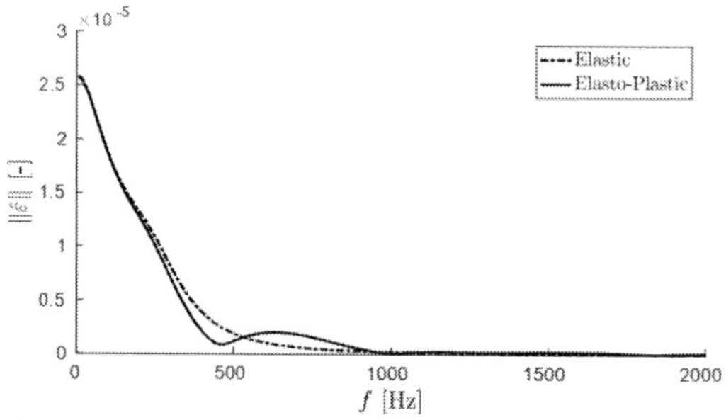

(b) Amplitude spectra.

Figure 3: Simulated strain signals at $x = a$ for an elastic and an elasto-plastic pile.

4 Quantification

In order to quantify the plastic deformation, the energy balance is considered

$$E_0 = E_w + E_p + E_{\text{loss}}, \tag{3}$$

where E_0 is the input energy of the hammer, E_w the wave energy that passed through the considered cross-section, E_p the energy lost in plastically deforming the material and E_{loss} denotes the losses in sound radiation and material damping. Each contribution to the energy balance is now further discussed.

The input energy, E_0, is determined by the aforementioned hammer model [3] or by scaling the linear response of a low energy impact. The energy that passed through

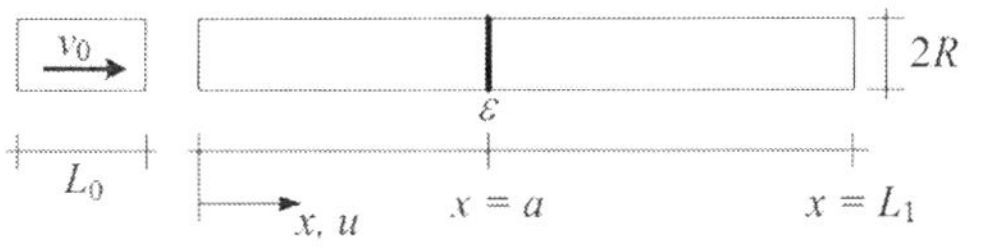

Figure 4: The modelled set-up of the experiment of Kolsky and Douch. [7]

the cross-section, E_w, is defined as the integral of the energy flux over time,

$$E_w = \int_{t=t_0}^{t=t_1} F v \, dt,$$

which can be rewritten in terms of axial strain ε, using the relations $F = A\sigma$, $\sigma = E\varepsilon$ and $v = \frac{\sigma}{\sqrt{E\rho}}$ resulting in

$$E_w = \frac{AE^2}{\sqrt{E\rho}} \int_{t=t_0}^{t=t_1} \varepsilon^2 \, dt, \tag{4}$$

in which A is the area of the cross-section. The plastic work, E_p, is defined as the area under the stress-plastic strain diagram integrated over the distance between the impact and the sensor location,

$$E_p = A \int_{x=0}^{x=a} \int_{\varepsilon=0}^{\varepsilon=\varepsilon_p} \sigma_y \, d\varepsilon \, dx = A\sigma_y \int_{x=0}^{x=a} \varepsilon_p \, dx.$$

The evaluation of the integral of ε_p over x gives the total plastic deformation u_p. This results in the final expression

$$E_p = A\sigma_y u_p. \tag{5}$$

Provided that the other losses in the pile, E_{loss}, are negligibly small, the permanent axial displacement, u_p, can be computed with

$$u_p = \frac{E_0 - E_w}{A\sigma_y}. \tag{6}$$

This equation gives a simple relation to quantify the plastic deformation caused by an impact load based on a strain measurement at a certain distance from the impact point. Naturally, it is expected that this expression will give an upper bound estimation of the expected plastic strain due to the fact that E_{loss} is assumed equal to zero.

5 Numerical example

As an example of the proposed method, the experiments of Kolsky and Douch [7] are considered. In their paper, a cylindrical copper specimen of length L_0 is impacted with initial velocity v_0 onto a stationary cylindrical copper rod with length L_1 as shown in Figure 4. Both the specimen and the rod have the same diameter. Dimensions and material properties are listed in Table 1. From the initial velocity $v_0 = 14.02$ m/s (= 46 ft/s), it follows that the initial energy $E_0 = 1/2mv_0^2 = 36.8$ J.

Since only the shape of the resulting stress wave is reported in [7], the stress wave propagation is simulated using the model introduced in Section 2. Equations 1 are

Table 1: Parameters used for the simulation of the experiment of Kolsky and Douch. [7]

Parameter	Value
L_0	0.1524 m
L_1	1.2192 m
R	9.525 mm
a	0.5476 m
E	129.8 GPa
ν	0.34
ρ	8960 kg/m^3
σ_y	210 MPa

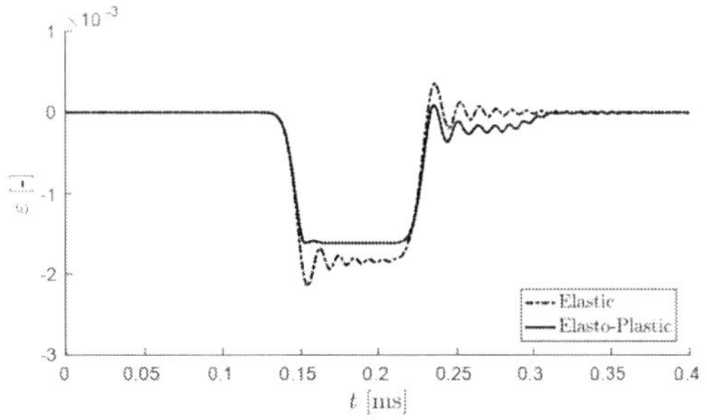

(a) Time signals.

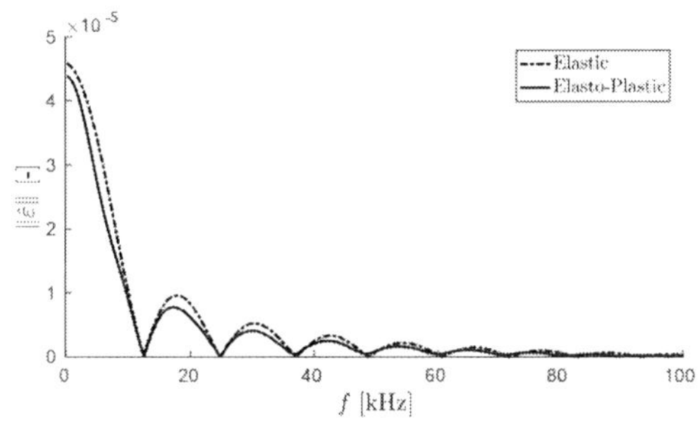

(b) Amplitude spectra.

Figure 5: Simulated strain signals at $x = 0.548$ m for the elastic and the elasto-plastic rod.

spatially discretised with the Finite Element Method in FEniCS [8] and an explicit Newmark scheme [9] is used for the time integration. The spatial resolution and time step are $\Delta x = 2.74 \cdot 10^{-3}$ m and $\Delta t = 1.0 \cdot 10^{-7}$ s, respectively. Equations 2 are satisfied with the help of a return mapping algorithm [6].

Axial strain signals for the elastic and the elasto-plastic simulation at $x = 0.548$ m are shown in Figure 5. Just as reported in [7], a plateau and a tail are present in the shape of the stress wave for the elasto-plastic case, which indicates that plastic deformation occurred. By integrating the square of the strain signal, the energy that passed the cross-section is determined: $E_w \approx 29.0$ J. With the help of Equation 6, the permanent deformation computed equals $u_p = 0.130$ mm.

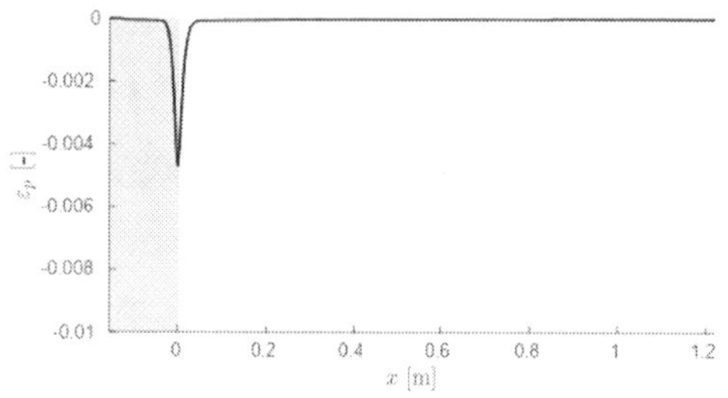

Figure 6: Plastic axial strain along the length of the rod. The shaded grey area is the length of the impact specimen L_0.

From the simulation, the permanent deformation can also be determined by integrating the plastic strain along the length of the rod up to the location of the sensor. The plastic strain profile is presented in Figure 6. Integrating the plastic strain results in $u_p = \int \varepsilon_p \, dx = 0.131$ mm, which is in good agreement with the value found using the wave energy. This simple example shows how the plastic deformation can be quantified with the proposed method.

6 Conclusions and future work

By considering a one-dimensional rod model with an elastic-perfectly plastic constitutive relation, it is shown that the occurrence of plastic deformation caused by an impact load can be detected from a local measurement a certain distance from the impact point, since the shape of the stress wave is changed by the physically non-linear behaviour of the system. Furthermore, the plastic deformation can be quantified by the same measurement with the help of an energy balance, relating the input energy and the wave energy to the amount of permanent axial displacement. A numerical example simulating an impact experiment shows the effectiveness of the proposed plasticity detection and quantification method.

Future work will focus on generalising the constitutive relation to incorporate hardening behaviour and on improving the modelling of the monopile with a cylindrically symmetrical shell theory which will also include hammer-pile interaction.

Acknowledgements

This research is part of the EUROS programme, which is supported by NWO domain Applied and Engineering Sciences and partly funded by the Dutch Ministry of Economic Affairs.

References

[1] WindEurope, Tech. rep., WindEurope (2017), *The European Offshore Wind Industry. Key Trends and Statistics 2016*

[2] F. Rausche, M. Nagy, S. Webster, L. Liang, *CAPWAP and Refined Wave Equation Analyses for Driveability Predictions and Capacity Assessment of Offshore Pile Installations*, in *ASME 2009 28th International Conference on Ocean, Offshore and Arctic Engineering* (American Society of Mechanical Engineers, 2009), pp. 375–383

[3] A.J. Deeks, M.F. Randolph, International Journal for Numerical and Analytical Methods in Geomechanics **17**, 279 (1993)

[4] G.P. DeVault, Journal of the Mechanics and Physics of Solids **13**, 55 (1965)

[5] K.F. Graff, *Wave Motion in Elastic Solids* (Dover Publications, Mineola, N. Y., 1991)

[6] J.C. Simo, T.J. Hughes, *Computational Inelasticity*, Vol. 7 (Springer Science & Business Media, 2006)

[7] H. Kolsky, L.S. Douch, Journal of the Mechanics and Physics of Solids **10**, 195 (1962)

[8] A. Logg, K.A. Mardal, G.N. Wells et al., *Automated Solution of Differential Equations by the Finite Element Method* (Springer, 2012), ISBN 978-3-642-23098-1

[9] N.M. Newmark, Journal of the Engineering Mechanics Division **85**, 67 (1959)

Tree root detection from ground surface vibration measurements

Michał K. Kalkowski[1,2,*], *Jennifer M* Muggleton[1], and *Emiliano* Rustighi[1]

[1] *Institute of Sound and Vibration Research, Faculty of Engineering and the Environment, University of Southampton SO17 1BJ, UK*

[2] *Department of Mechanical Engineering, Imperial College, London SW7 2AZ, UK*

Abstract. Rapid development of urban infrastructure in past decades together with a relatively recent growth of awareness of its impact on the natural environment result in an increased interest in non-destructive ground interrogation methods. Tree root damage is a very well known issue in civil engineering and can emerge as road surface fracture, building foundations disintegration or pipe penetration, among others. In this paper we investigate the feasibility of using a vibroacoustic method for tree root mapping. The core of the idea is that the mechanical waves induced by an excitation mechanism acting on the tree trunk propagate to the roots and then radiate into the surrounding soil. Owing to that, the response measured at the ground surface contains the contribution of waves radiating from roots and can be used for mapping their extent. In this paper, we report a set of field experiments on a 'purpose-built' root-trunk model buried underground. These preliminary results both demonstrate the technique and shed light on related challenges and limitations.

1 Introduction

Trees play a vital role in the carbon cycle, through oxygen synthesis and biomass production. However, in the era of rapid urbanisation and climate changes, tree ecosystems are endangered. On one hand it is very much desired that cities contain large green areas, but at the same time, intense exploitation of city space is a threat for trees. Particularly challenging aspect is the coexistence of tree root networks with city infrastructure. Roots' growth often results in an unexpected damage to pipe networks, road surfaces and building foundations. On the other hand, roots can be severely damaged by building works. It is essential that the location and growth direction of living roots can be monitored periodically, to allow for an informed construction planning and infrastructure maintenance.

There have been a number of attempts to develop remote root mapping techniques, but probably the only approaches that received considerable attention are based on the geotechnical penetrating radar (GPR) [1] and resistivity imaging [2]. Despite promising results reported, several limitations remain, including a complicated measurement setup (resistivity), sensitivity to weather/moisture conditions, common artefacts, or data processing requiring highly expert algorithms and/or operator [3–6].

In this paper, we investigate the possibility of using vibration to detect tree roots from ground surface vibration measurements. Conceptually, this research originates from vibroacoustic water pipe mapping which our research group has pioneered for past decades [7, 8]. The approach is based on sensing waves re-radiated by externally excited underground structures. In the considered case, the excitation is applied to the tree trunk. Wave energy travels down to the root network, is re-radiated into the surrounding ground and eventually propagates to the ground surface, where it can be sensed. This technique has proved itself successful for locating water pipes, as reported in the aforementioned references.

In this contribution, we present an experimental setup involving a mock up trunk-root system, which is used to investigate how root-like structures radiate waves to the ground surface. This is followed by some preliminary measured results from two types of sensor arrays; it is noted that we analyse 'bare' dynamic responses in this work and do not employ any advanced signal processing. Finally, we draw a number of observations directing current ant future work on this topic.

2 Field test rig set-up

Given the complexity of the problem, it was concluded that a fundamental physical insight may be gained from a controlled experiment. We built a mock up trunk-root system, which was equipped with a number of sensors and buried in the ground. In this following paragraphs, the rig used for our investigation is described in more detail.

2.1 Root

The root was represented by its simplest physical model – an exponentially tapered wooden bar ($\approx$ 2 m long with radius varying exponentially from 0.0945 m to 0.0035 m). Arborists indicate that the change of diameter of the main roots in the vicinity of the trunk can be characterised by an

*e-mail: m.kalkowski@imperial.ac.uk

 https://doi.org/10.1051/matecconf/201814815004

Figure 1. Trunk-root assembly before installation.

exponential profile and is called a 'zone of rapid taper' [9]. Compared to real roots, our setup involved two significant approximations; firstly, we neglected curvature typical for real roots and secondly, the specimen had a non-realistic material structure. Tree roots are cylindrically orthotropic; we used a wooden specimen, but owing to the complexity associated with sourcing such a large block of wood, our model was a glued assembly of several parts which did not preserve the cylindrical symmetry. Nevertheless, we believed that such a configuration is representative for the purpose of the investigation. Prior to the experiment, we measured wave characteristics of a free root in the laboratory and confronted them with a wave propagation model, obtaining very good agreement [10].

2.2 Trunk

The trunk was represented by a piece of timber ($0.1 \times 0.225 \times 1.5$ m) which was screwed and epoxy-bonded to the root after the former had been installed in the ground. The trunk hosted a shaker and an accelerometer placed next to the shaker for the purpose of recording input acceleration. The complete root-trunk assembly before installation is shown in Fig. 1.

2.3 On-root sensors

Our interest was in measuring both the ground surface response and the response at the root surface. For this purpose, we equipped the root with a number of in-house assembled MEMS accelerometers. The choice of the sensors had a two-fold motivation. Firstly, there was a need for perfectly waterproof sensors which can operate buried in the soil under different environmental conditions. Secondly, it seemed appealing to verify whether low-cost sensors can be used for scientific investigation. We used widely available GY-61 boards equipped with tri-axial ADXL335 accelerometer chip. After changing respective capacitors to widen the bandwidth, the boards were placed in 3D-printed enclosures, facilitating installation on the root and potted using an epoxy-based compound. The sensors were installed at two opposite sides of the root and at nine locations along its axis, giving 18 tri-axial accelerometers buried underground altogether (visible in Fig. 1.

2.4 Field installation

The mock-up trunk-root system was installed at a research site at Chilworth, owned by the University of Southampton. The root was buried at the depth of 0.5 m, with sand lining in the immediate vicinity of the root (to ensure better contact with the ground and protect the sensors) (Fig. 2). The trench was filled back with soil and compacted successively using a trench rammer. During installation it was discovered that the soil was very inhomogeneous at that site, with a large number of stones and hair-thin roots (Fig. 2 (a)). Such structure, although likely to occur in practical settings, surely is an additional challenge for our investigation. At first, a small access hole was left around the large end of the root. This was used for a series of measurements aimed at characterising waves in a buried exponential structure as such, which are not reported in this paper. After these measurements were taken, the trunk was attached using a number of heavy-duty screws and epoxy, as depicted in Fig. 2 (b).

2.5 Ground surface sensors

The response at the ground surface was recorded using SM-24 tri-axial geophones (see Fig. 2 (d)) arranged in two kinds of array (see Fig. 3). The first was a rectangular array with eight locations along the root axis and nine locations perpendicular to it. The second was a circular array, with seven geophones arranged radially, at approximately every 30 degrees around the trunk. The former configuration is typical for pipe measurements reported above, but the latter seems to correspond better to the current application context.

3 Experimental procedure

Our first investigation was aimed at verifying whether the signatures of the root are present in the ground surface response to trunk excitation. For this reason a number of frequency response functions were recorded using a PROSIG P8020 system. As the site was located in a business park, very close to a compressor station and other facilities, measured ground vibration responses were found to be very noisy. To address the low signal-to-noise issue, we used stepped sine excitation and extracted the magnitude and phase at every frequency step from Fourier series

Figure 2. Field installation: (a) inhomogeneous structure of the soil and root installation; (b) trunk attachment; (c) shaker on the trunk; (e) ground surface sensors.

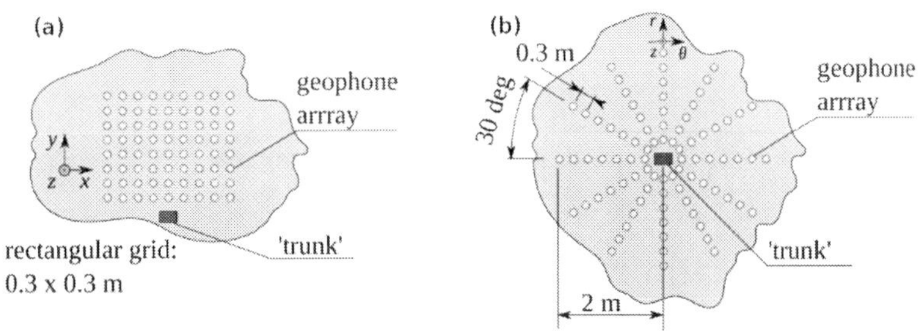

Figure 3. Two array configurations used in the experiment: (a) rectangular; (b) circular.

coefficients. Limited by a number of geophones available, the recording needed to be repeated a number of times to populate the array. However, given that measurements for a single array were taken very close in time with the same excitation setup and under approximately constant environmental conditions, we did not expect them to introduce any noticeable variability. For both geophone arrays, the trunk was excited along three orthogonal excitation directions using a Wilcoxon Research (WR) F4 shaker with a dedicated WR amplifier.

4 Results

Wave-based buried utility mapping uses unwrapped phase maps to identify the objects of interest, as reported in [7, 8]. Adopting an appropriate phase unwrapping scheme is an important factor associated with the method. For our considered problems of tree roots, we used a number of spatial approaches, but none of them was found to bring clear results. Consequently, in this paper we only present some magnitude graphs, in which root signatures can be observed. While analysing ground vibration response maps, one needs to use some criterion for the choice of frequencies to look at. Our first, rather simplistic choice, was to focus on peaks of the frequency response functions averaged over the whole array.

4.1 Rectangular array

Two sets of illustrative results – for the axial (y) and transverse (x) excitation – recorded with the rectangular array are shown in Fig. 4. The signatures of the root are clearly visible in the total superimposed response u_{total} (vector sum of all directional components), but for the axial excitation the signature is shifted with respect to its actual location. A possible factor contributing to this is that the soil just above the root was disturbed and mechanically compacted, which possibly introduced further macro-scale inhomogeneity into the soil. One remaining concern is that the response at the opposite side to the trunk is unknown. Given the type of excitation, it is expected that the response would visually resemble a dipole, which may misguide the interpretation. This is expected to be addressed with the circular array.

The chosen response to transverse excitation is slightly more convincing – the x-component of the response is clearly dominated by waves re-radiated from the root. Both graphs indicate the potential of vibration-based methods for tree root mapping.

4.2 Circular array

Using a circular array showed that, the blue spot around 40 Hz is, indeed, the signature of the root (see Fig. 5). Interestingly, at higher frequencies, it becomes more focused, clearly aligning with the line of the root. As expected, moving towards shorter wavelenghts increases the resolution of detection (u_r response in Fig. 5).

Illustrative responses for the two other directions are shown in Fig. 6. For transverse excitation the root is visible at 57 Hz (u_θ, u_{total}), but it appears at an incorrect angle.

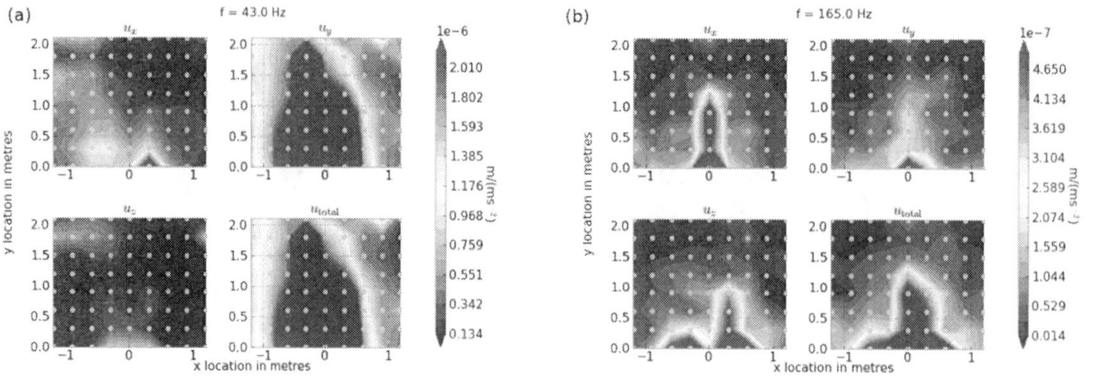

Figure 4. Illustrative magnitude maps for the rectangular array; the root is located along the y-direction, at $x = 0$: (a) axial excitation at 43 Hz; (b) transverse excitation at 165 Hz.

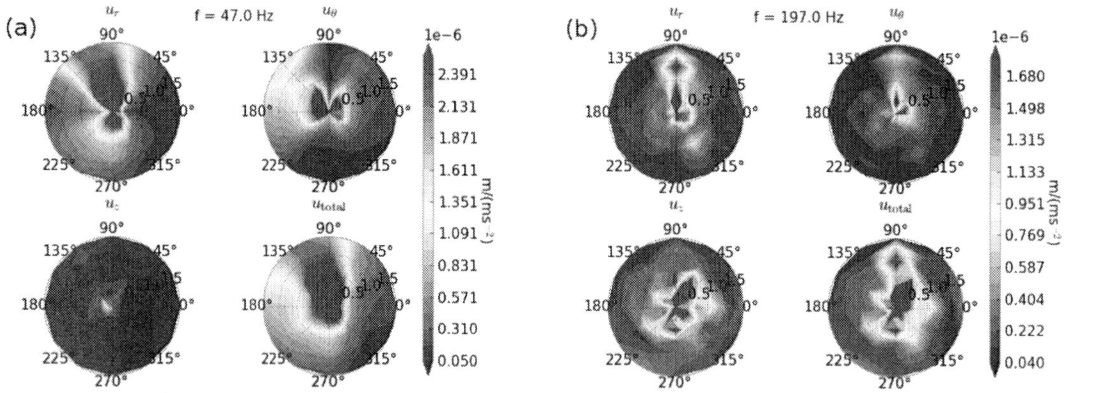

Figure 5. Illustrative magnitude maps for the circular array with axial excitation; the root is located along the 90 deg direction: (a) at 47 Hz; (b) at 197 Hz.

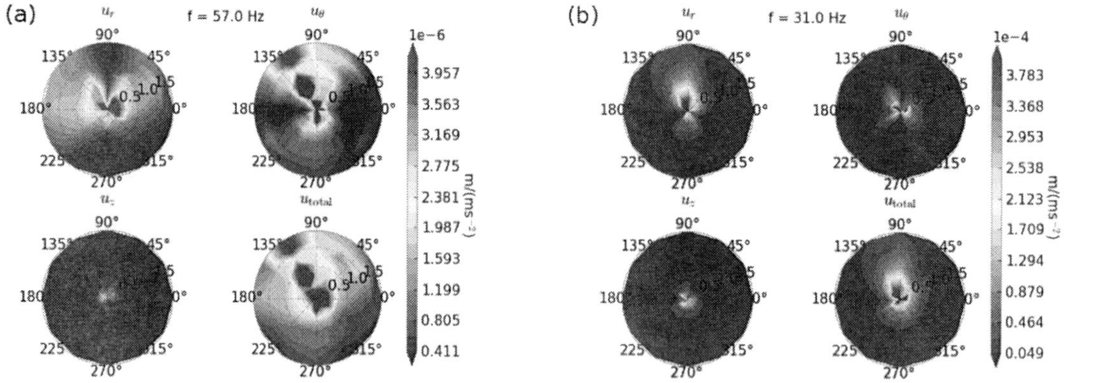

Figure 6. Illustrative magnitude maps for the circular array with transverse and vertical excitation; the root is located along the 90 deg direction: (a) transverse, at 57 Hz; (b) vertical at 31 Hz.

It is possibly related to the modified soil structure after trenching, but needs further investigation. The low frequency response to vertical forcing indicates the location of the root correctly for both u_{total} and u_r.

5 Discussion and conclusions

Illustrative results presented above prove vibration-based root mapping feasible. From all presented figures we clearly observe implications of the challenge which is inherent to the method as such – ground surface signals are a superposition of waves excited directly at the point where the excitation source is attached and waves re-radiated from the buried structures. Separating the two is often necessary to reveal the radiated component and it is not a straightforward task, given that the soil is rarely homogeneous. It is complicated even further by strong acoustic coupling between the soil and the root (owing to the material properties).

Whilst it is expected that advanced signal processing on the array readings would greatly enhance the performance of this method, analysing 'raw' ground surface responses as maps is to some extent attractive for its intuitiveness and simplicity. We note, that in this contribution only preliminary results are reported. Several important issues remain to be addressed in our future publications include application of array signal processing techniques and eliminating the effect of waves directly excited in the ground. Following from those, the method would need to be tested for more realistic configurations including more root branches.

Nevertheless, our contribution shows that tree roots leave a distinct signature on ground surface vibration responses, when excited at the trunk. Moreover, their location can be inferred preliminarily without advanced signal processing. Our hope is that after further developments, the method will be implemented in arboreal practice.

Acknowledgement

The support provided by the EPSRC (under grant EP/K021699/1: 'Assessing the Underworld') is gratefully acknowledged.

References

[1] L. Guo, J. Chen, X. Cui, B. Fan, H. Lin, Plant and Soil **362**, 1 (2013)

[2] M. Amato, B. Basso, G. Celano, G. Bitella, G. Morelli, R. Rossi, Tree Physiology **28**, 1441 (2008)

[3] R.W. Mankin, S.N. Johnson, D.V. Grinev, P.J. Gregory, Root Feeders: An Ecosystem Perspective. CABI Publishing, Wallingford, UK pp. 20–32 (2008)

[4] Y. Hirano, M. Dannoura, K. Aono, T. Igarashi, M. Ishii, K. Yamase, N. Makita, Y. Kanazawa, Plant and Soil **319**, 15 (2009)

[5] L.F. Ow, E.K. Sim, Plant Biosystems - An International Journal Dealing with all Aspects of Plant Biology **146**, 288 (2012)

[6] C.V.M. Barton, K.D. Montagu, Tree Physiology **24**, 1323 (2004)

[7] J.M. Muggleton, M.J. Brennan, Y. Gao, Journal of Applied Geophysics **75**, 54 (2011)

[8] J.M. Muggleton, E. Rustighi, Géotechnique Letters **3**, 137 (2013)

[9] B.F. Wilson, *The Growing Tree* (Univ of Massachusetts Press, 1984), ISBN 0-87023-424-2

[10] M.K. Kalkowski, J.M. Muggleton, E. Rustighi, Journal of Sound and Vibration **390**, 67 (2017)

Reflection of acoustic waves from the boundary or layer of two-phase medium

D.A. Gubaidullin[1], D.D. Gubaidullina[1], and Yu.V. Fedorov[1,*]

[1] Institute of Mechanics and Engineering KazSC RAS, Department of Continuum Mechanics, 420111, Kazan, Russia

Abstract. The inclined incidence of the acoustic wave on a layer of gas-droplet mixture or bubbly liquid of finite thickness is theoretically investigated. In the case of the incidence of the low-frequency acoustic wave to interface between the pure gas and aerosol or to interface between pure liquid and bubbly liquid the basic laws of reflection and transmission of a wave are established. This circumstance allows us to evaluate the transmission and reflection coefficients, depending on the volume content of inclusions and the angle of incidence of the acoustic wave. In particular, for the interface between pure gas and aerosol analytical expressions of the critical angle of wave incidence at which reflection coefficient becomes zero are obtained, i.e. thus there is a complete passage of the acoustic wave through the interface. It is established that the increase of the angle of incidence of the acoustic wave on the boundary or layer of aerosol causes: first, either to increase or to decrease of the reflection coefficient at low frequencies, and second, to appearance of additional minima depending on the reflection coefficient from frequency of disturbances related to the difference of speed of sound and density of the medium.

1 Introduction

The problem of wave propagation in complex layered media containing layers of bubbly liquid or suspension of gas with particles is topical. This is due to both natural phenomena, such as, for example, near-bottom or subsurface layers containing gas bubbles in seas and lakes [1–5], layers of fog, dusty gas, containing in atmosphere [6–8], and engineering processes [9, 10]. The design and parameters of a combined panel consisting of two glasses or polycarbonates with a gap between them filled with a bubbly fluid is described in [10]; this panel is used for the simultaneous information protection in the acoustic, optical, and electromagnetic technical channels.

Among experimental studies we notice [11], where the two-layer water - water with air bubbles - air medium was considered and the reflection coefficient was determined in the case of the normal acoustic-wave incidence from the direction of the water. In [12] the two-layer water - water with air bubbles - water medium was considered and the coefficient of transmission through a bubble screen was experimentally estimated. In [13] the propagation of a pulsed disturbance through a three-layer specimen placed in a fluid and consisting of a bubbly gel layer bounded by polycarbonate layers was experimentally investigated. The comparison of the theory with the experimental data obtained in [13] was made in [14] and good agreement was found to exist.

The main approaches to an investigation of the acoustics of bubbly fluids and gas-particle suspension can be found in the well known books [15–18], while paper [19] gives an overview of certain publications.

The problem of small disturbance propagation in a liquid with gas-vapor bubbles in some or another formulation was studied in [20–33].

In this study reflection of acoustic waves from the boundary or layer of two-phase medium is investigated.

1.1 Reflection and transmission coefficients

To determine the transmission and reflection from two-layer medium coefficients, a diagram of which is shown in fig. 1, the following method of calculations is used. According to [34], the result of the reflection and transmission of a plane monochromatic wave $\exp i(\omega t - Kz)$ from multi-layer media will be plane waves $R \exp i(\omega t + Kz)$ and $T \exp i(\omega t - Kz)$, where R and T – reflection and transmission of the waves coefficients, which are defined through layers impedances Z_j and input impedances of layers borders Z_j^{in}. For a two-layer medium reflection and transmission coefficients are of the form

$$R = \frac{Z_2^{in} - Z_1}{Z_2^{in} + Z_1}, T = \prod_{j=1}^{2} \left(\frac{Z_j^{in} + Z_j}{Z_j^{in} + Z_{j+1}} \exp iK_j s_j \right) \quad (1)$$

where

$$Z_j = \rho_j \frac{\omega}{K_j}, Z_2^{in} = \frac{Z_3 - iZ_2 \tan(K_2 s)}{Z_2 - iZ_3 \tan(K_2 s)} Z_2$$

$$i^2 = -1, j = 1, 2, 3$$

[*] e-mail: kopperfildd@yandex.ru

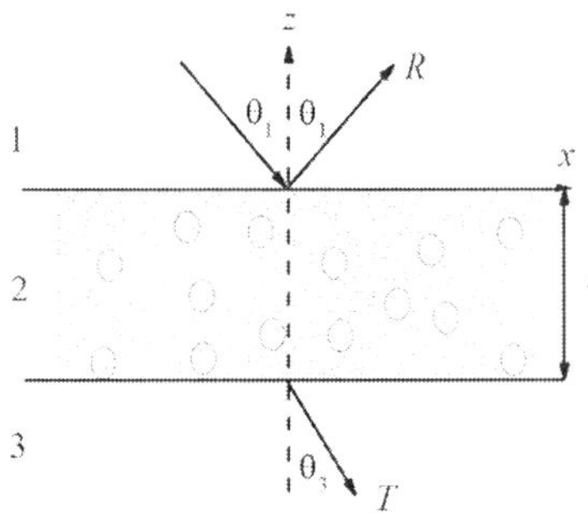

Figure 1. Scheme of the media

Here s – the thickness of the bubble layer, K – wave number, ω – frequency of disturbances, ρ – density of the layer. For a homogeneous layer the wave number is defined as $K_j = \omega/C_j$, where C_j – the speed of sound in the j-th layer.

1.2 Complex wave number for bubbly liquid

Since in the second layer bubble liquid will be considered, i.e., non- homogeneous and absorbing medium, to determine, in general, a complex wave number $K_2 = K_{bl}$ the relation obtained in [21] will be used. This dependence of a complex wave number on the disturbances frequency is given by

$$\left(\frac{K_{bl}}{\omega}\right)^2 = \frac{1}{C_f^2} + \frac{3\alpha_2\alpha_1\rho_1 Q}{3\gamma_2 p_1 - QS} \tag{2}$$

where

$$Q = 1 + \phi, \phi = 3(\gamma_2 - 1)\frac{y\coth y - 1}{y^2}, y = \left(-\frac{i\omega a^2}{\kappa_2}\right)^{1/2}$$

$$S = \frac{i\omega a^2 h\rho_1}{1 + ht}, h = \frac{4\nu_1}{a^2} - i\omega, t = \frac{a}{C_1\alpha_2^{1/6}}$$

Here, the subscript 1 refers to the parameters of the liquid phase, index 2 – to the parameters of the bubbles. The parameters α determine the volume content, ρ_1 – liquid density, ρ_2 – density of gas in bubbles, p_1 – pressure, C_f – frozen velocity, is almost equal to the speed of sound C_1 in pure liquid, γ – the adiabatic index, κ – coefficient of the gas thermal conductivity, ν_1 – the kinematic viscosity of the fluid, a – radius of bubbles.

Let us briefly explain the derivation of equation (2). A system of differential equations that determines the propagation of an acoustic wave in the medium under consideration is recorded. Then the system of equations is linearized. The solution of the obtained linear system of equations in the form of progressive waves is considered: $A_\phi \exp(i(Kx - \omega t))$, where x – coordinate, A_ϕ–amplitude, K – wave number, t–time. The dispersion equation (2) is obtained from the condition of the existence of uniqueness of a non-trivial solution for a linear system of equations. A detailed derivation of this relation is presented in [21].

1.3 Complex wave number for gas-drops suspension

Similarly an expression of the wave number for a gas-drops suspension, which has a view [18] is derived:

$$\left(\frac{C_1 K_{gl}}{\omega}\right)^2 = V(\omega)D(\omega), V(\omega) = 1 + \frac{m}{1 - i\omega\tau_v^*} \tag{3}$$

where

$$D(\omega) = 1 + (\gamma_1 - 1)\frac{t_d}{1 + t_d}, t_d = \frac{mc_{p2}/c_{p1}}{1 - i\omega\tau_T^*}$$

$$\tau_v^* = \tau_v\left(1 + \frac{1-i}{\sqrt{2}}\left(\omega\tau_{\mu_1}\right)^{\frac{1}{2}}\right)^{-1}, \tau_v = \frac{2\rho_2 a^2}{9\mu_1}, \tau_{\mu_1} = \frac{\rho_1 a^2}{\mu_1}$$

$$m = \frac{\alpha_2\rho_2}{\alpha_1\rho_1}, \tau_T^* = \tau_{T2}^* + \frac{mc_{p2}}{c_{p1}}\tau_{T1}^*, \tau_{T1}^* = \frac{\alpha_1\tau_{\lambda_1}}{3\alpha_2(1 + z_1)}$$

$$z_1 = \frac{1-i}{\sqrt{2}}\left(\omega\tau_{\lambda_1}\right)^{\frac{1}{2}}, \tau_{T2}^* = \frac{\tau_{\lambda_2}}{3}\frac{3z_2 - (3 + z_2^2)\tanh z_2}{z_2^2(\tanh z_2 - z_2)}$$

$$z_2 = \frac{1-i}{\sqrt{2}}\left(\omega\tau_{\lambda_2}\right)^{\frac{1}{2}}, \tau_{\lambda_j} = \frac{a^2}{\kappa_j}, \kappa_j = \frac{\lambda_j}{\rho_j c_{pj}}, j = 1, 2$$

The subscript 1 refers to the parameters of the carrier phase, index 2 – to the parameters of the droplets. The parameters α determine the volume content, ρ – density, C_1 – sound velocity, ν – the kinematic viscosity, λ – coefficient of thermal conductivity, a – radius of drops.

2 Results of the calculations

Let's consider the air mixtures with drops of water and water with air bubbles and calculate the phase velocity in these mixtures. The figures 2 a, b show the results of comparing of dependencies of the phase velocity on the disturbance frequency $f = \omega/2\pi$ for different values of the volume content of droplets (a) and bubbles (b). The calculations are carried by using formulas (2), (3). As for a gas suspension and for a bubble liquid in a large range of low frequencies, the phase velocity is not changed and possesses a certain equilibrium value, which depends on the volume content of the inclusions. With this increase in the volume content of inclusions, this equilibrium value decreases. At high frequencies, the phase velocity for these two media tends to the speed of sound in the carrier phase.

Analytic expressions of the equilibrium velocity for a gas suspension and a bubble liquid are obtained by passing to the limit $\omega \to 0$ in formulas (2),(3) and are written accordingly

$$C_{bl}^e = \left(\frac{1}{C_1^2} + \frac{\rho_1\alpha_1\alpha_2}{p_1}\right)^{-\frac{1}{2}} \tag{4}$$

$$C_{gl}^e = C_1\left(\frac{\gamma_e}{\alpha_1(1 + m)\gamma_1}\right)^{\frac{1}{2}}, \gamma_e = \frac{c_{p1} + mc_{p2}}{c_{p1}/\gamma_1 + mc_{p2}} \tag{5}$$

Let us consider the fall of a low-frequency acoustic wave at different angles to the interface between pure air and a mixture of air with water droplets on the air side when the speed of sound in the gas-droplet medium takes

 https://doi.org/10.1051/matecconf/201814815005

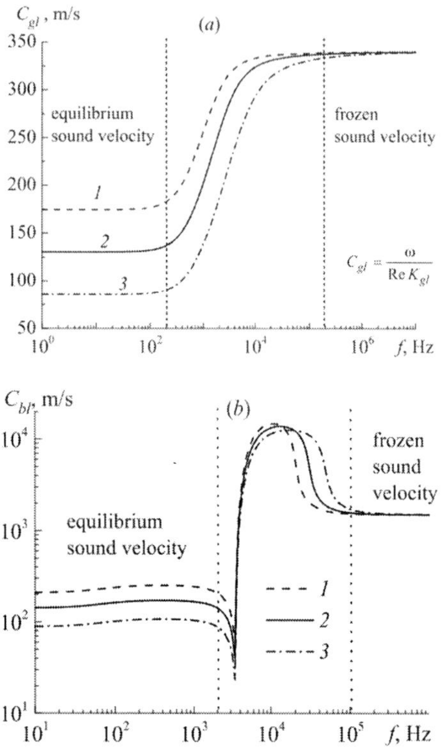

Figure 2. Dependences of phase velocity on the frequency of perturbation for mixture of air with water drops (a) and water with air bubbles (b) at various volume contents of inclusions: 1 - 0.0023, 2 - 0.005, 3 - 0.013; radius of drops - 5 mkm, radius of bubbles - 1 mm

on an equilibrium value C_{gl}^{e} (5). The reflection and transmission coefficients are written as the following formulas.

$$R = \frac{Z_{gl} - Z_1}{Z_{gl} + Z_1}, Z_{gl} = \frac{\rho_{gl} C_{gl}^{e}}{\cos\theta_{gl}}, Z_1 = \frac{\rho_1 C_1}{\cos\theta_1} \quad (6)$$

$$\theta_{gl} = \arcsin\left(\frac{C_{gl}^{e}}{C_1}\sin\theta_1\right), \rho_{gl} = \alpha_1\rho_1 + \alpha_2\rho_2 \quad (7)$$

$$T - R = 1 \quad (8)$$

The figures 3 a and b show the dependencies of the reflection coefficient on the volume content of water droplets (a) and the angle of incidence of the acoustic wave (b). Lets pay attention to the fact that for a value of the angle θ_1= 0, the dependence of the reflection coefficient on the volume content of droplets (Figure 3 a) is monotonic, and the acoustic wave begins to be reflected only at the values of the volume content of the droplets $\alpha_2 > 10^{-3}$. However, if we increase the angle of incidence, then the dependence $|R(\alpha_2)|$ becomes nonmonotonic, and the wave begins to be reflected from the boundary already when α_2<0.001. Another important conclusion is that for certain values of the angle of incidence and the volume content of drops, the reflection coefficient has zero values (Figs. 3 a and b). In this case, the wave passes completely (without reflection) through the interface. Below is the equation for the critical angle at which this situation is realized.

$$\sin\theta_1^* = \sqrt{\frac{\left(\rho_{gl} C_{gl}^{e}\right)^2 - \rho_1^2 C_1^2}{\left(C_{gl}^{e}\right)^2 \left(\rho_{gl}^2 - \rho_1^2\right)}} \quad (9)$$

It should be noted that the extremum of the reflection coefficient is possible only if the density of the gas droplet medium differs from the density of pure gas $\rho_{gl} \neq \rho_1$.

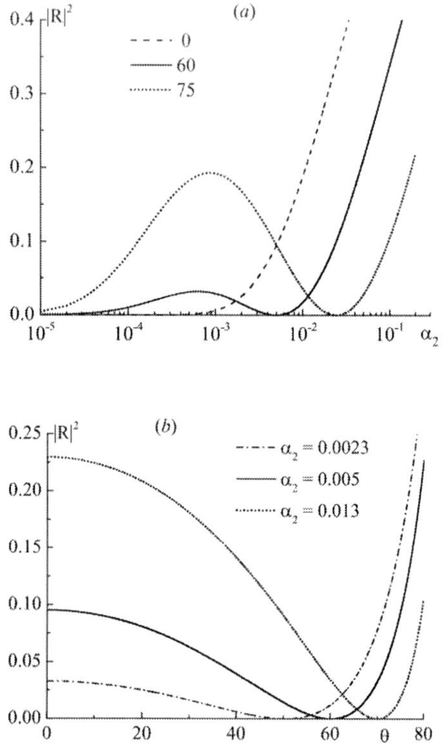

Figure 3. Dependences of reflection coefficient on volume content of drops at various angle of incidence (a) and on angle of incidence at various volume content of drops (b)

The figures 4 and 5 show the results of comparing of the dependencies of the reflection coefficient on the disturbance frequency $f = \omega/2\pi$. Curve 1 is constructed for the medium air / gas-droplet mixture. Comparison of this curve 1 with theoretical data 4 is shown. The curves 2, 3 correspond to the media air / gas-drop mixture / air and air / gas-droplet / water mixtures, respectively. The radius a of the water droplets was 5 μm, the volume content α_2=0.005, the acoustic wave fell from the side of air at an angle θ_1=0, the thickness of the gas-droplet mixture layer s=0.3 m. The nature of the curve 1 of the reflection coefficient corresponds to the character of the curve of sound speed in the gas droplet mixture. At frequencies up to 200 Hz and above 20 000 Hz, the reflection coefficient, like the speed of sound, does not change its equilibrium value and is approximately 0.1 and 0.44, respec-

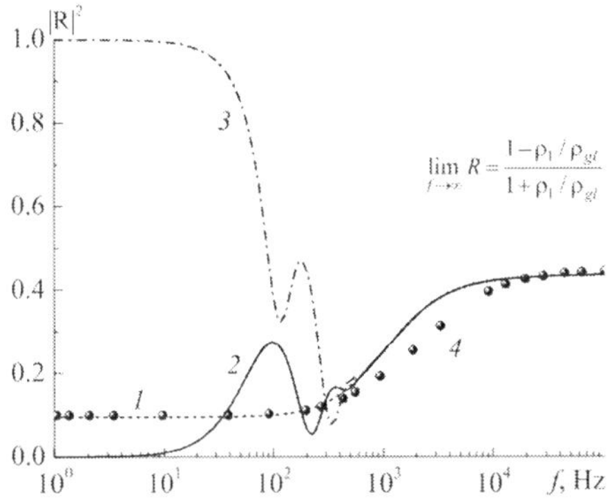

Figure 4. Results of calculations of the reflection coefficient for media: 1 – air / air with water drops, 2 – air / air with water drops / air, 3 – air / air with water drops / water, 4 – theoretical data [16]; the thickness of the layer s is 0.3 m, a = 5 μm, α_2 = 0.005

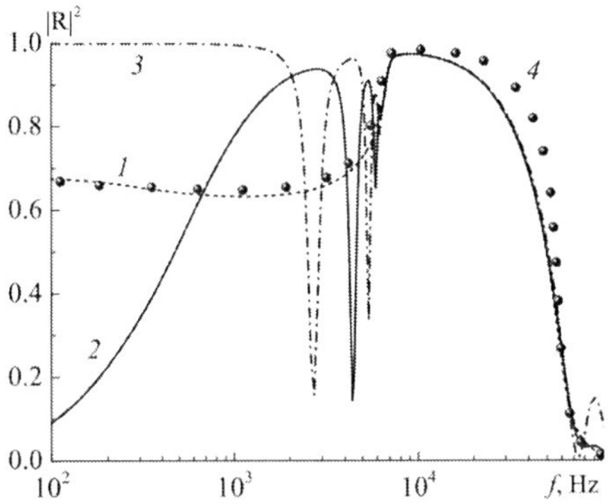

Figure 5. Results of calculations of the reflection coefficient for media: 1 – water / bubbly liquid, 2 – water / bubbly liquid / water, 3 – water / bubbly liquid / air, 4 – theoretical data [16]; the thickness of the layer s is 0.015 m, a = 0.5 mm, α_2 = 0.005

tively. At frequencies f > 600 Hz, the curves 2, 3 go to curve 1. This means that the influence of the third medium decreases and the reflection coefficient is of the same nature as if the wave fell on the interface between the air and the gas-droplet mixture. An analogous character of the reflection coefficient is also observed for the water/bubble liquid boundary.

In the example of interface of the medium air-gas suspension, the effect of the angle of incidence of the wave on the dependencies of the reflection coefficient on the disturbance frequency is shown in the figure 6. At frequencies up to 200 Hz, where the sound speed C_{gl} assumes an equilibrium value C^e_{gl}, the nature of the reflection coefficient corresponds to the nature of the reflection coefficient in the low-frequency region calculated by us earlier. It is seen that for certain angles of incidence of the wave, a nonmonotonic dependence of the reflection coefficient on the frequency of the disturbances is observed. You can evaluate the position of extremums using the following formula.

$$\min|R(\omega)| \Rightarrow \sin^2\theta_1 = \frac{(\sigma^2-1)n^2+m^4}{m^4-1} -$$

$$-\frac{\sqrt{(m^4(1+\sigma^2)^2-4\sigma^2)n^4+2m^4(\sigma^2-1)n^2+m^4}}{m^4-1} \quad (10)$$

$$\sigma = \frac{\Im K_{gl}}{\Re K_{gl}}, n = \frac{C_1}{C_{gl}}, m = \frac{\rho_{gl}}{\rho_1} \quad (11)$$

If $\omega \to 0$ then $\sigma \to 0$, $n \to C_1/C^e_{gl}$ and formula (10) $\to$ (9). If $\omega \to \infty$ then $\sigma \to 0$, $n \to 1$ and $\theta_1 \to \pi/2$. From the estimates, we come to the conclusion that the nonmonotonicity of the reflection coefficient from the disturbance frequency appear when the angle of incidence of the wave is in the interval $(\theta_1^*, \pi/2)$ and the density of the gas droplet medium ρ_{gl} is bigger than the density of the pure gas ρ_1.

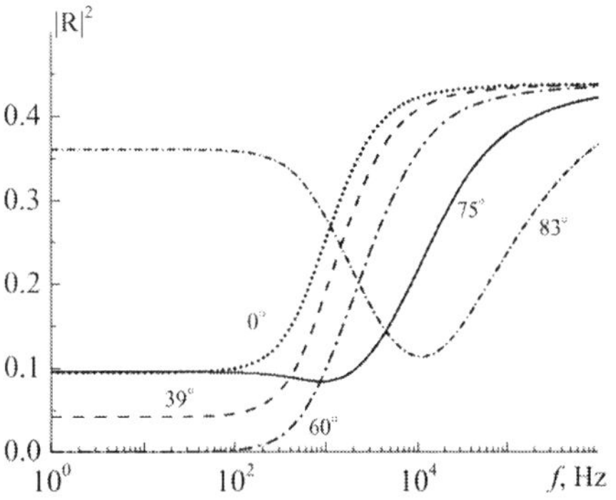

Figure 6. Dependencies of reflection coefficient on the frequency at various angle of incidence of wave for interface "air – air with water drops"

Conclusion

1. At normal incidence of the acoustic wave, the nature of the reflection coefficient at low frequencies for the layers containing the gas-droplet mixture or the bubble liquid does not qualitatively differ, but at high frequencies the reflection coefficient for the bubble liquid layer tends to zero, and for the gas-suspension layer with volume contents of inclusions $\alpha_2 > 0.2 \times 10^{-3}$ has a finite nonzero value.

2. With an inclined incidence of the acoustic wave, the difference is at low frequencies, namely, the reflection coefficient for the boundary or the layer of the gas-droplet mixture can both increase and decrease, while for the bubble liquid the reflection coefficient only increases with increasing angle of incidence of the wave.

3. For the boundary or the layer of the gas-droplet mixture, a minimum of the reflection coefficient may appear, depending on the frequency of the disturbances in a certain range of the angle of incidence of the wave and at certain values of the volume content of droplets, which is associated mainly with the difference in the densities of the gas-droplet mixture and pure gas.

Acknowledgment

This work was supported by the Russian Science Foundation (grant 15-11-20022).

References

[1] V.Sh. Shagapov, V.V. Sarapulova, Acoustical Physics **61**, 37 (2015)

[2] D.A. Gubaidullin, Yu.V. Fedorov, Fluid Dynamics **52**, 107 (2017)

[3] V.A. Gusev, O.V. Rudenko, Acoustical Physics **61**, 152 (2015)

[4] D.A. Gubaidullin, Yu.V. Fedorov, Doklady Physics **61**, 499 (2016)

[5] D.A. Gubaidullin, D.D. Gubaidullina, Yu.V. Fedorov, J. Phys.: Conf. Ser. **789**, 012016 (2017)

[6] J.E. Cole, R.A. Dobbins, J. Atmos. Sci. **28**, 202 (1971)

[7] G.A. Davidson, J. Atmos. Sci. **32**, 2201 (1975)

[8] R. Ishii, H. Matsuhisa, J. Fluid Mech. **130**, 259 (1983)

[9] V. Leroy, A. Strybulevich, M. Lanoy, F. Lemoult, A. Tourin, J.H. Page, Phys. Rev. B. **91**, 020301 (2015)

[10] O.B. Zelmanskiy, S.N. Petrov, A.A. Kazeka, Doklady BGUIR **78**, 30 (2013)

[11] K. Lee, B.K. Choi and S.W. Yoon, Journal of the Korean Physical Society **40**, 256 (2002)

[12] T.M. Tien, *Sound propagation through a bubble screen at finite gas-volume fraction* (Taiwan: National Cheng Kung University, 2001)

[13] V. Leroy, A. Strybulevych, J.H. Page, M.G. Scanlon, J. Acoust. Soc. Amer. **123**, 1931 (2008)

[14] R.I. Nigmatulin, D.A. Gubaidullin, A.A. Nikiforov, Doklady Physics **59**, 286 (2014)

[15] R.I. Nigmatulin, *Dynamics of multiphase media* (Moscow: Nauka, 1987)

[16] S. Temkin, *Suspension acoustics: an introduction to the physics of suspensions* (Cambridge: Cambridge University Press, 2005)

[17] V.E. Nakoryakov, B.G. Pokusaev, I.R. Shreiber, *Wave Dynamics of Gas- and Vapor - Liquid Media* (Moscow: Energoatomizdat, 1990)

[18] D.A. Gubaidullin, *Dynamics of two-phase gas-vapor-droplet media* (Kazan: Kazan mathematical society, 1998)

[19] A.Yu. Varaksin, High Temperature **51**, 377 (2013)

[20] R.N. Gafiyatov, D.A. Gubaidullin, A.A. Nikiforov, Fluid Dynamics **48**, 366 (2013)

[21] D.A. Gubaidullin, Yu.V. Fedorov, Journal of Applied Mathematics and Mechanics **77**, 532 (2013)

[22] D.A. Gubaidullin, R.N. Gafiyatov, J. Phys.: Conf. Ser. **567**, 012020 (2014)

[23] D.A. Gubaidullin, Yu.V. Fedorov, Acoustical Physics **62**, 179 (2016)

[24] T.G. Leighton, J. Jiang, K. Baik, J. Acoust. Soc. Amer. **131**, 2413 (2012)

[25] D.A. Gubaidullin, Yu.V. Fedorov, Fluid Dynamics **50**, 61 (2015)

[26] A. Prosperetti, Annu. Rev. Fluid Mech. **49**, 221 (2017)

[27] A. Prosperetti, Interface Focus. 20140024 (2015) doi:10.1098/rsfs.2015.0024

[28] D. Fuster, F. Montel, J. Fluid Mech. **779**, 598 (2015)

[29] D.A. Gubaidullin, D.D. Gubaidullina, Yu.V. Fedorov, J. Phys.: Conf. Ser. **567**, 012011 (2014)

[30] D.A. Gubaidullin, A.A. Nikiforov, R.N. Gafiyatov, High Temperature **53**, 240 (2015)

[31] D.A. Gubaidullin, A.A. Nikiforov, R.N. Gafiyatov, J. Phys.: Conf. Ser. **669**, 012019 (2016)

[32] D.A. Gubaidullin, R.N. Gafiyatov, J. Phys.: Conf. Ser. **789**, 012019 (2017)

[33] D.A. Gubaidullin, D.D. Gubaidullina, Yu.V. Fedorov, J. Phys.: Conf. Ser. **669**, 012011 (2016)

[34] L.M. Brekhovskikh, O.A. Godin, *Acoustics of Layered Media* (Moscow: Nauka, 1989)

Interaction acoustic waves with a layered structure containing layer of bubbly liquid

Damir Gubaidullin[1,*], *Anatolii* Nikiforov[1]

[1]Institute of Mechanics and Engineering, Kazan Science Center, Russian Academy of Sciences, 420111 Kazan, Russia

Abstract. The results of a theoretical study of the effect of a bubble layer on the propagation of acoustic waves through a thin three-layered barrier at various angles of incidence are presented. The barrier consists of a layer of gel with polydisperse air bubbles bounded by layers of polycarbonate. It is shown that the presence of polydisperse air bubbles in the gel layer significantly changes the transmission and reflection of the acoustic signal when it interacts with such an obstacle for frequencies close to the resonant frequency of natural oscillations of the bubbles. The frequency range is identified where the angle of incidence has little effect on the reflection and transmission coefficients of acoustic waves.

1 Introduction

Gas bubbles are the most acoustically active, natural and widespread formations in the liquid. The presence of a small amount of bubbles significantly increases the compressibility of the medium, while the density of the medium remains close to the density of the liquid. In addition, the oscillating bubble medium has strong dissipative properties, the resulting dissipative losses are mainly associated with interfacial heat transfer. Such special properties of bubble liquids make it possible to use bubble screens as ultrathin anti-hydrolocation coatings [1]. At present, air bubble layers are already being used to shield noise from anthropogenic activities during underwater drilling operations, noise from vibrations of other technical devices, and pile driving in coastal zones [2]. Acoustic properties of the near-bottom bubble layers are actively studied, including in the case of an oblique incidence of an acoustic wave [3]. The main aspects of the mechanics and thermophysics of bubbles and bubble liquids are presented in [4-6]. Various results of theoretical and experimental studies of the dynamics of acoustic waves in bubble media, including media containing bubble layers, are presented in [7-18]. Experimental work [11] presents the results of distortion of an acoustic signal when it falls along the normal to a flat sample consisting of two layers of polycarbonate and a layer containing an industrial gel with polydisperse bubbles.

In this paper we investigate the effect of a bubble layer on the passage of an acoustic signal through a three-layered barrier at various angles of incidence.

2 Theory

The present analysis is based on a dispersion relation which is well established in the case of propagation of pressure pulses of the small amplitude through the multilayer sample containing a layer of liquid with polydisperse gas bubbles (e.g.[9,12])

$$\left(\frac{K_*}{\omega}\right)^2=\frac{1}{C_f^2}+\frac{3\alpha_{g0}\alpha_{l0}\rho_{l0}\langle Q\rangle}{3\gamma_g p_{l0}-\langle QS\rangle}, \qquad (1)$$

$$C_f=C_l/\alpha_{l0},\quad S=i\omega a^2 h_a\rho_{l0}/(1+h_a t_a),$$

$$h_a=-i\omega+4\nu_l/a^2,\quad t_a=a/(C_l(\alpha_{g0})^\beta),\quad \beta=\frac{1}{6},$$

$$Q=1+3(\gamma_g-1)[y\cdot\coth(y)-1]/y^2,\quad y=\sqrt{-i\omega a^2/\kappa_g},$$

$$\langle Q\rangle=\frac{1}{\rho_g}\int_{a_{\min}}^{a_{\max}}N(a)g(a)Q\,da,\quad \rho_g=\int_{a_{\min}}^{a_{\max}}N(a)g(a)\,da,$$

$$\langle QS\rangle=\frac{1}{\rho_g}\int_{a_{\min}}^{a_{\max}}N(a)g(a)QS\,da,\quad g(a)=\frac{4}{3}\pi a^3\rho_g$$

in which ρ_{l0}, p_{l0} are the density and pressure of the unperturbed liquid, α_{g0} is the bubble volume content ($\alpha_{l0}=1-\alpha_{g0}$), C_l the speed of sound in liquid, ν_l the coefficient of the kinematic viscosity of liquid, ρ_{g0} the density of the gas in bubbles, γ_g the adiabatic index of the gas in bubbles, $N(a)$ the distribution function of bubbles over the radii a ($a_{\min}\le a\le a_{\max}$), κ_g the coefficient of thermal diffusivity of the gas, and c_p the specific heat capacity of the gas in bubbles at constant pressure. This dependence of the wave number $K_*=K_{**}+iK$ on the frequency ω allows us to determine the attenuation coefficient K and the phase velocity $C_p=\omega/K_{**}$.

* Corresponding author: damirgubaidullin@gmail.com

According to [8], the size distribution function $N(a)$ is determined as

$$N(a)=\frac{n_t}{\sqrt{2\pi}\varepsilon}\exp\left(-\frac{(\ln(a/a_0))^2}{2\varepsilon^2}\right) \tag{2}$$

in which $a_0 = 0.08$ mm, $\varepsilon = 0.06$, $n_t = 50$ are the distribution parameters [8].

The pulse perturbation of the pressure is given using the imaginary part of the Morlet wavelet with the spike frequency ω_s

$$p'(0,t)=\mathrm{Im}\left(\exp(-i2\pi\omega_s(t-\tau))\exp\left(\frac{-(t-\tau)^2}{2\delta^2}\right)\right) \tag{3}$$

Parameter τ determines the location, and δ is the spike width.

The fast Fourier transform (FFT) method is used for the calculation of the dynamics of small-amplitude pulse perturbations. In accordance with the representation of the Fourier analysis, the arbitrary spatial-temporal pulse can be written in the form of the following integral:

$$y(x,t)=\int_{-\infty}^{\infty}Y(\omega)\exp[i(K_*x-\omega t)]d\omega \tag{4}$$

This relation determines the inverse Fourier transform of the spectral function $Y(\omega)$. The function $Y(\omega)$ is found from the given initial signal

$$Y(\omega)=\frac{1}{2\pi}\int_{-\infty}^{\infty}y(0,t)\exp(i\omega t)d\omega \tag{5}$$

This relation determines the direct transform, the function $y(0,t)$.

The methods of the discrete Fourier transform are used for the numerical calculation of the dynamics of pulse pressure perturbation p'. Pulse p' is presented in the form of the finite sun of harmonic waves

$$p'(x,t)=\sum_{m=0}^{L-1}z_m\exp[i(K_*(\omega_m)x-\omega_m t)] \tag{6}$$

$$p'(0,t)=\sum_{m=0}^{L-1}z_m\exp(i\omega_m t) \tag{7}$$

In accordance with (5) for the determination of $p'(x,t)$, it is necessary to find coefficients z_m and the relation (6) should be fulfilled. FFT algorithms were used to find coefficients z_m and to determine $p'(x,t)$. Expanding the initial pulse into the discrete Fourier series and summing its harmonic components in the new position, it is possible to obtain the transformed pulse formed under the action of the dispersion and dissipation.

The following calculation technique is used for the calculation of the interaction of the acoustic signal in water with the multilayer object containing a layer of the bubbly liquid. According to [13], the result of the passage of the plane monochromatic wave $\sim\exp(iKx-\omega t)$ through the multilayer object is plane wave $\sim W\exp(iKx-\omega t)$ where W is the transmission coefficient of the wave, which is determined in terms of impedances of layers Z_i and input impedances of boundaries of layers Z_i^{in}. For a multilayer object consisting of n layers, the transmission coefficient W and the reflection coefficient R have the form

$$W=\frac{2Z_1}{Z_1+Z_2}\prod_{j=2}^{n}\left(\frac{Z_j^{in}+Z_j}{Z_j^{in}+Z_{j+1}}\exp(iK_jd_j\cos\theta_j)\right) \tag{8}$$

$$R=\frac{Z_n^{in}-Z_{n+1}}{Z_n^{in}+Z_{n+1}} \tag{9}$$

$$Z_j=\rho_j\frac{\omega}{K_j\cos\theta_j},$$

$$Z_j^{in}=\rho_j\frac{Z_{j-1}^{in}-iZ_j\mathrm{tg}(K_jd_j\cos\theta_j)}{Z_j-iZ_{j-1}^{in}\mathrm{tg}(K_jd_j\cos\theta_j)}Z_j,\quad Z_1^{in}=Z_1,$$

in which d_j, K_j are the thickness and wave number of the j-th layer, θ_j is the angle formed by the direction of propagation of the wave in the layer with the normal to the boundaries of the layer. The wave number for a layer with a bubble liquid is determined using the dispersion relation (1).

3 Results

The calculations were carried out for a multilayered media, as shown in Fig. 1. Similarly to [8], a multilayered object consisting of two identical layers of polycarbonate of a thickness of $d_2 = d_4 = 1.6$ mm a layer between which ($d_3 = 1$ mm) is filled with a gel with air bubbles is used. This object is placed into a water between a source of the acoustic signal (piezoelectric transducer) and a hydrophone.

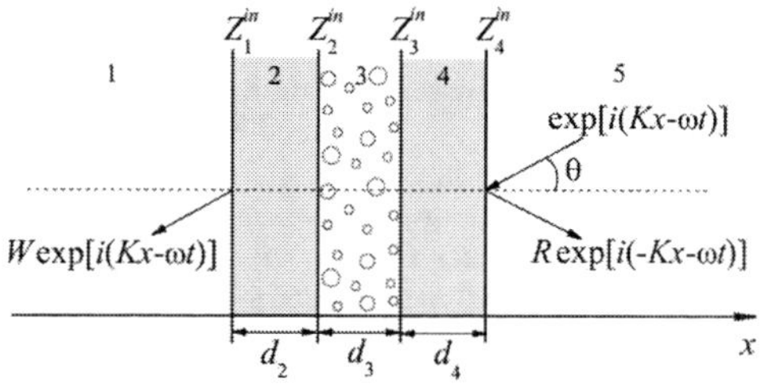

Fig. 1. Multilayered media scheme.

Using the inverse Fourier transform, the complex amplitudes of the harmonic components of the initial pressure pulse $p'(0,t)$ are calculated, where the boundary between media 5 (water) and layer 4 is taken as the position $x = 0$. Then, the amplitudes of the harmonic components of the pulse passing through the obstacle to the media 1 (water) are determined, i.e. in water ($x = d_2 + d_3 + d_4$). The change in pressure as a function of time in the new position is calculated using a direct Fourier transform.

In Fig. 2 and Fig. 3 show the results of the calculation of the dependences of the reflection coefficient modulus and the transmission coefficient modulus on the dimensionless frequency f/f_M ($f = \omega/2\pi$), where f_M is the Minnaert resonant frequency

$$f_M = \frac{1}{2\pi a_0}\sqrt{\frac{3\gamma_g p_0}{\rho^{\circ}_{l0}}} \qquad (10)$$

The radius a_0 is a parameter of the distribution function used (2). For the given conditions, $f_M = 40.8$ kHz. The calculations are performed both for a three-layer barrier containing a bubble gel layer (curves *I* and *II*) and for a barrier containing a bubble-free gel layer, i.e. at $\alpha_g = 0$ (curves *III* and *IV*). Accordingly, curves *I* and *III* are calculated at $\theta = 0$, *II* and *IV* – at $\theta = \pi/6$.

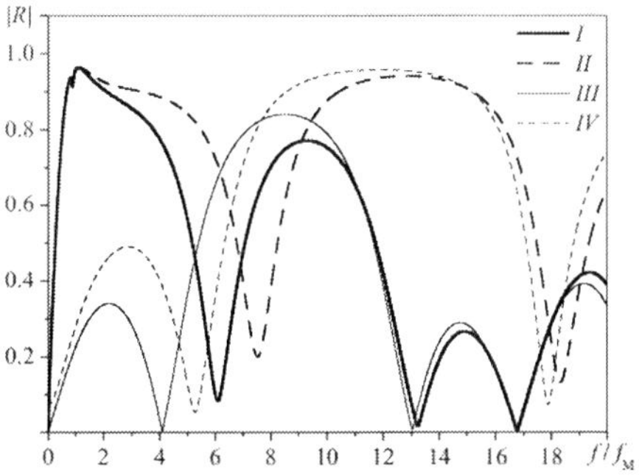

Fig. 2. Dependence of the reflection coefficient modulus on the dimensionless frequency.

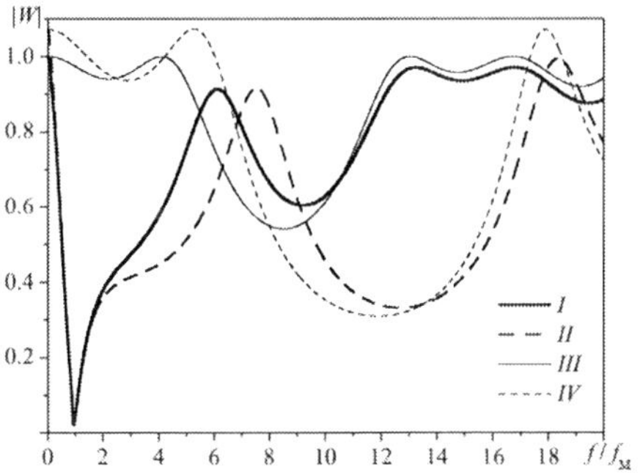

Fig. 3. Dependence of the transmission coefficient modulus on the dimensionless frequency.

The dependence *R* on Fig. 2 is largely determined by the effects of interference of waves reflected on individual boundaries [13]. It is not difficult to calculate that the minima of curve III in Fig. 2 correspond to frequencies determined by the relations previously obtained in [14]

$$f\left(\frac{d_P}{C_P} + \frac{d_l}{C_l}\right) = \frac{1}{4}n,\ (n = 1,3,5,...) \qquad (11)$$

$$f\left(\frac{d_l}{C_l}\right) = \frac{1}{2}n,\ (n = 1,2,3,...) \qquad (12)$$

Here, the subscript P denotes the parameters of the polycarbonate layer ($d_P = d_2 = d_4$), the subscript *l* denotes the parameters of the bubble-free gel.

For curves *I* and *II* in Fig. 2 and Fig. 3, i.e. when there are air bubbles in the gel, a significant difference from curves *III* and *IV*, when there are no bubbles in the gel, is achieved in the frequency range $f/f_M < 5$. This range includes two frequency-domain regions characteristic of the bubble medium-the low-frequency region and the band of acoustic opacity [4, 11]. In the low-frequency region ($f \le f_M$), the effect of the sound dispersion is determined by interphase temperature nonequilibrium and the value of the phase velocity is much smaller C_l. In the band of acoustic opacity ($f_M < f < f_C$, $f_C = f_M\sqrt{1 + \frac{\rho^{\circ}_{l0}\alpha_{g0}C_l^2}{\gamma_g p_0}} \approx 5f_M$), the values of the phase velocity and the attenuation coefficient assume anomalously high values. For frequencies $2 < f/f_M < 5$ the influence of the angle of incidence is weak, the character of the curves $|R|$ and $|W|$ varies little. For frequencies $f/f_M < 2$ curves *I* and *II* practically do not differ, i.e in this frequency range the effect of the angle of incidence on $|R|$ and $|W|$ is not essential. The minimum $|W|$ and maximum $|R|$ are reached in the region of the resonance frequency of the bubbles ($f/f_M = 1$), i.e. there is almost complete reflection ($|R| \approx 0.95$). For frequencies $f/f_M > 5$, the phase velocity C_p in a gel with bubbles tends to its asymptotic value – the frozen velocity of sound $C_f = C_l/\alpha_l$ which differs little from the speed of sound in a pure gel [4]. Reduced density of bubble gel $\rho = \rho_l + \rho_g = \rho^{\circ}_l\alpha_l + \rho^{\circ}_g\alpha_g \approx \rho^{\circ}_l\alpha_l \approx \rho^{\circ}_l$. Thus, for high frequencies $f/f_M > 5$, the acoustic properties of the bubble gel layer and the bubble-free gel layer differ little.

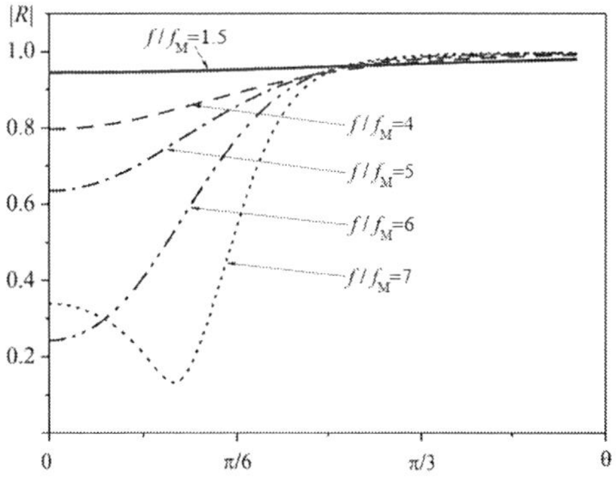

Fig. 4. The angular dependence of the reflection coefficient modulus on the barrier at various dimensionless frequencies.

In Fig. 4 and Fig. 5, the dependences of the modules of the coefficients of reflection and transmission of acoustic waves are plotted against the angle of incidence.

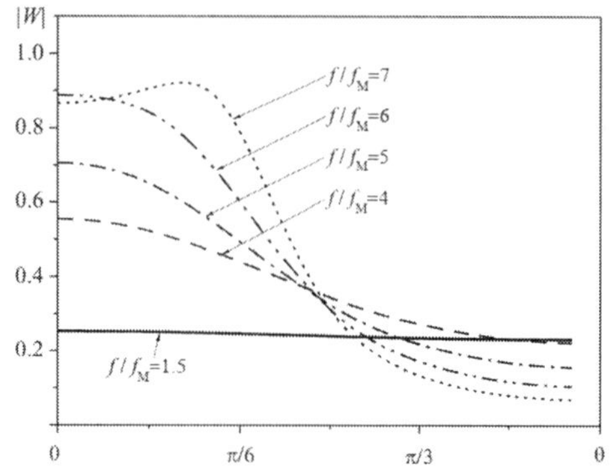

Fig. 5. The angular dependence of the transmission coefficient modulus on the barrier at various dimensionless frequencies.

As noted above, for frequencies $f/f_M < 2$ the influence of the angle of incidence on $|R|$ and $|W|$ is practically absent. In the frequency range $2 < f/f_M < 5$ the value of the angle of incidence already has some effect on $|R|$ and $|W|$ but the curves still remain monotonic. For frequencies $f/f_M > 5$ where the effect of bubbles in the gel on the acoustic properties of the layer does not appear, the angular dependences $|R|$ and $|W|$ become nonmonotonic. But such a character of the curves $|R|$ and $|W|$ is determined only by the laws of reflection and transmission of waves in a multilayer media [13] and will differ little from the form of the curves calculated for the object with a layer of bubble-free gel.

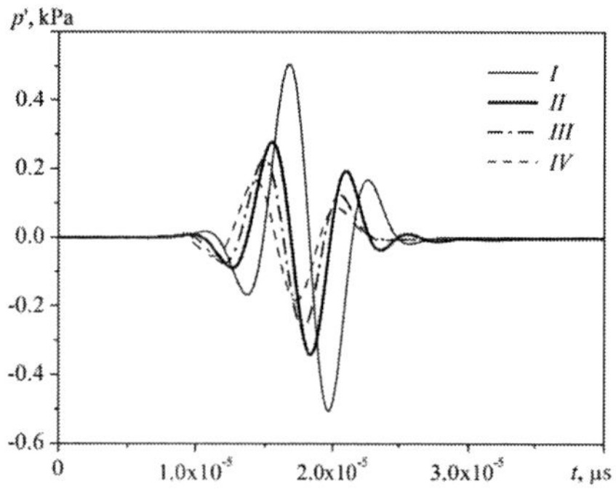

Fig. 6. Pulsed pressure perturbation in water containing a multilayered object.

The form of the pressure pulse after its passage through the three-layer object at various angles of incidence onto the obstacle is shown in Fig. 6, and in Fig. 7 shows the corresponding dimensionless amplitude spectrum of the Fourier transform. The number of harmonics of the FFT subroutine is selected from the motion condition of the pulse in the purified water (when there is no object) without distortion. Lines *I* – the initial pressure pulse (3), lines *II*, *III* and *IV* are calculated for the transmitted signal at $\theta = 0$, $\theta = \pi/6$ and $\theta = \pi/4$, respectively.

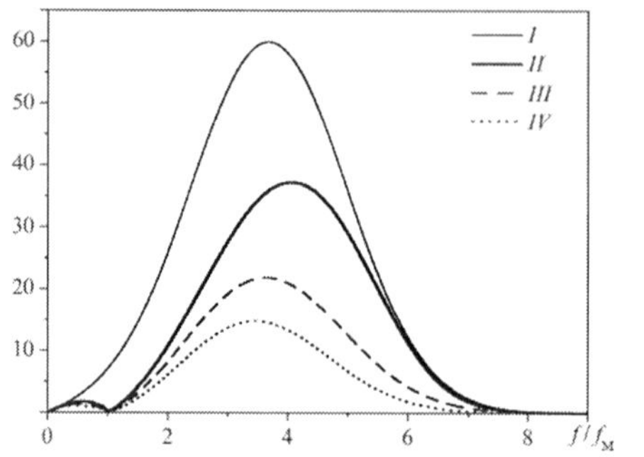

Fig. 7. Amplitude spectrum of Fourier transform.

As the angle of incidence increases, the attenuation of the pulse passing through the object increases. The main carrier frequency (the spike frequency) of the considered pulse is $f_s = 0.15$ MHz ($f_s/f_M = 3.8$). If we compare curves *II-IV* in Fig. 6 and the corresponding amplitude spectra in Fig. 7, it is obvious that the amplitude of the low-frequency harmonics ($f/f_M < 2$) with a change in the slope angle remains practically unchanged. However, the original signal is set in such a way that the major part of its harmonics is in the frequency region $2 < f/f_M < 5$. This determined the dependence of the attenuation of the transmitted pressure pulse on the angle of incidence. Attenuation of the pressure passing through the object, whose amplitude spectrum will be in the frequency range $f/f_M < 2$ will not depend on the angle of incidence. For a pressure pulse whose amplitude spectrum will be in the frequency range $f/f_M > 5$ the effect of bubbles in the gel layer will not appear.

4 Conclusions

The interaction of a small-amplitude pressure pulse with a thin three-layer object at various angles of incidence is theoretically studied. It is shown that the presence of polydispersed air bubbles in the gel layer substantially changes the transmission and reflection of acoustic waves for frequencies less than or close to the resonance frequency oscillations of the bubbles with characteristic mean radius of the bubble size distribution function and for frequencies of the acoustic opacity band. A range of frequencies less than the doubled resonant frequency is revealed, where the dependence of the transmission of acoustic waves through an three-layer object on the angle of incidence is not significant. In this frequency range, the transmission and reflection of acoustic waves will be influenced by the dimensions of the bubbles, their volume content and their thermophysical properties.

The study was performed by a grant from the Russian Science Foundation (project No. 15-11-20022).

References

1. V. Leroy, A. Strybulevych, M. Lanoy, F. Lemoult, A. Tourin, J.H. Page, Phys. Rev. B. **91**, 2, 020301 (2015)
2. B. Würsig, C.R. Jr. Greene, T.A. Jefferson, Mar. Environ. Res. **49**, 1, 79-93 (2000)
3. V.A. Gusev, O.V. Rudenko, Acoust. Phys. **61**, 2, 152-164 (2015)
4. R.I. Nigmatulin, *Dynamics of Multiphase Media* (Hemisphere, New York, 1990)
5. S. Temkin, *Suspension Acoustics: An Introduction to the Physics of Suspensions* (Cambridge Univ. Press, New York, 2005)
6. T.G. Leighton, *The Acoustic Bubble* (Academic, London, 1994)
7. D.A. Gubaidullin, A.A. Nikiforov, High Temp. **48**, 2, 170–175 (2010)
8. V. Leroy, A. Strybulevych, J.H. Page, M.G. Scanlon, J. Acoust. Soc. Am. **123**, 4, 1931–1940 (2008)
9. D.A. Gubaidullin, Yu.V. Fedorov, J. Appl. Math. Mech. **77**, 5, 532-540 (2013)
10. R.I. Nigmatulin, D.A. Gubaidullin, A.A. Nikiforov, Dokl. Phys. **59**, 6, 286–288 (2014)
11. V.Sh. Shagapov, V.V. Sarapulova, Acoust. Phys. **61**, 1, 37-44 (2015)
12. D.A. Gubaidullin, A.A. Nikiforov, High Temp. **55**, 1, 95–100 (2017)
13. L.M. Brekhovskikh, *Waves in Layered Media* (Academic, New York, 1976)
14. D.L. Folds, C.D. Loggins, J. Acoust. Soc. Am. **62**, 5, 1102–1109 (1977)

Small- Scale Drilling Test Rig For Investigation of Axial Excitation On The Drilling Process

A. Austefjord,[1] S. Blaylock,[1] I. Forster,[1 *], M. Sheehan,[1] C. Wright[1]

1: National Oilwell Varco, Conroe, Texas, United States

Abstract. This paper describes the design, construction and operation of a small-scale drilling rig for the purpose of investigation of the effect of axial excitation on the drilling process. The rig is bench top in size and has been designed to drill small rock samples, whilst at the same allowing axial excitation to be induced into the drilling process. The rig has been designed to drill the rock without any drilling fluids – so allowing improved observation of the chip generation process. Additionally, the drilling weight on bit is applied via masses, so allowing greater representation of the dynamic behavior of the drilling process – i.e. capturing more natural frequencies. The results from the rig have been obtained over two frequency ranges – low frequency (0-50 Hz) and high frequency (50-250 Hz). Results show that improved rate of penetration is obtained with axial excitation – with low and high frequency optima occurring. These optima can be related to the behavior of the string in the two frequency ranges – in the low frequency range, the entire string acts in unison; whereas at high frequency, only the bit/rock system is active. As a result, it is concluded that for low frequency operation, only information about the drill string is required to optimize performance; whereas for high frequency operation, information about the bit/rock system is required to optimize performance. Observation of the chip generation process via high speed video has shown that during axial excitation, regular shaped bricks are ejected when compared with the typical wedge- shaped chips that are normally ejected during the drilling process. It is concluded that, during the axial excitation process, the chips are being ejected via a levering action, so allowing a more efficient and quicker process. MIT [1] provided background classes, project guidance and project review as part of an NOV/MIT advanced technology program. Larger scale lab tests and/or field tests are required to verify/validate these conclusions.

1 Introduction

With respect to National Oilwell Varco (NOV) and NOV customers, of particular interest has been vibrations that aid the drilling process. Current tools that meet this requirement are tools that aid the drilling process by inducing axial excitation into the drill bit – either at medium frequency or at low frequency. There are also potential future tools operating at higher and very low frequencies.

There have been many published works in this field – however the most practical recent approaches, that have the greatest relevance for NOV, have been from the University of Aberdeen [2] and the University of Newfoundland [3].

Using the current axial excitation tools that are available as a basis for the study, on the small to medium scale, this has presented a low risk strategy to get a minimum viable product (MVP). The project was carried out on the as an innovation exercise, in order to determine if useful information could be obtained on such a fast track project.

2 Test Rig Description

The basic approach in the test rig design and its operation is that any modelling and small-scale tests should be concurrent and interactive and should cover all areas of the drill string – i.e. drilling bit (including bit/rock interaction), BHA and upper string.

It has been appropriate, given the broad subject matter, to cover one aspect of drilling vibration in isolation initially, but with a view to considering, and combining, other aspects in the future.

The small-scale tests and modelling have been initially discrete spring/mass based, with a view to develop into distributed mass/stiffness systems.

The modelling tools to achieve this have been math package based - MATHCAD/MATLAB for example [4] - and/or FEA based – i.e. ANSYS [5]. Lab tests have been initially in the vertical orientation, but could be developed into other orientations.

Rig test and mathematical modelling derived from testing can be used as a design tool for up- scale to full size down hole tools.

* Corresponding author: ian.forster@nov.com

The test rig is shown in figure 1.

Figure 1 – Test rig

Small-scale rig – pros

- Low cost.
- Fast build.
- Fail fast philosophy.
- Low risk.
- Small, light- weight – can be located virtually anywhere.
- Low voltage – minimal safety issues.
- Better to have it, than to be without it.
- Even if it fails, lessons will be learned – to the extent of achieving success at the next iteration.
- Allows NOV to be in a better position to ask the right questions – i.e. it helps to have a good idea of the answers first.
- Will allow any literature search to be more focused – i.e. will complement a literature search.

Small-scale rig – cons

- Fails – see above.

The main design consideration has been the application of weight on bit via masses, rather than by force/actuator – so enabling the dynamic behavior of a typical drill string to be captured – i.e. more natural frequencies. Another design consideration is the use of standard, off the shaft readily available components – for example, using standard masonry bits. Dry cutting has also been adopted, to aid visualization of the chip generation process.

The small-scale approach has been adopted in other work [6].

3 Test Rig Modelling

Two independent modelling approaches were adopted – running concurrently with the test rig design and operation: the first being a MATHCAD- based spring mass model approach; and the second ANSYS FEA-based spring model approach. The former allows more readily varied parameters and model adjustment; whereas the latter allows good visualization/animation of results.

The MATHCAD results in figure 2 show that the rig has two main natural frequencies – so justifying the mass based weight on bit application approach.

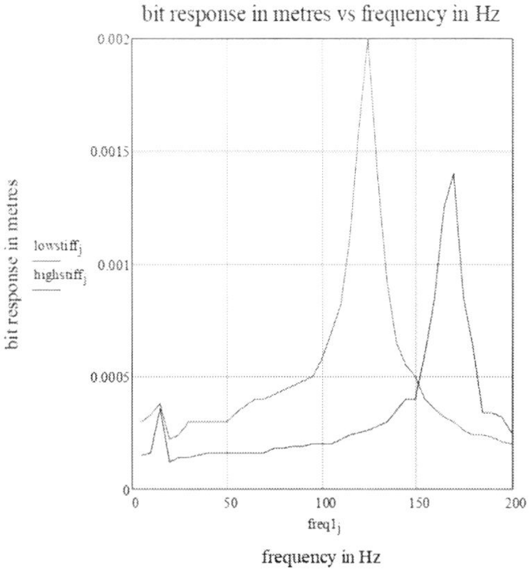

Figure 2 – MATHCAD rig model frequency response

The results are presented for two rock stiffness's – and indicate that the low frequency response is largely independent of rock stiffness – whereas the high frequency response is not. This high frequency result agrees with other published work (reference 1).

A further conclusion can be drawn from the ANSYS simulations – that the low frequency response is dependent on the drill string (with the drill string acting in unison); whereas the high frequency response is dependent on the bit rock system in isolation.

4 Test Rig Results

Test rig results presented are for soft rock and hard rock – sandstone and limestone. Two frequency ranges were considered – 0-50 Hz; and 50-250 Hz. High speed video was used to capture the chip generation process for the hard rock case.

Soft Rock

For the soft rock results in figures 3 and 4, the largest percent gain in rate of penetration (ROP) was shown to coincide with the best low and high frequency rig

* Corresponding author: ian.forster@nov.com

responses.

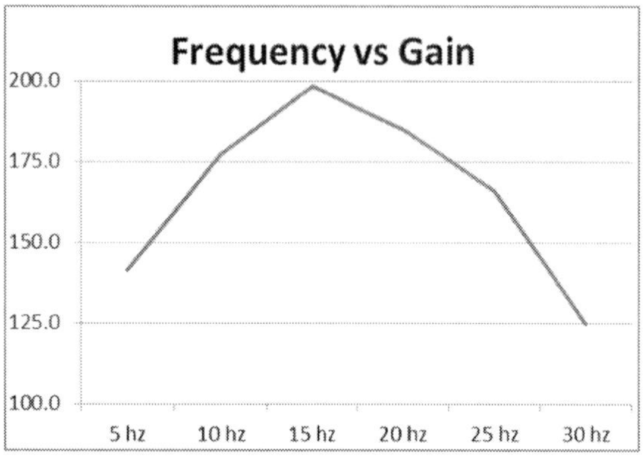

Figure 3 – Test rig low frequency percent ROP gains

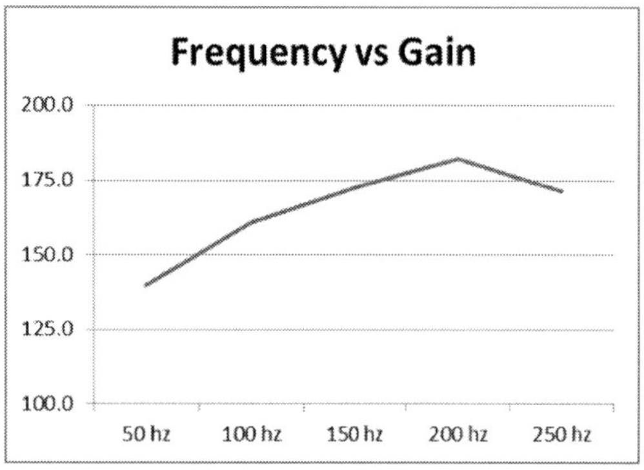

Figure 4 – Test rig high frequency percent ROP gains

<u>Hard Rock</u>

With the harder rock at low frequency and an upgraded rig, the same peak ROP frequency was repeated – for small and larger bits; however, there was a very low frequency secondary peak - not see in the first results – figures 5 and 6. This latter feature may have been as a result of the excitation frequency synchronizing with the chip generation frequency, and may have not been apparent on the pre-upgrade rig - due to energy dissipation. Additionally, it could also be due, in part, to rig structural resonance (again, only apparent after rig upgrade). This justifies further investigation.

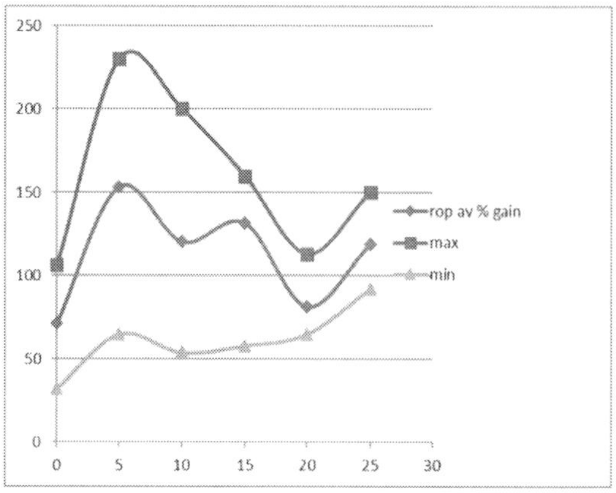

Figure 5 – Test rig low frequency small bit percent ROP gains

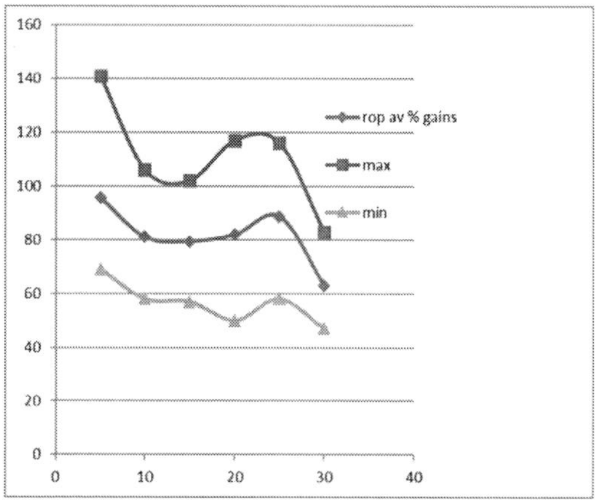

Figure 6 – Test rig low frequency large bit percent ROP gains

<u>High Speed Video</u>

High speed video test were conducted over a variety of frequencies – from 5 hz up to 150 hz.
However, the 100 hz result is presented here – as the migration of cuttings across the rock during the 100 Hz test, lended itself well to analysis of the debris field via microscope – figures 7-10. Note that the calutated normal drilling (non excitation) chip generation frequency based on depth of cut and rpm would be of the order of 100 Hz. Observed chip generation frequency was of the order of the natural frequency of the string – suggesting that this is an important driver.

Any issues with dry cutting were also investigated – i.e. a back to back test was run with and without dust extraction applied – and no observable difference was noted in tersms of ROP. However, further invetigation is still justified.

From the microscope photos, the chip size and shape matched the expected size and shape based on excitation frequency, rpm and bit size.

* Corresponding author: ian.forster@nov.com

A proposal for the chip generation process during axial excitation is given – suggesting a levering action is used to generate the chips – analogous to a jack hammer – resulting in lower energy required to fail the rock (in bending and tension) – figure 11.

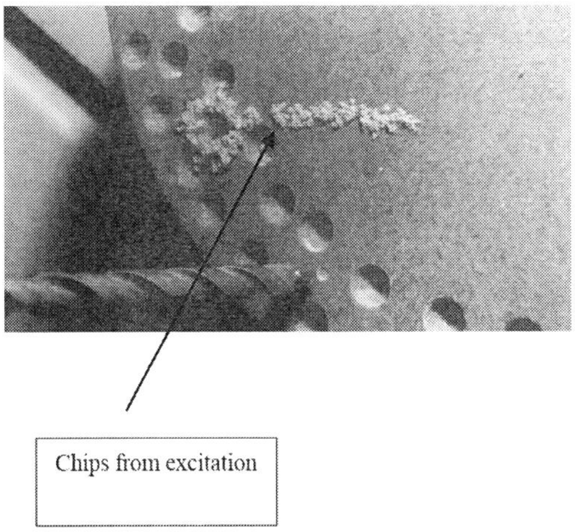

Figure 7 – debris field image

Figure 8 – debris field microscope image

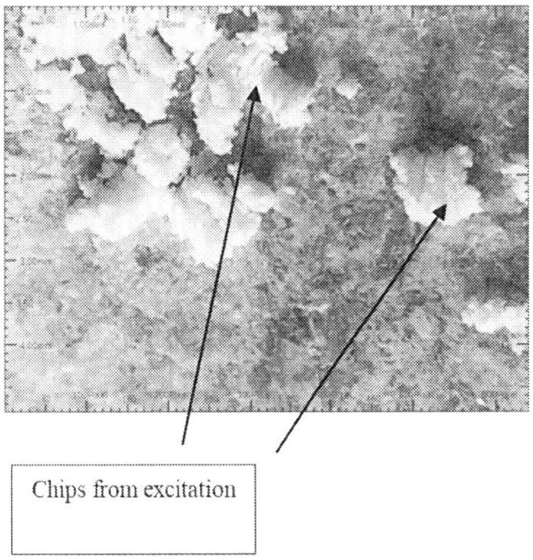

Figure 9 – debris field microscope image

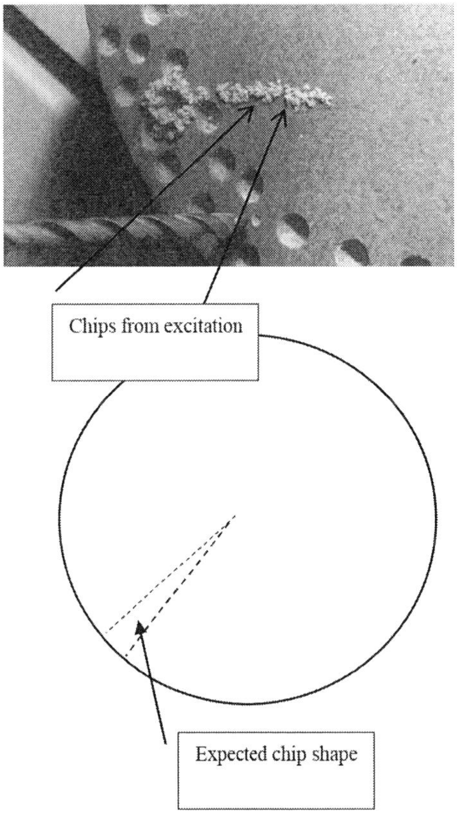

Figure 10 – expected chip shape

- Chip tensile failure and ejection – earlier due axial upward movement – chip size depends on frequency and rpm - chip is effectively levered out– smaller chip and small amount of

* Corresponding author: ian.forster@nov.com

material needing to be compressed/crushed – chip size more controlled

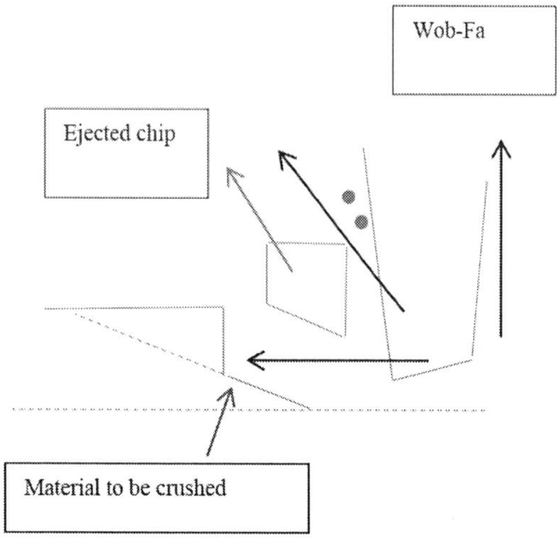

Note chip failure is mainly tensile – requiring less energy as the rock will have low tensile strength compared to compressive strength
Note also that lower frequency excitation will tend to form larger chips than at high frequency – at a given rpm

Figure 11 – excitation chip generation

5 Results Summary

1. Very low frequency benefits – possible synchronization with chip generation and/or rig structural resonance – quasi-static.
2. Low frequency peak ROP gain at 20-25 Hz coincident with string resonance.
3. High frequency peak ROP gain coincident with bit rock system resonance.
4. Larger bit tests - results variability offset.
5. Dry testing beneficial for visualization, effect on results checked but further investigation still required.
6. Mechanical specific energy (MSE) results mirrored ROP during excitation– but mitigated due to increased torque.
7. When electrical energy included in MSE formulation – increased by a factor of 9.
8. MSE suggests that a mechanical system would be preferable for excitation.

6 Conclusions

1. The small-scale rig approach has been validated.
2. Low and high frequency tools have been identified – and a possible very low frequency tool.
3. Low frequency tools are current NOV tools – and therefore the results can be used for optimization and modelling development.
4. The chip generation results can feed into bit/cutter design.
5. High frequency tools are possible for the future.

7 Further Work

1. Further tests using hard rock
2. Further upgrades – higher data rate data acquisition for higher frequency tests.
3. Investigation of varying rock/support stiffness.
4. String modelling development with respect to NOV tools.
5. Further High- speed video of bit rock interaction
6. Higher frequency tests – up to 1 KHZ and higher
7. Inclination tests
8. Further investigate the large variations in repeated, identical tests.
9. Very low frequency investigation with respect to synchronization with normal chip generation rate.
10. Cloned rigs to investigate specific phenomena in more detail – possibly at other establishments.
11. Investigation on any disadvantages of dry testing over wet testing.

Acknowledgements

1. ICOEV 2017
2. NOV
3. MIT
4. Project team members

Glossary

NOV – National Oilwell Varco
MIT – Massachusetts Institute of Technology
MSE – mechanical specific energy
RPM – revs per minute
ROP – rate of penetration
MVP – minimum viable product

* Corresponding author: ian.forster@nov.com

FEA – finite element analysis
WOB – weight on bit
Fa – axial excitation force

References

1 MIT/NOV – Advanced Technology Program – (2015-2016)
2 University of Aberdeen – publications relating work carried out on, and associated with, the Resonance Enhanced Drilling Project – (2009-2017)
3 University of Newfoundland - publications relating work carried out on, and associated with, axial excitation during drilling
4 MATHCAD – v14
5 ANSYS - v14.5
6 NOV – publications related to work carried out on, and associated with, the use of small-scale test rigs – (2009-2017)
7 G. Hareland et al., JCPT – paper number 2007-082 – Cutting Efficiency of a Single PDC Cutter on Hard Rock – (2007)

* Corresponding author: ian.forster@nov.com

Stick-slip and Torsional Friction Factors in Inclined Wellbores

Ulf Jakob F. Aarsnes[1,2,⋆] and *Roman J.* Shor[3,⋆⋆]

[1] *International Research Institute of Stavanger (IRIS), Oslo, Norway*
[2] *DrillWell - Drilling and well centre for improved recovery, Stavanger, Norway*
[3] *University of Calgary, Department of Chemical and Petroleum Engineering, Calgary, Canada*

Abstract. Stick slip is usually considered a phenomenon of bit-rock interaction, but is also often observed in the field with the bit off bottom. In this paper we present a distributed model of a drill string with an along-string Coulomb stiction to investigate the effect of borehole inclination and borehole friction on the incidence of stick-slip. This model is validated with high frequency surface and downhole data and then used to estimate static and dynamic friction.

A derivation of the torsional drill string model is shown and includes the along-string Coulomb stiction of the borehole acting on the string and the 'velocity weakening' between static and dynamic friction. The relative effects of these two frictions is investigated and the resulting drillstring behavior is presented. To isolate the effect of the along-string friction from the bit-rock interaction, field data from rotational start-ups after a connection (with bit off bottom) is considered. This high frequency surface and downhole data is then used to validate the surface and downhole behavior predicted by the model.

The model is shown to have a good match with the surface and downhole behavior of two deviated wellbores for depths ranging from 1500 to 3000 meters. In particular, the model replicates the amplitude and period of the oscillations, in both the topside torque and the downhole RPM, as caused by the along-string stick slip. It is further shown that by using the surface behavior of the drill-string during rotational startup, an estimate of the static and dynamic friction factors along the wellbore can be obtained, even during stick-slip oscillations, if axial tension in the drillstring is considered. This presents a possible method to estimate friction factors in the field when off-bottom stick slip is encountered, and points in the direction of avoiding stick slip through the design of an appropriate torsional start-up procedure without the need of an explicit friction test.

1 Introduction

Exploration and production of oil and gas in the deep subsurface, where hydrocarbon reservoirs are found at depths between 2,000 and 36,000 feet [1], requires that a narrow borehole, between 4 and 24 inches in diameter, be drilled using a slender drill-string through a varied downhole environment and along an often snaking wellpath. Drill string vibrations, and their negative consequences on ROP and equipment, are a well known phenomenon when drilling for hydrocarbons. In particular, the torsional oscillations known as stick slip, which are considered to be some of the most prevalent vibrations, are to be avoided. These stick-slip oscillations are characterized by a series of stopping – "sticking" – and releasing – "slipping" – events of the bit.

Significant literature exists which seeks to explain the incidence of stick slip through various implementations of bit-rock interaction and various complexities of drill-string dynamic models, summarized in [2]. Many models considers the drill-string as a lumped mass, representing the bottom hole assembly (BHA) inertia, and a torsional spring, representing the drill-string stiffness [3, 4], while more realistic, higher order, models have also been considered [5–7]. Stick-slip is then introduced as a self-excited vibration (see example in Fig. 1, right hand side) by inducing instability through a bit-rock interaction as a *velocity weakening* frictional force (Stribeck-like effect) [8–10], or through a regenerative effect [11–13] causing an apparent velocity weakening effect.

A key factor in common with the above referenced literature is that the bit-rock interaction is used to explain the cause of stick slip. A consequence of this is that the models are unable to explain the occurrence of stick-slip with the bit off bottom (where there is no bit-rock interaction). Off-bottom stick slip is a well known phenomenon from the field, and when mentioned in literature is hypothesized to be caused by a negative difference between static and dynamic *along-string* Coulomb-type friction [14–16]. This is an important phenomena as it indicates that non-linear frictional forces along the drill-string (and not just the bit rock interaction), in deviated or horizontal wells, plays a significant role in the torsional oscillatory behavior of drill-strings. This is also a qualitatively different mechanism for causing stick slip from the self-excited type

⋆e-mail: ujfa@iris.no
⋆⋆e-mail: roman.shor@ucalgary.ca

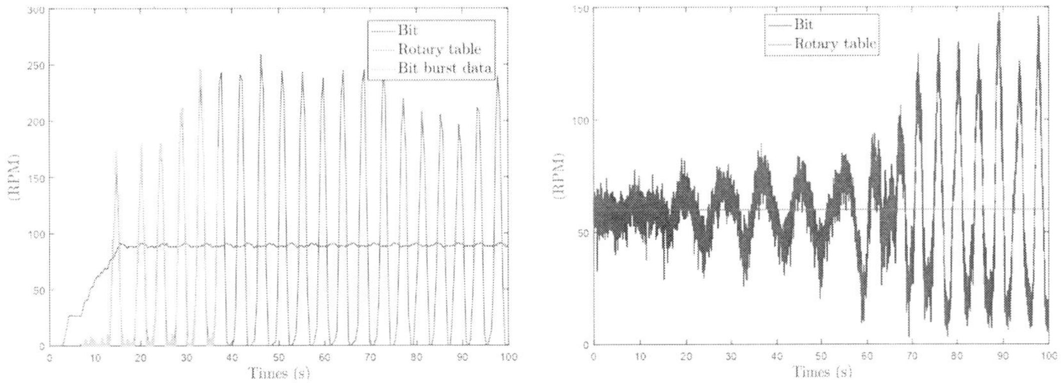

Figure 1: Two different kinds of stick-slip from two different wells. In the present paper we consider the type shown left, and not the self-excited type shown right.

caused by a velocity weakening effect in the bit-rock interaction, as is exemplified in Fig. 1:

- Self-excited vibrations are can occur when a slight change in e.g. bit-rock parameters causes the system to bifurcate and the resulting instability causes growing torsional oscillations to result in a stick-slip limit cycle. Shown right in Fig. 1.
- Off-bottom stick slip is cause by an initial exogenous shock, such as a rotation start-up, which pushes the drill string into activating the static Coulomb friction in the along string pipe–wellbore interaction. Shown left in Fig. 1.

Hence, models which only incorporate the bit rock interaction as the cause of torsional stick slip fail to explain off-bottom stick slip vibrations, as observed in field data after connections and in back-reaming operations. This potential cause of stick-slip, have not received nearly the same analysis or attention in the literature, in particular in the context of occurrence, characterization and avoidance.

2 Model

In this section we derive the model that will be used to recreate and analyze the torsional vibrations. To facilitate its relative simplicity, the formulation requires certain approximations, the main assumptions used are the following

- The bit is assumed to be off bottom and no bit-rock interaction is modeled.
- The torsional motion of the drill string is the dominating dynamic. No lateral or axial motion is assumed.
- The transition from static to dynamic Coulomb friction is modeled as a jump, i.e., the Stribeck curve is assumed negligible.
- The effects of along-string cuttings distribution on the friction is assumed to be homogenous.
- The effect of the pressure differential, inside and outside the drill string, on the bending moment is not represented and is assumed to be negligible.

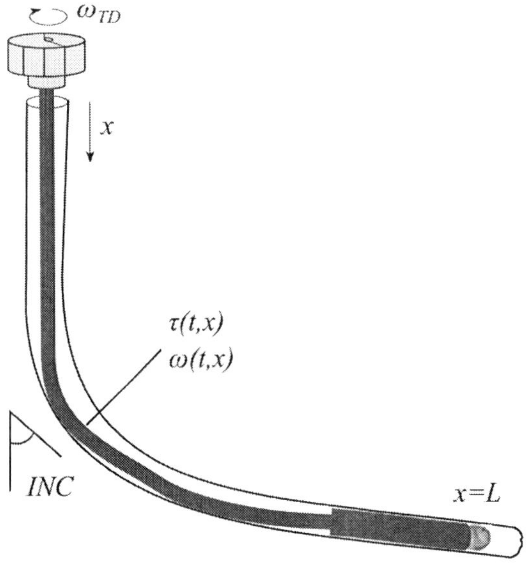

Figure 2: Schematic indicating the distributed drill string lying in deviate borehole.

2.1 Torsional dynamics of the drill string

We use a distributed model, similar to [17–19] and described in detail in [20], where we consider only the torsional dynamics of the drill string. For angular motion, angular velocity and torque are denoted as as $\omega(t,x)$, $\tau(t,x)$, respectively, with $(t,x) \in [0,\infty) \times [0,L]$. See Fig. 2 for a schematic indicating locations. For an infinitesimal element dx, as shown in Fig. 3, the torque is found as the shear strain, or twist per unit length. Letting ϕ denote the angular displacement in the string s.t. $\frac{\partial\phi(t,x)}{\partial t} = \omega(t,x)$, we have $\tau(t,x) = JG(\phi(t,x) - \phi(t,x+dx))/dx$, where J is the polar moment for inertia and G is the shear modulus.

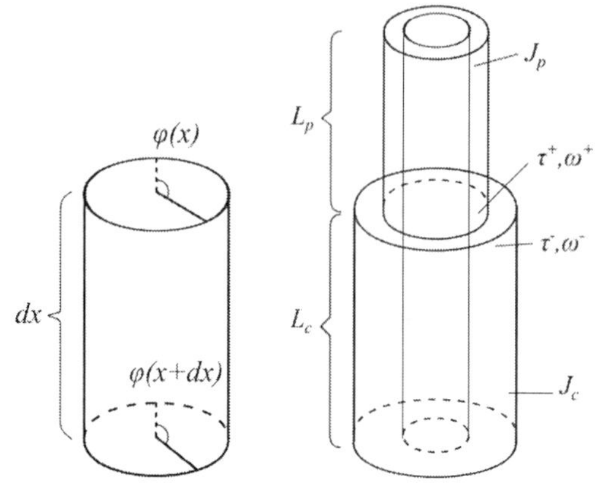

Figure 3: Infinitesmal drill string element (left) and the collar-Pipe transition (right).

Hence the equations for the angular motion are given by

$$\frac{\partial \tau(t,x)}{\partial t} + JG\frac{\partial \omega(t,x)}{\partial x} = 0 \tag{1}$$

$$J\rho\frac{\partial \omega(t,x)}{\partial t} + \frac{\partial \tau(t,x)}{\partial x} = S(\omega,x), \tag{2}$$

where the source term S is modeled as

$$S(\omega,x) = -k_t \rho J\omega(t,x) - \mathcal{F}(\omega,x), \tag{3}$$

where k_t is a damping constant representing the viscous shear stresses and $\mathcal{F}(\omega)$ is a differential inclusion, to be described, representing the Coulomb friction between the drill string and the borehole. The viscous shear stress coefficient k_t represnts the combined damping effects of the viscous shear of the drilling mud and the rolling contact between drill string and the cuttings bed.

The lowermost section of the drill string is typically made up of drill collars which may have a great impact on the drill string dynamic due to their added inertia. In particular, the transition from the pipes to collars in the drill string will cause reflections in the traveling waves due to the change in characteristic line impedance [17].

We split the drill string into a pipe section with polar moment of inertia and lengths J_p, L_p and a collar section with the same parameters given as J_c, L_c. We use τ^-, ω^- to denote the strain and velocity at the top of the drill collar and τ^+, ω^+ at the bottom of the pipe, see Fig. 3. Enforcing the boundary conditions $\omega^+ = \omega^-$ and $\tau^+ = -\tau^-$ at the transition.

2.2 Coulomb friction as an inclusion

The Coulomb friction is modeled as an inclusion

$$\begin{cases} \mathcal{F}(\omega,x) = F_d(x), & \omega > \omega_c, \\ \mathcal{F}(\omega,x) \in [-F_c(x), F_c(x)], & |\omega| < \omega_c, \\ \mathcal{F}(\omega,x) = -F_d(x), & \omega < -\omega_c, \end{cases} \tag{4}$$

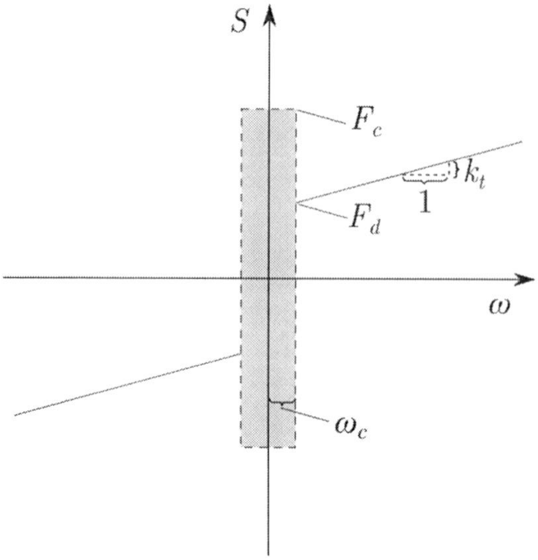

Figure 4: Schematic illustrating the four parameters determining the friction: the coulomb friction parameters ω_c, F_c, F_d and the viscous friction coefficient k_t, with the shaded region indicating the region of static torque, and the red curve the dynamic torque.

where ω_c is the threshold on the angular velocity where the Coulomb friction transitions from static to dynamic, F_d is the dynamic Coulomb torque, F_c is the static Coulomb torque, and $\mathcal{F}(\omega) \in [-F_c, F_c]$ denotes the inclusion where

$$\mathcal{F}(\omega,x) = -\frac{\partial \tau(t,x)}{\partial x} - k_t\rho J\omega(t,x) \in [-F_c(x), F_c(x)], \tag{5}$$

and take the boundary values $\pm F_c(x)$ if this relation does not hold. We define the non-dimensional coefficient $f_{rat} = F_c/F_d$ to help characterize the magnitude of the oscillations. The shape of the friction source term is illustrated in Fig. 4.

Using the torque model of [21], we first derive the profile of the normal force $F_N(x)$ which is then used to express the torque.

The normal force is expressed in terms of the tension profile σ_e, assuming a planar well and torsional rotation of the drill string, the tension gradient is given as [21]:

$$\frac{\partial \sigma_e}{\partial x} = W_b \cos\theta, \tag{6}$$

where $W_b(x) = gA(x)(\rho - \rho_{mud})$ is the buoyed weight per meter. That is, we have

$$\sigma_e(x) = \int_L^x W_b \cos\theta(\xi) d\xi \tag{7}$$

The normal force profile, F_N, is then found as

$$F_N(x) = \left|\sigma_e(x)\frac{\partial \theta}{\partial x} + W_b \sin(\theta)\right|, \tag{8}$$

and the maximum Coulomb torque

$$F_c(x, \mu) = \mu r_o(x) F_N(x), \tag{9}$$

where μ is the static Cuolomb friction factor. This friction factor is determined by factors such as wether the section is cased or openhole, as well as the lithology of the surrounding formation [14].

2.3 Boundary condition

At the topside boundary, the top drive is actuated by a motor torque, τ_m, controlled by a PI controller to a desired velocity set-point ω_{SP}:

$$e = \omega_{SP} - \omega_{TD} \tag{10}$$

$$I_e = \int_0^t e(\xi)\mathrm{d}\xi \tag{11}$$

$$\tau_m = k_p e + k_i I_e, \tag{12}$$

with a proportional gain, k_p and an integral gain, k_i.

The topdrive has the inertia J_{TD} and hence satisfies the dynamics

$$\frac{\partial \omega_{TD}}{\partial t} = \frac{1}{J_{TD}}\left(\tau_m - \tau_0\right), \tag{13}$$

and finally, the angular velocity at the top of the drill string is given as $\omega_0 = \omega_{TD}$.

2.3.1 Characterizing oscillations

The period of the oscillations is a combination of the time required to build the twist needed to overcome the static Coulomb friction, and the torsional fundamental period of the drill string.

The required change in twist at the top-drive, $\Delta\phi(0)$, see [20] for details, enables us to find the build-up time of the cycle as

$$t_{bu} = \frac{\Delta\phi(0)}{\omega_{TD}}. \tag{14}$$

When the drill string is free, i.e. there is no build up time, the drill string oscillation will be dominated by the fundamental period of the free drill string, which is dependent on the drill pipe and collar length and the relative impedance $\tilde{Z}$, [18]. Similarly, as the drill string breaks free, the trajectory will follow close to one fundamental period before getting stuck again.

That is, denoting the fundamental period of the drill string as t_{fp}, the period of the limit cycle is

$$t_p = t_{bu} + t_{fp}. \tag{15}$$

This allows us to heuristically distinguish between two different types of oscillations:

1. The oscillation is dominate by the fundamental period of the drill string: $t_{bu} < t_{fp}$. The torque trend resembles that of free oscillations.
2. The oscillation is dominated by the torque build up. A characteristic sawtooth pattern of the torque is obtained: $t_{bu} > t_{fp}$.

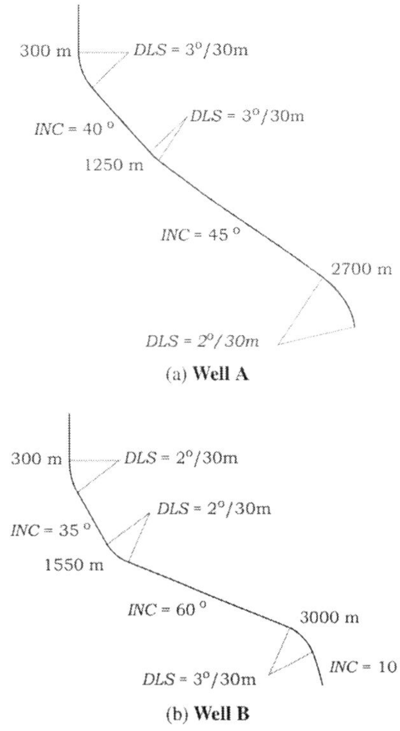

Figure 5: Wellbore surveys of the two field wells

3 Model comparison with field data

To validate the modeling approach taken in the present work, a simulation study was undertaken to compare the behavior of the model to that of recorded field data. Two field comparisons are presented, with one exhibiting inerita dominated oscillations and the other exhibiting stick dominated oscillations, as categorized in [20].

3.1 Simulation study case description

Field data for two deviated wells, the surveys of which are shown in Figures 5a and 5b, are considered. Rotational data – rotary rpm and torque – is recorded at 100 Hz and provides a clear picture of the rpm, both set-points and realized values, used and during the drilling process and the torsional drill-string response.

The torque from the variable frequency drive is controlled with a PI controller with known gains of k_p = 38 kNm per m/s and k_i = 100 kNm per m/s^2, see (10)–(13). In all cases presented herein, the drill-string starts at rest with zero torque at the surface. However, the stored torsional energy within the drill-string is not known. In both cases, the drill-string design is a simple directional assembly which is simplified to a 230 meter 5¾" OD BHA and monodiameter drillpipe to the surface. Downhole rpm and vibration data was collected for drilling performance improvement and control system verification and included

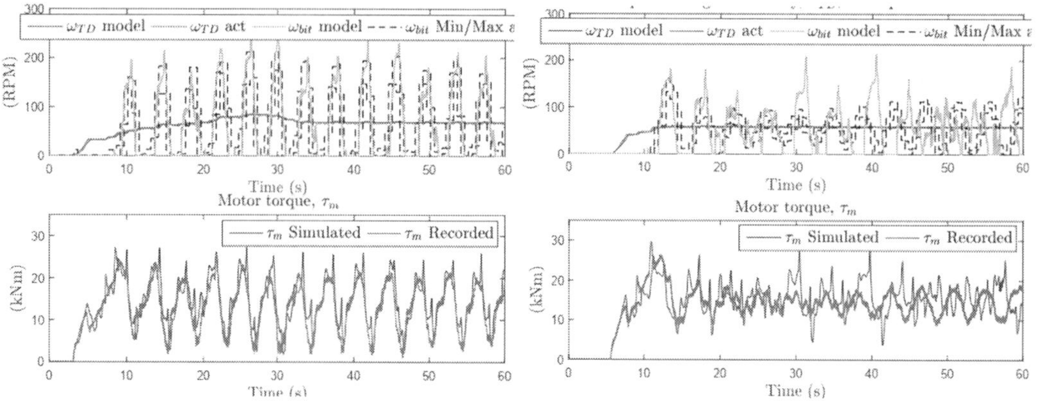

Figure 6: Left: Recorded and simulated drill-string response at a bit depth of 1,733 m in **Well A**, using fitting parameters: $\mu = 0.34, f_{rat} = 0.55, \omega_c = 19$ (RPM). Right: Recorded and simulated drill-string response at a bit depth of 2,235 m in **Well A**, using fitting parameters: $\mu = 0.35, f_{rat} = 0.45, \omega_c = 16$(RPM).

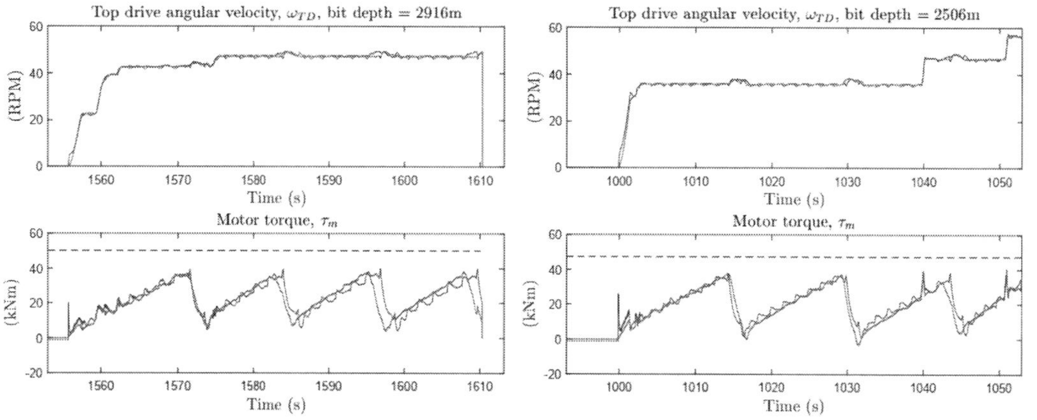

Figure 7: Left: Recorded and simulated drill-string response at a bit depth of 2,916 m in **Well B**. Right: Recorded and simulated drill-string response at a bit depth of 2,2506 m in **Well B**.

continuous low frequency (0.5Hz) data as well as occasional burst sequences of high frequency data (125 Hz). For Well B, downhole data was sampled continuously at 125 Hz and then processed to record minimum, maximum and mean rpm for the 0.5 second sampling window.

3.2 Well A: Inertia dominated oscillations

Recorded field data and a model fit for two different depths, 1,733 m and 2,235 m are shown in Figures 6 and 6. In each figure, the top plot shows the surface (in red) and downhole (dashed) recorded data as well as the modeled data (in blue and yellow, respectively). The bottom plot shows surface torque, with recorded data in red and simulated data in blue. Only the Coulomb friction parameters changes between the two model fits, which is expected as the friction coefficient is expected to change as new rock is encountered and the side force is distribution on the drill string is changed.

In Figure 6 the max torque peak is repeated throughout, and the period is regular, indicating that the full max torque is reached in this stick slip limit cycle. This is not the case in Figure 6, however, as can be seen in the irregular max torques. The lower-amplitude oscillations that occur after this initial peak is due to the torque contribution from the static Coulomb friction at the upper-most kick-off point not being activated (see Fig. 5b), i.e. only the Coulomb friction in the lower section of the well contributes to the stick slip limit cycle, thus lowering the amplitude.

3.3 Well B: Stick dominated oscillations

The field data from **Well B** comes from a deeper section with a shallower profile – the tangent section of the well is 60° instead of the 45° as in **Well A**, which results in a higher break-off torque. This, coupled with the fact that a lower RPM is used, causes **Well B** to exhibit stick dominated oscillations, where the majority of the limit cycle is spent building up the required break off torque. Downhole data was not available for this well, however, a fit of the surface data may still be made and is shown in Figure 7 and 7. Here the torque build up can be clearly seen, which is comparatively slow relative to the torque release.

3.4 Remarks on the model comparison

The model fit is found to be very sensitive and the quality of fit varied from dataset to dataset. This is due to a number of factors:

- Pre-existing torsional stress state, the amount of twist existing in the drillstring, is not known at each connection event. Thus the total number of twists needed to break static friction is not known. This is very difficult to both model and quantify in field data since there is no method to to know the number of twists in the drillstring from the available recorded field data.
- The model only accounts for torsional state within the drillstring and does not take into account the axial stress state, hence once the drillstring begins moving axially, the model assumptions are violated and deviation between the field and simulated data appears.

4 Conclusions and further work

A distributed Coulomb friction model has been presented which is parameterized using three parameters: the static and dynamic friction, F_c and F_d, or similarly, the coefficient of friction, μ, and the ratio between static and dynamic friction, f_{rat}, along the wellbore and the cutoff angular friction, ω_c. The model is applied to two datasets and the surface and downhole behavior is found to match closely with model predictions in most cases. Furthermore, once axial motion is introduced, a key assumption of the presented model - that only torsional dynamics are considered - is violated.

Acknowledgment

This work was supported by the Research Council of Norway, ConocoPhillips, Det norske oljeselskap, Lundin, Statoil and Wintershall through the research center DrillWell (203525/O30) at IRIS and through the Schulich School of Engineering at the University of Calgary.

The authors also wish to acknowledge Shell International Exploration and Production for allowing use of their data and Sicco Dwars for his support of the effort. Finally, the authors wish to acknowledge Eric Cayeux at IRIS for helpful input to this work.

References

[1] S. Stokka, E. Cayeux, D. Gardner, S. Kragset, H.P. Lohne, E. Randeberg, H.J. Skadsem, B. Aas, H. Kyllingstad, T. Larsen et al., *Long Reach Well Concept*, in *IADC/SPE Drilling Conference and Exhibition* (Forth Worth, 2016)

[2] R.J. Shor, M. Pryor, E. van Oort, *Drillstring Vibration Observation, Modeling and Prevention in the Oil and Gas Industry*, in *ASME 2014 Dynamic Systems and Control Conference* (2014)

[3] J.J. Bailey, I. Finnie, Transactions of the American Society of Mechanical Engineers pp. 122–127 (1960)

[4] D. Dashevskiy, J. Rudat, L. Pohle, *Model-Based Stability Analysis of Torsional Drillstring Oscillations*, in *Proceedings of 2011 IEEE International Conferencce on Control Applications* (Denver, CO, 2011), ISBN 9781457710636

[5] R.I. Leine, D.H. van Campen, W.J.G. Keultjes, Journal of Vibration and Acoustics **124**, 209 (2002)

[6] K. Nandakumar, M. Wiercigroch, Journal of Sound and Vibration **332**, 2575 (2013)

[7] X. Liu, N. Vlajic, X. Long, G. Meng, B. Balachandran, *State-Dependent Delay Influenced Drill String Dynamics and Stability Analysis*, in *Volume 7B: 9th International Conference on Multibody Systems, Nonlinear Dynamics, and Control* (ASME, 2013), p. V07BT10A065, ISBN 978-0-7918-5597-3

[8] J. Brett, SPE Drilling Engineering **7**, 168 (1992)

[9] P.E. Pastusek, D. Ertas, L. Wang, J.R. Bailey, Proceedings of 2013 SPE / IADC Drilling Conference and Exhibition (2013)

[10] M. Kapitaniak, V. Vaziri, J. Páez, K. Nandakumar, M. Wiercigroch, International Journal of Mechanical Sciences **101-102**, 324 (2015)

[11] T. Richard, C. Germay, E. Detournay, Comptes Rendus Mécanique **332**, 619 (2004)

[12] T. Richard, C. Germay, E. Detournay, Journal of Sound and Vibration **305**, 432 (2007)

[13] C. Germay, N. van de Wouw, H. Nijmeijer, R. Sepulchre, SIAM Journal on Applied Dynamical Systems **8**, 527 (2009)

[14] J. Brett, A. Beckett, C. Holt, D. Smith, SPE Drilling Engineering **4**, 223 (1989)

[15] G. Halsey, A. Kyllingstad, T. Aarrestad, D. Lysne, *Drillstring Vibrations: Comparison Between Theory and Experiments on a Full-Scale Research Drilling Rig*, in *SPE/IADC Drilling Conference* (Society of Petroleum Engineers, 1986), IADC/SPE 14760, pp. 311–321, ISBN 9781555636074

[16] D. Zhao, S. Hovda, S. Sangesland, Journal of Petroleum Science and Engineering **In press** (2016)

[17] U.J.F. Aarsnes, O.M. Aamo, Journal of Sound and Vibration **360**, 239 (2016)

[18] U.J.F. Aarsnes, N. van de Wouw, in review (2017)

[19] C. Germay, V. Denoël, E. Detournay, Journal of Sound and Vibration **325**, 362 (2009)

[20] U.J.F. Aarsnes, R.J. Shor, Journal of Petroleum Science and Engineering (accepted) (2017)

[21] M. Sheppard, C. Wick, T. Burgess, SPE Drilling Engineering **2**, 344 (1987)

Monitoring of drilling conditions using the Hilbert-Huang transformation

Piotr Wolszczak[1,2,*], *Grzegorz* Litak[1,3,**] and *Marek* Dziuba[4,***]

[1] *Department of Automation, Lublin University of Technology, Nadbystrzycka 36, 20-618 Lublin, Poland*
[2] *Department of Technology and Life Sciences, The State School of Higher Education in Zamość, Pereca 2, 22-400 Zamość, Poland*
[3] *Department of Process Control, AGH University of Science and Technology, Mickiewicza 30, 30-059 Kraków, Poland*
[4] *J.G. SERVICE, ul. Kuźnicza 4, 21-040 Świdnik, Poland*

Abstract. The article presents the results of design and monitoring the drilling process. Vibroacoustic sensors were used to observe spindle vibrations. These signals were subjected to a Huang decomposition and a Fourier transform. Results for various conditions were studied and classified with help of Fourier spectra and the envelope curves. Using the additional results of numerical simulations sources of vibration were identified. We considered four different types of drilling which were diversified in terms of geometrical parameters of blades. The application of Hilbert transform enable to find nonlinear characteristics via the deflection profile of resonance backbone curves.

1 Introduction

The recent development of materials and new technologies used in the construction of machines require improvements of tools used for machining [1]. The improvement consists of applying wear resistant coatings and adjusting the geometry of cutting tools but also certain machining parameters and identification of dynamical system response [2, 3]. This trend can be seen in aerospace and automotive industries, where the use of ultra-light and durable materials are important in construction [4].

In many industrial areas, the desire to reduce manufacturing cost, shorten manufacturing time, or even the time for implementing the production of a new type of material, e.g. composite material, Inconel and stainless steel, is important [4–6]. Modern computer tools for simulating the operations of machining tools and measuring paths with various sensors are applied to this end [3, 7]. Numerical methods and research tools applied in laboratories are being increasingly applied in tool-rooms. Computer technologies allow new methods and algorithms to control and evaluate of machining tools' operation.

In this note, we apply empirical mode decomposition in accordance with Huang [9] and Hilbert [8] algorithms designed for the assessment of machining tools' operation. As a result of Hilbert-Huang decomposition, information on the amplitude and frequency of integral signals composing the analysed signal in current time is obtained [12]. The evaluated empirical modes characterize phenomena emerging during aggressive machining, which induces vibrations in the machine-material-tool system (m-m-t system). Isolation of integral systems with regards to their frequency and amplitude enables information on the tool's vibrations while the individual cutting edges penetrate the material to be obtained. Such information allows for machining parameters to be selected and optimized, the tool's working conditions, including cooling, lubrication and wear, to be monitored.

The heat generated during the cutting process is less well evacuated because the drill is always surrounded by workpiece material. There are no entering phases of cutting edge into the material (for shallow holes, without breaking the chip).

Modeling of the cutting process takes into account the types of materials and geometries of workpiece and tool, and cutting speeds and chip load. Cutting coefficients are determined during experimental cutting or using mechanistic models. These coefficients are used to predict cutting forces, torque, power and limits of the chatter stability. In order to estimate the stress, strain and temperature distribution in the tool and chip is used micro mechanics of metal cutting. Beyond material properties like elastic modulus, yield stress and fracture toughness, the friction can also be important[13, 14].

An analysis of the slip line field with numerical modeling can provide detailed information about the cutting zones. However, due to the deformation, rates of strain and temperature these numerical modeling of cutting process is complicated and complex. For the prediction of reaction in parts and tools depend on cutting conditions developed an Arbitrary Lagrangian-Eulerian Finite Element (ALE-FE) [13, 15].

*e-mail: p.wolszczak@pollub.pl
**e-mail: g.litak@pollub.pl
***e-mail: mdziuba@jgservice.pl

Among the significant size of the cut affecting the service life of tools can mentioned the flow stress and average forces per tooth. The finite element (FE) model used in this study is based on the commercial FE software "Thirdwave AdvantEdge!". This software conducts calculations based on six-nodded quadratic triangular elements. The tool drills through the workpiece at a constant speed and constant feed rate. Simulations were started in a state after entering the blade into the material [16, 17].

2 Materials and apparatus

Example signals were collected during extensive studies conducted in cooperation with a producer of special cutting tools. The studies pertained to the adaptation of tools' geometry and working parameters to the type of the machined material, and requirements associated with tools' durability and manufacturing capacity they were intended for.

The object of the study were 6 mm-diameter drills tested on steel. In the course of the study, vibrations of the drilling machine's spindle caused by forces impacting the tool's cutting edges in the machining process were recorded and analysed. The signal of the tool vibration during drilling process recorded by means of Hansford Sensor HS-102 measuring system with the range of 16 kHZ and sensitivity of 100 mV/g. Vibroacoustic sensors provide the acceleration signal, which is then filtered by means of Buttherworth's high-pass filter to eliminate signal's drift and obtain an average value of zero. The sensors were attached to the spindle perpendicularly to the tool's axis by means of magnets (X and Y directions). Signals were recorded by means of National Instruments data acquisition card type NI USB6341 equipped with a 16-bit converter of 500 kS/s update rate. The recorded acceleration signal underwent consecutive integration operations in order for characteristics of speed and sensors' shift to be determined. Such characteristics provide information on forces influencing the spindle. Due to the ambiguity of signals' interpretation obtained from the vibroacoustic sensors, the study was comparative in character. Drills of various materials, types of coatings and geometry of teeth were compared in twos. Studies were commissioned by the producer of special cutting tools (adapted to the requirements of manufacturing process of the ordering party) in order to evaluate advantages of changes introduced in improved tools.

Example characteristics analysed further in the paper were recorded in the course of 6mm drills' work, which were intended to cut series of holes with depth of 15mm.

We use decomposition methods of empirical algorithms developed by Hilbert and Huang to assess the working conditions of cutting tools. As a result of Huang decomposition, and Hilbert transform information about the amplitude and frequency of the signal components forming the analyzed signal is obtained. As a consequence, signal components are extracted and information about the amplitude and frequency is obtained. The resulting components (modes) characterize phenomena occurring during an aggressive type of machining, during which induced vibration is a complex tooling system (machine, grip, object, tool). Separation of the component signals, because of their frequency and amplitude information can be confirmed on the tool vibrates during insertion of the material of the panels. This information allows the selection of cutting parameters, and their optimize and monitor the con-

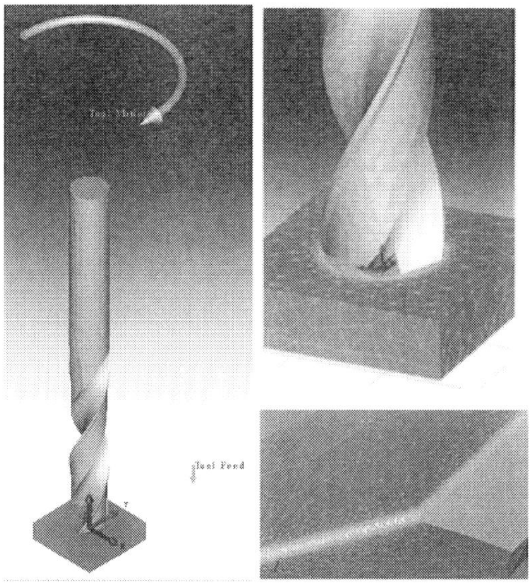

Figure 1. 3D simulation: solid model of drilling (a), mesh (b) and zoom of cutting edge with roundness (c).

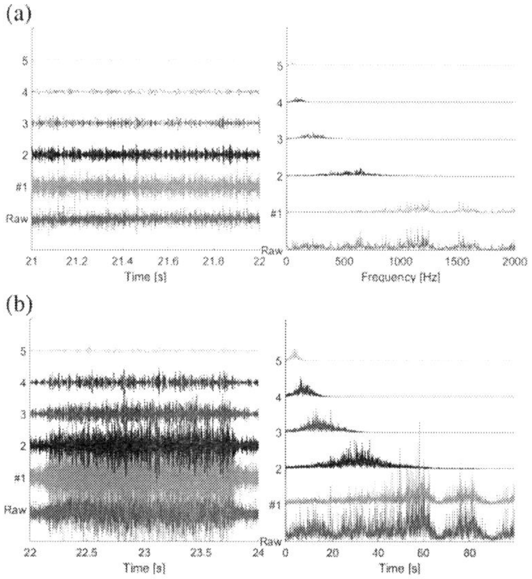

Figure 2. Huang's empirical mode decomposition of the vibration signal recorded during (a) positioning passes without machining (channel 1, command G0, time=1 s) and (b) drilling work pass (channel 1, command G1, time=2 s) into modes (left side) and FFT spectra of mod Huang (right side).

ditions of work tools, including cooling, lubrication and wear of tools [3, 4, 10, 11].

In the analysis by Hilbert-Huang one performs the so-called signal decomposition into empirical modes (Huang decomposition): $F_x^1(t)$, $F_x^2(t)$, ..., $F_x^m(t)$ [4]:

$$F_x(t) = \sum_{j=1}^{m} F_x^j(t) + r_m \tag{1}$$

where r_m is a truncation error after estimation of successive m modes. Each next empirical j mode is defined after subtracting average of maximum and minimum values interpolated by cubic splines of the local envelope $F_x^{j-1}(t)$. Note that the first mode $F_x^1(t)$ is obtained from the original signal $F_x(t) = F_x^0(t)$ and the Huang decomposition procedure [9]. The Hilbert transform is obtained from the original signal $g(t) = F_x^j(t)$ as follows:

$$\hat{g}(t) = \frac{1}{\pi}\int_{-\infty}^{\infty} \frac{g(\tau)}{t-\tau} d\tau. \tag{2}$$

Finally, a new complex function z(t):

$$z(t) = g(t) + i\hat{g}(t), \tag{3}$$

provide information about the instantaneous amplitude, A, phase, Θ, and frequency $f = 2\pi(d\Theta/dt)$.

3 Results and Discussion

The simulation encompassed the sliding motion pass of two cutting edge through the material. The material has shape of cone of a cone matching the cutting edge of drill. The analysis of phenomena occurring in machining zone encompasses tools' FEM simulations. Figure 1 presents an example of geometrical and working condition choosing including different radius of cutting edge rounding. Consequently, the 3D simulations encompassed the sliding motion pass of one cutting edge into the material.

On the basis of 3D machining models, force characteristics impacting the tool were determined. Results of simulations are presented in Tab. 1. Low values of maximum force and standard deviations (values in parentheses) in X, Y and Z directions should increase tool life. The lowest values of maximum force and standard deviations characterized drill with R =10 μm.

After approval of changes to the credit card's credentials, there are a series of prototype tools that are subjected to testing at a nearby position in the vibration sensors. Comparative tests have been carried out, monitoring the working conditions of cutters divided into groups differing in the rounding radius of the blade. Five tools were prepared in both groups. During testing at the Manufacturer's Factory, stainless steel material. The drilling depth was 15mm. The treatment was carried out at spindle rotation S =2000 rpm.

Experimental results are presented in Fig. 2. Namely, the comparison of decomposition of the signal recorded during tools' passes (command G1), when the tool cuts the material with positioning passes (command G0), when the spindle spins and moves without the contact between the tool and material. The isolation of signal's components (vibrations) associated with the drive system of the machine allows them to be isolated from the registered signal in the course of the tool's work, and the analysis of only these components which were induced by phenomena resulting from machining itself. Furthermore, Fig. 2a presents an example characteristic of a signal recorded without drilling (Fig. 2b, plot labeled "raw"). The recorded interval was of 1 second, with sampling frequency of 4kHz. The left side plots outline the recorded signal and consecutive modes obtained by Huang's decomposition algorithm. On the right side, the corresponding spectra obtained by Fast Fourier Transform are collected. Finally, Fig. 2a features clear cut frequency intervals characterizing individual Huang's modes (components of the tool's vibration signal). High spectra were recorded to higher frequencies and outlined in modes 1 and 2. These frequencies result from phenomena occurring in machining zone including: the way the material is cut by the edges, head's vibrations (tools of small diameter), change of drilling depth in individual passes through the material in circular motion. Due to the was the Fast Fourier Transform (FFT) is estimated, component signals induced by one type of the phenomena may induce peaks in the spectrum in various locations. However, the featured characteristics demonstrate the utility of Huang's decomposition in monitoring the conditions of machining. It is also worth to notice that the truncation error r_m (Eq. 1) was negligibly small after first 5 modes. Fig. 2 show that the Huang modes from 1 to 5 decrease considerably.

Hilbert-Huang empirical mode decomposition is based on isolating composing signals on the basis of their frequency and amplitudes. Values of the amplitudes characterising the component signals enables the sources of the generated signals to be determined and the analysis of their root causes to be conducted [3, 4]. In case of FFT, the selection of signals is based upon frequencies. An additional difficulty in interpretation is associated with intervals of composite signals which were identified as being multiple from the original frequency, and which form a separate concentration on the spectrum chart. FFT analysis revolves around the quantitative analysis with regards to frequency, while Hilbert-Huang decomposition enables values of amplitudes and frequency in time to be established. Such a feature is of particular utility in the evaluation of machining in real time in industrial conditions.

Figure 3 presents example of 1.75 s interval for tool's passes (record of spindle's vibrations) in the course of a single operation during the tool's pass (a) and 1s interval for a positioning pass without cutting (b). They underwent Huang empirical mode decomposition, and subsequently Hilbert spectral analysis. Values of instantaneous frequencies (vertical axis) and amplitudes (colour of the point) in 6 initial Hilbert-Huang's modes were presented by means of points. Due to the overlap of results on the chart the point's numerical force are invisible in frequency (time) plane. Therefore, their force in individual areas may be assessed by evaluating neighbouring concentrations' results.

The intervals featured in 3 are significantly different with regards to Huang modes' amplitudes of higher fre-

Table 1. Maximum values (and standard deviations) of FEM results analysis of tools with the corner radius R =(0, 10, 20, 30 [μm])

Corner radius [μm]	0	10	20	30
Force X [N]	579(199)	522(137)	1556(421)	5638(1943)
Force Y [N]	640(169)	412(102)	595(151)	8091(1877)
Force Z [N]	1574(177)	1468(195)	2052(255)	2382(456)
Power [W]	3827(701)	4563(929)	4062(907)	10834(2395)
Torque [Nm]	9(1.6)	10(2.1)	9(2.1)	24.63(5.44)
Temperature [C]	1224(42)	935(42)	1277(47)	1347(61)
Tool Stress [MPa]	3727(117)	1566(98)	1596(89)	1626(122)
Contact Pressure [MPa]	5745(258)	8069(383)	10205(504)	9996(662)
Contact Shear [MPa]	2804(207)	2432(156)	2468(164)	2417(180)
Deflection X [mm]	0.044(0.007)	0.022(6.5e-3)	0.018(0.016)	2.59e-5(7.9e-3)
Deflection Y [mm]	0.014(4e-3)	0.015(4.8e-3)	0.009(4.2e-3)	4.42e-6(8.27e-3)
Deflection Z [mm]	0.002(6e-4)	0.002(5e-4)	0.004(5e-4)	3.06e-5(1.05e-3)

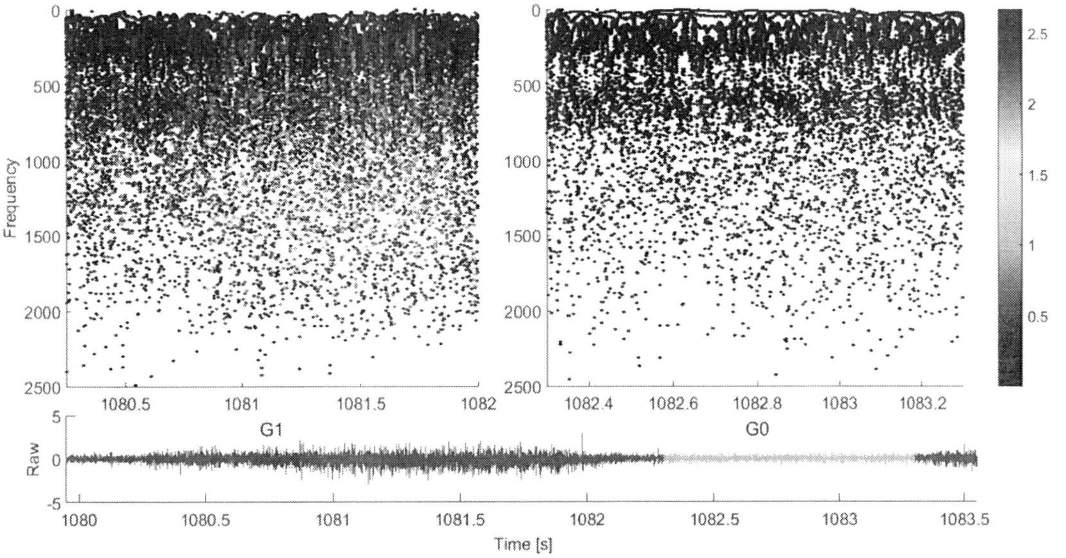

Figure 3. Sequence of frequencies from Hilbert-Huang's transform of example passes of tool No. 8 (group of tools with rounding edge of the blade R = 10 μm) in the course of a) drilling (command G1, left drawing) and positioning pass (command G0, right drawing), featured in the bottom channel. The color scale in the upper channel denotes the amplitude size.

quencies, over 1100Hz. An increase in the amplitude in the course of a tool's pass (Fig. 3a) in relation to vibrations originating from the machine (ranging from 50 to 1500Hz) may also be noticed. Such an analysis enables vibrations induced by the machine's drive system to be distinguished from vibrations generated by the cutting tool.

Fig. 4 presents the Magnitude (Frequency) of Hilbert-Huang's instantaneous decompositions in the course of the end mill's pass and spindle's positioning pass. Such a projection enables differences between passes to be determined. Vibrations' areas presented in Fig. 4b are characterize the machine's work in the course of positioning passes, above the machined material, in the same direction the tool's pass was made (Fig. 4a). The area visible in fig. (b) has two higher(elevated) areas with maximums in 400 and 1400 Hz. Similar concentrations are visible in chart (a). However, the amplitudes increased approximately 1,5-fold for frequencies 400HZ and 3-fold for wide area around 100Hz by forces originating from the tool's work. The remaining points present in fig. 4a result from the material being cut by cutting edges.

Initial evaluation of machining quality comes down to the observation of the points' concentration above the limit line of approx. 0.06 on the amplitude axis of accelerations registers by means of vibroacoustic sensors and decomposed by Hilbert-Huang's decomposition. High amplitude values characterize values of forces impacting the tool's edges. When selecting machining parameters, reduction of amplitudes is desirable.

Fig. 5 presents the analysis of vibrations' sources included in Huang's individual modes, conducted on the basis of frequency spectra. It may be conducted on the basis https://doi.org/10.1051/mateccconf/201814816003

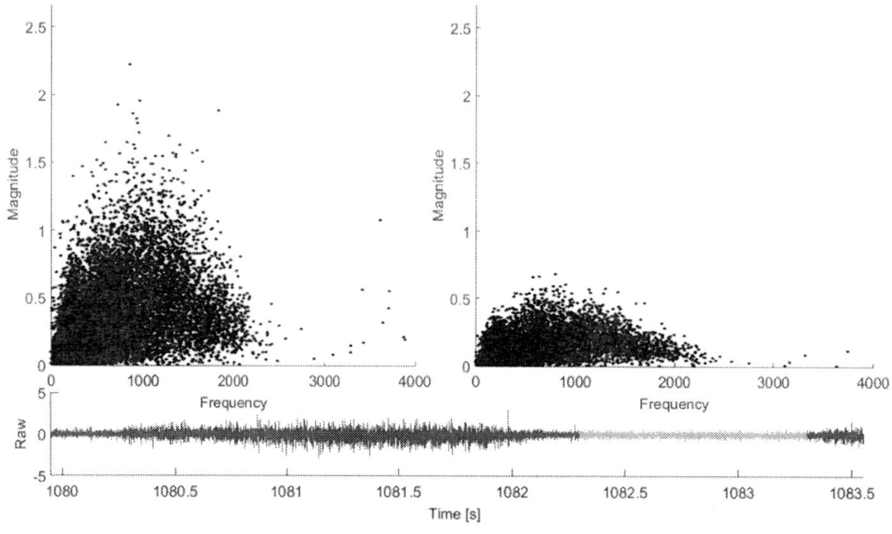

Figure 4. Magnitude versus frequency plot of Hilbert-Huang's instantaneous decompositions results for example passes of tool No. 8 (group of tools with rounding edge of the blade $R = 10\,\mu$m) in the course of drilling (command G1, left drawing) and positioning pass (command G0, right drawing).

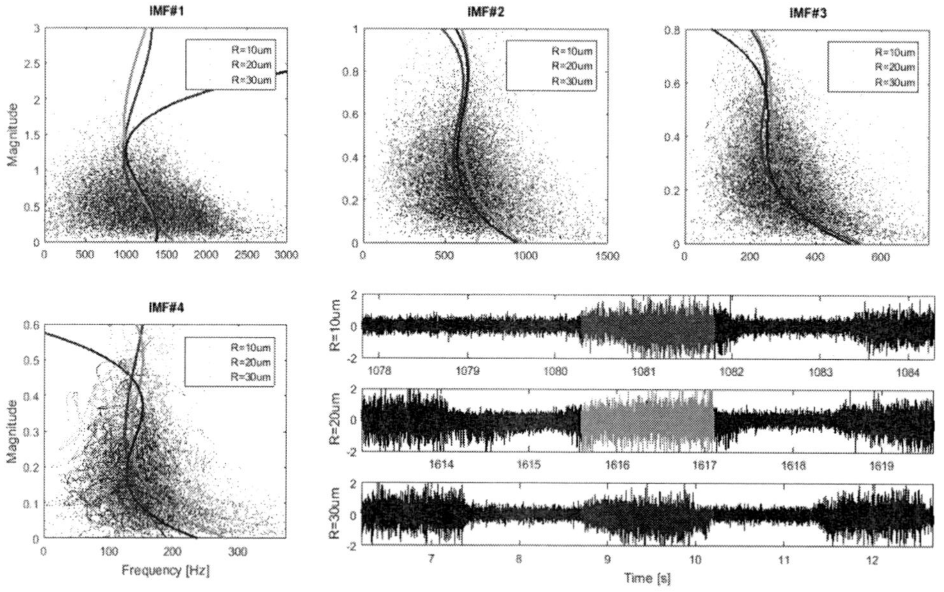

Figure 5. Relationship of amplitude and frequency of Hilbert-Huang empirical mode decomposition in example passes of three types of drills (with various parameters R of the tool corner radius, indicated). The deflection of the backbone of the modal (first four modes: INF#1-4) resonance curves. The fitting polynomial backbone lines are plotted exceeding the estimated clouds of estimated points in purpose to show the differences in nonlinear responses. Note, the system nonlinearities are signalled as backbone deflections in the particular modes.

of Amplitude (Frequency) charts. Relationship of amplitude and frequency of Hilbert-Huang empirical mode decomposition in example passes of three types of drills with different radius of corners, adequately 10, 20 and 30 [μm]. The backbones of the resonances obtained from the analysis indicate the nonlinearities in the physical process. Interestingly, both of nonlinearities are present in the empirical mode characteristics: hardening and softening. The most transparent are first three modes. Additionally, the backbones of the first and third modes are sensitive to the parameter, R, representing the type of drills, while the second mode seems to be independent.

4 Summary

When monitoring machining conditions with vibration sensors, recorded signals are distorted by components generated by the drive system and machine tool's mechanisms. The application of Hilbert-Huang empirical mode decomposition enables the extraction of vibrations' signals with regards to their amplitude and frequency.

An example decomposition of signals recorded by means of vibroacoustic sensors was made during drilling. FTT spectral analysis of Huang's modes was employed in order to establish frequencies characterizing known vibration sources such as: spindle's rotational speed, number of teeth on the cutting tool, and tool's vibration frequency determined during numerical simulations. Component signals generated in machining zone were isolated on Hilbert-Huang modes Amplitude (frequency) chart.

Hilbert-Huang decomposition enables amplitudę and frequency values to be determined in time. Such a feature is of particular significance for evaluating machining conditions in real time. The Huang's lower modes contained components of higher frequencies can be used during the real-time monitoring of the cutting.

The backbone shapes in the frequency-amplitude distribution specify uniformity of tool working. A vertical shape of backbone is characterized for oscillations with a lower content of nonlinear components. Nonlinear components are related to the cutting process dynamics and reflect larger amplitude chatter oscillations. They are most frequently undesirable, so we indicate such a tool with lower durability.

Note that the results of signal decomposition for different tools are very close to each other. The shapes of mode resonance backbones for all of them have similar shapes. This is expected phenomenon because the tools were nearly identical (same number of teeth, values of angles and other outside the rounded edge). Huang decomposition allows to detect subtle differences in the work of geometrically improved tools. It is also possible to observe the effects of such phenomena like slip-stick motion and contact losses.

References

[1] Altintas Y., *Manufacturing automation. Metal cutting mechanics, machine tool vibration, and CNC design* (Cambridge, New York 2000)

[2] Sen A.S., Litak G., Syta A., Rusinek R., Meccanica **48**, 738-789 (2013)

[3] Wolszczak P., Litak G., Łygas K., Adv. Sci. Technol. Res. J. **11**, 96–103 (2017)

[4] Litak G., Kecik K., Rusinek R., Latin American Journal of Solids and Structures **10**, 133-140 (2013)

[5] Lajmert P., Journal of Machine Engineering **10**, 39-49 (2010)

[6] Litak G., Rusinek R., Meccanica **47**, 1517-1526 (2012)

[7] Rusinek, R., Wiercigroch, M., Wahi, P., Int. J. Bif. Chaos **24**, 1450115 (2014)

[8] Hilbert D., *Grundzuege einer allgemeinen Theorie der linearen Integralgleichungen* (Chelsea Pub. Co., New York 1953)

[9] Huang N.E. Shen Z., Long S.R., et al., *The empirical mode decomposition and the Hilbert spectrum for nonlinear and non-stationary time series analysis* (Proceedings of Royal Society London A454, 1998) 903-993

[10] Rusinek, R., Lajmert, P., Kecik, K., Kruszynski, B., Warminski, J., Int. J. Mechan. Scien. **99**, 196-207 (2015)

[11] Liu, C.F., Zhu, L.d., Ni, C.B., Int. J. Adv. Manuf. Tech **91** 3339-3348 (2017)

[12] Feldman M., Mechanical Systems and Signal Processing **25**, 735-802 (2011)

[13] Altintas Y., Mach. Sci. Technol. **4**, 445–478 (2000)

[14] Williams J.G., Patel Y., Interface Focus **6** (2016).

[15] Movahhedy, M. S. Gadala and Y. Altintas, Machining Science and Technology **4**, 15-42 (2000)

[16] Kadirgama K et al., JMES **1**, 37-46 (2011)

[17] Third Wave Systems,*AdvantEdge FEM User's Manual* (Minneapolis, 2014) 340-344

Finite element modelling of the dynamics of groovy ball bearings

Olamide Ajala[1,*], *Ekaterina* Pavlovskaia[1,**], and *Marian* Weircigroch[1,***]

[1] *Centre for Applied Dynamics Research, School of Engineering, University of Aberdeen, UK*

Abstract. Geometry modified thrust bearings exposed to rolling and sliding contact are subjected to wear and localized frictional heating caused by relative slip between the two sliding surfaces. This leads to a rise in temperature, thermal stresses and changes in the elastic and plastic strength and physical properties of the material. The changes in the properties in turn alter the stress state, the displacement field, life and reliability of the bearing. Hence, a finite-element model is created to study the dynamics of groovy thrust bearing. In this paper, Hertz contact theory and numerical method are used to simulate the dynamics and kinematics of groovy ball bearings. The optimal loading parameters are identified in this study based on the analysis of the system responses and properties. The results of the numerical analysis and validation are presented. The numerical analysis proves the concept of transforming rotational motion into axial oscillation and demonstrates the capabilities of the numerical simulation to accurately model the dynamics of the groovy ball bearing.

1 Introduction

Drilling through hard rock has been a restraining factor to economically develop significant hydrocarbon reserves in several geological settings. The high cost to explore and develop these reserves is of considerable importance. Thus, Resonance Enhanced Drilling (RED), was developed at the University of Aberdeen to overcome this challenge [1]. This current research is focused on contact interaction between rolling element bearings (REB), which serve as the main mechanical component of RED module designed to operate under high frequency dynamic rotary action and static preload. This study, aims to develop an ultimate tool for optimal design of the dynamic exciter of RED module [2].

2 Finite element contact mechanics of groovy bearing

The exciter is modelled in Abaqus, as a solid continuum body represented by a 3-D finite-element mesh, with encastre and displacement boundary conditions on the cage and race respectively to simulate the exciter operation under dynamic load. First order, 8-node linear brick, hourglass control, reduced integration, 3-D stress-displacement elements and second order tetrahedral 3-D modified elements (C3D10M) were adopted (see figure 1). A total of 346,557 elements were used to model the assembly in Abaqus. All the parts of the modelled exciter, excluding the balls, are drawn using SolidWorks and subsequently imported into Abaqus, to ensure accurate geometry representation.

* e-mail: olamide.sherifah.ajala@abdn.ac.uk
** e-mail: e.pavlovskaia@abdn.ac.uk
*** e-mail: m.wiercigroch@abdn.ac.uk

The balls are constrained in translation in both lateral and longitudinal directions and the top race is constrained in all degrees of freedom (DOF) with exception of translation and rotation in the vertical direction. The Bottom race is constrained in all DOF with the exception of rotation in the vertical direction, while the spring is constrained in all DOF with exception of the translation in vertical direction. The connection between the disc and bearing is modelled using a spring element with spring stiffness of 6.67 kN/mm. The geometry and excitation loads for the selected bearing is presented in table 1.

The contact condition is simulated by adapting general contact algorithm and finite-sliding contact evolution discretisation to avoid missing contacts. The interactions between the bearing elements are defined with a normal hard pressure overclosure relationship and a penalty constraint enforcement formulation to avoid penetration of the contacting surface.

The model considers both frictionless and isotropic Coulomb friction formulations to simulate the lubricant behaviour of the bearing. Static instabilities in the model, are controlled by adjusting the initial contact states, the external excitation sequence and stiffness damping for numerical accuracy.

3 Numerical analysis and experimental verification

The groovy bearing described in this paper adopts a standard thrust ball bearing in which the driven track is modified as N number of sinusoid over which the rolling elements pass to periodically increase and decrease axial distance between the rotation element and base element while rotating, thereby creating an oscillatory axial motion

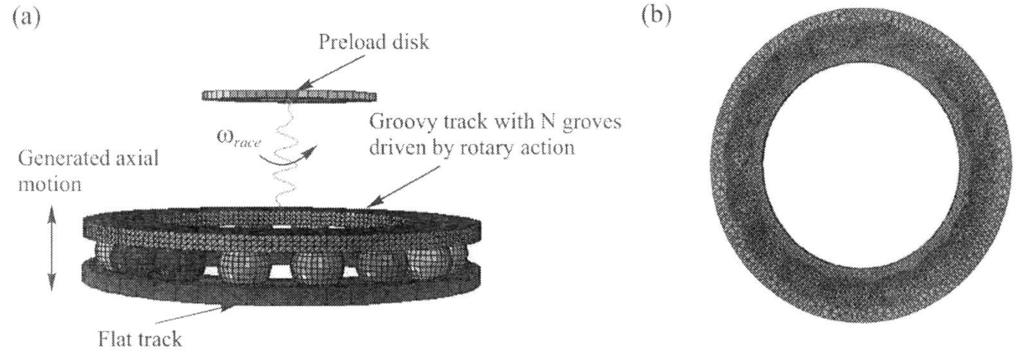

Figure 1. Schematic of the numerical model with C3D10M elements. (a) Main component of the modelled exciter; (b) Close up view of the groovy top race, with N grooves.

on the rotation element relative to the base element. This arrangement makes the dynamic analysis of the contact interaction between the rolling elements challenging from modelling perspective.

To obtain a bounded solution for adequate analysis of the model characteristics, the stability limit (Δt_{min}) of the assembly is estimated in terms of the highest eigenvalue (ω_{max}) of the assembly and the fraction of critical damping (ξ) in the highest mode as shown in Eq. (1) [4]:

$$\triangle\, t_{min} \leq \frac{2}{\omega_{max}}(\sqrt{1+\xi^2}-\xi). \quad (1)$$

To evaluate the accuracy of the numerical simulation, the energy balance of the assembly is reviewed for the entire model considering the energy balance equation in Eq. (2). For an accurate finite element analysis, the artificial energy (E_{VD} and E_{FD}) of the model should be less than 10 percent of the real energy (E_{KE} and E_I) of the system and the total energy of the system be constant E_T (i.e law of conservation of energy). Problems indicators when reviewing the energy balance of the model are excessive artificial strain energy due to hourglassing of the finite elements and a large change in the total energy E_T due to conflicting constraints and exceeding the stability limit of the model:

$$E_I + E_{VD} + E_{FD} + E_{KE} = E_T. \quad (2)$$

3.1 Model characteristics and analysis

Dimension and properties of the bearing analysed are listed in table 1. Using the configuration of the FE model described in previous section (see Sect. 2), series of analysis is performed on the groovy bearing. First, the model is subjected to preload of 6.4 kN, that is, downward force is applied to the preload disc shown in figure 1.

Once the assembly reaches stable value of preload, angular velocity is applied to the top plate of the bearing (one with the sinusoidal profile). As the nominal rotational velocity of 100 rad/s is reached, a steady state periodic solution is observed when looking into the vertical displacement of the top plate of the bearing, as shown in figure 2 (a). Consequently, when looking into the force generated by the exciter see figure 2 (b), it is visible that the force oscillates around the preload value, with peak-to-peak amplitude of around 8 kN.

The energy balance (see figure 2 (c)) of the model is very good. The artificial strain energy is 0.07 percent of the internal energy of the system, indicating no hourglassing of the model and element mesh size is adequate while the total energy of the system is constant (3 J). The observed internal energy in figure 2 (c) is the sum of recoverable elastic strain energy in the assembly through out the simulation and the observed artificial energy includes all energy stored in the hourglass (excessive deformation) resistance. Large values of artificial strain energy will indicate that mesh refinement or other changes to the mesh are necessary in the model. Figure 2 (c), indicates an accurate explicit dynamic analysis where no numerical instabilities occurs in the assembly and thus clearly proves capabilities of the FE model to accurately simulate the characteristics of the modelled exciter.

Table 1. Groovy bearing parameters

Parameters	Values
Number of rows	1
Number of rolling elements	16
Contact angle	90 degrees
Pitch diameter	115.00 mm
Bore diameter	90.00 mm
Ball diameter	17.00 mm
Bearing width	8.00 mm
Race diameter	140.00 mm
Race curvature factor	0.52
Preload	6400 N
Rotation	100 rad/s
Density	7850 kg/m^3
Elastic modulus	2.1 e+11 Mpa

https://doi.org/10.1051/matecconf/201814816004

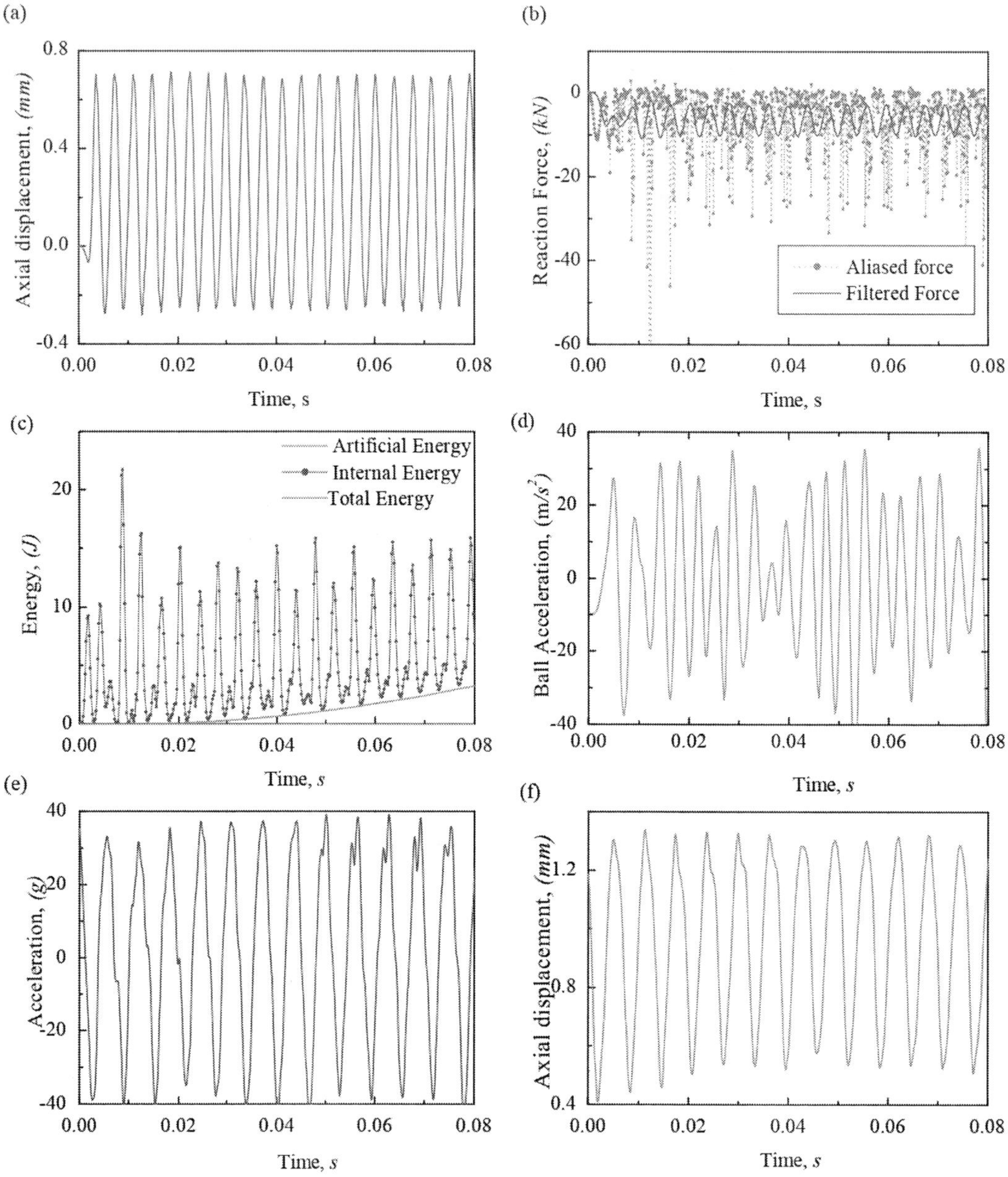

Figure 2. Time history of response of the exciter numerical model and RED experimental result. (a) Axial oscillation of the exciter; (b) Reaction force of exciter; (c) Energy history of the model; (d) Tangential acceleration of the rolling element of FE model; (e) Generated acceleration of the assembly from experiment; (f) Generated displacement of the assembly from experiment.

3.2 Experimental verification

In order to validate the numerical model, the results (axial displacement, reaction torque, force and acceleration) of the finite element analysis have been compared against available RED experimental results in [1]. A typical example of this, is the comparison of the acceleration and displacement from the FE model with experimental result.

Looking at figure 2 (e), it is observed that the measured experimental acceleration of the system oscillates at 40 m/s^2, as the balls interact with the cage pockets and the race of the bearing, this is in good agreement with the acceleration obtained from the numerical model presented in figure 2 (d).

Furthermore, the measured experimental displacement (see figure 2 (f)), shows a 1 mm peak-peak axial oscillation of the assembly, an exact match to the numerical result presented in figure 2 (a). The results of the experimental validation clearly shows that the numerical model accurately reflects the mechanics of the exciter analysed.

4 Summary and Future Work

Modelling of the dynamics groovy ball bearing is undertaken in this study, and the results of the numerical analysis and validation are presented. A parametric study of the developed models has been performed considering different modulus, preload value, amplitude of wave on the groovy race, contact load and other system parameters. The performance characteristics, capabilities and limitation of the developed model under varying excitation loads and initial conditions have been considered.

The next stage of this study, is development of a low dimensional analytical model for optimum design of groovy ball bearing, to improve the performance of the bearing over its service life and geometry optimization of the design.

In conclusion, a finite element contact mechanics bearing model is established based on a contact algorithm suited to high-precision elastic bodies and well-established for rolling element contact mechanics. Appropriate boundary conditions to simulate the mechanics of the bearing being modelled are adopted and load dependent rolling element contact for race-ball and cage-ball interactions is solved using appropriate algorithm in Abaqus.

The numerical model fully captures all the important groovy bearing geometry and demonstrates the capability of the assembly to produce 1 mm peak-to-peak axial oscillation under applied excitation frequency. The accuracy of the developed model is validated via FEM energy balance equation and experimental result.

Acknowledgements

This work is performed using Maxwell High Performance Computing (HPC) cluster of the University of Aberdeen IT Service (www.abdn.ac.uk/stanet/research/research-computing), provided by Dell Inc. and supported by Alces Software.

References

[1] M. Wiercigroch, M. Kapitaniak, and V. Vaziri. RED: Revolutionary Drilling Technology for Hard Rock Formations, SPE-184665-MS, Hague, (2017).
[2] M. Wiercigroch, M. Kapitaniak, V. Vaziri and N. Yari. *RED Rotary Actuator*. Patent No. WO 2016142537 A3 (2016).
[3] O. C. Zienkiewicz. *The finite Element Method*. McGraw-Hill, (1977).
[4] B. S. Simulia. *Abaqus Analysis Users Guide* Vol I-IV. Providence, RI, USA, (2012).

Proportional-Derivative Control of Stick-Slip Oscillations in Drill-Strings

Wei Lin[1,2], *Yang* Liu[2,*]

[1]School of Mechatronic Engineering, Southwest Petroleum University, Chengdu, 610500, China
[2]College of Engineering Mathematics and Physical Sciences, University of Exeter, North Park Road, Exeter, EX4 4QF, UK

Abstract. Stick-slip oscillation in drill-string is a universal phenomenon in oil and gas drilling. It could lead to the wear of drill bit, even cause catastrophic failure of drill-strings and measurement equipment. Therefore, it is crucial to study drilling parameters and develop appropriate control method to suppress such oscillation. This paper studies a discrete model of the drill-string system taking into account torsional degree-of-freedom, drill-string damping, and highly nonlinear friction of rock-bit interaction. In order to suppress the stick-slip oscillation, a new proportional-derivative controller, which can maintain drill bit's rotation at a constant speed, is developed. Numerical results are given to demonstrate its efficacy and robustness.

1 Introduction

For conventional rotary drilling, drill-string is a principal component for the entire drilling system. As shown in Fig.1 (a), a typical drilling system is composed of a rotary table, a slender drill pipe, a drill collar, and a drill bit. The drill-string could run several kilometres deep, but the diameter of the drill pipe does not exceed 0.3 metres. When drill pipes transmit the drive torque from rotary table to drill bit, deformations in drill pipes are very complex, especially considering the constraints of the borehole. The slenderness of the drill pipe makes it prone to exhibit undesired oscillations, including bit bouncing, lateral oscillation, torsional oscillation, and whirling motion. In oil and gas drilling, stick-slip behavior exists in 50% of drilling time [1], and the whipping and high rotation of the drill bit may cause bit bouncing and whirling motion [2]. Moreover, formation change is also very complex during drilling procedure, so the interactive model between rock and bit is highly nonlinear [3]. In addition, the interaction between drill-strings and borehole also makes drilling system uncertain [4].

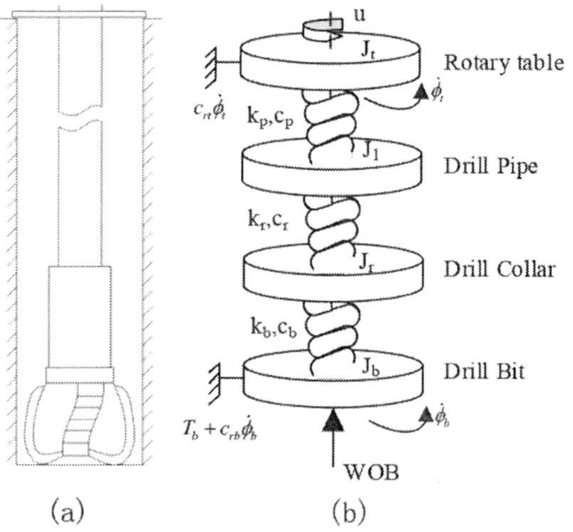

Fig. 1. The schematics of (a) the drill-string and (b) the simplified physical model of the drill-string system

There are vast of existing literatures using lumped-parameters model to study the drill-string dynamics, e.g.

---
* Corresponding author: y.liu2@exeter.ac.uk

 https://doi.org/10.1051/matecconf/201814816005

[5-7]. Balakumar [5] developed a lumped-parameter model to study the coupled axial, torsion, and lateral dynamics of a drill-string dynamics, including stick-slip and delay effect. Canudas-de-Wit et al. [6] proposed a lumped parameter model to use the weight on bit as an additional variable for suppress stick-slip oscillation. Liping Tang [7] proposed a lumped torsional pendulum model of the drilling system, and studied the stick-slip dynamics and negative damping effect. However, most of the dynamics model could not be guaranteed to suppress the stick-slip oscillation steadily and appropriate to the parameters changing. Since this paper proposed a proportional-derivative (PD) controller to suppress stick-slip oscillations using the lumped-mass model.

The remaining sections of this paper are organized as follows. In Section 2, a discrete dynamic model is developed for the drill-string system, and a new PD controller is studied. In Section 3, simulation results are presented to validate the effectiveness of the controller. Finally, conclusions are drawn in Section 4.

2 Drill-String Model and Proportional-Derivative Controller

The drilling system, which is illustrated in Fig.1 (a), is comprised of rotary table, along with drill pipe, drill collar and drill bit. In order to simplify the system, a discrete drill-string model is developed for analytical and numerical study. The mechanical schematic of drill-string system shown in Fig.1 (b) is considered as four disks, each with rotational inertia properties. The drill-string is connected by massless springs, which have torsional stiffness and torsional damping. The rotary table is driven by a motor fixed on the rotary table with control torque output. The interaction between drill bit and rock is a highly nonlinear friction.

Making use of the dynamic theory, the government equation of motion for the discrete system [8] can be written as

$$J\ddot{\phi}+C\dot{\phi}+K\phi+T=U \tag{1}$$

Where Φ is the rotational position of lumped mass, J is the rotational inertia matrix, C is the torsional damping matrix, and K is the torsional stiffness matrix, T is the torque of friction between drill bit and rock, and U is the control torque applied on the rotary table. According to the finite element method, matrixed (J, C, and K) are comprised as follows.

$$J=\begin{bmatrix} J_t & 0 & 0 & 0 \\ 0 & J_1 & 0 & 0 \\ 0 & 0 & J_r & 0 \\ 0 & 0 & 0 & J_b \end{bmatrix}$$

$$C=\begin{bmatrix} c_{rt}+c_p & -c_p & 0 & 0 \\ -c_p & c_p+c_r & -c_r & 0 \\ 0 & -c_r & c_r+c_b & -c_b \\ 0 & 0 & -c_b & c_{rb}+c_b \end{bmatrix}$$

$$K=\begin{bmatrix} k_p & -k_p & 0 & 0 \\ -k_p & k_p+k_r & -k_r & 0 \\ 0 & -k_r & k_r+k_b & -k_b \\ 0 & 0 & -k_b & k_b \end{bmatrix}$$

The frictional torque for the rock-bit contact is modelled as a combination of Stribeck [9] and the Karnopp's [10] model described as

$$T_b=\begin{cases} T_r & if\,|\dot{\phi}_b|<\zeta\; and\, |T_r|\le T_s, \\ T_s\,\mathrm{sgn}(T_r) & if\,|\dot{\phi}_b|<\zeta\; and\, |T_r|>T_s, \\ \mu_d R_b W_{ob}\,\mathrm{sgn}(\dot{\phi}_b) & if\,|\dot{\phi}_b|>\zeta. \end{cases} \tag{2}$$

In the sticking phase$\left(|\dot{\Phi}_b|<\zeta \text{ and } |T_r|<T_s\right)$, the bit velocity is less than a small positive constant ξ, and the torque T_r is less or equals to the static friction torque T_s, where $T_r=c_b(\dot{\Phi}_t-\dot{\Phi}_b)+k_b(\Phi_t-\Phi_b)$ and $T_s=\mu_s W_{ob} R_b$, where μ_s and μ_d are the static friction coefficient and dynamic friction coefficient, respectively. W_{ob} is the weight on bit (WOB), R_b is the bit radius.

In the process of stick to slip$\left(|\dot{\Phi}_b|<\zeta \text{ and } |T_r|>T_s\right)$, the drill bit velocity is still less than the small constant ξ, but the reaction torque T_r is greater than the static friction torque T_s and the drill bit start to move.

In the slip phase$\left(|\dot{\Phi}_b|>\zeta\right)$, the drill bit starts to rotate, and the friction torque includes the effects of drill bit radius, the WOB, and the dynamic friction coefficient [11].

For the purpose of control stick-slip oscillation, a new proportion-differential controller is designed.

$$u=k_1(\phi_t-\phi_b)+k_2(\dot{\phi}_d-\dot{\phi}_t)+k_3(\dot{\phi}_d-\dot{\phi}_b) \tag{3}$$

Where u is the control torque applied on rotary table, k_1 is proportion coefficient, k_2 and k_3 is the differential coefficient, $\dot{\Phi}_d$ is the desired angular velocity, $\dot{\Phi}_t$ is the rotary table velocity, and $\dot{\Phi}_b$ is the drill bit velocity.

https://doi.org/10.1051/matecconf/201814816005

3 Simulation Results

Numerical studies were conducted with the earlier described discrete system model by applying the PD controller. The geometry and physical parameters for the drill-string are given in Table 1. In the follow subsections, the stick-slip oscillation, control of stick-slip, and the respond to WOB oscillation are brought forth.

Table 1. Geometry and physical parameters for the drill-string

Parameter	Value
J_t	910 kg • m^2
J_l	2800 kg • m^2
J_r	750 kg • m^2
J_b	450 kg • m^2
c_p	150 N • m • s/rad
c_r	190 N • m • s/rad
c_b	180 N • m • s/rad
c_{rt}	410 N • m • s/rad
c_{rb}	80 N • m • s/rad
k_p	700 N • m • s/rad
k_r	1080 N • m/rad
k_b	910 N • m/rad
μ_s	0.8
μ_d	0.45
WOB	30 kN
R_b	0.15 m
v_d	6 rad/s
ζ	10^{-6}
k_1	500
k_2	300
k_3	200

Fig.2 shows the simulation result without controller. The figure presents that both angular velocities of the rotary table and the drill bit change repeatedly with the constant torque. Moreover, in Fig.2 (a), the peak speed of the drill bit is about 3.5 rad/s, which is much greater than the rotary table velocity. As is show in Fig.2 (b), the angular acceleration of drill bit also changes between positive and negative. In this rapidly changes of angular speed and angular acceleration, the tooth of the drill bit will be more likely to wear out than it is in constant speed. Therefore, it is necessary to study the stick-slip oscillation and suppress it.

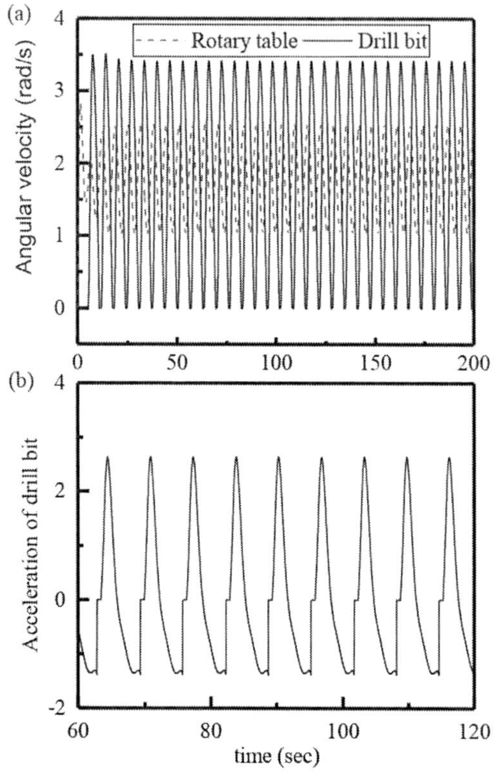

Fig. 2. Time histories of (a) angular velocities of the rotary table (red dash line) and the drill bit (black solid line) with WOB = 30 kN, and u = 3450 Nm.

The simulation result using the PD controller is shown in Fig.3. It can be observed from Fig.3 (a) that the stick-slip is suppressed when the controller is switched on at t = 33 seconds, and both the angular velocities of rotary table and the drill bit are stabilized at about t = 200 seconds. At the same time, the maximum amplitude control torque is about 8.5 kN • m, as seen in Fig.3 (b). In Fig.4, we can see that drill bit velocity converges at 5.26 rad/s, which is a little less than the desired velocity 6 rad/s. The result for torsional oscillation in Fig.3 and Fig.4 indicate that the PD controller can suppress stick-slip oscillation and keep the velocity on a constant desired velocity.

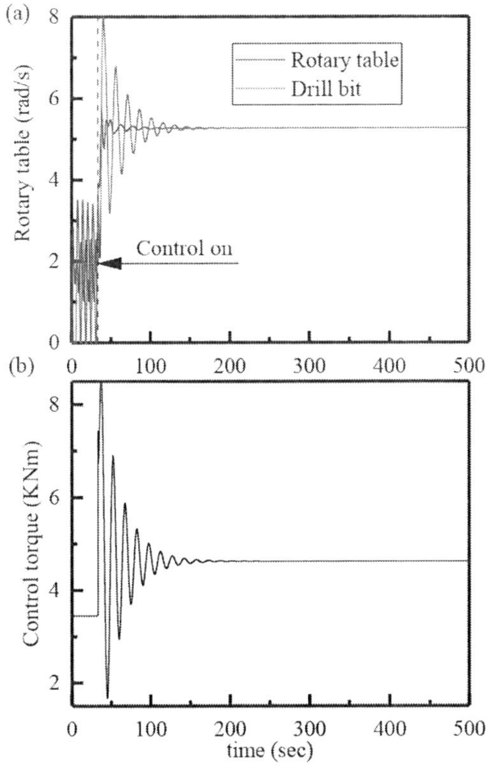

Fig. 3. Time histories of (a) angular velocities of rotary table (red solid line) and drill bit (black solid line), and (b) the control torque using the PD controller with WOB = 30 kN.

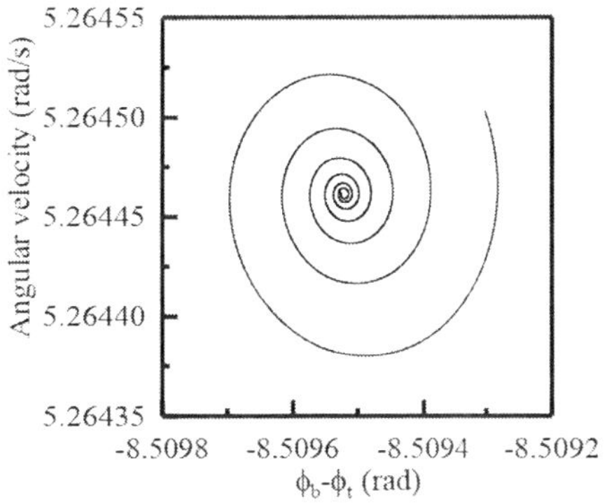

Fig.4 Trajectory of the drill-string on the phase plane (Φ_b - Φ_t) by using the PD controller with WOB = 30 kN

To study the stability of PD controller to parameter sensitive, three different WOB values are applied on the drill bit, and the results are shown in Fig. 5 and Fig. 6. It can be observed that the angular velocity can be able to stabilize the system with WOB changing. When WOB decreases from 30 kN to 20 kN, and 20 kN to 10 kN, the angular velocity has a small fluctuation, and then stay at a constant speed. Trajectory of the drill-string on the phase plane (Φb - Φt) by using the PD controller with different WOB are presented in Fig.6. It is clearly to describe the changing of the velocity with the angular displacement difference between drill bit and rotary table. For the PD controller, it has a proper estimation of unknown physical parameters for the switching control.

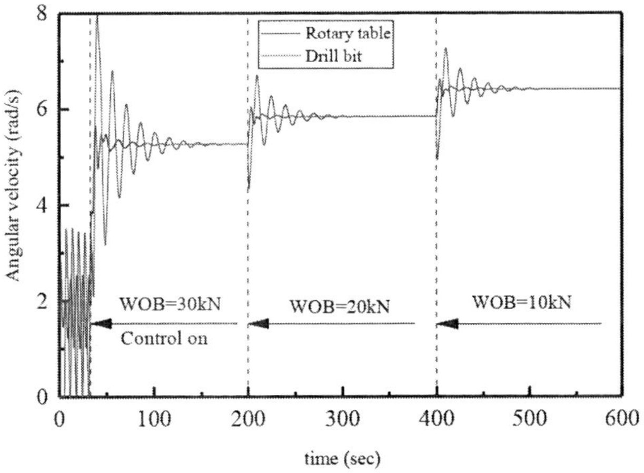

Fig.5 Angular velocities of rotary table (red solid line) and drill bit (blue solid line) by using PD controller with WOB = 30 kN, 20 kN, and 10 kN.

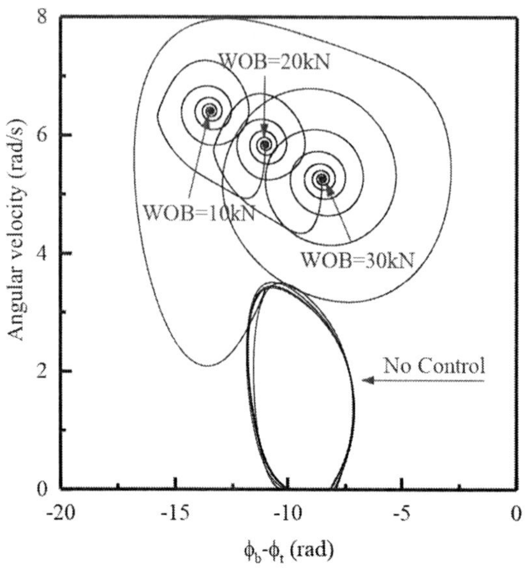

Fig.6 Trajectory of the drill-string on the phase plane (Φ_b - Φ_t) by using the PD controller with WOB = 30 kN, 20 kN, and 10 kN

4 Conclusions

This paper studied the stick-slip oscillations in drill-strings by using a discrete model with torsional degree-of-freedom from a control point of view. The model has one control input and multi-degree-of-freedom to be controlled, considering a series of drill pipes, drill collars, and the highly nonlinear friction for rock-bit interaction. A new PD controller was developed to suppress stick-slip oscillations and track desired rotary speed under variations of other drilling parameters, such as WOB.

Future work will focus on improving controller's accuracy and verify through experimentation. Borehole constraints and axial bit movement will also be considered in the discrete model.

Reference

1. J. F. Breu. The genesis of torsional drillstring vibrations. SPE Drilling Engineering. 168-174, (1992)
2. E. M. Navarro and R. Suarez. Practical approach to modelling and controlling stick-slip oscillations in oilwell drillstrings. Proc of IEEE Int Conf on Control Applications, 1454-1460, (2004)
3. D. M. William. Nonlinear drill string dynamics. PhD Dissertation, the University of Tulsa, (1996)
4. J. Páez Chávez, V. Vaziri Hamaneh, and M. Wiercigroch. Modeling and experimental verification of an asymmetric Jeffcott rotor with radial clearance. Journal of Sound and Vibration, **334**:86-97, (2015)
5. X. Liu, N. Viajic, X. Long, G. Meng, and B. Balachandran. Nonlinear motion of a flexible rotor with a drill bit: stick-slip and delay effects. Nonliear Dyn, **72**:61-77, (2013)
6. C. Canudas-de-Wit, F. R. Rubio, and M. A. Corchero. D-OSKIL: a new mechanism for controlling stick-slip oscillations in oil well drillstrings. IEEE Trans. Control Systems Technology. **16**:1177-1191, (2008)
7. L. Tang, X. Zhu, X. Qian, and C. Shi. Effects of weight on bit on torsional stick-slip vibration of oilwell drill string. Journal of Mechanical Science and Technology, **31**:4590-4507, (2017)
8. Y. Liu. Suppressing stick-slip oscillations in underactuated multibody drill-strings with parametric uncertainties using sliding-mode control. IET Control Theory Appl., **9**:91–102, (2015)
9. R. Stribeck. Die wesentlichen Eigenschaften der Gleitund rollenlager – the key qualities of sliding and roller bearings. Zeitschrift des vereines deutscher ingenieure, **46**:1342-1348, (1902)
10. D. Karnopp. Computer simulation of stick-slip friction in mechanical dynamic systems. ASME J. Dyn. Sys., **107**:100-103, (1985)
11. Y. Liu. Sliding-Mode control of an underactuated oilwell drill-string with parametric uncertainty. UKACC International conference on control, 597-602 (2014)

Calibrated FEM modelling of rock cutting with PDC cutter

Nina Yari[1,*], *Marcin* Kapitaniak[1], *Vahid* Vaziri[1], *Lifeng* Ma[2], and *Marian* Wiercigroch[1]

[1]Centre for Applied Dynamics Research, University of Aberdeen, AB24 3UE, Scotland, UK
[2]S&V Lab, Department of Engineering Mechanics, Xi'an Jiaotong University, 710049 China

Abstract. Complexity of bit-rock interactions and formation fracture makes rock drilling a challenging task from modelling perspective. A finite element model is developed which can plausibly capture the rock fractures while it is capable of evaluating drilling forces for various values of Rate of Penetration (ROP), bit angular velocity and cutter configuration. Firstly, experiments on a single cutter rig are performed with various drilling parameters to be compared to the proposed FE model. Prior to detailed numerical simulation, a rigorous computation test on material model characterization and the element size and type is conducted to ensure feasible and reliable simulation results. The cutting forces are analysed for different values of WOB, rake angle, initial depth of cut and cutter speed for a single-cutter linear cutting. The formation fragmentation process and computed results are comparable to laboratory tests.

1 Introduction

Polycrystalline diamond compact (PDC) bits have been used in in drilling industry from early 1970 [4, 5]. PDC bits consist of a set of PDC cutters on a tungsten carbide body which shear the rock [4, 5]. Rock drilling modelling is still a challenge due to the rock material nature and complexity of rock-bit interaction. In order to enhance the efficiency of drilling and improve the drill bit performance, a calibrated model which is able to estimate the drilling forces would be highly beneficial.

In Oil and Gas drilling industry, an axial force is exerted on the bit due to the drill-string weight which is called weight on bit (*WOB*) while the bit is rotated with an angular velocity (ω) creating a torque on bit (*TOB*). In order to investigate drilling forces on each PDC cutter, a linear single cutter rig was designed and built in drilling lab within Centre for Applied Dynamics Research (CADR) at University of Aberdeen. Linear single cutter rig has been used for a long time by many researchers [3, 7, 9] to model rock-cutter interaction. Similar tests have been performed to assess the unconfined compressive strength of the rock at very small depth of cuts [9].

Numerical simulation methods such as Discrete Element Method (DEM), Finite Element Method (FEM), Finite Difference Method (FEM) and Boundary Element Method (BEM) have been widely used to model rock fracturing and rock-cutter interaction [6, 8]. The aim of this study is to develop a Finite Element model utilizing ABAQUS CAE which can simulate the rock fracturing and predict the cutting forces in different operating parameters such as the cutter back rake angle, depth of cut and cutter velocity. The model assists in understanding the stress and strain field in the region of rock involved in drilling accompanied by how the cracks are formed and create a chip in the formation. The FE models are then tested against the experimental results.

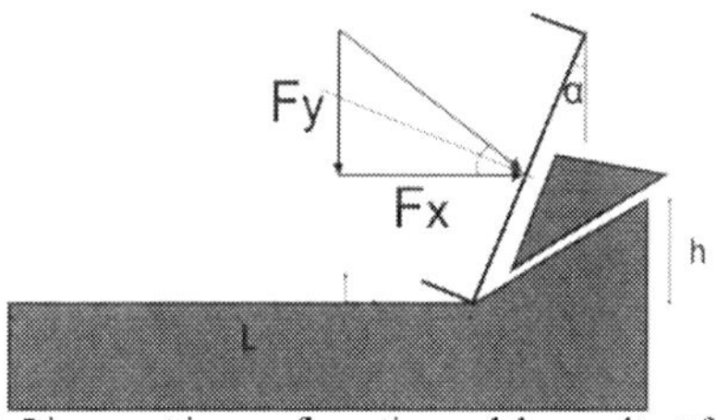

Fig. 1. Linear cutting configuration and the resultant forces.

2 Experiments

This section briefly introduces the linear cutting rig followed by test results.

2.1 Experimental setup

The linear single cutter rig is a metal shaper machine which is modified to suit for rock cutting. It consists of a stationary table which accommodates the rock, a single PDC cutter (3mm diameter) which is attached to a three dimensional force sensor moving along the rock length to cut a groove. The test is conducted under kinematic control which means the depth of cut and the cutting velocity is constant. Figure 1 represents linear cutting parameters. The cutter velocity in direction of the cutting is equal to rω, where r is the distance of the cutter from the bit centre. Depth of cut (h) is defined as the bit rate

* Corresponding author: nina.yari@abdn.ac.uk

https://doi.org/10.1051/matecconf/201814816006

of penetration per revolution ($2\pi \times ROP/\ \omega$). Interfacial frictional angle (ψ) is the angle between the resultant cutting force and the resolved force normal to the cutter [5].

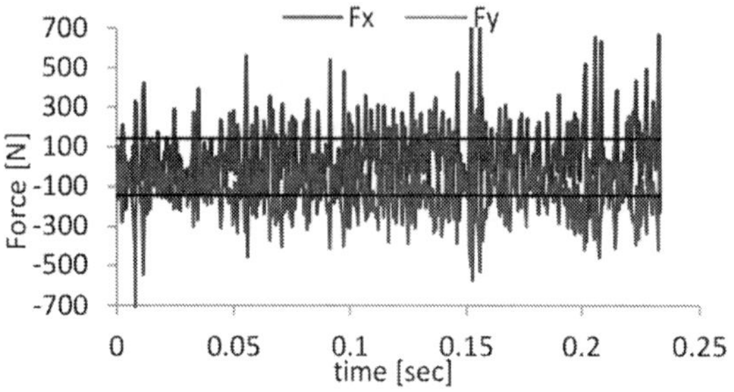

Fig. 2. Experimental results for sandstone cutting with 20 ° back rake angle, 1.3mm DOC and 500mm/sec cutting velocity

2.2 Experimental results

The corresponding cutting and normal force values for the experiment of a 3mm single PDC cutter with 20° back rake angle cutting a groove of 1.3 mm at 500mm/sec velocity in a sandstone sample are represented in Figure 2. The tests are performed for the same cutter at different values of depth of cut and the resultant cutting forces are measured which are demonstrated in Figure 3.

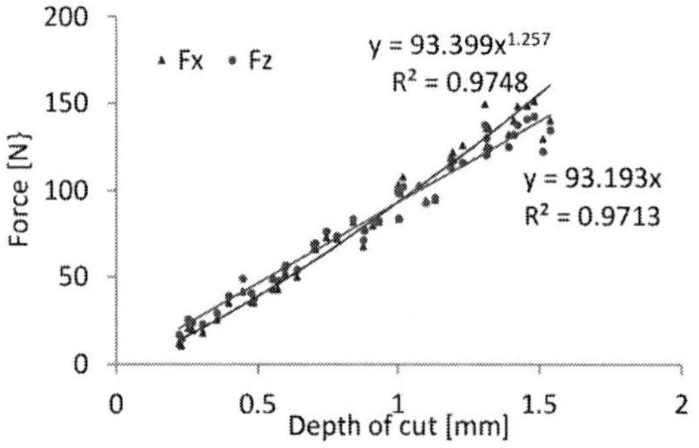

Fig.3. Resultant force values for sandstone with PDC bit with 20 ° back rake angle at various depth of cut values.

3 Numerical models

Finite Element Code, explicit ABAQUS CAE is utilized for modelling the rock cutting. The rock is modelled as a 2D rectangular deformable solid part and the cutter is a discrete rigid element. The mesh size and type were selected after a careful mesh sensitivity analysis. The bottom part of the rock is fully constrained in all directions while the sides are constrained in Z direction and the cutter is free to move in X direction. The cutter's size is 3mm and its back rake angle is 20°. The contact between the cutter and rock selected as cutter surface to rock node contact to capture the fractures and the friction coefficient is 0.1.

3.1 Material model

Drucker-Prager and Ductile Damage are used as Plasticity and damage models consecutively. Drucker-Prager plasticity model represents a pressure dependant yield [1] which makes it a suitable fit for modelling rock plasticity as rock becomes stronger at higher pressure values. Ductile damage is defined as a damage phenomenon by nucleation, growth and coalescence of microscopic voids [1]. In microscopic scale, the rock consists of voids and the cracks are formed by similar phenomenon. In Ductile Damage criterion, the equivalent plastic strain at the damage initiation is a function of strain rate and stress triaxiality η, which is the ratio of mean stress to Von-Mises stress [1]:

$$\omega_D = \int \frac{d\bar{\varepsilon}^{pl}}{\bar{\varepsilon}_D^{pl}(\eta,\dot{\bar{\varepsilon}}^{pl})} = 1 \qquad (1)$$

Where $\bar{\varepsilon}^{pl}$ is equivalent plastic strain at the onset of damage and $\dot{\bar{\varepsilon}}^{pl}$ is its rate and ω_D is the damage parameter which increases with plastic deformation [1]. As soon as the damage initiated, the stiffness is degraded which is identified by damage evolution law. In ABAQUS, the damage evolution is governed by [1]:

$$\sigma = (1 - D)\bar{\sigma} \qquad (2)$$

Where D is damage variable, σ is stress tensor at any time and $\bar{\sigma}$ is undamaged stress tensor. Published data on triaxial experiments on Vosges Sandstone [2] along with some experimental results were used as material model parameters.

3.2 Mesh size sensitivity analysis

The results of FE models are highly sensitive to mesh size and type. After a careful analysis, it is found that triangular elements are better representative of rock heterogeneity compared to quadrilateral elements. In order to select element size for rock, the model was run with various numbers of elements and the resultant force values were compared to each other (Figure 4). As displayed in Figure 4, the results cutting force value is converged when 48542 elements are used where the size of each element in cutting zone is 0.05mm. As the high stress region in rock cutting is at the vicinity of the contact, the mentioned element size is used in top part of the formation where cutting takes place and coarser elements forms the rest of the rock.

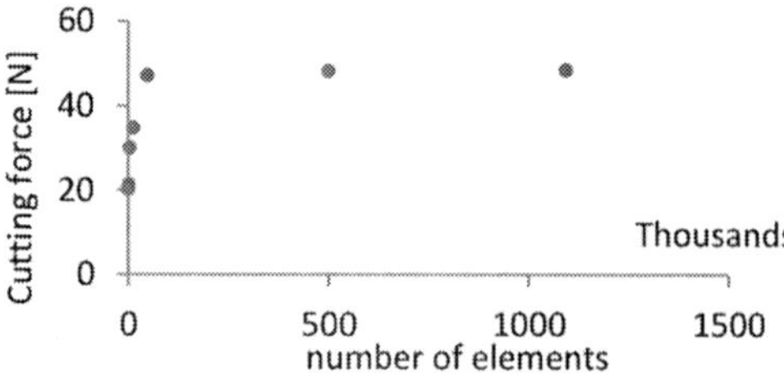

Fig. 4. Cutting forces from FE model for different rock element numbers

3.3 Numerical model results

3.3.1 Fixed depth of cut

Figure 5 shows the stress field of the FE model when a 3mm cutter with 20° back rake angle is cutting a groove of 1.4mm along the rock with 500mm/sec velocity. The force responses of the numerical simulation for the same cutter with 0.7mm depth of cut are demonstrated in Figure 6.

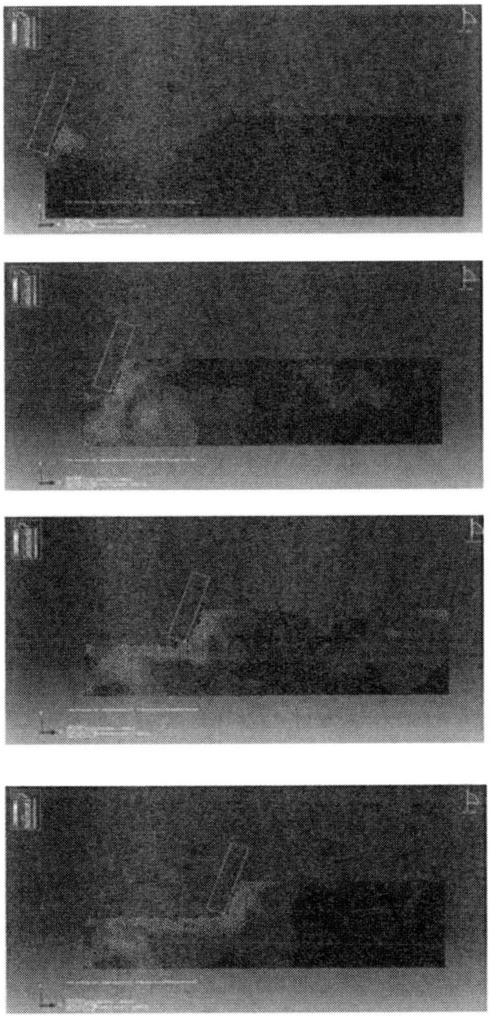

Fig. 5. Stress field evolution for a 2D FEM model of rock cutting with a single cutter for cutting velocity of 500mm/sec, 20° back rake angle and 1.4mm depth of cut

3.3.2 Unfixed depth of cut

In real drilling condition depth of cut is not fixed throughout the cutting process and it is increasing by time. As a result, there would be a velocity in Z direction as well. Consequently, work done (W), volume of cut (V), cutting velocities (v_x& v_y) and arear of cut (A) are calculated as:

$$W = F_x dl + F_z dh \quad (1)$$

$$V = A_{cutting} \times L \quad (2)$$

$$v_x = \frac{dl}{dt},\ v_z = \frac{dh}{dt} \quad (3)$$

$$A_{cutting} = \left[\arccos\left(\frac{r-h}{r}\right) r^2 - \sqrt{r^2-(r-h)^2}(r-h)\right] \quad (4)$$

Where h is depth of cut and L is the length of the rock in X direction.

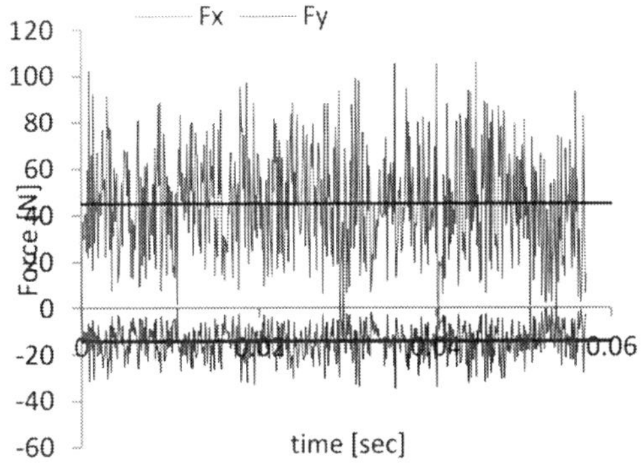

Fig. 6. Computer cutting force values for sandstone with PDC bit with 20 ° back rake angle, 0.7mm depth of cut and 500mm/sec cutting velocity

Assuming that cutting efficiency (η) is a parameter equal to:

$$\eta = \frac{cutting\ volume\ rate}{work\ rate} \quad (5)$$

So:

$$\eta = \frac{dV/_{dt}}{dw/_{dt}} = \frac{dV}{dw} = f(r,\ v_x, v_z) = g\left(r, \frac{v_z}{v_x}\right) = g\left(r,\ \frac{dh}{dl}\right) \quad (6)$$

Because r is fixed:

$$\eta = \eta\left(\frac{dh}{dl}\right) \quad (7)$$

Consequently, at a ratio of horizontal velocity to vertical velocity, the efficiency of drilling is optimum.

The model was run for various quantities of velocity ratios and Figure 7 is an example of the stress field and computed force response when v_z/v_x is 0.2.

Figures 8 and 9 demonstrate FEM computed forces and efficiencies for various velocity ratios. It is shown that for each test condition, there is an optimum velocity ratio which enhances the cutting efficacy.

(a)

(b)

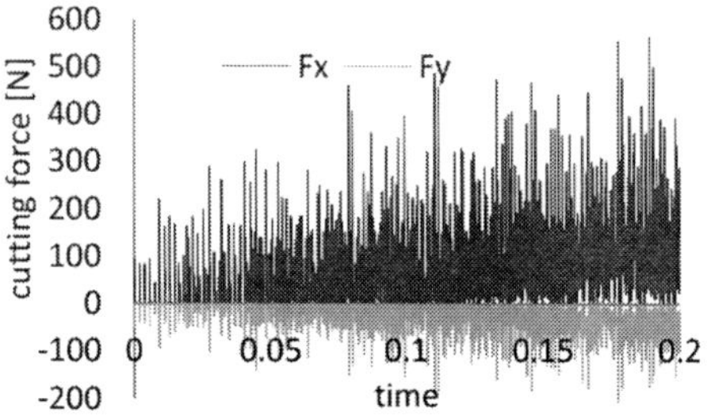

Fig. 7. (a) Stress field 2D FEM model of rock cutting with a 3mm single cutter with 20° back rake when v_x=400mm/sec and v_z=80mm/sec, (b) computed cutting force vs cutter advance

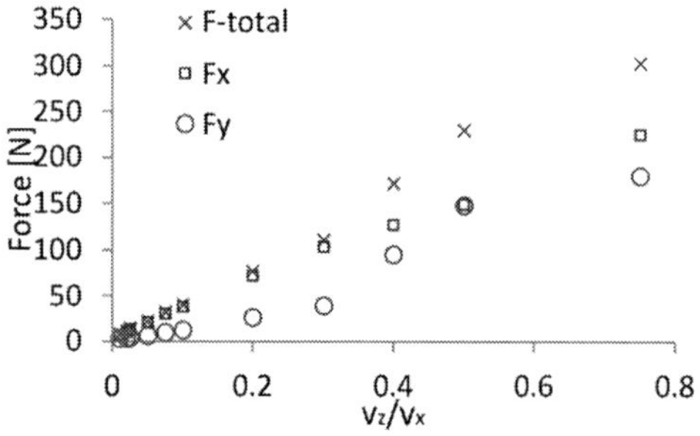

Fig. 8. Computed cutting forces for 2D FEM model of rock cutting with a 3mm single cutter with 20° back rake for unfixed depth of cut vs velocity ratio

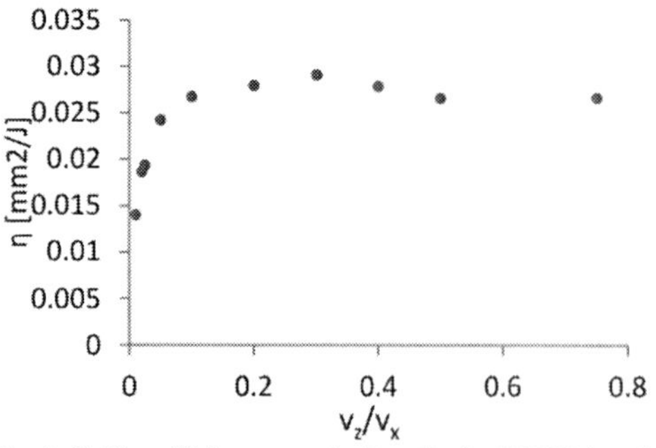

Fig. 9. Cutting efficiency vs velocity ratio for 2D FEM model of rock cutting with a 3mm single cutter with 20° back rake for unfixed depth of cut

4 Conclusion

In this study, a drilling process has been studied with both experimental and numerical simulations. To compare to the proposed numerical model, experiments on a single cutter rig have been first performed with various drilling parameters. Then a finite element model has been developed, and the cutting forces have been analysed for different values of WOB, rake angle, initial depth of cut and cutter speed for a single-cutter linear cutting. To evaluate the drilling process, a new parameter, the drilling efficiency, has been proposed. Initial numerical simulations demonstrated that this parameter can be employed to optimize the drilling parameters. The proposed parameter will be analysed further by being used for complex drilling.

References

1. ABAQUS documentation. USA: Dassault Systems.
2. P. Bésuelle, J. Desrues, S. Raynaud , Experimental characterisation of the localisation phenomenon inside a vosges sandstone in a triaxial cell, *Int. J. Rock Mech. Min. Sci.*, 37 (2000)
3. D. Che, W.L. Zhu, K.F. Ehmann, Chipping and crushing mechanisms in orthogonal rock cutting, *Int. J. Mech. Sci.*, **119** (2016)
4. E. Detournay, P. Defourny, A phenomenological model for the drilling action of drag bits, (1992)
5. E. Detournay, T. Richard, M. Shepherd, Drilling response of drag bits: Theory and experiment, *Int. J. Mech. Sci.*, (2008)
6. M. C. Jamie, Y. Zhou, J. Kin, I. K. Gamwo, Finite element modelling of rock cutting and its fragmentation process, *Int. J. Rock Mech. Min. Sci.*, **80** (2015)
7. Y. Nishimatsu, The mechanics of rock cutting, *International Journal of Rock Mechanics and Mining Sciences and Geomechanics Abstracts*, (1972)
8. L. M. Pradeep, Influence of cutter velocity, friction coefficient and rake angle on the formation of discontinuous rock fragments during rock cutting process, *Int. J. Adv. Manuf. Tech.*, (2016)
9. T. Richard, F. Dargain, E. Poyol, E. Detournay, Rock strength determination from scratch tests, *Eng. Geol.*, (2012)

HELICAL BUCKLING OF DRILL-STRINGS

Marcin Kapitaniak[1,*], *Vahid* Vaziri[1], and *Marian* Wiercigroch[1]

[1]*Centre for Applied Dynamics Research, School of Engineering, University of Aberdeen, Aberdeen, AB24 3UE, Scotland, UK*

Abstract. This paper presents experimental and numerical investigations of the complex drill-string vibration, with particular attention to the helical buckling of drill-strings and its influence on the system dynamics. The analysis is conducted using a novel experimental drilling rig, capable of exhibiting of all major types of drill-string vibration, including torsional, axial and lateral modes, which utilizes real industrial drill-bits and rock samples. A good qualitative and quantitative agreement between the experiments and numerical model is demonstrated.

1 Introduction

The drill-strings used for oil and gas extraction, due to their length reaching up to several kilometers, have a very low bending stiffness compared to the torsional one, what makes them impossible to be driven it in a completely straight configuration. Thereby the drill-string will buckle under its own weight and as the rotation from the top is applied, will take a helical shape due to the constraint in a form of the borehole. Due to the multiple contact points between the drill-string and the borehole, as well as the helical shape the equivalent properties of the structure are different than for the straight configuration. This can have a significant influence on the system dynamics, leading for example to well known stick-slip phenomenon at the drill-bit rock interface, which in turn can result in reduced efficiency of the drilling process, increased wear of the equipment and subsequent catastrophic failures [1]. The aim of this paper is to present the experimentally calibrated model, that takes into account helical buckling of drill-strings and provides further capabilities and understanding comparing to other models [2–4].

2 Aberdeen drilling rig

In order to gain a deeper understanding of drill-string vibration, a novel experimental drilling rig [2] has been constructed at the University of Aberdeen, which is capable of reproducing stick-slip, whirling and bit-bounce phenomena. In this paper we focus on the torsional buckling of the drill-string and demonstrate that it can be a cause of stick-slip oscillations. In our experimental facility, the drill-string is driven from top by a 3kW electric motor which can provide up to 1032 rpm. Depending on the chosen configuration either rigid or flexible shafts are used to transmit rotary motion to the Bottom Hole Assembly (BHA) section and ultimately to the drill-bit. The BHA is made of a heavy steel shaft, held in transversal direction using a loose bearing, as shown schematically in Fig. 1 (a). An axial static force or a Weight on Bit (WOB) is realized by placing steel disks on the top of

*e-mail: marcin.kapitaniak@abdn.ac.uk

https://doi.org/10.1051/matecconf/201814816007

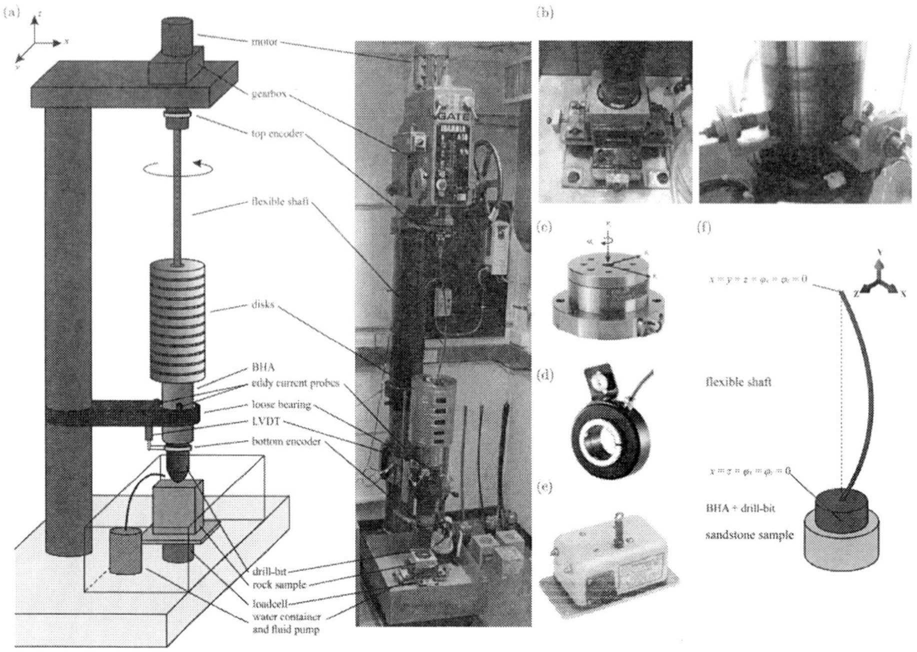

Figure 1. (a) Schematic diagram and photograph of the experimental setup. Main components of the system are: electric motor, flexible or rigid shafts, disks, the BHA, drill-bit and rock samples. (b) Drill-bit rock interface and the BHA with position of the eddy current probes, (c) 4-component load cell (d) top and bottom encoders and (e) LVDT. (f) Configuration of the FE model for pre-buckled configuration, showing all the boundary conditions applied at the top and bottom nodes.

the BHA providing WOB within the range 0.93 to 2.79 kN. A variety of different sensors are used in the experimental setup, which allow to conduct detailed measurements of the most important variables of the drilling process. These include two incremental rotary quadrature encoders (Fig. 1 (d)) to measure top and bottom angular velocity, an axial position transducer (Fig. 1 (e)) to observe Rate of Penetration (ROP) and most importantly two eddy current probes to measure the lateral position of the BHA within the range 0-5mm (see Fig. 1 (b)). The most advanced sensor in our setup is the four component dynamometer (Kistler 9272), placed directly below the rock sample, which allows to measure the axial force F_z, reaction torque T_z, and two forces acting in transversal directions F_x and F_y as depicted in Figure 1 (c). The ranges of measurement for Torque on Bit (TOB) and WOB are: ± 200 Nm and -5-20 kN respectively. For more details about the experimental rig, please refer to [5–7].

3 Experiments and Modelling

In our work we perform experimental observations of the stick-slip phenomenon, with the drill-string in the buckled configuration and on this basis calibrate a Finite Element (FE) model of the rig. Note that in the experimental analysis we represent the drill-string section using flexible shaft of very low bending to torsional stiffness ratio. As can be seen in Fig. 1(f), the FE model of the experimental setup consists of two cylinders, one representing the drill-bit and the other the rock sample, as well as the flexible shaft set in the prebuckled configuration, which allows us to observe helical buckling condition. As the flexible shaft exhibits different properties depending on the direction, it is required to describe it as an orthotropic material, that has different properties along 3 principal axes. This means 9 independent material constants are required to form elasticity matrix **D** to describe the stress-strain relationship. One of the critical steps in the modelling of drill-string dynamics is the mathematical description of the bit-rock interactions. Quantitatively, these interactions are represented in terms of TOB versus bit velocity curves, which in turn allow us to compute equivalent friction coefficient (slip-rate dependent) that captures the main phenomena observed in our drilling tests, namely, cutting and friction between the drill-bit and the rock sample. In the FE model, we assign following friction coefficient between the disc representing the drill-bit and the BHA and the one representing the rock sample: $\mu = \mu_k + (\mu_s - \mu_k)e^{-d_c r\dot{\varphi}_b} + \mu_{str} r\dot{\varphi}_b$, where μ_k, μ_s, μ_{str}, d_c, r and $\dot{\varphi}_b$ represent the kinetic, static, Stribeck friction coefficient, the decay rate, drill-bit radius and angular velocity respectively. For calibration details of the FE model of the flexible shaft and the drill-bit rock interactions please refer to [6].

An example of experimental time history and phase portrait (marked in red), clearly showing stick-slip behaviour, is shown in top panel in Fig. 2 (c), which is compared with the response from calibrated FE model of the assembly (in green) below. Note that the same top excitation is applied in experiment and the FE model (black curve), which assigns following inertia properties to the disc representing the BHA (m = 148.579 kg and J = 0.8416 kgm^2), elastic properties of the flexible shaft (Young moduli: $E_1 = E_2$ = 3.64 GPa, E_3 = 1.50 GPa, shear moduli: G_{12} = 0.02 GPa, $G_{13} = G_{23}$ = 18.16 GPa, Poisson ratios: $\nu_{12} = \nu_{13} = \nu_{23} = 0.30$) and equivalent friction coefficients ($\mu_s = 0.1280$, $\mu_k = 0.0874$, $\mu_{str} = 0.0085$, $d_c = 2.2047$), that correspond to experimental values, as identified in [6]. A very good qualitative and quantitative agreement is achieved, which is particulary visible when comparing shapes of the flexible shaft during different stages of the stick-slip motion, corresponding to experiment and the FE model (See front and top views shown in Figs 1 (a) and (b) respectively). Note, that there are small differences, as the model predicts slightly smaller amplitude of stick-slip vibrations, while keeping the same period. At the same time, when comparing the phase portraits for both cases, we observe that the sticking interval is also slightly smaller than the one observed in experiment, which might be caused by the simplification of cutting process using an equivalent friction approach.

Additionally, we observe experimentally the variation in WOB and in TOB, which can be directly related to the sticking phases (points A-B) visible in angular velocity of the drill-bit, as shown in left hand side panels of Figs 2 (d)-(f). This behaviour is confirmed in the FE model with corresponding time histories shown in the right panels (Figs 2 (d)-(f)), which provides further validation for the model's calibration. As shown in Fig. 2 (e), in both the experimental and numerical results, there is a drop in WOB exactly at a point where sticking phase begins (point A). This suggests an upward force acting on the drill-bit, due to helical deformation of the flexible shaft and elastic energy accumulated. As for the TOB, its variations happen mainly due to the changes of the velocity of the drill-bit and frictional/cutting characteristics of the drill-bit-rock interface. Looking closely on the time history of TOB, we observe that exactly on the onset of sticking phase (point A) a clear sudden increase of TOB is visible, indicat-

https://doi.org/10.1051/matecconf/201814816007

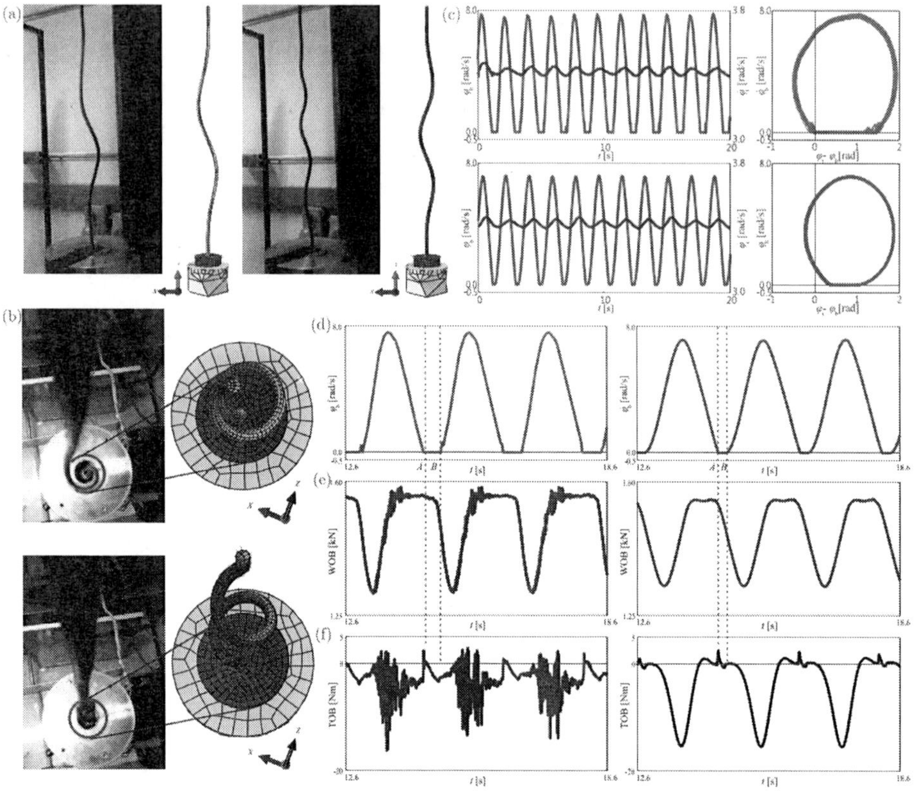

Figure 2. (a)-(b) Comparison of shape of the flexible shaft during stick-slip oscillations for experiment and FE model for prebuckled configuration (d_{db} = 0.0381mm) and WOB of 1.45 kN (front and top views respectively). (c) Corresponding time histories and phase portraits for experiment (in red) and FE model (in green). (d)-(f) A detailed view into time histories of bit angular velocity $\dot{\varphi}_b$, WOB and TOB respectively (Left and right panels depict experimental and numerical responses respectively).

ing reaching a static torque. During the sticking phase, TOB decreases to the initial value by the time slip phase occurs again (point B).

4 Conclusions

In this paper, we focus on the experimental verification of the developed FE model of the drilling rig introduced in [7, 6]. The drill-string (represented by the flexible shaft in the experiment) is driven in a buckled configuration, which allows us to investigate effects of helical buckling on stick-slip motion. After careful calibration of the FE model, we are able to achieve the qualitative and quantitative agreement between experiments and the model. The developed model is capable of predicting accurately the shape of the flexible shaft during different phases of stick-slip phenomena as well as changes in TOB and WOB, as shown in detail in Figs. 2 (a)-(f). In view of positive verification of the developed FE model, we can conclude that the model applicability has been demonstrated, which will allow for further investigation into the complexity of helical buckling phenomenon.

References

[1] A.M Chevallier, N.P Politis, M.L Payne, and P.D. Spanos, Shock and Vibration Digest, **35(2)**, 85–103 (2003)

[2] H. Al-Naser and Y.A. Khulief, Finite Elements in Analysis and Design, **41(13)**, 1270–1288 (2005)

[3] C. Liao, B. Balachandran, M. Karkoub, and Y.L. Abdel-Magid, Journal of Vibration and Acoustics, **133(4)**, (2011)

[4] E. M. Navarro-López and D. Cortés, Journal of Sound and Vibration, **307**, 152–171 (2007)

[5] M. Kapitaniak, V. Vaziri Hamaneh, J. Páez Chávez, and M. Wiercigroch, International Journal of Mechanical Sciences, **101-102**, 324–337 (2015)

[6] M. Kapitaniak, PhD thesis, University of Aberdeen, (2015)

[7] M. Kapitaniak, V. Vaziri Hamaneh, J. Páez Chávez, and M. Wiercigroch, Mechanical Systems and Signal Processing, **100**, 454–465 (2018)

 https://doi.org/10.1051/matecconf/201814816008

Suppression of Drill-String Stick-Slip Vibration

Vahid Vaziri[1,⋆], *Marcin* Kapitaniak[1,⋆⋆], and *Marian* Wiercigroch[1,⋆⋆⋆]
[1] *Centre for Applied Dynamic Research, University of Aberdeen, UK*

Abstract. We investigate experimentally and numerically suppression of drill-string torsional vibration while drilling by using a sliding mode control. The experiments are conducted on the novel experimental drilling rig developed at the University of Aberdeen [1] and using PDC commercial drill-bits and real rock-samples. A mathematical model of the experimental setup which takes into account the dynamics of the drill-string and the driving motor, is proposed. Then a sliding mode control method is employed to suppress stick-slip oscillations. The experimental and numerical results considering a time delay of the actuator are in a close agreement. Stick-slip vibration is eliminated and significant reduction in vibration amplitude has been observed when using the controller.

1 Introduction

During a downhole drilling process excessive vibrations can occur, which in most cases have a negative impact on the effectiveness of the process and the drilling equipment [2]. Such vibration may lead to an accelerated wear and premature damage of the expensive drilling equipment. Often different dynamic effects such as bit-bounce, stick–slip, forward and backward whirls may appear in the drilling process. Recently several attempts have been made to replicate these vibrations in academic laboratories. However, in most studies the cutting process is simulated by a friction between two disks [3].

In this work we focus on suppression of stick-slip phenomenon while drilling, which nature is still not well understood and can lead to a catastrophic failure of the drill-stings. Therefore, firstly, a 2-DOF lumped mass model is developed using the torque-on-bit curves, which capture both the frictional and cutting components of the drill-bit rock interactions [4]. Then, a sliding surface is defined and consequently a sliding mode controller is developed which accommodates the parameters uncertainties. The experimental and numerical results demonstrate the predictive capabilities of the mathematical models and the power of the controller to eliminate the stick-slip vibrations in the given conditions.

2 Drill-String Experimental Stand

A novel experimental drilling facility has been designed and built at the University of Aberdeen, capable of reproducing all major types of drill-string vibration [4–7]. It allows to investigate nonlinear behaviour between the drill-bit and the formation and to introduce different control methods of vibration suppression. Unlike typical other research experimental academic facilities, our rig is configured in such a way that the cutting process is undertaken us-

⋆e-mail: vahid.vaziri@abdn.ac.uk
⋆⋆e-mail: marcin.kapitaniak@abdn.ac.uk
⋆⋆⋆e-mail: m.wiercigroch@abdn.ac.uk

ing real commercial drill-bits and rock samples. The rig shown in Fig.1 is equipped with a variety of different sensors and transducers. Angular positions are measured by two quadrature encoders having 500 pulse per revolution, where axial motion of the drill-bit is captured by a P1010 position transducer attached to the Bottom Hole Assembly. Horizontal and vertical forces as well as torque coming from the bit to the rock are measured by a 4-component Kistler dynamometer placed under the rock sample. All the voltage signals are sent to a NI PCIe-6321 data acquisition card which has multiple analogue input (16-Bit, 250 kS/s) and output (900 kS/s) channels and four 32bit counters/timers. Accessing to 4 counters, allows us to precisely synchronise two encoders. This is important as then the drill-pipe twist can be calculated accurately. Furthermore, the card high performance allows high frequency data sampling up to 30 kHz. This card is controlled by a LabVIEW programme with custom built graphical interface allowing to monitor responses of the system and to present time histories of variables and phase portraits in real-time. This program also sends the command to the top AC motor through the NI card and the ABB frequency convertor. This convertor can work in speed or torque control modes which control the velocity or torque provided by the motor. Different speed or torque nonlinear control methods can be implemented in the LabVIEW program such as a sliding mode control in Fig.1.

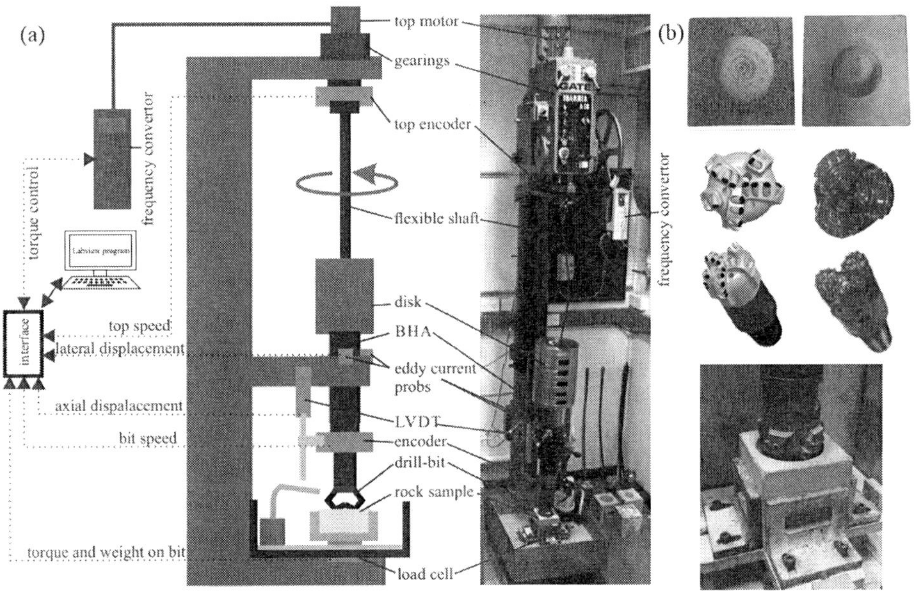

Figure 1. (a) Schematic diagram of the experimental setup, (middle) photograph of the experimental rig and (b) examples of rock samples and drill-bits used in the setup. Main components of the system are: sensors (top and bottom encoders, eddy current probes, LVDT and 4-component load cell), electric motor, flexible or rigid shafts, disks, the BHA, drill-bit and rock samples.

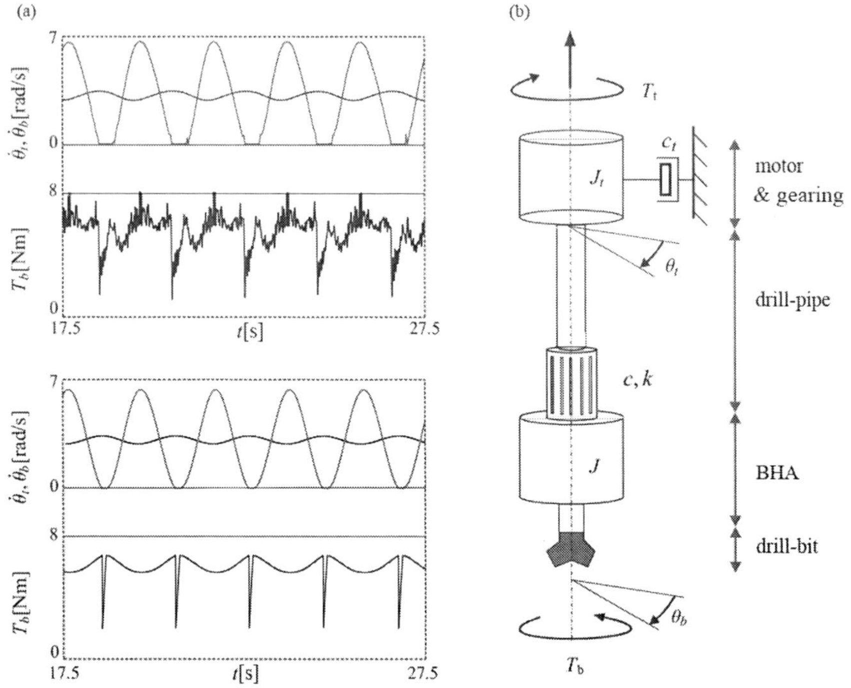

Figure 2. (a) An example of stick-slip oscillations occurring in the experimental rig for W_b = 1.79 kN and 1.5 inch pre-buckled flexible shaft. The time histories of the angular velocities at the bit, $\dot{\theta}_b$, and the surface, $\dot{\theta}_t$ together with TOB obtained in experimental studies (top) and calculated using proposed model (bottom). (b) A physical model of a 2-DOF lump mass torsional system. The viscous damping property of the motor and gearing system and the visco-elasto properties of the pipe are given by c_t, c and k, respectively. The reactive torque acting on the system during drilling is represented by T_b.

3 Mathematical Model and Sliding Model Control

The most used class of models for capturing uncoupled torsional vibration consists of several parallel disks, rotating around their common axis and connected to each other by torsional spring and damper. Top disk in all these models represents the rotary table and the bottom disk represents the drill-bit. The bit-rock interaction is modeled by the velocity dependent resistive torque acting on the bottom disk. This class of models has been widely employed in studies focusing in drill-string torsional vibration. The number of disks varies in those studies, for example 2 disks [8], 3 disks [9], 4 disks [10] and even 18 disks [11]. In the current work a 2-disk model has been employed (see Fig.2(b)) to model the CADR experimental drilling rig, following the work presented in [8]. The viscous damping property of the motor and gearing system and the visco-elasto properties of the drill-pipe are given by c_t, c and k, respectively. The reactive torque acting on the system during drilling is represented by T_b. In this model, the frequency convertor is set in the torque control mode and the top velocity can be calculated from the equation of motion.

https://doi.org/10.1051/matecconf/201814816008

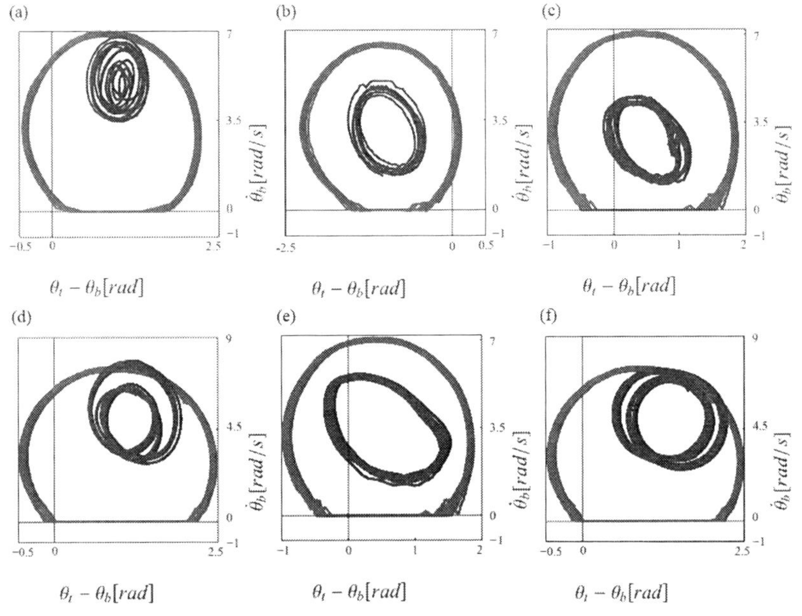

Figure 3. Phase portraits of the drilling experiments using sliding-mode controller. The uncontrolled stick-slip trajectories and the controlled limit cycles are shown in red and blue respectively. The controller achieves (a) 47.86% (b) 59.26% (c) 51.52% (d) 57.58% (e) 66.72% (f) 64.72% reduction in vibration.

A response is considered of the stick-slip oscillations of the drill-bit, for WOB value of W_b = 1.76 kN, and a 1.5 inch pre-buckled flexible shaft and T_t = 39.57 Nm torque generated by motor. Fig.2(a) presents its (top) experimental and (bottom) numerical time history. As can be seen, the top velocities in experiment and simulation (black curves) have sinusoidal vibrations, and the drill-bit velocities (red and green curves) experience stick-slip oscillations of almost constant amplitude. The numerical results were obtained using the model shown in Fig.2(b) with parameters identified as k = 10.00 Nm/rad, c = 0.005 Nms/rad, J_t = 13.93 kg m^2 and c_t = 11.38 Nms/rad. A TOB model is used for the TOB formulation, with corresponding parameters as developed in the previous work [4]. TOB recorded in the experiment and modelled can be seen in Fig.2(b) (blue curves). There is excellent agreement between the experimental observations and performed simulation.

After observing and modelling the stick-slip vibration, we focus on the suppression methods. To control stick-slip, one of the most commonly encountered vibration while drilling for oil and gas exploration, a sliding surface can be defined as

$$s = (\dot{\theta}_t - \omega_d) + \lambda \int_0^t (\dot{\theta}_t - \omega_d) d\tau + \lambda \int_0^t (\dot{\theta}_t - \dot{\theta}_b) d\tau, \quad (1)$$

where θ_t and θ_b are angular positions of the rotary table and the drill-bit respectively, ω_d is the desired angular velocity and λ is a positive control parameter. Based on the defined surface, then a new sliding-mode controller has been successfully designed and applied to lead the system to the sliding surface (s=0), which leads the system to its equilibrium ($\dot{\theta}_t = \dot{\theta}_t = \omega_d$) [5]. In presence of the delay in the motor, the controller has been tested in several different cases. Fig.3 presents phase portraits of the drilling experiments using sliding-mode controller. The uncontrolled stick-slip trajectories and the controlled limit cycles are shown in red and blue respectively. The controller achieves (a) 47.86% (b) 59.26% (c) 51.52% (d) 57.58% (e) 66.72% (f) 64.72% reduction in vibration.

4 Conclusion

In this paper we have presented a lumped mass model for the drilling rig shown in Fig.1. By using identified model parameters in the preformed simulation, an excellent match between experiment and simulation is achieved capturing the stick-slip vibration. The next step of this study was to adapt a sliding-mode control method and applied to the obtained model in order to eliminate the stick-slip vibration observed in the drilling rig and simulation. The controller implemented in the experiment is successful in eliminating the stick-slip vibration in the presence of a delay and dead-zone in the actuator. Several experimental results were presented in this paper with a variety of the estimated parameters applied to the controller. A significant reduction in vibration amplitude is observed when the controller is applied.

Taking into consideration the positive experimental results reported in this paper, we can conclude that a robust mathematical model and a sliding-mode controller have been developed, which proved to predict accurately the responses of the analyzed experimental setup and eliminate the stick-slip oscillations and reduce it to torsional vibration in presence of delay in the actuator.

References

[1] M. Wiercigroch, Modelling and Analysis of BHA and Drill-string Vibrations R&D Project sponsored by the BG Group (2010)

[2] P. Spanos, A. Chevallier, N. Politis, M. Payne, The Shock and Vibration Digest **35**, 81 (2003)

[3] P.A. Patil, C. Teodoriu, Journal of Petroleum Science and Engineering **112**, 227 (2013)

[4] M. Kapitaniak, V. Vaziri, J. Páez Chávez, K. Nandakumar, M. Wiercigroch, International Journal of Mechanical Sciences **101-102**, 324 (2015)

[5] V. Vaziri, Ph.D. thesis, University of Aberdeen (2015)

[6] M. Kapitaniak, Ph.D. thesis, University of Aberdeen (2015)

[7] M. Kapitaniak, V. Vaziri, J. Páez Chávez, M. Wiercigroch, Mechanical Systems and Signal Processing **100**, 454 (2018)

[8] E. Navarro-López, R. Suárez, IEEE Conference on Control Applications pp. 1454–1460 (September 2004)

[9] E. Navarro-López, E. Licéaga-Castro, Chaos, Solitons and Fractals **41**, 2035 (2009)

[10] Y. Liu, IET Control Theory & Applications **9**, 91 (2015)

[11] T. Vromen, C.H. Dai, N. Van De Wouw, T. Oomen, P. Astrid, H. Nijmeijer, IFAC Proceedings Volumes **48**, 266 (2015)

https://doi.org/10.1051/matecconf/201814816009

Numerical Application of a Stick-Slip Control and Experimental Analysis using a Test Rig

Leonardo D. Pereira[1,*], *Bruno* Cayres[1,2,**], and *Hans I.* Weber[1,***]

[1] *Dept. of Mechanical Engineering, PUC-Rio, Brazil*
[2] *Dept. of Mechanical Engineering, CEFET/RJ, Brazil*

Abstract. Part of the process of exploration and development of an oil field consists of the drilling operations for oil and gas wells. Particularly for deep water and ultra deep water wells, the operation requires the control of a very flexible structure which is subjected to complex boundary conditions such as the nonlinear interactions between drill bit and rock formation and between the drill string and borehole wall. Concerning this complexity, the stick-slip phenomenon is a major component related to the torsional vibration and it can excite both axial and lateral vibrations. With these intentions, this paper has the main goal of confronting the torsional vibration problem over a test rig numerical model using a real-time conventional controller. The system contains two discs in which dry friction torques are applied. Therefore, the dynamical behaviour were analysed with and without controlling strategies.

1 General Introduction

1.1 Oilwell drilling and test-rig model overview

The actual organizational model of oil and gas industry is no longer sustainable with oil prices below $50 a barrel. Christopher *et al.* (2016) [1] describes what they call the "potentially game-changing disruptions that may lead oil and gas companies to rethink their operating models fundamentally". Then, to address the industry challenges, this paper focuses its efforts on modelling conventional drilling operations using a reduced scale set-up model.

Briefly, the drilling process involves rock failure by a rotating **drill bit**. To rotate this drill bit from the top-end position (surface), the drilling rig's power source is used to turn a **rotary table**. A mechanism, consisting of **master bushing** and **kelly** is used to transmit the rotation from the rotary table (or **top drive**) to the **drill string**. The drill string is the subsurface component through which torque is transmitted from the surface to the bottom of the drilling system (downhole). It consists of connected lengths of **drill pipes**, the **bottom hole assembly (BHA)**, and the **drill bit**. The BHA is the portion of the drill string between the drill pipe and the drill bit. It is made up, primarily, of **drill collars (DC)** and **heavy weight drill pipe (HWDP)**.

Previous studies have identified three types of vibrations that may occur during drilling operations: **Torsional, Lateral** and **Axial**, as cited in [2–4].

In order to reproduce the phenomena inherent to the torsional vibrations present in the rotating drilling system a reduced-scale test rig set-up model (shown in Fig. 1) was developed at the PUC-Rio Dynamics and Vibration Laboratory. The set-up model consists of an electrical and a mechanical subsystems. Electrical subsystem represents the drilling rig's power source and the rotary table (or top drive), while the mechanical system represents the drill pipes, BHA and drill bit.

The test rig consists of three degrees of freedom: disc 1 (R1), disc 2 (R2) and the DC-motor inertia. Also, it contains an electric system which provides torque to the mechanical system.

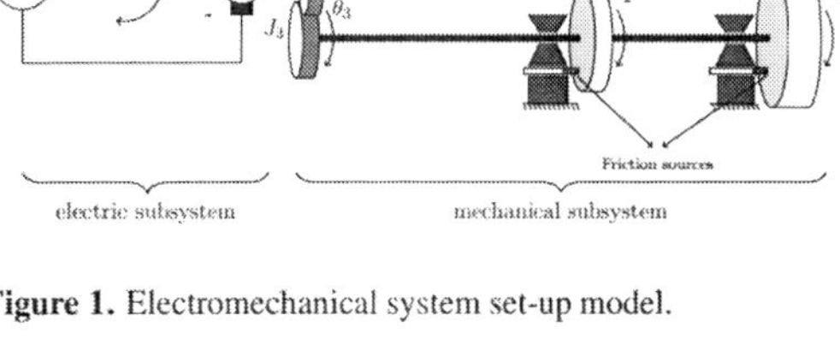

Figure 1. Electromechanical system set-up model.

The electric subsystem is modelled as a voltage source connected in series with a resistor and an inductor, providing torque T_m. The angular velocity Ω_m imposed by T_m is eight times greater than the angular velocity Ω_3 transmitted to the mechanical subsystem due to the transmission factor $\eta = 8 : 1$.

The system presents two resistive torque sources. At disc 1 (R1), the friction means the bit/rock friction interaction. The intermediate friction source (at disc 2) may

*e-mail: leopereiradias@yahoo.com.br
**e-mail: bruno.cayres@cefet-rj.br
***e-mail: hans@puc-rio.br

represent a friction raise from the interaction between drill pipes and borehole wall.

1.2 Motivation and objectives

Since vibration problems were detected and identified in drilling operations, several approaches have been suggested, both in industry and literature, to model and control these vibrations. Some of these approaches were conducted to the surface system (electric subsystem). Most of them dealt with the torsional behaviour and the suppression of the stick-phase of the stick-slip oscillations [5]. Moreover, the nonlinearity and uncertainty of the system parameter can be modelled in the control design processes. These facts are the motivation of this article.

Therefore, the scope of this paper is to minimize the torsional vibration problem of the test-rig set-up model using a control strategy. Specific objectives address:

- Open-loop analysis of the electrical and mechanical subsystems.
- Application of a control strategy using different velocity inputs in a closed-loop system.
- Control of the torsional vibration considering the nonlinearity due to friction interaction with the first and second inertia.
- Improvement on a developed experimental set-up for future verification and validation of the models.

1.3 Methodology

The methodology required to evaluate the performance of the control system is based on five steps [6]:

1. The modelling of the mechanical and mathematical representation of the test-rig set-up model.
2. An open-loop analysis is performed on a simplified model using a linear DC motor and a three degrees-of-freedom (DOF) system. In this system, a friction model capable of generating torsional vibrations is proposed acting on the first and second inertia, i.e., Stribeck friction.
3. The closed-loop analysis starts assuming a set of parameters.
4. Two situations are analysed concerning the control strategies:
   (a) Only few parameters such as Motor Torque, J3 RPM, and Normal 1 are known.
   (b) All the parameters (such as Normal forces and all angular velocities) are known.

Initially, the Proportional and Integral (PI) is the control strategy used in this comparison.

2 Preliminary concepts

2.1 Open-loop system

Generally, there are two open-loop system models: controlled or not controlled. In other words, for an open-loop analysis in a drilling system or in the proposed test rig set-up, the process may be equipped with a controller plus an actuator (Fig. 2a) or just with an actuator driven by an input signal Fig. 2b. For the purpose of this paper, the open-loop system will be considered without the controller.

(a) Open-loop system with feedforward control

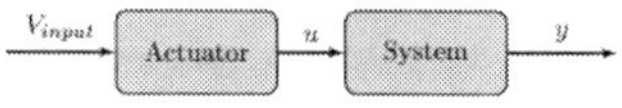

(b) Open-loop system without feedforward control

Figure 2. Block diagrams of open-loop systems.

The S_{input} is the input signal of the controller in the block diagram of Fig. 2a. The V_{input} and u_c are the voltage input signal of the actuator, the u is the torque input signal of the drilling system. The y is the output angular velocity of the system.

2.2 Closed-loop system

For this paper the closed-loop analysis of the test rig set-up will be performed with disturbances and without disturbances. Furthermore, the closed-loop control system, also known as feedback control system, is represented by the block diagram shown in Fig. 3.

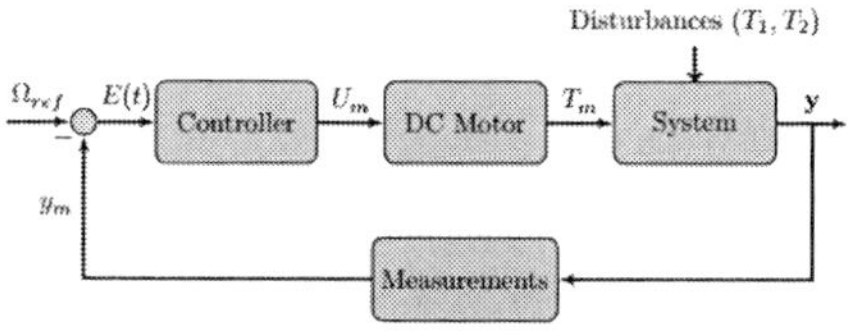

Figure 3. General block diagrams of a closed-loop system.

The Ω_{ref} is the reference signal of the system to be compared to the measured angular velocity y_m in the block diagram of Fig. 3 and generates the error signal $E(t)$. The U_m is the voltage input signal of the actuator, the T_m is the torque input signal of test rig set-up. The **y** is the output angular velocities of test rig set-up.

2.3 Automatic control systems

The control strategy (PI controller) used in the current paper is a well known industrial closed-loop controller, chosen according to their control actions, as explained in more details in [6].

2.3.1 Proportional and integral controller (PI)

The principle used in the SoftTorque control system (PI controller) is based on the proportional and integral control theory. So, based on the general block diagram of Fig. 3, the mathematical control law has the form:

$$\mathbf{u_c}(t) \;=\; k_p\, e(t) + k_i \int_0^t e(t)dt \qquad (1)$$

where u_c is the control signal, y_m is the controlled variable, r is the manipulated variable and $e(t)$ is the control error ($e(t) = [r(t) - y_m(t)]$). k_p is the proportional gain. k_i is the integral gain and can be represented as $k_i = \frac{k_p}{T_i}$. T_i is the integral time constant.

3 Mathematical modelling of the test rig set-up model

A simple schematic of the reduced scale test rig set-up model is illustrated in Fig. 1.

In this case, the torsional set-up model is divided in two dynamic subsystems: the *electrical system* and the *mechanical system*. The first part (electrical), is represented by a DC motor, delivering torque on the third inertia. The motor dynamics has a considerable impact on the full dynamical behaviour. On the other hand, the mechanical part consists of three inertia. Both subsystems are physically connected to perform the torsional dynamics and will be investigated separately.

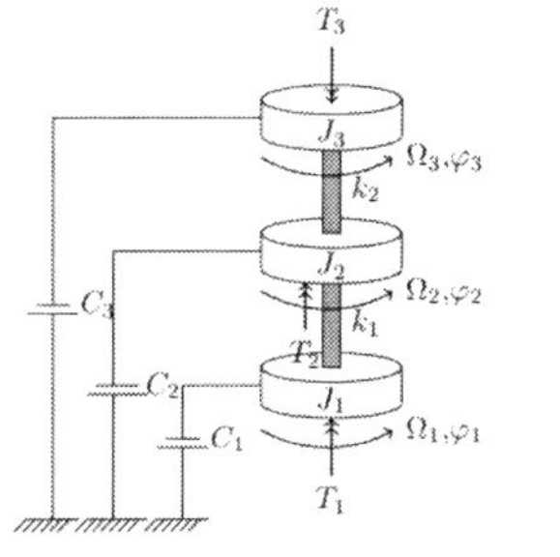

Figure 4. A test rig set-up modelled as a torsional pendulum.

The corresponding equations of motion of the test rig set-up model are represented by the model of torsional dynamics in Eq. 2.

$$\begin{vmatrix} J_1 & 0 & 0 \\ 0 & J_2 & 0 \\ 0 & 0 & J_3 \end{vmatrix} \begin{vmatrix} \dot{\Omega}_1 \\ \dot{\Omega}_2 \\ \dot{\Omega}_3 \end{vmatrix} + \begin{vmatrix} C_1 & 0 & 0 \\ 0 & C_2 & 0 \\ 0 & 0 & C_3 \end{vmatrix} \begin{vmatrix} \Omega_1 \\ \Omega_2 \\ \Omega_3 \end{vmatrix} + \begin{vmatrix} k_1 & -k_1 & 0 \\ -k_1 & k_1 + k_2 & -k_2 \\ 0 & -k_2 & k_2 \end{vmatrix} \begin{vmatrix} \varphi_1 \\ \varphi_2 \\ \varphi_3 \end{vmatrix} = \begin{vmatrix} -T_1 \\ -T_2 \\ T_3 \end{vmatrix} \qquad (2)$$

where φ_1, φ_2 and φ_3 are the angular displacements of the three discs, Ω_1, Ω_2 and Ω_3 are the angular velocities measured at the three discs. J_1, J_2 and J_3 are the equivalent mass moment of inertia of the three discs. C_1, C_2 and C_3 are the equivalent viscous damping coefficients, k_1 and k_2 is the equivalent torsional stiffness of the wire, and T_1 and T_2 are the nonlinear functions representing the resistive torque at discs 1 and 2. T_3 is the supplied torque to the mechanical system.

The behaviour of the DC electronic motor has its dynamical principles represented by two linear models: (1) the mechanical equation of free-body diagram of the rotor, and (2) the electric circuit equation of the armature.

The equation of motion of the motor dynamics is presented as:

$$L\,\dot{I} + R\,I + V_{emf} = U_m \qquad (3)$$

$$V_{emf} = k_e\,\eta\,\Omega_3 \qquad (4)$$

$$T_m = k_t\,I - C_m\,\eta\,\Omega_3 - T_{ff} - J_m\,\eta\,\Omega_3 \qquad (5)$$

where $T_3 = \eta\,T_m$ due to the gearbox factor. I is defined as armature current, R and L are electrical resistance and electrical inductance, respectively. U_m is the motor input voltage. V_{emf} represents the back-electromotive force (back-emf) related to the angular velocity Ω_3. k_e and k_t parameters are respectively the electromotive force constant and the motor torque constant. T_{ff} represents a internal friction torque of the DC motor.

3.1 Friction torque modelling

For this reason, the Eq. 6 describes how the resistive torque at the drill bit is modelled based on the friction law presented in [7, 8].

$$T_i = N_i\, r_i \left[\mu_k + (\mu_s - \mu_k) \cdot e^{-vb|\Omega_i|}\right] \cdot \operatorname{sign}(\Omega_i) \qquad (6)$$

where $i = 1, 2$, N_i is the normal force, also constant in the model, and r_i is the distance between the centre of the disc and the normal contact point (see Fig. 1). μ_s and μ_k are static and kinetic friction coefficient. The friction model curve is illustrated in Fig. 5.

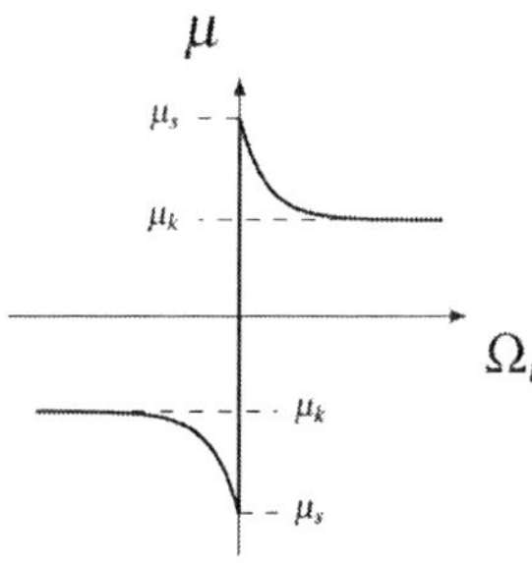

Figure 5. Stribeck friction model.

4 Stick-slip severity (SSS)

The **Stick-Slip Severity (SSS)** is a severity criterion that evaluates the vibration amplitudes in relation to the nominal velocity. Therewith, **SSS** is performed for the used PI control strategy. In this paper, the **SSS** may be presented as a 2D and 3D grey scale colour-map that plots the relation between $N1$ (normal at R1) and nominal angular velocity (RPM at DC motor), exposing limits where the previous defined torsional vibration (stick-slip) becomes severe according to the criterion.

$$\mathbf{SSS} = \left(\frac{\Omega_1^{max} - \Omega_1^{min}}{2\,\Omega_{ref}} \right) \cdot 100\% \qquad (7)$$

where Ω_1^{max} and Ω_1^{min} are respectively the maximum and minimum angular velocity of the disc 1. Ω_{ref} is the reference angular velocity.

5 Analysis of the results

An optimization method is used to evaluate the effectiveness of the open-loop analysis in control design, as presented in the flowchart schematic in Fig. 6. This procedure considers the subsystems separately. Tab. 1 contains the parameters of the test rig set-up.

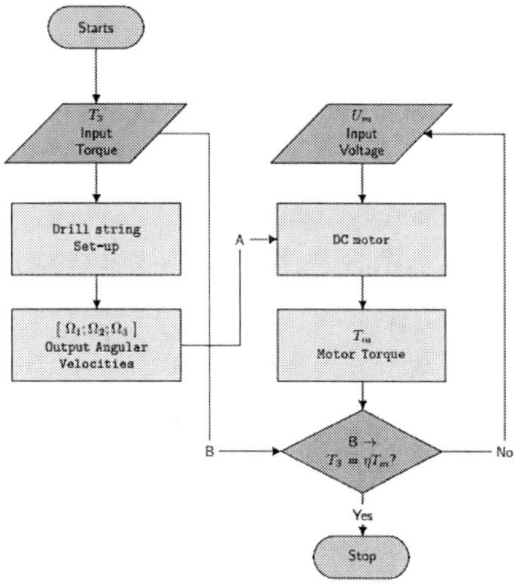

Figure 6. The iterative open-loop flowchart used in the analysis.

Multiple step responses $[\Omega_3, \Omega_2, \Omega_1]$ of the drill string system are shown in Figs. 7a and 7b. Remembering that in the open-loop analysis there is no need to satisfy any control criteria or to make consistent conclusions about steady state errors, because there is no pre-defined setpoint velocity (Ω_{ref}).

After substituting the parameters of Tab. 1 into the Eq. 2 and 5, a summary of the open-loop angular velocity responses of the test rig set-up and the DC motor systems, given in respect to time when the torque varies, are displayed in Tab. 2. Moreover, for each constant torque from 0.01 $N \cdot m$ to 0.30 $N \cdot m$ and the associated results, it is defined a Number (**Num.**) varying from 1 to 15, respectively, in the first column of the Tab. 2.

5.1 Open-loop results

A *Simulink* modelling is designed to simulate the dynamic of the mechanical part of test rig set-up, in other words, without a DC motor actuator. The simulations are performed to evaluate the system behaviour when a range of input torques are disturbed by a set of N1 and N2. The white zone indicates the zone where an input torque is effective to keep the angular velocity at a constant value. The black zone indicates the zone where the input torque

Table 1. Simulation parameters values.

Parameter	Description	Value	Unit
J_1	Disc 1 (R1)	0.0289	$kg\,m^2$
J_2	Disc 2 (R2)	0.0197	$kg\,m^2$
J_m	DC motor inertia	$0.37\ 10^{-3}$	$kg\,m^2$
C_1	R1 damping	0.0134	Nms/rad
C_2	R2 damping	0.0067	Nms/rad
k_1	R1-R2 stiffness	0.6441	Nm/rad
k_2	R2-R3 stiffness	0.1762	Nm/rad
R	Electrical resistance	0.33	Ohm
L	Electrical inductance	$1.1\ 10^{-3}$	H
k_e	Electromotive force constant	$6.02\ 10^{-2}$	$V/rad/s$
k_t	Motor torque constant	0.12	Nm/Amp
C_m	Motor speed constant	$1.91\ 10^{-4}$	Nms/rad
T_f	Internal friction torque	0.1	Nm
η	Transmission factor	8	–
r_1	N1 contact point	0.047	m
r_2	N2 contact point	0.047	m
μ_s	Static coefficient	0.461	–
μ_k	kinetic coefficient	0.370	–
v_b	velocity weakening parameter	1.5	–
k_i	integral gain	1.4891	–
k_p	proportional gain	0.4408	–

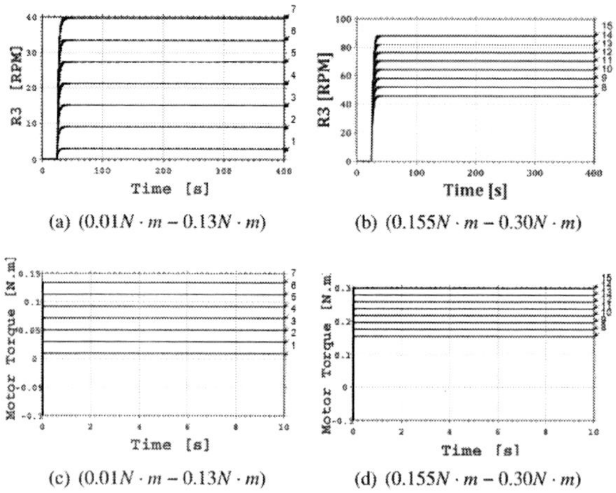

(a) $(0.01N \cdot m - 0.13N \cdot m)$ (b) $(0.155N \cdot m - 0.30N \cdot m)$

(c) $(0.01N \cdot m - 0.13N \cdot m)$ (d) $(0.155N \cdot m - 0.30N \cdot m)$

Figure 7. Angular velocity and torque responses.

Table 2. Test rig set-up and DC Motor open-loop results.

	Discs 1 and 2		DC Motor
Num.	T_3^{input}	$[\Omega_1 \& \Omega_2]^{settled}$	U_m
1	$0.0100 N{\cdot}m$	2.9499 rpm	0.1865 V
2	$0.0307 N{\cdot}m$	9.0605 rpm	0.5018 V
3	$0.0514 N{\cdot}m$	15.1710 rpm	0.8171 V
4	$0.0721 N{\cdot}m$	21.2816 rpm	1.1324 V
5	$0.0929 N{\cdot}m$	27.3921 rpm	1.4477 V
6	$0.1136 N{\cdot}m$	33.5026 rpm	1.7630 V
7	$0.1343 N{\cdot}m$	39.6132 rpm	2.0783 V
8	$0.1550 N{\cdot}m$	45.7237 rpm	2.3936 V
9	$0.1757 N{\cdot}m$	51.8343 rpm	2.7089 V
10	$0.1964 N{\cdot}m$	57.9448 rpm	3.0242 V
11	$0.2171 N{\cdot}m$	64.0554 rpm	3.3395 V
12	$0.2379 N{\cdot}m$	70.1659 rpm	3.6548 V
13	$0.2586 N{\cdot}m$	76.2765 rpm	3.9701 V
14	$0.2793 N{\cdot}m$	82.3870 rpm	4.2854 V
15	$0.3000 N{\cdot}m$	88.4976 rpm	4.6006 V

is not sufficient to suppress the resistive torque generated by the normal forces (N1 and N2).

The Hopf supercritical bifurcation diagrams of the test rig system are assumed in a non-controlled system for a set of N1 and N2 0 N and 10 N, as illustrated in Figs. 8a and 8b and Figs. 9a and 9b, respectively.

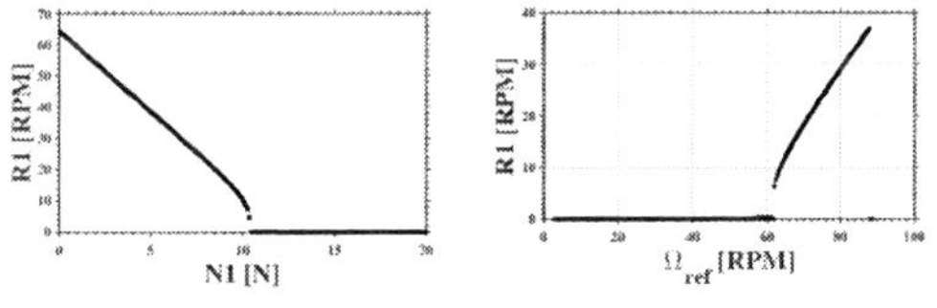

(a) Ω_{ref} = 64.05 rpm (b) $N1$ = 10 N

Figure 8. Bifurcation diagrams for $N2$ = 0 N.

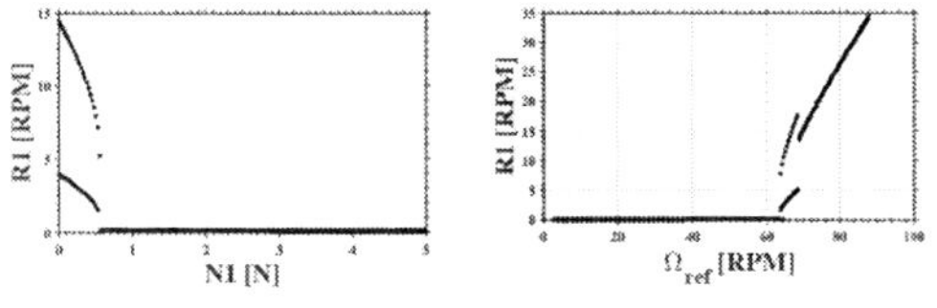

(a) Ω_{ref} = 64.05 rpm (b) $N1$ = 0.50 N

Figure 9. Bifurcation diagrams for $N2$ = 10 N.

5.2 Closed-loop results: PI controller

Now, a *Simulink* modelling is designed to simulate the dynamic test rig set-up with a DC motor actuator. The torsional vibration map of the test rig system when subjected to PI controller is illustrated in Figs. 10a and 10b. It is noticed that the system reaches amplitudes of vibration that are 4x (200%) the nominal velocity. The white zone indicates the zone where a PI controller is effective. For best understanding see [9].

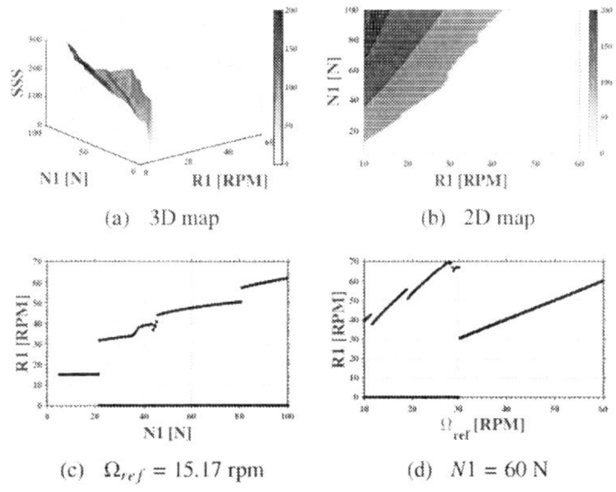

(a) 3D map (b) 2D map

(c) Ω_{ref} = 15.17 rpm (d) $N1$ = 60 N

Figure 10. SSS and Bifurcation diagrams for PI ($N2$ = 0 N).

In this case, it is possible to say (see Fig. 10c) that, for a constant angular velocity ($\Omega_{ref} = 15.17\ rpm$), it presents transition region from equilibrium point to periodic solution. It is a Hopf supercritical bifurcation diagram and may show the amplitude of vibration increasing when different $N1$ are applied to the system PI controller, as illustrated in Fig. 10c. Also, using Ω_{ref} as nonlinear control parameter, Fig. 10d, for $N1 = 60\ N$, the periodic solution becomes equilibrium point around $\Omega_{ref} = 30\ rpm$.

In order to analyse the influence of intermediate friction on the dynamic responses and the effectiveness zone of PI controller, $N2 \neq 0$, Figs. 11a and 11b present the **SSS** maps. The vibration zone is clearly increased compared to Figs. 10a and 10b. Also, the bifurcation points have changed observing Figs. 10c and 10d comparing to 11c and 11d.

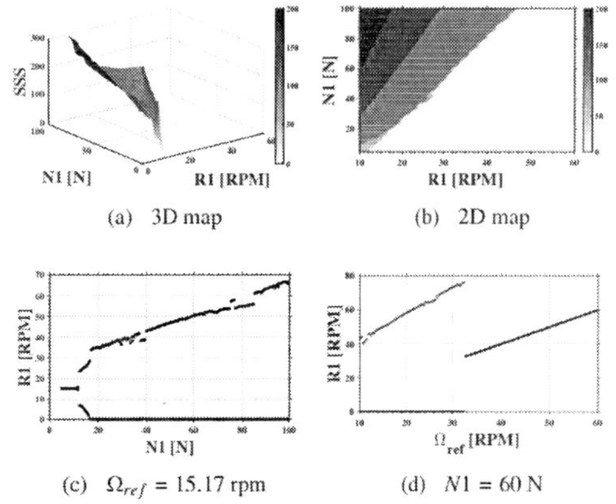

(a) 3D map (b) 2D map

(c) Ω_{ref} = 15.17 rpm (d) $N1$ = 60 N

Figure 11. SSS and Bifurcation diagrams for PI ($N2$ = 10 N).

Figs. 11b and 11c still show two nonlieanr jumps before 20 *rpm* when $N1$ is varied. Following the same reasoning, Fig. 12 illustrates large variations of **SSS**. However, for all three $N2$ values, the greater influence occurs for low values of Ω_{ref} where it may be observed a decreasing of the effective zone of the PI controller.

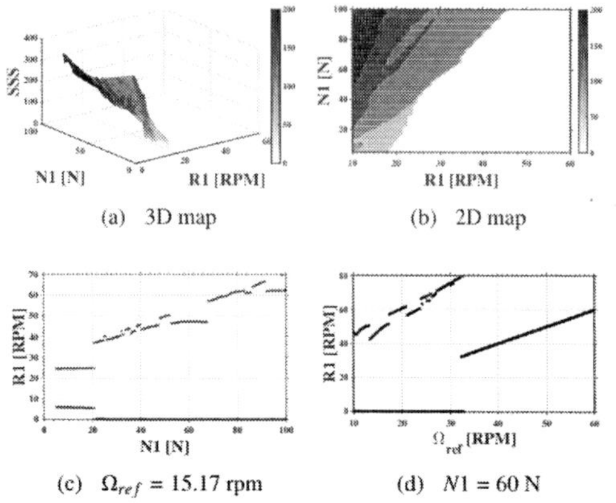

(a) 3D map (b) 2D map

(c) Ω_{ref} = 15.17 rpm (d) $N1$ = 60 N

Figure 12. SSS and Bifurcation diagrams for PI ($N2$ = 20 N).

Finally, a comparison between numerical and experimental values is performed. The system is settled to 30 and 50 *rpm* and the both transient are observed in Figs. 13a and 13b, respectively.

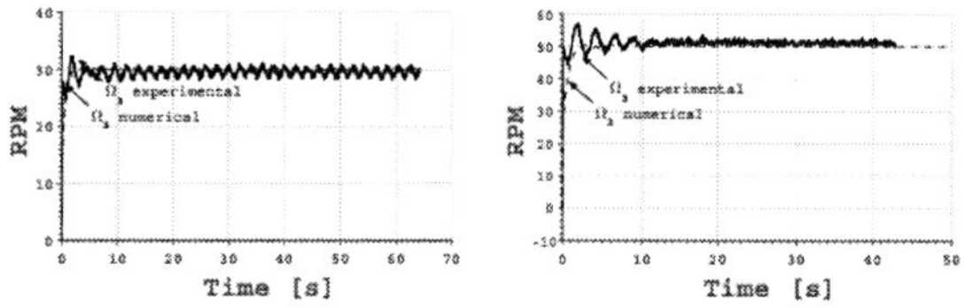

(a) 30 *rpm* (b) 50 *rpm*

Figure 13. Numerical and Experimental Ω_3 comparison.

For this test, both data converge to the settled angular velocity and the step response characteristics of both systems show the following performance indicators (see Tabs. 3 and 4).

6 Conclusions

The drilling system provides a myriad of problems to investigate. Controlling torsional vibration of this kind of system is challenging.

It is possible to observe that the PI controller is effective in a certain zone. By varying the intermediate friction torque T_2 (via $N2$), the controllable zone changes mainly for low nominal velocities Ω_{ref}. PI controller has proven to be very ineffective when the system is in torsional vibration. This is because PI controller is a linear controller,

Table 3. Attributes of the test rig set-up. (30 rpm).

Attributes	Ω_3 Exp.	Ω_3 Num.
Rise time [$\tau_r(s)$]	1.3207	1.4841
Settling time [$\tau_s(s)$]	N/A	1.8692
Peak time [$\tau_p(s)$]	1.9462	6.0147
Overshoot [% OS]	8.9790	0.4908

Table 4. Attributes of the test rig set-up. (50 rpm).

Attributes	Ω_3 Exp.	Ω_3 Num.
Rise time [$\tau_r(s)$]	1.2246	1.4793
Settling time [$\tau_s(s)$]	N/A	1.8646
Peak time [$\tau_p(s)$]	1.8598	6.0165
Overshoot [% OS]	12.0614	0.4919

but also has shown to restrain torsional vibrations for high nominal velocities.

In a first approach, this methodology is important to understand the test rig system and which controller (or combination of controllers) may be applied in experimental analysis.

References

[1] C. Handscom, S. Sharabur, J. Woxholth, **The oil and gas organization of the future**, Online; accessed 08-Feb (2017), http://www.mckinsey.com/industries/oil-and-gas/our-insights/the-oil-and-gas-organization-of-the-future

[2] F. Abdulgalil, H. Siguerdidjane, **PID Based on Sliding Mode Control for Rotary Drilling System**, in *IEEE/EUROCON The International Conference on "Computer as a Tool"* (Serbia & Montenegro, Belgrade, 2005), Vol. 2, pp. 262–265

[3] A.H. Suleiman, Tech. rep., Universiti Teknologi MARA (UiTM), Selangor Darul Ehsan, Malaysia (2006)

[4] E.M. Navarro-López, **Bit-sticking Phenomena in a Multidegree-of-freedom Controlled drill string**, in *Exploration & Production* (2010), Vol. 8 of 2, pp. 70–75

[5] M.B. Saldivar, S. Mondié, J.J. Loiseau, **Reducing Stick-Slip Oscillations in Oilwell drill strings**, in *6th International Conference on Electrical Engineering, Computing Science and Automatic Control* (Toluca de Lerdo, Mexico, 2009), pp. 1–6

[6] L.D. Pereira, M.Sc. Thesis, Pontifícia Universidade Catolica do Rio de Janeiro, Rio de Janeiro, Brazil (2017)

[7] T.G.M. Vromen, Ph.D. thesis, Eindhoven University of Technology, Eindhoven, The Netherlands (2015)

[8] J. Rudat, D. Dashevskiy, **Development of an innovative model-based stick/slip control system**, in *SPE/IADC Drilling Conference and Exhibition* (Society of Petroleum Engineers, 2011)

[9] B. Cayres, M.Sc. Thesis, Pontifícia Universidade Catolica do Rio de Janeiro, Rio de Janeiro, Brazil (2013)

Dynamic Analysis of a Drillstring-Riser System Drilling in Deep Water

Maolin Liao*, *Yingcao* Zhou, *Yinao* Su, *Zhilong* Lian, *Hongwei* Jiang
CNPC Drilling Research Institute, Beijing, China, 102206

Abstract. A mathematical model of a drillstring-riser system is developed to analyze the system dynamics when drilling in deep water. For the proposed model, a riser and a drillstring are combined as a pipe-in-pipe structure, along which the interactions between the riser and the drillstring are introduced by a series of spring-friction units. Numerical simulations of the developed mathematical model are carried out by Abaqus in which the finite element method is adopted. Comparing with the riser deflection calculated by previously published models which exclude the influences of the drillstring-riser interactions, the maximal riser deflection obtained by using the proposed drillstring-riser system is relatively small. Namely, the drillstring-riser interactions restrict the riser deflection. This finding indicates that, in a given safety margin of the riser deflection, the drillstring-riser system has stronger capacity of anti-deflection. Additionally, the influence of sea depth is investigated; the dynamic response of a drillstring-riser system drilling in deep water is obviously different from the corresponding condition drilling in shallow water. In order to maintain the drillstring-riser system work in a safe condition, the operations, such as thickening the riser wall, enhancing the control of the top drift, removing a part of buoyancy modules, decreasing WOB, and increasing the top tension of the riser, are suggested to be applied.

1 Introduction

The main difference of offshore drilling from onshore drilling is that drilling rigs and drilling fluids reach the wellhead through a riser system whose length is the same as the sea depth [1]. The top side of a riser system is connected with a drilling platform or a drilling vessel, and the bottom side of a riser system joins with a blowout preventer (BOP) which is installed on the wellhead [2]. A drillstring system rotates inside the riser system. Under the effects of the ocean loads, the drillstring-riser system deflects, and the corresponding contact and friction occur intermittently along the whole pipe-in-pipe structure.

The investigations of the riser dynamics can be traced back to 1970s. Burke built up the static and dynamic models of a riser system for 800ft sea depth [3]. Simmonds used the finite difference method to solve Burke's differential equations, and analyzed the distribution of riser deflection [4]. Still based on Burke's model, Krolikowski and Gay applied an improved linearization technique to analyze the frequency domain of a riser system [5]. Dareing and Huang discussed the variations of the natural frequencies for the low order modes of a riser system as adjusting system parameters [6]. Kirk et al. analyzed the influences of top tension on the variation of frequency domain when the riser system was subjected to periodic excitation. They also provided the bending moment distribution of a riser system under different order resonances [7]. Patel and Sarohia applied the finite element method to analyze riser dynamics. In their work, the ocean loads were set as regular, and the hydrodynamic damping was linearized [8]. Bueno and Morooka designed an interaction model to calculate the drillstring-riser contact force [9]. Yazdchi and Crisfeld applied the Reissner-Simo beam rather than the Euler-Bernoulli beam to consider the hydrodynamic effect on the shear deformation of a riser system [10]. Nair and Baddour divided a riser system into pieces which were connected by a series of spring-damper units [11]. Chen et al. analyzed the design factors of a riser system [12]. Wang et al. did the coupled dynamic analysis of a riser system under combined forcing and parametric excitation [13]. Meanwhile, aiming at the minimum variance of the bottom flexible joint deflection angle, they optimized the top tension of the riser system [14]. Based on the theoretical and experimental studies, Zhou et al. found 'one third effect' for a riser system drilling in deep water. Specifically, when a riser system works in shear flows, its maximal deflection is explored around 1/3 sea depth [15]. Liu et al. developed an experimental system for testing the riser system working in deep water, their system was designed based on the system parameters of the 6th generation deep water drilling platform, HYSY-981 [16].

Summarily, the previous investigations mainly focused on the dynamic analysis of the riser system. Both Reissner-Simo beam and Euler-Bernoulli beam were used for developing the mathematical models. Morrison equations were applied to describe the loads caused by sea wave and sea current. Several models considered the influences of the flow of drilling fluid, and a few models

* Corresponding author: liaomldr@cnpc.com.cn

discussed the effects of the drillstring vibrations. The developed models were solved by finite difference methods or finite element methods. Both time domain analysis and frequency domain analysis were carried out. The influences of the system parameters, such as top tension, water depth, platform drift, and buoyancy factor, etc., on the riser dynamics were analyzed. According to the literature review, a dynamic model about the drillstring-riser system which takes the drillstring-riser interactions into consideration needs to be developed, and a detailed dynamic analysis of the developed model should be carried out, both of which are benefit for feasibility and safety of offshore drilling in complicated deep water.

This paper is organized as follows. Section 2 introduces the mathematical modeling of a drillstring-riser system. Section 3 conducts the numerical simulation of the developed mathematical model using finite element method in Abaqus. In Section 4, the parameter sensitivity is analyzed. The influences of sea depth and design parameters on the dynamic responses of a drillstring-riser system are analyzed successively. Concluding remarks are provided in Section 5.

2 Mathematical modeling of a drillstring-riser system

The dynamic model of a drillstring-riser system consists of a pipe-in-pipe deformation model (see Fig.1 (a)) and a drillstring-riser interaction model (see Fig.1 (b)). The pipe-in-pipe structure deforms under the effects of the ocean loads from its outside. The interaction model describes the contact force and friction between a drillstring and a riser along the pipe-in-pipe structure. Two independent coordinate systems are set to describe the riser system (xyz) and the drillstring system (x'y'z'), respectively. The correlations between the two coordinate systems connect the two models together to form a complete dynamic model for the investigated drillstring-riser system.

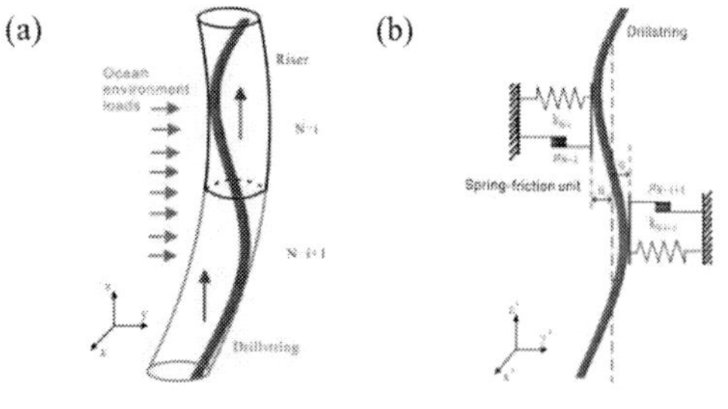

Fig. 1. (a) Pipe-in-pipe deformation model and (b) drillstring-riser interaction model.

Primarily, to describe the ocean loads around the pipe-in-pipe deformation model, Ekman drift theory [2] is adopted to calculate the flow velocity of sea current as

$$u_c = \begin{cases} K_w v_m (z - L + H_c)/H_c + v_t (z/L)^{1/7} & (L \le H_c) \\ v_t (z/L)^{1/7} & (L > H_c) \end{cases} \tag{1}$$

where the friction depth is calculated as

$$H_c = \frac{7.6 v_w}{\sqrt{\sin\varphi}} \tag{2}$$

Airy wave theory [2] is applied to calculate the velocity and the acceleration of sea wave, which are listed in Eq.3 and Eq.4, respectively.

$$u_w = \pi h_w / T_w e^{2\pi(z-L)/L_w} \cos(\omega t) \tag{3}$$

$$a_w = 2\pi^2 h_w / T_w{}^2 e^{2\pi(z-L)/L_w} \sin(\omega t) \tag{4}$$

According to the obtained parameters of sea wave and sea current, Morison equation is used to calculate the drag force and the inertial force caused by the ocean loads [3]. Meanwhile, considering the shear force, tension, gravity, and inertia force of a riser system, the differential equation about the dynamic behavior of a riser system can be described as Eq.5 as

$$\frac{\partial^2}{\partial z^2}\left[EI(z)\frac{\partial^2 y}{\partial z^2}\right] + P(z)\frac{\partial^2 y}{\partial z^2} + (1-B_f)m(z)g\frac{\partial y}{\partial z} + m(z)\frac{\partial^2 y}{\partial t^2} +$$
$$\frac{1}{2}C_d D\rho\left(u_w + u_c - \frac{\partial y}{\partial t}\right)\left|u_w + u_c - \frac{\partial y}{\partial t}\right| +$$
$$\frac{1}{4}D^2\pi\rho\left[C_m a_w - (C_m - 1)\frac{\partial^2 y}{\partial t^2}\right] = 0$$

(5)

In order to further introduce the effects of the drillstring-riser interactions into the mathematical model, a series of spring-friction units are set along the pipe-in-pipe structure in the physical model. A Heaviside function, $H(y - y' - G)$, is considered to check whether the drillstring-riser interaction happens at certain position.

$$H(y - y' - G) = \begin{cases} 0 & (y - y' < G) \\ 1 & (y - y' \ge G) \end{cases} \tag{6}$$

When $y - y' \ge G$, the drillstring-riser interaction happens; hence the contact force and friction are triggered, and the corresponding calculation equation is shown in Eq.7.

$$F_{s-r} = K(z)(y - y' - G) + \mu(z)K(z)(y - y' - G) \tag{7}$$

Based on Eq.5 and Eq.7, the differential equation about the dynamic behavior of a riser system can be described as Eq.8 as

$$H(y - y' - G)[K(z)(y' - y - G) + \mu(z)K(z)(y' - y - G)] +$$
$$\frac{\partial^2}{\partial z^2}\left[EI(z)\frac{\partial^2 y}{\partial z^2}\right] + P(z)\frac{\partial^2 y}{\partial z^2} + (1-B_f)m(z)g\frac{\partial y}{\partial z} + m(z)\frac{\partial^2 y}{\partial t^2} +$$
$$\frac{1}{2}C_d D\rho\left(u_w + u_c - \frac{\partial y}{\partial t}\right)\left|u_w + u_c - \frac{\partial y}{\partial t}\right| +$$
$$\frac{1}{4}D^2\pi\rho\left[C_m a_w - (C_m - 1)\frac{\partial^2 y}{\partial t^2}\right] = 0 \tag{8}$$

The boundary conditions and the initial conditions for the riser system can be described as Eq.9 and Eq.10, respectively.

$$y(0,t) = 0;\ \ EI(0)\frac{\partial^2 y(0,t)}{\partial z^2} = k_b\frac{\partial y(0,t)}{\partial z};$$

$$y(L,t) = S(t);\ EI(L)\frac{\partial^2 y(L,t)}{\partial z^2} = k_t\frac{\partial y(L,t)}{\partial z} \tag{9}$$

$$y(z,0) = 0;\ \frac{\partial y(z,0)}{\partial t} = 0 \tag{10}$$

Similar to Eq.8, the corresponding differential equation about the dynamic behavior of the drillstring system can be expressed as Eq.11 as

$$H(y - y' - G)[K(z)(y' - y - G) + \mu(z)K(z)(y' - y - G)] + \frac{\partial^2}{\partial z'^2}\left[EI'(z')\frac{\partial^2 y'}{\partial z'^2}\right] + P'(z')\frac{\partial^2 y'}{\partial z'^2} + m'(z')g\frac{\partial y'}{\partial z'} + m'(z')\frac{\partial^2 y'}{\partial t^2} = 0 \tag{11}$$

The boundary conditions and the initial conditions of the drillstring system can be described as Eq.12 and Eq.13, respectively.

$$y'(0,t) = A\sin(\Omega t);\ \frac{\partial y'(0,t)}{\partial z'} = 0;$$
$$y'(L,t) = S(t);\ \frac{\partial y'(L,t)}{\partial z'} = 0 \tag{12}$$

$$y'(z',0) = 0;\ \frac{\partial y'(z',0)}{\partial t} = 0 \tag{13}$$

3 Numerical simulation of a drillstring-riser system

Due to the introduction of the drillstring-riser interactions in the proposed mathematical model, repeated iterations are needed to confirm the impact positions, and the iterations increase the calculation complexity, hence decrease the calculation efficiency. Under such circumstances, the finite element analysis software, Abaqus, was considered to carry out numerical simulation. Since, in Abaqus, a number of interaction models are available to simulate a variety of interactions, such as contact, friction, etc.

To build the corresponding finite element model of the proposed mathematical model in Abaqus, the Timoshenko beam element, B31, was chosen to construct both the riser system and the drillstring system. The edge-to-edge contact model which belongs to the general-contact module was adopted to simulate the drillstring-riser interactions. Since the riser system connects with the drilling platform at its top side by an upper flexible joint and connects with the BOP at its bottom side with a lower flexible joint. The connector element, CONN3D2 which includes a joint-rotation combination, was used to describe the upper and lower flexible joints. Additionally, the ocean loads were set by the Aqua module. Aqua calculates ocean loads based on Morison equation which agrees with the method applied in the mathematical model. After the completion of the finite element model, the dynamic analysis was carried out. According to Ekman drift theory, the current velocity was set to be decreased exponentially from 1.35m/s to 0 m/s from the sea level to the mud line. Additionally, in order to exclude the disturbances of transient dynamic behaviors, the simulation duration was set as 100 periods of sea wave, and only the last 20 periods were chosen for the analyses of dynamic responses, to secure the simulated system reaches its dynamic stable state.

The simulation parameters about drillstring, riser, and fluids are listed in Table.1. The result of a numerical simulation case is shown in Fig.2. In subplot (a), the drillstring-riser system in its dynamic stable state is displayed, where the system lateral deflection is enlarged 10 times. The red part, which locates around 1/3 sea depth, marks the maximal deflection. Subplot (b) shows the drillstring-riser interaction, the inner pipe represents the drillstring and the outer pipe represents the riser. The red curve, which is the central line of the pipe-in-pipe structure, is applied to observe the dynamic response of this structure in real-time. Subplot (c) compares the riser deflection distributions between a riser system and a drillstring-riser system, in which, the drillstring-riser interaction is demonstrated to be significant for restricting the riser deflection. Therefore, to precisely simulate the dynamic responses of a deep water drilling system, it is necessary to take the drillstring-riser interactions into considerations.

Table 1. List of simulation parameters

drillstring	material density (kg/m^3)	elastic modulus (GPa)	external diameter (m)	thickness (m)	top tension ratio	contact stiffness (kN/m)	friction coefficient	amplitude of drillbit (m)
	7850	210	0.1492	0.00917	1.0	18000	0.05	0.02
riser	material density (kg/m^3)	elastic modulus (GPa)	external diameter (m)	thickness (m)	stiffness of upper joint (N·m/rad)	stiffness of lower joint (N·m/rad)	top tension ratio	buoyancy factor
	7850	210	0.5334	0.0191	5000000	7300000	1.2	0.9
fluids	seawater density (kg/m^3)	drag coefficient	inertial coefficient	sea wave amplitude (m)	sea wave period (s)	wind speed (m/s)	factor of wind speed	density of drilling fluid (kg/m^3)
	1050	0.7	2.0	3.0	9.0	5.0	0.03	1200

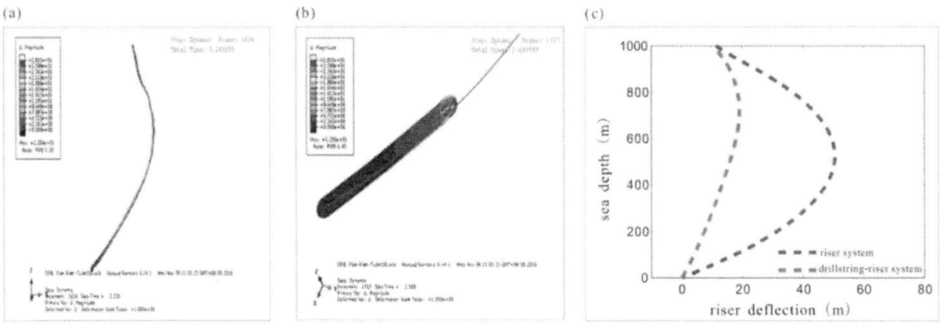

Fig.2. Numerical simulation using Abaqus. (a) Dynamic stable state of a drillstring-riser system, (b) the drillstring-riser interaction, and (c) comparison of the riser deflection distributions between a riser system and a drillstring-riser system

4 Parameter sensitivity analysis

The numerical simulation conducted in Section 3 demonstrates the feasibility of dynamic analysis of a drillstring-riser system by using Abaqus. Therefore, in this section, a series of parameter sensitivity analyses of the drillstring-riser system will be carried out based on the developed finite element model in Section 3.

4.1 Influences of sea depth

The dynamic response of a drillstring-riser system submerged in deep water is expected to be different from that submerged in shallow water. In order to fully understand the influences of sea depth, numerical simulations for three different sea depths, 200m, 1000m, and 2000m were conducted, the simulation results are shown in Fig.3. Specifically, Subplots (a), (b), (c) show the distributions of the riser deflection, riser bending moment, and riser Mises stress, respectively. Subplots (d), (e), (f) show the distributions of the drillstring deflection, drillstring bending moment, and drillstring Mises stress, respectively. Subplots (g), (h), (i) show the rotation angle distribution of the riser system, and the time histories of the upper and low flexible joints, respectively.

In subplots (a) and (d), the deflections of the riser and the drillstring are similar; both of their maximal deflections locate around 1/3 sea depth. The maximal deflection increases as the increase of sea depth; however, the ratio between the maximal deflection and the sea depth decreases as the increase of sea depth. Comparing subplots (b) and (e), the bending moment of the riser is much larger than that of the drillstring; since strong ocean loads act on the riser. As shown in subplot (b), in the shallow water (200m), the bending moment varies considerably along the whole riser; however, as the sea depth increases to 1000m or 2000m, the large bending moments are observed only within the top 10% sea area. This is because the ocean loads attenuate exponentially as the increase of sea depth. In addition, the maximal bending moment decreases as the increase of sea depth. This is because the riser length extends as the water depth, and a stronger gravity helps the riser keep vertical. Similar to the bending moment, in deep water, the large Mises stress of the riser also concentrates on the top 10% sea area, see subplot (c); since its Mises stress is mainly determined by the high bending stress. As a comparison, for the drillstring working inside the riser, its bending moment is small, and its Mises stress is mainly determined by tension stress; hence its Mises stress distributes approximately linearly, see subplot (f).

As shown in subplot (g), the rotation angle of the riser varies nonlinearly from its top to bottom. In subplot (h), the rotation angle of the upper flexible joint shows periodic fluctuation which agrees with the periodic motion of sea wave. In subplot (i), the periodic fluctuation of the rotation angle of the lower flexible joint is only observed for the shallow water case (200m). As the sea depth increases to 1000m or 2000m, the influences of ocean loads on the lower flexible joint become weak; hence, the rotation angle of the lower flexible joint remains dynamic stable state.

4.2 Influences of design parameters

The influences of sea depth have been demonstrated to be significant; however, it is a precondition rather than a design parameter which can be controlled or adjusted. According to the developed dynamic model, the design parameters for a drillstring-riser system mainly include riser thickness, top drift, buoyancy factor, weight on bit (WOB), and top tension of riser.

In order to analyze the parameter sensitivity, a standard study case with 2000m sea depth was simulated primarily, and then the design parameters were modified successively to compare with the dynamic responses of the standard case. All the simulation results are displayed in Fig.4. Specifically, Subplots (a), (b), (c) show the distributions of the riser deflection, riser bending moment, and riser Mises stress, respectively. Subplots (d), (e), (f) show the distributions of the drillstring deflection, drillstring bending moment, and drillstring Mises stress, respectively. Subplots (g), (h), (i) show the rotation angle distribution of the riser system, and the time histories of the upper and low flexible joints, respectively.

According to the comparisons in Fig.4, all the investigated design parameters are proved to be able to

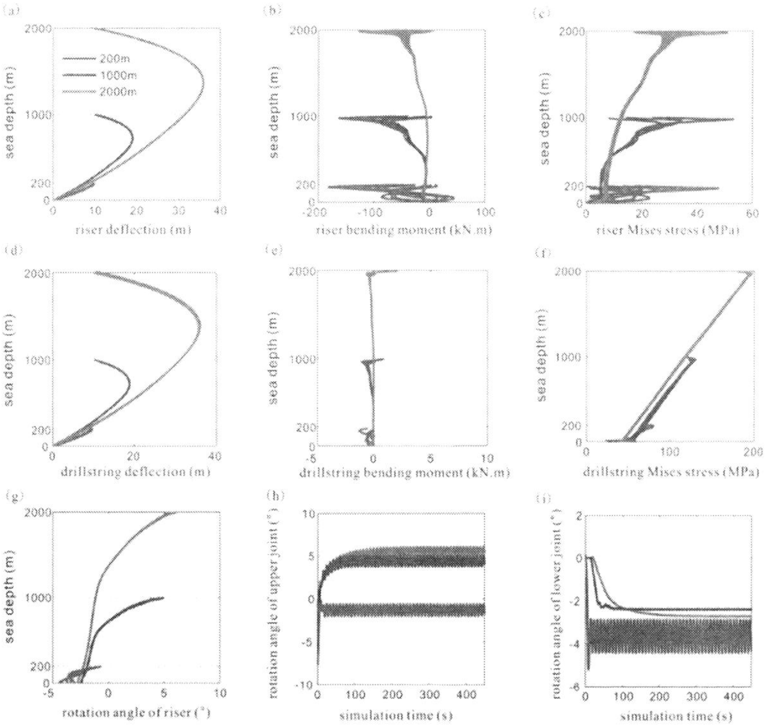

Fig.3. Comparison of dynamic responses of drillstring-riser systems working in different sea depths.

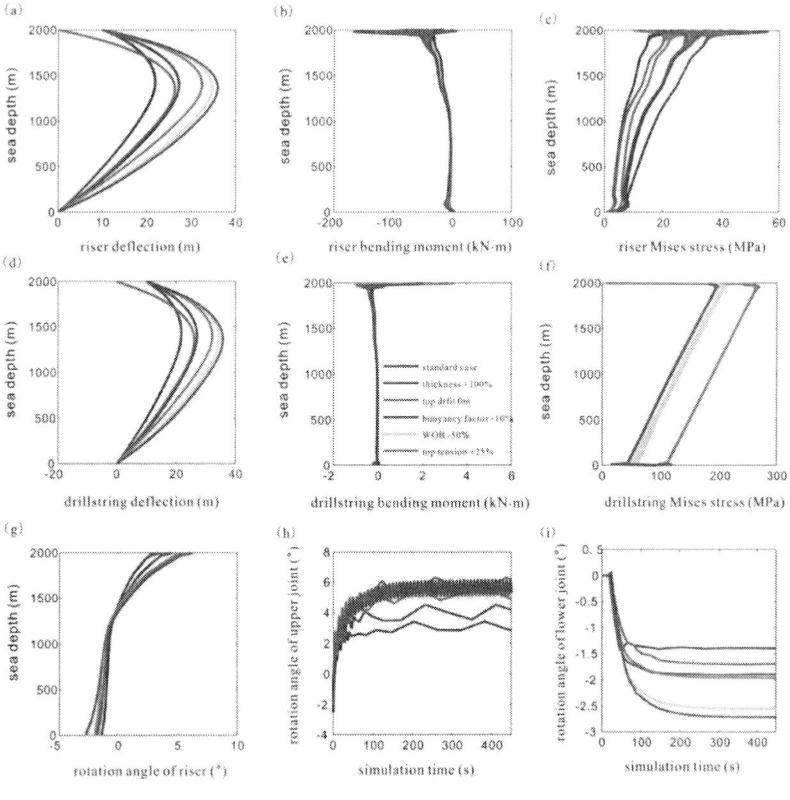

Fig.4. Comparison of dynamic responses of drillstring-riser systems working with different design parameters

affect the dynamic responses of the drillstring-riser system. Therefore, in order to keep the drillstring-riser system work in a safe condition, the following strategies can be applied, including thickening the riser wall, enhancing the control of the top drift, removing a part of buoyancy modules, decreasing WOB, and increasing the top tension of the riser.

5 Concluding remarks

In the presented work, the mathematical model of a drillstring-riser system was developed. The drillstring-riser interactions were introduced into the model by a series of spring-friction units along the pipe-in-pipe structure. The drillstring-riser interactions were proved to restrict the system deflection. This finding indicates that, with the same design parameters, a deep water drilling system is capable of working in tougher ocean environment.

The dynamic response of a drillstring-riser system drilling in deep water was obviously different from its response when drilling in shallow water. Specifically, in shallow water, both the bending moment and the Mises stress of the riser varied considerably along the whole system; however, as the sea depth increases to deep levels, the large bending moment and Mises stress were only observed within the top 10% sea area.

According to the sensitivity analyses of the design parameters, in order to keep the drillstring-riser system work in a safe condition, the operations, such as thickening the riser wall, enhancing the control of the top drift, removing a part of buoyancy modules, decreasing WOB, and increasing the top tension of the riser, were suggested to be applied.

Nomenclature

K_w: factor of wind speed	v_w: wind speed (m/s)	v_t: current speed (m/s)
z: distance from mud line (m)	L: water depth (m)	φ: geographic latitude (°)
h_w: amplitude of sea wave (m)	T_w: period of sea wave (s)	L_w: sea wave length (m)
ω: circular frequency of sea wave (rad/s)	C_d: drag coefficient	C_m: inertial coefficient
D: external diameter of riser (m)	ρ: sea-water density (kg/m^3)	EI: flexural stiffness of riser ($N\cdot m^2$)
m: mass of riser per unit length (kg)	B_f: buoyancy factor	P: tension of riser (N)
EI': flexural stiffness of drillstring ($N\cdot m^2$)	P': tension of drillstring (N)	g: gravitational acceleration (m/s^2)
m': mass of drillstring per unit length (kg)	K: contact stiffness (N/m)	μ: friction coefficient
S: top drift of riser (m)	A: amplitude of drillbit (m)	Ω: frequency of drillbit (rad/s)
k_b: rotational stiffness of the bottom flexible joint ($N\cdot m/rad$)	k_t: rotational stiffness of the top flexible joint ($N\cdot m/rad$)	G: initial gap between riser and drillstring (m)

References

[1] W. Jiang, Deepwater Drilling and Completion Technology: Theory and Practice [M]. Beijing: Petroleum Industry Press, 2015.
[2] S. Zhou, Q. Liu. Theory and Application Research of Deepwater Drilling Riser System Mechanical Behavior [M]. Beijing: Petroleum Industry Press, 2016.
[3] B.G. Burke, An Analysis of Marine Risers For Deep Water [C]. In Offshore Technology Conference, Houston, Texas, USA, 1973.
[4] D.G. Simmonds, Dynamic Analysis Of The Marine Riser [C]. In Society of Petroleum Engineers, Aberdeen, Scotland, UK, 1980.
[5] L.P. Krolikowski, T.A. Gay, An Improved Linearization Technique For Frequency Domain Riser Analysis [C]. In Offshore Technology Conference, Houston, Texas, USA, 1980.
[6] D.W. Dareing, T. Huang, Natural Frequencies of Marine Drilling Risers [J]. Journal of Petroleum Technology, 1976,28(7):813-818.
[7] C.L. Kirk, E.U. Etok, M.T. Cooper, Dynamic and Static Analysis of a Marine Riser [J]. Applied Ocean Research, 1979, 1(3):125-135.
[8] M.H. Patel, S. Sarohia, K.F. Ng, Finite-Element Analysis of the Marine Riser [J]. Engineering Structures, 1984, 6:175-184.
[9] R.C.S. Bueno, C.K. Morooka, Analysis Method for Contact Forces Between Drillstring-Well-Riser [C]. In International Petroleum Conference and Exhibition of Mexico, Veracruz, Mexico, 1994.
[10] M. Yazdchi, M.A. Crisfeld, Nonlinear Dynamic Behaviour of Flexible Marine Pipes and Risers [J]. International Journal for Numerical Methods in Engineering, 2002, 54:1265-1308.
[11] W.R. Nair, R.E. Baddour, Three-Dimensional Dynamics of a Flexible Marine Riser Undergoing Large Elastic Deformations [J]. Multibody System Dynamics, 2003, 10:393-423.
[12] Chang Y, Chen G, Xu L, et al. Influential Factors for the Design of Ultra-Deepwater Drilling Risers [J]. Petroleum Exploration and Development, 2009, 36(4):523-528.
[13] Y. Wang, D. Gao, J. Fang, Coupled Dynamic Analysis of Deepwater Drilling Riser under Combined Forcing and Parametric Excitation [J]. Journal of Natural Gas Science and Engineering, 27:1739-1747, 2015.
[14] Y. Wang, D. Gao, J. Fang, Optimization Analysis of the Riser Top Tension Force in Deepwater Drilling: Aiming at the Minimum Variance of Lower Flexible Joint Deflection Angle [J]. Journal of Petroleum Science and Engineering, 146:149-157, 2016.
[15] S. Zhou, Q. Liu, W. Jiang, et al. The Discovery of One Third Effect for Deep Water Drilling Riser: based on the Theoretical and Experimental Study of Deformation Characteristics of Deep Water Drilling Riser by Ocean Currents [J]. China Offshore Oil & Gas, 2013, 25(6):1-7.

[16] L. Mao, Q. Liu, S. Zhou, et al. Deep Water Drilling Riser Mechanical Behavior Analysis Considering Actual Riser String Configuration [J]. Journal of Natural Gas Science and Engineering, 2016, 33:240-254.

AUTHOR INDEX

AUTHOR INDEX

AUTHOR INDEX